Quantum Mechanics

This book has developed from the author's four decades of zealous teaching of classical mechanics, quantum theory, and atomic physics at premium IITs in India, and from his lectures on these subjects delivered for the popular National Programme on Technology Enhanced Learning (NPTEL) and SWAYAM PRABHA online channels, which are effective initiatives of the Government of India.

Quantum Mechanics: Formalism, Methodologies, and Applications covers the current curriculum requirements of most courses offered in different programs of physics. It will be particularly useful for masters and PhD students taking core courses on quantum mechanics, atomic, molecular, and optical physics, condensed matter, and nuclear and particle physics. It will also be useful to students learning quantum information science and quantum computing. It discusses a wide range of topics beginning with a pedagogical formulation of quantum mechanics including vector space formalism, matrix mechanics, path integrals, and also relativistic quantum mechanics. Quantum mechanics of many-electron atoms is discussed, and applications in spectroscopy and quantum collisions are covered. Topics like the optical and the reciprocity theorems, Eisenbud–Wigner–Smith scattering time delay, and an introduction to quantum computing and teleportation are also included. These topics represent major milestones in the advances in our understanding of atomic physics and condensed matter. They have propelled revolutions in nanosciences and nanotechnologies, atomic clocks and ultrafast dynamics, and quantum information science and quantum computing. *Quantum Mechanics: Formalism, Methodologies, and Applications* would serve a modern graduate curriculum very well, since it provides a rapid but gentle ramp-up from the basics of quantum mechanics to the methodologies of its applications in the frontiers technology.

P. C. Deshmukh is Mentor and Convener, Center for Atomic, Molecular, and Optical Sciences and Technologies (Joint Initiative of IIT Tirupati and IISER Tirupati), Tirupati; Adjunct Professor of Physics, IIT Tirupati; and Adjunct Professor of Physics, Dayananda Sagar University, Bengaluru. He was formerly Professor of Physics at IIT Madras, IIT Mandi, and IIT Tirupati. He obtained his PhD from Nagpur University and worked for his post-doctoral research at the University of Aarhus, University of Notre Dame, and Georgia State University. In addition, he has taught at the University of Western Ontario.

He is the author of *Foundations of Classical Mechanics* (Cambridge University Press, 2019) and an editor of *Quantum Collisions and Confinement of Atomic and Molecular Species, and Photons* (Springer, 2019). P. C. Deshmukh's research group is one of the major contributors to the study of attosecond time-delay in atomic photoionization.

Quantum Mechanics

Formalism, Methodologies, and Applications

P. C. Deshmukh

CAMBRIDGE
UNIVERSITY PRESS

CAMBRIDGE
UNIVERSITY PRESS

Shaftesbury Road, Cambridge CB2 8EA, United Kingdom

One Liberty Plaza, 20th Floor, New York, NY 10006, USA

477 Williamstown Road, Port Melbourne, vic 3207, Australia

314 to 321, 3rd Floor, Plot No.3, Splendor Forum, Jasola District Centre, New Delhi 110025, India

103 Penang Road, #05–06/07, Visioncrest Commercial, Singapore 238467

Cambridge University Press is part of Cambridge University Press & Assessment, a department of the University of Cambridge.

We share the University's mission to contribute to society through the pursuit of education, learning and research at the highest international levels of excellence.

www.cambridge.org
Information on this title: www.cambridge.org/9781316512258

First published 2023

Printed in India by Nutech Print Services, New Delhi 110020

A catalogue record for this publication is available from the British Library

ISBN 978-1-316-51225-8 Hardback
ISBN 978-1-009-44655-6 Paperback

To my teachers with much gratitude
and
to my students with best wishes

Contents

Figures

Foreword

Since its formulation during the first part of the twentieth century, quantum mechanics has fascinated everybody who has tried to grasp it. Classical physics regarded the world to be deterministic; it claimed that if we just knew everything with enough precision, we should be able to predict what will happen tomorrow. Laws of nature, however, can be best explained by quantum mechanics, which is very different from classical physics: only the probability of a certain outcome of an experiment can be predicted. There is also a fundamental limit to how precisely certain pairs of physical quantities can be measured: when we improve the precision of measurement of one quantity, we lose it on another! Even more mind-boggling is the concept of entanglement. Two entangled particles can travel far from each other and still have a connection so that measurements on one of them immediately forces the other into a specific quantum state, regardless of the distance between them. Through the history of quantum mechanics, accomplished scholars and students alike have found this hard to accept, and argued that the theory must not be complete, that we are still waiting for its final version. Nevertheless, quantum mechanics has been proven to be a very successful theory. As far as we know, its predictions are all correct and technologies based on quantum mechanics are nowadays used everywhere: the smart electronic devices in our pockets, the energy efficient LED lamps, and the solar panels that harvest sunlight – deep inside they function because of the laws of nature explained by quantum mechanics.

It is often said that it is not possible to really understand quantum mechanics. This might be true, but with enough effort it is certainly possible to learn to master its machinery and use it to explain physical phenomena and develop new technology. Professor Pranawa Deshmukh writes in this book: "Quantum theory may shock and confuse us, but it is a successful theory of the physical world. It is cast in a mathematical framework which must be learned with patience and rigour."

As a university teacher I know that the first course in quantum mechanics brings something special to many students. While classical physics seems to be completely settled and just for new generations to learn, quantum mechanics comes with surprises, riddles, and philosophical discussions. This book acknowledges this fascination; it does not compromise with the mathematical tools needed to be able to use the theory. Starting with the question of measurement and the wave–particle duality, the book continues along a path that takes the reader from the solutions of a particle in a box, through the harmonic oscillator and to the hydrogen atom, but it does not stop there. A whole chapter is devoted to many-electron systems. Here the necessity to approximate the many-problem is elaborated and several common approximations are explained. Advanced topics such as the path integral formulation of quantum mechanics, the role of symmetry, and the relativistic equation of Dirac are also covered and carefully integrated with the more traditional textbook content. The book comes with many solved problems, invaluable for the serious students and each chapter of the book starts with an historical photograph and an interesting quote from an important scientist, which nicely sets the theme for the following pages.

This is a broad and modern approach to the subject. It includes material for a longer graduate course that will prepare students in many subfields of physics. Its strength is the coverage of important applications of quantum mechanics, often subjects of contemporary research. Ample space has been provided, for example, to the use of transitions from one quantum state to another. Radiation waves emitted or absorbed during such transitions are fingerprints of the particular atom or molecule and can be used to analyse the element composition and temperature of distant stars and galaxies. Even the atmosphere of extra-terrestrial planets can today be studied with such spectroscopic methods. Collisions, involving both particles and photons, are treated with equal care and include a discussion on how time-reversal symmetry relates the seemingly different processes of particle scattering and photoionization.

The discussion on the measurement problem does not stop with the debate between Bohr and Einstein in the 1930s. This discussion was in fact just the starting point for the field of quantum information science. The inequality that was formulated by John Bell took the question of realism from being a philosophical issue to speculate on, to a question that could be settled with experiments. Bell's inequality is here discussed in detail, as is quantum teleportation and the theoretical foundations of quantum computing. Today, 'entanglement' is not just something to be puzzled about, but a reality that can be utilized in quantum computers. The in-depth discussion on these issues underlines a contemporary approach to the subject taken in this textbook.

Eva Lindroth
Professor of Physics
Stockholm University
Stockholm, Sweden

Preface

A graduate course on quantum mechanics is a daunting task – for both students and teachers. Students come for such a course with a fair amount of background in classical physics, *classical* in the sense that it is time-tested. They are familiar with the works of Newton, Lagrange, Hamilton, Euler, etc. In this scheme, an object's physical state is described by its position q and momentum p, and temporal evolution by Hamilton's equations of motion for the time rates $\dot{q}$ and $\dot{p}$. Their experience with classical physics entrenches their faith in it and builds their intuition, but they must now be taught that a physical theory that requires simultaneous knowledge of position and momentum is *fundamentally untenable*. Students must now settle with the fact that classical mechanics 'works' only when it is a very good approximation to a *more appropriate* theory of Nature, which is quantum mechanics. The foundational principles of quantum mechanics *conflict* with those of classical physics, causing confusion and doubt. Overcoming the resulting befuddlement involves learning what seems like an *abstract* formalism, which nonetheless turned out to be an unassailable theory of practical value. It changed our lives in the last century with quantum devices, and is now all set to take another leap into the second quantum revolution. It ushers in mind-blowing technology driven by entanglement and quantum computing.

The route to diligent applications of quantum mechanics begins with a *shock*. Students must grapple with formidable challenges on the path to comprehending consequential principles in a mystified territory. They have to develop proficiency in new methodologies involving abstract mathematics before they can see for themselves that quantum theory simply *works*; *nothing succeeds like success*. They can then use the theory to propel the frontiers of sciences, engineering, and technology. Amid this bewilderment, a graduate course in quantum mechanics is as romantic as it is challenging. One must learn to see beyond the corners of your vision, acquire rigorous capabilities in mathematics, enjoy luminous discourses between brilliant minds, cultivate an inventiveness to develop new technology that impacts human life, and understand the cosmos. *Quantum mechanics: formalism, methodologies, and applications* is a vast subject, very young compared to classical physics, but a very rich field to which some of the most outstanding intellectuals have made dazzling contributions during the past hundred odd years.

Some of my colleagues were surprised when a little over three years ago Cambridge University Press published my *Foundations of Classical Mechanics* (*FoCM*). They had expected that I would write a book on *quantum mechanics*, not *classical*. *FoCM* sets the stage for the present book: concepts, vocabulary, and notations employed in it are frequently referred to in *Quantum Mechanics: Formalism, Methodologies, and Applications*. This book has grown out of joyful improvisations I have labored over four decades to provide graduate students with a rapid, but gentle, ramp-up from foundational principles of quantum theory to advances in its practices of contemporary interests. We emphasise that there aren't two sets of laws of Nature, one for the microscopic and the other for the macroscopic world. In Chapter 1, we jump without much ado into the vector space formulation of quantum mechanics with a brief comment on the incompatibility of measurement of position and momentum using the Heisenberg microscope. We discuss the *complete set of compatible observables* and proceed to deliberate on Heisenberg's principle of uncertainty, and also on the Schrödinger equation. Chapter 1 also compares, and contrasts, the energy–time uncertainty with that in position–momentum. We underscore the fact that a system has *discrete* or *continuum* energy eigenstates depending on the boundary conditions on the Schrödinger equation.

The sequencing of topics in this book is perhaps a bit untraditional, but purposefully so. Immediately after introducing foundational principles in Chapter 1, we introduce in Chapter 2 Feynman's path integral formulation, along with a discussion on the geometrical phase, because of the importance of the *phase* of the wavefunction. The mindboggling Aharonov–Bohm effect is also discussed in this chapter. Chapter 2 also includes a commentary on why classical mechanics works at all, when and where it does. Chapter 3 is primarily dedicated to simple one-dimensional problems, whose applications go as far as laying the foundations of nanoscience, but it also includes a relatively new method to solve quantum mechanical problems using the Lambert W function, developed by S. R. Valluri and Kenneth Roberts.

The *shock* from the simultaneous immeasurability of position and momentum is accentuated by that of the impossibility of determining orthogonal components of angular momentum. We appreciate the role of symmetry and conservation laws in formulating laws of nature, and enter an analysis of the angular momentum in considerable detail in Chapter 4. In Chapter 5, we use it to understand quantum mechanics of the hydrogen atom from the standpoint of the geometrical, and also the dynamical, symmetry of the Coulomb interaction. Discrete bound states spectrum and the continuum eigenstates of the hydrogen atom are both discussed.

Approximation methods are dealt with in Chapter 6, but degenerate perturbation theory is deferred to Chapter 8 on Stark–Lu Surdo, Zeeman, and hyperfine spectroscopies. Perturbative interpretation of relativistic effects is discussed in the context of the Foldy–Wouthuysen transformations of the fully relativistic Dirac Hamiltonian in Chapter 7, which also presents a decoupling of the radial and the angular parts of the relativistic 4-component wavefunction – notwithstanding the presence of *odd* operators in the Dirac Hamiltonian. We stress that there aren't two laws, a *relativistic* law for particles moving at high speeds, and another *nonrelativistic* for those at low speeds. A fundamental particle *even at rest* has an intrinsic 'spin' angular momentum (discussed in Chapters 4 and 7, in particular), which requires *relativistic quantum mechanics* for its interpretation. It is this property that makes a particle a Fermion or a Boson.

The many-electron self-consistent field (Hartree–Fock) theory of the atomic structure is detailed in Chapter 9. In Chapter 10, on scattering theory, pedagogical treatment of the partial-waves analysis is boosted to explicate the role of time-reversal symmetry which connects solutions of quantum collisions with those from photoionization/photodetachment spectroscopy. This chapter discusses the optical theorem, reciprocity theorem, Eisenbud–Wigner–Smith time delay, Born approximations, Green function methods, etc.

We accentuate the fact that it is misleading to say that quantum theory is mind-boggling and counter-intuitive; rather, it is Nature which is – quantum theory describes it correctly. *Apparently* strange phenomena occur in Nature, not in theories. To an uneducated intuition, they appear strange. One requires a reinterpretation of *reality*, which paves the way to quantum computing and to the *second* quantum revolution. Chapter 11 provides an introduction to quantum computing, teleportation, and dense coding.

Quantum mechanics is a vast subject. We attempt to lay down a strong foundational formalism, move up to exemplify intricate methodologies, and exhibit significant applications which impact advances in engineering and technology. We hope this approach will help students and researchers to gain confidence in deploying crucial quantum tools resourcefully. We celebrate the interlacing of mathematics with the physical laws of Nature and hope that the coverage of each topic is satisfactory. An attempt has been made to maintain the presentation simple by focusing on the main ideas, and relegating some details to either a few problems at the end of each chapter, or to an appendix. Chapter 5 (on the non-relativistic hydrogen atom) has five appendices, 5A–5E, to provide necessary details with regard to the continuum eigenfunctions of the hydrogen atom, and about the symmetry group of the Coulomb potential, which accounts for the degeneracy of its discrete eigenstates. There also are five appendices (A–E) at the *end* of the book. These include a brief commentary on the role of *discrete symmetries*, a summary of Schrödinger, Heisenberg, and Dirac pictures of quantum mechanics, a brief account of the spherical harmonics, a short introduction to *second* quantization, and a shorter introduction to the Variational Quantum Eigensolver to simulate a many-electron system using *qubits*. Readers would benefit by regarding the appendices and the end-of-chapter problems as *vital* and *integral* content of the subject matter.

The contents of this book provide a compilation of my lecture notes prepared for a number of courses I had the opportunity to teach over four decades at the Indian Institute of Technology Madras, at the Indian Institute of Technology Mandi, at the Indian Institute of Technology Tirupati, and at the Indian Institute of Science Engineering and Research Tirupati. The course contents also benefited from video-lecture courses I had the opportunity to deliver for the *NPTEL* (Physics - Select/ Special Topics in Atomic Physics - YouTube, https://www.youtube.com/playlist?list=PLbMVogVj5 nJQAcTv17ETSh5-GNDbAo6BM, and Physics - Special/Select Topics in the Theory of Atomic Coll - YouTube, https://www.youtube.com/playlist?list=PLbMVogVj5nJSdsqPcC1J9SmCuKg5DIUwn) and for SWAYAM PRABHA (Special/Select Topics in Classical and Quantum Physics - YouTube, https://www.youtube.com/playlist?list=PLJoALJA_KMOAbZCaNzL28v8zqa0mpewf7). Parts of the contents of Chapter 11 have also been taught at the Dayananda Sagar University, Bengaluru. I am indebted to each of these institutions for giving me an opportunity to teach their students which has been an amazing learning experience for me.

An advance copy of the unedited pre-final complete manuscript has been read by Dr. Eva Lindroth, Dr. G. Baskaran, Dr. Anatoli Kheifets, Dr. Eugene Kennedy, and Dr. Pietro Decleva. I gratefully acknowledge their criticism and comments. I have benefited and learned much from my research collaborators over the years; in particular from Dr. P. L. Khare, Dr. C. Mande, Dr. Jan Linderberg, Dr. Walter R. Johnson, Dr. Steven T. Manson, Dr. Vojislav Radojevic, Dr. Valery Dolmatov, Dr. Himadri Chakraborthy, Dr. Anatoli Kheifets, Dr. Kenneth Roberts, and Dr. S. R. Valluri. They have helped me with many different aspects of quantum mechanics. The residual errors and gaps are strictly due to my inability to grasp various subtleties and complexities. Abundant and gainful contributions to my excursions in teaching and learning quantum mechanics have been made by my graduate students: N. Shanthi, R. Padma, E. W. B. Dias, H. R. Varma, Tanima Banerjee, M. Ummal Momeen, S. Sunil Kumar, G. B. Pradhan, Jobin Jose, Manas R. Parida, N. M. Murthy, K. Sindhu, Akash Singh Yadav, G. Aarthi, Ankur Mandal, Soumyajit Saha, and Sourav Banerjee.

I wish to thank the Niels Bohr Archive (specially Mr. Robert James Sunderland), the American Institute of Physics (specially Mr. Max Howell), and the International Business Machines (specially Ms. Benita Naidu) for permitting me to use some of their photographs in this book. The staff at Cambridge University Press has been amazingly helpful. Taranpreet Kaur, Vaishali Thapliyal, Qudsiya Ahmad, and Ankush Kumar have helped me at every stage. The meticulous and patient handling of the production process by Aniruddha De and Vikash Tiwari has been outstanding. I am very grateful to all of them.

A number of graduate students and colleagues have helped me prepare the contents of this book. In particular, the *end-of-chapter* problems have been compiled by Nishita Manohar Hosea, Bharath Manchikodi, Rasheed Sheik (IIT Mandi), Saumyashree Baral, Aiswarya Rajendran (IIT Patna), Aliasgar Musani, Shreyas Suresh, Jeyasitharam J. (IIT Tirupati), and Abheek Roy (IISERT). Dr. R. Padma painstakingly read through the complete work and helped improve the presentation. Thanks also to Pranav Sharma who made some useful suggestions. I acknowledge fruitful criticism from a few colleagues, who took the trouble of reading through an early draft of one or the other chapter, and advised critical improvements. In particular, I wish to thank Dr. Dilip Kumar Singh Angom, Dr. James Libby, Dr. S. Sunil Kumar, Dr. Sourav Banerjee, Dr. Jobin Jose, Dr. S. Aravinda, Dr. Aarthi Ganesan, Dr. Soumyajit Saha, Dr. Hari R. Varma, Dr. Srinivasa Prasannaa, and Dr. Kenneth Roberts. Constant support and encouragement by Dr. K. N. Satyanarayana, Dr. K. N. Ganesh, Dr. C. Vijayan, Dr. G. Aravind, Dr. Sivarama Krishnan, Dr. B. Koteswara Rao, Dr. Reetesh Gangwar, Dr. Premachandra Sagar, Dr. K. N. B. Murthy, Dr. A. Srinivas, Dr. Uday Kumar Reddy, and Dr. K. Vijaya Kumar is specially acknowledged. I wish to thank Dr. Pruthul Desai, Dr. Aniket Joglekar, and Dr. Ranjan Modak for their generous help in the final stages of the book production. My sister, Pradnyatai, and her husband, Balkrishna Tambe, accommodated me in their home during my numerous visits to Atlanta over four decades. These visits greatly facilitated my research collaboration with Dr. Steven T. Manson. Primary support came from my family: Sudha, Wiwek, Pradnya, and Aditi. They sustained, endured, and powered me.

February 14, 2023

P. C. Deshmukh
[a] Mentor and Convener,
Center for Atomic, Molecular, and Optical Sciences and Technologies
(Joint Initiative of IIT Tirupati and IISER Tirupati)
and Adjunct Professor of Physics, Indian Institute of Technology Tirupati, Tirupati
[b] Adjunct Professor of Physics, Dayananda Sagar University, Bengaluru

1

Description of a Physical System

Paul Dirac and Werner Heisenberg. B464, https://arkiv.dk/en/vis/5940636. Courtesy: Niels Bohr Archive.

The measure of greatness in a scientific idea is the extent to which it stimulates thought and opens up new lines of research.

—P. A. M. Dirac

From our day-to-day experience, we develop our notion of reality. However, our perception of physical properties, such as position and momentum, needs refinement to describe natural phenomena correctly. In this chapter we introduce fundamental principles of the quantum theory that does so with enduring cogency. The laws of nature that govern the functioning of the physical universe cannot be accounted for using classical physics of Newton, Lagrange, and Hamilton. Foundational principles and mathematical structure of quantum theory are introduced in this chapter.

1.1 Quantum versus Classical Physics Theories

In classical physics, the mechanical state of a system is represented by a point in the position–velocity phase space, or equivalently in the position–momentum phase space. The entire theoretical formalism of Newtonian–Lagrangian–Hamiltonian mechanics is based on this Galilean conjecture. The classical hypothesis seems appropriate in a large number of physical situations that concern our day-to-day experiences. With this ansatz, the temporal evolution of a system is described by the trajectory of the point in the phase space. The trajectory is obtained from the equation of motion that governs the time dependence of position and velocity, represented respectively by q and $\dot{q}$ (or alternatively the time dependence of position and momentum, represented respectively by q and p). The alternative equations of motion – Newton's, Lagrange's, and Hamilton's – are equivalent. Their applicability must be reconciled with the upper limit on accuracy with which *conjugate* physical properties of an object, viz. *position* and *momentum*, are simultaneously measured. There is, however, an *inverse* relation between the accuracy of simultaneous measurement of these two properties. The *more* accurately you measure either, the *less* accurately can you measure the other. The act of measuring either of the conjugate properties requires an observation.

Observations resulting in accurate measurements of conjugate variables are, however, not compatible with each other. Heisenberg's principle of uncertainty is the quantitative expression of this law of nature. It is expressed as a rigorous mathematical inequality that is neatly written in a compact form. Nonetheless, we refrain from advancing its mathematical expression too soon. It has no classical analogue. It cannot be written in any terms of what we are familiar with from classical mechanics. *Symbols* for position and momentum that are used to express the uncertainty principle are also found in classical mechanics, but they have a *new meaning* in quantum theory. Heisenberg's principle of uncertainty and also the Schrödinger equation will be introduced in Section 1.4 after the required notation is formally construed in Sections 1.2 and 1.3.

Classical and quantum physics are mathematical models which both aim at describing the state of a physical system and its temporal evolution. There certainly are not *two* laws of nature, one for large objects and the other for small. Many, but not all, macroscopic phenomena are fairly well accounted for by classical physics. If one comes across a claim that classical laws work for macroscopic objects, it must be understood only in the sense that in several macroscopic events, classical laws are very good *approximations* to quantum laws. In the present section, a few salient features of classical physics will be briefly recapitulated. There are, of course, many excellent books on classical mechanics. The formalism and notation used in Reference [1] provides a well-suited platform for topics developed in the present book. Classical physics has two alternative and equivalent formulations. One of these is based on the principle of causality and determinism. Its backbone is the linear stimulus–response hypothesis advanced by Isaac Newton. Newtonian dynamics is based on Galileo's identification of the constancy of momentum of an object as determined only by its *initial* mechanical state. The stimulus (i.e., force) that *changes* the momentum of an object is exactly equal to the *rate* at which the change in momentum occurs. The application of a force on an object therefore imparts to it an acceleration that is *directly* proportional to the force itself. Newtonian formalism is therefore a linear stimulus–response theory.

It will be argued in the next chapter that the alternative formalism of classical physics based on the *principle of variation* is in some sense more powerful than Newton's method, although the two approaches produce equivalent results. In it, the notion of force is not used. Instead, methods of variational calculus are used. The principle of variation stipulates that a mechanical system is described by its *Lagrangian*, $L(q,\dot{q})$. The essential premise of this formulation is therefore the same as that of Newton's method, since the Lagrangian is given in terms of the generalized position q and the generalized velocity $\dot{q}$. Its simplest form is

$$L(q,\dot{q}) = f_1(\dot{q}^2) + f_2(q), \tag{1.1}$$

where f_1 and f_2 are suitable functions respectively of velocity and position. The dependence on the square of the velocity rather than on its first power is prompted by the isotropy of space. The equation of motion of the system is then determined on the basis of an ansatz that the definite integral over time from the initial time t_i to the final time t_f, i.e.,

$$S = \int_{t_i}^{t_f} L(q,\dot{q})\,dt, \tag{1.2}$$

called *action*, is an extremum. The choice

$$L(q,\dot{q}) = T(\dot{q}^2) - V(q) = \frac{1}{2}m\dot{q}^2 - V(q), \tag{1.3}$$

with the requirement that action is stationary, i.e.,

$$\delta S = 0, \tag{1.4}$$

leads one [1] to the equation of motion as the necessary and sufficient condition that describes motion:

$$-\frac{\partial V}{\partial q} = \frac{\partial L}{\partial q} = \frac{d}{dt}\frac{\partial L}{\partial \dot{q}}. \tag{1.5}$$

In Eq. 1.3, T is the kinetic energy of the object, V its potential energy, and m its mass. That Eq. 1.3 is the right choice for the Lagrangian is seen on recognizing that it reproduces Newton's second law, since

$$\text{force} = -\frac{\partial V}{\partial q} \tag{1.6a}$$

$$\text{and} \quad \frac{d}{dt}\frac{\partial L}{\partial \dot{q}} = \frac{dp}{dt}, \tag{1.6b}$$

where p is the generalized momentum *defined* as the partial derivative of the Lagrangian with respect to $\dot{q}$.

Equation 1.5 is the Lagrange's equation of motion.

Instead of representing the state of a system by $(q,\dot{q})$ or $L(q,\dot{q})$, one may of course represent it by (q,p), or for that matter by a function $H(q,p)$ of the same.

The necessary and sufficient conditions for action to be an extremum turn out to be given by two first-order equations of motion,

$$\dot{p} = -\frac{\partial H(q,p)}{\partial q} \tag{1.7a}$$

and

$$\dot{q} = \frac{\partial H(q,p)}{\partial p}, \tag{1.7b}$$

$$\text{with } H(q,p) = T + V = \frac{p^2}{2m} + V(q). \tag{1.8}$$

Equation 1.7a, b are known as the Hamilton equations of motion, and H (Eq. 1.8) as the Hamiltonian.

Funquest: What constants of integration are required to solve (i) Newton's, (ii) Lagrange's, and (iii) Hamilton's equations of motion? How are these constants to be obtained?

Time-reversal symmetry of the equations of motion (Newton's, Lagrange's, or Hamilton's) ensures that we can not only predict from its solution the mechanical state of the system anytime in the future but also determine what it would have been at any time in the past. We advance a caution here that time-reversal symmetry has a very different connotation in quantum mechanics. We shall discuss it later.

Funquest: Show that (i) Newton's equation of motion, (ii) Lagrange's equation of motion, and (iii) Hamilton's equation of motion are symmetric under time-reversal $t \rightarrow -t$.

Classical mechanics is a study of the pair $(q, \dot{q})$ or equivalently of (q, p). The classical equations of motion require the initial conditions $(q_i, \dot{q}_i)$ or (q_i, p_i) at the initial time t_i. Lack of knowledge of either of the two physical properties in each of these pairs makes it impossible to determine the mechanical state of a system at an arbitrary time using the equations of motion.

A *pair of measurements* is necessary to determine $(q_i(t=0), \dot{q}_i(t=0))$ or $(q_i(t=0), p_i(t=0))$. Likewise, the correctness of the solution at an arbitrary time t to the equation of motion can also be verified only by a *measurement* of the *pair*, $(q(t), \dot{q}(t))$, or $(q(t), p(t))$.

A fundamental question therefore arises: Is simultaneous *accurate* measurement of both members of the pair $(q(t), p(t))$ possible at all, at any instant t? To address this question, we consider the measurement of position and momentum of an electron (Fig. 1.1). The experiment involves shining light of wavelength λ on the electron to observe it through the lens of a microscope. This apparatus is called as the *Heisenberg microscope*. An analysis of this experiment would lead us to the quantum theory, which supersedes classical physics. Our discussion of this experiment requires acquaintance with two brilliant advances that were made in the early days of quantum physics. Historical accounts of concurrent progress in quantum theory and atomic physics are as fascinating and intriguing as romantic. Exciting accounts of these advances are available [2, 3, 4]. We restrict ourselves only to two rather significant milestones used in Eq. 1.9a,b that would be employed in the analysis of the experiment with the Heisenberg microscope. The first is Planck's hypothesis, made in December 1899, about the corpuscular nature of energy in an electromagnetic field, succinctly stated as $E = \mathrm{h}\nu$, ν being the frequency of an electromagnetic wave. Planck described his hypothesis as an *act of desperation* and struggled hard for many years later to negate his own theory. Planck's postulate however gained robust support in Einstein's Nobel Prize–winning explanation (in 1905) of the 1887 experiments, conducted by Hertz and Lenard, in which they discovered the photoelectric effect. The corpuscular nature of the quantum of energy in the electromagnetic field was advanced by Einstein in this work. It was firmly established afterward, in 1924, in the *statistical* description of the electromagnetic field by S. N. Bose. The constant h is commonly named after Planck, but considering the necessary and decisive contributions made by Einstein and Bose that established its significance, it is more appropriately called as *Planck–Einstein–Bose* (PEB) constant. The second significant milestone we shall employ in Eq. 1.9a is *de Broglie's hypothesis* of *wave–particle dualism*. It associates a wave having a wavelength

$$\lambda = \frac{\mathrm{h}}{p} \tag{1.9a}$$

with every particle, p being the particle's momentum. Using these two landmark advances, we see that a photon of energy

$$E = h\nu = \frac{h}{\lambda}c = pc, \tag{1.9b}$$

carries a momentum p. This is well known from the observation of comet tails, which are always pointed *away* from sunlight. The momentum carried by an electromagnetic field is well described by the Poynting vector (Chapter 13 of Reference [1]). It is taken full advantage of in the laser cooling of atoms.

We will now use (a) the Planck–Einstein–Bose quantum of energy of a photon and (b) the de Broglie wave–particle duality hypothesis, to discuss the Heisenberg microscope experiment to determine the position of an electron. Our limited objective in deliberating on this experiment is to demonstrate that the assumption we make in classical physics that the state of a system is represented by a point in the position–momentum phase space cannot withstand scrutiny by an experiment. It would thus expose the inadequacy of classical mechanics and prepare us to look for an alternative theory, which turns out to be the quantum theory.

An electron (assumed to be initially at rest) would be seen through a microscope's lens (Fig. 1.1). This requires a photon to be scattered by the electron along OB, after maximal momentum transfer to the electron, or along OF, after minimal momentum transfer to the electron, or of course at some intermediate angle determined by the conservation of momentum between the photon and the electron.

For minimum momentum transfer, we have

$$p = \frac{h}{\lambda} = \frac{h}{\lambda'}\sin\theta + m_e v'_x, \tag{1.10a}$$

and for maximum momentum transfer, we have

$$p = \frac{h}{\lambda} = \frac{h}{\lambda''}(-\sin\theta) + m_e v''_x. \tag{1.10b}$$

In Eq. 1.10a and Eq. 1.10b m_e stands for the electron's mass and v_x' and v_x'' are its x-components of the velocity.

Accordingly, the least momentum gained by the scattered electron is

$$p_{\text{least}} = \frac{h}{\lambda} - \frac{h}{\lambda'}\sin\theta = m_e v'_x, \tag{1.11a}$$

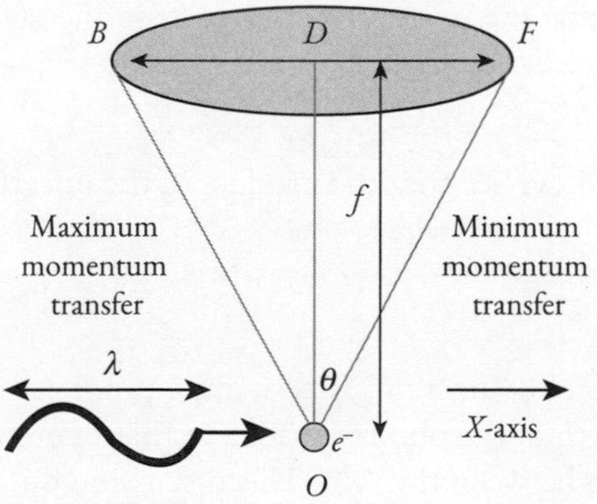

Fig. 1.1 The *Heisenberg microscope* is an apparatus we employ to discuss a thought experiment that demonstrates that accuracy in the measurement of an object's position is inversely related to that in a concurrent measurement of its momentum.

and the most momentum gained by it is

$$p_{\text{most}} = \frac{h}{\lambda} + \frac{h}{\lambda''}\sin\theta = m_e v_x''. \tag{1.11b}$$

The electron gains momentum from the probing photon, and the change in momentum along the x-axis is therefore

$$\Delta p_{\text{electron}} = \frac{h}{\lambda} \mp \frac{h}{\lambda}\sin\theta \approx \frac{h}{\lambda}(1 \mp \theta). \tag{1.12a}$$

This range would be minimal for small angles, and hence the momentum uncertainty is

$$\Delta p \approx \frac{2h}{\lambda}\theta. \tag{1.12b}$$

In order to minimize uncertainty in momentum, λ must therefore be large. Any attempt to minimize this uncertainty therefore comes at a price, since it would require a photon having a large wavelength, but a photon with larger wavelength would result in a greater uncertainty in position. Now, the resolving power (RP) of a microscope is

$$RP = \frac{1}{D} \simeq \frac{\sin\theta}{\lambda}, \tag{1.13a}$$

$$\text{or } \sin\theta \approx \frac{\lambda}{D}, \tag{1.13b}$$

$$\text{i.e., } \theta \approx \frac{\lambda}{D}. \tag{1.13c}$$

The location of the electron can at best be placed between $-f\tan\theta$ and $+f\tan\theta$, where

$$\tan\theta = \frac{\frac{1}{2}D}{f}, \tag{1.14}$$

f being the focal length of the microscope's lens (Fig. 1.1).

$$\text{Hence, } \frac{f}{D} = \frac{1}{2\tan\theta} \approx \frac{1}{2\theta}. \tag{1.15}$$

The uncertainty in locating the position of the electron therefore is

$$\Delta x \approx 2f\tan\theta \approx 2f\theta \approx 2f\frac{\lambda}{D} \simeq \frac{\lambda}{\theta}. \tag{1.16}$$

From Eqs. 1.12 and 1.16, we see that the product of the uncertainty in position with that in momentum is

$$\Delta p \Delta x \approx \frac{2h}{\lambda}\theta \times \frac{\lambda}{\theta} \simeq 2h, \tag{1.17}$$

which is of the order of the PEB constant. It is small, but not zero! The position uncertainty increases with the wavelength of the photon used, and that in momentum decreases! There is thus an inverse relationship between the two. This discussion on the Heisenberg microscope is mostly qualitative, but it brings out an essential limitation of classical physics. It exposes its intrinsic inconsistency, since the formalisms of Newton's, Lagrange's, and Hamilton's mechanics *all* depend on the *simultaneous* knowledge of position and momentum.

Physicists ran also into many other limitations of classical physics. These included the spectral intensity distribution of the blackbody radiation that was alluded to in Planck's hypothesis, analysis of atomic spectra, specific heat of solids, and so on. Readers should consult other sources for details on these topics. The way forward would be as romantic as challenging. A new formalism had to be developed which must admit the impossibility of simultaneous accurate measurement of position and momentum of a particle. These physical properties could not be represented merely by two numbers; a number (having dimension L) for the position of a particle, and another (having dimension MLT^{-1}) for momentum. Instead, it had to be stipulated that each particle would be represented by a wave–particle duality. The new theory that was developed using this scheme was called *wave mechanics*. It was developed in the 1920s by de Broglie and Schrödinger. An alternative theory that provided a viable replacement of classical physics was called *matrix mechanics*. It was developed by Heisenberg, Born, and Jordan around the same time. In matrix mechanics, position and momentum were represented by non-commuting matrices, and not merely by single numbers. Wave mechanics and matrix mechanics were raised on revolutionary ideas that would bring about an upheaval which impacted not just physics but nearly every single aspect of human life.

1.2 Hilbert Space Description of a Physical System

The experiment in Fig. 1.1 requires us to concede that representing the mechanical state of a system by a point in the phase space by its position and momentum is intrinsically flawed. This scheme must be replaced by a new theory. How about representing the state of a system by a vector in the Hilbert space? One may well ask why one should represent the state of a system by a vector. Well, what if it turns out to be useful in determining physical properties of the system? Quantum physics is the theory that answers this question magnificently! Quantum physics is the only theory that not only accounts for physical properties, but all of its predictions have come out to be correct! From the properties of fundamental particles to those of atoms, molecules, and condensed matter, it explains everything; nothing else does! Quantum theory may shock and confuse us, but it is a successful theory of the physical world. It is cast in a mathematical framework that must be learned with patience and rigor. At some point on this journey, we figure out *how* it works, if not *why* it works. Questions about the laws of nature reduce to *what the laws are* and *how they account for natural phenomena*, and not *why are the laws what they are*. The success of quantum theory is a compelling reason to admit that the theory is good, *extremely good*. It is a mathematical framework, and what makes it immensely appropriate to describe the laws of nature is "a wonderful gift that we neither understand nor deserve," as Wigner would say [5]. We therefore proceed to learn the mathematical formulation of quantum physics and hope that after putting in sufficient effort, we will discover how physical properties of a system can be extracted from this theory. More than a century of scintillating success of quantum theory is a strong argument to expect this hope to be very well placed.

Funquest: If, after patiently and thoroughly studying the mathematical framework described in the next few chapters, you discover that quantum theory accounts for most of the observable properties of the physical universe, what arguments would you have *against* the ansatz that the physical state of a system is described by a vector in the Hilbert space?

Intellectual giants like Bohr, Einstein, Feynman, and many others have all conceded, in telling ways, that quantum theory is shocking and mindboggling. Feynman went as far as saying that "nobody understands quantum mechanics." Quantum theory's success, and our quest to determine the laws of nature, is however a sufficient reason to explore, and hopefully master, its ansatz and methods. At a fundamental level, the ansatz of quantum theory has two components. That the state of the system is represented by a vector is the first of these. The other is how physical dynamical variables, such as position q, momentum p, angular momentum $\vec{j}$, energy E, etc., are represented. In quantum theory, these are represented by the operators $q_{op}, p_{op}, \vec{j}_{op}$, and the Hamiltonian H_{op}, respectively. These operators operate on the vectors (operands) in the Hilbert space. However, most often, the subscript "*op*" on the operators is omitted; the context would tell us if the symbol we are referring to corresponds to their quantum connotations. There, of course, are many physical dynamical variables of interest, over and above the four mentioned. As in classical mechanics, we would be interested in the results of measurements of dynamical variables, each of which would be represented by an appropriate operator.

We must interpret the result of a measurement in terms of the *operator algebra* that we have barely begun to develop. Nowhere have we referred to the size or mass of the physical system under investigation. We must therefore expect the new formalism to be applicable to all objects, whether microscopic or large. Quantum theory is based on a set of ansatz, a set of mathematical rules, which initially appear to be abstract and unrelated to the questions about physical properties of matter and energy. Yet it would turn out that it is just these mathematical structures that provide the *best* description of the laws of nature, albeit only *after* the mathematical framework is developed further.

We therefore begin with an abstract formalism of the quantum theory and rely on your motivation to patiently learn the rules of the game, and begin to have fun playing it. You will discover as you go along how beautifully the quantum theory connects to physical measurements and to *tangible* properties of the physical universe. The *Hilbert* space is the mathematical space in which the physical state vectors reside. It was named by John von Neumann after his Guru, David Hilbert. A vector that represents the physical state of a system at time t, if its state is known at an initial time t_0, is denoted as $|\gamma, t_0 : t\rangle$. The label γ must appropriately *designate* the quantum state. It is actually a *set* of one or more physical properties of the system that are *measurable* by an observer and that characterize the physical properties of the system. Temporal evolution of the system is then described by its derivative with respect to time, i.e., by $\frac{\partial}{\partial t}|\gamma, t_0 : t\rangle$. In Section 1.4, we shall see that it is given by the Schrödinger equation

$$i\hbar\frac{\partial}{\partial t}|\gamma, t_0 : t\rangle = H|\gamma, t_0 : t\rangle, \tag{1.18}$$

where H is the Hamiltonian operator, also to be introduced in Section 1.4. The symbol $|\ \rangle$ represents a vector, called a "ket." We shall soon define a mathematical object that will be represented by $\langle\ |\ \rangle$, which looks like a *bracket*. Hence the terminology *ket* for $|\ \rangle$ and *bra* for $\langle\ |$. This terminology was introduced by Dirac, and the notation we have used is named after him. $\langle\ |$ is also a vector in a space that is called Dual Conjugate (D.C.) Hilbert vector space, defined through a one-to-one correspondence with the ket Hilbert space:

$$\forall\ |\ \rangle \underset{\text{D.C.}}{\overset{\substack{\text{one}\\\text{to}\\\text{one}}}{\leftrightarrow}} \langle\ |. \tag{1.19}$$

The space of ket vectors $|\ \rangle$ is referred to as the *direct* space and that of bra vectors $\langle\ |$ as the *dual conjugate* (D.C.) space. The set of measurable properties represented by γ cannot include both members of the pair (q,p); in Section 1.1 we have already found that position and momentum are not simultaneously measurable. The label γ can therefore only be a set of physical properties whose measurements are *compatible* with each other. A physicist is interested in whatever *most* that can be learned about the target of her/his curiosity. Hence, we are interested in determining the *Complete Set of Compatible Observables* that would constitute the set γ. Dirac abbreviated this as CSCO, which would also stand for *Complete Set of Commuting Operators*. By the end of this chapter, this dual meaning of CSCO would become clear.

Vectors in the Hilbert space can be added to each other, weighted by *complex* numbers α_i to get new vectors,

$$|x_{\text{resultant}}\rangle = \alpha_1|x_1\rangle + \alpha_2|x_2\rangle. \tag{1.20a}$$

The corresponding, one-to-one, relation in the D.C. space is

$$\langle x_{\text{resultant}}| = \alpha_1^*\langle x_1| + \alpha_2^*\langle x_2|. \tag{1.20b}$$

The asterisk employed here denotes complex conjugation. We shall work with an N-dimensional linear vector space, spanned by N linearly independent base vectors $\{|x_1\rangle,|x_2\rangle,..,|x_N\rangle\}$. This basis is identified by the criterion that for *complex* numbers $\alpha_i, i = 1,2,..,N$,

$$\alpha_1|x_1\rangle + \alpha_2|x_2\rangle + .. + \alpha_N|x_N\rangle = |0\rangle, \tag{1.21a}$$

if, and only if, *each* $\alpha_i = 0$.

Since the correspondence between the direct space and the D.C. space is one-to-one, one must expect a linearly independent basis $\{\langle x_1|,\langle x_2|,..,\langle x_N|\}$ for the D.C. space such that

$$\alpha_1^*\langle x_1| + \alpha_2^*\langle x_2| + .. + \alpha_N^*\langle x_N| = 0, \tag{1.21b}$$

if and only if *each* $\alpha_i^* = 0$.

Observe that complex conjugate coefficients are employed in Eq. 1.20b. An arbitrary vector in the Hilbert space can always be written as a linear superposition of a complete set of base vectors. We now define products of vectors in the direct and the D.C. spaces. Three types of products can be defined:

i. an inner product, represented by a bra-ket $\langle\ |\ \rangle$, which is a scalar,
ii. an outer product, represented by a ket-bra $|\ \rangle\langle\ |$, which is an operator, and
iii. a direct product, also called as tensor product, $|\ \rangle|\ \rangle$, of two ket vectors, which belong to two *disjoint* vector spaces. We shall use these in Chapter 4.

For the time being, we shall mostly work with the scalar (i.e., "inner") product and with the outer product. Without even explicitly stating it, we have already employed *addition* of vectors in the Hilbert space. For now we will work with a finite dimensional Hilbert space, i.e., one for which N is finite. When required, especially when we work with continuum/scattering states, we shall introduce techniques to extend the mathematical machinery to address an infinite dimensional Hilbert space.

The Hilbert space is a *metric* space. An example of a metric space is the familiar n-dimensional Euclidean space, $R^{(n)}$. It is an ordered set of n-tuples $(x_1, x_2, x_3, ..., x_n)$ such that a *measure* of distance between two points in the space can be defined by employing a suitable criterion. In $R^{(3)}$, the distance between two Cartesian points (x_1, x_2, x_3) and (x'_1, x'_2, x'_3) is

$$d = d\{\vec{r}, \vec{r}'\} = d\{(x_1, x_2, x_3), (x'_1, x'_2, x'_3)\} = \sqrt{\left\{\sum_{i=1}^{3}(x_i - x'_i)^2\right\}}, \tag{1.22a}$$

$$\text{i.e., } d = \sqrt{(\vec{r} - \vec{r}')\cdot(\vec{r} - \vec{r}')}, \tag{1.22b}$$

which is, of course, the Pythagoras theorem. The notion of such a measure is readily extended to vector spaces of higher dimensions, such as the non-Euclidean *flat* space–time continuum of the Special Theory of Relativity, and even the *curved* space–time of the General Theory of Relativity (Chapters 2, 13, 14 of Reference [1]). The function d that can be appropriately defined is called a "metric" on the vector space. We see that the metric is defined with respect to the scalar product. The generalization of this idea to an n-dimensional linear Hilbert space leads us to the *inner* product of two vectors, $|a\rangle$ and $|b\rangle$, denoted by $\langle b|a\rangle$. With reference to any three vectors $|a\rangle, |b\rangle, |c\rangle$ and arbitrary complex numbers λ, μ, κ, the inner product has the following properties:

i. $$\langle b|a\rangle = \langle a|b\rangle^*, \tag{1.23a}$$

ii. $$\langle a|(\lambda b + \mu c)\rangle = \langle a|(\lambda b)\rangle + \langle a|(\mu c)\rangle = \lambda\langle a|b\rangle + \mu\langle a|c\rangle, \tag{1.23b}$$

and

iii. $$\langle(\lambda a + \mu b)|\kappa c\rangle = \langle(\lambda a)|\kappa c\rangle + \langle(\mu b)|\kappa c\rangle = \lambda^* \kappa\langle a|c\rangle + \mu^* \kappa\langle b|c\rangle. \tag{1.23c}$$

The inner product of a ket vector with itself, i.e., with its dual conjugate bra vector, is $\langle a|a\rangle$, and is defined to be essentially a non-negative quantity, i.e.,

$$\langle a|a\rangle \geq 0. \tag{1.24}$$

Equation 1.24 makes it possible to define a *real norm* of a vector as

$$\|a\| = \sqrt{\langle a|a\rangle}. \tag{1.25}$$

A vector whose norm is equal to unity is said to be *normalized*. Two vectors $|a\rangle, |b\rangle$ are said to be *orthogonal* if their inner product is zero. There is a significant parallel between the algebra of the linear vector Hilbert space and that of $R^{(3)}$. The difference however is not only in its possibly different, finite or infinite, dimensionality but also in the use of complex conjugation employed above.

An arbitrary vector in an N-dimensional Hilbert space is expressible as a linear combination of a complete set of *linearly independent* base vectors $\{|e_1\rangle, |e_2\rangle, ..., |e_N\rangle\}$, not necessarily normalized, nor orthogonalized. It can of course be expressed as a linear superposition of base vectors in an *orthonormal* basis $\{|u_1\rangle, |u_2\rangle, ..., |u_N\rangle\}$ in which the linearly independent basis set vectors are orthogonal and also normalized:

$$|\gamma\rangle = \sum_{i=1}^{N} a_i |e_i\rangle, \text{ and also} \tag{1.26a}$$

$$|\gamma\rangle = \sum_{i=1}^{N} b_i |u_i\rangle. \tag{1.26b}$$

An orthonormal basis is one which is both orthogonal and normalized (see Solved Problem P1.1). The base vectors in this case satisfy the property

$$\forall i, j: \quad \langle i | j \rangle = \delta_{ij}, \tag{1.27}$$

the right-hand side being the Kronecker-δ. The symbol "u" that we used in Eq. 1.26b is now dropped in Eq. 1.27 to make the notation more compact.

Now, just like the classical Kepler–Newton two-body gravitational problem has solutions describing different trajectories such as ellipse, parabola, or hyperbola (see Chapter 8 of Reference [1]), the quantum theory for a particle provides for alternative solutions for different energy (E) states. For $E < 0$, we have bound states; for $E > 0$, we have continuum states; and for $E = 0$, we have barely bound (or barely free) states. This would become clear as we develop the formalism further. For now, it is worth mentioning that the orthonormalizaton criterion (Eq. 1.27) for continuum states will have to be somewhat modified. Continuum states are used a lot in quantum collision theory.

We now introduce *operators* that operate on vectors in the Hilbert space. These are characterized further as linear, antilinear, unitary, anti-unitary, Hermitian, non-Hermitian, etc., depending on their special attributes. First, we introduce linear operators. These are a set of operators such that any two of these, say A_{op} and B_{op}, satisfy the following properties:

i. $\left(A_{op} + B_{op}\right)|x\rangle = A_{op}|x\rangle + B_{op}|x\rangle,$ (1.28a)

ii. $\left(\lambda A_{op}\right)|x\rangle = \lambda\left(A_{op}|x\rangle\right),$ λ being an arbitrary scalar (complex number), (1.28b)

iii. $\left(A_{op}B_{op}\right)|x\rangle = A_{op}\left(B_{op}|x\rangle\right),$ (1.28c)

iv. $(A_{op}B_{op})C_{op} = A_{op}(B_{op}C_{op}) = A_{op}B_{op}C_{op}$ (associative property), (1.28d)

but the order in which the operators appear in the multiplication must be preserved unless a pair of operators commute.

The null operator and the unit operator are defined *respectively* by the following relations:

i. $O_{op}|x\rangle = 0,$ (1.29a)

and

ii. $1_{op}|x\rangle = |x\rangle,$ (1.29b)

where $|x\rangle$ is an arbitrary vector in the Hilbert space.

An operator Λ_{op} is said to be the inverse, A_{op}^{-1}, of A_{op} if, for an arbitrary operand $|x\rangle$ in the Hilbert space, $\Lambda_{op}A_{op}|x\rangle = |x\rangle = A_{op}\Lambda_{op}|x\rangle.$ (1.29c)

Furthermore, an operator is said to be singular if its inverse does not exist; else it is called non-singular. At this point, we will take half a step back and discuss the mechanism to *label* the Hilbert state vectors. A label needs to be inserted in the ket vector $|\ \rangle$ in order to designate it unambiguously. Since the vector is meant to represent the physical state of a system, it can be labeled only by one (or more) of its *observable*, *measurable*, properties. When we measure a physical property denoted (say) by A of a state represented by the ket vector $|\ \rangle$, the act of measurement is represented by an operator A_{op}, which performs a mathematical operation on the vector. The result of the measurement of the observable physical property in question is, say, "a." This is denoted in quantum theory by A_{op} operating on $|\ \rangle$, giving us a new vector $|\ \rangle'$. If the original vector was in a state that is called as an eigenstate of that measurement, the new vector $|\ \rangle'$ is *proportional* to the original ket. The proportionality that is employed is just the result "a" of the measurement. We may therefore write the previous statement as a mathematical equation,

$$A_{op}|\ \rangle = |\ \rangle' = a|\ \rangle. \tag{1.30a}$$

Equation 1.30a is known as an eigenvalue equation. The ket $|\ \rangle$ is called as an eigenket of the operator A_{op}, and the proportionality "a" is called as the eigenvalue of A_{op}. The result of the measurement, "a," is obviously a physical property of the system. It can therefore be used to designate it. Accordingly, we insert the label "a" in the ket vector, and write it as $|a\rangle$. Having chosen the label to designate the ket vector, we can write the eigenvalue equation as

$$A_{op} \underset{\text{eigenstate}}{|a\rangle} = \underset{\text{labels eigenstate}}{\overset{\text{eigenvalue}}{a}} \underset{\text{eigenstate}}{|a\rangle}, \tag{1.30b}$$

$$\text{i.e., } A_{op}|a\rangle = a|a\rangle. \tag{1.30c}$$

Of particular importance in the context of measurable physical properties are operators that are Hermitian. A Hermitian operator $\hat{\Omega}$ is defined considering the effect of an operator in the direct space and the corresponding relation in the dual conjugate space, given by

$$\langle\alpha|\hat{\Omega}^{\dagger} \overset{\substack{\text{dual}\\ \text{conjugate}}}{\leftrightarrow} \hat{\Omega}|\alpha\rangle. \tag{1.30d}$$

The operator $\hat{\Omega}^{\dagger}$ is called as the Hermitian adjoint of $\hat{\Omega}$. Self-adjoint operators are said to be Hermitian operators; i.e., if $\hat{\Omega}$ is a Hermitian operator, then

$$\hat{\Omega}^{\dagger} = \hat{\Omega}, \tag{1.30e}$$

$$\text{and } \langle\beta|\hat{\Omega}|\alpha\rangle = \langle\beta|\left(\hat{\Omega}|\alpha\rangle\right) = \left(\langle\alpha|\hat{\Omega}^{\dagger}\right)|\beta\rangle^{*} = \langle\alpha|\hat{\Omega}^{\dagger}|\beta\rangle^{*} = \langle\alpha|\hat{\Omega}|\beta\rangle^{*}. \tag{1.30f}$$

In quantum theory, physical states represented by a vector $|a\rangle$, and another that is proportional to it, such as $\alpha|a\rangle$, are essentially the same. They differ only in their length,

$$\sqrt{\langle\langle a|\alpha^{*}||\alpha|a\rangle\rangle} = \sqrt{|\alpha|^{2}\langle a|a\rangle} = |\alpha| \times \|a\|. \tag{1.31}$$

A vector scaled by a complex number thus differs from the original one only in its normalization (see Solved Problem P1.1b). Measurement is central to physics, and quantum theory accounts for the results of a measurement exhaustively. If a physical property A is measured yet again, the result of the measurement would again be "*a*." In other words, a physical system in an eigenstate of an operator (measurement) *remains* in the same state. Repeated measurements of the same property return the same result of the measurement, which is the eigenvalue of the corresponding operator. A fascinating experiment carried out by Otto Stern and Walther Gerlach in 1922 illustrates this correspondence. It would also demonstrate how the recovery of an eigenvalue of property A in *repeated* experiments is marred by an intermediate measurement of an *incompatible* property B.

Apparatus for the Stern–Gerlach (SG) experiment is schematically shown in Fig. 1.2a. Results of this experiment would shock you, since the outcome of this experiment has no classical analogue. Silver atoms darting out of an oven are passed through an inhomogeneous magnetic field, say along the z-axis. The atoms carry zero net electric charge, and hence from classical electrodynamics [1], one would expect that they would traverse undeviated through the SG apparatus, producing a single spot at the centre of the detector screen. Instead, one actually finds two spots, one above the centre along the z-axis and the other symmetrically below it. The splitting of the silver atom beam into two components must be interpreted using relativistic quantum theory, which accounts for intrinsic "spin" angular momentum and concomitant dipole magnetic moment of an electron. The natural suspicion that "spin" is something like the rotation of the earth (or a gyroscope) about its own axis is misleading. "Spin" of a particle has no classical analogue. We regard "spin" as a meaningless name of a meaningful property of the particle. In other words, one could have given some other name to this intrinsic physical property, but the one that has stuck is "spin." It emerges naturally from the relativistic formulation of quantum mechanics (Chapter 7).

The 47 electrons in an atom of silver impart a net angular momentum to the atom (which is effectively from the unpaired electron in the valence shell of the atom). The associated magnetic dipole moment of the silver atom then tends to get aligned along, or opposite to, the inhomogeneous magnetic field in the apparatus. The atomic beam thus gets split, producing two spots on the detector. The SG experiment turned out to be a historic one since

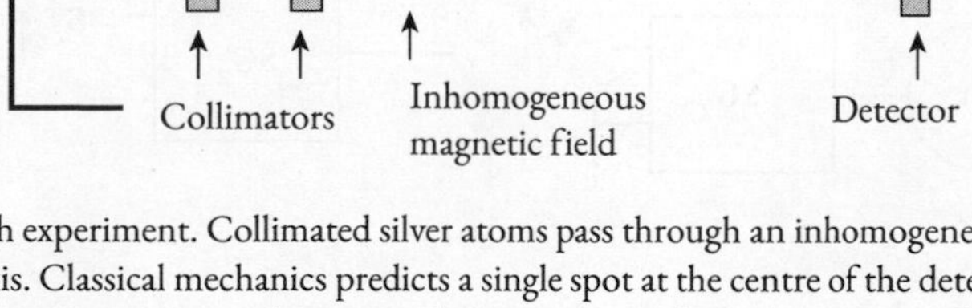

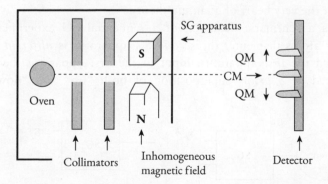

Fig. 1.2a Stern–Gerlach experiment. Collimated silver atoms pass through an inhomogeneous magnetic field along the z-axis. Classical mechanics predicts a single spot at the centre of the detector in line with the collimated beam. Quantum mechanics predicts *two* separated spots, as are *actually observed* in the SG experiment. One spot is above the centre, and the other is symmetrically below it.

it established the "spin" angular momentum. The separation of the beam of silver atoms into two components was a startling revelation demonstrating the inadequacy of classical electrodynamics.

Further inquiry into the results of SG experiments reveals *supplemental* characteristics about measurements that are inexplicable by classical physics. We might construe that the dipole magnetic moment vector of silver atoms would be fully along (or opposite to) the direction of the magnetic field. A single measurement is however insufficient to provide three components of the dipole magnetic moment vector. We therefore require additional measurements with the magnetic field oriented along orthogonal directions to get a complete description of the vector magnetic moment. Mindboggling laws of nature (discussed later in Chapter 4) totally unknown to classical physics however thwart our ability to unambiguously determine orthogonal components of angular momentum (and of the associated dipole magnetic moment). First-time readers of quantum theory would rejoice how a meticulous analysis of the SG experiment would make it possible for quantum theory to grow on them.

The state of silver atoms from the oven that *enter* the SG magnetic field requires *a priori* description as a *superposition* of *two* states corresponding to dipole moments parallel and anti-parallel to the magnetic field. We shall denote these two states respectively by arrows $\uparrow$ and $\downarrow$. This superposition is expressed in quantum mechanics as

$$|\psi\rangle = a_{\uparrow}|\uparrow\rangle + a_{\downarrow}|\downarrow\rangle, \tag{1.32}$$

where $a_{\uparrow}$ and $a_{\downarrow}$ are coefficients of the respective *pure* eigen-components $|\uparrow\rangle$ and $|\downarrow\rangle$, akin to Eq. 1.26b, since the base vectors $|\uparrow\rangle$ and $|\downarrow\rangle$ are orthogonal. The experimental set-up in Fig. 1.2a schematically illustrates the separation of these two components, producing the unexpected two spots on the detector.

Furthermore, another fundamental property in nature is revealed by *sequential* augmentation of the SG experiment. The necessary improvisations are depicted in Figs. 1.2b–d and discussed below. They demonstrate a *different* fundamental property that is related to compatibility, or lack of it, of two or more measurements. The phenomena we discuss now are at the very heart of quantum theory.

Figure 1.2b is a schematic depiction of a sequential SG experiment. Observe that the state of silver atoms that enter the *second* SG apparatus is *different* from that of the atoms that entered the *first* apparatus. Input to the first apparatus, being directly from the oven, consisted of *a priori* superposition of both "up" and "down" components.

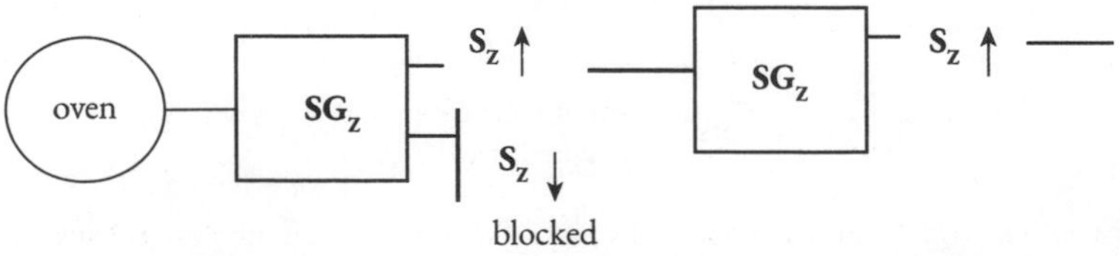

Fig. 1.2b This experiment shows that a mixed state collapses into an eigenstate by a measurement. Furthermore, a system that is produced in an eigenstate (in this case by the first SG apparatus) remains in the same eigenstate if the same measurement is repeated.

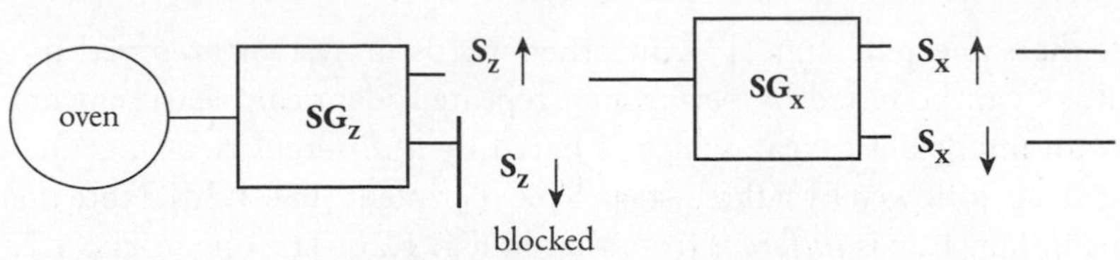

Fig. 1.2c *Selected* output of the first SG experiment in which the magnetic field is along the *z*-axis is input to a second SG apparatus in which the magnetic field is *orthogonal* to that in the first.

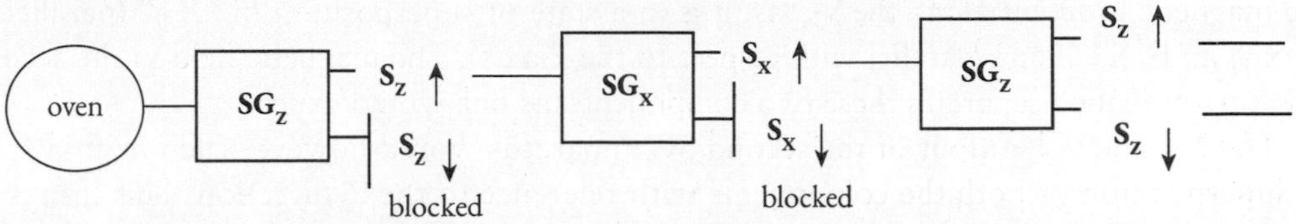

Fig. 1.2d This sequential SG experiment shows that measurements of Z- and X- components of the magnetic moment are not compatible with each other.

The magnetic field of the first SG apparatus *measures* the component of the magnetic moment along the Z-direction. This experiment has a congruous expression in quantum theory: a system in *a priori* state of superposition *collapses* into an eigenstate of the measurement. The collapse of an *a priori* mixed state into an eigenstate is therefore to be understood as a result of this measurement. The *probability* of getting "up" or "down" result is proportional to the *square* of the modulus of the coefficient of corresponding states in the superposition. Since the two probabilities are equal, the coefficient $a_\uparrow$ and $a_\downarrow$ in Eq. 1.32 are both $\frac{1}{\sqrt{2}}$. This is a typical representative of a very general interpretation of an arbitrary state in quantum theory, which is expressed as a linear superposition of orthonormal base vectors. For a normalized state, the sum of the modulus square of coefficients in an orthonormal basis adds up to unity:

$$\sum_{i=1}^{N} |b_i|^2 = 1, \tag{1.33}$$

where b_i are coefficients in Eq. 1.26b. The first SG apparatus reveals this result. This apparatus separates the two components. In the two-stage SG experiment shown in Fig. 1.2b, the ↓ component is blocked. Only the ↑ component is admitted into an *identical* second SG apparatus in which the magnetic field is also along the *z*-axis, just as in the first SG apparatus. Output of the second apparatus can therefore only be the ↑ component. This result reveals another characteristic feature of the measurement process. The input to the second SG apparatus being pure "up" state, the coefficient of $|\uparrow\rangle$ state in the input to the second SG apparatus is unity. It is guaranteed that the state is essentially "up" with reference to the *z*-axis. The output of the second SG apparatus is therefore always $|\uparrow\rangle$; it is the only possibility. The "up" branch output of the first SG apparatus *prepares* a pure $|\uparrow\rangle$ state that is input to the second SG apparatus. Subsequent measurements, *no matter how many*, would yield always the same pure state $|\uparrow\rangle$. In other words, a system *prepared* in a pure state remains in it, as can be tested by performing repeated identical measurements. This result however has an important caveat, which is bared by a *different* two-stage SG experiment, shown in Fig. 1.2c, followed by a three-stage SG experiment (Fig. 1.2d). Note that the second SG apparatus in Fig. 1.2c is *different* from that in Fig. 1.2b. The one in Fig. 1.2c produces a magnetic field that is orthogonal (along x-axis) to the one in the first SG apparatus, which is along the z-axis. The output of the first SG apparatus is a pure "up" state, but only with reference to the magnetic field along the z-axis. With respect to a field that would measure the magnetic moment along the x-axis, it is in a state of superposition of "X↑" (parallel to x-axis) and "X↓" (anti-parallel with respect to the x-axis). The magnetic field in the second SG apparatus then separates these two components, as one would expect.

The X↑ and X↓ output of the second SG apparatus would however each be in a state of superposition of both the components with reference to the Z-direction. This brings us to the caveat that was mentioned earlier about the *sustainability* of a pure state in repeated measurements. The caveat is that the repeated measurements must not be interspersed by any interaction of the system in an experiment that would measure a component of the magnetic moment along an orthogonal direction. Even an inadvertent interaction of the system that (directly or indirectly) amounts to such an experiment destroys the purity of the select state; it tosses the system back into a state of superposition. The three-stage SG experiment shown in Fig. 1.2d illustrates this. The outcome of this experiment may actually shock you, unless your intuition is by now overhauled by quantum theory. The $S_X\downarrow$ component from stage-2 is blocked and $S_X\uparrow$ passed through the third SG apparatus. The magnetic field in the stage-3 apparatus is once again along the z-axis. The $S_X\uparrow$ is a pure state with respect to measurement of the magnetic dipole moment along the x-axis, but it is in a mixed (superposed) state of $S_Z\uparrow$ and $S_Z\downarrow$. Measurement of the magnetic dipole moment along the z-axis in the third SG apparatus therefore returns both the "up" and "down" possibilities, again with equal probabilities, just as the very first SG apparatus did! This is surprising, even shocking, since the Z↓ component was blocked between the first and the second stage apparatus, but it reappears after the stage-3 apparatus. There is absolutely no classical analogue for the re-appearance of the Z↓ component. Strange as it may well seem, that is just how nature behaves. Quantum theory accounts for it correctly, since it employs the principle of *a priori* superposition of pure states, unless the system is *prepared* in a pure state by a measurement.

The intervention by measurement of the component of the magnetic dipole moment along x-axis (Figs. 1.2c, d) therefore imposes a stringent condition on the caveat that repeated measurements of the component along the z-axis yield the same result: the condition is that repeated measurements are not interspersed by an incompatible measurement. Comments made in the paragraph right after Eq. 1.31 assert the essential conclusions from the sequential SG experiments. We infer that an eigenstate of measurement of component of magnetic moment along the z-axis is *not* an eigenstate of measurement of component along x-axis. Since measurements are represented by operators, the incompatible measurements correspond to non-commuting operators. A maximal description of a state of the system therefore is in terms of eigenvalues of the complete set of commuting operators. These eigenvalues correspond to the results of measurements of the corresponding complete set of compatible observables. Both the sets are handily abbreviated as CSCO [5].

We now discuss other intriguing aspects of the Hilbert vector space formalism. Consider an N-dimensional linear vector space and a *second* Hilbert space that is M-dimensional. The N-dimensional space may be considered to be spanned by an orthonormal basis $\{|e_j\rangle; j = 1,2,..,N\}$, and the M-dimensional space by the orthonormal basis $\{|f_i\rangle; i = 1,2,3,..,M\}$. Each basis being *complete* is an important property. It means that an arbitrary vector in that particular vector space can be written as a superposition of the base vectors.

It is possible to transform a vector

$$|V\rangle = \sum_{i=1}^{N} x_i |e_i\rangle \equiv \underbrace{(x_1, x_2, .., x_N)}_{\text{components}} \tag{1.34a}$$

in the N-dimensional vector space into a vector

$$|W\rangle = \sum_{k=1}^{M} y_k |f_k\rangle \equiv \underbrace{(y_1, y_2, .., y_M)}_{\text{components}} \tag{1.34b}$$

in the M-dimensional vector space. Note that the vectors $|V\rangle$ and $|W\rangle$ are uniquely specified by their respective components $(x_1, x_2, .., x_N)$ and $(y_1, y_2, .., y_M)$. Consider the transformation

$$A|V\rangle = |W\rangle, \tag{1.35}$$

where A is the operator that effects this transformation. This transformation is linear. It is represented using complex coefficients A_{ij}, with $i = 1,2,..,M$ and $j = 1,2,..,N$ as:

$$\begin{array}{l} A_{11}x_1 + A_{12}x_2 + \ldots\ldots + A_{1N}x_N = y_1 \\ A_{21}x_1 + A_{22}x_2 + \ldots\ldots + A_{2N}x_N = y_2 \\ \ldots\ldots\ldots\ldots\ldots\ldots\ldots\ldots\ldots\ldots\ldots \\ A_{M1}x_1 + A_{M2}x_2 + .. + A_{MN}x_N = y_M \end{array}. \tag{1.36a}$$

The above family of equations can be written in a compact matrix notation:

$$\begin{bmatrix} A_{11} & A_{12} & A_{13} & \cdots & A_{1N} \\ A_{21} & A_{22} & A_{23} & \cdots & A_{2N} \\ . & . & . & \cdots & . \\ . & . & . & \cdots & . \\ . & . & . & \cdots & . \\ A_{M1} & A_{M2} & A_{M3} & \cdots & A_{MN} \end{bmatrix} \begin{bmatrix} x_1 \\ x_2 \\ .. \\ .. \\ .. \\ .. \\ .. \\ x_N \end{bmatrix} = \begin{bmatrix} y_1 \\ y_2 \\ .. \\ .. \\ .. \\ .. \\ .. \\ y_M \end{bmatrix}. \tag{1.36b}$$

Representing the matrices by *fat* letters, we may write the matrix equation using rectangular matrices as

$$\mathbb{A}_{M\times N}\mathbb{X}_{N\times 1} = \mathbb{Y}_{M\times 1}, \tag{1.36c}$$

or simply as $\mathbb{A}\mathbb{X} = \mathbb{Y}$. (1.36d)

Using slim letters, we can write the same relation as

$$AX = Y. \tag{1.36e}$$

The slim letter A is used for the matrix $\mathbb{A}$ in Eq. 1.36e, but the *same* symbol (A) represents an operator in Eq. 1.35. It is the *context* in which a symbol is used that determines its meaning. *Hence, there can be no confusion about what a letter symbol represents. In a given context, it is always unambiguous.*

Funquest: Consider the case $M = N$. Given that the basis set $\left\{\left|e_j\right\rangle; j = 1,2,..,N\right\}$ used in Eq. 1.34 is orthonormal, can you show that $A_{ij} = \langle i|A|j\rangle$?

It should now be clear that the matrix $\mathbb{A}$ *represents* the operator A in Eq. 1.35. When the ket vector is not an eigenvector of A_{op}, the result of operating on the ket by operator A_{op} gives a new vector $|\ \rangle'$. This vector is *not* proportional to the original ket. It can of course be written as a superposition of the complete set of base vectors. Having obtained the result of measurement represented by the operator A_{op}, one would like to get more information about the system. One could measure a different property, say B, which would be represented by the operator B_{op} acting on $|a|a\rangle\rangle$. If the measurement of B leaves the system in a pure eigenstate of the previous measurement of property A, then the measurement B is *compatible* with the measurement A. *The measurements A and B can then be made any number of times and in any order*. Results of the previous measurement would remain unaffected. While the measurement of A renders the system in an eigenstate of A, that of B renders it in an eigenstate of B. When measurements A and B are *compatible*, the operators that represent them have *simultaneous eigenstates*. In essence, compatible measurements are represented by operators which *commute* with each other:

$$[A,B]_- = AB - BA = 0. \tag{1.37}$$

The quantity $[A, B]_-$ is called as the commutator. For commuting properties it is zero. Often, the subscript "−" on the commutator is dropped. In contrast, an anti-commutator is written with a "+" subscript:

$$[R,S]_+ = RS + SR. \tag{1.38}$$

We shall need both the commutator and the anti-commutator in quantum theory.

We have already seen that measurements of the Z- and X- components of the intrinsic dipole magnetic moment of an electron described in the SG experiment (Fig. 1.2d) are *not* compatible with each other. The primary reason for this is that the intrinsic angular momentum of the electron has components that do not commute with each other. Representing their measurements by operators S_z and S_x, we see that

$$\left[S_z,S_x\right]_- \neq 0. \tag{1.39a}$$

We will learn in Chapter 4 that

$$\left[S_z, S_x\right]_- = i\hbar S_y. \tag{1.39b}$$

This would turn out to be a *defining* attribute of angular momentum in quantum theory, as we shall learn in Chapter 4. As you read further, you would gain increasing confidence that quantum theory provides a rigorous mathematical framework that correctly accounts for physical observations of nature. It is a theory to which no exception has been found. The *assumption* made in classical mechanics that position and momentum together specify the state of a system is not tenable since their measurements are not compatible with each other. We have already seen this in Section 1.1 in the discussion on measurements with the Heisenberg microscope. Accordingly, position and momentum must be represented by operators that do not commute. We shall discuss the operator description of position and momentum in the next section.

1.3 Position, Translation Displacement, and Momentum Operators

The position of an object in the usual three-dimensional Euclidean space is typically specified by three independent degrees of freedom that indicate the object's *translational displacement* from a reference point, which is the origin of a reference frame. With reference to such a frame, we therefore represent the measurement of position of an object by the position operator $\vec{x}_{op}$. The measurement returns the position $\vec{x}'$ of the target object, which is the eigenvalue employed to label the position eigenstate as $|\vec{x}'\rangle$. This measurement is then represented by an eigenvalue equation for the position operator:

$$\vec{x}_{op} \underbrace{|\vec{x}'\rangle}_{\text{eigenstate}} = \underbrace{\overset{\text{eigenvalue}}{\vec{x}'}}_{\text{labels eigenstate}} \underbrace{|\vec{x}'\rangle}_{\text{eigenstate}}, \tag{1.40a}$$

$$\text{i.e., } \vec{x}_{op}|\vec{x}'\rangle = \vec{x}'|\vec{x}'\rangle. \tag{1.40b}$$

Equation 1.40a,b must be understood in the spirit of Eqs. 1.30. A measurement resulting in position information is depicted in Fig. 1.3. The tiny detector measures the position accurately; it is considered here to have an infinitesimal size. Equation 1.40a,b is the position eigenvalue equation.

A position *translational displacement operator* $\tau\left(\overrightarrow{dx'}\right)$ would operate on a position eigenvector (i.e., eigenstate), and give you a new vector, which is also an eigenstate of the position operator, but belonging to a *different* eigenvalue. The position *label* for the latter would be *displaced* from the previous one by an amount $\overrightarrow{dx'}$:

$$\tau\left(\overrightarrow{dx'}\right)|\,\vec{x'}\rangle = |\,\vec{x}' + \overrightarrow{dx'}\rangle. \tag{1.41}$$

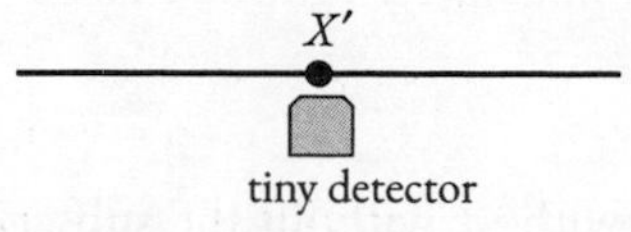

Fig. 1.3 The position operator represents measurement of position of the target object.

Equation 1.41 expresses the response of a *position eigenstate* to the *translational displacement* operator. The new state is an eigenstate of the position operator but belongs to a different eigenvalue, and hence it is *not* an eigenstate of the *translational displacement operator*. We now ask how an arbitrary state vector $|\gamma\rangle$ (which may not be a position eigenket) would respond to the translational displacement operator $\tau\left(\overrightarrow{dx'}\right)$. We express this operation as

$$|\gamma\rangle \rightarrow |\gamma\rangle' = \tau\left(\overrightarrow{dx'}\right)|\gamma\rangle. \tag{1.42}$$

The label γ is what we call as an *appropriate* label, or a *good* label, that describes the arbitrary state. What makes the label *appropriate* and *good* is that it represents a property (or a set of properties) that is measurable. The result of a measurement is what is used to label it. The eigenvalue equation (Eq. 1.30a,b,c) represents this. If several properties can be measured without disturbing each other, then they constitute a set of compatible observables, represented in quantum theory by commuting operators. γ is thus a *good* label if it comes from the result of a measurement of an observable. It can also represent a set of labels coming from the complete set of compatible observables, which would provide eigenvalues of a complete set of commuting operators. It is called as a *good label*, or commonly as *good quantum number*.

We consider an N-dimensional Hilbert space to be spanned by an orthonormal basis $\left\{|u_i\rangle; i = 1,2,..,N\right\}$ in which the vector $|\gamma\rangle$ would be a superposition as described in Eq. 1.26b.

Its *projection* on a base vector, say the jth one, would be

$$\langle u_j|\gamma\rangle = \sum_{i=1}^{N} b_i \langle u_j|u_i\rangle = \sum_{i=1}^{N} b_i \delta_{ji} = b_j, \tag{1.43}$$

which provides a measure of the *probability amplitude* that the arbitrary state $|\gamma\rangle$ is in the jth pure state.

The modulus square,

$$b_j b_j^* = |b_j|^2, \tag{1.44}$$

gives the corresponding probability. It will become increasingly clear that probability and statistics play a fundamental role in quantum theory. The total probability that the system $|\gamma\rangle$ is in one or the other of the N base states is unity; we therefore have

$$1 = \sum_{j=1}^{N} |b_j|^2 = \sum_{j=1}^{N} b_j b_j^* = \sum_{j=1}^{N} \langle u_j|\gamma\rangle\langle\gamma|u_j\rangle = \sum_{j=1}^{N} \langle\gamma|u_j\rangle\langle u_j|\gamma\rangle = \left\langle \gamma \left| \left(\sum_{j=1}^{N} |u_j\rangle\langle u_j| \right) \right| \gamma \right\rangle. \tag{1.45}$$

We know that when the state $|\gamma\rangle$ is *normalized*,

$$1 = \langle\gamma|\gamma\rangle. \tag{1.46}$$

From Eqs. 1.45 and 1.46, it immediately follows that the unit operator can be resolved as

$$1_{op} = 1 = \sum_{j=1}^{N} |u_j\rangle\langle u_j|. \tag{1.47}$$

The unit operator is written with or without the subscript "*op*." For an arbitrary vector $|\gamma\rangle$, we have

$$|\gamma\rangle = 1|\gamma\rangle = \sum_{j=1}^{N} |u_j\rangle\langle u_j||\gamma\rangle = \sum_{j=1}^{N} |u_j\rangle\langle u_j|\gamma\rangle. \tag{1.48a}$$

It is therefore easy to recognize that the operator

$$\Lambda_j = |u_j\rangle\langle u_j| \tag{1.48b}$$

projects out the jth component $|u_j\rangle$ out of $|\gamma\rangle$ with weight $\langle u_j|\gamma\rangle$. The operator defined in Eq. 1.48b is therefore called as the *projection operator*. Equation 1.47 represents the *resolution of unity* in a finite N-dimensional Hilbert space. The position eigenspace is infinite dimensional; the discrete sum must be therefore replaced by integration over whole space, represented by $\int d^3x[\text{integrand}]$. The resolution of unity therefore takes the following form:

$$1_{op} = \int d^3\vec{x}|\vec{x}\rangle\langle\vec{x}|. \tag{1.49a}$$

When the unit operator operates on a position eigenket $|\vec{x}''\rangle$, we get

$$\begin{aligned} 1_{op}|\vec{x}''\rangle &= \int_{\substack{\text{whole}\\\text{space}}} d^3\vec{x}'|\vec{x}'\rangle\langle\vec{x}'||\vec{x}''\rangle \\ &= \int_{\substack{\text{whole}\\\text{space}}} d^3\vec{x}'|\vec{x}'\rangle\langle\vec{x}'|\vec{x}''\rangle \\ &= \int_{\substack{\text{whole}\\\text{space}}} d^3\vec{x}'|\vec{x}'\rangle\delta^3(\vec{x}'-\vec{x}'') = |\vec{x}''\rangle, \end{aligned} \tag{1.49b}$$

wherein Dirac-δ integration is carried out (see, for example, Eq. 11.41, 11.42a,b in [1]).

The response to the translational displacement operator of an arbitrary state $|\gamma\rangle$ is then given, on insertion of the unit operator, by

$$|\gamma\rangle = \tau(\overrightarrow{dx'})\left[\int d^3\vec{x}|\vec{x}\rangle\langle\vec{x}|\right]|\gamma\rangle - \int d^3\vec{x}\left\{\tau(\overrightarrow{dx'})|\vec{x}\rangle\right\}\langle\vec{x}|\gamma\rangle, \tag{1.49c}$$

$$\text{i.e., } |\gamma\rangle' = \int d^3\vec{x}|\vec{x}+\overrightarrow{dx'}\rangle\langle\vec{x}|\gamma\rangle = \int d^3\vec{x}''|\vec{x}''\rangle\langle\vec{x}''-\overrightarrow{dx'}|\gamma\rangle = \int d^3\vec{x}|\vec{x}\rangle\langle\vec{x}-\overrightarrow{dx'}|\gamma\rangle, \tag{1.49d}$$

with $\vec{x} = \vec{x}'' - \overrightarrow{dx'}$. The integration label is of course dummy; it is over the whole space. However, depending on the dummy variable being represented by $\vec{x}$ or $\vec{x}''$, the symbol in the integrand must be consistently shifted. The factor $\langle\vec{x}|\gamma\rangle$ in the integrand in Eq. 1.49c is the *coordinate representation* of the state $|\gamma\rangle$. It is called as a *wavefunction*. It can be written in familiar notation as

$$\psi_\gamma(\vec{x}) = \langle\vec{x}|\gamma\rangle. \tag{1.50}$$

We have identified the subscript γ as a good quantum number, which describes the quantum state. Most often, the wavefunction $\psi_\gamma(\vec{x})$ is an analytical function; it is continuous and possesses continuous derivatives. Its analytical properties will become clear in Section 1.4 when we shall gather experience with the Schrödinger equation. The integrand $\langle\vec{x}-\overrightarrow{dx'}|\gamma\rangle$ in Eq. 1.49d is likewise a wavefunction, but its argument is subjected to a translation displacement:

$$\left\langle \vec{x} - \overrightarrow{dx'} \middle| \gamma \right\rangle = \psi_\gamma\left(\vec{x} - \overrightarrow{dx'}\right) = \psi_\gamma(\vec{x}) - \overrightarrow{dx'} \cdot \vec{\nabla}\psi_\gamma(\vec{x}), \tag{1.51}$$

$\vec{\nabla}\psi$ being the gradient of the wavefunction.

We now examine the translational displacement operator $\tau\left(\overrightarrow{dx'}\right)$ further. Our motivation to introduce it is to effect the transformation indicated in Eq. 1.41. It would transform a position eigenstate $|\,\overrightarrow{x'}\rangle$ to a new position eigenstate $|\,\overrightarrow{x'} + \overrightarrow{dx'}\rangle$, but if the original vector $|\,\overrightarrow{x'}\rangle$ was normalized, we do not expect the new vector $|\,\overrightarrow{x'} + \overrightarrow{dx'}\rangle$ to have a different norm. In other words, the norm of the state vector would be preserved:

$$\left\langle \overrightarrow{x'} + \overrightarrow{dx'} \middle| \overrightarrow{x'} + \overrightarrow{dx'} \right\rangle = \left\langle \overrightarrow{x'} \middle| \overrightarrow{x'} \right\rangle. \tag{1.52}$$

The bra vector $\left\langle \overrightarrow{x'} + \overrightarrow{dx'} \right|$ in the D.C. space can now be identified as the dual conjugate vector of $|\,\overrightarrow{x'} + \overrightarrow{dx'}\rangle = \tau\left(\overrightarrow{dx'}\right)|\,\overrightarrow{x'}\rangle$. In the algebra of Hilbert space and its D.C. space, which we have introduced earlier, the dual conjugate of $\tau\left(\overrightarrow{dx'}\right)|\,\overrightarrow{x'}\rangle$ is written as

$$\left\langle \overrightarrow{x'} \middle| \tau^\dagger\left(\overrightarrow{dx'}\right) = \left\langle \overrightarrow{x'} + \overrightarrow{dx'} \right|. \tag{1.53}$$

The symbol $\dagger$ used as a superscript in Eq. 1.53 is read as "dagger." In general, for any operator Ω, $\Omega^\dagger$ is read as "Ω dagger." The operator $\Omega^\dagger$ is called as "adjoint of Ω," or more fully as "Hermitian adjoint of Ω." We see that preservation under translation of the norm of the state vector necessitates that

$$\left\langle \overrightarrow{x'} + \overrightarrow{dx'} \middle| \overrightarrow{x'} + \overrightarrow{dx'} \right\rangle = \left\langle \overrightarrow{x'} \middle| \tau^\dagger\left(\overrightarrow{dx'}\right)\tau\left(\overrightarrow{dx'}\right) \middle| \overrightarrow{x'} \right\rangle = \left\langle \overrightarrow{x'} \middle| \overrightarrow{x'} \right\rangle, \tag{1.54}$$

$$\text{i.e., } \tau^\dagger\tau = 1. \tag{1.55}$$

Essentially, the result of Eq. 1.55 means that the translational displacement operator must have the property that its adjoint is exactly equal to its inverse:

$$\tau^\dagger = \tau^{-1}. \tag{1.56}$$

The property described by Eq. 1.56 is called *unitarity*. An operator that satisfies this property is called as a *unitary* operator. Every unitary operator necessarily preserves the norm of the vector on which it operates. Without more ado, we now summarize uncomplicated properties of the translational displacement operator:

i. Unitarity: $\tau\left(\overrightarrow{dx'}\right)^\dagger \tau\left(\overrightarrow{dx'}\right) = 1$; i.e., $\tau\left(\overrightarrow{dx'}\right)^\dagger = \tau\left(\overrightarrow{dx'}\right)^{-1}$, (1.57a)

ii. $\tau\left(\overrightarrow{dx''}\right)\tau\left(\overrightarrow{dx'}\right) = \tau\left(\overrightarrow{dx'} + \overrightarrow{dx''}\right)$, (1.57b)

iii. $\tau\left(-\overrightarrow{dx'}\right) = \tau\left(\overrightarrow{dx'}\right)^{-1}$, (1.57c)
and

iv. $\lim_{dx' \to 0} \tau\left(\overrightarrow{dx'}\right) = 1.$ (1.57d)

The arguments $\overrightarrow{dx'}$ and $\overrightarrow{dx''}$ in these equations both represent infinitesimal translational displacements. The property (i) has already been discussed and justified. Properties (ii), (iii), and (iv) are incontrovertible.

Now, $\tau\left(\overrightarrow{dx'}\right)$ has the form $1 - \vec{K} \cdot \overrightarrow{dx'}$, K being Hermitian (see Solved Problem P1.2).

Having studied the effects of the position operator and the translational displacement operator on position eigenkets, we now inquire what would happen if they operate in a different order; i.e., we ask whether or not they commute. For an arbitrary position eigenket $\left|\vec{x'}\right\rangle$, we see that

$$\vec{x}\left(1 - i\vec{K}\bullet\overrightarrow{dx'}\right)\left|\vec{x'}\right\rangle = \vec{x}\tau\left(\overrightarrow{dx'}\right)\left|\vec{x'}\right\rangle = \vec{x}\left|\vec{x'} + \overrightarrow{dx'}\right\rangle = \left(\vec{x'} + \overrightarrow{dx'}\right)\left|\vec{x'} + \overrightarrow{dx'}\right\rangle, \tag{1.58a}$$

and
$$\begin{aligned}\left(1 - i\vec{K}\bullet\overrightarrow{dx'}\right)\vec{x}\left|\vec{x'}\right\rangle &= \tau\left(\overrightarrow{dx'}\right)\vec{x}\left|\vec{x'}\right\rangle \\ &= \tau\left(\overrightarrow{dx'}\right)\vec{x}'\left|\vec{x'}\right\rangle \\ &= \vec{x}'\tau\left(\overrightarrow{dx'}\right)\left|\vec{x'}\right\rangle \\ &= \vec{x}'\left|\vec{x'} + \overrightarrow{dx'}\right\rangle.\end{aligned} \tag{1.58b}$$

Hence, on ignoring the second-order term in infinitesimal translational displacement,

we see that $$\left[\vec{x}, \tau\left(\overrightarrow{dx'}\right)\right]_{-}\left|\vec{x'}\right\rangle = \overrightarrow{dx'}\left|\vec{x'} + \overrightarrow{dx'}\right\rangle \simeq \overrightarrow{dx'}\left|\vec{x'}\right\rangle. \tag{1.59a}$$

Accordingly, to first order the commutator $$\left[\vec{x}, \tau\left(\overrightarrow{dx'}\right)\right]_{-} = \overrightarrow{dx'} = \overrightarrow{dx'}1_{op}, \tag{1.59b}$$

i.e., $$\left[\vec{x}, \left(1 - i\vec{K}\bullet\overrightarrow{dx'}\right)\right]_{-} = \overrightarrow{dx'}1_{op}, \tag{1.59c}$$

1_{op} being the unit operator.

Therefore, $$\left[\vec{x}, \vec{K}\bullet\overrightarrow{dx'}\right]_{-} = i\overrightarrow{dx'}, \tag{1.59d}$$

i.e., $$\vec{x}\left(\vec{K}\bullet\overrightarrow{dx'}\right) - \left(\vec{K}\bullet\overrightarrow{dx'}\right)\vec{x} = i\overrightarrow{dx'} = i\overrightarrow{dx'}1_{op}. \tag{1.59e}$$

The translational displacement $\overrightarrow{dx'}$ and the orthonormal basis $\left\{\hat{e}_j, j = 1,2,3\right\}$ are both arbitrary. Without any lack of generality, we can choose the former to be along one of the basis unit vectors, say the jth. Equation 1.59e then reduces to

$$\vec{x}K_j dx' - K_j dx'\vec{x} = i\hat{e}_j dx' = i\hat{e}_j dx'1_{op}. \tag{1.60}$$

Projecting now the above equation on the Cartesian ℓ^{th} axis,

$$\hat{e}_\ell\bullet\vec{x}K_j dx' - K_j dx'\hat{e}_\ell\bullet\vec{x} = i\hat{e}_\ell\bullet\hat{e}_j dx' = i\hat{e}_\ell\bullet\hat{e}_j dx'1_{op}, \tag{1.61}$$

i.e., $$x_\ell K_j dx' - K_j dx' x_\ell = i\delta_{\ell j} dx'1_{op}. \tag{1.62}$$

Canceling now the common arbitrary magnitude of the infinitesimal displacement on both sides, we get a commutator that has far-reaching consequences in the quantum theory. It is

$$\left[x_\ell, K_j\right] = i\delta_{\ell j}1_{op}, \text{ for } \ell, j = 1,2,3. \tag{1.63}$$

The astute reader would have recognized that the operator that would generate translations in homogeneous space must be the one that corresponds to the classical dynamically conjugate of position; it must therefore be the linear momentum. These are well-known results, intimately connected with one of the most beautiful theorems in physics, namely Noether's theorems (Chapter 1 of Reference [1]). One knows that the momentum that is canonically conjugate to a coordinate is conserved, which is a manifestation of the deep connection between symmetry and a conservation law. The linear momentum is the generator of translations in homogeneous space. This relationship is best described by the generating function for the following canonical transformations:

$$\vec{x} \rightarrow \vec{x}_{\text{new}} = \vec{X} = \vec{x} + d\vec{x}, \tag{1.64a}$$

$$\vec{p} \rightarrow \vec{p}_{\text{new}} = \vec{P}. \tag{1.64b}$$

This transformation is effected by the generating function [6]

$$F(\vec{x},\vec{P}) = \vec{x}\bullet\vec{P} + \vec{p}\bullet d\vec{x}. \tag{1.65a}$$

We now map the classical generating function (Eq. 1.65a) with that in quantum theory, given by

$$\tau\left(\overrightarrow{dx'}\right) = 1 - i\vec{K}\bullet\overrightarrow{dx'}, \tag{1.65b}$$

$\vec{K}$ being a Hermitian operator as found in the Solved Problem P1.2.

The first term in Eq. 1.65b is obviously dimensionless, so the dimension of the operator $\vec{K}$ must be L^{-1}. The choice

$$\vec{K} = \frac{\vec{p}}{\hbar} \tag{1.65c}$$

would seem perfect, since it not only has the required dimension but also provides a beautiful mapping of the generators by comparing

$$\tau\left(\overrightarrow{dx'}\right) = 1 - i\frac{\vec{p}}{\hbar}\bullet\overrightarrow{dx'} \tag{1.65d}$$

with Eq. 1.65a.

The choice $\hbar$ to divide the linear momentum might appear to be arbitrary, but it is a natural one since the PEB constant is the *fundamental* quantum of action. Further confidence in this particular choice would come when we shall discuss Eq. 1.85 in Section 1.4. It follows immediately from Eq. 1.63 that

$$\left[x_{\ell},p_{j}\right] = i\hbar\delta_{\ell j}1_{op} \text{ for } \ell,j = 1,2,3. \tag{1.66a}$$

More generally, representing the position operator by q_{op} and momentum by p_{op}, we have

$$\left[q_{op},p_{op}\right]_{-} = i\hbar 1_{op}. \tag{1.66b}$$

We may drop the subscripts in Eq. 1.66b and simplify the notation to write the same commutator as

$$[q,p]_{-} = i\hbar, \tag{1.66c}$$

wherein the unit operator on the right-hand side of Eq. 1.66c is of course present, invisibly. We continue to use the same symbols (like "q" and "p" in Eq. 1.66c) in quantum theory as we did in classical physics, but with *new* meanings. The operators q and p now respectively represent generalized coordinate and the generalized momentum that is conjugate to it. This must *always* be remembered. Equations 1.66a–c involve operators, although the symbols used therein appear even in classical physics. These relations are at the very foundation of the famous inequality that we shall discuss in Section 1.4, known as the Heisenberg principle of uncertainty.

The result in Eqs. 1.66a–c is completely in line with the discussion in Section 1.2 on the Heisenberg microscope. It tells us that the operators for position and momentum do not commute, and measurements of the corresponding observables are not compatible with each other. Furthermore, by operating on an arbitrary function $f(x)$ of x, it is easy to see that the commutator

$$\left[x, \frac{d}{dx}\right]_- = -1. \tag{1.67}$$

It therefore follows by comparing Eq. 1.66 and Eq. 1.67 that

$$p_x = -i\hbar\frac{d}{dx}, \text{ and } \vec{p} = -i\hbar\vec{\nabla}. \tag{1.68a}$$

Hence, for arbitrary states $|\alpha\rangle$ and $|\beta\rangle$,

$$\begin{aligned} \langle\beta|\vec{p}|\alpha\rangle &= \int_{\substack{\text{whole}\\ \text{space}}} d^3\vec{r}\,\langle\beta|\vec{r}\rangle\left(-i\hbar\vec{\nabla}\right)\langle\vec{r}|\alpha\rangle \\ &= \int_{\substack{\text{whole}\\ \text{space}}} d^3\vec{r}\,\psi_\beta^*(\vec{r})\left(-i\hbar\vec{\nabla}\right)\psi_\alpha(\vec{r}) \\ &= -i\hbar\int_{\substack{\text{whole}\\ \text{space}}} d^3\vec{r}\,\psi_\beta^*(\vec{r})\vec{\nabla}\psi_\alpha(\vec{r}). \end{aligned} \tag{1.68b}$$

Conservation of linear momentum in homogeneous space is the very basis for Newton's third law (Chapter 1 of Reference [1]). It is therefore only natural that the quantum momentum *operator* involves the gradient operator; after all it is the gradient of a function that relates the value of a continuous function at neighbouring points in homogeneous space.

The relationship between the position operator and the momentum operator is a very general one; it is applicable to operators corresponding to any pair of canonically conjugate generalized position and generalized momentum. However, there is an exception. The canonically conjugate relationship between energy and time has to be handled differently. There is no operator for *time* in quantum theory, though there is one for *time delay* [7]. This has an important bearing on the temporal evolution of the wavefunction (Eq. 1.50). Typically, it is a function of both space and time, written as

$$\psi_\gamma(\vec{x},t) = \langle\vec{x}|\gamma,t\rangle. \tag{1.69}$$

Its time-evolution is given by $\frac{\partial}{\partial t}\left\{\psi_\gamma(\vec{x},t)\right\}$. Since the generalized momentum *conjugate* to time is energy, we should expect the operator $\frac{\partial}{\partial t}$ to be *related* to the operator for energy.

It is instructive to appreciate these connections. Just as linear momentum is conserved in homogeneous space, energy is conserved from past to present to future, i.e., as time changes. We must *not* however hasten to develop an analogue of Eq. 1.66 involving time and energy, since there is no operator for time in quantum theory. The operator for energy is called as the Hamiltonian operator, introduced in the next section.

1.4 Time-Evolution Operator, Schrödinger Equation, and Heisenberg Principle of Uncertainty

In this section, we shall introduce two of the most celebrated aspects of quantum theory, namely the Heisenberg principle of uncertainty and the Schrödinger equation. This will require the mathematical formalism introduced in the previous section to be developed further. The investment in mathematical abstraction turns out to be a smart one, as it leads us to understand the laws of nature. As fascinating events in the twentieth century have revealed, experiments confirmed this, leading to phenomenal advances in technology that revolutionized life on earth. Paul Adrien Maurice Dirac had this to say about developing the mathematical framework:

> *If you are receptive and humble, mathematics will lead you by the hand. Again and again, when I have been at a loss how to proceed, I have just had to wait until I have felt the mathematics led me by the hand. It has led me along an unexpected path, a path where new vistas open up, a path leading to new territory, where one can set up a base of operations, from which one can survey the surroundings and plan future progress.*

The mathematical structure of matrix algebra is very well suited to develop quantum theory in which operators play a fundamental role. You have already discovered in answering the *Funquest* after Eq. 1.36 how the matrix

$$\mathbb{A} = \left[\langle i|A_{op}|j\rangle\right] = \left[\langle i|A|j\rangle\right];\ i,j = 1,2,..,N \tag{1.70}$$

represents the operator A_{op}. In Eq. 1.70, the operator is deliberately written both with the subscript "*op*" and without it. You must get used to this equivalence and interpret the term correctly from its context. In fact, even the left-hand side is often written only as a slim letter A; it would be understood from the context whether it is a symbol for an operator or the dynamical variable it represents, or for the matrix that represents the operator.

For many important applications in quantum theory, we make use of operators known as *linear operators*. An operator A is said to be a *linear operator* if for arbitrary ket vectors $|\alpha\rangle$ and $|\beta\rangle$ in the Hilbert space, and for arbitrary complex numbers c_α and c_β, the following defining criterion holds:

$$A\left[c_\alpha|\alpha\rangle + c_\beta|\beta\rangle\right] = c_\alpha A|\alpha\rangle + c_\beta A|\beta\rangle. \tag{1.71}$$

We also define *Hermitian adjoint* $A^\dagger$ of an operator A (often simply called *adjoint* of A), relying on the *one-to-one* correspondence between the ket vector "$A|\alpha\rangle$" in the direct space and its dual conjugate "$\langle\alpha|A^\dagger$":

$$A|\alpha\rangle \overset{\text{DC}}{\leftrightarrow} \langle\alpha|A^{\dagger}. \tag{1.72}$$

We have in fact already used the Hermitian adjoint of the translational displacement operator (Eq. 1.53). Now, the dual conjugate of the vector $AB|\alpha\rangle$, i.e., $A\left(B|\alpha\rangle\right)$, is $\left(\langle\alpha|B^{\dagger}\right)A^{\dagger}$, i.e., $\langle\alpha|B^{\dagger}A^{\dagger}$. It therefore follows that the *adjoint of a product of two operators* is the product of their adjoints in the reverse order:

$$(AB)^{\dagger} = B^{\dagger}A^{\dagger}. \tag{1.73}$$

A simple way of recognizing this important result is to note that the result of the operation on a ket vector by an operator gives another ket vector for which the D.C. relations are

$$\left|B|\alpha\rangle\right\rangle \overset{\text{DC}}{\leftrightarrow} \left\langle\langle\alpha|B^{\dagger}\right|, \tag{1.74}$$

and $$A\left|B|\alpha\rangle\right\rangle \overset{\text{DC}}{\leftrightarrow} \left\langle\langle\alpha|B^{\dagger}\right|A^{\dagger}. \tag{1.75}$$

The inverse of an operator, the unit operator, and the null operator have already been defined in Eq. 1.29.

If, for two operators X and Y,

$$X|\alpha\rangle = Y|\alpha\rangle, \tag{1.76}$$

for an arbitrary operand ket $|\alpha\rangle$, then the two operators are said to be *equal.*

Addition of operators is

i. commutative: $A + B = B + A$, (1.77a)

ii. associative: $(A + B) + C = A + (B + C) = A + B + C$. (1.77b)

We have already seen that the operation on a ket vector by an operator gives another ket vector in the Hilbert space. You also know that when the original ket is in an eigenstate of the measurement represented by the operator, the new ket is simply a scalar (complex number) multiple of the original; if not, it can always be expressed as a linear superposition of a complete set of linearly independent basis that spans the Hilbert space. While a new *ket* results from the operation on a ket by an operator, a new *operator* can be constructed from a vector in the direct space and another in the adjoint (D.C.) space. $|\alpha\rangle\langle\beta|$ is an *operator* that operates on a ket-vector $|\ \rangle$, essentially on the right, to get a new $|\ \rangle$, and operates on a bra-vector $\langle\ |$, essentially on the left, to get a new $\langle\ |$. We already know that $\langle\alpha|\beta\rangle$ is called as the *inner* product; $|\alpha\rangle\langle\beta|$ is called as the *outer* product. An operator placed on the right of $|\ \rangle$ vector is illegal $\left(\xcancel{|\alpha\rangle A}\right)$, and an operator placed on the left of $\langle\ |$ vector is also incorrect $\left(\xcancel{A\langle\alpha|}\right)$. The direct product of two ket vectors which belong to the same Hilbert space, which you may write as $|\alpha\rangle|\beta\rangle$, is also prohibited; but if the two ket vectors belong to different Hilbert spaces (such as the Hilbert spaces of two independent particles), which you may write as $|\alpha\rangle|b\rangle$, then it is admissible, and is called as *direct* product or *tensor* product. The direct product of vectors will be extensively used in Chapters 4 and 11. It should now be clear that in the term $|\alpha\rangle\langle\beta|\gamma\rangle$, the operator $|\alpha\rangle\langle\beta|$ operates on $|\gamma\rangle$ from its left, and in the term $\langle\gamma|\alpha\rangle\langle\beta|$, the operator $|\alpha\rangle\langle\beta|$ operates on $\langle\gamma|$ from its right. The associative property of multiplication applies to all legal multiplications:

$$|\alpha\rangle\langle\beta|\gamma\rangle = (|\alpha\rangle\langle\beta|)|\gamma\rangle = |\alpha\rangle(\langle\beta|\gamma\rangle). \tag{1.78}$$

Funquest: Show that $(|\alpha\rangle\langle\beta|)^{\dagger} = |\beta\rangle\langle\alpha|$.

A mathematical property every physical observable has is called its *hermiticity*. Physical properties that are observable are represented by operators that are necessarily *Hermitian*. This is a necessary, and not sufficient, property that characterizes operators which represent observables. It is named after Charles Hermite (1822–1901); the moon's north crater is named after him. The defining criterion for an operator Ω to be identified as a Hermitian operator is its *self-adjoint* property:

$$\Omega^{\dagger} = \Omega. \tag{1.79}$$

Position, momentum, angular momentum, energy, etc., are physical observables, and they are represented by Hermitian operators. We have already introduced in the previous section operators for position and momentum. The operator for angular momentum will be introduced in Chapter 4. The operator for energy, called as the Hamiltonian, is introduced later in this chapter. There are of course many observables, and each is represented by a unique Hermitian operator.

In the remaining part of this section, we introduce two celebrated expressions of quantum theory: (i) Heisenberg principle of uncertainty and (ii) Schrödinger equation. These require probabilistic interpretation of the expansion of a state vector in a complete set of linearly independent basis that spans the Hilbert space (Eq. 1.26), with expansion coefficients that are in general complex numbers (Eq. 1.43 and Eq. 1.44). From the sequential SG experiment (Figs. 1.2a–d), we have learned that (i) measurement of an observable of a system in *a priori* unknown state results in *collapse* from the *unknown superposition* of basis states into a *known* eigenstate, (ii) repeated measurements of the same observable of a system that is known *a priori* to be in an eigenstate leaves the system invariant in the *same* eigenstate and (iii) measurement of a second observable that is not *compatible* with the previous one throws an eigenstate (of the first measurement) into a superposition state.

Funquest: Can you take a pause here, revisit the discussion on the SG experiments (Figs. 1.2a–d) and consolidate in your mind the three points (i)–(iii) mentioned earlier?

What makes quantum theory so fascinating and incredibly useful is its intimacy with measurements and observations. This ground reality, illustrated by the sequential SG experiments, remains at the very heart of quantum theory; it permeates through its every application.

Funquest: Do commuting operators share *all* of their eigenvectors?
Also, is it possible for two non-commuting operators to have a common eigenvector?

We now consider an arbitrary state $\left|\alpha, t_0\right\rangle$ and ask what the state would be at a *different* time t. We may express the latter by $\left|\alpha, t_0 : t\right\rangle$. We are therefore seeking to determine the time evolution operator $U(t, t_0)$ such that

$$\left|\alpha, t_0 : t\right\rangle = U(t, t_0)\left|\alpha, t_0\right\rangle. \tag{1.80}$$

In order to determine an explicit form that the time evolution operator would have, we summarize its essential *desirable* properties:

i. It must preserve the norm of the state vector. Hence,

$$\left\langle \alpha, t_0 : t \middle| \alpha, t_0 : t\right\rangle = \left\langle \alpha, t_0 \middle| \alpha, t_0\right\rangle, \tag{1.81a}$$

i.e., $$\left\langle \alpha, t_0 \middle| U(t, t_0)^\dagger U(t, t_0) \middle| \alpha, t_0\right\rangle = \left\langle \alpha, t_0 \middle| \alpha, t_0\right\rangle, \tag{1.81b}$$

i.e., the time evolution operator must be unitary:

$$1 = U(t, t_0)U(t, t_0)^\dagger = U(t, t_0)^\dagger U(t, t_0), \tag{1.81c}$$

i.e., $$U(t, t_0)^\dagger = U(t, t_0)^{-1}. \tag{1.81d}$$

ii. The temporal evolution of a state from a certain reference instant t_0 until a later time t_2 must be the same as if it is attained in two segments: from the start t_0 to an intermediate instant t_1, with $t_0 < t_1 < t_2$, followed by time evolution from t_1 to t_2. Hence,

$$U(t_2, t_0) = U(t_2, t_1)U(t_1, t_0), \; t_1 \text{ being } \textit{any} \text{ intermediate instant of time.} \tag{1.82}$$

iii. Finally, one must expect that the limiting value of the time evolution operator as the time-interval from start to finish shrinks to zero must be the unit operator:

$$\lim_{\delta t \to 0} U(t_0 + \delta t, t_0) = 1_{op}. \tag{1.83}$$

The three properties we have listed here are similar to the properties given by Eqs. 1.57a, 1.57b, and 1.57d, which we considered for the translational displacement operator. The analogue of Eq. 1.57c is however omitted since *time-reversal* has a subtle meaning in quantum theory [7]. The time-reversal operator is *not* to be mistaken as inverse of the unitary time-evolution operator (Eq. 1.80). The time-reversal operator is *anti-unitary*; it is discussed in Appendix A. The properties described by Eq. 1.81 through Eq. 1.83 are satisfied (as can be readily verified using the method used in Solved Problem P1.2) if the time evolution operator is considered to have the following form:

$$U(t_0 + \delta t, t_0) = 1 - i\Omega\delta t, \tag{1.84}$$

Ω being a Hermitian operator, whose dimension must be T^{-1}. Since energy and time are canonically conjugate to each other, one would expect that Ω is related to the operator for energy, i.e., to H, the Hamiltonian (Eq. 1.7). If you allow your intuition to assist you, a heuristic approach would lead you to recognize that the operator Ω can be identified as

$$\Omega = \frac{H}{\hbar}, \tag{1.85}$$

and accordingly, the time-evolution operator becomes

$$U(t_0 + \delta t, t_0) = 1 - \frac{i}{\hbar}H\delta t. \tag{1.86a}$$

For evolution from an initial time t_i to a final time t_f, the time evolution operator is obtained by dividing the time interval $(t_i < t_f)$ into n infinitesimal sub-intervals, and let $n \to \infty$. The time-evolution operator is thus given by

$$U\left(t_f, t_i\right) = \lim_{n\to\infty}\left(1 - \frac{i}{\hbar}\frac{t_f - t_i}{n}H\right)^n = \exp\left(-\frac{i}{\hbar}H\left(t_f - t_i\right)\right). \tag{1.86b}$$

The choice of angular momentum in the denominator in Eq. 1.85 is obvious, just as it was in the denominator of Eq. 1.65. What is obscure however is the fact that it is $\hbar$ (PEB constant divided by 2π), since dimensional analysis alone would accommodate an arbitrary multiple of the same. The specific *value* of the common denominator in Eq. 1.65 and Eq. 1.85 is a smart choice. In Chapter 2, using Feynman's path integral approach to quantum mechanics, we will see how $\hbar$ plays a pivotal role in providing backward integration of classical mechanics in quantum theory as a limiting case.

Now, from Eq. 1.82 and Eq. 1.86,

$$U(t + \delta t, t_0) = U(t + \delta t, t)U(t, t_0) = \left(1 - \frac{i}{\hbar}H\delta t\right)U(t, t_0), \tag{1.87a}$$

i.e.,
$$\frac{U(t + \delta t, t_0) - U(t, t_0)}{\delta t} = -\frac{i}{\hbar}HU(t, t_0). \tag{1.87b}$$

On seeking the limit $\delta t \to 0$, we obtain the time-rate at which the time-evolution operator varies:

$$i\hbar\frac{\partial U}{\partial t} = HU(t, t_0). \tag{1.88}$$

Operating now on an arbitrary state vector $\left|\alpha, t_0\right\rangle$,

$$i\hbar\frac{\partial U}{\partial t}\left|\alpha, t_0\right\rangle = HU(t, t_0)\left|\alpha, t_0\right\rangle, \tag{1.89a}$$

i.e.,
$$i\hbar\frac{\partial}{\partial t}\left[U(t, t_0)\left|\alpha, t_0\right\rangle\right] = H\left[U(t, t_0)\left|\alpha, t_0\right\rangle\right], \tag{1.89b}$$

or
$$i\hbar\frac{\partial}{\partial t}\left|\alpha, t_0 : t\right\rangle = H\left|\alpha, t_0 : t\right\rangle. \tag{1.90}$$

Equation 1.90 states that the time-rate at which the state vector $\left|\alpha, t_0 : t\right\rangle$ evolves is given by the operation on it by the Hamiltonian operator multiplied by $\frac{1}{i\hbar}$. It is called as Schrödinger equation for the state vector $\left|\alpha, t_0 : t\right\rangle$. Its coordinate representation (defined by Eq. 1.50) is

$$i\hbar\frac{\partial}{\partial t}\left\langle\vec{x}\middle|\alpha, t_0 : t\right\rangle = H\left\langle\vec{x}\middle|\alpha, t_0 : t\right\rangle, \tag{1.91a}$$

i.e.,
$$i\hbar\frac{\partial\psi_\alpha(\vec{x}, t)}{\partial t} = H\psi_\alpha(\vec{x}, t), \tag{1.91b}$$

which is the Schrödinger equation for the wavefunction. Equation 1.91a is written in the *Dirac* notation, and Eq. 1.91b in the *de Broglie–Schrödinger* notation.

The classical viewpoint was that physical state of a system is represented by (q,p), and its temporal evolution is given by the Hamilton's equations (Eqs. 1.7a,b). Since (q,p) are not simultaneously measurable accurately, classical mechanics is replaced by quantum mechanics in which the system is represented by a state vector $\left|\alpha,t_0 : t\right\rangle$ in the Hilbert space, or equivalently by its coordinate representation, namely the wavefunction. Its evolution with the passage of time is given by the time-derivative (Eqs. 1.91a,b) of the state vector (or equivalently that of the wavefunction). In analogy with the term "equation of motion" that describes the Hamilton (or Newton, or Lagrange) equations, the Schrödinger equation is often referred to as the quantum equation of motion for a physical system. Replacement of classical mechanics by a new theory is demanded by the fact that simultaneous accurate description of *pairs* of some physical properties (e.g., (q,p), (S_x,S_y)) is not possible since their measurements are not compatible with each other. A physicist's effort to describe nature must not run into internal inconsistency. Quantum physics provides a successful replacement of classical physics. The Schrödinger equation circumvents the problem faced in classical mechanics faced in the analysis of Heisenberg microscope's experiment. It does not ask for simultaneous information about position and momentum. Instead, it works with the state vector. This may look odd, but using this approach (a) one can describe the physical state of a system in nature and predict its temporal evolution and (b) develop control over the physical properties for human benefit. In order to see how quantum theory connects to concrete properties of physical systems, one must wait till the new formalism is further developed. At this point, we have only gotten our feet wet in the sands of the beach of an ocean in which we must swim confidently!

We have learned from the SG experiment that a measurement induces the system to collapse into an eigenstate of a measurement. There are of course various eigenstates present in the *a priori* superposition state. Into which particular eigenstate the system would collapse on measurement is probabilistic. Statistical considerations are fundamental to quantum theory. The very notion of probability in quantum theory is however intrinsically *different* from that in classical physics. Classical physics being deterministic, statistics enters only because we often have to deal with too much information to process and *average* properties are sufficient to address our concerns. However, laws of nature are not describable by classical mechanics. Quantum mechanics describes them correctly, but it employs statistical description even for a single particle – in fact, even for no-particle (vacuum)! Strange as this may sound, we must allow our acquaintance with quantum theory to grow on us till we can use it rewardingly. Our first step in this direction is to define *expectation value* of an observable/operator. Using it, we shall define *dispersion* of that observable/operator. Hereafter, we shall always bear in mind that quantum operators represent physical observables; we shall always expect quantum theory to connect intimately with measurements, experiments, and observations. Later, we shall consider *transition* of a physical system from an initial state to a final state. An interaction that effects this transition would also be represented by appropriate operators. For now, we consider operators that represent observable physical properties of a system and proceed to define their *expectation values*.

The operator A_{op} for an observable A would often be written without the subscript "*op*." The expectation value of A when the system is in an arbitrary state $\left|\alpha\right\rangle$ is written as $\left\langle A\right\rangle$ or $\left\langle\alpha\middle|A\middle|\alpha\right\rangle$ and is defined below. We consider the Hilbert space to be spanned by a complete set of orthonormal base vectors that are eigenvectors of the operator A: $A\left|a_i\right\rangle = a_i\left|a_i\right\rangle$; $i = 1,2,\ldots$

The basis $\{|a_i\rangle;\ i = 1,2,\ldots\}$ is thus an eigenbasis of the operator A. The expectation value of the operator A in an arbitrary state $|\alpha\rangle$ is *defined* sandwiching twice the resolution of the unit operator as

$$\langle A\rangle = \langle\alpha|A|\alpha\rangle = \sum_{a'}\sum_{a''}\langle\alpha|a'\rangle\langle a'|A|a''\rangle\langle a''|\alpha\rangle, \tag{1.92a}$$

i.e.,
$$\langle A\rangle = \sum_{a'}\sum_{a''}\langle\alpha|a'\rangle a''\langle a'|a''\rangle\langle a''|\alpha\rangle, \tag{1.92b}$$

or
$$\begin{aligned}\langle A\rangle &= \sum_{a'}\sum_{a''}\langle\alpha|a'\rangle a''\delta_{a'a''}\langle a''|\alpha\rangle \\ &= \sum_{a'}\langle\alpha|a'\rangle\langle a'|\alpha\rangle a' \\ &= \sum_{a'}\left|\langle\alpha|a'\rangle\right|^2 a' \\ &= \sum_{a'}\left|\langle a'|\alpha\rangle\right|^2 a'.\end{aligned} \tag{1.92c}$$

We see that the right-hand side of Eq. 1.92c is a sum of all eigenvalues a', each weighted by the modulus square $\left|\langle a'|\alpha\rangle\right|^2$. Equation 1.92c has a classical analogue. For example, in a class of 100 students, if the probability that a student has height h_i is p_i, then the average of the heights of all students is $\sum_i p_i h_i$. This is just the kind of sum we have in Eq. 1.92c. The term *expectation value* is thus absolutely apt; it is the average value. The interpretation of weight factors $\left|\langle a'|\alpha\rangle\right|^2$ in Eq. 1.92c as *probability* is in accordance with our earlier discussion on Eq. 1.44, the complex coefficients $\langle a'|\alpha\rangle$ being the probability amplitudes. The eigenvalues a' are returned as results of measurements of A with probability $\left|\langle a'|\alpha\rangle\right|^2$. This interpretation of probability (known as the *Born interpretation*, after Max Born) is of pivotal importance in using *Bell inequality* to interpret *reality*. This will be discussed in Chapter 11.

The SG experiment is an illustration of a measurement process; it seeks to measure the component of the intrinsic magnetic moment of the atoms passing through the SG apparatus. The separation of the input atomic beam in the SG apparatus of Fig. 1.2a into two equally intense components is an experimental manifestation of the fact that the two components, "up" and "down," of the magnetic moment are returned as a result of their measurement, with each having an equal probability of ½. Input to the second SG apparatus (Fig. 1.2b) consists purely of the "up" component, and hence its measurement returns the same value with probability unity, i.e., with certainty. This one-to-one correspondence between experiments and theory makes quantum physics incredibly useful, elegant, and beautiful. Having defined the *expectation value* of an operator, we now define its *dispersion*, as

$$\left\langle(\Delta A)^2\right\rangle = \left\langle\left(A - \langle A\rangle\right)^2\right\rangle, \tag{1.93a}$$

i.e., $$\left\langle(\Delta A)^2\right\rangle = \left\langle A^2 - A\langle A\rangle - \langle A\rangle A + \langle A\rangle^2\right\rangle, \tag{1.93b}$$

i.e., $$\left\langle(\Delta A)^2\right\rangle = \left\langle A^2 - 2A\langle A\rangle + \langle A\rangle^2\right\rangle, \tag{1.93c}$$

i.e., $$\left\langle(\Delta A)^2\right\rangle = \langle A^2\rangle - 2\langle A\rangle\langle A\rangle + \langle A\rangle^2 = \langle A^2\rangle - \langle A\rangle^2. \tag{1.93d}$$

The *dispersion* of an observable (i.e., the operator that represents it) is also called as *mean square deviation*, and alternatively also as *variance*. We note that it is state-dependent and vanishes if the system is in an eigenstate of the operator. Quantitative limits of accuracy of simultaneous measurements of position and momentum can be stated in terms of their *dispersion*. This limit would appear as Heisenberg's uncertainty principle. Its satisfactory expression requires an understanding of dispersion (Eq. 1.93) of an observable. We consider two Hermitian operators, A and B, corresponding to observations of two physical properties. For each, we shall now consider the difference operators,

$$\Delta A = A - \langle A \rangle, \tag{1.94a}$$

and $$\Delta B = B - \langle B \rangle. \tag{1.94b}$$

We now consider two vectors $|a\rangle$ and $|b\rangle$, obtained by operating ΔA and ΔB on an arbitrary vector $|\ \rangle$. We have not labeled the arbitrary vector operand; being arbitrary, the label does not matter. However, we must label the vectors that *result* from the operations on $|\ \rangle$ by the *difference* operators. Thus,

$$|a\rangle = \Delta A|\ \rangle, \tag{1.95a}$$

and $$|b\rangle = \Delta B|\ \rangle. \tag{1.95b}$$

The operand $|\ \rangle$ on the right-hand side of Eqs. 1.95a and 1.95b may be arbitrary, but it is the *same* in both the cases. Using the Schwarz inequality (Solved Problem P1.3a) for the vectors in Eqs. 1.95a and 1.95b, we see that

$$\langle\ |\Delta A^\dagger \Delta A|\ \rangle \langle\ |\Delta B^\dagger \Delta B|\ \rangle \geq \left|\langle\ |\Delta A^\dagger \Delta B|\ \rangle\right|^2. \tag{1.96a}$$

Using hermiticity of the operators A and B, we have

$$\langle\ |(\Delta A)^2|\ \rangle \langle\ |(\Delta B)^2|\ \rangle \geq \left|\langle\ |\Delta A \Delta B|\ \rangle\right|^2. \tag{1.96b}$$

Now, $$\Delta A\Delta B = \frac{1}{2}(\Delta A\Delta B + \Delta B\Delta A + \Delta A\Delta B - \Delta B\Delta A) = \frac{1}{2}[\Delta A, \Delta B]_+ + \frac{1}{2}[\Delta A, \Delta B]_-, \tag{1.97}$$

where $[\ ,\ ]_+$ is the anti-commutator and $[\ ,\ ]_-$ is the commutator.

Furthermore,

$$\begin{aligned}[\Delta A, \Delta B]_- &= [A - \langle A \rangle, B - \langle B \rangle]_- \\ &= AB - \cancel{A\langle B \rangle} - \cancel{\langle A \rangle B} + \cancel{\langle A \rangle\langle B \rangle} - BA + \cancel{B\langle A \rangle} + \cancel{\langle B \rangle A} - \cancel{\langle B \rangle\langle A \rangle},\end{aligned}$$

i.e., $$[\Delta A, \Delta B]_- = [A - \langle A \rangle, B - \langle B \rangle]_- = [A, B]_-. \tag{1.98}$$

Hence, $$\Delta A\Delta B = \frac{1}{2}[\Delta A, \Delta B]_+ + \frac{1}{2}[A, B]_-. \tag{1.99}$$

Furthermore, we see that

$$\begin{aligned}[A, B]_-^{\ \dagger} &= (AB - BA)^\dagger \\ &= (AB)^\dagger - (BA)^\dagger \\ &= B^\dagger A^\dagger - A^\dagger B^\dagger \\ &= BA - AB = -[A, B]_-,\end{aligned} \tag{1.100}$$

i.e., $[A, B]_-$ is anti-Hermitian.

Let us now examine the Eq. 1.99. Its right-hand side consists of two terms, the first of which $\left(\frac{1}{2}[\Delta A, \Delta B]_+\right)$ is Hermitian, and the second $\left(\frac{1}{2}[A, B]_-\right)$ is anti-Hermitian. Hence, on taking the expectation value of all the terms, we see that the expectation value of the left-hand side is a sum of

i. the expectation value of the first term on the right-hand side, which is necessarily real, and
ii. the expectation value of the second term on the right-hand side, which is necessarily imaginary (see Exercise E1.2).

We therefore get

$$\langle \Delta A \Delta B \rangle = \underbrace{\frac{1}{2}\langle [\Delta A, \Delta B]_+ \rangle}_{\text{real number}} + \underbrace{\frac{1}{2}\langle [A, B]_- \rangle}_{\text{imaginary number}}. \tag{1.101}$$

Its modulus square is

$$|\langle \Delta A \Delta B \rangle|^2 = \frac{1}{4}|\langle [\Delta A, \Delta B]_+ \rangle|^2 + \frac{1}{4}|\langle [A, B]_- \rangle|^2. \tag{1.102}$$

Using Eq. 1.96 (obtained from Schwarz inequality) together with the result in Eq. 1.102, we arrive at a fundamental relation:

$$\langle (\Delta A)^2 \rangle \langle (\Delta B)^2 \rangle \geq |\langle \Delta A \Delta B \rangle|^2 \geq \frac{1}{4}|\langle [A, B]_- \rangle|^2, \tag{1.103a}$$

i.e., dropping the subscript '*minus sign*' after the commutator to simplify the notation,

$$\langle (\Delta A)^2 \rangle \langle (\Delta B)^2 \rangle \geq \frac{1}{4}|\langle [A, B] \rangle|^2. \tag{1.103b}$$

The relations displayed in Eq. 1.103a and Eq. 1.103b are known as the *principle of uncertainty*. The relationship between the left-hand side with that on the right is a *combination of an inequality and an equality. It is extremely important to remember this*! The uncertainty relations tell us that the product of the dispersion of two observables A and B, represented by their quantum operators, is always *greater than or equal to* $\frac{1}{4}|\langle [A, B] \rangle|^2$. The product of the two dispersions is zero if and only if the two observables are compatible; i.e., if the operators that represent them commute. Using now the square root of the dispersion of the two observables,

$$\delta A = \sqrt{\langle (\Delta A)^2 \rangle} \tag{1.104a}$$

and $$\delta B = \sqrt{\langle (\Delta B)^2 \rangle}, \tag{1.104b}$$

the *principle of uncertainty* may be written in terms of the square roots of the dispersions:

$$\delta A \ \delta B \geq \frac{1}{2}|\langle [A, B] \rangle|. \tag{1.105}$$

The form in Eq. 1.105 is rather commonly found in literature. Its left-hand side must be understood in terms of the square roots of the dispersions of the two observables A and B.

The special case of Eq. 1.105 when the two observables in question are the *position* and *momentum* is of great interest. Observations of position and momentum are not compatible. Their commutator is given by Eq. 1.66. It follows from Eq. 1.66 and Eq. 1.105 that

$$\delta q \delta p \geq \frac{\hbar}{2}. \tag{1.106}$$

It is in the form expressed in Eq. 1.106 that the principle of uncertainty is most widely called as *Heisenberg's principle of uncertainty*. The uncertainties δq and δp are the square roots of *dispersion* respectively of the two conjugate variables position and momentum. The uncertainties must therefore be understood essentially in terms of expectation values and dispersion relations. Assigning any meaning to these terms using classical ideas would be hazardous. The quantum uncertainty relations (Eq. 1.66 and correspondingly Eq. 1.106) hold between dynamical variables (operators) that are canonically conjugate to each other. The canonical momentum $\vec{p}_c$ must be replaced by the operator

$$\vec{p}_c \to -i\hbar\vec{\nabla}, \tag{1.107}$$

so that Eqs. 1.65 and 1.66 remain compatible with each other. Considering the canonically conjugate relationship between energy and time it becomes pertinent to ask if an energy–time uncertainty relation similar to Eq.1.66 and Eq. 1.106 holds. It is a weighty quantum curio that indeed one can obtain an energy–time uncertainty relation similar to Eq. 1.106, even if *only* in appearance. This relationship is

$$\delta E \delta t \geq \frac{\hbar}{2}, \tag{1.108}$$

in which the uncertainty δE in energy, and δt in time, is however *defined* differently. Because of this difference, explained below, the Eq. 1.108 for energy–time only *looks* like Eq. 1.106, but is *fundamentally* different from it. We note that Eq. 1.106 is built on the basis of Eq. 1.103 and Eq. 1.104 which employ operators for both the canonically conjugate variables. However, there is *no* operator for time (see Reference [7], *and* references therein).

Prior to discussing the genesis and the meaning of energy–time uncertainty relation (Eq. 1.108), we shall first discuss an important connection between the wavefunction (Eq. 1.69) and physical properties of a system. The wavefunction of our interest is the solution to the Schrödinger equation (Eq. 1.91). Being a differential equation, the nature of the solutions depends on the boundary conditions. We shall first discuss this for *bound* states of a system. A bound state is characterized by the property that (i) its wavefunction vanishes asymptotically (i.e., as $r \to \infty$) and (ii) in the asymptotic region the spatial gradient of the wavefunction is finite. We shall learn in Chapter 3 that the asymptotic properties of a wavefunction place a constraint on the eigenvalues of the Hamiltonian, viz. the energy spectrum must be discrete. In contrast, unbound states have a continuum of energy eigenvalues. The continuum eigenfunctions do not vanish asymptotically; they have an oscillatory (e.g., sinusoidal) behaviour in the asymptotic region. A property of utmost importance from the physical point of view is the *Max Born interpretation of the wavefunction*. For bound-state wavefunctions, we have

$$\iiint_{\text{ws}} \psi_\gamma(\vec{x},t)^* \psi_\gamma(\vec{x},t)\, dV = \iiint_{\text{ws}} \langle \gamma,t|\vec{x}\rangle\langle\vec{x}|\gamma,t\rangle\, dV = N, \tag{1.109}$$

where N is finite. The letters "ws" below the integration symbol remind us that integration is over "whole space." The whole space integral being finite, such wavefunctions are called "square integrable." After all, it is the modulus *square* of the complex wavefunction that is being integrated. Now, since a solution to a differential equation multiplied by a constant is also a solution to the same equation, it is convenient to define the "normalized wavefunction":

$$\psi_\gamma^{(N)}(\vec{x},t) = \frac{1}{\sqrt{N}}\psi_\gamma(\vec{x},t), \tag{1.110}$$

such that $$\iiint\limits_{ws} \psi_\gamma^{(N)}(\vec{x},t)^* \psi_\gamma^{(N)}(\vec{x},t)\,dV = \iiint\limits_{ws} \left|\psi_\gamma^{(N)}(\vec{x},t)\right|^2 dV = 1. \tag{1.111a}$$

A wavefunction that satisfies Eq. 1.111a is said to be "normalized." For any bound-state wavefunction, since an appropriate multiplication factor can always be found, normalization is always possible. Unless otherwise stated, we shall presume that our bound-state wavefunction is normalized. Accordingly, we shall hereafter omit the superscript (N) in Eq. 1.111a, i.e., we shall assume that $\psi_\gamma(\vec{x},t) \leftrightarrow \psi_\gamma^{(N)}(\vec{x},t)$. Thus,

$$\iiint\limits_{ws} \left|\psi_\gamma(\vec{x},t)\right|^2 dV = \sum_{i=1}^{\infty}\left|\psi_\gamma(\vec{x}_i,t)\right|^2 \delta V_i = 1. \tag{1.111b}$$

Since probability of an event is a number between 0 and 1, Max Born proposed that $\left|\psi_\gamma(\vec{x}_i,t)\right|^2 \delta V_i$ is the *probability* of finding a particle represented by the wave function $\psi_\gamma(\vec{x}_i,t)$ in the volume element δV_i at $\vec{x}_i$. Accordingly, $\left|\psi_\gamma(\vec{x},t)\right|^2$ is interpreted as probability *density* whose integration over *whole* space is 1. Only a particle confined in a finite region having impenetrable walls would give unity for the right-hand side of Eq. 1.111b in integration over a finite volume. However, as we shall discuss in Chapter 3, penetrability is determined by boundary conditions in a given situation, and it defies classical physics. Born interpretation is therefore very important because it compares and contrasts classical and quantum descriptions of a particle. In classical physics, there is never any ambiguity about where a particle exists in space. This corpuscular idea breaks down; it is replaced by the wavefunction. The physical system we are talking about can no longer be considered to be sharply located in any specific isolated region of physical space; it must be considered to be spread out. Its occurrence in any region of space can only be interpreted in terms of the probability $\left|\psi_\gamma(\vec{x}_i,t)\right|^2 \delta V_i$ mentioned earlier, in accordance with the Born interpretation of its wavefunction. Continuum wavefunctions however are not square integrable. For example, in the Solved Problem P1.10, we discuss momentum eigenstates that are *not* square integrable. They are normalized using a different procedure, called Dirac-δ normalization. It is obvious from Eq. 1.111 that the physical dimension of probability density

$$\rho_\gamma(\vec{x},t) = \left|\psi_\gamma(\vec{x},t)\right|^2 \tag{1.112}$$

is inverse volume. Hence, the dimension of the wavefunction ψ itself is $L^{-\frac{3}{2}}$.

Funquest: Determine the physical dimension of a wavefunction in a (i) one-dimension and (ii) two-dimension problem.

We have considered the integral of probability density over whole space that must add up to unity, the maximum probability. It is the scaling by $\frac{1}{\sqrt{N}}$ in Eq. 1.110 that has enabled setting the maximum probability as unity, in line with the usual statistical interpretation of probability. The multiplicative process described in Eq. 1.110 is called *normalization* of the wavefunction (see also Eq. 1.31). Once the solution to the Schrödinger equation is obtained, it can always be *normalized* using this procedure so that its modulus square is interpreted as probability density.

We now return to the perplexing energy–time uncertainty relation. As mentioned earlier, "time" is *not* an observable; it cannot be represented by a self-adjoint operator. We dodge the philosophical query about just what "time" is. We treat it only as a parameter and proceed to discuss the genesis of the energy–time uncertainty (Eq. 1.108). To appreciate the energy–time uncertainty, we first consider physical states of a system known as *stationary states*. These are also called as energy eigenkets. They are defined to be eigenstates of an operator A, which commutes with the Hamiltonian H:

$$[A,H]_{-} = 0. \tag{1.113}$$

The temporal evolution of the eigenkets from time t_0 to t of the operator A is described by

$$|a',t\rangle = U(t,t_0)|a',t_0\rangle = e^{-i\frac{H(t-t_0)}{\hbar}}|a',t_0\rangle = e^{-i\frac{E_{a'}(t-t_0)}{\hbar}}|a',t_0\rangle, \tag{1.114a}$$

since eigenkets of A are also eigenkets of the Hamiltonian. Considering $t_0 = 0$, we have the temporal evolution of the stationary states described as

$$|a',t\rangle = e^{-i\frac{E_{a'}t}{\hbar}}|a',0\rangle. \tag{1.114b}$$

The projection of the state $|a',t\rangle$ on the initial pure state $|a',t_0\rangle$ is the inner product

$$\langle a',t=0|a',t\neq 0\rangle = c_{a'}(t) = e^{-i\frac{E_{a'}t}{\hbar}}. \tag{1.115}$$

Eq. 1.115 provides a measure of the probability amplitude that a system in a pure state at t_0, when left to itself, remains in the same state at a different time t. The phase factor preserves the norm of the state. Equation 1.115 thus provides a temporal correlation between a pure state at t_0 with that at a different time t. The reason energy eigenkets (Eq. 1.115a,b) are called as *stationary states* is that the expectation value of any operator B in such a state is independent of time *regardless* of whether or not B commutes with H. This can be easily verified, since

$$\langle B\rangle_{\substack{at\\ t>0}} = \langle a',t|B|a',t\rangle = \langle a',0|e^{+i\frac{E_{a'}}{\hbar}t}Be^{-i\frac{E_{a'}}{\hbar}t}|a',0\rangle = \langle a',0|B|a',0\rangle = \langle B\rangle_{\substack{at\\ t=0}}. \tag{1.116}$$

The time-dependent coefficient $c_{a'}(t) = e^{-i\frac{E_{a'}t}{\hbar}}$ (Eq. 1.114) provides a measure of *correlation* of a pure stationary energy eigenket state at time $t \neq 0$ to that at $t = 0$. It is merely a phase factor, called as the *dynamical phase* of the stationary state. Now, instead of considering the system to be in a pure energy eigenstate at $t = 0$, we consider it to be in a *mixed* state, i.e., in a superposition of various different energy eigenkets. In this case, the expectation value of an operator B would not be invariant. It would be given by

$$\langle B\rangle_{\substack{at \\ t>0}} = \sum_i\sum_j \langle\langle a_i,t|c_i|B|c_j|a_j,t\rangle\rangle = \sum_i\sum_j c_i^* c_j e^{-\left(\frac{E_j-E_i}{\hbar}\right)t}\langle a_i,0|B|a_j,0\rangle. \tag{1.117}$$

Unlike the previous case, the expectation value in a mixed state changes with time. We now inquire what would be the nature of such a temporal correlation, if the state at t_0 is represented by a mixed state, rather than by a pure state. In this case,

$$|\alpha,t_0\rangle = \sum_i c_i(t_0)|a_i\rangle. \tag{1.118a}$$

At a different time, we must express this state as a different superposition, with time-dependent coefficients, i.e.,

$$\begin{aligned}|\alpha,t;t\neq t_0\rangle &= \sum_j c_j(t;t\neq t_0)|a_j\rangle \\ &= \sum_j c_j(t_0)e^{-i\frac{E_{a_j}(t-t_0)}{\hbar}}|a_j\rangle \\ &= \sum_j c_j(t_0)e^{-i\frac{E_{a_j}\Delta t}{\hbar}}|a_j\rangle,\end{aligned} \tag{1.118b}$$

where $\Delta t = t - t_0$.

The temporal correlation between $|\alpha,t\rangle$ and $|\alpha,t_0\rangle$ is then given by the inner product between Eq. 1.118b and Eq. 1.118a:

$$\tilde{C}(t_0,t) = \langle\alpha,t_0|\alpha,t\rangle = \sum_i\sum_j c_i^*(t_0)c_j(t)\langle a_i|a_j\rangle,$$

i.e., $$\tilde{C}(t_0,t) = \sum_i\sum_j c_i^*(t_0)c_j(t_0)e^{-i\frac{E_{a_j}\Delta t}{\hbar}}\langle a_i|a_j\rangle = \sum_i |c_i|^2 e^{-i\frac{E_{a_j}\Delta t}{\hbar}}. \tag{1.119}$$

The modulus square of each of the coefficients in this temporal correlation is modulated by time-dependent sine and cosine functions. The sum in Eq. 1.119 has its maximum value at $t = t_0$, i.e., $\Delta t = 0$. In this case,

$$\tilde{C}(t_0,t_0) = \sum_i |c_i|^2 = 1. \tag{1.120}$$

As time changes, the temporal correlation $\tilde{C}(t_0,t) = \langle\alpha,t_0|\alpha,t\rangle$ diminishes with time, because of the oscillatory sinusoidal terms in Eq. 1.119. When the number of states is large, the sum in Eq. 1.119 may be represented as an integral:

$$\tilde{C}(t_0,t) = \sum_i |c_i|^2 e^{-i\frac{E_{a_j}\Delta t}{\hbar}} \rightarrow \int_{\substack{\text{all} \\ \text{energies}}} dE\rho(E)|f(E)|^2 e^{-i\frac{E\Delta t}{\hbar}}, \tag{1.121}$$

wherein $\rho(E)$ is the density of energy eigenstates, giving $dE\rho(E)$ as the number of states at energy E, and $f(E)\big]_{E=E_i} \leftrightarrow c_i$. Corresponding to the overall normalization (Eq. 1.120), we must have

$$\int dE\rho(E)|f(E)|^2 = 1. \tag{1.122}$$

Consider now the case when the integrand $\rho(E)|f(E)|^2$ has a significant value in an energy interval ΔE. In such a case,

$$\tilde{C}(t) \simeq \int\limits_{\substack{\Delta E \\ \text{about } E_0}} dE\rho(E)|f(E)|^2 e^{-i\frac{E\Delta t}{\hbar}} = e^{-i\frac{E_0\Delta t}{\hbar}} \int\limits_{\substack{\Delta E \\ \text{about } E_0}} dE\rho(E)|f(E)|^2 e^{-i\frac{(E-E_0)\Delta t}{\hbar}}. \tag{1.123}$$

The integrand $\rho(E)|f(E)|^2 e^{-i\frac{(E-E_0)\Delta t}{\hbar}}$ oscillates rapidly with increasing value of t, and results in cancellation which would make the correlation term $\tilde{C}(t)$ diminish from unity, which was its value at t_0. As $\Delta t \to 0$, we already know that this correlation term is unity. The question we now ask is how large should Δt be so that a departure of the correlation term from unity would be observed? This would depend inversely on the energy width ΔE over which $\rho(E)|f(E)|^2$ is significant. If ΔE is large, $E - E_0$ would take a wide range of values. Even for a small value of Δt the sinusoidal oscillations would be then significant, and result in a reduction from unity of the correlation term. On the other hand, if ΔE is narrow then the cancellation would require a longer time interval. This reciprocal relationship between energy and time is written as

$$\Delta E \Delta t \simeq \hbar. \tag{1.124}$$

It is referred to energy–time uncertainty. It is best referred to as the *Mandelstam–Tamm uncertainty relation* [8], and not as Heisenberg energy–time uncertainty. However, the latter term is sometimes used. To see a *closer* connection of the energy–time uncertainty relation with Eq. 1.108, we consider an observable Ω which does not explicitly depend on time. Then, the rate of change with time of its expectation value is

$$\frac{d}{dt}\langle\Omega\rangle = \frac{d}{dt}\langle\psi|\Omega|\psi\rangle = \left\langle\frac{\partial\psi}{\partial t}|\Omega|\psi\right\rangle + \left\langle\psi\left|\frac{\partial\Omega}{\partial t}\right|\psi\right\rangle + \left\langle\psi|\Omega|\frac{\partial\psi}{\partial t}\right\rangle, \tag{1.125a}$$

$$\text{i.e., } \frac{d}{dt}\langle\Omega\rangle = \frac{d}{dt}\langle\psi|\Omega|\psi\rangle = \left\langle\frac{\partial\psi}{\partial t}|\Omega|\psi\right\rangle + \left\langle\psi|\Omega|\frac{\partial\psi}{\partial t}\right\rangle, \tag{1.125b}$$

$$\text{i.e., } \frac{d}{dt}\langle\Omega\rangle = \frac{1}{-i\hbar}\langle\psi|H\Omega|\psi\rangle + \frac{1}{i\hbar}\langle\psi|\Omega H|\psi\rangle = \frac{1}{i\hbar}\langle\psi|[\Omega,H]|\psi\rangle = \frac{i}{\hbar}|\langle[H,\Omega]\rangle|. \tag{1.125c}$$

Using Eq. 1.93 for dispersion

$$\text{We have } \left\langle(\Delta\Omega)^2\right\rangle = \left\langle(\Omega - \langle\Omega\rangle)^2\right\rangle, \tag{1.126a}$$

$$\text{Likewise, } \left\langle(\Delta H)^2\right\rangle = \left\langle(H - \langle H\rangle)^2\right\rangle. \tag{1.126b}$$

Using now the inequality in Eq. 1.103,

$$\begin{aligned}\left\langle(\Delta H)^2\right\rangle\left\langle(\Delta\Omega)^2\right\rangle \geq \frac{1}{4}|\langle[H,\Omega]\rangle|^2 &= \left\{\frac{1}{2i}\langle[H,\Omega]\rangle\right\}^2 \\ &= \left\{\frac{1}{2i}\frac{\hbar}{i}\frac{d}{dt}\langle\Omega\rangle\right\}^2 \\ &= \left\{\frac{\hbar}{2}\right\}^2\left\{\frac{d}{dt}\langle\Omega\rangle\right\}^2.\end{aligned} \tag{1.127}$$

Taking square root of both sides,

$$\delta H \delta \Omega \geq \left\{\frac{\hbar}{2}\right\}\left\{\left|\frac{d}{dt}\langle\Omega\rangle\right|\right\}. \tag{1.128}$$

We now *define* uncertainties δE and δH by the following relations:

$$\delta E \overset{\text{definition}}{=} \delta H, \tag{1.129a}$$

$$\text{and } \delta t \overset{\text{definition}}{=} \frac{\delta\Omega}{\left(\left|\dfrac{d\langle\Omega\rangle}{dt}\right|\right)}. \tag{1.129b}$$

Using Eq. 1.129 in Eq. 1.128, we get $\delta E \delta t \geq \dfrac{\hbar}{2}$, (1.130)

which offers an interpretation of Eq. 1.108. Note however that our *definition* of δt is such that the square root of the dispersion is to be understood as

$$\delta\Omega = \left(\left|\frac{d\langle\Omega\rangle}{dt}\right|\right)\delta t. \tag{1.131}$$

In other words, δt is the time-interval over which $\langle\Omega\rangle$ changes by one unit of standard deviation. We therefore see that the energy–time uncertainty is fundamentally different from the position–momentum uncertainty, since δt refers to a time-interval defined in Eq. 1.129b. We are no longer talking about compatibility of measurements of canonically conjugate variables. The energy–time uncertainty relation is therefore similar to Heisenberg's uncertainty principle, but also different from it in a subtle way.

The uncertainty relations have a classical analogue in the Fourier representations of functions. Consider, for example, the Fourier representation of a one-dimensional wavefunction:

$$\phi(x,t) = \int_{-\infty}^{+\infty} a(k)e^{i(kx-\omega t)}dk. \tag{1.132a}$$

Often, $a(k)$ may have a significant amplitude in a small range Δk centred about some particular value, such as k_0. We then have a "wave packet":

$$\phi(x,t) \approx \int_{\Delta k} a(k)e^{i(kx-\omega t)}dk, \tag{1.132b}$$

in which the range of integration is restricted, but we must remember that this is, after all, only an *approximation* to Eq. 1.132a. In a dispersive medium, we have

$$\omega = \omega(k) = \omega(k_0) + \left[\frac{d\omega}{dk}\right]_{k_0}(k-k_0) + O\left\{(k-k_0)^2\right\}. \tag{1.133}$$

The last term here represents collectively all terms of *order* $(k-k_0)^2$ and higher. Ignoring these terms, the wave packet now is expressed as

$$\phi(x,t) = \int_{\Delta k} a(k)e^{i\left[kx-\omega(k_0)t-k\left[\frac{d\omega}{dk}\right]_{k_0}t+k_0\left[\frac{d\omega}{dk}\right]_{k_0}t\right]}\left\{e^{+ik_0x}\times e^{-ik_0x}\right\}dk, \tag{1.134a}$$

$$\text{i.e., } \phi(x,t) = \left[\int_{\Delta k} a(k) e^{i(k-k_0)[x-v_g t]}\, dk \right] e^{+i(k_0 x - \omega_0 t)} = \tilde{A} e^{+i(k_0 x - \omega_0 t)} \tag{1.134b}$$

$$\text{where } \omega_0 = \omega(k)\big]_{k=k_0}, \tag{1.135a}$$

$$v_g = \left. \frac{d\omega(k)}{dk} \right]_{k=k_0}, \tag{1.135b}$$

$$\text{and } \tilde{A} = \int_{\Delta k} a(k) e^{i(k-k_0)[x-v_g t]}\, dk. \tag{1.135c}$$

Eq.1.134 represents a wave of length $\lambda_0 = \frac{2\pi}{k_0}$ and frequency $\nu_0 = \frac{\omega_0}{2\pi}$, but whose amplitude $\tilde{A}$ is a function of both space and time. The dependence on space and time is through the combination term $(x - v_g t)$. In other words, $\tilde{A}$ is not the same at every point along the x-axis, but where $(x - v_g t) = \eta$ is a constant. Equation 1.134 therefore essentially represents a "wave packet" that propagates in space such that $\delta\eta = 0 = (\delta x - v_g \delta t)$. The wave packet consists of a group of waves having wave numbers k in the range Δk. This group moves at the group velocity v_g. The components in this wave packet corresponding to different values of $k = \frac{2\pi}{\lambda_k} = \frac{2\pi\nu_k}{c_k} = \frac{\omega_k}{c_k}$ move at different individual phase velocities c_k. The wave packet is thus not localized at a particular point on the x-axis. It has a spread over a range Δx. To localize the wave packet, the spread Δk needs to be large. Δk and Δx therefore have a reciprocal relationship:

$$\Delta k \Delta x \approx 1. \tag{1.136a}$$

Scaled by $\hbar$, we can immediately obtain a relation that would be similar to the Heisenberg uncertainty principle, between momentum and position. Relations Eq. 1.132 exist between *any* pair $(\phi(x), a(k))$ of Fourier transforms; they relate to expressing a function in direct space to its Fourier resolution in the corresponding reciprocal space. Instead of the x-space and its reciprocal k-space, if we had considered the frequency-domain and its reciprocal time-domain, we would therefore obtain a similar relation,

$$\Delta\nu\Delta t \approx 1. \tag{1.136b}$$

On scaling Eq. 1.132b by $\hbar$, we would essentially obtain Eq. 1.124. For those who may wish to study the energy–time uncertainty further, it must however be added that it has also raised some controversies (see, for example, Reference [9]).

For *stationary* eigenstates (Eq. 1.113 to Eq. 1.115) of the Schrödinger equation, the wavefunction from Eq. 1.69, $\psi_\alpha(\vec{x},t)$, is factorable into a purely space-dependent part and a purely time-dependent part:

$$\psi_\alpha(\vec{x},t) = \psi_\alpha(\vec{x}) T(t). \tag{1.137}$$

Using the above form in the Schrödinger equation (Eq. 1.191b), we obtain

$$\frac{H\psi(\vec{x})}{\psi(\vec{r})} = \frac{i\hbar}{T(t)}\frac{\partial T}{\partial t} = E, \tag{1.138}$$

where E is the constant of separation. We are then led to two separate equations, one for the time-alone part, and the other for the space-alone part:

$$i\hbar\frac{\partial T}{\partial t} = ET(t), \tag{1.139}$$

$$\text{and } H\psi(\vec{x}) = E\psi(\vec{x}). \tag{1.140}$$

It is readily seen that the solutions of Eq. 1.139 have the form as in Eq. 1.114a,b, which provide the dynamical phase (Eq. 11.113 to Eq. 1.115):

$$T(t) = c\exp\left(-i\frac{E}{\hbar}t\right). \tag{1.141}$$

The Eq. 1.140 is the *time-independent Schrödinger equation*. We see that E is its energy eigenvalue. The Hamiltonian consists of the sum of the kinetic energy operator and the potential energy operator which we have used in Eq. 1.118. The kinetic energy operator would involve the Laplacian $\vec{\nabla}\bullet\vec{\nabla}$ because the momentum operator is given by Eq. 1.168a. The time-independent Schrödinger equation therefore is a second order differential equation whose general solution would depend on the form of the potential energy operator. The solution would consist of two arbitrary constants to be determined from the physical boundary conditions for each particular system.

The counterintuitive idea of representing observables by matrices proposed by Heisenberg [10], Born and Jordan [11], and consolidated in the famously known *three-man paper* [12], turned out to provide a brilliant surpass of classical physics. Products of these matrices would commute for compatible observables, and *not commute* for those incompatible. Matrix mechanics employed a viewpoint and articulation that was at dramatic variance from Schrödinger's wave mechanics [13]. However, the two methods turned out to be equivalent. The discerning work using vector spaces by Dirac [14] enabled the synthesis of *matrix mechanics* and *wave mechanics* into what is now called *quantum mechanics* [15]. A captivating account of this fusion, which mostly took place during the months of March and April 1926, is available in References [16] and [17]. A third approach to quantum mechanics, independent but equivalent to the other two was developed by Richard P. Feynman, a little over two decades later. It is called as the *path integral approach*. Feynman's method is introduced in the next chapter. In addition to the Schrödinger's, Heisenberg's, and Feynman's formulations of quantum mechanics, there are other, equivalent, formulations that we shall not discuss here, but only refer the readers to Reference [18] for further reading.

Solved Problems

P1.1(a): From a non-orthogonal linearly independent basis $\{|e_1\rangle, |e_2\rangle, \ldots, |e_N\rangle\}$ which spans an N-dimensional Hilbert space, obtain an orthogonal linearly independent basis $\{|u_1\rangle, |u_2\rangle, \ldots, |u_N\rangle\}$ such that for every i, j, with $i \neq j$, $\langle u_j | u_i\rangle = 0$.

Solution:

We begin with $|u_1\rangle = |e_1\rangle$ and construct the next basis vector $|u_2\rangle$ using a complex number c_{12} as a mixing coefficient as $|u_2\rangle = |e_2\rangle + c_{12}|u_1\rangle$ such that it would be orthogonal to $|u_1\rangle$. This condition leads us to determine the mixing coefficient c_{12} that would make $|u_2\rangle$ orthogonal to $|u_1\rangle$. The requirement $\langle u_1|u_2\rangle = 0$ is satisfied when $0 = \langle u_1 \mid e_2\rangle + c_{12}\langle u_1 \mid u_1\rangle$ that is, when $c_{12} = -\frac{\langle e_1 \mid e_2\rangle}{\langle e_1 \mid e_1\rangle}$.

Thus, $|u_2\rangle = |e_2\rangle - \frac{\langle e_1 \mid e_2\rangle}{\langle e_1 \mid e_1\rangle}|u_1\rangle$. Essentially, we remove the component of $|e_2\rangle$ along $|e_1\rangle$ so that the residual vector would be orthogonal to $|u_1\rangle = |e_1\rangle$. We continue the application of this strategy to get the remaining base vectors to get an orthogonal basis. The next basis vector $|u_3\rangle$ must be orthogonal to both $|u_1\rangle$ and $|u_2\rangle$. Hence, we construct it as $|u_3\rangle = |e_3\rangle + c_{13}|u_1\rangle + c_{23}|u_2\rangle$ and look for mixing coefficients c_{13} and c_{23} such that $\langle u_1|u_3\rangle = 0$ and also $\langle u_2|u_3\rangle = 0$. Orthogonality of $|u_1\rangle$ and $|u_3\rangle$ requires $\langle u_1 \mid e_3\rangle + c_{13}\langle u_1 \mid u_1\rangle + c_{23}\langle u_1 \mid u_2\rangle = 0$, of which the last term is already zero.

We must therefore have $c_{13} = -\frac{\langle u_1 \mid e_3\rangle}{\langle u_1 \mid u_1\rangle} = -\frac{\langle e_1 \mid e_3\rangle}{\langle e_1 \mid e_1\rangle}$. Similarly, by requiring the inner product $\langle u_2 \mid u_3\rangle$ to be zero, we get $c_{23} = -\frac{\langle u_2 \mid e_3\rangle}{\langle u_2 \mid u_2\rangle}$ and hence $|u_3\rangle = |e_3\rangle - \frac{\langle u_1 \mid e_3\rangle}{\langle u_1 \mid u_1\rangle}|u_1\rangle - \frac{\langle u_2 \mid e_3\rangle}{\langle u_2 \mid u_2\rangle}|u_2\rangle$. Continuing like this, we have $|u_j\rangle = |e_j\rangle + \sum_{i,j=1;i\neq j}^{N-1} c_{ij}|u_i\rangle$ with mixing complex coefficients $c_{ij} = -\frac{\langle u_i \mid e_j\rangle}{\langle u_i \mid u_i\rangle}$.

P1.1(b): Normalize the orthogonal basis vectors so obtained; i.e., for every i, $\langle u_i|u_i\rangle = 1$.

Solution:

The basis $\{|u_1\rangle, |u_2\rangle, \ldots, |u_N\rangle\}$ we obtained in part (a) is orthogonal, but the base vectors are not normalized. Normalization is achieved by dividing each vector by its norm (Eq. 1.25). We may denote the normalized basis by placing the caret (also called as circumflex) grapheme ^ to indicate that the normalized vector $|\hat{u}_i\rangle$ has unit magnitude. The technique described above is called 'Gram–Schmidt orthonormalization' (or 'orhogonalization').

P1.2: Show that if the translational infinitesimal displacement operator is considered to have the general form $\tau(\overrightarrow{dx'}) = 1 - i\vec{K}\bullet\overrightarrow{dx'}$, where each component K_x, K_y, K_z of the vector operator $\vec{K}$ is Hermitian (i.e., $\vec{K}^\dagger = \vec{K}$), then all properties described in Eq. 1.57a–d are satisfied.

Solution:

We have $\tau(\vec{dx}') = 1 - i\vec{K}\bullet\vec{dx}'$ and $\tau^\dagger(\vec{dx}') = 1 + i\vec{K}^\dagger\bullet\vec{dx}'$.

Therefore, $\tau(\vec{dx}')\tau^\dagger(\vec{dx}') = (1 - i\vec{K}\bullet\vec{dx}')(1 + i\vec{K}^\dagger\bullet\vec{dx}') \simeq 1 + i(\vec{K}^\dagger - \vec{K})\bullet\vec{dx}' = 1$, with $\vec{K}^\dagger = \vec{K}$ (hermiticity).

We have ignored the quadratic term in the infinitesimal displacement. The result $\tau(\vec{dx}')\tau^\dagger(\vec{dx}') = 1$ is a signature of unitarity:

(a) $\tau^{-1}(d\vec{x}') = \tau^{\dagger}(d\vec{x}')$. (Eq. 1.57a)

(b) $\tau(d\vec{x}') = 1 - i\vec{K}\bullet d\vec{x}'$ and $\tau(dx'') = 1 - i\vec{K}\bullet d\vec{x}''$.

Therefore, $\tau(d\vec{x}')\tau(d\vec{x}'') = (1 - i\vec{K}\bullet d\vec{x}')(1 - i\vec{K}\bullet d\vec{x}'') \simeq 1 - i\vec{K}\bullet(d\vec{x}' + d\vec{x}'') = \tau(d\vec{x}' + d\vec{x}'')$ (Eq. 1.57b).

(c) $\tau(-d\vec{x}') = 1 + i\vec{K}\bullet d\vec{x}' = \tau^{\dagger}(d\vec{x}') = \tau^{-1}(d\vec{x}')$ (Eq. 1.57c).

(d) $\lim_{dx'\to 0} \tau(d\vec{x}') = \lim_{dx'\to 0} (1 - i\vec{K}\bullet d\vec{x}') = 1$ (Eq. 1.57d).

P1.3(a): For any two vectors $|a\rangle$ and $|b\rangle$ in the Hilbert space, prove that $\langle a|a\rangle\langle b|b\rangle \geq \langle b|a\rangle\langle a|b\rangle$. This inequality is known as *Schwarz inequality*.

Solution:

For a complex number λ and vectors $|a\rangle$ and $|b\rangle$, we have the dual conjugate vector $\langle a - \lambda b|$ corresponding to the vector $|a - \lambda b\rangle = |a\rangle - \lambda|b\rangle$. From Eq. 1.24, we know that the inner product is non-negative: $\langle a - \lambda b|a - \lambda b\rangle \geq 0$, i.e., $\langle a|a\rangle - \lambda\langle a|b\rangle - \lambda^*\langle b|a\rangle + \lambda^*\lambda\langle b|b\rangle \geq 0$. Since λ is an arbitrary complex number, we may freely choose it. Consider $\lambda = \frac{\langle b|a\rangle}{\langle b|b\rangle}$. With this choice, the above expression reduces to $\langle a|a\rangle - \frac{\langle b|a\rangle}{\langle b|b\rangle}\langle a|b\rangle \geq 0$, which ensures that $\langle b|b\rangle\langle a|a\rangle - \langle b|a\rangle\langle a|b\rangle \geq 0$, which is the Schwarz inequality. Note that it includes the possibility that the left-hand side is equal to zero; it cannot however be negative.

P1.3(b): For any two vectors $|a\rangle$ and $|b\rangle$ in the Hilbert space, prove that $\|a + b\| \leq \|a\| + \|b\|$. This inequality is known as the *triangle inequality*.

Solution:

Consider the square of the norm of the sum of two vectors:

$\|a + b\|^2 = \langle a + b|a + b\rangle = \langle a|a\rangle + \langle b|a\rangle + \langle a|b\rangle + \langle b|b\rangle = \|a\|^2 + 2\mathrm{Re}\langle a|b\rangle + \|b\|^2$. Since for any complex number z, $|z| \geq \mathrm{Re}(z)$, it follows that $\|a + b\|^2 \leq \|a\|^2 + 2|\langle a|b\rangle| + \|b\|^2$. Using now the Schwarz's inequality, we have $2|\langle a|b\rangle| \leq 2\|a\| \times \|b\|$ and hence $\|a + b\|^2 \leq \|a\|^2 + 2\|a\|\ \|b\| + \|b\|^2$. Taking the square roots of both left- and the right-hand sides, we get $\|a + b\| \leq \|a\| + \|b\|$, which is the *triangle inequality*. It draws its name from the fact that in a triangle, the length of any side is less than or equal to the sum of the lengths of the other two sides.

P1.4: Determine the commutator $[\vec{r}, \vec{p}]$, corresponding to the position–momentum uncertainty, of the position vector operator $\vec{r}$ with the momentum vector operator $\vec{p}$.

Solution:

The required commutator is $(\vec{r}\vec{p} - \vec{p}\vec{r})$; it is a difference of two dyadic operators.

For two vector operators $\vec{A} = A_x\hat{e}_x + A_y\hat{e}_y + A_z\hat{e}_z$ and $\vec{B} = B_x\hat{e}_x + B_y\hat{e}_y + B_z\hat{e}_z$, $\vec{A}\vec{B}$ is a dyadic, which may be written in matrix form as $\mathbb{A}\mathbb{B}^T = \begin{pmatrix} A_x \\ A_y \\ A_z \end{pmatrix}\begin{pmatrix} B_x & B_y & B_z \end{pmatrix} = \begin{pmatrix} A_xB_x & A_xB_y & A_xB_z \\ A_yB_x & A_yB_y & A_yB_z \\ A_zB_x & A_zB_y & A_zB_z \end{pmatrix}$.

The unit dyadic is $\mathfrak{I} = \hat{e}_x\hat{e}_x + \hat{e}_y\hat{e}_y + \hat{e}_z\hat{e}_z$,

where $\hat{e}_x\hat{e}_x = \begin{pmatrix} 100 \\ 000 \\ 000 \end{pmatrix}$, $\hat{e}_y\hat{e}_y = \begin{pmatrix} 000 \\ 010 \\ 000 \end{pmatrix}$ and $\hat{e}_z\hat{e}_z = \begin{pmatrix} 000 \\ 000 \\ 001 \end{pmatrix}$.

The unit dyadic therefore is $\mathfrak{I} = \begin{pmatrix} 100 \\ 010 \\ 001 \end{pmatrix} = \begin{pmatrix} \hat{e}_x\hat{e}_x & 0 & 0 \\ 0 & \hat{e}_y\hat{e}_y & 0 \\ 0 & 0 & \hat{e}_z\hat{e}_z \end{pmatrix}$.

$$\text{Hence, } \left[\vec{r},\vec{p}\right]_- = [x\hat{e}_x + y\hat{e}_y + z\hat{e}_z, p_x\hat{e}_x + p_y\hat{e}_y + p_z\hat{e}_z]_- = \begin{Bmatrix} [x,p_x]\hat{e}_x\hat{e}_x + [x,p_y]\hat{e}_x\hat{e}_y + [x,p_z]\hat{e}_x\hat{e}_z + \\ [y,p_x]\hat{e}_y\hat{e}_x + [y,p_y]\hat{e}_y\hat{e}_y + [y,p_z]\hat{e}_y\hat{e}_z + \\ [z,p_x]\hat{e}_z\hat{e}_x + [z,p_y]\hat{e}_z\hat{e}_y + [z,p_z]\hat{e}_z\hat{e}_z \end{Bmatrix}$$

$$\text{i.e., } \left[\vec{r},\vec{p}\right] = \begin{Bmatrix} i\hbar\hat{e}_x\hat{e}_x + 0 + 0 + \\ 0 + i\hbar\hat{e}_y\hat{e}_y + 0 + \\ 0 + 0 + i\hbar\hat{e}_z\hat{e}_z \end{Bmatrix} = i\hbar\left\{\hat{e}_x\hat{e}_x + \hat{e}_y\hat{e}_y + \hat{e}_z\hat{e}_z\right\} = i\hbar\mathfrak{I}.$$

P1.5: Show that the coordinate representation of the resultant vector obtained from the operation by the momentum operator $\vec{p}$ on an arbitrary state vector $|\alpha\rangle$ is equal to $(-i\hbar)$ times the gradient of the coordinate representation of the state $|\alpha\rangle$.

Solution:

We must show that $\langle\vec{x}|\vec{p}|\alpha\rangle = -i\hbar\vec{\nabla}\langle\vec{x}|\alpha\rangle$. Let us first operate on an arbitrary state $|\alpha\rangle$ by the infinitesimal displacement operator (Eq. 1.65d)

$$\left(1 - \frac{i}{\hbar}\vec{p}\bullet\vec{ds}\right)|\alpha\rangle = \underline{\tau\left(\vec{ds}\right)\iiint d^3x'\left|\vec{x}'\right\rangle}\left\langle\vec{x}'\right||\alpha\rangle = \iiint \underline{d^3x'\left|\vec{x}' + \vec{ds}\right\rangle}\left\langle\vec{x}'\middle|\alpha\right\rangle.$$

$$\text{Therefore, } \left(1 - \frac{i}{\hbar}\vec{p}\bullet\vec{ds}\right)|\alpha\rangle = \iiint d^3x''\left|\vec{x}''\right\rangle\left\langle\vec{x}'' - \vec{ds}\middle|\alpha\right\rangle = \iiint d^3x''\left|\vec{x}''\right\rangle\left[\left\langle\vec{x}''\middle|\alpha\right\rangle - \left(\vec{\nabla}''\left\langle\vec{x}''\middle|\alpha\right\rangle\right)\bullet\vec{ds}\right].$$

Cancelling the vector $|\alpha\rangle$ on both sides of the equation, $\left(\frac{i}{\hbar}\vec{p}\bullet\vec{ds}\right)|\alpha\rangle = \iiint d^3x''\left|\vec{x}''\right\rangle\left(\vec{\nabla}''\left\langle\vec{x}''\middle|\alpha\right\rangle\right)\bullet\vec{ds}.$

Taking the coordinate representation of the above vector we get:

$$\left\langle\vec{x}\right|\left(\frac{i}{\hbar}\vec{p}\bullet\vec{ds}\right)|\alpha\rangle = \left\langle\vec{x}\right|\iiint d^3x''\left|\vec{x}''\right\rangle\left(\vec{\nabla}''\left\langle\vec{x}''\middle|\alpha\right\rangle\right)\bullet\vec{ds} = \iiint d^3x''\left\langle\vec{x}\middle|\vec{x}''\right\rangle\left(\vec{\nabla}''\left\langle\vec{x}''\middle|\alpha\right\rangle\right)\bullet\vec{ds},$$

$$\text{i.e., } \left\langle\vec{x}\right|\left(\frac{i}{\hbar}\vec{p}\bullet\vec{ds}\right)|\alpha\rangle = \iiint d^3x''\delta^3\left(\vec{x} - \vec{x}''\right)\left(\vec{\nabla}''\left\langle\vec{x}''\middle|\alpha\right\rangle\right)\bullet\vec{ds}, \text{ or } \langle\vec{x}|\vec{p}|\alpha\rangle\bullet\vec{ds} = \left(-i\hbar\vec{\nabla}\right)\langle\vec{x}|\alpha\rangle\bullet\vec{ds}.$$

Since $\vec{ds}$ is an arbitrary infinitesimal displacement in homogeneous space, we conclude that $\langle\vec{x}|\vec{p}|\alpha\rangle = -i\hbar\vec{\nabla}\langle\vec{x}|\alpha\rangle$.

P1.6: Show that the operator $\frac{d}{dx}$ is anti-Hermitian.

Solution:

We consider the matrix element of the one-dimensional momentum operator in bound-state functions $\phi_1(x)$ and $\phi_2(x)$.

$$\left\langle \phi_1 \middle| \left(-i\hbar\frac{d}{dx}\right) \middle| \phi_2 \right\rangle = -i\hbar\int_{-\infty}^{\infty} dx \phi_1(x)^* \frac{d\phi_2(x)}{dx} = -i\hbar\left[\phi_1(x)^*\phi_2(x)\right]_{-\infty}^{\infty} - (-i\hbar)\int_{-\infty}^{\infty} dx \left(\frac{d\phi_1(x)^*}{dx}\right)\phi_2(x),$$

i.e., $$\left\langle \phi_1 \middle| \left(-i\hbar\frac{d}{dx}\right) \middle| \phi_2 \right\rangle = -i\hbar\left\{\left[\phi_1(x)^*\phi_2(x)\right]_{-\infty}^{\infty} - \int_{-\infty}^{\infty} dx \left(\frac{d\phi_1(x)^*}{dx}\right)\phi_2(x)\right\},$$

or $$\left\langle \phi_1 \middle| \left(-i\hbar\frac{d}{dx}\right) \middle| \phi_2 \right\rangle = i\hbar\int_{-\infty}^{\infty} dx \left(\frac{d\phi_1(x)^*}{dx}\right)\phi_2(x) = i\hbar\left\langle \frac{d\phi_1}{dx} \middle| \phi_2 \right\rangle = \left\langle -i\hbar\frac{d\phi_1}{dx} \middle| \phi_2 \right\rangle,$$ on using Eq. 1.23c.

i.e., $\langle \phi_1 | p_x \phi_2 \rangle = \langle p_x \phi_1 | \phi_2 \rangle \Rightarrow \boxed{p_x = p_x^\dagger}$.

In other words, $\left(-i\hbar\frac{d}{dx}\right)^\dagger = \left(-i\hbar\frac{d}{dx}\right)$. The momentum operator is Hermitian.

i.e., $(+i\hbar)\left(\frac{d}{dx}\right)^\dagger = -i\hbar\frac{d}{dx} \Rightarrow \boxed{\left(\frac{d}{dx}\right)^\dagger = -\frac{d}{dx}}$.

P1.7(a): Prove that eigenvalues of a Hermitian operator are *real*.

Solution:

For an operator Ω and a vector $|\alpha\rangle$, the dual conjugate vector of $\langle\alpha|\Omega$ is $|\alpha\rangle\Omega^\dagger$, where $\Omega^\dagger$ is the Hermitian adjoint of the operator Ω. From the properties of linear operators that operate on vectors in a Hilbert space, we have $\langle\beta|\Omega|\alpha\rangle = \langle\alpha|\Omega^\dagger|\beta\rangle^*$. For a Hermitian operator, this relation becomes $\langle\beta|\Omega|\alpha\rangle = \langle\alpha|\Omega|\beta\rangle^*$. Consider the eigenvalue equation $\Omega|\omega'\rangle = \omega'|\omega'\rangle$. Its dual conjugate relation is $\langle\omega''|\Omega^\dagger = \langle\omega''|\omega''^*$, where the asterisk denotes complex conjugation. Taking the inner product of $\Omega|\omega'\rangle$ with $\langle\omega''|$, we get

$$\langle\omega''|\Omega|\omega'\rangle = \omega'\langle\omega''||\omega'\rangle = \omega'\langle\omega''|\omega'\rangle. \tag{P1.7a.1}$$

For a Hermitian operator (defined by $\Omega^\dagger = \Omega$), the relation $\langle\omega''|\Omega^\dagger = \langle\omega''|\omega''^*$ becomes $\langle\omega''|\Omega = \omega''^*\langle\omega''|$. Taking the inner product of the last equation with $|\omega'\rangle$,

we get $\langle\omega''|\Omega|\omega'\rangle = \omega''^*\langle\omega''|\omega'\rangle$. (P1.7a.2)

Subtracting Eq. P1.7a.2 from Eq. P1.7a.1, we get $0 = (\omega' - \omega''^*)\langle\omega''|\omega'\rangle$.

When $\omega' = \omega''$, $\langle\omega''|\omega'\rangle \neq 0$ and hence $\omega' = \omega'^*$, and the eigenvalues are manifestly real.

P1.7(b): Prove that eigenkets of a Hermitian operator belonging to different eigenvalues are orthogonal.

Solution:

Continuing from the solution to P1.7a, we see that when $\omega' \neq \omega''$, we must have $\langle\omega''|\omega'\rangle = 0$; i.e., the vectors $|\omega'\rangle$ and $|\omega''\rangle$ are orthogonal.

P1.8: Prove that for two commuting Hermitian operators A and B, if *at least* one of the two operators (say A) has non-degenerate eigenstates, the two operators have a common eigenbasis; i.e., both the operators have a diagonal representation in the common eigenbasis.

Solution:

The operator A has non-degenerate eigenstates.

$$A|a_i\rangle = a_i|a_i\rangle$$

$$BA|a_i\rangle = a_iB|a_i\rangle$$

The operator on the left-hand side can also be written as AB, hence

$$A\left(B|a_i\rangle\right) = a_i\left(B|a_i\rangle\right);$$

i.e., $\left(B|a_i\rangle\right)$ is also an eigenvector of A belonging to the same eigenvalue. Since the operator A is non-degenerate, $B|a_i\rangle$ and $|a_i\rangle$ can differ only in their respective normalization; the vectors $\left(B|a_i\rangle\right)$ and $|a_i\rangle$ represent the same physical state in the Hilbert space; they are proportional to each other:

$B|a_i\rangle = b_i|a_i\rangle$ (proportionality from the above argument),

$B|a_i\rangle = b_i|a_i\rangle = b_i|a_i, b_i\rangle$ (recognition as an eigenvalue equation).

The earlier arguments hold for every eigenvector of A. We see that every eigenvector of A is also an eigenvector of B, and hence both the operators A and B have a diagonal representation in a common eigenbasis.

P1.9: For commuting Hermitian operators A and B, *both* of which having degenerate eigenstates, prove that every basis in which A is diagonal does not provides a diagonal representation of the operator B. Prove also that one can, nonetheless, find a basis in which both A and B are simultaneously diagonal.

Solution:

Let the eigenvalue a_i of the operator A be d_i-fold degenerate, where d_i is greater than 1 for at least one value of i. For the sake of our discussion, let us assume that a_1 and a_m are two-fold degenerate and all other eigenvalues are non-degenerate.

The matrix representation of the operator A in its eigenbasis $\{|a_1^1\rangle, |a_1^2\rangle, \ldots, |a_m^1\rangle, |a_m^2\rangle\}$ is

$$\mathbb{A} = \begin{bmatrix} a_1 & & & & & \\ & a_1 & & & & \\ & & \ddots & & & \\ & & & \ddots & & \\ & & & & a_m & \\ & & & & & a_m \end{bmatrix}$$

, but the matrix would be essentially the same if the first base vectors in the eigenbasis are interchanged. The actual representation depends on how the base vectors are ordered. We have chosen an ordering in which the two base vectors which both belong to the eigenvalue a_1 are placed *adjacent* to each other in the first two positions, and the two base vectors which both belong to the eigenvalue a_m are placed *adjacent* to each other in the last two positions. A different ordering would make the matrix look different. The degeneracy of the ith eigenstate may be denoted as n_i-fold.

Consider an n_i-dimensional vector space spanned by $\left\{ \left| a_i, \alpha_j \right\rangle ; i = 1,2,..,m; j = 1,2,..,n_i \right\}$ consisting of n_i linearly independent functions which all belong to the same eigenvalue a_i. The matrix elements of the operator B would be diagonal in the index i, but not in j. The matrix representation of the operator B would, in general, have a *block-diagonal* form

$$\mathbb{B} = \begin{bmatrix} \boxed{b_1} & & & \\ & \boxed{b_2} & & \\ & & \ddots & \\ & & & \boxed{b_k} \end{bmatrix}.$$

However, each block $\boxed{b_i}$ can be individually diagonalized. If we diagonalize each and every individual block $\boxed{b_i}$ for $i = 1,2,..,k$, then the basis functions that span each block are transformed to a new basis, but even in the new basis the operator A would continue to remain diagonal. The eigenvalues would remain the same, on account of the degeneracy with respect to the eigenvalue a_i. Both the operators A and B are now diagonal in the transformed basis, but prior to the transformation, only A was diagonal. *In other words, for commuting operators A and B, one can find a basis in which both A and B are simultaneously diagonal, but not every basis in which A is diagonal provides a diagonal representation of the operator B.* The base vectors in the transformed basis can then be labeled by two indices $\left| a_i, b_i \right\rangle$ and the set of eigenvectors denoted by $\left\{ \left| a_i, b_i \right\rangle \right\}$ will have at least one of the two labels different. However, if the operator B also has degenerate eigenstates in the subspace of eigenvectors of A belonging to the eigenvalue a_i, then that vector is not uniquely specified by just the two quantum labels a_i and b_i. We can then look for a *third* compatible measurement corresponding to which we have an operator C that could provide a *distinct* good quantum label/number. One could continue the process till we find the CSCO (the complete set of commuting operators corresponding to the complete set of compatible observables). For example, the eigenvectors of the *non-relativistic* hydrogen atom (Chapter 5) are simultaneous eigenvectors of three commuting operators $\left\{ H, L^2, L_z \right\}$ which provide the complete set, albeit in the non-relativistic domain. The eigenvalues $\left\{ n, \ell, m \right\}$ respectively of the three operators provide good quantum numbers, at least one of which is unique in each set. The simultaneous eigenvectors of the operators $\left\{ H, L^2, L_z \right\}$ will be discussed further in Chapter 5.

P1.10(a): Show that the position representation of a momentum state vector is not square integrable.

Solution:

Consider the action of the position operator on an arbitrary state vector $\left| \gamma \right\rangle$:

$$\vec{r}_{op} \left| \gamma \right\rangle = \vec{r}_{op} \iiint_{ws} d^3\vec{r}' \left| \vec{r}' \right\rangle \left\langle r' \right| \left| \gamma \right\rangle = \iiint_{ws} d^3\vec{r}' \, \vec{r}_{op} \left| \vec{r}' \right\rangle \left\langle \vec{r}' \middle| \gamma \right\rangle = \iiint_{ws} d^3\vec{r}' \, \vec{r}' \left| \vec{r}' \right\rangle \left\langle \vec{r}' \middle| \gamma \right\rangle. \text{ Hence,}$$

$$\left\langle \vec{r} \middle| \vec{r}_{op} \middle| \gamma \right\rangle = \left\langle \vec{r} \right| \iiint_{ws} d^3\vec{r}' \, \vec{r}' \left| \vec{r}' \right\rangle \left\langle \vec{r}' \middle| \gamma \right\rangle = \iiint_{ws} d^3\vec{r}' \, \vec{r}' \left\langle \vec{r} \middle| \vec{r}' \right\rangle \left\langle \vec{r}' \middle| \gamma \right\rangle = \iiint_{ws} d^3\vec{r}' \, \vec{r}' \delta^3 \left(\vec{r} - \vec{r}' \right) \left\langle \vec{r}' \middle| \gamma \right\rangle = \vec{r} \left\langle \vec{r} \middle| \gamma \right\rangle.$$

The function $\langle \vec{r}|\gamma\rangle = \psi_\gamma(\vec{r})$ is the wavefunction in the position representation. Now we consider the momentum state vectors $|\vec{p}\rangle$; these are eigenvectors of the momentum operator:

$$\vec{p}_{op}|\vec{p}\rangle = \vec{p}|\vec{p}\rangle, \text{ i.e., } \left(-i\hbar\vec{\nabla}\right)|\vec{p}\rangle = \vec{p}|\vec{p}\rangle.$$

Hence, $\langle \vec{r}|\left(-i\hbar\vec{\nabla}\right)|\vec{p}\rangle = \left(-i\hbar\vec{\nabla}\right)\langle \vec{r}|\vec{p}\rangle = \left(-i\hbar\vec{\nabla}\right)\psi_{\vec{p}}(\vec{r})$,

where $\langle \vec{r}|\vec{p}\rangle = \psi_{\vec{p}}(\vec{r})$ is the coordinate representation of the momentum state vector $|\vec{p}\rangle$.

In one-dimension, say corresponding to the Cartesian X coordinate, we have:

$$\langle x|\left(-i\hbar\frac{d}{dx}\right)|p\rangle = \left(-i\hbar\frac{d}{dx}\right)\psi_p(x); \text{ i.e., } \left(-i\hbar\frac{d}{dx}\right)\psi_p(x) = p\psi_p(x).$$

Hence, $\dfrac{d}{dx}\ell n\psi_p(x) = \dfrac{\left(\dfrac{d\psi_p(x)}{dx}\right)}{\psi_p(x)} = \dfrac{ip}{\hbar}$, or $\ell n\psi_p(x) = \dfrac{ip_x x}{\hbar} + c_x$.

Therefore, $\psi_p(x) = e^{\frac{ipx}{\hbar}}e^{c_x} = A_x e^{\frac{ipx}{\hbar}}$.

The integral $\iiint\limits_{ws} d^3\vec{r}'\,\psi_p(\vec{r}')^*\,\psi_p(\vec{r}')$ is not finite; hence, the momentum eigenfunction is not square integrable. In particular, if a physical system is in a state that is an eigenvector of the position operator, the position dispersion is zero but its product with that of the momentum nonetheless has a minimum value on account of Eq. 1.105. This ties up with the fact that the momentum eigenfunctions are *not* square integrable.

P1.10(b): Normalize the coordinate representation of the momentum state vector $\langle \vec{r}|\vec{p}\rangle = \psi_{\vec{p}}(\vec{r}) = Ae^{\frac{i}{\hbar}\vec{p}\cdot\vec{r}}$, using the Dirac-$\delta$ function.

Solution:

First, we show this for the one-dimensional case.

$$\langle p|p'\rangle = \delta(p-p') = \langle p|\left\{\int_{-\infty}^{\infty} dx|x\rangle\langle x|\right\}|p'\rangle = \int_{-\infty}^{\infty} dx\langle p|x\rangle\langle x|p'\rangle = |A_x|^2\int_{-\infty}^{\infty} dxe^{\frac{i(p'-p)x}{\hbar}}.$$

Using now $\mu = \dfrac{p-p'}{\hbar}$ in the definition of the Dirac-δ function, we get

$$\delta\left(\frac{p-p'}{\hbar}\right) = \hbar\delta(p-p') = \frac{1}{2\pi}\int_{-\infty}^{\infty} d\omega e^{\frac{i}{\hbar}(p-p')\omega} = \frac{1}{2\pi}\frac{\delta(p-p')}{|A_x|^2}.$$

Therefore, $|A_x|^2 = \dfrac{1}{2\pi\hbar}$, and hence $\psi_p(x) = \dfrac{1}{\sqrt{2\pi\hbar}}e^{\frac{ip_x x}{\hbar}}$. The Dirac-$\delta$ normalized 3-dimensional momentum state eigenvector in the position representation therefore is:

$$\psi_{\vec{p}}(\vec{r}) = \frac{1}{(2\pi\hbar)^{\frac{3}{2}}}e^{\frac{i\vec{p}\cdot\vec{r}}{\hbar}}.$$

Exercises

E1.1: Prove the *Jacobi identity* for three operators *A*, *B*, *C*:

$$[A,[B,C]]+[B,[C,A]]+[C,[A,B]]=0.$$

E1.2: Prove that eigenvalues of an anti-Hermitian operator $\mho$ (defined by $\mho^\dagger=-\mho$) can only be purely imaginary or zero.

E1.3: Determine the eigenvectors and eigenvalues of an operator Ω, which satisfies the equation $\Omega^2-5\Omega+6=0$. Is the operator Hermitian?

E1.4: The Schrödinger equation of a particle having mass *m* is $\left(\frac{p^2}{2m}-b\nabla_p^2\right)\psi(\vec{p},t)=i\hbar\frac{\partial\psi(\vec{p},t)}{\partial t}$. Write the corresponding equation in coordinate space.

E1.5: Determine if $\left\langle\frac{dV(x)}{dx}\right\rangle\neq\left[\frac{dV(x)}{dx}\right]_{x=\langle X\rangle}$. Comment on the result.

E1.6: Show that $[x,p^n]_-=i\hbar np^{n-1}$.

E1.7: Show that $[p,x^n]_-=-i\hbar nx^{n-1}$.

E1.8: Determine the uncertainty product $\delta x\delta k$ for a wave packet given by $\psi(x)=\int_{-\infty}^{+\infty}\eta(k)e^{ikx}dk$, where $\eta(k)=\eta$ for $k_0-(\Delta k/2)<k<k_0+(\Delta k/2)$, and $\eta(k)=0$ otherwise.

E1.9: Determine (a) $\langle x^n\rangle$ and (b) $\langle p\rangle$ for a Gaussian function, $\psi(x)=\left(\frac{1}{\sigma\sqrt{\pi}}\right)^{\frac{1}{2}}e^{-x^2/2\sigma^2}$ (*n* is a whole number and σ is the "standard deviation"). Comment on the cases when *n* is an even integer and when it is odd.

E1.10: Determine if $\frac{d\langle\vec{r}\times\vec{p}\rangle}{dt}=\iiint dV\psi^*\{\vec{r}\times(-\vec{\nabla}V)\}\psi$ where $\vec{r},\vec{p}$ and *V* are respectively the position, momentum, and potential-energy operators.

E1.11: A system is prepared in an eigenstate $|\omega_i\rangle$ of the observable Ω. A different measurement Ξ, which is incompatible with Ω is now carried out on the system originally prepared in the state $|\omega_i\rangle$. What is the probability that the system lands in the eigenstate $|\sigma_k\rangle$ of the operator Ξ?

E1.12: This problem is similar to E1.11, but with a crucial difference. As in E1.11, the system is first prepared in an eigenstate $|\omega_i\rangle$ of the observable Ω. A different measurement Λ, which is *not* compatible with Ω, nor with Ξ, is carried out on the system that was originally prepared in the eigenstate $|\omega_i\rangle$ of Ω. The measurement of Λ is now followed by the measurement Ξ. What is the probability that the system that was originally prepared in the eigenstate $|\omega_i\rangle$ lands in the eigenstate $|\sigma_k\rangle$ of the operator Ξ, remembering that a measurement of Λ was carried out between the measurements Ω and Ξ? Is the result the same as that of E1.11?

E1.13: (a) Does the average momentum of a wave packet that represents a free particle change with time? (b) Does the average position of the wave packet for a free particle move at a constant velocity even if the wave packet may get deformed with passage of time?

E1.14: If an eigenvalue ω of the operator Ω is degenerate, then show that an arbitrary linear combination of the linearly independent eigenfunctions corresponding to the degenerate eigenvalue remains an eigenstate even if the wave packet may get deformed with passage of time.

E1.15: A bead of mass w slides on a frictionless straight wire of length L between two rigid walls. Determine the energy levels of this system? Show explicitly that the wave functions of this system belonging to different eigenvalues are mutually orthogonal. Also, determine the relative probabilities for various energy eigenstates to be occupied when the bead is exactly at the centre of the wire.

References

[1] P. C. Deshmukh, *Foundations of Classical Mechanics* (Cambridge University Press, 2019).

[2] G. Gamow, *Thirty Years that Shook Physics: The Story of Quantum Theory* (Dover, 1966).

[3] M. M.Gördesli, *The History of Quantum Physics* (Amazon, 2017).

[4] J. Mehra and H. Rechenberg, *The Historical Development of Quantum Theory*, Vol. 1 through 6 (Springer, 1982).

[5] P. A. M. Dirac, *Principles of Quantum Mechanics* (Oxford University Press, 1930).

[6] Herbert Goldstein, John Safko, and Charles P. Poole, *Classical Mechanics* (Pearson, 2011).

[7] P. C. Deshmukh and S. Banerjee, Time Delay in Atomic and Molecular Collisions and Photoionisation/Photodetachment. *Int. Rev. in Phys. Chem.* 40:1, 127–153 (2020).

[8] L. Mandelstam and I. Tamm, The Uncertainty Relation between Energy and Time in Non-relativistic Quantum Mechanics. In: B. M. Bolotovskii, V. Y. Frenkel, and R. Peierls (eds.), *Selected Papers* (Springer, 1991). https://doi.org/10.1007/978-3-642-74626-0_8.

[9] Jan Hilgevoord, Time in Quantum Mechanics. *Amer. Jour. of Phys.* 70, 301 (2002).

[10] W. Heisenberg, Über Quantentheoretische Umdeutung Kinematischer und Mechanischer Beziehungen. *Zeits. f. Physik*. 33, 879 (1925).

[11] M. Born and P. Jordan, On Quantum Mechanics, *Zeits. f. Physik*. 34, 858 (1925).

[12] M. Born, W. Heisenberg, and P. Jordan, Zur Quantenmechanik II. Z. *Phys.* 35, 557–615 (1926). https://doi.org/10.1007/BF01379806.

[13] E. Schrödinger, Quantisierung als Eigenwertproblem. *Ann. d. Physik*. 79, 361 (1926).

[14] P. A. M. Dirac, The Fundamental Equations of Quantum Mechanics, *Proc. Roy. Soc.* A109, 642–653 (1925); A110, 561 (1926).

[15] William A. Fedak and Jeffrey J. Prentis, The 1925 Born and Jordan Paper 'On Quantum Mechanics'. *Am. J. Phys.* 77:2, 128 (2009).

[16] Carl Eckart, Operator Calculus and the Solution of the Equations of Quantum Dynamics. *Phys. Rev.* 28, 711–726 (1926).

[17] B. L. Van der Waerden, (1973). From Matrix Mechanics and Wave Mechanics to Unified Quantum Mechanics. In: J. Mehra (ed.), *The Physicist's Conception of Nature*, pp. 276–293 (Dordrecht:Reidel Publishing Company, 1973). Reprinted in *Notices of the Am. Math. Soc.*, 44:3, 323–328 (1997).

[18] Daniel F. Styer, Miranda S. Balkin, Kathryn M. Becker, Matthew R. Burns, Christopher E. Dudley, Scott T. Forth, Jeremy S. Gaumer, Mark A. Kramer, David C. Oertel, Leonard H. Park, Marie T. Rinkoski, Clait T. Smith, and Timothy D. Wotherspoon, Nine Formulations of Quantum Mechanics. *Amer. Jour. of Phys.* 70, 288 (2002).

2

Path Integral Formulation of Quantum Mechanics

We choose to examine a phenomenon which is impossible, absolutely impossible, to explain in any classical way, and which has in it the heart of quantum mechanics. In reality, it contains the only mystery.

—Richard P. Feynman

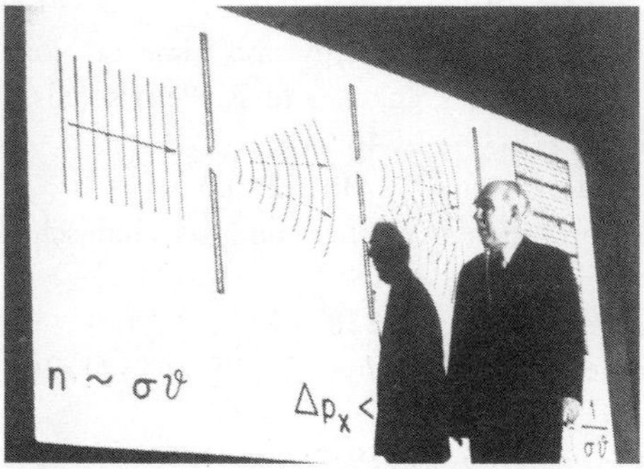

Niels Bohr lecturing on the Young's double slit experiment. B129, https://arkiv.dk/en/vis/5902686. Courtesy: Niels Bohr Archive.

In this chapter we show that a reformulation of quantum mechanics, namely Feynman's path integral approach, is equivalent to the earlier ones, due to Schrödinger and Heisenberg. We shall introduce and apply it in this chapter to interpret the Aharonov–Bohm effect, which is considered to be a "quantum wonder." It underscores the importance of electromagnetic potentials. We shall find that in addition to the *dynamical phase* introduced in Chapter 1, another angle, namely the *geometric phase* (Pancharatnam–Berry phase), is an important physical property of the wavefunction whose consequences are measurable.

2.1 Propagator: Propagation of a Quantum of Knowledge

Notwithstanding the equivalence of wave mechanics and matrix mechanics, Schrödinger and Heisenberg did not appreciate each other's methodologies [1]. In 1926, Schrödinger said: "*I knew of [Heisenberg's] theory, of course, but I felt discouraged, not to say repelled, by the methods of transcendental algebra, which appeared difficult to me, and by the lack of visualizability.*" Heisenberg's dislike for Schrödinger's wave mechanics was equally intense. In the same year, in a letter to Pauli, he wrote "*The more I think about the physical portion of Schrödinger's theory, the more repulsive I find it.... What Schrödinger writes about the visualizability of his theory 'is probably not quite right,' in other words it's crap.*" In the mean time, Paul A. M. Dirac, who had made path-breaking contributions [2] in that eventful year (1926) to the formulation of quantum mechanics, had drawn a penetrating analogy [3] between classical and quantum mechanics.

Dirac's remark drew the attention of Richard P. Feynman, who used it to build the path integral approach to quantum mechanics based on the variational principle.

Feynman was introduced to the principle of extremum action [4] rather early, while in high school. He reveals in his book [5] that "*my Physics teacher ... Mr. Bader ... told me something ... absolutely fascinating.... Every time the subject comes up, I work on it ... the principle of least* (rather, 'extremum') *action.*" This principle provides an alternative platform to build classical mechanics without employing the Newtonian principle of causality and determinism – i.e., without using the idea of a stimulus being the cause of a mechanical system's departure from equilibrium. The principle of variation is an *ansatz* that motion of a mechanical system occurs along a path in the configuration space that *extremizes* "action," which is the definite integral $\int_{t_i}^{t_f} L(q,\dot{q})dt$ of the system's Lagrangian, $L(q,\dot{q}) = T - V$. T is the system's kinetic energy and V its potential energy, described respectively in terms of velocity $\dot{q}$ and position q [4]. As we shall see below, "action" provides an essential measure of a phase angle (in units of $\hbar$) in terms of which Feynman's path integral formulation of the quantum law of nature is developed.

Funquest: Is the Lagrangian for a system unique? If two Lagrangians $L(q,\dot{q},t)$ and $L'(q,\dot{q},t)$ lead to the same equation of motion, what can be the difference between the two?

We shall now present the fundamental *ansatz* of Feynman's approach to quantum mechanics. It would appear as if it has come from nowhere, out of the blue; that is exactly what makes it an *ansatz*. Such is the power of a genius mind that can *postulate* a rule, that can be tested and verified. We shall first state the ansatz, no matter how far-fetched it might seem, and then proceed to analyse and interpret it. In doing so, we essentially follow the spirit of Feynman's admiration for Dirac. Said Feynman, "*He* (Dirac) *had the courage to simply guess at the form of an equation, and try to interpret it afterwards.*" Fittingly, we state the ansatz about the propagator in a mathematical form. We shall interpret it later. It would turn out that Feynman's ansatz produces a theory that is totally equivalent to matrix mechanics and wave mechanics. It was like re-inventing quantum mechanics almost two decades later, but with a different opening!

For simplicity, we consider motion in one dimension (along the x-axis of a Cartesian coordinate reference frame) of a system of mass m. The ansatz of the path integral method states that the *kernel* (which contains the crux of the central idea) that provides the probability amplitude $K(x_f,t_f;x_0,t_0)$ for a system to propagate (i.e., evolve) from a state $|x_0,t_0\rangle$ (at the initial time $t_i = t_0$) to the final state $|x_f,t_f\rangle$ (at the end time t_f) is given by

$$K(x_f,t_f;x_0,t_0) = \langle x_f|\ \exp\left[-i\frac{H(t_f-t_0)}{\hbar}\right]|x_0\rangle = \int_{x_0}^{x_f} D[x(t)]exp\left[\frac{i}{\hbar}\int_{t_0}^{t_f} dtL(x,\dot{x})\right], \quad (2.1a)$$

i.e.,

$$\langle x_f|\ \exp\left[-i\frac{H(t_f-t_0)}{\hbar}\right]|x_0\rangle = \lim_{N\to\infty}\int_{-\infty}^{\infty} dx_1 \ldots \int_{-\infty}^{\infty} dx_{N-1}\left(\sqrt{\frac{m}{i2\pi\hbar\Delta t}}\right)^N \exp\left[\frac{i}{\hbar}\int_{t_0}^{t_f} dtL(x,\dot{x})\right]. \quad (2.1b)$$

In writing Eq. 2.1b, we have only expanded the short notation:

$$\int_{x_0}^{x_f} D[x(t)] \equiv \lim_{N\to\infty} \int_{-\infty}^{\infty} dx_1 \ldots \int_{-\infty}^{\infty} dx_{N-1} \left(\frac{m}{i2\pi\hbar\Delta t}\right)^{N/2}. \tag{2.2}$$

We shall now unscramble the ansatz stated in Eq. 2.1. In doing so, we shall also unravel the source of the cryptic term $\left(\sqrt{\dfrac{m}{i2\pi\hbar\Delta t}}\right)^N$ in Eq. 2.2. After this, we shall provide a physical description of an experimental situation using the above ansatz. Dirac had pointed out [3] that the kernel, which tells us how a wavefunction propagates from $\psi(x_0,t_0)$ to $\psi(x_f,t_f)$, was *analogous* to $\exp\left[\dfrac{i}{\hbar}\int_{t_0}^{t_f} dt L(x,\dot{x})\right]$. The phase angle (in units of $\hbar$) of the complex kernel is expressed in terms of the classical "action" $S = \int_{t_i}^{t_f} dt L(x,\dot{x})$, whose extremum for alternative paths between x_0 and x_f in the configuration space produces the classical trajectory. Feynman expanded Dirac's plan into the ansatz stated in Eq. 2.1. In order to appreciate Feynman's ansatz, we first segment the time-evolution operator (Eq. 1.86a,b), as a product of N factors, each being $\exp\left[-\dfrac{i}{\hbar} H\Delta t\right]$, and let $N \to \infty$, i.e.,

$$U(t_f,t_i) = \lim_{N\to\infty}\left(1 - \frac{i}{\hbar}\frac{t_f - t_i}{N} H\right)^N = \exp\left[-\frac{i}{\hbar} H(t_f - t_i)\right]$$

$$= \overbrace{\exp\left[-\frac{i}{\hbar} H\Delta t\right]\exp\left[-\frac{i}{\hbar} H\Delta t\right]\ldots\exp\left[-\frac{i}{\hbar} H\Delta t\right]}^{N\ factors}. \tag{2.3}$$

The kernel

$$K(x_f,t_f;x_0,t_0) = \langle x_f | U(t_f,t_i = t_0) | x_0\rangle \tag{2.4a}$$

is then given by

$$\langle x_f | \exp\left[-i\frac{H(t_f - t_0)}{\hbar}\right] | x_0\rangle = \int_{-\infty}^{\infty} dx_1 \ldots \int_{-\infty}^{\infty} dx_{N-1} \left\{\begin{array}{r} \langle x_f | \exp\left[-\frac{i}{\hbar} H\Delta t\right] | x_{N-1}\rangle \times \\ K_{N-1,N-2} \ldots \times \\ \langle x_2 | \exp\left[-\frac{i}{\hbar} H\Delta t\right] | x_1\rangle\langle x_1 | \\ \exp\left[-\frac{i}{\hbar} H\Delta t\right] | x_0\rangle \end{array}\right\}, \tag{2.4b}$$

with the understanding that we would be interested in the limit $N \to \infty$.

In Eq.2.4b, we have used the short notation $K_{N-1,N-2}$ for the sub-kernel from $(N-2)^{th}$ step to $(N-1)^{th}$. Also, we have inserted the resolution of the unit operator $\int_{-\infty}^{\infty} dx_i |x_i\rangle\langle x_i|$, for $i = 1,2,\ldots,(N-1)$, between each consecutive step. We can now write a typical sub-kernel by employing the resolution of the unit operator in the momentum space as

$$\left\langle x_j \middle| e^{-i\frac{H\Delta t}{\hbar}} \middle| x_{j-1} \right\rangle = \int_{-\infty}^{\infty} dp_{j-1} \left\langle x_j \middle\| p_{j-1} \right\rangle \left\langle p_{j-1} \middle| e^{-i\frac{H\Delta t}{\hbar}} \middle| x_{j-1} \right\rangle. \tag{2.5}$$

We hasten to add that the subscript $(j-1)$ on the momentum over which integration is carried out is mute (since it is the dummy variable that gets integrated out), but it is notionally useful. Even if redundant and clumsy, it is advantageous for bookkeeping the *N* sub-kernels.

The first factor in the integrand of Eq. 2.5 is nothing but the coordinate representation of the momentum eigenstate (see Solved Problem P1.9 in Chapter 1). It is given by

$$\left\langle x_j \middle| p_{j-1} \right\rangle = \frac{1}{\sqrt{2\pi\hbar}} e^{\frac{i}{\hbar} p_{j-1} x_j}. \tag{2.6}$$

To determine the second factor $\left\langle p_{j-1} \middle| e^{-i\frac{H\Delta t}{\hbar}} \middle| x_{j-1} \right\rangle$ in the integrand, we first note that neither the position eigenstate vector $\left| x_{j-1} \right\rangle$ nor the momentum state eigenvector $\left\langle p_{j-1} \right|$ is an eigenstate of the Hamiltonian *H*. We shall now take advantage of operator algebra, recognizing that

$$\left\langle p_{j-1} \middle| e^{-i\frac{H\Delta t}{\hbar}} \middle| x_{j-1} \right\rangle = \left\langle p_{j-1} \middle| e^{-i\left(\frac{p_{op}^2}{2m} + V_{op}\right)\frac{\Delta t}{\hbar}} \middle| x_{j-1} \right\rangle = \overparen{\left\langle p_{j-1} \middle| e^{-i\left(\frac{p_{op}^2}{2m}\right)\frac{\Delta t}{\hbar}}} \overparen{e^{-i(V_{op})\frac{\Delta t}{\hbar}} \middle| x_{j-1} \right\rangle}. \tag{2.7}$$

In the last term, we have highlighted (by placing arcs above two factors) the parts $\left\langle p_{j-1} \right| e^{-i\left(\frac{p_{op}^2}{2m}\right)\frac{\Delta t}{\hbar}}$ and $e^{-i(V_{op})\frac{\Delta t}{\hbar}} \left| x_{j-1} \right\rangle$, to point out that we can now use the *eigenvalue equation* for the position operator to determine $e^{-i(V_{op})\frac{\Delta t}{\hbar}} |x\rangle$, and that for the momentum operator to determine $\langle p | e^{-i\left(\frac{p_{op}^2}{2m}\right)\frac{\Delta t}{\hbar}}$. In particular, we use the fact that

$$e^{-i(V_{op})\frac{\Delta t}{\hbar}} \left| x_{j-1} \right\rangle = e^{-iV_{j-1}\frac{\Delta t}{\hbar}} \left| x_{j-1} \right\rangle, \tag{2.8a}$$

V being the potential energy of the system when it is in position eigenstate; V_{j-1} is the potential energy when the particle is in the state $\left| x_{j-1} \right\rangle$. Likewise,

$$\left\langle p_{j-1} \right| e^{-i\left(\frac{p_{op}^2}{2m}\right)\frac{\Delta t}{\hbar}} = \left\langle p_{j-1} \right| e^{-i\left(\frac{p_{j-1}^2}{2m}\right)\frac{\Delta t}{\hbar}}, \tag{2.8b}$$

p_{j-1}^2 is the square of the momentum when the particle is in the state $\left| p_{j-1} \right\rangle$. Using Eqs. 2.8a,b in Eq. 2.7, we get

$$\begin{aligned} \left\langle p_{j-1} \middle| e^{-i\frac{H\Delta t}{\hbar}} \middle| x_{j-1} \right\rangle &= \overparen{\left\langle p_{j-1} \middle| e^{-i\left(\frac{p_{j-1}^2}{2m}\right)\frac{\Delta t}{\hbar}}} \overparen{e^{-i(V_{j-1})\frac{\Delta t}{\hbar}} \middle| x_{j-1} \right\rangle} \\ &= \left\langle p_{j-1} \middle| e^{-i\frac{E_{j-1}\Delta t}{\hbar}} \middle| x_{j-1} \right\rangle = \left\langle p_{j-1} \middle| x_{j-1} \right\rangle e^{-i\frac{E_{j-1}\Delta t}{\hbar}}. \end{aligned} \tag{2.9}$$

On the left-hand side of Eq. 2.9, we have the matrix element of the Hamiltonian operator and on the right-hand side we have a product of two complex functions. An eigenvalue equation for the Hamiltonian could not be used, but equality of the two sides of Eq. 2.9 results from Eq. 2.7 and Eq. 2.8a,b.

It now follows that

$$\left\langle p_{j-1}\right| e^{-i\frac{H\Delta t}{\hbar}}\left|x_{j-1}\right\rangle = \frac{1}{\sqrt{2\pi\hbar}} e^{-i\frac{p_{j-1}x_{j-1}}{\hbar}} e^{-i\frac{E_{j-1}\Delta t}{\hbar}} = \frac{1}{\sqrt{2\pi\hbar}} e^{-\frac{i}{\hbar}\left(p_{j-1}x_{j-1}+E_{j-1}\Delta t\right)}. \tag{2.10}$$

Eq. 2.5 and Eq. 2.10 now give

$$\left\langle x_{j}\right| e^{-i\frac{H\Delta t}{\hbar}}\left|x_{j-1}\right\rangle = \int_{-\infty}^{\infty} dp_{j-1}\left(\frac{1}{\sqrt{2\pi\hbar}} e^{\frac{i}{\hbar}p_{j-1}x_j}\right)\left(\frac{1}{\sqrt{2\pi\hbar}} e^{-\frac{i}{\hbar}\left(p_{j-1}x_{j-1}+E_{j-1}\Delta t\right)}\right)$$

$$= \int_{-\infty}^{\infty} \frac{dp_{j-1}}{2\pi\hbar} e^{\frac{i}{\hbar}p_{j-1}\left(x_j-x_{j-1}\right)} e^{-\frac{i}{\hbar}E_{j-1}\Delta t}. \tag{2.11}$$

Using this result in Eq. 2.4, the kernel becomes

$$\left\langle x_f\right| e^{-i\frac{H\left(t_f-t_0\right)}{\hbar}}\left|x_0\right\rangle = \lim_{N\to\infty}\int_{-\infty}^{\infty} dx_1 \ldots \int_{-\infty}^{\infty} dx_{N-1}\int_{-\infty}^{\infty}\frac{dp_0}{2\pi\hbar}\ldots\int_{-\infty}^{\infty}\frac{dp_{N-1}}{2\pi\hbar}\, e^{\frac{i}{\hbar}\sum_{j=1}^{N}\left[p_{j-1}\frac{\left(x_j-x_{j-1}\right)}{\Delta t}-E_{j-1}\right]\Delta t}. \tag{2.12}$$

A typical integral we must now evaluate, to solve Eq. 2.12, has the following form:

$$\left\langle x_{j}\right| e^{-i\frac{H\Delta t}{\hbar}}\left|x_{j-1}\right\rangle = \frac{1}{2\pi\hbar}\int_{-\infty}^{\infty} dp_{j-1} e^{i\frac{p_{j-1}\left(x_j-x_{j-1}\right)}{\hbar}} e^{-i\frac{E_{j-1}\Delta t}{\hbar}} = \frac{1}{2\pi\hbar} e^{-i\frac{\Delta t}{\hbar}V\left(x_{j-1}\right)}\int_{-\infty}^{\infty} dp\; e^{-ap^2} e^{bp}, \tag{2.13a}$$

with $a = \frac{i\Delta t}{2m\hbar}$; $b = \frac{i}{\hbar}\left(x_j - x_{j-1}\right)$. (2.13b)

Solving for the Gaussian integral (Solved Problem P2.1), it immediately follows that

$$\left\langle x_{j}\right| e^{-i\frac{H\Delta t}{\hbar}}\left|x_{j-1}\right\rangle = \frac{1}{2\pi\hbar} e^{-i\frac{\Delta t}{\hbar}V\left(x_{j-1}\right)} e^{\left\{\frac{\left[\frac{i}{\hbar}\left(x_j-x_{j-1}\right)\right]^2}{4\left(\frac{i\Delta t}{2m\hbar}\right)}\right\}}\sqrt{\frac{\pi}{\left(\frac{i\Delta t}{2m\hbar}\right)}}, \tag{2.14a}$$

i.e., $$\left\langle x_{j}\right| e^{-i\frac{H\Delta t}{\hbar}}\left|x_{j-1}\right\rangle = \sqrt{\frac{m}{i2\pi\hbar\Delta t}}\, e^{+i\frac{\Delta t}{\hbar}\left\{\frac{m}{2}\left[\frac{\left(x_j-x_{j-1}\right)}{\Delta t}\right]^2-V\left(x_{j-1}\right)\right\}} = \sqrt{\frac{m}{i2\pi\hbar\Delta t}}\, e^{+i\frac{\Delta t}{\hbar}L_j}, \tag{2.14b}$$

where the Lagrangian $L_j = \frac{m}{2}\left(\frac{x_j - x_{j-1}}{\Delta t}\right)^2 - V\left(x_{j-1}\right)$. (2.15)

The kernel in Eq. 2.12 is therefore given by

$$\left\langle x_f\right| e^{-i\frac{H\left(t_f-t_0\right)}{\hbar}}\left|x_0\right\rangle = \lim_{N\to\infty}\int_{-\infty}^{\infty} dx_1 \ldots \int_{-\infty}^{\infty} dx_{N-1}\left[\sqrt{\frac{m}{i2\pi\hbar\Delta t}}\right]^N e^{-\frac{i}{\hbar}\sum_{j=1}^{N}V\left(x_{j-1}\right)\Delta t}\, e^{\left[\frac{i}{\hbar}\frac{m\Delta t}{2}\sum_{j=1}^{N}\left(\frac{\left(x_j-x_{j-1}\right)}{\Delta t}\right)^2\right]}. \tag{2.16}$$

Recognizing now the limiting value of the sum (as the time intervals $\Delta t \to 0$) as an integral, we have

$$\lim_{\substack{N\to\infty \\ \Delta t\to 0}}\left[\frac{i}{\hbar}\Delta t\sum_{j=1}^{N}\left\{\frac{m}{2}\left(\frac{\left(x_j-x_{j-1}\right)}{\Delta t}\right)^2-V\left(x_{j-1}\right)\right\}\right]=\frac{i}{\hbar}\int_{t_0}^{t_f}dtL(x,\dot{x})=\frac{i}{\hbar}S\left[x(t)\right]. \tag{2.17}$$

Accordingly, the kernel (propagator) becomes

$$\begin{aligned}\left\langle x_f\right|e^{-i\frac{H\left(t_f-t_0\right)}{\hbar}}\left|x_0\right\rangle &=\lim_{N\to\infty}\int_{-\infty}^{\infty}dx_1\ldots\int_{-\infty}^{\infty}dx_{N-1}\left[\frac{m}{i2\pi\hbar\Delta t}\right]^{N/2}e^{\frac{i}{\hbar}\int_{t_0}^{t_f}dtL(x,\dot{x})}\\ &=\int_{x_0}^{x_f}D\left[x(t)\right]e^{\frac{i}{\hbar}\int_{t_0}^{t_f}dtL(x,\dot{x})},\end{aligned} \tag{2.18}$$

which is just the original ansatz in Eq. 2.1.

Funquest: In what other alternative ways can you evaluate the Gaussian integral?

Feynman's ansatz turned out to be a *déjà vu*, for he showed that it essentially reproduced the same results as from the Schrödinger equation. The ansatz in Eq. 2.1 incorporates the revolutionary idea in quantum mechanics which Einstein would regard as "spooky action at a distance." Entanglement resulting from the superposition of alternatives, no matter what distance separates them, is well integrated in Feynman's inclusion of summation over all paths (all histories). The tenet of *entanglement* is a mind-boggling cardinal signature of quantum theory. Its interpretation was hotly debated by Einstein and Bohr. We shall visit salient features of this debate, and its resolution, in Chapter 11. For now, we examine how it succeeds in interpreting how nature functions.

Having unscrambled the ansatz in Eq. 2.18, we shall now proceed to interpret it. We shall be led to the conclusion that the propagator involves a sum over *all* histories, i.e., a sum over *all* paths. A *gedanken* Young's interference experiment described now would explain this. We consider an experiment in which a particle travels from its source (shown by a "star" in Fig. 2.1) to a point on the detector screen. We shall refer to this particle as FP (*Feynman particle*). It would represent both a particle and a wave as understood in classical physics. The FP, which begins its journey at the source and ends on a detector screen, must be considered to have traveled through every *possible* path in between, no matter how improbable. The inclusion of every possible path consists of paths even with multiple way points – wherever they may be – such as at the North pole, or for that matter in the Andromeda galaxy! We illustrate these alternative paths in Figs. 2.2(a–d). One would of course wonder how the shortest path between the source and the detector can interfere with a path that has Mars and Andromeda galaxy as hopping intermediary points. At the speed of light it would take ~2.54 million years to get to Andromeda, and another ~2.54 million years to get back. In comparison, if the experiment is performed in a laboratory in which the shortest distance between the source and the detector is just a few meters, it would take only a few nanoseconds at the speed of light. This may seem to defy *common sense*, but it is in agreement with quantum theory, which correctly describes the *laws of nature*. After all, speed (momentum) of a particle is not knowable along with its path (position)! As Niels Bohr could say, physics may not answer all the questions, but it tells us what the right questions to be asked are.

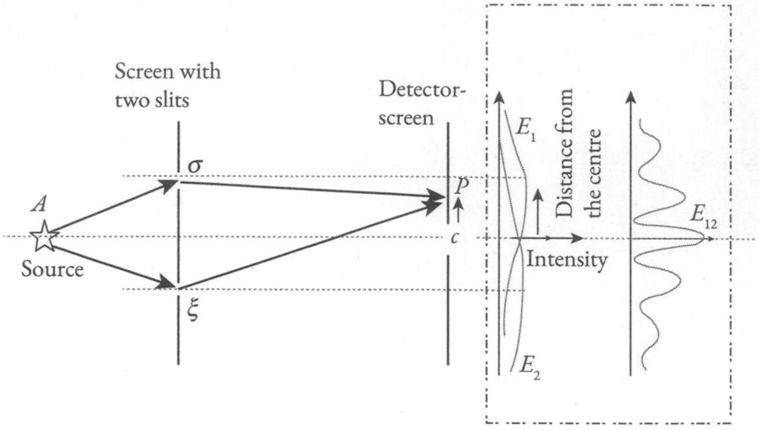

Fig. 2.1 The intensity vs. distance from centre C of the detector-screen is shown by the envelop E_1 when slit σ alone is open and by the envelop E_2 when slit ξ alone is open. It is shown by the envelop E_{12} when both the slits are open.

Figure 2.1 shows various routes a FP can take, beginning its journey at the source A and ending at a point P on the detector-screen. If we were guided by intuition based on *classical* physics, we would ask whether the FP would go through slit σ *or* through slit ξ. However, nature is best described by quantum laws, not classical. Feynman proposed that every possible path, no matter what the way points are, must *all* be considered dead evenly. This is a strange idea from the point of view of common intuition. It needs a new interpretation that contradicts judgment based on classical theory. Additionally, if we are required to use the same vocabulary as in classical physics, then it can only be done by assigning a new meaning to those words. When a FP hits the detector-screen on the right, the location of the point P at which it hits the screen is registered. In the language of Chapter 1, the FP *collapses* at the moment of its detection into a *position eigenstate*. The number of FPs recorded at each point on the detector-screen provides a measure of the intensity (probability) of the FPs that arrive at that point. We are interested in the analysis of the distribution of the number of hits at different points on the detector-screen over a sufficiently long exposure.

The result of the intensity distribution pattern on the detector-screen is mind-boggling. The bright and dark fringes produced in Young's double-slit experiment symbolize the hallmark of quantum description of laws of nature. Nature permits us to employ the "*either* σ *or* ξ" consideration *if and only if an observation is performed that determined exactly which of the two slits was actually traversed through*. Such an observation requires an improvisation of the apparatus. The simplest arrangement to achieve this can be made by introducing a shutter that would block one or the other slit. One could then conclusively claim that the FP would have traversed essentially through the open slit, not the one that is shut. Alternatively, one may keep both slits open but contrive an inclusion in the apparatus of a light/particle source *behind* the entrance screen. This would collide with the FP after it entered the apparatus and thereby determine exactly which of the two alternative paths was taken by the FP. An experiment in which this information is obtained is therefore referred to as a "*which way*" (abbreviated as WW) experiment (discussed further in Chapter 3). If an observation is *not* made to determine which slit was traversed through, i.e., if a WW experiment is not performed, then interference pattern shown by the envelop E_{12} results on the screen. It can be accounted for only by considering entry through the slit σ *and* through the slit ξ, not one *or* the other. When both the

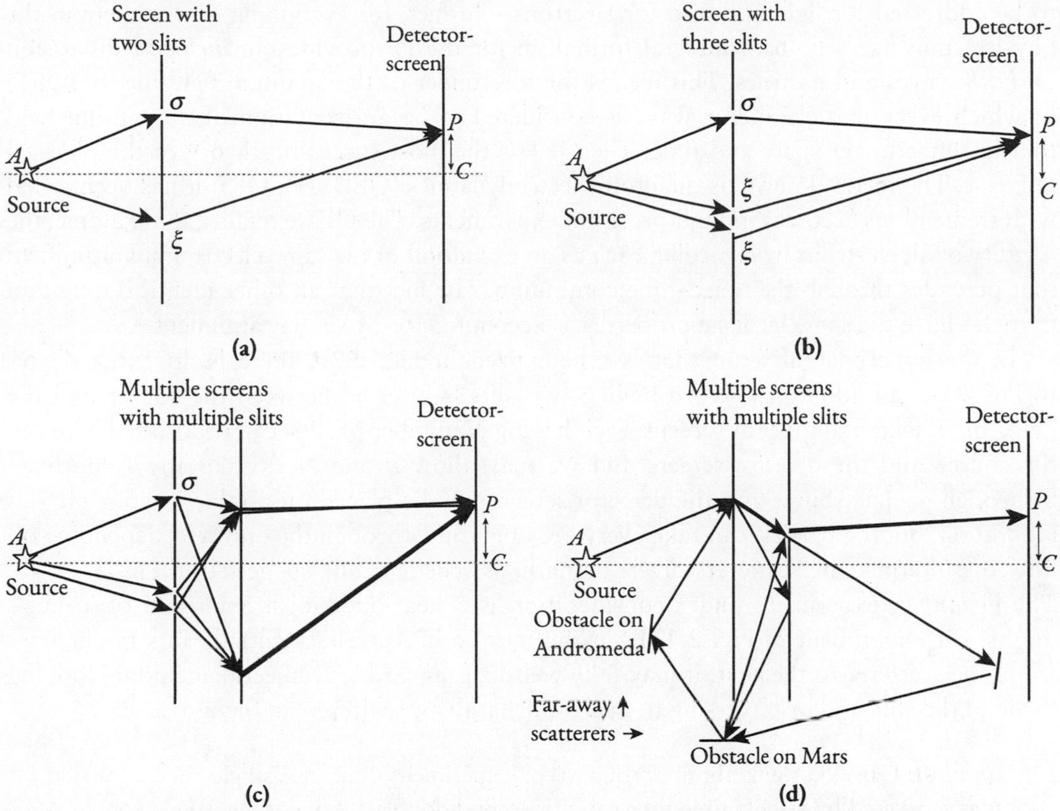

Fig. 2.2 The four panels in this figure, (a) through (d), show a *gedanken* Young's interference experiment in which the experimental setup allows increasingly zigzag paths, no matter how bizarre they seem to be, which FP can take from its source to a destination point on the detector-screen.

slits are open, the intensity distribution is not the one with a double-hump envelop (Fig. 2.1) as one might expect from the sum of E_1 and E_2. Instead, it turns out to be the interference pattern consisting of maxima and minima (bright and dark fringes) indicated by the envelop E_{12}.

Such an experiment was first performed and explained by Thomas Young (1773–1829) in 1801, using a light source. Young's explanation was constructed surmising that light consists of waves. The wave nature of light is very well described by the electrodynamics theory of James Clerk Maxwell (see, for example, Chapters 12 and 13 in Reference [4]). Young's use of the wave theory of light was in contradiction with the corpuscular theory developed by Newton (1642–1726/27). It was, however, the corpuscular (particle) theory of light that Einstein (1879–1955) employed in 1905 to provide a satisfactory explanation of the photoelectric effect. The wave (Young–Maxwell) *or* particle (Newton–Einstein) nature of light was at the centre of a debate that both frustrated and inspired some brilliant minds in the early part of the twentieth century. They put together an astounding notion of wave–particle duality which holds for both electrons and light. The experiments of Thomson (1856–1940) and Millikan (1868–1953) had conjectured that electrons had a corpuscular character. However, intensity distribution observed in experiments conducted in 1927 by Davisson (1881–1958) and Germer (1896–1971), in which they fired slow electrons at a crystalline nickel target, could be accounted for only using wavelike properties. The wave–particle controversy needed to be addressed for light and also for electrons – in fact, for everything that is there in the physical universe. The path integral formalism for the FP provides an *inclusive* discussion on *both* waves and particles. This idea is the forerunner of the quantum field theory (QFT) in which every particle and/or wave is considered to be an excitation of a quantum field that is defined over space and time. The QFT is the most successful theory of the physical universe. The theory known as quantum electrodynamics (QED) is a QFT; it has been tested with tremendous accuracy in high-precision experiments. The QFT explains, for example, the *identity* of all electrons by regarding each as an excitation of the same electron quantum field that pervades through the space–time continuum. The fact that all other identical quantum particles have the same intrinsic properties is accounted for by similar arguments.

In the experimental setup that is schematized in Fig. 2.2b, there is an extra slit S_3. In Fig. 2.2c, an additional screen having two slits is inserted between the source and the detector. In fact, n number of screens, each having N number of slits, can be inserted between the source and the detector-screen, and we may allow n and N to take any value from 0 through ∞. The source and the detector-screen may be placed on a table in your physics laboratory, but the particle can take every possible route through the universe, depending on how the intermediate scatterers divert its path, as schematically suggested in Fig. 2.2(a–d). The *gedanken* experiment under consideration is a near-outlandish extension of Young's double-slit experiment of Fig. 2.1. Even though the FP travels essentially in straight lines from one scatterer to the next, it may take weird zigzag paths, even eerie, including looping through the slits and/or traversing to nearby, or faraway, scatterers in the universe.

Funquest: Can you think about a Young-type experiment done at a single-slit but at *two times* separated by a short time-interval (attosecond double-slit experiment)?

The intensity at a point P on the detector-screen that would be observed is determined by the *modulus square* of the complex amplitude of the propagator. It is obtained from the *sum* (represented by the integral on the right-hand side of Eq. 2.18) of complex amplitudes over *every possible path* (i.e., every possible history), however wacky the alternative path may be. Each path contributes essentially the same *magnitude* (modulus) of the complex amplitude. The alternative paths differ only in their *phases*, which are determined by the *action* – the integral of the Lagrangian – in units of $\hbar$. Feynman's ansatz (Eq. 2.1 and Eq. 2.18) is now interpreted in terms of a sum over all the histories, including all creepy ones such as those depicted in Fig. 2.2d. Toward the end of Chapter 1, we had taken note of the synthesis of *matrix mechanics* and *wave mechanics* into *quantum mechanics*. The underpinning frameworks of the two schemes are vastly dissimilar, and neither has even a grain of resemblance with Feynman's ansatz (Eq. 2.18). In the next section, we shall find that the results from Eq. 2.18 and Eq. 1.86 are, nevertheless, equivalent.

2.2 Sum Over Infinite Histories

Classical physics had met its failure. Both wave mechanics and matrix mechanics accounted for physical phenomena in nature even if their formalisms seemed counterintuitive, and also vastly different from each other. Feynman's ansatz (Eq. 2.18) provides a mathematical kernel whose modulus square would provide a measure of the probability for the propagation $|x_0,t_0\rangle \to |x_f,t_f\rangle$. The kernel itself is defined (Eq. 2.18) in terms of an infinite sum (an integral) over every possible path (i.e., sum over all histories). Feynman introduced his ansatz in the 1940s by when quantum mechanics, based on wave and matrix mechanics, was well established for over two decades. We shall therefore refer to quantum mechanics based on Eq. 1.86 as *conventional* quantum mechanics. Now, we demonstrate that prediction from Feynman's ansatz for the time evolution of a state agrees exactly with that from conventional quantum mechanics [5, 6].

From *conventional* quantum mechanics, i.e., using Eq. 1.86, we have

$$\left\langle x_f \right| \exp\left[-i\frac{H\left(t_f - t_0\right)}{\hbar}\right] \left| x_0 \right\rangle = \left\langle x_f \right| \exp\left[-i\frac{H\Delta t}{\hbar}\right] \left| x_0 \right\rangle, \tag{2.19}$$

where we set the zero of the time scale at t_0. Equation 2.19 represents the propagator for evolution of a system from the state $\left|x_0, t_0\right\rangle$ to $\left|x_f, t_f\right\rangle$ over a time interval $\Delta t = t_f - t_0$. We shall consider a free particle. As explained below, this would pose absolutely no challenge to the *generality* of the result we shall obtain. Inserting the resolution of the unit operator in the momentum space, we get

$$\left\langle x_f \right| e^{-i\frac{H(t_f-t_0)}{\hbar}} \left| x_0 \right\rangle = \int_{-\infty}^{\infty} dp \left\langle x_f \left| e^{-\frac{i}{\hbar}\frac{p_x^2}{2m}(t_f-t_0)} \right| p \right\rangle \left\langle p \middle| x_0 \right\rangle = \int_{-\infty}^{\infty} dp \left\langle x_f \middle| p \right\rangle \left\langle p \middle| x_0 \right\rangle e^{-\frac{i}{\hbar}\frac{p_x^2}{2m}(t_f-t_0)}, \tag{2.20a}$$

i.e.,

$$\left\langle x_f \right| e^{-i\frac{H(t_f-t_0)}{\hbar}} \left| x_0 \right\rangle = \int_{-\infty}^{\infty} dp \frac{1}{\sqrt{2\pi\hbar}} e^{i\frac{px_f}{\hbar}} \frac{1}{\sqrt{2\pi\hbar}} e^{-i\frac{px_0}{\hbar}} e^{-\frac{i}{\hbar}\frac{p^2}{2m}(t_f-t_0)}$$

$$= \frac{1}{2\pi\hbar}\int_{-\infty}^{\infty} dp e^{\frac{i}{\hbar}p(x_f-x_0)} e^{-\frac{i}{\hbar}\frac{p^2}{2m}(t_f-t_0)}. \tag{2.20b}$$

Using now the result of the Solved Problem P2.1 for the Gaussian integral, we get

$$\left\langle x_f \right| e^{-i\frac{H(t_f-t_0)}{\hbar}} \left| x_0 \right\rangle = \left[\sqrt{\frac{m}{2\pi\hbar i\left(t_f - t_0\right)}}\right]\left[e^{i\frac{m\left(x_f-x_0\right)^2}{2\hbar\left(t_f-t_0\right)}}\right] = \left[N_{t_f,t_0}\right]\left[e^{i\theta}\right]. \tag{2.21}$$

This result is simple: the probability amplitude for the evolution $\left|x_0, t_0\right\rangle \to \left|x_f, t_f\right\rangle$ is given simply by a normalization constant N_{t_f,t_0} which depends on (a) the time difference $\Delta t = t_f - t_0$ (other than constants like the PEB constant "h" and the mass "m" of the particle) and (b) a phase factor θ. Feynman had the intuition and courage (inspired by that of Dirac) to stipulate that the path integral *ansatz* (Eq. 2.18) would yield exactly the same result (Eq. 2.21) as we got from *conventional* quantum mechanics. We now demonstrate that indeed it does. Toward this goal, we first recognize that the *definite time integral* in the space-integrand of Eq. 2.18 is the limit of a sum over infinite number $(N; N \to \infty)$ of time-steps. Hence,

$$\left\langle x_f \right| e^{-i\frac{H(t_f-t_0)}{\hbar}} \left| x_0 \right\rangle = \lim_{N\to\infty} \int_{-\infty}^{\infty} dx_1 \ldots \int_{-\infty}^{\infty} dx_{N-1} \left[\frac{m}{i2\pi\hbar\Delta t}\right]^{N/2} e^{\frac{i}{\hbar}\Delta t\left[\sum_{j=1}^{N}\frac{m}{2}\left(\frac{x_j-x_{j-1}}{\Delta t}\right)^2\right]}. \tag{2.22}$$

In order to render our notation compact, we introduce a dimensionless variable

$$y_j = x_j\sqrt{\frac{m}{2\hbar\Delta t}}, \tag{2.23a}$$

with

$$dx_j = \sqrt{\frac{2\hbar\Delta t}{m}}dy_j. \tag{2.23b}$$

This gives

$$\left\langle x_f\right| e^{-i\frac{H(t_f-t_0)}{\hbar}}\left|x_0\right\rangle = \lim_{N\to\infty}\left[\sqrt{\frac{2\hbar\Delta t}{m}}\right]^{N-1}\int_{-\infty}^{\infty}dy_1\ldots\int_{-\infty}^{\infty}dy_{N-1}\left[\frac{m}{i2\pi\hbar\Delta t}\right]^{N/2} e^{\left[\frac{i}{\hbar}\Delta t\sum_{j=1}^{N}\frac{m}{2}\frac{2\hbar\Delta t}{m}\left(\frac{(y_j-y_{j-1})}{\Delta t}\right)^2\right]}, \tag{2.24a}$$

$$\text{i.e., } \left\langle x_f\right| \exp\left[-i\frac{H(t_f-t_0)}{\hbar}\right]\left|x_0\right\rangle = \lim_{N\to\infty}\left[\sqrt{\frac{2\hbar\Delta t}{m}}\right]^{N-1}\int_{-\infty}^{\infty}dy_1\ldots\int_{-\infty}^{\infty}dy_{N-1}\left[\frac{m}{i2\pi\hbar\Delta t}\right]^{N/2} \exp\left[i\left\{\sum_{j=1}^{N}(y_j-y_{j-1})^2\right\}\right]. \tag{2.24b}$$

The results of the Solved Problem P2.2 and Eq. 2.24 now can be combined to give

$$\left\langle x_f\right| e^{-i\frac{H(t_f-t_0)}{\hbar}}\left|x_0\right\rangle = \left[\frac{m}{i2\pi\hbar(t_f-t_0)}\right]^{1/2} e^{\text{im}\left[\frac{(x_f-x_0)^2}{2\hbar(t_f-t_0)}\right]} = \left[N_{t_f,t_0}\right]\left[e^{i\theta}\right], \tag{2.25}$$

since $N\Delta t = t_f - t_0$. It must be stressed that the normalization constant in Eq. 2.25 is completely determined by the time interval $(t_f - t_0)$ *alone* (apart from universal constants). Equation 2.25 is exactly the same result (Eq. 2.21) that we had obtained from *conventional* quantum mechanics. We have established equivalence of conventional quantum mechanics and the path integral method. Even if we have shown this result only for free particles, it is no limitation; *between* multiple scattering interventions (as in Fig. 2.2d), the FP travels essentially as a free particle. The $N \to \infty$ hops in Feynman's method (Eq. 2.22) accommodate every alternative route from the source to the detector, via arbitrary way points anywhere in the universe. The equivalence of Schrödinger's wave mechanics, Heisenberg's matrix mechanics, and Feynman's path integral approach is thus on a robust foundation. In some situations, one formulation of quantum mechanics may, however, have an advantage over another. We shall come across such an example in the consideration of degeneracy of eigenstates in the discrete spectrum of hydrogen atom (Chapter 5). In this case, the operator formalism of quantum mechanics would be found to have a decisive advantage over Schrödinger's differential equation method. Notwithstanding such differences, the three alternative formulations essentially converge to the same theory. The reason for this confluence *cannot* be found in *any* classical mechanism.

We can now address a nagging oddity that may have troubled you: how come Newton's laws work in a vast majority of physical applications in our day-to-day experience, even if they require simultaneous knowledge of position and momentum? There is no threshold *size* for classical laws to be applicable. Any suggestion that there is a different theory for microscopic objects than there is for macroscopic objects is misleading. Quantum

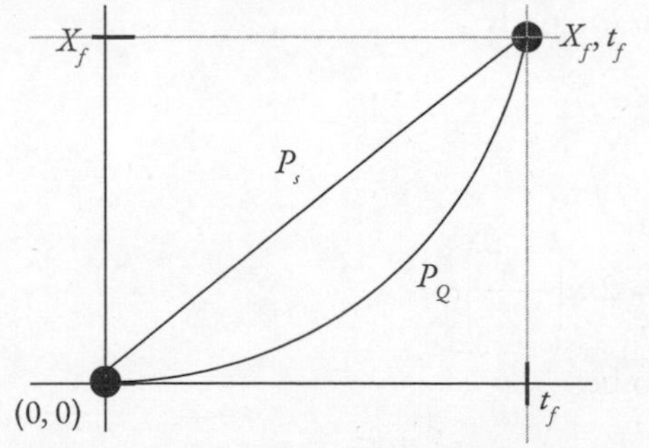

Fig. 2.3 Schematic diagram showing one-dimensional motion along the (vertical) x-axis against the (horizontal) time axis. A free particle's motion from ($x_i = 0$, $t_i = 0$) to (x_f, t_f). Newton's method provides the correct path P_S (straight line). Consideration of variations in the quadratic path P_Q explains why the classical trajectory turns out to be correct.

laws are universal. It is the success of classical laws, when and where they apply, that ought to baffle. In terms of Feynman's path integral approach to quantum mechanics, we shall be able to appreciate the success of classical laws, i.e., in those special domains that they seem to work. We shall now identify a simple criterion that determines the limit of applicability of classical mechanics. Figure 2.3 schematically illustrates one-dimensional motion of an object moving from $\left(x_i = 0, t_i = 0\right)$ to $\left(x_f, t_f\right)$ in an interaction-free region. It is obvious that the Newton–Lagrange–Hamilton method would give us a straight trajectory, labeled P_S, in this figure. This is readily validated using path integrals. Using Feynman's method, we compare *variations* to the straight path P_S and also to the quadratic path P_Q (Fig. 2.3). Not just P_Q, but infinite number of paths can be considered, regardless of how inelegant they may be. We shall see that when *action* is much larger than the PEB constant h, the phase factor in Eq. 2.18 cancels out [7] in all but a few paths. The path that survives corresponds to the classical trajectory, when *action* is much larger than h. This is just the path that makes *action* stationary, thereby making classical law applicable.

First, we determine *action* corresponding to the straight line

$$x(t) = xt = \left(\frac{x_f - x_i}{t_f - t_i}\right)t = \left(\frac{x_f}{t_f}\right)t \tag{2.26a}$$

from x_0 to x_f. It is

$$S_S = \int_0^{t_f} dt \frac{1}{2} m \left(\frac{x_f}{t_f}\right)^2 = \frac{1}{2} m \frac{x_f^2}{t_f}. \tag{2.26b}$$

The subscript S in Eq. 2.26b refers to the "straight" path. The Feynman propagator for this path is

$$\langle x_f | e^{-i\frac{H(t_f - t_0)}{\hbar}} | x_0 \rangle_S = \int_{x_0}^{x_f} D[x(t)] e^{\frac{i}{\hbar} S_S[x(t)]}. \tag{2.26c}$$

The phase angle in this integrand is

$$\theta_S = \frac{m x_f^2}{2\hbar t_f}. \tag{2.26d}$$

Likewise, for the quadratic path

$$x_Q(t) = \left(\frac{x_f}{t_f^{\,2}}\right)t^2, \tag{2.27a}$$

the Lagrangian is

$$L_Q = \frac{1}{2}m4\left(\frac{x_f}{t_f^{\,2}}\right)^2 t^2 = 2m\left(\frac{x_f^{\,2}}{t_f^{\,4}}\right)t^2,$$

and *action* turns out to be

$$S_Q = \int_0^{t_f} dtL_Q = \int_0^{t_f} dt2m\left(\frac{x_f^2}{t_f^4}\right)t^2 = \frac{2}{3}\frac{mx_f^2}{t_f}. \tag{2.27b}$$

The propagator for the quadratic path therefore is

$$\left\langle x_f\right| e^{-i\frac{H(t_f-t_0)}{\hbar}} \left|x_0\right\rangle_Q = \int_{x_0}^{x_f} D\left[x(t)\right]e^{\frac{i}{\hbar}S_Q[x(t)]}, \tag{2.27c}$$

and the phase in its integrand is

$$\theta_Q = \frac{2}{3}\frac{mx_f^2}{\hbar t_f}. \tag{2.27d}$$

The reason that nature prefers the straight path P_S over the quadratic path P_Q is *not* just that $\theta_Q > \theta_S$. To describe nature's preference appropriately, we must investigate what happens to the phases for *alternative* paths when the path is *varied*. For an equitable comparison, we subject both the linear and the quadratic paths to essentially the same variation. We therefore consider a *variation* represented by

$$t \to t\left(1 + \varepsilon\frac{t - t_f}{t_f}\right), \tag{2.28}$$

where $\varepsilon \ll 1$ is a tiny perturbation parameter. Accordingly, the new paths are

$$x_{SV}(t) = \left(\frac{x_f}{t_f}\right)\left(t + \frac{\varepsilon t}{t_f}\left(t - t_f\right)\right), \tag{2.29}$$

$$\text{and } x_{QV}(t) = \left(\frac{x_f}{t_f^2}\right)\left\{t\left(1 + \varepsilon\frac{t - t_f}{t_f}\right)\right\}^2 \tag{2.30}$$

respectively, instead of Eq. 2.26a and Eq. 2.27a. The subscript V in Eqs. 2.29 and 2.30 reminds us of the *variation* indicated in Eq. 2.28, and the subscripts S and Q designate the straight and the quadratic paths respectively. It is straightforward to see that the *action* corresponding to the path SV is

$$S_{SV} = \frac{m}{2}\frac{x_f^{\,2}}{t_f}\left(1 + \frac{\varepsilon^2}{3}\right) = S_S\left(1 + \frac{\varepsilon^2}{3}\right). \tag{2.31}$$

Likewise, the *action* corresponding to the path QV is

$$S_{QV} = \frac{2}{3}\frac{mx_f^{\,2}}{t_f}\left(1 + \frac{\varepsilon}{2} + O(\varepsilon^2)\right) = S_Q\left(1 + \frac{\varepsilon}{2} + O(\varepsilon^2)\right). \tag{2.32}$$

In Eq. 2.32, $O(\varepsilon^2)$ represents terms of the order of ε^2, or weaker. To first order in the variation parameter (ε), the phase factors therefore are

$$\theta_{SV} \simeq \frac{S_S}{\hbar}, \tag{2.33a}$$

and

$$\theta_{QV} \simeq \frac{S_Q}{\hbar}\left(1+\frac{\varepsilon}{2}\right). \tag{2.33b}$$

The variation in *action* due to the alterations in the original path therefore is

$$(\delta S)_S = \left(\frac{\partial S_{SV}}{\partial \varepsilon}\right)(\varepsilon) = \frac{2}{3}S_S\varepsilon^2 \simeq 0, \tag{2.34a}$$

for the original straight line path and

$$(\delta S)_Q = \left(\frac{\partial S_{QV}}{\partial \varepsilon}\right)(\varepsilon) = \frac{1}{2}S_Q\varepsilon \neq 0, \tag{2.34b}$$

for the original quadratic path, to *first* order in ε. Recall that the Feynman propagator is a complex number consisting of a constant times $e^{\frac{i}{\hbar}S}$, where S is *action* and $\hbar$ is the PEB constant divided by 2π. Thus, when action is much greater than $\hbar$, paths that are nearby to the straight path P_S reinforce each other, whereas paths that are in the proximity of the quadratic paths cancel each other. The complex propagator may be represented by a vector $\vec{V}$ in a plane, which in the Cartesian coordinate system (notation from Reference [4]) is

$$\vec{V} = V_x\hat{e}_x + V_y\hat{e}_y, \tag{2.35a}$$

or, in the plane–polar coordinate system as (from Eq. 2.1 or Eq. 2.18)

$$\vec{V} = \hat{e}_\rho e^{i\varphi}\sqrt{V_x^2 + V_y^2} = \hat{e}_\rho e^{i\frac{S}{\hbar}}\sqrt{V_x^2 + V_y^2}. \tag{2.35b}$$

Feynman propagators along different paths contribute essentially the same *magnitude*, but due to different *action* along various paths, vectors that represent them have different directions. When $S \gg \hbar$, a small variation in the classical path (P_S in Fig. 2.3) causes practically no change in the phase angle $\varphi = \frac{S}{\hbar}$, since action is not affected to first order in the variation ε (Eq. 2.34a). The corresponding propagator vectors point in the same direction; they add up coherently. However, for small variations to the paths such as P_Q (Fig. 2.3), the phase changes significantly (Eq. 2.34b). The corresponding vectors would then have the same magnitude but different directions. The corresponding vector representation of the summation of propagators would be akin to polygonal addition of vectors (Fig. 2.4), resulting in null displacement. Only the path P_S contributes and explains why the Newton–Lagrange–Hamilton method is successful in a large number of cases. We note that *action* scales as mass of the particle. Hence, as mass decreases, $\frac{S}{\hbar}$ becomes increasingly significant and hence the path integral method (i.e., quantum mechanics) would not give the same results as classical mechanics. One must then abandon classical physics and employ quantum theory. The mass of a particle therefore matters, but not by itself; it is in combination with the value of $\hbar$ that it does.

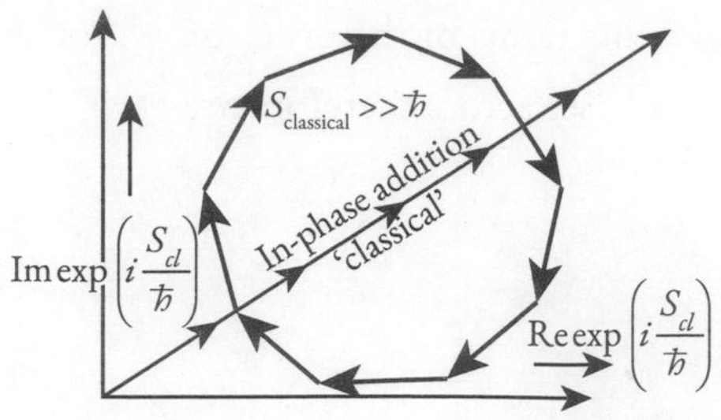

Fig. 2.4 When $S \gg \hbar$, paths nearby to the classical paths add up in phase; those away from the same cancel each other.

Funquest: Is it correct to say that quantum mechanics is required only for particles with (a) less mass or (b) less volume?

We now see why classical mechanics succeeds in many day-to-day experiences, even if it rests on the impossible simultaneous knowledge of position and momentum. The essence of the more appropriate quantum theory is stealthily present in it, through the approximation that is applicable when $S \gg \hbar$.

2.3 Quantum Superposition using Path Integrals

Entanglement resulting from *a priori* superposition of every possible alternative is beautifully exhibited in Young's double-slit experiment, whether with light or with particles. Its analysis defies classical mechanics. We investigate results of this experiment in the next section using Feynman's path integrals.

The well-known double-slit experiment performed by Thomas Young (1773–1829) – *a child prodigy, often described as the last man who knew everything* – is commonly endorsed as one of the most beautiful experiments in physics. It promoted the wave theory of light in the nineteenth century. Wave–particle duality in de Broglie–Schrödinger's quantum mechanics inspired efforts to perform a similar experiment with electrons. We do not recount here these outstanding advances but urge the readers to go through References [8, 9] to survey experiments (including double-slit interference with a single electron) to find exciting literature on this fundamental topic. We shall assume that the reader is familiar with the interpretation of the interference fringes in this experiment based on the wave theory. Detailed analysis of the fringes is easily found in most books on optics that account for the intensity distribution in terms of the modulus square of the *sum of the complex amplitudes* of waves from the slit σ and from slit ξ (Fig. 2.1), arriving at a point on the detector. The complex amplitude of each of the two waves consists of two real parameters: (a) amplitude and (b) phase of the respective waves along each alternative path. Use of complex amplitudes is essential; separate algebras of two real numbers would not work. In-phase arrival of the two waves results in constructive interference (bright fringes), and opposite-phase arrival in destructive interference (dark fringes). Essentially, this is caused by the difference in the path lengths of the two alternative paths. If the two path lengths differ by an integer multiple $(n\lambda)$ of the wavelength λ (n being an integer), the two waves interfere constructively. On the other hand, if the path difference is $\left(n+\frac{1}{2}\right)\lambda$ we get destructive interference.

The probability that FP propagates from an initial state $|x_0,t_0\rangle$ to the final state $|x_f,t_f\rangle$ is given by the modulus square of the kernel $\langle x_f | e^{-i\frac{H(t_f-t_0)}{\hbar}} |x_0\rangle$. As discussed in Section 2.1, it is given by

$$P(f,0) = |K(f,0)|^2 = \left| \sum_j^{\text{all paths}} N_{f,0} e^{\frac{i}{\hbar}S_j(f,0)} \right|^2 \tag{2.36}$$

where $K(f,0)$ is the said kernel, in which "0" represents the point of origin of FP and "f" represents a destination point on the detector. We are interested in determining the net intensity distribution on the detector screen. We shall now apply Eq. 2.36 to understand the fringes in the double-slit experiment, following the discussion in Reference [10]. As seen in Fig. 2.5a, we must consider two alternative paths, one through the slit σ and the other through the slit ξ. We shall denote the *action* determined along the path via slit σ by S_σ and that along the path via slit ξ by S_ξ. The sum over paths indicated in Eq. 2.36 is then the sum over paths σ and ξ. *After* we analyse the interference pattern using path integrals, we shall discuss an improvisation of the experiment using an electric current through a solenoid (Fig. 2.5b) that is placed behind the two slits (Fig. 2.5c).

For now, we discuss the interference experiment in Fig. 2.5a. For the path via slit σ, we have

$$K_\sigma(f,0) = N_{f,0}\exp\left[\frac{i}{\hbar}S_\sigma(f,0)\right], \tag{2.37a}$$

$$S_\sigma(f,0) = S_\sigma(f,\sigma) + S_\sigma(\sigma,0), \tag{2.37b}$$

and integration over time must be carried out as a sum over two segments corresponding to "0" to σ, and from σ to "f," i.e.,

$$\int_{t_0}^{t_f} \equiv \int_{t_0}^{t_\sigma} + \int_{t_\sigma}^{t_f}\ ;\ t_0 \le t_\sigma \le t_f. \tag{2.37c}$$

For the path via the slit ξ, we have

$$K_\xi(f,0) = N_{f,0}\exp\left[\frac{i}{\hbar}S_\xi(f,0)\right], \tag{2.38a}$$

$$S_\xi(f,0) = S_\xi(f,\xi) + S_\xi(\xi,0). \tag{2.38b}$$

In this case we need a sum over the two segments from "0" to ξ and from ξ to "f."

The kernel in Eq. 2.36 therefore is

$$K(f,0) = K_\sigma(f,\sigma)K_\sigma(\sigma,0) + K_\xi(f,\xi)K_\xi(\xi,0). \tag{2.39}$$

In-between scatterers, the FP propagates as a free particle. Hence, we use Eq. 2.25 to determine each of the four kernels: $K_\sigma(f,\sigma), K_\sigma(\sigma,0), K_\xi(f,\xi)$, and $K_\xi(\xi,0)$.

From symmetry, we see that

$$K_\sigma(\sigma,0) = K_\xi(\xi,0) = K_0, \text{ say,} \tag{2.40a}$$

and therefore

$$K(f,0) = K_0\{K_\sigma(f,\sigma) + K_\xi(f,\xi)\}. \tag{2.40b}$$

Also, from symmetry, the time interval for the paths from the source "0" to each of the two slits σ and ξ is the same:

$$\Delta t_{0,\sigma} = \Delta t_{0,\xi}. \tag{2.41a}$$

We must therefore conclude that

$$\Delta t_{\sigma,f} = \Delta t_{\xi,f}. \tag{2.41b}$$

To our minds that are commonly tuned to classical physics, the result in Eq. 2.41b could seem to be counterintuitive, since the *path* lengths "d" from the two slits to the destination point "f" are clearly different: $d(\sigma,f) \neq d(\xi,f)$. If we were to use ideas from classical physics, we would infer that the FP must travel at different momenta (speeds) for Eq. 2.41b to hold. We have, however, already agreed that momentum is not measurable simultaneously with position, hence unknown in the present consideration. Furthermore, the normalization constant is determined solely by time intervals, hence

$$N_{\sigma,f} = N_{\xi,f}, \text{ say } N. \tag{2.41c}$$

The kernel in Eq. 2.40b therefore becomes

$$K(f,0) = K_0\left\{Ne^{i\theta_\sigma} + Ne^{i\theta_\xi}\right\} = K_0N\left\{e^{i\theta_\sigma} + e^{i\theta_\xi}\right\}, \tag{2.42}$$

where θ_σ and θ_ξ are the phases in Feynman's ansatz, given in terms of action determined along respective paths. The probability (Eq 2.36) that determined the intensity produced by FP at the point "f" on the detector is

$$P(f,0) = |K(f,0)|^2 = K^*(f,0)K(f,0). \tag{2.43}$$

Now, let

$$|K_0N| = \sqrt{I}, \tag{2.44a}$$

and the difference in phase angle to be represented by

$$\delta = (\theta_\sigma - \theta_\xi). \tag{2.44b}$$

We immediately get

$$P[(f,0)] = 2I(1 + \cos\delta), \tag{2.45a}$$

giving us a beautiful result:

$$0 \leq P[(f,0)] \leq 4I. \tag{2.45b}$$

We see that

$$P(f,0) = 4I \text{ when the phase difference is } \delta = 0, 2\pi, 4\pi, \ldots \text{ (bright fringes)}, \tag{2.46a}$$

and

$$P(f,0) = 0 \text{ when the phase difference is } \delta = \pi, 3\pi, 5\pi, \ldots \text{ (dark fringes)}. \tag{2.46b}$$

While accounting for interference fringes we have not employed the language of Young's waves, nor the de Broglie wave hypothesis for particles. We have essentially used the path integral approach. We now show that it corresponds exactly to conclusions from wave theory.

For brevity, let us represent the distance $d(\sigma,f)$ by d_σ and $d(\xi,f)$ by d_ξ. Then,

$$\exp\left(i\theta_\sigma\right) = \exp\left(i\frac{md_\sigma^{\,2}}{2\hbar\tau}\right). \tag{2.47a}$$

and

$$\exp\left(i\theta_\xi\right) = \exp\left(i\frac{md_\xi^{\,2}}{2\hbar\tau}\right), \tag{2.47b}$$

where τ is essentially the *same time* that the FP would take to traverse the distance from either slit to the destination point f.

Hence,

$$\exp(i\delta) = \exp\left(i\left(\theta_\xi - \theta_\sigma\right)\right) = \exp\left(i\frac{m\left(d_\xi + d_\sigma\right)\left(d_\xi - d_\sigma\right)}{2\hbar\tau}\right),$$

$$\text{or } \exp(i\delta) = \exp\left(i\frac{m\bar{d}\Delta}{\hbar\tau}\right) = \exp\left(i\frac{m\bar{u}\Delta}{\hbar}\right) = \exp\left(i\frac{\bar{p}\Delta}{\hbar}\right) = \exp\left(i\frac{2\pi\Delta}{\bar{\lambda}}\right), \tag{2.48}$$

with $\bar{u} = \frac{\bar{d}}{\tau}$ representing the average *speed* of a *de Broglie particle* (Eq. 1.9a). The path difference is $\Delta = d_\xi - d_\sigma$. We have represented the average distance between the two slits and destination point f as $\bar{d}$. We have employed de Broglie hypothesis that the wavelength of the wave is given by the ratio of the PEB constant h to the particle's momentum. The phase difference between the waves (de Broglie and/or Young) from the two slits when they arrive at the destination point "f" therefore is $\delta = \frac{2\pi\Delta}{\bar{\lambda}}$, and the corresponding path difference between d_σ and d_ξ is $\Delta = \frac{\delta}{2\pi} \times \bar{\lambda}$. Thus, when $\delta = n2\pi$ (n being an integer), the path difference between the two waves would be $\Delta = n\bar{\lambda}$, resulting in constructive interference (bright fringes). When $\delta = \left(n + \frac{1}{2}\right)2\pi$, the path difference would be $\Delta = \left(n + \frac{1}{2}\right)\bar{\lambda}$, resulting in destructive interference (dark fringes). We have found that the correspondence between wave mechanics and Feynman's path integral approach is completely satisfactory.

It is an amazing feature of nature that the fringes appear after sufficiently long exposures to the particles, even when the source yield is weakened so that there is only one particle at a time in the apparatus [8, 9]. In the next section, we extend Feynman's path integral approach to discuss a rather subtle feature of quantum physics, viz., the geometrical phase, also known as Pancharatnam–Berry phase.

2.4 Path Integral Analysis of the Pancharatnam–Berry Phase in the Aharonov–Bohm Effect

As an interesting application of the Feynman path integral method, we now consider the Pancharatnam–Berry phase [11, 12] in the Aharonov–Bohm effect [13, 14]. This is an important phenomenon in quantum electrodynamics and *gauge field theory*.

The primary methodology is based on an approximation technique often employed in quantum mechanics called as the *adiabatic approximation*. This section aims at providing an illustration of the adiabatic method, introduce the geometrical phase, provide a fascinating application of Feynman's path integral approach, and discuss a phenomenon – the Aharonov–Bohm effect – which is characteristically a quantum effect providing insights into the nature of the electromagnetic potentials. Approximation methods in quantum mechanics (including the 'adiabatic' approximation) are discussed in Chapter 6.

The basic phenomenon of the Pancharatnam–Berry phase was studied by Pancharatnam in the context of optics. A more general formulation is due to Berry; it is also referenced as Aharonov–Anandan phase [15]. For reasons that would become clear in this section, it is called *geometrical phase*. The Aharonov–Bohm effect has been described as an enigmatic quantum wonder [16] since it demonstrates an effect of a field that is not even there! We illustrate it by a tiny improvisation of the Young's double-slit experiment (Fig. 2.5a) that has been discussed in Section 2.3.

We consider *nearly* the same experiment as described in Fig. 2.1 (also shown in Fig. 2.5a), but with a small alteration. We install a long solenoid (Fig. 2.5b) between, and immediately behind, the two slits (Fig. 2.5c). The circle seen between the two slits in Fig. 2.5c provides a top view of the solenoid in the schematic diagram. The experiment is done with an electron gun at the source in a region of electromagnetic vacuum. The only electromagnetic field in the apparatus would be a constant magnetic field that can be set up *inside* the solenoid by passing a steady current through its coil. The solenoid is impenetrable. Electrons traveling from the source to the screen must *circumvent* the solenoid, through the region where electromagnetic fields $\left(\vec{E},\vec{B}\right)$ are both strictly zero. Classical Lorentz force on the electrons therefore vanishes. However strange as it may seem, the interference fringes on the screen get *displaced* (Fig. 2.5c) when a current is passed through the solenoid. This is an intriguing mystery called as the Aharonov–Bohm effect. Note that the magnetic vector potential $\vec{A}$ is nonzero, even if electromagnetic fields $\left\{\vec{E},\vec{B}\right\}$ are zero, along the possible path of electron trajectories. Theories from classical physics have us believe that electromagnetic potentials $\left\{\phi,\vec{A}\right\}$ are mere mathematical conveniences; that Newtonian dynamics results essentially from the Lorentz force we express in terms of the fields $\left\{\vec{E},\vec{B}\right\}$. However, we shall see that the potential $\vec{A}$ plays a decisive role in the displacement of fringes in Aharonov–Bohm effect.

Electron dynamics must be described in terms of temporal evolution of a complex state vector in the Hilbert space. The system Hamiltonian in the Aharonov–Bohm effect involves the electromagnetic gauge potentials $\left\{\phi,\vec{A}\right\}$, which can be adiabatically controlled externally by regulating the current through the solenoid. We represent the Hamiltonian as

$$H = H\left(R(t)\right), \tag{2.49a}$$

where R stands for an externally controlled time-dependent parameter that can be changed adiabatically. After a sufficient quasi-static change of the external parameter(s), the Hamiltonian can be brought back over a *closed* cycle to its original value:

$$H\left(R(t+T)\right) = H\left(R(t)\right). \tag{2.49b}$$

The actual value T of the time taken for the Hamiltonian to return to its original value would not be of great significance in our discussion. The changes must, however, be slow

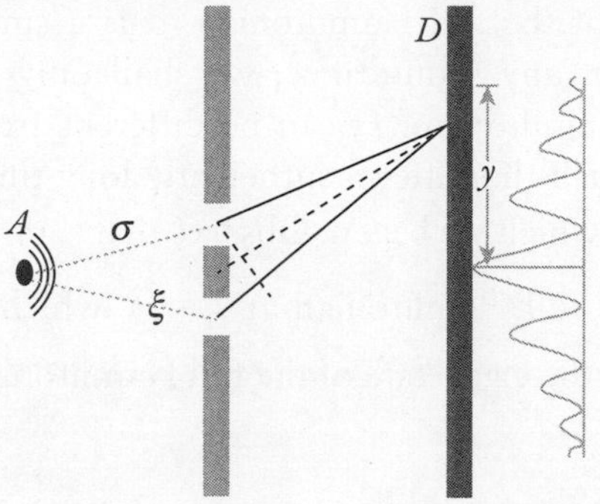

Fig. 2.5a Phase difference between alternative paths produces interference through superposition.

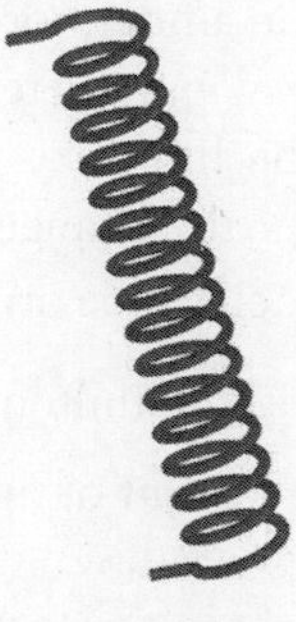

Fig. 2.5b Solenoid that is placed in a modified setup to conduct the Aharonov–Bohm experiment in an improvised Young's double-slit experiment.

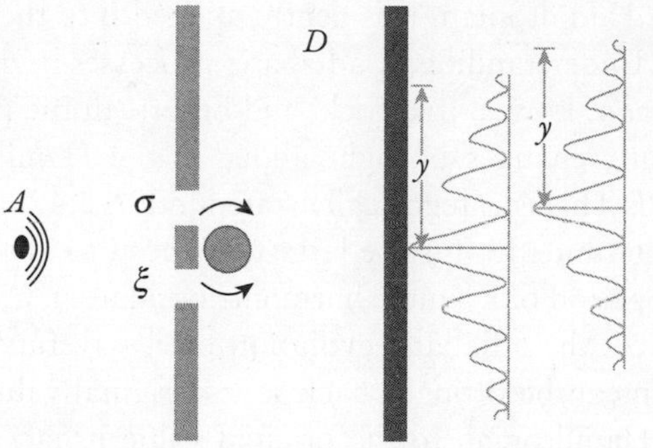

Fig. 2.5c The Aharonov–Bohm experiment is essentially a Young's double-slit experiment in which a solenoid is placed right behind the slits and in between them. The solenoid is impenetrable, so electron trajectories can only skirt the solenoid on either side. The shaded circle right behind the slits shows the top view of the solenoid.

enough to preserve adiabiticity. In such a case, if the state function of the system was originally the nth eigenstate Ψ_n when the change in the external parameters was triggered, then it would continue to be in the same state, except for a phase factor. T must be much larger than any internal time scale that may characterize the system. We ask: "How would the system wavefunction evolve with time if the parameters in the Hamiltonian are varied adiabatically around some closed cycle, as per Eq. 2.49b?" To address this question, we consider first a system described by a full Hamiltonian, that develops from a seed value H_0, on which a gently changing part H_1 rides. Changes to the Hamiltonian can be controlled in a quasi-static manner, practically infinitely slowly. Let us write the Schrödinger equation for the Hamiltonian H_0:

$$H_0\Phi_0 = E_0\Phi_0. \tag{2.50}$$

The full Hamiltonian H develops over time from the seed Hamiltonian H_0, as may be represented by the relation

$$H = \left(H_0 + \exp(\alpha t)H_1\right). \tag{2.51}$$

This relation underscores adiabatic development of the full Hamiltonian. α is a small positive notional parameter. Its actual value is not of any significance; we shall only be interested in the limit $\alpha \to 0$ which gives the full Hamiltonian. H can be different from the original, but the changes introduced can also be such that after a sufficiently long time, H is brought back to the original Hamiltonian H_0 with which we began. Observe that $\lim_{t\to-\infty} H$ gives us the seed Hamiltonian H_0, while $\lim_{t\to 0} H$ is the full Hamiltonian at $t = 0$, which is some reference instant of time. In an adiabatic process, the eigenstate of the full Hamiltonian H can be represented by

$$\frac{|\psi_0\rangle}{\langle\Phi_0|\psi_0\rangle} = \lim_{\alpha\to 0} \frac{U_\alpha(0,-\infty)|\Phi_0\rangle}{\langle\Phi_0|U_\alpha(0,-\infty)|\Phi_0\rangle}, \tag{2.52}$$

where $U_\alpha(0,-\infty)$ is the operator responsible for the temporal evolution of the state we began with. In an adiabatic evolution, the eigenstate Φ_0 of H_0 the system begins with adapts itself continuously to the new Hamiltonian. It is gently nudged into the corresponding eigenstate of the full Hamiltonian. Understanding of adiabatic processes in quantum mechanics comes from the works of Gellman, Low, Born, Fock, and others. In the present context, *H returns to H_0* through quasi-static changes brought about in the Hamiltonian by some external parameters $R(t)$ in time T. This is categorically stated in Eq. 2.49b. In practice, T only has to be sufficiently long, not infinite. It must be large compared to any internal time scale of the system (such as the time period of a simple harmonic oscillator). Quasi-static variation of the Hamiltonian would enable the adiabatic evolution of a wavefunction. The parameter that changes the Hamiltonian can be brought back to its original value in the parameter space. This would restore the Hamiltonian to the original value it had at the start. The question we now ask is: If changes in the parameters are brought about *slowly* over a *closed cycle* (Eq. 2.49b), what would be the net change in the eigenfunctions? Since the Hamiltonian returns to the original value, we expect that only the phase of the complex wavefunction be affected. The phase accumulated by the wavefunction would consist of two components:

- a *dynamical time-dependent phase* θ that represents the temporal evolution of a stationary eigenstate. It is associated with the eigenvalue of the time-evolution operator to which an energy eigenket belongs (Eqs. 1.114a,b).
- a *geometrical phase* γ that depends on the path in the parameter space along which the parameters are adiabatically varied to change the Hamiltonian.

We therefore write the time-dependent wavefunction as

$$\psi_n(x,t) = \psi_n(x)\exp\left[-\frac{i}{\hbar}\int_0^t E_n(t')dt'\right]\exp[i\gamma_n(t)] = \psi_n(x)\exp[i\theta_n(t)]\exp[i\gamma_n(t)]. \tag{2.53}$$

The angle

$$\theta_n(t) = -\frac{1}{\hbar}\int_0^t E_n(t')dt' \tag{2.54}$$

is the *dynamical phase*. If a measurement of energy is made at time t', the result would be $E_n(t')$, and the system's state function would continue to be the same n^{th} eigenstate. The other phase angle γ in Eq. 2.53 depends on time, but its time dependence is fundamentally different from that of θ. *Primarily*, γ is determined by *geometry*, i.e., by the path in the parameter space along which the externally controlled parameters in the Hamiltonian can be changed in a full cycle so that the Hamiltonian can return to what it was at the beginning of these changes. This is achieved by bringing back the parameter set to its original start values in one cycle, resulting in Eq. 2.49. The angle γ is also called as the Pancharatnam–Berry phase. It is this term that is of central consideration in the present section.

That changes in the Hamiltonian are adiabatic is central to our discussion. The system must, however, interact with the agency that *changes* the set of parameters on which the Hamiltonian depends, resulting in mild non-adiabiticity, even if only *weak*. The time dependence of an eigenstate is therefore represented by the following expression, consisting of a sum of two terms:

$$\psi_n(\vec{r},t) = \psi_n(\vec{r},t)\exp\left(i\theta_n(t)\right)\exp\left(i\gamma_n\right) + \varepsilon\sum_{m\neq n} C_m(t)\psi_m(\vec{r},t). \tag{2.55}$$

In Eq. 2.55,

- the *first* term on the right-hand side of Eq. 2.55 has already been discussed in the context of Eq. 2.53. In this term, the total phase is the sum of the *dynamical* phase $\theta_n(t)$ and the geometric phase γ.
- the *second* term includes a scaling parameter ε, which would be exactly zero if the process under consideration is strictly adiabatic. A tiny nonzero ε represents a minute departure from adiabiticity. This departure from adiabiticity results from the (weak) coupling between the system and the agency that changes the parameters R. The non-adiabiticity would result in a superposition with other eigenstates ψ_m belonging to different energies E_m, with $m \neq n$. The weak non-adiabatic mixing with other energy states represents the fact that energy is not conserved.

The time-dependent Schrödinger equation that we must therefore solve is

$$H\left(R(t)\right)\psi(\vec{r},t) = i\hbar\frac{\partial}{\partial t}\psi(\vec{r},t), \tag{2.56a}$$

$$\text{i.e., } H\left(R(t)\right)\psi_n(\vec{r},t) = i\hbar\frac{\partial}{\partial t}\left[\psi_n(\vec{r},t)e^{i\theta_n(t)}e^{i\gamma_n(t)} + \varepsilon\sum_{m\neq n} C_m(t)\psi_m(\vec{r},t)\right], \tag{2.56b}$$

i.e.,

$$\begin{bmatrix} \cancel{E_n(t)\psi_n(\vec{r},t)e^{i\theta_n(t)}e^{i\gamma_n(t)}} \\ +\varepsilon\sum_{m\neq n} c_m(t)E_m(t)\psi_m(\vec{r},t) \end{bmatrix} = i\hbar\begin{bmatrix} \left\{\frac{\partial}{\partial t}\psi_n(\vec{r},t)\right\}e^{i\theta_n(t)}e^{i\gamma_n(t)} \\ \cancel{+\psi_n(\vec{r},t)\left(-\frac{i}{\hbar}E_n\right)e^{i\theta_n(t)}e^{i\gamma_n(t)}} \\ +i\psi_n(\vec{r},t)e^{i\theta_n(t)}e^{i\gamma_n(t)}\frac{\partial\gamma_n}{\partial t} \\ +\varepsilon\sum_{m\neq n}\left\{\frac{\partial c_m(t)}{\partial t}\right\}\psi_m(\vec{r},t) \\ +\varepsilon\sum_{m\neq n}\left\{c_m(\vec{r},t)\frac{\partial\psi_m(t)}{\partial t}\right\} \end{bmatrix}. \tag{2.56c}$$

On moving the term for n to the left-hand side, and writing all the other terms on the right, we get

$$\frac{\partial \psi_n(\vec{r},t)}{\partial t} + i\psi_n(\vec{r},t)\frac{\partial \gamma_n}{\partial t} = -\varepsilon e^{-i\theta_n(t)} e^{-i\gamma_n(t)} \sum_{m\neq n} \left\{ \begin{array}{l} c_m(\vec{r},t)\dfrac{\partial \psi_m(t)}{\partial t} + \dfrac{\partial c_m(t)}{\partial t}\psi_m(\vec{r},t) \\ -\dfrac{i}{\hbar} c_m(t) E_m(t) \psi_m(\vec{r},t) \end{array} \right\}. \tag{2.56d}$$

Both the terms on the left-hand side involve changes induced by the alteration in *parameters* on which the Hamiltonian depends; these terms are of first order in ε. On the right-hand side, the first term inside the bracket is of second order, which we shall drop. Also, to first order, $e^{-i\gamma_n(t)} \simeq 1$. We therefore get

$$\frac{\partial \psi_n(\vec{r},t)}{\partial t} + i\psi_n(\vec{r},t)\frac{d\gamma_n}{dt} = -\varepsilon e^{-i\theta_n(t)} \sum_{m\neq n} \left\{ \dot{c}_m(t) - \frac{i}{\hbar} c_m(t) E_m(t) \right\} \psi_m(\vec{r},t). \tag{2.57}$$

On multiplying Eq. 2.57 by ψ_n^* and integrating over whole space, we now find that

$$\int dV \psi_n^*(\vec{r},t)\frac{\partial \psi_n(\vec{r},t)}{\partial t} + i\dot{\gamma}_n = 0, \tag{2.58a}$$

since the wavefunctions for $m \neq n$ are orthogonal.

$$\text{Hence, } \dot{\gamma}_n = i\int dV \psi_n^*(\vec{r},t)\frac{\partial \psi_n(\vec{r},t)}{\partial t} = i\left[\int dV \psi_n^*(\vec{r},t)\frac{\partial \psi_n(\vec{r},t)}{\partial R}\right]\frac{dR}{dt}. \tag{2.58b}$$

Integrating,

$$\gamma_n = i\int_0^t dt' \left\{ \iiint_{\substack{whole\\ space}} \psi_n^* \frac{\partial \psi_n}{\partial R} \right\} \frac{dR}{dt'} = i\int_{R_i}^{R_f} \left\{ \iiint_{\substack{whole\\ space}} \psi_n^* \frac{\partial \psi_n}{\partial R} \right\} dR. \tag{2.59}$$

In writing this result, we have now taken cognizance of the fact, that the primary variation responsible for change in the wavefunction, is the change in the parameter(s) on which the Hamiltonian depends. If we now apply a cyclic change in the parameter(s) such that

$$H(R+T) = H(R);\ R_{\text{final}} = R_{\text{initial}};\ \delta R = 0, \tag{2.60}$$

$$\text{then we find that } \gamma_n(T) = i\oint \left\{ \iiint_{\substack{whole\\ space}} \psi_n^* \frac{\partial \psi_n}{\partial R} \right\} dR = \oint \left\{ i\left\langle \psi_n \middle| \frac{\partial \psi_n}{\partial R} \right\rangle \right\} dR = 0. \tag{2.61}$$

This result may disenchant you, since the Pancharatnam–Berry phase has turned out to be zero in this case. However, when the Hamiltonian depends on a *set of N parameters*, not on just a single parameter, then we run into a very interesting situation. In this case, R would stand for a set,

$$R \equiv \{R_1, R_2, R_3, R_4, \ldots, R_N\}, \tag{2.62}$$

and we must replace the operator $\dfrac{\partial}{\partial R}$ by the gradient operator $\vec{\nabla}_R$ in the N-dimensional parameter space. The line integral from Eq. 2.61 in the parameter space now becomes

$$\gamma_n(T) = \oint \left\{ i\left\langle \psi_n \middle| \vec{\nabla} \psi_n \right\rangle \right\} \cdot \overrightarrow{dR}, \tag{2.63}$$

and in general, the result of Eq. 2.63 is not zero. Its value depends on the path of integration in the parameter space. It does not depend on the total time taken (unlike the dynamical phase given in Eq. 2.51), as long as the process can be considered to be adiabatic. From the first term $\psi_n(\vec{r},t)\exp[i\theta_n(t)]\exp[i\gamma_n(t)]$ in Eq. 2.55, it is obvious that if the Pancharatnam–Berry phase were to be imaginary, it would correspond to attenuation, not to phase. Preservation of the norm in the parameter space, however, guarantees that

$$0 = \vec{\nabla}_R \langle \Psi_n(\vec{r},t) | \Psi_n(\vec{r},t) \rangle = \left\langle \left\{ \vec{\nabla}_R \Psi_n(\vec{r},t) \right\} \middle| \Psi_n(\vec{r},t) \right\rangle + \left\langle \Psi_n(\vec{r},t) \middle| \left\{ \vec{\nabla}_R \Psi_n(\vec{r},t) \right\} \right\rangle. \quad (2.64)$$

The right-hand side is a sum of a complex number and its complex conjugate, and can be zero only if the real part is zero. Having now concluded that the integrand in Eq. 2.63 must be imaginary, we conclude that the left-hand side of that equation must be real. It therefore corresponds to a phase, and not attenuation. Furthermore, we note that if the eigenfunctions are real, so would be their gradients, and the integrand in Eq. 2.63 would be real. However, we just concluded that it must be imaginary. The only way a number can be both real and imaginary is when it is zero, i.e., the geometric phase must vanish whenever the eigenfunctions are real.

We now remind ourselves of an ansatz that the Lagrangian for a particle having mass m and charge e is

$$L = \frac{1}{2}mv^2 - e\phi(\vec{r},t) + \frac{e}{c}\vec{v}\bullet\vec{A}(\vec{r},t). \quad (2.65a)$$

It produces the correct equation of motion (see Eq. 2.70, below) for the said charge particle. In this Lagrangian, $\{\phi(\vec{r},t), \vec{A}(\vec{r},t)\}$ are the electromagnetic scalar and vector potentials experienced by the charged particle, $\vec{v}$ is its velocity, and c the speed of light. Note that we have employed the Gaussian-cgs system in which the physical dimensions of electric charge are $M^{\frac{1}{2}}L^{\frac{3}{2}}T^{-1}$. In this system, the dimensions of the electric field $\vec{E}$ and the magnetic field $\vec{B}$ are both the same, $M^{\frac{1}{2}}L^{-\frac{1}{2}}T^{-1}$ and the dimensions of both the vector potential $\vec{A}$ and the scalar potential ϕ are $M^{\frac{1}{2}}L^{\frac{1}{2}}T^{-1}$.

We first write the Lagrangian using Cartesian degrees of freedom $\{q_1, q_2, q_3\} = \{x, y, z\}$:

$$L = \frac{1}{2}m\left(\sum_{j=1}^{3} v_j^2\right) - e\phi(\vec{r},t) + \frac{e}{c}\left(\sum_{j=1}^{3} v_j A_j(\vec{r},t)\right). \quad (2.65b)$$

The ith component (i = 1,2,3) of the canonically conjugate generalized momentum $\vec{p}_c$ therefore is

$$p_{c,i} = \frac{\partial L}{\partial \dot{q}_i} = \frac{\partial L}{\partial v_i} = mv_i + \frac{e}{c}A_i(\vec{r},t) = p_{k,i} + \frac{e}{c}A_i. \quad (2.66a)$$

Note that in addition to the kinetic momentum $\vec{p}_k = m\vec{v} = m\dot{\vec{q}}$, the *canonically conjugate generalized momentum* $\vec{p}_c$ includes $\frac{e}{c}$ times the vector potential $\vec{A}$.

Thus, $\vec{p}_c = \vec{p}_k + \frac{e}{c}\vec{A} = m\vec{v} + \frac{e}{c}\vec{A} = m\dot{\vec{q}} + \frac{e}{c}\vec{A}$ (2.66b)

Hence,

$$\frac{d}{dt}\frac{\partial L}{\partial \dot{q}_i} = m\dot{v}_i + \frac{e}{c}\frac{dA_i(\vec{r},t)}{dt} = m\dot{v}_i + \frac{e}{c}\left\{\sum_{j=1}^{3}\dot{q}_j\frac{\partial A_i(\vec{r},t)}{\partial q_j} + \frac{\partial A_i(\vec{r},t)}{\partial t}\right\}$$

$$= \dot{p}_{ki} + \frac{e}{c}\left(\vec{v}\bullet\vec{\nabla} + \frac{\partial}{\partial t}\right)\vec{A}. \tag{2.67a}$$

From Lagrange's equation, we know that

$$\dot{p}_{c,i} = \frac{\partial L}{\partial q_i} = -e\frac{\partial \phi}{\partial q_i} + \frac{e}{c}\frac{\partial}{\partial q_i}\left(\vec{v}\bullet\vec{A}\right). \tag{2.67b}$$

Hence, $m\dot{v}_i + \frac{e}{c}\left(\vec{v}\bullet\vec{\nabla} + \frac{\partial}{\partial t}\right)A_i = -e\frac{\partial \phi}{\partial q_i} + \frac{e}{c}\frac{\partial}{\partial q_i}\left(\vec{v}\bullet\vec{A}\right)$

or $$m\dot{v}_i = -e\left(\vec{\nabla}\phi\right)_i + \frac{e}{c}\frac{\partial A_i}{\partial t} + \frac{e}{c}\left\{\left(\vec{v}\bullet\frac{\partial \vec{A}}{\partial q_i}\right) - \left(\vec{v}\bullet\vec{\nabla}\right)A_i\right\}. \tag{2.68}$$

It is now easy to see that the term in the curly bracket is nothing but the ith component of $\vec{v}\times\vec{B} = \vec{v}\times\left(\vec{\nabla}\times\vec{A}\right)$:

$$\vec{v}\bullet\frac{\partial \vec{A}}{\partial q_i} - \left(\vec{v}\bullet\vec{\nabla}\right)A_i = \sum_{j=1}^{3}\left(v_j\frac{\partial A_j}{\partial q_i} - v_j\frac{\partial A_i}{\partial q_j}\right) = \sum_{j=1}^{3}v_j\left(\frac{\partial A_j}{\partial q_i} - \frac{\partial A_i}{\partial q_j}\right)$$

$$= \left\{\vec{v}\times\left(\vec{\nabla}\times\vec{A}\right)\right\}_i = \left(\vec{v}\times\vec{B}\right)_i. \tag{2.69}$$

Since we are using the Cartesian coordinate system, it is instructive to develop a compact notation using the Levi-Civita symbol, ε_{ijk}, where i, j, k represent the three Cartesian components. As is well known, the Levi-Civita symbol is antisymmetric in every pair of its indices and has the following properties: $\varepsilon_{123} = \varepsilon_{231} = \varepsilon_{312} = +1$ and $\varepsilon_{132} = \varepsilon_{321} = \varepsilon_{213} = -1$ and all *other* $\varepsilon_{ijk} = 0$. Using the Levi-Civita symbol, we can write the components of the cross-product $\vec{H} = \vec{F}\times\vec{G}$ of two vectors as $H_i = \varepsilon_{ijk}F_jG_k \equiv \sum_{j=1}^{3}\sum_{k=1}^{3}\varepsilon_{ijk}F_jG_k$, in which the Einstein summation notation is employed.

Thus,

$$\left(\vec{v}\times\vec{B}\right)_i = \left\{\vec{v}\times\left(\vec{\nabla}\times\vec{A}\right)\right\}_i = \varepsilon_{ijk}v_j\left(\vec{\nabla}\times\vec{A}\right)_k = \varepsilon_{ijk}v_j\left\{\varepsilon_{pmk}\frac{\partial A_p}{\partial q_m}\right\}$$

$$= \sum_{j=1}^{3}v_j\left\{\sum_{p=1}^{3}\sum_{m=1}^{3}\left(\sum_{k=1}^{3}\varepsilon_{ijk}\varepsilon_{pmk}\right)\frac{\partial A_p}{\partial q_m}\right\}.$$

Since $\sum_{k=1}^{3}\varepsilon_{ijk}\varepsilon_{pmk} = \delta_{ip}\delta_{jm} - \delta_{im}\delta_{jp}$, we get $\left(\vec{v}\times\vec{B}\right)_i = \sum_{j=1}^{3}v_j\left(\frac{\partial A_j}{\partial q_i} - \frac{\partial A_i}{\partial q_j}\right)$.

Collecting now expressions for all the three components ($i = 1, 2, 3$) of Eqs. 2.68 and 2.69, we get the celebrated Lorentz force equation:

$$\dot{\vec{p}}_k = m\ddot{\vec{q}} = e\left[\left(-\vec{\nabla}\phi\right) + \frac{1}{c}\frac{\partial \vec{A}}{\partial t} + \frac{1}{c}\left(\vec{v}\times\vec{B}\right)\right] = e\left[\vec{E} + \frac{1}{c}\left(\vec{v}\times\vec{B}\right)\right]. \tag{2.70}$$

As seen in Fig. 2.5c, the trajectories of the electrons, no matter whether they pass from slit σ or slit ξ (or both!) lie totally *outside* the solenoid, and hence in a region where the magnetic field $\vec{B} = \vec{0}$. We are now ready to figure out the mystery behind the Aharonov–Bohm effect. This is essentially a quantum effect, and hence we must describe the state of the system by a wavefunction, and its temporal evolution by the Schrödinger equation. We therefore proceed to set up the Schrödinger equation for the charged particle described by the Lagrangian in Eq. 2.65. The classical Hamilton's principal function [4] is

$$H = \dot{\vec{q}}_c\bullet\vec{p}_c - L = \frac{\vec{p}_k}{m}\bullet\vec{p}_c - \frac{\vec{p}_k\bullet\vec{p}_k}{2m} + e\phi - \frac{e}{c}\frac{\vec{p}_k}{m}\bullet\vec{A}, \tag{2.71a}$$

$$\text{i.e., } H = \frac{1}{m}\left\{\vec{p}_c - \frac{e}{c}\vec{A}\right\}\bullet\vec{p}_c - \frac{\left\{\vec{p}_c - \frac{e}{c}\vec{A}\right\}\bullet\left\{\vec{p}_c - \frac{e}{c}\vec{A}\right\}}{2m} + e\phi - \frac{e}{c}\frac{\left\{\vec{p}_c - \frac{e}{c}\vec{A}\right\}}{m}\bullet\vec{A}. \tag{2.71b}$$

$$\text{Hence, } H = \frac{1}{2}\frac{\vec{p}_c\bullet\vec{p}_c}{m} - \frac{e}{2mc}\vec{A}\bullet\vec{p}_c - \frac{e}{2mc}\vec{p}_c\bullet\vec{A} + \frac{e^2}{2mc^2}\vec{A}\bullet\vec{A} + e\phi. \tag{2.71c}$$

On replacing now the classical canonical dynamical variables by operators $\vec{q}_c \to \vec{q}_{operator}$ and $\vec{p}_c \to -i\hbar\vec{\nabla}$ (see Eq. 1. 107), we get:

$$H = \frac{\left[\vec{p}_c - \frac{e}{c}\vec{A}\right]^2}{2m} + e\phi \underset{operator}{\overset{quantization}{\longrightarrow}} \frac{\left[\left(-i\hbar\vec{\nabla}\right) - \frac{e}{c}\vec{A}\right]^2}{2m} + e\phi. \tag{2.72}$$

While we are at it, we consider the consequences of the quadratic terms in this Hamiltonian, since the operators $\vec{\nabla}$ and $\vec{A}$ do not commute. The Hamiltonian therefore is

$$H = \frac{1}{2m}\left(-\hbar^2\vec{\nabla}^2 + i\hbar\frac{e}{c}\left(\vec{A}\bullet\vec{\nabla} + \vec{\nabla}\bullet\vec{A}\right) + e^2A^2\right) + e\phi. \tag{2.73}$$

The operator $\left(\vec{A}\bullet\vec{\nabla}\right)$ does not commute with the *operator* $\left(\vec{\nabla}\bullet\vec{A}\right)$. We examine how their sum operates on an arbitrary function f of space coordinates.

$$\left(\vec{\nabla}\bullet\vec{A}(\vec{r},t) + \vec{A}(\vec{r},t)\bullet\vec{\nabla}\right)f = \vec{\nabla}\bullet\left(\vec{A}(\vec{r},t)f\right) + \vec{A}(\vec{r},t)\bullet\vec{\nabla}f, \tag{2.74a}$$

i.e.,

$$\left(\vec{\nabla}\bullet\vec{A}(\vec{r},t) + \vec{A}(\vec{r},t)\bullet\vec{\nabla}\right)f = \left(\vec{\nabla}\bullet\vec{A}(\vec{r},t)\right)f + \vec{A}(\vec{r},t)\bullet\vec{\nabla}f + \vec{A}(\vec{r},t)\bullet\vec{\nabla}f. \tag{2.74b}$$

The first term involves the scalar divergence of the vector potential; it is no longer the *operator* $\left(\vec{\nabla}\bullet\vec{A}\right)$. We cannot freely choose the curl of the vector potential, since it is already fixed by its equality to the magnetic field. We can, however, use the gauge freedom to choose the Coulomb gauge in which the vector potential is solenoidal.

$$\text{Hence, } \left(\vec{\nabla}\bullet\vec{A}(\vec{r},t) + \vec{A}(\vec{r},t)\bullet\vec{\nabla}\right)f = 2\vec{A}(\vec{r},t)\bullet\vec{\nabla}f. \tag{2.74c}$$

The Hamiltonian operator then becomes

$$H = \left(-\frac{\hbar^2\vec{\nabla}^2}{2m} + \frac{i\hbar e}{mc}\vec{A}\bullet\vec{\nabla} + \frac{e^2}{2mc^2}A^2 \right) + e\phi, \tag{2.75}$$

and it may be used in the Schrödinger equation.

An interesting question must now be addressed. When you study the quantum physics of charged particles in electromagnetic fields $(\vec{E},\vec{B})$, how would the solution to the Schrödinger equation be affected if the electromagnetic potentials in the Hamiltonian (Eq. 2.72) are expressed using a different gauge? We know that the fields $(\vec{E},\vec{B})$ are invariant under the gauge freedom

$$\vec{A}(\vec{r},t) \to \vec{A}'(\vec{r},t) = \vec{A}(\vec{r},t) + \vec{\nabla}\chi(\vec{r},t), \tag{2.76a}$$

along with $\phi(\vec{r},t) \to \phi'(\vec{r},t) = \phi(\vec{r},t) - \frac{1}{c}\frac{\partial\chi(\vec{r},t)}{\partial t}$. (2.76b)

The dimensions of $\vec{A}$ and ϕ being the same, we see that dimensions of the gauge function $[\chi]$ are $M^{\frac{1}{2}}L^{\frac{3}{2}}T^{-1}$. In Eqs. 2.76a,b, $\chi(\vec{r},t)$ is an arbitrary (but well behaved) scalar field that we shall refer to as the gauge function. We now address our concern: *In what way is the solution to the quantum equation of motion, namely the Schrödinger equation, sensitive to the gauge freedom indicated in Eq. 2.76a,b*? Surely, we must expect the wavefunction to be susceptible to the choice of the gauge employed to describe the electromagnetic potentials appearing in the Hamiltonian. Yet, in some sense, it might *seem* that nothing would change, since under the transformations $(\phi,\vec{A}) \to (\phi',\vec{A}')$, the electromagnetic fields $(\vec{E},\vec{B})$ remain invariant. We therefore expect the solution to the Schrödinger equation to change, but by no more than by a phase factor. This phase factor must, however, be *fundamentally different* from the dynamic phase of Eq. 2.54. The gauge-related phase that now enters our analysis is the Pancharatnam–Berry phase factor. We must expect this phase to involve the gauge function $\chi(\vec{r},t)$ in some way. It turns out, *as you will find in the Solved Problem P2.3*, that the new (dimensionless) geometric phase angle is $\frac{e\chi(\vec{r},t)}{\hbar c}$. We shall see that if $\psi(\vec{r},t)$ is the solution of the Schrödinger equation in which the Hamiltonian is expressed in terms of the potentials $(\phi,\vec{A})$, then

$$\psi'(\vec{r},t) = \exp\left(i\frac{e}{\hbar c}\chi(\vec{r},t)\right)\psi(\vec{r},t) \tag{2.77}$$

is the solution when the Hamiltonian is expressed in terms of the potentials $(\phi',\vec{A}')$. It must be stressed that $\chi(\vec{r},t)$ is essentially the very same gauge function that we had employed in the gauge transformations (Eq. 2.76) of the electromagnetic potentials. The gauge function depends on the local space and time $(\vec{r},t)$. The transformations in Eq. 2.76a,b and in Eq. 2.77 go essentially in tandem. Now, we know that the following Schrödinger equation (Eq. 1.91b) involves the Hamiltonian described in terms of the potentials $(\phi,\vec{A})$. Its solution is the wavefunction $\psi(\vec{r},t)$:

$$i\hbar\frac{\partial}{\partial t}\psi(\vec{r},t) = \left[\frac{1}{2m}\left(-i\hbar\vec{\nabla} - \frac{e\vec{A}(\vec{r},t)}{c}\right)^2 + e\phi(\vec{r},t)\right]\psi(\vec{r},t). \tag{2.78}$$

The Schrödinger equation in which the Hamiltonian is set up in terms of the gauge-transformed potentials $\left(\phi',\vec{A}'\right)$ has a *different* solution. In this case, the wavefunction is $\psi'(\vec{r},t)$:

$$i\hbar\frac{\partial}{\partial t}\psi'(\vec{r},t) = \left[\frac{1}{2m}\left(-i\hbar\vec{\nabla} - \frac{e\vec{A}'(\vec{r},t)}{c}\right)^2 + e\phi'(\vec{r},t)\right]\psi'(\vec{r},t). \tag{2.79}$$

The transformations given in Eqs. 2.76a,b and Eq. 2.77 connect the set $\left\{\phi(\vec{r},t),\vec{A}(\vec{r},t),\psi(\vec{r},t)\right\}$ with $\left\{\phi'(\vec{r},t),\vec{A}'(\vec{r},t),\psi'(\vec{r},t)\right\}$. We shall soon find that the geometric phase factor (in Eq. 2.77) accounts for displacement of fringes in the Aharonov–Bohm effect.

Local gauge symmetry of the Schrödinger equation is ensured by gauge transformations of the electromagnetic potentials and the existence of the electric charge e. It appears explicitly in the transformation of the wavefunction (Eq. 2.77). The FPs (electrons, in the present case) can reach the detector only by circumventing the impenetrable solenoid. The magnetic field $\vec{B}$ outside the solenoid is strictly zero. The electromagnetic vector potential *outside* the solenoid is obtainable from its circulation about a closed path:

$$\oint\vec{A}'\cdot\overrightarrow{dr} = \oint\left(\vec{A}+\vec{\nabla}\chi\right)\cdot\overrightarrow{dr} = \oint\vec{A}\cdot\overrightarrow{dr} + \oint\vec{\nabla}\chi\cdot\overrightarrow{dr} = \oint\vec{A}\cdot\overrightarrow{dr} + \oint d\chi = \oint\vec{A}\cdot\overrightarrow{dr}. \tag{2.80}$$

It is essentially the same, regardless of the potential being expressed as $\vec{A}$ or by the gauge transformed $\vec{A}'$. Using the Kelvin–Stokes theorem (see, for example, section 10.4 in Reference [4]), the circulation is equal to a surface integral over an open surface (bounded by the path of integration of the line integral) of the curl of the vector potential, which is just the magnetic field $\vec{B}$:

$$\oint\vec{A}'\cdot\overrightarrow{dr} = \oint\vec{A}\cdot\overrightarrow{dr} = \iint\left(\vec{\nabla}\times\vec{A}\right)\cdot\overrightarrow{dS} = \iint\vec{B}\cdot\overrightarrow{dS} = \Phi_{enclosed}. \tag{2.81}$$

$\Phi_{enclosed}$ is the gauge-independent flux enclosed by the path along which the circulation is determined. Consider the integral over a circular closed path of radius $\rho > s$ around the solenoid in a plane orthogonal to the solenoid's axis, s being the solenoid's radius. The axis of the solenoid is chosen to define the z-axis of a Cartesian coordinate system. The vector potential in the apparatus is given by $\vec{A} = A\hat{e}_\varphi$, $\hat{e}_\varphi$ being the unit vector along the direction in which the azimuthal angle φ of a cylindrical polar coordinate system (see Chapter 2 in [4]) increases.

From Eq. 2.81, $A2\pi\rho = \Phi = B\pi a^2$, and hence $\vec{A} = \frac{\Phi}{2\pi\rho}\hat{e}_\varphi$. (2.82)

Using the Hamiltonian from Eq. 2.75, the Schrödinger equation (Eq. 1.140) is

$$\frac{1}{2m}\left(-\hbar^2\vec{\nabla}^2 + 2i\frac{e}{c}\hbar\vec{A}\cdot\vec{\nabla} + \frac{e^2}{c^2}A^2\right)\psi(\vec{r}) = E\psi(\vec{r}) \tag{2.83}$$

can be most conveniently dealt with using the cylindrical coordinate system. Exploiting symmetry, we need to work with only a one-dimensional differential equation. We use the fact that

$$2i\frac{e\hbar}{c}\hbar\vec{A}\cdot\vec{\nabla}\psi(\vec{r}) = 2i\frac{e\hbar}{c}\frac{\Phi}{2\pi\rho}\frac{1}{\rho}\frac{\partial\psi(\vec{r})}{\partial\varphi} = i\frac{e\hbar}{c}\frac{\Phi}{\pi\rho^2}\frac{\partial\psi(\vec{r})}{\partial\varphi}, \tag{2.84}$$

which now requires us to solve the following equation:

$$\frac{1}{2m}\left(-\hbar^2\frac{1}{\rho^2}\frac{d^2}{d\varphi^2}+i\frac{e\hbar}{c}\frac{\Phi}{\pi\rho^2}\frac{d}{d\varphi}+\frac{e^2}{c^2}\frac{\Phi^2}{4\pi\rho^2}\right)\psi(\varphi) = E\psi(\varphi). \tag{2.85}$$

The electric potential ϕ in our apparatus is thankfully zero, otherwise we would have had another symbol in Eq. 2.85, which would *sound* (not look) like Φ (flux) and φ (azimuthal angle). Introducing

$$F = \frac{e\Phi}{2c\pi\hbar} \tag{2.86a}$$

and

$$\eta = \frac{2ma^2E}{\hbar^2} - F^2, \tag{2.86b}$$

Eq. 2.85 can be written in a compact form:

$$\frac{d^2\psi}{d\varphi^2} - 2iF\frac{d\psi}{d\varphi} + \eta\psi = 0. \tag{2.87}$$

Its solution is $\psi = Ae^{i\mu\varphi}$, (2.88)

with $\mu = F \pm \sqrt{F^2+\eta} = \dfrac{e\Phi}{2c\pi\hbar} \pm \sqrt{\dfrac{2ma^2E}{\hbar^2}}$. (2.89)

In writing Eq. 2.89, we have used Eq. 2.86b. Considering a single-valued character of the wavefunction ψ, we note that its value at angles φ and $\varphi + 2\pi$ must be the same. This is a *periodic boundary condition*. Hence, μ must be an integer "n", i.e.,

$$\frac{e\Phi}{2c\pi\hbar} \pm \sqrt{\frac{2ma^2E}{\hbar^2}} = n. \tag{2.90a}$$

It follows that $E = \dfrac{\hbar^2}{2ma^2}\left(n - \dfrac{e\Phi}{2c\pi\hbar}\right)^2$, with $n = 0, \pm1, \pm2, \pm3, ...$ (2.90b)

Three interesting observations can now be made:

a. The energy spectrum is discrete. This has resulted from the boundary condition on the Schrodinger equation and a distinctive feature of quantum mechanics.
b. If the flux were zero, the energy levels for the positive and the negative integers would be *degenerate*, which is to say that *linearly independent* eigenfunctions exist, which belong to the *same* eigenvalue. The degeneracy in the present case has resulted from the use of the periodic boundary condition we have employed. In Chapter 3, you will learn that usually a one-dimensional bound-state problem does *not* have degenerate eigenfunctions; however, the boundary conditions alluded

to there would *not* be periodic; rather you will have the wavefunction *vanish* asymptotically. The removal of degeneracy when the potential is *changed* even if slightly is a general feature of *perturbation theory*, which we shall discuss in detail in Chapter 6. In the present case, the change in the potential is due to the current in the solenoid.

c. If the value of $\frac{e\Phi}{2c\pi\hbar}$ happens to be an integer, as can be controlled by the current through the solenoid, then the energy eigenspectrum of Eq. 2.90b would be essentially unchanged.

It is now convenient to write the eigen-energies with positive and negative integers separately:

$$E_{+} = \frac{1}{2m}\left(\frac{\hbar}{a}\right)^2\left(n - \frac{e\Phi}{2c\pi\hbar}\right)^2 \quad ; \; n = 0,1,2,3,.... \tag{2.90c}$$

$$E_{-} = \frac{1}{2m}\left(\frac{\hbar}{a}\right)^2\left(n + \frac{e\Phi}{2c\pi\hbar}\right)^2 \quad ; \; n = 0,1,2,3,.... \tag{2.90d}$$

We observe that for the same n, $E_{+} < E_{-}$.

We can now appreciate the physical reason for the displacement of the interference pattern on the detector on passing a current through the solenoid in the Aharonov–Bohm experiment (Fig. 2.5c). The components of the electron beam that get split in going through the slits σ and ξ pass around the two sides of the solenoid. On one side of the solenoid, the path is along the vector potential, and on the other, it is opposite to it. Thus, the FPs going around the solenoid on opposite sides pick up different phases from the magnetic vector potential $\vec{A}$. At the detector these two components recombine, generating the interference pattern with recalibrated phases. In Eq. 2.63, we have seen that the Pancharatnam–Berry phase resulting from the electromagnetic vector potential in the Schrödinger equation is

$$\gamma_n(T) = \oint\left\{i\left\langle\psi_n\middle|\vec{\nabla}\psi_n\right\rangle\right\}\bullet d\vec{R} = \iint\vec{\nabla}\times\left\{i\left\langle\psi_n\middle|\vec{\nabla}\psi_n\right\rangle\right\}\bullet d\vec{S}. \tag{2.91}$$

We immediately recognize the conformity,

$$\left\{i\frac{\hbar c}{e}\left\langle\psi_n\middle|\vec{\nabla}\psi_n\right\rangle\right\} \leftrightarrow \left[\vec{A}\right]. \tag{2.92}$$

We now use the result of the Solved Problem P2.4 to write the solution to the one-dimensional Schrödinger equation as

$$\psi_0 = \exp\left[\frac{i}{\hbar}\int\vec{p}_k\bullet\vec{dr}\right] \rightarrow \exp\left[\frac{i}{\hbar}\int\left(\vec{p}_c - \frac{e}{c}\vec{A}\right)\bullet\vec{dr}\right] = \exp\left[\frac{i}{\hbar}\int\left(\vec{p}_c\bullet\vec{dr} - \frac{e}{c}\vec{A}\bullet\vec{dr}\right)\right]. \tag{2.93a}$$

i.e.,

$$\psi_0 = \exp\left[\frac{i}{\hbar}\int\vec{p}_k\bullet\vec{dr}\right] = \exp\left[\frac{i}{\hbar}\int\vec{p}_c\bullet\vec{dr}\right]\exp\left[-\frac{i}{\hbar}\frac{e}{c}\int\vec{A}\bullet\vec{dr}\right]. \tag{2.93b}$$

Comparing Eq. 2.93 with Eq. 2.77, we can now recognize the Aharonov–Bohm phase picked up by the electrons in traveling around the solenoid on its either side. From Fig. 2.5c, we see that the electron trajectory on one side of the solenoid would be along the unit vector $\hat{e}_\varphi$ of the cylindrical polar coordinate system we have employed, whereas that on the other side would be opposite to it. Accordingly, the integrand $\vec{A}\cdot\overrightarrow{dr}$ in Eq. 2.93 is positive for the trajectory on the semi-circular path on one side of the solenoid and negative on the other. Carrying out the definite integral between limits corresponding to diametrically opposite points M and N on the semi-circular path,

$$\chi(\vec{r}) = -\int_M^N \vec{A}\cdot\overrightarrow{dr} = -\int_M^N \left(\frac{\Phi}{2\pi\rho}\hat{e}_\varphi\right)\cdot\left(\rho d\varphi\hat{e}_\varphi\right) = -\frac{\Phi}{2\pi}\int_M^N d\varphi = -\frac{\Phi}{2\pi}(\pm\pi) = \mp\frac{\Phi}{2}. \tag{2.94}$$

The term $(\pm\pi)$ in Eq. 2.94 is the result of the integration $\int d\varphi$ over the *semi-circular* paths skirting the solenoid from either side. The Pancharatnam–Berry phase requires integration over a full cycle, i.e., over the complete circle (indicated here by $\oint$) around the solenoid. Hence,

$$\gamma_n(T) = \oint\left\{i\left\langle\psi_n\middle|\vec{\nabla}\psi_n\right\rangle\right\}\cdot\overrightarrow{dR} = \frac{e}{\hbar c}\oint\vec{A}\cdot\overrightarrow{dR} = \frac{e\Phi}{\hbar c}. \tag{2.95}$$

The Pancharatnam–Berry phase has turned out to be such that in accordance with the statement "(c)" made right after Eq. 2.90b, the energy eigenspectrum remains essentially unchanged. This phase *difference* has resulted from the trajectories of FPs (electrons) on the two sides of the solenoid. It is caused in a region where the magnetic field $\vec{B}$ is zero, but the vector potential $\vec{A}$ is not. Control on the fringes is achieved essentially through tuning the Pancharatnam–Berry phase resulting from the integral $\frac{e}{\hbar c}\oint\vec{A}\cdot\overrightarrow{dR}$ in Eq. 2.95. The final phase has the flux and not the vector potential. The formalism is therefore gauge invariant. It is the *difference* in the Pancharatnam–Berry phase picked up by the FPs that skirt the impenetrable solenoid from its two sides which modulates the *net* phase in the paths through slits σ and ξ (Fig. 2.5c). It has resulted from Feynman's ansatz (Eq. 2.1). Aharonov–Bohm effect is thus essentially an illustration of the Pancharatnam–Berry phase. Feynman's path integral approach to quantum mechanics offers a direct and simple interpretation of the shift in the interference pattern seen in the Aharonov–Bohm effect. Control on the interference pattern in the Aharonov–Bohm experiment by a magnetic field $\vec{B}$ in a region – *where it isn't* – is a wonder [15] that can be accounted for only by quantum theory. It should be pointed out that this effect was predicted in an earlier study, by Ehrenberg and Siday [17]. It can therefore be called as the Ehrenberg–Siday–Aharonov–Bohm effect. Commonly, it is known as the Aharonov–Bohm effect. While we have discussed it only as a *thought* experiment, it has certainly been realized in the laboratory (see, for example, References [18, 19]).

Funquest: Has a gravitational Aharnov–Bohm effect been studied? *You may come back to this question after you go through the Solved Problem P8.2 of Chapter 8.*

Figure 2.5c provides a schematic representation of the Aharonov–Bohm effect. It shows a displacement of interference fringes due to different geometric phases picked up in the Feynman paths of the electrons going around the solenoid on its two sides. A numerical simulation [20] of this phenomenon is presented in Fig. 2.6.

i. $\gamma_n(T) = 0$

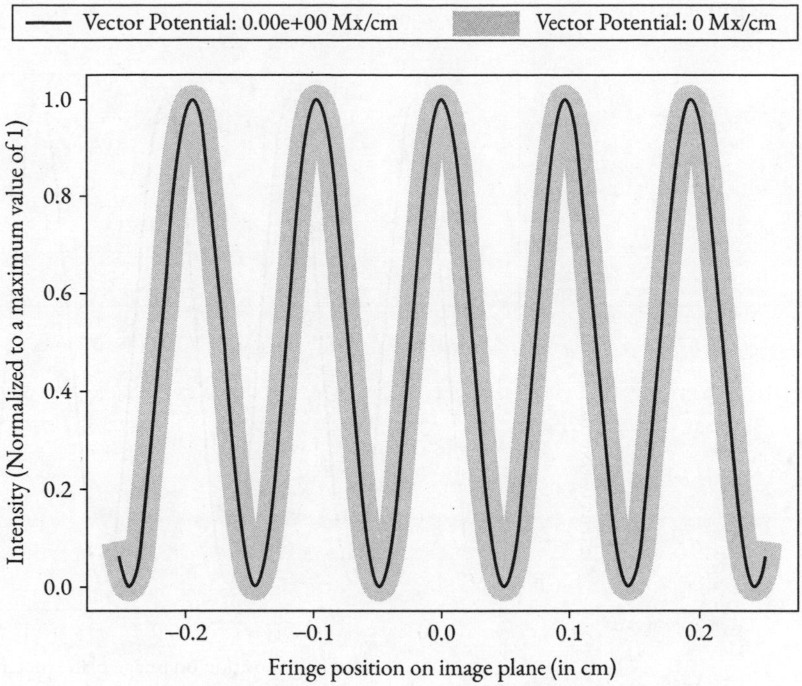

ii. $\gamma_n(T) = \frac{\pi}{2}$

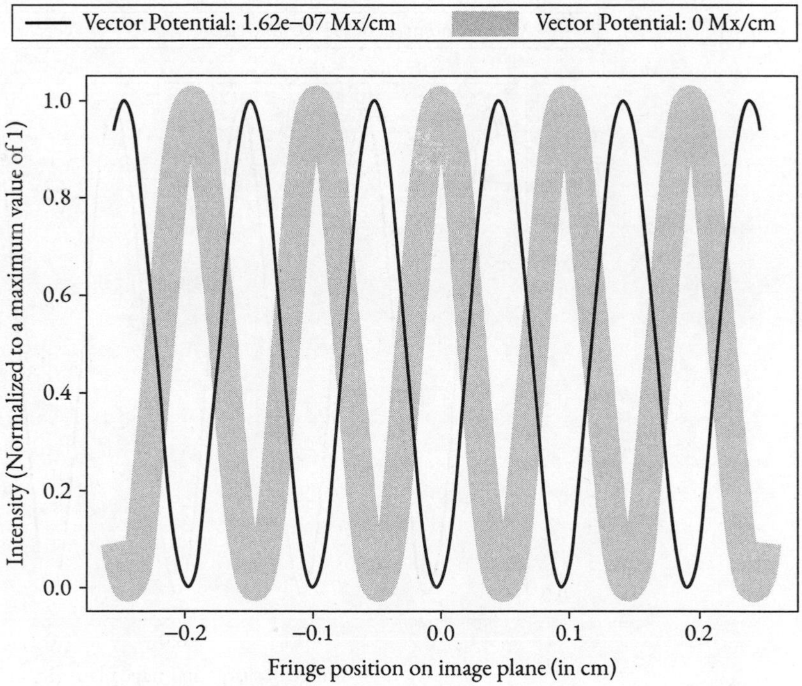

iii. $\gamma_n(T) = \pi$

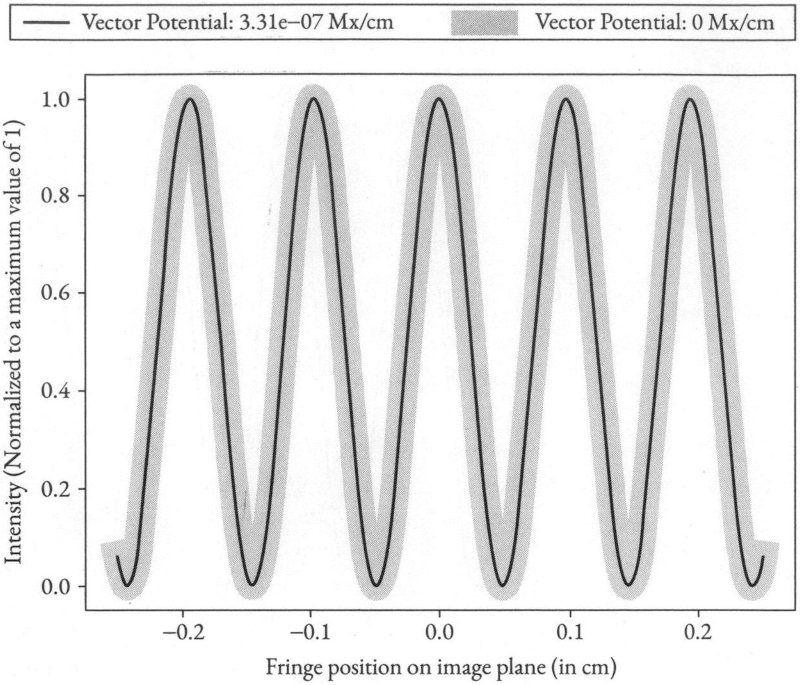

iv. $\gamma_n(T) = \frac{3}{2}\pi$

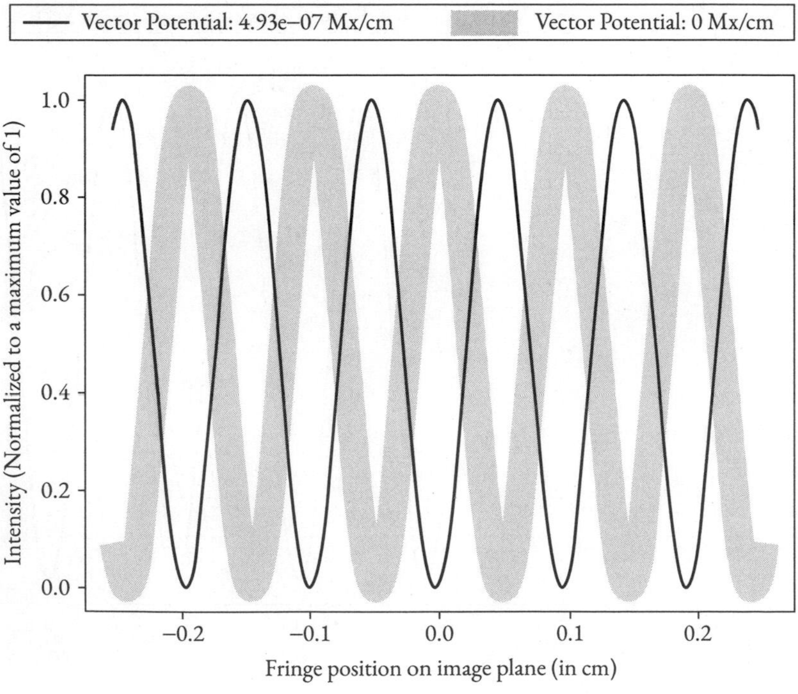

v. $\gamma_n(T) = 2\pi$

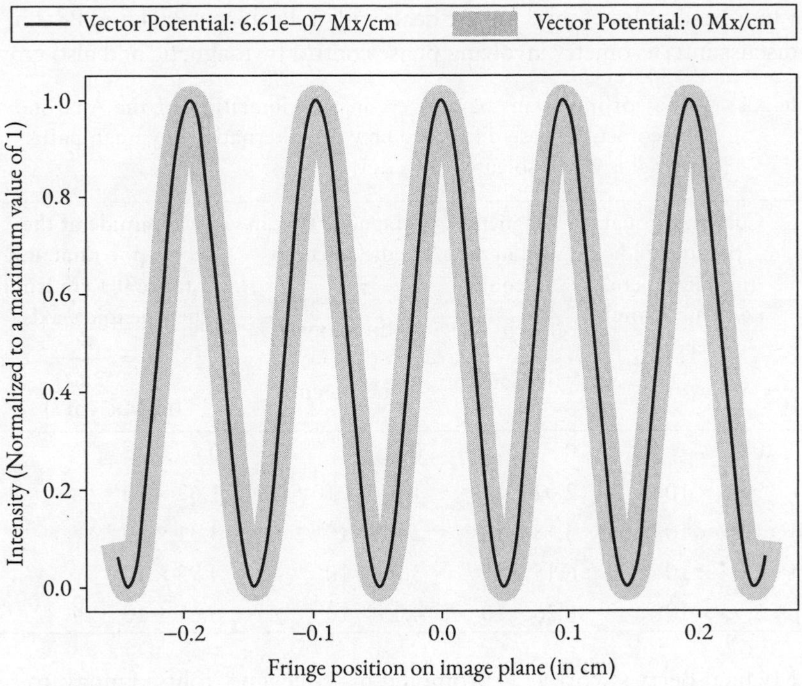

Fig. 2.6 Displacement of the fringes due to the Aharonov–Bohm effect respectively for the five cases tabulated in Table 2.1. There is a lateral displacement as the geometric phase increases. The thick curve for "no current" is included in each to provide reference for the relative displacement. The similarity of the pattern in (i), (iii), and (v) is only apparent as the displaced fringes fall on top of adjacent ones as the geometric phase changes. The same is true for (ii) and (iv).

Interference fringes are obtained in this simulation using the following parameters for the Young's double-slit experiment:

Distance between the two slits	: 0.25 cm
Slit width	: 0.01 cm
Distance between the slits and the detector screen	: 200 cm
Energy of the monoenergetic electrons from the source	: 1.63×10^{-18} erg

The interference pattern produced due to quantum entanglement between two paths of electron(s) through the slits σ and ξ is shown by a thick curve in light gray color in Fig. 2.6. The simulation program then provides for an input choice of the electron current through a solenoid (Fig. 2.5b) of length 2 cm and having 100 turns, uniformly packed. The diameter of the solenoid is 0.08 cm.

Figure 2.6 shows shift in the fringes corresponding to the current through the solenoid set to be (a) 0 statampere, (b) 9.66×10^2 statampere, (c) 1.97×10^3 statampere, (d) 2.94×10^3 statampere, and (e) 3.95×10^3 statampere. Gaussian CGS system of units has been used. Magnetic field inside the solenoid, magnitude of vector potential outside the solenoid at a distance of 0.1 cm from its axis, and corresponding values of flux, and of the geometric phase, are presented in Table 2.1. Experimental study of the geometrical phase in the Aharonov–Bohm effect provides a laboratory method to appreciate Feynman's path integral approach to quantum mechanics.

Two more examples of this kind are dealt with in Problems P8.1 and P8.2 in Chapter 8, in which we discuss interferometry involving phase control by magnetic and also gravitational fields.

Table 2.1 Values of important parameters in the simulation of the Aharonov–Bohm effect due to the geometric phase difference between alternative Feynman paths of the electron(s) from either side of the solenoid (Fig. 2.5)

Sr. No.	Current through the solenoid (in statampere, i.e., statcoulomb per sec)	Magnetic field in the solenoid $B = \mu\frac{N}{L}I$ (in Gauss)	Magnetic flux in the solenoid $\Phi = \pi r^2_{solenoid} \times B$ (in maxwell (Mx), i.e., G·cm²)	Magnitude of the vector potential at a distance 0.1 cm from the solenoid's axis. $A = \frac{\Phi}{2\pi\rho}$ (in Mx. cm^{-1})	Geometric phase difference between the Feynman paths from either side of the solenoid. $\gamma_n(T) = \frac{e\Phi}{\hbar c}$
(i)	0	0	0	0	0
(ii)	9.66×10^2	2.024×10^{-5}	1.017×10^{-7}	1.62×10^{-7}	$\pi/2$
(iii)	1.97×10^3	4.13×10^{-5}	2.07×10^{-7}	3.31×10^{-7}	π
(iv)	2.94×10^3	6.15×10^{-5}	3.09×10^{-7}	4.93×10^{-7}	$3\pi/2$
(v)	3.95×10^3	8.26×10^{-5}	4.15×10^{-7}	6.61×10^{-7}	2π

Michael Berry's work [12] prompted his colleague, John Hannay, to search for a classical analogue of the Berry's phase. Hannay [21] inquired if a classical Hamiltonian is parametrically varied adiabatically and brought back to its original value, would the system return exactly to its state that it had where it started, or would it pick up a geometric phase similar to the Pancharatnam–Berry phase? A classic example of this is the Foucault pendulum (see, for example, Chapter 3 of Ref.[4]). In a local frame of reference, the pendulum swings in a vertical plane that can be defined with reference to its orientation with respect to the North (or South) pole. The Foucault pendulum is a nonholonomic system, since the plane of oscillation does not return to the original after the earth completes one cycle of rotation about its axis.

The plane of oscillation of the Foucault pendulum does not return to its original orientation (with reference to the laboratory in which it is placed) over the time period taken by the earth to complete one rotation about its axis. The physical situation is akin to carrying the laboratory in which the Foucault pendulum experiment is carried out over a closed loop (left panel in Fig. 2.7) on the surface of a sphere, with the understanding that the laboratory actually encircles the earth's axis of rotation along the laboratory's latitude on the earth. The solid angle $\delta\Omega$ subtended at the earth's centre by a surface element δS enclosed by the closed loop is $\delta\Omega = \frac{\delta S}{r^2}$, r being the earth's radius. This is clearly independent of the shape of the surface. As the earth completes one rotation about its axis, the pendulum's temporal evolution is determined not merely by its internal time (dynamics) but also by the adiabatic changes brought about by geometry as the earth takes a much longer time compared to the pendulum's time period. Over one rotation of the earth, the solid angle is essentially just what the latitude subtends at the earth's centre and is given by (using spherical polar coordinates)

$$\Omega = 2\pi[-\cos\theta]_0^{\theta_0} = 2\pi(1-\cos\theta_0) = 2\pi(1-\sin\lambda), \tag{2.96}$$

where λ is the latitude.

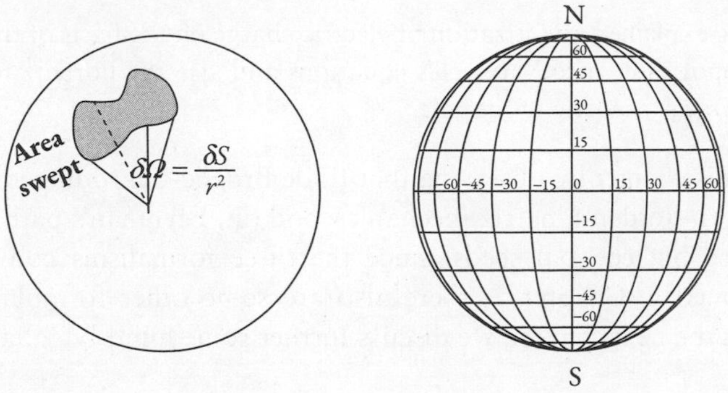

Fig. 2.7 The figure on the left shows a closed curve on a sphere along which a pendulum can be parallel and transported adiabatically. The figure on the right shows that a pendulum set in motion, say somewhere in the northern hemisphere of the earth, would begin its oscillation in a certain plane to which it would not return when the earth completes one full rotation about its North–South axis.

The angle through which the plane of oscillation shifts over one rotation of the earth about its axis is therefore $2\pi \sin \lambda$. The time period for the plane of oscillation of the Foucault pendulum that would return to the original orientation with reference to the laboratory is therefore

$$T = \frac{2\pi}{\omega \sin \lambda}, \tag{2.97}$$

where ω is the circular (angular) frequency of the earth's rotation about its axis. The angle between orientations of the plane is the classical analogue of the Pancharatnam–Berry phase. It is usually called as the "Hannay angle" in the context of classical processes. We have determined the periodicity (Eq. 2.97) of the Foucault pendulum purely from geometry, i.e., merely from the consideration of the Hannay angle (Eq. 2.96). It can also be obtained *artificially* using *dynamics*, albeit only on using *pseudo*-forces (centrifugal and Coriolis) in a rotating frame of reference (see, for example, Eq. 3.34, Chapter 3 in Reference [4]). Reference [22] summarizes some important applications of the Hannay angle, including analysis of semi-rigid bodies, in which a cycle of *internal* twists can produce a new orientation without any external torques. An example of this is the *topological transition* inside a cat "dropped", safely but freely, with its feet up under gravity. With no external torque available to the cat to turn around, the cat lands safely on its feet – giving it nine lives, so to say! In the absence of any external torque it is essentially a geometric topological adjustment that lands the cat safely. The same principle is exploited by an athlete, a pole-vaulter for example, in clearing a high-raised bar. She/he is able to flex the body to go over a higher bar by adjusting the body shape to clear the best height even as her/his centre of mass may actually pass below the bar! One can only expect the underlying physics to be of great importance in the understanding of topological phase transitions. Manipulation of the geometric phase is important also for the understanding of nanoscopic physics [23], to provide stability, address decoherence in a quantum computer [24, 25], and to control properties of polarized light [26]. On the side, it is also interesting to observe that if one carried out the circulation of the vector potential on the closed curve shown in Fig. 2.7, then it would correspond to a *radial* outward magnetic flux through the enclosed surface, as would originate from Dirac's monopole at the centre! Readers may wish to search for specialized literature [27] on this topic.

Funquest: Dirac explained quantization of electric charge on the basis of the existence of a magnetic monopole. Re-write Maxwell's equations and also the Lorentz force law taking into account Dirac's magnetic monopole.

We conclude this chapter by observing that (i) de Broglie–Schrödinger's wave mechanics, (ii) Heisenberg–Born–Jordan's matrix mechanics, and (iii) Feynman's path integral approach have vastly diverse but coequal seeds, since the three formalisms converge to quantum theory. As mentioned in Chapter 1, there also are some other formulations of quantum mechanics [28]. In the next chapter, we discuss further some mind-boggling consequences of quantum theory.

Solved Problems

P2.1: Determine the Gaussian integral, $J(a,b) = \int_{-\infty}^{\infty} dx \exp(-ax^2 + bx)$.

Solution:

First, we evaluate the integral $I(a) = \int_{-\infty}^{\infty} dx \exp(-ax^2)$. The function that is integrated has the well-known Gaussian bell-shape, symmetric about $x = 0$. Using plane polar coordinates $\sqrt{x^2 + y^2} = \rho$ and $\frac{y}{x} = \tan^{-1}\varphi$ with $0 \le \rho < \infty$ and $0 \le \varphi < 2\pi$, we note that

$$I^2(a) = \int_{-\infty}^{\infty} dx\, e^{-ax^2} \int_{-\infty}^{\infty} dy\, e^{-ay^2} = \int_{-\infty}^{\infty} dx \int_{-\infty}^{\infty} dy\, e^{-a\left(x^2+y^2\right)} = 2\pi \int_0^{\infty} \rho e^{-a\rho^2} d\rho = \frac{\pi}{a}, \text{ we conclude that } I(a) = \sqrt{\frac{\pi}{a}}.$$

Hence, $J(a,b) = \int_{-\infty}^{\infty} dx e^{-ax^2+bx} = \int_{-\infty}^{\infty} dx e^{\left[\frac{b^2}{4a} - a\left(x - \frac{b}{2a}\right)^2\right]} = e^{\frac{b^2}{4a}} \int_{-\infty}^{\infty} dx e^{-a\left(x - \frac{b}{2a}\right)^2} = e^{\frac{b^2}{4a}} \sqrt{\frac{\pi}{a}}$, seen easily on transferring the integration over x to that over $y = \left(x - \frac{b}{2a}\right)$.

P2.2: Evaluate the right-hand side of Eq. 2.24 for $N = 2$ and develop a generalization of the result as N increases. Our eventual interest is of course in the limit $N \to \infty$.

Solution:

For $N = 2$, we reach t_f from t_0 in two steps:

$$\langle x_f | \exp\left[-i\frac{H(t_f - t_0)}{\hbar}\right] |x_0\rangle = \langle x_f | \exp\left[-i\frac{H(t_f - t_0)}{\hbar}\right] |x_1\rangle\langle x_1| \exp\left[-i\frac{H(t_f - t_0)}{\hbar}\right] |x_0\rangle.$$

The integral over y_1 is

$$\int_{-\infty}^{\infty} dy_1 e^{i\left[\sum_{j=1}^{N=2}\left(y_j - y_{j-1}\right)^2\right]} = \int_{-\infty}^{\infty} dy_1 e^{i\left[\left(y_2-y_1\right)^2 + \left(y_1 - y_0\right)^2\right]} = e^{i\left(y_2^2 + y_0^2\right)} \int_{-\infty}^{\infty} dy_1 e^{\left(2iy_1^2 - 2i\left(y_0+y_2\right)y_1\right)} = e^{i\left(y_2^2+y_0^2\right)} e^{\frac{\left\{-2i\left(y_0+y_2\right)\right\}^2}{4(-2i)}} \sqrt{\frac{i\pi}{2}},$$

i.e., $$\int_{-\infty}^{\infty} dy_1 e^{i\left[\sum_{j=1}^{N=2}\left(y_j - y_{j-1}\right)^2\right]} = e^{i\frac{\left(y_2-y_0\right)^2}{2}} \sqrt{\frac{i\pi}{2}}.$$

Next, we shall need, for $N = 3$, integration over y_2:

$$\sqrt{\frac{i\pi}{2}}\int_{-\infty}^{\infty} dy_2 e^{i\frac{(y_2-y_0)^2}{2}} e^{i(y_3-y_2)^2} = \sqrt{\frac{i\pi}{2}}\int_{-\infty}^{\infty} dy_2 e^{i\left[(y_3-y_2)^2+\frac{(y_2-y_0)^2}{2}\right]} = \sqrt{\frac{i\pi}{2}}\sqrt{\frac{i2\pi}{3}}e^{i\left[\frac{(y_3-y_0)^2}{3}\right]} = \sqrt{\frac{(i\pi)^2}{3}}e^{i\left[\frac{(y_3-y_0)^2}{3}\right]}.$$

Continuing for N steps, we get

$$\left\langle x_f \right| e^{-i\frac{H(t_f-t_0)}{\hbar}} \left| x_0 \right\rangle = \lim_{N\to\infty}\left[\frac{m}{i2\pi\hbar\Delta t}\right]^{N/2}\left[\sqrt{\frac{2\hbar\Delta t}{m}}\right]^{N-1}\sqrt{\frac{(i\pi)^{N-1}}{N}}e^{i\left\{\frac{(y_N-y_0)^2}{N}\right\}},$$

$$\text{i.e., } \left\langle x_f \right| e^{-i\frac{H(t_f-t_0)}{\hbar}} \left| x_0 \right\rangle = \lim_{N\to\infty}\left[\frac{m}{i2\pi\hbar N\Delta t}\right]^{1/2} e^{im\left[\frac{(x_N-x_0)^2}{2\hbar N\Delta t}\right]}.$$

P2.3: Prove that the Schrödinger equation (Eq. 2.79) for the gauge-transformed electromagnetic potential $\left\{\phi'(\vec{r},t),\vec{A}'(\vec{r},t)\right\}$ follows from that for $\left\{\phi(\vec{r},t),\vec{A}(\vec{r},t)\right\}$ with $\psi'(\vec{r},t)$ given by Eq. 2.77 involving the gauge function $\chi(\vec{r},t)$.

Solution:

From Eq. 2.79:

$$i\hbar\frac{\partial}{\partial t}\psi'(\vec{r},t) = i\hbar\frac{\partial}{\partial t}\left\{\exp\left(i\frac{e}{\hbar c}\chi(\vec{r},t)\right)\psi(\vec{r},t)\right\},$$

$$\text{i.e., } i\hbar\frac{\partial}{\partial t}\psi'(\vec{r},t) = \exp\left(i\frac{e}{\hbar c}\chi(\vec{r},t)\right)\left(-\frac{e}{c}\frac{\partial\chi(\vec{r},t)}{\partial t} + i\hbar\frac{\partial}{\partial t}\right)\psi(\vec{r},t), \tag{P2.3.1}$$

The right-hand side of Eq. 2.79 is

$$\frac{1}{2m}\left(\begin{matrix}-i\hbar\vec{\nabla}\\ -\frac{e\vec{A}'(\vec{r},t)}{c}\end{matrix}\right)^2\psi'(\vec{r},t) = \frac{1}{2m}\left[\begin{matrix}\left(\frac{e}{c}\right)^2\exp\left[i\frac{e}{\hbar c}\chi(\vec{r},t)\right]\left\{\vec{\nabla}\chi(\vec{r},t)\right\}^2\psi(\vec{r},t) - \frac{i\hbar e}{c}\exp\left[i\frac{e}{\hbar c}\chi(\vec{r},t)\right]\left\{\vec{\nabla}^2\chi(\vec{r},t)\right\}\psi(\vec{r},t)\\ -2\frac{i\hbar e}{c}\exp\left[i\frac{e}{\hbar c}\chi(\vec{r},t)\right]\left\{\vec{\nabla}\chi(\vec{r},t)\right\}\left\{\vec{\nabla}\psi(\vec{r},t)\right\} - \hbar^2\exp\left[i\frac{e}{\hbar c}\chi(\vec{r},t)\right]\left\{\vec{\nabla}^2\psi(\vec{r},t)\right\}\\ +\frac{i\hbar e}{c}\left\{\vec{\nabla}\bullet\vec{A}(\vec{r},t)\right\}\exp\left[i\frac{e}{\hbar c}\chi(\vec{r},t)\right]\psi(\vec{r},t)\\ -\left(\frac{e}{c}\right)^2\exp\left[i\frac{e}{\hbar c}\chi(\vec{r},t)\right]\vec{A}\bullet\left\{\vec{\nabla}\chi(\vec{r},t)\right\}\psi(\vec{r},t) + \frac{i\hbar e}{c}\exp\left[i\frac{e}{\hbar c}\chi(\vec{r},t)\right]\vec{A}\bullet\left\{\vec{\nabla}\psi(\vec{r},t)\right\}\\ +\frac{i\hbar e}{c}\exp\left[i\frac{e}{\hbar c}\chi(\vec{r},t)\right]\left\{\vec{\nabla}^2\chi(\vec{r},t)\right\}\psi(\vec{r},t)\\ -\left(\frac{e}{c}\right)^2\exp\left[i\frac{e}{\hbar c}\chi(\vec{r},t)\right]\left\{\vec{\nabla}\chi(\vec{r},t)\right\}^2\psi(\vec{r},t) + \frac{i\hbar e}{c}\exp\left[i\frac{e}{\hbar c}\chi(\vec{r},t)\right]\left\{\vec{\nabla}\chi(\vec{r},t)\right\}\left\{\vec{\nabla}\psi(\vec{r},t)\right\}\\ -\left(\frac{e}{c}\right)^2\exp\left[i\frac{e}{\hbar c}\chi(\vec{r},t)\right]\vec{A}\bullet\left\{\vec{\nabla}\chi(\vec{r},t)\right\}\psi(\vec{r},t)\\ +\frac{i\hbar e}{c}\exp\left[i\frac{e}{\hbar c}\chi(\vec{r},t)\right]\vec{A}\bullet\left\{\vec{\nabla}\psi(\vec{r},t)\right\} - \left(\frac{e}{c}\right)^2\exp\left[i\frac{e}{\hbar c}\chi(\vec{r},t)\right]\left\{\vec{\nabla}\chi(\vec{r},t)\right\}^2\psi(\vec{r},t)\\ +\frac{i\hbar e}{c}\exp\left[i\frac{e}{\hbar c}\chi(\vec{r},t)\right]\left\{\vec{\nabla}\chi(\vec{r},t)\right\}\left\{\vec{\nabla}\psi(\vec{r},t)\right\}\\ +\left(\frac{e}{c}\right)^2\left\{\begin{matrix}\vec{A}^2(\vec{r},t)\psi(\vec{r},t) + \vec{A}\bullet\left\{\vec{\nabla}\chi(\vec{r},t)\right\}\psi(\vec{r},t)\\ +\left\{\vec{\nabla}\chi(\vec{r},t)\right\}\bullet\vec{A}(\vec{r},t)\psi(\vec{r},t)\\ +\left\{\vec{\nabla}\chi(\vec{r},t)\right\}^2\psi(\vec{r},t)\end{matrix}\right\}\exp\left[i\frac{e}{\hbar c}\chi(\vec{r},t)\right]\end{matrix}\right],$$

(P2.3.2a)

i.e.,

$$\left(-i\hbar\vec{\nabla}-\frac{e\vec{A}'(\vec{r},t)}{c}\right)^2\psi'(\vec{r},t)=\frac{1}{2m}\exp\left[i\frac{e}{\hbar c}\chi(\vec{r},t)\right]\begin{bmatrix}-\hbar^2\vec{\nabla}^2\psi(\vec{r},t)+\frac{i\hbar e}{c}\left\{\vec{\nabla}\bullet\vec{A}(\vec{r},t)\right\}\psi(\vec{r},t)\\+2\frac{i\hbar e}{c}\vec{A}\bullet\left\{\vec{\nabla}\psi(\vec{r},t)\right\}\\+\left(\frac{e}{c}\right)^2\vec{A}^2(\vec{r},t)\psi(\vec{r},t)\end{bmatrix}. \quad \text{(P2.3.2b)}$$

Furthermore,

$$\left[e\phi'(\vec{r},t)\right]\psi'(\vec{r},t)=\left[e\left(\phi(\vec{r},t)-\frac{1}{c}\frac{\partial\chi(\vec{r},t)}{\partial t}\right)\right]\exp\left[i\frac{e}{\hbar c}\chi(\vec{r},t)\right]\psi(\vec{r},t). \quad \text{(P2.3.3)}$$

Therefore,

$$\begin{bmatrix}\frac{1}{2m}\left(-i\hbar\vec{\nabla}-\frac{e\vec{A}'(\vec{r},t)}{c}\right)^2\\+e\phi'(\vec{r},t)\end{bmatrix}\psi'(\vec{r},t)=\exp\left[i\frac{e}{\hbar c}\chi(\vec{r},t)\right]\begin{bmatrix}\frac{1}{2m}\left\{-i\hbar\vec{\nabla}-\frac{e}{c}\vec{A}(\vec{r},t)\right\}^2\\+e\phi(\vec{r},t)-\frac{e}{c}\frac{\partial\chi(\vec{r},t)}{\partial t}\end{bmatrix}\psi(\vec{r},t) \quad \text{(P2.3.4)}$$

Hence,

$$\begin{bmatrix}\frac{1}{2m}\left(-i\hbar\vec{\nabla}-\frac{e\vec{A}'(\vec{r},t)}{c}\right)^2\\+e\phi'(\vec{r},t)\end{bmatrix}\psi'(\vec{r},t)=\exp\left[i\frac{e}{\hbar c}\chi(\vec{r},t)\right]\begin{pmatrix}i\hbar\frac{\partial}{\partial t}\\-\frac{e}{c}\frac{\partial\chi(\vec{r},t)}{\partial t}\end{pmatrix}\psi(\vec{r},t), \quad \text{(P2.3.5)}$$

or $$\left[e\phi'(\vec{r},t)\right]\psi'(\vec{r},t)=\exp\left[i\frac{e}{\hbar c}\chi(\vec{r},t)\right]\left[e\phi(\vec{r},t)-\frac{e}{c}\frac{\partial\chi(\vec{r},t)}{\partial t}\right]\psi(\vec{r},t). \quad \text{(P2.3.6)}$$

Therefore, $$i\hbar\frac{\partial}{\partial t}\psi'(\vec{r},t)=\left[\frac{1}{2m}\left(-i\hbar\vec{\nabla}-\frac{e\vec{A}'(\vec{r},t)}{c}\right)^2+e\phi'(\vec{r},t)\right]\psi'(\vec{r},t). \quad \text{(P2.3.7)}$$

P2.4: In a one-dimensional Schrödinger equation for a particle whose motion is restricted to the Cartesian *x*-axis, show using a semi-classical argument that the approximate solution is given by the wavefunction $\psi_o \approx \exp\left[\frac{i}{h}\int p(x)dx\right]$.

Solution:

The Schrödinger equation is: $\left[\frac{1}{2m}\left(-i\hbar\frac{d}{dx}\right)^2+V(x)\right]\psi(x)=E\psi(x)$.

Using what is referred to as *semi*-quantization $p(x)\leftrightarrow\sqrt{2m\{E-V(x)\}}$,

$$p_x\psi_0\approx\left[\sqrt{2m\{E-V(x)\}}\right]\psi_0;\ \text{i.e.,}\ \left(-i\hbar\frac{d}{dx}\right)\psi_0\simeq\left[\sqrt{2m\{E-V(x)\}}\right]\psi_0.$$

Thus, $\frac{d\psi_0}{\psi_0}\approx\frac{i}{\hbar}\left[\sqrt{2m\{E-V(x)\}}\right]dx$; or $ln(\psi_0)=\frac{i}{\hbar}\int\left[\sqrt{2m\{E-V(x)\}}\right]dx$.

Hence, $\psi_0 \approx \exp\left[\frac{i}{\hbar}\int\left[\sqrt{2m\{E-V(x)\}}\right]dx\right] = \exp\left[\frac{i}{\hbar}\int p(x)dx\right]$.

The approximation alluded to here is discussed more fully in Section 6.1 of Chapter 6 (in particular, see Eq. 6.43).

P2.5: Show that the transformation of the electromagnetic potentials $\{\vec{A}(\vec{r},t),\phi(\vec{r},t)\}$ to $\{\vec{A}'(\vec{r},t),\phi'(\vec{r},t)\}$, where $\vec{A}'(\vec{r},t)=\vec{A}(\vec{r},t)+\vec{\nabla}\Lambda(\vec{r},t)$ and $\phi'(\vec{r},t)=\phi(\vec{r},t)-\frac{\partial\Lambda(\vec{r},t)}{\partial t}$ leave the electromagnetic field $\{\vec{E}(\vec{r},t),\vec{B}(\vec{r},t)\}$ invariant, $\Lambda(\vec{r},t)$ being an arbitrary (but well-behaved) scalar function.

Solution:

From Maxwell's equations, we have

$\vec{\nabla}\times\vec{E}(\vec{r},t) = -\frac{\partial\vec{B}(\vec{r},t)}{\partial t} = -\frac{\partial\left[\vec{\nabla}\times\vec{A}(\vec{r},t)\right]}{\partial t} = -\vec{\nabla}\times\frac{\partial\vec{A}(\vec{r},t)}{\partial t}$ since the derivative operators with respect to space and time variables commute. Thus, $\vec{\nabla}\times\left[\vec{E}(\vec{r},t)+\frac{\partial\vec{A}(\vec{r},t)}{\partial t}\right]=\vec{0}$. It is therefore the combined vector $\left\{\vec{E}(\vec{r},t)+\frac{\partial\vec{A}(\vec{r},t)}{\partial t}\right\}$ that constitutes an irrotational vector field, which we can always write as $-\vec{\nabla}\phi(\vec{r},t)$, since the gradient of a scalar function is essentially irrotational. In other words, $\vec{E}(\vec{r},t) = -\left[\vec{\nabla}\phi(\vec{r},t)+\frac{\partial\vec{A}(\vec{r},t)}{\partial t}\right]$. Furthermore, the magnetic induction field $\vec{B}(\vec{r},t)$, being solenoidal, is derivable from a vector potential $\vec{A}(\vec{r},t)$ as its curl. However, the gradient of any scalar point function is essentially irrotational, and therefore the magnetic induction field $\vec{B}(\vec{r},t)$ is derivable also from a different potential, $\vec{A}'(\vec{r},t)=\vec{A}(\vec{r},t)+\vec{\nabla}\Lambda(\vec{r},t)$, as its curl. If we therefore change the vector potential from $\vec{A}(\vec{r},t)$ to $\vec{A}'(\vec{r},t)=\vec{A}(\vec{r},t)+\vec{\nabla}\Lambda(\vec{r},t)$, then the electric field becomes $\vec{E}(\vec{r},t) = -\left[\vec{\nabla}\phi(\vec{r},t)+\frac{\partial\{\vec{A}(\vec{r},t)+\vec{\nabla}\Lambda\}}{\partial t}\right]$, i.e.,

$\vec{E}(\vec{r},t) = -\left[\vec{\nabla}\left\{\phi(\vec{r},t)+\frac{\partial\Lambda}{\partial t}\right\}+\frac{\partial\vec{A}(\vec{r},t)}{\partial t}\right]$. A transformed scalar field, given by $\phi'(\vec{r},t)=\phi(\vec{r},t)+\frac{\partial\Lambda(\vec{r},t)}{\partial t}$, would therefore leave the electric intensity field invariant. We thus see that the gauge transformations $\vec{A}(\vec{r},t)\to\vec{A}'(\vec{r},t)=\vec{A}(\vec{r},t)+\vec{\nabla}\Lambda(\vec{r},t)$ and $\phi(\vec{r},t)\to\phi'(\vec{r},t)=\phi(\vec{r},t)-\frac{\partial\Lambda(\vec{r},t)}{\partial t}$ leave the electromagnetic field $\{\vec{E}(\vec{r},t),\vec{B}(\vec{r},t)\}$ invariant.

P2.6: Analysis using path integrals often require solutions of Gaussian integrals of the form $\iiint d^3\vec{q}\exp\left(\frac{i}{\hbar}(\alpha\vec{q}^2+\vec{q}'\bullet\vec{q})\right)$. In P2.1, we have already considered an integral of this type. Such integrals can be evaluated by reducing them to a form $I=\int\exp(-\alpha(z+c)^2)dz$.

Evaluate the integral I using Cauchy's contour integration carried out in the complex z plane and show that

$$\int_{-\infty}^{\infty} \exp(-\alpha(z+c)^2)dz = \sqrt{\frac{\pi}{\alpha}}; \quad \alpha, c \in \mathbb{C}; \Re\{t\} > 0.$$

Solution:

Let the complex number c be given by $c = a + ib$.

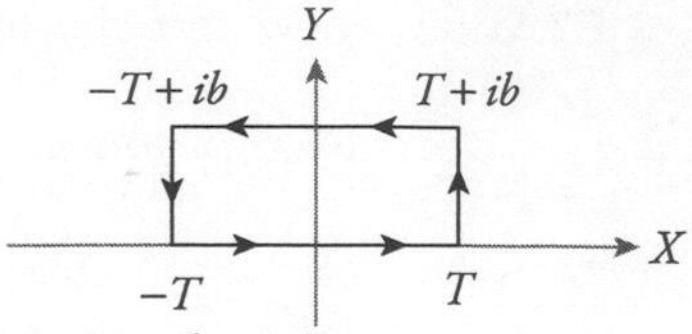

Complex t-plane: $t = x + iy$

$$f(z) = f(x+iy) = \exp\left(-\alpha\begin{pmatrix} x+iy \\ +a+ib \end{pmatrix}^2\right)$$

$$f(z) = f(x+iy) = \exp\left(-\alpha\begin{pmatrix} (x+a) \\ +i(y+b) \end{pmatrix}^2\right)$$

Let us consider the path integral

$$I = \int \exp(-\alpha(z+c)^2)dz = \begin{bmatrix} \int_{-T}^{+T} f(x)dx + i\int_{y=0}^{y=b} f(T+iy)dy \\ + \int_{x=T}^{x=-T} f(x+ib)dx + i\int_{b}^{0} f(-T+iy)dy \end{bmatrix}$$

In the limit $T \to \infty$, and using the result of P2.1, we get

$$0 = \int_{-T}^{+T} f(x)dx + \int_{x=T}^{x=-T} f(x+ib)dx = \sqrt{\frac{\pi}{\alpha}} + \int_{x=T}^{x=-T} f(x+ib)dx$$

$$\int_{x=-\infty}^{x=\infty} f(x+ib)dx = \sqrt{\frac{\pi}{\alpha}}$$

Changing the variable $x = z + a = z + c - ib$ corresponding to $x + ib = z + c$,

$$\int_{-\infty}^{\infty} f(z+c)\,dz = \int_{-\infty}^{\infty} \exp(-\alpha(z+c)^2)dz = \sqrt{\frac{\pi}{\alpha}}.$$

P2.7: Determine the multidimensional Gaussian integral $Z = \int d^n x \exp\left(-\frac{1}{2}x^T \mathbb{A} x\right)$, where $d^n x \equiv \prod_i dx_i$, and $\mathbb{A}$ is a symmetric matrix with eigenvalues a_i.

Solution:

First, we write $y = \mathbb{M}^T x$, where $\mathbb{M}$ is an orthogonal matrix that diagonalizes $\mathbb{A} : \mathbb{D} = \mathbb{M}^T \mathbb{A} \mathbb{M} = \text{diag}(a_1, a_2 .. a_n)$.

Then, $Z = \int d^n y \exp\left(-\frac{1}{2}y^T D y\right) = \prod_i \int dy_i \exp\left(-\frac{1}{2}a_i y_i^2\right)$.

Each of the integrals in the product can be evaluated independently. Using the solution to P2.1, we see that $Z = \prod_i \sqrt{\frac{2\pi}{a_i}} = \sqrt{\frac{(2\pi)^n}{\det|\mathbb{A}|}}$.

P2.8: Obtain the propagator $K(x_a t_a; x_b t_b) = \int_{x_a}^{x_b} Dx(t) \exp\left(\frac{iS(x(t))}{\hbar}\right)$ for a particle in a potential $V = a + bx + cx^2 + d\dot{x} + ex\dot{x}$.

Solution:

Let each path be represented by $x(t) = x_{cl}(t) + y(t)$; x_{cl} being the classical path and y denoting a departure from it. Hence, $\dot{x}(t) = \dot{x}_{cl}(t) + \dot{y}(t)$.

Slicing the time interval into N parts, $x_i \equiv x(t) = x_{cl}(t_i) + y_i$. With $dx_i = dy_i$, we get

$$\int_{x_a}^{x_b} Dx(t) = \int_{y(0)=0}^{y(t)=0} Dy(t), \text{ and the propagator is } K(x_a t_a; x_b t_b) = \int_{y(0)=0}^{y(t)=0} Dy(t)\exp\left(\frac{i}{\hbar}S(x_{cl}(t) + y(t))\right).$$

Expanding S_{cl} in a Taylor series about x_{cl}, we get $S(x_{cl} + y) = \int_0^t L(x_{cl} + y, \dot{x}_{cl} + \dot{y})dt'$,

$$\text{i.e., } S(x_{cl} + y) = \int_0^t \left[L(x_{cl}, \dot{x}_{cl}) + \underbrace{\left(\left.\frac{\partial L}{\partial x}\right|_{x_{cl}} y + \left.\frac{\partial L}{\partial \dot{x}}\right|_{x_{cl}} \dot{y} \right)}_{=0} + \frac{1}{2}\left(\left.\frac{\partial^2 L}{\partial x^2}\right|_{x_{cl}} y^2 + 2\left.\frac{\partial^2 L}{\partial x \partial \dot{x}}\right|_{x_{cl}} y\dot{y} + \left.\frac{\partial^2 L}{\partial \dot{x}^2}\right|_{x_{cl}} \dot{y}^2 \right) \right] dt'.$$

$$\text{Hence, } K(x_a t_a; x_b t_b) = \exp\left(\frac{i}{\hbar}S_{cl}\right)\int_0^0 \exp\left[\frac{i}{\hbar}\int_0^t \left(\frac{1}{2}m\dot{y}^2 - cy^2 - ey\dot{y}\right)dt'\right] Dy(t'').$$

P2.9: Determine Feynman's path integral for a particle moving in a simple harmonic oscillator potential $u(x) = \frac{1}{2}kx^2$.

Solution:

The Lagrangian is $L = T - U = L = \frac{m}{2}(\dot{x}^2 - \omega^2 x^2)$ with $\omega = \sqrt{k/m}$. This can be seen as a special case of P2.8 with $c = m\omega^2$ and $a = b = d = e = 0$. This gives $F(T) = \int Dy(t)\exp\left[\frac{i}{\hbar}\int_0^T \frac{m}{2}(\dot{y}^2 - \omega^2 y^2)\right]$ wherein $F(t)$ appears in the final step of the solution to P2.8. Let the various paths $y(t)$ be represented as Fourier series: $y(t) = \sum_{n=1}^{\infty} a_n \sin\left(\frac{n\pi t}{T}\right)$.

The requirements that $y(t_a) = 0 = y(t_b)$ and $\dot{y}(t_a) = 0 = \dot{y}(t_b)$ imply that $y(t)$ and $\dot{y}(t)$ are periodic functions in T. We shall consider $\int Dy(t) \to \frac{1}{A(\varepsilon)^N}\int da_1 \int da_2 \ldots \int da_N$ and then take the limit $N \to \infty$. Accordingly, $F(T) = \frac{1}{A(\varepsilon)^N}\int da_1 \ldots \int da_N \exp\left[\frac{im}{2\hbar}\frac{T}{2}\sum_{n=1}^{N}\left(\left(\frac{n\pi}{T}\right)^2 - \omega^2\right)a_n^2\right]$.

Recognizing the Gaussian integrals, we have $\int \frac{da_n}{A}\exp\left[\frac{imT}{4\hbar}f_n(\omega,T)a_n^2\right] = \sqrt{\frac{2}{Tf_n(\omega,T)}}$ with $f_n(\omega,T) = \frac{n^2\pi^2}{T^2} - \omega^2$. There are no linear terms in a_n; hence, the result of the N integrations is a product of the independent terms: $F(T) \propto \prod_{n=1}^{N}\frac{1}{\sqrt{f_n(\omega,T)}}$.

Now, expanding $\sqrt{f_n(\omega,T)}$, we have $\prod_{n=1}^{N}\frac{1}{\sqrt{f_n(\omega,T)}} = \prod_{n=1}^{N}\left(\frac{n^2\pi^2}{T^2}\right)^{-1/2} \times \prod_{n=1}^{N}\left(1 - \frac{\omega^2 T^2}{n^2\pi^2}\right)^{-1/2}$.

The first factor is independent of ω; it can be absorbed in the normalization factor. The second factor in the limit $N \to \infty$ is $\lim_{N\to\infty} \prod_{n=1}^{N} \left(1 - \frac{\omega^2 T^2}{n^2 \pi^2}\right)^{-1/2} = \sqrt{\frac{\omega T}{\sin \omega T}}$.

To obtain the normalization factor C, we take the limit $\lim_{\omega \to 0} F(T)$, which gives $C = \sqrt{m / 2i\pi\hbar T}$.

Thus, the required kernel $K(x_a t_a; x_b t_b) = \sqrt{\frac{m\omega}{2i\pi\hbar \sin \omega T}} \exp\left(\frac{i}{\hbar} S_d\right)$.

Exercises

E2.1: (a) Use Eq. 2.61 to determine the geometric phase change when the width of an infinitely deep square well expands adiabatically from a_1 to a_2.

(b) What would be the dynamical phase if the expansion of the width occurs at a constant rate?

(c) If the well then contracts back to its original size, what is the Berry's phase for the cycle?

E2.2: Consider a Dirac-δ potential well $V(x) = -\alpha\delta(x)$ supporting a single bound state $(\alpha > 0)$ given by $\psi(x) = \frac{\sqrt{m\alpha}}{\hbar} \exp\left(-\frac{m\alpha|x|}{\hbar^2}\right)$. Determine the change in the geometric phase when α, the strength of the potential, is adiabatically increased from α_1 to α_2. What would be the dynamical phase if this adiabatic increase in the strength of the potential occurs at a constant rate? If you wish, you may put this problem on hold till you go through Chapter 3 (especially the Solved Problem P3.6).

E2.3: Show that the geometric phase vanishes if the wave function $\psi_n(t)$ is real. What happens if $\psi_n(t)$ is trivially complex, i.e., $\psi_n'(t) = e^{i\phi_n} \psi_n(t)$?

E2.4: The statistical mechanical partition function Z can be expressed as a propagator when written with an imaginary time argument. Such a transformation is called Wick's rotation: $\tau = i(t - t_0)$; $\frac{dx}{d\tau} = \frac{dx}{dt}\frac{dt}{d\tau} = i\frac{dx}{dt}$ for a continuous parameter x. The partition function for a Hamiltonian H is $Z = \int \langle x | \exp(-\beta H) | x \rangle dx$, where $\beta = i\left(\frac{t_b - t_a}{\hbar}\right)$. This allows us to reinterpret a quantum mechanical system evolving in imaginary time as a classical statistical mechanics system. Make use of the imaginary time transformation and obtain the partition function for a harmonic oscillator in terms of the path integral. Furthermore, determine its energy levels.

E2.5: From the eigenfunctions and the eigenvalues we can easily construct the propagator for the harmonic oscillator: $K(x_a, t_a; x_b t_b) = \sum_n \psi_n(x) \psi_n^*(x) \exp\left(-\frac{i}{\hbar} E_n (t_b - t_a)\right)$. Now consider the reverse process, for the case of the oscillator. Set $x_a = x_b = t_a = 0$ and $t_b = t$. (Take the propagator from Problem E2.10. The $\exp\left(\frac{i}{\hbar} S_d\right)$ term will vanish when $x_a = x_b = t_a = 0$.) Obtain the energy levels of the oscillator by expanding both sides of the given expression. Are these the only eigenvalues of the harmonic oscillator? If not, what happened to the missing eigenvalues?

E2.6: Consider the propagator expression give in the previous problem (**E2.5**). As in the previous case, expand both sides of the expression, but this time set $x_a = x_b$, and $t_a = 0, t_b = t$. Obtain E_0, E_1, $|\psi_0(0)|^2$ and $|\psi_1(0)|^2$.

E2.7: Eq. 2.25 gives the propagator for a free particle. Show by substitution that the free particle kernel satisfies the differential equation $i\hbar\frac{\partial K}{\partial t_b} = -\frac{\hbar^2}{2m}\frac{\partial^2 K}{\partial x_b^2}$.

E2.8:

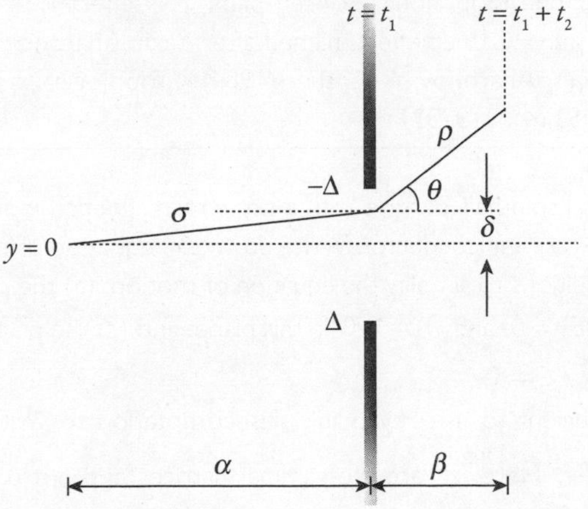

Consider a slit of width 2Δ as shown in the adjacent figure. Let $y = 0$ be the midpoint of the slit. At time $t = 0$ a particle starts from the origin and "strikes" the slit at a distance δ from the mid-point after a time interval t_1. After "striking" the slit, the particle gets diffracted and is found at a distance y from the origin after a time t_2. Determine the wave amplitude for the particle at some time $t \geq t_1 + t_2$ and express it in terms of σ and ρ.

E2.9: In problem **E2.8**, we determined the wave amplitude for a particle striking a single slit. Expand the exponent in the integral and simplify it to express it in terms of the two Fresnel integrals:

$$C(\zeta) = \int_0^{\zeta} \cos\left(\frac{\pi}{2}Z^2\right)dZ \text{ and } S(\zeta) = \int_0^{\zeta} \sin\left(\frac{\pi}{2}Z^2\right)dZ.$$

(*Hint:* Write σ in terms of α and δ and ρ in terms of β, y and δ. You can simplify the resulting expression by completing the squares and separating δ parts from non-δ parts. You will then be able to write the exponent as $e^{iB}e^{iA(\delta-\delta_0)^2}$. By defining $A(\delta - \delta_0)^2 \equiv i\frac{\pi}{2}Z^2$ and $Z_\pm \equiv \left(\frac{2A}{\pi}\right)^{1/2}(\Delta \pm \delta_0)$, you'll be able to express the original integral in terms of $C(\zeta)$ and $S(\zeta)$.

E2.10: For Fraunhofer diffraction, the source should be at infinity, i.e., $\sigma \to \infty$, $t_1 \to \infty$. Apply these conditions in the expression for the wave amplitude obtained in problem **E2.9**, and obtain the expression for Fraunhofer diffraction far away from the slit ($\Delta \ll y$).

E2.11: Consider an electron at the origin subjected to a time-varying magnetic field of a constant magnitude but changing direction. Let the magnetic field vector with magnitude B_0 sweep out a cone of angle α on the surface of a sphere of radius $r = B_0$ at a constant angular velocity ω. Obtain the Berry's phase for one cycle. The eigenstate that represents the spin-up along $\vec{B}(t)$ is $\chi_+ = \begin{pmatrix} \cos(\theta/2) \\ e^{i\varphi}\sin(\theta/2) \end{pmatrix}$.

E2.12: In **E2.11**, instead of an electron, consider a charged particle having spin 1. Can you intuitively guess what the Berry's phase will be? Solve for the Berry's phase and verify that your guess is indeed correct (or wrong in case you guessed incorrectly.)

NOTE: The problems E2.13, E2.14, and E2.15 together describe the motion of a spin-half neutral particle with nonzero magnetic moment in the presence of an external electric field, moving in the region where $\nabla \bullet E = 0,$ in a closed path. Surprisingly, this particle picks up a nonzero geometrical phase when it completes one loop along a closed path. This effect is a dual to Aharonov–Bohm effect and is called Aharonov–Casher effect, named after Yakir Aharonov and Aharon Casher who first described it in 1984 (Y. Aharonov; A. Casher (1984). "Topological quantum effects for neutral particles." *Phys. Rev. Lett.* 53 (4): 319–321).

E2.13: Consider a neutral spin-half particle with a nonzero magnetic dipole moment, traveling in the field of a current-carrying conductor. Write down the equation of motion for a particle. Apply the following conditions to simplify the equation of motion: (a) the particle is confined to the *xy* plane; (b) $E_z = \partial E_z = 0$ and $\partial \psi_z = 0$ in this plane; and (c) the particle travels a closed path in a region where $\vec{\nabla} \bullet \vec{E} = 0.$

(Note: It is convenient to use Feynman's slashed notation to write the equation of motion: $\partial\!\!\!/ = \gamma_\mu \partial^\mu \equiv \gamma_\mu \dfrac{\partial}{\partial x_\mu}.$ Here γ_μ are the gamma matrices and not to be confused with Berry's phase $\gamma_n(T)$.)

E2.14: Rewrite the equation of motion obtained in the previous problem (**E2.13**), i.e., $\left[\partial\!\!\!/' + m + \mu\left(\gamma_1 E_1 + \gamma_2 E_2\right)\gamma_4\right]\psi = 0,$ such that ψ can be expressed as an eigenstate of the Hamiltonian.

(Hint: Rewrite $m + \mu\left(\gamma_1 E_1 + \gamma_2 E_2\right)\gamma_4$ as $m\sigma_3 - i\vec{\sigma}' \bullet \vec{\nabla} - s\mu\vec{\sigma}' \bullet \vec{A}',$ where $\vec{\sigma}' \equiv \left(\sigma_1, \sigma_2\right)$ and σ_i are the Pauli matrices (see Solved Problem P4.3, Chapter 4), and $\vec{A}' \equiv \left(E_2, -E_1\right).$ Then you can write the wave function in terms of $\vec{A}'$.)

E2.15: Use the wave function $\psi(\vec{r})$ obtained in problem **E2.14**, in Eq. 2.61 to obtain the Berry's phase when the particle completes a complete loop in the closed path that was described in **E2.13**, condition (c).

References

[1] https://history.aip.org/history/exhibits/heisenberg/p08.htm.

[2] P. A. M. Dirac, The Fundamental Equations in Quantum Mechanics. *Proc. Roy. Soc.* A109, 642–653 (1925); A110, 561 (1926).

[3] P. A. M. Dirac, *The Lagrangian in Quantum Mechanics* in Physikalische Zeitschrift der Sowjetunion, Band3, Heft 1, pp. 64–72 (1933).

[4] P. C. Deshmukh, *Foundations of Classical Mechanics* (Cambridge University Press, New Delhi, 2019).

[5] R. P. Feynman and A. R. Hibbs, *Path Integral Approach to Quantum Mechanics* (McGraw-Hill, New York, 1965).

[6] A. O. Barut and S. Basri, Path Integrals and Quantum Interference. *Am. J. Phys.* 60 (10) 896–899 (1992).

[7] John S. Townsend, *A Modern Approach to Quantum Mechanics* (University Science Books, Sausalito, CA, 2010).

[8] Peter Rodgers, *The Double Slit Experiment* (PhysicsWorld, 1 September 2002).

[9] A. Tonomura, J. Endo, T. Matsuda, T. Kawasaki, and H. Ezawa, Demonstration of Single-electron Build Up of an Interference Pattern. *Am. J. Phys.* 57:2, 117–120 (1989).

[10] A. O. Barut and S. Basri, Path Integrals and Quantum Interference. *Am. J. Phys.* 60:10, 896–899 (1992).

[11] S. Pancharatnam, Generalized Theory of Interference, and Its Applications. Part I. Coherent Pencils. *Proc. Indian Acad. Sci. A.* 44:5, 247–262 (1956).

[12] Michael Victor Berry, Quantal Phase Factors Accompanying Adiabatic Changes. *Proc. Roy. Soc. A*, 392, 45–57 (1984), https://doi.org/10.1098/rspa.1984.0023.

[13] Y. Aharonov and D. Bohm, Significance of Electromagnetic Potentials in Quantum Theory. *Phys. Rev.* 115 (3) 485–491 (1959).

[14] Y. Aharonov and J. Anandan, Phase Change during a Cycle of Quantum Evolution. *Phys. Rev. Lett.* 58, 1593 (1987).

[15] D. J. Fernandez, L. M. Nietot, M. A. del Olmo and M. Santander, Aharonov-Anandan Geometric Phase for Spin-1/2 Periodic Hamiltonians. *J. Phys. A: Math. Gen.* 25, 5151 (1992).

[16] https://www.newscientist.com/article/mg20627596-600-quantum-wonders-the-field-that-isnt-there/.

[17] W. Ehrenberg, and R. E. Siday, The Refractive Index in Electron Optics and the Principles of Dynamics. *Proceedings of the Physical Society. Series B.* 62:1, 8–21 (1949).

[18] R. G. Chambers, Shift of an Electron Interference Pattern by Enclosed Magnetic Flux. *Phys. Rev. Lett.* 5, 3 (1960).

[19] A. Tonomura et al., Evidence for Aharonov-Bohm Effect with Magnetic Field Completely Shielded from Electron Wave. *Phys. Rev. Lett.* 56, 792–795 (1986).

[20] P. C. Deshmukh, Subham Ghosh, Uday Kumar, C. Hareesh, and G. Araving, A Primer on Path Integrals, Aharonov–Bohm Effect and the Geometric Phase, *The Physics Educator* 4 (1) 2250005 World Scientific Publishing Company, DOI: 10.1142/S2661339522500056 (2022).

[21] J. H. Hannay, Angle Variable Holonomy in Adiabatic Excursion of an Integrable Hamiltonian. *J. Phys. A: Math. Gen.* 18, 221 (1985).

[22] J. M. Robbins, The Hannay Angle, Thirty Years On. *J. Phys. A: Math. Theor.* 49, 431002 (3pp) (2016).

[23] Sushanta Dattagupta, Quantum Phase and Its Measurable Attributes `a la Aharonov–Bohm Effect. *Resonance* p.949, September 2018.

[24] L.-M. Duan, J. I. Cirac, and P. Zoller, Geometric Manipulation of Trapped Ions for Quantum Computation. *Science* 292:1, 1695 (2001).

[25] Vlatko Vedral, *Geometric Phases and Topological Quantum Computation* (2002). https://arxiv.org/abs/quant-ph/0212133.

[26] Rajaram Nityananda, The Interference of Polarised Light. *Resonance* p.309, April 2013.

[27] P. Bandyopadhyay, Anisotropic Spin System, Quantized Dirac Monopole and the Berry Phase. *Proc. R. Soc. A.* 467, 427–438 (2011).

[28] Daniel F. Styer, Miranda S. Balkin, Kathryn M. Becker, Matthew R. Burns, Christopher E. Dudley, Scott T. Forth, Jeremy S. Gaumer, Mark A. Kramer, David C. Oertel, Leonard H. Park, Marie T. Rinkoski, Clait T. Smith, and Timothy D. Wotherspoon, 9 Formulations of Quantum Mechanics *Amer. Jour. of Phys.* 70, 288 (2002).

3

Probability Tangles and Eigenstates of One-dimensional Potentials

Max and Heidi Born at Bengaluru, India, c. 1937 . https://en.wikipedia.org/wiki/Max_Born. Prof. B.S. Madhava Rao family archives, courtesy of Mrs Usha Ramaiah.

The belief that there is only one truth, and that oneself is in possession of it, is the root of all evil in the world.

—Max Born

In this chapter, we build on the principles of quantum mechanics introduced in the previous two chapters to apply the same to physical situations. First, we shall address simple one-dimensional physical problems. These would involve single particles interacting with a potential.

3.1 Probability Current Density

We have learned in Chapter 1 that if $|\gamma\rangle$ represents an energy eigenstate, then it is *stationary* and its time dependence is described by $\exp\left(-i\frac{E}{\hbar}t\right)$ (Eqs. 1.114a,b and Eq. 1.115). The contribution of this term to the integrand in Eq. 1.109 and in Eq. 1.111a,b is essentially unity, no matter what the value of time t is. In this case, the probability density becomes independent of time. However, in other cases, the probability density $\rho_\gamma(\vec{x},t)$, and hence the *probability* $\left|\psi_\gamma(\vec{x}_i,t)\right|^2 \delta V_i$ of finding a particle represented by the state vector $|\gamma\rangle$ within a volume element δV_i about the position $\vec{x}_i$, is time dependent. It then becomes necessary to determine

$$\frac{\partial}{\partial t}\left\{\rho_\gamma(\vec{x},t)\right\} = \frac{\partial}{\partial t}\left\{\psi_\gamma(\vec{x},t)^* \psi_\gamma(\vec{x},t)\right\}. \quad (3.1)$$

The partial derivative with respect to time of the product of the wavefunction and its complex conjugate requires $\frac{\partial}{\partial t}\psi_\gamma(\vec{x},t)$, which is available from the time-dependent Schrödinger equation. Also required is $\frac{\partial}{\partial t}\psi_\gamma(\vec{x},t)^*$, which is readily obtainable from its complex conjugation.

The result of the Solved Problem P3.1 at the end of this chapter, viz.,

$$\frac{\partial\rho(\vec{r},t)}{\partial t}+\vec{\nabla}\cdot\vec{j}(\vec{r},t)=0, \tag{3.2}$$

has exactly the same form as the *equation of continuity* in fluid dynamics and also in electrodynamics (Eq. 10.26 and Eq. 13.71c in Reference [1]). In the above equation,

$$\vec{j}(\vec{r},t)=\frac{-i\hbar}{2m}\left\{\psi^*(\vec{r},t)\vec{\nabla}\psi(\vec{r},t)-\psi(\vec{r},t)\vec{\nabla}\psi^*(\vec{r},t)\right\}=\frac{\hbar}{m}\mathrm{Im}\left\{\psi^*(\vec{r},t)\vec{\nabla}\psi(\vec{r},t)\right\}, \tag{3.3a}$$

or writing the same compactly, with the arguments implicit,

$$\vec{j}=\frac{-i\hbar}{2m}\left\{\psi^*\vec{\nabla}\psi-\psi\vec{\nabla}\psi^*\right\}=\frac{\hbar}{m}\mathrm{Im}\left\{\psi^*\vec{\nabla}\psi\right\}. \tag{3.3b}$$

In terms of the momentum operator,

$$\vec{j}=\frac{1}{2m}\left\{\psi^*\vec{p}\psi-\psi\vec{p}\psi^*\right\}=\frac{1}{m}\mathrm{Re}\left(\psi^*\vec{p}\psi\right), \tag{3.3c}$$

and it is called as the probability current density vector. Its integral over the whole space gives us

$$\iiint_{\text{whole space}}\vec{j}(\vec{r},t)dV=\frac{1}{m}\iiint_{\text{whole space}}\left\{\psi^*(\vec{r},t)\vec{p}\psi(\vec{r},t)\right\}dV=\frac{\langle\vec{p}\rangle_t}{m}, \tag{3.4}$$

where $\langle\vec{p}\rangle_t$ is the expectation value of the momentum operator at the instant t. The dimensions of the probability current density vector are $L^{-2}T^{-1}$. Notwithstanding the above relations, the probability current density vector at a point is not directly measurable. It would not be appropriate to equate it to the average particle flux at a point at a specific instant of time, though such an interpretation can sometimes be useful.

An electric current involves the motion of charges. Charge, however, cannot be considered localized as modeled in classical physics. Instead, it must be considered spread out in space according to a quantum statistical distribution. We are therefore interested in determining how the probability density of an electric charge would vary with time. We recall (from Eq. 2.66) that the kinetic momentum $(\vec{p}_k)$ is related to the canonically conjugate generalized momentum $(\vec{p}_c)$ by

$$\vec{p}_k=m\vec{v}=\vec{p}_c-\frac{e}{c}\vec{A}, \tag{3.5a}$$

and the corresponding Hamiltonian for a charged particle in an electromagnetic field is

$$H=\frac{\left[\vec{p}_c-\frac{e}{c}\vec{A}\right]^2}{2m}+e\phi\ \underset{\text{operator}}{\overset{\text{quantization}}{\rightarrow}}\ \frac{\left[\left(-i\hbar\vec{\nabla}\right)-\frac{e}{c}\vec{A}\right]^2}{2m}+e\phi. \tag{3.5b}$$

It is then prudent to define a *charge density* that is proportional to the probability density (Eq. 3.1) as

$$\rho_\gamma^{(e)}(\vec{x},t) = e\left|\psi_\gamma(\vec{x},t)\right|^2. \tag{3.6}$$

Variation in charge density is akin to a flow of charges, thereby constituting an electric current. These relations between Max Born's probability density and the quantum current density vector therefore represent a conservation principle *aka* Gauss' law (see, for example, Chapter 10 in Reference [4]). It is remarkable that identical expressions appear in classical fluid dynamics, electrodynamics, and quantum mechanics. In this chapter, we shall find interesting engineering and technology applications of quantum probability, probability current, and the equation of continuity. Physical phenomena in nature, for example the working of a scanning tunneling microscope are mind-boggling but are accounted for by quantum mechanics. The idea of a delocalized particle represented by a wavefunction, whose modulus square at a point represents the probability *density* of finding it there, is alien to classical theory. Only a measurement produces a definite result into which the system collapses. This result is one of the eigenstates of the position operator. The probabilistic wavefunction dictates the *chance* that a measurement would produce a particular result.

The already mysterious probabilistic interpretation gets even more bizarre in the consideration of superposition of wavefunction: If $\psi_1(\vec{r},t)$ and $\psi_2(\vec{r},t)$ are two eigenstates of a system, then the system may also be in a state of their linear superposition

$$\psi(\vec{r},t) = c_1\psi_1(\vec{r},t) + c_2\psi_2(\vec{r},t), \tag{3.7}$$

where c_1 and c_2 are complex numbers in general. Essentially, this superposition means that only a measurement can tell if the system is in the state with subscript "1" or "2." The result could be either of these, but nothing else. The probability that the measurement would result in state "1" is $\left|c_1\right|^2$, and in state "2" it is $\left|c_2\right|^2$. The sum of the two probabilities is equal to unity for *normalized* $\psi(\vec{r},t)$. The corresponding probability density is

$$\psi^*(\vec{r},t)\psi(\vec{r},t) = \left\{c_1{}^*\psi_1{}^*(\vec{r},t) + c_2{}^*\psi_2{}^*(\vec{r},t)\right\}\left\{c_1\psi_1(\vec{r},t) + c_2\psi_2(\vec{r},t)\right\}, \tag{3.8a}$$

$$\text{i.e., } \psi^*\psi = \left|c_1\right|^2\left|\psi_1\right|^2 + 2\,\mathrm{Re}\left(c_1{}^*c_2\psi_1{}^*\psi_2\right) + \left|c_2\right|^2\left|\psi_2\right|^2. \tag{3.8b}$$

The first and the last term in Eq. 3.8b constitute the *sum* of two probability densities I_1 and I_2. The middle term in Eq. 3.8b has no analogue in classical physics. It arises due to interference, described only by quantum mechanics. It involves not only the product of coefficients of the two wavefunctions but also that of the two complex wavefunctions themselves. The interference term contributes at locations only where overlap of the two functions occurs. The sum of the three terms can be both greater than or less than $I_1 + I_2$ by an amount given by $\left|2\,\mathrm{Re}\left(c_1{}^*c_2\psi_1{}^*\psi_2\right)\right|$. We discussed interference in Chapter 2 in the case of Young's double-slit experiment, with the slits σ and ξ providing two alternative states from paths '1' and '2'. This is already an astonishing result having no classical analogue. What makes the working of nature an even deeper mystery is that the interference vanishes if we add an auxiliary experiment that would determine if the system is in state "1" or "2." The auxiliary experiment is a *measurement* on the system because of which the system collapses into one of the eigenstates. Such a measurement destroys the interference, but prior to the measurement, the system can only be described by a wavefunction that must include superposition of *all* the eigenstates. How nature determines whether a measurement is arranged so that interference would be destroyed is a controvertible question. Need we personify nature? Our goal is to *describe* nature; quantum theory does so correctly, classical physics does not. We do *not* ask *why* the laws of nature are what they are. Rather, in all modesty, we seek to know just *what* the laws of nature are. We want to learn *how* to *describe* the state of a system and *how* it evolves with time. If we answer this correctly, we would "understand" nature at least to the extent we possibly could. From this discernment, we not merely gratify our academic curiosity but also use that knowledge to engineer new devices and advance technology.

The alternative slits σ or ξ that we discussed in Chapter 2, or the state "1" or "2" of Eq. 3.7, are mutually exclusive in classical physics. Nature is, however, described more appropriately by considering them as inclusive possibilities, with only a measurement inducing a result. Superposition of the two alternatives is representative of a general principle in quantum theory: Until a measurement is made, a system must be described by a superposition that is *inclusive* of all possible outcomes of that measurement. The number of possibilities could be many, say N, which may be finite or infinite, and also discrete or continuous. For example, a function $\psi_\gamma(\vec{r}) = \langle \vec{r}|\gamma\rangle$ requires for its description a *superposition* of an infinite number of momentum eigenfunctions:

$$\psi_\gamma(\vec{r}) = \langle \vec{r}|\gamma\rangle = \iiint_{\text{ws}} d^3\vec{p}\langle \vec{r}|\vec{p}\rangle\langle \vec{p}|\gamma\rangle = \frac{1}{(2\pi\hbar)^{\frac{3}{2}}}\iiint_{\text{ws}} d^3\vec{p}\langle \vec{p}|\gamma\rangle e^{\frac{i}{\hbar}\vec{p}\cdot\vec{r}}$$

$$= \frac{1}{(2\pi\hbar)^{\frac{3}{2}}}\iiint_{\text{ws}} d^3\vec{p}\tilde{\psi}_\gamma(\vec{p}) e^{\frac{i}{\hbar}\vec{p}\cdot\vec{r}}. \tag{3.9a}$$

This integration is over the *w*hole of the momentum *s*pace, which is the space reciprocal (in units of $\hbar$ since $p = \hbar k = \hbar\frac{1}{\lambda}$) to the usual Euclidean three-dimensional space we are used to. The function $\tilde{\psi}_\gamma(\vec{p})$ that appears in the integrand of Eq. 3.9 is the Fourier transform of $\psi(\vec{r})$. Reciprocally, the momentum space wavefunction $\tilde{\psi}_\gamma(\vec{p})$ is

$$\tilde{\psi}_\gamma(\vec{p}) = \langle \vec{p}|\gamma\rangle = \iiint_{\text{ws}} d^3\vec{r}\langle \vec{p}|\vec{r}\rangle\langle \vec{r}|\gamma\rangle = \frac{1}{(2\pi\hbar)^{\frac{3}{2}}}\iiint_{\text{ws}} d^3\vec{r}\langle \vec{r}|\gamma\rangle e^{-\frac{i}{\hbar}\vec{p}\cdot\vec{r}}$$

$$= \frac{1}{(2\pi\hbar)^{\frac{3}{2}}}\iiint_{\text{ws}} d^3\vec{r}\psi_\gamma(\vec{r}) e^{-\frac{i}{\hbar}\vec{p}\cdot\vec{r}}. \tag{3.9b}$$

The localization of the system at the position $\vec{r}$ represented in Eq. 3.9a requires a superposition of all possible momentum states, since the integration is over the *w*hole of the momentum *s*pace. This is strictly in line with the principle of uncertainty. If the superposition (i.e., integration) is limited to a subset, the localization at $\vec{r}$ is compromised, which results in an uncertainty in position. Likewise, the localization in momentum space at $\vec{p}$ requires a superposition of <u>all</u> position states, as indicated in Eq. 3.9b.

If the whole space can, however, be factored into a part that is accessible to the physical system under consideration, separated from the rest by impenetrable barriers, then the position space gets bounded to a finite region. The superposition must exclude the inaccessible position eigenfunctions, thereby diminishing the position uncertainty. This would result in increasing the uncertainty in momentum. The momentum would then not be defined sharply as in Eq. 3.9b. Instead, we would need a wavepacket in the momentum space resulting from superposition of a number of momentum eigenfunctions. Mutual uncertainty is built into the description of $\tilde{\psi}_\gamma(\vec{p})$ and $\psi_\gamma(\vec{r})$, which are the Fourier transforms of each other (Eq. 1.132). Eq. 1.106 provides a precise relationship of this uncertainty, since position and momentum measurements are not compatible with each other. Consequences of this relation go *way beyond* the classical property (Eq. 1.136) of Fourier transforms.

Next, we discuss bizarre consequences of a measurement resulting in the collapse of a superposition state into an eigenstate. A "which way (WW) experiment" described here illustrates this collapse. It would highlight the quantum signature of physical laws. A further improvisation of this experiment, called "delayed choice which way experiment" would provide additional revelations of nature's strange ways. These experiments leave no suspicion that the *classical* reciprocal relationship exhibited by Fourier transforms is grossly inadequate to describe nature.

The principle of superposition, whether it involves a finite number of states (as in Eq. 3.7) or an infinite number of states (as in Eq. 3.9), is one of the prime signatures of quantum mechanics. Built into it is a counterintuitive description of the natural state of a system to consist of an *inclusion of all alternatives*, such as "1" and "2" in Eq. 3.7. Classical physics regards these as strictly exclusive; either of these excludes the other. Feynman's sum over histories is an equivalent statement of quantum mechanics, as we have reasoned in Chapter 2; it requires sum over all histories. Nature is correctly described by the quantum description, of which the principle of superposition is a characteristic expression. In every physical situation that provides for alternative pathways, the outcome of the interference term in Eq. 3.8 arises. The phenomenology that we discuss here is characteristic of any situation in which alternatives "1" and "2" are available; the double-slit experiment represents *all* of such situations. An example of a very different situation to which the analysis that follows would apply equally well is that of an excited atom that can possibly return to its ground state either via a radiative transition or via a nonradiative transition. Experimental observations on such a system can be understood only in terms of interference between the two alternative pathways.

From the classical point of view, this is counterintuitive. What makes it more mysterious is that the interference disappears when a measurement is made to determine which of the available alternatives is chosen by nature. All that is needed for the system to collapse into an eigenstate from the superposition is the mere *provision* in the physical apparatus of a any tangible contrivance that has the capacity to *determine* the alternative. Such a provision may be deliberately installed, or it may be present inadvertently. The WW experiment touched upon in Chapter 2 in the context of Young's double-slit experiment exemplifies this facet. The interference appears even when the particle density is as low as having only a single particle in the apparatus at a time [2, 3]. We shall now discuss this in further detail.

Figure 3.1 is a schematic diagram of the interference fringes obtained in Young's double-slit experiment. It is not a snapshot of an actual experiment; hence, the scale, size, brightness, etc., of the fringes would not be of any concern for our discussion. Our focus is on the principle of superposition that is behind the appearance of constructive and destructive interference (Eq. 3.13). In Figs. 3.2a and 3.2b, we present a slight improvisation of the double-slit experiment. In the experiments described by these figures, we incorporate a "which path marker." The two states "1" and "2" in Eq. 3.13 are represented by the two slits in this figure. In Figure 3.2a there are two WW markers, viz., (i) shutter and (ii) polarizer. The shutter is a WW marker. When either of the two shutters is closed, the only path that remains accessible

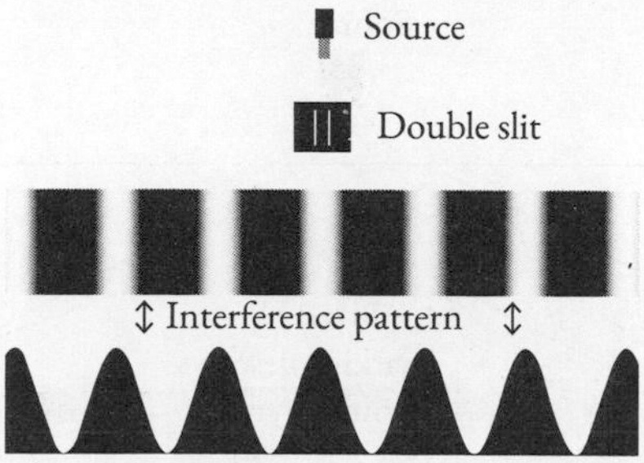

Fig. 3.1 Schematic representation of interference between alternatives resulting in bright and dark fringes.

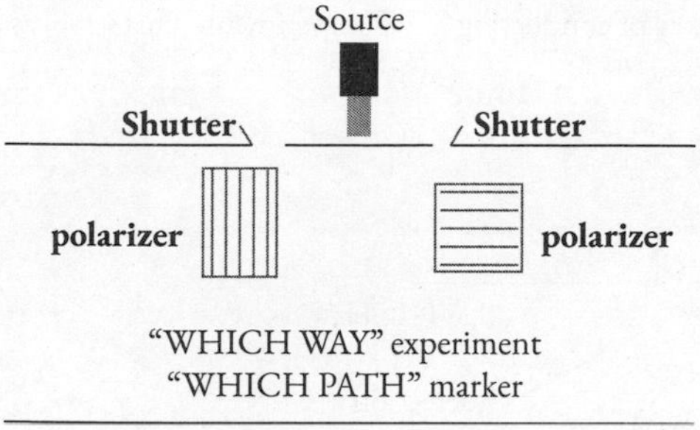

Fig. 3.2a A WW marker destroys interference. The detector screen is represented in this figure by the straight line at the bottom.

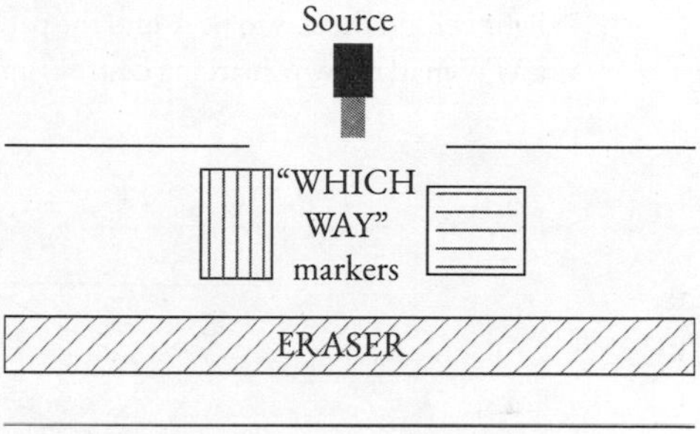

Fig. 3.2b In this arrangement, an additional polarizer is placed in the apparatus whose polarization axis is at 45° angle with respect to that of the filter. This additional polarizer erases the WW information and restores interference.

is the one that is open, thus marking the WW. Similarly, polarizers placed behind the slits in Fig. 3.2a act as filters. Thereby they serve as WW markers. Polarizable particles pass through them only if their polarization is along the axis of polarization of the filter.

Figure 3.2 brings out an astonishing quality of natural phenomena. This figure is essentially schematic, but it shows that an eraser restores interference. To the uninitiated mind, this is bizarre, since erasing occurs well "after" (from the point of view of perceptions nurtured by classical physics) the particle(s) has/have entered through either/both slits. The arrangement in Fig. 3.3 shows another way of conducting the WW experiment. Figure 3.3a shows that WW information can be obtained by shining light on the slits to determine WW the particles traversed. Figure 3.3b shows the insertion of an eraser of that information. The lens recombines the rays scattered from the two slits thereby destroying the WW information. While determination of WW information destroys interference, erasing that information restores it. What makes this even more abstruse is the fact that the eraser restores interference even when placed very far – even kilometers or light years away. This is unfathomable from the viewpoints of classical causality, which has us believe that an eraser placed far away cannot affect the intensity distribution pattern on the detector.

The experimental arrangements in Figs. 3.2 and 3.3 are referred to as WW experiments. A WW experiment destroys interference. In Figs. 3.2b and 3.3b, the quantum eraser destroys information obtained in a WW experiment. The eraser restores the interference. The arrangement in Fig. 3.3c is called as *delayed choice* WW experiment. The eraser restores

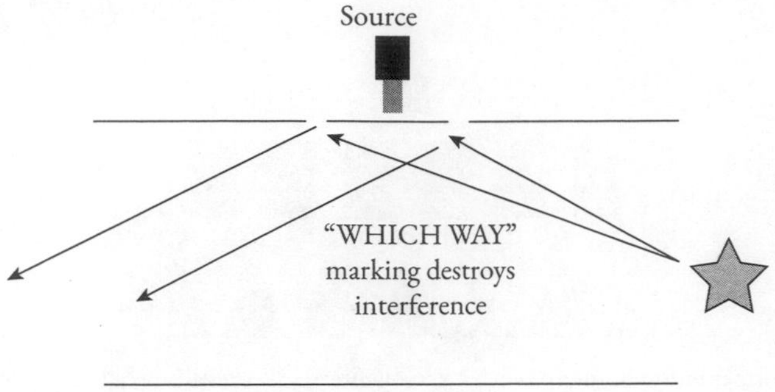

Fig. 3.3a As in Fig. 3.2, the detector screen is represented in this figure by the straight line at the bottom. Shining light behind the slits to peek into the path is another way of conducting a WW experiment. This serves as a WW marker. WW marking destroys interference.

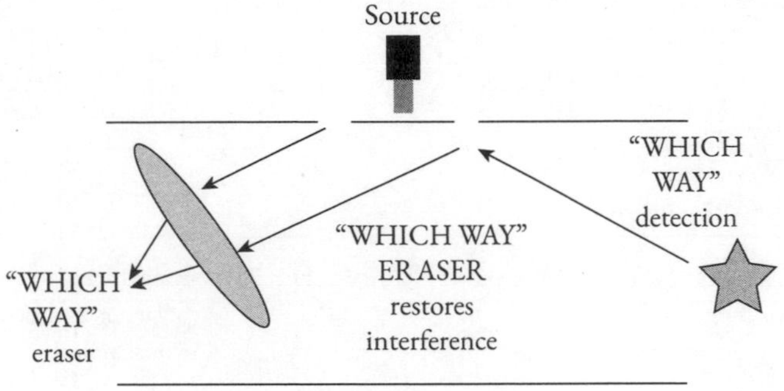

Fig. 3.3b A lens would recombine the scattered light from the two slits and erase the WW information. The quantum eraser restores interference.

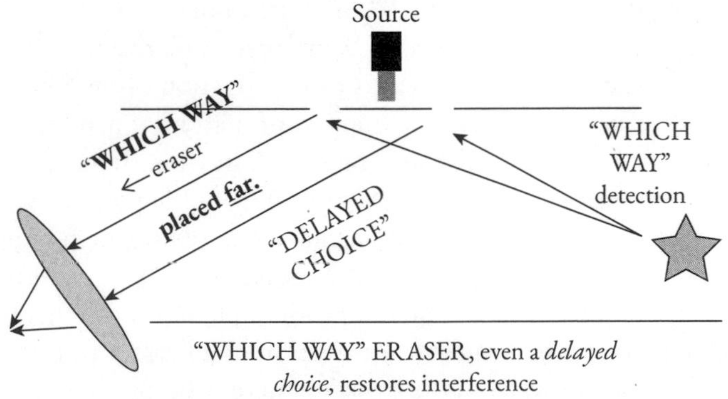

Fig. 3.3c In this setup, the eraser is placed far away. However, regardless of where the eraser is placed, close by or far away, the mere provision of an eraser restores interference.

interference even when placed very far. The provision of marker and eraser determines the intensity distribution on the detector screen. A marker destroys interference fringes, and the eraser restores them. The experimental situations described in Fig. 3.1, 3.2a,b, and 3.3a,b,c are essentially schematic, but such experiments have been actually performed in a laboratory [4, 5], with the outcome described above. No matter how bizarre the principle of superposition and the role of measurement are from the point of view of classical physics, interpretation of these experiments from the standpoint of quantum physics is straightforward. The mere provision of WW *markers* necessitates *expansion of the Hilbert space* to include the *marker states*. Denoting the alternative paths by $|1\rangle$ and $|2\rangle$ and the marker states by $|M_1\rangle$ and $|M_2\rangle$, the system's state vector in the expanded Hilbert space is

$$|\psi\rangle = \frac{1}{\sqrt{2}}|1\rangle|M_1\rangle + \frac{1}{\sqrt{2}}|2\rangle|M_2\rangle. \tag{3.10}$$

The marker states $|M_1\rangle$ and $|M_2\rangle$ are exclusive; they are orthogonal:

$$\langle M_2|M_1\rangle = 0. \tag{3.11}$$

Due to this orthogonality, just the availability of the "WW" marker in the apparatus destroys the interference. This is independent of how close or how far the markers are located. Likewise, regardless of how close or how far the eraser is located, it eliminates the marker states from the Hilbert space and restores interference. *Superposition of Hilbert space vectors must include marker states when present, unless an eraser annuls their consequences.* This principle fittingly accounts for all similar natural phenomena. Interference ensues when both alternatives are accessible, but a WW measurement/observation renders the system's collapse into one of the eigenstates and eliminates interference. This is extremely satisfying from the point of view of having a good theory of nature. After all, a theory of the physical universe seeks to account for what the laws of nature are, and how the physical states evolve with time. The apparent surreptitious mechanism of information exchange between the markers and the slit(s) supposedly *after* the particles have passed through the slits is understood if we are cautious in how we pose our inquiry. Our perception of what happened *before* and *after* the particles pass through the slits may well be totally off the track. The slits and the markers *together* constitute the system *at each instant of time*. This requires for its description an expanded Hilbert space (see Eq. 3.10). The temporal sequencing of events as we may have in our minds therefore simply does not apply. An inquiry on what mechanism enables communication between the slits and the markers is not the right question to be asked, as Bohr might have said. The alternatives in a superposition are intrinsically *entangled*. They remain so until an observation is made on the system at which instant the system collapses into one of the eigenstates. Such a description of nature is clearly beyond classical physics. Quantum laws of nature come across as counterintuitive and stealthy only because we try to interpret nature from the viewpoint of classical physics. One can only expect therefore that quantum theory was hotly debated, even as it continued to succeed. In Chapter 11, we shall peek into the debate, and its resolution, between two of quantum theory's greatest creators, Niels Bohr and Albert Einstein. In the next section, we shall see that the probabilistic interpretation of the wavefunction, no matter how weird it seems, has enabled physicists, material scientists, and engineers to fabricate devices with incredible capabilities for advances in technology.

3.2 Tunneling: *Over*-the-Barrier and *Under*-the-Barrier

Even though disagreements on fundamental principles continued (and continue), quantum theory has kept getting firmly entrenched as the theory of nature, supported by the proverbial precept that *nothing succeeds like success*. An early triumph of quantum theory was the explanation of transmission of matter (*matter-waves*) past potential barriers in regions considered energetically forbidden in classical physics. Experimental verification of these phenomena enabled engineering of technological devices for societal benefits. In this section, we shall discuss how Max Born's probabilistic interpretation of the wavefunction (Eq. 1.111 and Eq. 1.112) and that of the current density vector (Eq. 3.2) expound travel past obstacles that are classically baffling. We shall first describe rather unpretentious *one-dimensional* posers, which deliver unimaginable revelations. Notwithstanding their simplicity, surprising applications of one-dimensional problems are found in a wide range of phenomena such as the understanding of stellar nucleo-synthesis, radioactivity, and field emission (also called *Fowler–Nordheim tunneling*). The physical situations we now address come under the general class of *quantum tunneling*.

We have seen in Chapter 2 (specifically, in the discussion of Fig. 2.4) that classical paths correspond to those for which action is much larger than the PEB constant. It is therefore not surprising that the Schrödinger equation reduces to the Hamilton–Jacobi equation of classical physics in the limit $h \to 0$. This is is discussed in Section 6.1 of Chapter 6. There is of course no mechanism to seek the reverse; i.e., there is no way we can get a quantum feature from any kind of *extrapolation* of classical physics. Tunneling is an intriguing phenomenology in nature, but the term *tunneling* now enters a new domain where its meaning has no classical analogue. When a tunnel is dug into a mountain for a motor road or a train track to pass, classical physics completely accounts for all the associated physical conditions. Nature, however, has a counterintuitive property, which requires quantum theory. In our discussion of these properties, absolutely *nothing* would be *scooped* out of the barrier ("mountain", in our example). We shall, however, continue to use the term "tunneling" to describe the passage of matter across a potential barrier. The term "barrier" is strictly inappropriate; it is only a pointer to how it is described in classical physics. In the actual physical universe, there are no finite sized barriers that can confine an object to a limited region of space. Concomitant nonclassical behavior includes *reflection* of a particle by an obstacle even if the energy of the particle is *greater* than the height of an obstruction. Max Born's interpretation of the wavefunction accounts for tunneling in a straightforward manner, described below.

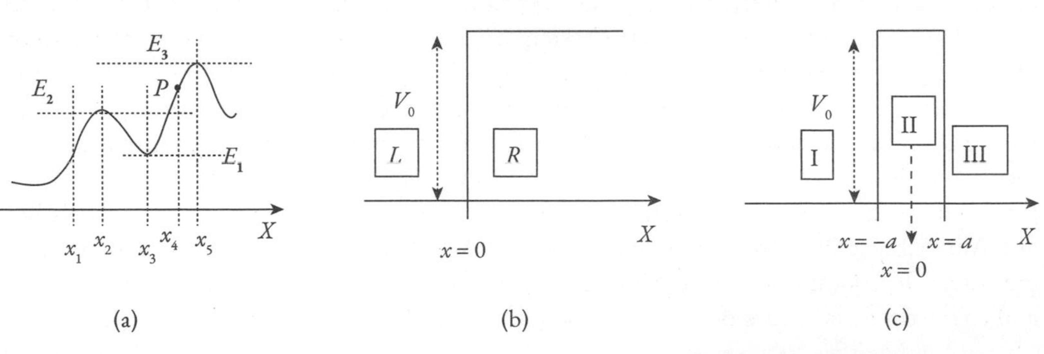

Fig. 3.4 The analysis of one-dimensional motion along the x-axis is different in classical mechanics and in quantum mechanics. (a) A particle on a one-dimensional roller coaster. (b) A step potential. The potential is zero for $x < 0$ and it is V_0 for $x > 0$, all the way as $x \to \infty$. (c) A potential that is zero for $x < -a$ (region I), V_0 in region between $x = -a$ and $x = +a$ (region II), and zero again in region $x > a$ (region III).

First, we consider one-dimensional motion on a roller coaster (Fig. 3.4a) from the perspective of classical physics. We shall regard the roller coaster to be frictionless. A particle

P at x_4 moving to the left would certainly overcome the potential hump at x_2 and keep moving left, toward x_1. Newtonian physics does not consider the possibility that this particle would reverse the direction of its motion at x_2, and on taking a U-turn begin to move to its right. Likewise, if this particle has energy at x_4 that is *less* than E_3 and happens to be moving toward its right, classical mechanics would not expect it to go over the hump at x_5 and continue its rightward motion. However, conclusions we arrive at on applying conservation of energy using classical mechanics are challenged by nature. Quantum mechanics offers a straightforward explanation of the nonclassical behavior. Below, we illustrate how quantum mechanics accounts for natural phenomena by considering two simple one-dimensional physical potentials. These are schematically shown in Fig. 3.4b, which is a step potential, and in Fig. 3.4c, the rectangular potential barrier.

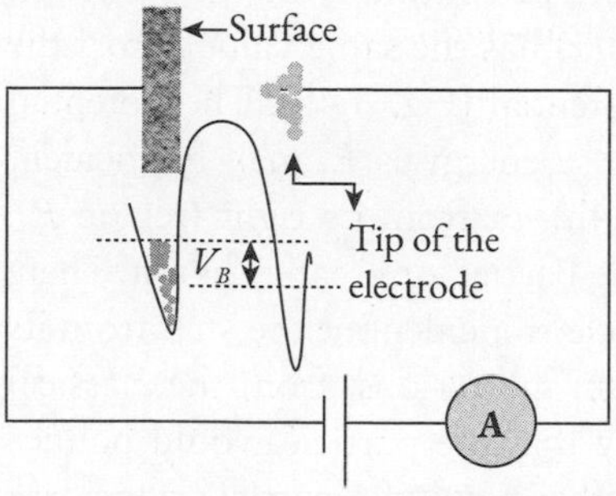

Fig. 3.5a The inset of this figure shows a schematic representation of the potential function.

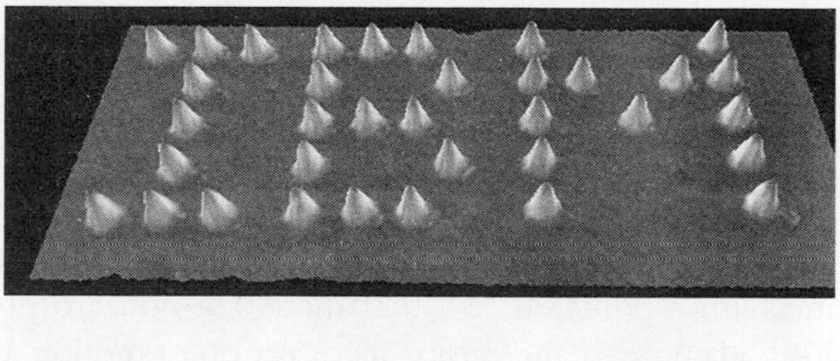

Fig. 3.5b An early example of the control on repositioning of atoms on a surface by moving around the microtip electrode is shown here. A team of researchers [9] in 1989 moved around 35 atoms of xenon to spell "IBM". Reprint courtesy of IBM Corporation ©.

A physical realization of the potential barrier (Fig. 3.4c) is illustrated in the electrical dc circuit shown in Fig. 3.5a. For the electrical circuit to complete, electrons would need to flow through vacuum between the surface and the microtip of the electrode shown schematically in this figure. The microtip shown is grossly magnified; it is actually so tiny that it tapers down to atomic size. The vacuum between the surface and the tip represents a huge potential barrier that must be overcome by the electrons to complete the circuit. The inset of the figure is a schematic display of this potential barrier. Electrons (represented in this inset figure by tiny filled black circles) on the surface must overcome the potential barrier to go across a stretch of vacuum to flow into the microtip of the electrode to complete the dc circuit. A bias voltage V_B is applied to control the current, but there is no mechanism we know from classical physics that would allow the electrons to move across the potential barrier raised by the gap between the surface and the microtip electrode. Nature, however, allows a current to flow; it is measured in an ammeter. This unexpected phenomenon is explained by quantum theory. By moving the microtip across the surface to be studied, one can probe the detailed physical structure of the surface at an atomic level. This is the principle behind the scanning tunneling microscope, commonly abbreviated as STM [8]. The atoms on the surface can even be repositioned by controlling the movement of the microtip electrode. The STM was invented by Gerd Binnig and Heinrich Rohrer for which they were awarded jointly the 1986

Nobel Prize in physics. A team of researchers in 1989 succeeded in moving around 35 atoms of xenon [9] on a substrate of nickel crystal to spell "IBM" (Fig. 3.5b). The fundamental principles involved in such processes were at the heart of Feynman's visionary talk "Plenty of Room at the Bottom" to the American Physical Society on December 29, 1959. As is history now, this talk triggered the development of nanoscience and nanotechnology.

Application of the Schrödinger equation to one-dimensional problems (Figs. 3.4 and 3.5) would explain the functioning of an STM. From Eq. 3.2 we know that for stationary states, the equation of continuity is a statement of conservation of "flux." For $\vec{j}$ to be solenoidal, the wavefunction and its gradient must be continuous, i.e., both $\psi(x)$ and $\psi'(x)=\frac{d\psi}{dx}$ are continuous at the boundaries of regions where the potential changes, such as at the step (Fig. 3.4b), or at the edges of a rectangular barrier (Fig. 3.4c). The continuity of wavefunction and that of its first derivative with respect to x ensures that $j(x)$ has the same value *across* the edge(s) of the potential. First, we shall consider the step potential (Fig. 3.4b). The potential step is located at $x = 0$ and its height is V_0. A particle whose energy is E, may be incident at the potential step either from its left (region L in Fig. 3.4b), or from its right (region R). The energy E of the particle may be less than the height of the barrier, or greater than it. There are thus different physical situations to consider. If the particle is incident at the step from its *right*, arriving at the step with energy $E > V_0$, it would seem mysterious, from the classical mechanics point of view, that there is a nonzero probability that the particle would bounce back at the step and *reverse* its direction of motion. On the other hand, if a particle is incident on the potential step from region L, moving rightward with energy $E < V_0$, it would be equally mysterious if there would be any nonzero probability that the particle would be found in region R. The predictions of classical mechanics are, however, not correct. Nature is best described by quantum mechanics. To understand nature's surprises, we must use Schrödinger's equation. (Eq. 1.140), with the Hamiltonian consisting of the sum of the kinetic energy operator and the potential energy operator, as appropriate for the regions L and R in the present situation. Clearly, in the region L, the Hamiltonian H consists of only the kinetic energy operator, but in the region R, H has its sum with the potential energy operator.

Let us consider a particle to be incident on the potential step from region L, moving rightward with energy $E < V_0$. It is straightforward to see that the potential being different in regions L and R, the most *general* solution for the two regions of the Schrödinger equation is

$$\text{for region } \boxed{L} \quad \psi_{Left}(x,t) = e^{i(k_0x-\omega t)} + \text{R}e^{-i(k_0x+\omega t)}, \tag{3.12a}$$

$$\text{and for region } \boxed{R} \quad \psi_{Right}(x,t) = Te^{i(kx-\omega t)}, \tag{3.12b}$$

where k_0 and k are defined by the following relations:

$$\boxed{L} \; E = \frac{\hbar^2 k_0^{\,2}}{2m}, \tag{3.13a}$$

$$\boxed{R} \; E - V_0 = \frac{\hbar^2 k^2}{2m}. \tag{3.13b}$$

In the region L, which is the one on the *left* side of the step potential, the solution is a superposition of a wave traveling toward the right, and another toward the left, the latter with a relative *coefficient* R (for 'reflection'). The use of the coefficient R in Eq. 3.12a to denote '*reflection*', and also to denote the 'right side of the barrier' surely would cause no confusion; the context defines it clearly. In the *region R*, which is to the *right* side of the step potential, we have only a wave moving *toward the right*, as there is no possible interaction in that region which can reverse the particle's motion. This component is the one that is transmitted across the barrier, and hence weighted by a coefficient T (for '*transmission*'). The coefficients R and T can be determined from the boundary conditions imposed by the equation of continuity, which requires the value of $j(x)$ determined from the solutions (Eqs. 3.12a,b) in regions L and R to be equal. As we have already seen, this condition is satisfied by requiring the wavefunction and its derivative to be both continuous at the boundary of the potential, which in this case is at $x = 0$. Furthermore, since $E < V_0$, k must be imaginary, and we write it as

$$k = i\kappa, \tag{3.14}$$

where κ is real and positive.

The time-independent solution in region R therefore is

$$\psi_{Right}(x) = Te^{-\kappa x}. \tag{3.15}$$

The solutions have been written inclusive of the time-dependent *dynamical phase* (Eq. 1.114; Eq. 1.115) in which $\omega = \frac{E}{\hbar}$. The solutions can be written as a superposition of any basis pair of linearly independent functions. Using the base pair of traveling waves, one moving toward right and the other moving toward left, is particularly insightful, as we shall see shortly. Of course, any other linearly independent basis pair of functions, such as odd and even eigenfunctions of parity (e.g., sine and cosine functions), can also be used.

As noted above, the Eq. 3.12b has only a wave traveling toward the right. From the continuity of wavefunction (boundary condition "1," which we shall refer to as B.C. 1) and that of its derivative (which we shall refer to as B.C. 2) at $x = 0$, we see that

$$\boxed{\text{B.C. 1}}\quad \psi_I(x \to 0_-) = \psi_{II}(x \to 0_+) \Rightarrow 1 + R = T, \tag{3.16a}$$

and $$\boxed{\text{B.C. 2}}\quad \psi_I'(x \to 0_-) = \psi_{II}'(x \to 0_+) \Rightarrow ik_0(1 - R) = -\kappa T. \tag{3.16b}$$

It immediately follows that the coefficients of the traveling waves are

$$T = \frac{2k_0}{k_0 + i\kappa} \tag{3.17a}$$

and $$R = \frac{k_0 - i\kappa}{k_0 + i\kappa}. \tag{3.17b}$$

When $E \ll V_0$, κ is large and $R \to -1$ and the wavefunction goes to zero at the step.

We leave it to the reader to work out an exercise to determine the coefficient R when a particle with $E > V_0$ is incident on the step potential from the region L. The result is

$$R = \frac{\sqrt{E} - \sqrt{E - V_0}}{\sqrt{E} + \sqrt{E - V_0}} = \frac{\left(\sqrt{\frac{E}{E - V_0}}\right) - 1}{\left(\sqrt{\frac{E}{E - V_0}}\right) + 1}. \tag{3.18}$$

It approaches the value "0," which we get from the (*incorrect*) classical theory, but only in the limit $E \gg V_0$. Occurrence of such phenomena in nature is a massive endorsement of the inapplicability of classical physics, and of the validity of quantum theory. This phenomenon is called *over-the-barrier reflection*. It is interpreted as tunneling in the momentum space [6, 7]. The result in Eq. 3.18 is akin to that for the classical optical reflection coefficient when the refraction index μ changes *sharply*:

$$R = \frac{\mu - 1}{\mu + 1}. \tag{3.19a}$$

Therefore, we have the correspondence

$$\mu = \sqrt{\frac{E}{E - V_0}}. \tag{3.19b}$$

Going back to the case when the particle incident on the step potential *from its left* has energy E *less* than the hight of the step potential, we see that the current density J in the region L can be written as the sum of an ingoing (moving rightward, toward the step) and an outgoing (moving leftward, away from the step) components. From Eq. 3.3, we see that

$$J_x^{L\overset{(\text{incoming})}{\rightarrow}} = \left(\frac{\hbar k_0}{m}\right), \tag{3.20a}$$

$$J_x^{L\overset{(\text{reflected})}{\leftarrow}} = \left(\frac{\hbar k_0}{m}\right)|R|^2 = \left(\frac{\hbar k_0}{m}\right), \text{ since } |R|^2 = 1, \tag{3.20b}$$

$$\text{and } J_x^{R(\text{transmitted})} = 0. \tag{3.20c}$$

Essentially, there is no current in the region L, since the incident and reflected components cancel each other exactly, nor is there any current in the region R. After all, the current density vector is solenoidal for stationary states. The nonzero *probability density* in the region R represents an *evanescent* wave in that region whose amplitude decays exponentially (Eq. 3.15). It is not a worry, since the current is zero in that region; reflection being total in region L:

$$\left|J_x^{L\overset{(\text{incoming})}{\rightarrow}}\right| = \left|J_x^{L\overset{(\text{reflected})}{\leftarrow}}\right|. \tag{3.21}$$

The term *evanescent* has a perplexing allusion; it acknowledges the queer presence of a wave that does not even exist. To decrypt this peculiar condition, we recall that *transport of energy* through a region of space is an essential property that a wave has (see, for example, section 4.3 in [1]). There is certainly nonzero amplitude of the wavefunction (Eq. 3.15) in region R, but the entire flux is reflected back (Eq. 3.20a,b,c) into the region L. This is a characteristic property of all wave phenomena. It is well known for electromagnetic waves.

An interesting situation develops when the raised step in Fig. 3.4b does not extend all the way to $x \to \infty$. Figure 3.4c depicts such a scenario. It shows a potential that drops to zero after a certain width, say 2a. Such a potential is called as rectangular barrier, because in classical physics it would be concluded that the potential poses an obstruction to motion of a particle across it. The width 2a being finite, it is not large enough to provide for the exponentially decaying wavefunction (Eq. 3.15) to vanish. Application of the equation of continuity at the boundary of the potential at $x = a$ would require the continuity of wavefunction ψ and its derivative $\psi' = \frac{d\psi}{dx}$. These boundary conditions produce an unforeseen outcome. We had seen that the current in both regions L and R across the *step potential* vanishes. It is zero everywhere. The rectangular barrier with finite width, however, produces a nonzero current everywhere even for particles having energy that is *less* than the height V_0 of the barrier. This phenomenon is called *tunneling*, even if nothing is physically *scooped out* as in a classical tunnel for vehicular tracks through mountain tunnels. The appearance of nonzero current across a barrier has no classical analogue, and the term "tunneling" must be understood essentially in a nonclassical sense. Equally startling is the appearance of a *reflected* current although the energy of a particle that is incident at the barrier is *greater* than the height of the barrier. We have been nonplussed earlier once before by a similar situation - while dealing with the step potential. The end result for the current is independent of direction of incidence, whether from the left or the right side of the step potential. Independence with respect to the direction of incidence from either side of the rectangular barrier persists. Specifically, we have

$$\left| J^{\begin{pmatrix}\text{step}\\ \text{potential}\end{pmatrix}} \right|_{\substack{\text{everywhere,}\\ \text{both in regions L \& R}}} = 0, \tag{3.22a}$$

and
$$\left| J^{\begin{pmatrix}\text{rectangular}\\ \text{barrier}\end{pmatrix}} \right|_{\substack{\text{everywhere,}\\ \text{in each region I, II, and III.}}} \neq 0. \tag{3.22b}$$

Analysis of the rectangular potential barrier proceeds the same way as for the step potential. Again, any basis pair of functions can be used to express the most general solution of the second-order differential equation, but employing the basis of traveling waves is particularly useful. The fact that the probability current density vector is solenoidal in the case of stationary states goes along with the continuity (at the positions of discontinuity of the potential) of the wavefunction and its spatial gradient. This property requires relationships between the coefficients of waves traveling to the left and to the right. These relations would give us, for the rectangular potential barrier, the reflection and the transmission coefficients (similar to Eqs. 3.16 and 3.17).

In the schematic representation of the one-dimensional rectangular potential barrier (Fig. 3.4, right-most panel), we shall consider a particle to be incident on the barrier from region I, traveling left to right. This corresponds to the *boundary condition* $G = 0$ (Eq. 3.23c) as there can be no traveling wave moving from right to left in region III. In regions I and III, the entire energy of the particle is essentially kinetic, since the potential in these regions is zero. Solution to the Schrödinger equation (Eq. 1.140) therefore is given below by Eq. 3.23a, Eq. 3.23b, and Eq. 3.23c, respectively, for the three regions I, II, and III:

$$\text{Region } \boxed{\text{I}} \quad \psi_{\text{I}}(x,t) = e^{i(kx-\omega t)} + Re^{-i(kx+\omega t)}, \tag{3.23a}$$

$$\text{Region } \boxed{\text{II}} \quad \psi_{\text{II}}(x,t) = Ae^{i(qx-\omega t)} + Be^{-i(qx+\omega t)}, \tag{3.23b}$$

$$\text{Region } \boxed{\text{III}} \quad \psi_{\text{III}}(x,t) = Te^{i(kx-\omega t)} + \not{G}e^{-i(kx+\omega t)}, \tag{3.23c}$$

$$\text{where } k = \frac{\sqrt{2mE}}{\hbar} \tag{3.24a}$$

$$\text{and } q = \frac{\sqrt{2m\left(E - V_0\right)}}{\hbar}. \tag{3.24b}$$

We see that q^2 is positive or negative depending on $E > V_0$ or $E < V_0$. When q^2 is positive, the kinetic energy in region II is reduced. The situation is similar to electromagnetic waves that travel at a reduced speed $v = \frac{c}{\mu}$ when they enter a dielectric medium having a refractive index

μ, c being the speed of light in vacuum. The analogy must be used with caution, since an energy eigenstate is an eigenstate of momentum also only in free space. This is of course not the case in region II, where the particle interacts with a potential $V(x)$. The operators for the potential energy and that for the kinetic energy do not commute with each other. When applying the boundary conditions stemming from the solenoidal property of the probability current density vector at the edges of the potential at $x = -a$ and $x = +a$, the dynamical phase drops out. When q^2 is negative, it is useful to introduce a positive real number κ such that

$$q = i\kappa. \tag{3.25}$$

In this case, on using the solenoidal property of the probability current density and the associated boundary conditions at $x = \pm a$, we get:

$$T = e^{-2ika}\frac{(2kq)}{2kq\cos(2qa) - i(q^2 + k^2)\sin(2qa)}, \tag{3.26a}$$

$$\text{i.e., } T = e^{-2ika}\frac{(2ki\kappa)}{2ki\kappa\cos(2i\kappa a) - i(-\kappa^2 + k^2)\sin(2i\kappa a)}, \tag{3.26b}$$

$$\text{or } T = e^{-2ika}\frac{(2k\kappa)}{2k\kappa\cosh(2\kappa a) - i(k^2 - \kappa^2)\sinh(2\kappa a)}. \tag{3.26c}$$

The probability of transmission is given by its modulus square,

$$|T|^2 = \frac{(2k\kappa)^2}{(2k\kappa)^2 + (k^2 + \kappa^2)^2\sinh^2(2\kappa a)}. \tag{3.27}$$

For $\kappa a \gg 1$, $\sinh(2\kappa a) \to \left(\frac{1}{2}e^{2\kappa a}\right)$, and hence

$$|T|^2 \approx \frac{(2k\kappa)^2}{(2k\kappa)^2 + (k^2 + \kappa^2)^2\left(\frac{1}{2}e^{2\kappa a}\right)^2} = \frac{(4k\kappa)^2 e^{-4\kappa a}}{(k^2 + \kappa^2)^2}. \tag{3.28a}$$

Writing the barrier width as $d = 2a$, we see that at a given energy of the particle and for a particular barrier height, the *tunneling probability* goes as

$$|T|^2 \to e^{-2\kappa d}. \tag{3.28b}$$

The same result can be drawn using an approximation called as the WKBJ (Wentzel–Kramers–Brilluin–Jeffreys) approximation, which we discuss in Chapter 6 (see Solved Problem P6.1). When the energy of the incident particle is greater than the height of the barrier, the nonzero probability of *reflection above the barrier* (tunneling in the momentum space) is of importance, as mentioned already in the case of the step potential. A few physical examples of this very fascinating phenomenon are discussed in [7].

One of the first successes of the quantum theory of tunneling was the explanation, by George Gamov in 1929 [10], of the inverse relation between the half-life of radioactive nuclei and the energy of the α particle that is emitted. Such a relation was empirically known earlier as the Geiger–Nuttall law. The α particle is a bound state (with binding energy of ~24.2 MeV) of two neutrons and two protons. The Coulomb barrier it must overcome to escape from the nucleus is obtainable from the potential $\frac{2(Z-2)e^2}{4\pi\varepsilon_0 r}$ determined at a distance corresponding to the nuclear radius, where Z is the atomic number of the parent nucleus, e the proton charge, and r the distance from the center of the nucleus. After its escape, the α particle's energy is essentially kinetic; it is obtained from scattering experiments. A schematic representation of the decay potential barrier that must be overcome by the α particle in the decay of $^{238}U_{92}$ is shown in Fig. 3.6. The α particle escapes with ~4.2 MeV of essentially kinetic energy. Gamov combined the predictions of the tunneling probability (Eq. 3.28a,b) with some imaginative casting of the α decay of a radioactive nucleus using a semiclassical model, as discussed below.

If we consider N rectangular strips into which the potential function can be broken down (Fig. 3.6b), then the tunneling probability would be proportional to

$$|T|^2 = \tau = \tau_1\tau_2..\tau_N, \tag{3.29}$$

$\tau_i = |T_i|^2$ being the probability of tunneling across the ith rectangular strip.

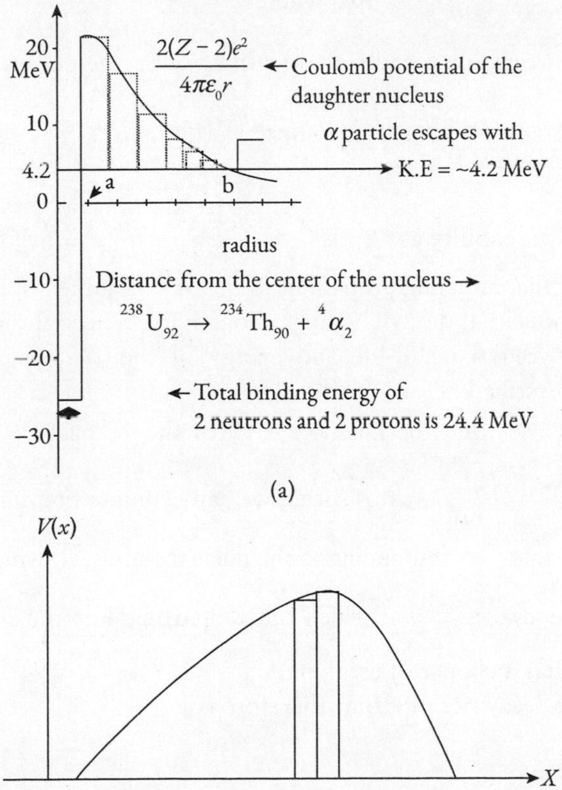

An arbitrarily shaped potential can be broken down into small rectangular strips over its entire domain.

(b)

Fig. 3.6 In α decay, radioactive $^{238}U_{92}$ nucleus decays into $^{234}Th_{90}$ and an α particle. From being bound inside the nucleus to a continuum state with a specific kinetic energy of ~4.2 MeV, the α particle must pass through the Coulomb barrier shown in (a).

Thus, $\ell n|T|^2 = \ell n\ \tau = \ell n\tau_1 + \ell n\tau_2 + .. + \ell n\tau_N.$ (3.30)

Using Eq. 3.28, we see that

$$\ell n|T|^2 \simeq \sum_{i=1}^{N} \ell n|T_i|^2 = -2\sum_{i=1}^{N} \kappa_i d_i = -2\int dx\kappa(x) \simeq -2\frac{\sqrt{2m}}{\hbar} \int_{\text{barrier width}} dx\sqrt{(V(x) - E)}. \quad (3.31)$$

Therefore, $|T|^2 \simeq Ce^{-2\int_{\text{barrier width}} dx\sqrt{\frac{2m}{\hbar^2}(V(x)-E)}}$. (3.32)

In the case of α decay, we recognize that our one-dimensional integration variable is the radial distance r that corresponds to the distance of the escaping particle from the center of the nucleus (Fig. 3.5). It is useful to introduce a parameter γ, defined by

$$\gamma = \frac{\sqrt{2m}}{\hbar}\int_a^b dr\sqrt{(V(r) - E)} = \frac{\sqrt{2mE}}{\hbar}\int_a^b dr\sqrt{\frac{V(r)}{E} - 1} = \sqrt{\frac{2m}{\hbar^2 E}\frac{Z_1Z_2e^2}{4\pi\varepsilon_0}}\, g(\rho), \quad (3.33a)$$

with $g(\rho) = \left[\cos^{-1}\sqrt{\rho} - \sqrt{\rho(1-\rho)}\right]$, where $\rho = \frac{V(b)}{V(a)} = \frac{E}{V(a)}$, (3.33b)

and the limits of integration a and b are identified with reference to Fig. 3.6a.

We can therefore write $\ell n|T|^2 \simeq -2\int_a^b dr\sqrt{\frac{2m}{\hbar^2}(V(r) - E)} = -2\gamma,$ (3.34)

and the tunneling probability as $P = |T|^2 \simeq e^{-2\int_a^b dr\sqrt{\frac{2m}{\hbar^2}(V(r)-E)}} = e^{-2\gamma}.$ (3.35)

Gamov connected the tunneling probability (Eq. 3.41) obtained by using the Schrödinger equation with the exponential decay law for radioactivity to account for the Geiger–Nuttall relationship [11] between the half-life and energy of the α particle emitted using the semiclassical model described below.

From Fig. 3.6a, we see that the kinetic energy of the α particle is $E = \frac{Z_1Z_2e^2}{4\pi\varepsilon_0 b}$, and therefore its speed is $v = \sqrt{\frac{2E}{m}}$. At this speed, we can estimate the time δt taken by the α particle to clear the distance corresponding to the nuclear radius, R, which would be $\delta t = \frac{R}{v}$. This gives us a frequency $\nu = \frac{1}{\delta t}$ at which the α particle hits the surface of the nucleus, which is just when it has a chance of escaping.

The probability of decay per unit time therefore is

$$\tilde{P} = \frac{\mathrm{v}}{R}|T|^2 \simeq \frac{\mathrm{v}}{R}e^{-2\gamma}. \quad (3.36)$$

When we multiply $\tilde{P}$ by $N(t)$, the number of radioactive nuclei at a given instant of time, we shall get the rate at which population of radioactive nuclei would diminish, given by the familiar exponential law. Hence,

$$\frac{dN}{dt} = -N\tilde{P} = -N\nu|T|^2 \simeq -N\nu e^{-2\gamma}, \quad (3.37)$$

from which we immediately obtain $N(t) = N(\text{at } t = 0)e^{-\nu|T|^2 t}$, giving (3.38)

$$N\left(t_{half-life}\right) = N(\text{at } t = 0)e^{-\nu|T|^2 t_{half-life}} = \frac{N(\text{at } t = 0)}{2}. \tag{3.39a}$$

It therefore follows that $e^{\nu|T|^2 t_{half-life}} = 2$, (3.39b)

$$\text{and hence } t_{half-life} = \frac{\ell n2}{\nu e^{-2\gamma}} = \frac{\ell n2}{\nu}e^{2\gamma} = (\ell n2)R\sqrt{\frac{m}{2E}}e^{\frac{2}{\hbar}\sqrt{\frac{2m}{E}}\frac{Z_1Z_2e^2}{4\pi\varepsilon_0}g(\rho)}. \tag{3.40a}$$

We note that it is independent of $N(\text{at } t = 0)$. This result is often written in the following equivalent form:

$$\ell nt_{half-life} = \ell n\left[R(\ell n2)\right] + \frac{1}{2}\ell n\left[\frac{m}{2E}\right] + \frac{2}{\hbar}\sqrt{\frac{2m}{E}}\frac{Z_1Z_2e^2}{4\pi\varepsilon_0}g(\rho). \tag{3.40b}$$

This corresponds to the relation between $\ell nt_{half-life}$ and $\sqrt{\frac{1}{E}}$ that was empirically known earlier. Gamov's explanation of this relationship is fascinating, especially since it was seen as one of the first validations of the probabilistic interpretation of the wavefunction that is employed in quantum tunneling phenomena. Tunneling seems to be bizarre, but it occurs in many different natural phenomena and is satisfactorily accounted for by quantum mechanics. Examples include nuclear fission, radioactive decay, photosynthesis, and superconducting quantum interference device magnetometers. Futuristic quantum computers are under construction employing superconducting qubits; the underlying phenomenology makes use of tunneling. With advances in technology, it is now possible even to determine the actual time taken for tunneling [12].

For the discussion on tunneling, we considered scattering of an incident particle by a potential > 0. On the other hand, if we have a particle in a region of space where the potential is < 0, we essentially would have a *particle in a rectangular well.* The well represents an attractive rather than a repulsive potential. In the next section, we consider an attractive potential in one dimension. It accounts for many natural phenomena.

3.3 Discrete Bound States of an Attractive Square Well and the One-Dimensional Harmonic Oscillator

A one-dimensional attractive square well is a marvelous example of an apparently simple model that is capable of providing an accurate description of a physical system having advanced applications in technology. A peculiar feature of the quantum dot technology is that one can *design* materials to have specific desired electronic and optical properties. Essentially, one can actually *fabricate* an *artificial atom* in the laboratory to possess opto-electronic properties [13] almost at will. This is achieved by controlling the size of the quantum dot during its preparation. The size of a quantum dot is roughly between two and ten nanometers; consisting of ~10 to ~50 atoms. A quantum dot is an object that has characteristic features both of a single atom and of condensed matter. Hence, several physical properties of a quantum dot are best described by an atomic model, and some other properties by that of bulk condensed matter. Size of the quantum dot can be controlled using a *top-down* or a *bottom-up* method of preparation. In the former method, one slashes down bulk matter to a required size; in the latter, one prepares a quantum dot by assembling just

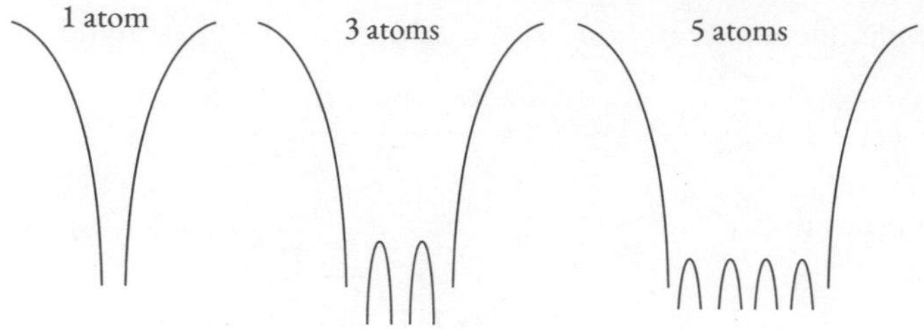

Fig. 3.7a Atomic potentials in a nanostructure.

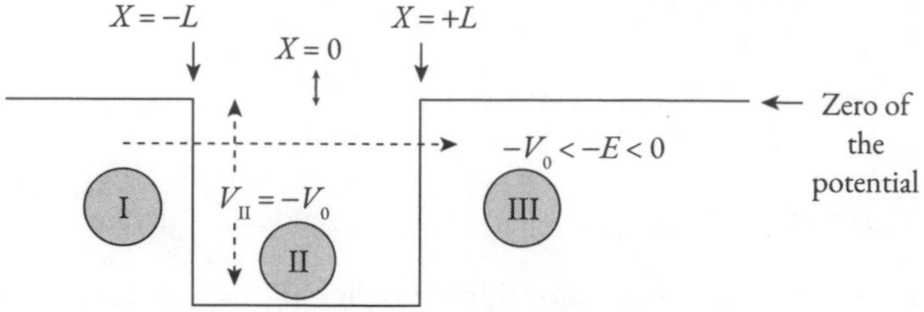

Fig. 3.7b One-dimensional potential well having a finite depth.

as many atoms as would be required to optimize its intended size. Figure 3.7a shows the progression of potential function for an assembly of atoms. The potential between adjacent nuclei practically flattens out. It is schematically depicted in Fig. 3.7b, which provides an excellent representation of the effective potential. The mathematical problem that this model potential poses is essentially that of a *particle in a rectangular square well*. In Fig. 3.7b, a one-dimensional well of width $2L$ is set up symmetrically about $x = 0$. The solution is obtained from quantum mechanics, by solving the Schrödinger equation (Eq. 1.140) for the three regions I $(x < -L)$, II $(-L < x < +L)$, and III $(L < x)$. Subsequently, boundary conditions enforced by the equation of continuity that requires continuity of the wavefunction and of its derivative at $x = \pm L$ would determine the nature of the physical solutions.

We represent energy by the symbol $E > 0$ and consider a particle of energy $-E$ such that $-V_0 < -E < 0$. To simplify our notation, we introduce α and β defined by

$$\alpha^2 = \left(\frac{2m}{\hbar^2}\right)\left(V_0 - E\right), \tag{3.41a}$$

$$\beta^2 = \frac{2mE}{\hbar^2}. \tag{3.41b}$$

We note that α and β are both positive. The general solution to the Schrödinger equation in the basis of *traveling* waves (considering implicit time-dependence of a stationary state given by $\exp\left(-i\frac{E}{\hbar}t\right)$) for the three regions are written in Eqs. 3.42a,b,c:

$$\left.\left[\frac{-\hbar^2}{2m}\frac{d^2}{dx^2}\right]\psi(x) = -E\psi(x)\right\}_{\boxed{I}} \quad \boxed{E > 0} \leftrightarrow \text{Solution: } \psi_I(x) = \cancel{A}e^{-\beta x} + De^{\beta x} = De^{\beta x}, \tag{3.42a}$$

$$\left.\left[\frac{-\hbar^2}{2m}\frac{d^2}{dx^2} - V_0\right]\psi(x) = -E\psi(x)\right\}_{\boxed{II}} \leftrightarrow \text{Solution: } \psi_{II}(x) = Ae^{-i\alpha x} + Be^{i\alpha x}, \tag{3.42b}$$

and

$$\left.\left[\frac{-\hbar^2}{2m}\frac{d^2}{dx^2}\right]\psi(x) = -E\psi(x)\right\}_{\boxed{III}} \leftrightarrow \text{Solution: } \psi_{III}(x) = Ce^{-\beta x} + \cancel{G}e^{\beta x} = Ce^{-\beta x}. \quad (3.42c)$$

The coefficients F and G (respectively in Eq. 3.42a and Eq. 3.42c) are set equal to zero since there can be no wave traveling from left to right in region I, nor from right to left in region III. Essentially, we are considering the particle in the one-dimensional box, not scattering by an attractive potential. It is in the latter case that one would have a particle-wave in region I moving from left to right toward the potential, or one moving right to left toward the potential in region III. One may expect that the solutions to the scattering problem bear a relation to those of the particle in a box. Indeed, such a relationship exists; it is known as the Levinson's theorem [13]. For now, we focus on the particle in a one-dimensional attractive square well.

Continuity of the wavefunction $\psi(x)$ and its derivative $\psi'(x) = \dfrac{d\psi}{dx}$ at $x = +L$ gives

$$A(i\alpha)e^{-i\alpha L} + B(i\alpha)e^{i\alpha L} = C(i\alpha)e^{-\beta L}, \quad (3.43a)$$

$$\text{and } A(-i\alpha)e^{-i\alpha L} + B(i\alpha)e^{i\alpha L} = -\beta Ce^{-\beta L}. \quad (3.43b)$$

Adding Eq. 3.43a and Eq. 3.43b, we get

$$2B(i\alpha)e^{i\alpha L} = (i\alpha - \beta)Ce^{-\beta L}, \quad (3.44a)$$

and subtracting Eq. 3.43b from Eq. 3.43a gives

$$2A(i\alpha)e^{-i\alpha L} = (i\alpha + \beta)Ce^{-\beta L}. \quad (3.44b)$$

Likewise, continuity of the wavefunction $\psi(x)$ and its derivative $\psi'(x) = \dfrac{d\psi}{dx}$ at $x = -L$ gives

$$A(i\alpha)e^{+i\alpha L} + B(i\alpha)e^{-i\alpha L} = D(i\alpha)e^{-\beta L} \quad (3.45a)$$

$$\text{and } A(-i\alpha)e^{+i\alpha L} + B(i\alpha)e^{-i\alpha L} = D\beta e^{-\beta L}. \quad (3.45b)$$

Adding Eq. 3.45a and Eq. 3.45b, we get

$$2B(i\alpha)e^{-i\alpha L} = D(i\alpha + \beta)e^{-\beta L}, \quad (3.46a)$$

and subtracting Eq. 3.45b from Eq. 3.45a gives

$$2A(i\alpha)e^{i\alpha L} = D(i\alpha - \beta)e^{-\beta L}. \quad (3.46b)$$

Now, taking the ratio of Eq. 3.44a to Eq. 3.46b, we get

$$\frac{2B(i\alpha)e^{i\alpha L}}{2A(i\alpha)e^{i\alpha L}} = \frac{C(i\alpha - \beta)e^{-\beta L}}{D(i\alpha - \beta)e^{-\beta L}}, \text{ which implies that } \frac{B}{A} = \frac{C}{D} = \varepsilon, \text{ say.} \quad (3.47a)$$

Similarly, taking the ratio of Eq. 3.44b to Eq. 3.46a, we get

$$\frac{2A(i\alpha)e^{-i\alpha L}}{2B(i\alpha)e^{-i\alpha L}} = \frac{C(i\alpha + \beta)e^{-\beta L}}{D(i\alpha + \beta)e^{-\beta L}}, \text{ which implies that } \frac{1}{\varepsilon} = \frac{A}{B} = \frac{C}{D}. \quad (3.47b)$$

From Eq. 3.47a and Eq. 3.47b, we conclude that $\varepsilon = \dfrac{1}{\varepsilon}$, or $\varepsilon = \pm 1$. There are thus two types of solutions, those with $\varepsilon = +1$ (i.e., $A = B$ and $C = D$) and those with $\varepsilon = -1$ (i.e., $A = -B$ and $C = -D$). Thus, the solutions are

For $\varepsilon = +1 \rightarrow \psi_I(x) = Ce^{\beta x}$, (3.48a)

$$\psi_{II}(x) = A\left(e^{-i\alpha x} + e^{i\alpha x}\right) = 2A\cos(\alpha x), \quad (3.48b)$$

$$\psi_{III}(x) = Ce^{-\beta x}. \quad (3.48c)$$

For $\varepsilon = -1 \rightarrow \psi_I(x) = -De^{\beta x}$, (3.49a)

$$\psi_{II}(x) = A\left(e^{-i\alpha x} - e^{i\alpha x}\right) = -2iA\sin(\alpha x), \quad (3.49b)$$

$$\psi_{III}(x) = De^{-\beta x}. \quad (3.49c)$$

We find that the two types of solutions with $\varepsilon = +1$ and $\varepsilon = -1$ correspond, respectively, to $\psi(x)$ being *even* and *odd*. Considering the symmetry of the potential about $x = 0$, we could have certainly used a basis of parity eigenstates instead of the traveling wave. Any complete linearly independent basis can be used, but some choices bring out the physical essence more lucidly. This point would be clarified if you are introduced to the "quantum defect theory" [15]. For the *even* case $(\varepsilon = +1)$, matching the wavefunction and its derivative at $x = -L$ gives

$$Ce^{\beta L} = 2A\cos(-\alpha L) = 2A\cos(\alpha L), \text{ i.e., } Ce^{\beta L} = 2A\cos(-\alpha L) = 2A\cos(\alpha L), \quad (3.50a)$$

and

$$\beta Ce^{\beta L} = -2\alpha A\sin(-\alpha L) = 2\alpha A\sin(\alpha L). \quad (3.50b)$$

Dividing Eq. 3.50b by Eq. 3.50a, we get

$$\beta = \alpha\tan(\alpha L), \text{ or } \beta L = \alpha L\tan(\alpha L). \quad (3.51)$$

To develop compact relations, we put $\beta L = u$ and $\alpha L = v$. (3.52)

From Eq. 3.40, we have $u^2 + v^2 = \dfrac{2mV_0 L^2}{\hbar^2}$. (3.53)

Equation 3.51 now takes the following form for the even $(\varepsilon = +1)$ solutions:

$$u = v\tan(v). \quad (3.54a)$$

For the odd solutions $(\varepsilon = -1)$, on applying the boundary conditions, instead of Eq. 3.53 we get

$$u = -v\cot(v). \quad (3.54b)$$

Boundary conditions at $x = +L$ essentially give relations same as in Eqs. 3.54a,b. Solutions to the attractive square well problem therefore require simultaneous solutions to the relationship expressed in Eq. 3.53 (which is essentially an equation to a circle of radius $R = L\sqrt{\dfrac{2mV_0}{\hbar^2}}$) and the (u, v) relationships involving transcendental functions in Eqs. 3.54a,b. The simultaneous applicability of the *pair* of relationships (Eq. 3.53 *and* Eq. 3.54a for "even," and Eq. 3.53 *and* Eq. 3.54b for "odd") between (u, v) places constraints on what values energy can have. Energy eigenvalues can therefore only be discrete. This is a characteristic feature of solutions of bound states of any potential. Bound-state wavefunctions go to zero in the asymptotic region. Furthermore, the number of bound-state solutions depends on the product L^2V_0, which is therefore called as the *strength parameter* of the potential. Thus, if one is to make an artificial rectangular square well, one could increase the number of bound states by deepening the well, or by increasing its width. The dependence on depth is linear, while that on width is quadratic. It is therefore more efficient to increase the number of bound states by widening the potential. Since transcendental functions have to be dealt with, most of the literature on this topic employs numerical or graphical methods. We shall discuss in the next section a method, developed rather recently by S. R. Valluri and Kenneth Roberts [16], which makes use of a geometric-analytical method using complex functions. In the present section, we illustrate the solutions using the graphical method.

It is now clear that solutions of the rectangular square well potential occur at the intersection of the family of circles corresponding to $u^2 + v^2 = \frac{2mV_0 L^2}{\hbar^2} = R^2$, with the family of curves that represent $u = v\tan(v)$ for the "even" solutions (Fig. 3.8a) and with curves represented by $u = -v\cot(v)$ for "odd" solutions (Fig. 3.8b). Circles with increasing radii correspond to increasing strength of the potential. One can immediately see that the number of intersections (i.e., number of bound states) of the curves increases with the *strength* L^2V_0 of the potential. For $R = 1$, there is only one "even" bound state. The first "odd" bound state requires a larger strength of the potential; it occurs only at $R = 3$. At this strength, the number of "even" states, however, continues to remain only one. At $R = 5$ we have two bound states of both "even" and "odd" types, a total of four bound states.

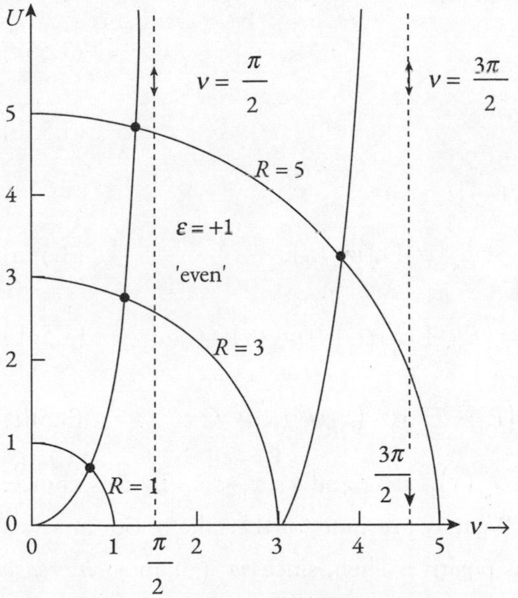

Fig. 3.8a "Even" solutions occur at the intersection of curves represented by $u = v\tan(v)$ and $u^2 + v^2 = \frac{2mV_0L^2}{\hbar^2} = R^2$.

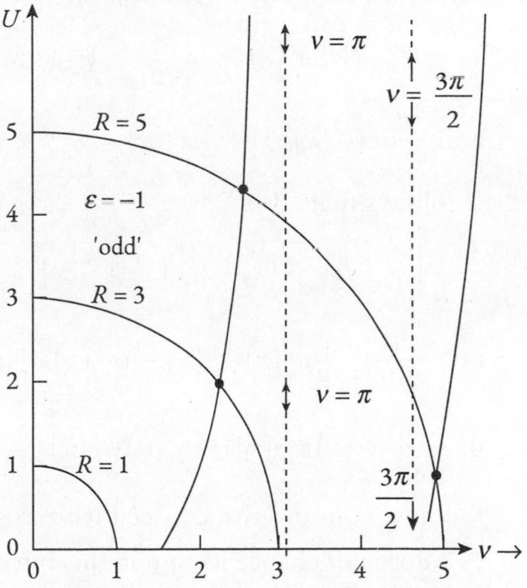

Fig. 3.8b "Odd" solutions occur at the intersection of curves represented by $u = -v\cot(v)$ and $u^2 + v^2 = \frac{2mV_0L^2}{\hbar^2} = R^2$.

The solutions we have obtained are in conformity with some general characteristics of quantum mechanics in one dimension. First, we observe that the number of solutions must be finite when the strength of the potential $\left(L^2V_0\right)$ is finite. Consider now the property of *degeneracy*. If two linearly independent eigenfunctions of the Schrödinger equation exist that belong to the same eigenvalue, the corresponding energy state is said to be *degenerate*. In the *bound-state spectra* of *one-dimensional* potentials, the energy states are essentially *nondegenerate*. This is an important result (see the Solved Problem P3.3; however, also contrast it from the degeneracy discussed in Chapter 2 in the context of Eq. 2.90b). We see that the eigen-energies of the rectangular square well provide an excellent illustration of the non-degeneracy of bound states in one-dimensional problems. However, we must forewarn that the Coulomb potential in the hydrogen atom has a degenerate bound-state energy spectrum that cannot be accounted [17, 18] for by the radial one-dimensional Schrödinger equation. This is an involved problem, which we shall discuss in Chapter 5.

The solutions of the attractive square well potential illustrate another general property of one-dimensional bound states. We have already recognized that the number of bound states is finite and limited by the strength of the well. Let us now arrange the energy eigenvalues of the one-dimensional Schrödinger equation in an increasing order, $E_1 < E_2 < E_3 < \dots < E_{n-1} < E_n < E_{n+1} < \dots$ and identify the corresponding *nondegenerate* eigenfunctions as $\psi_1, \psi_2, \psi_3, \dots, \psi_{n-1}, \psi_n, \psi_{n+1}, \dots\dots$ Then, for $m < n$, ψ_n has a zero between two zeros of ψ_m. This is readily established, as shown below, directly from the Schrödinger equation, which we first write separately for m and n:

$$\frac{-\hbar^2}{2m}\psi_m''(x) + V(x)\psi_m(x) = E_m\psi_m(x), \tag{3.55a}$$

and $$\frac{-\hbar^2}{2m}\psi_n''(x) + V(x)\psi_n(x) = E_n\psi_n(x). \tag{3.55b}$$

It follows that

$$\frac{-\hbar^2}{2m}\left[\psi_n(x)\psi_m''(x) - \psi_m(x)\psi_n''(x)\right] = \left(E_m - E_n\right)\psi_n(x)\psi_m(x), \tag{3.56a}$$

i.e., $$\frac{-\hbar^2}{2m}\frac{d}{dx}\left[\psi_n(x)\psi_m'(x) - \psi_m(x)\psi_n'(x)\right] = \left(E_m - E_n\right)\psi_n(x)\psi_m(x), \tag{3.56b}$$

or $$\frac{d}{dx}\left[\psi_n(x)\psi_m'(x) - \psi_m(x)\psi_n'(x)\right] = \frac{2m}{\hbar^2}\left(E_n - E_m\right)\psi_n(x)\psi_m(x). \tag{3.56c}$$

We now consider two consecutive zeros of $\psi_m(x)$, at x_1 and x_2, $x_1 < x_2$. By this choice, $\psi_m(x)$ does not change its sign in this interval. We may presume for the sake of our analysis (without any loss of generality) that this sign is positive. Then, since $\psi_m(x)$ must *increase* from the zero at x_1, bend over, and again become zero at x_2. We conclude that $\psi_m'\left(x_1\right) > 0$ and $\psi_m'\left(x_2\right) < 0$. Integrating now Eq. 3.56c between x_1 and x_2, we get

$$\left\{\begin{matrix} \psi_n\left(x_2\right)\psi_m'\left(x_2\right) - \psi_m\left(x_2\right)\psi_n'\left(x_2\right) \\ -\psi_n\left(x_1\right)\psi_m'\left(x_1\right) + \psi_m\left(x_1\right)\psi_n'\left(x_1\right) \end{matrix}\right\} = \frac{2m}{\hbar^2}\left(E_n - E_m\right)\int_{x_1}^{x_2} dx\ \psi_n(x)\psi_m(x), \tag{3.57a}$$

$$\text{i.e., } \psi_n\left(x_2\right)\psi_m'\left(x_2\right) - \psi_n\left(x_1\right)\psi_m'\left(x_1\right) = \frac{2m}{\hbar^2}\left(E_n - E_m\right)\int_{x_1}^{x_2} dx\ \psi_n(x)\psi_m(x), \tag{3.57b}$$

since $\psi_m\left(x_1\right) = 0 = \psi_m\left(x_2\right)$.

Now, if $\psi_n(x)$ is *assumed* to have no zero between x_1 and x_2, it can also be assumed to be positive throughout the interval, invoking exactly the same argument as we did to ascertain that $\psi_m(x)$ is positive. This makes the right-hand side of Eqs. 3.57a,b to be positive definite. However, the left-hand side becomes essentially negative, resulting in a contradiction. The assumption we made must therefore be incorrect. In other words, $\psi_n(x)$ must have a zero in the interval; it must go through a node. It is very instructive to consider a limiting case of the square well in which the walls of the well become impenetrable. This is modeled by dividing the one-dimensional space into three regions of space, I $(x = \pm L)$, II $(-L < x < L)$, and III $(x > L)$ such that in regions I and III, the energy is infinite (unphysical). The wavefunction must therefore reside only in the region II; it must go to zero at the boundaries $x = \pm L$ of the well. The derivative of the wavefunction would not be continuous across the walls, and this is acceptable *when* the jump in the potential is infinite (*not* otherwise).

The bottom of the well now sets the zero of the energy scale. Essentially, we have a particle in a one-dimensional infinitely deep square well.

In region II, the Schrödinger equation is

$$\frac{-\hbar^2}{2m}\frac{d^2\psi(x)}{dx^2} = E\psi(x) = \frac{\hbar^2k^2}{2m}\psi(x). \tag{3.58}$$

We can write the general solution in any base pair of linearly independent functions. We write it in the basis $\left(\cos(kx), \sin(kx)\right)$:

$$\psi(x) = A\cos(kx) + B\sin(kx). \tag{3.59}$$

For $\psi(x = L) = 0$, we must have one of the two possibilities:

i. $B = 0$ *and* $\cos(kL) = 0$, which would require $k = \dfrac{n\pi}{2L} = k_n$ with $n = 1,3,5,7,\ldots.$

ii. $A = 0$ *and* $\sin(kL) = 0$, which would require $k = \dfrac{n\pi}{2L} = k_n$ with $n = 2,4,6,8,\ldots.$

The normalized eigenfunctions therefore are

$$\psi_n(x) = \frac{1}{\sqrt{L}}\cos\left(k_n x\right), \text{ with } n = 1,3,5,7,\ldots \tag{3.60a}$$

$$\psi_n(x) = \frac{1}{\sqrt{L}}\sin\left(k_n x\right), \text{ with } n = 2,4,6,8,\ldots \tag{3.60b}$$

Now, the $\sin\left(k_n x\right)$ function goes through a zero $(n-1)$ times between $k_n x = 0$ and $k_n x = n\pi$, i.e., between $x = 0$ and $x = \dfrac{n\pi}{k_n} = n\pi \times \dfrac{2L}{n\pi} = 2L$, which is the width of the well. The same is true for the cosine function. Hence, $\psi_n(x)$ has $(n-1)$ nodes within the well. We have considered only positive integers, since negative integers only give us the same function, merely scaled by ± 1. The energy of the nth state is

$$E_n = \frac{\hbar^2 k_n^2}{2m} = \frac{h^2 n^2 \pi^2}{32mL^2}, \tag{3.61a}$$

and the minimum energy it can have is

$$E_1 = \frac{\hbar^2 k_1^2}{2m} = \frac{h^2 \pi^2}{32mL^2}. \tag{3.61b}$$

This is called as the zero point energy. Classical physics predicts the minimum energy to be zero, but nature defies this.

The minimum energy is accounted for correctly by quantum mechanics, since the uncertainty in the particle's location in the box bears a relation to that in momentum, and hence in energy. Now, a node reduces the uncertainty in position of the particle by telling us where the particle *cannot* be. As a result of this, the uncertainty in momentum increases, thereby increasing the particle's energy. It is therefore only natural that the number of nodes *increases* with energy. The energy of the system would approach zero in the classical limit $L \to \infty$. The spacing between the discrete energy states increases with n (Fig. 3.9). The energy eigenvalues are nondegenerate, as we certainly expect. The results we have obtained for the infinitely deep square well illustrate two rather general theorems [19] that are valid for the Schrödinger equation. The first of these is called as the *separation theorem*, which states that zeros of linearly independent solutions alternate. The other is the *comparison theorem*, which is essentially the result we already established, that for $m < n$, ψ_n has a zero between two zeros of ψ_m. In the larger study of differential equations, these theorems are discussed more fully in the Sturm–Liouville theory.

The functionality of the *separation theorem* and the *comparison theorem* that we have seen earlier in the case of the infinitely deep square well persists in the finite square well solutions (Eq. 3.48, for $\varepsilon = +1$, and Eq. 3.49, for $\varepsilon = -1$). However, while the infinite square well supports an infinite number of bound states, the finite square well supports only a finite number of bound states which is restricted by the strength of the potential L^2V_0.

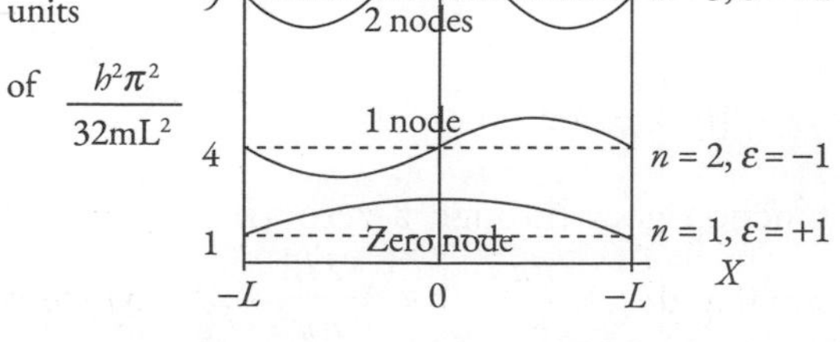

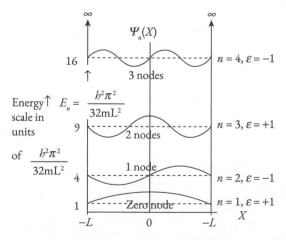

Fig. 3.9 Energy levels and corresponding wavefunctions of a particle confined in an impenetrable potential barrier. Observe the nodal structure in conformity with the *separation* theorem and the *comparison* theorem discussed in the text.

The energy eigenvalues of the finite square well are given by the points of intersection of the curves shown in Fig. 3.8a,b. In the limit $V_0 \to \infty$, these would correspond to Eqs. 3.61a,b. Corresponding to the strength of the potential indexed by $R = \frac{L}{\hbar}\sqrt{2mV_0}$ belonging to the range $\left(n^{(e)} - 1\right)\pi < R \leq n^{(e)}\pi$, there are $n^{(e)}$ number of even $(\varepsilon = +1)$ bound-state wavefunctions of the finite square well. Likewise, when $\left(n^{(o)} - \frac{1}{2}\right)\pi < R \leq \left(n^{(o)} + \frac{1}{2}\right)\pi$, there are $n^{(o)}$ number of odd $(\varepsilon = -1)$ bound-state wavefunctions. The wavefunctions do not go to zero at the edges of the well at $x = \pm L$, unlike the case for the infinite well. Instead, the wavefunctions tunnel through the walls of the well into the classically forbidden regions I and III where they have an evanescent presence, decaying exponentially. Inside the well, the wavefunctions of the finite square well are no longer pure sine or pure cosine functions; instead, they require a superposition of the two.

We conclude this section by pointing out that the optical properties of a quantum dot are determined by the spacing between these energy states, which in turn are inversely related to the size of the quantum dot (Fig. 3.10a), just as we have seen for the square well. The natural energy states of a quantum dot are discrete, the quantum dot being a system that is confined in all dimensions.

The discrete energy states of the quantum dot are atom-like, but the conglomeration of atoms has some bulk-like behavior. Consequently, the optical properties are determined by the energy states of an exciton (Fig. 3.10b), which is a bound state of an electron that is raised into the dielectric material's conduction band with the hole it leaves behind in the valence band. The effective mass of the electron (μ_e) in the conduction band, that of the hole (μ_h) in the valence band, and the energy spacing between the discrete levels determine the color of the quantum dot. The result is known as the Brus equation [20]:

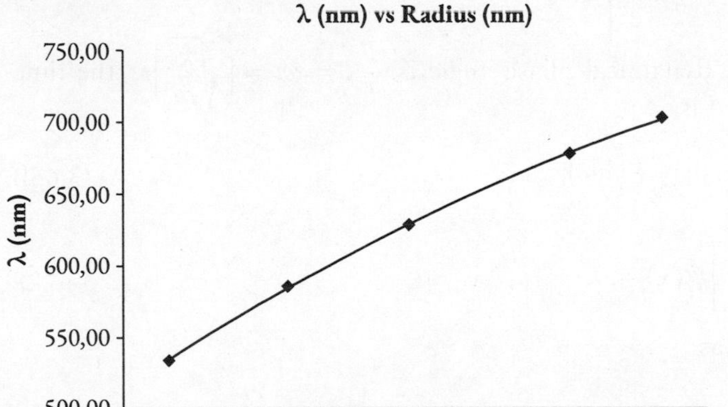

Fig. 3.10a Optical properties of a quantum dot are related to its size.

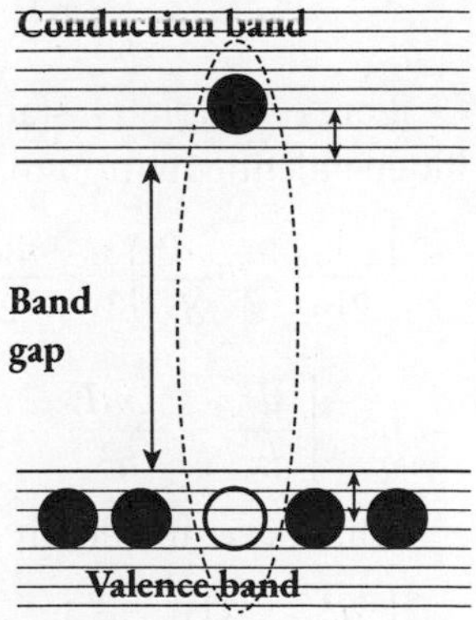

Fig. 3.10b An exciton is a bound state of an electron and a hole.

$$\Delta E = E_{gap} + \frac{\hbar^2}{8\mu\rho^2}, \tag{3.62a}$$

where $\mu = \dfrac{\mu_e \mu_h}{\mu_e + \mu_h}$. (3.62b)

We shall now obtain the eigenstates of the one-dimensional simple harmonic oscillator.

An arbitrarily shaped one-dimensional potential $U(x)$ at a point x near a point of stable equilibrium at x_0, such as the one shown in Fig. 3.4a, can be expanded near x_0 as follows:

$$U(x) = U(x_0) + \left.\frac{\partial U}{\partial x}\right|_{x_0}(\delta x) + \frac{1}{2!}\left.\frac{\partial^2 U}{\partial x^2}\right|_{x_0}(\delta x)^2 + \frac{1}{3!}\left.\frac{\partial^3 U}{\partial x^3}\right|_{x_0}(\delta x)^3 + \ldots, \tag{3.63a}$$

$$U(x) = U(x_0) + \frac{1}{2}k(\delta x)^2 + O(\delta x^3), \tag{3.63b}$$

where $\delta x = x - x_0$ and $k = \left.\dfrac{\partial^2 U}{\partial x^2}\right|_{x_0}$ is usually called as the spring constant. A large number of problems can be cast in terms of the harmonic oscillator, or variants and improvisations thereof such as the anharmonic oscillator, the double-oscillator, and coupled oscillators. With the zero of the potential scale set at x_0, which can also be chosen to be the zero of the x scale, we have the harmonic oscillator potential, $U(x) = \dfrac{1}{2}kx^2$. The Lagrangian for the harmonic oscillator is

$$L(x, \dot{x}) = \frac{1}{2}m\dot{x}^2 - \frac{1}{2}kx^2 \tag{3.64a}$$

and the Hamiltonian is $H(x,p) = \dfrac{p^2}{2m} + \dfrac{1}{2}kx^2$. (3.64b)

The Schrödinger equation for the oscillator is

$$i\hbar\frac{\partial\psi(x,t)}{\partial t} = \left[\frac{1}{2m}\left(-i\hbar\frac{\partial}{\partial x}\right)^2 + \frac{1}{2}kx^2\right]\psi(x,t). \tag{3.65a}$$

For a stationary state whose dynamical phase angle is $\dfrac{E}{\hbar}t = \omega t = \left(\sqrt{\dfrac{k}{m}}\right)t$, the time-independent equation to be solved is

$$\left[\frac{1}{2m}\left(-i\hbar\frac{d}{dx}\right)^2 + \frac{m\omega^2}{2}x^2\right]\psi(x) = E\psi(x), \tag{3.65b}$$

i.e., $$\left[\frac{d^2}{dx^2} + \frac{2mE}{\hbar^2} - \frac{m^2\omega^2}{\hbar^2}x^2\right]\psi(x) = 0, \tag{3.65c}$$

which is equivalently written as

$$\left[\frac{d^2}{dy^2} + \frac{2E}{\hbar\omega} - y^2\right]\psi(y) = 0 \tag{3.65d}$$

with $y = \left(\sqrt{\dfrac{m\omega}{\hbar}}\right)x.$ (3.66)

For *large distances*, the differential equation approximately is

$$\left[\frac{d^2}{dy^2} + \frac{m^2\omega^2}{\hbar^2} y^2\right]\psi_{y\to\infty}(y) \simeq 0. \tag{3.67a}$$

and is easy to solve.

Its solution is $\psi_{y\to\infty}(y) = Ae^{-\frac{y^2}{2}} + Be^{+\frac{y^2}{2}}$, (3.67b)

but we must consider B = 0 in order to reject unphysical states at large distances. The solution for *all* distances must therefore be a *modification* of that at large distance, and may be written as

$$\psi(y) = H(y)e^{-\frac{y^2}{2}}, \tag{3.68}$$

where $H(y)$ must satisfy

$$\left[\frac{d^2}{dy^2} - 2y\frac{d}{dy} + (\gamma - 1)\right]H(y) = 0, \tag{3.69}$$

where $\gamma = \frac{2E}{\hbar\omega}$. (3.70)

Equation 3.69 is amenable to a power series solution

$$H(y) = y^s \sum_{k=0}^{\infty} a_k y^k = \sum_{k=0}^{\infty} a_k y^{k+s} \tag{3.71}$$

with $s \geq 0$ to avoid divergence at $y = 0$. Using Eq. 3.71 in Eq. 3.69 gives us

$$\sum_{k=0}^{\infty}\left[(k+s)(k+s-1)a_k y^{k+s-2} - \{2(k+s) - (\gamma - 1)\}a_k y^{k+s}\right] = 0. \tag{3.72a}$$

The coefficient of the lowest power of y in Eq. 3.72a must vanish, and hence $s(s-1)a_0 = 0$.

With $a_0 \neq 0$, we note that we must have $s = 0$ or $s = 1$. We shall consider $s = 0$; we shall find out very soon that consideration of $s = 1$ would turn out to be unnecessary. Equation 3.72a then gives us

$$\sum_{n=0}^{\infty}\left[n(n-1)a_n y^{n-2} - \{2n - (\gamma - 1)\}a_n y^n\right] = 0. \tag{3.72b}$$

Coefficient of each power of y must vanish independently. The coefficient in the first term that would correspond to the same exponent of y as in the second term would have n replaced by $n + 2$.

Hence, $(n+2)(n-1)a_{n+2} - \{2n - (\gamma - 1)\}a_n = 0$, (3.73a)

i.e., $a_{n+2} = \frac{\{2n - (\gamma - 1)\}}{(n+2)(n-1)}a_n$. (3.73b)

The coefficients of the $(n + 2)$ term is determinable from that of the nth term; hence all the coefficients are determinable from a_0 and a_1. The solutions thus have power series in even and odd powers of y:

Even: $H_e(y) = a_0 + a_2 y^2 + a_4 y^4 + \ldots$ (3.74a)

and

Odd: $H_o(y) = a_1 y + a_3 y^3 + a_5 y^5 + ...$ (3.74b)

The odd and even solutions are linearly independent. Having found two linearly independent solutions, there being no more, the consideration of $s = 1$ becomes redundant. The general solution to the Schrödinger equation for the harmonic oscillator therefore is a linear superposition of $H_e(y)e^{-\frac{y^2}{2}}$ and $H_0(y)e^{-\frac{y^2}{2}}$. Successive terms in the solution appear in the ratio $\frac{a_{n+2}}{a_n}\frac{y^{n+2}}{y^n}$, which goes to zero for large n:

$$\frac{a_{n+2}}{a_n}\frac{y^{n+2}}{y^n} = \frac{\{2n-(\gamma-1)\}}{(n+2)(n-1)}y^2 \underset{n\to\infty}{\rightarrow} \frac{2n}{n^2}y^2 \rightarrow \frac{2}{n}y^2 \rightarrow 0, \tag{3.75}$$

hence the series converges. This behavior is the same as that of another power series,

$$e^{x^2} = \sum_{\substack{n=0,\\ n:even}}^{\infty} \frac{x^n}{\left(\frac{n}{2}\right)!} = \sum_{\substack{n=0,\\ n:even}}^{\infty} \frac{2x^n}{n!}. \tag{3.76}$$

The net solution $\psi(y) = H(y)e^{-\frac{y^2}{2}} \rightarrow e^{y^2}e^{-\frac{y^2}{2}} \rightarrow e^{+\frac{y^2}{2}}$ would blow up at large distances, unless the infinite power series is required to terminate at some value of n. From Eqs. 3.73a,b, we see that the series would terminate for those values of γ for which

$$2n = \gamma - 1 = \frac{2E}{\hbar\omega} - 1, \text{ i.e., } E_n = \left(n + \frac{1}{2}\right)\hbar\omega. \tag{3.77}$$

Physically acceptable solutions therefore occur for integer values of n, including $n = 0$ at which there is a zero-point energy of $\frac{1}{2}\hbar\omega$. The condition in Eq. 3.77 renders the functions $H(y)$ as polynomials $H_n(y)$ for $n = 0,1,2,3,,,,$. These polynomials are known as Hermite polynomials, after the French mathematician Charles Hermite (1822–1901). He also proved that eigenvalues of self-adjoint operators are real and hence the operators represented by such matrices are called as Hermitian operators. On the side, we mention that a crater at the moon's North pole is also named after him. A rather common way of defining the Hermite polynomials is

$$H_n(y) = (-1)^n e^{y^2} \frac{d^n}{dy^n} e^{-y^2}. \tag{3.78}$$

The polynomials in Eq. 3.78 satisfy the Hermite differential equation

$$\left[\frac{d^2}{dy^2} - 2y\frac{d}{dy} + 2n\right]H_n(y) = 0. \tag{3.79}$$

Their orthogonality property is stated as

$$\int_{-\infty}^{\infty} dy H_n(y) H_m(y) e^{-y^2} = \delta_{nm} n! 2^n \sqrt{\pi}, \tag{3.80}$$

which enables us to normalize the wavefunctions of the oscillator.

The energy spectrum is obviously discrete, and successive energy levels are equally spaced at an energy difference of $\hbar\omega$. Table 3.1 lists some of the lowest few eigenfunctions. In this table,

Table 3.1 The harmonic oscillator wavefunctions; probability amplitudes and probability densities; and the uncertainty product $\delta x \delta p$ for the ground state ($n = 0$) and the first three excited states, with $n = 1, 2, 3$. In the sketches for the probability amplitudes and the probability densities, the vertical dashed lines show the classical turning points. Spacing between the turning points increases with n since the energy increases, but the probability amplitude and the probability density are nonzero even beyond the turning points, in the classically forbidden regions.

n	Hermite polynomial $H_n(y)$	Normalized wavefunction $\psi_n(y) = N\psi_n(y)e^{-\frac{y^2}{2}}$	Probability amplitude $\psi_n(y)$	Probability density $\lvert\psi_n(y)\rvert^2$	Uncertainty product $\delta x \delta p$ $\frac{\hbar}{2}(2n+1)$
3	$a_1 y + a_3 y^3$	$\left[\left(\frac{\alpha}{\pi}\right)^{1/4}\frac{1}{\sqrt{3}}\right](2y^3 - 3y)e^{-\frac{y^2}{2}}$	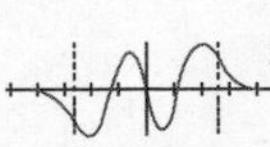	$n = 3$ 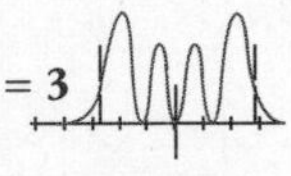	$\frac{7\hbar}{2}$
2	$a_0 + a_2 y^2$	$\left[\left(\frac{\alpha}{\pi}\right)^{1/4}\frac{1}{\sqrt{2}}\right](2y^2 - 1)e^{-\frac{y^2}{2}}$	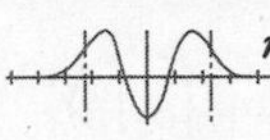	$n = 2$ 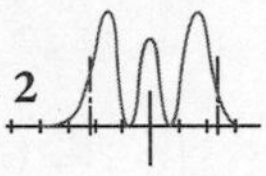	$\frac{5\hbar}{2}$
1	$a_1 y$	$\left[\left(\frac{\alpha}{\pi}\right)^{1/4}\sqrt{2}\right]ye^{-\frac{y^2}{2}}$	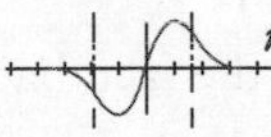	$n = 1$ 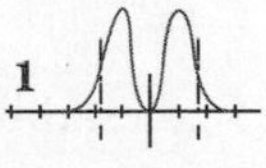	$\frac{3\hbar}{2}$
0	a_0	$\left(\frac{\alpha}{\pi}\right)^{1/4}e^{-\frac{y^2}{2}}$	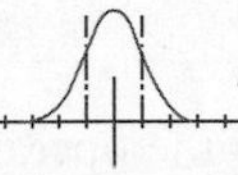	$n = 0$ 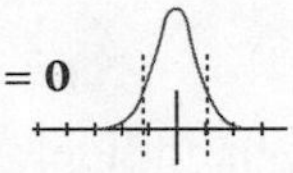	$\frac{\hbar}{2}$

$\alpha = \frac{m\omega}{\hbar}$ and y is from Eq. 3.66. Determination of the normalization constants N and the uncertainty products $\delta x \delta p$ (*determined by using Eqs. 1.104a,b*) is left as an exercise.

It is instructive now to define a non-Hermitian operator

$$c = \sqrt{\frac{m\omega}{2\hbar}}\left(x + \frac{ip}{m\omega}\right), \tag{3.81a}$$

and its adjoint

$$c^{\dagger} = \sqrt{\frac{m\omega}{2\hbar}}\left(x - \frac{ip}{m\omega}\right), \tag{3.81b}$$

in terms of the position and momentum operators.

From the primary commutation relation between position and momentum, it is easy to see that $\left[c, c^{\dagger}\right]_{-} = 1$ (3.82a)

and also that $c^{\dagger}c = \frac{H}{\hbar\omega} - \frac{1}{2}$, (3.82b)

$$\text{i.e., } H = \hbar\omega\left(c^{\dagger}c + \frac{1}{2}\right) = \hbar\omega\left(N + \frac{1}{2}\right), \tag{3.83a}$$

where $N = c^{\dagger}c$ (3.83b)

is a self-adjoint operator.

As usual, we designate eigenkets of N by its eigenvalues, which we label as n:

$$N|n\rangle = n|n\rangle \tag{3.84}$$

From Eq. 3.83, we see that the H and N are linearly related, hence $|n\rangle$ are energy eigenkets and

$$H|n\rangle = \left(n + \frac{1}{2}\right)\hbar\omega|n\rangle,$$

and the energy eigenvalues are $E_n = \left(n + \frac{1}{2}\right)\hbar\omega,$ (3.85)

just as in Eq. 3.77, if we could ascertain that $n = 0,1,2,3,....$ This is achieved by recognizing that

$$[N,c]_- = [c^{\dagger}c,c]_- = c^{\dagger}[c,c]_- + [c^{\dagger},c]_- c = -c, \tag{3.86a}$$

likewise $[N,c^{\dagger}]_- = c^{\dagger},$ (3.86b)

and hence $N\{c^{\dagger}|n\rangle\} = (n+1)\{c^{\dagger}|n\rangle\},$ (3.86c)

and $N\{c|n\rangle\} = (n-1)\{c|n\rangle\}.$ (3.86d)

Equations 3.86c,d establish the fact that the operators $c^{\dagger}$ and c produce new eigenkets of N with eigenvalues shifted by ± 1. Recognizing therefore that

$$c|n\rangle = \tilde{c}|n-1\rangle,$$

where on the left-hand side c is the operator given in Eq. 3.81 and $\tilde{c}$ on the right-hand side is a scalar to be determined.

Since $n = \langle n|N|n\rangle = \langle n|c^{\dagger}c|n\rangle = \tilde{c}^{*}\tilde{c} = |\tilde{c}|^2$, and $\langle n|c^{\dagger}c|n\rangle \geq 0$. i.e., $n \geq 0$, we choose the positive phase convention and identify $\tilde{c}$ as $\sqrt{n}$. Hence,

$$c|n\rangle = \sqrt{n}|n-1\rangle, \tag{3.87a}$$

and likewise, $c^{\dagger}|n\rangle = \sqrt{n+1}|n+1\rangle.$ (3.87b)

The smallest value of n is zero, and it can increase in steps of unity by applying $c^{\dagger}$, which is therefore called as the creation operator. Correspondingly, c is called as the annihilation (or destruction) operator, which lowers an energy eigenket from $|n\rangle$ to $|n-1\rangle$. Zero being the lowest attainable number, we have

$$c|0\rangle = 0. \tag{3.88}$$

Finally, the operator defined by Eq. 3.83b is called as the "number" operator. The names attributed to the operators $c^{\dagger}$, c, and N would appeal to everyone, for they appropriately describe their roles with regard to the manipulation of the number n. Their usage in the analysis of the quantum mechanics of the harmonic oscillator is a precursor to a very powerful method that is most useful in many-body theory, known as *second* quantization. A brief introduction to 'second quantization' is presented in Appendix D.

In the next section, we discuss a *relatively new method* to obtain solutions of the Schrödinger equation.

3.4 Lambert W Solution to the Quantum One-Dimensional Square Well Problem

We have seen that the solution to the Schrödinger equation for the finite square well involves transcendental functions (Eqs. 3.54a,b). It is widely argued in literature that the transcendental equations must be solved using either numerical or graphical methods, such as the one we employed in Fig. 3.8. Exact solutions, which employ contour integration in the complex plane, are available [21]. In the present section, we discuss a *geometric-analytic* method, which makes use of conformal mapping $w \to z = we^{w}$ between two complex domains to obtain solutions to the quantum finite depth square well. A technique that does not use contour integration was developed rather recently [16] by S. R. Valluri and Kenneth Roberts. This method is based on the Lambert W function [22, 23]. In the present section, we illustrate the application of this method to the quantum finite depth square well problem following the treatment in Reference [16]. The primary results we discuss are essentially the same as we obtained in the previous section using the graphical method. It is, however, instructive to develop acquaintance with the Lambert W technique, especially since it helps examine solutions in response to some changes in a physical parameter of the problem, thereby providing control on material properties [24, 25]. The Lambert W method employs conformal mapping, but does not use contour integration.

Using Eq. 3.44a and Eq. 3.47b,

$$2\frac{B}{A}A(i\alpha)e^{i\alpha L} = \varepsilon 2A(i\alpha)e^{i\alpha L} = (i\alpha - \beta)Ce^{-\beta L}, \tag{3.89a}$$

and from Eq. 3.44b, we have $2A(i\alpha)e^{-i\alpha L} = (i\alpha + \beta)Ce^{-\beta L}$. (3.89b)

Dividing Eq. 3.89a by Eq. 3.89b, we get

$$(\pm 1)e^{2i\alpha L} = \frac{(i\alpha - \beta)}{(i\alpha + \beta)}, \text{ or } (\mp 1)e^{2i\alpha L} = \frac{(\beta - i\alpha)}{(i\alpha + \beta)}. \tag{3.90a}$$

$$\text{i.e., } (\mp 1)e^{2iv} = \frac{\beta L - i\alpha L}{\beta L + i\alpha L} = \frac{u - iv}{u + iv} = \frac{(u - iv)^2}{u^2 + v^2}. \tag{3.90b}$$

Hence, $(\mp 1)R^2 e^{2iv} = (u - iv)^2$, or $\left(\pm\sqrt{\mp 1}\right)R = (u - iv)e^{-iv}$, (3.91)

where R is the radius of the circles in Fig. 3.8. The four complex roots of unity $\left(\pm\sqrt{\mp 1}\right)$ in Eq. 3.91 are

$$\gamma = +1, +i, -1, -i. \tag{3.92}$$

Thus, the symbol γ represents four alternative values. Beginning with any of these values, its inverse, or its negative, or its complex conjugate is also represented by γ.

Thus, $w^{*}e^{-iv} = (u - iv)e^{-iv} = \gamma R$ (3.93a)

and $we^{+iv} = (u + iv)e^{+iv} = \gamma R$, (3.93b)

where $w = u + iv$ (3.94)

is a complex number in the w plane. In the case of the finite attractive square well, the condition

$$ww^* = (u + iv)(u - iv) = u^2 + v^2 = \frac{2mV_0 L^2}{\hbar^2} = R^2 \tag{3.95}$$

describes a circle in the complex-w plane whose radius is determined by the strength of the potential. $\gamma = \pm i$ represent two rays for the positive and negative imaginary axis (Fig. 3.11a) in the complex z plane. These correspond to the "even" $(\varepsilon = +1)$ solutions (Fig. 3.8a). Likewise, $\gamma = \pm 1$ represent two rays for the positive and negative real axis (Fig. 3.11a) and correspond to "odd" $(\varepsilon = +1)$ solutions (Fig. 3.8b). We now consider a mapping on the complex plane consisting of complex numbers $z = x + iy$ to $w = u + iv$ such that

$$W(z)e^{W(z)} = we^w = z. \tag{3.96}$$

This has countable infinite number of solutions.

Each solution is denoted by $W_k(z)$, where k is an integer. The symbol $W(z)$ represents the principal branch, or some other branch, depending on the context. Each k corresponds to a particular branch of the Lambert W function. The branches with $k = 0$ and $k = -1$ are real values for any z. The branch $k = 0$ is called as the principal branch. The range of all other branches excludes the real axis. $W_0(z)$ contains positive values in its range. $W_{-1}(z)$ includes $\left(-\infty, \frac{1}{e}\right]$ in its closure. The set of all $W_k(z)$ for integer k is called as the Lambert W function, even if it is multivalued.

The multi-branch inverse mapping of this relationship is provided by the Lambert $W(z)$ function. A solution of this equation is one of the branches of $w \to z = we^w$ between the complex z and the complex-w planes. Since logarithm is the inverse of $f(w) = e^w$, the inverse function of we^w is often called as *product logarithm*. Each branch is a single valued complex function. For a given complex number z, the equation $we^w = z$ has a countable infinite number of solutions. The axial rays in the complex z plane exclude the origin. These rays (Fig. 3.11a) map into the Lambert W curves in the complex w plane. Thus, we seek to obtain the set $W(S) \equiv W(S_r) \cup W(S_i)$ with $z = We^W$ as a family of various curves in the w plane for various branches k. The intersections of these curves with a curve that describes a constraint imposed by some physical parameter, such as that imposed by Eq. 3.70, provide the solutions of the problem of interest. Likewise, a curve in the w plane that describes some constraint maps back into a curve in the z plane, which may well have multiple loops. The intersections of these curves with the axial rays in the z plane correspond to the solutions of interest. We shall discuss the *finite attractive square well* further to illustrate this.

At $z = \frac{-1}{e}$, we have two roots of $We^w = z$ which are $W_0\left(-\frac{1}{e}\right) = -1$ and $W_{-1}\left(-\frac{1}{e}\right) = -1$.

The phase e^{+iv} rotates the complex W counter-clockwise by an angle $|v| \leq R$. The condition $we^{+iv} = \gamma R$ does not allow rotation into an arbitrary quadrant. Since γR is purely real or purely imaginary, the rotation must land on one of the coordinate axes in the complex w plane. Only certain rotations are therefore permitted.

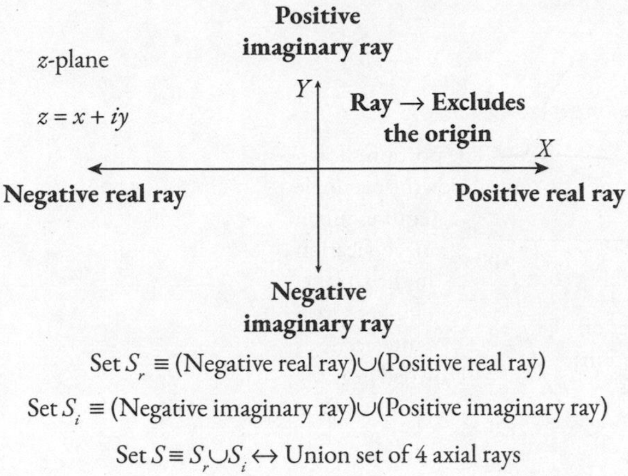

Fig. 3.11a The complex z plane.

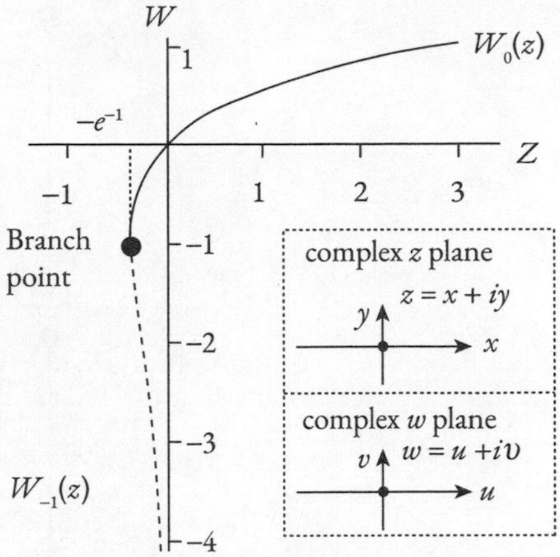

Fig. 3.11b The real branches W_0 and W_{-1} of the Lambert W function. W_0 is the principal branch.

Now, we have $z = we^{w} = (u+iv)e^{(u+iv)} = e^{u}(u+iv)e^{iv} = e^{u}(u+iv)(\cos v + i\sin v)$,

$$(3.97a)$$

i.e., $z = e^{u}\gamma R = e^{u}\left\{(u\cos v - v\sin v) + i(u\sin v + v\cos v)\right\}$. (3.97b)

The set S_i (Fig. 3.11a) consists of the two rays along the positive and negative imaginary axes. For these two rays, z is positive imaginary $(\gamma = +i)$ or negative imaginary $(\gamma = -i)$, and the real part of Eq. 3.97b would be zero, which means that

$$u = v\tan v, \tag{3.98a}$$

which is just the necessary condition to have a bound state of the finite square well that we discussed in Section 3.3 (Eq. 3.54 and Fig. 3.8a, corresponding to the "even" solution, having $\varepsilon = +1$). Likewise, for the set S_r, we have two rays along the positive and negative real axis, and the imaginary part of Eq. 3.97b is zero. This provides the condition

$$u = -v\cot v, \tag{3.98b}$$

which is the second necessary condition to have a bound state (Eq. 3.55 and Fig. 3.8b, corresponding to the "odd" solution, having $\varepsilon = -1$). Of course, the conditions expressed in Eqs. 3.98a,b are only "necessary" but not "sufficient". In addition to these, for sufficiency we must take into account the physical parameters of the problem that determine the *strength* of the attractive well. The latter determines the radius of the circle in the w plane described by Eq. 3.53 (and Fig. 3.8b). This circle that may be labeled as P in the w plane must therefore be mapped into the z plane. We denote this mapping as $Q = Pe^{P}$. The set Q consists of points that describe the multi-loop self-intersecting closed curve in the z plane. The Q curve is shown in Fig. 3.12. The points of intersection of this curve with the set of rays $S_r \cup S_i$ then provide the complete solution to the finite square well problem.

Instead of mapping the circle in the w plane into the z plane, we can also map the axial rays of the z plane to the w plane to obtain the solutions. Mapping of the axial rays yields

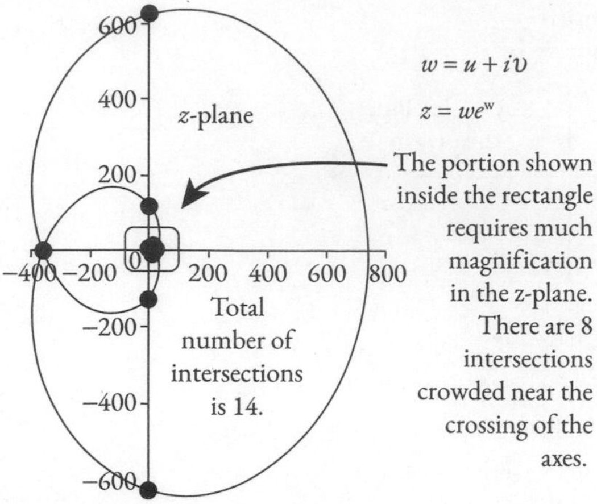
600
400
200
−200
−400
−600
−400
−200
0
200
400
600
800
z-plane
$w = u + iv$
$z = we^{w}$
The portion shown inside the rectangle requires much magnification in the z-plane. There are 8 intersections crowded near the crossing of the axes.
Total number of intersections is 14.

Fig. 3.12 Image of the R-circle in w plane gives closed multi-loop self-intersecting curve, which is referred to as the Q-curve in the z plane. Solutions to the FSW (finite square well) problem are given by the points of intersection of the Q curve in the z plane with the coordinate axes. On successive magnification shown in the three panels below of the portion in the rectangle near the origin, one can see that the Q curve crosses the axes 14 times.

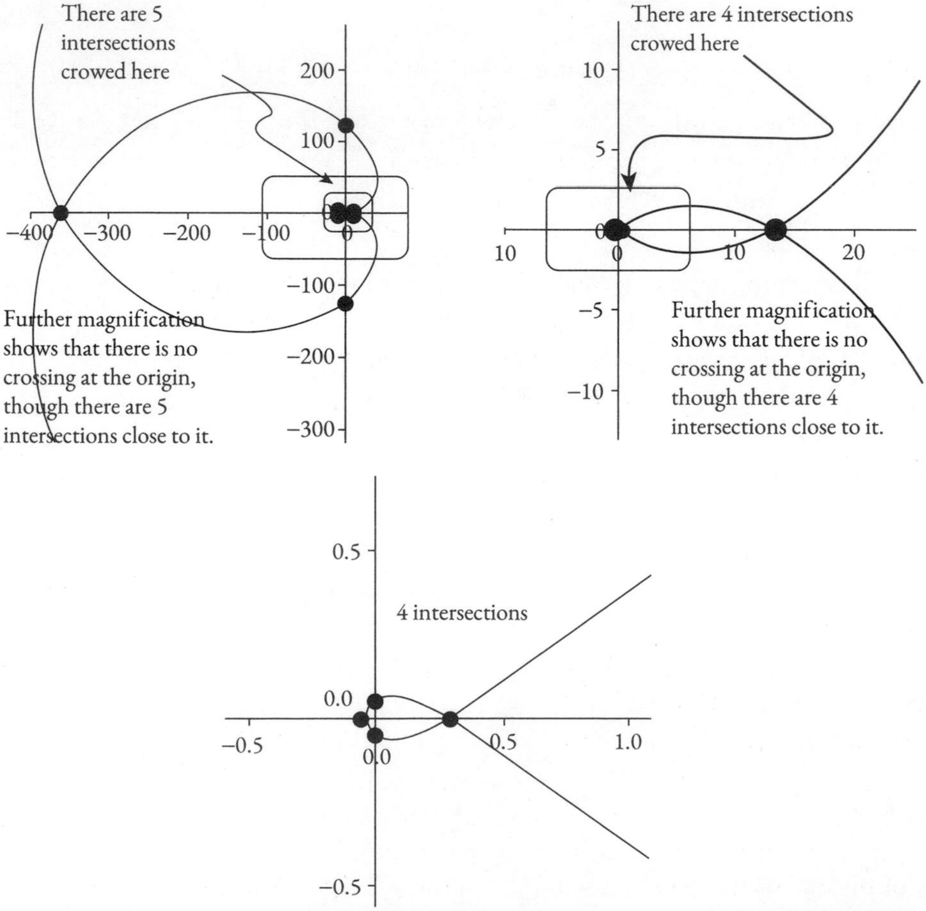
There are 5 intersections crowed here
200
100
−100
−200
−300
−400
−300
−200
−100
0
Further magnification shows that there is no crossing at the origin, though there are 5 intersections close to it.
There are 4 intersections crowed here
10
5
−5
−10
10
0
10
20
Further magnification shows that there is no crossing at the origin, though there are 4 intersections close to it.
0.5
4 intersections
0.0
−0.5
−0.5
0.0
0.5
1.0

a family of curves which are the different branches of the Lambert W functions. The set of these curves is labeled as $W(S) \equiv W(S_r) \bigcup W(S_i)$. Each of these curves is labeled as $w(k, \text{ray})$ to specify the kth branch, and the specific ray specified. Solutions to the finite square well problem are then given by the intersections of $W(k, \text{ray})$ with the circle in the w plane which represents the particular strength of the potential. The labeling of the curves in the w plane is as per the convention described in References [14, 19]. The intersections of the family of curves $W(S)$ in the w plane by the circle $|w| = R$ are shown in Fig. 3.13, which provide the solutions to the finite square well problem. Only the branches with $k = 0, -1$ intersect with the $R = 5$ circle. Solutions in the 1st quadrant have $(u > 0, v > 0)$. There are a total of 14 intersections of the Q curve in the z plane with the coordinate axes (Fig. 3.12). Correspondingly, there are 14 intersections of the circle $|w| = R$ in the w plane with family of P curves, labeled as $w(k, \text{branch})$ in Fig. 3.13, obtained from mapping the z plane coordinate axes rays into the w plane.

The mapping $w \leftrightarrow z$ is conformal. Hence, the angles which the points of intersection of the $|w| = R$ circle with the Lambert family of curves for various branches make in the complex w plane (Fig. 3.12), and angles which the points of intersection of the multi-loop self-intersecting Q-curve make with the coordinate axes in the complex z plane (Fig. 3.13), belong essentially to the same set of 14 angles, tabulated in Table 3.2.

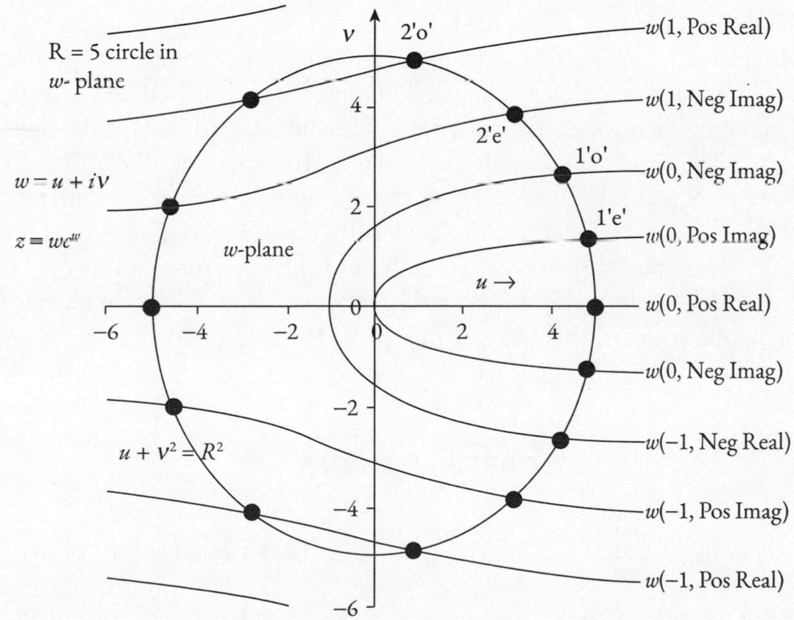

Fig. 3.13 Maps in the w plane of the axes-rays in the z plane, using the Lambert W function. The notation $w(k, \text{ray})$ which labels the curves in the w plane specifies the kth branch, and the specific ray specified. The four solutions in the quadrant for $(u > 0, v > 0)$ are labeled as 1'e', 1'o', 2'e', and 2'o' corresponding to the 1st and 2nd "even" $(\varepsilon = +1)$ and "odd" $(\varepsilon = -1)$ solutions of Fig. 3.8.

Table 3.2 Angles corresponding to the conformal mapping of the points of intersection in Figs. 3.12 and 3.13.

Crossing on the real axis X in the z plane	Crossing on the imaginary axis Y in the z plane
0.000	0.264
0.546	0.875
1.377	2.735
2.179	3.548
3.142 (i.e., μ)	5.408
4.105	6.019
4.906	–
5.737	–

In this table the four angles in the range $0 < \theta < \frac{\pi}{2}$ correspond to the four solutions in the quadrant for $(u > 0, v > 0)$. These are labeled as 1'e', 1'o', 2'e', and 2'o' in Fig. 3.13. They correspond to the 1st and 2nd "even" $(\varepsilon = +1)$ and "odd" $(\varepsilon = -1)$ solutions of Fig. 3.8. A similar *geometric analytical* method to obtain solutions of the time-independent Schrödinger equation for bound states of a particle in a one-dimensional finite *double*-square well potential has also been developed [24].

The technique described in this section have some very interesting applications. For example, it has been used to suggest how to minimize thermal conductivity to augment the figure of merit of thermoelectric materials [25] and also to study the eigenvalues of nanoribbons [26]. The model potentials discussed in this chapter illustrate properties of nature which demand quantum theory. These include tunneling, over the barrier tunneling, appearance of evanescent wavefunctions in classically forbidden region, etc. Understanding of phenomena based on probabilistic interpretation of the wavefunction has led to exciting advances [8, 9, 16, 25–27] in technology. There are optical (i.e., electromagnetic) analogues of the phenomena discussed in this chapter that provide further connects between quantum theory and optical and electronic properties.

Solved Problems

P3.1: Show that $\frac{\partial \rho(\vec{r},t)}{\partial t} + \vec{\nabla}\cdot\vec{j}(\vec{r},t) = 0$, where $\vec{j}(\vec{r},t) = \frac{\hbar}{m}\mathrm{Im}\left\{\psi^*(\vec{r},t)\vec{\nabla}\psi(\vec{r},t)\right\}$, or for short, $\vec{j} = \frac{\hbar}{m}\mathrm{Im}\left(\psi^*\vec{\nabla}\psi\right)$. "Im" stands for "Imaginary part of".

Solution:

From the time-dependent Schrödinger equation, we have

$$\frac{\partial \rho}{\partial t} = \psi^*\frac{\partial \psi}{\partial t} + \psi\frac{\partial \psi^*}{\partial t} = \psi^*\left\{\frac{-i}{\hbar}H\psi\right\} + \psi\left\{\frac{i}{\hbar}H^\dagger\psi^*\right\} = \frac{i}{\hbar}\left\{\psi\left(\frac{-\hbar^2}{2m}\vec{\nabla}\cdot\vec{\nabla}\right)\psi^* - \psi^*\left(\frac{-\hbar^2}{2m}\vec{\nabla}\cdot\vec{\nabla}\right)\psi\right\},$$

i.e., $\frac{\partial\rho}{\partial t} = \frac{-i\hbar}{2m}\left\{\psi\left(\vec{\nabla}\cdot\vec{\nabla}\right)\psi^* - \psi^*\left(\vec{\nabla}\cdot\vec{\nabla}\right)\psi\right\}$,

where the kinetic energy operator is $T = \frac{\left(-i\hbar\vec{\nabla}\right)\cdot\left(-i\hbar\vec{\nabla}\right)}{2m} = T^\dagger$ and the potential energy operator is $V = V^\dagger$. The potential energy term drops out. We therefore see that

$$\frac{\partial\rho(\vec{r},t)}{\partial t} = \vec{\nabla}\cdot\left\{\frac{-i\hbar}{2m}\left(\psi\vec{\nabla}\psi^* - \psi^*\vec{\nabla}\psi\right)\right\} = \vec{\nabla}\cdot\left\{-\vec{j}(\vec{r},t)\right\},$$

with $\vec{j}(\vec{r},t) = \frac{i\hbar}{2m}\left\{\psi(\vec{r},t)\vec{\nabla}\psi^*(\vec{r},t) - \psi^*(\vec{r},t)\vec{\nabla}\psi(\vec{r},t)\right\} = \frac{\hbar}{m}\mathrm{Im}\left\{\psi^*(\vec{r},t)\vec{\nabla}\psi(\vec{r},t)\right\}$.

We have used (a) $\left(\vec{\nabla}\psi\right)\cdot\left(\vec{\nabla}\psi^*\right) = \left(\vec{\nabla}\psi^*\right)\cdot\left(\vec{\nabla}\psi\right)$ and (b) $z - z^* = 2i\times\mathrm{Im}(z)$, for an arbitrary complex number z. In the present case, we have identified z as $\psi^*\vec{\nabla}\psi$.

P3.2: Show for a particle having charge e that the equation of continuity is satisfied with the probability charge current density vector given by

$$\vec{j}^{(e)} = \frac{1}{2m}\left\{\psi^*\vec{p}_c\psi - \psi\vec{p}_c\psi^* - 2\frac{e}{c}\vec{A}(\psi^*\psi)\right\}.$$

Solution:

Using the Hamiltonian (from Eq. 2.73)

$$H = \frac{1}{2m}\left\{-\hbar^2\vec{\nabla}^2 + \frac{i\hbar e}{c}\left(\vec{\nabla}\cdot\vec{A} + \vec{\nabla}\cdot\vec{A}\right) + \frac{e^2}{c^2}\vec{A}^2\right\} + e\phi$$

in the time-dependent Schrödinger equation we get

$$\frac{\partial\rho}{\partial t} = \frac{-i}{2\hbar m}\left\{-\hbar^2\psi^*\vec{\nabla}^2\psi + \hbar^2\psi\vec{\nabla}^2\psi^*\right\} - \frac{i}{2\hbar m}\left\{\frac{2i\hbar e}{c}\psi^*\left(\vec{A}\cdot\vec{\nabla}\right)\psi + \frac{2i\hbar e}{c}\psi\left(\vec{A}\cdot\vec{\nabla}\right)\psi^*\right\},$$

i.e., $\frac{\partial\rho}{\partial t} = \frac{-i\hbar}{2m}\left\{\psi\vec{\nabla}\cdot\vec{\nabla}\psi^* - \psi^*\vec{\nabla}\cdot\vec{\nabla}\psi\right\} + \frac{1}{2m}\frac{2e}{c}\left\{\psi^*\left(\vec{A}\cdot\vec{\nabla}\right)\psi + \psi\left(\vec{A}\cdot\vec{\nabla}\right)\psi^*\right\}$,

or $\frac{\partial\rho}{\partial t} = \vec{\nabla}\cdot\left\{\frac{-i\hbar}{2m}\left(\psi\vec{\nabla}\psi^* - \psi^*\vec{\nabla}\psi\right)\right\} + \frac{1}{2m}\frac{2e}{c}\vec{\nabla}\cdot\left[\vec{A}(\psi^*\psi)\right]$.

Hence, $\frac{\partial\rho}{\partial t} = -\vec{\nabla}\cdot\vec{j}^{(e)}$, where $\vec{j}^{(e)} = \frac{1}{2m}\left[\psi\vec{p}_c\psi^* - \psi^*\vec{p}_c\psi - \frac{2e}{c}\vec{A}(\psi^*\psi)\right]$ and $\vec{p}_c = -i\hbar\vec{\nabla}$.

P3.3: Prove that *none* of the energy eigenvalues in the *bound-state spectrum* of a *one-dimensional* potential is *degenerate*.

Solution:

Assume two linearly independent solutions, $\psi_1(x)$ and $\psi_2(x)$, which belong to the same eigenvalue in a one-dimensional Schrödinger equation:

$$\frac{1}{2m}\left(-i\hbar\frac{d}{dx}\right)^2\psi_1(x) + V(x)\psi_1(x) = E\psi_1(x), \text{ and } \frac{1}{2m}\left(-i\hbar\frac{d}{dx}\right)^2\psi_2(x) + V(x)\psi_2(x) = E\psi_2(x).$$

First, we multiply the differential equation for ψ_1 by ψ_2 and that for ψ_2 by ψ_1. Next, taking the difference of the resulting equations, we get:

$\frac{-\hbar^2}{2m}\left(\psi_2\psi_1'' - \psi_1\psi_2''\right) = 0,$ i.e., $\psi_2\psi_1'' - \psi_1\psi_2'' = 0,$ or $\frac{d}{dx}\left(\psi_2\psi_1' - \psi_1\psi_2'\right) = 0.$ We have denoted each differentiation with respect to x by placing a prime on the function. It now follows that $\psi_2(x)\psi_1'(x) - \psi_1(x)\psi_2'(x) = K,$ a constant, whose value of course cannot depend on the value of the coordinate x. Since we know that for physically bound states, $\psi_{1,2}(x \to \pm\infty) = 0,$ we conclude easily that the constant K must be zero. Hence, $\frac{\psi_2'}{\psi_2} = \frac{\psi_1'}{\psi_1},$ i.e., $\ln\psi_2 = \ln\psi_1 + C,$ or $\psi_2 = e^C\psi_1.$ We had begun with the *assumption* that ψ_2 and ψ_1 are linearly independent, but we now get a result that is obviously contrary to that! We therefore conclude that *in a one-dimensional problem, none of the energy values of the bound-state spectrum is degenerate*. In Chapter 2, we did come across a degeneracy in a one-dimensional problem n which we have *discrete* energy states (Eq. 2.90b); however the boundary condition there was the *periodic boundary condition*, whereas in the present case it is different; the wavefunction goes to zero in the asymptotic region.

P3.4: Use the coordinate representation of Eq. 3.88 to obtain the ground state wavefunction of the simple harmonic oscillator.

Solution:

Using Eq. 3.88 and Eq. 3.81a, we see that

$$\langle x'|c|0\rangle = \sqrt{\frac{m\omega}{2\hbar}}\langle x'|x + \frac{ip}{m\omega}|0\rangle = \sqrt{\frac{m\omega}{2\hbar}}\langle x'|x + \frac{\hbar}{m\omega}\frac{d}{dx}|0\rangle = 0.$$

i.e., $\langle x'|x + \frac{\hbar}{m\omega}\frac{d}{dx}|0\rangle = 0$ or $\langle x'|x|0\rangle + \frac{\hbar}{m\omega}\langle x'|\frac{d}{dx}|0\rangle = 0;$ i.e., $x'\langle x'|0\rangle + \frac{\hbar}{m\omega}\frac{d}{dx'}\langle x'|0\rangle = 0.$

In the de-Broglie-Schrödinger notation, we have a differential equation

$x'\psi_0(x') + \frac{\hbar}{m\omega}\frac{d}{dx'}\psi_0(x') = 0.$ The normalized solution to this differential equation is the same, not surprisingly, as that for the ground state wavefunction of the harmonic oscillator listed in Table 3.1 for $n = 0$. It is important to observe that the differential equation we solved is *not* the Schrödinger equation, but it of course is based on the quantum mechanics of the position and momentum operators. Wavefunctions for the excited states are also obtained in the same manner. We shall use this idea again in Chapter 5 to obtain eigenfunctions of the hydrogen atom without using the Schrödinger equation.

P3.5: Rewrite the space-part of the solutions (Eqs. 3.23a,b,c) in the three regions in Fig. 3.4c as $\psi_I(x) = A\exp(ikx) + B\exp(-ikx);$ $\psi_{II}(x) = C\exp(-\kappa x) + D\exp(\kappa x);$ and $\psi_{III}(x) = F\exp(ikx) + G\exp(-ikx).$ *Determine* the 2×2 matrix $\mathbb{M}_{2\times2}$ *such that* $\begin{bmatrix} A \\ B \end{bmatrix} = \mathbb{M}\begin{bmatrix} F \\ G \end{bmatrix}.$

Solution:

The equation of continuity for the probability current density vector requires that the wavefunction and its derivative are both continuous at the boundary of the regions I and II, and at the boundary of the regions II and III. This condition translates to

$$\begin{bmatrix} A \\ B \end{bmatrix} = \frac{1}{2}\begin{bmatrix} \left(1+\frac{i\kappa}{k}\right)\exp(\kappa a + ika) & \left(1-\frac{i\kappa}{k}\right)\exp(-\kappa a + ika) \\ \left(1-\frac{i\kappa}{k}\right)\exp(\kappa a - ika) & \left(1+\frac{i\kappa}{k}\right)\exp(-\kappa a - ika) \end{bmatrix}\begin{bmatrix} C \\ D \end{bmatrix}$$

and

$$\begin{bmatrix} C \\ D \end{bmatrix} = \frac{1}{2}\begin{bmatrix} \left(1-\frac{ik}{\kappa}\right)\exp(\kappa a + ika) & \left(1+\frac{ik}{\kappa}\right)\exp(\kappa a - ika) \\ \left(1+\frac{ik}{\kappa}\right)\exp(-\kappa a + ika) & \left(1-\frac{ik}{\kappa}\right)\exp(-\kappa a - ika) \end{bmatrix}\begin{bmatrix} F \\ G \end{bmatrix}.$$ Therefrom it follows that

$$\mathbb{M} = \begin{bmatrix} \left(\cosh 2\kappa a + \frac{i\gamma_-}{2}\sinh 2\kappa a\right)\exp(2ika) & \frac{i\gamma_+}{2}\sinh 2\kappa a \\ -\frac{i\gamma_+}{2}\sinh 2\kappa a & \left(\cosh 2\kappa a - \frac{i\gamma_-}{2}\sinh 2\kappa a\right)\exp(-2ika) \end{bmatrix} \text{ with } \gamma_\pm = \frac{\kappa}{k} \pm \frac{k}{\kappa}.$$

In Chapter 10 (Problem P10.7) you will see how this is related to the S-matrix of collision theory. As one might expect, you may verify that the determinant $|\mathbb{M}| = 1$.

When $G = 0$, in the limit that of having a strong barrier $(\kappa a \gg 1)$ you would find that the probability of transmission (Eq. 3.27; Eq. 3.28a) provides a measure of the tunneling probability.

P3.6: Show that an attractive Dirac-δ one-dimensional potential has a single *bound* state and determine its energy eigenvalue.

Solution:

Let us denote the potential as $V(x) = -\lambda\delta(x); \lambda > 0$.

The Schrödinger equation is $-\frac{\hbar^2}{2m}\frac{d^2\psi(x)}{dx^2} - \lambda\delta(x)\psi(x) = E\psi(x)$, with $E < 0$ for *bound* states.

The general solution is $\psi(x) = A\exp(\mu x) + B\exp(-\mu x)$ with $\mu = \sqrt{\left(-\frac{2mE}{\hbar^2}\right)}$. Therefore, we must have $\psi_I(x) = A\exp(\mu x); x < 0$ and $\psi_{II}(x) = B\exp(-\mu x); x > 0$ to prevent the wavefunction from blowing up. Continuity of the wavefunction requires $A = B$. In other words, $\psi(x) = A\exp(-\mu|x|)$. The derivative of the wavefunction, $\psi' = \frac{d\psi}{dx}$, cannot be continuous at $x = 0$ since the potential blows up at that point.

Integrating the Schrödinger equation, $-\frac{\hbar^2}{2m}\int_{-\varepsilon}^{\varepsilon} dx\frac{d^2\psi(x)}{dx^2} - \lambda\int_{-\varepsilon}^{\varepsilon} dx\delta(x)\psi(x) = E\int_{-\varepsilon}^{\varepsilon} dx\psi(x)$,

i.e., $-\frac{\hbar^2}{2m}\left[\frac{d\psi(x)}{dx}\right]_{-\varepsilon}^{\varepsilon} - \lambda\int_{-\varepsilon}^{\varepsilon} dx\delta(x)\psi(x) = E\int_{-\varepsilon}^{\varepsilon} dx\psi(x)$.

In the limit $\varepsilon \to 0$, $-\frac{\hbar^2}{2m}\left[\frac{d\psi(x)}{dx}\right]_{-\varepsilon}^{\varepsilon} - \lambda\psi(x = 0) = 0$, i.e., $-\frac{\hbar^2}{2m}\left[\frac{d\psi(x)}{dx}\right]_{-\varepsilon}^{\varepsilon} - \lambda A = 0$.

$$-\frac{\hbar^2}{2m}\left[\frac{d\psi(x)}{dx}\right]_{-\varepsilon}^{\varepsilon} = \frac{\hbar^2}{m}A\mu\lim_{\varepsilon\to 0}\exp(-\mu\varepsilon) = \frac{\hbar^2}{m}A\mu.$$

Hence, $\frac{\hbar^2}{m}A\mu = A\lambda$ or $\mu = \frac{m}{\hbar^2}\lambda$. The bound-state energy of the Dirac-δ attractive potential therefore is unique, given by $E = -\frac{m}{2\hbar^2}\lambda^2$. Note that the wavefunction is normalizable (square integrable) given by $\psi(x) = \frac{\sqrt{m\lambda}}{\hbar}\exp\left(-\frac{m\lambda}{\hbar^2}|x|\right)$.

Supplementary Note: Please Revisit P2.12 (Chapter 2).

P3.7: Determine the scattering states (i.e., with $E > 0$) of an (i) attractive and (ii) repulsive Dirac-δ one-dimensional potential.

Solution:

The general solution for $x \neq 0$ is:

$$\psi_I(x) = A\exp(ikx) + B\exp(-ikx); x < 0 \text{ and } \psi_{II}(x) = C\exp(ikx) + D\exp(-ikx); x > 0.$$

Since the wavefunction must be continuous at $x = 0$, we must have $A + B = C + D$.

Form the limiting behavior of the derivative: $-\frac{\hbar^2}{2m}(ik)\left[(C - D) - (A - B)\right] = \lambda(C + D)$.

Wavefunctions for scattering states $E > 0$ are not square integrable, so the normalization condition does not provide assistance toward determining the four unknowns A, B, C, D from just two equations we have with us. For a particle incident from the left, there would be a possibility of tunneling and also of reflection, but after tunneling there is nothing in the region $x > 0$ to reverse its direction of motion; this boundary condition is equivalent to choosing $D = 0$. (*In Chapter 10, we shall interpret this as outgoing wave boundary condition.*) That reduces the number of unknowns from four to three and we get $\frac{B}{A} = -\frac{k_0}{k_0 + ik}$ and $\frac{C}{A} = \frac{ik}{k_0 + ik}$ with $k_0 = \frac{m\lambda}{\hbar^2}$.

The reflection coefficient is $R = \frac{|B|^2}{|A|^2} = \frac{k_0^2}{k_0^2 + k^2} = \frac{E_0}{E_0 + E}$ tunneling (i.e., transmission) coefficient is $T = 1 - R = \frac{k^2}{k_0^2 + k^2} = \frac{E}{E_0 + E}$, where $E_0 = \frac{\hbar^2 k_0^2}{2m} = \frac{m\lambda^2}{2\hbar^2}$. The result being quadratic in λ holds good for both attractive and repulsive Dirac-δ potentials, but of course the repulsive potential does not hold any bound state.

P3.8: This problem is based on Reference [24]. You would use the Lambert W method discussed in Section 3.4 to obtain solutions for a one-dimensional Double Square Well Potential (DSWP) shown in the adjacent figure.

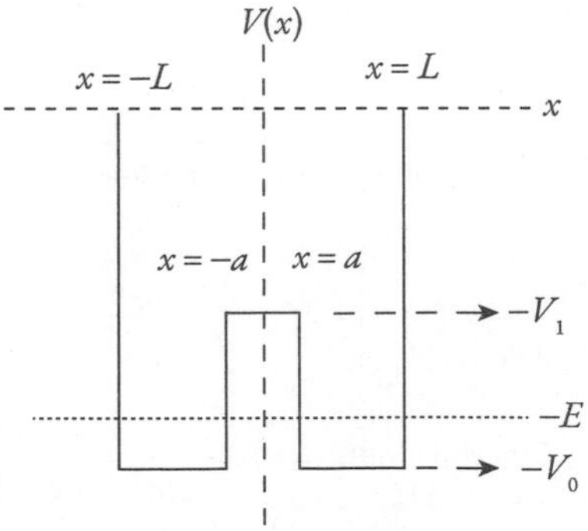

The outer walls ($x < -L$ and $x > L$) are at zero potential. Two symmetric wells having depth $-V_0$ are in *regions 1* ($-L < x < -a$) and region 3 ($a < x < L$). The middle barrier (*region 2*, $-a < x < a$)

has a finite depth $-V_1$, with $(-V_0) < (-V_1)$. Consider a particle with energy $-E$ in a bound state such that $(-V_0) < (-E) < (-V_1)$.

Note that V_0, E, V_1 are defined to be intrinsically positive. Consider E to be large, so that the potential barriers at $x = \pm L$ are very large. Using the equation of continuity at the boundaries of the potential, eliminate the coefficients B and C first, and then also A_1 and A_2 (Hint: First obtain the ratio $\frac{A_1}{A_2}$) and obtain relations between α, β, γ.

Solution:

Due to the symmetry of the potential, the Schrödinger wavefunction is either even or odd; i.e., $\psi(-x) = \varepsilon\psi(x)$ with $\varepsilon = \pm 1$.

Region "2": $\psi_2(x) = B_1 \exp(-\beta x) + B_2 \exp(\beta x)$, (P3.8.1)

i.e., $\psi_2(x) = \varepsilon B \exp(-\beta x) + B \exp(\beta x)$, (P3.8.2)

with $\beta^2 = \frac{2m}{\hbar^2}(E - V_1)$, since the region 2 straddles the origin and the wavefunction is odd or even.

Region "3": $\psi_3(x) = A_1 \exp(-i\alpha x) + A_2 \exp(+i\alpha x)$, with $\alpha^2 = \frac{2m}{\hbar^2}(V_0 - E)$. (P3.8.3)

Region $x > L$: $\psi_r(x) = C \exp(-\gamma x)$ with $\gamma = \sqrt{\frac{2mE}{\hbar^2}}$. (P3.8.4)

We assign the value "zero" to the coefficient of the component $\exp(+\gamma x)$ as per the boundary conditions; otherwise, it would blow up as $x \to \infty$.

Note that each of the α, β, γ is positive. The offset of the energy $-E$ from $-V_0$ and from $-V_1$ is important for the determination of the bound state. The parameters α, β provide a measure of this offset. From the equation of continuity (see Eq. 3.2 and Eq. 3.3), the wavefunction and its derivative must be continuous.

$\psi_3(a) = \psi_2(a)$, therefore

$A_1 \exp(-i\alpha a) + A_2 \exp(i\alpha a) = \varepsilon B \exp(-\beta a) + B \exp(\beta a)$. (P3.8.5)

Also, $\psi_3'(a) = \psi_2'(a)$, therefore

$-i\alpha A_1 \exp(-i\alpha a) + i\alpha A_2 \exp(i\alpha a) = -\varepsilon\beta B \exp(-\beta a) + \beta B \exp(\beta a)$. (P3.8.6)

Again, since $\psi_3(L) = \psi_r(L)$,

$A_1 \exp(-i\alpha L) + A_2 \exp(i\alpha L) = C \exp(-\gamma a)$. (P3.8.7)

Likewise, $\psi_3'(L) = \psi_r'(L)$, and hence

$-i\alpha A_1 \exp(-i\alpha L) + i\alpha A_2 \exp(i\alpha L) = -\gamma C \exp(-\gamma a)$. (P3.8.8)

Now, $\frac{i\alpha * (Eq.P3.8.5) - Eq.P3.8.6}{i\alpha * (Eq.P3.8.5) + Eq.P3.8.6}$ eliminates the coefficient B and gives

$$\frac{A_1}{A_2}\exp(-i2\alpha a) = \frac{(i\alpha - \beta)\exp(\beta a) + \varepsilon(i\alpha + \beta)\exp(-\beta a)}{(i\alpha + \beta)\exp(\beta a) + \varepsilon(i\alpha - \beta)\exp(-\beta a)}. \quad \text{(P3.8.9)}$$

Likewise, $\dfrac{i\alpha * (Eq.P3.8.7) - Eq.P3.8.8}{i\alpha * (Eq.P3.8.7) + Eq.P3.8.8}$ eliminates the coefficient B, and gives

$$\frac{A_1}{A_2}\exp(-i2\alpha L) = \frac{(i\alpha + \gamma)}{(i\alpha - \gamma)} = -\left[1 + 2\frac{i\alpha}{\gamma} + 2\left(\frac{i\alpha}{\gamma}\right)^2 + ...\right]. \quad \text{(P3.8.10)}$$

The series on the right-hand side of the above equation converges for $V_1 > \frac{V_0}{2}$.

Using Eq. P.3.8.9 and Eq. 3.8.10, we eliminate the coefficients A_1 and A_2 to get

$$\frac{(i\alpha + \gamma)}{(i\alpha - \gamma)}\exp(+i2\alpha(L - a)) = \frac{(i\alpha - \beta)\exp(\beta a) + \varepsilon(i\alpha + \beta)\exp(-\beta a)}{(i\alpha + \beta)\exp(\beta a) + \varepsilon(i\alpha - \beta)\exp(-\beta a)}. \quad \text{(P3.8.11)}$$

P3.9: Use the limit $\gamma \to$ large. Use (a) new dimensional variables $u = \beta a$ and $v = \alpha a$ and (b) $u^2 + v^2 = R^2$ and solve for energy eigenvalues of the DSWP.

Solution:

In the limit $\gamma \to$ large, from Eq. P3.8.10, we get $\frac{A_1}{A_2}\exp(-i2\alpha L) = -1$, using which in Eq. P.3.8.9,

$$\text{we get } \exp(+i2\alpha(L - a)) = -\frac{(i\alpha - \beta)\exp(\beta a) + \varepsilon(i\alpha + \beta)\exp(-\beta a)}{(i\alpha + \beta)\exp(\beta a) + \varepsilon(i\alpha - \beta)\exp(-\beta a)}. \quad \text{(P3.9.1)}$$

$$\text{Now, } u^2 + v^2 = R^2 = (\alpha^2 + \beta^2)a^2 = \frac{2ma^2}{\hbar^2}(V_0 - V_1). \quad \text{(P3.9.2)}$$

Introducing $f = \dfrac{L - a}{a}$ from Eq. P3.9.1 we get

$$\exp(+i2fv) = -\frac{(iv - u)\exp(u) + \varepsilon(iv + u)\exp(-u)}{(iv + u)\exp(u) + \varepsilon(iv - u)\exp(-u)}. \quad \text{(P3.9.3)}$$

The term f is determined by the specific geometry of the DSWP. In Reference [24], Eq. 3.9.2 is referred to as "*Radial Equation*" (RE) and Eq. 3.9.3 as "*Structural Equation*" (SE). The locus of points $w \equiv (u,v)$ in the complex-w plane $(w = u + iv)$ satisfies the SE and constitute a set of curves called as the *Lambert lines* (or "Lamlines"), referred in P3.10. The RE provides descriptions of circles in the w plane. The curves generated by SE have complicated shapes. Their intersections provide bound-state energies.

P3.10: Use the *geometric-analytic* method discussed in Section 3.4 to obtain the energy eigenvalues of DSWP.

Solution:

The RE and SE discussed in P3.9 each defines a family of curves in the complex-w plane $(w = u + iv)$ which is generated by the locus of the point (u,v). In order to use the Lambert W method (Section 3.4), we need the map $(w = u + iv) \to (z = x + iy)$. Specifically, our objective is to determine the image $z \equiv (x,y)$ of the locus of $w \equiv (u,v)$, which satisfies the SE. The multiply branched Lambert W function is the inverse map $(z = x + iy) \to (w = u + iv)$. The Lambert lines (referenced in P3.9) are the images of the axial rays in the complex $z \equiv (x,y)$ plane.

From the SE (Eq. P3.9.3), we have

$$(iv+u)\exp(u)\exp(+i2fv)+\varepsilon(iv-u)\exp(-u)\exp(+i2fv)=-(iv-u)\exp(u)-\varepsilon(iv+u)\exp(-u).$$

Hence, multiplying throughout by $\exp(-ifv)$, we get

$$\begin{bmatrix}(u+iv)\exp(u+ifv)\\ +\varepsilon(-u+iv)\exp(-u+ifv)\end{bmatrix}=\begin{bmatrix}(u-iv)\exp(u-ifv)\\ -\varepsilon(u+iv)\exp(-(u+ifv))\end{bmatrix}. \quad \text{(P3.10.1)}$$

First, we consider a map $w \to z$ such that $(x,y) \equiv z = Y(w) \equiv Y(u,v) = (u+iv)\exp(u+ifv)$. In this map, Y is described as a function of two real variables (u,v). The result of this map would give the two real variables $z \equiv (x,y)$. The mapping $z = Y(w) = Y(u,v)$ is not conformal for $f \neq 1$, but nonetheless it sustains important features including (i) intersections of curves in the two planes and (ii) relative magnitudes of the angles of intersection.

We now restate Eq. P3.10.1 as

$$Y(w)+\varepsilon Y(-w^*)=Y(w^*)+\varepsilon Y(-w), \quad \text{(P3.10.2a)}$$

i.e., $Y(u,v)+\varepsilon Y(-u,v)=Y(u,-v)+\varepsilon Y(-u,-v)$. (P3.10.2b)

Now, $(v) \rightleftarrows (-v)$ is equivalent to $(w) \rightleftarrows (w^*)$,

$(u,v) \rightleftarrows (-u,-v)$ is $(w) \rightleftarrows (-w)$,

and $(u,v) \rightleftarrows (-u,v)$ is $(w) \rightleftarrows (-w^*)$.

Therefore, we consider a second map

$$z=G(w)=G(u,v)=Y(u,v)-\varepsilon Y(-u,-v), \quad \text{(P3.10.3a)}$$

i.e., $z=G(w)=Y(w)-\varepsilon Y(-w)$. (P3.10.3b)

Accordingly, Eq. P3.10.2 becomes $z=G(w)=G(w^*)=\left[G(w)\right]^*$, (P3.10.4)

i.e., $G(w)$ is real.

The map $z=G(w)$ identifies the relation $G(w)=\left[G(w)\right]^*$ as the SE which provides the solution to the DSWP one-dimensional quantum problem. $\varepsilon=\pm 1$ provides the eigenvalues for the even and odd states. This map transforms the circles in the *w* plane to multiloop curves in the *z* plane. Intersections of these loops with the real axis provide simultaneous solutions of the RE and SE, providing the eigenvalues of the DSWP.

For a detailed discussion on P3.8, P3.9, and P3.10, please refer to Reference [24].

P3.11: In Section 3.3, we considered a quantum dot. In this problem, we consider a particle confined in a three-dimensional slab. Such a particle is confined in what is called a *quantum well*. Depending on the symmetry of the potential in the Hamiltonian, a two- or three-dimensional problem can sometimes be reduced to a one-dimensional problem.

Solution:

The particle is in a potential given by $V(\vec{r}) = 0$, for $0 < z < a$; $0 < x < L_x$; $0 < y < L_y$ and infinite elsewhere. The width of the well is 'a', and the *other* two lengths L_x, L_y are significantly longer. Boundary conditions on the wavefunction are

$$\psi(0,y,z) = 0,\ \ 0 \le y \le L_y,\ \ 0 \le z \le a,$$

$$\psi(L_x,y,z) = 0,\ \ 0 \le y \le L_y,\ \ 0 \le z \le a,$$

$$\psi(x,0,z) = 0,\ \ 0 \le x \le L_x,\ \ 0 \le z \le a,$$

$$\psi(x,L_y,z) = 0,\ \ 0 \le x \le L_x,\ \ 0 \le z \le a,$$

$$\psi(x,y,0) = 0,\ \ 0 \le x \le L_x,\ \ 0 \le y \le L_y,$$

$$\psi(x,y,a) = 0,\ \ 0 \le x \le L_x,\ \ 0 \le y \le L_y.$$

The Schrödinger equation inside the box for a carrier in a semiconductor having effective mass μ is

$$-\frac{\hbar^2}{2\mu}\left(\frac{\partial^2}{\partial x^2} + \frac{\partial^2}{\partial y^2} + \frac{\partial^2}{\partial z^2}\right)\psi = E\psi.$$

Considering a separable wavefunction of the form $\psi(x,y,z) = \psi_x(x)\psi_y(y)\psi_z(z)$, the equation to be solved is

$$\frac{1}{\psi_x}\left(-\frac{\hbar^2}{2\mu}\right)\frac{d^2\psi_x}{dx^2} + \frac{1}{\psi_y}\left(-\frac{\hbar^2}{2\mu}\right)\frac{d^2\psi_y}{dy^2} + \frac{1}{\psi_z}\left(-\frac{\hbar^2}{2\mu}\right)\frac{d^2\psi_z}{dz^2} = E.$$

For this relation to hold, each term must be equal to a constant. Instead of having the three-dimensional problem, we now have *three* one-dimensional problems.

Hence $-\frac{\hbar^2}{2\mu}\frac{d^2\psi_x}{dx^2} = E_x\psi_x$; $\quad -\frac{\hbar^2}{2\mu}\frac{d^2\psi_y}{dy^2} = E_y\psi_y$; $\quad -\frac{\hbar^2}{2\mu}\frac{d^2\psi_z}{dz^2} = E_z\psi_z$.

In the region $0 \le x \le L_x, \psi_x(x) = \frac{1}{L_x}\exp(ik_x x)$ with $k_x = \frac{2n_x\pi}{L_x}$, $n_x = 0, \pm 1, \pm 2, \ldots$;

In the region $0 \le y \le L_y$, we have $\psi_y(y) = \frac{1}{L_y}\exp(ik_y y)$ with $k_y = \frac{2n_y\pi}{L_y}$, $n_y = 0, \pm 1, \pm 2, \ldots$;

In the region $0 \le z \le a$, we have $\frac{d^2\psi_z}{dz^2} + k_z^2\psi_z = 0$

with $k_z^2 = \frac{2\mu E_z}{\hbar^2}$ and $\psi_z(z) = A_z\exp(ik_z z) + B_z\exp(-ik_z z)$.

Applying the boundary conditions $\psi_z(0) = \psi_z(a) = 0$, we have $\psi_z(0) = A_z + B_z$, i.e., $A_z = -B_z$ and $\psi_z(z) = A_z\left(\exp(ik_z z) - \exp(-ik_z z)\right) = 2iA_z\sin(k_z z)$.

Applying now the second boundary condition, we get $\psi_z(a) = 2iA_z\sin(k_z a) = 0$, i.e., $A_Z = 0$ (which corresponds to a trivial solution), or $k_z a = n_z\pi$, i.e., $k_z = \frac{n_z\pi}{a}$ with $n_z = 1, 2, 3, \ldots$

Therefore, $\psi_z(z) = \sqrt{\frac{2}{a}}\sin\left(\frac{n_z\pi}{a}z\right)$.

The total wavefunction therefore is $\psi(x,y,z) = \frac{1}{\sqrt{L_x L_y}}\sqrt{\frac{2}{a}}\sin\left(\frac{n_z\pi}{a}z\right)\exp(ik_x x)\exp(ik_y y)$.

The total energy is $E = \frac{\hbar^2}{2\mu}\left(k_x^2 + k_y^2\right) + \frac{\hbar^2 n_z^2}{2\mu a^2}$.

P3.12: Model the wavefunction of a carrier (having effective mass μ) in a semiconductor nanowire (*quantum wire*) by "particle in a cylindrical potential" and solve its Schrödinger equation by using the method of separation of variables and reducing the problem to a one-dimensional problem.

Solution:

We shall use the cylindrical polar coordinates (ρ,φ,z) described in Chapter 2 of Reference [1]. The potential is:

$$V(\rho,\varphi,z) = \begin{cases} 0, & \text{for } 0 \le \rho \le a,\ 0 < z < L_z \\ \infty, & \text{elsewhere} \end{cases}$$ with $L_z \gg a$, the radius of the cylindrical wire. The Schrödinger equation with zero-potential is: $-\frac{\hbar^2}{2\mu}\left(\frac{1}{\rho}\frac{\partial}{\partial\rho}\left(\rho\frac{\partial}{\partial\rho}\right)+\frac{1}{\rho^2}\frac{\partial^2}{\partial\varphi^2}+\frac{\partial^2}{\partial z^2}\right)\psi = E\psi,$

i.e., $\left(\rho\frac{\partial}{\partial\rho}\left(\rho\frac{\partial}{\partial\rho}\right)+\frac{\partial^2}{\partial\varphi^2}+\rho^2\frac{\partial^2}{\partial z^2}\right)\psi + (k\rho)^2\psi = 0$, with $k^2 = \frac{2\mu E}{\hbar^2}$.

Separating the z-dependence of the wavefunction, we have $\psi(\rho,\varphi,z) = \xi(\rho,\varphi)\chi(z)$, hence

$$\frac{1}{\xi(\rho,\varphi)\rho}\frac{\partial}{\partial\rho}\left(\rho\frac{\partial}{\partial\rho}\right)+\frac{1}{\xi(\rho,\varphi)\rho^2}\frac{\partial^2\xi(\rho,\varphi)}{\partial\varphi^2}+k^2 = -\frac{1}{\chi(z)}\frac{d^2\chi(z)}{dz^2},$$

and both sides must be equal to a constant, say k_z^2. Accordingly,

$$\frac{1}{\xi\rho}\frac{\partial}{\partial\rho}\left(\rho\frac{\partial}{\partial\rho}\right)+\frac{1}{\xi\rho^2}\frac{\partial^2\xi}{\partial\varphi^2}+k^2 = k_z^2 \text{ and } \frac{d^2\chi}{dz^2}+k_z^2\chi = 0.$$

Solution to the z-equation is $\chi(z) = c_1\exp(ik_z z) + c_2\exp(-ik_z z)$, and applying the boundary conditions $\chi(0) = 0 = \chi(L_z)$, the function is $\chi(z) = \frac{1}{\sqrt{L_z}}\exp(ik_z z)$.

Now, to solve the differential equation for $\xi(\rho,\varphi)$ by the method of separation of variables, we write $\xi(\rho,\varphi) = \zeta(\rho)\eta(\varphi)$ and get the following *two* one-dimensional differential equations:

$$\frac{1}{\zeta(\rho)}\left[\rho^2\frac{d^2\zeta(\rho)}{d\rho^2}+\rho\frac{d\zeta(\rho)}{d\rho}\right]+\left[(k\rho)^2-(k_z\rho)^2\right] = m^2 \quad\text{and}\quad -\frac{1}{\eta(\varphi)}\frac{d^2\eta(\varphi)}{d\varphi^2} = m^2,$$ where m is a constant. The solution to the φ-equation is $\eta(\varphi) = \eta_1\exp(im\varphi) + \eta_2\exp(-im\varphi)$. From the two linearly independent solutions, we may choose either. With $\eta(\varphi) = \eta_1\exp(im\varphi)$ and applying periodic boundary conditions $\eta(\varphi) = \eta(\varphi + 2\pi)$ which ensure uniqueness of the wavefunction, we get $\eta_1\exp(im\varphi) = \eta_1\exp(im\varphi + 2\pi)$, which gives $\eta(\varphi) = \eta_1\exp(im\varphi)$ with $m = 0,\pm1,\pm2,\ldots$

The radial differential equation now is $\left[\rho^2\frac{d^2\zeta(\rho)}{d\rho^2}+\rho\frac{d\zeta(\rho)}{d\rho}\right]+\left[(\kappa\rho)^2-m^2\right]\zeta(\rho) = 0,$

where $\kappa = k^2 - k_z^2$. Changing the variable $\rho \to \omega = \kappa\rho$ along with $\frac{d}{d\omega} \equiv \frac{d}{d(\kappa\rho)} \equiv \frac{d}{\kappa d\rho}$,

i.e., $\frac{d}{d\rho} \equiv \kappa\frac{d}{d\omega}$,

we get $\left[\omega^2\frac{d^2\zeta(\omega)}{d\omega^2}+\omega\frac{d\zeta(\omega)}{d\omega}\right]+\zeta(\omega)(\omega^2-m^2) = 0.$

The solution to the above differential equation is a linear combination of the Bessel function of first kind (J_m) and of the second kind (Y_m): $\zeta(\rho) \Leftrightarrow \zeta(\omega) = \zeta_1 J_m(\omega) + \zeta_2 Y_m(\omega) \equiv \zeta_1 J_m(\omega)$ since $Y_m(\omega)$ diverge as $\omega \to 0$ and we must choose $\zeta_2 = 0$.

Hence, $\zeta(\rho) = \zeta_1 J_m(\omega) = \zeta_1 J_m\left(\frac{\alpha_{m,n}}{a}\rho\right)$, where $\alpha_{m,n}$ denote the roots of the Bessel function.

The total wavefunction is $\psi(\rho,\varphi,z) = NJ_m\sqrt{\frac{2}{L_z}}\left(\frac{\alpha_{m,n}}{a}\rho\right)\exp(im\varphi)\sin\left(\frac{n_z\pi}{L_z}z\right)$ and the energy is $E = \frac{\hbar^2}{2\mu}\left(\frac{\alpha^2_{m,n}}{a^2} + \frac{n_z^2\pi^2}{L_z^{\,2}}\right)$.

Exercises

E3.1: An electron with energy 2 eV is incident on a rectangular barrier of height 4 eV. The transmission probability of the electron obtained at the other end of the barrier is found to be 10^{-3}. Determine the width of the barrier.

E3.2: Consider the STM (Fig. 3.5a). Consider the wavefunction for an electron in the gap between a sample surface and the tip of the electrode.

(a) Determine the transmission probability for an electron having energy 2 eV from the surface of a sample to the tip of the electrode through a potential barrier of length 0.5 *nm* and height 4 eV.

(b) By what percentage does the current drop if the tip of the electrode is moved from $\ell_1 = 0.45\ nm$ to $\ell_2 = 0.55\ nm$ from the sample surface?

E3.3: Consider tunneling across a thin potential barrier of height V_0 and width $2a$, with the condition that $a \ll \frac{\hbar}{\sqrt{2mE}}$. Determine the tunneling probability across two such barriers whose centers are separated by a distance *d*. Furthermore, analyze resonance conditions with respect to (a) energy and (b) barrier separation *d*.

E3.4: Determine the eigenfunctions and eigenvalues of a one-dimensional repulsive periodic square wells $V(x+L) = V(x)$ of width 'a' and depth V_0. Furthermore, solve also for the case when the potential is a Dirac-δ periodic potential. Comment on the *band structure* of energies.

E3.5: Determine the eigenfunctions and eigenvalues of a one-dimensional attractive periodic square wells $V(x+L) = V(x)$ of width 'a' and depth V_0. Comment on the *band structure* of energies.

E3.6: Determine the potential energy function experienced by a particle in a one-dimensional space if its mass is *m*, energy is zero, and the particle's wavefunction is $Nx\exp\left(-\frac{x^2}{c^2}\right)$, where *N* and *c* are constants.

E3.7: (a) Determine $\langle x\rangle, \langle x^2\rangle, \langle p_x\rangle, \langle p_x^{\,2}\rangle$ for the ground state and the lowest two excited states of a one-dimensional simple harmonic oscillator. (b) Determine the uncertainty product $\delta x \delta p_x$ when the oscillator is in its ground state and in its lowest two excited states.

E3.8: A particle in a one-dimensional potential is in mixed state ψ that is a superposition of its ground state ψ_0 and the first excited state ψ_1. Let $\Psi = \frac{1}{\sqrt{2}}\left(\psi_0 + \psi_1\right)$. Presume that ψ_0 and ψ_1 are both normalized. Determine (a) the expectation value of its Hamiltonian and (b) the uncertainty in energy.

E3.9: For the particle described in E3.8, determine the average position $\langle x \rangle$ and the average momentum $\langle p_x \rangle$.

E3.10: Prove that the reflection and transmission coefficients in the problem of tunneling across a one-dimensional potential barrier are the *same* regardless of whether the particle is incident from the left or from the right.

E3.11: What are the consequences on the wave function and energy of a particle if a constant potential is added in the Schrödinger's equation?

E3.12: Determine the group velocity of the Gaussian wavepacket $\psi(x,t) = \int_{-\infty}^{\infty} dk \exp\left[-\frac{\left(k-k_0\right)^2}{a}\right] \exp\left[i(kx-\omega t)\right]$ and show that the wavepacket expands as it travels.

E3.13: Determine the momentum probability distribution for the ground state of a particle in an infinitely deep one-dimensional square well potential.

REFERENCES

[1] P. C. Deshmukh, *Foundations of Classical Mechanics* (Cambridge University Press, New Delhi, 2019).

[2] Peter Rodgers, *The Double Slit Experiment* (PhysicsWorld, September 1, 2002).

[3] A. Tonomura, J. Endo, T. Matsuda, T. Kawasaki, and H. Ezawa, Demonstration of Single-electron Build Up of an Interference Pattern. *Am. J. Phys.* 57:2, 117–120 (1989).

[4] Yoon-Ho Kim, R. Yu, S. P. Kulik, Y. H. Shih, and Marlan Scully, A Delayed Choice Quantum Eraser. *Phys. Rev. Lett.* 84:1, 1–5 (2000).

[5] Rachel Hillmer and Paul Kwiat, A Do-It-Yourself Quantum Eraser *Scientific American*, pp. 90–95 (May 2007).

[6] R. H. Dicke and J. P. Wittke, *Introduction to Quantum Mechanics* (p. 48, Addison-Wesley, 1960).

[7] R. L. Jaffe, Reflection above the Barrier as Tunnelling in Momentum Space. *Amer. J. Phys.* 78:6, 620–623 (2010).

[8] Scanning Tunnelling Microscope. https://en.wikipedia.org/wiki/Scanning_tunneling_microscope.

[9] D. M. Eigler and E. K. Schweizer, Positioning Single Atoms with a Scanning Tunnelling Microscope. *Nature* 344:6266, 524–526 (1990). doi:10.1038/344524a0. ISSN 0028-0836.

[10] George Gamow, Zur Quantentheorie der Atomzertrümmerung. *Z. Phys.* 52, 510 (1929).

[11] C. Qi, A. N. Andreyev, M. Huyse, R. J. Liotta, P. Van Duppen, and R. Wyss, On the Validity of the Geiger–Nuttall Alpha-decay Law and Its Microscopic Basis. *Phys. Lett. B* 734:27, 203–206 (2014).

[12] Ramón Ramos, David Spierings, Isabelle Racicot, and Aephraim M. Steinberg, Measurement of the Time Spent by a Tunnelling Atom within the Barrier Region. *Nature* 583, 529–532 (2020).

[13] R. Ashoori, Electrons in Artificial Atoms. *Nature* 380, 559 (1996). https://doi.org/10.1038/380559b0.

[14] Marcel Wellner, Levinson's Theorem (an Elementary Derivation). *Amer. J. Phys.* 32, 787 (1964). https://doi.org/10.1119/1.1969857.

[15] M. J. Seaton, Quantum Defect Theory. *Reports on Progress in Physics.* 46:2, 167 (1983).

[16] Kenneth Roberts and S. R. Valluri, Tutorial: The Quantum Finite Square Well and Lambert W Function. *Can. J. Phys.*, 95, 105–110 (2017).

[17] P. C. Deshmukh, Aarthi Ganesan, N. Shanthi, Blake Jones, James Nicholson, and Andrea Soddu, Accidental Degeneracy of the Hydrogen Atom Is Not an Accident. *Canadian Journal of Physics*, 10.1139/cjp-2014-0300.

[18] P. C. Deshmukh, Aarthi Ganesan, Sourav Banerjee, and Ankur Mandal, Accidental Degeneracy of the Hydrogen Atom and Its Non-accidental Solution in Parabolic Coordinates. *Can. J. Phys.* 99:10 853–860 (2021). https://doi.org/10.1139/cjp-2020-0258.

[19] M. Moriconi, Nodes of Wavefunctions. *Am. J. Phys.* 75:3, 284–285 (2007).

[20] L. E. Brus, Electron–Electron and Electron–Hole Interactions in Small Semiconductor Crystallites: The Size Dependence of the Lowest Excited Electronic State. *Journal of Chemical Physics* 80:9, 4403 (1984). doi:10.1063/1.447218.

[21] C. E. Siewert, Explicit Results for the Quantum-Mechanical Energy States Basic to a Finite Square-Well Potential. *J. Math. Phys.* 19, 434 (1978).

[22] S. R. Valluri, D. J. Jeffrey, and R. M. Corless, Some Applications of the Lambert W Functions to Physics. *Can. J. Phys.* 78, 823–830 (2000).

[23] R. M. Corless, G. H. Gonnet, D. E. G. Hare, D. J. Jeffrey, and D. E. Knuth, On the Lambert W Function. *Adv. Comput. Math.*, 5:1, 329–359 (1996).

[24] Narola Harsh Bharatbhai, P. C. Deshmukh, Roberty B. Scott, Ken Roberts, and S. R. Valluri Lambert W Function Methods in Double Square Well and Waveguide Problems. *Journal of Physics Communications* 4, 065001 (2 June 2020).

[25] Aakash Yadav, P. C. Deshmukh, Ken Roberts, N. M. Jisrawi, and SR Valluri, An Analytic Study of the Wiedemann-Franz Law and the Thermoelectric Figure of Merit. *Journal of Physics Communications* 3, 105001 (2019).

[26] L. Prabhat Reddy, Sibibalan Jeevanandam, P. C. Deshmukh, Aude Maignan, Najeh Jisrawi, Ken Roberts, and S. R. Valluri, A Study on the Electronic Properties of Graphene Nanoribbons using the Offset Logarithm Function. *Materials Today: Proceedings* (2021).

[27] S. E. Kocabas, G. Veronis, D. A. B. Miller, and S. Fan, Transmission Line and Equivalent Circuit Models for Plasmonic Waveguide Components. *Phys. Rev. B*, 79, 035120 (2009).

4

Angular Momentum

Hendrik "Henk" Brugt Gerhard Casimir (1909–2000) with Neils Bohr. B174, https://arkiv.dk/en/vis/5904586. Courtesy: Niels Bohr Archive.

Also known for "Casimir effect" *physical attraction between conducting plates due to quantum fluctuations.*

In Chapter 1 we discussed the role of linear momentum as the generator of translational displacement in homogeneous space. Likewise, it is instructive to explore the role of angular momentum as the generator of rotational displacements in isotropic space. Not surprisingly, analogies drawn from classical physics are not adequate to describe physical particles and fields. One must therefore redefine *angular momentum* to describe appropriately the physical properties in nature by quantum mechanics.

4.1 Definition and Properties of Angular Momentum

Kepler orbits of planets around the sun signify our intuitive understanding of angular momentum of an object about a point (Fig. 4.1). It interprets angular momentum as *moment of the momentum*, expressed as the cross product of the position vector of an object with its instantaneous linear momentum. We recognize it to represent a physical quantity that is

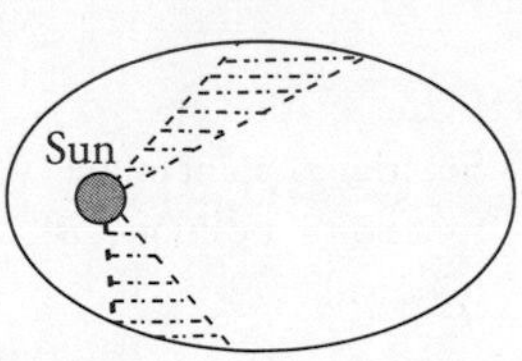

Fig. 4.1 Our perception of angular momentum is built on the basis of classical models such as the Kepler–Newton planetary orbits.

conserved in isotropic space. The classical definition of angular momentum is however not sustainable, and not just because simultaneous measurements of position and momentum are incompatible. Physical properties of particles and fields have properties that have no classical analogues. Quantum mechanics acknowledges the intrinsic incompatibility in observations of some physical properties and provides a defensible alternative theory to describe nature. We shall therefore be able to persist with the classical interpretation of angular momentum as the generator of rotation in isotropic space only up to a certain point of our analysis; beyond it, we shall require a worthier mathematical model that is robust enough to describe nature acceptably.

We consider an object having an arbitrary shape $f_\alpha(\vec{r})$ and subject it to a rotation about an axis through it, such as the one schematically shown in Fig. 4.2a. On turning it through an infinitesimal angle $\overrightarrow{\delta\phi}$, the shape transforms to $g_\alpha(\vec{r})$. We denote the rotation by an operator $U_R\left(\overrightarrow{\delta\phi}\right)$ and express this operation as

$$U_R\left(\overrightarrow{\delta\phi}\right)f_\alpha(\vec{r}) = g_\alpha(\vec{r}). \tag{4.1}$$

With reference to a Cartesian coordinate system, a rotation relocates a point whose position vector is $\vec{r}$ to $\vec{r}_R$ (Fig. 4.2b):

$$\vec{r}_R = \vec{r} + \overrightarrow{\delta r} = \vec{r} + \left(\overrightarrow{\delta\phi} \times \vec{r}\right). \tag{4.2}$$

Equation 4.1 tells us that the function $g_\alpha(\vec{r})$ has the same value at the new orientation as the function $f_\alpha\left(R^{-1}\vec{r}\right)$, i.e., as $f_\alpha\left(\vec{r} - \overrightarrow{\delta r}\right)$, R being the *operator* corresponding to the said rotation. We shall refer to such a description of transformation of the function as *active*. In *passive* description, an equivalent inverse operation on the coordinate system is employed (Fig. 4.2c), leaving the function itself unchanged. The active and the passive descriptions are equivalent. The former is convenient when static displacements are of most significance, but the latter is more suitable when dynamics is under consideration.

Thus,

$$U_R\left(\overrightarrow{\delta\phi}\right)f_\alpha(\vec{r}) = g_\alpha(\vec{r}) = f_\alpha\left(R^{-1}\vec{r}\right) = f_\alpha(\vec{r} - \delta\vec{r}). \tag{4.3}$$

An arbitrary state vector $|\alpha\rangle$ would transform under the rotation operator to a new vector $|\alpha\rangle_R$ in the Hilbert space. We describe this operation as

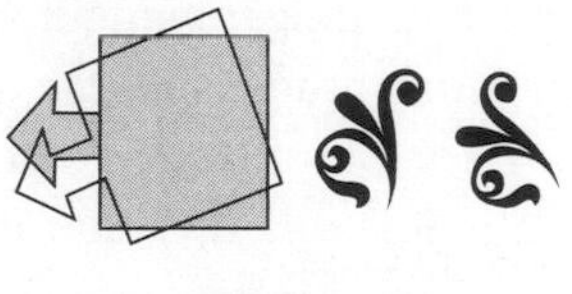

Fig. 4.2a Rotation of an arbitrary function $f(\vec{r})$.

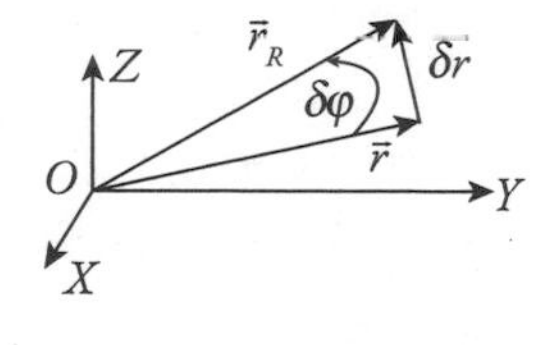

Fig. 4.2b In the rotation of a function, a point in the function at position $\vec{r}$ is relocated to $\vec{r}_R$.

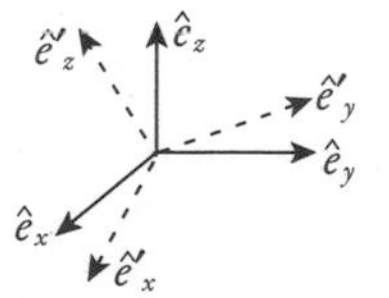

Fig. 4.2c The rotation of a function can also be described from a passive point of view in which the coordinate frame itself is rotated in the opposite rotation.

$$|\alpha\rangle \to |\alpha\rangle_R = U_R\left(\overrightarrow{\delta\phi}\right)|\alpha\rangle = U_R\left(\overrightarrow{\delta\phi}\right)\underset{ws}{\iiint} dV|\vec{r}\rangle\langle\vec{r}|\alpha\rangle = \underset{ws}{\iiint} dV\left\{U_R\left(\overrightarrow{\delta\phi}\right)|\vec{r}\rangle\right\}\psi_\alpha(r), \quad (4.4a)$$

where "ws" denotes that the integration is to be carried out over the "whole space."

Hence, $$U_R\left(\overrightarrow{\delta\phi}\right)|\alpha\rangle = \underset{ws}{\iiint} dV\left|\vec{r}+\overrightarrow{\delta r}\right\rangle\psi_\alpha(r) = \underset{ws}{\iiint} dV'|\vec{r}'\rangle\psi_\alpha\left(\vec{r}'-\overrightarrow{\delta r}\right), \quad (4.4b)$$

where $\vec{r}' = \vec{r} + \overrightarrow{\delta r}$.

Therefore,

$$U_R\left(\overrightarrow{\delta\phi}\right)|\alpha\rangle = \underset{ws}{\iiint} dV'|\vec{r}'\rangle\left\{\psi_\alpha(\vec{r}') - \overrightarrow{\delta r}\boldsymbol{\cdot}\vec{\nabla}\psi(\vec{r}')\right\} = \underset{ws}{\iiint} dV'|\vec{r}'\rangle\left\{1-\overrightarrow{\delta r}\boldsymbol{\cdot}\vec{\nabla}\right\}\psi_\alpha(\vec{r}'), \quad (4.4c)$$

i.e., $$U_R\left(\overrightarrow{\delta\phi}\right)|\alpha\rangle = \underset{ws}{\iiint} dV'|\vec{r}'\rangle\left\{1-\frac{i}{\hbar}\left(\overrightarrow{\delta\varphi}\times\vec{r}\right)\boldsymbol{\cdot}\vec{p}\right\}\langle\vec{r}'|\alpha\rangle$$

$$= \underset{ws}{\iiint} dV'|\vec{r}'\rangle\left\{1-\frac{i}{\hbar}\left(\vec{\ell}\boldsymbol{\cdot}\overrightarrow{\delta\varphi}\right)\right\}\langle\vec{r}'|\alpha\rangle, \quad (4.4d)$$

since in a scalar triple product $\vec{A}\boldsymbol{\cdot}\vec{B}\times\vec{C}$ of three vectors, the positions of the scalar product and the vector product can be interchanged.

We therefore recognize that the operator for infinitesimal rotation is

$$U_R^{(\vec{\ell})}\left(\overrightarrow{\delta\phi}\right) - 1 - \frac{i}{\hbar}\left(\vec{r}\times\vec{p}\right)\boldsymbol{\cdot}\overrightarrow{\delta\varphi} = 1 - \frac{i}{\hbar}\vec{\ell}\boldsymbol{\cdot}\overrightarrow{\delta\varphi}, \quad (4.5a)$$

and that for a finite rotation it is

$$U_R^{(\vec{\ell})}\left(\theta\hat{\theta}\right) = \lim_{n\to\infty}\left(1-\frac{i}{\hbar}\vec{\ell}\boldsymbol{\cdot}\left(\frac{\theta\hat{\theta}}{n}\right)\right)^n = \exp\left(-\frac{i}{\hbar}\vec{\ell}\boldsymbol{\cdot}\theta\hat{\theta}\right). \quad (4.5b)$$

We have identified the vector operator

$$\vec{\ell} - \vec{r}_{\text{operator}}\times\vec{p}_{\text{operator}} - \vec{r}\times\left(-i\hbar\vec{\nabla}\right) \quad (4.6)$$

as the *orbital angular momentum operator*. Its interpretation links the *operator*

$$\vec{\ell} = \sum_{i=1}^{3}\ell_i\hat{e}_i = \ell_x\hat{e}_x + \ell_y\hat{e}_y + \ell_z\hat{e}_z \quad (4.7)$$

to the three generators ℓ_x, ℓ_y, ℓ_z of rotations about orthogonal Cartesian axes X, Y, and Z in the ordinary three-dimensional Euclidean space R^3. We have employed $\vec{r}$ and $-i\hbar\vec{\nabla}$ to respectively represent the position operator and the momentum operator (in the coordinate representation). The superscript (ℓ) we have placed in Eq. 4.5 is tentative. Most often, we will suppress it. It is used here only to distinguish it from a similar operator $U_R^{(\vec{s})}\left(\overrightarrow{\delta\phi}\right)$ to be soon introduced, which would correspond to the *spin* angular momentum. Using the notation we have developed so far, we represent the operators for rotations about the X, Y, and Z Cartesian axes respectively by

$$U_R(\phi_x) = 1 - \frac{i}{\hbar}\ell_x\phi_x - \frac{\ell_x^2{\phi_x}^2}{2\hbar^2} + \ldots, \quad (4.8a)$$

$$U_R(\phi_y) = 1 - \frac{i}{\hbar}\ell_y\phi_y - \frac{\ell_y^2\phi_y^2}{2\hbar^2} + \ldots, \tag{4.8b}$$

and $U_R(\phi_z) = 1 - \frac{i}{\hbar}\ell_z\phi_z - \frac{\ell_z^2\phi_z^2}{2\hbar^2} + \ldots.$ (4.8c)

Now, an "active" rotation of a point through an angle φ about the Z-axis of a Cartesian coordinate system would turn its position vector

$$\vec{r} = x\hat{e}_x + y\hat{e}_y + z\hat{e}_z \tag{4.9a}$$

to $\vec{r}_R = (x\cos\varphi - y\sin\varphi)\hat{e}_x + (x\sin\varphi + y\cos\varphi)\hat{e}_y + z\hat{e}_z.$ (4.9b)

The angle φ is the same as employed in the polar coordinate systems, and hence positive for counter-clockwise rotation in the XY-plane. We represent the transformation

$$\vec{r}_R = \mathbb{R}_z(\varphi)\vec{r} \tag{4.10a}$$

using a matrix equation

$$\begin{bmatrix} x \\ y \\ z \end{bmatrix} \to \begin{bmatrix} x_R \\ y_R \\ z_R \end{bmatrix} = \begin{bmatrix} \cos\varphi & -\sin\varphi & 0 \\ \sin\varphi & \cos\varphi & 0 \\ 0 & 0 & 1 \end{bmatrix} \begin{bmatrix} x \\ y \\ z \end{bmatrix} = \mathbb{R}_z(\varphi) \begin{bmatrix} x \\ y \\ z \end{bmatrix}, \tag{4.10b}$$

where $\mathbb{R}_z(\varphi)$ is the transformation matrix. This transformation should be contrasted with inversion/reflection which is also a symmetry operation. For example, reflection of a point whose Cartesian coordinates are (x, y, z) in the ZX-plane gives us new coordinates $(x_{P,ZX}, y_{P,ZX}, z_{P,ZX})$ of the image point:

$$\begin{bmatrix} x \\ y \\ z \end{bmatrix} \to \begin{bmatrix} x_{P,ZX} \\ y_{P,ZX} \\ z_{P,ZX} \end{bmatrix} = \begin{bmatrix} 1 & 0 & 0 \\ 0 & -1 & 0 \\ 0 & 0 & 1 \end{bmatrix} \begin{bmatrix} x \\ y \\ z \end{bmatrix} = \mathbb{P}_{ZX}(\varphi) \begin{bmatrix} x \\ y \\ z \end{bmatrix}. \tag{4.10c}$$

The matrix $\mathbb{R}$ in Eq. 4.10b and the matrix $\mathbb{P}$ in Eq. 4.10c are characteristic respectively of a typical *rotation* matrix and an *inversion/reflection* matrix, also called as *parity* matrix. Both are orthogonal unimodular matrices, but their determinants have opposite signs. While $|\mathbb{R}|$ is always +1, $|\mathbb{P}|$ is always −1. The difference in sign of these two determinants has consequences on the symmetry group generated by the set of rotation operators and that of inversion/reflection (i.e., parity) operators.

For rotations through small angles, $\xi \to 0$. We can expand the trigonometric functions in Eq. 4.10b in powers of ξ. Ignoring terms in ξ^n with $n \geq 3$, we see that

$$\mathbb{R}_z(\xi) = \begin{bmatrix} 1 - \frac{\xi^2}{2} & -\xi & 0 \\ \xi & 1 - \frac{\xi^2}{2} & 0 \\ 0 & 0 & 1 \end{bmatrix}. \tag{4.11a}$$

Likewise,

$$\mathbb{R}_x(\xi) = \begin{bmatrix} 1 & 0 & 0 \\ 0 & 1-\frac{\xi^2}{2} & -\xi \\ 0 & \xi & 1-\frac{\xi^2}{2} \end{bmatrix}, \tag{4.11b}$$

and

$$\mathbb{R}_y(\xi) = \begin{bmatrix} 1-\frac{\xi^2}{2} & 0 & \xi \\ 0 & 1 & 0 \\ -\xi & 0 & 1-\frac{\xi^2}{2} \end{bmatrix}. \tag{4.11c}$$

It is easy to see using these matrix representations that rotations about orthogonal coordinate axes do not commute. For example,

$$\left[\mathbb{R}_x(\xi), \mathbb{R}_y(\xi)\right]_- = \begin{bmatrix} -\cancel{\frac{\xi^4}{2}} & -\xi^2 & 0 \\ \xi^2 & -\cancel{\frac{\xi^4}{2}} & 0 \\ 0 & 0 & 0 \end{bmatrix}$$

$$= \begin{bmatrix} 1-\cancel{\frac{\xi^4}{2}} & -\xi^2 & 0 \\ \xi^2 & 1-\cancel{\frac{\xi^4}{2}} & 0 \\ 0 & 0 & 1 \end{bmatrix} - \begin{bmatrix} 1 & 0 & 0 \\ 0 & 1 & 0 \\ 0 & 0 & 1 \end{bmatrix} = \mathbb{R}_z(\xi^2) - \mathbb{I}, \tag{4.12}$$

where $\mathbb{I}$ is the 3×3 unit matrix corresponding to a rotation through 0 angle about an arbitrary axis. Terms in ξ^4, are ignorable, they are crossed out. That rotations through infinitesimal angular displacements about orthogonal axes do not commute is an expression of the fact that the generators of rotations about orthogonal axes do not commute. This result foreshadows the emergence of a fundamental relationship, which provides an essential criterion to *define* the angular momentum.

Replacing Cartesian alphabetical indexing by numerals, it immediately follows (from the solution to the Solved Problem P4.1) that

$$\left[\ell_i, \ell_j\right] = \varepsilon_{ijk} i\hbar \ell_k;\ i, j, k = 1, 2, 3, \tag{4.13a}$$

where we have used the Levi-Civita symbol.

$$\left.\begin{aligned} \varepsilon_{123} = \varepsilon_{231} = \varepsilon_{312} = +1, \\ \varepsilon_{132} = \varepsilon_{321} = \varepsilon_{213} = -1, \\ \text{and all other } \varepsilon_{ijk} = 0 \end{aligned}\right\}. \tag{4.13b}$$

We can obtain the commutation relation (Eq. 4.13a) also by using Eq. 4.6 and employing Eq. 1.66a. Doing so is easiest using the Cartesian coordinate system, but it is instructive to ensure that the result is independent of the choice of a coordinate system. One can verify this result using other coordinate systems, such as the cylindrical polar and the spherical polar coordinate systems. After all, Eq. 1.66b and Eq. 1.66c hold for any coordinate system for canonically conjugate coordinates and momenta.

Now, the scalar product of two vector *operators* $\vec{F}$ and $\vec{G}$ is

$$\vec{F}\bullet\vec{G} = \left(\sum_{i=1}^{3} F_i G_i\right) = \left(\sum_{i=1}^{3} G_i F_i\right) + \left(\sum_{i=1}^{3}[F_i, G_i]_-\right) = \vec{G}\bullet\vec{F} + \left(\sum_{i=1}^{3}[F_i, G_i]_-\right), \tag{4.14a}$$

and their vector cross product $\vec{H} = \vec{F}\times\vec{G}$ has components given by

$$H_i = \varepsilon_{ijk} F_j G_k \equiv \sum_{j=1}^{3}\sum_{k=1}^{3} \varepsilon_{ijk} F_j G_k = -\left(\vec{G}\times\vec{F}\right)_i + \varepsilon_{ijk}\left[F_j, G_k\right]_-, \text{ for } i = 1,2,3. \tag{4.14b}$$

It is instructive to discern the somewhat perplexing result that for the operators $\vec{r}$ (position) and $\vec{p}$ (momentum),

$$\vec{r}\bullet\vec{p} \neq \vec{p}\bullet\vec{r}, \tag{4.14c}$$

$$\text{and } \vec{r}\times\vec{p} = -\vec{p}\times\vec{r}. \tag{4.14d}$$

Using Eqs. 4.13a,b and Eqs. 4.14a,b, we obtain the following *nonclassical* expression:

$$\vec{\ell}\times\vec{\ell} = i\hbar\vec{\ell}. \tag{4.15}$$

That the cross product of $\vec{\ell}$ with itself does not vanish must no longer surprise us; after all, it is a quantum vector *operator*, not merely a classical vector.

From the Solved Problem P4.2 we see that

$$\left[\vec{\ell}, \vec{\ell}\right] = -i\hbar\vec{\ell}\times\Im, \tag{4.16}$$

where $\Im$ is the unit dyadic that was used in the Solved Problem P1.4. Again, not surprisingly, this is a manifestly nonclassical expression.

The term "orbital" angular momentum has originated from the classical notion of *moment of momentum*. However, measurements of position and momentum being incompatible, the classical notion of an orbit is not sustainable. We shall continue to use the expression "*orbital* angular momentum," but *not* the classical depiction of an orbit. The latter is no more meaningful since its description simultaneously requires accurate knowledge of both position *and* momentum.

Equations 4.13, 4.15, and 4.16 are equivalent. Representation of the quantum orbital angular momentum vector operator by these expressions is satisfactory; they employ quantum operators, not classical dynamical variables. Description of angular momentum by these operators is nonetheless *insufficient* to describe *all* of the physical properties of elementary particles, which have an additional *independent* and *intrinsic* source of angular momentum, called as the *spin* angular momentum. The vector spaces of the spin and the orbital angular momentum have an essentially empty intersection.

The spin angular momentum has its roots in another groundbreaking discovery that was made in the early part of the 20th century by Albert Einstein, in the same year (1905) that he accounted for the photoelectric effect. It came from Einstein's bold reconciliation with the constancy of the speed of light in all inertial frames of reference (predicted by Maxwell's theory of electrodynamics) as a law of nature (see, for example, Chapters 12 and 13 in Reference [1]). Now, the Schrödinger formulation of quantum mechanics is essentially nonrelativistic; it considers space and time intervals as *independent* invariants under transformations between different inertial coordinate frames. In Chapter 7, we shall see that a relativistic formulation of quantum mechanics naturally leads to the notion that elementary particles have an intrinsic property, named "spin." This property has no analogue in Newtonian physics. It fortuitously corresponded to a classical gyroscope model that George Eugene Uhlenbeck and Samuel Goudsmit had proposed in 1925 for the analysis of the spectrum of an atom placed in a magnetic field. They had proposed *ad hoc* that an electron has a spin, similar to a rotating top. Dirac's relativistic quantum mechanics provides for the newly discovered intrinsic property, which cannot be accounted for by the gyroscopic model. We continue to use the term "spin" to describe this property, but we shall discredit the classical undertone of rotation of a particle about an axis. The description of angular momentum associated with the spin of elementary particles requires relativistic quantum mechanics. It is a measurable intrinsic physical property of the particle. We represent it by an operator " $\vec{s}$," which satisfies *all* the properties as those in Eqs. 4.13, 4.15, and 4.16.

The operators $\vec{\ell}$ and $\vec{s}$ operate on operands in their own respective disjoint vector spaces. Classical models such as the Kepler orbits (which describe *orbital* angular momentum) and a gyroscope rotation (that describes *spin* angular momentum) have only a limited utility and entail a risk of being misleading. We replace these models by the quantum mechanical *definition* of angular momentum given below. We shall represent *both* the "spin" and the "orbital" angular momentum by the common symbol

$$\vec{j} = \sum_{k=1}^{3} j_k \hat{e}_k. \tag{4.17}$$

In a given context if it is important to emphasize that the angular momentum under consideration is "orbital," we shall use the symbol $\vec{\ell}$, and when it is "spin," we shall use the symbol $\vec{s}$. Having abandoned the Keplerian-orbit and gyroscopic-spin models, we *define* angular momentum $\vec{j}$ by the following characteristic properties:

$$\left[j_i, j_j\right] = \varepsilon_{ijk} i\hbar j_k;\ i,j,k = 1,2,3 \text{ and } i \to j \to k \to i, \tag{4.18a}$$

or, equivalently $\vec{j} \times \vec{j} = i\hbar\vec{j}$, (4.18b)

or, equivalently $\left[\vec{j}, \vec{j}\right] = -i\hbar\vec{j} \times \mathfrak{I}$. (4.18c)

This definition of the angular momentum is in sharp departure with how it is defined in classical mechanics [1,2]. The equivalent characteristics defining relations for the angular momentum operator $\vec{j}$ given in Eqs. 4.18a,b,c are similar to those for $\vec{\ell}$, but now the spin $\vec{s}$ is also included. As we study this topic further, an understanding of the orbital and spin angular momenta would grow on us and provide freedom from the outmoded and incongruous classical cartoons.

It is now left as an exercise to show that the commutator $\left[\vec{j}\cdot\hat{u},\vec{j}\cdot\vec{j}\right]_-$ (where $\hat{u}$ is a unit vector pointed in an arbitrary direction in the isotropic space R^3) vanishes; i.e.,

$$\left[\vec{j}\cdot\hat{u},\vec{j}\cdot\vec{j}\right]=\left[j_u,j^2\right]=0. \tag{4.19a}$$

We can always choose a Cartesian coordinate system whose Z-axis is along the direction of $\hat{u}$; i.e., the unit vector $\hat{e}_z$ is the same as $\hat{u}$. Without any loss of generality, we can then write the above relation as

$$\left[j_z,j^2\right]=0. \tag{4.19b}$$

Measurements corresponding to j^2 and j_z are compatible. Commuting operators have the same eigenkets, labeled by the eigenvalues of both, in arbitrary order. We represent the hitherto unknown eigenvalues of the Hermitian operators j_z and j^2 respectively by $m\hbar$ and $f(j)\hbar^2$. The advantage in doing so is obvious; appropriate dimensions of the angular momentum and of its square are yanked out respectively in units of $\hbar$ and $\hbar^2$. The residual parts of the two eigenvalues are left in the real numbers m and $f(j)$, which are to be now determined. The eigenvectors of j^2 and j_z are therefore labeled respectively by j and m. We call j as the *angular momentum quantum number*. We must remember that the Z-axis of our coordinate system is directed in an *arbitrary* direction $\hat{u}$. As such, the quantum physical system under our present consideration is not placed in a magnetic field. In Chapter 8 we shall consider an atom in a steady magnetic field whose direction in a laboratory would be chosen to define $\hat{e}_z$. The quantum number m would play a rather important role in this case, in anticipation of which it is called as the *magnetic quantum number*. The axis along $\hat{u}$ is called as the axis of quantization, even in the absence of any magnetic field, and the component $\vec{j}\cdot\hat{u}$ therefore defines the operator j_z, regardless of whether a magnetic field is present or not.

Having now identified the *good* quantum numbers j and m having properties to be yet uncovered, we have the following eigenvalue equations:

$$j_z\left|j,m\right\rangle=\hbar m\left|j,m\right\rangle \tag{4.20a}$$

and $$j^2\left|j,m\right\rangle=\hbar^2 f(j)\left|j,m\right\rangle. \tag{4.20b}$$

It certainly does not matter whether the eigenkets are labeled as $\left|j,m\right\rangle$ or $\left|m,j\right\rangle$. Most often, we write them as $\left|jm\right\rangle$, without a coma separating them. They can also be written as $\left|\lambda jm\right\rangle$, where λ would represent any other (one or more) eigenvalue(s) of some other operator(s), which may also commute with both $\left(j_z,j^2\right)$. We shall work with a complete set $\left\{\left|jm\right\rangle\right\}$ of eigenbasis in which the base vectors are linearly independent, orthogonalized, and normalized. Hence,

$$\left\langle j,m\right|j_z\left|j',m'\right\rangle=\hbar m\delta_{jj'}\delta_{mm'} \tag{4.21a}$$

and $$\left\langle j,m\right|j^2\left|j',m'\right\rangle=\hbar^2 f(j)\delta_{jj'}\delta_{mm'}. \tag{4.21b}$$

We now introduce two operators

$$j_\pm=j_x\pm ij_y, \tag{4.22a}$$

(where $i=\sqrt{-1}$), called "step up" (with + sign) and "step down" (with – sign) operator, with $j_x = \vec{j} \bullet \hat{e}_x$ (4.22b)

and $j_y = \vec{j} \bullet \hat{e}_y$. (4.22c)

The operators $j_\pm$ are also called as ladder operators. The nomenclature would soon become clear. It is easy to see that

$$\left[j^2, j_\pm\right] = 0 \tag{4.23a}$$

and $\left[j_z, j_\pm\right] = \pm\hbar j_\pm$, (4.23b)

from which it follows that

$$\left\langle jm \mid j_z j_+ - j_+ j_z \mid j''m''\right\rangle = \hbar\left\langle jm \mid j_+ \mid j''m''\right\rangle. \tag{4.24a}$$

Inserting the resolution of unity we get

$$\sum_{j'm'}\langle jm|j_z|j'm'\rangle\langle j'm'|j_+|j''m''\rangle - \sum_{j'm'}\langle jm|j_+|j'm'\rangle\langle j'm'|j_z|j''m''\rangle = \hbar\left\langle jm \mid j_+ \mid j''m''\right\rangle. \tag{4.24b}$$

The matrix element $\langle jm|j_z|j'm'\rangle$ is zero unless $j = j'$ and $m = m'$, and $\langle j'm'|j_+|j''m''\rangle$ is zero unless $j' = j''$. Hence,

$$\left\{\begin{aligned}&\sum_{j'm'}\left(\delta_{mm'}\delta_{jj'}\langle jm|j_z|j'm'\rangle\right)\left(\langle j'm'|j_+|j''m''\rangle\delta_{j'j''}\right)\\ &-\sum_{j'm'}\left(\langle jm|j_+|j'm'\rangle\delta_{jj'}\right)\left(\delta_{m'm''}\delta_{j'j''}\langle j'm'|j_z|j''m''\rangle\right)\end{aligned}\right\} = \hbar\left\langle jm \mid j_+ \mid j''m''\right\rangle\delta_{jj''}. \tag{4.24c}$$

Nontrivial solution therefore requires $j = j' = j''$. Summing over the magnetic quantum number on the left-hand side,

$m\hbar\langle jm|j_+|jm''\rangle - m''\hbar\langle jm|j_+|jm''\rangle = \hbar\left\langle jm \mid j_+ \mid jm''\right\rangle$, we get (4.24d)

or $(m - m'' - 1)\langle jm|j_+|jm''\rangle = 0$. (4.24e)

The necessary condition for $\langle jm|j_+|jm''\rangle \neq 0$ therefore is

$$m = m'' + 1. \tag{4.25a}$$

Likewise, the necessary condition for $\langle jm|j_-|jm''\rangle \neq 0$ is

$$m = m'' - 1. \tag{4.25b}$$

Equations 4.25a and 4.25b explain the nomenclature 'ladder operators' for j_+ and j_-; they are respectively also called, more meaningfully, as the "step up" and "step down" operators. We now study their action on the angular momentum eigenvectors $|jm\rangle$. Toward that end, we examine the matrix elements $\langle j,m+1|j_+|jm\rangle$ and $\langle j,m|j_-|j,m+1\rangle$.

Let $\langle j,m+1|j_+|jm\rangle = \alpha_m\hbar$. (4.26a)

The relation adjoint to the above is

$$\langle j,m|j_-|j,m+1\rangle = \alpha_m^*\hbar. \tag{4.26b}$$

It follows from Eq. 4.26a that $j_+|j,m\rangle = \alpha_m\hbar|j,m+1\rangle$, (4.27a)

and from Eq. 4.26b it follows that $j_-|j,m+1\rangle = \alpha_m^*\hbar|j,m\rangle$. (4.27b)

Furthermore, since $[j_+, j_-] = 2\hbar j_z$, (4.28a)

we get $\langle j,m|j_+j_-|jm\rangle - \langle jm|j_-j_+|j,m\rangle = 2\hbar\langle j,m|j_z|j,m\rangle$, (4.28b)

i.e., $\alpha^*_{m-1}\alpha_{m-1} - \alpha_m\alpha^*_m = 2m$, (4.28c)

i.e., $|\alpha_{m-1}|^2 - |\alpha_m|^2 = 2m$. (4.28d)

This is a first-order linear difference equation. It is easy to verify that its solution is given by the roots of

$$|\alpha_m|^2 = k_j - m(m+1) = k_j - m^2 - m, \tag{4.29a}$$

$$\text{or, } |\alpha_{m-1}|^2 = k_j - (m-1)(m) = k_j - m^2 + m. \tag{4.29b}$$

In Eqs. 4.29a,b, k_j is a constant that depends on the angular momentum quantum number j, but not on the magnetic quantum number m. Subtraction of Eq. 4.29a from Eq. 4.29b gives the right-hand side of Eq. 4.28d, as it of course should. The left-hand side of Eq. 4.29a is strictly nonnegative, and so is the term m^2 on the right-hand side. We see that 'm' being a real number, the right-hand side may go negative for *large* positive values of 'm', and also for its *large* negative values. To avoid contradiction with the nonnegative left-hand side, 'm' must therefore be restricted to a range

$$m_2 < m \leq m_1. \tag{4.30}$$

The step-up operator would therefore *not* yield $|j,m_1+1\rangle$ from $|j,m_1\rangle$, and the step-down operator would *not* yield $|j,m_2\rangle$ from $|j,m_2+1\rangle$, thereby confining the basis $\{|j,m\rangle\}$ between $|j,m_2+1\rangle$ and $|j,m_1\rangle$. We express these limits by writing

$$\langle j,m_1+1|j_+|j,m_1\rangle = \alpha_{m_1}\hbar = 0 \tag{4.31a}$$

and $\langle j,m_2|j_-|j,m_2+1\rangle = \alpha^*_{m_2}\hbar = 0$. (4.31b)

Hence, using $\alpha_{m_1} = 0$ and $\alpha^*_{m_2} = 0$ in Eqs. 4.29a,b,
we get

$$k_j - m_1^2 - m_1 = 0 \tag{4.32a}$$

and $k_j - m_2^2 - m_2 = 0$. (4.32b)

We see that m_1 and m_2 are the roots of the quadratic equation

$$k_j - m^2 - m = 0. \tag{4.33}$$

Therefore,

$$m_1 = -\frac{1}{2} + \frac{1}{2}\sqrt{1+4k_j} \tag{4.34a}$$

and $m_2 = -\dfrac{1}{2} - \dfrac{1}{2}\sqrt{1+4k_j}$. (4.34b)

We note that $m_2 = -m_1 - 1$. (4.35)

The step-down operator can never yield the value m_2 (from Eq. 4.31b). The magnetic quantum number is therefore restricted to the range

$$-m_1 \le m \le m_1. \tag{4.36a}$$

Without any loss of generality, we *re-title* m_1 as j. The above range therefore is

$$-j \le m \le j. \tag{4.36b}$$

We thus identify the magnetic quantum number "m" to be a real number, which may vary from $+j$ to $-j$ in steps of unity, since the step-up and the step-down operators shift it by "1" at a time. Accordingly, the difference $2j$ between $+j$ and $-j$ can only be zero, or it can be a positive integer. In other words, j can only be one of the following: $j \to 0, \frac{1}{2}, 1, \frac{3}{2}, 2, \frac{5}{2}, 3, \frac{7}{2}, 4, \ldots.$

We can now determine the value of the constant k_j, since we must have
$k_j - m_1(m_1+1) = 0,$

which gives $k_j = j(j+1)$. (4.37)

Finally, since $\langle jm|j^2|jm\rangle = \left\langle jm\left|\frac{1}{2}(j_+j_- + j_-j_+) + j_z^2\right|jm\right\rangle,$ (4.38a)

we get $\hbar^2 f(j) = \frac{1}{2}\langle jm|j_+j_-|jm\rangle + \frac{1}{2}\langle jm|j_-j_+|jm\rangle + m^2\hbar^2,$ (4.38b)

i.e., $\hbar^2 f(j) = \frac{1}{2}\hbar^2|\alpha_{m-1}|^2 + \frac{1}{2}\hbar^2|\alpha_m|^2 + m^2\hbar^2,$ (4.38c)

i.e., $f(j) = \frac{1}{2}\left(k_j - m^2 + m\right) + \frac{1}{2}\left(k_j - m^2 - m\right) + m^2 = k_j = j(j+1).$ (4.39)

Having found out what the real numbers m and $f(j)$ in Eqs. 4.16a,b can be, we rewrite these equations, but this time with eigenvalues that are formally determined:

$$j_z|j,m\rangle = \hbar m|j,m\rangle \tag{4.40a}$$

and $j^2|j,m\rangle = \hbar^2 j(j+1)|j,m\rangle,$ (4.40b)

with the angular momentum quantum number j being one of the following,

$$j = 0, \frac{1}{2}, 1, \frac{3}{2}, 2, \frac{5}{2}, 3, \frac{7}{2}, 4, \ldots,$$

and for each value of j, the magnetic quantum number m taking $(2j+1)$ possible values, given by $m = (-j, -j+1, \ldots, j-1, j)$, i.e., $-j \le m \le +j$ in steps of 1. We know from Eqs. 4.31a,b that the ladder operators cannot shift the magnetic quantum number above the maximum $m = j$ nor diminish it below the minimum $m = -j$. j_+ can however operate on $|j, m = -j\rangle$ and raise the magnetic quantum number by unity every successive time it operates until we get $|j, m = +j\rangle$. Likewise, j_- would operate on $|j, m = +j\rangle$ and diminish the magnetic quantum number down the ladder, one step at a time, until we get $|j, m = -j\rangle$. The matrix elements of the step-up and step-down operators at the extreme highest and lowest steps are zero (Eqs. 4.31a,b). We shall now proceed to determine the nonzero intermediate values $\langle j, m+1|j_+|jm\rangle$ and $\langle j, m-1|j_-|jm\rangle$. The result is obtainable using Eqs. 4.29a,b. On employing a sign convention of using the positive square root of their left-hand sides, it is

$$\langle j, m+1|j_+|jm\rangle = +\hbar\sqrt{j(j+1) - m(m+1)} = +\hbar\sqrt{(j-m)(j+m+1)} \tag{4.41a}$$

and $\langle j,m-1|j_-|jm\rangle = +\hbar\sqrt{j(j+1)-m(m-1)} = +\hbar\sqrt{(j+m)(j-m+1)}$, (4.41b)

i.e., more compactly,

$$\langle j,m\pm 1|j_\pm|jm\rangle = +\hbar\sqrt{j(j+1)-m(m\pm 1)} = +\hbar\sqrt{(j\mp m)(j\pm m+1)}. \tag{4.41c}$$

The sign convention employed here is due to Condon and Shortley [3]. We can therefore express the effect of the ladder (shift) operators on angular momentum eigenstates in a compact form:

$$j_\pm|j,m\rangle = +\hbar\sqrt{j(j+1)-m(m\pm 1)}\,|j,m\pm 1\rangle = +\hbar\sqrt{(j\mp m)(j\pm m+1)}\,|j,m\pm 1\rangle. \tag{4.42}$$

The angular momentum vector operator $\vec{j}$ for spin-half particles therefore has a matrix structure. Using the matrix representations we have obtained in Solved Problem P4.3a, we see that it is given by

$$[\vec{j}] = \frac{\hbar}{2}[\vec{s}] = \frac{\hbar}{2}\left[s_x\hat{e}_x + s_y\hat{e}_y + s_z\hat{e}_z\right] = \frac{\hbar}{2}\begin{bmatrix} \hat{e}_z & \hat{e}_x - i\hat{e}_y \\ \hat{e}_x + i\hat{e}_y & -\hat{e}_z \end{bmatrix}. \tag{4.43}$$

We have used *non*relativistic quantum mechanics and angular momentum algebra to get the angular momentum quantum number j to be 0, or a positive integer, or a positive half integer. The electron spin angular momentum vector operator $\vec{s}$ however does not come from the Schrödinger equation; but the orbital angular momentum $\vec{\ell}$ does, with $\ell = 0,1,2,3,...$ (integers, not half-integer). The intrinsic spin half of an electron comes naturally from relativistic quantum mechanics using the Dirac equation. The electron spin has an interesting history. In 1924, Pauli had proposed a new "two-valued" quantum number for electrons to account for the filling of electron shells in successive atoms in the periodic table, following the spirit of the *aufbau* (which means "build up") principle. At that point, it was not attributed to any physical property, such as angular momentum, let alone the *spin* angular momentum. It was motivated by the need for a fourth quantum number, which had to be double-valued, to apply the *aufbau* principle and formulate the *exclusion principle* that no two electrons in an atom can have the same set of four quantum numbers. The spin angular momentum was postulated, as mentioned earlier, by Uhlenbeck and Goudsmit, but without any physical basis. Uhlenbeck referred to it as *abracadabra* (magic) and Goudsmit said that "*it was a kind of numerology ... it is a miracle that we arrived at the correct expressions which later could be derived by quantum mechanics.*" Ironically, Wolfgang Pauli had dismissed the idea of half-integer quantum numbers, though later he developed the algebra of 2 × 2 matrices (now named after him) to represent the spin-half operators. Discussion on the natural emergence of electron spin must wait until Chapter 7. For now, we expand the coordinate space to include the two-dimensional spin space. An arbitrary state $\chi(\zeta)$ in this space is called as the spin wavefunction, or just a spinor. We shall denote the spin coordinate by ζ and represent the spinor by a column matrix having two rows:

$$\chi(\zeta) \equiv \begin{pmatrix} c_1 \\ c_2 \end{pmatrix} \equiv c_1\begin{pmatrix} 1 \\ 0 \end{pmatrix} + c_2\begin{pmatrix} 0 \\ 1 \end{pmatrix} \equiv c_1|\alpha\rangle + c_2|\beta\rangle \equiv c_1|\uparrow\rangle + c_2|\downarrow\rangle \equiv c_1|+\rangle + c_2|-\rangle. \tag{4.44}$$

The pair of vectors $\{|\uparrow\rangle,|\downarrow\rangle\}$, or equivalently the pair $\{|+\rangle,|-\rangle\}$, represents the complete basis of angular momentum states $|j,m\rangle$ corresponding to the spin $j = s = \frac{1}{2}$ for which the magnetic quantum number can take two values $m=+\frac{1}{2}$ and $m=-\frac{1}{2}$, represented by up and down arrows, or + and – signs. In Eq. 4.44, we have employed equivalent notations (Eq. 1.32 through Eq. 1.36) to represent the physical state of the spin system by vectors in a Hilbert space or by column matrices. The complete wavefunction for an electron, inclusive of the spin, is

$$\psi(\vec{r})\to\psi(q)=\psi(\vec{r},\zeta)=\psi(\vec{r})\chi(\zeta)=\psi(\vec{r})\begin{pmatrix}c_1\\c_2\end{pmatrix}. \tag{4.45}$$

$\psi(q)$ is also called as spin-orbital in which $\psi(\vec{r})$ is the orbital part and $\chi(\zeta)$ is the spinor, i.e., the spin part. The spin space is a mathematical space that is disjoint from the usual three-dimensional space R^3. We therefore introduce a rotation operator similar to that in Eq. 4.5, but replacing $\vec{\ell}$ by $\vec{s}$. The orbital angular momentum vector operator $\vec{\ell}$ was associated with $\vec{r}\times\left(-i\hbar\vec{\nabla}\right)$, but the spin angular momentum vector operator $\vec{s}$ is defined only by Eqs. 4.14a,b,c. Instead of Eq. 4.5 the rotation operator corresponding to the spin angular momentum is

$$U_R^{(\vec{s})}\left(\overrightarrow{\delta\phi}\right)=1-\frac{i}{\hbar}\vec{s}\cdot\overrightarrow{\delta\varphi}. \tag{4.46}$$

Superscript $(\vec{s})$ is added on the rotation operator on the left-hand side of Eq. 4.46 to denote “spin”, only to highlight its difference from Eq. 4.5 corresponding to the orbital angular momentum. To appreciate the role of the spin angular momentum, we take up the case of an electron in an atom of hydrogen. The eigenvalues in Eq. 4.40b for the orbital angular momentum are

$$j\to\ell=0,1,2,3,..., \tag{4.47a}$$

whereas in the case of the electron spin angular momentum the eigenvalue corresponds to the particle's *intrinsic* property

$$j\to s=\frac{1}{2}. \tag{4.47b}$$

Both the orbital and the spin angular momenta would be represented by $\vec{j}$, whose components are the generators of infinitesimal rotations effected by the operator

$$\left(U_R^{(\vec{\ell})}\left(\overrightarrow{\delta\phi}\right),U_R^{(\vec{s})}\left(\overrightarrow{\delta\phi}\right)\right)\to U_R^{(\vec{j})}\left(\overrightarrow{\delta\phi}\right)\equiv U_R\left(\overrightarrow{\delta\phi}\right)=1-\frac{i}{\hbar}\vec{j}\cdot\overrightarrow{\delta\varphi}. \tag{4.48a}$$

Superscripts in Eqs. 4.5 and 4.48a will be usually suppressed in the rest of the book; the context in a given situation would carry that information. In the case of both the orbital and the spin angular momentum operators, we will represent a rotation through an *infinitesimal* angle $\delta\xi=\lim_{N\to\infty}\frac{\xi}{N}$ about an axis along the unit vector $\hat{\xi}$ by the operator

$$U_R^{(\vec{j})}\left(\hat{\xi}\delta\xi\right)=1-\frac{i}{\hbar}\vec{j}\cdot\hat{\xi}\delta\xi. \tag{4.48b}$$

A finite rotation results from repeated applications of infinitesimal ones. Accordingly, we represent the operator for a *finite* angle ξ about $\hat{\xi}$ by the operator

$$U_R^{(\vec{j})}\left(\hat{\xi}\xi\right)=\lim_{n\to\infty}\left(1-\frac{i}{\hbar}\vec{j}\bullet\left(\frac{\hat{\xi}\xi}{n}\right)\right)^n=\exp\left(-\frac{i}{\hbar}\vec{j}\bullet\hat{\xi}\xi\right). \tag{4.48c}$$

We have denoted this angle as $\left(\hat{\xi}\xi\right)$ rather than as $\left(\vec{\xi}\right)$ since a *finite* rotation is not a vector (see, for example, Chapter 2 of Reference [1]). To simplify the notation, we rewrite Eq. 4.48c as

$$U_R^{(\vec{j})}\left(\hat{e}_z\xi\right)=\exp\left(-\frac{i}{\hbar}j_z\xi\right), \tag{4.48d}$$

where we have chosen a Cartesian coordinate system with $\hat{e}_z=\hat{\xi}$, along the axis of quantization. The operator $\vec{j}\bullet\hat{\xi}$ is Hermitian, and it is referred to as the *generator* of rotation about an axis along $\hat{\xi}$.

The difference between Eq. 4.47a and Eq. 4.47b has important consequences on the rotation properties of the operators $U_R^{(\vec{\ell})}\left(\xi\hat{\xi}\right)$ and $U_R^{(\vec{s})}\left(\xi\hat{\xi}\right)$. We see that corresponding to a rotation through $\xi=2\pi$ about $\hat{e}_z=\hat{\xi}$, the eigenvalue $\exp\left(-\frac{i}{\hbar}2\pi\left(m\hbar\right)\right)$ of the rotation operator would be +1 when m is an integer, but it is −1 when m is half-integer. In the case of spin-half angular momentum, we therefore need a rotation through $\xi=4\pi$ to get the identity, whereas in the case of the orbital angular momentum $(\ell=0,1,2,3,..)$, we need a rotation through $\xi=2\pi$ to return the identity.

The above difference between $U_R^{(\vec{\ell})}$ and $U_R^{(\vec{s})}$ is because of a primary difference between the orbital and the spin angular momenta. Components of the orbital angular momentum are generators of rotations about three mutually orthogonal coordinate axes passing through the origin of the usual three-dimensional Euclidean space $\mathbb{R}^3$. The rotations leave one point in $\mathbb{R}^3$ fixed, which is the origin of the coordinate system. The rotation operators constitute a symmetry group represented by *orthogonal* matrices. The rotation group is a Lie group (named after Sophus Lie); the parameters of the product of two elements in this group are continuous functions of the parameters of the factor elements. This property enables the collection of all elements corresponding to finite rotations, including the identity constitute a mathematical group. In general, length-preserving transformations are represented by orthogonal matrices whose determinant is ±1. The value −1 of the determinant corresponds to inversions (or, reflection/parity) and +1 to rotations. The rotation group is denoted by $SO(3)$, but more generally, the symmetry group for rotations (including in higher dimensions) is denoted by $SO(N)$. In this notation, "O" stands for the nature of the transformations being *orthogonal*, and *S* stands for the *special* choice for the value of the determinant of these matrices being +1. This choice corresponds to pure rotations and excludes inversion/parity. For rotations in $\mathbb{R}^3$, $N=3$ corresponds to the number of generators of the group $SO(3)$, given by $\frac{N(N-1)}{2}=\frac{3(3-1)}{2}$, which are the three orthogonal components ℓ_x,ℓ_y,ℓ_z of the orbital angular momentum vector $\vec{\ell}$ (Eq. 4.7). These generators had primarily entered our analysis in the context of rotation (Eq. 4.3) of an arbitrary function $f(\vec{r})$, which may even be the Schrödinger wavefunction $\psi(\vec{r})$. We now proceed to examine the consequences of referring the complete wavefunction

(Eq. 4.45), inclusive of the spin, to a rotated coordinate system (Fig. 4.2). Let us consider a spin-orbital $\psi_\gamma(q)$ in the quantum state $|\gamma\rangle$:

$$\psi_\gamma(q) = \psi_\gamma(\vec{r},\zeta) = \langle \vec{r},\zeta \mid \gamma\rangle = \langle \vec{r},\zeta \mid \gamma,\pm\rangle \leftrightarrow \begin{bmatrix} \langle \vec{r},\zeta \mid \gamma,+\rangle \\ \langle \vec{r},\zeta \mid \gamma,-\rangle \end{bmatrix} = \begin{bmatrix} u_+(\vec{r}) \\ u_-(\vec{r}) \end{bmatrix}. \tag{4.49a}$$

Subjecting the spin-orbital to a rotation (such as in Figs. 4.2a,b) results in the transformation of the spin-orbital $\psi_\gamma(\vec{r},\zeta) \to \psi_\gamma(\vec{r}\,',\zeta)$ with

$$\psi_\gamma(q') = \langle q' \mid \gamma\rangle = \begin{bmatrix} u_+(\vec{r}\,') \\ u_-(\vec{r}\,') \end{bmatrix} = \mathbb{U}\begin{bmatrix} u_+(\vec{r}) \\ u_-(\vec{r}) \end{bmatrix} = \begin{bmatrix} \xi & \eta \\ \lambda & \mu \end{bmatrix}\begin{bmatrix} u_+(\vec{r}) \\ u_-(\vec{r}) \end{bmatrix}. \tag{4.49b}$$

In this relationship, $\mathbb{U}$ is a unitary transformation matrix. The four complex elements (ξ,η,λ,μ) of the matrix $\mathbb{U}$ have eight real numbers, but unitarity, expressed by the matrix equation

$$\mathbb{U}\mathbb{U}^\dagger = \mathbb{U}^\dagger\mathbb{U} = \begin{bmatrix} 1 & 0 \\ 0 & 1 \end{bmatrix}, \tag{4.50a}$$

provides four relationships between them. Furthermore, unimodularity, i.e., the fact that we consider those unitary transformations for which the determinant

$$|\mathbb{U}| = 1, \tag{4.50b}$$

provides a fifth relationship. That leaves us with a set of three real numbers, which define an arbitrary rotation. Typically, these correspond to the three Euler angles that we are familiar with from the description of rotations of a rigid body (see, for example, Chapter 7 in Reference [1]). The set of all possible transformations $\mathbb{U}$ constitutes the group SU(2), which is a set of all unitary unimodular 2 × 2 matrix operators. It is homomorphic, rather than isomorphic, with the SO(3) group, on account of the fact that a rotation through 4π is required in the case of spin-half angular momentum. We shall highlight this feature again, toward the end of this section.

We shall now study the coordinate representation of orbital angular momentum states in the eigenvector space of the operators (ℓ^2,ℓ_z). We write this as

$$\langle \hat{r}|\ell m\rangle = \langle \theta\varphi|\ell m\rangle = Y_{\ell,m}(\theta,\varphi), \tag{4.51}$$

where (θ,φ) are the usual spherical polar coordinates (see, for example, Chapter 2 of Reference [1]), and the functions $Y_{\ell,m}(\theta,\varphi)$ are called as spherical harmonics, with $\ell = 0,1,2,3,4,\ldots$ etc.; and for each value of the orbital angular momentum quantum number ℓ, we have $2\ell+1$ values of the magnetic quantum number, $m = -\ell,-\ell+1,..,\ell-1,\ell$. The spherical harmonics are solutions of a differential equation and constitute a complete set of orthogonal functions to express square-integrable, and normalizable functions of the polar coordinates (θ,φ). We present salient properties of the differential equation for spherical harmonics in Appendix C. These include some important properties of the Legendre polynomials, which we shall use subsequently, and in various other chapters also.

Let us now examine the coordinate representation of the state obtained from operating ℓ^2 on an arbitrary state vector $|\alpha\rangle$. Toward that end, we first recognize an operator identity for two vector operators (workout unsolved Problem E4.1). It is only a tad bit tedious to establish. It is

$$\left(\vec{F}\times\vec{G}\right)\cdot\left(\vec{F}\times\vec{G}\right)=\left\{F^2G^2-\left(\vec{F}\cdot\vec{G}\right)^2\right\}-\left\{\begin{matrix}\sum_j\sum_k F_j\left[F_j,G_k\right]G_k-\sum_j\sum_k F_j\left[F_k,G_k\right]G_j\\+\sum_j\sum_k F_j\left[F_k,G_j\right]G_k+\sum_j\sum_k F_jF_k\left[G_k,G_j\right]\end{matrix}\right\}. \quad (4.52a)$$

Putting $\vec{F}=\vec{r}$ and $\vec{G}=\vec{p}$, in Eq. 4.52a, it follows that

$$\ell^2=r^2p^2-\left(\vec{r}\bullet\vec{p}\right)^2+i\hbar\left(\vec{r}\bullet\vec{p}\right). \quad (4.52b)$$

Hence,

$$\langle\vec{r}|\ell^2|\alpha\rangle=\langle\vec{r}|r^2p^2|\alpha\rangle-\langle\vec{r}|\left(\vec{r}\bullet\vec{p}\right)^2|\alpha\rangle+i\hbar\langle\vec{r}|\left(\vec{r}\bullet\vec{p}\right)|\alpha\rangle. \quad (4.53a)$$

i.e., $\langle\vec{r}|\ell^2|\alpha\rangle=r^2\langle\vec{r}|p^2|\alpha\rangle-\langle\vec{r}|\left(\vec{r}\bullet\frac{\hbar}{i}\vec{\nabla}\right)^2|\alpha\rangle+i\hbar\langle\vec{r}|\left(\vec{r}\bullet\frac{\hbar}{i}\vec{\nabla}\right)|\alpha\rangle,$

or $\langle\vec{r}|\ell^2|\alpha\rangle=r^2\langle\vec{r}|p^2|\alpha\rangle+\hbar^2\langle\vec{r}|\left\{\left(r\frac{\partial}{\partial r}\right)\left(r\frac{\partial}{\partial r}\right)+\left(r\frac{\partial}{\partial r}\right)\right\}|\alpha\rangle.$ (4.53b)

Since for an arbitrary function $f(r)$, $\left\{\left(r\frac{\partial}{\partial r}\right)\left(r\frac{\partial}{\partial r}\right)+\left(r\frac{\partial}{\partial r}\right)\right\}f(r)=\left\{2r\frac{\partial}{\partial r}+r^2\frac{\partial^2}{\partial r^2}\right\}f(r),$

we get $\langle\vec{r}|\ell^2|\alpha\rangle=r^2\langle\vec{r}|p^2|\alpha\rangle+\hbar^2\langle\vec{r}|\left\{\left(2r\frac{\partial}{\partial r}\right)+\left(r^2\frac{\partial^2}{\partial r^2}\right)\right\}|\alpha\rangle.$ (4.53c)

We have used the fact that $\langle\vec{r}|$ is an eigenvector of the position operator operating from its right. Using now the result of the Solved Problem P1.5 (Chapter 1), we get

$$\langle\vec{r}|\ell^2|\alpha\rangle=r^2\langle\vec{r}|p^2|\alpha\rangle+\hbar^2 2r\frac{\partial}{\partial r}\langle\vec{r}|\alpha\rangle+\hbar^2r^2\frac{\partial^2}{\partial r^2}\langle\vec{r}|\alpha\rangle. \quad (4.54)$$

We now observe the following connection between the kinetic energy operator

$$T=\frac{\vec{p}\bullet\vec{p}}{2m}=-\frac{\hbar^2}{2m}\vec{\nabla}\bullet\vec{\nabla}=-\frac{\hbar^2\vec{\nabla}^2}{2m}, \quad (4.55a)$$

and that for the square of the orbital angular momentum,

$$\frac{1}{2mr^2}\langle\vec{r}|\ell^2|\alpha\rangle-\frac{\hbar^2}{mr}\frac{\partial}{\partial r}\langle\vec{r}|\alpha\rangle-\frac{\hbar^2}{2m}\frac{\partial^2}{\partial r^2}\langle\vec{r}|\alpha\rangle=\langle\vec{r}|\frac{\vec{p}\bullet\vec{p}}{2m}|\alpha\rangle=-\frac{\hbar^2}{2m}\langle\vec{r}|\nabla^2|\alpha\rangle. \quad (4.55b)$$

We now remind ourselves that in the spherical polar coordinates, the gradient operator is given by

$$\vec{\nabla}=\hat{e}_r\frac{\partial}{\partial r}+\hat{e}_\theta\frac{1}{r}\frac{\partial}{\partial\theta}+\hat{e}_\varphi\frac{1}{r\sin\theta}\frac{\partial}{\partial\varphi} \quad (4.56a)$$

and the Laplacian by

$$\begin{aligned}\nabla^2&=\frac{1}{r^2}\frac{\partial}{\partial r}\left(r^2\frac{\partial}{\partial r}\right)+\frac{1}{r^2}\left\{\frac{1}{\sin\theta}\frac{\partial}{\partial\theta}\left(\sin\theta\frac{\partial}{\partial\theta}\right)+\frac{1}{\sin^2\theta}\frac{\partial^2}{\partial\varphi^2}\right\}\\&=\left(\frac{\partial^2}{\partial r^2}+\frac{2}{r}\frac{\partial}{\partial r}\right)+\frac{1}{r^2}\left\{-\frac{\ell^2}{\hbar^2}\right\}.\end{aligned} \quad (4.56b)$$

Using the above relations, you would now recognize the following polar representation of the operator for the square of the orbital angular momentum:

$$\ell^2 = -\hbar^2 \left\{ \frac{1}{\sin\theta} \frac{\partial}{\partial\theta} \left(\sin\theta \frac{\partial}{\partial\theta} \right) + \frac{1}{\sin^2\theta} \frac{\partial^2}{\partial\varphi^2} \right\}. \tag{4.57a}$$

It is left as an exercise for you to see that we would get exactly the same result (Eq. 4.57a) if we were to take the square of the expression (Eq. 4.6) of the orbital angular momentum operator and use Eq. 4.56a for the gradient operator. Using the same method, you can also discover the following polar representation of the operator $\vec{\ell} \bullet \hat{e}_z = \ell_z$:

$$\ell_z = -i\hbar \frac{\partial}{\partial\varphi}. \tag{4.57b}$$

The spherical harmonics mentioned in Eq. 4.51 are simultaneous eigenfunctions of the orbital angular momentum vector operators in Eqs. 4.57a,b, expressed in the polar representation.

Using the method demonstrated in the Solved Problem P4.3, we can obtain the $(2j+1)\times(2j+1)$ dimensional matrix representations of the angular momentum operators j^2, j_z (whether orbital or spin, both) in the $\{|j,m\rangle\}$ basis, with $j = 0, \frac{1}{2}, 1, \frac{3}{2}, 2, \frac{5}{2}, 3, \frac{7}{2}, \ldots$; and for each value of j the magnetic quantum number m takes $2j+1$ values, $m = -j, -j+1, .., j-1, j$. Likewise, we now discuss the $(2j+1)\times(2j+1)$ dimensional matrix representation of the rotation operator $\exp\left(-\frac{i}{\hbar}\left(\vec{j} \bullet \hat{e}_z\right)\xi\right)$ defined in Eq. 4.48c. It is given by

$$\left\langle j'm' \left| \exp\left(-\frac{i}{\hbar}\left(\vec{j} \bullet \hat{e}_z\right)\xi\right) \right| jm \right\rangle = \delta_{jj'} \mathfrak{D}^{(j)}_{m'm}(R) = \left\langle jm' \left| \exp\left(-\frac{i}{\hbar}\left(\vec{j} \bullet \hat{e}_z\right)\xi\right) \right| jm \right\rangle. \tag{4.58a}$$

The Kronecker-δ in the above equation underscores the fact that the rotation operator $\exp\left(-\frac{i}{\hbar}\left(\vec{j} \bullet \hat{\xi}\right)\xi\right)$ commutes with j^2 no matter what the direction of $\hat{\xi}$ is, hence $U_R j^2 |jm\rangle$ and $j^2 U_R |jm\rangle$ correspond essentially to the same state.

Thus, $\langle jm' | U_R | jm \rangle = \mathfrak{D}^{(j)}_{m'm}(R) = \left\langle jm' \left| \exp\left(-\frac{i}{\hbar}\left(\vec{j} \bullet \hat{e}_z\right)\xi\right) \right| jm \right\rangle$. (4.58b)

The scalar functions $\mathfrak{D}^{(j)}_{m'm}(R)$ are known as Wigner Functions, in honor of Eugene Paul Wigner (1902–1995; 1963 Physics Nobel Laureate). They constitute an *irreducible representation of the rotation operator* (**IRRO**) U_R and are laid out as matrix *elements* of a block-diagonal matrix as follows:

$$[\mathfrak{D}(R)]_{IRRO} = \left[\left[\mathfrak{D}^{(j)}_{m'm}(R) \right]_{(2j+1)\times(2j+1)} \right]$$

$$= \begin{bmatrix} \left[\mathfrak{D}^{(j=0)}_{m'm}(R)\right]_{1\times1} & [0]_{1\times2} & [0]_{1\times3} & [0] \\ [0]_{2\times1} & \left[\mathfrak{D}^{\left(j=\frac{1}{2}\right)}_{m'm}(R)\right]_{2\times2} & [0]_{2\times3} & [0] \\ [0]_{3\times1} & [0]_{3\times2} & \left[\mathfrak{D}^{(j=1)}_{m'm}(R)\right]_{3\times3} & [0] \\ [0] & [0] & [0] & [\ldots] \end{bmatrix}. \tag{4.59}$$

This matrix is an irreducible representation; it is not possible to further diagonalize the block-matrices into smaller blocks. Note that the diagonal blocks are square matrices of increasing dimensions as the value of j increases. Their dimensionality $(2j+1)\times(2j+1)$ is indicated in Eq. 4.59 as a subscript on each block. The rectangular matrices $[0]_{p\times q}$ are null matrices of appropriate number of rows and columns as required to lay out the block matrices made up of the Wigner functions along the diagonal of the IRRO. The set of all rotation matrices $\{U_R(\xi\hat{\xi})\}$ for each j and all rotations $R\leftrightarrow(\xi\hat{\xi})$ through the angle ξ about the axis $\hat{\xi}$ constitute a group. This group of matrices represents the group SO(3) mentioned earlier. Each of the three generators of this group, namely the operators $j_i; i=1,2,3$ corresponding to three mutually orthogonal axes, commutes with the square of the angular momentum operator, j^2. An operator that commutes with *each* generator of a group is called the *Casimir operator* for that group, after Hendrik Casimir (1909–2000). Casimir first proved that one operator that commutes with the operators for all the generators of a Lie group can be built from a bilinear combination of the generators. The number of Casimir operators for a group (i.e., the number of operators that commute with each of its generators) is the same as the *rank* of the group. This theorem [4,5] is due to Racah. The only Casimir operator for the Rank-1 group SO(3) is j^2.

The matrix corresponding to zero rotation provides the identity, and rotation through the angle $-\xi$ about the same axis provides the inverse operation corresponding to rotation through $+\xi$. We express the "closure" property of the group corresponding to any two rotations R_i and R_j by the following relation:

$$\left[\mathfrak{D}^{(j)}_{m'm}(R_iR_j)\right]_{(2j+1)\times(2j+1)}=\left[\sum_{m''=-j}^{+j}\mathfrak{D}^{(j)}_{m'm''}(R_i)\mathfrak{D}^{(j)}_{m''m}(R_j)\right]_{(2j+1)\times(2j+1)}. \tag{4.60}$$

We also note that

$$\mathfrak{D}^{(j)}_{m'm}(R^{-1})=\mathfrak{D}^{(j)*}_{mm'}(R), \tag{4.61a}$$

which expresses the unitary character of the rotation matrices. It is a consequence of the primary fact that rotation operators themselves are unitary:

$$U(R^{-1})=U^{-1}(R)=U^{\dagger}(R). \tag{4.61b}$$

An arbitrary rotation is expressible in terms of three rotation matrices corresponding to the Euler angles φ,θ,ψ (see, for example, Eq. 7.52 in Reference [1]):

$$\mathbb{A}(\varphi,\theta,\psi)=\mathbb{R}_{Z''}(\psi)\mathbb{R}_{X'}(\theta)\mathbb{R}_{Z}(\varphi). \tag{4.62}$$

Wigner functions corresponding to such an arbitrary rotation therefore is

$$\mathfrak{D}^{(j)}_{m'm}(R:\varphi,\theta,\psi)=\left\langle jm'\left|\exp\left(-\frac{i}{\hbar}\left(\vec{j}\bullet\hat{e}_Z\right)\varphi\right)\exp\left(-\frac{i}{\hbar}\left(\vec{j}\bullet\hat{e}_{X'}\right)\theta\right)\times\right.\right.$$
$$\left.\left.\times\exp\left(-\frac{i}{\hbar}\left(\vec{j}\bullet\hat{e}_{Z''}\right)\psi\right)\right|jm\right\rangle \tag{4.63a}$$

i.e., $$\mathfrak{D}^{(j)}_{m'm}(R:\varphi,\theta,\psi)=\exp(-im'\varphi)\left\langle jm'\left|\exp\left(-\frac{i}{\hbar}j_{X'}\theta\right)\right|jm\right\rangle\exp(-im\psi), \tag{4.63b}$$

or $$\mathfrak{D}^{(j)}_{m'm}(R:\varphi,\theta,\psi)=\exp(-im'\varphi)\,d^{jx'}_{m'm}(\theta)\exp(-im\psi), \tag{4.63c}$$

where $d^{jx'}_{m'm}(\theta) = \mathfrak{D}^{(j)}_{m'm}(R:0,\theta,0)$. (4.63d)

Note that the lowercase $d^{jx'}_{m'm}(\theta)$ is only a special case of the Wigner $\mathfrak{D}^{(j)}_{m'm}(R:\varphi,\theta,\psi)$. Using now the fact that a rotation operator acting on an angular momentum state is expressible as a linear combination of the complete set of vectors in the same-j space

$$\exp\left(-\frac{i}{\hbar} j_{X'}\theta\right)|jm\rangle = U(R:\theta_{X'})|jm\rangle = \sum_{m''=-j}^{j} \mathfrak{D}^{(j)}_{m''m}(R:\theta_{X'})|jm''\rangle, \tag{4.64a}$$

we can therefore write for a rotation

$$U_R|jm\rangle = \sum_{m'=-j}^{j} |jm'\rangle\langle jm'|U_R|jm\rangle = \sum_{m'=-j}^{j} \mathfrak{D}^{(j)}_{m'm}(R)|jm'\rangle. \tag{4.64b}$$

$$\text{Hence, } \mathfrak{D}^{(j)}_{m'm}(R:\varphi,\theta,\psi) = \exp(-i\{m'\varphi + m\psi\})\langle jm'\left|\left\{\sum_{m''=-j}^{j} \mathfrak{D}^{(j)}_{m''m}(R:\theta_{X'})|jm''\rangle\right\}\right.. \tag{4.64c}$$

In carrying out such rotation analysis, one must remember that the orthogonality relation $\langle jm'|jm''\rangle = \delta_{m'm''}$ can be used only when the magnetic quantum number in the $\langle \text{bra}|$ vector and that in the $|\text{ket}\rangle$ vector refer to essentially the *same* axis of quantization. In Eqs. 4.64a,b, note that the summation is over the first index. We have already seen that in the case of the integer angular momentum (example: orbital angular momentum, ℓ), the coordinate representation of the angular momentum states $|\ell m\rangle$ is very useful. The relation corresponding to 4.64a for the spherical harmonics is

$$\langle\hat{r}|\ell m\rangle = \sum_{m'=-\ell}^{\ell} \mathfrak{D}^{(\ell)}_{m'm}(R^{-1})\langle\hat{r}_R|\ell m'\rangle, \tag{4.64d}$$

$$\text{i.e., } Y_{\ell m}(\theta\varphi) = \sum_{m'=-\ell}^{\ell} \mathfrak{D}^{(\ell)}_{m'm}(R^{-1})Y_{\ell m'}(\theta_R\varphi_R) = \sum_{m'=-\ell}^{\ell} \mathfrak{D}^{(\ell)}_{mm'}(R)^* Y_{\ell m'}(\theta_R\varphi_R), \tag{4.64e}$$

where the polar angles are as shown in Fig. 4.3.

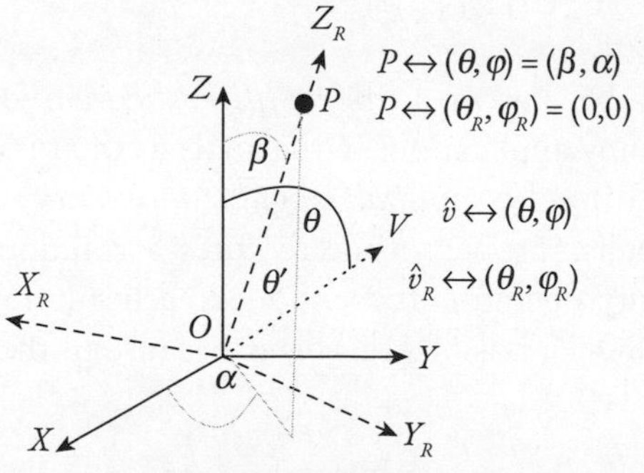

Fig. 4.3 With a common origin O, we have a Cartesian coordinate frame of reference (X_R,Y_R,Z_R) that is rotated with respect to (X,Y,Z). The polar and azimuthal coordinates of an *arbitrary point on the line OV* along the direction given by the unit vector $\hat{v} = \hat{v}_R$ are (θ,φ) with respect to the frame (X,Y,Z) and (θ_R,φ_R) with respect to (X_R,Y_R,Z_R). The polar coordinates of a point P on the Cartesian axis Z_R are $(\theta,\varphi) = (\beta,\alpha)$ with respect to the frame (X,Y,Z) and $(\theta_R,\varphi_R) = (0,0)$ with respect to (X_R,Y_R,Z_R). We shall denote a unit vector along the axis Z_R by $\hat{u}$.

Note that $(\theta\varphi)$ and $(\theta_R\varphi_R)$ are the polar coordinates of essentially the same point which is along the unit vector $\hat{v}$, but the latter are with respect to a rotated frame of reference. The polar coordinates of a point P on the Z_R axis are $(\theta,\varphi)=(\beta,\alpha)$ with respect to the frame (X,Y,Z) and $(\theta_R,\varphi_R)=(0,0)$ with respect to (X_R,Y_R,Z_R). We know that the square of the orbital angular momentum operator commutes with one of its components, which could be any direction, such as along Z, or along Z_R,

i.e., we have $\left[\ell^2,\ell_Z\right]=0=\left[\ell^2,\ell_{Z_R}\right]$. (4.65)

For the point P,

$$Y_{\ell m}(\beta,\alpha)=\sum_{m'=-\ell}^{\ell}\mathfrak{D}_{mm'}^{(\ell)\,*}(R)Y_{\ell m'}(0,0)=\sum_{m'=-\ell}^{\ell}\mathfrak{D}_{mm'}^{(\ell)\,*}(R)\sqrt{\frac{2\ell+1}{4\pi}}\delta_{m'0}=\mathfrak{D}_{m0}^{(\ell)\,*}(R)\sqrt{\frac{2\ell+1}{4\pi}}, \quad (4.66)$$

since for every (ℓ,m'), $Y_{\ell m'}(0,0)=\sqrt{\frac{2\ell+1}{4\pi}}\delta_{m'0}$. (4.67)

From Eq. 4.66 it follows that $\mathfrak{D}_{m0}^{(\ell)\,*}(R)=\sqrt{\frac{4\pi}{2\ell+1}}\,Y_{\ell m}(\beta\alpha)$. (4.68)

Equation 4.64e for $m=0$ gives

$$Y_{\ell 0}(\theta\varphi)=\sum_{m'=-\ell}^{\ell}\mathfrak{D}_{0m'}^{(\ell)\,*}(R)Y_{\ell m'}(\theta_R\varphi_R)=\sum_{m'=-\ell}^{\ell}\mathfrak{D}_{m'0}^{(\ell)}(R)Y_{\ell m'}(\theta_R\varphi_R), \quad (4.69a)$$

i.e., $Y_{\ell 0}(\theta\varphi)=\sum_{m=-\ell}^{\ell}\sqrt{\frac{4\pi}{2\ell+1}}\,Y_{\ell m}^{*}(\theta\varphi)Y_{\ell m}(\theta_R\varphi_R)$. (4.69b)

Therefore, $\sqrt{\frac{2\ell+1}{4\pi}}P_\ell\left(\cos\theta_R\right)=\sum_{m=-\ell}^{\ell}\sqrt{\frac{4\pi}{2\ell+1}}Y_{\ell m}^{*}(\beta,\alpha)Y_{\ell m}(\theta\varphi)$. (4.69c)

In other words, $P_\ell\left(\cos\theta_R\right)=\frac{4\pi}{2\ell+1}\sum_{m=-\ell}^{\ell}Y_{\ell m}^{*}(\beta,\alpha)Y_{\ell m}(\theta\varphi)$, (4.70a)

i.e., $P_\ell\left(\hat{u}\bullet\hat{v}\right)=\frac{4\pi}{2\ell+1}\sum_{m=-\ell}^{\ell}Y_{\ell m}^{*}(\hat{u})Y_{\ell m}(\hat{v})$. (4.70b)

The result in Eqs. 4.70a,b is known as the *addition theorem for spherical harmonics*. It is an important one; it has many applications in the analysis of atomic structure and dynamics.

The rotations generated by the *spin-half angular momentum*, introduced in the Solved Problem P4.3, have a peculiar property: one requires a rotation through 4π rather than 2π to represent the identity operator, as mentioned earlier (after Eq. 4.50b). The rotation algebra in this case is not SO(3), it is SU(2). This is because of the fact the rotation operator (Eqs. 4.48a–d) in this case is

$$U_R\left(\theta\hat{\theta}\right)=e^{-\frac{i}{\hbar}\theta\hat{\theta}\bullet\vec{s}}=\exp\left(-\frac{i}{\hbar}\theta\hat{\theta}\bullet\frac{1}{2}\hbar\vec{\sigma}\right)=\exp\left(-\frac{i}{2}\theta\hat{\theta}\bullet\vec{\sigma}\right), \quad (4.71)$$

and hence for rotation through 2π, we get

$$U_R\left(2\pi\hat{\theta}\right)=\exp\left(-i\pi\hat{\theta}\bullet\vec{\sigma}\right)=\cos\left(\pi\begin{bmatrix}1&0\\0&-1\end{bmatrix}\right)-i\sin\left(\pi\begin{bmatrix}1&0\\0&-1\end{bmatrix}\right)=-1\begin{bmatrix}1&0\\0&1\end{bmatrix}, \quad (4.72a)$$

and $U_R\left(4\pi\hat{\theta}\right)=\begin{bmatrix}1 & 0\\ 0 & 1\end{bmatrix}=1_{2\times 2}^{\text{operator}}$. (4.72b)

An experiment in which the 4π rotation of a spin-half particle is verified is discussed in the context of the Solved Problem P8.1 of Chapter 8.

4.2 Addition of Angular Momenta: Clebsch–Gordan Coefficients

An elementary particle can have two independent sources of angular momentum, orbital and spin. We describe a particle's state by a vector in the Hilbert space, but the orbital and spin angular momenta being independent of each other, the Hilbert space of the orbital angular momentum is disjoint from that of the spin. It is always good to keep it at the back of our mind that the terms *orbital* and *spin* have no classical analogues. We describe the *net* (generally called as *total*) angular momentum of the particle by the vector operator

$$\vec{j}=\vec{\ell}\oplus\vec{s}=\vec{\ell}+\vec{s}. \tag{4.73}$$

For the sake of simplicity, we shall continue to use the symbol + to denote the *addition* of two angular momenta. However, the symbol $\oplus$ in Eq. 4.73 emphasizes that this is not merely the addition of two classical vectors; it is a provision to examine the combined effect of two independent sources of angular momentum, both described by Eqs. 4.18a,b,c. What is *termed* as *addition* of angular momenta in Eq. 4.73 is the study of this combined effect. The addition in Eq. 4.73 is also called "s-o", for *spin–orbit, coupling*.

There are other situations in which the addition of angular momentum is important, not just those which involve the spin and the orbital angular momentum of an elementary particle. It could be the combined effect of spin angular momentum of one particle and that of another, or the orbital angular momentum of one particle with that of a second particle, or for that matter, the spin of one particle and the orbital angular momentum of another particle (called "s-o-o", for *spin–other–orbit coupling*). In general, for two independent angular momentum vector operators $\vec{j}_1$ and $\vec{j}_2$, we define their "sum" as

$$\vec{j}=\vec{j}_1\oplus\vec{j}_2=\vec{j}_1+\vec{j}_2. \tag{4.74}$$

The combined vector space of the sum $\vec{j}$ is the so-called *tensor product space* of the vectors $\{|j_1m_1\rangle\}$ and $\{|j_2m_2\rangle\}$. The "addition" in Eq. 4.74 is neither an addition of two numbers nor that of two classical vectors; likewise, the product space is neither the inner (scalar) product $\langle a|b\rangle$ (which is a bra-ket) nor the outer product $|a\rangle\langle b|$ (which is a ket-bra). The inner and the outer product are both defined using a vector in the direct space and another in its dual conjugate space. The product space of the combined angular momentum vector operator $\vec{j}$ is different from both the inner and the outer product since it involves disjoint vector spaces of the two independent angular momenta. Vectors in the product space are obtained as tensor (or direct) product of vectors from the two independent vector spaces. We write the direct product of vectors in two independent Hilbert spaces as

$$|j_1m_1\rangle\otimes|j_2m_2\rangle=|j_1m_1\rangle\times|j_2m_2\rangle=|j_1m_1j_2m_2\rangle=|(j_1j_2)m_1m_2\rangle\equiv|m_1m_2\rangle. \tag{4.75}$$

The direct (i.e., tensor) product is denoted by $\otimes$, or simply by $\times$. In Eq. 4.75, we have presented alternative and equivalent notations in common use to write the direct product of vectors that belong to two disjoint Hilbert spaces of the operators $\vec{j}_1$ and $\vec{j}_2$. The dimensionality of the vector space of $\vec{j}_1$ is $(2j_1+1)$, since m_1 takes $2j_1+1$ values $-j_1, -j_1+1, .., j_1-1, j_1$. Likewise, the dimensionality of the vector space of $\vec{j}_2$ is $(2j_2+1)$. We therefore expect the dimensionality of their tensor product space to be $(2j_1+1)\times(2j_2+1)$, and indeed it is so. The direct product space is an eigenspace of the net angular momentum (Eq. 4.74); i.e., it is spanned by the eigenvectors of $\vec{j}\bullet\vec{j}$ and $\vec{j}\bullet\hat{u}$ with

$$[\vec{j}\bullet\vec{j}, \vec{j}\bullet\hat{u}] = 0. \tag{4.76}$$

All of the defining characteristic properties (Eqs. 4.18a,b,c) of the angular momentum operators are satisfied by the sum defined in the Eq. 4.73. The commuting operators in Eq. 4.76 therefore have a diagonal representation in the eigenbasis $\{|(j_1 j_2)\, jm\rangle\}$ of the coupled angular momentum, in which the range of values of j is determined by j_1 and j_2, and for each j the magnetic quantum number m takes $2j+1$ values from $-j$ to $+j$ in steps of unity. The eigenbasis of the coupled angular momentum $\{|(j_1 j_2)\, jm\rangle\}$ and the basis of the tensor product of the uncoupled factor states $\{|(j_1 j_2) m_1 m_2\rangle\}$ are both *orthonormal* and *complete*. It is natural to expect that an arbitrary vector in the direct product space is expressible as a linear superposition of the base vectors consisting of the tensor product states $\{|m_1 m_2\rangle\}$:

$$|(j_1 j_2)\, jm\rangle = \sum_{m_1=-j_1}^{j_1} \sum_{m_2=-j_2}^{j_2} |(j_1 j_2) m_1 m_2\rangle \langle (j_1 j_2) m_1 m_2 | (j_1 j_2)\, jm\rangle. \tag{4.77a}$$

In brief notation, we write this result as

$$|jm\rangle = \sum_{m_1=-j_1}^{j_1} \sum_{m_2=-j_2}^{j_2} |m_1 m_2\rangle \langle m_1 m_2 | jm\rangle. \tag{4.77b}$$

We interpret Eqs. 4.77a,b in two completely equivalent ways:

a. resolution of the unit operator

$$1 = \sum_{m_1=-j_1}^{j_1} \sum_{m_2=-j_2}^{j_2} |m_1 m_2\rangle \langle m_1 m_2| \equiv \sum_{m_1=-j_1}^{j_1} \sum_{m_2=-j_2}^{j_2} |(j_1 j_2) m_1 m_2\rangle \langle (j_1 j_2) m_1 m_2| \tag{4.78}$$

in the composite factor space may be considered to have operated on the direct product vector that you see on the left-hand side of Eqs. 4.77a,b.

b. expansion of the vector on the left-hand side of Eqs. 4.77a,b in the complete set of base vectors $\{|(j_1 j_2) m_1 m_2\rangle\} \equiv \{|m_1 m_2\rangle\}$ of the composite factor space; the expansion coefficients being $\langle (j_1 j_2) m_1 m_2 | (j_1 j_2)\, jm\rangle \equiv \langle m_1 m_2 | jm\rangle$.

Similarly, we can use superposition of a complete set of base vectors $\{|jm\rangle\}$ in the direct product space (which are simultaneous eigenvectors of $\vec{j}\bullet\vec{j}$ and j_z) to express every base vector $|(j_1 j_2) m_1 m_2\rangle \equiv |m_1 m_2\rangle$ in the product of factor spaces:

$$|(j_1 j_2) m_1 m_2\rangle = \sum_{j=j_{\min}}^{j_{\max}} \sum_{m=-j}^{j} |(j_1 j_2)\, jm\rangle \langle (j_1 j_2)\, jm | (j_1 j_2) m_1 m_2\rangle. \tag{4.79a}$$

In the short notation, this relation is

$$\left|m_1m_2\right\rangle = \sum_{j=j_{\min}}^{j_{\max}} \sum_{m=-j}^{j} \left|jm\right\rangle\left\langle jm\middle|m_1m_2\right\rangle. \tag{4.79b}$$

Equations 4.79a,b also have alternative equivalent interpretations, coming from the recognition of the unit operator

$$1 = \sum_{j=j_{\min}}^{j_{\max}} \sum_{m=-j}^{j} \left|jm\right\rangle\left\langle jm\right| \equiv \sum_{m_1=-j_1}^{j_1} \sum_{m_2=-j_2}^{j_2} \left|(j_1j_2)\,jm\right\rangle\left\langle(j_1j_2)\,jm\right|, \tag{4.80}$$

and identifying $\left\langle(j_1j_2)\,jm\middle|(j_1j_2)\,m_1m_2\right\rangle \equiv \left\langle jm\middle|m_1m_2\right\rangle$ as expansion coefficients of $\left|m_1m_2\right\rangle$ in the basis $\left\{\left|jm\right\rangle\right\}$. The expansion coefficients $\left\langle(j_1j_2)m_1m_2\middle|(j_1j_2)\,jm\right\rangle = \left\langle m_1m_2\middle|jm\right\rangle$ in Eqs. 4.77a,b and $\left\langle(j_1j_2)\,jm\middle|(j_1j_2)m_1m_2\right\rangle = \left\langle jm\middle|m_1m_2\right\rangle$ in Eqs. 4.79a,b are called as the Clebsch–Gordan (or "CG") coefficients, named after Alfred Clebsch (1833–1872) and Paul Albert Gordan (1837–1912). They would seem to be complex conjugates of each other, but in the Condon and Shortley convention mentioned earlier, they turn out to be real numbers, as would be soon explained. The works of Clebsch and Gordan are older than most developments in quantum mechanics; they are credited for their underlying seminal contributions to algebraic geometry. We shall now begin to use the shorter notation $\left\{\left|m_1m_2\right\rangle\right\}$ instead of $\left\{\left|(j_1j_2)m_1m_2\right\rangle\right\}$ and $\left\{\left|jm\right\rangle\right\}$ instead of $\left\{\left|(j_1j_2)\,jm\right\rangle\right\}$. First, we note that the CGCs (i.e., Clebsch–Gordan coefficients) are zero unless $m = m_1 + m_2$. To prove that, we observe from Eq. 4.74 that

$$j_z\left|jm\right\rangle = \left(j_{1z} + j_{2z}\right) \sum_{m_1=-j_1}^{j_1} \sum_{m_2=-j_2}^{j_2} \left|m_1m_2\right\rangle\left\langle m_1m_2\middle|jm\right\rangle, \tag{4.81a}$$

$$\text{i.e., } \sum_{m_1=-j_1}^{j_1} \sum_{m_2=-j_2}^{j_2} \left|m_1m_2\right\rangle\left\langle m_1m_2\middle|j_z\middle|jm\right\rangle = \left\{\begin{aligned} & j_{1z} \sum_{m_1=-j_1}^{j_2} \sum_{m_2=-j_2}^{j_2} \left|j_1m_1\right\rangle\left|j_2m_2\right\rangle\left\langle m_1m_2\middle|jm\right\rangle \\ & +j_{2z} \sum_{m_1=-j_1}^{j_1} \sum_{m_2=-j_2}^{j_2} \left|j_1m_1\right\rangle\left|j_2m_2\right\rangle\left\langle m_1m_2\middle|jm\right\rangle \end{aligned}\right\}, \tag{4.81b}$$

$$\text{i.e., } \hbar \sum_{m_1=-j_1}^{j_1} \sum_{m_2=-j_2}^{j_2} \left|m_1m_2\right\rangle m\left\langle m_1m_2\middle|jm\right\rangle = \hbar\left\{\begin{aligned} & \sum_{m_1=-j_1}^{j_2} \sum_{m_2=-j_2}^{j_2} m_1\left|m_1m_2\right\rangle\left\langle m_1m_2\middle|jm\right\rangle \\ & + \sum_{m_1=-j_1}^{j_1} \sum_{m_2=-j_2}^{j_2} m_2\left|m_1m_2\right\rangle\left\langle m_1m_2\middle|jm\right\rangle \end{aligned}\right\}, \tag{4.81c}$$

$$\text{i.e., } \sum_{m_1=-j_1}^{j_1} \sum_{m_2=-j_2}^{j_2} \left|m_1m_2\right\rangle\left\{m\left\langle m_1m_2\middle|jm\right\rangle\right\} = \sum_{m_1=-j_1}^{j_2} \sum_{m_2=-j_2}^{j_2} \left|m_1m_2\right\rangle\left\{\left(m_1+m_2\right)\left\langle m_1m_2\middle|jm\right\rangle\right\}. \tag{4.81d}$$

Both sides of Eq.4.81d are superpositions of essentially the same linearly independent base vectors; hence, the coefficients of the corresponding vectors on the two sides must be equal, i.e.,

$$m\left\langle m_1m_2\middle|jm\right\rangle = \left(m_1+m_2\right)\left\langle m_1m_2\middle|jm\right\rangle, \tag{4.82a}$$

$$\text{or } \left(m_1+m_2-m\right)\left\langle m_1m_2\middle|jm\right\rangle = 0. \tag{4.82b}$$

We therefore conclude that a CGC must be zero unless $m_1 + m_2 = m$. This is a very useful property of the CGCs. As seen subsequently, it immediately helps us determine the range $j_{\min} \le j \le j_{\max}$ that we employed in the summations in Eqs. 4.79a,b and Eq. 4.80. Toward that, let us consider the CGC $\langle m_1 m_2 | jm \rangle$ with $m = m_{\max} = j$ and $m_1 = m_{1,\max} = j_1$. For this coefficient, we must have $m_2 = m - m_1 = j - j_1$. Now, we know that $-j_2 \le m_2 \le j_2$, i.e., $-j_2 \le j - j_1 \le j_2$ and hence $(j_1 - j_2) \le j \le (j_1 + j_2)$. However, we could have executed this reasoning with the indices "1" and "2" interchanged, which would have given us $(j_2 - j_1) \le j \le (j_1 + j_2)$. We must therefore conclude that

$$|j_1 - j_2| \le j \le (j_1 + j_2);\ \text{i.e.,}\ j_{\min} = |j_1 - j_2| \text{ and } j_{\max} = j_1 + j_2. \tag{4.83}$$

The above inequality is called as the *triangle inequality* for reasons that become clear from Fig. 4.4.

Let us tentatively assume that $j_1 > j_2$. Since for each value of j, m takes $2j + 1$ values, the dimensionality d of the $\{|jm\rangle\}$ eigenbasis of the coupled angular momentum is given by

$$d = \sum_{j=j_1-j_2}^{j_1+j_2} (2j+1) = \begin{bmatrix} \{(2j_{\min}+1)\} + \{2(j_{\min}+1)+1\} + \{2(j_{\min}+2)+1\} + \dots \\ +\{2(j_{\max}-2)+1\} + \{2(j_{\max}-1)+1\} + \{(2j_{\max}+1)\} \end{bmatrix}. \tag{4.84}$$

From the triangle inequality $(j_1 - j_2) \le j \le (j_1 + j_2)$, it is clear that j can take a total of $(2j_2 + 1)$ values between $j_{\min} = (j_1 - j_2)$ and $j_{\max} = (j_1 + j_2)$. Therefore, the number of curly brackets $\{\ \}$ on the right-hand side of Eq. 4.84 is $(2j_2 + 1)$. Simplifying,

$$\begin{aligned} d &= \sum_{j=j_1-j_2}^{j_1+j_2} (2j+1) \\ &= \begin{bmatrix} (2(j_1-j_2)+1) + \{2((j_1-j_2)+1)+1\} + \{2((j_1-j_2)+2)+1\} + \dots \\ +\{2((j_1+j_2)-2)+1\} + \{2((j_1+j_2)-1)+1\} + (2(j_1+j_2)+1) \end{bmatrix}, \end{aligned} \tag{4.85a}$$

$$\begin{aligned} \text{i.e., } d &= \sum_{j=j_1-j_2}^{j_1+j_2} (2j+1) \\ &= \begin{bmatrix} \{(2j_1+1) \cancel{-2j_2}\} + \{(2j_1+1) \cancel{-2j_2+2}\} + \{(2j_1+1) \cancel{-2j_2+4}\} + \dots \\ +\{(2j_1+1) \cancel{+2j_2-4}\} + \{(2j_1+1) \cancel{+2j_2-2}\} + \{(2j_1+1) \cancel{+2j_2}\} \end{bmatrix}. \end{aligned} \tag{4.85b}$$

On canceling pairs of terms placed symmetrically away from the ends of the outermost (rectangular) bracket above, because they have opposite signs, we find ourselves adding $(2j_1 + 1)$ repeatedly $(2j_2 + 1)$ times. Therefore,

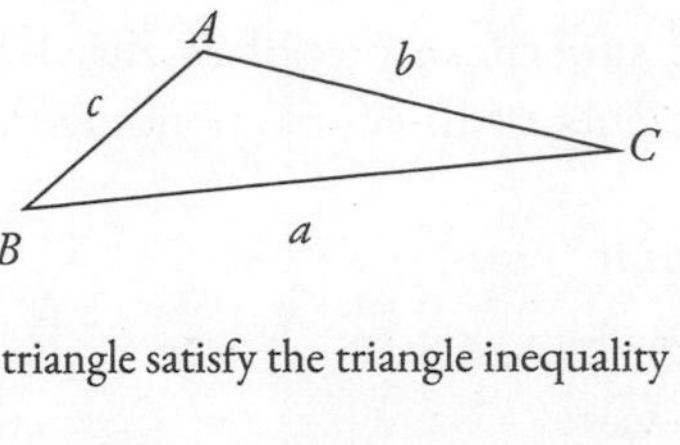

Fig. 4.4 Lengths of the three sides of a triangle satisfy the triangle inequality $|c - a| \le b \le (a + c)$.

$$d=\sum_{j=j_1-j_2}^{j_1+j_2}(2j+1)=(2j_1+1)\times(2j_2+1). \tag{4.86}$$

Assuming $j_2>j_1$ would give us exactly the same result, we would have only ended up adding $(2j_2+1)$ a total of $(2j_1+1)$ times. There is therefore no loss of generality in the assumption $j_1>j_2$, which we had tentatively made earlier. We of course knew the result of Eq. 4.86 from the comments between Eq. 4.75 and Eq. 4.76, since the disjoint factor spaces of the two angular momenta j_1 and j_2 have dimensions $(2j_1+1)$ and $(2j_2+1)$ respectively.

The CGCs are derivable from a single-seed coefficient for which we employ the Condon and Shortley convention. This is easy to see by using the ladder operators for the two angular momenta under addition to produce the composite angular momentum (Eq. 4.72), from which it follows that

$$j_\pm|jm\rangle=j_{1\pm}\sum_{m_1=-j_1}^{j_1}\sum_{m_2=-j_2}^{j_2}|m_1m_2\rangle\langle m_1m_2|jm\rangle+j_{2\pm}\sum_{m_1=-j_1}^{j_1}\sum_{m_2=-j_2}^{j_2}|m_1m_2\rangle\langle m_1m_2|jm\rangle. \tag{4.87}$$

Equation 4.87 is in fact a pair of two equations, one for the plus sign and the other for the minus sign. We shall continue parallel analysis of both the relations.

Using Eq. 4.42, and canceling $\hbar$ on both sides of the equation, we get

$$\sqrt{(j\mp m)(j\pm m+1)}\,|j,m\pm 1\rangle = \left[\begin{array}{l}\displaystyle\sum_{m_1=-j_1}^{j_1}\sum_{m_2=-j_2}^{j_2}\sqrt{j_1(j_1+1)-m_1(m_1\pm 1)}\,|j_1,m_1\pm 1\rangle|j_2m_2\rangle\langle m_1m_2|jm\rangle\\ \displaystyle\sum_{m_1=-j_1}^{j_1}\sum_{m_2=-j_2}^{j_2}\sqrt{j_2(j_2+1)-m_2(m_2\pm 1)}\,|j_1m_1\rangle|j_2,m_2\pm 1\rangle\langle m_1m_2|jm\rangle\end{array}\right]. \tag{4.88}$$

It is now convenient to employ new auxiliary summation indices, with primes,

$$m_1'=m_1\pm 1 \text{ and } m_2'=m_2\pm 1. \tag{4.89a}$$

Correspondingly, we have $m_1'\mp 1=m_1$ and $m_2'\mp 1=m_2$. (4.89b)

We can now rewrite the result of Eq. 4.86 in terms of the primed indices:

$$\sqrt{(j\mp m)(j\pm m+1)}\,|j,m\pm 1\rangle = \left[\begin{array}{l}\displaystyle\sum_{m_1'=-j_1}^{j_1}\sum_{m_2=-j_2}^{j_2}\sqrt{j_1(j_1+1)-(m_1'\mp 1)(m_1')}\,|m_1',m_2\rangle\langle m_1'\mp 1,m_2|jm\rangle\\ \displaystyle\sum_{m_1=-j_1}^{j_1}\sum_{m_2'=-j_2}^{j_2}\sqrt{j_2(j_2+1)-(m_2'\mp 1)(m_2')}\,|m_1,m_2'\rangle\langle m_1,m_2'\mp 1|jm\rangle\end{array}\right].$$

Dropping now the prime on the dummy summation indices, we get

$$\sqrt{(j\mp m)(j\pm m+1)}\,|j,m\pm 1\rangle = \sum_{m_1=-j_1}^{j_1}\sum_{m_2=-j_2}^{j_2}\left[\begin{array}{l}\sqrt{j_1(j_1+1)-(m_1\mp 1)(m_1)}\,|j_1m_1\rangle|j_2m_2\rangle\langle m_1\mp 1,m_2|jm\rangle\\ +\sqrt{j_2(j_2+1)-(m_2\mp 1)(m_2)}\,|j_1m_1\rangle|j_2m_2\rangle\langle m_1,m_2\mp 1|jm\rangle\end{array}\right].$$

Inserting now, on the right hand side, the unit operator resolved in the basis of the tensor product of the factor uncoupled states, we have

$$\sum_{m_1=-j_1}^{j_1}\sum_{m_2=-j_2}^{j_2}|m_1m_2\rangle\begin{Bmatrix}\langle m_1m_2|j,m\pm1\rangle\\ \times\sqrt{(j\mp m)(j\pm m+1)}\end{Bmatrix}$$

$$=\begin{bmatrix}\sum_{m_1=-j_1}^{j_1}\sum_{m_2=-j_2}^{j_2}\sqrt{j_1(j_1+1)-(m_1\mp1)(m_1)}|m_1m_2\rangle\langle m_1\mp1,m_2|jm\rangle\\ \sum_{m_1=-j_1}^{j_1}\sum_{m_2=-j_2}^{j_2}\sqrt{j_2(j_2+1)-(m_2\mp1)(m_2)}|m_1m_2\rangle\langle m_1,m_2\mp1|jm\rangle\end{bmatrix}.$$

We have superposition in essentially the same linearly independent basis on both the left- and the right-hand sides of the equation, guaranteeing that the coefficients of the corresponding terms (i.e., terms referenced to the same base vectors) are equal. Hence,

$$\langle m_1m_2|j,m\pm1\rangle\sqrt{(j\mp m)(j\pm m+1)}$$
$$=\begin{bmatrix}\langle m_1\mp1,m_2|jm\rangle\sqrt{j_1(j_1+1)-(m_1\mp1)(m_1)}\\ +\langle m_1,m_2\mp1|jm\rangle\sqrt{j_2(j_2+1)-(m_2\mp1)(m_2)}\end{bmatrix}. \tag{4.90a}$$

i.e., $\langle m_1m_2|j,m\pm1\rangle\sqrt{(j\mp m)(j\pm m+1)}$

$$=\begin{bmatrix}\langle m_1\mp1,m_2|jm\rangle\sqrt{(j_1\pm m_1)(j_1\mp m_1+1)}\\ +\langle m_1,m_2\mp1|jm\rangle\sqrt{(j_2\pm m_2)(j_2\mp m_2+1)}\end{bmatrix}. \tag{4.90b}$$

Equations 4.90a,b are a pair of relations, known as the recursion relations. All the CGCs can be determined using the recursion relations from just *one* seed CGC. A convenient choice, by convention, is to have the CGC corresponding to the maximum possible values of the magnetic quantum numbers m and m_1 to be equal to +1; i.e.,

$$\langle m_1m_2|jm\rangle\leftrightarrow\langle m_1=j_1,m_2=m-m_1=j-j_1|j,m=j\rangle\leftrightarrow\langle j_1,j-j_1|j,j\rangle=+1. \tag{4.91}$$

This choice, together with the Condon and Shortley sign convention used in the CGC recursion relations, guarantees that all the CGCs are essentially real numbers. A *not-so-uncommon* source of error in angular momentum algebra occurs from mixing up formalisms in which the phase conventions are different; one must therefore watch out for the phase conventions very carefully. Since atomic physics mostly involves electron dynamics, among physical properties of greatest importance is the electron spin. As mentioned earlier, this property of the electron results from the fact that laws of nature are relativistic and hence correctly described only by a relativistic theory. Dirac's formulation of relativistic quantum mechanics accounts for the electron spin; we shall discuss this in Chapter 7. Here we only acknowledge that the spin angular momentum of an electron is $s=\frac{1}{2}$. We often need to employ the *s–o* coupling. For this case, $j_1=\ell$, the orbital angular momentum quantum number of an electron, and $j_2=s$, the spin angular momentum quantum number. We therefore pay special attention to the case when one has an arbitrary angular momentum quantum number j_1 and the other is $j_2 = \frac{1}{2}$. In this case, when m and m_1 take their respective largest values $m = j$ and $m_1 = j_1$, we get from recursion relation (Eq. 4.90, with the *lower* sign)

$$\langle m_1 m_2 | j, m-1 \rangle \sqrt{(j+m)(j-m+1)} = \begin{bmatrix} \langle m_1+1, m_2 | jm \rangle \sqrt{(j_1-m_1)(j_1+m_1+1)} \\ +\langle m_1, m_2+1 | jm \rangle \sqrt{(j_2-m_2)(j_2+m_2+1)} \end{bmatrix}. \quad (4.92a)$$

The CGC on the left-hand side can be nonzero only when $m_2 = (m-1) - m_1$. Thus, when $m_2 = j - 1 - j_1$ if $m_1 = j_1$ and $m = j$ whence the first term on the right-hand side goes to zero, giving

$$\langle j_1, j-1-j_1 | j, j-1 \rangle \sqrt{2j} = \langle j_1, j-j_1 | jj \rangle \sqrt{(j_2 - j + 1 + j_1)(j_2 + j - j_1)}. \quad (4.92b)$$

Furthermore, from the recursion relation for the CGCs (Eqs. 4.90a,b, upper sign), we get

$$\langle m_1 m_2 | j, m+1 \rangle \sqrt{(j-m)(j+m+1)} = \begin{bmatrix} \langle m_1-1, m_2 | jm \rangle \sqrt{(j_1+m_1)(j_1-m_1+1)} \\ +\langle m_1, m_2-1 | jm \rangle \sqrt{(j_2+m_2)(j_2-m_2+1)} \end{bmatrix}. \quad (4.93a)$$

The CGC on the left-hand side would be nonzero only when $m_2 = (m+1) - m_1$. Thus, when $m_1 = j_1$ and $m = j-1$ (i.e., not the highest, but the next), we shall have $m_2 = (j-1+1) - m_1 = j - j_1$, and hence

$$\langle j_1, j-j_1 | j, j \rangle \sqrt{2j} = \begin{bmatrix} \langle j_1 - 1, j - j_1 | j, j-1 \rangle \sqrt{2j_1} \\ +\langle j_1, j - j_1 - 1 | j, j-1 \rangle \sqrt{(j_2 + j - j_1)(j_2 - j + j_1 + 1)} \end{bmatrix}. \quad (4.93b)$$

Using Eq. 4.92b, we can determine $\langle j_1, j-1-j_1 | j, j-1 \rangle$ from $\langle j_1, j-j_1 | jj \rangle$. The latter is the seed CGC corresponding to the maximum values of the magnetic quantum numbers in the respective ranges for j_1 and j. The value and phase of this coefficient are chosen as per the Condon and Shortley convention; it is set equal to +1. Once the two coefficients $\langle j_1, j-j_1 | jj \rangle$ and $\langle j_1, j-1-j_1 | j, j-1 \rangle$ are fixed, we can use Eq. 4.93b to determine $\langle j_1 - 1, j - j_1 | j, j-1 \rangle$. Subsequently, continuing the use of the recursion relations enables the calculation of *all* the CGCs. The CGCs are presented most efficiently as unitary matrices having the dimension $(2j_1+1) \times (2j_2+1)$. The CGCs satisfy the following self-evident orthogonality conditions:

$$\delta_{m_1 m_1'} \delta_{m_2 m_2'} = \sum_{j=|j_1-j_2|}^{j_1+j_2} \sum_{m=-j}^{j} \langle m_1' m_2' | jm \rangle \langle jm | m_1 m_2 \rangle \quad (4.94a)$$

and $$\delta_{jj'} \delta_{mm'} = \sum_{m_1=-j_1}^{j_1} \sum_{m_2=-j_2}^{j_2} \langle jm | m_1 m_2 \rangle \langle m_1 m_2 | j'm' \rangle. \quad (4.94b)$$

We illustrate the construction of the CGC matrix for the case of coupling $j_1 = \frac{1}{2}$ and $j_2 = \frac{1}{2}$. The composite vector space is of dimension $d = \left\{2\left(\frac{1}{2}\right)+1\right\} \times \left\{2\left(\frac{1}{2}\right)+1\right\} = 4$. The factor basis is

$$\{|m_1,m_2\rangle\} = \left\{\left|\frac{1}{2},\frac{1}{2}\right\rangle, \left|\frac{1}{2},-\frac{1}{2}\right\rangle, \left|-\frac{1}{2},\frac{1}{2}\right\rangle, \left|-\frac{1}{2},-\frac{1}{2}\right\rangle\right\} \tag{4.95a}$$

and the composite basis is

$$\{|j,m\rangle\} = \{|0,0\rangle, |1,-1\rangle, |1,0\rangle, |1,1\rangle\}. \tag{4.95b}$$

We present the 4×4 matrix of the CGCs $\mathbb{C}_{4\times 4} = [\langle m_1 m_2 | jm\rangle]$ next in which we label the columns by $|jm\rangle$ and the rows by $\langle m_1 m_2|$.

$$\mathbb{C}_{4\times4}\left(j_1 = \frac{1}{2}, j_2 = \frac{1}{2}\right) = \begin{bmatrix} \boxed{\downarrow \langle m_1 m_2|/|jm\rangle \rightarrow} & |1,1\rangle & |1,0\rangle & |0,1\rangle & |1,-1\rangle \\ \left\langle\frac{1}{2},\frac{1}{2}\right| & 1 & 0 & 0 & 0 \\ \left\langle\frac{1}{2},-\frac{1}{2}\right| & 0 & \frac{1}{\sqrt{2}} & \frac{1}{\sqrt{2}} & 0 \\ \left\langle-\frac{1}{2},\frac{1}{2}\right| & 0 & \frac{1}{\sqrt{2}} & -\frac{1}{\sqrt{2}} & 0 \\ \left\langle-\frac{1}{2},-\frac{1}{2}\right| & 0 & 0 & 0 & 1 \end{bmatrix}. \tag{4.96a}$$

The table of CGCs to couple an arbitrary angular momentum j_1 with $j_2 = \frac{1}{2}$ is presented in Eq. 4.96b:

$$\mathbb{C}_{2\times2}\left(j_1 : arbitrary, j_2 = \frac{1}{2}\right) = \begin{bmatrix} \boxed{\downarrow \langle m_1 m_2|/|jm\rangle \rightarrow} & \left|\frac{1}{2},\frac{1}{2}\right\rangle & \left|\frac{1}{2},-\frac{1}{2}\right\rangle \\ \left\langle j_1+\frac{1}{2},m\right| & \sqrt{\dfrac{j_1+m+\frac{1}{2}}{2j_1+\frac{1}{2}}} & \sqrt{\dfrac{j_1-m+\frac{1}{2}}{2j_1+\frac{1}{2}}} \\ \left\langle j_1 1\frac{1}{2},m\right| & -\sqrt{\dfrac{j_1-m+\frac{1}{2}}{2j_1+\frac{1}{2}}} & \sqrt{\dfrac{j_1+m+\frac{1}{2}}{2j_1+\frac{1}{2}}} \end{bmatrix}. \tag{4.96b}$$

The CGC matrix (Eq. 4.96a) has a block-diagonal unitary character. The largest block has dimensionality 2×2 when at least one (or both) of the angular momenta in the coupling is $j = \frac{1}{2}$. In the Eq. 4.96a, both j_1 and j_2 have that value, whereas in the Eq. 4.96b, only j_2 is $\frac{1}{2}$ and j_1 is arbitrary.

As another example, we present $\mathbb{C}_{6\times 6}\left(j_1 = 1, j_2 = \frac{1}{2}\right)$:

$$\mathbb{C}_{6\times 6}\left(j_1=1, j_2=\frac{1}{2}\right)$$

$$= \begin{bmatrix} \boxed{\downarrow \langle m_1 m_2 | / | jm\rangle \rightarrow} & \left|\frac{3}{2},\frac{3}{2}\right\rangle & \left|\frac{3}{2},\frac{1}{2}\right\rangle & \left|\frac{1}{2},\frac{1}{2}\right\rangle & \left|\frac{3}{2},-\frac{1}{2}\right\rangle & \left|\frac{1}{2},-\frac{1}{2}\right\rangle & \left|\frac{3}{2},-\frac{3}{2}\right\rangle \\ \left\langle 1,\frac{1}{2}\right| & 1 & 0 & 0 & 0 & 0 & 0 \\ \left\langle 1,-\frac{1}{2}\right| & 0 & \sqrt{\frac{1}{3}} & \sqrt{\frac{2}{3}} & 0 & 0 & 0 \\ \left\langle 0,\frac{1}{2}\right| & 0 & \sqrt{\frac{2}{3}} & -\sqrt{\frac{1}{3}} & 0 & 0 & 0 \\ \left\langle 0,-\frac{1}{2}\right| & 0 & 0 & 0 & \sqrt{\frac{2}{3}} & \sqrt{\frac{1}{3}} & 0 \\ \left\langle -1,\frac{1}{2}\right| & 0 & 0 & 0 & \sqrt{\frac{1}{3}} & -\sqrt{\frac{2}{3}} & 0 \\ \left\langle -1,-\frac{1}{2}\right| & 0 & 0 & 0 & 0 & 0 & 1 \end{bmatrix} \quad (4.97)$$

In this case also the largest block has a dimensionality 2×2, since one of the two angular momenta being coupled is $j_2=\frac{1}{2}$. Block diagonal and unitary character of the CGC matrix dictates the arrangements of rows and columns. Together with an important theorem about the matrix elements $\mathfrak{D}^{(j)}_{m'm}(R)$ (Wigner functions) of the rotation operator (Eqs. 4.58a,b), the CGCs provide an often-used relationship known as the Clebsch–Gordan series. The theorem we just mentioned states that *the matrix element of the rotation operator in the direct (i.e., tensor) product states is equal to the product of the matrix elements of the rotation operator in the "factor" states*. To prove this theorem, we consider the response of a direct product state $|j_1m_1j_2m_2\rangle$ to the rotation operator. Both the vectors $|j_1m_1\rangle$ and $|j_2m_2\rangle$ must undergo rotations in response to this process; hence,

$$U_R|j_1m_1j_2m_2\rangle = U_R|j_1m_1\rangle|j_2m_2\rangle = \sum_{m_1=-j_1}^{j_1}\mathfrak{D}^{(j_1)}_{m_1'm_1}(R)\left|j_1m_1'\right\rangle\sum_{m_2=-j_2}^{j_2}\mathfrak{D}^{(j_2)}_{m_2'm_2}(R)\left|j_2m_2'\right\rangle, \quad (4.98a)$$

$$\text{i.e., } U_R|j_1m_1j_2m_2\rangle = \sum_{m_1'=-j_1}^{j_1}\sum_{m_2'=-j_2}^{j_2}\mathfrak{D}^{(j_1)}_{m_1'm_1}(R)\mathfrak{D}^{(j_2)}_{m_2'm_2}(R)\left|j_1m_1'\right\rangle\left|j_2m_2'\right\rangle. \quad (4.98b)$$

Projecting the above vector on $\left\langle j_1m_1''\,j_2m_2''\right|$, we get

$$\left\langle j_1m_1''\,j_2m_2''\right|U_R|j_1m_1j_2m_2\rangle = \sum_{m_1'=-j_1}^{j_1}\sum_{m_2'=-j_2}^{j_2}\mathfrak{D}^{(j_1)}_{m_1'm_1}\mathfrak{D}^{(j_2)}_{m_2'm_2}\left\langle j_1m_1''\,j_2m_2''\middle|j_1m_1'\,j_2m_2'\right\rangle, \quad (4.99a)$$

$$\text{i.e., } \left\langle j_1m_1''\,j_2m_2''\right|U_R|j_1m_1j_2m_2\rangle = \sum_{m_1'=-j_1}^{j_1}\sum_{m_2'=-j_2}^{j_2}\mathfrak{D}^{(j_1)}_{m_1'm_1}(\mathrm{R})\mathfrak{D}^{(j_2)}_{m_2'm_2}(\mathrm{R})\delta_{m_1'm_1''}\delta_{m_2'm_2''} \quad (4.99b)$$

$$\text{or, } \left\langle j_1m_1''\,j_2m_2''\right|U_R|j_1m_1j_2m_2\rangle = \mathfrak{D}^{(j_1)}_{m_1''m_1}(R)\mathfrak{D}^{(j_2)}_{m_2''m_2}(R). \quad (4.99c)$$

In other words, $\left\langle j_1 m_1'' j_2 m_2'' \middle| U_R \middle| j_1 m_1 j_2 m_2 \right\rangle = \left\langle j_1 m_1'' \middle| U_R \middle| j_1 m_1 \right\rangle \left\langle j_2 m_2'' \middle| U_R \middle| j_2 m_2 \right\rangle$, (4.99d)

thus proving the theorem. We now insert (twice), in the matrix element on the left-hand side of Eq. 4.99d, the resolution of the unit operator in the eigenspace of the coupled angular momentum, as shown below:

$$\left\langle m_1'' m_2'' \middle| U_R \middle| m_1 m_2 \right\rangle$$

$$= \left\langle m_1'' m_2'' \right| \left\{ \sum_{j=|j_1-j_2|}^{j_1+j_2} \sum_{m=-j}^{j} |jm\rangle\langle jm| \right\} U_R \left\{ \sum_{j'=|j_1-j_2|}^{j_1+j_2} \sum_{m'=-j'}^{j'} |j'm'\rangle\langle j'm'| \right\} |m_1 m_2\rangle. \tag{4.100a}$$

i.e., $\left\langle m_1'' m_2'' \middle| U_R \middle| m_1 m_2 \right\rangle$

$$= \sum_{j=|j_1-j_2|}^{j_1+j_2} \sum_{m=-j}^{j} \sum_{j'=|j_1-j_2|}^{j_1+j_2} \sum_{m'=-j'}^{j'} \left\langle m_1'' m_2'' \middle| jm \right\rangle \left\langle jm \middle| U_R \middle| j'm' \right\rangle \left\langle j'm' \middle| m_1 m_2 \right\rangle. \tag{4.100b}$$

We know from the discussion on Eqs. 4.58a,b that $\left\langle jm \middle| U_R \middle| j'm' \right\rangle = \mathfrak{D}^{(j)}_{mm'}(R)\,\delta_{jj'}$; hence, using Eq. 4.99c, we get

$$\mathfrak{D}^{(j_1)}_{m_1'' m_1}(\mathrm{R})\,\mathfrak{D}^{(j_2)}_{m_2'' m_2}(\mathrm{R})$$

$$= \sum_{j=|j_1-j_2|}^{j_1+j_2} \sum_{m=-j}^{j} \sum_{j'=|j_1-j_2|}^{j_1+j_2} \sum_{m'=-j'}^{j'} \left\langle m_1'' m_2'' \middle| jm \right\rangle \left\langle j'm' \middle| m_1 m_2 \right\rangle \delta_{jj'} \mathfrak{D}^{(j)}_{mm'}(\mathrm{R}). \tag{4.100c}$$

Contracting now the summation over j' using the Kronecker-δ, we have

$$\mathfrak{D}^{(j_1)}_{m_1'' m_1}(R)\,\mathfrak{D}^{(j_2)}_{m_2'' m_2}(R) = \sum_{j=|j_1-j_2|}^{j_1+j_2} \sum_{m=-j}^{j} \sum_{m'=-j}^{j} \left\langle m_1'' m_2'' \middle| jm \right\rangle \left\langle jm' \middle| m_1 m_2 \right\rangle \mathfrak{D}^{(j)}_{mm'}(R). \tag{4.101a}$$

It is important to keep track of which index is summed over, and where it is placed. The positions of the scalar factors $\left\langle m_1'' m_2'' \middle| jm \right\rangle$, $\left\langle jm' \middle| m_1 m_2 \right\rangle$, and $\mathfrak{D}^{(j)}_{mm'}(R)$ are interchangeable; they are mere coefficients/numbers and hence commute. The CGCs, being real, may also be seen with labels in the $|\ \rangle$ and $\langle\ |$ swapped.

Using the property stated in Eq. 4.82b, we see that

$$\mathfrak{D}^{(j_1)}_{m_1'' m_1}\,\mathfrak{D}^{(j_2)}_{m_2'' m_2} = \sum_{j=|j_1-j_2|}^{j_1+j_2} \sum_{m=-j}^{j} \sum_{m'=-j}^{j} \left\langle m_1'' m_2'' \middle| jm \right\rangle \delta_{m,m_1''+m_2''|} \left\langle jm' \middle| m_1 m_2 \right\rangle \delta_{m',m_1+m_2|} \mathfrak{D}^{(j)}_{mm'}, \tag{4.101b}$$

or

$$\mathfrak{D}^{(j_1)}_{m_1'' m_1}\,\mathfrak{D}^{(j_2)}_{m_2'' m_2} = \sum_{j=|j_1-j_2|}^{j_1+j_2} \left\langle m_1'' m_2'' \middle| j, m_1'' + m_2'' \right\rangle \left\langle j, m_1 + m_2 \middle| m_1 m_2 \right\rangle \mathfrak{D}^{(j)}_{m_1''+m_2'',\, m_1+m_2}. \tag{4.101c}$$

The expression in Eq. 4.101a is called as the Clebsch–Gordan (CG) series. The condition in Eqs. 4.82a,b for the CGC to be nonvanishing is represented in Eq. 4.101b by the two occurrences of Kronecker-δ. The CGCs being real numbers in the phase convention we have adopted, the CG series is often found in the literature in alternative and equivalent forms using $\langle jm|m_1 m_2\rangle = \langle m_1 m_2|jm\rangle$. A special case of Eq. 4.101c is of great interest. This is the one we get when $m_1 = 0$ and $m_2 = 0$. After we make this choice, let us relabel m_1'' as m_1, and m_2'' as m_2. On doing so, we get

$$\mathfrak{D}^{(\ell_1)}_{m_1 0}\mathfrak{D}^{(\ell_2)}_{m_2 0} = \sum_{\ell=|\ell_1-\ell_2|}^{\ell_1+\ell_2} \left\langle m_1 m_2 \middle| \ell, m_1 + m_2 \right\rangle \left\langle \ell, 0 \middle| 00 \right\rangle \mathfrak{D}^{(\ell)}_{m_1+m_2,0}. \tag{4.101d}$$

Now, using Eq. 4.68 for each of the $\mathfrak{D}$ functions that appears above, and the fact that the complex conjugate of a product of two complex numbers is equal to the product of the complex conjugates of the original numbers, we get an often-used relationship for the spherical harmonics:

$$Y_{\ell_1 m_1}(\hat{u})Y_{\ell_2 m_2}(\hat{u}) = \sum_{\ell=|\ell_1-\ell_2|}^{\ell_1+\ell_2}\sqrt{\frac{(2\ell_1+1)(2\ell_2+1)}{4\pi(2\ell+1)}}\langle m_1 m_2|\ell, m_1+m_2\rangle\langle \ell,0|00\rangle Y_{\ell,m_1+m_2}(\hat{u}). \tag{4.102}$$

Together with the orthogonality relations for the spherical harmonics, the above result has many applications, especially in the evaluation of integrals of products of three spherical harmonics, known as *Gaunt integrals*:

$$\iint Y_{\ell_3 m_3}{}^{*}(\hat{u})Y_{\ell_1 m_1}(\hat{u})Y_{\ell_2 m_2}(\hat{u})d\Omega = \iint \sum_{\ell=|\ell_1-\ell_2|}^{\ell_1+\ell_2}\left\{\begin{array}{l}\sqrt{\dfrac{(2\ell_1+1)(2\ell_2+1)}{4\pi(2\ell+1)}}\\ \times\langle m_1 m_2|\ell, m_1+m_2\rangle\\ \times\langle \ell,0|00\rangle\end{array}\right\} Y_{\ell_3 m_3}{}^{*}(\hat{u})Y_{\ell,m_1+m_2}(\hat{u})d\Omega, \tag{4.103a}$$

$$\iint Y_{\ell_3 m_3}{}^{*}(\hat{u})Y_{\ell_1 m_1}(\hat{u})Y_{\ell_2 m_2}(\hat{u})d\Omega = \sum_{\ell=|\ell_1-\ell_2|}^{\ell_1+\ell_2}\left\{\begin{array}{l}\sqrt{\dfrac{(2\ell_1+1)(2\ell_2+1)}{4\pi(2\ell+1)}}\\ \times\langle m_1 m_2|\ell, m_1+m_2\rangle\\ \times\langle \ell,0|00\rangle\end{array}\right\}\delta_{\ell,\ell_3}\delta_{m_1+m_2,m_3}, \tag{4.103b}$$

$$\iint Y_{\ell_3 m_3}{}^{*}(\hat{u})Y_{\ell_1 m_1}(\hat{u})Y_{\ell_2 m_2}(\hat{u})d\Omega = \sqrt{\frac{(2\ell_1+1)(2\ell_2+1)}{4\pi(2\ell_3+1)}}\langle m_1 m_2|\ell_3, m_3\rangle\langle \ell_3,0|00\rangle. \tag{4.103c}$$

We now have with us many of the important tools required to study quantum theory of the atomic structure, which has a dominant central field (spherical) symmetry. All this analysis is required even before we examine the simplest atom, that of hydrogen, which we shall take up in the next chapter. Before we plunge into further details, we should point out that the angular momentum coupling constants, namely the CG coefficients, are often employed in the literature using some alternative and equivalent forms. Among the most prominent commonly used forms is the so-called Wigner-3*j* symbol, defined by the following relationship:

$$\begin{pmatrix} j_1 & j_2 & j_3 \\ m_1 & m_2 & m_3 \end{pmatrix} = (-1)^{j_1-j_2+m}[j]^{-\frac{1}{2}}\langle (j_1 j_2)m_1 m_2|(j_1 j_2)jm\rangle, \tag{4.104a}$$

where $|(j_1 j_2)j_3 m_3\rangle = |(j_1 j_2)jm\rangle$ and $[j]=2j+1$. By choosing $m_3=-(m_1+m_2)=-m$, the 3*j* symbol is

$$\begin{pmatrix} j_1 & j_2 & j_3 \\ m_1 & m_2 & m_3 \end{pmatrix} = (-1)^{j_1 - j_2 + m} [j]^{-\frac{1}{2}} \langle (j_1 j_2) m_1 m_2 | (j_1 j_2) j, -(m_1 + m_2) \rangle. \tag{4.104b}$$

Note that the sum of the three magnetic quantum numbers in the Wigner $3j$ symbol must be zero for the $3j$ symbol itself to have a nonzero value. The Wigner $3j$ symbols have convenient symmetry properties. In particular, changing the columns in a cyclic order leaves the Wigner $3j$ symbol invariant:

$$\begin{pmatrix} j_1 & j_2 & j_3 \\ m_1 & m_2 & m_3 \end{pmatrix} = \begin{pmatrix} j_2 & j_3 & j_1 \\ m_2 & m_3 & m_1 \end{pmatrix} = \begin{pmatrix} j_3 & j_1 & j_2 \\ m_3 & m_1 & m_2 \end{pmatrix}, \tag{4.105a}$$

Interchanging adjacent columns affects the phase of the $3j$ symbol in the following manner:

$$\text{and } -1^{(j_1+j_2+j_3)} \begin{pmatrix} j_1 & j_2 & j_3 \\ m_1 & m_2 & m_3 \end{pmatrix} = \begin{pmatrix} j_2 & j_1 & j_3 \\ m_2 & m_1 & m_3 \end{pmatrix} = \begin{pmatrix} j_1 & j_3 & j_2 \\ m_1' & m_3 & m_2 \end{pmatrix} = \begin{pmatrix} j_3 & j_2 & j_1 \\ m_3 & m_2 & m_1 \end{pmatrix}. \tag{4.105b}$$

Also, reversal of the sign of all the three magnetic quantum numbers affects the phase as follows:

$$\begin{pmatrix} j_1 & j_2 & j_3 \\ m_1 & m_2 & m_3 \end{pmatrix} = (-1)^{(j_1+j_2+j_3)} \begin{pmatrix} j_1 & j_2 & j_3 \\ -m_1 & -m_2 & -m_3 \end{pmatrix}. \tag{4.105c}$$

The orthogonality properties (corresponding to Eqs. 4.94a,b) of the Wigner $3j$ symbols are also conveniently stated:

$$\sum_{j=|j_1-j_2|}^{j_1-j_2} \sum_{m=-j}^{j} (2j+1) \begin{pmatrix} j_1 & j_2 & j_3 \\ m_1 & m_2 & m_3 \end{pmatrix} \begin{pmatrix} j_1 & j_2 & j_3 \\ m_1' & m_2' & m_3 \end{pmatrix} = \delta_{m_1 m_1'} \delta_{m_2 m_2'} \tag{4.106a}$$

$$\text{and } \sum_{m_1=-j_1}^{j_1} \sum_{m_2=-j_2}^{j_3} (2j+1) \begin{pmatrix} j_1 & j_2 & j \\ m_1 & m_2 & m \end{pmatrix} \begin{pmatrix} j_1 & j_2 & j' \\ m_1 & m_2 & m' \end{pmatrix} = \delta_{jj'} \delta_{mm'}. \tag{4.106b}$$

Tables of the CGCs are easily accessible on the internet, but some of you may enjoy coding the following general formula, also due to E. P. Wigner, in which the summation over κ must include all integer values of κ such that the argument of none of the factorials goes negative:

$$\langle (j_1 j_2) m_1 m_2 | (j_1 j_2) jm \rangle$$

$$= \left[\begin{array}{l} \dfrac{(2j+1)(j+j_1-j_2)!(j-j_1+j_2)!(j_1+j_2-j)!(j+m)!(j-m)!}{(j_1+j_2+j+1)!(j_1-m_1)!(j_1+m_1)!(j_2-m_2)!(j_2+m_2)!} \\ \times \displaystyle\sum_{\kappa} \left\{ \begin{array}{l} (-1)^{\kappa+j_2+m_2} \\ \times \dfrac{(j_2+j_3+m_1-\kappa)!(j_1-m_1+\kappa)!}{\kappa!(j-j_1+j_2-\kappa)!(j+m-\kappa)!(\kappa+j_1-j_2-m)!} \\ \times \delta_{m,m_1+m_2} \end{array} \right\} \end{array} \right]. \tag{4.107}$$

One can also build angular momentum vector coupling algebra for adding more than two angular momentum vectors. Thus, we can construct coupling coefficients for coupling $\vec{j}_1 \oplus \vec{j}_2 \oplus \vec{j}_3$. One must consider the alternatives $\vec{j}_1 \oplus (\vec{j}_2 \oplus \vec{j}_3) = \vec{j}_1 \oplus \vec{j}_{23}$ and $(\vec{j}_1 \oplus \vec{j}_2) \oplus \vec{j}_3 = \vec{j}_{12} \oplus \vec{j}_3$. The coupling schemes become complex in this case, and one employs the so-called Racah coupling coefficients and the Wigner $6j$ symbols. An example of coupling four angular momenta is that of a pair of electrons. Each electron has its orbital angular momentum as well as spin angular momentum. We denote these by $\vec{\ell}_1$, $\vec{\ell}_2$, $\vec{s}_1$, and $\vec{s}_2$ for the two electrons. In this case, we have alternative coupling schemes. We may couple $\vec{\ell}_1 \oplus \vec{\ell}_2 = \vec{L}$ and $\vec{s}_1 \oplus \vec{s}_2 = \vec{S}$, and finally $\vec{L} \oplus \vec{S} = \vec{J}_{LS(RS)}$. The subscript LS(RS) stands for the name of this coupling scheme, viz., LS-coupling or Russel–Saunders coupling scheme. Alternatively, we may couple $\vec{\ell}_1 \oplus \vec{s}_1 = \vec{j}_1$ and $\vec{\ell}_2 \oplus \vec{s}_2 = \vec{j}_2$, and finally $\vec{j}_1 \oplus \vec{j}_2 = \vec{J}_{jj(\text{relativistic})}$; this coupling scheme is called as the jj-coupling or relativistic coupling scheme. In this case, Wigner $9j$ symbols are employed in the angular momentum coupling schemes. We refer the readers to other sources [6, 7, 8] for advanced treatments on angular momentum coupling algebra.

4.3 Irreducible Tensor Operators

The central theme we are carrying in this chapter is hinged on the study of *rotations* under which *angular momentum* is conserved. We have known from classical physics that the response to *rotation* of a physical property determines its predominant character; viz. its tensor attributes. Tensors of rank 0 remain invariant under rotation of a coordinate system; these are scalars. Likewise, tensors of rank 1 are vectors; they are defined by how their components transform under the rotation of a coordinate system. Other physical quantities, such as electric susceptibility of a material, or moment of inertia of an object, are tensors of rank 2. *All* physical quantities are tensors of various ranks. They are further classified as polar or axial tensors depending on specific laws, which govern the transformation properties of their components. One commonly encounters contravariant (polar) and covariant (axial) vectors (see, for example, Chapter 2 in Reference [1]). In the previous two sections, we have deliberated on the role of the Wigner $\mathfrak{D}$ functions in describing rotations of angular momentum vectors in their Hilbert spaces. Physical quantities being represented by operators, we now examine how various *operators* respond to rotations.

Let us consider a vector $|\alpha\rangle$ that is transformed to $|\alpha_R\rangle$ under the influence of a rotation R effected by a unitary operator U_R, then we have

$$U_R|\alpha\rangle = |\alpha_R\rangle. \tag{4.108}$$

In order to consider how an operator transforms under corresponding rotation, we shall first introduce an *irreducible tensor operator* (ITO) of rank k, defined as a *family* of $(2k+1)$ operators $T_{-k}^{(k)}, T_{-k+1}^{(k)}, T_{-k+2}^{(k)}, \ldots, T_{k-2}^{(k)}, T_{k-1}^{(k)}, T_k^{(k)}$. Some of these have a special name; ITO of rank zero is called as a scalar operator, that of rank 1 as a vector operator, etc. The operator itself is defined by how its $(2k+1)$ components transform under rotations. Using Wigner $\mathfrak{D}$ functions (Eqs. 4.58a,b), we describe the response of a component $T_k^{(k)}$ of an ITO to a rotation effected by the rotation operator U_R by

$$T_k^{(k)} \to \left[T_k^{(k)}\right]_R = U_R T_q^{(k)} U_R^\dagger = \sum_{q'=-k}^{k} \mathfrak{D}_{q'q}^{(k)}(R) T_{q'}^{(k)}. \tag{4.109}$$

Note that this law of transformation under rotation of the spherical components of an ITO is the same as that for angular momentum eigenvectors given in Eq. 4.64b. In particular, a tensor operator whose rank is 0 has only a single component, $T_0^{(k=0)}$.

Under rotation, it remains invariant, just as the angular momentum vector $|j=0,m=0\rangle$. $T_0^{(k=0)}$ is therefore referred to as a scalar operator. Likewise, an ITO of rank 1 is a vector operator. The transformation law given above is in line with what we would expect; it guarantees that the expectation value of a rotated operator in rotated states (denoted by subscript R in the next equation) is equal to that of the original operator in the original states:

$$\left\langle \alpha_R \middle| \left[T_k^{(k)}\right]_R \middle| \alpha_R \right\rangle = \left\langle \alpha_R \middle| U_R T_q^{(k)} U_R^\dagger \middle| \alpha_R \right\rangle = \left\langle \alpha \middle| U_R^\dagger U_R T_q^{(k)} U_R^\dagger U_R \middle| \alpha_R \right\rangle = \left\langle \alpha \middle| T_q^{(k)} \middle| \alpha \right\rangle. \tag{4.110}$$

For a rotation through an *infinitesimal* angle ξ (so that we may ignore terms that are quadratic or smaller in ξ) about an axis along $\hat{\xi}$, we have

$$\left[T_k^{(k)}\right]_R = \left(1 - \frac{i}{\hbar}\vec{j}\cdot\hat{\xi}\xi\right) T_q^{(k)} \left(1 + \frac{i}{\hbar}\vec{j}\cdot\hat{\xi}\xi\right) = \sum_{q'=-k}^{k} \mathfrak{D}_{q'q}^{(k)}\left(R:\hat{\xi}\xi\right) T_{q'}^{(k)}, \tag{4.111a}$$

$$\text{i.e.,}\ \left(1 - \frac{i}{\hbar}\vec{j}\cdot\hat{\xi}\xi\right) T_q^{(k)} \left(1 + \frac{i}{\hbar}\vec{j}\cdot\hat{\xi}\xi\right) = \sum_{q'=-k}^{k} \left\langle kq' \middle| 1 - \frac{i}{\hbar}\left(\vec{j}\cdot\hat{\xi}\right)\xi \middle| kq \right\rangle T_{q'}^{(k)}, \tag{4.111b}$$

since we know from Eq. 4.58b that

$$\mathfrak{D}_{m'm}^{(j)}(R) = \left\langle jm' \middle| U_R \middle| jm \right\rangle = \left\langle jm' \middle| \exp\left(-\frac{i}{\hbar}\left(\vec{j}\cdot\hat{e}_z\right)\xi\right) \middle| jm \right\rangle. \tag{4.112}$$

Simplifying just as we did in the Solved Problem P4.1, and identifying $\hat{e}_z = \hat{\xi}$, we get a result that has exactly the same *form* as Eq. 4.13a, or its more general avatar, Eq. 4.18a.

$$\left[j_z, T_q^{(k)}\right] = \hbar q T_q^{(k)}. \tag{4.113a}$$

$$\text{Likewise, } \left[j_\pm, T_q^{(k)}\right] = \pm\hbar\sqrt{(k\mp q)(k\pm q+1)}\,T_{q\pm1}^{(k)}. \tag{4.113b}$$

As discussed subsequently, it is of far-reaching consequence that the commutation properties of the components of an ITO with angular momentum operators $\left(j_z, j_\pm\right)$ have exactly the same *form* as the effect of angular momentum operators on the eigenvectors $|jm\rangle$, given in Eq. 4.40 and Eq. 4.42. For a vector to be classified as a vector operator, it must satisfy Eqs. 4.113a,b. A vector operator is therefore an irreducible spherical tensor operator of rank 1. Its spherical components $\left\{V^{(1)}: V_{-1}^{(1)}, V_0^{(1)}, V_1^{(1)}\right\}$ have the following relationship with the Cartesian components $\left\{\vec{V}: V_x, V_y, V_z\right\}$:

$$V_1^{(1)} = -\frac{V_x + iV_y}{\sqrt{2}};\ V_0^{(1)} = V_z;\ V_{-1}^{(1)} = \frac{V_x - iV_y}{\sqrt{2}}. \tag{4.114}$$

The result of rotation of a vector $\vec{A} = \sum_{i=1}^{3} \hat{e}_i A_i$ is expressible in terms of a superposition of the original components. Each component A_i of the vector transforms to a new one, and the transformation law is

$$A_i \to A_{i,R} = \sum_{j=1}^{3} R_{ij} A_j. \tag{4.115}$$

The corresponding vector operator $\hat{\vec{A}} = \sum_{i=1}^{3} \hat{e}_i \hat{A}_i$ would have components that follow a similar transformation law:

$$\hat{A}_i \to U_R{}^\dagger \hat{A}_i U_R = \hat{A}_{i,R} = \sum_{j=1}^{3} R_{ij} \hat{A}_j, \tag{4.116a}$$

The transformation in Eq. 4.116a can be written for all the three components in a compact matrix form:

$$\left[\hat{\vec{A}}\right]_R = U_R^\dagger\left[\hat{A}\right]U_R = U_R^\dagger\begin{bmatrix}\hat{A}_1\\\hat{A}_2\\\hat{A}_3\end{bmatrix}U_R = \begin{bmatrix}\hat{A}_{1,R}\\\hat{A}_{2,R}\\\hat{A}_{3,R}\end{bmatrix} = \begin{bmatrix}R_{11} & R_{12} & R_{13}\\R_{21} & R_{22} & R_{23}\\R_{31} & R_{32} & R_{33}\end{bmatrix}\begin{bmatrix}\hat{A}_1\\\hat{A}_2\\\hat{A}_3\end{bmatrix} = \mathbb{R}\left[\hat{\vec{A}}\right]. \quad (4.116b)$$

The expectation value of the components of the vector operator therefore follows the same transformation law as that of the components of the vector:

$$\langle\hat{A}_i\rangle \to \langle\hat{A}_i\rangle_R = \sum_{j=1}^{3} R_{ij}\langle\hat{A}_j\rangle. \quad (4.117a)$$

Note that the transformation written in Eq. 4.116a ensures that the expectation value of a rotated component in the rotated states is equal to that of the unrotated component in the unrotated states:

$$\langle\alpha_R|\hat{A}_{i,R}|\alpha_R\rangle = \langle\alpha|U_R^\dagger\hat{A}_{i,R}U_R|\alpha\rangle = \langle\alpha|U_R^\dagger U_R\hat{A}_i U_R^\dagger U_R|\alpha\rangle = \langle\alpha|\hat{A}_i|\alpha\rangle. \quad (4.117b)$$

From the solution to the Solved Problem P4.6, we see that the transformation law in Eqs. 4.116a,b is completely equivalent to the following commutation rules

$$\left[\hat{A}_i, j_i\right] = 0;\ i = 1,2,3, \quad (4.118a)$$

$$\left[\hat{A}_i, j_j\right] = i\hbar\hat{A}_k;\ (i,j,k) = (1,2,3);\ i \to j \to k \to i, \quad (4.118b)$$

and $\left[\hat{A}_i, j_k\right] = -i\hbar\hat{A}_j;\ (i,j,k) = (1,2,3);\ i \to j \to k \to i.$ (4.118c)

The angular momentum itself is a vector operator defined by Eqs. 4.118a,b,c, which are equivalent to the transformation law of the components of a vector operator defined by Eqs. 4.116a,b. Another important learning in this section is the fact that angular momentum eigenvectors and the ITOs have exactly the same response to rotations; both are described using the same Wigner $\mathfrak{D}_{rs}^{(j)}(R)$ functions.

Before we conclude this section, we specifically visit angular momentum coupling between orbital angular momentum of an electron and its intrinsic spin angular momentum, since it is of special interest in atomic physics. One must couple the spherical harmonics (Appendix C) and the spin states, which we used in Eqs. 4.43 through 4.45, and in the Solved Problem P4.4. The spin states are known as spinors, $\chi_{\frac{1}{2},m_s}(\zeta)$, which are the two-components of electron spin functions:

$$\chi_{\frac{1}{2},m_s=\frac{1}{2}}(\zeta) = \begin{bmatrix}1\\0\end{bmatrix} \quad (4.119a)$$

and $\chi_{\frac{1}{2},m_s=-\frac{1}{2}}(\zeta) = \begin{bmatrix}0\\1\end{bmatrix}.$ (4.119b)

The coupling of the orbital angular momentum states, viz., the spherical harmonics $Y_{\ell m_\ell}(\hat{r})$ (Appendix C) and the electron spinors, is represented by functions known as the *vector spherical harmonics*. These are defined using the CGCs as

$$\Omega_{jm} = \sum_{m_s=-\frac{1}{2}}^{\frac{1}{2}} Y_{\ell m_\ell}(\hat{r})\chi_{\frac{1}{2},m_s}(\zeta)\left\langle \ell, m_\ell, \frac{1}{2}, m_s \middle| \left(\ell, \frac{1}{2}\right) j, m \right\rangle. \tag{4.120}$$

Summing over the two values $m_s = \frac{1}{2}, -\frac{1}{2}$ in Eq. 4.120, and knowing that the CGC is zero unless $m_\ell + m_s = m$, we see that

$$\Omega_{j\ell m} = \begin{bmatrix} Y_{\ell,\left(m_\ell = m+\frac{1}{2}\right)}(\hat{r})\chi_{\frac{1}{2},-\frac{1}{2}}(\zeta)\left\langle \ell, m+\frac{1}{2}, \frac{1}{2}, -\frac{1}{2} \middle| \left(\ell, \frac{1}{2}\right) j, m \right\rangle + \\ Y_{\ell,\left(m_\ell = m-\frac{1}{2}\right)}(\hat{r})\chi_{\frac{1}{2},\frac{1}{2}}(\zeta)\left\langle \ell, m-\frac{1}{2}, \frac{1}{2}, \frac{1}{2} \middle| \left(\ell, \frac{1}{2}\right) j, m \right\rangle \end{bmatrix}. \tag{4.121a}$$

Inserting now the spin states from Eqs. 4.119a,b, we see that

$$\Omega_{jm} = \begin{bmatrix} Y_{\ell\left(m_\ell = m+\frac{1}{2}\right)}(\hat{r})\begin{bmatrix}0\\1\end{bmatrix}\left\langle \ell, \left(m_\ell = m+\frac{1}{2}\right), \frac{1}{2}, -\frac{1}{2} \middle| \left(\ell, \frac{1}{2}\right) j, m \right\rangle + \\ Y_{\ell\left(m_\ell = m-\frac{1}{2}\right)}(\hat{r})\begin{bmatrix}1\\0\end{bmatrix}\left\langle \ell, \left(m_\ell = m-\frac{1}{2}\right), \frac{1}{2}, \frac{1}{2} \middle| \left(\ell, \frac{1}{2}\right) j, m \right\rangle \end{bmatrix}. \tag{4.121b}$$

The vector spherical harmonics therefore are represented by a 2 × 1 matrix, which is

$$\text{i.e., } \Omega_{jm} = \begin{bmatrix} Y_{\ell,\left(m_\ell = m+\frac{1}{2}\right)}(\hat{r})\left\langle \ell, m+\frac{1}{2}, \frac{1}{2}, -\frac{1}{2} \middle| \left(\ell, \frac{1}{2}\right) j, m \right\rangle \\ Y_{\ell,\left(m_\ell = m-\frac{1}{2}\right)}(\hat{r})\left\langle \ell, m-\frac{1}{2}, \frac{1}{2}, \frac{1}{2} \middle| \left(\ell, \frac{1}{2}\right) j, m \right\rangle \end{bmatrix}. \tag{4.121c}$$

The value of the CGCs in the vector spherical harmonics depends on whether $j = \ell + \frac{1}{2}$, or $j = \ell - \frac{1}{2}$. These can be easily determined from Eq. 4.94b.

For $j = \ell + \frac{1}{2}$,

$$\left\langle \ell, \left(m_{\ell'} = m - \frac{1}{2}\right), \frac{1}{2}, \frac{1}{2} \middle| \left(\ell, \frac{1}{2}\right)\left(j = \ell + \frac{1}{2}\right), m \right\rangle = \sqrt{\frac{j+m}{2j}}, \tag{4.122a}$$

$$\text{and } \left\langle \ell, \left(m_{\ell'} = m + \frac{1}{2}\right), \frac{1}{2}, -\frac{1}{2} \middle| \left(\ell, \frac{1}{2}\right)\left(j = \ell + \frac{1}{2}\right), m \right\rangle = \sqrt{\frac{j-m}{2j}}, \tag{4.122b}$$

and for $j = \ell - \frac{1}{2}$,

$$\left\langle \ell, \left(m_{\ell'} = m - \frac{1}{2}\right), \frac{1}{2}, \frac{1}{2} \middle| \left(\ell, \frac{1}{2}\right)\left(j = \ell - \frac{1}{2}\right), m \right\rangle = -\sqrt{\frac{j-m+1}{2j+2}}, \tag{4.123a}$$

$$\text{and } \left\langle \ell, \left(m_{\ell'} = m + \frac{1}{2}\right), \frac{1}{2}, -\frac{1}{2} \middle| \left(\ell, \frac{1}{2}\right)\left(j = \ell - \frac{1}{2}\right), m \right\rangle = \sqrt{\frac{j+m+1}{2j+2}}. \tag{4.123b}$$

$$\text{Hence, for } j=\ell+\frac{1}{2},\ \Omega_{jm}^{+}=\begin{bmatrix}\sqrt{\dfrac{j+m}{2j}}\,Y_{\left(\ell=j-\frac{1}{2}\right),\left(m_\ell=m-\frac{1}{2}\right)}(\hat{r})\\ \sqrt{\dfrac{j-m}{2j}}\,Y_{\left(\ell=j-\frac{1}{2}\right),\left(m_\ell=m+\frac{1}{2}\right)}(\hat{r})\end{bmatrix}, \tag{4.124a}$$

$$\text{and for } j=\ell-\frac{1}{2},\ \Omega_{jm}^{-}=\begin{bmatrix}\sqrt{\dfrac{j-m+1}{2j+2}}\,Y_{\left(\ell=j+\frac{1}{2}\right),\left(m_\ell=m-\frac{1}{2}\right)}(\hat{r})\\ -\sqrt{\dfrac{j+m+1}{2j+2}}\,Y_{\left(\ell=j+\frac{1}{2}\right),\left(m_\ell=m+\frac{1}{2}\right)}(\hat{r})\end{bmatrix}. \tag{4.124b}$$

The eigenstates of $\left(j^2, j_z\right)$ with $\vec{j}=\vec{\ell}+\vec{s}$ are also eigenstates of

$$\vec{\Lambda}=2\vec{\ell}\bullet\vec{s}=2\sum_{i=1}^{3}\ell_i s_i=2\ell_z s_z+2\left(\ell_x s_x+\ell_y s_y\right)=2\ell_z s_z+\left(\ell_+ s_-+\ell_- s_+\right), \tag{4.125}$$

since

$$\ell_\pm=\left(\ell_x\pm i\ell_y\right), \tag{4.126a}$$

$$\text{and } s_\pm=\left(s_x\pm is_y\right). \tag{4.126b}$$

It turns out to be fruitful to introduce an operator

$$K=-\left(1_{2\times2}+\vec{\sigma}\bullet\vec{\ell}\right), \tag{4.127}$$

for which we can write an eigenvalue equation

$$K\Omega_{j\ell m}(\hat{r})=\kappa\,\Omega_{j\ell m}(\hat{r}) \tag{4.128a}$$

$$\text{and } \left(\vec{\sigma}\bullet\vec{\ell}\right)\Omega_{j\ell m}(\hat{r})=\hbar(1+\kappa)\Omega_{j\ell m}(\hat{r}), \tag{4.128b}$$

$$\text{or equivalently as } K\Omega_{\kappa m}(\hat{r})=\kappa\Omega_{\kappa m}(\hat{r}). \tag{4.129a}$$

$$\text{and } \left(\vec{\sigma}\bullet\vec{\ell}\right)\Omega_{\kappa m}(\hat{r})=\hbar(1+\kappa)\Omega_{\kappa m}(\hat{r}). \tag{4.129b}$$

$$\text{The eigenvalues are } \kappa=-\ell-1 \text{ for } j=\ell+\frac{1}{2} \tag{4.130a}$$

$$\text{and } \kappa=\ell \text{ for } j=\ell-\frac{1}{2}. \tag{4.130b}$$

In other words,

$$\kappa=\pm\left(j+\frac{1}{2}\right) \text{ corresponds to } j=\ell\mp\frac{1}{2}. \tag{4.131}$$

The response of the spherical spinor to the operator $\vec{\sigma}\bullet\hat{r}$ is of great importance. In Chapter 7, we shall identify (Eq. 4.131) to provide a *good* quantum number to characterize the relativistic wavefunction of the hydrogen atom.

From the properties of the Pauli matrices there are two important results that we shall often employ, both of which come directly from the Solved Problem P4.4. These are

$$\left(\vec{\sigma}\bullet\hat{r}\right)\left(\vec{\sigma}\bullet\hat{r}\right)=1 \tag{4.132}$$

and $\left(\vec{a}\bullet\vec{\sigma}\right)\left(\vec{b}\bullet\vec{\sigma}\right)=\left(\vec{a}\bullet\vec{b}\right)1_{op}+i\vec{a}\times\vec{b}\bullet\vec{\sigma}.$ (4.133)

Under parity, the sign of $\left(\vec{\sigma}\bullet\hat{r}\right)$ changes, hence its effect on the vector spherical harmonics is

$$\left(\vec{\sigma}\bullet\hat{r}\right)\Omega_{\kappa,m}\left(\hat{r}\right)=-\Omega_{-\kappa,m}\left(\hat{r}\right). \tag{4.134}$$

The response of a two-components wavefunction $f\left(r\right)\Omega_{\kappa m}\left(\hat{r}\right)$, which is separable in its radial and angular part to $\left(\vec{\sigma}\bullet\vec{p}\right)$ (using Eqs. 4.131, 4.132, and 4.133), is

$$\left(\vec{\sigma}\bullet\vec{p}\right)f\left(r\right)\Omega_{\kappa m}\left(\hat{r}\right)=\frac{1}{r^2}\left(\vec{\sigma}\bullet\vec{r}\right)\left(\vec{\sigma}\bullet\vec{r}\right)\left(\vec{\sigma}\bullet\vec{p}\right)f\left(r\right)\Omega_{\kappa m}\left(\hat{r}\right)$$

$$=\frac{1}{r^2}\left(\vec{\sigma}\bullet\vec{r}\right)\left(\vec{r}\bullet\vec{p}+i\vec{\sigma}\bullet\vec{r}\times\vec{p}\right)f\left(r\right)\Omega_{\kappa m}\left(\hat{r}\right),$$

i.e., $$\left(\vec{\sigma}\bullet\vec{p}\right)f\left(r\right)\Omega_{\kappa m}\left(\hat{r}\right)=\left(\vec{\sigma}\bullet\hat{r}\right)\left(\hat{r}\bullet\left(-i\hbar\vec{\nabla}\right)+\frac{i\vec{\sigma}\bullet\vec{\ell}}{r}\right)f\left(r\right)\Omega_{\kappa m}\left(\hat{r}\right)$$

$$=\left[-i\hbar\frac{d}{dr}f\left(r\right)\right]\left[\left(\vec{\sigma}\bullet\hat{r}\right)\Omega_{\kappa m}\left(\hat{r}\right)\right]-\frac{i\hbar\left(1+\kappa\right)}{r}f\left(r\right)\left[\left(\vec{\sigma}\bullet\hat{r}\right)\Omega_{\kappa m}\left(\hat{r}\right)\right],$$

where we have used the operator $\frac{d}{dr}$ instead of $\frac{\partial}{\partial r}$ since the operand $f\left(r\right)$ is a function of the radial distance alone. We therefore get

$$\left(\vec{\sigma}\bullet\vec{p}\right)f\left(r\right)\Omega_{\kappa m}\left(\hat{r}\right)=-i\left(\hbar\frac{d}{dr}f\left(r\right)\left[\left(-1\right)\Omega_{-\kappa m}\left(\hat{r}\right)\right]+\frac{\hbar\left(1+\kappa\right)}{r}f\left(r\right)\left[\left(-1\right)\Omega_{-\kappa m}\left(\hat{r}\right)\right]\right),$$

i.e., $$\left(\vec{\sigma}\bullet\vec{p}\right)f\left(r\right)\Omega_{\kappa m}\left(\hat{r}\right)=i\hbar\left(\frac{df\left(r\right)}{dr}+\frac{\left(1+\kappa\right)f\left(r\right)}{r}\right)\Omega_{-\kappa m}\left(\hat{r}\right). \tag{4.135}$$

In Chapter 7, we shall discuss the relativistic (Dirac) equation for the hydrogen atom where we will use the above relations to separate the radial part from the angular part of the relativistic wavefunction.

4.4 Wigner–Eckart Theorem; Spectroscopic Selection Rules

We have seen that Eq. 4.109 and Eq. 4.111 provide equivalent definitions of an ITO. We have also seen that the response of an angular momentum vector $\left|kq\right\rangle$ to rotations is identical to that of a component $T_q^{(k)}$ of an ITO. When we refer to an angular momentum state $\left|kq\right\rangle$, q is a magnetic quantum number, coming from the eigenvalue $q\hbar$ of j_z, and k is the angular momentum quantum number, coming from the eigenvalue $\hbar^2k\left(k+1\right)$ of the commuting operator j^2. On the other hand, when we refer to the component $T_q^{(k)}$, q is the index for the component of an ITO of rank k. We now ask if we can define meaningfully a composite physical quantity by *coupling two ITOs*, $X^{(k_1)}$ and $W^{(k_2)}$, using the very same CGCs $\left\langle\left(k_1k_2\right)q_1q_2\middle|\left(k_1k_2\right)kq\right\rangle\equiv\left\langle q_1q_2\middle|kq\right\rangle$ employed in the coupling $\vec{j}=\vec{j}_1\oplus\vec{j}_2$ of two angular momenta. In analogy with Eqs. 4.75a,b, we write such a composite construct as

$$T_{\mu}^{(\Omega)} = \sum_{q_1=-k_1}^{k_1} \sum_{q_2=-k_2}^{k_2} X_{q_1}^{(k_1)} \; W_{q_2}^{(k_2)} \left\langle (k_1 k_2) q_1 q_2 \middle| (k_1 k_2) \Omega\mu \right\rangle$$

$$= \sum_{q_1=-k_1}^{k_1} \sum_{q_2=-k_2}^{k_2} X_{q_1}^{(k_1)} \; W_{q_2}^{(k_2)} \langle q_1 q_2 | \Omega\mu \rangle. \tag{4.136a}$$

Guided by the analogy we have invoked, we *expect* the construct $T_q^{(k)}$ to be also an ITO with

$$q = -k, -k+1, .., k-1, k, \text{ for each value of } k, \tag{4.136b}$$

whose range is $k = |k_1 - k_2|, |k_1 - k_2| + 1, .., k_1 + k_2 - 1, k_1 + k_2$. (4.136c)

That Eqs. 4.136a,b,c *indeed* define a composite ITO is readily established by subjecting it to either of the two equivalent *defining criteria* given in Eq. 4.109 and Eq. 4.111. Validation of this claim is left as an exercise; we recommend that *both* the alternative criteria are employed in this exercise workout. Going further, we now construct a vector that is composed using the same recipe as in Eq. 4.136, but this time around using $|k_2 q_2\rangle$ instead of $W_{q_2}^{(k_2)}$:

$$\left|(k_1 k_2) \Omega\mu\right\rangle = \sum_{q_1=-k_1}^{k_1} \sum_{q_2=-k_2}^{k_2} X_{q_1}^{(k_1)} |k_2 q_2\rangle \left\langle (k_1 k_2) q_1 q_2 \middle| (k_1 k_2) \Omega\mu \right\rangle, \tag{4.137a}$$

i.e., in shorter notation, $\left|(k_1 k_2) \Omega\mu\right\rangle = \sum_{q_1=-k_1}^{k_1} \sum_{q_2=-k_2}^{k_2} X_{q_1}^{(k_1)} |k_2 q_2\rangle \langle q_1 q_2 | \Omega\mu \rangle,$ (4.137b)

with $q = -k, -k+1, .., k-1, k$, for each value of k,

whose range is $k = |k_1 - k_2|, |k_1 - k_2| + 1, .., k_1 + k_2 - 1, k_1 + k_2$.

The relation that is the inverse of Eqs. 4.137a,b is

$$X_{q'}^{(k_1)} |k_2 m'\rangle = \sum_{J=|k_1-k_2|}^{k_1+k_2} \sum_{M=-J}^{+J} |JM\rangle \langle JM | q'm' \rangle. \tag{4.138}$$

As you would expect, the vector $|\Omega\mu\rangle$ constructed as above is an angular momentum eigenstate; i.e., it is a simultaneous eigenvector of j^2 and j_z with eigenvalues $\hbar^2 \Omega(\Omega+1)$ and $\mu\hbar$ respectively. We validate this claim by examining the response of $\left|(k_1 k_2) \Omega\mu\right\rangle$ to a rotation U_R:

$$U_R |\Omega\mu\rangle = U_R \sum_{q_1=-k_1}^{k_1} \sum_{q_2=-k_2}^{k_2} X_{q_1}^{(k_1)} |k_2 q_2\rangle \langle q_1 q_2 | \Omega\mu \rangle. \tag{4.139}$$

We recognize that

$$U_R |\Omega\mu\rangle = \sum_{q_1=-k_1}^{k_1} \sum_{q_2=-k_2}^{k_2} \left(U_R X_{q_1}^{(k_1)} U_R^{\dagger} \right) \left(U_R |k_2 q_2\rangle \right) \langle q_1 q_2 | \Omega\mu \rangle. \tag{4.140a}$$

Using Eq. 4.58 (or Eq. 4.64) and Eq. 4.111, we see that

$$U_R |\Omega\mu\rangle = \sum_{q_1=-k_1}^{k_1} \sum_{q_2=-k_2}^{k_2} \left(\sum_{q_3=-k_1}^{k_1} \mathfrak{D}_{q_3 q_1}^{(k_1)} (R) X_{q_3}^{(k_1)} \right) \left(\sum_{q_4=-k_2}^{k_2} \mathfrak{D}_{q_4 q_2}^{(k_2)} (R) |k_2 q_4\rangle \right) \langle q_1 q_2 | \Omega\mu \rangle, \tag{4.140b}$$

i.e.,

$$U_R\left|\Omega\mu\right\rangle=\sum_{q_1=-k_1}^{k_1}\sum_{q_2=-k_2}^{k_2}\sum_{q_3=-k_1}^{k_1}\sum_{q_4=-k_2}^{k_2}\left\{\mathfrak{D}_{q_3q_1}^{(k_1)}\mathfrak{D}_{q_4q_2}^{(k_2)}\right\}\left\{X_{q_3}^{(k_1)}\left|k_2q_4\right\rangle\right\}\left\langle q_1q_2\middle|\Omega\mu\right\rangle. \tag{4.140c}$$

Using now the CG series (Eq. 4.101) and Eq. 4.138, we get

$$U_R\left|\Omega\mu\right\rangle=\left[\begin{array}{l}\sum_{q_1=-k_1}^{k_1}\sum_{q_2=-k_2}^{k_2}\sum_{q_3=-k_1}^{k_1}\sum_{q_4=-k_2}^{k_2}\\ \left\{\sum_{k=|k_1-k_2|}^{k_1+k_2}\sum_{m=-k}^{k}\sum_{m'=-k}^{k}\left\langle q_3q_4\middle|km\right\rangle\left\langle km'\middle|q_1q_2\right\rangle\mathfrak{D}_{mm'}^{(k)}(R)\right\}\\ \left\{\sum_{J=|k_1-k_2|}^{k_1+k_2}\sum_{M=-J}^{+J}\left|JM\right\rangle\left\langle JM\middle|q_3q_4\right\rangle\right\}\end{array}\right]\left\langle q_1q_2\middle|\Omega\mu\right\rangle. \tag{4.140d}$$

Rearranging the terms now to recognize readily the orthogonality relations for the CGCs, we get

$$U_R\left|\Omega\mu\right\rangle$$

$$=\left[\begin{array}{l}\sum_{q_1=-k_1}^{k_1}\sum_{q_2=-k_2}^{k_2}\sum_{q_3=-k_1}^{k_1}\sum_{q_4=-k_2}^{k_2}\left\{\sum_{k=|k_1-k_2|}^{k_1+k_2}\sum_{m=-k}^{k}\sum_{m'=-k}^{k}\mathfrak{D}_{mm'}^{(k)}(R)\right\}\\ \left\{\sum_{J=|k_1-k_2|}^{k_1+k_2}\sum_{M=-J}^{+J}\left\langle JM\middle|q_3q_4\right\rangle\left\langle q_3q_4\middle|km\right\rangle\right\}\end{array}\right]\left\langle km'\middle|q_1q_2\right\rangle\left\langle q_1q_2\middle|\Omega\mu\right\rangle\left|JM\right\rangle.$$

Now summing over q_1, q_2, q_3, and q_4,

$$U_R\left|\Omega\mu\right\rangle=\sum_{k=|k_1-k_2|}^{k_1+k_2}\sum_{m=-k}^{k}\sum_{m'=-k}^{k}\left(\delta_{Jk}\delta_{Mm}\right)\left(\delta_{k\Omega}\delta_{m'\mu}\right)\mathfrak{D}_{mm'}^{(k)}(R)\left|JM\right\rangle=\sum_{m=-\Omega}^{\Omega}\mathfrak{D}_{m\mu}^{(\Omega)}(R)\left|\Omega m\right\rangle. \tag{4.141}$$

The vector $\left|(k_1k_2)\Omega\mu\right\rangle$ defined by Eq. 4.137 is an angular momentum eigenstate since it transforms just as per the *defining criterion* for such a state. One can verify further that it satisfies the expected responses to angular momentum operators:

$$j_z\left|(k_1k_2)\Omega\mu\right\rangle=\hbar\mu\left|(k_1k_2)\Omega\mu\right\rangle, \tag{4.142a}$$

and $j_\pm\left|(k_1k_2)\Omega\mu\right\rangle=\hbar\sqrt{\Omega(\Omega+1)-\mu(\mu\pm1)}\left|(k_1k_2)\Omega,\mu\pm1\right\rangle.$ (4.142b)

We are now well equipped to introduce a theorem known for its importance in various applications of group theory and quantum mechanics, especially in the domain of spectroscopy. This theorem is known after E. P. Wigner (1902–1995) and Carl Henry Eckart (1902–1973). The Wigner–Eckart theorem analyzes a crucial physical quantity, $\left\langle j'm'\middle|T_q^{(k)}\middle|jm\right\rangle$, which is the probability amplitude for $T_q^{(k)}\left|jm\right\rangle$ to be in the state $\left|j'm'\right\rangle$; it is essentially the transition probability amplitude for an interaction represented by the operator $T_q^{(k)}$ to induce a transition from the state $\left|j,m\right\rangle$ to $\left|j',m'\right\rangle$, which is given by the matrix element $\left\langle j'm'\middle|T_q^{(k)}\middle|jm\right\rangle$. Transition of a system from one quantum state to another is a fundamental process of importance in spectroscopy (Chapter 8) and in collision physics (Chapter 10).

From Eq. 4.113a, we have

$$\langle j'm' \| [j_\pm, T_q^{(k)}] \| jm\rangle = \hbar\sqrt{(k \mp q)(k \pm q+1)}\langle j'm' | T_{q\pm1}^{(k)} | jm\rangle, \tag{4.143a}$$

$$\text{i.e., } \langle j'm' | j_\pm T_q^{(k)} | jm\rangle - \langle j'm' | T_q^{(k)} j_\pm | jm\rangle = \hbar\sqrt{(k \mp q)(k \pm q+1)}\langle j'm' | T_{q\pm1}^{(k)} | jm\rangle. \tag{4.143b}$$

Also, the adjoint relation of $j_\pm | j,m\rangle = \hbar\sqrt{(j \mp m)(j \pm m+1)}\; | j,m \pm 1\rangle$ is

$$\langle jm | j_\pm^\dagger = \langle jm | j_\mp = \hbar\sqrt{(j \mp m)(j \pm m+1)}\langle j,m \pm 1|, \tag{4.144a}$$

$$\text{i.e., } \langle j'm' | j_\mp = \hbar\sqrt{(j' \mp m')(j' \pm m'+1)}\langle j',m' \pm 1|, \tag{4.144b}$$

$$\text{or, } \langle j'm' | j_\pm = \hbar\sqrt{(j' \pm m')(j' \mp m'+1)}\langle j',m' \mp 1|. \tag{4.144c}$$

Thus,

$$\langle j'm' | j_\pm T_q^{(k)} | jm\rangle - \langle j'm' | T_q^{(k)} j_\pm | jm\rangle = \hbar\sqrt{(k \mp q)(k \pm q+1)}\langle j'm' | T_{q\pm1}^{(k)} | jm\rangle. \tag{4.145}$$

Hence,

$$\hbar\langle j'm' | T_{q\pm1}^{(k)} | jm\rangle\sqrt{(k \mp q)(k \pm q+1)} = \left\{\begin{matrix} \hbar\langle j',m' \mp 1 \| T_q^{(k)} | jm\rangle\sqrt{(j' \pm m')(j' \mp m'+1)} \\ -\hbar\langle j'm' | T_q^{(k)} | \; | j,m \pm 1\rangle\sqrt{(j \mp m)(j \pm m+1)} \end{matrix}\right\}. \tag{4.146a}$$

Eq. 4.146a has exactly the same form as the recursion relation (Eq. 4.90) for the CGCs, which we *rewrite* in the following form:

$$\langle j'm' | m,q \pm 1\rangle\sqrt{(k \mp q)(k \pm q+1)} = \left\{\begin{matrix} \langle j',m' \mp 1 | mq\rangle\sqrt{(j' \pm m')(j' \mp m'+1)} \\ -\langle\langle j'm' | m \pm 1,q\rangle\sqrt{(j \mp m)(j \pm m+1)} \end{matrix}\right\}. \tag{4.146b}$$

In Eq. 4.146a and Eq. 4.146b, we have a pair of linear equations of the type

$$\sum_j a_{ij} x_j = 0 \tag{4.147a}$$

$$\text{and } \sum_j a_{ij} y_j = 0. \tag{4.147b}$$

Such a pair of simultaneous linear equations with corresponding coefficients guarantees that for every j, k we must have

$$\frac{x_j}{x_k} = \frac{y_j}{y_k} \tag{4.148a}$$

and therefore, for every index j, $x_j = \rho y_j$, (4.148b)

where ρ is a constant.

We therefore identify the ratio $\dfrac{\langle j',m' \mp 1 | T_q^{(k)} | jm\rangle}{\langle j',m' \mp 1 | mq\rangle} = \rho$ (4.149)

to be a constant, i.e., independent of the magnetic quantum numbers. This ratio is independent of the *orientation* of the axis of quantization; i.e., it is independent of geometry.

Thus, $\langle j',m'\mp 1|T_q^{(k)}|jm\rangle = \rho\langle j',m'\mp 1|mq\rangle,$ (4.150)

or, equivalently, $\langle j'm'|T_q^{(k)}|jm\rangle = \dfrac{\langle j'\|T_q^{(k)}\|j\rangle}{\sqrt{2j'+1}}\times\langle j'm'|mq\rangle \equiv pp\times gp,$ (4.151)

where "pp" stands for "physical part" and "gp" for "geometrical part." The result stated in Eq. 4.151 is called as the Wigner–Eckart theorem. Concurrently, the theorem *defines* the physical part as the *reduced matrix element* which has been written with the double-bar. It may be noted that the factor $\sqrt{2j'+1}$ in the denominator is absorbed within the definition of the reduced matrix element by some authors.

For an interaction represented by an ITO of rank 1, i.e., $k = 1$, we have from the triangle law

$$|j-k| \le j' \le (j+k);\ \text{i.e.,}\ |j-1| \le j' \le (j+1) \tag{4.152a}$$

and also

$$|j-j'| \le k \le (j+j');\ \text{i.e.,}\ |j-j'| \le 1 \le (j+j'). \tag{4.152b}$$

Thus, a transition induced by a vector operator (ITO of rank 1) from a state $|jm\rangle$ to $|j'm'\rangle$ can take place if the change in the angular momentum quantum number is either 0, or +1, or –1. That the quantum number must change only according to this rule for a transition to be possible is a selection rule; unless it is satisfied, the transition cannot occur under the interaction chosen to be represented by the ITO we considered. This selection rule is summarily written as $\delta j = 0,\pm 1$ when $k = 1$. The sub-rule $\delta j = 0$ however has an exception: a transition from $j = 0$ to $j' = 0$ cannot take place, since Eq. 4.152b requires $1 \le (j+j')$, which cannot be satisfied for the transition $(j=0)\rightarrow(j'=0)$. One can also see that a vector operator cannot bring about a transition between spherical harmonics having the same parity. *The rule that parity-conserving transitions are forbidden (for an interaction represented by a vector operator) is known after Otto Laporte*. The Laporte's rule is discussed later again in Chapter 8 (see the discussion on Eqs. 8.83a,b), where we also discuss its violation. Finally, we observe that the vectors $|jm\rangle$ may of course be labeled by any additional quantum number(s) if there are other measurements (for example, energy) that are compatible with those of (j^2, j_z). Inclusive of the possible additional good quantum number(s), we denote the vectors under consideration by $|\gamma, j, m\rangle$ and $|\gamma', j's, m'\rangle$, and the statement of the Wigner–Eckart theorem takes an appropriately modified form, but the physical essence of course remains the same.

Solved Problems

P4.1: In Eq. 4.12, we have seen that the matrix representations of the rotation operators for rotation through an infinitesimal angle ξ about the *X*- and *Y*-axis of the Cartesian coordinate system do not commute. Migrate this result from the rotation matrices to rotation operators. Use terms up to ξ^2 in Eqs. 4.8a,b,c to replace the matrices in Eq. 4.12 and determine if the generators of rotations about orthogonal axes commute.

Solution:

The commutator in Eq. 4.12 with the operators in Eqs. 4.8a,b,c replacing the matrices is

$$\left[1-\frac{i}{\hbar}\xi\ell_x-\frac{\xi^2\ell_x^2}{2\hbar^2},1-\frac{i}{\hbar}\xi\ell_y-\frac{\xi^2\ell_y^2}{2\hbar^2}\right]=\not{1}-\frac{i}{\hbar}\xi^2\ell_z-\frac{\xi^4\ell_z^2}{2\hbar^2}-\not{1},$$ since the term in ξ^4 is ignorable. We now expand the commutator on the left-hand side and equate the coefficients of *corresponding* powers of the infinitesimal angle ξ on the two sides of the equation. There being no term on the right-hand side of the above relation in the zeroth and the first power of ξ, the corresponding coefficients on the left-hand side turn out to be zero, as can be verified. From the coefficient of ξ^2, we get

$$\frac{1}{\hbar^2}\left[\ell_x,\ell_y\right]=\frac{i}{\hbar}\ell_z.$$

We find that the generators of rotations about orthogonal coordinate axes do not commute. Essentially, we have discovered Eq. 4.13a.

P4.2: Determine the commutator $\left[\vec{\ell},\vec{\ell}\right]_-$.

Solution:

$$[\vec{\ell},\vec{\ell}]=[\ell_x\hat{e}_x+\ell_y\hat{e}_y+\ell_z\hat{e}_z,\ell_x\hat{e}_x+\ell_y\hat{e}_y+\ell_z\hat{e}_z].$$

i.e., $$[\vec{\ell},\vec{\ell}]=\left\{\begin{matrix}[\ell_x,\ell_x]\hat{e}_x\hat{e}_x+[\ell_x,\ell_y]\hat{e}_x\hat{e}_y+[\ell_x,\ell_z]\hat{e}_x\hat{e}_z+\\ [\ell_y,\ell_x]\hat{e}_y\hat{e}_x+[\ell_y,\ell_y]\hat{e}_y\hat{e}_y+[\ell_y,\ell_z]\hat{e}_y\hat{e}_z+\\ [\ell_z,\ell_x]\hat{e}_z\hat{e}_x+[\ell_z,\ell_y]\hat{e}_z\hat{e}_y+[\ell_z,\ell_z]\hat{e}_z\hat{e}_z\end{matrix}\right\}=\left\{\begin{matrix}i\hbar\ell_z\hat{e}_x\hat{e}_y-i\hbar\ell_y\hat{e}_x\hat{e}_z\\ -i\hbar\ell_z\hat{e}_y\hat{e}_x+i\hbar\ell_x\hat{e}_y\hat{e}_z\\ +i\hbar\ell_y\hat{e}_z\hat{e}_x-i\hbar\ell_x\hat{e}_z\hat{e}_y\end{matrix}\right\},$$

i.e., $$[\vec{\ell},\vec{\ell}]=i\hbar\left\{\left(\ell_z\hat{e}_x-\ell_x\hat{e}_z\right)\hat{e}_y+\left(\ell_x\hat{e}_y-\ell_y\hat{e}_x\right)\hat{e}_z+\left(\ell_y\hat{e}_z-\ell_z\hat{e}_y\right)\hat{e}_x\right\}.$$

Now, $\vec{\ell}\times\hat{e}_x=\left(\ell_x\hat{e}_x+\ell_y\hat{e}_y+\ell_z\hat{e}_z\right)\times\hat{e}_x=-\ell_y\hat{e}_z+\ell_z\hat{e}_y$, and there are two other similar relations that we get on making cyclic changes $x\to y\to z\to x$.

Therefore, $[\vec{\ell},\vec{\ell}]=i\hbar\left\{\left(-\vec{\ell}\times\hat{e}_y\right)\hat{e}_y+\left(-\vec{\ell}\times\hat{e}_z\right)\hat{e}_z+\left(-\vec{\ell}\times\hat{e}_x\right)\hat{e}_x\right\}=-i\hbar\vec{\ell}\times\left(\hat{e}_x\hat{e}_x+\hat{e}_y\hat{e}_y+\hat{e}_z\hat{e}_z\right)$, which is Eq. 4.16.

P4.3: Find the matrices representing the operators $j^2,j_\pm,j_x,j_y,j_z$ in the eigenbasis of angular momentum commuting operators for (a) $j=\frac{1}{2}$ and (b) $j=1$.

Solution:

(a) For $j=\frac{1}{2}$, it is obvious that the eigenbasis of $\left(j^2,j_z\right)$ is $\left\{\left|j=\frac{1}{2},m=\frac{1}{2}\right\rangle,\left|j=\frac{1}{2},m=-\frac{1}{2}\right\rangle\right\}$,

i.e., $\left\{\left|\frac{1}{2},\frac{1}{2}\right\rangle,\left|\frac{1}{2},-\frac{1}{2}\right\rangle\right\}$. Hence, the matrix representation of the operator j^2 is

$$\left[j^2\right]=\begin{bmatrix}\left\langle\frac{1}{2},\frac{1}{2}\middle|j^2\middle|\frac{1}{2},\frac{1}{2}\right\rangle & \left\langle\frac{1}{2},\frac{1}{2}\middle|j^2\middle|\frac{1}{2},-\frac{1}{2}\right\rangle\\ \left\langle\frac{1}{2},-\frac{1}{2}\middle|j^2\middle|\frac{1}{2},\frac{1}{2}\right\rangle & \left\langle\frac{1}{2},-\frac{1}{2}\middle|j^2\middle|\frac{1}{2},-\frac{1}{2}\right\rangle\end{bmatrix}=\hbar^2\frac{3}{4}\begin{bmatrix}1&0\\0&1\end{bmatrix},$$

since $j^2\left|\frac{1}{2},\pm\frac{1}{2}\right\rangle=\hbar^2\frac{1}{2}\left(\frac{1}{2}+1\right)\left|\frac{1}{2},\pm\frac{1}{2}\right\rangle.$

Likewise, the matrix representation of the operator j_z is

$$[j_z]=\begin{bmatrix}\left\langle \frac{1}{2},\frac{1}{2}\middle|j_z\middle|\frac{1}{2},\frac{1}{2}\right\rangle & \left\langle \frac{1}{2},\frac{1}{2}\middle|j_z\middle|\frac{1}{2},-\frac{1}{2}\right\rangle \\ \left\langle \frac{1}{2},-\frac{1}{2}\middle|j_z\middle|\frac{1}{2},\frac{1}{2}\right\rangle & \left\langle \frac{1}{2},-\frac{1}{2}\middle|j_z\middle|\frac{1}{2},-\frac{1}{2}\right\rangle\end{bmatrix}=\hbar\frac{1}{2}[\sigma_z], \text{ where } [\sigma_z]=\begin{bmatrix}1 & 0\\ 0 & -1\end{bmatrix}.$$

Furthermore, the matrix representation of $j_\pm$ is

$$[j_\pm]=\begin{bmatrix}\left\langle \frac{1}{2},\frac{1}{2}\middle|j_\pm\middle|\frac{1}{2},\frac{1}{2}\right\rangle & \left\langle \frac{1}{2},\frac{1}{2}\middle|j_\pm\middle|\frac{1}{2},-\frac{1}{2}\right\rangle \\ \left\langle \frac{1}{2},-\frac{1}{2}\middle|j_\pm\middle|\frac{1}{2},\frac{1}{2}\right\rangle & \left\langle \frac{1}{2},-\frac{1}{2}\middle|j_\pm\middle|\frac{1}{2},-\frac{1}{2}\right\rangle\end{bmatrix}=\begin{bmatrix}0 & \left\langle \frac{1}{2},\frac{1}{2}\middle|j_\pm\middle|\frac{1}{2},-\frac{1}{2}\right\rangle \\ \left\langle \frac{1}{2},-\frac{1}{2}\middle|j_\pm\middle|\frac{1}{2},\frac{1}{2}\right\rangle & 0\end{bmatrix}.$$

Now, using Eq. 4.38, $[j_+]=\hbar\begin{bmatrix}0 & 1\\ 0 & 0\end{bmatrix}$ and $[j_-]=\hbar\begin{bmatrix}0 & 0\\ 1 & 0\end{bmatrix}$.

Finally, since $j_x=\frac{1}{2}(j_+ + j_-)$ and $j_y=\frac{1}{2i}(j_+ - j_-)$, we get

$[j_x]=\frac{\hbar}{2}[\sigma_x]$ with $[\sigma_x]=\begin{bmatrix}0 & 1\\ 1 & 0\end{bmatrix}$, and $[j_y]=\frac{\hbar}{2}[\sigma_y]$ with $[\sigma_y]=\begin{bmatrix}0 & -i\\ i & 0\end{bmatrix}$.

The matrices $[\sigma_x],[\sigma_y],[\sigma_z]$ defined above are simply written as $\sigma_x,\sigma_y,\sigma_z$ for short and are called as the Pauli matrices. Together with the 2×2 unit matrix $\begin{bmatrix}1 & 0\\ 0 & 1\end{bmatrix}$, they constitute a complete basis of matrices in terms of which an arbitrary 2×2 matrix can be written as a linear superposition. The spin-half angular momentum is then represented by the operator $\vec{j}=\vec{s}=\frac{\hbar}{2}\vec{\sigma}$, where the Cartesian components of $\vec{\sigma}$ are the three Pauli matrices.

(b) For $j=1$, the eigenbasis of (j^2, j_z) is $\{|1,1\rangle,|1,0\rangle,|1,-1\rangle\}$. Having identified the basis, it is now easy to follow the same procedure as for $j=\frac{1}{2}$ and show that for $j=1$ we shall have

$$j^2=2\hbar^2\begin{bmatrix}1 & 0 & 0\\ 0 & 1 & 0\\ 0 & 0 & 1\end{bmatrix},\ j_x=\frac{\hbar}{\sqrt{2}}\begin{bmatrix}0 & 1 & 0\\ 1 & 0 & 1\\ 0 & 1 & 0\end{bmatrix},\ j_y=\frac{\hbar}{\sqrt{2}}\begin{bmatrix}0 & i & 0\\ -i & 0 & i\\ 0 & -i & 0\end{bmatrix}, \text{ and } j_z=\hbar\begin{bmatrix}1 & 0 & 0\\ 0 & 0 & 0\\ 0 & 0 & -1\end{bmatrix}.$$

P4.4: Prove that (a) $\sigma_i\sigma_j=\delta_{ij}1_{op}+i\varepsilon_{ijk}\sigma_k$ and (b) for any two vectors $\vec{a}$ and $\vec{b}$, $(\vec{a}\bullet\vec{\sigma})(\vec{b}\bullet\vec{\sigma})=(\vec{a}\bullet\vec{b})1_{op}+i\vec{a}\times\vec{b}\bullet\vec{\sigma}$

Solution:

For any two operators *R*, *S*, we have $RS=\frac{1}{2}RS-\frac{1}{2}SR+\frac{1}{2}RS+\frac{1}{2}SR=\frac{1}{2}[R,S]_-+\frac{1}{2}[R,S]_+$.

Now, $\vec{a}\bullet\vec{b}=\sum_{j=1}^{3}a_jb_j\equiv a_jb_j=a_ib_j\delta_{ij}$, and the *k*th component of the cross product of the two vectors is $(\vec{a}\times\vec{b})_\ell=a_jb_k\varepsilon_{jk\ell}$. Subscripts minus and plus above denote commutator and anti-commutator respectively.

Therefore, $\sigma_j\sigma_k = \frac{1}{2}\left[\sigma_j,\sigma_k\right]_- + \frac{1}{2}\left[\sigma_j,\sigma_k\right]_+ = \frac{1}{2}\left(2i\varepsilon_{jk\ell}\sigma_\ell\right) + \frac{1}{2}\left(\delta_{jk}2\times 1\right) = i\varepsilon_{jk\ell}\sigma_\ell + \delta_{jk}1$, establishing (a) and

$$\left(\vec{a}\bullet\vec{\sigma}\right)\left(\vec{b}\bullet\vec{\sigma}\right) = a_j b_k \sigma_j\sigma_k = i\varepsilon_{jk\ell}a_j b_k \sigma_\ell + \delta_{jk}a_j b_k 1_{op} = i\left(\vec{a}\times\vec{b}\right)_\ell \sigma_\ell \equiv i\left(\vec{a}\times\vec{b}\right)\bullet\vec{\sigma} + \left(\vec{a}\bullet\vec{b}\right)1_{op},$$ proving (b).

P4.5: Show that for the orbital-angular-momentum eigenfunctions, the smallest possible value of the angle between $\vec{\ell}$ and the Z-axis obeys the relation $\cos^2\theta = \ell/(\ell+1)$. As ℓ increases, does this smallest possible angle increase or decrease?

Solution:

The orbital angular momentum $\vec{\ell}$ precesses about the Z-axis. The smallest possible angle between $\vec{\ell}$ and Z would correspond to the *largest* possible eigenvalue of ℓ_z, which is $\ell\hbar$. Hence, $\cos\theta = \frac{\ell\hbar}{\left|\vec{\ell}\right|\hbar} = \frac{\ell}{\sqrt{\ell(\ell+1)}}$ or $\cos^2\theta = \frac{\ell^2}{\ell(\ell+1)} = \frac{\ell}{(\ell+1)}$.

As ℓ increases, $\cos^2\theta \to 1$ and the angle θ becomes smaller and tends to zero.

P4.6: Determine the commutation relations $\left[\hat{A}_i, j_k\right]; i,k = 1,2,3$ between Cartesian components of a vector operator $\hat{\vec{A}} = \sum_{i=1}^{3}\hat{e}_i\hat{A}_i$ and the angular momentum vector operator by subjecting the components $\hat{A}_i; i = 1,2,3$ to a rotation $U_R = \exp\left(1 - \frac{i}{\hbar}\vec{\eta}\bullet\vec{j}\right)$. Consider only infinitesimal rotations; ignore quadratic and all smaller terms in the rotation angle.

Solution:

For *infinitesimal* rotations, the rotation matrix would only be infinitesimally different from the unit matrix. We therefore write the rotation matrix as

$$\begin{bmatrix} R_{11} & R_{12} & R_{13} \\ R_{21} & R_{22} & R_{23} \\ R_{31} & R_{32} & R_{33} \end{bmatrix} - \begin{bmatrix} 1 & 0 & 0 \\ 0 & 1 & 0 \\ 0 & 0 & 1 \end{bmatrix} + \begin{bmatrix} 0 & \eta_{12} & \eta_{13} \\ \eta_{21} & 0 & \eta_{23} \\ \eta_{31} & \eta_{32} & 0 \end{bmatrix} = \begin{bmatrix} 1 & \eta_{12} & \eta_{13} \\ \eta_{21} & 1 & \eta_{23} \\ \eta_{31} & \eta_{32} & 1 \end{bmatrix}, \tag{P4.6.1}$$

where the η_{ij}'s are infinitesimally small numbers.

Thus, $R_{ij} = \delta_{ij} + \eta_{ij}$. (P4.6.2)

We know that the rotation group $SO(3)$ is orthogonal, the orthogonality being expressed as

$$\delta_{ik} = \sum_{j=1}^{3} R_{ij}R_{kj} = \sum_{j=1}^{3}\left(\delta_{ij} + \eta_{ij}\right)\left(\delta_{kj} + \eta_{kj}\right),$$

$$\text{i.e., } \delta_{ik} = \sum_{j=1}^{3}\left(\delta_{ij}\delta_{kj} + \delta_{ij}\eta_{kj} + \eta_{ij}\delta_{kj} + \eta_{ij}\eta_{kj}\right) = \delta_{ik} + \eta_{ki} + \eta_{ik} + O\left(\eta^2\right). \tag{P4.6.3}$$

The elements of the matrix $\left[\eta_{ij}\right]$ therefore must have the antisymmetric property $\eta_{ki} = -\eta_{ik}$.

We therefore write the rotation matrix as

$$\mathbb{R} = \begin{bmatrix} 1 & -\eta_3 & \eta_2 \\ \eta_3 & 1 & -\eta_1 \\ -\eta_2 & \eta_1 & 1 \end{bmatrix}. \tag{P4.6.4}$$

Using Eqs. 4.114a,b to subject each component of a vector *operator* to an infinitesimal rotation,

$$\hat{A}_i \to \hat{A}_{i,R} = \left(1 + \frac{i}{\hbar}\vec{\eta}\cdot\vec{j}\right)\hat{A}_i\left(1 - \frac{i}{\hbar}\vec{\eta}\cdot\vec{j}\right), \tag{P4.6.5}$$

$$\text{i.e., } \hat{A}_i \to \hat{A}_{i,R} = \left(1 + \frac{i}{\hbar}(\eta_1 j_1 + \eta_2 j_2 + \eta_3 j_3)\right)\hat{A}_i\left(1 - \frac{i}{\hbar}(\eta_1 j_1 + \eta_2 j_2 + \eta_3 j_3)\right),$$

$$\text{i.e., } \hat{A}_i \to \hat{A}_{i,R} = \begin{bmatrix} \left(\hat{A}_i - \frac{i}{\hbar}\eta_1\hat{A}_i j_1 - \frac{i}{\hbar}\eta_2\hat{A}_i j_2 - \frac{i}{\hbar}\eta_3\hat{A}_i j_3\right) \\ +\frac{i}{\hbar}\eta_1 j_1\left(\hat{A}_i - \frac{i}{\hbar}\eta_1\hat{A}_i j_1 - \frac{i}{\hbar}\eta_2\hat{A}_i j_2 - \frac{i}{\hbar}\eta_3\hat{A}_i j_3\right) \\ +\frac{i}{\hbar}\eta_2 j_2\left(\hat{A}_i - \frac{i}{\hbar}\eta_1\hat{A}_i j_1 - \frac{i}{\hbar}\eta_2\hat{A}_i j_2 - \frac{i}{\hbar}\eta_3\hat{A}_i j_3\right) \\ +\frac{i}{\hbar}\eta_3 j_3\left(\hat{A}_i - \frac{i}{\hbar}\eta_1\hat{A}_i j_1 - \frac{i}{\hbar}\eta_2\hat{A}_i j_2 - \frac{i}{\hbar}\eta_3\hat{A}_i j_3\right) \end{bmatrix}. \tag{P4.6.6}$$

Ignoring terms of $O(\eta^2)$,

$$\hat{A}_i \to \hat{A}_{i,R} = \hat{A}_i - \frac{i}{\hbar}\eta_1\left[\hat{A}_i, j_1\right] - \frac{i}{\hbar}\eta_2\left[\hat{A}_i, j_2\right] - \frac{i}{\hbar}\eta_3\left[\hat{A}_i, j_3\right], \tag{P4.6.7}$$

and now setting $(\eta_1, \eta_2, \eta_3) \to (\eta_i, \eta_j, \eta_k)$:

$$\hat{A}_i \to \hat{A}_{i,R} = \hat{A}_i - \frac{i}{\hbar}\eta_i\left[\hat{A}_i, j_i\right] - \frac{i}{\hbar}\eta_j\left[\hat{A}_i, j_j\right] - \frac{i}{\hbar}\eta_k\left[\hat{A}_i, j_k\right]. \tag{P4.6.8}$$

Using Eqs. 1.116a,b and Eq. P4.6.4, we see that

$$\hat{A}_i \to \hat{A}_{i,R} = \sum_{j=1}^{3} R_{ij}\hat{A}_j = \hat{A}_i - \eta_k\hat{A}_j + \eta_j\hat{A}_k. \tag{P4.6.9}$$

The required result follows by comparing coefficients of the corresponding infinitesimal rotation angles in Eq. P4.6.8 and Eq. P4.6.9.

P4.7: A spin-less particle is in a state represented by the wavefunction $\psi = K(x + y + 2z)\exp(-\alpha r)$, where *x*, *y*, *z* are Cartesian coordinates, *r* is the distance from the origin in the spherical polar coordinates, and *K* and α are constants. Determine (a) the angular momentum of the particle, (b) the expectation value of the *Z*-component of the orbital angular momentum of the particle, and (c) the probability that a measurement of ℓ_z results in the eigenvalue $\hbar$.

Solution:

The given wave function in spherical polar coordinates is $\psi = Kr(\sin\theta\cos\varphi + \sin\theta\sin\varphi + 2\cos\theta)\exp(-\alpha r)$.

Resolved into a product of radial and angular parts, the wavefunction is $\psi(r,\theta,\varphi) = \psi(r)\psi(\theta,\varphi)$ $= N_r r e^{-\alpha r}\left[N_s(\sin\theta\cos\varphi + \sin\theta\sin\varphi + 2\cos\theta)\right]$, where N_r and N_s are the normalization constants respectively for the radial and angular parts. Accordingly, we have

$$N_s^2 \int_{\theta=0}^{\pi}\int_{\varphi=0}^{2\pi} (\sin\theta\cos\varphi + \sin\theta\sin\varphi + 2\cos\theta)^2 \sin\theta d\theta d\varphi = 1.$$

Rewriting $\cos\varphi = \frac{1}{2}(\exp(i\varphi) + \exp(-i\varphi))$ and $\sin\varphi = \frac{1}{2i}(\exp(i\varphi) - \exp(-i\varphi))$, we get $N_s = \frac{1}{\sqrt{8\pi}}$.

Hence, $\psi(\theta,\varphi)=\frac{1}{\sqrt{8\pi}}\left(-\frac{1}{2}(1-i)\sqrt{\frac{8\pi}{3}}Y_1^1+\frac{1}{2}(1+i)\sqrt{\frac{8\pi}{3}}Y_1^{-1}+2\sqrt{\frac{4\pi}{3}}Y_1^0\right)$. The given wave function therefore corresponds to orbital angular momentum state with $\ell=1$.

(a) The total orbital angular momentum of the particle is therefore given by

$$\sqrt{\langle \ell^2\rangle}=\sqrt{\ell(\ell+1)}\hbar=\sqrt{2}\hbar.$$

(b) The z-component of the angular momentum is given by

$$\langle\psi|\ell_z|\psi\rangle=N_s^2\left[\frac{1}{2}\frac{8\pi}{3}\hbar\left(Y_1^1\right)^2+\frac{1}{2}\frac{8\pi}{3}\hbar\left(Y_1^{-1}\right)^2+4\frac{4\pi}{3}(0)\left(Y_1^0\right)^2\right]=0.$$

(c) The probability of finding the result of measurement of ℓ_z to be $\hbar$ is

$$P=\left|\langle(m=+\hbar)|\psi(\theta,\varphi)\rangle\right|^2=\frac{1}{8\pi}\frac{1}{2}\frac{8\pi}{3}=\frac{1}{6}.$$

Exercises

E4.1: Prove that $\left(\vec{F}\times\vec{G}\right)\bullet\left(\vec{F}\times\vec{G}\right)=\left\{F^2G^2-\left(\vec{F}\bullet\vec{G}\right)^2\right\}-\left\{\begin{matrix}\sum_j\sum_k F_j\left[F_j,G_k\right]G_k-\sum_j\sum_k F_j\left[F_k,G_k\right]G_j\\ \sum_j\sum_k F_j\left[F_k,G_j\right]G_k+\sum_j\sum_k F_jF_k\left[G_k,G_j\right]\end{matrix}\right\}$.

E4.2: Show for a particle having charge e and a magnetic moment $\vec{\mu}_s=-g_s\mu_B\frac{\vec{s}}{\hbar}$ that the equation of continuity is satisfied with the probability charge current density vector given by

$$\vec{j}^{(e,\mu_s)}=\frac{1}{2m}\left\{\psi^*\vec{p}_c\psi-\psi\vec{p}_c\psi^*-2\frac{e}{c}\vec{A}(\psi^*\psi)\right\}+\frac{\mu_s c}{s}\vec{\nabla}\times(\psi^*\vec{s}\psi),$$

E4.3: Show that $\ell_x=i\hbar\left(\sin\varphi\frac{\partial}{\partial\theta}+\cot\theta\cos\varphi\frac{\partial}{\partial\varphi}\right)$, $\ell_y=-i\hbar\left(\cos\varphi\frac{\partial}{\partial\theta}-\cot\theta\sin\varphi\frac{\partial}{\partial\varphi}\right)$ and $\ell_z=i\hbar\frac{\partial}{\partial\varphi}$.

E4.4: Determine the expectation values $\langle\ell_x\rangle$ and $\langle\ell_x^2\rangle$ for a system that is in the simultaneous eigenstate of $\vec{\ell}^2$ and ℓ_z.

E4.5: Calculate the values of the Wigner 3j symbols (a) $\begin{pmatrix}\frac{3}{2} & \frac{3}{2} & 2\\ \frac{1}{2} & -\frac{1}{2} & 0\end{pmatrix}$ and (b) $\begin{pmatrix}\frac{5}{2} & \frac{3}{2} & 2\\ \frac{3}{2} & -\frac{1}{2} & 1\end{pmatrix}$.

E4.6: Consider a state $|\psi\rangle$ which is a superposition of states $|\psi\rangle$ = $|\ell=1,m\rangle$ given by $|\psi\rangle=C_0|1\ 0\rangle+C_{-1}|1\ -1\rangle+C_1|1\ 1\rangle$.

Find a direction $\hat{n}$ such that this state is an eigen state of the operator $\hat{n}\bullet\vec{\ell}$. Express the coefficients C_i in terms of the angles θ,φ that define the direction $\hat{n}$.

Write down the expression for the eigen vectors of ℓ_x and ℓ_y.

E4.7: Prove that $\left[\vec{\ell}\bullet\vec{\ell},\vec{p}\bullet\vec{p}\right]=0$.

E4.8: Prove that $\left[\vec{\ell}\bullet\vec{\ell},\vec{r}\bullet\vec{r}\right]=0$.

E4.9: Show that the average values of $\langle \ell_x \rangle$ and $\langle \ell_y \rangle$ are equal to zero if a system is an eigenstate of ℓ_z. Furthermore, determine the uncertainty $\delta \ell_i = \sqrt{\langle \ell_i^2 \rangle - \langle \ell_i \rangle^2}$ in that state for i = 1, 2, 3. Check if your result is in line with the uncertainty principle (Eq. 1.105).

E4.10: An electron's state is described by the wave function $\psi = \frac{1}{\sqrt{4\pi}}\left(e^{i\varphi}\sin\theta + \cos\theta\right)g(r)$, where $\int_0^\infty |g(r)|^2 r^2 dr = 1$ (spherical polar coordinates used). (a)What are the possible results of a measurement of ℓ_z? (b) What is the probability of obtaining each of the possible results in part a? (c) What is the expectation value of ℓ_z?

E4.11: The spin functions in matrix form for an electron in a basis in which s_z is diagonal are $\begin{pmatrix} 1 \\ 0 \end{pmatrix}$ and $\begin{pmatrix} 0 \\ 1 \end{pmatrix}$ with eigenvalues equal to $\pm \frac{1}{2}$, respectively. Determine the normalized column matrix eigenfunction of s_y with an eigenvalue $-\frac{1}{2}$.

E4.12: Determine the expectation value of the magnetic dipole moment of an electron in the state $\left| \ell \frac{1}{2}; j\, m \right\rangle$.

E4.13: For a particle with spin $s = 1$, determine the eigenstates of the operator $\hat{n} \cdot \vec{s}$ along the arbitrary direction $\hat{n} = \sin\theta\left(\hat{e}_x \cos\varphi + \hat{e}_y \sin\varphi\right) + \cos\theta\hat{e}_z$.

E4.14: A beam of particles is subjected to measurements of $\vec{\ell}^2, \ell_z$. The results of the measurements are the values $\ell = 0, m = 0$ and $\ell = 1, m = -1$ with probabilities $\frac{3}{4}$ and $\frac{1}{4}$ respectively. (a) Write the mixed state of the beam immediately *before* the measurement. (b) Obtain the possible outcomes and the corresponding probabilities when the particles in the beam with $\ell = 1, m = -1$ are separated out and subjected to a measurement of ℓ_x.

E4.15: The angular momentum ladder operators $\ell_\pm$ are defined as $\ell_\pm = \ell_x \pm i\ell_y$. Using the expressions for ℓ_x and ℓ_y you obtained in E4.3, obtain the expression for $\ell_\pm$. Furthermore, calculate $Y_{3,\pm 1}(\theta,\phi)$ using $\ell_\pm$, given $Y_{30}(\theta,\phi) = \sqrt{\frac{7}{16\pi}}\left(5\cos^3\theta - 3\cos\theta\right)$.

E4.16: Using the Pauli matrices σ_x, σ_y and σ_z, we may write (a) $\exp(-i\alpha\sigma_x) = A\cos\alpha + B\sin\alpha$ and (b) $\exp(i\alpha\sigma_x)\sigma_z \exp(-i\alpha\sigma_x) = C\cos(2\alpha) + D\sin(2\alpha)$. Determine *A*, *B*, *C*, and *D*.

E4.17: The unitary operator that generates a rotation by φ about an arbitrary axis $\hat{n}$ in spinor space is given as $U = \exp\left(\frac{i}{\hbar}\varphi\hat{n}\cdot\vec{S}\right)$. Show that $U = \cos(\varphi/2) + i\hat{n}\cdot\vec{\sigma}\sin(\varphi/2)$, where $\vec{S} = \frac{\hbar}{2}\left[\sigma_x\hat{e}_1 + \sigma_y\hat{e}_2 + \sigma_z\hat{e}_3\right]$ and σ_i are the Pauli matrices.

E4.18: Using the expression for *U* from the previous problem, show that $U\vec{\sigma}U^\dagger = \hat{n}\left(\hat{n}\cdot\vec{\sigma}\right) - \hat{n}\times\left(\hat{n}\times\vec{\sigma}\right)\cos\varphi + \left(\hat{n}\times\sigma\right)\sin\varphi$.

References

[1] P. C. Deshmukh, *Foundations of Classical Mechanics* (Cambridge University Press, New Delhi, 2019).

[2] H. Goldstein, C. P. Poole, and J. Safko, *Classical Mechanics* 3rd Edition (Pearson, 2014).

[3] E. U. Condon and G. H. Shortley, *The Theory of Atomic Spectra* (Cambridge University Press, 1970).

[4] Giulio Racah, Sulla caratterizzazione delle rappresentazioni irriducibili dei gruppi semisemplici di Lie. *Lincei-Rend. Sc. fis. mat. e nat* 8, 108–112 (1950).

[5] E. G. Beltrametti and A.Blasi, On the Number of Casimir Operators Associated with Any Lie group. *Physics Letters*, 20:1, 62–64 (1966). https://doi.org/10.1016/0031-9163(66)91048-1.

[6] A. R. Edmonds, *Angular Momentum in Quantum Mechanics* (Princeton University Press, 1957).

[7] Lawrence Biedenharn and J. D. Louck, *Angular Momentum in Quantum Physics: Theory and Applications* (Cambridge University Press, 2014).

[8] W. R. Johnson, *Atomic Structure Theory: Lectures on Atomic Physics* (Springer, 2007).

5

The Non-relativistic Hydrogen Atom

Niels Bohr. B34, https://arkiv.dk/en/vis/5900707. Courtesy: Niels Bohr Archive.

One must always do what one really cannot.

—Niels Bohr

A groundbreaking event in the development of the quantum theory of atomic structure is the remarkable discovery by Johann Jacob Balmer in 1884. He figured out that Anders Jonas Angstrom's measurements of the wavelengths $\lambda = 4101.2, 4340.1, 4860.74$, and 6562.10 Å of the four prominent lines in the spectrum of the hydrogen atom fit the formula $\lambda = b\left(\frac{n^2}{n^2 - 4}\right)$, where $b = 3645.6$ Å, and $n = 3, 4, 5, 6$. This is an outstanding illustration of the arithmetic genius that Balmer was, being an arithmetic teacher in a girls' school. Balmer's formula was independently rediscovered by Johannes Robert Rydberg six years later, and rationalized in the planetary model of the *old* quantum theory proposed by Niels Bohr in 1913. Orbits, however, are intangible, position and momentum measurements not being compatible. Solution to the Schrödinger equation with appropriate boundary conditions satisfactorily account for the Balmer–Rydberg–Bohr formula $\left(\frac{1}{n^2}; n = 1,2,3,..\right)$, explaining the discrete eigenenergies of the hydrogen atom. In Section 5.1, we discuss the bound-state solutions of the Schrödinger equation for the hydrogen atom. However, the Schrödinger equation does *not* fully account for the degeneracy of the energy levels of (even the nonrelativistic) hydrogen atom. This bemusing condition is addressed in the Sections 5.2 and 5.3 of this chapter, and the continuum eigenfunctions are discussed in Section 5.4.

5.1 Eigenvalues and Eigenfunctions of the Hydrogen Atom

As the very first atom in the periodic table, the hydrogen atom is the simplest one. It is the prototype of one-electron central field atomic systems (including ions) for which *exact* analytical solution to the Schrödinger equation can be obtained. The electron–proton two-particles Schrödinger equation for the hydrogen atom is

$$i\hbar\frac{\partial\Psi(\vec{r}_e,\vec{r}_p,t)}{\partial t}=\left[\frac{(-i\hbar\vec{\nabla}_e)^2}{2m_e}+\frac{(-i\hbar\vec{\nabla}_p)^2}{2m_p}+V(|\vec{r}_e-\vec{r}_p|)\right]\Psi(\vec{r}_e,\vec{r}_p,t), \tag{5.1}$$

where $\vec{r}_e$ and $\vec{r}_p$ are respectively the notional position vectors in a laboratory frame of reference of the electron and the proton, and m_e and m_p their respective masses. The gradient operators $\vec{\nabla}_e$ and $\vec{\nabla}_p$ seek space derivatives with respect to the electron and the proton coordinates respectively, and the interaction potential energy operator depends on the relative distance between the two particles. The center of mass of the two particles is located at

$$\vec{R}=\frac{m_e\vec{r}_e+m_p\vec{r}_p}{m_e+m_p}, \tag{5.2a}$$

and the reduced mass

$$\mu=\frac{m_e m_p}{m_e+m_p}\approx m_e \tag{5.2b}$$

which is located at the position vector $\vec{r}=\vec{r}_e-\vec{r}_p$ with respect to the center of mass. The Schrödinger equation can be rewritten as

$$i\hbar\frac{\partial\Psi(\vec{R},\vec{r},t)}{\partial t}=\left[\frac{(-i\hbar\vec{\nabla}_R)^2}{2M}+\frac{(-i\hbar\vec{\nabla})^2}{2\mu}+V(r)\right]\Psi(\vec{R},\vec{r},t), \tag{5.3a}$$

$$\text{where } \Psi(\vec{R},\vec{r},t)=\left\{F(\vec{R})\exp\left(-\frac{i}{\hbar}E_M t\right)\right\}\left\{\psi(\vec{r})\exp\left(-\frac{i}{\hbar}Et\right)\right\}. \tag{5.3b}$$

The function $F(\vec{R})$ describes the potential-free center of mass in a stationary state having energy E_M, and $\psi(\vec{r})$ describes the solution of the reduced mass in a potential $V(r)=V(|\vec{r}|)$. The temporal evolution of the stationary state of the center of mass is described by its dynamic phase $\exp\left(-\frac{i}{\hbar}E_M t\right)$ and that of the reduced mass by $\exp\left(-\frac{i}{\hbar}Et\right)$. The time-independent wave function for the center of mass satisfies the free-particle Schrödinger equation

$$\frac{(-i\hbar\vec{\nabla}_R)^2}{2M}F(\vec{R})=E_M F(\vec{R}). \tag{5.4}$$

Since the mass of the electron $m=m_e\simeq\mu$, the time-dependent Schrödinger equation for the electron in the hydrogen atom is described by

$$i\hbar\frac{\partial}{\partial t}\psi(\vec{r},t)=\left[\frac{p^2}{2m}+V(|\vec{r}|)\right]\psi(\vec{r},t) \tag{5.5a}$$

$$\text{with } \psi(\vec{r},t)=\psi(\vec{r})\exp\left(-\frac{i}{\hbar}Et\right), \tag{5.5b}$$

$\psi(\vec{r})$ being the stationary state whose temporal evolution is described by the dynamical phase $\theta = \frac{Et}{\hbar}$ (Eq. 1.115). The space part $\psi(\vec{r})$ is the solution of the time-independent Schrödinger equation, which from Eq. 1.140 is known to be

$$\left[\frac{\left(-i\hbar\vec{\nabla}\right)^2}{2\mu} + V(r)\right]\psi_E(\vec{r}) = E\psi_E(\vec{r}). \tag{5.6}$$

We have labeled the wave function $\psi_E(\vec{r}) = \langle\vec{r}|E\rangle$ by the energy, but there also are other good quantum numbers that emerge from measurements that are compatible with the measurement of energy, as we shall find shortly.

We observe that while the classical Hamiltonian for a central field potential is

$$H_{classical} = \frac{p_{r,\text{classical}}^2}{2\mu} + \frac{\ell_{classical}^2}{2\mu r^2} + V(r), \tag{5.7}$$

$$\text{where } p_{r,classical} = \left[\frac{1}{r}(\vec{r}\bullet\vec{p})\right]_{\text{classical}}, \tag{5.8}$$

quantization of the *radial momentum* requires *symmetrization*, since position and momentum operators do not commute; their measurements are not mutually compatible. The quantum operator for the radial momentum therefore (as established in the Solved Problem P5.1) is

$$p_r^{(\text{quantum})} = \frac{1}{2}\left[\frac{1}{r}(\vec{r}\bullet\vec{p}) + (\vec{p}\bullet\vec{r})\frac{1}{r}\right]_{\text{quantum}} = -i\hbar\left(\frac{\partial}{\partial r} + \frac{1}{r}\right). \tag{5.9}$$

The radial momentum operator defined in Eq. 5.9 provides a real expectation value (Hermiticity). Furthermore, we must replace $\ell_{classical}^2$ in the classical Hamiltonian (Eq. 5.7) by the operator ℓ^2 from Eq. 4.57a. Accordingly, the Schrödinger equation for the hydrogen atom becomes

$$\left[\frac{1}{2\mu}\left(p_r^{\,2} + \frac{\ell^2}{r^2}\right) + V(r)\right]\psi_E(\vec{r}) = E\psi_E(\vec{r}), \tag{5.10}$$

where p_r is the quantum radial momentum operator (Eq. 5.9) and ℓ^2 is the square of the quantum orbital angular momentum vector operator. The operators $\{H, \ell^2, \ell_z\}$ commute with each other. The eigenvalues of these operators provide a set of three "good" quantum numbers for the electron's quantum states; they can be obtained from measurements that are compatible with each other; allowing measurements of these three physical properties in arbitrary order. The quantum state of the hydrogen atom is thus labeled by $\{E, \ell, m\}$, and we write the wave function as $\psi_{E\ell m}(\vec{r}) = \langle\vec{r}|E\ell m\rangle$. The relativistic hydrogen atom (Chapter 7) introduces another quantum number resulting from the electron's intrinsic spin angular momentum. In the present chapter, we work only with the nonrelativistic hydrogen atom. The spherical symmetry of the Hamiltonian enables us to write the wave function as a product of a radial part and an angular part:

$$\psi(\vec{r}) = R(r)Y(\hat{r}) = R(r)Y(\theta,\varphi). \tag{5.11}$$

Using Eq. 5.11 in the Schrödinger Eq. 5.6 (or Eq. 5.10), we get

$$\frac{1}{R}\frac{d}{dr}\left(r^2\frac{d}{dr}\right)+\frac{2\mu r^2}{\hbar^2}\{E-V(r)\}=-\frac{1}{Y}\left(\frac{1}{\sin\theta}\frac{\partial}{\partial\theta}\left(\sin\theta\frac{\partial Y}{\partial\theta}\right)+\frac{1}{\sin^2\theta}\frac{\partial^2 Y}{\partial\varphi^2}\right). \tag{5.12}$$

Since the radial distance and the polar coordinates are independent degrees of freedom, both sides of Eq. 5.12 can be set to be equal to λ, which must be independent of (r,θ,φ). λ is the constant of separation; it separates the differential equation for the function $Y(\theta,\varphi)$

viz., $$\frac{1}{\sin\theta}\frac{\partial}{\partial\theta}\left(\sin\theta\frac{\partial Y}{\partial\theta}\right)+\frac{1}{\sin^2\theta}\frac{\partial^2 Y}{\partial\varphi^2}+\lambda Y=0, \tag{5.13}$$

from the one-dimensional differential equation for the function $R(r)$,

viz., $$\frac{1}{r^2}\frac{d}{dr}\left(r^2\frac{dR(r)}{dr}\right)+\left[\frac{2\mu}{\hbar^2}\{E-V(r)\}-\frac{\lambda}{r^2}\right]R(r)=0. \tag{5.14}$$

> ***Funquest:*** Determine the commutator $\left[r,p_r^{(q)}\right]_-$ of the radial coordinate operator with the radial momentum operator. The superscript q used here denotes '*quantum*'.

Note that partial derivative operators $\frac{\partial}{\partial\theta}$ and $\frac{\partial}{\partial\varphi}$ are used in Eq. 5.13, since $Y=Y(\theta,\varphi)$, whereas the total derivative operator $\frac{d}{dr}$ is used in Eq. 5.14, which has only one independent variable. From the previous chapter (see Eq. 4.57a), we see that Eq. 5.13 is essentially the eigenvalue equation for ℓ^2. Its solutions are the spherical harmonics $Y_{\ell m}(\theta,\varphi)=Y_{\ell m}(\hat{r})$ belonging to the eigenvalues

$$\ell(\ell+1)=\lambda, \tag{5.15a}$$

with $\ell=0,1,2,3,4,5,....,$ (5.15b)

and m taking a total of $(2\ell+1)$ values for each value of ℓ, from $-\ell$ to $+\ell$ in steps of unity;

i.e., $m=-\ell,-\ell+1,-\ell+2,....,\ell-2,\ell-1,\ell.$ (5.16)

The spherical harmonics are discussed in Appendix C. We therefore now proceed to discuss the radial solutions.

On identifying the constant of separation (Eq. 5.15a), the one-dimensional radial Eq. 5.10 becomes

$$\frac{1}{r^2}\frac{d}{dr}\left(r^2\frac{dR(r)}{dr}\right)+\left[\frac{2\mu}{\hbar^2}\{E-V(r)\}-\frac{\ell(\ell+1)}{r^2}\right]R(r)=0, \tag{5.17a}$$

or $$\frac{d}{dr}\left(r^2\frac{dR}{dr}\right)-\ell(\ell+1)R+\frac{2\mu}{\hbar^2}r^2\left[E-V(r)\right]R=0. \tag{5.17b}$$

We can also write Eq. 5.17a,b as

$$\frac{1}{r^2}\frac{d}{dr}\left(r^2\frac{dR(r)}{dr}\right)+\frac{2\mu}{\hbar^2}\left[E-V_{eff}(r)\right]R(r)=0, \tag{5.17c}$$

$$\text{or,} \left(R'' + \frac{2}{r}R'\right) + \frac{2\mu}{\hbar^2}\left[E - V_{eff}(r)\right]R(r) = 0, \tag{5.17d}$$

with an effective one-dimensional potential along the radial distance

$$V_{eff}(r) = V(r) + \frac{\hbar^2}{2\mu}\frac{\ell(\ell+1)}{r^2} = -\frac{e^2}{r} + \frac{\hbar^2}{2\mu}\frac{\ell(\ell+1)}{r^2}. \tag{5.18}$$

The effective potential thus depends on the value of the orbital angular momentum quantum number ℓ, which may certainly be zero in a special case. The effective potential $V_{eff}(r)$ consists of the *attractive* Coulomb potential $-\frac{e^2}{r}$ and the *repulsive* part $+\frac{\hbar^2}{2m}\frac{\ell(\ell+1)}{r^2}$. The physical potential in the electron–proton pair system that makes up the hydrogen atom is of course only the Coulomb attraction $-\frac{e^2}{r}$, but the effective potential $V_{eff}(r) = -\frac{e^2}{r} + \frac{\hbar^2}{2m}\frac{\ell(\ell+1)}{r^2}$ has resulted from the mathematical artifact of reducing the three dimensional problem to a differential equation only in one dimension. The repulsive part $\frac{\hbar^2}{2m}\frac{\ell(\ell+1)}{r^2}$ is therefore a *pseudo*-potential. Like the pseudo-forces in a rotating frame of reference (see Chapter 3 of Reference [5]), it has *real* effects. Its role in the one-dimensional radial Schrödinger equation is similar to that of the centrifugal force in a *rotating* frame of reference. It is therefore called as the *centrifugal potential*, or the *centrifugal barrier* (since it is repulsive). It determines the radial functions, and consequently many atomic properties. Some observable physical properties influenced by the centrifugal barrier are discussed in this chapter. Notice that the radial differential equation (Eqs. 5.17a,b,c) does not contain the magnetic quantum number m; it certainly depends on the orbital angular momentum quantum number ℓ. We know from Chapter 4 that for a spherically symmetric Hamiltonian, the axis of quantization can be oriented in any direction. Since the energy E parameter appears only in the radial equation and not in the angular equation, we may expect the energy states to be degenerate with respect to $m = -\ell, -\ell+1, \ldots, \ell-1, \ell$. This $(2\ell+1)-$ *fold* degeneracy would be well accounted for by *geometry* of the Hamiltonian in the Euclidean space R^3, since the axis of quantization can be oriented in an arbitrary direction, the potential being spherically symmetric. The degeneracy of the bound-state energy eigenstates of the hydrogen atom is, however, not fully accounted for by the spherical symmetry of the Hamiltonian in R^3; a subtle aspect associated with it will be discussed in Sections 5.2 and 5.3. Meanwhile, we take cognizance of the fact that the spherical harmonics are completely determined by the quantum numbers (ℓ, m).

The second order radial differential equation has solutions that depend on the boundary conditions at $r = 0$ and at $r \to \infty$. Now, when the potential $V(r)$ is everywhere finite, the radial function $R(r)$ must also be finite everywhere and well behaved, including at the origin $r = 0$. In the asymptotic region $(r \to \infty)$, the radial functions for $E < 0$ (bound states) must go to zero, whereas for $E > 0$ states, the wavefunctions are oscillatory. We shall see shortly that in the former case the energy eigenvalues can only be *discrete*, but in the latter case they are *continuous*. One must therefore remember that discreteness of the energy spectrum is a property of *bound*-state $(E < 0)$ solutions; it is *not* the *signature* of quantum theory.

The radial differential equation (Eqs. 5.17a,b,c) is somewhat difficult to solve. We employ a commonly used strategy to obtain its solution. First, we introduce an auxiliary function $\chi(r)$, such that

$$R(r) = \frac{\chi(r)}{r}, \tag{5.19a}$$

and recognize that it must satisfy

$$-\frac{\hbar^2}{2\mu}\frac{d^2\chi_{E,\ell}(r)}{dr^2} + \left\{V(r) + \frac{\hbar^2}{2\mu}\frac{\ell(\ell+1)}{r^2}\right\}\chi_{E,\ell}(r) = E\chi_{E,\ell}(r). \tag{5.19b}$$

We seek a power series solution,

$$R(r) = r^s\sum_{i=0}^{\infty} a_i r^i, \tag{5.20}$$

whose dominant term in the region $r \to 0$ is obviously r^s. Thus, near the origin, we get

$$-\frac{\hbar^2}{2\mu}s(s-1)r^{s-2} + \left\{V(r) + \frac{\hbar^2}{2\mu}\frac{\ell(\ell+1)}{r^2}\right\}r^s = Er^s, \tag{5.21a}$$

$$\text{i.e., } \frac{\hbar^2}{2\mu}\left\{\ell(\ell+1) - s(s-1)\right\}r^s + r^2V(r)r^s = Er^{s+2}. \tag{5.21b}$$

Hence, for potentials which satisfy $\lim_{r\to 0} r^2V(r) = 0,$ (5.22)

the behavior of the function $\chi_{E,\ell}(r \to 0)$ close to the origin is described by the following relation, regardless of the value of the energy E, whether positive (continuum) or negative (discrete):

$$\frac{\hbar^2}{2\mu}\left\{\ell(\ell+1) - s(s-1)\right\}r^s = 0, \tag{5.23a}$$

$$\text{i.e., } \ell(\ell+1) - s(s-1) = 0. \tag{5.23b}$$

Funquest: Sketch the physical coulomb potential and also the effective radial potential $V_{eff}(r) = V(r) + \frac{\hbar^2}{2\mu}\frac{\ell(\ell+1)}{r^2}$ on the same graph paper for $\ell = 0,1,2,3$ and compare them.

We find that solutions exist for $s = -\ell$ or for $s = (\ell + 1)$. We can, however, accept only the solution that can be normalized, the normalization condition being

$$1 = \int_0^{\infty} r^2 dr\left|R(r)\right|^2 = \int_0^{\infty} dr\left|\chi(r)\right|^2. \tag{5.24}$$

This integrand diverges at the origin for $\chi = r^{-\ell}$ for $\ell \geq 1$, and this makes the solution $s = -\ell$ unacceptable. For $\ell = 0$, χ becomes a constant, and the normalization integral is finite. Nonetheless, the solution is unacceptable as it would make the radial function $R(r \to 0) = \frac{r^{-\ell}}{r} = \frac{1}{r}$, for $\ell = 0$. This would give, on insertion in the radial Schrodinger equation,

$$\left(\vec{\nabla}\cdot\vec{\nabla}\right)\frac{1}{r} = -4\pi\delta^3(\vec{r}), \tag{5.25}$$

from the kinetic energy operator in the Hamiltonian. We consider potentials that do not have a $\delta-$*function* singularity at the origin, and hence the possibility $\chi = r^{-\ell}$ must be discarded for all values of ℓ, including $\ell = 0$. Thus, the small-r behaviour of the function $\chi_{E,\ell}(r \to 0) \to r^{\ell+1}$ is the only solution that is physically admissible. Correspondingly, the radial wave functions are given by

$$R(r \to 0;\text{for } \underline{\text{both}}\ E < 0 \text{ and } E > 0) \to r^{\ell}. \tag{5.26}$$

The function $R_{\ell}(r, r \to 0) \to r^{\ell}$ is sketched in Fig. 5.1 (apart from the normalization) for the lowest few values of the orbital angular momentum quantum number. The probability of finding the electron between r and $r + dr$ is proportional to $r^2|R(r)|^2$, i.e., to $r^2 r^{2\ell} = r^{2(\ell+1)}$ in the small-r region.

The radial wave functions go rapidly to zero with increasing value of the orbital angular momentum quantum number, in accordance with the real effect of the centrifugal barrier (pseudo-potential) mentioned earlier. This property also impacts, for example, the shape of the atomic photoionization cross-section, especially if it is from a bound state with a high value of the orbital angular momentum quantum number.

The small-r $(r \to 0)$ behavior for the $E > 0$ and the $E < 0$ states is similar; but it is totally different in the asymptotic $(r \to \infty)$ region depending on the sign of the energy eigenvalue. For $E > 0$, the asymptotic behavior of the radial functions is oscillatory, whereas for $E < 0$ the radial functions vanish asymptotically. Their differences are significant also in the intermediate region. We now write the physical potential $V(r)$ in the Schrödinger equation explicitly as the Coulomb potential $-\dfrac{Ze^2}{r}$. The solutions will be applicable not just for the hydrogen atom, but for all hydrogenic atomic systems, including the positronium, and all hydrogen-like ions. Accordingly, we get

$$-\frac{\hbar^2}{2\mu}\frac{d^2\chi_{E,\ell}(r)}{dr^2} - \frac{Ze^2}{r}\chi_{E,\ell}(r) + \frac{\hbar^2}{2\mu}\frac{\ell(\ell+1)}{r^2}\chi_{E,\ell}(r) = E\chi_{E,\ell}(r). \tag{5.27}$$

We shall now discuss the negative energy solutions (bound states). We parameterize distance by rescaling the distance from the origin by introducing

$$\rho = \left(\sqrt{\frac{8\mu|E|}{\hbar^2}}\right)r, \tag{5.28}$$

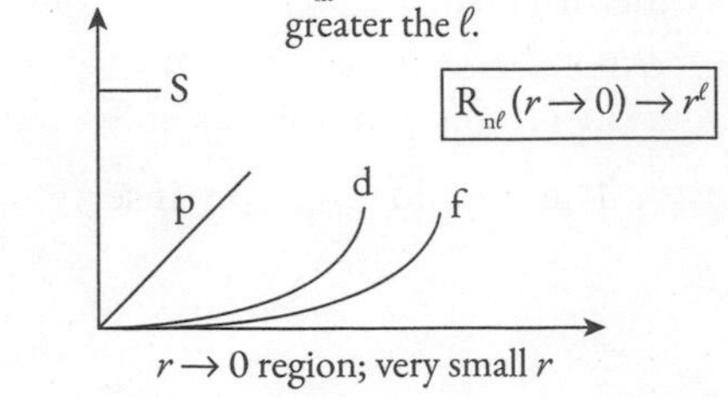

Fig. 5.1 Schematic depiction of the small-r $(r \to 0)$ behavior of the *radial* wave functions of the hydrogen atom.

which gives

$$\frac{d^2\chi_{E,\ell}(\rho)}{d\rho^2} - \frac{\ell(\ell+1)}{\rho^2}\chi_{E,\ell}(\rho) + \frac{\Omega}{\rho}\chi_{E,\ell}(\rho) = \frac{1}{4}\chi_{E,\ell}(\rho) \tag{5.29}$$

$$\text{with } \Omega = \frac{Ze^2}{\hbar}\sqrt{\frac{\mu}{2|E|}}. \tag{5.30}$$

The differential equation Eq. 5.29 is easy to solve in the asymptotic limit $\rho \to \infty$, since in this limit it simplifies to

$$\frac{d^2\chi_{E,\ell}(\rho)}{d\rho^2} - \frac{1}{4}\chi_{E,\ell}(\rho) = 0, \tag{5.31}$$

whose solution is

$$\chi_{E,\ell}(\rho;\rho \to \infty) = Ce^{-\frac{\rho}{2}} + De^{+\frac{\rho}{2}}. \tag{5.32}$$

The second term here is incompatible with the normalization condition (Eqs. 1.111a,b). Therefore, the solution to Eq. 5.29 is

$$\chi_{E,\ell}(\rho;0 \le \rho < \infty) = \rho^{\ell+1}e^{-\frac{\rho}{2}}F_{\ell}(E,\rho). \tag{5.33}$$

The first two factors in this solution provide the correct description of the solution in the regions $\rho \to 0$ and $\rho \to \infty$ respectively. The third (i.e., the last) factor must satisfy the following differential equation:

$$\rho\frac{d^2F_{E,\ell}(\rho)}{d\rho^2} + \{2\ell + 2 - \rho\}\frac{dF_{E,\ell}(\rho)}{d\rho} + (\Omega - \ell - 1)F_{E,\ell}(\rho) = 0. \tag{5.34}$$

The advantage in the strategy we adopted to factor the function $\chi_{E,\ell}$ in Eqs. 5.19a,b is now manifest: the differential equation Eq. 5.34 for $F_{E,\ell}(\rho)$ is amenable to a power series solution,

$$F_{E,\ell}(\rho) = \sum_{k=0}^{\infty} c_k\rho^k. \tag{5.35}$$

Thus,

$$\chi_{E,\ell}(\rho;0 \le \rho < \infty) = \rho^{\ell+1}e^{-\frac{\rho}{2}}\sum_{k=0}^{\infty} c_k\rho^k. \tag{5.36}$$

If $c_0 = 0$, then in the limit $\rho \to 0$ the leading term in the summation $\sum_{k=0}^{\infty} c_k\rho^k$ would be $c_1\rho^1$ making $\chi_{E,\ell}(\rho \to 0) \to \rho^{\ell+2}$ instead of $\rho^{\ell+1}$. Hence, $c_0 \ne 0$. Inserting the power series (Eq. 5.35) in the differential equation for $F_{E,\ell}(\rho)$, we get

$$\sum_{k=2}^{\infty} k(k-1)c_k\rho^{k-2} + \sum_{k=1}^{\infty} k(2\ell+2)c_k\rho^{k-2} + \sum_{k=0}^{\infty}(-k + \Omega - \ell - 1)c_k\rho^{k-1} = 0. \tag{5.37}$$

Putting $(k - 1) = j$ in the first two terms,

$$\sum_{j=1}^{\infty} j(j+1)c_{j+1}\rho^{j-1} + \sum_{j=0}^{\infty}(j+1)(2\ell+2)c_{j+1}\rho^{j-1} + \sum_{k=0}^{\infty}(-k+\Omega-\ell-1)c_k\rho^{k-1} = 0. \tag{5.38a}$$

The summation index is of course dummy, hence we may write

$$\sum_{k=1}^{\infty} k(k+1)c_{k+1}\rho^{k-1} + \sum_{k=0}^{\infty}\{(k+1)(2\ell+2)c_{k+1} + (-k+\Omega-\ell-1)c_k\}\rho^{k-1} = 0. \quad (5.38b)$$

We can now extend the summation in the first term to include the $k = 0$ term; after all it only adds *zero* to the summation. Hence,

$$\sum_{k=0}^{\infty}\{k(k+1)c_{k+1} + (k+1)(2\ell+2)c_{k+1} + (-k+\Omega-\ell-1)c_k\}\rho^{k-1} = 0, \quad (5.38c)$$

$$\text{i.e., } c_{k+1}\{k(k+1) + (k+1)(2\ell+2)\} = (k-\Omega+\ell+1)c_k. \quad (5.38d)$$

Observe that for large values of k,

$$\frac{c_{k+1}}{c_k} \underset{k \to \text{large}}{\to\to} \frac{k+\ell+1-\Omega}{(k+1)(k+2\ell+2)} \underset{k \to \text{large}}{\to\to} \frac{1}{k}. \quad (5.39)$$

This ratio of successive coefficients suggests that the power series $F_{E,\ell}(\rho)$ from Eq. 5.35 behaves as $\rho^n e^\rho$ for finite n. The function from Eq. 5.36 would then go as $\rho^{n+\ell+1}e^{+\frac{\rho}{2}}$; it diverges and blows up for very large values of k, *unless* for some integer $k = n_r$ the series terminates. This would make further (larger) powers of ρ irrelevant and prevent the series from unphysical divergence. From Eq. 5.39 we see that the condition for series termination is

$$\Omega = n_r + \ell + 1, \quad (5.40)$$

where n_r is an integer, $n_r = 0,1,2,3,\ldots$ (5.41)

From Eq. 5.30 and Eq. 5.40 we see that

$$\Omega = \frac{Ze^2}{\hbar}\sqrt{\frac{\mu}{2|E|}} = n_r + \ell + 1 = n, \quad (5.42)$$

which must be an integer $n \geq 1$, since *both* n_r and ℓ are greater than or equal to zero.

The functions $F_{E,\ell}(\rho) = \sum_{k=0}^{n_r} c_k\rho^k$ are *polynomials* of degree n_r, known as the Laguerre polynomials after Edmond Nicolas Laguerre (1834–1886). The Laguerre functions belong to the family of confluent hypergeometric functions [4], denoted by $F(-n+\ell+1, 2\ell+2, \rho)$, discussed in Appendix 5A, at the end of this chapter. The Laguerre differential equation is

$$x\frac{d^2y}{dx^2} + (1-x)\frac{dy}{dx} + ny(x) = 0, \quad (5.43a)$$

where n is an integer. Its solutions are the Laguerre polynomials (see Appendix 5A), conveniently written as a series

$$L_n(x) = \sum_{m=0}^{n}(-1)^m \frac{n!}{(n-m)!}\frac{x^m}{m!m!}. \quad (5.43b)$$

In particular,

$$L_{n=0}(x) = 1, \quad (5.44a)$$

$$L_{n=1}(x) = -x + 1, \tag{5.44b}$$

$$L_{n=2}(x) = \frac{1}{2!}(x^2 - 4x + 2), \tag{5.44c}$$

$$L_{n=3}(x) = \frac{1}{3!}(-x^3 + 9x^2 - 18x + 6), \text{ etc.} \tag{5.44d}$$

They can be written using the *Rodrigue's formula*:

$$L_n(x) = \frac{e^x}{n!}\frac{d^x}{dx^n}(x^n e^{-x}). \tag{5.45}$$

From Eq. 5.42, it follows that

$$|E| = \frac{Z^2 e^4}{\hbar^2}\frac{\mu}{2n^2}, \tag{5.46}$$

$$\text{i.e., } E = E_n = -\frac{\mu Z^2 e^4}{2\hbar^2}\frac{1}{n^2} = -\frac{Z^2 e^2}{2a_0 n^2}, \tag{5.47}$$

$$\text{where } a_0 = \frac{\hbar^2}{\mu e^2}, \tag{5.48}$$

the radius of the first "orbit" in Bohr's planetary model of the atom. The Bohr model was based on an *assumption* that the centripetally accelerated electrons stay in "stationary" (i.e., "stable," without radiating energy) orbits having energies given by the same semi-empirical formula that was obtained earlier by Balmer and Rydberg. The energy eigenvalues (Eq. 5.47) of the Schrödinger equation provide a theoretical basis for the energy eigen-spectrum that was only *explained away* in the *old* quantum theory. It is schematically shown in Fig. 5.2. Wave mechanics, however, provides not merely the energy eigenvalues but also the wave functions from which other physical properties of the hydrogen atom are obtainable. The radial wave functions are listed in Table 5.1 for the lowest few states. The functions are scaled in this table by the normalization constant $N_{E,\ell}$.

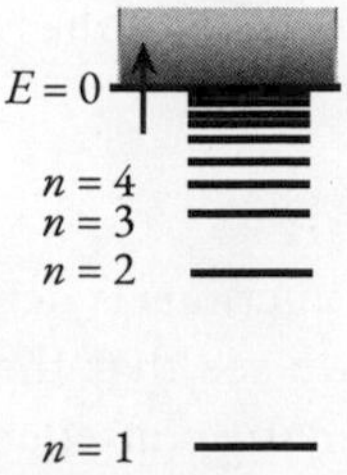

Fig. 5.2 Schematic diagram of the energy levels of the hydrogen atom modeled by the Schrödinger equation. The bound states energy eigenvalues are given by Eq. 5.47. The energy eigenvalues turn out to be exactly the same as in the Rydberg-Balmer semi-empirical model, and also the Bohr-model of the old quantum theory. Bound states ($E < 0$) have a discrete spectrum, and unbound states ($E > 0$) have a continuum.

Table 5.1 Lowest few normalized radial wave functions of the hydrogen atom. The spectroscopic notation $s, p, d, f, g,\ldots$ respectively for $\ell = 0, 1, 2, 3, 4, \ldots$ is commonly used.

Sr. No.	$n = n_r + \ell + 1$	ℓ	n_r	Spectroscopic notation	$R_{E,\ell}(r) \leftrightarrow R_{n,\ell}(r)$ $E_n = -\frac{Z^2 e^2}{2a_0 n^2}$
1	1	0	0	1s	$R_{1s}(r) = 2\left(\frac{Z}{a_0}\right)^{\frac{3}{2}} e^{-\frac{Zr}{a_0}}$
2	2	0	1	2s	$R_{2s}(r) = 2\left(\frac{Z}{2a_0}\right)^{\frac{3}{2}}\left(1-\frac{Zr}{2a_0}\right)e^{-\frac{Zr}{2a_0}}$
3	2	1	0	2p	$R_{2p}(r) = \frac{1}{\sqrt{3}}\left(\frac{Z}{2a_0}\right)^{\frac{3}{2}}\frac{Zr}{a_0}e^{-\frac{Zr}{2a_0}}$
4	3	0	2	3s	$R_{3s}(r) = 2\left(\frac{Z}{3a_0}\right)^{\frac{3}{2}}\left\{1-\frac{2Zr}{3a_0}+\frac{2(Zr)^2}{27a_0^{\ 2}}\right\}e^{-\frac{Zr}{3a_0}}$
5	3	1	1	3p	$R_{3p}(r) = \frac{4\sqrt{2}}{9}\left(\frac{Z}{3a_0}\right)^{\frac{3}{2}}\left\{1-\frac{Zr}{6a_0}\right\}\left(\frac{Zr}{a_0}\right)e^{-\frac{Zr}{3a_0}}$
6	3	2	0	3d	$R_{3d}(r) = \frac{2\sqrt{2}}{27\sqrt{5}}\left(\frac{Z}{3a_0}\right)^{\frac{3}{2}}\left(\frac{Zr}{a_0}\right)^2 e^{-\frac{Zr}{3a_0}}$
..	..	..	..	..	..

Funquest: Won't you like to test the recursion relation for Laguerre polynomials? Check it out for $n = 2$.

The number of nodes (other than the zero at the origin) each radial function has is

$$n_r = n - \ell - 1. \tag{5.49}$$

The nodal structure of the radial functions is determined by the Laguerre polynomials (see Appendix 5A). From Eq. 5.49, we see that the radial functions 1s, 2p, 3d, 4f, etc., are *nodeless*. This property has a dramatic consequence on collision and photoionization dynamics of atoms. For example, photoionization from atomic states often vanishes (or at least nearly so) at some energies above the ionization threshold as a result of the nodes in the radial function. The presence of one or more node makes the radial wavefunction positive in some radial segments, and negative in the other. The overlap between the initial state radial function with that of the final state radial function (in an atomic transition resulting from absorption of a photon) is therefore positive over some radial segment(s) and negative over other segments, over the radial distance $0 \leq r < \infty$. If the cancellation is complete, the net contribution may go to zero at the corresponding energy of the electromagnetic radiation. At that energy, the transition probability would then drop to zero, called as the *Cooper minimum* (Fig. 5.3). However, if the initial state radial function is nodeless, there can be no Cooper minimum.

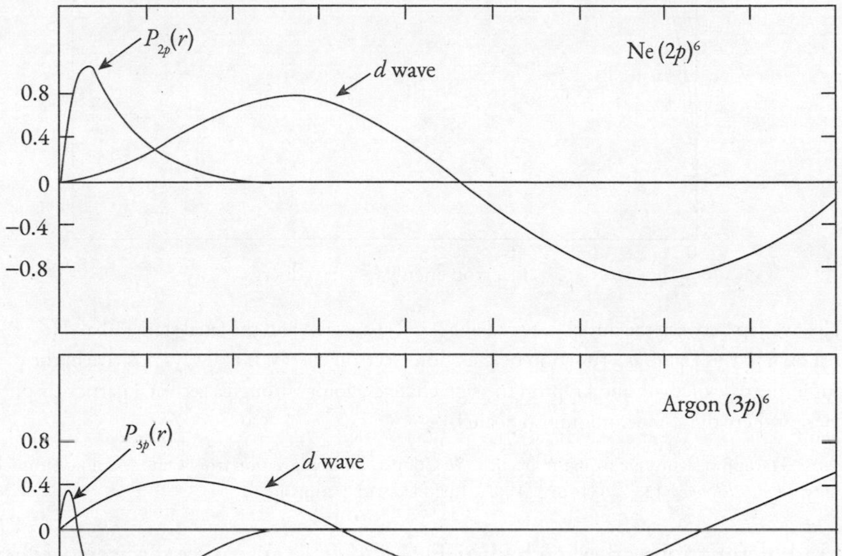

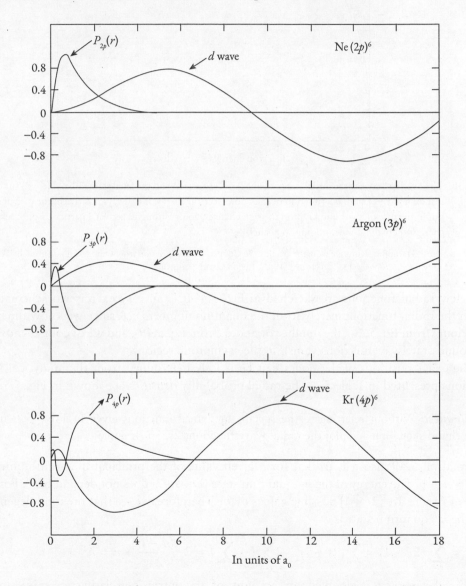

Fig. 5.3a The outermost $2p$ radial function in neon is nodeless, the argon $3p$ has one node, and the krypton $4p$ has two nodes. Also shown in each panel is the continuum d radial wavefunction at an energy that is just enough for ionization of the atom to take place on absorption of a photon. The photoelectron energy at the ionization threshold is $\epsilon \simeq 0$.

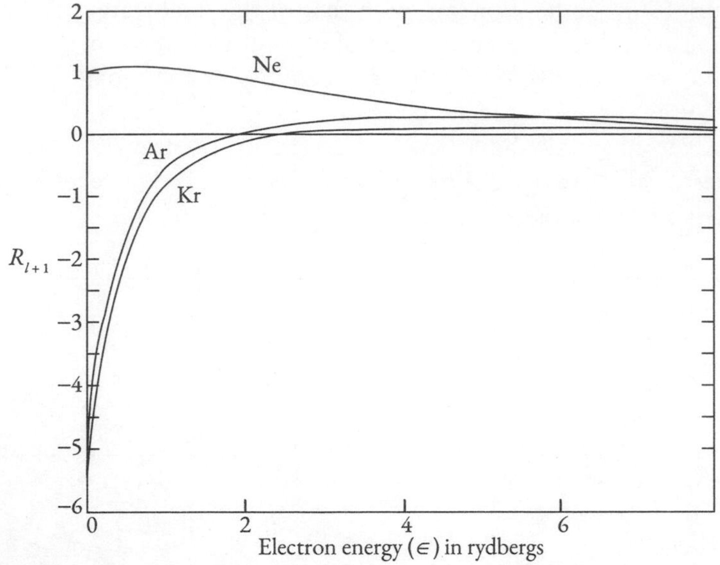

Fig. 5.3b The overlap between the initial state radial p ($\ell = 1$) function and the final state radial d ($\ell = 2$) function is shown in this figure as a function of the photoelectron energy. It is always positive in the case of neon, but in the case of argon and krypton the sign changes, going through a zero at a particular photoelectron energy where the Cooper minimum occurs.

Source: The above figures are from the article J. W. Cooper, Photoionization from Outer Atomic Subshells. A Model Study, *Physical Review* 128:2, 681–693 (1962). Included with permission.

A few radial functions are sketched in Fig. 5.4. It is an easy exercise to schematically sketch the radial functions merely from the quantum numbers n, ℓ, since we know the $r \to 0$ behaviour (from Eq. 5.26), the number of nodes (from Eq. 5.49), and we of course know that the bound-state wavefunctions vanish in the asymptotic region.

The exact positions of the nodes can be readily determined from the analytical radial functions presented in Table 5.1. The radial probability *densities* are shown in Fig. 5.5.

Funquest: Without referring to the specific analytical functions given in Table 5.1, use freehand sketching and plot the radial wave functions 2s, 3p, 4f and 5s.

Equation 5.49 also tells us that for a given value of the principal quantum number n, $\ell_{\max} = n - 1$. The energy of the nth quantum state (Eq. 5.47) does not depend on ℓ. For each value of ℓ there are $(2\ell + 1)$ possible values of m. Therefore, we see that the total degeneracy d of the nth quantum state is

$$d = \sum_{\ell=0}^{\ell_{\max}} (2\ell + 1) = \sum_{\ell=0}^{n-1} (2\ell + 1) = 2\sum_{l=0}^{n-1} \ell + \sum_{l=0}^{n-1} 1 = 2\frac{n(n-1)}{2} + n = n^2 . \tag{5.50}$$

This degeneracy is doubled on account of the intrinsic (relativistic, spin) angular momentum of an electron, which we shall discuss in the Chapter 7. The degeneracy of the energy eigenstate determines the *aufbau* filling of the orbitals in the atoms in the periodic table, as per the Pauli exclusion principle. It is the Fermi-Dirac statistics that validates the

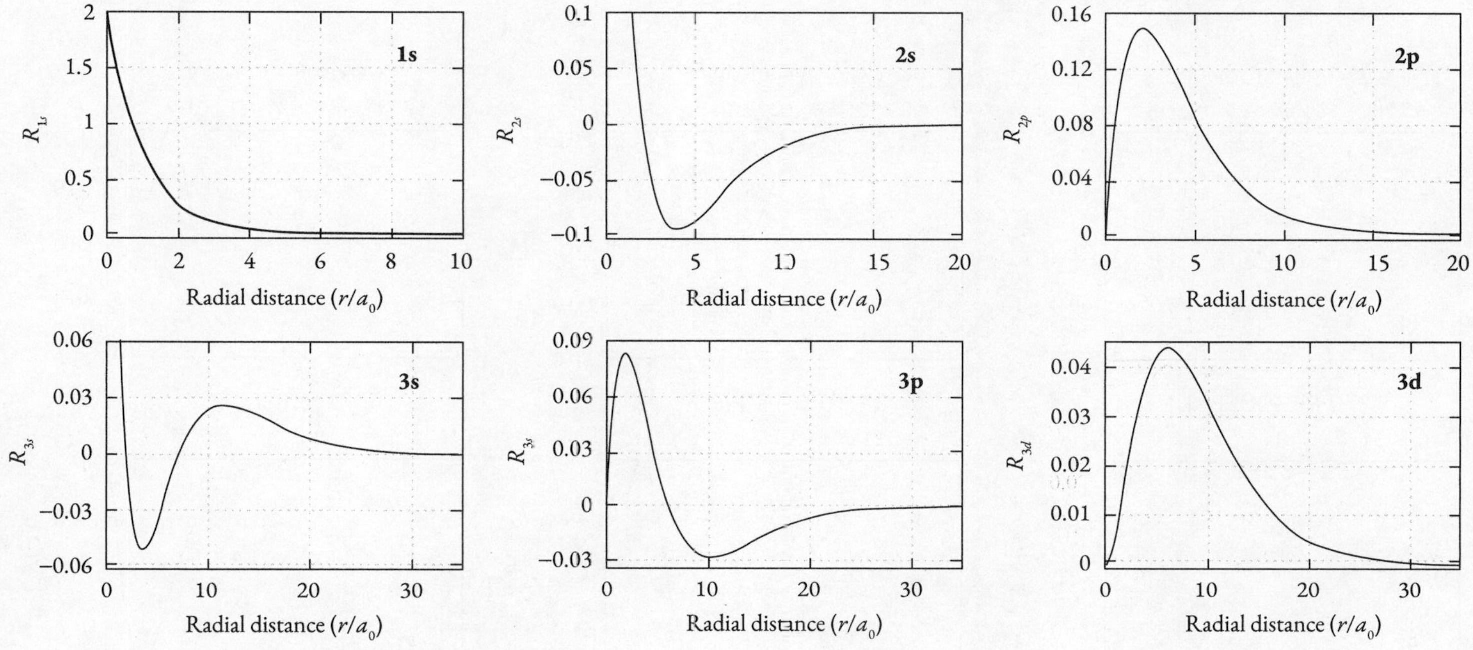

Fig. 5.4 Illustrative radial *wave functions* of the hydrogen atom. Figures plotted by Nishi[redacted]a Manohar Hosea.

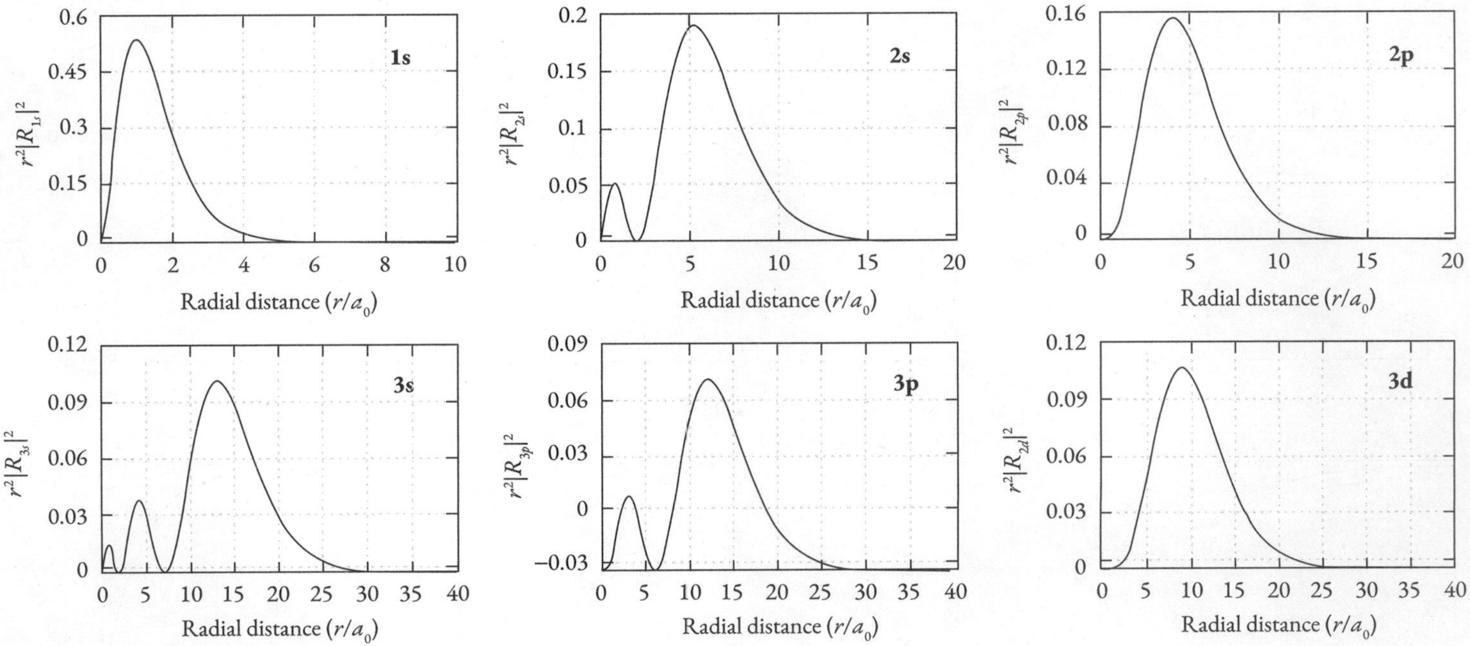

Fig. 5.5 Plots of radial probability *densities* corresponding to the functions shown in Fig. 5.4. Figures plotted by Nishita Manohar Hosea.

exclusion principle. As mentioned earlier, the $(2\ell + 1) - fold$ degeneracy with respect to the linearly independent functions for $m = -\ell, -\ell + 1, ..., \ell - 1, \ell$ is satisfactorily accounted for by the spherical symmetry of the Hamiltonian (since the axis of quantization can be oriented in any arbitrary direction when the potential has central field symmetry). The $n^2 - fold$ degeneracy (Eq. 5.50), however, begs an explanation, since Eqs. 5.17a,b,c ought to have nondegenerate eigenvalues (see the Solved Problem P3.3). The degeneracy of the nth state with respect to different values of the orbital angular momentum $\ell = 0,1,2,..,\ell_{max} = n - 1$ is responsible for a total of n^2 linearly independent eigenfunctions of the nonrelativistic Hamiltonian belonging to the same energy eigenvalue $\left(-\frac{Z^2e^2}{2a_0n^2}\right)$. This degeneracy is *not* accounted by the geometry of the Hamiltonian in R^3; The $n^2 - fold$ degeneracy is therefore commonly called as *accidental* degeneracy of the hydrogen atom. The Schrödinger equation makes this degeneracy look mysterious! In the next section, we discuss the physical reasons responsible for the $n^2 - fold$ degeneracy.

Funquest: Determine the commutators $\left[r^2, p_r^{(q)}\right]_-$ and $\left[r^{-1}, p_r^{(q)}\right]_-$.

5.2 THE "ACCIDENTAL" DEGENERACY IS NOT AN ACCIDENT!

The generators of rotations corresponding to the spherical symmetry of the Hamiltonian are the three components of the angular momentum. On this basis alone, the symmetry group of the Hamiltonian for the hydrogen atom would be identified as SO(3), as discussed in the previous chapter. The hydrogen atom, however, has an additional symmetry; this scarce feature stems from an interesting manifestation of the intimate connection between symmetry and conservation laws [1, 2, 3], famously known as the Noether's theorem. The additional symmetry comes from the exact *form* of the interaction between the electron and the proton which goes as *one-over-distance* in the entire space. The one-over-distance form of the interaction also occurs in the gravitational Kepler–Newton planetary model of the solar system. The physical quantity that is conserved in the Kepler–Newton model is the fixation of the major axis of the elliptic orbits, represented by the constancy of the Laplace–Runge–Lenz (LRL) vector (see, for example, Chapter 8 of References [5], or [6]) also called as the eccentricity vector, shown in Fig. 5.6.

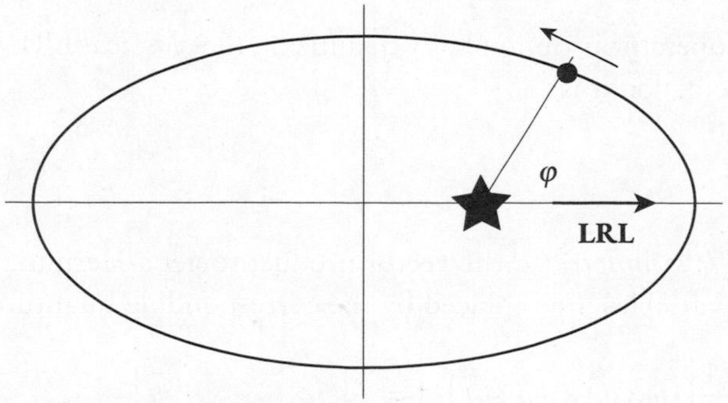

Fig. 5.6 The classical Laplace–Runge–Lenz vector points from the focus to the perigee. It is a constant vector for the one-over-distance potential. It keeps the major axis fixed; prevents its precession.

The spherical symmetry of the interaction *alone* fixes only the plane of the classical orbit due to the onservation of angular momentum. It does not prevent the *precession* of the orbit *within* that plane. That the major axis of the Kepler–Newton ellipse does not precess is a manifestation of a *conserved physical entity*, associated with which is a *symmetry* resulting from the specific *form* of the potential in the Hamiltonian. Being governed by the specific form of the interaction (viz., one-over-distance) it is called as the *dynamical* symmetry. It belongs to the *pair* (symmetry, conservation laws) in the Noether's theorem. As a result of the rotational *and* the *additional* dynamical symmetry, the symmetry group of the hydrogen atom is SO(4), not SO(3). The condition that the interaction has the exact $\frac{1}{r}$ form over the whole space $0 \le r < \infty$ is *strict* for the SO(4) symmetry to hold.

On account of the SO(4) symmetry, the Schrödinger equation for the hydrogen atom is *superintegrable*. As we shall see below, a comprehensive analysis of the quantum mechanics of the hydrogen atom on the basis of its SO(4) symmetry accounts for the n^2-degeneracy of its quantum states. The superintegrability enables the application of the method of *separation of variables* to solve the Schrödinger equation not only in the spherical polar coordinate system but also in other coordinate systems. We shall discuss the solutions in the parabolic coordinate system in Section 5.3. In the present section, we focus on obtaining the Casimir operators (defined in the previous chapter, after Eq. 4.59) for the SO(4) group. Application of the Casimir operators would resolve the *n^2-fold* degeneracy, which would no longer remain a mystery. It would be rationalized on the basis of the complete symmetry of the Hamiltonian and not [3] have to be referred to as *accidental*.

We have already seen that the operators $\{H, \ell^2, \ell_z\}$ commute with each other. The Hamiltonian being *superintegrable*, there exists an additional algebraically independent operator that commutes with it, albeit in the *subspace* of bound states $(E < 0)$. The number of generators of SO(4) is $\frac{4(4-1)}{2} = 6$. These include the three components $\{\ell_x, \ell_y, \ell_z\}$ of the angular momentum vector operator and another three components $\{A_x, A_y, A_z\}$ of the Pauli–Laplace–Runge–Lenz vector operator (or simply as the *Pauli–Lenz* vector operator), which is defined as

$$\vec{A} = \frac{1}{2\mu}\left[\left(\vec{p} \times \vec{\ell}\right) - \left(\vec{\ell} \times \vec{p}\right)\right] - \kappa\frac{\vec{r}}{r}. \tag{5.51}$$

The Pauli–Lenz operator is obtained by quantizing the *classical* LRL vector [5-7]. For the electron–proton interaction it is

$$\vec{A} = \frac{\vec{p} \times \vec{\ell}}{\mu} - \kappa\hat{e}_r \tag{5.52}$$

with $\kappa = e^2$. Pauli [7] *symmetrized* the vector product to get a Hermitian operator since the classical dynamical variables are replaced by the corresponding quantum operators:

$$\left(\vec{p} \times \vec{\ell}\right)_{\text{classical}} \rightarrow \frac{1}{2}\left[\left(\vec{p} \times \vec{\ell}\right) + \left(\vec{p} \times \vec{\ell}\right)^{\dagger}\right] = \frac{1}{2}\left[\vec{p} \times \vec{\ell} - \vec{\ell} \times \vec{p}\right]_{\substack{\text{quantum}\\ \text{operators}}}. \tag{5.53}$$

The symmetrizing methodology is identical to that applied in obtaining the quantum radial momentum operator (Eq. 5.9). We need $\vec{p} \times \vec{\ell}$ and $\vec{\ell} \times \vec{p}$ to construct the Pauli–Lenz operator with $\vec{\ell} = \vec{r} \times \vec{p}$ and $\vec{p} = \vec{r} \times \left(-i\hbar\vec{\nabla}\right)$. Now, the components of an operator

$$\vec{F} = \vec{C} \times \vec{D} \tag{5.54a}$$

are given by

$$F_i = \sum_j \sum_k \varepsilon_{ijk} C_j D_k \equiv \varepsilon_{ijk} C_j D_k \tag{5.54b}$$

$$\text{with } \varepsilon_{ijk} = \begin{cases} 1 & \text{if } (ijk) \text{ is cyclic} \\ -1 & \text{if } (ijk) \text{ is non-cyclic} \\ 0 & \text{if any of the two indices are equal} \end{cases} . \tag{5.55}$$

Accordingly [8], we get

$$\vec{\ell}_i = \left(\vec{r} \times \vec{p}\right)_i = \sum_j \sum_k \varepsilon_{ijk} x_j p_k \equiv \varepsilon_{ijk} x_j p_k = -i\hbar\varepsilon_{ijk} x_j \frac{\partial}{\partial x_k}, \tag{5.56}$$

$$\left(\vec{p} \times \vec{\ell}\right)_i = \sum_j \sum_k \varepsilon_{ijk} p_j \ell_k \equiv \varepsilon_{ijk} p_j \ell_k, \tag{5.57}$$

$$\text{and } \left(\vec{\ell} \times \vec{p}\right)_i = \sum_j \sum_k \varepsilon_{ijk} \ell_j p_k = \varepsilon_{ijk} \ell_j p_k. \tag{5.58}$$

The Einstein convention of summation over the double index is used here. Using Eq. 5.56 in Eq. 5.58 for the jth component of the orbital angular momentum operator, we get

$$\left(\vec{\ell} \times \vec{p}\right)_i = \sum_j \sum_k \sum_m \sum_n \varepsilon_{ijk} \varepsilon_{jmn} x_m p_k p_n. \tag{5.59}$$

Further, using the commutator that represents the quantum uncertainty principle

$$\left[x_m, p_k\right]_- = i\hbar 1_{op} \delta_{mk} \tag{5.60}$$

in Eq. 5.59, we get

$$\left(\vec{\ell} \times \vec{p}\right)_i = \sum_j \sum_k \sum_m \sum_n \varepsilon_{ijk} \varepsilon_{jmn} \left(i\hbar\delta_{mk} + p_k x_m\right) p_n. \tag{5.61}$$

Hence,

$$\left(\vec{\ell} \times \vec{p}\right)_i = i\hbar \sum_j \sum_k \sum_n \varepsilon_{ijk} \varepsilon_{jkn} p_n + \sum_j \sum_k \varepsilon_{ijk} p_k \left(\sum_m \sum_n \varepsilon_{jmn} x_m p_n \right), \tag{5.62a}$$

$$\text{i.e., } \left(\vec{\ell} \times \vec{p}\right)_i = i\hbar \sum_j \sum_k \sum_n \varepsilon_{ijk} \varepsilon_{jkn} p_n + \sum_j \sum_k \varepsilon_{ijk} p_k \ell_j, \tag{5.62b}$$

$$\text{or } \left(\vec{\ell} \times \vec{p}\right)_i = i\hbar \sum_j \sum_k \sum_n \varepsilon_{ijk} \varepsilon_{jkn} p_n - \sum_j \sum_k \varepsilon_{ikj} p_k \ell_j. \tag{5.62c}$$

Now, using $\sum_j \sum_k \varepsilon_{jki}\varepsilon_{jkn} = 2\delta_{in}$,

$$\left(\vec{\ell} \times \vec{p}\right)_i = i\hbar \sum_j \sum_k \sum_n \varepsilon_{jki}\varepsilon_{jkn} p_n - \left(\vec{p} \times \vec{\ell}\right)_i = 2i\hbar p_i - \left(\vec{p} \times \vec{\ell}\right)_i. \tag{5.62d}$$

Since this relation holds for all the three components, we get the following vector identity:

$$\vec{\ell} \times \vec{p} = 2i\hbar\vec{p} - \left(\vec{p} \times \vec{\ell}\right). \tag{5.63}$$

Using now Eq. 5.63 in Eq. 5.52, we get the following *equivalent* form of the Pauli–Lenz vector operator:

$$\vec{A} = \frac{1}{\mu}\left[\vec{p} \times \vec{\ell} - i\hbar\vec{p}\right] - \kappa\frac{\vec{r}}{r}. \tag{5.64}$$

It is now instructive to construct a scalar operator

$$\vec{A} \bullet \vec{A} = \left(\frac{1}{\mu}\left[\left(\vec{p} \times \vec{\ell}\right) - i\hbar\vec{p}\right] - \kappa\frac{\vec{r}}{|r|}\right) \bullet \left(\frac{1}{\mu}\left[\left(\vec{p} \times \vec{\ell}\right) - i\hbar\vec{p}\right] - \kappa\frac{\vec{r}}{|r|}\right). \tag{5.65}$$

In Appendix 5B it is shown that

$$\vec{A}^2 = \frac{2H}{\mu}\left(\vec{\ell}^2 + \hbar^2\right) + \kappa^2. \tag{5.66}$$

We shall now use the following four commutation properties:

(a) $\left[\vec{A}, H\right] = 0,$ (5.67)

which means that the Pauli–Lenz vector operator is a constant of motion. This property is established in Appendix 5C.

(b) $\left[A_i, \ell_j\right] = i\hbar\varepsilon_{ijk}A_k$ (5.68)

- an important property, shown in Appendix 5D.

(c) $\left[\ell_i, \ell_j\right] = i\hbar\varepsilon_{ijk}\ell_k$ (5.69)

- already discussed at length in the previous chapter.

(d) $\left[A_i, A_j\right] = -2i\frac{\hbar}{\mu}H\varepsilon_{ijk}\ell_k$ (5.70)

- demonstrated in Appendix 5E.

The commutators (a), (b), (c), and (d) *together* enable us to conclude that the three components of the Pauli–Lenz vector operator $\vec{A}$, along with the three components of the angular momentum operator $\vec{\ell}$, provide a *closed* algebra over the *subspace* of the Hilbert space of eigenvectors belonging to a *particular bound-state* energy eigenvalue E. The Hamiltonian operator on the right-hand side of the commutator (d) restricts the algebra of $\vec{\ell}, \vec{A}$ to be *closed* only over the *subspace* of the Hilbert space of eigenvectors belonging to a *particular* bound-state energy eigenvalue $E < 0$. The three components of the angular momentum operator $\vec{\ell}$, together with the three components of the Pauli–Lenz vector operator $\vec{A}$, constitute a set of six generators of the Lie Group SO(4) of rank 2. It is now useful to introduce an operator proportional to $\vec{A}$, defined by

$$\vec{A}' = \sqrt{\frac{-\mu}{2E}}\vec{A}. \tag{5.71a}$$

Note that since we have considered only the bound states $(E < 0)$, the scaling in Eq. 5.71a is not by an imaginary number. We shall denote $\sqrt{-\frac{2E}{\mu}}$ by p_s, giving

$$\vec{A}' = \frac{1}{p_s}\vec{A}. \tag{5.71b}$$

We shall now proceed to determine the Casimir operators [9] for SO(4). It is convenient to employ *natural index* for the components of the angular momentum and those of the *scaled* Pauli–Lenz vector operator. This is done by indexing the three components of the orbital angular momentum operator as $\vec{\ell} \to \{\ell_{23}, \ell_{31}, \ell_{12}\}$ and those of the scaled Pauli–Lenz operator as $\vec{A}' \to \{\ell_{14}, \ell_{24}, \ell_{34}\}$. The *Casimir Operators* for a Lie group commute with all of its generators. As mentioned in Chapter 4, from Racah's theorem we know that the number of Casimir operators for a group is equal to the rank of the group. One operator that is a suitable *bilinear* combination of the generators, can always be found that commutes with *all* the generators. The group SO(4) being of rank 2 has two Casimir operators, which are

$$C_1 = I^2 + K^2 \tag{5.72a}$$

and

$$C_2 = I^2 - K^2, \tag{5.72b}$$

wherein two *new* operators $\vec{I}$ and $\vec{K}$ are introduced, in terms of the angular momentum and the (scaled) Pauli–Lenz operator, by the following relations:

$$\vec{I} = \frac{1}{2}\left(\vec{\ell} + \vec{A}'\right) \tag{5.73a}$$

and

$$\vec{K} = \frac{1}{2}\left(\vec{\ell} - \vec{A}'\right). \tag{5.73b}$$

That the operators defined in Eqs. 5.72a,b are Casimir operators for SO(4) can be easily seen by verifying that

$$\forall i = 1,2,3,\ \left[C_1, \ell_i\right] = 0, \text{ and also } \left[C_1, A_i\right] = 0, \tag{5.74a}$$

and likewise

$$\forall i = 1,2,3,\ \left[C_2, \ell_i\right] = 0, \text{ and also } \left[C_2, A_i\right] = 0. \tag{5.74b}$$

It is likewise easy to verify that

$$\left[I_x, I_y\right]_- = i\hbar I_z, \text{ etc., for cyclic (x,y,z)} \rightarrow \text{(y,z,x)} \rightarrow \text{(z,x,y)} \tag{5.75a}$$

$$\text{and } \left[K_x, K_y\right]_- = i\hbar K_z, \text{ etc., for cyclic (x,y,z)} \rightarrow \text{(y,z,x)} \rightarrow \text{(z,x,y).} \tag{5.75b}$$

Equations 5.75a,b satisfy the *defining* relations (Eq. 4.18a) for angular momentum. Both $\vec{I}$ and $\vec{K}$ are therefore angular momentum vector operators. They are independent; all components of $\vec{I}$ commute with all components of $\vec{K}$. Furthermore, both $\vec{I}$ and $\vec{K}$ commute with the Hamiltonian. These operators are the *sum* and the *difference* of an axial vector operator $\left(\vec{L}\right)$ and a polar vector operator $\vec{A}'$. Hence, they do not have well-defined parity (discussed in Chapter 2 of Reference [5]) eigenstates. This has as important consequence on the hydrogen atom spectrum, discussed below. From Chapter 4, we know that the eigenvalues of I^2 and K^2 are respectively $\hbar^2 i(i+1)$ and $\hbar^2 k(k+1)$.

The eigenvalue c_2 of the Casimir operator C_2 is zero, since this operator turns out to be equal to the scalar operator $\vec{A}\bullet\vec{L}$ which is identically zero (Solved Problem P5.4). Accordingly, from Eq. 5.72b we see that the eigenvalues i and k must be *equal*, with

$$i,k = 0, \frac{1}{2}, 1, \frac{3}{2}, 2, \frac{5}{2}, 3, \ldots. \tag{5.76}$$

Now, from Eq. 5.72a, we see that the eigenvalue of the first Casimir operator is

$$c_1 = 2\hbar^2 i(i+1) = 2\hbar^2 k(k+1). \tag{5.77}$$

Using Eqs. 5.72 and 5.73, we have

$$C_1 = \frac{1}{4}\left(\vec{\ell}^2 + \vec{\ell}\bullet\vec{A}' + \vec{A}'\bullet\vec{\ell} + \vec{A}'^2\right) + \frac{1}{4}\left(\vec{\ell}^2 - \vec{\ell}\bullet\vec{A}' - \vec{A}'\bullet\vec{\ell} + \vec{A}'^2\right), \tag{5.78a}$$

$$\text{i.e., } C_1 = \frac{1}{4}\left(\vec{\ell}^2 + \vec{A}'^2\right) + \frac{1}{4}\left(\vec{\ell}^2 + \vec{A}'^2\right) = \frac{1}{2}\left(\vec{\ell}^2 + \vec{A}'^2\right). \tag{5.78b}$$

To determine C_1, we note that

$$\vec{A}'^2 = \left(\sqrt{\frac{-\mu}{2E}}\vec{A}\right)\bullet\left(\sqrt{\frac{-\mu}{2E}}\vec{A}\right) = \frac{\mu}{2E}\vec{A}\bullet\vec{A}. \tag{5.79}$$

The eigenvalue c_1 of the first Casimir operator C_1 is related to that of $\vec{A}\bullet\vec{A}$. We see that

$$C_1 = \frac{1}{2}\left(\vec{\ell}^2 + \vec{A}'^2\right) = \frac{1}{2}\vec{\ell}^2 + \frac{1}{2}\left(\frac{-\mu}{2E}\right)\vec{A}^2 = \frac{1}{2}\vec{\ell}^2 + \frac{1}{2}\left(\frac{-\mu}{2E}\right)\left\{\frac{2H}{\mu}\left(\vec{\ell}^2 + \hbar^2\right) + \kappa^2\right\},$$

$$\text{i.e., } C_1 = \frac{1}{2}\vec{\ell}^2 - \frac{1}{2}\vec{\ell}^2 - \frac{1}{2}\hbar^2 - \frac{1}{2}\left(\frac{\mu}{2E}\right)\kappa^2, \tag{5.80}$$

from which it follows that the eigenvalue of the first Casimir operator is

$$c_1 = -\left(\frac{\mu}{4E}\right)\kappa^2 - \frac{1}{2}\hbar^2 = 2i(i+1)\hbar^2, \tag{5.81a}$$

$$\text{i.e.,} \left\{2i(i+1)+\frac{1}{2}\right\}\hbar^2 = -\frac{\mu\kappa^2}{4E}, \tag{5.81b}$$

and hence,

$$E = -\frac{\mu\kappa^2}{2\hbar^2\{4i(i+1)+1\}} = -\frac{\mu\kappa^2}{2\hbar^2(2i+1)^2} = -\frac{\mu\kappa^2}{2\hbar^2 n^2}, \tag{5.82}$$

$$\text{where } n = 2i+1 \tag{5.83}$$

must be an integer, having values

$$n = 1,2,3,4,...., \tag{5.84}$$

on account of the fact that the only possible values of i are $0,\frac{1}{2},1,\frac{3}{2},2,\frac{5}{2},3,...$

Eq. 5.82 is just the usual Balmer–Rydberg–Bohr expression for the energy of a discrete bound energy level in the hydrogen atom. We had obtained it earlier (Eq. 5.47) by applying boundary conditions on the solution to the radial Schrödinger equation. However, in obtaining Eq. 5.82, the Schrödinger equation has *not* been used; it has been obtained using the SO(4) algebra of the angular momentum and the Pauli–Lenz quantum operators. One may recall that by exploiting the properties of the classical LRL vector we can obtain the complete equation to the Kepler–Newton elliptical orbit *without* solving a differential equation of motion (see, for example, Chapter 8 of Reference [5]). The derivation of Eq. 5.82 *without* employing the Schrödinger equation is a quantum analogue of that result.

We will see now that the SO(4) algebra properly accounts for the n^2 – *fold* degeneracy, which Schrödinger's wave mechanics would only describe as *accidental*. The eigenvalues i_z and k_z of the operators defined in Eqs. 5.73a,b can *each* take $2i+1$ (or equivalently $2k+1$) values. We immediately see that the degeneracy of the level is

$$(2k+1)\times(2k+1) = (2i+1)\times(2i+1) = n^2. \tag{5.85}$$

It is not "accidental" [3]; it has resulted naturally from the SO(4) symmetry of the hydrogen atom's strict-Coulomb potential, described by $\frac{1}{r}$; $0 \le r < \infty$. This *degeneracy* is the quantum analogue of the constancy of the Laplace–Runge–Lenz vector which remains fixed along the major axis, preventing it from precessing.

In Table 5.2 are presented the atomic degenerate orbitals for the lowest few bound states. A remarkable feature of the results tabulated here is the fact that states with odd and even parity are degenerate (e.g., 2*s* and 2*p*). This is due to the fact that the vector operators $\vec{I}$ and $\vec{K}$ are *pseudo*-angular-momentum operators, being made up from the sum and the difference of an axial and a polar vector operator.

Funquest: Is the angular momentum vector polar or axial? Justify the term "pseudo-angular momentum" used above for the operators $\vec{I}$ and $\vec{K}$.

Table 5.2 The n^2 – *fold* degeneracy for the SO(4) symmetry of the hydrogen atom is accounted for by the different eigenvalues of I_z and K_z. The subscripts in the last column denote the magnetic quantum number m, for each value of l. For $n = 5$, the 25 degenerate orbitals are indicated using spectroscopic notation.

i	$2i$	$n = 2i + 1$	n^2	*atomic orbitals*
0	0	1	1	$\{1s\}$
$\frac{1}{2}$	1	2	4	$\{2s, 2p_{-1}, 2p_0, 2p_{+1}\}$
1	2	3	9	$\left\{\begin{matrix} 3s, 3p_{-1}, 3p_0, 3p_{+1} \\ 3d_{-2}, 3d_{-1}, 3d_0, 3d_{+1}, 3d_{+2} \end{matrix}\right\}$
$\frac{3}{2}$	3	4	16	$\left\{\begin{matrix} 4s, 4p_{-1}, 4p_0, 4p_{+1} \\ 4d_{-2}, 4d_{-1}, 4d_0, 4d_{+1}, 4d_{+2} \\ 4f_{-3}, 4f_{-2}, 4f_{-1}, 4f_0, 4f_{+1}, 4f_{+2}, 4f_{+3} \end{matrix}\right\}$
2	4	5	25	$\{5s(1), 5p(3), 5d(5), 5f(7), 5g(9)\}$

The algebra of the SO(4) operators provides not merely the eigenvalues and accounts for their degeneracies; but also provide the eigenfunctions, *without* using the Schrödinger equation. Let us denote the ground state $(n = 1)$ of the hydrogen atom by the ket vector $|1\rangle$ in the Hilbert space. For $n = 1$ we have $i = 0 = k$. Therefore,

$$\vec{I}|1\rangle = 0, \tag{5.86a}$$

and also $\vec{K}|1\rangle = 0.$ (5.86b)

Hence, $\vec{\ell}|1\rangle = \left(\vec{I} + \vec{K}\right)|1\rangle = 0,$ (5.87a)

and likewise $\vec{A}'|1\rangle = \left(\vec{I} - \vec{K}\right)|1\rangle = 0.$ (5.87b)

We therefore conclude that

$$\vec{A}|1\rangle = \left[\frac{1}{2\mu}\left(\vec{p} \times \vec{\ell} - \vec{\ell} \times \vec{p}\right) - \kappa\hat{\rho}\right]|1\rangle = 0. \tag{5.88}$$

Now, the ith component of $\left(\vec{\ell} \times \vec{p}\right)$ is

$$\left(\vec{\ell} \times \vec{p}\right)_i = \sum_{j=1}^{3}\sum_{k=1}^{3} \in_{ijk} \ell_j p_k \equiv \in_{ijk} \ell_j p_k = \in_{ijk} \in_{jlm} \mathrm{r}_l \left(\mathrm{p}_m p_k\right) = \left(- \in_{jik}\right) \in_{jlm} \left(r_l p_k\right) p_m, \tag{5.89a}$$

$$\text{i.e., } \left(\vec{\ell} \times \vec{p}\right)_i = -\in_{jik}\in_{jlm}\left(i\hbar\delta_{lk} + p_k r_l\right)p_m, \tag{5.89b}$$

on account of commutation property of position and momentum operators.

Using the equivalent form of the Pauli–Lenz vector operator

$$\vec{A} = \frac{1}{\mu}\left[\left(\vec{p} \times \vec{\ell}\right) - i\hbar\vec{p}\right] - \kappa\hat{r}, \tag{5.90}$$

it follows from Eq. 5.88 that

$$\left(\vec{p} \times \vec{\ell} - i\hbar\vec{p} - \mu\kappa\hat{r}\right)|1\rangle = 0, \tag{5.91a}$$

i.e., using Eq. 5.87a,

$$\left(i\hbar\vec{p} + \mu\kappa\hat{r}\right)|1\rangle = 0. \tag{5.91b}$$

Eq. 5.91b is a *first*-order *quantum differential equation.* It is remarkable that it has *not* been obtained from the Schrödinger equation; rather, it has been obtained using the Pauli–Lenz vector operator. Inbuilt in it are all the consequences of quantum theory although it is independent of the Schrödinger equation. While the Schrödinger equation has the Laplacian operator in it, the Eq. 5.91b contains the gradient operator and admits the solution

$$\psi_{1,0}(\vec{r}) = \frac{1}{\sqrt{\pi a_0^{\,3}}}e^{-r/a_0}, \tag{5.92}$$

which is the familiar ground state of the hydrogen atom (Table 5.1). The algebra of the Pauli–Lenz vector operator also provides the excited states [10] following a similar technique (see the Solved Problem P5.6).

In Chapter 1, we argued that Schrödinger's wave mechanics and Heisenberg's matrix mechanics are equivalent expressions of quantum theory. Yet the former does not account for the hydrogen atom's degeneracy while the latter does, when used implicitly in the formulation based on the Pauli–Lenz operator, as we have just seen. Another offshoot of the SO(4) symmetry is the fact that the Schrödinger equation for the hydrogen atom can be solved using the method of separation of variables in more than one coordinate system. We discuss this in the next section.

5.3 Superintegrability and Non-Accidental Solutions in Parabolic Coordinates

We have seen that a study of the SO(4) symmetry of the Coulomb Hamiltonian negates the impression that the degeneracy of the bound-state solutions is an accident. We now proceed to examine a concomitant issue in the context of supersymmetric quantum mechanics (SUSYQM), of which the SO(4) symmetry of the hydrogen atom is a prominent example. Studies in this area started long back, and the works of Fock [11] and Sommerfeld [12] are among the earliest ones. First, we recall from classical mechanics that for a dynamical physical property $O(q,p)$, if for every orbit in the phase space and at *every* pair $\left(t_1,t_2\right)$ of instants of time,

$$O\left(q(t_1),p(t_1)\right) = O\left(q(t_2),p(t_2)\right), \tag{5.93}$$

then O is said to be an *integral of motion*. The dynamical property O is a *constant*. Every constant cannot, however, be considered to be an integral of motion, as it may not satisfy Eq. 5.93 for all points in the phase space. An *Isolating Integral* is an integral of motion which *restricts* the phase space available to a dynamical system. A system with n degrees of freedom would have a $2n$ dimensional phase space $\left(q_1,q_2,..,q_n,p_1,p_2,..,p_n\right)$. However, if the number of isolating integrals is λ, the trajectory of the system in the phase space would be found in the $(2n - \lambda)$ dimensional manifold subspace of the original phase space. In other words, the isolating integrals *isolate* the available manifold accessible for the evolution of the system and for the system's trajectories to wander into. There also are, in some cases which are not of much interest, non-isolating integrals of motion. They satisfy Eq. 5.93 but do not reduce the dimensionality of the trajectory. The classical Hamilton–Jacobi equation is solvable in more than one coordinate system when isolating integrals of motion can be found.

Consider the Hamiltonian for a system having n degrees of freedom given by $H\left(q_1,q_2,...,q_n,p_1,p_2,...p_n\right)$. For a quantum system, the arguments of the Hamiltonian are operators. This system is integrable [**13–15**] if there exist $(n-1)$ algebraically independent linear operators $\left\{X_\alpha;\alpha = 1,2,..,(n-1)\right\}$ which all commute with the Hamiltonian, and also with each other. In other words, the following relations hold:

$$\left[H,X_\alpha\right]_- = 0 \text{ for } \left\{X_\alpha;\alpha = 1,2,..,(n-1)\right\}, \tag{5.94a}$$

$$\text{and } \left[X_\beta,X_\alpha\right]_- = 0\left\{\forall\ X_\alpha,X_\beta\ ;\ \alpha,\beta = 1,2,..,(n-1)\right\}. \tag{5.94b}$$

In the case of the hydrogen atom, we have $n = 3$ and $\left\{X_1,X_2\right\} \leftrightarrow \left\{\ell^2,\ell_z\right\}$. In some cases, there may exist *another* set of operators $\left\{Y_\gamma;\gamma = 1,2,..,k\right\}$, consisting of at least one, and a maximum of $(n-1)$ operators,

$$\text{i.e.,} 1 \le k \le (n-1), \tag{5.95}$$

such that they *also* commute with the Hamiltonian,

$$\text{i.e.,} \left[H,Y_\gamma\right]_- = 0 \text{ for} \left\{Y_\gamma;\gamma = 1,2,..,k\right\}, \tag{5.96}$$

and the set $\left\{H,\left\{X_\alpha;\alpha = 1,2,..,(n-1)\right\},\left\{Y_\gamma;\gamma = 1,2,..,k\right\}\right\}$ is algebraically linearly independent. The operators of the set $\left\{Y_\gamma;\gamma = 1,2,..,k\right\}$ may or may not commute with those in the set $\left\{X_\alpha;\alpha = 1,2,..,(n-1)\right\}$, nor with other operators within the same set $\left\{Y_\gamma;\gamma = 1,2,..,k\right\}$.

In this case, the quantum dynamical system is said to be *superintegrable*. It is called *minimally superintegrable* if $k = 1$ and *maximally superintegrable* if $k = (n-1)$. According to Bertrand's theorem, there exist (i) four or (ii) five *isolating* integrals in three dimensions. In the case of *four* isolating integrals, the trajectories are restricted to a two-dimensional surface. In the case of *five* isolating integrals, all finite trajectories are *closed*. This is the case to which the harmonic oscillator potential and the Newton–Kepler potential both belong. These potentials have *five isolating* integrals and are *maximally superintegrable*. The plane in which the Kepler orbit resides is fixed by the constancy of the orbital angular momentum. The constancy of its major axis is effected by the isolating integral resulting from the extra (dynamical) symmetry of the potential, which prevents the apogee and the perigee to advance on completion of each orbit.

There are three different classes of superintegrable systems [15]:

(i) Superseparable systems: these admit separability in at least two different coordinate systems. The constants of motion are linear or quadratic in the momenta.
(ii) Separable systems: solvable in only one coordinate system.
(iii) Nonseparable systems: Very few systems of this class are known. We only make a mention of this, and refer the reader to Reference [16] for details.

Sommerfeld [11] noted that when the dynamical system has additional *isolating* functionally independent integrals, the equation of motion is separable in *more than one* coordinate system. This is the case with dynamical systems that are *superintegrable*. The symmetry group of such systems includes a dynamical symmetry, such as the SO(4) symmetry of the hydrogen atom. *There is then an associated degeneracy of the energy levels*. For the hydrogen atom, it is the degeneracy with respect to the quantum number $\ell = 0,1,2,\ldots,(n-1)$. The hydrogen atom Hamiltonian commutes with the Pauli–Lenz operator within the subspace of the bound states $(E < 0)$. The three-dimensional isotropic linear harmonic oscillator (see the Solved Problem P5.11) is another example of such a system. Further details will be found, for example, in References [16, 17]. The underlying degeneracy of the superintegrable potentials compels classical orbits to be closed. An extraordinary property of both the Kepler and the harmonic oscillator potential is the fact that the orbits are not only closed, but they are *also* non-precessing. This manifests as constancy of the perihelion vectors for these orbits. For trajectories having the form of closed orbits with *one pair* of turning points, the perihelion vector is time independent, and therefore a true integral of the motion. The symmetry of the hydrogen atom is SO(4) and the Pauli–Lenz vector is an additional operator that commutes with the Hamiltonian within the subspace of the bound states. In the case of the three-dimensional isotropic harmonic oscillator, it is a second rank tensor, known as the *Fradkin tensor* which is a constant of motion.

In Section 5.1, we employed the spherical polar coordinate system to separate the functions of the variables (r,θ,φ). The functions of (θ,φ) are the spherical harmonics which we have studied in the previous chapter, and discussed in Appendix C. This part of the solution is the same for every central-field potential, no matter how the potential depends on the distance.

In analogy with the solvability of the classical Hamilton-Jacobi equation in an additional coordinate system, the Schrödinger equation for superintegrable systems can also be solved in more than one coordinate system. In Section 5.1 the solution to the Schrödinger equation for the hydrogen atom using separation of variables in the spherical polar coordinate system has been discussed. The solutions using the parabolic coordinate system is now presented below. Transformation relations between the basis set of unit vectors $\left\{\hat{e}_u,\hat{e}_v,\hat{e}_\varphi\right\}$ of the parabolic coordinates system and the basis set $\left\{\hat{e}_x,\hat{e}_y,\hat{e}_z\right\}$ of the Cartesian coordinate system can be easily obtained. We have used the notation from Chapter 2 of Reference [5], in which figures and expressions for the surfaces of constant u,v,φ are also presented. The three degrees of freedom in space are described by the parameters (u, v, φ) in the parabolic coordinates. These are related to the Cartesian coordinates by

$$x = \sqrt{uv}\cos\varphi,\ y = \sqrt{uv}\sin\varphi,\ z = \frac{1}{2}(u - v). \tag{5.97a}$$

The inverse relations are

$$u = r + z,\ v = r - z, \tag{5.97b}$$

φ being the usual azimuthal angle that is common to the cylindrical polar and also the spherical polar coordinate system. In this system of coordinates, the Coulomb potential is

$$V(r) = V(u + v) = -\frac{2\kappa}{u + v}, \tag{5.98a}$$

$$\text{with } \kappa = Ze^2. \tag{5.98b}$$

The time-independent Schrödinger equation for the hydrogen atom in the parabolic coordinates (*see the Solved Problem P5.7*) therefore takes the form [21]

$$-\frac{\hbar^2}{2\mu}\left\{\begin{array}{l}\left(\frac{4}{(u+v)}\right)\left[\frac{\partial}{\partial u}\left(u\frac{\partial}{\partial u}\right)+\frac{\partial}{\partial v}\left(v\frac{\partial}{\partial v}\right)\right]\psi(u,v,\varphi)+\\ \frac{1}{uv}\frac{\partial^2}{\partial\varphi^2}\psi(u,v,\varphi)\end{array}\right\}-\frac{2\kappa}{u+v}\psi(u,v,\varphi)=E\psi(u,v,\varphi). \tag{5.99}$$

Proposing the method of separation of variables, we write

$$\psi(u, v, \varphi) = f_1(u) f_2(v) \Phi(\varphi). \tag{5.100}$$

Dividing now Eq. 5.99 by $f_1(u)f_2(v)\Phi(\varphi)$ and subsequently multiplying by uv, we get

$$-\frac{\hbar^2}{2\mu}\left\{\left(\frac{4uv}{(u+v)}\right)\left[\frac{1}{f_1}\frac{\partial}{\partial u}\left(u\frac{\partial f_1}{\partial u}\right)+\frac{1}{f_2}\frac{\partial}{\partial v}\left(v\frac{\partial f_2}{\partial v}\right)\right]+\frac{1}{\Phi}\frac{\partial^2\Phi}{\partial\varphi^2}\right\}+\left(\frac{-2\kappa}{u+v}-E\right)uv=0. \tag{5.101}$$

Equation 5.101 has the form

$$F(u, v) + \Phi(\varphi) = 0, \tag{5.102}$$

from which we conclude that the functions $F(u, v)$ and $\Phi(\varphi)$ are independent of each other. It is now easy to see that we obtain essentially the same differential equation for $\Phi(\varphi)$ as in Appendix C. Accordingly,

$$\frac{1}{\Phi}\frac{\partial^2\Phi(\varphi)}{\partial\varphi^2} = -m^2, \tag{5.103}$$

where m is the magnetic quantum number.

The other differential equation to be solved is that for

$$F(u, v) = f_1(u) f_2(v) \tag{5.104a}$$

given by

$$\frac{4uv}{(u+v)}\left[\frac{1}{f_1}\frac{\partial}{\partial u}\left(u\frac{\partial f_1}{\partial u}\right)+\frac{1}{f_2}\frac{\partial}{\partial v}\left(v\frac{\partial f_2}{\partial v}\right)\right]+\frac{2\mu}{\hbar^2}\left[\frac{2\kappa}{u+v}+E\right]uv=m^2. \tag{5.104b}$$

Multiplying this equation by $\frac{u+v}{4uv}$, we get

$$\left[\frac{1}{f}\frac{\partial}{\partial u}\left(u\frac{\partial f_1}{\partial u}\right)+\frac{\mu E}{\hbar^2}\frac{u}{2}-\frac{m^2}{4}\frac{1}{u}\right]+\left[\frac{1}{f_2}\frac{\partial}{\partial v}\left(v\frac{\partial f_2}{\partial v}\right)+\frac{\mu E}{\hbar^2}\frac{v}{2}-\frac{m^2}{4}\frac{1}{v}\right]+\frac{\mu\kappa}{\hbar^2}=0. \tag{5.104c}$$

It is useful now to introduce new constants δ_1 and δ_2, such that

$$\frac{\mu\kappa}{\hbar^2}=\delta_1+\delta_2, \tag{5.105}$$

to separate the differential equations for u and v:

$$\frac{1}{f_1}\frac{d}{du}\left(u\frac{df_1}{du}\right)+\frac{\mu E}{\hbar^2}\frac{u}{2}-\frac{m^2}{4}\frac{1}{u}+\delta_1=0, \tag{5.106a}$$

and

$$\frac{1}{f_2}\frac{d}{dv}\left(v\frac{df_2}{dv}\right)+\frac{\mu E}{\hbar^2}\frac{v}{2}-\frac{m^2}{4}\frac{1}{v}+\delta_2=0. \tag{5.106b}$$

The above two differential equations are completely identical except for the two constants δ_1 and δ_2. They are solvable to yield physically acceptable solutions. We shall now restrict ourselves to the n^2-degenerate *bound-state* $(E<0)$ solutions and introduce a parameter $\beta=\left(\frac{-2\mu E}{\hbar^2}\right)^{1/2}$. Furthermore, we scale the variables u and v by using two new variables $\rho_1=\beta u$ and $\rho_2=\beta v$ to obtain the following differential equations for $f_1(\rho_1)$ and $f_2(\rho_2)$:

$$\frac{d^2f_1}{d\rho_1^2}+\frac{1}{\rho_1}\frac{df_1}{d\rho_1}+\left(\frac{-1}{4}+\frac{-m^2}{4\rho_1^2}+\frac{\lambda_1}{\rho_1}\right)f_1(\rho_1)=0, \tag{5.107a}$$

and

$$\frac{d^2f_2}{d\rho_2^2}+\frac{1}{\rho_2}\frac{df_2}{d\rho_2}+\left(\frac{-1}{4}+\frac{-m^2}{4\rho_2^2}+\frac{\lambda_2}{\rho_2}\right)f_2(\rho_2)=0, \tag{5.107b}$$

where $\lambda_1=\frac{\delta_1}{\beta}$ and $\lambda_2=\frac{\delta_2}{\beta}$.

Since we are dealing with bound-state solutions, $f_1(\rho_1)\to 0$ in the asymptotic region at the rate $e^{-\rho_1/2}$ and likewise $f_2(\rho_2)\to 0$ as $e^{-\rho_2/2}$. Close to the origin, for small values of ρ_1 and ρ_2, the functions behave as $\rho_1^{|m|/2}$ and $\rho_2^{|m|/2}$, respectively. Therefore, the solutions to Eqs. 5.107a and 5.107b are

$$f_1(\rho_1)=e^{-\rho_1/2}\rho_1^{|m|/2}\zeta_1(\rho_1), \tag{5.108a}$$

$$f_2(\rho_2) = e^{-\rho_2/2}\rho_2^{|m|/2}\zeta_2(\rho_2), \tag{5.108b}$$

where $\zeta_1(\rho_1)$ and $\zeta_2(\rho_2)$ must satisfy the differential equation

$$\left[\rho_1\frac{d^2}{d\rho_1^{\ 2}} + (|m|+1-\rho_1)\frac{d}{d\rho_1} + \left(\lambda_1 - \frac{1}{2}(|m|+1)\right)\right]\zeta_1(\rho_1) = 0, \tag{5.109a}$$

and $$\left[\rho_2\frac{d^2}{d\rho_2^{\ 2}} + (|m|+1-\rho_2)\frac{d}{d\rho_2} + \left(\lambda_2 - \frac{1}{2}(|m|+1)\right)\right]\zeta_2(\rho_2) = 0. \tag{5.109b}$$

Equations 5.109a and 5.109b are known as the Kummer–Laplace differential equation. Their solutions are

$$\zeta_1(\rho_1) = CL^{|m|}_{n_1+|m|}(\rho_1) = \tilde{C}_1F_1(-n_1,|m|+1,\rho_1) \tag{5.110a}$$

and $$\zeta_2(\rho_2) = CL^{|m|}_{n_2+|m|}(\rho_2) = \tilde{C}_2F_2(-n_2,|m|+1,\rho_2), \tag{5.110b}$$

where C *and* $\tilde{C}$ are constants, and the numbers $n_1 = \lambda_1 - \frac{1}{2}(|m|+1)$ and $n_2 = \lambda_2 - \frac{1}{2}(|m|+1)$ are positive integers, or zero.

Note that

$$\lambda_1 + \lambda_2 = n_1 + n_2 + |m| + 1 = n \tag{5.111}$$

is a nonzero integer. The wavefunctions of the hydrogen atom in parabolic coordinates therefore are specified by three '*good*' quantum numbers (n_1,n_2,m) similar to, *but also different from*, the quantum numbers (n,ℓ,m) in the spherical polar system. Only the magnetic quantum number is the same. The wavefunctions are given by

$$\psi_{n_1n_2m}(u,v,\varphi) = f_{1n_1|m|}(\rho_1)f_{2n_2|m|}(\rho_2)\frac{e^{im\varphi}}{\sqrt{2\pi}}, \tag{5.112a}$$

i.e., $$\psi_{n_1n_2m}(u,v,\varphi) = \left[\begin{array}{c}\dfrac{\sqrt{2}}{n^2}\left(\dfrac{Z}{a_\mu}\right)^{3/2}\left\{\dfrac{n_1!n_2!}{\left[(n_1+|m|)!(n_2+|m|)!\right]^3}\right\}^{1/2} \times \\ e^{-(\rho_1+\rho_2)/2}(\rho_1\rho_2)^{|m|/2}L^{|m|}_{n_1+|m|}(\rho_1)L^{|m|}_{n_2+|m|}(\rho_2)\dfrac{e^{im\varphi}}{\sqrt{2\pi}}\end{array}\right], \tag{5.112b}$$

with $\rho_1 = \beta u,\ \rho_2 = \beta v,\ n = n_1 + n_2 + |m| + 1.$ (5.113)

Eq. 5.112 provides the normalized eigenfunctions of the bound states of one-electron atoms (or ions) in parabolic coordinates. It is a primary result of the fact that the Coulomb potential has an additional symmetry, namely the dynamical symmetry that is (a) responsible for its superintegrability, and (b) makes its equation solvable in an additional coordinate system using the method of separation of variables.

Since $n = \lambda_1 + \lambda_2 = \dfrac{\delta_1 + \delta_2}{\beta} = \dfrac{\mu\kappa}{\hbar^2\beta}$, (5.114)

we get $\beta^2 = \dfrac{\mu^2\kappa^2}{\hbar^4 n^2} = \dfrac{\mu^2 Z^2 e^4}{\hbar^4 n^2}$ (5.115)

and $E = E_n = -\dfrac{\mu Z^2 e^4}{2\hbar^2 n^2}$. (5.116)

We thus recover the familiar discrete bound-state energy eigenvalues. The energy eigenvalue depends only on n, but there are different ways of combining n_1, n_2, and m to add up (Eq. 5.113) to get n. These different ways correspond to the degeneracy of the bound states. There are n ways to choose n_1 and n_2 when $m = 0$. For $|m| > 0$, the number of ways of choosing m are two, viz. $(\text{viz. } \pm|m|)$, and $n - |m|$ ways of choosing n_1 and n_2. Thus, the total degeneracy with energy corresponding to a particular value of n, is

$$n + 2\sum_{|m|=1}^{n-1}\left(n - |m|\right) = n + 2\left[n(n-1) - \frac{n(n-1)}{2}\right] = n^2, \tag{5.117}$$

which is exactly the $n^2 - \textit{fold}$ degeneracy that we had discussed in Sections 5.1 and 5.2. We have already discussed the pseudo-angular momentum vector operators $\vec{I}$ and $\vec{K}$ in Section 5.2. Corresponding to these operators, we now introduce ladder operators which we have used in Chapter 4:

$$I_\pm = I_x \pm iI_y, \tag{5.118a}$$

and

$$K_\pm = K_x \pm iK_y. \tag{5.118b}$$

The operators $\vec{I}$ and $\vec{K}$ satisfy the complete algebra of angular momentum operators and operate on the *isolating eigenspace* of H corresponding to the discrete bound $(E < 0)$ states. In this subspace, the physical states of the quantum system are simultaneous eigenvectors of (H, I_z, K_z). The *good* quantum numbers are (n, m_{Iz}, m_{Kz}), and the eigenstates are labeled as $|n, m_{Iz}, m_{Kz}\rangle$. Using the results of the algebra (Chapter 4) for the angular momentum ladder operators, we have

$$I_\pm|n, m_{Iz}, m_{Kz}\rangle = \hbar \times \sqrt{i(i+1) - m_{Iz}(m_{Iz} \pm 1)}\,|n, m_{Iz} \pm 1, m_{Kz}\rangle, \tag{5.119a}$$

and

$$K_\pm|n, m_{Iz}, m_{Kz}\rangle = \hbar \times \sqrt{k(k+1) - m_{Kz}(m_{Kz} \pm 1)}\,|n, m_{Iz}, m_{Kz} \pm 1\rangle. \tag{5.119b}$$

We know from the previous chapter that the ladder step-up operator cannot raise the corresponding magnetic quantum number any further if it already has its maximum value. Therefore,

$$I_+|n, n, n\rangle = 0 \text{ and } K_+|n, n, n\rangle = 0. \tag{5.120}$$

Using the chain rule to obtain partial derivative operators in the parabolic coordinate system from either the Cartesian or the spherical polar coordinate system, and using dimensionless variables

$$\xi = \frac{p_s}{\hbar} u \tag{5.121a}$$

and

$$\eta = \frac{p_s}{\hbar} v, \tag{5.121b}$$

the ladder operators are given by

$$I_+ = \hbar e^{i\varphi} \left(\sqrt{\xi} \frac{\partial}{\partial \xi} + \frac{i}{2\sqrt{\xi}} \frac{\partial}{\partial \varphi} + \frac{\sqrt{\xi}}{2} \right) \times \left(\sqrt{\eta} \frac{\partial}{\partial \eta} + \frac{i}{2\sqrt{\eta}} \frac{\partial}{\partial \varphi} - \frac{\sqrt{\eta}}{2} \right), \tag{5.122a}$$

$$I_- = \hbar e^{-i\varphi} \left(\sqrt{\xi} \frac{\partial}{\partial \xi} - \frac{i}{2\sqrt{\xi}} \frac{\partial}{\partial \varphi} - \frac{\sqrt{\xi}}{2} \right) \times \left(\sqrt{\eta} \frac{\partial}{\partial \eta} - \frac{i}{2\sqrt{\eta}} \frac{\partial}{\partial \varphi} + \frac{\sqrt{\eta}}{2} \right), \tag{5.122b}$$

$$K_+ = -\hbar e^{i\varphi} \left(\sqrt{\xi} \frac{\partial}{\partial \xi} + \frac{i}{2\sqrt{\xi}} \frac{\partial}{\partial \varphi} - \frac{\sqrt{\xi}}{2} \right) \times \left(\sqrt{\eta} \frac{\partial}{\partial \eta} + \frac{i}{2\sqrt{\eta}} \frac{\partial}{\partial \varphi} + \frac{\sqrt{\eta}}{2} \right), \tag{5.123a}$$

and

$$K_- = -\hbar e^{-i\varphi} \left(\sqrt{\xi} \frac{\partial}{\partial \xi} - \frac{i}{2\sqrt{\xi}} \frac{\partial}{\partial \varphi} + \frac{\sqrt{\xi}}{2} \right) \times \left(\sqrt{\eta} \frac{\partial}{\partial \eta} - \frac{i}{2\sqrt{\eta}} \frac{\partial}{\partial \varphi} - \frac{\sqrt{\eta}}{2} \right). \tag{5.123b}$$

Since $\ell_z = I_z + K_z$, the wavefunction $\psi_{n,n,n}$ corresponding to the highest value of m_{Iz} and also the highest value of m_{Kz} is represented by the "good" quantum numbers $|n,n,n\rangle$, which must have the form $F(\xi,\eta)e^{i2n\varphi}$. In other words,

$$\psi_{n,n,n} = N e^{-(\xi+\eta)/2} (\xi\eta)^n e^{i2n\varphi}, \tag{5.124}$$

where N is the normalization constant. It is determined by using the usual considerations of square integrability of the bound-state functions. It is found to be

$$|N| = \frac{(p_s \hbar)^{3/2}}{(2j)! \sqrt{(2j+1)\pi}}. \tag{5.125}$$

Other wave functions can be obtained using the ladder operators, and the following two identities:

$$\left(\sqrt{\xi} \frac{\partial}{\partial \xi} - \frac{i}{2\sqrt{\xi}} \frac{\partial}{\partial \varphi} + \frac{\sqrt{\xi}}{2} \right) \psi = \xi^{(1-m)/2} e^{\xi/2} \frac{\partial}{\partial \xi} \left(\xi^{m/2} e^{-\xi/2} \psi \right), \tag{5.126}$$

$$(I_-)^n \psi = \hbar^n e^{-in\varphi} \xi^{(n-m)/2} e^{\xi/2} \times \frac{\partial^n}{\partial \xi^n} \xi^{m/2} e^{-\xi/2} \eta^{(n-m)/2} e^{-\eta/2} \frac{\partial^n}{\partial \eta^n} \eta^{m/2} e^{\eta/2} \psi. \tag{5.127}$$

For further details, we refer the reader to References [18–20]. The basis set $\{|n, m_{Iz}, m_{Kz}\rangle\}$ corresponding to the solutions in the parabolic coordinate system and $\{|n,\ell,m\rangle\}$ corresponding to the spherical polar coordinate system are both complete; one may use either of these to describe a general state of Coulombic atomic systems. Each base vector in one system can therefore be obviously expressed as a superposition of base vectors of the other set. In particular,

$$|n,\ell,m\rangle = \sum_{m_{Iz}=-i}^{+i}\sum_{m_{Kz}=-k}^{+k} |m_{Iz}, m_{Kz}\rangle\langle m_{Iz}, m_{Kz}|n,\ell,m\rangle \tag{5.128a}$$

and

$$|n, m_{Iz}, m_{Kz}\rangle = \sum_{\ell=0}^{\infty}\sum_{m=-\ell}^{+\ell} |n,\ell,m\rangle\langle n,\ell,m|n, m_{Iz}, m_{Kz}\rangle. \tag{5.128b}$$

We immediately recognize the expansion coefficients as the Clebsch–Gordan coefficients that we have discussed in the previous chapter. The relations Eqs. 5.128a,b are known as *interbasis expansions*.

In the previous three sections, we have discussed the solutions for bound states $E < 0$. In this case, the energy eigenvalues turn out to be *discrete* on account of the boundedness of the wavefunctions which must vanish in the asymptotic region. The discreteness of the energy spectrum results from the *boundary conditions* in the asymptotic region, and not from the quantum mechanical equation of motion alone. Solutions of the very same Schrödinger equation with the very same Hamiltonian for $E > 0$ are unbound, they do not vanish asymptotically. These solutions are not discrete. On the contrary, they are continuous, discussed in the next section.

5.4 Continuum Eigenfunctions of the Coulomb Potential

In the previous three sections of this chapter, we studied the bound states $(E < 0)$ of the nonrelativistic hydrogen atom. In the present section we study the complementary part $(E > 0)$ of its quantum state space. The boundary condition in the asymptotic region now is different. The wavefunction for a positive energy state is unbound; it does not vanish asymptotically. We begin this analysis with the Schrödinger equation for a pair of charged particles Z_1e and Z_2e in an attractive Coulomb interaction of which the hydrogen atom is the classic example:

$$\left[\frac{(-i\hbar\vec{\nabla})^2}{2\mu} - \frac{(Z_1e)(Z_2e)}{r}\right]\psi(\vec{r}) = E\psi(\vec{r}), \tag{5.129}$$

$$\text{where } E = \frac{\hbar^2k^2}{2\mu} > 0. \tag{5.130}$$

$$\left[\frac{(-i\hbar\vec{\nabla})^2}{2\mu} - \frac{(Z_1Z_2e^2)}{r} - E\right]\psi(\vec{r}) = 0, \tag{5.131a}$$

$$\text{i.e.,} \left[\vec{\nabla}^2 + \frac{2(\mu Z_1 Z_2) k}{\hbar k r} \frac{e^2}{\hbar} + k^2 \right] \psi(\vec{r}) = 0, \tag{5.131b}$$

$$\text{or} \left[\vec{\nabla}^2 + k^2 + \frac{2\gamma k}{r} \right] \psi(\vec{r}) = 0, \tag{5.131c}$$

$$\text{where } \gamma = \frac{Z_1 Z_2 e^2 \mu}{\hbar^2 k} = \frac{Z_1 Z_2 e^2 \mu}{\hbar p}. \tag{5.132}$$

Note that the Coulomb interaction between the nucleus and the electron being attractive, $\gamma > 0$ represents the attractive interaction. We shall use parabolic coordinates (u, v, φ) as we did in Section 5.3 (Eq. 5.99). Following the *methods* in References [21, 22], we shall, however, represent the parabolic coordinate system by (z, ξ, φ), since

$$z = \frac{u - v}{2}, \tag{5.133a}$$

where z denotes the Cartesian degree of freedom along the z-axis,

$$\text{and } \xi = ikv = ik(r - z) = ikr(1 - \cos\theta) = ikr\left(2\sin^2\frac{\theta}{2}\right). \tag{5.133b}$$

Note that ik being a constant, the variable $\xi \equiv v$ (equivalent). In the coordinate system we have used earlier, a point in R^3 is completely specified by $(x, y, z) \equiv (u, v, \varphi) \equiv (u, \xi, \varphi)$. Furthermore, on the surface $(v, \varphi) \equiv (\xi, \varphi)$, the z coordinate completely specifies a point in space. The coordinates (z, ξ, φ) therefore *uniquely* specify the three degrees of freedom in the Euclidean space R^3. The advantage in choosing this coordinate system lies in the fact that the z-axis may be considered to be passing through the notional positions of the two interacting charged particles and recognize that the Coulomb interaction is symmetric with respect to the azimuthal angle φ. We shall therefore be able to seek a solution of the Schrödinger equation in a form that is independent of the azimuthal angle φ:

$$\psi(\vec{r}) = \psi(z, \xi) \neq \textit{function}(\varphi). \tag{5.134}$$

We now proceed to write the Laplacian operator in the Schrödinger equation (Eqs. 5.131a,b,c) using the coordinates (z, ξ, φ). We therefore seek to express the Cartesian partial derivative operators $\frac{\partial^2}{\partial x^2}, \frac{\partial^2}{\partial y^2}$, and $\frac{\partial^2}{\partial z^2}$ in the Laplacian in terms of the partial derivatives with respect to the parabolic coordinates $(z_p, \xi_p, \varphi_p) \equiv (z_p, v_p, \varphi_p)$, but we must remember that the z-coordinate z_p in the parabolic coordinate system involves $\xi_p \equiv v_p$, on account of Eqs. 5.133a,b. We shall use a slightly modified notation to interpret Eq. 5.133a as

$$z_p = \frac{u - v_c}{2}, \tag{5.135}$$

in which v_c is regarded as a constant. The *second* degree of freedom in our parabolic coordinate system is given by Eq. 5.133b and it cannot be treated as a *variable* in the description of the *first* degree of freedom, viz. z_p.

For clarity, let us use the subscript 'c' for Cartesian and 'p' for parabolic, since the 'z' coordinate of the parabolic system must be used with additional care. Instead of the partial derivative operators $\left(\frac{\partial}{\partial x_c}, \frac{\partial}{\partial y_c}, \frac{\partial}{\partial z_c}\right)$ that appear in the Laplacian $\left(\frac{\partial^2}{\partial x_c^{\,2}} + \frac{\partial^2}{\partial y_c^{\,2}} + \frac{\partial^2}{\partial z_c^{\,2}}\right)$, we must now employ the operators $\left(\frac{\partial}{\partial z_p}, \frac{\partial}{\partial \xi_p}, \frac{\partial}{\partial \varphi_p}\right)$, which are equivalent to $\left(\frac{\partial}{\partial z_p}, \frac{\partial}{\partial v_p}, \frac{\partial}{\partial \varphi_p}\right)$, in the description of the Laplacian.

Using the chain-rule of the partial derivative operators, we have

$$\frac{\partial}{\partial x_c} = \left(\frac{\partial z_p}{\partial x_c}\frac{\partial}{\partial z_p} + \frac{\partial v_p}{\partial x_c}\frac{\partial}{\partial v_p} + \frac{\partial \varphi_p}{\partial x_c}\frac{\partial}{\partial \varphi_p}\right) \equiv \frac{1}{2}\frac{x}{r}\frac{\partial}{\partial z_p} + \frac{x}{r}\frac{\partial}{\partial v} - \frac{\sin\varphi}{\sqrt{uv}}\frac{\partial}{\partial \varphi}, \tag{5.135a}$$

$$\frac{\partial}{\partial y_c} = \left(\frac{\partial z_p}{\partial y_c}\frac{\partial}{\partial z_p} + \frac{\partial v_p}{\partial y_c}\frac{\partial}{\partial v_p} + \frac{\partial \varphi_p}{\partial y_c}\frac{\partial}{\partial \varphi_p}\right) = \frac{1}{2}\frac{y}{r}\frac{\partial}{\partial z_p} + \frac{y}{r}\frac{\partial}{\partial v} + \frac{\cos\varphi}{\sqrt{uv}}\frac{\partial}{\partial \varphi}, \tag{5.135b}$$

$$\left(\frac{\partial}{\partial z}\right)_c = \left(\frac{\partial z_p}{\partial z}\frac{\partial}{\partial z_p} + \frac{\partial v_p}{\partial z_c}\frac{\partial}{\partial v_p} + \frac{\partial \varphi_p}{\partial z_c}\frac{\partial}{\partial \varphi_p}\right) = \frac{1}{2}\left(\frac{z}{r} + 1\right)\frac{\partial}{\partial z_p} + \left(\frac{z}{r} - 1\right)\frac{\partial}{\partial v}, \tag{5.135c}$$

In arriving at Eqs. 5.135a,b,c, we have recognized that

$$\frac{\partial v}{\partial x} = \frac{\partial(r - z)}{\partial x} = \frac{\partial r}{\partial x} = \frac{\partial\left(x^2 + y^2 + z^2\right)^{\frac{1}{2}}}{\partial x} = \frac{1}{2}\left(r^2\right)^{-\frac{1}{2}}(2x) = \frac{x}{r}, \tag{5.136a}$$

$$\text{and likewise } \left(\frac{\partial v}{\partial y} = \frac{y}{r}\right) \;\&\; \left(\frac{\partial v}{\partial z} = \frac{z}{r} - 1\right). \tag{5.136b}$$

The Laplacian in the coordinate system (z, v, φ) therefore is:

$$\vec{\nabla}\bullet\vec{\nabla} = \nabla^2 = \left[\begin{array}{l} \left(\frac{1}{2}\frac{x}{r}\frac{\partial}{\partial z_p} + \frac{x}{r}\frac{\partial}{\partial v} - \frac{\sin\varphi}{\sqrt{uv}}\frac{\partial}{\partial \varphi}\right)^2 \\ + \left(\frac{1}{2}\frac{y}{r}\frac{\partial}{\partial z_p} + \frac{y}{r}\frac{\partial}{\partial v} + \frac{\cos\varphi}{\sqrt{uv}}\frac{\partial}{\partial \varphi}\right)^2 + \left(\frac{1}{2}\left(\frac{z}{r} + 1\right)\frac{\partial}{\partial z_p} + \left(\frac{z}{r} - 1\right)\frac{\partial}{\partial v}\right)^2 \end{array}\right], \tag{5.137a}$$

or,

$$\nabla^2 = \frac{1}{\left(z_p + v\right)}\left\{\frac{\partial}{\partial z_p}\left[\frac{\left(2z_p + v\right)}{2}\frac{\partial}{\partial z_p}\right] + \frac{\partial}{\partial v}\left(2v\frac{\partial}{\partial v}\right) + \frac{\partial}{\partial \varphi}\left(\frac{\left(z_p + v\right)}{\left(2z_p + v_c\right)v}\frac{\partial}{\partial \varphi}\right)\right\}, \tag{5.137b}$$

which is equivalent to the Laplacian employed in Eq. 5.99 (see the Solved Problems P5.8 and P5.9).

As discussed earlier, the dynamical symmetry of the Schrödinger equation makes it separable in the parabolic coordinate system and we therefore seek its solution in the form

$$\psi(\vec{r}) = \psi(z, \xi) = e^{ikz}\chi(\xi). \tag{5.138}$$

Separability of the differential equation leads us to conclude that the function $\chi(\xi)$ must be a solution of the following differential equation:

$$\left[\xi\frac{d^2}{d\xi^2}+(1-\xi)\frac{d}{d\xi}-i\gamma\right]\chi(\xi)=0. \tag{5.139}$$

The general solution to the Schrödinger equation would be a superposition of two linearly independent solutions, one being $\chi_1(\xi)$ (say) and the other being $\chi_2(\xi)$, both of which would satisfy Eq. 5.139.

Let us now explore if for some parameter β, the form

$$\chi_1(\xi)=\xi^{-\beta} \tag{5.140}$$

is a solution of Eq. 5.139. Since the continuum $(E>0)$ solutions are of greatest interest in the asymptotic region $(r\to\infty)$, we shall specifically focus our interest in the limit $\xi\to\infty$ which represents the asymptotic region. However, note that $z=r$ makes $\xi=0$ on account of Eq. 5.97b. Using the form of the solution suggested by Eq. 5.140 in Eq. 5.139, we get

$$\left(\beta^2\xi^{-1}-\beta-i\gamma\right)\xi^{\beta}=0, \tag{5.141a}$$

which in the asymptotic limit gives

$$\left(-\beta-i\gamma\right)\xi^{\beta}\simeq 0, \tag{5.141b}$$

except for the singularity at $\xi=0$.

We therefore conclude that the choice

$$\beta=i\gamma \tag{5.142}$$

provides one solution of the Schrödinger equation:

$$\psi_1(\vec{r})=\exp(ikz)\chi_1(\xi)=\exp\left[i\left\{kz-\gamma\ln\left(k(r-z)\right)\right\}\right]. \tag{5.143}$$

Since

$$\chi_1(\xi)=\xi^{-i\gamma}=\left\{ik(r-z)^{-i\gamma}\right\}\equiv k(r-z)^{-i\gamma}, \tag{5.144a}$$

$$\ln\left(\xi^{-i\gamma}\right)=\ln\left\{k(r-z)^{-i\gamma}\right\}, \tag{5.144b}$$

$$-i\gamma\ln\xi=-i\gamma\ln\left\{k(r-z)\right\}, \tag{5.144c}$$

and hence $\chi_1(\xi)=\exp\left[-i\gamma\ln\left\{k(r-z)\right\}\right]$. (5.145)

We now examine if for some *other* value of the parameter β, we can have the second, linearly independent, solution

$$\psi_2(\vec{r})=\exp(ikz)\chi_2(\xi) \tag{5.146}$$

with $\chi_2(\xi)\simeq\xi^{\beta}e^{\xi}$. (5.147)

Putting Eq. 5.147 in Eq. 5.139, we get

$$\left[\beta^2\xi^{-1} + (\beta + 1 - i\gamma)\right]\xi^{\beta}e^{\xi} = 0. \tag{5.148}$$

Dropping the first term in the asymptotic region, we find that

$$\beta = -1 + i\gamma, \tag{5.149}$$

gives an acceptable solution:

$$\chi_2(\xi) = \frac{1}{\xi}\xi^{i\gamma}e^{\xi} = \frac{1}{ik(r-z)}\{ik(r-z)\}^{i\gamma}\, e^{\{ik(r-z)\}}. \tag{5.150}$$

Recognizing now that

$$\{k(r-z)\}^{i\gamma} = \left[e^{\ln\{k(r-z)\}}\right]^{i\gamma} = e^{i\gamma\ln\{k(r-z)\}}, \tag{5.151}$$

we get

$$\psi_2(\vec{r}) = e^{ikz}\frac{e^{\{ik(r-z)\}}}{k(r-z)}e^{i\gamma\ln\{k(r-z)\}} = \frac{e^{i\{kr+\gamma\ln\{k(r-z)\}\}}}{k(r-z)} = \frac{e^{ikr}}{k(r-z)}e^{i\gamma\ln\{k(r-z)\}}. \tag{5.152}$$

The general solution $\psi(\vec{r})$ to the Coulomb problem is a linear superposition of $\psi_1(\vec{r})$ and $\psi_2(\vec{r})$:

$$\psi(\vec{r}) \to e^{i\left[kz-\gamma\ln\{k(r-z)\}\right]} + f_c(k,\theta)\frac{e^{i\{kr+\gamma\ln\{k(r-z)\}\}}}{k(r-z)}. \tag{5.153}$$

Together with the inclusion of the *dynamical phase* (Eqs. 1.114, 1.115), the space and time dependent solution to the Coulomb problem is

$$\psi(\vec{r},t) \to e^{i\left[kz-\omega t-\gamma\ln\{k(r-z)\}\right]} + f_c(k,\theta)\frac{e^{i\{kr-\omega t+\gamma\ln\{k(r-z)\}\}}}{k(r-z)}. \tag{5.154}$$

The superposition expressed in Eq. 5.154 has a fascinating interpretation. We know that the surface of constant phase $\Phi_{plane} = kz - \omega t$ (which appears in the *first* term of Eq. 5.154) is a *plane wave* moving along the z-axis. Likewise, the phase $\Phi_{circular} = kr - \omega t$ (which appears in the *second* term) is constant on a spherical surface moving outward. The solution we have obtained thus represents scattering of a charged particle in a Coulomb potential. However, we also take note of the fact that there is an extra *phase shift* in both the plane-wave surface and the spherical-wave surfaces indicated by the logarithmic distance-dependent terms in the solutions. The additional logarithmic term is called as the *Coulomb phase shift*. It appears no matter how large the distance and makes it impossible to separate the incident wave from the scattered one. This is because of the fact that the *long-range* Coulomb interaction goes as $\frac{1}{r}$. Furthermore, when $\xi = 0$, i.e., when $\theta = 0$ (points along the polar axis) we have to ensure that our solution is regular along the polar axis. Towards that goal, we express $\chi(\xi)$ using its Laplace transform as an integral in the complex *t* space,

$$\chi(\xi) = \int_{t_1}^{t_2} e^{\xi t} f(t)\,dt, \tag{5.155}$$

and determine an appropriate path of integration in the complex t plane which would avoid the singularity. Using the Laplace transform, the differential equation for $\chi(\xi)$ is

$$\left[\xi\frac{d^2}{d\xi^2}+(1-\xi)\frac{d}{d\xi}-i\gamma\right]\int_{t_1}^{t_2}e^{\xi t}f(t)\,dt=0, \tag{5.156a}$$

$$\text{i.e.,}\ \int_{t_1}^{t_2}f(t)\left[\xi\frac{d^2\left(e^{\xi t}\right)}{d\xi^2}+(1-\xi)\frac{d\left(e^{\xi t}\right)}{d\xi}-i\gamma\left(e^{\xi t}\right)\right]dt=0, \tag{5.156b}$$

$$\text{or,}\ \int_{t_1}^{t_2}f(t)\left[\xi t^2+(1-\xi)t-i\gamma\right]e^{\xi t}dt=0, \tag{5.156c}$$

$$\text{or,}\ \int_{t_1}^{t_2}f(t)\left[(t-i\gamma)+t(t-1)\xi\right]e^{\xi t}dt=0. \tag{5.156d}$$

Therefore,

$$\int_{t_1}^{t_2}f(t)\left[t(t-1)\xi\right]e^{\xi t}dt+\int_{t_1}^{t_2}f(t)(t-i\gamma)e^{\xi t}dt=0. \tag{5.156e}$$

We now integrate the first of these two integrals by parts, by factoring the integrand as a product of $t(t-1)f(t)\xi$ and $e^{\xi t}$. Accordingly, we get

$$\left[f(t)\left[t(t-1)\cancel{\xi}\right]\frac{e^{\xi t}}{\cancel{\xi}}\right]_{t_1}^{t_2}-\int_{t_1}^{t_2}\left(\frac{d}{dt}\left\{f(t)\left[t(t-1)\cancel{\xi}\right]\right\}\right)\frac{e^{\xi t}}{\cancel{\xi}}dt+\int_{t_1}^{t_2}f(t)(t-i\gamma)e^{\xi t}dt=0, \tag{5.157}$$

$$\text{i.e.,}\ \left[t(t-1)f(t)e^{\xi t}\right]_{t_1}^{t_2}+\int_{t_1}^{t_2}\left[(t-i\gamma)f(t)-\frac{d}{dt}\left\{t(t-1)f(t)\right\}\right]e^{\xi t}dt=0. \tag{5.158}$$

Now, the extended complex t plane consists of all points in the t plane and the single point at infinity, which is essentially the same in all directions. If we *choose* a contour for the path of integration $\int_{t_1}^{t_2}dt$ in the complex t plane for the *second* term in Eq. 5.158, requiring that the *first* term $[..]_{t_1}^{t_2}$ (corresponding to the surface terms) vanishes, then setting the factor $\left[(t-i\gamma)f(t)-\frac{d}{dt}\left\{t(t-1)f(t)\right\}\right]$ in the integrand in Eq. 5.158 equal to zero provides us the solution, since Eq. 5.158 is then validated. We shall select the path for contour integration shortly. In anticipation of an appropriate path that would be satisfactory, we therefore examine the condition

$$\left[(t-i\gamma)f(t)-\frac{d}{dt}\left\{t(t-1)f(t)\right\}\right]=0. \tag{5.159}$$

On integrating this relation, we get

$$f(t)=At^{i\gamma-1}(1-t)^{-i\gamma}, \tag{5.160}$$

which gives $$\chi(\xi) = A\int_{t_1}^{t_2} e^{\xi t} t^{i\gamma-1}(1-t)^{-i\gamma}\, dt = A\int_{C} e^{\xi t} t^{i\gamma-1}(1-t)^{-i\gamma}\, dt. \tag{5.161}$$

The path C for the contour integration in the complex t plane must be chosen avoiding the branch points of the integrand $t = 0,1$ (Fig. 5.7a). Now the surface terms would vanish if $e^{\xi t} \to 0$. Since $\xi = ikr(1-\cos\theta)$, for $\cos\theta < 1$ we may write $\xi = i\kappa$ with $\kappa > 0$. This makes $e^{\xi t} = e^{i\kappa t} \to 0$ when the imaginary part of t goes to $+\infty$ $\left(\mathrm{Im}(t) \to +\infty\right)$ and the surface terms would vanish. A point to be taken cognizance of is that to avoid getting multiple-valuedness while choosing the path of integration, one must be mindful of the branch points.

Following the methods in References [4] and [22,23], we execute the integration by introducing a complex s plane (Fig. 5.7b) which is related to the complex t plane by

$$s = \xi t = i\kappa t. \tag{5.162}$$

This gives $e^{\xi t} = e^{s}$ which vanishes as $s(\text{real}) \to -\infty$ and ensures that the surface terms vanish, as required. *The path of integration in the complex s plane must therefore be chosen such that s goes to infinity along the negative real axis.*

The integral in Eq. 5.161 in the complex s plane is

$$\chi(\xi) = A\int_{\bar{C}} e^{s}\left(\frac{s}{\xi}\right)^{i\gamma-1}\left(1-\left(\frac{s}{\xi}\right)\right)^{-i\gamma}\frac{ds}{\xi} = A\int_{\bar{C}} e^{s} s^{i\gamma-1}(\xi-s)^{-i\gamma}\, ds = \chi(s). \tag{5.163}$$

In Appendix 5F, it is shown that the above contour integral for the asymptotic limit $\xi \to \infty$ is

$$\chi(\xi) = 2\pi i A\xi^{-i\gamma} g_1(\gamma)\left\{1 + e^{i\phi(k,\xi,\gamma)}(i\gamma)\frac{e^{\xi}}{\xi}\right\}, \tag{5.164}$$

where

$$e^{i\phi(k,\xi,\gamma)} = e^{2i\gamma \ln k(r-z)} e^{i\Theta(\gamma)}. \tag{5.165}$$

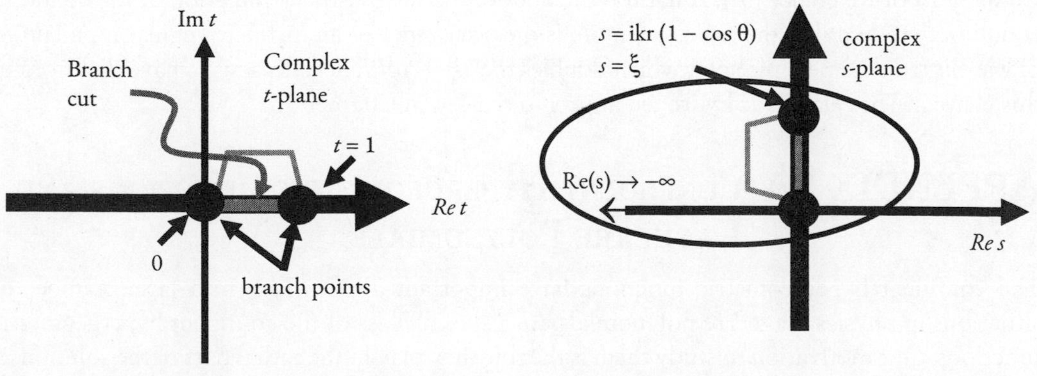

Fig. 5.7 a, b The left panel in this figure shows the branch points in the complex t plane and the right panel shows the same in the complex s plane. The flexed branch cut is also shown.

Using Eq. 5.133b, Eq. 5.164, and Eq. 5.165, the solution (Eq. 5.138) to the Schrödinger equation for the Coulomb problem then becomes

$$\psi(\vec{r}) = 2\pi i A e^{\gamma\frac{\pi}{2}} e^{-i\gamma\ln[k(r-z)]} g_1(\gamma)\left\{1 + e^{i\phi(k,\xi,\gamma)}(i\gamma)\frac{e^{\xi}}{\xi}\right\} e^{ikz}, \tag{5.167a}$$

$$\text{i.e., } \psi(\vec{r}) = 2\pi i A e^{\gamma\frac{\pi}{2}} g_1(\gamma)\left\{e^{i\{kz-\gamma\ln[k(r-z)]\}} + \gamma e^{i\gamma\ln k(r-z)} e^{i\Theta(\gamma)} \frac{e^{i\{kr+\gamma\ln k(r-z)\}}}{k(r-z)}\right\}. \tag{5.167b}$$

Comparing now Eq. 5.167b with Eq. 5.153, we see that

$$f_c(k,\theta) = \frac{\gamma e^{i\gamma\ln k(r-z)} e^{i\Theta(\gamma)}}{k(r-z)} = \frac{\gamma e^{i\gamma\ln k(r-z)} e^{i\Theta(\gamma)}}{k\left(2\sin^2\frac{\theta}{2}\right)}. \tag{5.168}$$

In quantum collision theory, which we shall discuss in Chapter 10 (in particular, see Eq.10.77), the modulus square of Eq. 5.168 is identified as the *differential scattering cross-section*:

$$\frac{d\sigma}{d\theta} = |f_c(k,\theta)|^2 \tag{5.169a}$$

and we see that

$$\frac{d\sigma}{d\theta} = \frac{\gamma^2}{4k^2\sin^4\frac{\theta}{2}}. \tag{5.169b}$$

It is interesting that this result is the same as the classical Coulomb–Rutherford formula (see, for example, appendix 8A of Chapter 8 in Reference [5]). The expression in Eq. 5.169a is well known in quantum collision physics; we shall discuss it in detail in Chapter 10, especially in Sections 10.2 and 10.3. In Section 10.4 we shall find that the result in Eq. 5.169b is borne out also in a commonly used method (known as the Born approximation) in the analysis of quantum collisions. Hydrogenic one-electron atomic systems (including ions) are among the very few quantum objects for which exact analytical solutions can be obtained. The Schrödinger equation is, however, not Lorentz covariant. Finiteness of the speed of light requires us to reformulate the quantum equation of motion in the non-Euclidean space-time continuum (see, for example, Chapter 13 of Reference [5]). A natural offshoot of the relativistic formulation of the quantum equation that describes the hydrogen atom is the occurrence of an intrinsic angular momentum of the electron, namely the spin, which doubles the $n^2 - fold$ degeneracy we have discussed in this chapter. The relativistic hydrogen atom is discussed in Chapter 7.

APPENDIX 5A Confluent Hypergeometric Functions and Laguerre Polynomials

The confluent hypergeometric functions have important applications in a large number of situations in physics. Laguerre polynomials are a special class of the confluent hypergeometric functions. Our motivation to study them is the role they play in the radial part of the solution of the Schrödinger equation for the hydrogen atom. In the context of Eq. 5.34, we shall follow the the spirit of the methods described in References [4,23]. We consider the differential equation

$$z\frac{d^2F}{dz^2}+(\gamma-z)\frac{dF}{dz}-\alpha F=0, \tag{5A.1}$$

for all finite points in the complex z plane. We admit arbitrary α but restrict γ such that $\gamma \neq 0$ and $\gamma \neq$ negative integer. We label the function F as $F(\alpha,\gamma,z)$ and inquire if we can find a function $\zeta(t)$ such that

$$f_1(z)=\int e^{tz}\zeta(t)\,dt \tag{5A.2a}$$

is a solution to Eq. 5A.1. In other words, we seek to determine the conditions on $\zeta(t)$ such that $f_1(z)$ a solution. The first and the second derivatives of Eq. 5A.2a are

$$\frac{df_1(z)}{dz}=\int e^{tz}t\zeta(t)\,dt, \tag{5A.2b}$$

and

$$\frac{d^2f_1(z)}{dz^2}=\int e^{tz}t^2\zeta(t)\,dt. \tag{5A.2c}$$

Putting Eqs. 5A.2a,b,c in Eq. 5A.1 gives

$$\int \zeta(t)e^{tz}\left\{zt^2+(\gamma-z)t-\alpha\right\}dt=0, \tag{5A.3a}$$

$$\text{i.e., } \int\zeta(t)(t^2-t)e^{tz}z\,dt+\int\zeta(t)e^{tz}(\gamma t-\alpha)\,dt=0. \tag{5A.3b}$$

$$\text{Since } \frac{d}{dt}\left(e^{tz}\right)=ze^{tz} \text{ and } d\left(e^{tz}\right)=ze^{tz}dt, \text{ we have} \tag{5A.4}$$

$$\int\left\{\zeta(t)(t^2-t)\right\}\left\{d\left(e^{tz}\right)\right\}+\int\zeta(t)e^{tz}(\gamma t-\alpha)\,dt=0, \tag{5A.5}$$

$$\text{i.e, } \left\{\zeta(t)(t^2-t)\right\}e^{tz}\Big]_{t_1}^{t_2}-\int\left[\frac{d}{dt}\left\{\zeta(t)(t^2-t)\right\}\right]e^{tz}+\int\zeta(t)e^{tz}(\gamma t-\alpha)\,dt=0. \tag{5A.6}$$

There are three terms in Eq. 5A.6; the first term is the difference of terms at the lower and the upper limits, and the second and the third terms are integrals. If we can find a contour over which the integration in the complex t plane gives the following result

$$-\int\left[\frac{d}{dt}\left\{\zeta(t)(t^2-t)\right\}\right]e^{tz}+\int\zeta(t)e^{tz}(\gamma t-\alpha)\,dt=0, \tag{5A.7a}$$

$$\text{or, } \int e^{tz}\left[\zeta(t)(\gamma t-\alpha)-\frac{d}{dt}\left\{\zeta(t)(t^2-t)\right\}\right]dt=0, \tag{5A.7b}$$

and makes the surface terms vanish, then we see that $f_1(z)$ is indeed the right solution. Now, Eq. 5A.7b would be satisfied if

$$\zeta(t)(\gamma t-\alpha)=\frac{d}{dt}\left\{\zeta(t)(t^2-t)\right\}, \tag{5A.8a}$$

i.e., if

$$(\gamma t - \alpha)dt = \frac{d\{\zeta(t)(t^2 - t)\}}{\zeta(t)}, \tag{5A.8b}$$

i.e., if

$$\frac{(\gamma t - \alpha)}{(t^2 - t)}dt = \frac{d\{\zeta(t)(t^2 - t)\}}{\zeta(t)(t^2 - t)} = d\left[\ln\{\zeta(t)(t^2 - t)\}\right] = \left\{\frac{\alpha}{t} + \frac{\gamma - \alpha}{t - 1}\right\}dt, \tag{5A.8c}$$

i.e., if

$$d\left[\ln\{\zeta(t)(t^2 - t)\}\right] = \left\{\frac{\alpha}{t} + \frac{\gamma - \alpha}{t - 1}\right\}dt. \tag{5A.8d}$$

Integrating,

$$\ln\{\zeta(t)(t^2 - t)\} = \int \frac{\alpha}{t}dt + \int \frac{\gamma - \alpha}{t - 1}dt = (\alpha \ln t) - (\gamma - \alpha)\ln(1 - t) + \tilde{k}, \tag{5A.9a}$$

i.e.,

$$\ln\{\zeta(t)(t^2 - t)\} = \ln(t^\alpha) + \ln(t - 1)^{(\gamma-\alpha)} + \ln e^{\tilde{k}} = \ln\left[e^{\tilde{k}} t^\alpha (t - 1)^{(\gamma-\alpha)}\right]. \tag{5A.9b}$$

Applying involution (i.e., exponentiating),

$$\zeta(t)t(t - 1) = e^{\tilde{k}} t^\alpha (t - 1)^{(\gamma-\alpha)}, \tag{5A.10}$$

$$\text{i.e. } \zeta(t) = e^{\tilde{k}} t^{\alpha-1} (t - 1)^{(\gamma-\alpha-1)} = k t^{\alpha-1} (t - 1)^{(\gamma-\alpha-1)}. \tag{5A.11}$$

We see that $f_1(z) = \int e^{tz} t^{\alpha-1} (t - 1)^{\gamma-\alpha-1}\, dt$ *provides a solution, if we choose the path of integration in the complex t plane to be such that*

$$V_1(t) = e^{tz} t^\alpha (t - z)^{\gamma-\alpha} \tag{5A.12}$$

provides surface terms which vanish in the complex t plane.

Introducing now a new function G such that

$$z^{1-\gamma}G(z) = F(z), \tag{5A.13}$$

we see that it must satisfy the differential equation

$$z\frac{d^2G}{dz^2} + (2 - \gamma - z)\frac{dG}{dz} - (\alpha - \gamma + 1)G = 0. \tag{5A.14}$$

Following the same indexing scheme, we thus identify a linearly independent particular integral $z^{1-\gamma}F(\alpha - \gamma + 1, 2 - \gamma, z)$. The general solution to the differential equation therefore is

$$\Phi(z) = c_1 F(\alpha, \gamma, z) + c_2 z^{1-\gamma} F(\alpha - \gamma + 1, 2 - \gamma, z). \tag{5A.15}$$

Negative integer values of γ can also be brought to the same form as Eq. 5A.14 and therefore do not require independent consideration. Exclusion of negative integer values of

μ therefore does not lead to any loss of generality. However, there is a singularity in the second term at $z = 0$. The integral representation that is now a candidate to be a solution is

$$f_2(z) = z^{1-\gamma}\int e^{tz} t^{\alpha-\gamma}(t-1)^{-\alpha}\,dt. \tag{5A.16}$$

Introducing a new variable s by defining

$$tz = s, \tag{5A.17}$$

we can transfer the integration over t to that over s, to get

$$f_2(z) = z^{1-\gamma}\int e^s s^{\alpha-\gamma}(s-z)^{-\alpha} z^{\gamma-\alpha} z^{\alpha}\frac{ds}{z} = \int e^s (s-z)^{-\alpha} s^{\alpha-\gamma} ds \equiv \int e^t (t-z)^{-\alpha} t^{\alpha-\gamma} dt, \tag{5A.18}$$

where, in the last step, we simply replace the dummy integration variable s by t. Applying the same method as for $f_1(z)$, *we find that the condition for* $f_2(z) = \int e^t (t-z)^{-\alpha} t^{\alpha-\gamma} dt$ *to be a solution is that we choose the path of integration in the complex t plane to be such that*

$$V_2(t) = e^t t^{\alpha-\gamma+1}(t-z)^{1-\alpha} \tag{5A.19}$$

provides vanishing surface terms. To address the singular points $t = 0$ and $t = z$ in the integrand of Eq. 5A.18, we employ the contour shown in Fig. 5A.1 which satisfactorily enables the surface terms V_2 to vanish. The integral in $f_2(z)$ given in Eq. 5A.18 evaluated along the contour C in Fig. 5A.1 does not have a singularity at $z = 0$; it is the same as the function $F(\alpha,\gamma,z)$, except perhaps for a constant factor.

For $z = 0$, the singular points $t = 0$ and $t = z$ in the integrand of Eq. 5A.18 are essentially the same.

Now, we shall use the Γ function defined by

$$\frac{1}{\Gamma(\gamma)} = \frac{1}{2\pi i}\int_C e^t t^{-\gamma} dt. \tag{5A.20}$$

With $F(\alpha,\gamma,0) = 1$, (5A.21)

we recognize that $F(\alpha,\gamma,z) = \dfrac{\Gamma(\gamma)}{2\pi i} f_2(z) = \dfrac{\Gamma(\gamma)}{2\pi i}\displaystyle\int e^t (t-z)^{-\alpha} t^{\alpha-\gamma} dt,$ (5A.22)

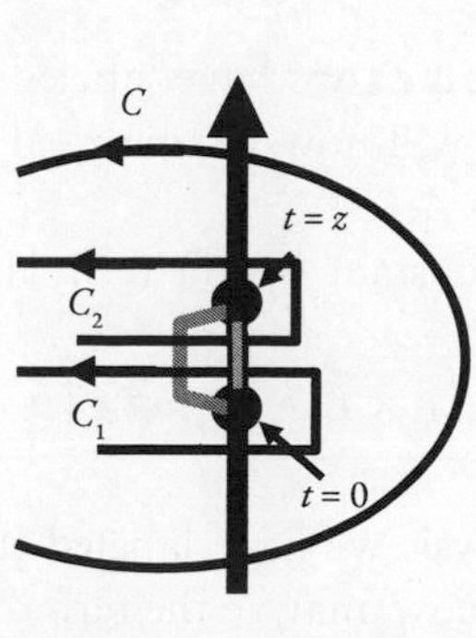

Fig. 5A.1 Contour chosen to avoid the singularities in the integrand in Eq. 5A.18. The flexed branch cut is also shown.

since for $z = 0$ it reduces to Eq. 5A.20. In Eq. 5A.22, if we replace $t \to t + z$, we see that

$$F(\alpha,\gamma,z) = \frac{\Gamma(\gamma)}{2\pi i}\int e^{z}e^{t}t^{-\alpha}(t+z)^{\alpha-\gamma}\,dt = e^{z}F(\gamma-\alpha,\gamma,-z). \tag{5A.23}$$

For $0 \le m < n$, $F(\alpha = -n, \gamma = m, z) = F(-n,m,z)$ are the generalized (or associated) Laguerre polynomials, and the subset $F(-n, m = 0, z) = F(-n,0,z)$ are (simply) called the Laguerre polynomials.

These are given by

$$L_n^m(z) = e^{z}z^{-m}\frac{d^n}{dz^n}\left(e^{z}z^{n+m}\right), \tag{5A.24a}$$

and the subset with $m = 0$ by

$$L_n(z) = L_n^0(z) = e^{z}\frac{d^n}{dz^n}\left(e^{z}z^{n}\right). \tag{5A.24b}$$

The confluent hypergeometric function $F(-n+\ell+1, \ell+1, \rho)$ is finite at $\rho = 0$ and does *not* diverge as $\rho \to \infty$ at a rate any more than a finite power of ρ. It provides a solution to the radial part of the Schrödinger equation for the hydrogen atom when $-n+\ell+1$ is zero or is a negative integer. If this condition is not satisfied, the solution diverges as e^{ρ}. Physical acceptable solutions for the bound states of the hydrogen atom occur for $-n+\ell+1 \le 0$ i.e., $\ell \le n-1$. The solutions are polynomials of degree $n-\ell-1$, and thus have $n-\ell-1$ nodes (see Eq. 5.49).

APPENDIX 5B SQUARE OF THE LRL-PAULI VECTOR OPERATOR

In this appendix, we show that $\vec{A}^2 = \frac{2H}{\mu}\left(\vec{\ell}^2 + \hbar^2\right) + \kappa^2$.

We know from Eq. 5.65 that

$$\vec{A}\bullet\vec{A} = \left(\frac{1}{\mu}\left[\left(\vec{p}\times\vec{\ell}\right) - i\hbar\vec{p}\right] - \kappa\frac{\vec{r}}{|r|}\right)\bullet\left(\frac{1}{\mu}\left[\left(\vec{p}\times\vec{\ell}\right) - i\hbar\vec{p}\right] - \kappa\frac{\vec{r}}{|r|}\right). \tag{5B.1}$$

i.e.

$$\vec{A}\bullet\vec{A} = \underbrace{\frac{1}{\mu^2}\left[\left(\vec{p}\times\vec{\ell}\right)-i\hbar\vec{p}\right]^2}_{(i)} \underbrace{-\frac{1}{\mu}\left[\left(\vec{p}\times\vec{\ell}\right)-i\hbar\vec{p}\right]\bullet\kappa\frac{\vec{r}}{r}}_{(ii)} \underbrace{-\frac{\kappa}{\mu}\frac{\vec{r}}{r}\bullet\left[\left(\vec{p}\times\vec{\ell}\right)-i\hbar\vec{p}\right]}_{(iii)} + \underbrace{\kappa^2\frac{\vec{r}^2}{r^2}}_{(iv)}. \tag{5B.2a}$$

On the right-hand side of Eq. 5B.2a we have labeled the four terms as (i), (ii), (iii) and (iv) for bookkeeping. These terms will now be analyzed separately and then assembled to reconstruct the scalar operator $\vec{A}\bullet\vec{A}$.

Apart from the multiplicative constant $\frac{1}{\mu^2}$ the term labeled as (i) in Eq. 5B.2a is

$$\left[\left(\vec{p}\times\vec{\ell}\right) - i\hbar\vec{p}\right]^2 = \underbrace{\left(\vec{p}\times\vec{\ell}\right)^2}_{A} \underbrace{- i\hbar\left(\vec{p}\times\vec{\ell}\right)\bullet\vec{p}}_{B} \underbrace{- i\hbar\vec{p}\bullet\left(\vec{p}\times\vec{\ell}\right)}_{C} \underbrace{- \hbar^2\vec{p}^2}_{D}. \tag{5B.2b}$$

Yet again to break up the analysis we have labeled the terms on the right-hand side of Eq. 5B.2b as A,B,C, and D. Let us now analyze the term "A":

$$[\text{A}]: \left(\vec{p}\times\vec{\ell}\right)^2 = \sum_{ijkmn}\varepsilon_{ijk}\varepsilon_{inm}p_j\ell_k p_n\ell_m = \sum_{jkmn}\left(\delta_{jn}\delta_{km} - \delta_{jm}\delta_{kn}\right)p_j\ell_k p_n\ell_m. \tag{5B.3}$$

Hence, $\left(\vec{p}\times\vec{\ell}\right)^2 = \sum_{jk}\left(p_j\ell_k p_j\ell_k - p_j\ell_k p_k\ell_j\right) = \sum_{jk} p_j\ell_k p_j\ell_k,$ (5B.4)

since $\sum_k \ell_k p_k = \sum_{kmn} \varepsilon_{kmn} x_m p_n p_k = 0.$ (5B.5)

Again, since $\ell_j = \left(\vec{r}\times\vec{p}\right)_j = \sum_{m=1}^{3}\sum_{n=1}^{3} \varepsilon_{jmn} x_m p_n,$ (5B.6)

we get $\left(\vec{p}\times\vec{\ell}\right)^2 = \sum_{jk} p_j{}^2\ell^2{}_k + i\hbar\sum_{jkm}\varepsilon_{kjm} p_j p_m \ell_k = \vec{p}^2\vec{\ell}^2.$ (5B.7)

The term "B" in Eq. 5B.2b is

[B]: $\left(\vec{p}\times\vec{\ell}\right)\bullet\vec{p} = \sum_{ijk}\varepsilon_{ijk} p_j \ell_k p_i = \sum_{ijk}\varepsilon_{ijk} p_j p_i \ell_k + i\hbar\sum_{ijkm}\varepsilon_{ijk}\varepsilon_{kim} p_j p_m = 2i\hbar\vec{p}^2,$ (5B.8)

since the sum

$$\sum_{ijk}\varepsilon_{ijk} p_j p_i \ell_k = 0, \tag{5B.9}$$

and $\sum_{k,l}\varepsilon_{hij}\varepsilon_{kim} = 2\delta_{jm}.$ (5B.10)

The third term "C" in Eq. 5B.2b is

[C]: $\vec{p}\bullet\left(\vec{p}\times\vec{\ell}\right) = \sum_{ijk}\varepsilon_{ijk} p_i p_j \ell_k = 0.$ (5B.11)

Hence, collecting the terms [A], [B], and [C] respectively from Eqs. 5B.3, 5B.8, and 5B.11, the term (i) on the right-hand side of Eq. 5B.2a becomes

$$\frac{1}{\mu^2}\left[\left(\vec{p}\times\vec{\ell}\right) - i\hbar\vec{p}\right]^2 = \frac{1}{\mu^2}\left[\vec{p}^2\vec{\ell}^2 - 2\left(i\hbar\right)^2\vec{p}^2 - \hbar^2\vec{p}^2\right] = \frac{1}{\mu^2}\left[\vec{p}^2\left(\vec{\ell}^2 + \hbar^2\right)\right]. \tag{5B.12}$$

The second term, (ii), in Eq. 5B.2a is

$$\underbrace{\frac{1}{\mu}\left[\left(\vec{p}\times\vec{\ell}\right) - i\hbar\vec{p}\right]\bullet\kappa\frac{\vec{r}}{r}}_{(ii)} = \underbrace{\frac{1}{\mu}\left(\vec{p}\times\vec{\ell}\right)\bullet\kappa\frac{\vec{r}}{r}}_{(ii.a)} - \underbrace{\frac{1}{\mu}i\hbar\vec{p}\bullet\kappa\frac{\vec{r}}{r}}_{(ii.b)}. \tag{5B.13}$$

We will now analyze the parts (ii.a) and (ii.b) from the right-hand side of Eq. 5B.13, of which *apart from the multiplier* κ the first part is:

(ii.a): $\frac{1}{\mu}\left(\vec{p}\times\vec{\ell}\right)\bullet\frac{\vec{r}}{r} = \frac{1}{\mu}\sum_{ijk}\varepsilon_{ijk} p_j \ell_k \frac{x_i}{r}.$ (5B.14)

Since $\ell_k \frac{x_i}{r} = \left[\ell_k, \frac{x_i}{r}\right] + \frac{x_i}{r}\ell_k,$ (5B.15a) for an arbitrary function $f(r)$, we have

$$\left[\ell_k, \frac{x_i}{r}\right] f(r) = \varepsilon_{klm}\left[x_l p_m, \frac{x_i}{r}\right] f(r). \tag{5B.15b}$$

$$\text{Thus, } \left[\ell_k, \frac{x_i}{r}\right] f(r) = -i\hbar\varepsilon_{klm} x_l \left[\partial_m, \frac{x_i}{r}\right] f(r) - i\hbar\varepsilon_{klm}\left[x_l, \frac{x_i}{r}\right]\partial_m f(r), \tag{5B.15c}$$

$$\text{i.e., } \left[\ell_k, \frac{x_i}{r}\right] f(r) = \varepsilon_{klm}\frac{x_l}{r}\left(-i\hbar\delta_{im}\right) f(r), \tag{5B.15d}$$

$$\text{or, } \left[\ell_k, \frac{x_i}{r}\right] f(r) = -i\hbar\varepsilon_{kli}\frac{x_l}{r} f(r). \tag{5B.15e}$$

$$\text{Therefore, } \ell_k \frac{x_i}{r} = i\hbar\varepsilon_{ilk}\frac{x_l}{r} + \frac{x_i}{r}\ell_k. \tag{5B.16}$$

Accordingly, the following expression can be employed to analyze the term (ii.a) of Eq. 5B.13 further:

$$\left(\vec{p}\times\vec{\ell}\right)\bullet\frac{\vec{r}}{r} = \sum_{ijk}\varepsilon_{ijk}p_j\left(i\hbar\varepsilon_{ilk}\frac{x_l}{r} + \frac{x_i}{r}\ell_k\right) = \sum_{ijk}\varepsilon_{ijk}p_j\left(i\hbar\varepsilon_{ilk}\frac{x_l}{r}\right) + \sum_{ijk}\varepsilon_{ijk}p_j\left(\frac{x_i}{r}\ell_k\right). \tag{5B.17}$$

Again, using Eq. 5B.10, we have

$$\left(\vec{p}\times\vec{\ell}\right)\bullet\frac{\vec{r}}{r} = 2\delta_{jl}i\hbar p_j\frac{x_l}{r} + \sum_{ijk}\varepsilon_{ijk}\left(p_j x_i\right)\frac{\ell_k}{r}, \tag{5B.18a}$$

$$\text{i.e., } \left(\vec{p}\times\vec{\ell}\right)\bullet\frac{\vec{r}}{r} = 2\delta_{jl}i\hbar p_j\frac{x_l}{r} + \sum_{ijk}\varepsilon_{ijk}\left(\left[p_j, x_i\right] + x_i p_j\right)\frac{\ell_k}{r}, \tag{5B.18b}$$

$$\text{or, } \left(\vec{p}\times\vec{\ell}\right)\bullet\frac{\vec{r}}{r} = 2\frac{i\hbar}{r}p_j x_j + \varepsilon_{ijk}i\hbar\delta_{ij}\frac{\ell_k}{r} + \varepsilon_{ijk}x_i p_j\frac{\ell_k}{r}. \tag{5B.18c}$$

We thus get for the term (ii.a) the following result:

$$\left(\vec{p}\times\vec{\ell}\right)\bullet\frac{\vec{r}}{r} = 2\frac{i\hbar}{r}p_j x_j + \sum_k\frac{\ell_k\ell_k}{r} = \frac{2i\hbar}{r}p_j x_j + \frac{\ell^2}{r}. \tag{5B.19}$$

We now analyze the terms (ii) and (iii) from Eq. 5B.2a *together*.

$$\underbrace{\left[\left(\vec{p}\times\vec{\ell}\right) - i\hbar\vec{p}\right]\bullet\frac{\vec{r}}{r}}_{(ii)} + \underbrace{\frac{\vec{r}}{r}\bullet\left[\left(\vec{p}\times\vec{\ell}\right) - i\hbar\vec{p}\right]}_{(iii)} = \underbrace{\frac{2i\hbar}{r}p_j x_j + \frac{\ell^2}{r}}_{(ii.a)} \underbrace{- i\hbar\vec{p}\bullet\frac{\vec{r}}{r}}_{(ii.b)} + \underbrace{\frac{\ell^2}{r}}_{(iii.a)} \underbrace{- i\hbar\frac{\vec{r}}{r}\bullet\vec{p}}_{(iii.b)}, \tag{5B.20a}$$

$$\text{i.e., } \underbrace{\left[\left(\vec{p}\times\vec{\ell}\right) - i\hbar\vec{p}\right]\bullet\frac{\vec{r}}{r}}_{(ii)} + \underbrace{\frac{\vec{r}}{r}\bullet\left[\left(\vec{p}\times\vec{\ell}\right) - i\hbar\vec{p}\right]}_{(iii)} = \underbrace{2i\hbar\vec{p}\bullet\frac{\vec{r}}{r} + \frac{\ell^2}{r}}_{(ii.a)} \underbrace{- i\hbar\vec{p}\bullet\frac{\vec{r}}{r}}_{(ii.b)} + \underbrace{\frac{\ell^2}{r}}_{(iii.a)} \underbrace{- i\hbar\frac{\vec{r}}{r}\bullet\vec{p}}_{(iii.b)}. \tag{5B.20b}$$

$$\text{Thus, } \underbrace{\left[\left(\vec{p}\times\vec{\ell}\right)-i\hbar\vec{p}\right]\bullet\frac{\vec{r}}{r}}_{(ii)}+\underbrace{\frac{\vec{r}}{r}\bullet\left[\left(\vec{p}\times\vec{\ell}\right)-i\hbar\vec{p}\right]}_{(iii)}=\frac{2\ell^2}{r}-i\hbar\frac{\vec{r}}{r}\bullet\vec{p}+i\hbar\vec{p}\bullet\frac{\vec{r}}{r}, \tag{5B.21a}$$

$$\text{i.e., } \underbrace{\left[\left(\vec{p}\times\vec{\ell}\right)-i\hbar\vec{p}\right]\bullet\frac{\vec{r}}{r}}_{(ii)}+\underbrace{\frac{\vec{r}}{r}\bullet\left[\left(\vec{p}\times\vec{\ell}\right)-i\hbar\vec{p}\right]}_{(iii)}=\frac{2\ell^2}{r}+\left(i\hbar\vec{p}\bullet\frac{\vec{r}}{r}-i\hbar\frac{\vec{r}}{r}\bullet\vec{p}\right). \tag{5B.21b}$$

For brevity, we now introduce an operator

$$\Gamma=\vec{p}\bullet\frac{\vec{r}}{r}-\frac{\vec{r}}{r}\bullet\vec{p}=\left\{\begin{matrix}(p_x\hat{e}_x+p_y\hat{e}_y+p_z\hat{e}_z)\bullet\frac{1}{r}(x\hat{e}_x+y\hat{e}_y+z\hat{e}_z)\\ -\frac{1}{r}(x\hat{e}_x+y\hat{e}_y+z\hat{e}_z)\bullet(p_x\hat{e}_x+p_y\hat{e}_y+p_z\hat{e}_z)\end{matrix}\right\}, \tag{5B.22a}$$

$$\text{i.e., } \Gamma=\left[p_x\frac{x}{r}-\frac{x}{r}p_x\right]+\left[p_y\frac{y}{r}-\frac{y}{r}p_y\right]+\left[p_z\frac{z}{r}-\frac{z}{r}p_z\right]. \tag{5B.22b}$$

$$\text{Thus, } \vec{p}\bullet\frac{\vec{r}}{r}-\frac{\vec{r}}{r}\bullet\vec{p}=\sum_{i=1}^{3}\left(p_i\frac{x_i}{r}-\frac{x_i}{r}p_i\right)=\sum_{i=1}^{3}\left[p_i,\frac{x_i}{r}\right]=\sum_{j=1}^{3}\sum_{i=1}^{3}\left[p_i,\frac{x_j}{r}\right]\delta_{ij}. \tag{5B.23}$$

Recognizing now that for an arbitrary function $f(r)$,

$$\left[p_i,\frac{x_j}{r}\right]f(r)=p_i\left(\frac{x_j}{r}f(r)\right)-\frac{x_j}{r}\left(p_if(r)\right), \tag{5B.24}$$

with $p_i=-i\hbar\frac{\partial}{\partial x_i}$,

$$\text{we find that } \Gamma=\vec{p}\bullet\frac{\vec{r}}{r}-\frac{\vec{r}}{r}\bullet\vec{p}=\sum_{j=1}^{3}\left[p_j,\frac{x_j}{r}\right]=\sum_{j=1}^{3}\sum_{i=1}^{3}\left[p_i,\frac{x_j}{r}\right]\delta_{ij}. \tag{5B.25}$$

$$\text{Again, } \left[p_i,\frac{x_j}{r}\right]f(r)=-i\hbar\frac{\partial}{\partial x_i}\left(\frac{x_j}{r}f(r)\right)-\frac{x_j}{r}\left((-i\hbar)\frac{\partial}{\partial x_i}f(r)\right),$$

$$\text{i.e., } \left[p_i,\frac{x_j}{r}\right]f(r)=-i\hbar\left(\frac{1}{r}\left(\frac{\partial}{\partial x_i}x_j\right)f(r)+x_j\left(\frac{\partial}{\partial x_i}\frac{1}{r}\right)f(r)+\frac{x_j}{r}\left(\frac{\partial}{\partial x_i}f(r)\right)\right)+\frac{x_j}{r}i\hbar\left(\frac{\partial}{\partial x_i}f(r)\right),$$

$$\text{or, } \left[p_i,\frac{x_j}{r}\right]f(r)=-i\hbar\left(\frac{\delta_{ij}}{r}-\frac{x_jx_i}{r^3}\right)f(r). \tag{5B.26}$$

$$\text{Hence, } \Gamma=\vec{p}\bullet\frac{\vec{r}}{r}-\frac{\vec{r}}{r}\bullet\vec{p}=-i\hbar\sum_{j=1}^{3}\sum_{i=1}^{3}\left(\frac{\delta_{ij}}{r}-\frac{x_jx_i}{r^3}\right)\delta_{ij}=-2i\hbar\frac{1}{r}. \tag{5B.27}$$

Accordingly, we can now write

$$\boxed{(ii)+(iii)}=\frac{2}{r}\vec{\ell}^2+(i\hbar)\left(-2i\hbar\frac{1}{r}\right)=\frac{2}{r}\left(\vec{\ell}^2+\hbar^2\right), \tag{5B.28}$$

and $\vec{A}\bullet\vec{A} = \frac{1}{\mu^2}\vec{p}^2\left(\vec{\ell}^2 + \hbar^2\right) - \frac{\kappa}{\mu}\frac{2}{r}\left(\vec{\ell}^2 + \hbar^2\right) + \kappa^2\frac{\vec{r}^2}{r^2},$ (5B.29a)

i.e., $\vec{A}\bullet\vec{A} = \frac{1}{\mu^2}\vec{p}^2\left(\vec{\ell}^2 + \hbar^2\right) - \frac{\kappa}{\mu}\frac{2}{r}\left(\vec{\ell}^2 + \hbar^2\right) + \kappa^2\frac{\vec{r}^2}{r^2} = \frac{2}{\mu}\left(\frac{\vec{p}^2}{2\mu} - \frac{\kappa}{r}\right)\left(\vec{\ell}^2 + \hbar^2\right) + \kappa^2.$ (5B.29b)

i.e., $\vec{A}^2 = \frac{2H}{\mu}\left(\vec{\ell}^2 + \hbar^2\right) + \kappa^2.$ (5B.29c)

APPENDIX 5C Commutation of the LRLP Vector Operator with the Hamiltonian

In this appendix, we prove the result in Eq. 5.67, viz., $\left[\vec{A}, H\right] = 0.$

From Eq. 5.65, we know that $\vec{A}_{Quantum} = \frac{1}{2\mu}\left[\left(\vec{p}\times\vec{\ell}\right) - \left(\vec{\ell}\times\vec{p}\right)\right] - \kappa\frac{\vec{r}}{|r|},$ (5C.1)

and the Hamiltonian is $\vec{H} = \frac{\vec{p}^2}{2\mu} - \frac{\kappa}{r}.$ (5C.2)

The commutator of the ith component of the quantum LRL-Pauli operator $\vec{A}$ with the Hamiltonian therefore is

$$\left[\vec{A}_i, H\right] = \frac{1}{2\mu}\left[\left(\vec{p}\times\vec{\ell}\right)_i, H\right] - \frac{1}{2\mu}\left[\left(\vec{\ell}\times\vec{p}\right)_i, H\right] - \left[\kappa\frac{x_i}{|r|}, H\right]. \quad (5C.3)$$

We shall refer to the three terms on the right-hand side of the above equation respectively as [A], [B], and [C] and analyze them individually.

[A]: $\frac{1}{2\mu}\left[\left(\vec{p}\times\vec{\ell}\right)_i, H\right] = \frac{1}{2\mu}\varepsilon_{ijk}\left[p_j\ell_k, H\right] = \frac{1}{2\mu}\varepsilon_{ijk}\left[p_j\ell_k H - Hp_j\ell_k\right].$ (5C.4a)

Hence, $\frac{1}{2\mu}\left[\left(\vec{p}\times\vec{\ell}\right)_i, H\right] = \frac{1}{2\mu}\varepsilon_{ijk}\left[p_j H\ell_k - Hp_j\ell_k\right] = \frac{1}{2\mu}\varepsilon_{ijk}\left[p_j, H\right]\ell_k.$ (5C.4b)

We have used the fact that each component of the orbital angular momentum vector operator commutes with the Hamiltonian.

In the last term in Eq. 5.C4b, we have the commutator of the jth component of the linear momentum operator with the Hamiltonian:

$$\left[p_j, H\right]_- = -\kappa\left[-i\hbar\frac{\partial}{\partial x_j}, \frac{1}{r}\right]_- = \kappa\left[i\hbar\frac{\partial}{\partial x_j}, \frac{1}{r}\right]_-. \quad (5C.5a)$$

On operating by this commutator on an arbitrary function of r, we get

$$\left[p_j, H\right]_- f(r) = i\hbar\kappa\left(\frac{1}{r}\frac{\partial f}{\partial x_j} + \left\{\frac{\partial(r^{-1})}{\partial x_j}\right\}f - \frac{1}{r}\frac{\partial f}{\partial x_j}\right) = -i\hbar\kappa\frac{x_j}{r^3}f(r), \quad (5C.5b)$$

hence

$$\left[\vec{p}_j, \vec{H}\right]_- = -i\hbar\kappa\frac{x_j}{r^3}. \quad (5C.5c)$$

Therefore, for the term [A], we get $\frac{1}{2\mu}\left[\left(\vec{p}\times\vec{\ell}\right)_i, H\right] = \frac{-i\hbar\kappa}{2\mu}\varepsilon_{ijk}\frac{x_j\ell_k}{r^3}$. (5C.6)

Likewise, for the term [B], we have

[B]: $\frac{1}{2\mu}\left[\left(\vec{\ell}\times\vec{p}\right)_i, H\right] = \frac{1}{2\mu}\varepsilon_{ijk}\left[\ell_j p_k, H\right] = \frac{-i\hbar\kappa}{2\mu}\varepsilon_{ijk}\frac{\ell_j x_k}{r^3}$. (5C.7)

Combining the terms [A] and [B],

$$\frac{1}{2\mu}\left\{\left[\left(\vec{p}\times\vec{\ell}\right)_i, H\right] - \left[\left(\vec{\ell}\times\vec{p}\right)_i, H\right]\right\} = \left\{\begin{matrix} \frac{-i\hbar\kappa}{2\mu}\varepsilon_{ijk}\frac{x_j\ell_k}{r^3} + \\ \frac{i\hbar\kappa}{2\mu}\varepsilon_{ijk}\frac{\ell_j x_k}{r^3} \end{matrix}\right\} = \frac{-i\hbar\kappa}{2\mu}\varepsilon_{ijk}\left(\frac{x_j\ell_k}{r^3} - \frac{\ell_j x_k}{r^3}\right). \quad (5C.8)$$

Now, $\ell_j = \sum_{m,n}\varepsilon_{jmn}x_m p_n$, (5C.9a)

$\ell_k = \sum_{m,n}\varepsilon_{kmn}x_m p_n$, and (5C.9b)

$$\frac{1}{2\mu}\left\{\left[\left(\vec{p}\times\vec{\ell}\right)_i, H\right] - \left[\left(\vec{\ell}\times\vec{p}\right)_i, H\right]\right\} = \frac{-i\hbar\kappa}{2\mu}\sum_{j,k}\varepsilon_{ijk}\sum_{m,n}\left(\varepsilon_{kmn}\frac{x_j x_m p_n}{r^3} - \varepsilon_{jmn}\frac{x_m p_n x_k}{r^3}\right). \quad (5C.10)$$

We can see that $\left[p_n, \frac{x_k}{r^3}\right]_- = -i\hbar\left(\frac{\delta_{kn}}{r^3} - 3\frac{x_k x_n}{r^5}\right)$, (5C.11a)

and hence $\frac{p_n x_k}{r^3} = \left\{-i\hbar\left(\frac{\delta_{kn}}{r^3} - 3\frac{x_k x_n}{r^5}\right) + \frac{x_k p_n}{r^3}\right\}$, (5C.11b)

using which in Eq. 5C.10 we get

$$\frac{1}{2\mu}\left\{\begin{matrix}\left[\left(\vec{p}\times\vec{\ell}\right)_i, H\right] - \\ \left[\left(\vec{\ell}\times\vec{p}\right)_i, H\right]\end{matrix}\right\} = \frac{-i\hbar\kappa}{2\mu}\sum_{j,k}\varepsilon_{ijk}\sum_{m,n}\left(\begin{matrix}\varepsilon_{kmn}\frac{x_j x_m p_n}{r^3} - \\ \varepsilon_{jmn}x_m\left\{-i\hbar\left(\begin{matrix}\frac{\delta_{kn}}{r^3} - \\ 3\frac{x_k x_n}{r^5}\end{matrix}\right) + \frac{x_k p_n}{r^3}\right\}\end{matrix}\right), \quad (5C.12a)$$

i.e.,

$$\frac{1}{2\mu}\left\{\begin{matrix}\left[\left(\vec{p}\times\vec{\ell}\right)_i, H\right] - \\ \left[\left(\vec{\ell}\times\vec{p}\right)_i, H\right]\end{matrix}\right\} = \left[\begin{matrix}\frac{-i\hbar\kappa}{2\mu}\sum_{j,k}\varepsilon_{ijk}\sum_{m,n}\left\{\varepsilon_{kmn}\frac{x_j x_m}{r^3} - \varepsilon_{jmn}\frac{x_m x_k}{r^3}\right\}p_n \\ -\frac{(i\hbar)^2\kappa}{2\mu}\sum_{j,k}\varepsilon_{ijk}\sum_{m,n}\varepsilon_{jmn}\left(\frac{x_m\delta_{kn}}{r^3}\right) \\ +\frac{(i\hbar)^2\kappa}{2\mu}\sum_{j,k}\varepsilon_{ijk}\sum_{m,n}\varepsilon_{jmn}\left(\frac{3x_m x_k x_n}{r^5}\right)\end{matrix}\right], \quad (5C.12b)$$

or

$$\frac{1}{2\mu}\left\{\begin{matrix}\left[\left(\vec{p}\times\vec{\ell}\right)_i, H\right]- \\ \left[\left(\vec{\ell}\times\vec{p}\right)_i, H\right]\end{matrix}\right\} = \left[\begin{matrix}\frac{(i\hbar)^2\kappa}{2\mu}\sum_{j,k}\varepsilon_{ijk}\sum_{m,n}\left(\varepsilon_{kmn}\frac{x_j x_m}{r^3} - \varepsilon_{jmn}\frac{x_m x_k}{r^3}\right)\frac{\partial}{\partial x_n} \\ -\frac{(i\hbar)^2\kappa}{2\mu}\sum_{j,k,m}\varepsilon_{ijk}\varepsilon_{jmk}\frac{x_m}{r^3} \\ +\frac{3(i\hbar)^2\kappa}{2\mu}\sum_{j,k,m,n}\varepsilon_{ijk}\varepsilon_{jmn}\frac{x_m x_k x_n}{r^5}\end{matrix}\right]. \quad (5C.12c)$$

The last term vanishes:

$$\sum_{j,k,m,n}\varepsilon_{ijk}\varepsilon_{jmn}\frac{x_m x_k x_n}{r^5} = \frac{1}{2}\left[\begin{matrix}\sum_{j,k,m,n}\varepsilon_{ijk}\varepsilon_{jmn}\frac{x_k x_m x_n}{r^5} + \\ \sum_{j,k,n,m}\varepsilon_{ijk}\varepsilon_{jnm}\frac{x_k x_n x_m}{r^5}\end{matrix}\right] = \frac{1}{2}\left[\begin{matrix}\sum_{j,k,m,n}\varepsilon_{ijk}\varepsilon_{jmn}\frac{x_k x_m x_n}{r^5} - \\ \sum_{j,k,n,m}\varepsilon_{ijk}\varepsilon_{jmn}\frac{x_k x_n x_m}{r^5}\end{matrix}\right] = 0, \quad (5C.13)$$

since $\left[x_k, x_n\right]_- = 0$.

Therefore,

$$\frac{1}{2\mu}\left\{\begin{matrix}\left[\left(\vec{p}\times\vec{\ell}\right)_i, H\right]- \\ \left[\left(\vec{\ell}\times\vec{p}\right)_i, H\right]\end{matrix}\right\} = \left[\begin{matrix}\frac{(i\hbar)^2\kappa}{2\mu}\sum_{j,k,m,n}\left\{\varepsilon_{ijk}\varepsilon_{kmn}\frac{x_j x_m}{r^3}\frac{\partial}{\partial x_n} - \varepsilon_{ijk}\varepsilon_{jmn}\frac{x_m x_k}{r^3}\frac{\partial}{\partial x_n}\right\} \\ -\frac{(i\hbar)^2\kappa}{2\mu}\sum_{j,k,m}\varepsilon_{ijk}\varepsilon_{jmk}\frac{x_m}{r^3}\end{matrix}\right]. \quad (5C.14)$$

Interchanging the dummy labels j and k and subsequently recognizing that $\varepsilon_{ikj} = -\varepsilon_{ijk}$, we see that

$$-\sum_{j,k,m,n}\varepsilon_{ijk}\varepsilon_{jmn}\frac{x_m x_k}{r^3}\frac{\partial}{\partial x_n} = -\sum_{k,j,m,n}\varepsilon_{ikj}\varepsilon_{kmn}\frac{x_m x_j}{r^3}\frac{\partial}{\partial x_n} = \sum_{k,j,m,n}\varepsilon_{ijk}\varepsilon_{kmn}\frac{x_m x_j}{r^3}\frac{\partial}{\partial x_n} \quad (5C.15)$$

and hence

$$\frac{1}{2\mu}\left\{\begin{matrix}\left[\left(\vec{p}\times\vec{\ell}\right)_i, H\right]- \\ \left[\left(\vec{\ell}\times\vec{p}\right)_i, H\right]\end{matrix}\right\} = \left[\begin{matrix}\frac{(i\hbar)^2\kappa}{\mu}\sum_{j,k,m,n}\varepsilon_{ijk}\varepsilon_{kmn}\frac{x_j x_m}{r^3}\frac{\partial}{\partial x_n} \\ -\frac{(i\hbar)^2\kappa}{2\mu}\sum_{j,k,m}\varepsilon_{ijk}\varepsilon_{jmk}\frac{x_m}{r^3}\end{matrix}\right]. \quad (5C.16)$$

Since $\sum_i \varepsilon_{ijk}\varepsilon_{imn} = \delta_{jm}\delta_{kn} - \delta_{jn}\delta_{km}$, (5C.17)

the sum in the first term on the right-hand side of 5C.14 is

$$\sum_{j,k,m,n}\varepsilon_{ijk}\varepsilon_{kmn}\frac{x_j x_m}{r^3}\frac{\partial}{\partial x_n} = \sum_{j,m,n}\left(\sum_k \varepsilon_{kij}\varepsilon_{kmn}\right)\frac{x_j x_m}{r^3}\frac{\partial}{\partial x_n} = \sum_{j,m,n}\left(\delta_{im}\delta_{jn} - \delta_{in}\delta_{jm}\right)\frac{x_j x_m}{r^3}\frac{\partial}{\partial x_n},$$

i.e., $$\sum_{j,k,m,n}\varepsilon_{ijk}\varepsilon_{kmn}\frac{x_j x_m}{r^3}\frac{\partial}{\partial x_n} = \sum_j \frac{x_j x_i}{r^3}\frac{\partial}{\partial x_j} - \sum_j \frac{x_j x_j}{r^3}\frac{\partial}{\partial x_i} = \frac{x_i}{r^3}\sum_j x_j\frac{\partial}{\partial x_j} - \frac{1}{r}\frac{\partial}{\partial x_i}, \quad (5C.18)$$

since $\sum_j x_j x_j = r^2$.

The sum in the second term on the right-hand side of 5C.14 is

$$-\sum_{j,k,m} \varepsilon_{ijk}\varepsilon_{jmk}\frac{x_m}{r^3} = +\sum_{j,k,m} \varepsilon_{jik}\varepsilon_{jmk}\frac{x_m}{r^3} = \sum_{k,m}\left(\sum_j \varepsilon_{jik}\varepsilon_{jmk}\right)\frac{x_m}{r^3} = \sum_{k,m}(\delta_{im}\delta_{kk} - \delta_{ik}\delta_{km})\frac{x_m}{r^3},$$

$$\text{i.e., } -\sum_{j,k,m} \varepsilon_{ijk}\varepsilon_{jmk}\frac{x_m}{r^3} = \left(\frac{3x_i}{r^3} - \frac{x_i}{r^3}\right) = \frac{2(i\hbar)^2\kappa}{\mu}\frac{x_i}{r^3} \tag{5C.19}$$

Using Eq. 5C.18 and 5C.19 in Eq. 5C.14, we get

$$\frac{1}{2\mu}\left\{\left[\left(\vec{p}\times\vec{\ell}\right)_i, H\right] - \left[\left(\vec{\ell}\times\vec{p}\right)_i, H\right]\right\} = \frac{(i\hbar)^2\kappa}{\mu}\left[\frac{x_i}{r^3} - \frac{1}{r}\frac{\partial}{\partial x_i} + \frac{x_i}{r^3}\sum_j x_j\frac{\partial}{\partial x_j}\right]. \tag{5C.20}$$

It is left as an exercise for the reader to show that the last term $-\left[\kappa\frac{x_i}{|r|}, H\right]$ in the commutator $\left[\vec{A}_i, H\right]$ has exactly the same terms as in Eq. 5C.19 but with opposite signs, thereby establishing the result, viz., $\left[\vec{A}_i, H\right] = 0$.

APPENDIX 5D Commutation Relations between LRLP and Angular Momentum Operators

In this appendix, we prove the result in Eq. 5.68, viz., $\left[A_i, \ell_j\right] = i\hbar\varepsilon_{ijk}A_k$.

We see shall determine the required commutator using Eq. 5.64. We have

$$\left[A_i, \ell_j\right]_- = \frac{1}{\mu}\left[\left(\vec{p}\times\vec{\ell}\right)_i, \ell_j\right] - \frac{1}{\mu}\left[\left(i\hbar\vec{p}\right)_i, \ell_j\right] - \left[\kappa\frac{x_i}{r}, \ell_j\right]. \tag{5D.1}$$

$$\text{Using } \left(\vec{p}\times\vec{\ell}\right)_i = \varepsilon_{ijk}\left(p_j\ell_k - p_k\ell_j\right), \tag{5D.2}$$

$$\left[A_i, \ell_j\right]_- = \begin{bmatrix} \frac{1}{\mu}\varepsilon_{ijk}\left(p_j\ell_k\ell_j - p_k\ell_j\ell_j - \ell_j p_j\ell_k + \ell_j p_k\ell_j\right) - \\ \frac{i\hbar}{\mu}\varepsilon_{ijk}\left(p_i\ell_j - \ell_j p_i\right) - \kappa\varepsilon_{ijk}\left(\frac{x_i}{r}\ell_j - \ell_j\frac{x_i}{r}\right) \end{bmatrix}. \tag{5D.3}$$

$$\text{Now, } p_j\ell_k\ell_j = p_j\ell_j\ell_k - i\hbar p_j\ell_i = \ell_j p_j\ell_k - i\hbar p_j\ell_i \text{ since } \left[\ell_j, p_j\right] = 0, \tag{5D.4a}$$

$$\text{and likewise } p_k\ell_j\ell_j = \ell_j p_k\ell_j - i\hbar p_i\ell_j, \text{ since } \left[\ell_j, p_k\right] = \varepsilon_{ijk}i\hbar p_i. \tag{5D.4b}$$

Therefore,

$$\left[A_i, \ell_j\right]_- = \begin{bmatrix} \frac{1}{\mu}\varepsilon_{ijk}\left(\ell_j p_j\ell_k - i\hbar p_j\ell_i - \ell_j p_k\ell_j + i\hbar p_i\ell_j - \ell_j p_j\ell_k + \ell_j p_k\ell_j\right) - \\ \frac{i\hbar}{\mu}\varepsilon_{ijk}\left(p_i\ell_j - \ell_j p_i\right) - \kappa\varepsilon_{ijk}\left(\frac{x_i}{r}\ell_j - \ell_j\frac{x_i}{r}\right) \end{bmatrix}, \tag{5D.5a}$$

$$\text{i.e., } \left[A_i, \ell_j\right]_- = \frac{1}{\mu}\varepsilon_{ijk}\begin{pmatrix} -i\hbar p_j \ell_i + \\ i\hbar p_i \ell_j \end{pmatrix} + \frac{i\hbar}{\mu}\varepsilon_{ijk}\begin{pmatrix} \ell_j p_i - \\ p_i \ell_j \end{pmatrix} - \kappa\varepsilon_{ijk}\begin{pmatrix} \frac{x_i}{r}\ell_j - \\ \ell_j \frac{x_i}{r} \end{pmatrix}, \tag{5D.5b}$$

$$\text{or, } \left[A_i, \ell_j\right]_- = \frac{i\hbar}{\mu}\varepsilon_{ijk}\begin{pmatrix} p_i \ell_j - \\ p_j \ell_i \end{pmatrix} + \frac{i\hbar}{\mu}\varepsilon_{ijk}\left(-i\hbar p_k\right) - \kappa i\hbar\varepsilon_{ijk}\frac{x_k}{r}. \tag{5D.5c}$$

$$\text{Hence, } \left[A_i, \ell_j\right]_- = i\hbar\varepsilon_{ijk}\left\{\frac{1}{\mu}\left(p_i \ell_j - p_j \ell_i - i\hbar p_k\right) - \kappa\frac{x_k}{r}\right\} = i\hbar\varepsilon_{ijk}A_k. \tag{5D.6}$$

APPENDIX 5E Commutation Relations between Components of the LRLP Vector Operator

In this appendix, we prove the result in Eq. 5.70, viz., $\left[A_i, A_j\right] = -2i\frac{\hbar}{\mu}H\varepsilon_{ijk}\ell_k$.

Using Eq. 5.64,

$$\left[A_i, A_j\right]_- = \begin{bmatrix} \left(\frac{1}{\mu}\left(\vec{p}\times\vec{\ell} - i\hbar\vec{p}\right)_i - \kappa\frac{x_i}{r}\right)\left(\frac{1}{\mu}\left(\vec{p}\times\vec{\ell} - i\hbar\vec{p}\right)_j - \kappa\frac{x_j}{r}\right) - \\ \left(\frac{1}{\mu}\left(\vec{p}\times\vec{\ell} - i\hbar\vec{p}\right)_j - \kappa\frac{x_j}{r}\right)\left(\frac{1}{\mu}\left(\vec{p}\times\vec{\ell} - i\hbar\vec{p}\right)_i - \kappa\frac{x_i}{r}\right) \end{bmatrix}_-, \tag{5E.1a}$$

i.e.,

$$\left[A_i, A_j\right]_- = \begin{bmatrix} \left(\frac{1}{\mu}\left(p_j\ell_k - p_k\ell_j - i\hbar p_i\right) - \kappa\frac{x_i}{r}\right)\left(\frac{1}{\mu}\left(p_k\ell_i - p_i\ell_k - i\hbar p_j\right) - \kappa\frac{x_j}{r}\right) - \\ \left(\frac{1}{\mu}\left(p_k\ell_i - p_i\ell_k - i\hbar p_j\right) - \kappa\frac{x_j}{r}\right)\left(\frac{1}{\mu}\left(p_j\ell_k - p_k\ell_j - i\hbar p_i\right) - \kappa\frac{x_i}{r}\right) \end{bmatrix}_-, \tag{5E.1b}$$

i.e.,

$$\left[A_i, A_j\right]_- = \left\{\begin{matrix} \frac{1}{\mu^2}\begin{bmatrix} \left(p_j\ell_k - p_k\ell_j - i\hbar p_i\right)\left(p_k\ell_i - p_i\ell_k - i\hbar p_j\right) - \\ \left(p_k\ell_i - p_i\ell_k - i\hbar p_j\right)\left(p_j\ell_k - p_k\ell_j - i\hbar p_i\right) \end{bmatrix} + \\ -\frac{\kappa}{\mu}\left[\frac{x_i}{r}\left(p_k\ell_i - p_i\ell_k - i\hbar p_j\right) + \left(p_j\ell_k - p_k\ell_j - i\hbar p_i\right)\frac{x_j}{r}\right] \\ +\frac{\kappa}{\mu}\left[\left(p_k\ell_i - p_i\ell_k - i\hbar p_j\right)\frac{x_i}{r} + \frac{x_j}{r}\left(p_j\ell_k - p_k\ell_j - i\hbar p_i\right)\right] \end{matrix}\right\}_-, \tag{5E.1c}$$

$$\text{i.e., } \left[A_i, A_j\right]_- = \left\{\begin{matrix} \frac{1}{\mu^2}[ab - ba] + \\ -\frac{\kappa}{\mu}\left[\frac{x_i}{r}b + a\frac{x_j}{r}\right] \\ +\frac{\kappa}{\mu}\left[b\frac{x_i}{r} + \frac{x_j}{r}a\right] \end{matrix}\right\}, \tag{5E.1d}$$

where $a = \left(p_j\ell_k - p_k\ell_j - i\hbar p_i\right)$, (5E.2a)

and $b = \left(p_k\ell_i - p_i\ell_k - i\hbar p_j\right)$. (5E.2b)

We have 30 terms on the right-hand side! The algebra is evidently laborious. Fortunately many terms drop out, and the terms that survive from $[ab - ba]$ part of Eq. 5E.1d are:

$$\frac{1}{\mu^2}\begin{pmatrix} p_j\ell_k p_k\ell_i - p_j\ell_k p_i\ell_k - p_k\ell_j p_k\ell_i + p_k\ell_j p_i\ell_k - \\ p_k\ell_i p_j\ell_k + p_k\ell_i p_k\ell_j + p_i\ell_k p_j\ell_k - p_i\ell_k p_k\ell_j \end{pmatrix} = \frac{1}{\mu^2}\begin{pmatrix} t_1 - t_2 - t_3 + t_4 - \\ t_5 + t_6 + t_7 - t_8 \end{pmatrix}. \tag{5E.3}$$

For bookkeeping, we have labeled *respective* terms above as $t_i, i = 1,2,..,7,8$.

We see that $t_1 = p_j\ell_k p_k\ell_i = p_j p_k \ell_k \ell_i = p_j p_k\left(i\hbar\ell_j + \ell_i\ell_k\right) = p_j p_k \ell_i \ell_k + i\hbar p_j p_k \ell_j$. (5E.4)

In obtaining Eq. 5E.4, we have used the familiar commutation properties

$$\left[\ell_k, p_k\right] = 0 \tag{5E.4a}$$

and

$$\left[\ell_k, \ell_i\right] = i\hbar\ell_j. \tag{5E.4b}$$

Also, we have

$$\left[p_j, p_k\right] = 0 \tag{5E.4c}$$

and

$$\left[\ell_i, p_j\right] = i\hbar p_k. \tag{5E.4d}$$

Using the commutation properties in Eq. 5E.4a–d, it follows that

$$t_1 = p_k\left(p_j\ell_i\right)\ell_k + i\hbar p_j p_k \ell_j = p_k\left(\ell_i p_j - i\hbar p_k\right)\ell_k + i\hbar p_j p_k \ell_j,$$

i.e., $t_1 = p_k\ell_i p_j\ell_k - i\hbar p_k p_k \ell_k + i\hbar p_j p_k \ell_j$, (5E.5a)

$t_4 = p_k\ell_j p_i\ell_k = i\hbar p_k p_i \ell_i + p_i\ell_k p_k\ell_j - i\hbar p_k p_k \ell_k$, (5E.5b)

$$t_6 = p_k\ell_i p_k\ell_j = p_k\left(p_k\ell_i - i\hbar p_j\right)\ell_j = p_k p_k \ell_i \ell_j - i\hbar p_k p_j \ell_j = p_k p_k\left(i\hbar\ell_k + \ell_j\ell_i\right) - i\hbar p_k p_j \ell_j,$$

i.e., $t_6 = i\hbar p_k p_k \ell_k + p_k p_k \ell_j \ell_i - i\hbar p_k p_j \ell_j = i\hbar p_k p_k \ell_k + p_k\left(\ell_j p_k - i\hbar p_i\right)\ell_i - i\hbar p_k p_j \ell_j$, (5E.5c)

$$t_7 = p_i\ell_k p_j\ell_k = p_i\left(p_j\ell_k - i\hbar p_i\right)\ell_k = p_i p_j \ell_k \ell_k - i\hbar p_i p_i \ell_k = p_j p_i \ell_k \ell_k - i\hbar p_i p_i \ell_k,$$

i.e., $t_7 = p_j\left(\ell_k p_i - i\hbar p_j\right)\ell_k - i\hbar p_i p_i \ell_k = p_j\ell_k p_i\ell_k - i\hbar p_j p_j \ell_k - i\hbar p_i p_i \ell_k$. (5E.5d)

Substituting Eq. 5E.5a–d in Eq. 5E.3, we see that
$-t_2$ in Eq. 5E.3 cancels the first term on the right-hand side of t_7 in Eq. 5E.5d,

$-t_3$ is cancelled by $p_k\ell_j p_k\ell_i$ that appears on the right-hand side of t_6,
t_5 is cancelled by the first term on the right-hand side of t_1
and $-t_8$ is cancelled by the second term on the right-hand side of t_4. After the cancelations, the residual terms give us:

$$\frac{1}{\mu^2}\begin{pmatrix} p_k\ell_i p_j\ell_k - i\hbar p_k p_k\ell_k + i\hbar p_j p_k\ell_j - p_j\ell_k p_i\ell_k - \\ p_k\ell_j p_k\ell_i + i\hbar p_k p_i\ell_i + p_i\ell_k p_k\ell_j - i\hbar p_k p_k\ell_k - \\ p_k\ell_i p_j\ell_k + i\hbar p_k p_k\ell_k + p_k\ell_j p_k\ell_i - i\hbar p_k p_i\ell_i - \\ i\hbar p_k p_j\ell_j + p_j\ell_k p_i\ell_k - i\hbar p_j p_j\ell_k - i\hbar p_i p_i\ell_k - \\ p_i\ell_k p_k\ell_j \end{pmatrix} = \frac{i\hbar}{\mu^2}\begin{pmatrix} -p_k^2\ell_k + p_j p_k\ell_j + p_k p_i\ell_i - \\ p_k^2\ell_k + p_k^2\ell_k - p_k p_i\ell_i - \\ p_k p_j\ell_j - p_j^2\ell_k - p_i^2\ell_k \end{pmatrix},$$

$$\text{i.e., } \frac{1}{\mu^2}\begin{pmatrix} p_k\ell_i p_j\ell_k - i\hbar p_k p_k\ell_k + i\hbar p_j p_k\ell_j - p_j\ell_k p_i\ell_k - \\ p_k\ell_j p_k\ell_i + i\hbar p_k p_i\ell_i + p_i\ell_k p_k\ell_j - i\hbar p_k p_k\ell_k - \\ p_k\ell_i p_j\ell_k + i\hbar p_k p_k\ell_k + p_k\ell_j p_k\ell_i - i\hbar p_k p_i\ell_i - \\ i\hbar p_k p_j\ell_j + p_j\ell_k p_i\ell_k - i\hbar p_j p_j\ell_k - i\hbar p_i p_i\ell_k - \\ p_i\ell_k p_k\ell_j \end{pmatrix} = -\frac{i\hbar}{\mu^2}\vec{p}^2\ell_k. \tag{5E.6}$$

The terms $-\left[\frac{x}{r}b + a\frac{y}{r}\right] + \left[b\frac{x}{r} + \frac{y}{r}a\right]$ in Eq. 5E.1d give us

$$-\left[\frac{x}{r}b + a\frac{y}{r}\right] + \left[b\frac{x}{r} + \frac{y}{r}a\right] = \begin{bmatrix} \frac{x_j}{r}p_j\ell_k - \frac{x_j}{r}p_k\ell_j - i\hbar\frac{x_j}{r}p_i - \\ p_j\ell_k\frac{x_j}{r} + p_k\ell_j\frac{x_j}{r} + i\hbar p_i\frac{x_j}{r} + \\ p_k\ell_i\frac{x_i}{r} - p_i\ell_k\frac{x_i}{r} - i\hbar p_j\frac{x_i}{r} - \\ \frac{x_i}{r}p_k\ell_i + \frac{x_i}{r}p_i\ell_k + i\hbar\frac{x_i}{r}p_j \end{bmatrix}. \tag{5E.7}$$

$$\text{Since } p_i\frac{x_j}{r} = \frac{x_j}{r}p_i + i\hbar\frac{x_j x_i}{r^3}, \tag{5E.8a}$$

$$\text{and } -p_j\frac{x_i}{r} = -\frac{x_i}{r}p_j - i\hbar\frac{x_i x_j}{r^3}, \tag{5E.8b}$$

it follows that

$$p_i\frac{x_j}{r} - p_j\frac{x_i}{r} + \frac{x_i}{r}p_j - \frac{x_j}{r}p_i = 0. \tag{5E.9}$$

Furthermore, in Eq. 5E.7, we may use

$$p_k\ell_i\frac{x_i}{r} = p_k\frac{x_i}{r}\ell_i = \frac{x_i}{r}p_k\ell_i + i\hbar\frac{x_i x_k}{r^3}\ell_i, \tag{5E.10a}$$

$$p_k \ell_j \frac{x_j}{r} = p_k \frac{x_j}{r} \ell_j = \frac{x_j}{r} p_k \ell_j - i\hbar \frac{x_j x_k}{r^3} \ell_j, \tag{5E.10b}$$

$$-p_i \ell_k \frac{x_i}{r} = -p_i \frac{x_i}{r} \ell_k - i\hbar p_i \frac{x_j}{r} = -\frac{x_i}{r} p_i \ell_k + i\hbar \left(\frac{1}{r} - \frac{x_i^2}{r^3} \right) \ell_k - i\hbar p_i \frac{x_j}{r}, \text{ and} \tag{5E.10c}$$

$$-p_j \ell_k \frac{x_j}{r} = -p_j \frac{x_j}{r} \ell_k - i\hbar p_j \frac{x_i}{r} = -\frac{x_j}{r} p_j \ell_k + i\hbar \left(\frac{1}{r} - \frac{x_j^2}{r^3} \right) \ell_k + +i\hbar p_j \frac{x_i}{r}. \tag{5E.10d}$$

The remaining terms in Eq. 5E.7 give us

$$\begin{bmatrix} p_k \ell_i \frac{x_i}{r} - p_i \ell_k \frac{x_i}{r} - \frac{x_i}{r} p_k \ell_i + \\ \frac{x_i}{r} p_i \ell_k - p_j \ell_k \frac{x_j}{r} + p_k \ell_j \frac{x_j}{r} + \\ \frac{x_j}{r} p_j \ell_k + \frac{x_j}{r} p_k \ell_j \end{bmatrix} = i\hbar \begin{pmatrix} \frac{x_i x_k}{r^3} \ell_i - \frac{x_j x_k}{r^3} \ell_j + \left(\frac{1}{r} - \frac{x_i^2}{r^3} \right) \ell_k + \\ \left(\frac{1}{r} - \frac{x_j^2}{r^3} \right) \ell_k + \frac{x_i}{r} p_j - \frac{x_j}{r} p_i \end{pmatrix} = R, \text{ say.} \tag{5E.11}$$

Now,

$$x_i \ell_i - x_j \ell_j = -i\hbar \begin{pmatrix} x_i \left(x_j \frac{\partial}{\partial x_k} - x_k \frac{\partial}{\partial x_j} \right) + \\ x_j \left(x_k \frac{\partial}{\partial x_i} - x_i \frac{\partial}{\partial x_k} \right) \end{pmatrix} = -i\hbar x_k \begin{pmatrix} x_j \frac{\partial}{\partial x_i} - \\ x_i \frac{\partial}{\partial x_j} \end{pmatrix} = -x_k \ell_k. \tag{5E.12}$$

Using now the fact that $x_i p_j - x_j p_i = \ell_k$, (5E.13)

The term R in Eq. 5E.11 becomes

$$R = i\hbar \left(-\frac{1}{r^3} \left(x_k^2 + x_i^2 + x_j^2 \right) \ell_k + \frac{2}{r} \ell_k + \frac{1}{r} \ell_k \right) = i\hbar \frac{2}{r} \ell_k. \tag{5E.14}$$

Finally, using Eq. 5E.6 and 5E.14, we get

$$\left[A_i, A_j \right]_- = -\frac{2i\hbar}{\mu} H \sum_k \varepsilon_{ijk} \ell_k, \tag{5E.15a}$$

$$\text{i.e., } \left[A_i, A_j \right]_- = -\frac{2i\hbar}{\mu} \left(\frac{\vec{p}^2}{2\mu} \right) \ell_k + \frac{\kappa}{\mu} \frac{2i\hbar}{r} \ell_k = -\frac{2i\hbar}{\mu} \left(\frac{\vec{p}^2}{2\mu} - \frac{\kappa}{r} \right) \ell_k. \tag{5E.15b}$$

We thus have the final result:

$$\left[A_x, A_y \right]_- = -\frac{2i\hbar}{\mu} \left(\frac{p^2}{2\mu} - \frac{\kappa}{r} \right) \ell_z = -\frac{2i\hbar}{\mu} H \ell_z. \tag{5E.16}$$

Other relations include those obtained by making cyclic changes $x \to y \to z \to x$. Further details are available in [25].

APPENDIX 5F Contour Integration in the Solution for the Continuum Coulomb Functions

In this Appendix we show how the Eq. 5.163,

$$\chi(\xi) = A\int_{\bar{C}} e^s \left(\frac{s}{\xi}\right)^{i\gamma-1} \left(1-\left(\frac{s}{\xi}\right)\right)^{-i\gamma} \frac{ds}{\xi} = A\int_{\bar{C}} e^s s^{i\gamma-1} (\xi - s)^{-i\gamma} \, ds \tag{5F.1}$$

leads to Eq. 5.164.

The integration in the complex s plane over the contour $\bar{C}$ (Fig. 5.6b) must let $\operatorname{Re}(s) \to -\infty$ to ensure that the surface terms are nullified.

We note that at $\xi = 0$, from Eq. 5.163, we have

$$\chi(\xi = 0) = A\int_{\bar{C}} e^s s^{i\gamma-1} (-1)^{-i\gamma} (s)^{-i\gamma} \, ds = (-1)^{-i\gamma} A\int_{\bar{C}} \frac{e^s}{s} ds = (-1)^{-i\gamma} A2\pi i. \tag{5F.2}$$

The branch points corresponding to $t = 0,1$ in the t plane are now at $s = 0, \xi$ (i.e., at $s = 0, i\kappa$) in the s plane. The second branch point is at a distance of κ right *directly above* the first branch point, i.e., directly in the direction of the positive imaginary axis in the complex s plane. These branch points are shown in Fig. 5.6b. The extended complex plane consists of all the points and a single point at infinity, which allows us to choose the contour $\bar{C}$ as

$$\int_{\bar{C}} ds(..) \equiv \int_{C_1} ds(..) + \int_{C_2} ds(..), \tag{5F.3a}$$

where the contours C_1 and C_2 are shown in Fig. 5F.1. The contour C_1 goes around the first branch point, and the C_2 around the second. Since our interest is in the asymptotic limit, we shall focus on the ratio $\frac{s}{\xi}$ so that we can develop an expansion in $\frac{s}{\xi}$ and retain the leading terms. Now, s changes on the contour over which the integral in Eq. 5F.1 is to be determined. Consider a typical value of s, say $s = s_0$ (Fig. 5F.1b). For the contour C_1, $s = -(s_0 \pm i\varepsilon)$ and $\frac{s}{\xi} = \frac{-(s_0 \pm i\varepsilon)}{\xi} = \frac{-(s_0 \pm i\varepsilon)}{i\kappa}$. In the asymptotic limit, $\xi \to \infty$, and therefore $\kappa \to \infty$ which makes $\frac{-(s_0 \pm i\varepsilon)}{i\kappa}$ rather small and $(\xi - s)$ can be expanded in terms of $\frac{s}{\xi}$. For the curve C_2, we have $i\kappa$ added, i.e., $s = -(s_0 \pm i\varepsilon) + i\kappa = -s_0 \mp i\varepsilon + i\kappa$ and $\frac{s}{\xi} = \frac{s}{i\kappa} = \left\{\frac{(-s_0 \mp i\varepsilon)}{i\kappa} + 1\right\}$, which is <u>*not*</u> *small*. Introducing a new variable $s' = s - \xi$ would take the contour C_2 to C_1 and give $\frac{s'}{\xi} = \frac{s - \xi}{i\kappa} = \left\{\frac{(-s_0 \mp i\varepsilon)}{i\kappa} + 1\right\} - \frac{\xi}{i\kappa} = \left\{\frac{(-s_0 \mp i\varepsilon)}{i\kappa}\right\}$, which is *small*, and we can now execute an expansion over $\frac{s'}{\xi}$.

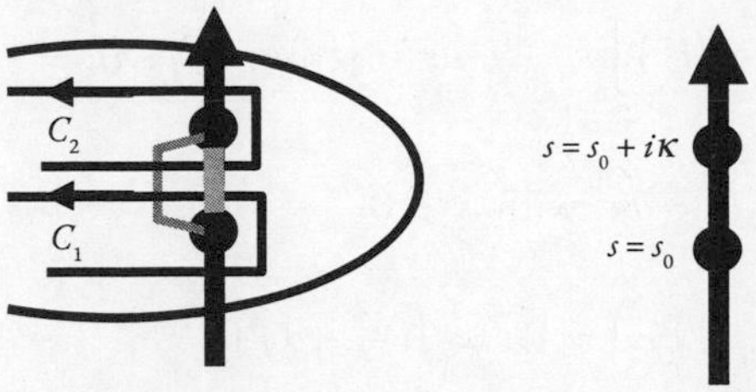

Fig. 5F.1 a, b The left panel (a) shows the choice of contour to address the branch points in the complex s plane. The flexed branch cut is also shown. The right panel (b) shows that for *each* point on the curve C_1, there is a corresponding point directly above it on C_2 at a distance $(i\kappa)$.

Recognizing now that $$\int_{\bar{C}} ds(..) \equiv \int_{C_1} ds(..) + \int_{C_1} ds'(..), \quad (5F.3b)$$

we proceed to evaluate the integral in Eq. 5F.1 for $\xi \to \infty$ to check out that we shall in fact indeed get the correct answer.

The integral takes the following form:

$$\chi(\xi) = A\left[\left\{\int_{C_1} e^s s^{i\gamma-1} \xi^{-i\gamma}\left(1-\frac{s}{\xi}\right)^{-i\gamma} ds\right\} + \left\{\int_{C_1} e^{s'+\xi} \xi^{i\gamma-1}\left(1+\frac{s'}{\xi}\right)^{i\gamma-1} (-s')^{-i\gamma} ds'\right\}\right], \quad (5F.4a)$$

$$\text{i.e., } \chi(\xi) = A\left[\xi^{-i\gamma}\left\{\int_{C_1} e^s s^{i\gamma-1}\left(1-\frac{s}{\xi}\right)^{-i\gamma} ds\right\} + \xi^{i\gamma-1} e^u (-1)^{-i\gamma}\left\{\int_{C_1} e^s\left(1+\frac{s}{\xi}\right)^{i\gamma-1} s^{-i\gamma} ds\right\}\right]. \quad (5F.4b)$$

In writing Eq. 5F.4b, we have merely extracted the factors independent of the integration variable and relabeled the (dummy) variable over which integration is carried out. In the asymptotic limit, we get

$$\chi(\xi) \underset{\frac{s}{\xi}\to 0}{=} A\left[(1\times\xi)^{-i\gamma}\left\{\int_{C_1} e^s s^{i\gamma-1} ds\right\} + (-1)e^{\xi}(1\times\xi)^{i\gamma-1}(-1)^{-i\gamma-1}\left\{\int_{C_1} e^s s^{-i\gamma} ds\right\}\right], \quad (5F.5a)$$

$$\text{i.e., } \chi(\xi) \approx A\left[\left\{\xi^{-i\gamma}\int_{C_1} e^s s^{i\gamma-1} ds\right\} - \left\{e^{\xi}(-\xi)^{i\gamma-1}\int_{C_1} e^s s^{-i\gamma} ds\right\}\right], \quad (5F.5b)$$

which can be written compactly as $$\chi(\xi) = 2\pi i A\left\{\xi^{-i\gamma} g_1(\gamma) - (-\xi)^{i\gamma-1} e^{\xi} g_2(\gamma)\right\}, \quad (5F.5c)$$

with $$2\pi i g_1(\gamma) = \int_{C_1} e^s s^{i\gamma-1} ds, \quad (5F.6a)$$

and $$2\pi i g_2(\gamma) = \int_{C_1} e^s s^{-i\gamma} ds. \quad (5F.6b)$$

Since $$2\pi i g_2(\gamma) = \int_{C_1} s^{-i\gamma} e^s ds = s^{-i\gamma} e^s\Big|_{-\infty-i\varepsilon}^{-\infty+i\varepsilon} - \int_{C_1}(-i\gamma) s^{-i\gamma-1} e^s ds, \quad (5F.7a)$$

we note that $2\pi i g_2(\gamma) = (i\gamma)\int\limits_{C_1} s^{-i\gamma-1}e^s ds = (i\gamma)g_1^*(\gamma)2\pi i,$ (5F.7b)

$$\xi = ikr(1-\cos\theta) = i\kappa \text{ with } \kappa > 0$$

$$\chi(\xi) = 2\pi i A\left\{\xi^{-i\gamma}g_1(\gamma) - (-\xi)^{i\gamma-1}e^{\xi}(i\gamma)g_1^*(\gamma)\right\}, \tag{5F.8a}$$

i.e., $\chi(\xi) = 2\pi i A\left\{\xi^{-i\gamma}g_1(\gamma) - (\xi^*)^{i\gamma-1}e^{\xi}(i\gamma)g_1^*(\gamma)\right\},$ (5F.8b)

or $\chi(\xi) = 2\pi i A\xi^{-i\gamma}g_1(\gamma)\left\{1 - \frac{(\xi^*)^{i\gamma}}{\xi^{-i\gamma}}\frac{g_1^*(\gamma)}{g_1(\gamma)}\frac{e^{\xi}(i\gamma)}{\xi^*}\right\}.$ (5F.8c)

Noting that $i\xi^* = i(-i\kappa) = \kappa = -i\xi,$ (5F.9a)

and also that $(\xi^*)^{i\gamma} = (-i\kappa)^{i\gamma} = (-i)^{i\gamma}[k(r-z)]^{i\gamma} = e^{\gamma\frac{\pi}{2}}e^{i\gamma\ln[k(r-z)]},$ (5F.9b)

and $(\xi)^{-i\gamma} = (i\kappa)^{-i\gamma} = (i)^{-i\gamma}[k(r-z)]^{-i\gamma} = e^{\gamma\frac{\pi}{2}}e^{-i\gamma\ln[k(r-z)]},$ (5F.9c)

we rewrite the solution as

$\chi(\xi) = 2\pi i A\xi^{-i\gamma}g_1(\gamma)\left\{1 + e^{i\phi(k,\xi,\gamma)}(i\gamma)\frac{e^{\xi}}{\xi}\right\},$ which is the Eq. 5.164, and (5F.10)

where $e^{i\phi(k,\xi,\gamma)} = \frac{(\xi^*)^{i\gamma}}{\xi^{-i\gamma}}\frac{g_1^*(\gamma)}{g_1(\gamma)} = e^{2i\gamma\ln[k(r-z)]}\frac{g_1{}^*(\gamma)}{g_1(\gamma)} = e^{2i\gamma\ln k(r-z)}e^{i\Theta(\gamma)},$ (5F.11)

with $\frac{g_1{}^*(\gamma)}{g_1(\gamma)} = e^{i\Theta(\gamma)}.$ (5F.12)

Solved Problems

P5.1: Show that the quantum operator for the radial momentum is not $-i\hbar\frac{\partial}{\partial r}$. Rather, show that it is $-i\hbar\left(\frac{\partial}{\partial r}+\frac{1}{r}\right)$, i.e., $-i\hbar\left(\frac{1}{r}\frac{\partial}{\partial r}\right)(r)$.

Solution:

From Eq. 5.9, we have $p_r^{(q)} = \frac{1}{2}\left[\frac{1}{r}(\vec{r}\bullet\vec{p}) + (\vec{p}\bullet\vec{r})\frac{1}{r}\right]_q$, where q stands for "quantum operator". Let us consider the coordinate representation of the state that results from the operation by $p_r^{(q)}$ on an arbitrary state $|\alpha\rangle$.

$$\langle\vec{r}|p_r^{(q)}|\alpha\rangle = \frac{1}{2}\langle\vec{r}|\hat{e}_r\bullet\vec{p}|\alpha\rangle + \frac{1}{2}\langle\vec{r}|\vec{p}\bullet\hat{e}_r|\alpha\rangle,$$

i.e., $\langle\vec{r}|p_r^{(q)}|\alpha\rangle = \frac{1}{2}\hat{e}_r\bullet\langle\vec{r}|(-i\hbar\vec{\nabla})|\alpha\rangle + \frac{1}{2}\iiint d^3\vec{r}'\langle\vec{r}|\vec{p}|\vec{r}'\rangle\bullet\langle\vec{r}'|\hat{e}_r|\alpha\rangle,$

i.e., $\langle \vec{r}|p_r^{(q)}|\alpha\rangle = \frac{1}{2}\hat{e}_r\bullet(-i\hbar\vec{\nabla})\langle \vec{r}|\alpha\rangle + \frac{1}{2}\iiint d^3\vec{r}'\langle \vec{r}|(-i\hbar\vec{\nabla})|\vec{r}'\rangle\bullet\langle \vec{r}'|\hat{e}_r|\alpha\rangle$,

i.e., $\langle \vec{r}|p_r^{(q)}|\alpha\rangle = \frac{-i\hbar}{2}\hat{e}_r\bullet\vec{\nabla}\psi_\alpha(\vec{r}) - \frac{i\hbar}{2}\iiint d^3\vec{r}'\langle \vec{r}|\vec{\nabla}|\vec{r}'\rangle\bullet\langle \vec{r}'|\hat{e}_r|\alpha\rangle$,

or, $\langle \vec{r}|p_r^{(q)}|\alpha\rangle = \frac{-i\hbar}{2}\frac{\partial\psi_\alpha(\vec{r})}{\partial r} - \frac{i\hbar}{2}\iiint d^3\vec{r}'\langle \vec{r}|\vec{\nabla}|\vec{r}'\rangle\bullet\langle \vec{r}'|\hat{e}_r|\alpha\rangle$.

Now, from the Solved Problem P1.5, we know that $\langle \vec{r}|\vec{p}|\alpha\rangle = -i\hbar\vec{\nabla}\langle \vec{r}|\alpha\rangle$,

hence $\langle \vec{r}|\vec{p}|\vec{r}'\rangle = -i\hbar\vec{\nabla}\langle \vec{r}|\vec{r}'\rangle = -i\hbar\vec{\nabla}\delta^3(\vec{r}-\vec{r}')$. Therefore,

$$\langle \vec{r}|p_r^{(q)}|\alpha\rangle = \left[\begin{array}{l}\frac{-i\hbar}{2}\frac{\partial\psi_\alpha(\vec{r})}{\partial r} - \\ \frac{i\hbar}{2}\iiint d^3\vec{r}'\vec{\nabla}\delta^3(\vec{r}-\vec{r}')\bullet\langle \vec{r}'|\hat{e}_r|\alpha\rangle\end{array}\right] = \left[\begin{array}{l}\frac{-i\hbar}{2}\frac{\partial\psi_\alpha(\vec{r})}{\partial r} - \\ \frac{i\hbar}{2}\vec{\nabla}\iiint d^3\vec{r}'\delta^3(\vec{r}-\vec{r}')\bullet\langle \vec{r}'|\hat{e}_r|\alpha\rangle\end{array}\right],$$

i.e., $\langle \vec{r}|p_r^{(q)}|\alpha\rangle = \frac{-i\hbar}{2}\frac{\partial\psi_\alpha(\vec{r})}{\partial r} - \frac{i\hbar}{2}\vec{\nabla}\bullet\langle \vec{r}|\hat{e}_r|\alpha\rangle = \frac{-i\hbar}{2}\frac{\partial\psi_\alpha(\vec{r})}{\partial r} - \frac{i\hbar}{2}\vec{\nabla}\bullet(\hat{e}_r\psi_\alpha(\vec{r}))$,

or, $\langle \vec{r}|p_r^{(q)}|\alpha\rangle = \frac{-i\hbar}{2}\frac{\partial\psi_\alpha(\vec{r})}{\partial r} - \frac{i\hbar}{2}\left[(\vec{\nabla}\bullet\hat{e}_r)\psi_\alpha(\vec{r}) + \hat{e}_r\bullet\vec{\nabla}\psi_\alpha(\vec{r})\right] = -i\hbar\left(\frac{\partial}{\partial r}+\frac{1}{r}\right)\psi_\alpha(\vec{r})$,

since $\vec{\nabla}\bullet\hat{e}_r = \frac{2}{r}$.

Hence, $p_r^{(q)} = -i\hbar\left(\frac{\partial}{\partial r}+\frac{1}{r}\right) \equiv \left(\frac{1}{r}\frac{\partial}{\partial r}\right)(r)$,

as $\left(\frac{1}{r}\frac{\partial}{\partial r}\right)rf(\vec{r}) = \left(\frac{\partial}{\partial r}+\frac{1}{r}\right)f(\vec{r})$ for arbitrary $f(\vec{r})$.

P5.2: Show that the quantum operator for the radial momentum $p_r^{(q)} = -i\hbar\left(\frac{\partial}{\partial r}+\frac{1}{r}\right)$ is Hermitian.

Solution:

Using the result of the Solved Problem P1.5, we have

$$\langle\beta|p_r^q|\alpha\rangle = \iiint d^3\vec{r}\langle\beta|\vec{r}\rangle\langle\vec{r}|p_r^q|\alpha\rangle = \iint d\Omega\int_{r=0}^{\infty} r^2dr\langle\beta|\vec{r}\rangle p_r^q\langle\vec{r}|\alpha\rangle.$$

Using further the result of P5.1,

$$\langle\beta|p_r^q|\alpha\rangle = \iint d\Omega\int_{r=0}^{\infty} r^2dr\psi_\beta^*(\vec{r})\left(\frac{\hbar}{i}\frac{1}{r}\frac{\partial}{\partial r}\right)(r)\psi_\alpha(\vec{r}) = \iint d\Omega\int_{r=0}^{\infty} r^2dr\psi_\beta^*(\vec{r})\left(\frac{\hbar}{i}\frac{1}{r}\frac{\partial}{\partial r}\right)(r\psi_\alpha(\vec{r}))$$

$\langle\beta|p_r^q|\alpha\rangle = \frac{\hbar}{i}\iint d\Omega\int_{r=0}^{\infty} dr(r\psi_\beta^*(\vec{r}))\frac{\partial}{\partial r}(r\psi_\alpha(\vec{r}))$. Integrating now the radial integral by parts,

$$\int_{r=0}^{\infty} dr(r\psi_\beta^*(\vec{r}))\frac{\partial}{\partial r}(r\psi_\alpha(\vec{r})) = r\psi_\beta^*(\vec{r})(r\psi_\alpha(\vec{r}))\Big|_0^\infty - \int_{r=0}^{\infty} dr\left[\frac{\partial}{\partial r}(r\psi_\beta^*(\vec{r}))\right](r\psi_\alpha(\vec{r})).$$

The first term on the right-hand side of this equation vanishes.

Hence, $\langle\beta|p_r^q|\alpha\rangle = \iint d\Omega \int_{r=0}^{\infty} dr\left[-\frac{\hbar}{i}\frac{\partial}{\partial r}\left(r\psi_\beta^*(\vec{r})\right)\right]\left(r\psi_\alpha(\vec{r})\right).$

Complex conjugation of this relation gives

$$\langle\beta|p_r^q|\alpha\rangle^* = \iint d\Omega \int_{r=0}^{\infty} dr\left(r\psi_\alpha^*(\vec{r})\right)\left[\frac{\hbar}{i}\frac{\partial}{\partial r}\left(r\psi_\beta(\vec{r})\right)\right],$$

i.e., $\langle\beta|p_r^q|\alpha\rangle^* = \iint d\Omega \int_{r=0}^{\infty} r^2 dr\left(\psi_\alpha^*(\vec{r})\right)\left[\frac{\hbar}{i}\frac{1}{r}\frac{\partial}{\partial r}\left(r\psi_\beta(\vec{r})\right)\right] = \langle\alpha|p_r^q|\beta\rangle.$

This equation is in conformity with the defining character of a Hermitian operator (Eq. 1.30f). In other words, $p_r^q = \left(p_r^q\right)^\dagger$.

P5.3: Determine the expectation values of the radial position operator $\langle r\rangle_{n\ell}$, for (a) 1s, (b) 3d, and (c) 4d radial orbitals. Is your answer the same as the value of r at which the radial probability density (Fig. 5.4) is a maximum? Comment on your findings.

Solution:

We shall develop a general strategy to determine $\langle r^s\rangle$. In the present problem, our interest is in $s = +1$ but in the next problem we shall determine $\langle r^s\rangle$ also with $s = -1, -2, -3$.

$\langle r^s\rangle = \langle R_{n\ell} | r^s | R_{n\ell}\rangle = \int_0^\infty R_{n\ell} r^s R_{n\ell} r^2 dr$ where the normalized radial wavefunction of the hydrogen atom is given by $R_{n\ell} = \left[\frac{1}{\left[\left(\frac{n}{2}\right)^{2\ell+3} 2n\frac{(n+\ell)!}{(n-\ell+1)!}\right]^{1/2}}\right] r^\ell e^{-r/n} L_{n-\ell-1}^{2\ell-1}$. The function $L_{n-\ell-1}^{2\ell-1}(2r/n)$ is the associated Laguerre polynomial. Let $X = \frac{2r}{n}$ therefore, $r = \frac{n}{2}X$. Also, let $n' = n - \ell - 1$ and $\alpha = 2\ell + 1$.

$$\text{Accordingly, } \langle r^s\rangle = N_{n\ell}^2\left(\frac{n}{2}\right)^{2\ell+3+s}\int_0^\infty e^{-X}L_{n'}^\alpha(X)L_{n'}^\alpha(X)X^{2\ell+2+s}dX = N_{n\ell}^2\left(\frac{n}{2}\right)^{2\ell+3+s} J_{n'n'}(s), \quad \text{(P5.3.1)}$$

wherein $J_{n'n'}(s)$ denotes the integral which can be evaluated using the generating function of the Laguerre polynomials.

The generating function is $G^\alpha(u,x) = \sum_n u^{n'}L_{n'}^\alpha = \frac{1}{(1-u)^{\alpha+1}}exp\left(\frac{-xu}{1-u}\right).$

Consider $G^\alpha(u,x)G^\alpha(t,x) = \sum_{n'm'} u^{n'}u^{m'}J_{n'm'}(s) = \sum_{n'm'} u^{n'}u^{m'}\int_0^\infty e^{-X}L_{n'}^\alpha(X)L_{m'}^\alpha(X)X^{2\ell+2+s}dX.$

Substituting the expression for the generating function, and simplifying, we get

$$G^\alpha(u,x)G^\alpha(t,x) = \int_0^\infty \frac{1}{[(1-u)(1-t)]^{\alpha+1}} exp\left(\frac{1-tu}{(1-u)(1-t)}\right)X^{2\ell+2+s}dX.$$

Let $Y = XA$, and define the integral $I = \frac{1}{[(1-u)(1-t)]^{\alpha+1}}\int_0^\infty \frac{1}{A^{2\ell+3+s}}Y^{2\ell+2+s}e^{-Y}dY.$

Hence,

$$I = \frac{1}{[(1-u)(1-t)]^{\alpha+1}} \int_0^{\infty} \frac{1}{A^{2\ell+3+s}} Y^{2\ell+2+s+1-1} e^{-Y} dY = \frac{[(1-u)(1-t)]^{1+s}}{(1-ut)^{2\ell+3+s}} (2\ell+s+2)! = I_{eq} + I_{neq}, \qquad \text{(P5.3.2)}$$

where I_{eq} contains the terms having equal powers of u and t, and I_{neq} contains the rest of the terms. The terms contained in I_{eq} are the only ones required to obtain $J_{n'n'}(s)$, and will be of the form $I_{eq} = \sum_k (ut)^k J_{n'n'}(s)$.

For $s = +1$, $\langle r^{+1} \rangle = N_{n\ell}^{\,2} \left(\frac{n}{2}\right)^{2\ell+3+1} J_{n'n'}(+1) = N_{n\ell}^{\,2} \left(\frac{n}{2}\right)^{2\ell+4} J_{n'n'}(+1)$, and the integral I can be written as

$$I = \frac{[(1-u)(1-t)]^2}{(1-ut)^{2\ell+4}} (2\ell+3)! = I_{eq} + I_{neq}.$$

The terms with equal powers of u and t in the numerator of equation (P5.3.2) are $1 + 4ut + (ut)^2$. Expanding the denominator of equation (in Eq. P5.3.2) as a binomial series, we get

$$I_{eq} = \left[1 + 4ut + (ut)^2\right] \sum_k \binom{2\ell+4+k-1}{k} (ut)^k = \sum_k \frac{\left[1 + 4ut + (ut)^2\right](2\ell+3+k)!}{k!} (ut)^k,$$

wherein, the notation $\binom{n}{r}$ denotes the binomial coefficient $\frac{n!}{(n-r)!r!}$. Now, expanding the summation and picking out $(ut)^{k=n-\ell-1}$ terms, we get

$$I_{eq} = \frac{(n+\ell)!}{(n-\ell-3)!} \left\{ \frac{(n+\ell+2)(n+\ell+1)}{(n-\ell-1)(n-\ell-2)} + 4\frac{(n+\ell+1)}{(n-\ell-2)} + 1 \right\} (u\iota)^{n-\ell-1}.$$

On multiplying and dividing by $(n-\ell-1)(n-\ell-2)$,

$$I_{eq} = \frac{(n+\ell)!}{(n-\ell-3)!} \left\{ \frac{(n+\ell+2)(n+\ell+1)}{(n-\ell-1)(n-\ell-2)} + 4\frac{(n+\ell+1)(n-\ell-1)}{(n-\ell-2)(n-\ell-1)} + \frac{(n-\ell-2)(n-\ell-1)}{(n-\ell-2)(n-\ell-1)} \right\} (ut)^{n-\ell-1},$$

$$I_{eq} = \frac{(n+\ell)!}{(n-\ell-1)!} \times 2\{3n^2 - \ell(\ell+1)\} (ut)^{n-\ell-1}.$$

Therefore, $J_{n'n'} = \frac{(n+\ell)!}{(n-\ell-1)!} \times 2\{3n^2 - \ell(\ell+1)\}$.

$$\langle r^{+1} \rangle_{n\ell} = N_{n\ell}^{\,2} \left(\frac{n}{2}\right)^{2\ell+4} J_{n'n'}(+1) = \frac{2(3n^2 - \ell(\ell+1))}{\left(\frac{n}{2}\right)^{2l+3} \frac{2n(n+\ell)!}{(n-\ell-1)!}} \left(\frac{n}{2}\right)^{2\ell+4} \frac{(n+\ell)!}{(n-\ell-1)!} = \frac{1}{2}[3n^2 - \ell(\ell+1)].$$

Rewriting $r = a_0 r$, we obtain $\langle r \rangle_{n\ell} = \frac{a_0}{2}[3n^2 - \ell(\ell+1)]$

$n = 1, \ell = 0, \langle r \rangle_{10} = \frac{3}{2}\frac{a_0}{Z}$	$n = 3, \ell = 2, \langle r \rangle_{32} = \frac{21}{2}\frac{a_0}{Z}$	$n = 3, \ell = 2, \langle r \rangle_{42} = 21\frac{a_0}{Z}$

An expectation value is the weighted average of all possible outcomes of a measurement, whereas the distance at which the radial probability density has its maximum is the most probable value of the measurement. The expectation value of the operator r is therefore always higher than the distance at which the radial probability density is maximum.

P5.4: Determine the expectation values of the operators, $\left\langle \frac{1}{r} \right\rangle_{n\ell}$ and $\langle r^2 \rangle_{n\ell}$, for (a) 1s, (b) 3d, and (c) 4d radial orbitals.

Solution:

For $s = -1$, Eq. P5.3.1 yields $\langle r^{-1} \rangle = N_{n\ell}^2 \left(\frac{n}{2}\right)^{2\ell+3-1} J_{n'n'}(s) = N_{n\ell}^2 \left(\frac{n}{2}\right)^{2\ell+2} J_{n'n'}(-1)$.

The integral I (in Eq. P5.3.2) becomes $\frac{1}{(1-ut)^{2\ell+2}}(2\ell+1)! = I_{eq} + I_{neq}$.

Expanding the terms with equal powers in a binomial series, we get

$$I_{eq} = \sum_{k=0}^{\infty} (ut)^k \binom{2\ell+1+k}{k}(2\ell+1)! = \sum_{k=0}^{\infty} (ut)^k \frac{(2\ell+1+k)!}{(2\ell+1)!k!}(2\ell+1)!.$$

Coefficients of $(ut)^{k=n'=n-\ell-1}$ give $J_{n'n'}(-1) = \frac{(n+\ell)!}{(n-\ell-1)} = I_{eq}$.

Hence,

$$\langle r^{-1} \rangle = N_{n\ell}^2 \left(\frac{n}{2}\right)^{2\ell+2} J_{n'n'} = \frac{1}{\left[\left(\frac{n}{2}\right)^{2\ell+3} 2n \frac{(n+\ell)!}{(n-\ell+1)!}\right]} \left(\frac{n}{2}\right)^{2\ell+2} \frac{(n+\ell)!}{(n-\ell-1)} = \frac{1}{n^2}.$$

Considering $r_p = a_0 r$, we get the result for the expectation value: $\left\langle \frac{1}{r} \right\rangle = \frac{Z}{a_0 n^2}$.

For $s = 2$, Eq. P5.3.1 gives $\langle r^{+2} \rangle = N_{n\ell}^{\ 2} \left(\frac{n}{2}\right)^{2\ell+3+2} J_{n'n'}(+2) = N_{n\ell}^{\ 2} \left(\frac{n}{2}\right)^{2\ell+5} J_{n'n'}(+2)$.

Thus, Eq. P5.3.2 becomes $\frac{[(1-u)(1-t)]^3}{(1-ut)^{2\ell+5}}(2\ell+4)! = I_{eq} + I_{neq}$.

Consideration of equal powers of u and t gives $I_{eq} = \left[1 + 9(ut) + 9(ut)^2 + (ut)^3\right] \sum_k \frac{(2\ell+4+k)!}{(2\ell+4)!k!}(ut)^k$.

Collecting the coefficients of powers of (ut) that equal $(n-\ell-1)$, we get $\langle r^2 \rangle_{n\ell} = \frac{n^2}{2}[5n^2 - 3\ell^2 - 3\ell + 1]$.

Finally, since $r_p = a_0 r$, we get $\langle r^2 \rangle_{n\ell} = \frac{n^2}{2}[5n^2 - 3\ell^2 - 3\ell + 1]\left(\frac{a_0}{Z}\right)^2$.

The following table gives the expectation values for the 1s, 3d, and 4d orbitals:

Orbital	$\left\langle \frac{1}{r} \right\rangle_{n\ell}$	$\langle r^2 \rangle$
1s	$\frac{Z}{a_0}$	$3\left(\frac{a_0}{Z}\right)^2$
3d	$\frac{1}{9}\frac{Z}{a_0}$	$126\left(\frac{a_0}{Z}\right)^2$
4d	$\frac{1}{16}\frac{Z}{a_0}$	$504\left(\frac{a_0}{Z}\right)^2$

P5.5: Determine the expectation values of the operators, $\left\langle \frac{1}{r^2} \right\rangle_{n\ell}$ and $\left\langle \frac{1}{r^3} \right\rangle_{n\ell}$, for (a) 1s, (b) 3d, and (c) 4d radial orbitals.

Solution:

Case 1: $\left\langle \frac{1}{r^2} \right\rangle$

Consider Eq. P5.3.1. With $s = -2$ we have $\langle r^{-2} \rangle = N_{n\ell}^2 \left(\frac{n}{2}\right)^{2\ell+3-2} J_{n'n'}(-2)$.

The integral I (Eq. P5.3.2) becomes $I = \frac{[(1-u)(1-t)]^{1-2}}{(1-ut)^{2l+3-2}}(2\ell - 2 + 2)! = I_{eq} + I_{neq}$.

Therefore, $I_{eq} + I_{neq} = \frac{[(1-u)(1-t)]^{-1}}{(1-ut)^{2\ell+1}}(2\ell)! = \frac{1}{(1-u)}\frac{1}{(1-t)}\frac{1}{(1-ut)^{2\ell+1}}(2\ell)!$.

The first and the second term on the left-hand side can be written as

$$\frac{1}{1-u} = \sum_{k_1=0}^{\infty} u^{k_1} \text{ and } \frac{1}{1-t} = \sum_{k_2=0}^{\infty} t^{k_2}.$$

Therefore, $\frac{1}{1-u}\frac{1}{1-t} = \sum_{k_1=0}^{\infty}\sum_{k_2=0}^{\infty} u^{k_1} t^{k_2}$.

Since we consider equal powers, $k_1 = k_2 = k$, $\frac{1}{1-u}\frac{1}{1-t} = \sum_{k}^{\infty} (ut)^k$.

Now, $\frac{1}{(1-ut)^{2\ell+1}} = \sum_{p=0}^{\infty} \binom{2\ell+1+p-1}{p} (ut)^p = \sum_{p=0}^{\infty} \binom{2\ell+p}{p} (ut)^p$.

Therefore, $I_{eq} = (2\ell)! \sum_{k,p} (ut)^{k+p} \binom{2\ell+p}{p}$.

Changing the order of summation,

$$I_{eq} = (2\ell)! \sum_{k,p} (ut)^{k+p} \binom{2\ell+p}{p} \sum_{r=0}^{\infty} \delta(r, k+p) = (2\ell)! \sum_{r} (ut)^r \left[\sum_{p=0}^{r} \binom{2\ell+p}{p} \right],$$

i.e., $I_{eq} = (2\ell)! \sum_{r} (ut)^r \binom{2\ell+r+1}{r} = (2\ell)! \sum_{r} (ut)^r \frac{(2\ell+r+1)!}{(2\ell+1)!r!}$.

Taking only the coefficients of $(ut)^{r=n-\ell-1}$: $J_{n'n'}(-2) = \frac{1}{(2\ell+1)}\frac{(n+\ell)!}{(n-\ell-1)!}$.

Therefore, $\langle r^{-2} \rangle = \frac{1}{\left[\left(\frac{n}{2}\right)^{2\ell+3} 2n \frac{(n+\ell)!}{(n-\ell+1)!} \right]} \left(\frac{n}{2}\right)^{2\ell+1} \frac{1}{(2\ell+1)}\frac{(n+\ell)!}{(n-\ell-1)!} = \frac{1}{n^3\left(\ell + \frac{1}{2}\right)}$.

Since, $r_{physical} = a_0 r$, $\left\langle \frac{1}{r^2} \right\rangle = \left(\frac{Z}{a_0}\right)^2 \frac{1}{n^3(\ell + 1/2)}$.

Case 2: $\left\langle \frac{1}{r^3} \right\rangle$

Consider Eq. P5.3.1 (with $s = -3$): $\langle r^{-3} \rangle = N_{n\ell}^2 \left(\frac{n}{2}\right)^{2\ell+3-3} J_{n'n'}(s) = N_{n\ell}^2 \left(\frac{n}{2}\right)^{2\ell} J_{n'n'}(-3).$

The integral I (Eq. 5.3.2) becomes

$$I = \frac{[(1-u)(1-t)]^{1-3}}{(1-ut)^{2\ell+3-3}}(2\ell - 3 + 2)! = I_{eq} + I_{neq} = \frac{1}{(1-u)^2}\frac{1}{(1-t)^2}\frac{1}{(1-ut)^{2\ell}}(2\ell - 1)!.$$

From the first two terms in this equation (considering equal powers in u and t),

$$\frac{1}{(1-u)^2}\frac{1}{(1-t)^2} = \sum_{k=0}^{\infty}\binom{k+1}{k}^2 (ut)^k.$$

The third term of the equation can be expressed as $\frac{1}{(1-ut)^{2\ell}} = \sum_{p=0}^{\infty}\binom{2\ell+p-1}{p}(ut)^p.$

Therefore, $I_{eq} = (2\ell-1)!\sum_{k=0}^{\infty}\binom{k+1}{k}^2 (ut)^k \sum_{p=0}^{\infty}\binom{2\ell+p-1}{p}(ut)^p.$

Changing the order of summation, we get $I_{eq} = (2\ell-1)!\sum_{k,p}(ut)^{k+p}(k+1)^2\binom{2\ell+p-1}{p}\sum_{r=0}^{\infty}\delta(r,k+p),$

or, $I_{eq} = (2\ell-1)!\sum_{r}(ut)^r\left[\sum_{p=0}^{r}(r-p+1)^2\binom{2\ell+p-1}{p}\right].$

To evaluate the term in the square brackets of this equation, we make use of the following three formulae

$$\sum_{p=0}^{r}\binom{m+p}{p} = \binom{m+r+1}{r},$$

$$\sum_{p=0}^{r}p\binom{m+p}{p} = (m+1)\binom{m+r+1}{r-1},$$

and $\sum_{p=0}^{r}p^2\binom{m+p}{p} = (m+1)(m+2)\binom{m+r+1}{r-2} + (m+1)\binom{m+r+1}{r-1}.$

For the case when $m = 2\ell - 1$ and $r + 1 = n - 1$, we get

$$I_{eq} = (2\ell-1)!\sum_{r}(ut)^r\left[\sum_{p=0}^{r}(r+1-p)^2\binom{2\ell+p-1}{p}\right],$$

i.e., $I_{eq} = (2\ell-1)!\sum_{r}(ut)^r\left[\sum_{p=0}^{r}(r+1)^2\binom{2\ell+p-1}{p} + p^2\binom{2\ell+p-1}{p} - 2(r+1)p\binom{2\ell+p-1}{p}\right].$

Considering now the coefficients of $(ut)^{n-\ell-1}$, we can write

$$J_{n'n'} = (2\ell-1)!\sum_{p=0}^{r}\left[(n-\ell)^2\binom{2\ell+p-1}{p} + p^2\binom{2\ell+p-1}{p} - 2(r+1)p\binom{2\ell+p-1}{p}\right],$$

which simplifies to $J_{n'n'} = \dfrac{(n+\ell)!}{(n-\ell-1)!}\dfrac{n}{4\ell(\ell+1/2)(\ell+1)}$.

Thus,

$$\langle r^{-3}\rangle = \frac{1}{\left[\left(\frac{n}{2}\right)^{2\ell+3} 2n\frac{(n+\ell)!}{(n-\ell+1)!}\right]}\left(\frac{n}{2}\right)^{2\ell}\frac{(n+\ell)!}{(n-\ell-1)!}\frac{n}{4\ell(\ell+1/2)(\ell+1)} = \frac{1}{n^3\ell(\ell+1/2)(\ell+1)}.$$

Since $r_{physical} = a_0 r$, we get $\left\langle \dfrac{1}{r^3}\right\rangle = \left(\dfrac{Z}{a_0}\right)^3 \dfrac{1}{n^3\ell(\ell+1/2)(\ell+1)}$.

The following table summarizes the results for the 1s, 3d, and 4d orbitals.

Orbital	$\left\langle \frac{1}{r^2}\right\rangle_{n\ell}$	$\left\langle \frac{1}{r^3}\right\rangle_{n\ell}$
1s	$2\left(\frac{Z}{a_0}\right)^2$	
3d	$\frac{2}{135}\left(\frac{Z}{a_0}\right)^2$	$405\left(\frac{Z}{a_0}\right)^3$
4d	$\frac{1}{160}\left(\frac{Z}{a_0}\right)^2$	$960\left(\frac{Z}{a_0}\right)^3$

P5.6: Using the Pauli–Lenz operator and the pseudo-angular momentum operators $\vec{I}$ and $\vec{K}$, obtain the 2s radial wave function of the hydrogen atom.

Solution:

We shall first determine the wavefunction of the 2p orbital.

We know that $\vec{I}|n=2\rangle = \frac{1}{2}|n=2\rangle$

and $\vec{K}|n=2\rangle = \frac{1}{2}|n=2\rangle$.

Hence, $\vec{\ell}|n=2,\ell=0\rangle = \left(\vec{I}+\vec{K}\right)|n=2,\ell=1\rangle = 1|n=2,\ell=1\rangle$

and $\vec{A}|n=2,\ell=1\rangle = \left(\vec{I}-\vec{K}\right)|n=2,\ell=1\rangle = 0$.

The Lenz operator $\vec{A}|n=2,\ell=1\rangle = \left[(\vec{p}\times\vec{\ell}) - i\hbar\vec{p}\right]|n=2,\ell=1\rangle - \mu\kappa\hat{r}|n=2,\ell=1\rangle$

becomes

$$0 = (\vec{p}\times\vec{\ell})|n=2,\ell=1\rangle - i\hbar\vec{p}|n=2,\ell=1\rangle - \mu\kappa\hat{r}|n=2,\ell=1\rangle.$$

Considering only the radial part of the wavefunction,

$$0 = -i\hbar\langle\vec{r}|\vec{p}|n=2,\ell=1\rangle - \mu\kappa\langle\vec{r}|\hat{r}|n=2,\ell=1\rangle.$$

Consider

$$\frac{d}{dr}\left(\frac{R_{21}}{r}\right)+\frac{\mu\kappa}{\hbar^2}\left(\frac{R_{21}}{r}\right)=\frac{r\frac{dR_{21}}{dr}-R_{21}}{r^2}+\frac{\kappa\mu}{\hbar^2}\frac{R_{21}}{r}=\frac{dR_{21}}{dr}-\frac{R_{21}}{r}+\frac{\kappa\mu}{\hbar^2}R_{21}=0$$

Performing integration,

$$\int\frac{dR_{21}}{R_{21}}=\int\left(\frac{1}{r}-\frac{\kappa\mu}{\hbar^2}\right)dr+c \text{ gives } \ln(R_{21})=\ln r-\frac{\kappa\mu}{\hbar^2}r+c.$$

Therefore, $R_{21}=C\times re^{-Zr/2a_0}$, where $C=\frac{1}{\sqrt{3}}\frac{Z}{a_0}\left(\frac{Z}{2a_0}\right)^{3/2}$.

Hence, $R_{21}=\frac{1}{\sqrt{3}}\left(\frac{Z}{2a_0}\right)^{3/2}\frac{Z}{a_0}re^{-Zr/2a_0}$.

Now, to determine the wavefunction of the 2s orbital, we use the recursive relation

$$\frac{R_{n,l-1}(r)}{r^{l-1}}=\frac{n}{\sqrt{n^2-l^2}}\left\{\left(\frac{2l+1}{Z}\right)la_0+\frac{la_0}{Z}r\frac{d}{dr}-r\right\}\frac{R_{nl}}{r^l}. \qquad \text{P5.6.1}$$

Hence,

$$R_{2,0}(r)=\frac{2}{\sqrt{3}}\frac{a_0}{Z}\left\{3+r\frac{d}{dr}-\frac{Z}{a_0}r\right\}\frac{R_{21}}{r}=\frac{2}{\sqrt{3}}\frac{a_0}{Z}\left\{3+r\frac{d}{dr}-\frac{Z}{a_0}r\right\}\frac{1}{\sqrt{3}}\left(\frac{Z}{2a_0}\right)^{3/2}\frac{Z}{a_0}e^{-Zr/2a_0}.$$

This equation simplifies to $R_{2,0}(r)=\frac{2}{3}\left(\frac{Z}{2a_0}\right)^{3/2}\left\{3-\frac{Zr}{2a_0}-\frac{Z}{a_0}r\right\}e^{-Zr/2a_0}$.

Therefore, $R_{2,0}(r)=2\left(\frac{Z}{2a_0}\right)^{3/2}\left(1-\frac{Zr}{2a_0}\right)e^{-Zr/2a_0}$.

P5.7: Show that in the parabolic coordinate system the Laplacian operator is given by

$$\nabla^2=\frac{4}{(u+v)}\left[\frac{\partial}{\partial u}\left(u\frac{\partial}{\partial u}\right)+\frac{\partial}{\partial v}\left(v\frac{\partial}{\partial v}\right)\right]+\frac{1}{uv}\frac{\partial^2}{\partial\varphi^2}.$$

Solution:

We employ inverse transformations:

$u=u(x,y,z), v=v(x,y,z)$ and $\varphi=\varphi(x,y,z)$.

We have $x^2+y^2=(2z+v)v$; $x^2+y^2=u(u-2z)$ and $\varphi=\tan^{-1}\left(\frac{y}{x}\right)$,

where $u=r+z$; $v=r-z$ and $\varphi=\tan^{-1}\left(\frac{y}{x}\right)$ and $r=\sqrt{x^2+y^2+z^2}=\frac{u+v}{2}$.

We see that

$$\frac{\partial u}{\partial x}=\frac{\partial r}{\partial x}+0=\frac{x}{r}, \qquad \frac{\partial v}{\partial x}=\frac{\partial r}{\partial x}-0=\frac{x}{r}, \qquad \frac{\partial\varphi}{\partial x}=\frac{-\sin\varphi}{\sqrt{uv}},$$

$$\frac{\partial u}{\partial y}=\frac{\partial r}{\partial y}+0=\frac{y}{r},\qquad \frac{\partial v}{\partial y}=\frac{\partial r}{\partial y}-0=\frac{y}{r},\qquad \frac{\partial\varphi}{\partial y}=\frac{\cos\varphi}{\sqrt{uv}},$$

$$\frac{\partial u}{\partial z}=\frac{z}{r}+1=\frac{u}{r},\qquad \frac{\partial v}{\partial z}=\frac{z}{r}-1=-\frac{v}{r},\qquad \frac{\partial\varphi}{\partial z}=0.$$

Now, $\frac{\partial}{\partial x}=\frac{\partial u}{\partial x}\frac{\partial}{\partial u}+\frac{\partial v}{\partial x}\frac{\partial}{\partial v}+\frac{\partial\varphi}{\partial x}\frac{\partial}{\partial\varphi}=\frac{x}{r}\frac{\partial}{\partial u}+\frac{x}{r}\frac{\partial}{\partial v}-\frac{\sin\varphi}{\sqrt{uv}}\frac{\partial}{\partial\varphi}$, hence

$$\frac{\partial^2}{\partial x^2}=\left[\begin{array}{l}\frac{1}{r}\frac{\partial}{\partial u}-\frac{x^2}{r^3}\frac{\partial}{\partial u}+\left(\frac{x}{r}\right)^2\frac{\partial^2}{\partial u^2}+\left(\frac{x}{r}\right)^2\frac{\partial^2}{\partial u\partial v}-\frac{x}{r}\frac{\sin\varphi}{\sqrt{uv}}\frac{\partial^2}{\partial u\partial\phi}+\frac{1}{r}\frac{\partial}{\partial v}-\frac{x^2}{r^3}\frac{\partial}{\partial v}\\ +\left(\frac{x}{r}\right)^2\frac{\partial^2}{\partial v^2}+\left(\frac{x}{r}\right)^2\frac{\partial^2}{\partial v\partial u}-\frac{x}{r}\frac{\sin\phi}{\sqrt{uv}}\frac{\partial^2}{\partial v\partial\phi}+\frac{\partial}{\partial x}\left(\frac{-\sin\varphi}{\sqrt{uv}}\frac{\partial}{\partial\varphi}\right)\end{array}\right]. \qquad \text{P5.7.1}$$

Similarly, $\frac{\partial}{\partial y}=\frac{y}{r}\frac{\partial}{\partial u}+\frac{y}{r}\frac{\partial}{\partial v}+\frac{\cos\varphi}{\sqrt{uv}}\frac{\partial}{\partial\varphi}$ and hence

$$\frac{\partial^2}{\partial y^2}=\left[\begin{array}{l}\frac{1}{r}\left(\frac{\partial}{\partial u}+\frac{\partial}{\partial v}\right)-\frac{y^2}{r^3}\left[\frac{\partial}{\partial u}+\frac{\partial}{\partial v}\right]+\left(\frac{y}{r}\right)^2\left[\frac{\partial^2}{\partial u^2}+\frac{\partial^2}{\partial v^2}\right]+2\left(\frac{y}{r}\right)^2\left[\frac{\partial^2}{\partial u\partial v}\right]\\ +\frac{y}{r}\frac{\cos\varphi}{\sqrt{uv}}\left[\frac{\partial^2}{\partial u\partial\varphi}+\frac{\partial^2}{\partial v\partial\varphi}\right]-\frac{y\cos\varphi}{2r\sqrt{uv}}\left[\frac{1}{u}\frac{\partial}{\partial\varphi}+\frac{1}{v}\frac{\partial}{\partial\varphi}\right]\\ -\frac{\sin\varphi\cos\varphi}{uv}\frac{\partial}{\partial\varphi}+\frac{\cos^2\varphi}{\left(\sqrt{uv}\right)^2}\frac{\partial^2}{\partial\varphi^2}\end{array}\right]. \qquad \text{P5.7.2}$$

Finally, $\frac{\partial}{\partial z}=\frac{\partial u}{\partial z}\frac{\partial}{\partial u}+\frac{\partial v}{\partial z}\frac{\partial}{\partial v}+\frac{\partial\varphi}{\partial z}\frac{\partial}{\partial\varphi}=\frac{u}{r}\frac{\partial}{\partial u}+\frac{-v}{r}\frac{\partial}{\partial v}$

and hence $\frac{\partial^2}{\partial z^2}=\frac{\partial}{\partial z}\left(\frac{u}{r}\frac{\partial}{\partial u}-\frac{v}{r}\frac{\partial}{\partial v}\right)$,

$$\text{i.e., }\frac{\partial^2}{\partial z^2}=\left[\begin{array}{l}\frac{\partial u}{\partial z}\frac{1}{r}\frac{\partial}{\partial u}-\frac{\partial v}{\partial z}\frac{1}{r}\frac{\partial}{\partial v}+\left(\frac{-u}{r^2}\right)\frac{\partial r}{\partial z}\frac{\partial}{\partial u}+\frac{v}{r^2}\frac{\partial r}{\partial z}\frac{\partial}{\partial v}\\ +\frac{u}{r}\left(\frac{u}{r}\frac{\partial}{\partial u}-\frac{v}{r}\frac{\partial}{\partial v}\right)\frac{\partial}{\partial u}-\frac{v}{r}\left(\frac{u}{r}\frac{\partial}{\partial u}-\frac{v}{r}\frac{\partial}{\partial v}\right)\frac{\partial}{\partial v}\end{array}\right].$$

Substituting now $z=\frac{u-v}{2}$ and using $\frac{\partial r}{\partial z}=\frac{z}{r}$, we get

$$\frac{\partial^2}{\partial z^2}=\left[\begin{array}{l}\frac{u}{r}\frac{1}{r}\frac{\partial}{\partial u}+\frac{v}{r}\frac{1}{r}\frac{\partial}{\partial v}-\frac{u}{r^2}\frac{z}{r}\frac{\partial}{\partial u}+\frac{v}{r^2}\frac{z}{r}\frac{\partial}{\partial v}\\ +\frac{1}{r^2}\left(u^2\frac{\partial^2}{\partial u^2}-uv\frac{\partial^2}{\partial u\partial v}\right)+\frac{1}{r^2}\left(v^2\frac{\partial^2}{\partial u^2}-uv\frac{\partial^2}{\partial u\partial v}\right)\end{array}\right],$$

$$\text{i.e., }\frac{\partial^2}{\partial z^2}=\left[\begin{array}{l}\frac{1}{r^2}\left(u\frac{\partial}{\partial u}+v\frac{\partial}{\partial v}\right)-\frac{u}{r^3}\left(\frac{u-v}{2}\right)\frac{\partial}{\partial u}+\frac{v}{r^3}\left(\frac{u-v}{2}\right)\frac{\partial}{\partial v}\\ +\frac{1}{r^2}\left(u^2\frac{\partial^2}{\partial u^2}+v^2\frac{\partial^2}{\partial u^2}\right)+\frac{-2uv}{r^2}\frac{\partial^2}{\partial u\partial v}\end{array}\right],$$

$$\text{or } \frac{\partial^2}{\partial z^2} = \begin{bmatrix} \frac{1}{r^2}\left(u\frac{\partial}{\partial u}+v\frac{\partial}{\partial v}\right)-\frac{1}{2r^3}\left(u^2\frac{\partial}{\partial u}+v^2\frac{\partial}{\partial v}\right) \\ +\frac{uv}{2r^3}\left(\frac{\partial}{\partial u}+\frac{\partial}{\partial v}\right)+\frac{1}{r^2}\left(u^2\frac{\partial^2}{\partial u^2}+v^2\frac{\partial^2}{\partial v^2}\right)+\frac{-2uv}{r^2}\frac{\partial^2}{\partial u\partial v} \end{bmatrix}. \quad \text{P5.7.3}$$

Using $x^2+y^2=uv$ and substituting $x=\sqrt{uv}\cos\varphi$ and $y=\sqrt{uv}\sin\varphi$, and the terms in $\frac{\partial^2}{\partial x^2}$ and $\frac{\partial^2}{\partial y^2}$:

$$\left(\frac{y}{r}\right)^2=\frac{uv\sin^2\varphi}{r^2} \text{ and } \left(\frac{x}{r}\right)^2=\frac{uv\cos^2\varphi}{r^2}.$$

Therefore, $\frac{x\sin\varphi}{r\sqrt{uv}}=\frac{y\cos\varphi}{r\sqrt{uv}}$.

Now, adding P5.7.1, P5.7.2, and P5.7.3, we get the Laplacian

$$\begin{pmatrix} \frac{\partial^2}{\partial x^2}+\frac{\partial^2}{\partial y^2} \\ +\frac{\partial^2}{\partial z^2} \end{pmatrix} = \begin{bmatrix} \frac{2}{r}\left(\frac{\partial}{\partial u}+\frac{\partial}{\partial v}\right)+\frac{1}{r^2}\left(u\frac{\partial}{\partial u}+v\frac{\partial}{\partial v}\right)-\frac{1}{2r^3}\left(v^2\frac{\partial}{\partial v}+u^2\frac{\partial}{\partial u}\right) \\ +\frac{x^2+y^2}{r^2}\left(\frac{\partial^2}{\partial u^2}+\frac{\partial^2}{\partial v^2}\right)+\frac{1}{r^2}\left[u^2\frac{\partial^2}{\partial u^2}+v^2\frac{\partial^2}{\partial v^2}\right]-\frac{uv}{2r^3}\left(\frac{\partial}{\partial v}+\frac{\partial}{\partial u}\right)+\frac{1}{uv}\frac{\partial^2}{\partial\varphi^2} \end{bmatrix}.$$

Collecting the coefficients of $\frac{\partial}{\partial u},\frac{\partial}{\partial v},\frac{\partial^2}{\partial u^2}$ and $\frac{\partial^2}{\partial v^2}$ together,

and separating the factor of $\frac{1}{2r^3}$ we have

$$\frac{1}{2r^3}\begin{bmatrix} \left(4r^2+2ru-u^2-uv\right)\frac{\partial}{\partial u}+\left(4r^2+2ru-v^2-uv\right)\frac{\partial}{\partial v} \\ +\left(2ruv+2ru^2\right)\frac{\partial^2}{\partial u^2}+\left(2ruv+2rv^2\right)\frac{\partial^2}{\partial v^2} \end{bmatrix}+\frac{1}{uv}\frac{\partial^2}{\partial\varphi^2}.$$

Using now $r=\frac{u+v}{2}$ and further simplifying gives us

$$\nabla^2=\frac{4}{(u+v)}\left[\frac{\partial}{\partial u}+\frac{\partial}{\partial v}+u\frac{\partial^2}{\partial u^2}+v\frac{\partial^2}{\partial v^2}\right]+\frac{1}{uv}\frac{\partial^2}{\partial\varphi^2}.$$

Furthermore, since $\frac{\partial}{\partial u}+u\frac{\partial^2}{\partial u^2}=\frac{\partial}{\partial u}\left(u\frac{\partial}{\partial u}\right)$,

we get the desired expression for the Laplacian in the parabolic coordinates (u,v,φ):

$$\nabla^2=\frac{4}{(u+v)}\left[\frac{\partial}{\partial u}\left(u\frac{\partial}{\partial u}\right)+\frac{\partial}{\partial v}\left(v\frac{\partial}{\partial v}\right)\right]+\frac{1}{uv}\frac{\partial^2}{\partial\varphi^2}.$$

P5.8: Obtain the expression for the Laplacian in (z_p,v,φ) coordinates.

Solution:

We recall that the z-coordinate in the Cartesian system represents an independent degree of freedom. While using the parabolic coordinate system (z,v,φ) we place a subscript p on z on account of Eq. 5.135.

Now,

$$\frac{\partial}{\partial x}=\frac{\partial z_p}{\partial x}\frac{\partial}{\partial z_p}+\frac{\partial v_p}{\partial x}\frac{\partial}{\partial v_p}+\frac{\partial \varphi}{\partial x}\frac{\partial}{\partial \varphi}=\frac{\partial}{\partial x}\left(\frac{u-v_c}{2}\right)\frac{\partial}{\partial z_p}+\frac{\partial v_p}{\partial x}\frac{\partial}{\partial v_p}+\frac{\partial \varphi}{\partial x}\frac{\partial}{\partial \varphi},$$

with $z_p=\left(\frac{u-v_c}{2}\right)$ for a *given* v_c, since the three independent degrees of freedom are represented by (z_p, v, φ);

i.e., the variation with respect to v must not be included in the variation with respect to the first degree of freedom u which is now replaced by z_p using Eq. 5.135.

Therefore, the operator $\frac{\partial}{\partial x}$ becomes

$$\frac{\partial}{\partial x}=\frac{1}{2}\frac{x}{r}\frac{\partial}{\partial z_p}+\frac{x}{r}\frac{\partial}{\partial v}-\frac{\sin\varphi}{\sqrt{uv}}\frac{\partial}{\partial \varphi}.$$

Similarly, $\frac{\partial}{\partial y}=\frac{1}{2}\frac{y}{r}\frac{\partial}{\partial z_p}+\frac{y}{r}\frac{\partial}{\partial v}+\frac{\cos\varphi}{\sqrt{uv}}\frac{\partial}{\partial \varphi}$

and $\frac{\partial}{\partial z}=\frac{1}{2}\left(\frac{z}{r}+1\right)\frac{\partial}{\partial z_p}+\left(\frac{z}{r}-1\right)\frac{\partial}{\partial v}.$

Now, to get the Laplacian we need $\frac{\partial^2}{\partial x^2}$, $\frac{\partial^2}{\partial y^2}$ and $\frac{\partial^2}{\partial z^2}$. Of these, the first second order operator is

$$\frac{\partial^2}{\partial x^2}=\frac{\partial}{\partial x}\left\{\frac{1}{2}\frac{x}{r}\frac{\partial}{\partial z_p}+\frac{x}{r}\frac{\partial}{\partial v}-\frac{\sin\varphi}{\sqrt{uv}}\frac{\partial}{\partial \varphi}\right\},$$

i.e.,

$$\frac{\partial^2}{\partial x^2}=\begin{bmatrix}\frac{1}{2r}\frac{\partial}{\partial z_p}+\frac{x}{2}\left(\frac{-x}{r^3}\right)\frac{\partial}{\partial z_p}+\frac{x}{2r}\frac{\partial^2}{\partial x\partial z_p}+\frac{1}{r}\frac{\partial}{\partial v}\\ +\frac{x}{2}\left(\frac{-x}{r^3}\right)\frac{\partial}{\partial v}+\frac{x}{r}\frac{\partial^2}{\partial x\partial z_p}-\frac{\partial}{\partial x}\left(\frac{\sin\varphi}{\sqrt{uv}}\frac{\partial}{\partial \varphi}\right)\end{bmatrix}.$$

Substituting for $\frac{\partial}{\partial x}$ from equation 5.135a, and simplifying we get

$$\frac{\partial^2}{\partial x^2}=\begin{bmatrix}\left(\frac{x}{2r}\right)^2\frac{\partial^2}{\partial z_p^2}+\left(\frac{1}{2r}-\frac{x^2}{2r^3}\right)\frac{\partial}{\partial z_p}+\left(\frac{x}{2r}\right)^2\frac{\partial^2}{\partial v^2}+\left(\frac{1}{r}-\frac{x^2}{r^3}\right)\frac{\partial}{\partial v}\\ +\left(\frac{x}{2r}\right)^2\frac{\partial^2}{\partial z_p\partial v}-\frac{x}{2r}\frac{\sin\varphi}{\sqrt{uv}}\frac{\partial^2}{\partial z_p\partial\varphi}-\frac{x}{r}\frac{\sin\varphi}{\sqrt{uv}}\frac{\partial^2}{\partial\varphi\partial v}-\frac{\partial}{\partial x}\left\{\frac{\sin\varphi}{\sqrt{uv}}\frac{\partial}{\partial\varphi}\right\}\end{bmatrix}. \quad \text{P5.8.1}$$

Similarly,

$$\frac{\partial^2}{\partial y^2}=\begin{bmatrix}\left(\frac{y}{2r}\right)^2\frac{\partial^2}{\partial z_p^2}+\left(\frac{1}{2r}-\frac{y^2}{2r^3}\right)\frac{\partial}{\partial z_p}+\left(\frac{y}{2r}\right)^2\frac{\partial^2}{\partial v^2}+\left(\frac{1}{r}-\frac{y^2}{r^3}\right)\frac{\partial}{\partial v}\\ +\left(\frac{y}{2r}\right)^2\frac{\partial^2}{\partial z_p\partial v}+\frac{y}{2r}\frac{c}{\sqrt{uv}}\frac{\partial^2}{\partial z_p\partial\varphi}+\frac{y}{r}\frac{\cos\varphi}{\sqrt{uv}}\frac{\partial^2}{\partial\varphi\partial v}-\frac{\partial}{\partial y}\left\{\frac{\sin\varphi}{\sqrt{uv}}\frac{\partial}{\partial\varphi}\right\}\end{bmatrix}, \quad \text{P5.8.2}$$

$$\text{hence } \frac{\partial^2}{\partial z^2} = \left[\begin{array}{l} \frac{1}{4}\left(1+\frac{z}{r}\right)^2 \frac{\partial^2}{\partial z_p^2} + \frac{1}{2r}\left(1-\frac{z^2}{r^2}\right)\frac{\partial}{\partial z_p} + \left(1-\frac{z}{r}\right)^2 \frac{\partial^2}{\partial v^2} \\ + \frac{1}{r}\left(1-\frac{z^2}{r^2}\right)\frac{\partial}{\partial v} + \left(\frac{z^2}{r^2}-1\right)\frac{\partial^2}{\partial z_p \partial v} \end{array}\right]. \qquad \text{P5.8.3}$$

Adding Eq. P5.8.1, Eq. P5.8.2, and Eq. P5.8.3, we get

$$\vec{\nabla}^2 = \left[\begin{array}{l} \left(\frac{x}{2r}\right)^2 \frac{\partial^2}{\partial z_p^2} + \left(\frac{1}{2r} - \frac{x^2}{2r^3}\right)\frac{\partial}{\partial z_p} + \left(\frac{x}{2r}\right)^2 \frac{\partial^2}{\partial v^2} + \left(\frac{1}{r} - \frac{x^2}{r^3}\right)\frac{\partial}{\partial v} + \left(\frac{x}{2r}\right)^2 \frac{\partial^2}{\partial z_p \partial v} \\ -\frac{x \sin\varphi}{2r\sqrt{uv}}\frac{\partial^2}{\partial z_p \partial \varphi} - \frac{x \sin\varphi}{r\sqrt{uv}}\frac{\partial^2}{\partial \varphi \partial v} - \frac{\partial}{\partial x}\left\{\frac{\sin\varphi}{\sqrt{uv}}\frac{\partial}{\partial \varphi}\right\} + \left(\frac{y}{2r}\right)^2 \frac{\partial^2}{\partial z_p^2} + \left(\frac{1}{2r} - \frac{y^2}{2r^3}\right)\frac{\partial}{\partial z_p} \\ +\left(\frac{y}{2r}\right)^2 \frac{\partial^2}{\partial v^2} + \left(\frac{1}{r} - \frac{y^2}{r^3}\right)\frac{\partial}{\partial v} + \left(\frac{y}{2r}\right)^2 \frac{\partial^2}{\partial z_p \partial v} + \frac{y}{2r}\frac{c}{\sqrt{uv}}\frac{\partial^2}{\partial z_p \partial \varphi} + \frac{y \cos\varphi}{r\sqrt{uv}}\frac{\partial^2}{\partial \varphi \partial v} \\ -\frac{\partial}{\partial y}\left\{\frac{\sin\varphi}{\sqrt{uv}}\frac{\partial}{\partial \varphi}\right\} + \left(\frac{u}{2r}\right)^2 \frac{\partial^2}{\partial z_p^2} + \frac{uv}{2r^3}\frac{\partial}{\partial z_p} + \left(\frac{v}{r}\right)^2 \frac{\partial^2}{\partial v^2} + \frac{uv}{r^3}\frac{\partial}{\partial v} + \frac{uv}{r^2}\frac{\partial^2}{\partial z_p \partial v} \end{array}\right].$$

Collecting and simplifying the coefficients of $\partial^2 / \partial z_p^2$, $\partial / \partial z_p$, $\partial^2 / \partial v^2$, and $\partial / \partial v$, we get $\vec{\nabla}^2$ in (z_p, v, φ) co-ordinates:

$$\vec{\nabla}^2 = \frac{r+z}{2r}\frac{\partial^2}{\partial z_p^2} + \frac{1}{r}\frac{\partial}{\partial z_p} + \frac{2v}{r}\frac{\partial^2}{\partial v^2} + \frac{2}{r}\frac{\partial}{\partial v} + \frac{1}{(r+z)v}\frac{\partial^2}{\partial \varphi^2}.$$

P5.9: Show that the form of the Laplacian obtained in the **P5.8** is equivalent to that in Eq. 5.137b, and hence correct.

Solution:

To show that this expression is equivalent to Eq. 5.137b, we make use of the relation $z_p = \frac{u - v_c}{2}$. Taking partial derivatives yields, $\frac{\partial}{\partial u} = \frac{1}{2}\frac{\partial}{\partial z_p}$. Plugging this in the equation for $\vec{\nabla}^2$ from **P5.8** we get

$$\vec{\nabla}^2 = \frac{u}{u+v}\frac{\partial^2}{\partial z_p^2} + \frac{2}{u+v}\frac{\partial}{\partial z_p} + \frac{4v}{u+v}\frac{\partial^2}{\partial v^2} + \frac{4}{u+v}\frac{\partial}{\partial v} + \frac{1}{uv}\frac{\partial^2}{\partial \varphi^2},$$

$$\text{i.e., } \vec{\nabla}^2 = \frac{4u}{u+v}\frac{\partial^2}{\partial u^2} + \frac{4}{u+v}\frac{\partial}{\partial u} + \frac{4v}{u+v}\frac{\partial^2}{\partial v^2} + \frac{4}{u+v}\frac{\partial}{\partial v} + \frac{1}{uv}\frac{\partial^2}{\partial \varphi^2}.$$

$$\text{Therefore, } \vec{\nabla}^2 = \frac{4}{u+v}\left\{\frac{\partial}{\partial u}\left(u\frac{\partial}{\partial u}\right) + \frac{\partial}{\partial v}\left(v\frac{\partial}{\partial v}\right)\right\} + \frac{1}{uv}\frac{\partial^2}{\partial \varphi^2}.$$

P5.10: The physical system formed by an antimuon (μ^+) and an electron constitutes an *exotic atom* called as the *muium*; it behaves like a light isotope of hydrogen. Determine the 'Bohr radius' and the ionization energy of the muonium. ($m_\mu = 207m_e$).

Solution:

The reduced mass of the muonium two-body system is $\mu = \frac{m_e m_\mu}{m_e + m_\mu} = \frac{m_e(207)m_e}{m_e + 207m_e} = \frac{207}{208}m_e$

$$= \left(1 - \frac{1}{208}\right)m_e \simeq \left(1 - \frac{1}{200}\right)m_e.$$

Hence,

$$\frac{1}{\mu} \simeq \left(1 + \frac{1}{200}\right)\frac{1}{m_e}.$$

The 'Bohr radius' is $a_0 = \frac{\hbar^2}{\mu e^2} = \frac{\hbar^2}{m_e e^2}\left(1 + \frac{1}{200}\right) = a_0(H)\left(1 + \frac{1}{200}\right)$,

where, $a_0(H)$ is the Bohr radius of hydrogen atom.

The ionization energy is $E(muonium) = \frac{\mu e^4}{2\hbar^2} = \frac{m_e e^4}{2\hbar^2}\left(1 - \frac{1}{200}\right) = E_I(H)\left(1 - \frac{1}{200}\right)$,

where $E_I(H)$ is the ionization energy of hydrogen atom. The muium is more suitable for the study of bound-state quantum electrodynamics (QED) than the hydrogen atom. The reason for this is that the muon and the antimuon is a lepton and has no internal structure, while the nucleus of the hydrogen atom is a proton (baryon) that has an internal structure. The antimuon and an electron is also sometimes called as *muonium*, but the latter term is rather used to describe the bound state of an antimuon and a muon.

P5.11: A particle in a three-dimensional isotropic harmonic oscillator potential is described by the Hamiltonian $H = \frac{p^2}{2m} + \frac{1}{2}m\omega^2 r^2$. Determine its energy eigenvalues and the degeneracy of each energy eigenvalue.

Solution:

(a) Using the Cartesian co-ordinate system, the Schrödinger equation is:

$$-\frac{\hbar^2}{2m}\left(\frac{\partial^2}{\partial x^2} + \frac{\partial^2}{\partial y^2} + \frac{\partial^2}{\partial z^2}\right)\psi(x,y,z) + \frac{1}{2}m\omega^2\left(x^2 + y^2 + z^2\right)\psi(x,y,z) = E\psi(x,y,z).$$

Using the separation of variables, $\psi(x,y,z) = X(x)Y(y)Z(z)$ and we have

$$\left(-\frac{\hbar^2}{2m}\frac{1}{X}\frac{\partial^2 X}{\partial x^2} + \frac{1}{2}m\omega^2 x^2\right) + \left(-\frac{\hbar^2}{2m}\frac{1}{Y}\frac{\partial^2 Y}{\partial y^2} + \frac{1}{2}m\omega^2 y^2\right) + \left(-\frac{\hbar^2}{2m}\frac{1}{Z}\frac{\partial^2 Z}{\partial z^2} + \frac{1}{2}m\omega^2 z^2\right) = E.$$

Each term on the left hand side is the Hamiltonian for a one-dimensional harmonic oscillator. Hence, writing $E = E_x + E_y + E_z$, the allowed eigen-energies are:

$$E_x = \left(n_x + \frac{1}{2}\right)\hbar\omega; \quad E_y = \left(n_y + \frac{1}{2}\right)\hbar\omega; \quad E_z = \left(n_z + \frac{1}{2}\right)\hbar\omega.$$

Therefore, $E_n = \left(n_x + n_y + n_z + \frac{3}{2}\right)\hbar\omega = \left(n + \frac{3}{2}\right)\hbar\omega.$

(b) The degeneracy is illustrated by the following table:

n	n_x	n_y	n_z	Number of ways in which $n_x + n_y + n_z = n$	Degeneracy
1	1	0	0	3	3
	0	1	0		
	0	0	1		
2	2	0	0	3	6
	0	2	0		
	0	0	2		
	1	1	0	3	
	1	0	1		
	0	1	1		
3	3	0	0	3	10
	0	3	0		
	0	0	3		
	2	1	0	6	
	2	0	1		
	1	2	0		
	1	0	2		
	0	2	1		
	0	1	2		
	1	1	1	1	

The degeneracy of the level E_n is essentially the number of different ways we can add n_x, n_y and n_z to get n. It is $d = \frac{(n+1)(n+2)}{2}$. It is not fully accounted for by the rotational invariance of the isotropic oscillator. The spherical symmetry would account only for the $(2\ell+1)-fold$ degeneracy. The extra degeneracy is owing to an additional conserved quantity [17].

Exercises

E5.1: Arrive at Eq. 5.3a from Eq. 5.1. Change the Schrödinger equation from the laboratory frame of reference to the center of mass and relative frame of reference.

E5.2: (a) Calculate the ground state energy of the hydrogen atom using SI units and convert the result to electron volts.

(b) Find the energy of Li^{2+} for $n = 2$, in eV.

E5.3: Verify that for large values of k, the ratio c_{k+1} / c_k in Eq. 5.39 in the same as the ratio of the coefficient of ρ^{k+1} to that of ρ^k in the power series of $\rho^n e^n$ for finite n.

E5.4: A positronium consists of an electron and a positron revolving about their common center of mass. Calculate in eV the ground state energy of positronium.

E5.5: For the ground state of hydrogen atom:

(a) Show that $\langle T\rangle + \langle V\rangle = E_n$.

(b) Find $\dfrac{\langle T\rangle}{\langle V\rangle}$.

(c) Determine the root-mean-square speed $\langle v^2\rangle^{1/2}$ of the electron using $\langle T\rangle$ and find the numerical value of $\dfrac{\langle v^2\rangle^{1/2}}{c}$, where c is the speed of light.

E5.6: Another solvable two-body problem by transforming to the center of mass coordinate system is the rigid rotor. Two particles of mass m are attached to the ends of a massless rigid rod of length a. The system is free to rotate in three dimensions about the fixed center.

(a) Verify that the moment of inertia is given by $I = \mu a^2$ using $\mu = \dfrac{m_1 m_2}{m_1 + m_2}$.

(b) Show that the allowed energies of the rigid rotor are $E_n = \dfrac{\hbar^2 n(n+1)}{2I}$.

(c) What are the normalized eigenfunctions for this system? What is the degeneracy of the nth energy level?

E5.7: (a) Show that the time derivative of the expectation value of some observable $\breve{Q}(x,p,t)$ is given by $\dfrac{d}{dt}\langle \breve{Q}\rangle = \dfrac{i}{\hbar}\langle [\breve{H},\breve{Q}]\rangle + \left\langle \dfrac{\partial \breve{Q}}{\partial t}\right\rangle$.

(b) Prove the three-dimensional virial theorem for stationary states using the result obtained in (a), where $\breve{Q} = \vec{r}\cdot\vec{p}$: $2\langle T\rangle = \langle \vec{r}\cdot\vec{\nabla}V\rangle$.

(c) Show that $\langle T\rangle = -E_n$; $\langle V\rangle = 2E_n$.

E5.8: The wavefunction of the hydrogen atom is given by $\psi_{n\ell m}(r,\theta,\varphi) = R_{n\ell}(r)Y_\ell^m(\theta,\varphi)$ where Y_ℓ^m is the spherical harmonic function. Prove the following relation between the spherical harmonic functions: $\sum\limits_{m=-\ell}^{m=+\ell} Y_\ell^m(\theta,\varphi)^* Y_\ell^m(\theta,\varphi) = \dfrac{2\ell+1}{4\pi}$.

E5.9: (a) For the hydrogen atom ground state, calculate the probability of finding the electron farther away from the nucleus ($r > 2a_0$)

(b) Find the probability of finding the electron in the classically forbidden region ($E < V$).

E5.10: What is the probability that an electron in the ground state of hydrogen will be found inside the nucleus?

E5.11: (a) Verify that the operator for the classical expression $\vec{p}\times\vec{\ell}$ is

$$\frac{1}{2}\left[(\vec{p}\times\vec{\ell}) + (\vec{p}\times\vec{\ell})\right] = \frac{1}{2}\left[(\vec{p}\times\vec{\ell}) - (\vec{\ell}\times\vec{p})\right].$$

(b) Show that the Runge–Lenz operator is conserved.

(c) Explain how conservation of the Runge–Lenz operator implies that the orbits are closed.

E5.12: Find the wavefunction of H-atom in parabolic coordinates for

(a) $n_1 = n_2 = m = 0$

(b) $n = 2$, $m = 0$

Express them in spherical coordinates.

E5.13: (a) Define $A_\pm = A_x \pm A_y$, use Eq. 5.118a and Eq. 5.118b to show that

$$\frac{\mu}{\hbar^2} A_- |n\ell m\rangle = \sum_{l'=l,l\pm1} |n\ell'\ell-1\rangle\langle n\ell'\ell-1|l_- - K_-|n\ell\ell\rangle\sqrt{\frac{2\mu E_n}{-\hbar^2}}.$$

(b) Find $\langle nl'l-1|l_-|nll\rangle$ and $\langle nl'l-1|K_-|nll\rangle$ using Eq. 5.119a and Eq. 5.119b.

E5.14: Arrive at the recursive relation Eq. P5.6.1, of $R_{n\ell}$ using the formula

$$\frac{\mu}{\hbar^2} A_- |n,\ell,\ell\rangle = \frac{Z}{na_0}\left\{\sqrt{\frac{2(n^2-(\ell+1)^2)}{(2\ell+1)(2\ell+3)}}|n,\ell+1,\ell-1\rangle - \sqrt{\frac{2\ell(n^2-\ell^2)}{(2\ell+1)}}|n,\ell-1,\ell-1\rangle\right\}.$$

E5.15: (a) Show that the expectation value of the electric dipole moment of an atom having N electrons in a state of well-defined parity vanishes.

(b) What is the dipole moment of the hydrogen atom in a (i) spherical (ii) parabolic eigenstate?

E5.16: (a) Find the most probable value of r for the ground state of hydrogen like atom.

(b) Show that the most probable values of r for $\ell = n-1$ states of hydrogen like atoms are $r = \frac{n^2 a_0}{Z}$.

References

[1] R. P. Feynman, *The Character of the Physical Law* ("Messenger Lectures" at Cornell University in 1964, published by Modern Library, 1994). http://www.youtube.com/watch?v=j3mhkYbznBk [link downloaded on August 15, 2013].

[2] E. P. Wigner, *Events, Laws of Nature, and Invariance Principles* ("Nobel Lecture" delivered on December 12, 1963). http://www.nobelprize.org/nobel_prizes/physics/laureates/1963/wigner-lecture.pdf [link downloaded on August 15, 2013].

[3] P. C. Deshmukh, Aarthi Ganesan, N. Shanthi, Blake Jones, James Nicholson, and A. Soddu, The "Accidental" Degeneracy of the Hydrogen Atom Is No Accident! *Canadian Journal of Physics*, 10.1139/cjp-2014-0300.

[4] L. D. Landau and E. M. Lifshitz, *Quantum Mechanics – Non-relativistic Theory* (3rd Edition, Pergamon Press Ltd, 1977).

[5] P. C. Deshmukh, *Foundations of Classical Mechanics* (Cambridge University Press, 2019).

[6] C. E. Burkhardt and J. J. Leventhal, Lenz Vector Operations on Spherical Hydrogen Atom Eigenfunctions. *American Journal of Physics* 72:8, 1013 (2004).

[7] W. Pauli, Perturbation of Hydrogen Degenerate Levels and SO (4). *Z. Phys*. 36, 336–363 (1926).

[8] W. Greiner and B. Mueller, *Quantum Mechanics – Symmetries* (Springer Verlag, 1992).

[9] H. B. G. Casimir, Rotation of a Rigid Body in Quantum Mechanics. *Proc. Roy. Soc. Amsterdam* 34, 844 (1931).

[10] A. Messiah, *Quantum Mechanics* (Dover Publications, 1999).

[11] V. A. Fock, On the Theory of the Hydrogen Atom. *Z. Phys*. 98:3–4, 145–154 (1935).

[12] A. Sommerfeld, *Atomic Structure and Spectral Lines* (Methuen, London, 1923).

[13] N. W. Evans, Superintegrability in Classical Mechanics. *Phys. Rev. A* 41:10, 5666–5676 (1990).

[14] Ian Marquette, Supersymmetry as a Method of Obtaining New Superintegrable Systems with Higher Order Integrals of Motion. *J. Math. Phys*. 50, 122102 (2009).

[15] M. A. Rodriguez and P. Winternitz, Quantum Superintegrability and Exact Solvability in n Dimensions. *Journal of Mathematical Physics* 43, 1309 (2002).

[16] M. F. Rañada, Higher Order Superintegrability of Separable Potentials with a New Approach to the Post–Winternitz System. *Journal of Physics A: Mathematical and Theoretical* 46:12, 125206 (2013).

[17] D. M. Fradkin, Existence of the Dynamics Symmetries O_4 and SU_3 for All Central Potential Problems. *Progress in Theoretical Physics* 37, 798–812 (1967).

[18] B. H. Bransden and C. J. Joachain, *Quantum Mechanics* (2nd Edition, Pearson Prentice Hall, 2000).

[19] G. F. Torres del Castillo and E. Navarro Morales, Bound States of the Hydrogen Atom in Parabolic Coordinates. *Revista Mexicana de Fisica* 54:6, 454–458 (2008).

[20] Davis Park, Relation between the Parabolic and Spherical Eigenfunctions of Hydrogen. *Zeitschrift fur Physik* 159, 155–157 (1960).

[21] P. C. Deshmukh, Aarthi Ganesan, Sourav Banerjee, and Ankur Mandal, The *Accidental* Degeneracy of the Hydrogen Atom and Its *Non-accidental* Solution Using Parabolic Coordinates. *Canadian Journal of Physics*, 2021. https://cdnsciencepub.com/doi/10.1139/cjp-2020-0258.

[22] G. B. Arfken and H. J. Waber, *Mathematical Methods for Physicists* (5th Edition, Academic Press, 2001).

[23] J. J. Sakurai and J. Napolitano, *Modern Quantum Mechanics* (2nd Edition, Addison-Wesley, 2011).

[24] K. Gottfried and T. Yan, *Quantum Mechanics: Fundamentals* (2nd Edition, Springer-Verlag, 2003).

[25] Walter Greiner and Berndt Müller, *Quantum Mechanics – Symmetries* (Springer-Verlag, 1984, 1992).

6

Approximation Methods

Painting of Stephen Hawking by Abhik Samui (IIT Madras). Included with kind permission from Abhik Samui

One of the basic rules of the universe is that nothing is perfect. Perfection simply doesn't exist..... Without imperfection, neither you nor I would exist
—Stephen Hawking (1942–2018)

Approximations are made in mathematics, physics, engineering, and life sciences for a variety of reasons such as (a) obtaining an exact solution is not possible and (b) an exact solution is not necessary, subject to a given context. Both of these factors are important in the development of approximations. We have already seen in Chapter 2 that out of the infinite alternative paths accessible to a system due to their entanglement/superposition, the classical path that makes *action* stationary results from the condition that *action* is much larger than the PEB constant. When this condition is satisfied, laws of classical physics provide an excellent approximation to quantum mechanics. In Chapter 2, we observed that the interpretation of the wavefunction as probability amplitude became the basis of the path integral formulation of quantum mechanics, in which the propagator's phase is the *action*. Considering this relationship, it is appropriate to wonder if the Schrödinger equation could have been *predicted* from classical mechanics, if only the wavefunction (Eq. 6.22, below) was introduced, even just as a *definition*. In Section 6.1, we demonstrate, in *hindsight*, how this could have been possible. Of course, historical development of quantum mechanics took place on a different course. We also demonstrate the inverse of this process, viz., how the classical equation of motion is obtainable as an approximation from quantum mechanics in the limit $h \to 0$. An approximation scheme often implemented to solve the Schrödinger equation, called the Wentzel–Kramers–Brilloiun–Jeffrey (WKBJ) method, is based on the *connections* between classical and quantum mechanics. It is introduced in Section 6.1. In Section 6.2, we discuss perturbation methods to obtain *approximate* solutions to unsolvable problems. Time-independent perturbation methods are discussed in Section 6.2, and time-dependent perturbation methods in Section 6.3. Major non-perturbative approaches to attack complex problems include the adiabatic approximation, the sudden approximation, and the method of variation. These are introduced in Section 6.4.

6.1 Classical Limit of the Schrödinger Equation and the WKBJ Approximation

Readers who are well conversant with the Hamilton–Jacobi (HJ) theory in classical mechanics [1] will be able to skip a significant part of the discussion in this section. It is, however, included here, even if only briefly, for the sake of completeness. The HJ formalism provides an efficient mechanism to solve the Hamilton's equations of motion by transforming the generalized coordinates and momenta (q_i, p_i) to a new set of generalized coordinates and momenta (Q_i, P_i), which make the solutions trivial. The dynamics gets included in the generating function that effects the transformation of the Hamiltonian

$$H(q_i, p_i, t) \to K(Q_i, P_i, t), \tag{6.1a}$$

where

$$Q_i = Q_i(q, p, t) \tag{6.1b}$$

$$\text{and } P_i = P_i(q, p, t). \tag{6.1c}$$

The transformed Hamiltonian K provides the equations of motion just as the original Hamiltonian H does; Goldstein [1] jocularly refers to it as "Kamiltonian." Just like

$$\dot{q}_i = \frac{\partial H}{\partial p_i}; \dot{p}_i = -\frac{\partial H}{\partial q_i}, \tag{6.2a}$$

the equations of motion [1,2] are given by

$$\dot{Q}_i = \frac{\partial K}{\partial P_i}; \dot{P}_i = -\frac{\partial K}{\partial Q_i}. \tag{6.2b}$$

The equivalence of Newton's formulation of classical dynamics which is based on the principle of causality and determinism, and the alternative framework based on the principle of variation, which leads to the Lagrange's and Hamilton's equations, is well known. Both of these formulations require simultaneous knowledge of position and momentum of an object that is under investigation. The HJ method is a strategy to make the application of Hamilton's equations solvable in a trivial manner, by transforming a Hamiltonian $H(q, p, t)$ to a new form $K(Q, P, t)$ in which the new coordinates and new momenta represent *equilibrium* points. In this method, the *new Hamiltonian is sought to be identically zero*, which incontestably results in the new coordinates, and the new momenta, *both* become constants in time, determined only by the initial conditions. The solutions to the equations of motion become trivial, and the inquiry is shifted to the generating function that enables the desired transformation. The Lagrangian for the physical system contains essential knowledge about the state of the system, which is independent of the representation in either of the two systems. Hence, we have

$$H = p_i \dot{q}_i - L \tag{6.3a}$$

$$\text{and } K = P_i \dot{Q}_i - L. \tag{6.3b}$$

A constant that scales the Lagrangian may, however, be involved, since the equations of motion will be insensitive to scaling by a constant. Motion occurs so as to make *action* stationary (extremum); i.e.,

$$\delta\int_{t_1}^{t_2} L dt = 0 = \delta\int_{t_1}^{t_2}\left(p_i\dot{q}_i - H(q,p,t)\right)dt \tag{6.4a}$$

or, equivalently $$\delta\int_{t_1}^{t_2} L dt = 0 = \delta\int_{t_1}^{t_2}\left(P_i\dot{Q}_i - K(Q,P,t)\right)dt. \tag{6.4b}$$

The generating function F connects the integrands in Eqs. 6.4a and 6.4b (in which the Einstein summation convention is implied), since a variation in its total derivative with respect to time vanishes at the limits of the temporal integration. Inclusive of a scale factor λ mentioned earlier, we see that

$$\lambda\left(p_i\dot{q}_i - H(q,p,t)\right) = \left(P_i\dot{Q}_i - K(Q,P,t)\right) + \frac{dF}{dt}. \tag{6.5a}$$

Transformations for which Eq. 6.5a holds are called "extended canonical transformations." The special case $\lambda = 1$ contains the essential physics for which, we have,

$$\left(p_i\dot{q}_i - H(q,p,t)\right) = \left(P_i\dot{Q}_i - K(Q,P,t)\right) + \frac{dF}{dt}. \tag{6.5b}$$

The transformations $(q,p)\overset{F}{\leftrightarrow}(Q,P)$ corresponding to $\lambda = 1$ are simply called "canonical transformations." The generating function F is most useful when it is determined by half of the variables (other than time t) from the new set (Q_i,P_i) and the other half from the old set (q_i,p_i). Considering

$$F = F_1(q,Q,t), \tag{6.6}$$

it follows that

$$p_i\dot{q}_i - H = P_i\dot{Q}_i - K + \frac{\partial F_1(q,Q,t)}{\partial q_i}\dot{q}_i + \frac{\partial F_1(q,Q,t)}{\partial Q_i}\dot{Q}_i + \frac{\partial F_1(q,Q,t)}{\partial t}, \tag{6.7a}$$

i.e., $$\left[p_i - \frac{\partial F_1(q,Q,t)}{\partial q_i}\right]\dot{q}_i - H = \left[P_i + \frac{\partial F_1(q,Q,t)}{\partial Q_i}\right]\dot{Q}_i - K + \frac{\partial F_1(q,Q,t)}{\partial t}. \tag{6.7b}$$

Since q and Q are independent of each other, coefficients of $\dot{q}_i$ and $\dot{Q}_i$ must vanish. We therefore get

$$p_i = \frac{\partial F_1(q,Q,t)}{\partial q_i} \tag{6.8a}$$

and $$P_i = -\frac{\partial F_1(q,Q,t)}{\partial Q_i}, \tag{6.8b}$$

which reduces Eq. 6.7b to

$$K = H + \frac{\partial F_1(q,Q,t)}{\partial t}. \tag{6.9}$$

An alternative generating function is

$$F = f_2(q,P,t) = F_2(q,P,t) - Q_i P_i. \tag{6.10}$$

In this case,

$$p_i \dot{q}_i - H = P_i \dot{Q}_i - K + \frac{dF}{dt} = P_i \dot{Q}_i - K + \frac{d}{dt}\left[F_2(q,P,t) - Q_i P_i\right], \tag{6.11a}$$

i.e.,

$$p_i \dot{q}_i - H = \cancel{P_i \dot{Q}_i} - K + \frac{\partial F_2(q,P,t)}{\partial q}\dot{q} + \frac{\partial F_2(q,P,t)}{\partial P}\dot{P} + \frac{\partial F_2(q,P,t)}{\partial t} - Q_i \dot{P}_i - \cancel{\dot{Q}_i P_i}, \tag{6.11b}$$

$$\text{i.e., } \left\{p_i - \frac{\partial F_2(q,P,t)}{\partial q}\right\}\dot{q}_i - H = -K + \left\{\frac{\partial F_2(q,P,t)}{\partial P_i} - Q_i\right\}\dot{P}_i + \frac{\partial F_2(q,P,t)}{\partial t}. \tag{6.11c}$$

Now, we get the following relations for the partial derivatives of the generating function

$$p_i = \frac{\partial F_2(q,P,t)}{\partial q} \tag{6.12a}$$

$$\text{and } Q_i = \frac{\partial F_2(q,P,t)}{\partial P_i}, \tag{6.12b}$$

and the new Hamiltonian is

$$K = H + \frac{\partial F_2(q,P,t)}{\partial t}. \tag{6.13}$$

The generating function that results in the transformed Hamiltonian to be zero is called the Hamilton's principal function:

$$K = H\left(q, \frac{\partial F_2}{\partial q}, t\right) + \frac{\partial F_2(q,P,t)}{\partial t} = 0. \tag{6.14}$$

The transformed Hamiltonian K being zero, Hamilton's equations written with the new coordinates and the new momenta have obvious solutions; they are at equilibrium, determined only by their initial values:

$$\left(Q_i(t), P_i(t)\right) = \left(Q_i(t=0), P_i(t=0)\right). \tag{6.15}$$

Two other types of generating function can be considered. The four canonical transformations are summarized in Table 6.1.

If we choose the Hamilton's principal function to be

$$S(q,P,t) = F_2(q,P,t), \tag{6.16}$$

the resulting relation is

$$H\left(q, \frac{\partial F_2}{\partial q}, t\right) + \frac{\partial S(q,P,t)}{\partial t} = 0, \tag{6.17}$$

called as the Hamilton–Jacobi (HJ) equation.

Table 6.1 Four types of canonical transformations

Generating function		Partial derivatives of the generating function
$[f_1]$	$F_1(q,Q,t)$	$p_i = \frac{\partial F_1(q,Q,t)}{\partial q_i}$; $P_i = -\frac{\partial F_1(q,Q,t)}{\partial Q_i}$
$[f_2]$	$F_2(q,P,t) - Q_i P_i$	$p_i = \frac{\partial F_2(q,P,t)}{\partial q}$; $Q_i = \frac{\partial F_2(q,P,t)}{\partial P_i}$
$[f_3]$	$F_3(p,Q,t) + q_i p_i$	$q_i = -\frac{\partial F_3(p,Q,t)}{\partial p_i}$; $P_i = -\frac{\partial F_3(p,Q,t)}{\partial Q_i}$
$[f_4]$	$F_4(p,P,t) + q_i p_i - Q_i P_i$	$q_i = -\frac{\partial F_4(p,P,t)}{\partial p_i}$; $Q_i = \frac{\partial F_4(p,P,t)}{\partial P_i}$

From Eqs. 6.12a,b (or, Table 6.1), we see that

$$p_i = \frac{\partial S(q,P,t)}{\partial q}; \quad Q_i = \frac{\partial S(q,P,t)}{\partial P_i}. \tag{6.18}$$

The first of the above equations for the Cartesian x-coordinate is

$$p_x = \frac{\partial S(x,P,t)}{\partial x}, \tag{6.19a}$$

and in three dimensions, in the vector form, it is

$$\vec{p} = \vec{\nabla} S. \tag{6.19b}$$

Solving the HJ equation for the Hamilton's principal function, $S(q,P,t)$, we see that

$$\frac{dS}{dt}(q,P,t) = \frac{\partial S}{\partial q}\dot{q} + \frac{\partial S}{\partial P}\cancel{\dot{P}} + \frac{\partial S}{\partial t} = p\dot{q} + \frac{\partial S}{\partial t} = p\dot{q} - H = L, \tag{6.20}$$

and its integral is essentially the *action*

$$S = \int L dt, \tag{6.21}$$

apart from an additive constant that is unimportant.

Funquest: What property of a medium through which light propagates would the symbol 'n' represent if the time taken for light to traverse from A to B is given by the condition that $T = \frac{1}{c}\int_A^B n ds$ is an extremum, c being the speed of light and the integration is over the path light takes.

Everything we discussed here is based on classical physics as was known in the 19th century. The introduction of a wavefunction through a *definition*

$$\psi(\vec{r},t) = e^{\frac{i}{\hbar}S(\vec{r},t)} \tag{6.22}$$

could have been attempted early on to discover, as shown below, the Schrödinger equation of quantum mechanics. Such an ansatz would have advanced the development of quantum mechanics. It can be conjectured now that such an initiative constituted the heart of the inkling proposed by Dirac, and developed further by Feynman, in the path integral formalism of quantum mechanics, as noted in Chapter 2. The wavefunction is interpreted as the probability amplitude that manifests as Feynman's path integral propagator in which the phase comes from the classical *action*. The reader may also study Reference [3] for an insightful discussion on this concept. The relationship between the wavefunction and the Feynman propagator (Chapter 2) can be expressed, with appropriate choice of 'zero' of time and space coordinates, as

$$K\left(\vec{r}_f,t_f;\vec{r}_i,t_i\right) = A(\vec{r},t)e^{\frac{i}{\hbar}S(\vec{r},t)}, \tag{6.23a}$$

$$\text{i.e.,}\quad \frac{K\left(\vec{r}_f,t_f;\vec{r}_i,t_i\right)}{A(\vec{r},t)} = e^{\frac{i}{\hbar}S(\vec{r},t)} = \psi(\vec{r},t). \tag{6.23b}$$

From Eq. 6.22, it follows that

$$S(\vec{r},t) = -i\hbar\ell n\psi, \tag{6.24}$$

and hence

$$\frac{i}{\hbar}\psi\frac{\partial S}{\partial x} = \frac{\partial\psi}{\partial x}\quad;\quad \frac{i}{\hbar}\psi\frac{\partial S}{\partial y} = \frac{\partial\psi}{\partial y}\quad;\quad \frac{i}{\hbar}\psi\frac{\partial S}{\partial z} = \frac{\partial\psi}{\partial z}\ ;\ \frac{i}{\hbar}\psi\frac{\partial S}{\partial t} = \frac{\partial\psi}{\partial t}. \tag{6.25}$$

Taking now the second derivative with respect to space coordinates, we see that

$$\frac{\partial^2\psi}{\partial x^2} = \frac{i}{\hbar}\frac{\partial\psi}{\partial x}\frac{\partial S}{\partial x} + \frac{i}{\hbar}\psi\frac{\partial^2 S}{\partial x^2}, \tag{6.26a}$$

$$\text{i.e.,}\quad \frac{\partial^2\psi}{\partial x^2} = \frac{i}{\hbar}\frac{\partial\psi}{\partial x}\frac{\partial S}{\partial x} + \frac{i}{\hbar}\psi\frac{\partial^2 S}{\partial x^2} = \frac{i}{\hbar}\left(\frac{i}{\hbar}\psi\frac{\partial S}{\partial x}\right)\frac{\partial S}{\partial x} + \frac{i}{\hbar}\psi\frac{\partial^2 S}{\partial x^2}, \tag{6.26b}$$

$$\text{or}\quad \frac{\partial^2\psi}{\partial x^2} = \frac{-1}{\hbar^2}\psi\left(\frac{\partial S}{\partial x}\right)^2 + \frac{i}{\hbar}\frac{\partial^2 S}{\partial x^2} = \frac{-1}{\hbar^2}\psi\left(\frac{\partial S}{\partial x}\right)^2, \tag{6.26c}$$

since we see from Eq. 6.19a that

$$\frac{\partial^2 S(x,P,t)}{\partial x^2} = \frac{\partial}{\partial x}\frac{\partial S}{\partial x} = \frac{\partial p_x}{\partial x} = 0, \tag{6.27}$$

position and momentum being independent of each other. Hence, using Eq. 6.19 and Eq. 6.26c, we get the following relations:

$$\left.\begin{aligned} -\hbar^2 \frac{\partial^2 \psi}{\partial x^2} &= \left(\frac{\partial S}{\partial x}\right)^2 \psi = p_x^{\,2}\psi \;; \\ -\hbar^2 \frac{\partial^2 \psi}{\partial y^2} &= \left(\frac{\partial S}{\partial y}\right)^2 \psi = p_y^{\,2}\psi \;; \\ -\hbar^2 \frac{\partial^2 \psi}{\partial z^2} &= \left(\frac{\partial S}{\partial z}\right)^2 \psi = p_z^{\,2}\psi \end{aligned}\right\}. \tag{6.28a}$$

In other words,

$$-\hbar^2\left[\frac{\partial^2}{\partial x^2}+\frac{\partial^2}{\partial y^2}+\frac{\partial^2}{\partial z^2}\right]\psi = \left[p_x^{\,2}+p_y^{\,2}+p_z^{\,2}\right]\psi, \tag{6.28b}$$

which essentially provides for the kinetic energy of the system. Adding potential energy V to get the total energy E represented by the Hamiltonian in the HJ equation, and using Eq. 6.24 to identify

$$\frac{\partial S}{\partial t} = -i\hbar\frac{1}{\psi}\frac{\partial \psi}{\partial t}, \tag{6.29}$$

we get

$$\left[H\left(q,\frac{\partial F_2}{\partial q},t\right)+\left(-i\hbar\frac{1}{\psi}\frac{\partial\psi}{\partial t}\right)\right]\psi = 0, \tag{6.30a}$$

$$\text{i.e., } H\psi = i\hbar\frac{\partial\psi}{\partial t}, \tag{6.30b}$$

which is essentially the Schrödinger equation. The introduction of the wavefunction in Eq. 6.22 is of course essential for this discussion, but there was no motivation in the 19th century to attempt this. The development of quantum mechanics therefore had to wait! The differential equation we have arrived at being linear is satisfied by both Eqs. 6.23a and 6.23b:

$$\left(-\frac{\hbar^2\vec{\nabla}^2}{2m}+V\right)\left(e^{\frac{i}{\hbar}S}\right) = i\hbar\frac{\partial\left(e^{\frac{i}{\hbar}S}\right)}{\partial t} \tag{6.31a}$$

$$\text{and } \left(-\frac{\hbar^2\vec{\nabla}^2}{2m}+V\right)\left(Ae^{\frac{i}{\hbar}S}\right) = i\hbar\frac{\partial\left(Ae^{\frac{i}{\hbar}S}\right)}{\partial t}, \tag{6.31b}$$

thereby reinforcing the interpretation of the probability amplitude wavefunction $A(\vec{r},t)\psi(\vec{r},t)$ as the Feynman propagator $K\left(\vec{r}_f,t_f;\vec{r}_i,t_i\right)$. It is equally instructive to see that the classical HJ equation is obtainable from the Schrödinger equation. The limit $h\rightarrow 0$ results in this, as demonstrated in the Solved Problem P6.1. For a stationary state, the Hamilton's principal function is separable in the space part and the time part:

$$S(\vec{r},t) = W(\vec{r}) - Et = \hbar\left(u(\vec{r}) - \omega t\right). \tag{6.32}$$

The function $W(\vec{r})$ is called as Hamilton's characteristic function. In wave theory, we know that a surface of constant phase $S(\vec{r},t)$ advances linearly with time and describes a wavefront. Since the Hamilton's principal function is a phase, we can now make the "*wave*" ↔ "*particle*" connection. From Eq. 6.19b and Eq. 6.32, we interpret the classical momentum as the gradient of Hamilton's characteristic function:

$$\vec{p}_{classical} = \vec{\nabla}S = \vec{\nabla}W. \tag{6.33}$$

Funquest: How would you obtain the wavefront of light using the Hamilton-Jacobi equation?

The classical momentum is pointed in the direction in which a particle moves, likened in the above relation to the direction normal to a wavefront. The trajectories of the particle are then orthogonal to the surface of constant value of the characteristic function W. This relationship between classical mechanics and wave mechanics is reminiscent of that between geometrical optics and wave optics.

The connections between classical and quantum mechanics outlined here inspire a method of obtaining approximate solutions for the stationary states of a quantum system in a semi-classical method known as the Wentzel–Kramers–Brilloiun–Jeffreys (WKBJ) approximation [4–7], salient features of which are outlined below; and further details can be found in References [8–10]. To illustrate the WKBJ method, we note that the dynamical phase of a stationary state is $\exp\left(-i\frac{E}{\hbar}t\right) = \exp(-i\omega t)$, and we consider a one-dimensional problem for which the time-independent Schrödinger equation is

$$\left[\frac{d^2}{dx^2} + \frac{2m(E - V(x))}{\hbar^2}\right]\psi(x) = \left[\frac{d^2}{dx^2} + k^2\right]\psi(x) = 0, \tag{6.34a}$$

$$\text{where } k^2 = \frac{2m(E - V(x))}{\hbar^2}. \tag{6.34b}$$

For $E > V(x)$, we define $k = k(x) = +\frac{1}{\hbar}\sqrt{2m(E - V(x))}$ and note that k is real. For $E < V(x)$, the same definition renders k imaginary. With regard to a potential that may change with position, we have explicitly admitted position dependence of the wave number $k = k(x)$. Let us write the one-dimensional wavefunction as

$$\psi(x) = A(x)\exp\left(\frac{i}{\hbar}S(x)\right). \tag{6.35}$$

This gives, with a *single prime* denoting the *first* derivative and *two primes* denoting the *second*,

$$\psi'(x) = A'\exp\left(\frac{i}{\hbar}S\right) + A\frac{i}{\hbar}S'\exp\left(\frac{i}{\hbar}S\right) = \left[A' + A\frac{i}{\hbar}S'\right]\exp\left(\frac{i}{\hbar}S\right) \tag{6.36}$$

$$\text{and } \psi''(x) = \left(A'' + 2\frac{i}{\hbar}A'S' + A\frac{i}{\hbar}S'' - A\frac{1}{\hbar^2}S'^2\right)\exp\left(\frac{i}{\hbar}S\right). \tag{6.37}$$

Using Eq. 6.35 and Eq. 6.37 in Eq. 6.34a, we get

$$\left\{A'' + 2\frac{i}{\hbar}A'S' + A\frac{i}{\hbar}S'' - \frac{1}{\hbar^2}AS'^2\right\}\cancel{\exp\left(\frac{i}{\hbar}S\right)} = -k^2A\cancel{\exp\left(\frac{i}{\hbar}S\right)}, \tag{6.38a}$$

$$\text{i.e., } \hbar^2 A'' + i\hbar^2 A'S' - AS'^2 + \hbar^2 k^2 A + i\hbar AS'' = 0. \tag{6.38b}$$

If we *tentatively* ignore the spatial dependence of the amplitude A, it follows that

$$-i\hbar S'' + S'^2 = \hbar^2 k^2 = p^2 = 2m(E - V). \tag{6.38c}$$

Equations 6.38b,c are straightforward consequences of the Schrödinger equation, but are nonlinear. The nonlinearity enables development of an approximation scheme by expanding the phase $S(x)$ as a power series in $\hbar$:

$$S = S_0 + \left(\frac{\hbar}{i}\right) S_1 + \left(\frac{\hbar}{i}\right)^2 S_2 + \ldots. \tag{6.39a}$$

Obtaining the first and the second derivatives of the phase and inserting in the Schrödinger equation we get

$$S' = S_0' + \left(\frac{\hbar}{i}\right) S_1' + \left(\frac{\hbar}{i}\right)^2 S_2' + \ldots,$$

$$\text{and } S'' = S_0'' + \left(\frac{\hbar}{i}\right) S_1'' + \left(\frac{\hbar}{i}\right)^2 S_2'' + \ldots.,$$

using which in Eq. 6.38c, we get

$$-i\hbar\left\{S_0'' + \left(\frac{\hbar}{i}\right) S_1'' + \left(\frac{\hbar}{i}\right)^2 S_2''\right\} + \left\{S_0' + \left(\frac{\hbar}{i}\right) S_1' + \left(\frac{\hbar}{i}\right)^2 S_2'\right\}^2 = 2m\left(E - V(x)\right). \tag*{6.39b}$$

Ignoring now terms of the order $O(\hbar^3)$ or smaller, we get

$$-i\hbar\begin{Bmatrix} S_0'' + \left(\frac{\hbar}{i}\right) S_1'' + \\ \left(\frac{\hbar}{i}\right)^2 S_2'' \end{Bmatrix} + \begin{Bmatrix} S_0'^2 + \left(\frac{\hbar^2}{i^2}\right) S_1'^2 + \\ 2\left(\frac{\hbar}{i}\right) S_0' S_1' + 2\left(\frac{\hbar}{i}\right)^2 S_0' S_2' \end{Bmatrix} = 2m\left(E - V(x)\right), \tag{6.40a}$$

$$\text{i.e., } \hbar^0 \begin{Bmatrix} S_0'^2 - \\ 2m\left(E - V(x)\right) \end{Bmatrix} + \hbar^1 \begin{Bmatrix} -iS_0'' + \\ \left(\frac{2}{i}\right) S_0' S_1' \end{Bmatrix} - \hbar^2 \begin{Bmatrix} S_1'' + \\ S_1'^2 + 2S_0' S_2' \end{Bmatrix} = 0. \tag{6.40b}$$

The coefficients of each power of $\hbar$ must be *independently* zero. Hence,

$$\left(-S_0'^2 + 2m(E - V)\right) = 0, \tag{6.41a}$$

$$\left(iS_0'' - \frac{2}{i} S_0' S_1'\right) = 0, \tag{6.41b}$$

$$\text{and } \left(S_1'' + S_1'^2 + 2S_0' S_2'\right) = 0, \ldots, \text{ etc.} \tag{6.41c}$$

As a first approximation, if we keep only the leading term in Eq. 6.39a, we get from Eq. 6.41a,

$$S_0' = \pm\sqrt{2m(E - V)} = \pm\hbar k(x), \tag{6.41d}$$

which is the momentum. Integrating this result, we get

$$S_0(x) = \pm\hbar\int^x k(x')dx' + \lambda_0. \tag{6.41e}$$

Ignoring S_1 in the first approximation, from Eqs. 6.41b and 6.41e we see that

$$S_0'' = \pm\hbar k'(x) = 0. \tag{6.41f}$$

Ignoring all but the leading term in Eq. 6.39a amounts to the neglect of $|\hbar S''|$ relative to $|S'^2|$ in Eq. 6.38c. The latter was obtained essentially from the Schrödinger equation. Applicability of this approximation therefore requires $\left|\frac{\hbar S''}{S'^2}\right| \ll 1$, i.e., $\left|\frac{d}{dx}\left(\frac{\hbar}{S'}\right)\right| \ll 1$; equivalently, we require $\left[\left|\frac{d}{dx}\left(\frac{1}{k(x)}\right)\right| = \left|\frac{k'}{k^2}\right| = \left|\frac{\hbar p'}{p^2}\right|\right] \ll 1$. The condition for the validity of this (WKBJ) approximation scheme therefore is

$$\left|\frac{d}{dx}\left(\frac{\lambda(x)}{2\pi}\right)\right| \ll 1. \tag{6.42}$$

The WKBJ approximation can be a very useful method to get approximate solutions to the Schrödinger equation for a stationary state when the potential energy $V(x)$ changes rather slowly. Then, the momentum of the particle would not change rapidly across the region of the potential. In particular, the potential must not change significantly over a distance of the *local* de Broglie wavelength. Furthermore, from Eq. 6.41b, we have $S_1' = \frac{i}{2}\frac{S_0''}{S_0'}$, integrating which we get

$$S_1(x) = \frac{i}{2}\left(\ell n S_0'\right) + \lambda_1 = \frac{i}{2}\left(\ell n \hbar k(x)\right) + \lambda_1 = \frac{i}{2}\left(\ell n \hbar + \ell n k(x)\right) + \lambda_1, \tag{6.43a}$$

$$\text{i.e., } S_1(x) = \frac{i}{2}\left(\ell n k(x)\right) = i\left(\ell n\left(k(x)\right)^{\frac{1}{2}}\right). \tag{6.43b}$$

We do not mention the constants of integration anymore, since they can be absorbed in the overall normalization of the wavefunction. On reconsidering the space dependence $A = A(x)$ of the amplitude, we notice that the *real part* and the *imaginary part* of Eq. 6.38b must be independently zero, hence

$$A'' - \frac{1}{\hbar^2}AS'^2 + k^2A = 0 \tag{6.44a}$$

$$\text{and } 2A'S' + AS'' = 0. \tag{6.44b}$$

Multiplication of Eq. 6.44b with the amplitude $A(x)$ must also result in zero, and hence

$$A[2A'S' + AS''] = 2AA'S' + A^2S'' = \frac{d}{dx}[A^2S'] = 0. \tag{6.44c}$$

From Eq. 6.44c we conclude that $(A^2 S')$ must be a *constant* real number. The amplitude of the wavefunction therefore is proportional to the *reciprocal* of the *square root* of the *gradient* of S, and hence (from Eq. 6.41d) to $\frac{1}{\sqrt{k}}$. This property, along with the form of the wavefunction in Eq. 6.35, would guide us in developing the form of the WKBJ wavefunction. Two common situations we encounter in solving one-dimensional potential problems (including the radial equation for a spherically symmetric potential) are represented in Figs. 6.1a and 6.1b. In these figures, t and T are the classical turning points for a particle with energy E. The slope $\frac{dV}{dx}$ is negative in Fig. 6.1a and positive in Fig. 6.1b. The initial condition determines which side of a turning point a particle would be confined according to the laws of *classical* physics. Nature, of course, is described correctly only by quantum theory and enables the particle's mobility described by its probability current past the classical turning points, as we have discussed in Chapter 3 and Solved Problem P6.2.

We present here salient features of the more detailed discussion in References [8–10]. The WKBJ wavefunction is given by Eq. 6.45a in the region $E < V$ and by Eq. 6.45b in the region $E > V$ for the case $\frac{dV}{dx} < 0$.

<u>Classically Forbidden (CF)</u>

$E < V \leftrightarrow x < t$

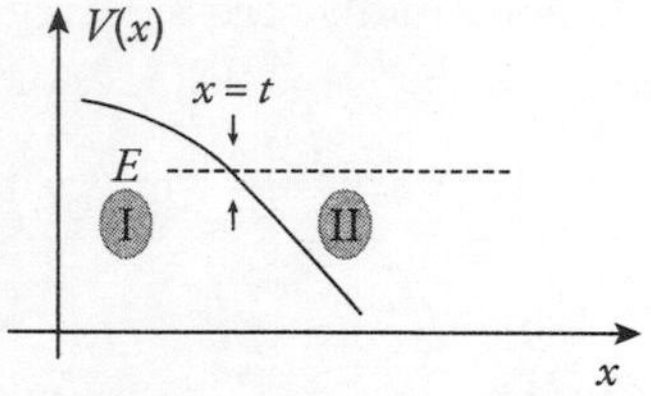

$$\psi_{CF}(x) \simeq \begin{bmatrix} \frac{A}{\sqrt{\kappa(x)}} \exp\left(-\int_t^x \kappa(x')dx'\right) + \\ \frac{B}{\sqrt{\kappa(x)}} \exp\left(\int_t^x \kappa(x')dx'\right) \end{bmatrix} \tag{6.45a}$$

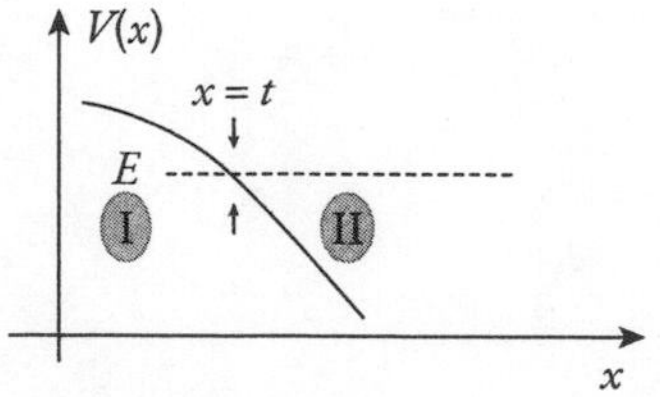

Fig. 6.1a Region I is classically forbidden (CF); region II is allowed. A particle described by classical physics can be found only in the region II, to the right of $x = t$, where $E > V$.

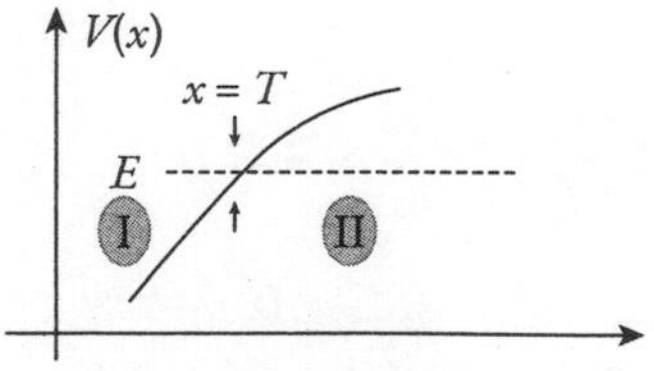

Fig. 6.1b Region I is classically allowed (CA); region II is forbidden. The classical turning point is at $x = T$.

Classically Allowed (CA)

$E > V \leftrightarrow x > t$

$$\boxed{\frac{dV}{dx} < 0}$$

$$\psi_{CA}(x) \simeq \left[\frac{C}{\sqrt{k(x)}} \exp\left(+i\int_x^t k(x')dx' \right) + \frac{D}{\sqrt{k(x)}} \exp\left(-i\int_x^t k(x')dx' \right) \right] \tag{6.45b}$$

Note that $\kappa(x) = ik(x)$. (6.46)

For the other case $\frac{dV}{dx} > 0$, Eq. 6.47a gives the solution for the CA region, and the solution for the CF region is given by Eq. 6.47b.

Classically Allowed (CA)

$E > V \leftrightarrow x < T$

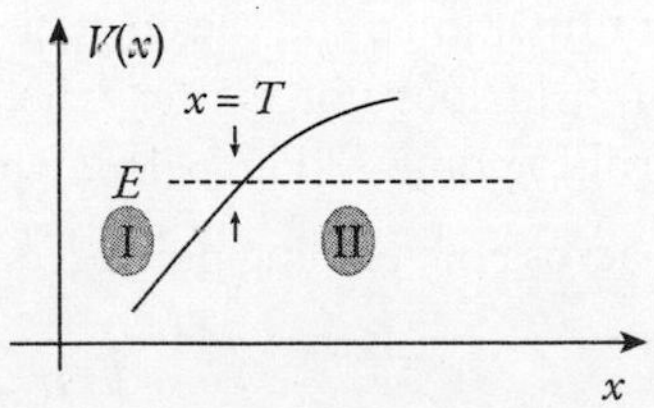

$$\psi_{CA}(x) \simeq \left[\frac{C}{\sqrt{k(x)}} \exp\left(+i\int_T^x k(x')dx' \right) + \frac{D}{\sqrt{k(x)}} \exp\left(-i\int_T^x k(x')dx' \right) \right] \tag{6.47a}$$

Classically Forbidden (CF)

$E < V \leftrightarrow x > T$

$$\boxed{\frac{dV}{dx} > 0}$$

$$\psi_{CF}(x) \simeq \left[\frac{A}{\sqrt{\kappa(x)}} \exp\left(-\int_T^x \kappa(x')dx' \right) + \frac{B}{\sqrt{\kappa(x)}} \exp\left(\int_T^x \kappa(x')dx' \right) \right] \tag{6.47b}$$

Both k and κ are real and positive. We have used a somewhat similar expression earlier in Chapter 2 to describe the wavefunction in the context of Feynman's path integral method (specifically, see Eq. 2.93 and the Solved Problem P2.4). The square root of the momentum that appears in the denominator (Eqs. 6.45a,b) of the wavefunction has two consequences. First, the probability density being given by the square of the amplitude, we see that it is less in those regions where the momentum is large. This interpretation is in line with what we expect from classical physics. The other consequence poses a difficulty; the wavefunction blows up as the momentum becomes small, i.e., at the turning points. It is then a challenge to determine the relationship between the coefficients A, B, C, and D in Eqs. 6.45a,b and in Eqs. 6.47a,b. The WKBJ solutions are therefore only asymptotically valid but not in the immediate vicinity of the classical turning points. Connecting the WKBJ wavefunctions across the turning points is therefore a challenge since the wavefunction blows up at both the turning points t and T. In order to circumvent the problem posed by the fact that the WKBJ wavefunction blows up at the classical turning points, a distinctive plan is needed.

A strategy to address the conundrum posed by the divergence of the wavefunction at the turning points is driven by examining the functions in the asymptotic regions. As is explained here, the WKBJ asymptotic solutions have the same forms as those of the solutions to the Airy differential equation. The Airy functions in the asymptotic regions are therefore employed to determine the connections of the WKBJ functions across the turning points. This approach is obviously *valid*, since the general solution to a second-order differential equation for a function $f(z)$ is expressible as a superposition of two linearly independent functions, and the coefficients in this superposition do not depend on the variable value of z; it does not matter whether $z \to 0$ or $z \to \infty$. We illustrate the method by considering the turning point T (Fig. 6.1b), where $\frac{dV}{dx} > 0$. We consider a linear approximation to the potential across the turning point T:

$$V(x) = Mx + \left(V_T - Mx_T\right) = V_T + M\left(x - x_T\right), \tag{6.48}$$

where M is the slope $M = \left.\frac{dV}{dx}\right]_T$. (6.49)

We see that in the present case (Fig. 6.1b) the slope M is positive. For such a linear potential, the Schrödinger equation can be written in a compact form

$$\frac{d^2\psi}{dz^2} - z\psi = 0. \tag{6.50}$$

On introducing a new dimensionless variable

$$z = \left(\frac{2\mu M}{\hbar^2}\right)^{1/3}\left(x - x_T\right) \tag{6.51}$$

corresponding to which

$$V(x) - E = M\left(x - x_T\right) = M\left(\frac{\hbar^2}{2mM}\right)^{1/3} z = M^{2/3}\left(\frac{\hbar^2}{2m}\right)^{1/3} z, \tag{6.52}$$

and $\frac{dz}{dx} = \left(\frac{2\mu M}{\hbar^2}\right)^{1/3}$. (6.53)

Accordingly, we have

$$k^2 = \frac{2\mu(|E-V|)}{\hbar^2} = \frac{2\mu}{\hbar^2} \times M^{2/3}\left(\frac{\hbar^2}{2\mu}\right)^{1/3} z = \left(\frac{2\mu M}{\hbar^2}\right)^{2/3} z, \tag{6.54a}$$

$$\text{and } k = \left(\frac{2\mu M}{\hbar^2}\right)^{1/3} z^{1/2}, \tag{6.54b}$$

which gives for its derivative with respect to x,

$$|k'| = \frac{dk}{dx} = \left(\frac{2\mu M}{\hbar^2}\right)^{1/3}\left(\frac{1}{2}z^{-1/2}\right)\left(\frac{2\mu M}{\hbar^2}\right)^{1/3} = \frac{1}{2}\left(\frac{2\mu M}{\hbar^2}\right)^{2/3}\left(z^{-1/2}\right). \tag{6.55}$$

The condition (Eq. 6.42) for the validity of the WKBJ approximation amounts to requiring $\frac{1}{2} \ll z^{3/2}$. Solutions to the differential equation Eq. 6.50 are known. They are called as Airy functions $(Ai(z), Bi(z))$. *Asymptotic* solutions to the Airy differential equation belong to two categories:

$$Ai(z; z \ll 0) \to Ai_{osc}(x; x \ll T) \to \frac{\Omega}{\sqrt{k}}\sin\int_x^T\left(k(x')dx' + \frac{\pi}{4}\right), \tag{6.56a}$$

$$Ai(z; z \gg 0) \to Ai_{exp}(x; x \gg T) \to \frac{\Omega}{2\sqrt{\kappa}}\exp\int_T^x -\kappa(x')dx', \tag{6.56b}$$

and

$$Bi(z; z \ll 0) \to Bi_{osc}(x; x \ll x_T) \to \frac{\Omega}{\sqrt{k}}\cos\int_x^{x_T}\left(k(x')dx' + \frac{\pi}{4}\right), \tag{6.57a}$$

$$Bi(z; z \gg 0) \to Bi_{exp}(x; x \gg x_T) \to \frac{\Omega}{\sqrt{\kappa}}\exp\int_{x_T}^x \kappa(x')dx'. \tag{6.57b}$$

In Eqs. 6.56a,b and Eqs. 6.57a,b, the position-independent parameter Ω is determined by the slope of the linear potential. We have subscripted the Airy functions by "osc" (for oscillatory) and "exp" for exponential. The exponential functions are a *continuation* in the CF region of the oscillatory functions in the CA region. Comparing Eqs. 6.47a,b with the asymptotic forms of the Airy functions (Eqs. 6.56a,b and Eqs. 6.57a,b), we get the following *connection formula*:

$$\left\{\begin{array}{c} \dfrac{2A}{\sqrt{k(x)}}\cos\left(\displaystyle\int_x^T k(x')dx' - \frac{\pi}{4}\right) - \dfrac{B}{\sqrt{k(x)}}\sin\left(\displaystyle\int_x^T k(x')dx' - \frac{\pi}{4}\right) \\ \Leftrightarrow \\ \dfrac{A}{\sqrt{\kappa(x)}}\exp\left(-\displaystyle\int_T^x \kappa(x')dx'\right) + \dfrac{B}{\sqrt{\kappa(x)}}\exp\left(\displaystyle\int_T^x \kappa(x')dx'\right) \end{array}\right\}. \tag{6.58a}$$

On carrying out a similar analysis using Eqs. 6.45a,b for the case when $\frac{dV}{dx} < 0$ (Fig. 6.1a) and comparing them with the asymptotic forms of the Airy functions, we get another *connection formula*:

$$\left\{\begin{array}{c}\frac{A}{\sqrt{\kappa(x)}}\exp\left(-\int_x^t \kappa(x')dx'\right)+\frac{B}{\sqrt{\kappa(x)}}\exp\left(\int_x^t \kappa(x')dx'\right)\\ \Leftrightarrow \\ \frac{2A}{\sqrt{k(x)}}\cos\left(\int_t^x k(x')dx'-\frac{\pi}{4}\right)-\frac{B}{\sqrt{k(x)}}\sin\left(\int_t^x k(x')dx'-\frac{\pi}{4}\right)\end{array}\right\}. \tag{6.58b}$$

The functions $\left(\sin\left(\theta-\frac{\pi}{4}\right),\cos\left(\theta-\frac{\pi}{4}\right)\right)$ in Eqs. 6.58a,b provide an alternative basis set to write a general solution that is alternatively written earlier in the basis $\left(\exp(i\theta),\exp(-i\theta)\right)$. The phase shift through the angle $\frac{\pi}{4}$ might seem strange, but it is perfectly legitimate as the difference is compensated readily by adjusting the coefficients of the base functions. Equations 6.58a,b are called *WKBJ connection formulas*. They make it possible to connect the WKBJ functions in the CA and CF regions across the classical turning points by taking advantage of the fact that the asymptotic forms of the Airy functions and the WKBJ functions are identical. The proportions of the mixing coefficients in the superposition are of significance. We note that in Eqs. 6.58a,b the coefficients are (2A,B) in the CF region and (A,B) in the CA region.

As an application of the WKBJ method, we consider a particle with energy E in a potential shown in Fig. 6.2.

From the connection formula at $x = t$, we find that

$$\frac{A}{\sqrt{\kappa(x)}}\exp\left(-\int_x^t \kappa(x')dx'\right)\Leftrightarrow\frac{2A}{\sqrt{k(x)}}\cos\left(\int_t^x k(x')dx'-\frac{\pi}{4}\right), \tag{6.59a}$$

and from the connection formula at $x = T$, we have

$$\frac{2A'}{\sqrt{k(x)}}\cos\left(\int_x^T k(x')dx'-\frac{\pi}{4}\right)\Leftrightarrow\frac{A'}{\sqrt{\kappa(x)}}\exp\left(-\int_T^x \kappa(x')dx'\right). \tag{6.59b}$$

We have taken care of employing different constants A and A' in Eqs. 6.59a and 6.59b. The connection formulas only connect each of them to the farthest sides, $x \to \pm\infty$, where the functions have an exponential behaviour. A relation between A and A' would, however, emerge from the fact that in the region II (Fig. 6.2), the physical solution must be unique: the solution that we get from the farthest left and from the farthest right must correspond to the same function in the region II. Hence,

$$A\cos\left(\int_t^x k(x')dx'-\frac{\pi}{4}\right)=A'\cos\left(\int_x^T k(x')dx'-\frac{\pi}{4}\right) \tag{6.60a}$$

$$\text{or, } A\sin\left(\int_t^x k(x')dx'+\frac{\pi}{4}\right)=A'\sin\left(\int_x^T k(x')dx'+\frac{\pi}{4}\right). \tag{6.60b}$$

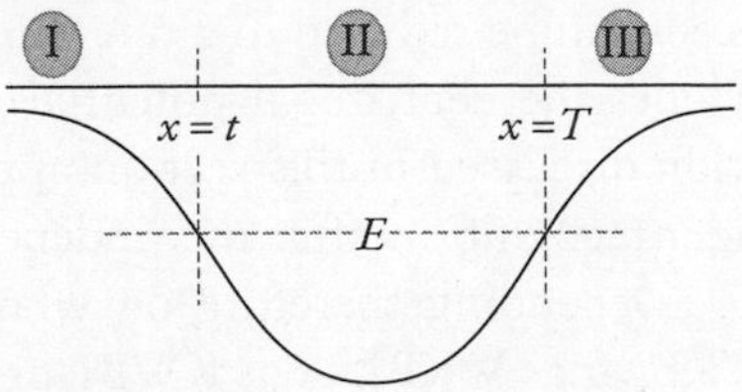

Fig. 6.2 A particle with energy E would be confined to region II, between the turning points t and T according to classical laws. The regions I and III are CF.

Recognizing now that the integration range $\int_t^x \square$ to be equivalent to $\int_t^T \square - \int_x^T \square$, we see that

$$\sin\left(\int_t^T k(x')dx' - \int_x^T k(x')dx' + \frac{\pi}{4}\right) = \frac{A'}{A}\sin\left(\int_x^T k(x')dx' + \frac{\pi}{4}\right),$$

$$\text{i.e., } \sin\left(\left(\int_t^T k(x')dx' + \frac{\pi}{2}\right) - \left(\int_x^T k(x')dx' + \frac{\pi}{4}\right)\right) = \frac{A}{A'}\sin\left(\int_x^T k(x')dx' + \frac{\pi}{4}\right). \tag{6.60c}$$

Furthermore, from the trigonometric identity

$$\sin(n\pi - \theta) = (-1)^{n-1}\sin\theta\,;\; n = 1,2,3,.., \tag{6.61}$$

it follows that

$$\left(\int_t^T k(x')dx' + \frac{\pi}{2}\right) = (n+1)\pi\,;\; n = 0,1,2,3,.. \tag{6.62}$$

$$\text{and } \frac{A}{A'} = (-1)^{n-1}. \tag{6.63}$$

Considering the integral over momentum p rather than the wavenumber k, if we carry out the integration over a *closed* path $t \to T \to t$, we get

$$\oint p(x')dx' = \left(n + \frac{1}{2}\right)\hbar. \tag{6.64}$$

The above relation gives the discrete energy states of a bound particle, and the particle has a nonzero probability of tunneling through the classical turning points into the nonclassical region. Equation 6.64 corresponds to the Bohr–Wilson–Sommerfeld (BWS) quantization rule of the *old* quantum theory, except for the factor $\frac{\hbar}{2}$, which is missing in the BWS rule. This rule was quite successful in the early days of quantum physics. One can easily check that it produces the correct energy spectrum of Bohr's *planetary* model of the hydrogen atom, or that of a particle in an impenetrable box, etc. One must, however, remember that although the result in Eq. 6.64 is an improvement over the BWS rule, it is itself the result of the *quasi-classical* WKBJ *approximation* to the Schrödinger equation. The WKBJ methodology can also be applied to solve the Schrödinger equation approximately for a central field problem since it can be reduced to a one-dimensional problem by defining an effective potential that includes the centrifugal potential (Eq. 5.18 in Chapter 5). The earliest use of the approximation discussed in this section appeared in the work of Jeffreys [4], but the method gained wider recognition after the (independent) papers by Wentzel [5], Kramers [6], and Brillouin [7]. Depending therefore on who your favorite author is, the method is referred to as the JWKB, or as WKBJ, or as KWBJ, or as BKWJ approximation. The approximation scheme is also known as the *semi-classical* (or *quasi-classical*) approximation, or as the *phase integral method*.

6.2 Time-Independent Perturbation Theory

Perturbation theory provides a systematic method to estimate errors due to the neglect of certain terms while solving physical problems. The method incorporates strategies to model corrections to a solvable, but incomplete, problem. Such methods were adopted even in classical physics. For example, in the 19th century, Jean Joseph Le Verrier predicted the existence and position of Neptune by mathematically analyzing discrepancies in the orbit of Uranus based on Kepler–Newton laws. Perturbative methods, however, go only as far as they go; the precession of the planet mercury could not be accounted for from the estimation of errors due to the precession of the equinoxes, gravitational tugs due to other planets, and even corrections due to the distribution of mass in the oblate shape of the sun. It needed, as shown in a rather simple introduction in Chapter 14 of [2], a totally new theory – the *general theory of relativity*. Perturbation theory, however, is a very powerful tool. Where it succeeds, it provides for a quantitative mechanism to determine consequences of the insolvable part V in the Hamiltonian.

There are very few examples of physical systems for which exact solutions of the Schrödinger equation can be obtained. These include the free particle, the particle in a box, under-the-barrier and over-the-barrier tunneling across a barrier, the harmonic oscillator, the atomic 1-electron system, a rigid rotor. The Hamiltonian of other real physical systems is usually more complicated, but it is often possible to express it as

$$H = H_0 + V, \tag{6.65}$$

where H_0 describes the *dominant* character of H and is solvable. Inclusion of the term V renders the problem mathematically intractable. The best that one can then do is to solve the Schrödinger equation with the Hamiltonian H_0 and make an attempt to estimate errors caused by the neglect of V. Perturbation theory is the methodology that enables us to achieve this. Several mathematicians and physicists, most notably Schrödinger himself, have contributed to the development of the methodology using techniques employed earlier in the field of optics by Rayleigh. One refers to the term H_0 in the *full Hamiltonian H* as the *unperturbed Hamiltonian* and V as the *perturbation*. H_0 must constitute the most significant part of the full Hamiltonian and for which the Schrödinger equation is *solvable*. Perturbation theory is applicable when certain conditions are satisfied, the most imperative of which is that the perturbation V is only a small part of the full Hamiltonian. There, however, are other requirements. In this chapter, we shall be mostly concerned with the development of perturbation theory for the bound states of a quantum system. The perturbation V consists of one or more terms that make the problem difficult. It may include terms that are intrinsic to the system, but neglected in order to make the problem solvable, or may contain terms that must be considered on the application of external fields in which a quantum system is placed; interactions with the external field can then be possibly considered as perturbations. It is important in perturbation theory that all terms that make a comparable change to the eigenfunctions and eigenvalues of the unperturbed Hamiltonian are included regardless of their origin, whether intrinsic to the physical system or triggered by the application of an external field.

A perturbation can modify the eigenvalues and eigenfunctions of the unperturbed Hamiltonian, it may remove a degeneracy in an eigenspectrum, and may do so partially or wholly, and it may even introduce transitions between different eigenstates.

Time-Independent Perturbation Theory of Non-Degenerate Eigenstates

To begin with, we shall consider perturbations to non-degenerate eigenstates of the unperturbed Hamiltonian. The eigenvalue equations for the stationary eigenstates of the Hamiltonian H_0 are

$$H_0\left|n^{(0)}\right\rangle = E_n^{(0)}\left|n^{(0)}\right\rangle. \tag{6.66}$$

The set of eigenvectors $\left\{\left|n^{(0)}\right\rangle\right\}$ is considered to consist of orthonormal vectors. When the perturbation is small (weak), we may assume that Hilbert space of the full Hamiltonian H is spanned by the complete eigenbasis of H_0,
i.e.,

$$\sum_m \left|m^{(0)}\right\rangle\left\langle m^{(0)}\right| = 1. \tag{6.67}$$

The superscript (0) on both the eigenvalues and the eigenvectors denote that they belong to the eigenspectrum of the unperturbed Hamiltonian. We now introduce a parameter λ called as an *order parameter* (for reasons to become clear soon) such that we can write the full Hamiltonian as

$$H = H_0 + \lambda V. \tag{6.68}$$

Obviously, only two values (i) $\lambda = 0$ and (ii) $\lambda = 1$ are of *physical* interest, the former to the case of the unperturbed system, and the latter to the perturbed one. In perturbation theory, the order parameter λ is a mathematical device. It is allowed to take a small value between zero and one. The ultimate results that we shall use would be independent of the magnitude of the order parameter, as will be seen shortly. Auxiliary introduction of the order parameter enables us to develop a powerful method of solving complicated problems without even requiring us to depend on its actual value; in fact, the auxiliary parameter would be eventually dispensed with.

The eigenvalue equation for the *full* Hamiltonian is

$$H|n\rangle = \left(H_0 + \lambda V\right)|n\rangle = E_n|n\rangle. \tag{6.69}$$

The eigenvectors $|n\rangle$ of the full Hamiltonian are also considered normalized:

$$\langle n|n\rangle = 1. \tag{6.70}$$

Perturbation theory makes the ansatz that the eigenvalues and the eigenvectors of the full Hamiltonian are expressible as a power series in the perturbation parameter $\lambda < 1$ as

$$E_n = \left\{\lambda^0 E_n^{(0)} + \lambda^1 E_n^{(1)} + \lambda^2 E_n^{(2)} + \lambda^3 E_n^{(3)} + ..\right\} = \sum_{i=0}^{\infty} \lambda^i E_n^{(i)} \tag{6.71}$$

and

$$|n\rangle = \lambda^0 \left|n^{(0)}\right\rangle + \lambda^1 \left|n^{(1)}\right\rangle + \lambda^2 \left|n^{(2)}\right\rangle + \lambda^3 \left|n^{(3)}\right\rangle + .. = \sum_{i=0}^{\infty} \lambda^i \left|n^{(i)}\right\rangle, \tag{6.72}$$

where $E_n^{(j)}$ is the jth order correction to the unperturbed eigenvalue and $\left|n^{(j)}\right\rangle$ is the jth order correction to the eigenvector. Since $\lambda < 1$, subsequent terms in the power series expansions in Eqs. 6.71 and 6.72 have diminishing importance. Perturbation theory rests on this essential property even as it must make the final results independent of the auxiliary order parameter λ. The ansatz (Eq. 6.67) about the completeness of the eigenvectors of the unperturbed Hamiltonian assures us that for every j,

$$\left|n^{(j)}\right\rangle = \sum_m \left|m^{(0)}\right\rangle\left\langle m^{(0)}\middle|n^{(j)}\right\rangle = \left|n^{(0)}\right\rangle\left\langle n^{(0)}\middle|n^{(j)}\right\rangle + \sum_{m\neq n} \left|m^{(0)}\right\rangle\left\langle m^{(0)}\middle|n^{(j)}\right\rangle. \tag{6.73}$$

From Eq. 6.72 and Eq. 6.73, it follows that

$$|n\rangle = \sum_{i=0}^{\infty} \lambda^i \left[\sum_m \left|m^{(0)}\right\rangle\left\langle m^{(0)}\right|\right]\left|n^{(i)}\right\rangle = \begin{bmatrix} \sum_{i=0}^{\infty} \lambda^i \left|n^{(0)}\right\rangle\left\langle n^{(0)}\middle|n^{(i)}\right\rangle + \\ \sum_{i=0}^{\infty} \lambda^i \left[\sum_{m\neq n} \left|m^{(0)}\right\rangle\left\langle m^{(0)}\right|\right]\left|n^{(i)}\right\rangle \end{bmatrix}, \tag{6.74a}$$

$$\text{i.e., } |n\rangle = \left|n^{(0)}\right\rangle \sum_{i=0}^{\infty} \lambda^i \left\langle n^{(0)}\middle|n^{(i)}\right\rangle + \sum_{m\neq n} \left|m^{(0)}\right\rangle \sum_{i=0}^{\infty} \lambda^i \left\langle m^{(0)}\middle|n^{(i)}\right\rangle, \tag{6.74b}$$

$$\text{or } |n\rangle = \left|n^{(0)}\right\rangle \begin{bmatrix} \left\langle n^{(0)}\middle|n^{(0)}\right\rangle + \\ \lambda\left\langle n^{(0)}\middle|n^{(1)}\right\rangle + \\ \lambda^2 \left\langle n^{(0)}\middle|n^{(2)}\right\rangle + .. \end{bmatrix} + \sum_{m\neq n} \left|m^{(0)}\right\rangle \begin{bmatrix} \cancel{\left\langle m^{(0)}\middle|n^{(0)}\right\rangle} + \\ \lambda\left\langle m^{(0)}\middle|n^{(1)}\right\rangle + \\ \lambda^2 \left\langle m^{(0)}\middle|n^{(2)}\right\rangle + .. \end{bmatrix}. \tag{6.74c}$$

On the right-hand side, $\left\langle m^{(0)}\middle|n^{(0)}\right\rangle = 0$ has been recognized on account of orthogonality of the eigenvectors of the unperturbed Hamiltonian; the corresponding term is therefore dropped. Hence,

$$|n\rangle = \left|n^{(0)}\right\rangle \begin{bmatrix} \left\langle n^{(0)}\middle|n^{(0)}\right\rangle + \\ \lambda\left\langle n^{(0)}\middle|n^{(1)}\right\rangle + \\ \lambda^2 \left\langle n^{(0)}\middle|n^{(2)}\right\rangle + .. \end{bmatrix} + \sum_{m\neq n} \left|m^{(0)}\right\rangle \begin{bmatrix} \lambda\left\langle m^{(0)}\middle|n^{(1)}\right\rangle + \\ \lambda^2 \left\langle m^{(0)}\middle|n^{(2)}\right\rangle + .. \end{bmatrix}. \tag{6.74d}$$

To determine the perturbed states, up to the jth order, we require the nonzero component of $\left|n^{(j)}\right\rangle$ not only on the parent state that is perturbed but also on other states that are perturbed. In other words, we require $\left\langle n^{(0)} \middle| n^{(j)}\right\rangle$, i.e., $\left\langle m^{(0)} \middle| n^{(j)}\right\rangle$ with $m = n$ and also $\left\langle m^{(0)} \middle| n^{(j)}\right\rangle$ with $m \neq n$. The eigenvalue equation for the full Hamiltonian (Eq. 6.50) can now be written using the power series expansions of the eigenvalues (from Eq. 6.71) and eigenvectors (from Eqs. 6.74a–d):

$$\left(H_0 + \lambda V\right)\left(\sum_{i=0}^{\infty} \lambda^i \left|n^{(i)}\right\rangle\right) = \left\{\begin{matrix} \lambda^0 E_n^{(0)} + \lambda^1 E_n^{(1)} + \\ \lambda^2 E_n^{(2)} + .. \end{matrix}\right\}\left(\sum_{i=0}^{\infty} \lambda^i \left|n^{(i)}\right\rangle\right). \tag{6.75a}$$

Equating coefficients of corresponding powers of λ, we get

$$\lambda^0 : H_0\left|n^{(0)}\right\rangle = E_n^{(0)}\left|n^{(0)}\right\rangle, \tag{6.75b}$$

$$\lambda^1 : H_0\left|n^{(1)}\right\rangle + V\left|n^{(0)}\right\rangle = E_n^{(0)}\left|n^{(1)}\right\rangle + E_n^{(1)}\left|n^{(0)}\right\rangle,$$

$$\text{i.e., } \left(H_0 - E_n^{(0)}\right)\left|n^{(1)}\right\rangle = \left(E_n^{(1)} - V\right)\left|n^{(0)}\right\rangle, \tag{6.75c}$$

$$\lambda^2 : H_0\left|n^{(2)}\right\rangle + V\left|n^{(1)}\right\rangle = \begin{bmatrix} E_n^{(0)}\left|n^{(2)}\right\rangle + E_n^{(1)}\left|n^{(1)}\right\rangle \\ + E_n^{(2)}\left|n^{(0)}\right\rangle \end{bmatrix},$$

$$\text{i.e., } \left(H_0 - E_n^{(0)}\right)\left|n^{(2)}\right\rangle = \left(E_n^{(1)} - V\right)\left|n^{(1)}\right\rangle + E_n^{(2)}\left|n^{(0)}\right\rangle, \tag{6.75d}$$

$$\lambda^3 : H_0\left|n^{(3)}\right\rangle + V\left|n^{(2)}\right\rangle = \begin{bmatrix} E_n^{(0)}\left|n^{(3)}\right\rangle + E_n^{(1)}\left|n^{(2)}\right\rangle + \\ E_n^{(2)}\left|n^{(1)}\right\rangle + E_n^{(3)}\left|n^{(0)}\right\rangle \end{bmatrix},$$

$$\text{i.e., } \left(H_0 - E_n^{(0)}\right)\left|n^{(3)}\right\rangle = \begin{bmatrix} E_n^{(1)} \\ -V \end{bmatrix}\left|n^{(2)}\right\rangle + \begin{bmatrix} E_n^{(2)}\left|n^{(1)}\right\rangle + \\ E_n^{(3)}\left|n^{(0)}\right\rangle \end{bmatrix}, \ldots \tag{6.75e}$$

In general, coefficients for the jth power of λ give

$$\lambda^j : \left(H_0 - E_n^{(0)}\right)\left|n^{(j)}\right\rangle = -V\left|n^{(j-1)}\right\rangle + \sum_{k=1}^{j} E_n^{(k)}\left|n^{(j-k)}\right\rangle. \tag{6.76}$$

The perturbed states also being normalized,

$$1 = \langle n|n\rangle = \sum_{i=0}^{\infty} \lambda^i \sum_{j=0}^{\infty} \lambda^{j*}\left[\sum_m \sum_{m'} \left\langle n^{(j)} \middle| m'^{(0)}\right\rangle\left\langle m'^{(0)} \middle\| m^{(0)}\right\rangle\left\langle m^{(0)} \middle| n^{(i)}\right\rangle\right], \tag{6.77a}$$

$$1 = \sum_{i=0}^{\infty} \lambda^i \sum_{j=0}^{\infty} \lambda^{j*} \sum_m \left\langle n^{(j)} \middle| m^{(0)}\right\rangle\left\langle m^{(0)} \middle| n^{(i)}\right\rangle = \sum_{i=0}^{\infty} \sum_{j=0}^{\infty} \lambda^i \lambda^{j*} \left\langle n^{(j)} \middle| n^{(i)}\right\rangle, \tag{6.77b}$$

$$1 = \left\{ \begin{matrix} \langle n^{(0)} | n^{(0)} \rangle + \lambda \left(\langle n^{(1)} | n^{(0)} \rangle + \langle n^{(0)} | n^{(1)} \rangle \right) + \\ \lambda^2 \left[\langle n^{(2)} | n^{(0)} \rangle + \langle n^{(1)} | n^{(1)} \rangle + \langle n^{(0)} | n^{(2)} \rangle \right] + .. \end{matrix} \right\}. \tag{6.77c}$$

Equating the coefficients of corresponding powers of λ, we get

$$\lambda^0 : \langle n^{(0)} | n^{(0)} \rangle = 1, \tag{6.78a}$$

$$\lambda^1 : \langle n^{(1)} | n^{(0)} \rangle + \langle n^{(0)} | n^{(1)} \rangle = 0, \tag{6.78b}$$

$$\lambda^2 : \langle n^{(2)} | n^{(0)} \rangle + \langle n^{(1)} | n^{(1)} \rangle + \langle n^{(0)} | n^{(2)} \rangle = 0, \tag{6.78c}$$

$$\lambda^j : \sum_{k=0}^{j} \langle n^{(j-k)} | n^{(k)} \rangle = 0. \tag{6.79}$$

The phase of these vectors is chosen so that $\langle n^{(0)} | n \rangle$ is real.

$$\langle n^{(0)} | n \rangle = \sum_i \lambda^i \langle n^{(0)} | n^{(i)} \rangle = \left[\begin{matrix} \langle n^{(0)} | n^{(0)} \rangle + \lambda \langle n^{(0)} | n^{(1)} \rangle + \\ \lambda^2 \langle n^{(0)} | n^{(2)} \rangle + \lambda^3 \langle n^{(0)} | n^{(3)} \rangle .. \end{matrix} \right]. \tag{6.80}$$

Since terms in different powers of λ must be considered independent, the requirement that $\langle n^{(0)} | n \rangle$ is real requires that for every j, $\langle n^{(0)} | n^{(j)} \rangle$ must be real, and hence

$$\langle n^{(0)} | n^{(j)} \rangle = \langle n^{(j)} | n^{(0)} \rangle. \tag{6.81}$$

In other words, $\langle n^{(j)} | n^{(0)} \rangle$ is essentially real.

Projecting Eq. 6.76 on $\langle n^{(0)} |$,

$$\langle n^{(0)} | \left(H_0 - E_n^{(0)} \right) | n^{(j)} \rangle = -\langle n^{(0)} | V | n^{(j-1)} \rangle + \sum_{k=1}^{j} E_n^{(k)} \langle n^{(0)} | n^{(j-k)} \rangle. \tag{6.82}$$

For $j \geq 1$, from Eq. 6.82, we get

$$0 = -\langle n^{(0)} | V | n^{(j-1)} \rangle + \sum_{k=1}^{j} E_n^{(k)} \langle n^{(0)} | n^{(j-k)} \rangle, \tag{6.83a}$$

i.e., separating the term for $k = j$ in the summation on the right,

$$0 = \left[\begin{matrix} -\langle n^{(0)} | V | n^{(j-1)} \rangle + \\ E_n^{(k)} \langle n^{(0)} | n^{(0)} \rangle + \sum_{k=1}^{j-1} E_n^{(k)} \langle n^{(0)} | n^{(j-k)} \rangle \end{matrix} \right] = \left[\begin{matrix} -\langle n^{(0)} | V | n^{(j-1)} \rangle + \\ E_n^{(j)} + \sum_{k=1}^{j-1} E_n^{(k)} \langle n^{(0)} | n^{(j-k)} \rangle \end{matrix} \right], \tag{6.83b}$$

$$\text{and hence } E_n^{(j)} = \left\langle n^{(0)} \middle| V \middle| n^{(j-1)} \right\rangle - \sum_{k=1}^{j-1} E_n^{(k)} \left\langle n^{(0)} \middle| n^{(j-k)} \right\rangle. \tag{6.84}$$

To determine $E_n^{(j)}$, we require $E_n^{(k)}; k = 1,2,..,(j-1)$ and also $\left| n^{(i)} \right\rangle; i = 1,2,..,(j-1)$.

We had noted earlier that to determine the perturbed vector states, up to the jth order, we require $\left\langle n^{(0)} \middle| n^{(j)} \right\rangle$, i.e., $\left\langle m^{(0)} \middle| n^{(j)} \right\rangle$ with $m = n$ and also $\left\langle m^{(0)} \middle| n^{(j)} \right\rangle$ with $m \neq n$. First, we shall determine $\left\langle m^{(0)} \middle| n^{(j)} \right\rangle; m \neq n$. Projecting Eq. 6.76 on $\left\langle m^{(0)} \right|$,

$$\text{for } m \neq n \left(E_m^{(0)} - E_n^{(0)} \right) \left\langle m^{(0)} \middle| n^{(j)} \right\rangle = -\left\langle m^{(0)} \middle| V \middle| n^{(j-1)} \right\rangle + \sum_{k=1}^{j} E_n^{(k)} \left\langle m^{(0)} \middle| n^{(j-k)} \right\rangle. \tag{6.85a}$$

Separating the $k = j$ term in the summation on the right-hand side,

$$m \neq n: \left\langle m^{(0)} \middle| n^{(j)} \right\rangle = -\frac{\left\langle m^{(0)} \middle| V \middle| n^{(j-1)} \right\rangle}{\left(E_m^{(0)} - E_n^{(0)} \right)} + E_n^{(j)} \frac{\cancel{\left\langle m^{(0)} \middle| n^{(0)} \right\rangle}}{\left(E_m^{(0)} - E_n^{(0)} \right)} + \sum_{k=1}^{j-1} E_n^{(k)} \frac{\left\langle m^{(0)} \middle| n^{(j-k)} \right\rangle}{\left(E_m^{(0)} - E_n^{(0)} \right)}, \tag{6.85b}$$

$$\left\langle m^{(0)} \middle| n^{(j)} \right\rangle = -\frac{\left\langle m^{(0)} \middle| V \middle| n^{(j-1)} \right\rangle}{\left(E_m^{(0)} - E_n^{(0)} \right)} + \sum_{k=1}^{j-1} E_n^{(k)} \frac{\left\langle m^{(0)} \middle| n^{(j-k)} \right\rangle}{\left(E_m^{(0)} - E_n^{(0)} \right)}, \tag{6.85c}$$

since $\left| m^{(0)} \right\rangle$ and $\left| n^{(0)} \right\rangle$ are orthogonal. Next, to determine $\left\langle n^{(0)} \middle| n^{(j)} \right\rangle$, we rewrite Eq. 6.79 by separating the terms for $k = 0$ and $k = j$ to get

$$\lambda^j: \left\langle n^{(j)} \middle| n^{(0)} \right\rangle + \left\langle n^{(0)} \middle| n^{(j)} \right\rangle + \sum_{k=1}^{j-1} \left\langle n^{(j-k)} \middle| n^{(k)} \right\rangle = 0. \tag{6.86a}$$

Recognizing now that the first two terms on the left hand side of the above equation are equal to each other (Eq. 6.81),
we get,

$$\left\langle n^{(0)} \middle| n^{(j)} \right\rangle = -\frac{1}{2} \sum_{k=1}^{j-1} \left\langle n^{(j-k)} \middle| n^{(k)} \right\rangle. \tag{6.86b}$$

Let us consider *first-order* perturbative corrections.
For $j = 1$, from Eq. 6.82, we get

$$\left\langle n^{(0)} \middle| \left(H_0 - E_n^{(0)} \right) \middle| n^{(1)} \right\rangle = -\left\langle n^{(0)} \middle| V \middle| n^{(0)} \right\rangle + E_n^{(1)} \left\langle n^{(0)} \middle| n^{(0)} \right\rangle = -\left\langle n^{(0)} \middle| V \middle| n^{(0)} \right\rangle + E_n^{(1)}. \tag{6.87}$$

Since $\left\langle n^{(0)} \right| H_0 = \left\langle n^{(0)} \right| E_n^{(0)}$, we find that

$$E_n^{(1)} = \left\langle n^{(0)} \middle| V \middle| n^{(0)} \right\rangle. \tag{6.88}$$

We see that the first-order correction to the energy eigenvalue is essentially the expectation value of the perturbation operator in the unperturbed state. This result is very commonly used in a large number of applications. We shall use it in the next two chapters. We do not need correction to the state vector to first order to obtain correction to energy to the first order. However, to determine many other quantum properties, we require the state vectors to be corrected due to the perturbation. We must therefore use Eq. 6.76d to obtain the state vectors. Truncating it at the first order term, we get

$$|n\rangle = \begin{bmatrix} |n^{(0)}\rangle\left[\langle n^{(0)}|n^{(0)}\rangle + \langle n^{(0)}|n^{(1)}\rangle\right] + \\ \sum_{m\neq n}|m^{(0)}\rangle\langle m^{(0)}|n^{(1)}\rangle \end{bmatrix} = \begin{bmatrix} |n^{(0)}\rangle\left[1 + \langle n^{(0)}|n^{(1)}\rangle\right] + \\ \sum_{m\neq n}|m^{(0)}\rangle\langle m^{(0)}|n^{(1)}\rangle \end{bmatrix}. \tag{6.89}$$

We must determine $\langle n^{(0)}|n^{(1)}\rangle$ and also $\langle m^{(0)}|n^{(1)}\rangle$ for $m \neq n$; the first of these $\langle n^{(0)}|n^{(1)}\rangle$ is readily obtained from Eq. 6.86b:

$$\langle n^{(0)}|n^{(1)}\rangle = -\frac{1}{2}\sum_{k=1,0}\langle n^{(j-k)}|n^{(k)}\rangle = -\frac{1}{2}\left[\langle n^{(1)}|n^{(0)}\rangle + \langle n^{(0)}|n^{(1)}\rangle\right] = -\langle n^{(0)}|n^{(1)}\rangle.$$

We therefore conclude that $\langle n^{(0)}|n^{(1)}\rangle = 0.$ (6.90a)

To determine $\langle m^{(0)}|n^{(1)}\rangle$ when $m \neq n$, we use Eq. 6.85c up to first order, which gives

$$\langle m^{(0)}|n^{(1)}\rangle = -\frac{\langle m^{(0)}|V|n^{(0)}\rangle}{\left(E_m^{(0)} - E_n^{(0)}\right)}. \tag{6.90b}$$

The perturbed state, up to first-order correction, therefore is

$$|n\rangle = |n^{(0)}\rangle + \sum_{m\neq n}|m^{(0)}\rangle\frac{\langle m^{(0)}|V|n^{(0)}\rangle}{\left(E_n^{(0)} - E_m^{(0)}\right)}. \tag{6.91}$$

Perturbation theory works on the premise that perturbation corrections are not very strong; we therefore see from Eq. 6.91 that a condition for the applicability of the perturbation theory is that $\langle m^{(0)}|V|n^{(0)}\rangle \ll \left(E_n^{(0)} - E_m^{(0)}\right)$.

Let us now consider the second-order perturbative corrections.

From Eq. 6.84, with $j = 2$,

$$E_n^{(2)} = \langle n^{(0)}|V|n^{(1)}\rangle - E_n^{(1)}\cancel{\langle n^{(0)}|n^{(1)}\rangle} = \sum_{m\neq n}\frac{\langle n^{(0)}|V|m^{(0)}\rangle\langle m^{(0)}|V|n^{(0)}\rangle}{\left(E_n^{(0)} - E_m^{(0)}\right)},$$

$$\text{i.e., } E_n^{(2)} = \sum_{m\neq n}\frac{\left|\langle n^{(0)}|V|m^{(0)}\rangle\right|^2}{\left(E_n^{(0)} - E_m^{(0)}\right)}. \tag{6.92}$$

Funquest: Do you think that the second order perturbation correction to energy of an eigenstate can ever be non-zero if the first-order correction is zero?

When $\left|n^{(0)}\right\rangle$ represents the ground state of the unperturbed Hamiltonian, $E_n^{(0)} < E_m^{(0)}$, and the denominator is negative for all values of m, while the numerator is always positive. The second-order energy correction to the ground state is therefore always negative. To obtain the perturbed state vector up to second-order corrections, we truncate Eq. 6.74d up to the second order to get

$$|n\rangle = \left|n^{(0)}\right\rangle \begin{bmatrix} \left\langle n^{(0)} \middle| n^{(0)} \right\rangle + \\ \left\langle n^{(0)} \middle| n^{(1)} \right\rangle + \\ \left\langle n^{(0)} \middle| n^{(2)} \right\rangle \end{bmatrix} + \sum_{m \neq n} \left|m^{(0)}\right\rangle \left[\left\langle m^{(0)} \middle| n^{(1)} \right\rangle + \left\langle m^{(0)} \middle| n^{(2)} \right\rangle \right],$$

$$\text{i.e., } |n\rangle = \left|n^{(0)}\right\rangle \left[1 + \left\langle n^{(0)} \middle| n^{(2)} \right\rangle \right] + \sum_{m \neq n} \left|m^{(0)}\right\rangle \left[\left\langle m^{(0)} \middle| n^{(1)} \right\rangle + \left\langle m^{(0)} \middle| n^{(2)} \right\rangle \right]. \tag{6.93}$$

To get the state vector up to order 2, we need the coefficients (a) $\left\langle n^{(0)} \middle| n^{(2)} \right\rangle$, $\left\langle m^{(0)} \middle| n^{(1)} \right\rangle$; $m \neq n$ (which we already have from Eq. 6.71b) and (b) $\left\langle m^{(0)} \middle| n^{(2)} \right\rangle$; $m \neq n$.

(a) $\left\langle n^{(0)} \middle| n^{(2)} \right\rangle = -\frac{1}{2} \left\langle n^{(1)} \middle| n^{(1)} \right\rangle = -\frac{1}{2} \sum_{m \neq n} \frac{\left|\left\langle m^{(0)} \middle| V \middle| n^{(0)} \right\rangle\right|^2}{\left(E_n^{(0)} - E_m^{(0)}\right)^2}$ (Eq. 6.86b used).

(b) To obtain $\left\langle m^{(0)} \middle| n^{(2)} \right\rangle$; $m \neq n$, we shall use Eq. 6.85c, with $j = 2$:

$$m \neq n: \left\langle m^{(0)} \middle| n^{(2)} \right\rangle = -\frac{\left\langle m^{(0)} \middle| V \middle| n^{(1)} \right\rangle}{\left(E_m^{(0)} - E_n^{(0)}\right)} + E_n^{(1)} \frac{\left\langle m^{(0)} \middle| n^{(1)} \right\rangle}{\left(E_m^{(0)} - E_n^{(0)}\right)},$$

$$\text{i.e., } m \neq n: \left\langle m^{(0)} \middle| n^{(2)} \right\rangle = -\frac{\left\langle m^{(0)} \middle| V \middle| n^{(1)} \right\rangle}{\left(E_m^{(0)} - E_n^{(0)}\right)} + \left\langle m^{(0)} \middle| V \middle| n^{0} \right\rangle \frac{\left\langle m^{(0)} \right|}{\left(E_m^{(0)} - E_n^{(0)}\right)} \left|n^{(1)}\right\rangle, \tag{6.94a}$$

i.e., $m \neq n$:

$$\left\langle m^{(0)} \middle| n^{(2)} \right\rangle = -\frac{\left\langle m^{(0)} \middle| V \middle| n^{(1)} \right\rangle}{\left(E_m^{(0)} - E_n^{(0)}\right)} + \sum_{m' \neq n} \frac{\left\langle n^{(0)} \middle| V \middle| n^{(0)} \right\rangle}{\left(E_m^{(0)} - E_n^{(0)}\right)} \left\langle m^{(0)} \middle| m'^{(0)} \right\rangle \frac{\left\langle m'^{(0)} \middle| V \middle| n^{(0)} \right\rangle}{\left(E_n^{(0)} - E_{m'}^{(0)}\right)}. \tag{6.94b}$$

Since $\left|n^{(1)}\right\rangle = \sum_{m' \neq n} \left|m'^{(0)}\right\rangle \frac{\left\langle m'^{(0)} \middle| V \middle| n^{(0)} \right\rangle}{\left(E_n^{(0)} - E_{m'}^{(0)}\right)}$,

$$-\frac{\left\langle m^{(0)} \middle| V \middle| n^{(1)} \right\rangle}{\left(E_m^{(0)} - E_n^{(0)}\right)} = -\sum_{m' \neq n} \frac{\left\langle m^{(0)} \middle| V \middle| m'^{(0)} \right\rangle}{\left(E_m^{(0)} - E_n^{(0)}\right)} \frac{\left\langle m'^{(0)} \middle| V \middle| n^{(0)} \right\rangle}{\left(E_n^{(0)} - E_{m'}^{(0)}\right)} = \sum_{m' \neq n} \frac{\left\langle m^{(0)} \middle| V \middle| m'^{(0)} \right\rangle \left\langle m'^{(0)} \middle| V \middle| n^{(0)} \right\rangle}{\left(E_m^{(0)} - E_n^{(0)}\right)\left(E_{m'}^{(0)} - E_n^{(0)}\right)}.$$

Hence, when $m \neq n$:

$$\left\langle m^{(0)} \middle| n^{(2)} \right\rangle = \sum_{m' \neq n} \frac{\left\langle m^{(0)} \middle| V \middle| m'^{(0)} \right\rangle \left\langle m'^{(0)} \middle| V \middle| n^{(0)} \right\rangle}{\left(E_m^{(0)} - E_n^{(0)}\right)\left(E_{m'}^{(0)} - E_n^{(0)}\right)} - \frac{\left\langle n^{(0)} \middle| V \middle| n^{(0)} \right\rangle \left\langle m^{(0)} \middle| V \middle| n^{(0)} \right\rangle}{\left(E_m^{(0)} - E_n^{(0)}\right)^2}. \tag{6.95}$$

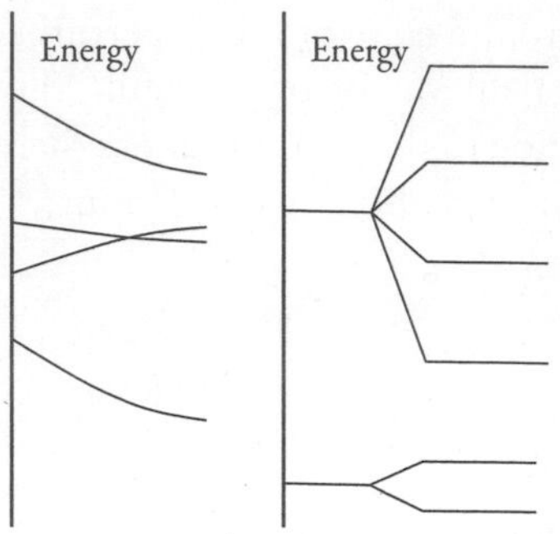

Fig. 6.3 As a result of a perturbation, the energy levels may change, and/or they may get split, in accordance with symmetry and conservation principles.

We will be applying the consequences of the results of perturbation theory in the following chapters. Chapter 7 would introduce the relativistic hydrogen atom based on *a priori* relativistic framework. Nonetheless, we shall also discuss if some of the relativistic effects are satisfactorily accounted for using perturbative methods using the non-relativistic form of the unperturbed Hamiltonian. In Chapter 8, we shall consider the effects of external fields using perturbation theory.

The energy eigenstates we considered here were non-degenerate. The energy levels of the quantum system often change due to a perturbation, as shown in the left panel of Fig. 6.3. When the energy states are degenerate, the degeneracy may be removed, and different linearly independent wavefunctions may belong to different eigenvalues, as shown in the right panel of Fig. 6.3. The degeneracy may be removed partly, or fully.

In the presence of degeneracy, a slight modification of the method discussed here is required. We will study it in the context of a specific physical problem, namely an atom having degenerate eigenstates placed in an electric field. We will see in Chapter 8 that the partial or total removal of degeneracy depends on how the perturbation term impacts the symmetry of the unperturbed Hamiltonian. The interaction perturbation term V considered in this section is independent of time, and the energy states were considered to be non-degenerate. In the next section we study the effects of a time-dependent perturbation.

The case of degenerate states will be treated in Chapter 8, where we illustrate the methodology in the context of Stark effect.

Funquest: Is it guaranteed that the $(n + 1)^{\text{th}}$ order perturbation correction to an energy eigenvalue is weaker than the nth order correction?

6.3 Time-Dependent Perturbation Theory and Atomic Photoionization

The Schrödinger equation describes the temporal evolution of a physical system described by a wavefunction. Dynamical variables are described by appropriate operators, as discussed in Chapter 1. The operators are independent of time; temporal evolution of the physical system is described by the time dependence of the wavefunction. This formulation is called as the *Schrödinger picture* of quantum mechanics. An alternative and equivalent formulation places the temporal evolution in the operators, rather than in the wavefunction, which sustain at all times their values at the initial time. Such a formulation is called as the *Heisenberg picture* of quantum mechanics. In a third *picture*, called as the *Dirac picture* or *Interaction picture*, both the wavefunctions and the operators are time dependent. The three *pictures* are discussed in Appendix B. Results of the three pictures are equivalent, and one can carry out transformations from one picture to another. The choice of using one picture or the other is driven by various factors that are often specific to a particular situation. In the time-dependent perturbation theory discussed in this section, a *time-dependent interaction* of a quantum system with a time-dependent applied field is treated as a *perturbation* operator in the Hamiltonian that is employed in the Schrödinger equation. Time-dependent perturbation theory describes the methodology to examine consequences of the perturbation term on the eigenstates of the unperturbed states.

We illustrate the methodology adopted in time-dependent perturbation theory using the example of the interaction between an atomic electron and an applied time-dependent electromagnetic field described by the four-component potential $\left(\vec{A}(\vec{r},t),\phi(\vec{r},t)\right)$. In this section, after developing the basic formalism of the time-dependent perturbation theory, we shall proceed to apply it to study the photoionization of atoms. The Hamiltonian for an atomic electron (charge: –e) in an electromagnetic field is, from Eq. 2.75,

$$H = \left(-\frac{\hbar^2\vec{\nabla}^2}{2m} - \frac{i\hbar e}{mc}\vec{A}\bullet\vec{\nabla} + \frac{e^2}{2mc^2}A^2\right) - e\phi, \tag{6.96}$$

and hence its Schrödinger equation is given by

$$i\hbar\frac{\partial\psi(\vec{r},t)}{\partial t} = \left[-\frac{\hbar^2\vec{\nabla}^2}{2m} - \frac{ie\hbar}{mc}\vec{A}\cdot\vec{\nabla} + \frac{e^2}{2mc^2}A^2 - \frac{Ze^2}{r}\right]\psi(\vec{r},t). \tag{6.97}$$

The ratio of the second term to the first term (which comes from the kinetic energy) is of the order of $\sim O\left(\frac{eA}{pc}\right)$. The ratio of the third (quadratic in the vector potential) to the second term is also of order ~ $O\left(\frac{eA}{pc}\right)$. Now, an order of magnitude estimate of $\frac{eA}{pc}$ can be obtained from the electric charge and momentum of an electron in an atomic orbit, and an estimate of the magnitude of the vector potential can be obtained from the average power per unit area delivered by an electromagnetic wave given by the Poynting vector (Eq. 13.43, Chapter 13 in [1]) corresponding to the temperature-dependent electromagnetic energy density in a cavity. Even at thousands of degrees Celsius, it turns out that $\frac{eA}{pc}$ is small, and hence the quadratic term $\frac{e^2A^2}{2mc^2}$ can be ignored. This approximation works satisfactorily up to the intensity of about a 100 W/cm^2. A 100 femtosecond laser pulse operating at 100 μJ and focused on a tiny 50 μm spot generates an electric field of the order of ~1 V/Å, which is comparable to the field strength inside an atom. In most situations neglect of the quadratic term is a very satisfactory approximation. We can therefore employ an approximation that is valid for weak fields and express the electron–field interaction term in the above Hamiltonian perturbatively as

$$H = H_0 + \lambda H', \tag{6.98}$$

with

$$H' = -\frac{ie\hbar}{mc}\vec{A}(\vec{r},t)\bullet\vec{\nabla} = \frac{e}{mc}\vec{A}(\vec{r},t)\bullet\vec{p}, \tag{6.99}$$

and λ is an order parameter of the perturbation theory, m its mass, and c is the speed of light. We consider the initial state of the quantum system to be one of the eigenstates of the unperturbed Hamiltonian H_0, identified as $\psi_0(\vec{r})\exp\left(-i\frac{E_0}{\hbar}t\right)$. The entire time dependence of the energy eigenstates is contained in the dynamical phase $e^{-i\frac{E_0}{\hbar}t}$. Other than this oscillatory time dependence, there is no other time dependence of the wavefunction. After the interaction of the atomic system with a time-dependent electromagnetic potential is switched on, the general solution to the Schrödinger equation with the full Hamiltonian (Eq. 6.98) is a *linear superposition of the unperturbed wavefunction*. This ansatz is satisfactory in the weak coupling limit. It works on the premise that the perturbation does not throw the system out of its original Hilbert space, but the time dependence of the general solution consists of two parts: the usual dynamical evolution $\exp\left(-i\frac{E_k}{\hbar}t\right)$ of the unperturbed wavefunctions, and an *additional* time dependence that results from the exposure of the atomic system to the time-dependent atom–field interaction. This *extra* time dependence is placed in the coefficients, $c_k(t)$. We write the wavefunction as

$$\Psi(\vec{r},t) = \sum_k c_k(t)\psi_k(\vec{r})\exp(-i\omega t), \tag{6.100}$$

which is a solution to the time-dependent Schrödinger equation

$$i\hbar\frac{\partial}{\partial t}\Psi(\vec{r},t) = H\Psi(\vec{r},t). \tag{6.101}$$

The main hypothesis in Eq. 6.100 is that even in the presence of a time-dependent interaction, the general solution of the Schrödinger equation is expressible in terms of the stationary eigenstates of the unperturbed Hamiltonian at each instant of time, by allowing the expansion coefficients to vary with time. Since these expansion coefficients provide a measure of the probability of finding the system in a particular eigenstate, the probability density now becomes time dependent. Without this time dependence, the probability density of a stationary energy eigenstate is independent of time. Time dependence of the coefficients in Eq. 6.100 therefore contains information about transition probabilities from an initial state to another, brought about by a time-dependent perturbation.

It follows from Eq. 6.100 and Eq. 6.101 that

$$i\hbar\frac{\partial}{\partial t}\left\{\sum_k c_k(t)\psi_k(\vec{r})\exp\left(-i\omega_k t\right)\right\} = \left[H_0 + \lambda H'\right]\left(\sum_k c_k(t)\psi_k(\vec{r})\exp\left(-i\omega_k t\right)\right), \tag{6.102}$$

i.e.,

$$\begin{bmatrix} i\hbar\left\{\sum_k \dot{c}_k(t)\psi_k(\vec{r})e^{-i\omega_k t}\right\} + \\ i\hbar\left\{\sum_k (-i\omega_k)c_k(t)\psi_k(\vec{r})e^{-i\omega_k t}\right\} \end{bmatrix} = \begin{bmatrix} H_0\left(\sum_k c_k(t)\psi_k(\vec{r})e^{-i\omega_k t}\right) + \\ \lambda H'\left(\sum_k c_k(t)\psi_k(\vec{r})e^{-i\omega_k t}\right) \end{bmatrix}, \tag{6.103}$$

in which equal terms on both sides of the equation are canceled. Projecting Eq. 6.103 on the final state $\langle f|$, and dividing both sides by $i\hbar$, we get

$$\dot{c}_f(t)\exp\left(-i\omega_f t\right) = \frac{1}{i\hbar}\left(\sum_k \left\{\lambda c_k(t)\right\}\langle f|H'|k\rangle \exp\left(-i\omega_k t\right)\right). \tag{6.104}$$

We now introduce, as in the previous section, a mathematical device, an auxiliary perturbation order parameter λ, and propose an expansion of the time-dependent coefficient in terms of its powers:

$$c_f(t) = \sum_{n=0}^{\infty} \lambda^n c_f^{(n)}(t) \tag{6.105}$$

with corresponding time derivatives given by

$$\dot{c}_f(t) = \sum_{n=0}^{\infty} \lambda^n \dot{c}_f^{(n)}(t). \tag{6.106}$$

The time dependence of the expansion coefficients in the superposition (Eq. 6.100) is a smart accommodation that allows us to obtain approximate solutions of the Schrödinger equation when a time-dependent interaction must be included in the Hamiltonian. This alludes to the interaction picture (Appendix B); temporal evolution of the system is described in the wavefunctions but in a way that enables a time-dependent perturbation operator to be introduced. The method works within the limits of its approximation, which requires weak coupling between the quantum system and the external time-dependent field. The perturbation operator parameter λ is considered to be a small dimensionless number, such that its increasing powers become progressively unimportant. From the physical point of view, only two values $\lambda = 0$, corresponding to zero interaction between the atom and the electric field, and $\lambda = 1$, which corresponds to the interaction being active, are of relevance. The mathematical device of employing the order parameter having a tiny nonzero value is a smart way of developing an approximation scheme to determine, as we shall see below, the *additional* time dependence of the coefficient $c_k(t)$ in Eq. 6.95.

Using Eqs. 6.105 and 6.106 in Eq. 6.104,

$$\sum_{n=0}^{\infty} \lambda^n \dot{c}_f^{(n)}(t) = \frac{1}{i\hbar}\sum_k \left\{\lambda \sum_{m=0}^{\infty} \lambda^m c_k^{(m)}(t)\right\}\langle f|H'|k\rangle \exp\left(+i\omega_{fk} t\right). \tag{6.107}$$

The zero-order coefficients do not depend on time. We can now equate the terms on the two sides of Eq. 6.107, which correspond to the *same* order in the perturbation parameter:

$$\dot{c}_f^{(s+1)}(t) = \frac{1}{i\hbar}\sum_k c_k^{(s)}\langle f|H'|k\rangle \exp\left(+i\omega_{fk} t\right). \tag{6.108}$$

In particular, to the first order, the time dependence of the coefficients in Eq. 6.106 is given by

$$\dot{c}_f^{(1)}(t) = \frac{1}{i\hbar}\sum_k c_k^{(0)}\langle f|H'|k\rangle \exp\left(+i\omega_{fk} t\right). \tag{6.109}$$

The perturbative interaction can be considered to be switched on at some reference time $t_0 = 0$ when the system is in the initial "pure" state $|i\rangle$, i.e., $c_k^{(0)} = \delta_{ki}$, (6.110)

giving, $\dot{c}_f^{(1)}(t) = \frac{1}{i\hbar}\langle f|H'|i\rangle \exp\left(+i\omega_{fi}t\right).$ (6.111)

On integrating and using the initial condition (Eq. 6.110) at time $t_0 = 0$, we get the value of the time-dependent coefficient of the final state $|f\rangle$ at a later time t:

$$c_f^{(1)}(t) = \frac{-i}{\hbar}\int_0^t dt'\langle f|H'(\vec{r},t')|i\rangle \exp\left(+i\omega_{fi}t'\right). \tag{6.112}$$

From elements of the principle of superposition in quantum physics, we know that $\left|c_f^{(1)}(t)\right|^2$ gives us the probability that the system would undergo a transition from the state $|i\rangle$ to the state $|f\rangle$ in time t from the instant that the atomic system is exposed to an applied electromagnetic oscillatory field. The right-hand side of Eq. 6.112 manifestly involves an integral over time, but there is also a space integral in the integrand $\langle f \mid H'(\vec{r},t') \mid i\rangle$, which we now analyze with the perturbation Hamiltonian given by

$$H' = \left(-\frac{ie\hbar}{mc}\right)\hat{\varepsilon}A_0(\Omega)\left\{e^{i\left(\vec{k}\bullet\vec{r}-\Omega t\right)} + e^{-i\left(\vec{k}\bullet\vec{r}-\Omega t\right)}\right\}\bullet\vec{\nabla}. \tag{6.113}$$

Hence the time-dependent coefficient is given by

$$c_f^{(1)}(t) = \left(-\frac{ie\hbar}{mc}\right)\left(\frac{-i}{\hbar}\right)\left\{\begin{array}{l}\int_0^t dt'\left\langle f\left|\hat{\varepsilon}A_0(\omega)\exp\left(i\left(\vec{k}\bullet\vec{r} - \Omega t'\right)\right)\bullet\vec{\nabla}\right|i\right\rangle e^{+i\omega_{fi}t'} + \\ \int_0^t dt'\left\langle f\left|\hat{\varepsilon}A_0(\omega)\exp\left(-i\left(\vec{k}\bullet\vec{r} - \Omega t'\right)\right)\bullet\vec{\nabla}\right|i\right\rangle e^{+i\omega_{fi}t'}\end{array}\right\}. \tag{6.114}$$

Now,

$$\int_0^t dt'\exp\left(\pm i\left(\omega \pm \omega_{fi}\right)t'\right) = \left.\frac{\exp\left(\pm i\left(\omega \pm \omega_{fi}\right)t'\right)}{\pm i\left(\omega \pm \omega_{fi}\right)}\right|_0^t = \frac{\exp\left(\pm i\left(\omega \pm \omega_{fi}\right)t\right) - 1}{\pm i\left(\omega \pm \omega_{fi}\right)}, \tag{6.115}$$

in which the numerator is of the order of unity and the denominator is the sum and difference $\left(\omega \pm \omega_{fi}\right)$. Close to the resonance frequency, considering the term that would dominate the right-hand side of Eq. 6.114, we see that

$$c_f^{(1)}(t) = \left(-\frac{eA_0(\omega)}{mc}\right)\times\left\langle f\left|\exp\left(i\vec{k}\bullet\vec{r}\right)\hat{\varepsilon}\bullet\vec{\nabla}\right|i\right\rangle\int_0^t dt'\exp\left(-i\left(\omega - \omega_{fi}\right)t'\right). \tag{6.116}$$

Hence, the transition probability $i \to f$ is given by

$$\left|c_f^{(1)}(t)\right|^2 = \left(\frac{eA_0(\omega)}{mc}\right)^2\left|\left\langle f\left|\exp\left(i\vec{k}\bullet\vec{r}\right)\hat{\varepsilon}\bullet\vec{\nabla}\right|i\right\rangle\right|^2 2F(t,\omega - \omega_{fi}), \tag{6.117}$$

where

$$2F(t,\omega - \omega_{fi}) = \left[\frac{\exp\left(-i\left(\omega - \omega_{fi}\right)t\right) - 1}{(-i)\left(\omega - \omega_{fi}\right)}\right]\left[\frac{\exp\left(-i\left(\omega - \omega_{fi}\right)t\right) - 1}{(-i)\left(\omega - \omega_{fi}\right)}\right]. \tag{6.118}$$

Introducing $\tilde{\omega} = \omega - \omega_{fi}$, we have

$$2F(t,\tilde{\omega}) = \frac{2[1-\cos\tilde{\omega}t]}{\tilde{\omega}^2} = \frac{2\left[2\sin^2\left(\frac{\tilde{\omega}t}{2}\right)\right]}{\tilde{\omega}^2}, \tag{6.119}$$

or

$$F(t,\tilde{\omega}) = \frac{2\sin^2\left(\frac{\tilde{\omega}t}{2}\right)}{\tilde{\omega}^2}. \tag{6.120}$$

Over a large time interval, it is shown in the Solved Problem P6.8, that the above function is a Dirac-δ distribution function (Fig. 6.4):

$$\lim_{t\to large} F(t,\tilde{\omega}) = \pi t\delta(\tilde{\omega}). \tag{6.121}$$

Accordingly, the transition probability is given by

$$\left|c_f^{(1)}(t)\right|^2 = \left(\frac{eA_0(\omega)}{mc}\right)^2 \left|\left\langle f\left|e^{i\vec{k}\bullet\vec{r}}\hat{\varepsilon}\bullet\vec{\nabla}\right|i\right\rangle\right|^2 \times 2\pi t\delta(\tilde{\omega}). \tag{6.122}$$

The transition rate (transition probability per unit time) is then given by

$$W_{fi} = \left(\frac{eA_0(\omega)}{mc}\right)^2 \left|\left\langle f\left|e^{i\vec{k}\bullet\vec{r}}\hat{\varepsilon}\bullet\vec{\nabla}\right|i\right\rangle\right|^2 \times 2\pi\delta(\tilde{\omega}). \tag{6.123}$$

The transition probability being experimentally measurable, the above relation connects a laboratory observable with a theoretical prediction. Recognizing its importance,

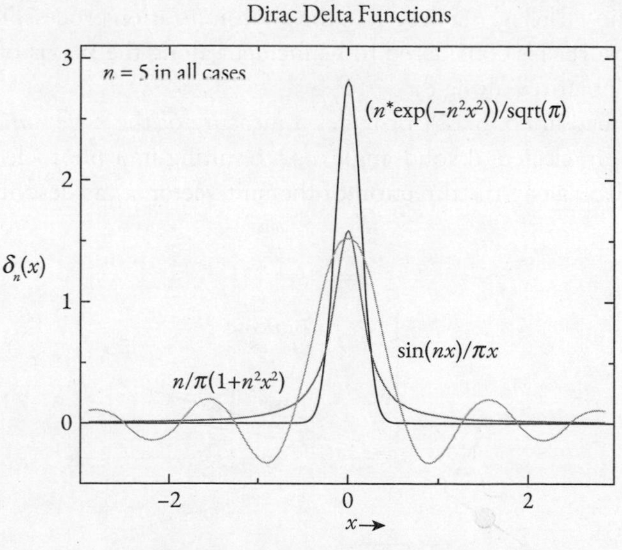

Fig. 6.4 The Dirac-δ "function" is in fact better known as a distribution, or a generalized function. It is narrow and sharply peaked. There are various representations of the Dirac-δ; three of these representations are shown in this figure. Figures plotted by Jobin Jose.

Fermi designated it as the *golden rule*, and it is commonly known as Fermi's golden rule, more completely as Fermi's golden rule number 2. The result was originally obtained by G. Wentzel. (Note: There is another result that is known as Fermi's golden rule number 1, which appears in the quantum theory of collisions.)

> ***Funquest:*** The method discussed in this section is semi-classical, since it treats the electromagnetic field using the classical Maxwell's theory even if the atom is treated using quantum mechanics. How should this model be improved if the electromagnetic field is also quantized?

In atomic spectroscopy, a quantity of great interest is the photoionization cross section. The term "cross section" comes from collision physics: its explanation would require considerable detour, so we shall only use the term as it is. We shall discuss it in Chapter 8 and then again, in much detail, in Chapter 10. Suffice it is for now that we recognize it to be a measure of the probability of transition from an initial state to a final state, $|i\rangle \to |f\rangle$, when electromagnetic radiation is absorbed by an atom. An atom undergoes such a transition under certain conditions; for example, the total energy and angular momentum must be conserved in the process. The energy lost from the electromagnetic field is gained by the atom that absorbs it. The physical quantity that corresponds to the photoionization cross section and determined by the transition probability is described by the ratio:

$$\frac{\text{Energy absorbed per unit time in } |i\rangle \to |f\rangle}{\textit{Energy} \text{ flux of the EM radiation}} = \frac{\text{Energy absorbed per unit time in } |i\rangle \to |f\rangle}{\textit{Energy} \text{ per unit area per unit time of the EM radiation}}. \tag{6.124}$$

The geometry shown in Fig. 6.5 describes the photoionization process. The electromagnetic radiation that is absorbed is considered to be incident along the x-axis of a Cartesian frame of reference and is polarized along $\hat{\varepsilon}$.

The ratio described in Eq. 6.124 provides a measure of the *differential cross section* for photoionization in an elemental solid angle $d\Omega$ resulting in a photoelectron's escape with momentum $\hbar\vec{k} = \hbar\vec{k}_f$, along the direction of the unit vector $\hat{k}_f$ as described in Fig. 6.5. It is given by

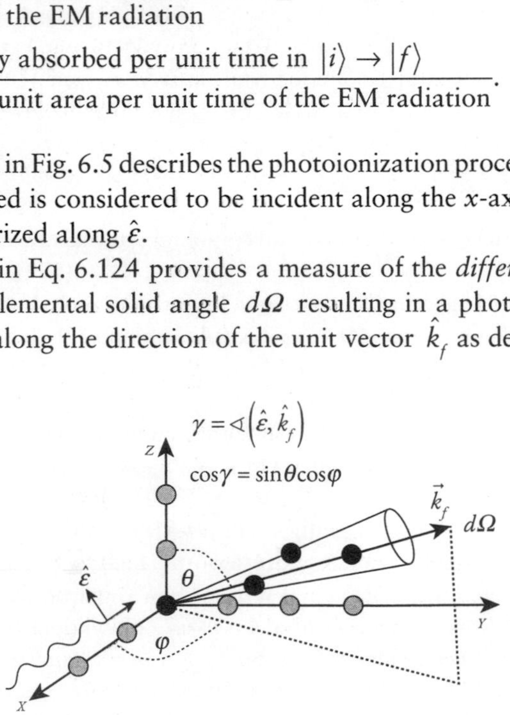

Fig. 6.5 The electromagnetic radiation is considered to be polarized along the direction $\hat{\varepsilon}$ and incident along the x-axis. The photoelectron is ejected along $\hat{\varepsilon}_r = \hat{k}_f$.

$$\left[\frac{d\sigma}{d\Omega}\right]_{\hat{k}_f}^{\hat{\varepsilon}} = \frac{\hbar\omega \times \left[W_{fi}\right]_{\hat{k}_f}^{\hat{\varepsilon}}}{I(\omega)}, \tag{6.125a}$$

$$\text{i.e.,} \quad \frac{d\sigma}{d\Omega} = \frac{\hbar\omega \times W_{fi}}{I(\omega)}. \tag{6.125b}$$

Equation 6.125b is essentially the same as Eq. 6.125a; only the notation is simplified, but it must be remembered that it is with respect to the polarization of the electromagnetic wave along $\hat{\varepsilon}$ and the photoelectron is ejected along $\hat{k}_f$. In the Eq. 6.125b, $\tilde{\omega} = \omega - \omega_{fi}$, and $I(\omega)$ is the intensity of the electromagnetic radiation at the circular frequency ω. Now, the vector potential of the electromagnetic field is

$$\vec{A}(\vec{r},t) = A_0(\omega)\hat{\varepsilon}\left[e^{i(\vec{k}\bullet\vec{r} - \omega t)} + e^{-i(\vec{k}\bullet\vec{r} - \omega t)}\right], \tag{6.126a}$$

and in the Gaussian-CGS system of units, the electric intensity vector is

$$\vec{E}(\vec{r},t) = -\vec{\nabla}\phi - \frac{1}{c}\frac{\partial\vec{A}}{\partial t}. \tag{6.126b}$$

In the Coulomb gauge, $\phi = 0$ and $\vec{\nabla} \bullet \vec{A} = 0$. The $\vec{H}$ field is

$$\vec{H}(\vec{r},t) = \vec{\nabla} \times \vec{A}. \tag{6.126c}$$

From the above relations, we can easily find the intensity of the electromagnetic field, which is just the average value of the Poynting vector (see, for example, Chapter 13 in [2]). Therefore,

$$I(\omega) = \left\langle\left|\vec{S}\right|\right\rangle = \left\langle\left|\frac{c}{4\pi}\vec{E} \times \vec{H}\right|\right\rangle = \frac{\omega^2}{2\pi c}A_0^2(\omega). \tag{6.127}$$

Using Eq. 6.125 and Eq. 6.127, we get

$$\frac{d\sigma}{d\Omega} = \frac{\hbar\omega\left(\frac{eA_0(\omega)}{mc}\right)^2 \left|\left\langle f\left|e^{i\vec{k}\bullet\vec{r}}\,\hat{\varepsilon} \bullet \vec{\nabla}\right|i\right\rangle\right|^2 2\pi\delta(\tilde{\omega})}{\frac{\omega^2}{2\pi c}A_0^2(\omega)}, \tag{6.128a}$$

$$\text{i.e.,} \quad \frac{d\sigma}{d\Omega} = \frac{4\pi^2\alpha\hbar^3}{m^2\omega}\left|\left\langle f\left|e^{i\vec{k}\bullet\vec{r}}\,\hat{\varepsilon} \bullet \vec{\nabla}\right|i\right\rangle\right|^2 \delta\left(E - E_{fi}\right). \tag{6.128b}$$

This ratio is called as the *differential cross section* for photoionization. We have considered energy conservation between the electromagnetic radiation field and the atom that absorbs it. The conservation of angular momentum involves certain conditions, known as spectroscopic selection rules. We shall consider the selection rules in Chapter 8. It can be easily verified that the dimensions of the ratio described by Eq. 6.125 and also Eqs. 6.128a,b are L^2. This exercise is, however, important since it justifies the name "cross section." The prefix differential relates to the fact it corresponds to the probability of atomic transitions resulting in photoelectron angle ejection, per unit solid angle, in the direction $\hat{k}_f$ (Fig. 6.5). The photoionization cross section is usually measured in terms of the unit Mb (Mega-barn), with 1 Mb = 10^{-18} cm^2. Essentially, the differential cross section in Eqs. 6.128a,b gives us a measure of the probability of photoionization into the solid angle $d\Omega$ about the photoelectron's escape direction $\hat{k}_f$. The differential cross section therefore has an angular distribution that is potentially, but not necessarily, isotropic. It is also, obviously, energy dependent. Transitions to a number of degenerate states are possible, and we must therefore estimate the number of such states. We shall use "box normalization" over a cubical box of length L (Fig. 6.6a), but our final result will be independent of L, the length of its side.

As a result of box normalization described in Fig. 6.6b,

$$E = \frac{\hbar^2 k^2}{2m} = \frac{\hbar^2}{2m}\left(k_x^2 + k_y^2 + k_z^2\right) = \frac{\hbar^2}{2m}\left(\frac{2\pi}{L}\right)^2\left(n_x^2 + n_y^2 + n_z^2\right) = \frac{2\pi^2\hbar^2}{mL^2}n^2. \tag{6.129}$$

The number of such states in the volume element shown in Fig. 6.6b is then given by

$$n^2 dn d\Omega = n^2\frac{dn}{dE}dEd\Omega = n^2\frac{dk}{dE}\frac{L^2}{4\pi^2}\frac{2\pi}{L}dEd\Omega = \left(\frac{L}{2\pi}\right)^3\left(\frac{mk}{\hbar^2}\right)dEd\Omega. \tag{6.130}$$

Each of these states can contribute to the transition from an initial bound state $|i\rangle$ to a final continuum state $|f\rangle$. Integrating over the possible energy states (Eq. 6.130), we get (from Eq. 6.128),

$$\frac{d\sigma}{d\Omega} = \int\left\{\frac{4\pi^2\alpha\hbar^3}{m^2\omega}\left|\left\langle f\left|\exp\left(i\vec{k}\bullet\vec{r}\right)\hat{\varepsilon}\bullet\vec{\nabla}\right|i\right\rangle\right|^2\rho(E)dE\delta\left(E - E_{fi}\right)\right\}, \tag{6.131a}$$

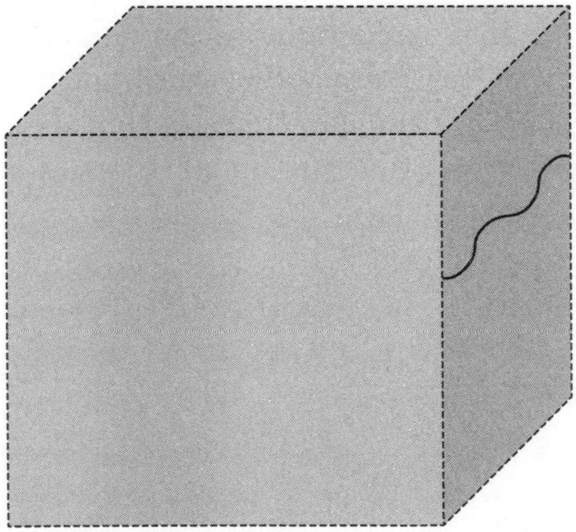

Fig. 6.6a Box normalization relies on determining that wavelengths can be sustained in the box since only an integer multiple of the wavelengths can do so. Therefore, the following relations hold: $n_x\lambda_x = L$; i.e., $n_x\frac{2\pi}{k_x} = L$, or $k_x = \frac{2\pi n_x}{L}$.

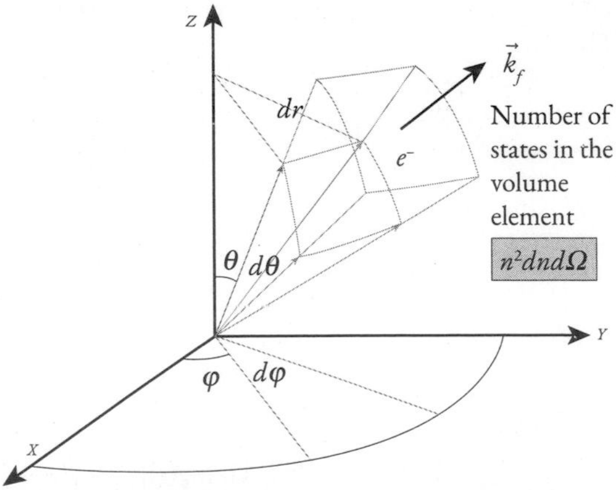

Fig. 6.6b Determination of the number of states into which the photoionization transition can take place.

where $\rho(E)$ stands for the *density of states*.

Accordingly,

$$\frac{d\sigma}{d\Omega} = \int \left\{ \frac{4\pi^2 \alpha \hbar^3}{m^2 \omega} \left| \left\langle f \left| \exp\left(i\vec{k} \bullet \vec{r}\right) \hat{\varepsilon} \bullet \vec{\nabla} \right| i \right\rangle \right|^2 \delta\left(E - E_{fi}\right) \left(\frac{L}{2\pi}\right)^3 \left(\frac{mk}{\hbar^2}\right) \right\} dE. \tag{6.131b}$$

Dirac-δ integration over energy gives

$$\left[\frac{d\sigma}{d\Omega}\right]_{\hat{k}_f}^{\hat{\varepsilon}} = \frac{4\pi^2 \alpha \hbar^3}{m^2 \omega_{fi}} \left| \left\langle f \left| \exp\left(i\vec{k} \bullet \vec{r}\right) \hat{\varepsilon} \bullet \vec{\nabla} \right| i \right\rangle \right|^2 \left(\frac{L}{2\pi}\right)^3 \left(\frac{mk}{\hbar^2}\right). \tag{6.131c}$$

The transition matrix element in this equation is

$$\begin{aligned} M &= \left\langle f \left| \exp\left(i\vec{k} \bullet \vec{r}\right) \hat{\varepsilon} \bullet \vec{\nabla} \right| i \right\rangle \\ &= \int \left[\left(\frac{1}{\sqrt{L^3}} \exp\left(-i\vec{k}_f \bullet \vec{r}\right) \right) \left(\exp\left(i\vec{k} \bullet \vec{r}\right) \hat{\varepsilon} \bullet \vec{\nabla} \right) \psi_i(\vec{r}) \right] dV, \end{aligned} \tag{6.132a}$$

i.e.,

$$\begin{aligned} M &= \frac{-i}{\sqrt{L^3}} \left[\hat{\varepsilon} \bullet \vec{k}_f \right] \int \left[\psi_i(\vec{r}) \exp\left(i\left(\vec{k} - \vec{k}_f\right) \bullet \vec{r} \right) \right] dV \\ &= \frac{-ik_f}{\sqrt{L^3}} [\cos\gamma] \int \left[\psi_i(\vec{r}) \exp\left(i\left(\vec{k} - \vec{k}_f\right) \bullet \vec{r} \right) \right] dV. \end{aligned} \tag{6.132b}$$

Equations 6.132a,b have a factor in $L^{-3/2}$, but we need its square in the differential cross section (Eqs. 6.131a–c), which renders our final result to be independent of the length L used for the *box normalization*. The integral in Eqs. 6.132a,b can be easily determined; it is proportional to the Fourier transform of the initial state wavefunction. For example, if the initial state function is the 1s wavefunction $\left[\frac{1}{\sqrt{\pi}}\left(\frac{Z}{a_0}\right)^{\frac{3}{2}} e^{-\frac{Zr}{a_0}}\right]$ of the electron in atomic hydrogen, the integral becomes

$$I = \int dV \psi_i(\vec{r}) e^{i\left(\vec{k} - \vec{k}_f\right) \bullet \vec{r}}, \tag{6.133}$$

i.e.,

$$I = \frac{1}{\sqrt{\pi}}\left(\frac{Z}{a_0}\right)^{\frac{3}{2}} \frac{8\pi\left(\frac{Z}{a_0}\right)}{\left[\left\{Z^2 + a_0^2\left|\vec{k} - \vec{k}_f\right|^2\right\}\left\{\frac{1}{a_0^2}\right\}\right]^2}. \tag{6.134}$$

Using Eq. 6.132b, Eq. 6.133, and Eq. 6.134 in Eq. 6.131c, we get

$$\left[\frac{d\sigma}{d\Omega}\right]_{\hat{k}_f}^{\hat{\varepsilon}} = \frac{32\alpha\hbar k_f^3}{m\omega_{fi}} \frac{Z^5 a_0^3 (\cos^2\gamma)}{\left[\left\{Z^2 + a_0^2\left|\vec{k} - \vec{k}_f\right|^2\right\}\right]^4}. \tag{6.135}$$

We now use an approximation, known as the *Born approximation*. This approximation is often used in the high-energy regime; we shall discuss it in Chapter 10 in the context of quantum collisions. The ionization potential (I.P.) is essentially the binding energy of the electron that is ejected, usually referred to as the *photoelectron*. Writing the energy of the photon that is absorbed as $h\nu$, we note that

$$h\nu = \frac{hc}{\lambda} = \frac{\hbar c}{\lambda\!\!\!^-} = \hbar kc = pc = \frac{\hbar^2 k_f^2}{2m} + I.P. \underset{\text{High Energy Approximation}}{\approx \rightarrow} \frac{\hbar^2 k_f^2}{2m}, \tag{6.136a}$$

$$\text{i.e.,} \quad \frac{k}{k_f} \simeq \frac{\hbar k_f}{2cm} = \frac{p_f}{2cm} = \frac{v_f}{2c} <<< 1. \tag{6.136b}$$

Thus, in the Born approximation,

$$\left[\frac{d\sigma}{d\Omega}\right]_{\hat{k}_f}^{\hat{\varepsilon}} = \frac{32\alpha\hbar}{m\omega_{fi}}\left(\frac{Z}{a_0 k_f}\right)^5 \frac{(\sin^2\theta\cos^2\phi)}{\left(1 - \frac{v_f}{c}\cos\theta\right)^4}. \tag{6.137}$$

For unpolarized light, we must use the average value

$$\langle\cos^2\varphi\rangle = \frac{1}{2\pi}\int_0^{2\pi}\cos^2\varphi\, d\varphi = \frac{1}{2},$$

which gives

$$\left[\frac{d\sigma}{d\Omega}\right]_{\hat{k}_f}^{unpolarized} \approx \frac{16\alpha\hbar}{m\omega_{fi}}\left(\frac{Z}{a_0 k_f}\right)^5 (\sin^2\theta)\left(1 + 4\frac{v_f}{c}\cos\theta\right). \tag{6.138}$$

The total photoionization cross section is then obtained by integrating the differential cross section over all the angles θ:

$$\sigma_{Total}^{unpolarized} = \int_{\theta=0}^{\pi}\int_{\varphi=0}^{2\pi}\left[\frac{d\sigma}{d\Omega}\right]_{\hat{k}_f}^{unpolarized}\sin\theta d\theta d\varphi = \frac{128\pi}{3m}\frac{\alpha\hbar}{\omega}\left(\frac{Z}{a_0}\right)^5\frac{1}{k_f^5}. \tag{6.139}$$

Furthermore,

$$k_f^5 = \left(k_f^2\right)^{5/2} = \left(\frac{2m}{\hbar}\right)^{5/2}\omega^{5/2}, \tag{6.140a}$$

and

$$\hbar\omega \approx \frac{\hbar^2 k_f^2}{2m}, \tag{6.140b}$$

hence,

$$\sigma_{Total}^{unpolarized} = \frac{128\pi}{3m}\alpha\hbar\left(\frac{2m}{\hbar}\right)^{-5/2}\left(\frac{Z}{a_0}\right)^5 \omega^{-7/2}. \tag{6.141}$$

Funquest: Can you list all factors (or, as many as you can think of) that may result in a departure of experimentally observed photoionization cross-section from the predictions of Eq. 6.141?

In the Born approximation, the photoionization cross section thus diminishes with energy as $E^{-7/2}$ and increases with the atomic number as Z^5. If one carries out similar analysis with other hydrogenic wavefunctions for higher principal quantum numbers n, we find that the photoionization cross section diminishes with the principal quantum number as n^{-3}. The Born approximation is very useful, but it does have limitations. A primary source of this limitation is that we have pretended that the electron that gets ejected on photoabsorption is independent of all other electrons. Strictly speaking, this is not correct. In a many-electron system, electrons have correlated dynamics. Broadly, electron correlations are of two kinds:

(a) Those which stem from the quantum statistical nature of the many-electron system, described by the Fermi–Dirac statistics. Correlations resulting from this property are known as the Fermi–Dirac, or exchange, or statistical correlations. We shall discuss these in Chapter 9.
(b) Additional correlations, namely the Coulomb correlations, which go beyond the scope of this book.

When we take into account all correlations, we shall discover that many corrections have to be made to the simple expression obtained in the Born approximation. As a result of these correlations, departures from the Born approximation become significant. One must concede that it only provides an exception rather than a rule [9, 10], but this statement really has to be understood in the context of the importance of correlations that modern studies are focused on. Other than that, let there be no doubt that the Born approximation provides an excellent account of the energy dependence of the photoionization cross section at high-enough photon energies. We shall come across further applications of the perturbation theory in subsequent chapters.

6.4 Non-Perturbative Methods

As Hawking said, nothing is perfect; we must live with imperfections. Exact solutions exist for very few model potentials. These are often called as toy problems. For most of the real problems, one can at best only look for approximate solutions. Other than the semi-classical approximations discussed in Section 6.1 and the perturbative methods discussed in Sections 6.2 and 6.3, there are other approximation methods. Notably, we mention the *adiabatic approximation*, the *sudden approximation*, and the *variational* methods. Each of these is a very powerful technique. The daunting task of introducing these methods in an introductory/intermediate text on quantum mechanics is rendered at least partially addressable by the fact that the adiabatic approximation has already been illustrated in Section 2.4 of Chapter 2, where it was used to study the Aharonov–Bohm effect and understand the geometric phase.

The terms *adiabatic* and *sudden* approximations refer to *time scales*; an energy interchange between the system and its environment, or lack of it, is *not* referenced. When the time scale over which a time-dependent perturbation effect must be considered is long compared to the internal time clock of a physical system, the adiabatic approximation may be useful. In the opposite case, the sudden approximation may be useful.

In an adiabatic process, the Hamiltonian changes gently in response to variations in some external parameters that control the Hamiltonian. Remind yourself of the discussion the geometrical phase that develops adiabatically in the Aharonov–Bohm effect (Chapter 2). The changes are essentially due to those in the external parameter space on which the Hamiltonian depends. We consider the physical system to be in one of its discrete bound eigenstates. The system evolves in response to changes in an external parameter that controls the system's Hamiltonian gently enough, so that if it is in the nth eigenstate of the Hamiltonian at the initial time t_i, it would remain in the nth eigenstate of the Hamiltonian at final time t_f. Consider, for example, a particle in an infinitely deep one-dimensional square well. Let us presume that it is in the nth odd eigenstate (Eq. 3.60b). Its wavefunction clearly depends on the width of the well. Now, consider the physical system to be adiabatically varied by controlling an external parameter that changes the width of the well. If the width changes gently, the system would continue to remain in the nth odd eigenstate of the changing Hamiltonian, even if the actual value of the width in the wavefunction is changing. Both the normalization constant and the *arguments* of the sine function (in the case of odd states) and of the cosine function (even states) change, but under the adiabatic approximation the system survives in the nth odd or even state – in whichever state it was at the initial time. One could give other examples of adiabatic variations, such as altering the spring constant of a simple harmonic oscillator that may be externally controlled to change the system wavefunctions. If the variation of the Hamiltonian is slow, i.e., adiabatic, then an oscillator in the nth eigenstate at the initial state would remain in the nth eigenstate of the changed Hamiltonian. A geometrical phase is picked up by the system in an adiabatic process if the Hamiltonian is varied slowly by controlling an external parameter cyclically. This gives the Pancharatnam–Berry phase for a quantum system. We have discussed this in Section 2.4 of Chapter 2.

If changes in the Hamiltonian are sudden, then change in the physical system's wavefunction is not so simple. Let us consider a quantum system in an initial state $|\psi_i\rangle$. The final state is then represented by a vector $|\psi_f\rangle$ that can of course be written as a superposition of the eigenbasis vectors $\left\{|\psi_n^f\rangle; n = 1,2,..\right\}$ of the *new* Hamiltonian. The ansatz of the sudden approximation is that the nth expansion coefficient in the superposition just mentioned is the projection of the initial state on the nth base vector in the eigenbasis of the new Hamiltonian:

$$|\psi_f\rangle = \left[\sum_n |\psi_n^f\rangle\langle\psi_n^f|\right]|\psi_i\rangle = \sum_n |\psi_n^f\rangle\left(\langle\psi_n^f|\psi_i\rangle\right). \tag{6.142}$$

Thus, the probability that the particle in the initial state $|\psi_i\rangle$ of the Hamiltonian $H(t_i)$ at time t_i will be found in one of the eigenstate $|\psi_n^f\rangle$ of the Hamiltonian $H(t_f)$ at time t_f is

$$P_n = \left|\langle\psi_n^f|\psi_i\rangle\right|^2. \tag{6.143a}$$

This expression must be contrasted with Eq. 6.132; the difference is due to the sudden approximation here as opposed to the first-order time-dependent perturbation theory that was applied in the previous section. The difference is essentially due to the time interval over which the perturbation is applied. The sudden approximation is required in situations when the Hamiltonian changes violently, suddenly, over a very short period. Consider the phenomenon of β-decay, for example. The probability that an atomic electron in the ith initial state wavefunction $\psi_i^{(Z)}(\vec{r})$ will be found, *after* the sudden emission of the β-particle from the nucleus, in the atomic state having wavefunction $\psi_f^{(Z+1)}(\vec{r})$ will be

$$P_{i\to f} = \left|\left\langle \psi_n^f \middle| \psi_i \right\rangle\right|^2 = \int_{r=0}^{\infty}\int_{\theta=0}^{\pi}\int_{\varphi=0}^{2\pi} r^2 \sin\theta d\theta d\varphi \psi_f^{(Z+1)}(r)^* \psi_i^{(Z)}(r)^*. \tag{6.143b}$$

Numerical calculations to determine transition probabilities under the sudden approximation also serve to test the quality of the wavefunctions [11] and determine electron shake-off probabilities following nuclear decay.

We now summarize salient features of the *variational method* that is often used to obtain approximate solutions. We have already come across even in classical physics the importance of the variational principle. We know very well that an efficacious formulation of classical mechanics based on the principle of variation in terms of the Lagrange's and Hamilton's equations [2] exists. In Chapter 2 we have seen that the variation method provides an elegant platform to develop even quantum mechanics using the Dirac–Feynman ansatz. We shall now discuss the application of the variational principle in obtaining approximate solutions to unsolvable problems.

Complex problems are not *exactly* solvable. No single approximation method is applicable in all such cases even to determine solutions that are at least moderately acceptable. Developing alternative approximation methods is therefore always a necessity. The method of variation provides a platform for one of the very successful approaches to solve complicated problems in a very satisfactory manner. It rests on the principle that the expectation value of a Hamiltonian H in an *arbitrary* state $|\psi\rangle$ is essentially greater than or equal to the *exact ground state energy* of the system:

$$E = \frac{\langle\psi|H|\psi\rangle}{\langle\psi|\psi\rangle} \geq E_0; \tag{6.144}$$

the equality holds when the state $|\psi\rangle$ is the ground state $|\psi_0\rangle$ itself. In order to establish this result, we first express the arbitrary state as a linear superposition of a *complete* set of eigenbasis $\{\phi_i; i = 0,1,2,..c\}$:

$$|\psi\rangle = \sum_{i=1}^{c} a_i \phi_i. \tag{6.145}$$

We have used the letter c to denote completeness, irrespective of whether the vector space of this system is finite or infinite dimensional. We choose the basis set to be the eigenbasis of the Hamiltonian, so that

$$H|\phi_i\rangle = E_i|\phi_i\rangle \text{ ; i=0,1,2,..} \tag{6.146}$$

Denoting the ground state by $i = 0$, we have $E_i > E_0$ for every $i \geq 1$. Thus, the norm of the state is

$$\langle\psi|\psi\rangle = \sum_{i=0}^{c}|a_i|^2 = |a_0|^2 + \sum_{i=1}^{c}|a_i|^2 \tag{6.147}$$

and the expectation value of the Hamiltonian in this state is

$$\langle\psi|H|\psi\rangle = \sum_{i=0}^{c}|a_i|^2 E_i = |a_0|^2 E_0 + \sum_{i=1}^{c}|a_i|^2 E_i. \tag{6.148}$$

In the second term, $E_i > E_0$ for every value of i, hence $\langle\psi|H|\psi\rangle > \left\{|a_0|^2 E_0 + \sum_{i=1}^{c}|a_i|^2 E_0 = E_0\langle\psi|\psi\rangle\right\}$, which proves Eq. 6.144. This method is most useful to make iterative corrections to determine the ground state of a system. When it is unknown, one can use a trial wavefunction – *any* trial function – to determine a *first estimate* of the expectation value of the Hamiltonian. From the above analysis, we know that, at best, with an enviable stroke of luck the trial function may be the correct ground state function. This would be verified by the fact that *any change* in the trial function that one may try would only *raise* the expectation value of the Hamiltonian. However, if the expectation value of the Hamiltonian diminishes, it would imply that the second guess is better than the previous. This feature enables us to develop an iterative procedure to revise the guess function till any subsequent revision of the trial function does not lower the expectation value of the Hamiltonian; instead, it raises the same. By making iterative corrections, this method helps us get as close as possible to the determination of the ground state. The variations in the trial functions are made by changing a set of parameters subject to appropriate physical constraints, such as preservation of the norm and orthogonality relations for the new (changed) functions.

SUPPLEMENTARY NOTES:

- *We illustrate the application of the method of variation in the Solved Problem P6.9. In Chapter 9, the variation method is discussed to determine the wavefunctions of a many-electron atom. It turns out to be a complicated problem that is attacked using the method of variation using a scheme that is called the Hartree–Fock self-consistent-field (HF SCF) method.*
- *Numerical approximations play a huge role in studying many physical problems. These are discussed at length in specialized books on numerical methods and computational physics. We restrict ourselves to illustrate some salient features of these approximations in the Solved Problems P6.10 and P6.11.*
- *Perturbation series sometimes does not converge, and other methods are also not guaranteed to be successful. Developing approximations therefore remains a continuing challenge.*

Solved Problems

P6.1: Show that the classical HJ equation can be obtained from the Schrödinger equation in the limit $\hbar \to 0$.

Solution:

The Schrödinger equation is $i\hbar\frac{\partial}{\partial t}\left\{A(\vec{r},t)e^{\frac{i}{\hbar}S(\vec{r},t)}\right\} = \left\{\frac{\left(-i\hbar\vec{\nabla}\right)^2}{2m} + V\right\}\left\{A(\vec{r},t)e^{\frac{i}{\hbar}S(\vec{r},t)}\right\}$,

$$\text{i.e., } i\hbar\left\{\frac{\partial A}{\partial t} + A\frac{i}{\hbar}\frac{\partial S}{\partial t}\right\}e^{\frac{i}{\hbar}S} = \left[\frac{-\hbar^2 e^{\frac{i}{\hbar}S}\vec{\nabla}^2 A}{2m} - \frac{i\hbar e^{\frac{i}{\hbar}S}\vec{\nabla}A\bullet\vec{\nabla}S}{2m} - \frac{\hbar^2\vec{\nabla}\bullet\left\{\left(A\frac{i}{\hbar}e^{\frac{i}{\hbar}S}\right)\left(\vec{\nabla}S\right)\right\}}{2m}\right]e^{\frac{i}{\hbar}S} + V\left(Ae^{\frac{i}{\hbar}S}\right),$$

$$\text{i.e., } i\hbar\left\{\frac{\partial A}{\partial t}+A\frac{i}{\hbar}\frac{\partial S}{\partial t}\right\}e^{\frac{i}{\hbar}S}=\left[\begin{array}{c}\dfrac{-\hbar^2 e^{\frac{i}{\hbar}S}\vec{\nabla}^2 A}{2m}-\dfrac{i\hbar e^{\frac{i}{\hbar}S}\vec{\nabla}A\bullet\vec{\nabla}S}{2m}-\\ \dfrac{\hbar^2\left\{\vec{\nabla}\left(A\frac{i}{\hbar}e^{\frac{i}{\hbar}S}\right)\bullet\left(\vec{\nabla}S\right)\right\}}{2m}-\dfrac{\hbar^2\left(A\frac{i}{\hbar}e^{\frac{i}{\hbar}S}\right)\left(\vec{\nabla}^2 S\right)}{2m}\end{array}\right]e^{\frac{i}{\hbar}S}+V\left(Ae^{\frac{i}{\hbar}S}\right),$$

i.e.,

$$i\hbar\left\{\frac{\partial A}{\partial t}+A\frac{i}{\hbar}\frac{\partial S}{\partial t}\right\}e^{\frac{i}{\hbar}S}=\left[\begin{array}{c}\dfrac{-\hbar^2 e^{\frac{i}{\hbar}S}\vec{\nabla}^2 A}{2m}-\dfrac{i\hbar e^{\frac{i}{\hbar}S}\vec{\nabla}A\bullet\vec{\nabla}S}{2m}-\dfrac{i\hbar e^{\frac{i}{\hbar}S}\left(\vec{\nabla}A\right)\bullet\left(\vec{\nabla}S\right)}{2m}-\\ \dfrac{\hbar^2 A\left(\frac{i}{\hbar}\right)^2 e^{\frac{i}{\hbar}S}\left(\vec{\nabla}S\right)\bullet\left(\vec{\nabla}S\right)}{2m}-\dfrac{\hbar^2\left(A\frac{i}{\hbar}e^{\frac{i}{\hbar}S}\right)\left(\vec{\nabla}^2 S\right)}{2m}\end{array}\right]e^{\frac{i}{\hbar}S}+V\left(Ae^{\frac{i}{\hbar}S}\right),$$

$$\text{i.e., } i\hbar\left\{\frac{\partial A}{\partial t}+A\frac{i}{\hbar}\frac{\partial S}{\partial t}\right\}e^{\frac{i}{\hbar}S}=\left[\begin{array}{c}\dfrac{-\hbar^2 e^{\frac{i}{\hbar}S}\vec{\nabla}^2 A}{2m}-\dfrac{i\hbar e^{\frac{i}{\hbar}S}\vec{\nabla}A\bullet\vec{\nabla}S}{m}+\\ \dfrac{Ae^{\frac{i}{\hbar}S}\left|\vec{\nabla}S\right|^2}{2m}-\dfrac{i\hbar\left(Ae^{\frac{i}{\hbar}S}\right)\left(\vec{\nabla}^2 S\right)}{2m}\end{array}\right]e^{\frac{i}{\hbar}S}+V\left(Ae^{\frac{i}{\hbar}S}\right),$$

$$\text{or } i\hbar\left\{\frac{\partial A}{\partial t}+A\frac{i}{\hbar}\frac{\partial S}{\partial t}\right\}=\left[\frac{-\hbar^2\vec{\nabla}^2 A}{2m}-\frac{i\hbar\vec{\nabla}A\bullet\vec{\nabla}S}{m}+\frac{A\left|\vec{\nabla}S\right|^2}{2m}-\frac{i\hbar A\left(\vec{\nabla}^2 S\right)}{2m}\right]+VA.$$

$$\text{Hence, } \left\{i\hbar\frac{\partial A}{\partial t}-A\frac{\partial S}{\partial t}\right\}=\left[\frac{-\hbar^2\vec{\nabla}^2 A}{2m}-\frac{i\hbar\vec{\nabla}A\bullet\vec{\nabla}S}{m}+\frac{A\left|\vec{\nabla}S\right|^2}{2m}-\frac{i\hbar A\left(\vec{\nabla}^2 S\right)}{2m}\right]+VA.$$

Now, taking the limit $\hbar \to 0$, we get $-A\dfrac{\partial S}{\partial t}-\dfrac{A\left|\vec{\nabla}S\right|^2}{2m}-VA\approx 0,$

i.e., $\dfrac{\partial S}{\partial t}+\left(\dfrac{p^2}{2m}+V\right)\approx 0$, which is just the "Hamilton–Jacobi equation" (Eq. 6.17).

P6.2: Determine the tunneling probability through a finite width rectangular potential barrier for a particle having energy $E < V_0$ (Fig. 3.4c from Chapter 3) using the WKBJ approximation. Consider the barrier width to be thin.

Solution:

An estimate of the tunneling probability is $T=\dfrac{\rho_{x=a}}{\rho_{x=-a}}$, where the numerator is the probability density of the wavefunction at $x = a$ and the denominator at $x = -a$, which respectively mark the end and the onset of the rectangular barrier (Fig. 3.4c). Using now Eq. 6.44b we get, since $\kappa(x')=\dfrac{2m\left(V(x')-E\right)}{\hbar^2}$,

$\psi_{x=a}(x) \simeq \psi_{x=-a}\exp\left(-\int \kappa(x')dx'\right) = \psi_{x=-a}\exp\left(-\int \sqrt{\frac{2m\left(V(x')-E\right)}{\hbar^2}}dx'\right) \approx \psi_{x=-a}\exp(-\kappa d)$. Here, d is the width of the barrier. Thus, $T = e^{-2\int_{-a}^{+a} dx\sqrt{\frac{2m}{\hbar^2}(V(x)-E)}} = e^{-2\gamma}$ (see Eqs. 3.28a,b). The tunneling probability was used by Gamow to explain α-decay, as discussed in Chapter 3.

P6.3: A mechanical one-dimensional linear harmonic oscillator having mass m whose motion is along the Cartesian x-axis has charge q. It is placed in a constant electric field $\vec{E} = E\hat{e}_x$; $E > 0$. Determine first-order corrections to its energy spectrum treating its interaction with the electric field perturbatively.

Solution:

The eigenspectrum of the harmonic oscillator has been discussed in Chapter 3 by solving the Schrödinger equation (Eq. 3.65). Discrete bound state energies of this system were found to be given by $E_n = \left(n+\frac{1}{2}\right)\hbar\omega$ (Eq. 3.77). The interaction of the oscillator with the applied electric field can be treated perturbatively, with the perturbation potential being $V(x) = -qEx$.

From Eq. 6.88, the first-order correction is

$$E_n^{(1)} = \left\langle n^{(0)}\middle|(-qEx)\middle|n^{(0)}\right\rangle = -qE\left\langle n^{(0)}\middle|x\middle|n^{(0)}\right\rangle = -qE\left\langle n^{(0)}\middle|\sqrt{\frac{\hbar}{2m\omega}}\left(c+c^{\dagger}\right)\middle|n^{(0)}\right\rangle,$$

i.e., $E_n^{(1)} = -qE\sqrt{\frac{\hbar}{2m\omega}}\left\langle n^{(0)}\middle|\left(c+c^{\dagger}\right)\middle|n^{(0)}\right\rangle = 0.$

We have used Eqs. 3.81a,b to express the position operator in terms of the creation and destruction operators of the harmonic oscillator used in Section 3.3. This result can be understood best by estimating the first-order correction using the coordinate representation of the state vectors of the oscillator. Using the wavefunctions, $E_n^{(1)} = -qE\int_{-\infty}^{+\infty} dx\left[\left|\psi_n(x)\right|^2 x\right]$. We know from Section 3.3 that the wavefunctions are either *even* or *odd* (Table 3.1, Chapter 3) so the parity of the integrand $\left[\left|\psi_n(x)\right|^2 x\right]$ is determined by the parity of x, which is odd. The integral over whole space therefore vanishes.

P6.4: Determine the *second*-order corrections to the energy spectrum of the harmonic oscillator dealt with in the Solved Problem P6.3.

Solution:

From Eq. 6.92, $E_n^{(2)} = \sum_{m\neq n}\frac{\left|\left\langle n^{(0)}\middle|(-qEx)\middle|m^{(0)}\right\rangle\right|^2}{\left(E_n^{(0)}-E_m^{(0)}\right)} = (qE)^2\sum_{m\neq n}\frac{\left|\left\langle n^{(0)}\middle|x\middle|m^{(0)}\right\rangle\right|^2}{\left(E_n^{(0)}-E_m^{(0)}\right)}$. This correction can now be determined using the creation and destruction operators (Eqs. 3.81 a,b).

$\left\langle n^{(0)}\middle|x\middle|m^{(0)}\right\rangle = -qE\sqrt{\frac{\hbar}{2m\omega}}\left(\sqrt{n+1}\langle n|m+1\rangle + \sqrt{n}\langle n|m-1\rangle\right)$. The result is $E_n^{(2)} = -\frac{q^2E^2}{2m\omega^2}$. It is peculiar that the second-order correction is nonzero even if the first-order correction vanishes. It is arising due to the fact that the electric field gradually displaces the equilibrium position of the oscillator, thereby generating a dipole moment. The induced dipole moment is $\vec{p} = q\vec{x}$ and its interaction with the applied field is $-\vec{p}\bullet\vec{E}$, which generates a correction that is quadratic in the magnitude of the electric field.

P6.5: The Coulomb potential $V(r) = -\frac{e^2}{r}$ of a hydrogen atom is perturbed by an additional potential $H' = bx^2$ (where b is a constant) to the Hamiltonian. Find the first-order perturbation correction to the ground state energy.

Solution:

$$E_0^{(1)} = \langle \psi_0 | H' | \psi_0 \rangle = \frac{b}{\pi a_0^3} \int_0^\infty r^2 e^{\frac{-2r}{a_0}} r^2 dr \int_0^\pi \sin^3\theta d\theta \int_0^{2\pi} \cos^2\varphi d\varphi = ba_0^2.$$

P6.6: Find the ground state energy for a particle of mass m moving in a one-dimensional box with rigid walls at $x = 0$ and $x = L$ by considering a parabolic trial function $\phi(x) = x(L - x)$.

Solution:

Given $\phi(x) = x(L - x)$.

$$\text{Normalization: } N^2 \int_0^L x^2(L-x)^2 dx = 1 \Rightarrow N^2\left[\frac{L^5}{3} + \frac{L^5}{5} - \frac{2L^5}{4}\right] = 1 \Rightarrow \boxed{N = \sqrt{\frac{30}{L^5}}}.$$

$$\text{Hence, } \phi(x) = \sqrt{\frac{30}{L^5}} x(L-x) \Rightarrow \frac{\partial \phi}{\partial x} = \sqrt{\frac{30}{L^5}}(L - 2x).$$

$$\therefore \langle T \rangle = \frac{\hbar^2}{2m}\int_0^L \left|\frac{\partial \phi}{\partial x}\right|^2 dx = \frac{\hbar^2}{2m} \times \frac{30}{L^5}\int_0^L (L-2x)^2 dx.$$

$$\text{Finally, we get: } \frac{15\hbar^2}{mL^5}\left[L^2x + \frac{4x^3}{3} - \frac{4Lx^2}{2}\right]_0^L = \frac{15\hbar^2}{mL^5} \times \frac{L^3}{3} = \frac{5\hbar^2}{mL^2}.$$

P6.7: Use the WKBJ method to determine the tunneling probability for a potential barrier $V(x) = \begin{cases} V_0 - \alpha x, & x > 0 \\ 0, & x < 0 \end{cases}$ [Hint: Refer to the Solved Problem P6.2].

Solution:

Tunneling probability $T = \exp\left(-2\int_{x_1}^{x_2} \gamma dx\right)$, where $\gamma^2 = \frac{2m}{\hbar^2}[V(x) - E]$.

$$E = V(x) \Rightarrow E = V_0 - \alpha x \Rightarrow x = \frac{V_0 - E}{\alpha}.$$

$$\text{Hence, } T = \exp\left(-2\int_0^{\frac{V_0-E}{\alpha}} \sqrt{V_0 - \alpha x - E} \times \frac{\sqrt{2m}}{\hbar} dx\right).$$

$$\text{Now, } \int_0^{\frac{V_0-E}{\alpha}} dx\sqrt{V_0 - \alpha x - E} = \frac{2}{3}\left[\frac{(V_0 - E - \alpha x)^{3/2}}{-\alpha}\right]_0^{\frac{(V_0-E)}{\alpha}} = \frac{2}{3\alpha}(V_0 - E)^{3/2}.$$

$$\text{Hence, } T = \exp\left[\frac{-4\sqrt{2m}}{3\alpha\hbar}(V_0 - E)^{3/2}\right].$$

P6.8: Prove that $\lim_{t\to\infty}\frac{1}{\pi t}F(t,\tilde{\omega}) = \delta(\tilde{\omega})$, where $F(t,\tilde{\omega}) = \frac{2\sin^2\left(\frac{\tilde{\omega}t}{2}\right)}{\omega^2}$ (see Eq. 6.120 and Eq. 6.121).

Solution:

We recollect that a sequence $s_m(x)$ is δ-convergent if $\lim_{m\to\infty}\int_{-\infty}^{\infty} s_m(x)f(x)dx = f(0)$ for all functions $f(x)$ that are sufficiently smooth in the domain $-\infty < x < \infty$. This allows the interpretation of the limit $\lim_{m\to\infty} s_m(x) = \delta(x)$.

Let $f(x) = \frac{1}{\pi}\frac{\sin^2 x}{x^2}$.

Observe that $\int_0^\infty ye^{-xy}dy = y\frac{e^{-xy}}{-x}\Big|_0^\infty - \int_0^\infty \frac{e^{-xy}}{-x}dy = 0 - 0 - \frac{e^{-xy}}{(-x)^2}\Big|_0^\infty = -\frac{e^{-xy}}{x^2}\Big|_0^\infty = \frac{1}{x^2}$.

Hence, $\int_0^\infty \frac{1}{x^2}\sin^2 x\,dx = \int_0^\infty\left(\int_0^\infty ye^{-xy}dy\right)\sin^2 x\,dx = \int_0^\infty\int_0^\infty ye^{-xy}\left(\frac{e^{ix}-e^{-ix}}{i2}\right)^2 dy\,dx$,

i.e., $\int_0^\infty \frac{1}{x^2}\sin^2 x\,dx = \frac{-1}{4}\int_0^\infty\int_0^\infty ye^{-xy}(e^{2ix}+e^{-2ix}-2e^{ix}e^{-ix})dy\,dx$. Hence,

$$\int_0^\infty \frac{1}{x^2}\sin^2 x\,dx = \frac{-1}{4}\int_0^\infty\int_0^\infty y\left[e^{-x(y+2i)}+e^{-x(y-2i)}-2e^{-xy}\right]dy\,dx = \frac{-1}{4}\int_0^\infty y\left[\frac{e^{-x(y+2i)}}{-(y+2i)}+\frac{e^{-x(y-2i)}}{-(y-2i)}-\frac{2e^{-xy}}{-y}\right]_{x=0}^{\infty}dy,$$

i.e.,

$$\int_0^\infty \frac{1}{x^2}\sin^2 x\,dx = \frac{-1}{4}\int_0^\infty y\left[0+0-0+\frac{1}{y+2i}+\frac{1}{y-2i}-\frac{2}{y}\right]dy = \frac{-1}{4}\int_0^\infty\left[\frac{y}{y+2i}+\frac{y}{y-2i}-2\right]dy,$$

or

$$\int_0^\infty \frac{1}{x^2}\sin^2 x\,dx = \frac{-1}{4}\int_0^\infty\left[\frac{y(y-2i)+y(y+2i)-2y^2-8}{y^2+4}\right]dy = \frac{-1}{4}\int_0^\infty\frac{-8}{y^2+4}dy = 2\int_0^\infty\frac{1}{y^2+4}dy.$$

Hence,

$$\int_0^\infty \frac{1}{x^2}\sin^2 x\,dx = \frac{2}{4}\int_0^\infty\frac{1}{\left(y/2\right)^2+1}dy = \int_0^\infty\frac{1}{z^2+1}dz,$$ wherein we have changed the variable now by introducing $\frac{y}{2} = z$ and $dy = 2dz$.

Since $\int\frac{1}{1+x^2}dx = \tan^{-1}x$, we get $\int_0^\infty\frac{1}{x^2}\sin^2 x\,dx = \tan^{-1}(\infty)-\tan^{-1}(0) = \frac{\pi}{2}$, we see that $\int_{-\infty}^{\infty} \frac{1}{x^2}\sin^2 x\,dx = \pi$ or $\int_{-\infty}^{\infty}\left(\frac{1}{\pi}\frac{1}{x^2}\sin^2 x\right)dx = 1.$

On writing $f_\varepsilon(\tilde{\omega}) = \frac{1}{\varepsilon} f\left(\frac{\tilde{\omega}}{\varepsilon}\right) = \frac{1}{\varepsilon}\frac{\sin^2\left(\frac{\tilde{\omega}}{\varepsilon}\right)}{\tilde{\omega}^2}$, we see that for $\varepsilon = \frac{2}{t}$, as $\varepsilon \to 0$, we have $\frac{2}{t} \to \infty$ and

hence $\lim_{t\to\infty} \frac{1}{\pi t} F(t,\tilde{\omega}) = \lim_{t\to\infty}\frac{1}{\pi t}\frac{2\sin^2\left(\frac{\tilde{\omega}t}{2}\right)}{\omega^2} = \delta(\tilde{\omega}).$

P6.9: Estimate the ground state energy of a system in a one-dimensional potential $V(x) = -b\delta(x)$ with $-a \le x \le a$ and $b > 0$ if its normalized variational trial wave function is $\psi = \frac{1}{\sqrt{a}}\cos\left(\frac{\pi x}{a}\right)$.

Solution:

Expectation values of the potential and kinetic energies are:

$$< V > = \int \psi^* V \psi dx = \frac{-b}{a}\int_{-a}^{a}\cos^2\left(\frac{\pi x}{a}\right)\delta(x)dx = \frac{-b}{a},$$

$$\text{and } < T > = \int \psi^* T \psi dx = \frac{-\hbar^2}{2m}\int \psi^* \frac{\partial^2 \psi}{\partial x^2}dx = \frac{\hbar^2}{2m}\frac{\pi^2}{4a^3}\int_{-a}^{a}\cos^2\left(\frac{\pi x}{a}\right) = \frac{\hbar^2\pi^2}{8ma^2}$$

Hence, the total energy $< E > = < T > + < V > = \frac{\hbar^2\pi^2}{8ma^2} - \frac{b}{a}.$

At the minimum, $0 = \frac{\partial E}{\partial a} \Rightarrow \frac{-2\pi^2\hbar^2}{8ma^3} + \frac{b}{a^2} = 0 \Rightarrow a = \frac{\pi^2\hbar^2}{4mb}$. Hence, the ground state energy is $E_{gs} = \frac{-2mb^2}{\pi^2\hbar^2}.$

P6.10: Numerical integration of an ordinary differential equation (ODE) is often carried out using the *forward Euler method*. It employs a first-order numerical methodology that requires the initial value to be available. Using the *forward Euler method*, obtain the radial part of the ground state wave function of the hydrogen atom. Comment on how your results with two different choices of the radial grid size (a) 0.1 and (b) 0.0001 compare with the analytic solution.

Solution:

For a general first-order differential equation $y'(t) = f(t, y(t))$ with initial condition $y(t_0) = y_0$ and grid size $h = \frac{t_n - t_0}{n}$, the forward Euler equation is $y_{n+1} = y_n + hf(t_n, y_n)$. Error associated with this approximation is of the order of h^2. To solve a second-order differential equation of the type $y''(t) = f(t, y, y')$, the equation must be converted into two first-order equations such as $y' = v$ and $v' = f(t, y, v)$. The two first-order equations are to be then solved using the forward Euler method simultaneously.

The radial part of Schrödinger equation for the hydrogen atom is written as

$$\frac{1}{r^2}\frac{d}{dr}\left(r^2\frac{dR(r)}{dr}\right)+\left[\frac{2\mu}{\hbar^2}\{E-V(r)\}-\frac{\ell(\ell+1)}{r^2}\right]R(r)=0, \quad \text{(P6.10.1)}$$

where $V(r)=\frac{e^2}{r}$. Using the auxiliary function $\chi_{E,\ell}(r)=rR(r)$, the radial part of the Schrödinger equation is given as

$$-\frac{\hbar^2}{2\mu}\frac{d^2\chi_{E,\ell}(r)}{dr^2}+\left\{V(r)+\frac{\hbar^2}{2\mu}\frac{\ell(\ell+1)}{r^2}\right\}\chi_{E,\ell}(r)=E\chi_{E,\ell}(r). \quad \text{(P6.10.2)}$$

This second-order differential equation could be converted into two first-order equations as $\chi_{E,\ell}'(r)=v$ and $v'=f(r,\ell,E)\chi_{E,\ell}$, with initial condition $\chi_{E,\ell_{r\to 0}}\approx r^{\ell+1}$ (see Eq. 5.26 in Chapter 5). To obtain the ground state wavefunction, let us take $\ell=0$ and solve equation P6.10.2 using the forward Euler equation. Table P6.10 shows the comparison of numerical solution obtained using the Euler equation using radial grid size h = 0.1 and h = 0.0001 and the analytical solution for the 1s electron wavefunction (from Table 5.1, Chapter 5):

$$\chi_{1s}(r)=2\left(\frac{1}{a_0}\right)^{\frac{3}{2}}re^{\frac{-r}{a_0}}.$$

Figure P6.10 shows the comparison of the radial wavefunction obtained using Euler equation (with radial grid size $h = 0.1$ and $h = 0.0001$) and analytical form. The finer grid with $h = 0.0001$ gives far better agreement with the analytical function than the coarse grid $h = 0.1$.

Table P6.10 Comparison of numerical and analytical solution of ground state of hydrogen atom. The numerical solutions are obtained using two different radial grid sizes.

Radial distance (a.u.)	**Analytical solution**	**Numerical solution**	
		(h = 0.0001)	**(h = 0.1)**
0.5	0.309	0.303	0.317
1	0.376	0.368	0.336
1.5	0.342	0.335	0.241
2	0.276	0.271	0.114
2.5	0.209	0.205	−0.017
3	0.153	0.149	−0.148
3.5	0.108	0.106	−0.289
4	0.075	0.073	−0.457
4.5	0.052	0.049	−0.678
5	0.035	0.034	−0.985

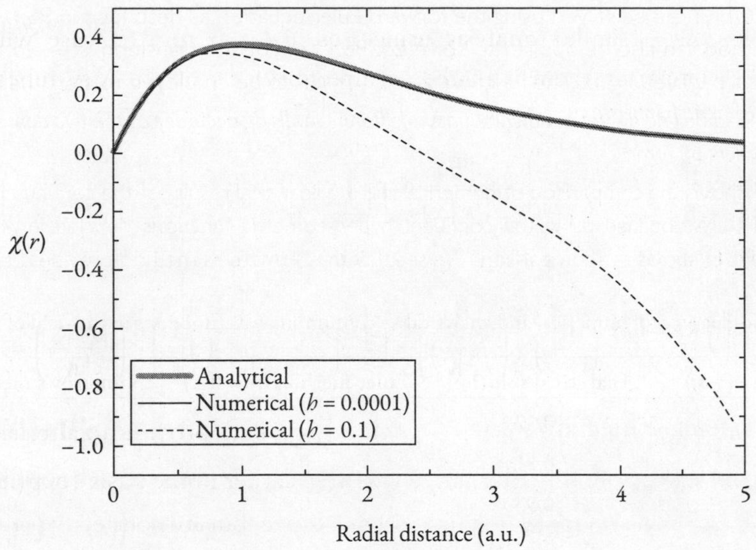

Fig. P6.10 Comparison of radial wavefunction obtained using the Euler equation ($h = 0.0001$: black solid, $h = 0.1$: black dashed) and analytical function (thick grey).

P6.11: Find the continuum electron wave function of H^+ ion with kinetic energy 1 a.u. using Euler and Numerov's method with radial grid size h = 0.1. Compare your result with the analytical solution.

Solution:

Refer to P6.10 regarding using forward Euler equation to solve a second-order differential equation. The error associated with the Euler equation approximation is of the order of h^2.

Numerov's method (also known as Cowell's method) is used for solving second-order differential equations. The error associated with this method is of the order of h^6.

For equations of the general form $y''(x) = f(x,y)y(x)$, (P6.11.1)

applying Numerov's method gives $W_{n+1} = 2W_n + h^2 f_n y_n - W_{n-1}$, (P6.11.2)

where $W_n = \left[1 - \frac{h^2}{12} f_n\right] y_n$ and $f_n = f(x_n, y_n)$. (P6.11.3)

For obtaining the continuum electron wave function of H^+ ion, consider the photoionization transition from $1s \rightarrow \varepsilon p$ of the H atom.

The auxiliary function $\chi_{E,l}(r) = rR(r)$ (Eq. 5.19a, Chapter 5) of the outgoing electron satisfies the Schrödinger equation

$$-\frac{\hbar^2}{2\mu}\frac{d^2\chi_{E,\ell}(r)}{dr^2} + \left\{V(r) + \frac{\hbar^2}{2\mu}\frac{\ell(\ell+1)}{r^2}\right\}\chi_{E,\ell}(r) = E\chi_{E,\ell}(r), \tag{P6.11.4}$$

where $E > 0$ and $V(r) = \frac{e^2}{r}$. The energy of the outgoing electron is taken as 1 a.u.

Equation P6.11.4 can be solved using the forward Euler equation as described in P6.10. Likewise, the Numerov's method can also be applied to solve Eq. P6.11.4. For the transition 1s → εp, the continuum wavefunction is a p wave ($\ell = 1$). The function $\chi_{E,\ell_{r\to 0}} \approx r^{\ell+1}$ is to be employed to start the iterations.

Note that the same grid discretization amounts to different levels of accuracy as the two numerical techniques have different orders of error.

Table P6.11 shows the comparison of numerical and the analytical results of the continuum wavefunction. Analytical results are obtained using the confluent hypergeometric functions detailed in Appendix 5A of Chapter 5. Fig. P6.11 shows the comparison. We see that the Numerov's method provides better result.

Table P6.11 Comparison of numerical and analytical εp continuum wavefunctions in the field of H^+ ion.

Radial distance (a.u.)	Analytical solution	Euler method (h = 0.1)	Numerov's method (h = 0.1)
0.5	0.32	0.25	0.33
1	0.81	0.71	0.85
1.5	0.93	0.92	0.97
2	0.52	0.61	0.54
2.5	−0.20	−0.12	−0.21
3	−0.80	−0.87	−0.84
3.5	−0.93	−1.19	−0.98
4	−0.54	−0.82	−0.56
4.5	0.17	0.08	0.18
5	0.78	1.04	0.82
5,5	0.96	1.52	1.01
6	0.72	1.32	0.75

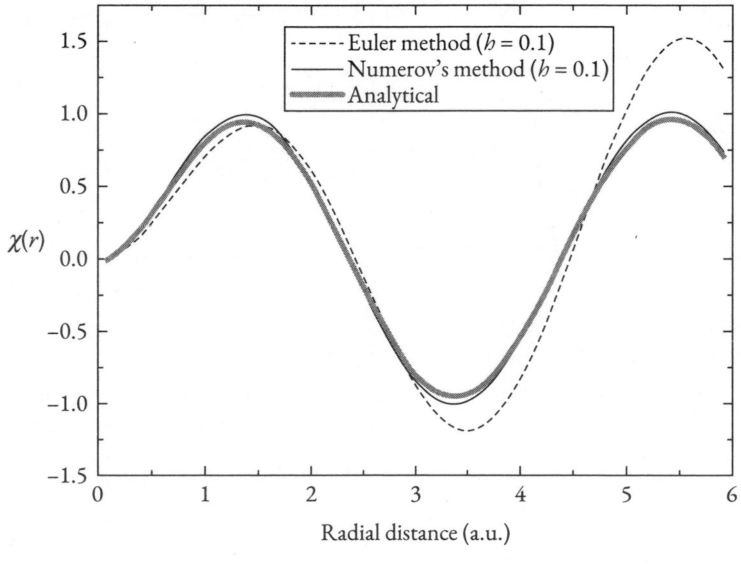

Fig. P6.11 Comparison of radial wavefunction obtained using the Euler (black dashed) and Numerov's method (black solid) and analytical function (thick grey).

Exercises

E6.1: A particle of mass m moves in the attractive central potential $V(r) = \frac{-g^2}{r^{3/2}}$, where g is a constant. Use the normalized trial wave function $\psi = \left(\frac{k^3}{8\pi}\right)^{1/2} \exp\left(\frac{-kr}{2}\right)$ to estimate the lowest energy state. [Use $\int_0^\infty x^n \exp(-ax)dx = \frac{n!}{a^{n+1}}$]

E6.2: Estimate the energy levels of a particle moving in the potential $V(x) = \begin{cases} \infty, & x < 0 \\ \frac{x}{a}, & x > 0 \end{cases}$ using WKBJ approximation.

E6.3: Using the first-order perturbation theory, determine the energy of the first three states of an infinite well of width a shown in the adjacent figure; note the shape of the bottom of the well since the portion AB has been sliced off.

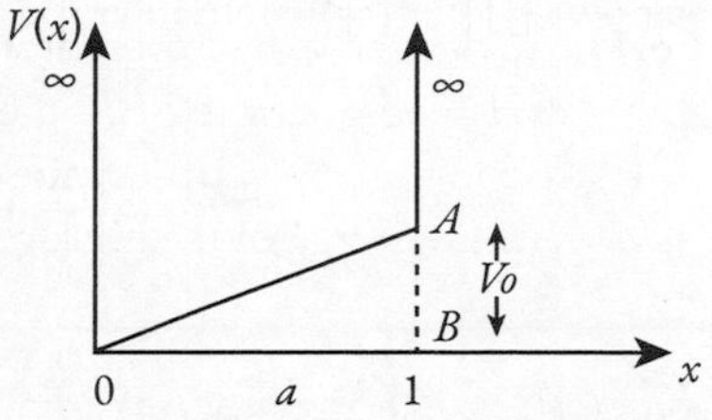

E6.4: A particle of mass m moves in a three-dimensional potential $V = \frac{1}{2}k(x^2 + y^2 + z^2 + 2\lambda xy)$, where λ is a small parameter. Find the ground state energy treating the term λkxy as a perturbation to first order.

E6.5: Obtain the general expression for the nth energy eigenvalue of a one-dimensional Hamiltonian $H = \frac{P^2}{2m} + \lambda x^4$ (where $\lambda > 0$ is a constant) using the WKBJ approximation.

E6.6: Consider the following potential $V = \begin{cases} \infty \text{ for } x < 0 \text{ and } x > \ell \\ -b \text{ for } 0 < x < \frac{\ell}{2} \\ 0 \text{ for } \frac{\ell}{2} < x < \ell \end{cases}$.

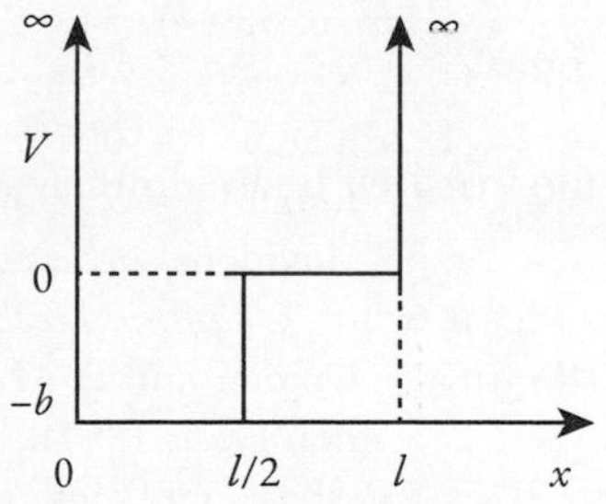

Treat the potential dip in the region x belongs to $\left[0, \frac{\ell}{2}\right]$ as a perturbation to a rigid box for the region x belongs to $[0, \ell]$. Obtain the eigenenergies using first-order perturbation theory.

E6.7: Find the ground state energy of a particle of mass m whose Hamiltonian is $H = \frac{-\hbar^2}{2m}\frac{d^2}{dx^2} + bx^4$ (b is a constant) using the trial wave function $\psi(x) = A\exp(-\alpha^2 x^2)$

(Use $\int_{-\infty}^{\infty} x^2 e^{-2\alpha^2 x^2} dx = \frac{1}{4\alpha^3}\sqrt{\frac{\pi}{2}}$ and $\int_{-\infty}^{\infty} x^4 e^{-2\alpha^2 x^2} dx = \frac{3}{16\alpha^5}\sqrt{\frac{\pi}{2}}$).

E6.8: Consider a particle of mass m that is bouncing vertically and elastically on a reflecting hard floor, where $V(z) = \begin{cases} mgz, & z > 0 \\ +\infty, & z \le 0 \end{cases}$. g is the local acceleration due to gravity. Use the WKBJ method to estimate the ground state energy of this particle. (Note: The potential examined here is of interest in obtaining the eigenspectrum of the quarkonium, which is a bound state of a quark and antiquark).

E6.9: A rigid rotor constrained to rotate in one plane has moment of inertia I about its axis of rotation and electric dipole moment μ (in the plane). This rotor is placed in a weak uniform electric field E, which is in the plane of rotation. Treating the electric field as a perturbation, find the first non-vanishing correction to the energy levels of the rotor.

[Hint: Use $\psi(\theta) = \frac{1}{\sqrt{2\pi}}\exp(im\theta); m = 0, \pm 1, \pm 2, ..$ and $E = \frac{m^2\hbar^2}{2I}$.]

E6.10: Find the best bound on the first excited state of the one-dimensional harmonic oscillator using the trial function $\psi(x) = Ax\exp(-ax^2)$.

E6.11: Consider a hydrogen atom in a weak electrostatic field $E\hat{e}_z$. The system's Hamiltonian is $H = H_0 + V$, where H_0 is the hydrogen atomic Hamiltonian and $V = -eEr\cos\theta$ is the perturbation due to the electrostatic field. Determine the first- and second-order correction to the ground state energy of the hydrogen atom (see also Chapter 8).

References

[1] H. Goldstein, C. P. Poole, and J. Safko, *Classical Mechanics* (3rd Edition, Pearson, 2014).

[2] P. C. Deshmukh, *Foundations of Classical Mechanics* (Cambridge University Press, New Delhi, 2019).

[3] J. H. Field, Derivation of the Schrödinger Equation from the Hamilton–Jacobi Equation in Feynman's Path Integral Formulation of Quantum Mechanics. *European Journal of Physics* 32:1, (2012). DOI: 10.1088/0143-0807/32/1/007.

[4] H. Jeffreys, On Certain Approximate Solutions of Linear Differential Equations of the Second Order. *Proc. London Math. Soc.* 2:23, 428 (1923).

[5] G. Wentzel, EineVerallgemeinerung der Quantenbedingungenfür die Zwecke der Wellenmechnik. Z. *Physik* 38:6–7, 518–529 (1926).

[6] H. A. Kramers, Wellenmechanik und halbzahlige Quantisierung. Z. *Physik* 39:10–11, 828–840 (1926). DOI: 10.1007/BF01451751.

[7] L. Brillouin, La mécanique ondulatorie de Schrödinger: une méthode générale de resolution par approximations successives. *Compt. Rend.* 183, 24 (1926).

[8] L. Schiff, *Quantum Mechanics* (3rd Edition, McGraw-Hill Book Company, 1968).

[9] L. D. Landau and E. M. Lifshitz, *Quantum Mechanics – Non-relativistic Theory* (3rd Edition, Pergamon Press, 1977).

[10] E. Merzbacher, *Quantum Mechanics* (3rd Edition, John Wiley & Sons, 1998).

[11] A. V. Tankhivale and C. Mande, Effective Z Values for Wave Functions in Multi-electron Atoms and Their Application to the Calculation of Shake-off Probability. *J. Phys. B: Atom. Mol. Phys.* 3:774 (1970).

7

The Relativistic Hydrogen Atom

Paul Dirac. B463, https://arkiv.dk/en/vis/5940625. Courtesy: Niels Bohr Archive.

If you are receptive and humble, mathematics will lead you by the hand....

Pure mathematics and physics are becoming ever more closely connected, ... Possibly, the two subjects will ultimately unify, ...

— Paul A. M. Dirac

A remarkable achievement of the 19th century was the formulation of the laws of electrodynamics by James Clerk Maxwell. It seemed incredible that the electromagnetic waves would propagate at a speed determined only by properties of vacuum irrespective of the state of motion of an observer. The speed of the electromagnetic waves is essentially constant in every inertial frame of reference. This seemed counterintuitive, since it required *no* adjustment of the relative difference in the motion of an observer and the observed waves. This phenomenology prompted Einstein to recognize that reconciliation of the constancy of speed of light in vacuum is possible by admitting space-contraction and time-dilation; this idea is formally structured in the special theory of relativity (STR; see Chapters 12, 13, and 14 of Reference [1] for a straightforward introduction to the theory of relativity). An immediate consequence of the STR is that neither a Euclidean space interval nor a time interval is separately Lorentz invariant. The Schrödinger equation is nonrelativistic; it is not Lorentz invariant. In this chapter, we develop relativistic quantum mechanics.

7.1 Klein–Gordan and the Dirac Equation

We consider a Cartesian frame of reference F' centered at O' with axes $O'X'$, $O'Y'$, $O'Z'$ respectively parallel to corresponding axes in another frame F centered at O whose axes are OX, OY, OZ. We consider the frame F' to be moving with respect to F at a constant velocity $\vec{v}$ along the x-axis as shown in Fig. 7.1. The invariant metric in the STR is the *space–time* interval between two events in the *non*-Euclidean space–time continuum. Using common notation (such as in [1]), the event intervals in the frames F and F' are related by the following well-known transformation relations:

$$x' = \gamma(x - vt) \quad ; \quad x = \gamma(x' + vt'), \tag{7.1a}$$

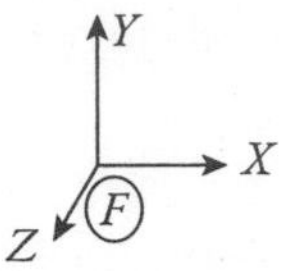

Origins O and O' of the two frames F and F' coincide at $t = 0$ and at $t' = 0$. F' moves with respect to F at a constant velocity $\vec{v}$.

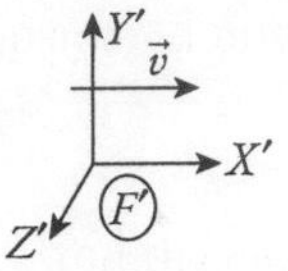

Fig. 7.1 Two Cartesian inertial frames of references with their respective axes parallel to each other and moving at a constant velocity $\vec{v}$ with respect to each other.

$$y' = y \quad ; \quad y = y', \tag{7.1b}$$

$$z' = z \quad ; \quad z = z', \tag{7.1c}$$

and

$$t' = \gamma\left(t - \frac{vx}{c^2}\right) \quad ; \quad t = \gamma\left(t' + \frac{vx'}{c^2}\right). \tag{7.1d}$$

In Eq. 7.1a–d, we have $\gamma = \dfrac{1}{\sqrt{1 - \dfrac{v^2}{c^2}}} = \dfrac{1}{\sqrt{1-\beta^2}}$, with $\beta = \dfrac{v}{c}$. (7.1e)

Consequences of the constancy of the speed of light in all inertial frames of reference include

(a) time dilation: $\Delta t = \dfrac{\Delta\tau}{\sqrt{1-\beta^2}} \geq \Delta\tau$, (7.2a)

(b) length contraction: $L' = L\sqrt{1-\beta^2} \leq L$, (7.2b)

(c) relativity of simultaneity; it is not absolute.

Contravariant $(x^0 \;\; x^1 \;\; x^2 \;\; x^3)$ and covariant coordinates $(x_0 \;\; x_1 \;\; x_2 \;\; x_3)$ will be used to denote events in the space–time continuum. The g-metric connects the same (again, using the notation from [1]):

$$a_\mu = g_{\mu\nu} a^\nu, \tag{7.3}$$

where the signature of the g-metric that we shall employ is $(1 \;\; -1 \;\; -1 \;\; -1)$. Thus, the interval between two events is given by

$$ds^2 = (dx^\mu) g_{\mu\upsilon} (dx^\upsilon) = \left[dx^0 \; dx^1 \; dx^2 \; dx^3\right] \begin{bmatrix} 1 & 0 & 0 & 0 \\ 0 & -1 & 0 & 0 \\ 0 & 0 & -1 & 0 \\ 0 & 0 & 0 & -1 \end{bmatrix} \begin{bmatrix} dx^0 \\ dx^1 \\ dx^2 \\ dx^3 \end{bmatrix}, \tag{7.4a}$$

$$\text{i.e., } ds^2 = (dx^0)(dx_0) + (dx^1)(dx_1) + (dx^2)(dx_2) + (dx^3)(dx_3). \tag{7.4b}$$

The invariance of the event interval is manifest from the following relation:

$$ds^2 = dx^\mu dx_\mu = dx^\mu g_{\mu\upsilon} dx^\upsilon = ds'^2 = dx'^\mu dx'_\mu = dx'^\mu g_{\mu\upsilon} dx'^\upsilon. \tag{7.4c}$$

One may expect a relativistic theory of quantum mechanics to be developed by quantizing the dynamical variables, i.e., by representing position and momentum by appropriate operators:

$$q \to q_{op}\ ;\ \ p \to p_{op}, \tag{7.5a}$$

and dynamical variables $F(q,p) \to F_{op} \to F_{op}\left(q_{op}, p_{op}\right).$ (7.5b)

However, we run into a difficulty since consideration of the momentum as mass times velocity involves the recognition of the velocity as $\vec{v} = \lim_{\delta t \to 0} \frac{\delta \vec{r}}{\delta t}$, but the numerator undergoes length contraction, and the denominator undergoes time dilation. Relativistic mechanics requires the introduction of *proper velocity*, defined by

$$\vec{\eta} = \text{proper velocity} = \frac{\text{proper length}}{\text{proper time}} = \frac{d\vec{r}}{d\left(t/\gamma\right)} = \gamma \frac{d\vec{r}}{dt}. \tag{7.6}$$

$\vec{\eta}$ gives three of the four components of the four-vector for *proper velocity* η^μ with $\mu = 0,1,2,3$. Thus,

$$\eta^0 = \frac{dx^0}{d(t/\gamma)} = \gamma \frac{dx^0}{dt} = \gamma \frac{d(ct)}{dt} = \gamma c, \tag{7.7a}$$

$$\eta^1 = \frac{dx^1}{d(t/\gamma)} = \gamma \frac{dx^1}{dt} = \gamma \frac{d(x)}{dt} = \gamma v_x, \tag{7.7b}$$

$$\eta^2 = \frac{dx^2}{d(t/\gamma)} = \gamma v_y, \tag{7.7c}$$

and $\eta^3 = \frac{dx^3}{d(t/\gamma)} = \gamma v_z.$ (7.7d)

The four-velocity then is

$$\eta^\mu = \left(\gamma c, \vec{\eta}\right) = \left(\gamma c, \gamma \vec{v}\right) = \gamma\left(c, \vec{v}\right). \tag{7.8}$$

One can see that the scalar product $\eta^\mu \eta_\mu$ is manifestly invariant since

$$\eta^\mu \eta_\mu = \eta^\mu g_{\mu\nu} \eta^\nu = \eta^{0^2} - \eta^{1^2} - \eta^{2^2} - \eta^{3^2} = c^2. \tag{7.9}$$

Defining now the *proper momentum* as mass times *proper velocity*, we have

$$p^0 = m\eta^0 = m\gamma c, \tag{7.10a}$$

$$p^1 = m\eta^1 = m\gamma v_x, \tag{7.10b}$$

$$p^2 = m\eta^2 = m\gamma v_y, \tag{7.10c}$$

and $p^3 = m\eta^3 = m\gamma v_z,$ (7.10d)

i.e., $p^\mu = m\eta^\mu = \left(\gamma mc, m\vec{\eta}\right) = \left(\gamma mc, \gamma \vec{v}\right) = \gamma m\left(c, \vec{v}\right).$ (7.11)

It is of course important to ascertain that the scalar product $p^{\mu}p_{\mu}$ is invariant. This is easily verified since

$$p^{\mu}p_{\mu} = p^{\mu}g_{\mu\nu}p^{\nu} = p^{0^2} - p^{1^2} - p^{2^2} - p^{3^2} = m^2\gamma^2c^2 - m^2\gamma^2v^2 = \frac{E^2}{c^2} - \vec{p}.\vec{p} = m^2c^2. \quad (7.12)$$

Funquest: Explain how the idea of 'simultaneity' breaks down considering the constancy of the speed of light in two inertial frames of reference moving with respect to each other.

Of course, one must remember that since STR establishes the equivalence between mass and energy, it is *pointless* to define a relativistic variable mass $m_{\text{relativistic}} = \gamma m$, where m is the rest-mass. The famous equation between energy and mass must be written as

$$E = \gamma mc^2 = mc^2\left(1 - \frac{v^2}{c^2}\right)^{-\frac{1}{2}} = mc^2\left(1 + \frac{1}{2}\frac{v^2}{c^2} + \frac{\frac{1}{2}\left(\frac{1}{2}+1\right)}{2!}\left(\frac{v^2}{c^2}\right)^2 + \ldots.\right), \quad (7.13a)$$

and not as $E = mc^2$ which is appropriate only for a particle at rest. (7.13b)

For a photon, the rest mass being zero, Eq. 7.13a gives an indeterminate quantity, thereby directing us to employ Eq. 7.12 (instead of Eq. 7.13a) to obtain

$$E = pc = \frac{h}{\lambda}c = h\frac{c}{\lambda} = h\nu, \quad (7.14)$$

where we have made use of the de Broglie relation between the momentum p and the wavelength λ. It may be mentioned that the confidence with which we know that the photon is massless is the same with which we know that the Coulomb interaction goes as $1/r$. One can see this by considering a variant of the Coulomb interaction, namely the screened Coulomb interaction. A departure from the Coulomb potential can be represented as $V(r) \sim \frac{e^{-\frac{r\mu c}{h}}}{r}$, where μ would be the photon mass. We see that $\mu \to 0$ goes with the strict Coulomb potential $V(r) \sim 1/r$.

We are now ready to quantize the system, i.e., we would use the operators $q \to q_{op}$ and $p \to p_{op} = -i\hbar\vec{\nabla}$. Thus, we expect to achieve quantization by introducing the following operators for the momentum:

$$p^{\mu} = i\hbar\partial^{\mu} = i\hbar\frac{\partial}{\partial x_{\mu}} \equiv i\hbar\left(\frac{\partial}{\partial ct}, -\vec{\nabla}\right) \quad (7.15a)$$

and $p_{\mu} = i\hbar\partial_{\mu} = i\hbar\frac{\partial}{\partial x^{\mu}} \equiv i\hbar\left(\frac{\partial}{\partial ct}, \vec{\nabla}\right).$ (7.15b)

Quantization of the position operator requires us to employ *operators*

$$q \to q_{op}, \quad (7.16a)$$

i.e., $(q^0, q^1, q^2, q^3) \to$ *operators*. (7.16b)

We therefore have $p^0 = \gamma mc = \frac{E}{c} \to \frac{1}{c} i\hbar \frac{\partial}{\partial t} = i\hbar \frac{\partial}{\partial (ct)}$, (7.17a)

and $\left(p^1, p^2, p^3\right) \equiv \left(p_x, p_y, p_z\right) \to \left(\vec{p}\right) \to \left(-i\hbar\vec{\nabla}\right)$. (7.17b)

By expressing all terms in the Eq. 7.12 as operators operating on a wavefunction ψ, we get

$$\left[\frac{E^2}{c^2} - \vec{p}.\vec{p} - m^2c^2\right]\psi = \left[-\frac{\hbar^2}{c^2}\frac{\partial^2}{\partial t^2} + \hbar^2\left(\vec{\nabla}\bullet\vec{\nabla}\right) - (mc)^2\right]\psi = 0, \tag{7.18a}$$

$$\text{i.e.,} \left[\frac{1}{c^2}\frac{\partial^2}{\partial t^2} - \left(\vec{\nabla}\bullet\vec{\nabla}\right) + \left(\frac{mc}{\hbar}\right)^2\right]\psi = 0. \tag{7.18b}$$

An operator called as *d'Alembert operator*, or simply as the "box" operator, is often used to write Eq. 7.18a,b. It is defined as

$$\Box = \left[\frac{1}{c^2}\frac{\partial^2}{\partial t^2} - \left(\vec{\nabla}\bullet\vec{\nabla}\right)\right] = \frac{\partial}{\partial x_\mu}\frac{\partial}{\partial x^\mu}, \tag{7.19}$$

using which we get

$$\left[\Box + \left(\frac{mc}{\hbar}\right)^2\right]\psi = 0, \tag{7.20}$$

known as the Klein–Gordon (KG) equation. It was developed by Klein [2] and Gordon [3] independently but around the same time (in 1926). It applies quantization (Eqs. 7.15a,b and Eqs. 7.16a,b) for the relativistic invariants in Eq. 7.12. It is thus a natural choice for a relativistic quantum equation. Apart from involving second-order space derivatives as in the Schrödinger equation, the KG equation has the second-order derivative with respect to time. It has an extra degree of freedom since its solution would require not only the initial time condition on the wavefunction but also on its first temporal derivative. The KG equation poses further difficulties with regard to the probability interpretation of the wavefunction (see the Solved Problem P7.5). Several important contributions to the study of the KG equation, including by Schrödinger himself, were attempted. However, we shall bypass these developments since it was soon realized that the KG equation is unsuitable to describe electrons. The relativistic quantum equation that describes the electron was developed by Paul Adrien Maurice Dirac (1902–1984). We shall discuss it now.

The Dirac equation [4–6] plays a fundamental role in accounting for atomic structure and dynamics. From the relativistic invariant in Eq. 7.12, we have

$$p^\mu p_\mu - m^2c^2 = p^0 p_0 - \vec{p}\bullet\vec{p} - m^2c^2 = \left(p^0\right)^2 - \vec{p}\bullet\vec{p} - m^2c^2 = 0. \tag{7.21}$$

When $\vec{p} = \vec{0}$, (7.22a)

Eq. 7.21 can be factored as

$$\left(p^0 + mc\right)\left(p^0 - mc\right) = 0, \tag{7.22b}$$

where either or both $(p^0 + mc)$ and $(p^0 - mc)$ would be zero. For the time being, let us consider

$$(p^0 - mc) = 0. \tag{7.23}$$

Funquest: What would happen if you chose the other factor (i.e., one with the + sign) to be zero instead of Eq. 7.23?

For the special case $\vec{p} = \vec{0}$, one may therefore develop a quantum equation by representing Eq. 7.23 as an operator, and determine the operand on which it acts that would represent the relativistic wavefunction. In Dirac's method, this objective is aimed at for the more general case, inclusive of $\vec{p} \neq \vec{0}$. We begin by asking if it is possible to factorize Eq. 7.21 as

$$p^\mu p_\mu - m^2c^2 = \left(\beta^\kappa\, p_\kappa + mc\right)\left(\gamma^\lambda\, p_\lambda - mc\right) = 0. \tag{7.24}$$

The motivation to explore the above form comes from its similarity with Eq. 7.22b, but it involves introducing eight unknown coefficients that would satisfy Eq. 7.24,

$$\beta^\kappa \equiv \{\beta^0, \beta^1, \beta^2, \beta^3\} \text{ and } \gamma^\lambda \equiv \{\gamma^0, \gamma^1, \gamma^2, \gamma^3\}, \tag{7.25}$$

to be determined.

Some respite can be found in the fact that such factorization requires

$$\left(\beta^\kappa\, p_\kappa + mc\right)\left(\gamma^\lambda\, p_\lambda - mc\right) = \beta^\kappa\gamma^\lambda p_\kappa p_\lambda - mc\beta^\kappa\, p_\kappa + mc\gamma^\lambda\, p_\lambda - m^2c^2, \tag{7.26a}$$

$$\text{or, } p^\mu p_\mu = \beta^\kappa\gamma^\lambda p_\kappa p_\lambda - mc\left(\beta^\kappa - \gamma^\kappa\right) p_\kappa. \tag{7.26b}$$

The left-hand side involves only quadratic terms in the momentum, so the coefficient of the linear term on the right-hand side must vanish. That reduces the number of unknowns from eight to four, since

$$\beta^\kappa = \gamma^\kappa. \tag{7.27a}$$

We then have

$$p^\mu p_\mu = \gamma^\kappa\gamma^\lambda p_\kappa p_\lambda. \tag{7.27b}$$

The challenge now is to figure out how the four terms on the left-hand side can be made equal to the 16 terms on the right-hand side:

$$\begin{bmatrix} (p^0)^2 \\ -(p^1)^2 \\ -(p^2)^2 \\ -(p^3)^2 \end{bmatrix} = \begin{bmatrix} \gamma^0\gamma^0 p_0 p_0 + \gamma^1\gamma^1 p_1 p_1 + \gamma^2\gamma^2 p_2 p_2 + \gamma^3\gamma^3 p_3 p_3 \\ +(\gamma^0\gamma^1 + \gamma^1\gamma^0)p_0 p_1 + (\gamma^0\gamma^2 + \gamma^2\gamma^0)p_0 p_2 \\ +(\gamma^0\gamma^3 + \gamma^3\gamma^0)p_0 p_3 + (\gamma^1\gamma^2 + \gamma^2\gamma^1)p_1 p_2 \\ +(\gamma^1\gamma^3 + \gamma^3\gamma^1)p_1 p_3 + (\gamma^2\gamma^3 + \gamma^3\gamma^2)p_2 p_3 \end{bmatrix}. \tag{7.27c}$$

It is now clear that if we can find $\gamma^\mu; \mu = 0, 1, 2, 3$, such that the equality in Eq. 7.27c is realized, then we can conclude that the factorization of Eq. 7.21 as suggested in Eq. 7.24 is indeed accomplishable. Dirac determined that this is achieved by realizing $\gamma^\mu; \mu = 0, 1, 2, 3$ to be the following matrices:

$$\gamma^0 = \begin{bmatrix} 1 & 0 & 0 & 0 \\ 0 & 1 & 0 & 0 \\ 0 & 0 & -1 & 0 \\ 0 & 0 & 0 & -1 \end{bmatrix}, \tag{7.28a}$$

$$\gamma^1 = \begin{bmatrix} 0 & 0 & 0 & 1 \\ 0 & 0 & 1 & 0 \\ 0 & -1 & 0 & 0 \\ -1 & 0 & 0 & 0 \end{bmatrix}, \tag{7.28b}$$

$$\gamma^2 = \begin{bmatrix} 0 & 0 & 0 & -i \\ 0 & 0 & i & 0 \\ 0 & i & 0 & 0 \\ -i & 0 & 0 & 0 \end{bmatrix}, \tag{7.28c}$$

$$\text{and } \gamma^3 = \begin{bmatrix} 0 & 0 & 1 & 0 \\ 0 & 0 & 0 & -1 \\ -1 & 0 & 0 & 0 \\ 0 & 1 & 0 & 0 \end{bmatrix}. \tag{7.28d}$$

The γ^μ matrices, also called as Dirac matrices, are made up of block matrices, the blocks being the Pauli matrices (from the Solved Problem P4.3) and/or the 2×2 unit matrix and the 2×2 null matrix:

$$\gamma^0 = \begin{bmatrix} 1_{2\times2} & 0_{2\times2} \\ 0_{2\times2} & -1_{2\times2} \end{bmatrix}, \tag{7.29a}$$

$$\gamma^i{}_{4\times4} = \begin{bmatrix} 0_{2\times2} & \sigma^i{}_{2\times2} \\ -\sigma^i{}_{2\times2} & 0_{2\times2} \end{bmatrix}. \tag{7.29b}$$

γ^0 is a Hermitian matrix, and the remaining three matrices $(\gamma^\mu; \mu = 1,2,3) \leftrightarrow \vec{\gamma}$ are anti-Hermitian. In order to get the relativistic quantum equation, we execute the following strategy:

- peel out one of the factors $(\gamma^\lambda\ p_\lambda - mc)$; $\lambda = 0,1,2,3$, in the Eq. 7.24,
- quantize momentum by using corresponding operators
  $p_\kappa = i\hbar\partial_\kappa = i\hbar\dfrac{\partial}{\partial x^\kappa} = i\hbar\left(\dfrac{\partial}{\partial ct}, \vec{\nabla}\right)$,
- and operate on what would be the relativistic wavefunction ψ,

to get

$$\left(\gamma^\kappa\ i\hbar\partial_\kappa - mc\right)_{4\times4} \psi_{4\times1} = 0_{4\times1}. \tag{7.30a}$$

The above relativistic quantum equation is called as the Dirac equation. It is obvious that the operand wavefunction ψ must have a matrix structure, consisting of four rows and one column.

The Einstein convention of summing over double index is applicable, so we may write the Dirac equation also as

$$\left(\gamma^0 p_0 + \gamma^1 p_1 + \gamma^2 p_2 + \gamma^3 p_3 - mc\right)\psi = 0, \tag{7.30b}$$

$$\text{i.e.,} \left\{ \begin{array}{l} \begin{bmatrix} 1 & 0 & 0 & 0 \\ 0 & 1 & 0 & 0 \\ 0 & 0 & -1 & 0 \\ 0 & 0 & 0 & -1 \end{bmatrix} p_0 + \begin{bmatrix} 0 & 0 & 0 & 1 \\ 0 & 0 & 1 & 0 \\ 0 & -1 & 0 & 0 \\ -1 & 0 & 0 & 0 \end{bmatrix} p_1 + \begin{bmatrix} 0 & 0 & 0 & -i \\ 0 & 0 & i & 0 \\ 0 & i & 0 & 0 \\ -i & 0 & 0 & 0 \end{bmatrix} p_2 \\ + \begin{bmatrix} 0 & 0 & 1 & 0 \\ 0 & 0 & 0 & -1 \\ -1 & 0 & 0 & 0 \\ 0 & 1 & 0 & 0 \end{bmatrix} p_3 - mc \begin{bmatrix} 1 & 0 & 0 & 0 \\ 0 & 1 & 0 & 0 \\ 0 & 0 & 1 & 0 \\ 0 & 0 & 0 & 1 \end{bmatrix} \end{array} \right\} \psi = 0, \tag{7.30c}$$

$$\text{or,} \begin{bmatrix} p_0 - mc & 0 & p_3 & p_1 - ip_2 \\ 0 & p_0 - mc & p_1 + ip_2 & -p_3 \\ -p_3 & -\left(p_1 - ip_2\right) & -(p_0 + mc) & 0 \\ -\left(p_1 + ip_2\right) & p_3 & 0 & -(p_0 + mc) \end{bmatrix} \begin{bmatrix} u_1 \\ u_2 \\ u_3 \\ u_4 \end{bmatrix} = 0, \tag{7.30d}$$

where the relativistic four-component wavefunction is written as

$$\psi_{4\times 1} = \begin{bmatrix} u_1 \\ u_2 \\ u_3 \\ u_4 \end{bmatrix}. \tag{7.31}$$

Dirac equation introduces a first-order time derivative operator, but we also note the fact that the momentum operators are only linear. The Dirac equation is also sometimes written in alternative but equivalent forms. It is rather common to use *Pauli representation* by defining matrix operators $\vec{\alpha}$ and β as

$$\gamma^\mu = \left\{\gamma^0, \gamma^1, \gamma^2, \gamma^3\right\} = \left\{\gamma^0, \vec{\gamma}\right\} = \left\{\beta, \beta\vec{\alpha}\right\}, \tag{7.32a}$$

$$\text{i.e., } \beta = \gamma^0 \text{ and } \vec{\alpha} = \beta^{-1}\vec{\gamma}, \tag{7.32b}$$

$$\text{or, in other words, } \beta = \begin{bmatrix} 1_{2\times2} & 0_{2\times2} \\ 0_{2\times2} & -1_{2\times2} \end{bmatrix} \text{ and } \alpha^i_{4\times4} = \begin{bmatrix} 0_{2\times2} & \sigma^i_{2\times2} \\ \sigma^i_{2\times2} & 0_{2\times2} \end{bmatrix}. \tag{7..32c}$$

The vector space of ψ has a matrix structure and the 2×2 blocks within the Dirac matrices operate on the upper 2×1 or the lower 2×1 part of the wavefunction. The Dirac vector space can thus be thought of as an amalgamation of the two Pauli vector spaces. One must remember, however, that $\gamma^{\mu=1,2,3}$, i.e., $\vec{\gamma} = \beta\vec{\alpha}$ has three components, like that of a vector. $\vec{\gamma}$ therefore represents operators that have both a vector and a matrix character. Each of the three-component matrices has off-diagonal elements. Hence, the two-component Pauli wavefunctions get scrambled on being operated by any of the three $\vec{\gamma}$ Dirac matrices. On the side, we note that 16 linearly independent matrices can be built from Dirac matrices (Table 7.1), which constitute Clifford algebra (see the Solved Problem P7.6).

Table 7.1 Sixteen linearly independent matrices that can be built from Dirac matrices.

$\Gamma^{(S)}$	$(\gamma^{\mu})^2 = (1_{4\times4}):\ \mu = 1,2,3,4$	1 matrix
$\Gamma^{(V)}$	$\gamma^{\mu} = \{\gamma^0,\gamma^1,\gamma^2,\gamma^3\} = \{\gamma^0,\vec{\gamma}\} = \{\beta,\beta\vec{\alpha}\}$	4 matrices
$\Gamma^{(T)}$	$\sigma_{\mu\nu} = i\gamma^{\mu}\gamma^{\nu};\ \mu \neq \nu$	6 matrices
$\Gamma^{(P)}$	$\gamma^5 = i\gamma^0\gamma^1\gamma^2\gamma^3 = \begin{bmatrix} 0_{2\times2} & 1_{2\times2} \\ 1_{2\times2} & 0_{2\times2} \end{bmatrix}$	1 matrix
$\Gamma^{(A)}$	$\gamma^5\gamma^{\mu}\ :\mu = 0,1,2,3$	4 matrixes

We shall now proceed to include the electromagnetic field in the relativistic quantum equation. We shall use the notation in Chapters 12 and 13 from Reference [1] to denote the electromagnetic potential

$$A^{\mu} \equiv \{A^0, A^1, A^2, A^3\} \equiv \{A^0, \vec{A}\} \equiv \{\phi, \vec{A}\}. \tag{7.33}$$

From Eq. 2.66b, we know that the *canonical* momentum $\vec{p}_c$ of a particle having an electric charge e in the presence of A^{μ} is different from its kinetic momentum $\vec{p}_k$:

$$\vec{p}_c = m\vec{v} + \frac{e}{c}\vec{A}(\vec{r},t) = p_k + \frac{e}{c}\vec{A}(\vec{r},t). \tag{7.34a}$$

The kinetic momentum is

$$\vec{\pi} = \vec{p}_k = \vec{p}_c - \frac{e}{c}\vec{A}(\vec{r},t), \tag{7.34b}$$

The four momentum is $\pi^{\mu} = (\pi^0,\vec{\pi}) = (p^0,\vec{p}_k) = \left(\gamma mc, \vec{p}_c - \frac{e}{c}\vec{A}\right)$, (7.35)

and quantization (i.e., employing 'operators') leads us to

$$p^{\mu} = \pi^{\mu} \rightarrow i\hbar\partial^{\mu} - \frac{e}{c}A^{\mu} \text{ for } \mu = 0,1,2,3. \tag{7.36}$$

The Dirac equation therefore becomes

$$\left(i\hbar\gamma^{\kappa}\partial_{\kappa} - \frac{e}{c}\gamma^{\kappa}A_{\kappa} - mc\right)_{4\times4}\psi_{4\times1} = 0_{4\times1}. \tag{7.37}$$

The result of the Solved Problem P7.1 (at the end of the chapter) is

$$i\hbar\frac{\partial}{\partial t}\psi_{4\times1} = \left(c\vec{\alpha}\bullet\vec{\pi} + \beta mc^2 + e\phi\right)_{4\times4}\psi_{4\times1} = H\psi_{4\times1} \tag{7.38}$$

$$\text{i.e., } i\hbar\frac{\partial}{\partial t}\begin{pmatrix}\tilde{\varphi}_{2\times1} \\ \tilde{\chi}_{2\times1}\end{pmatrix} = \left(c\vec{\alpha}\bullet\vec{\pi} + \beta mc^2 + e\phi\right)_{4\times4}\begin{pmatrix}\tilde{\varphi}_{2\times1} \\ \tilde{\chi}_{2\times1}\end{pmatrix}, \tag{7.39a}$$

where we have written the four-component wavefunction as $\psi_{4\times1} = \begin{pmatrix}\tilde{\varphi}_{2\times1} \\ \tilde{\chi}_{2\times1}\end{pmatrix}$. (7.39b)

The representation of the Dirac equation in Eq. 7.38 and in Eq. 7.39a is usually referred to as the *standard representation*. The operator H in Eq. 7.38 is the Dirac Hamiltonian. Simplifying Eq. 7.39, we get

$$i\hbar\frac{\partial}{\partial t}\begin{pmatrix}\tilde{\varphi}_{2\times1}\\ \tilde{\chi}_{2\times1}\end{pmatrix}=\begin{pmatrix} c\vec{\sigma}\bullet\vec{\pi}\begin{bmatrix}0_{2\times2} & 1_{2\times2}\\ 1_{2\times2} & 0_{2\times2}\end{bmatrix}+\begin{bmatrix}1_{2\times2} & 0_{2\times2}\\ 0_{2\times2} & -1_{2\times2}\end{bmatrix}mc^2\\ +\begin{bmatrix}1_{2\times2} & 0_{2\times2}\\ 0_{2\times2} & 1_{2\times2}\end{bmatrix}e\phi\end{pmatrix}_{4\times4}\begin{pmatrix}\tilde{\varphi}_{2\times1}\\ \tilde{\chi}_{2\times1}\end{pmatrix} \tag{7.40}$$

$$\text{or } i\hbar\frac{\partial}{\partial t}\begin{pmatrix}\tilde{\varphi}_{2\times1}\\ \tilde{\chi}_{2\times1}\end{pmatrix}=c\vec{\sigma}_{2\times2}\bullet\vec{\pi}\begin{pmatrix}\tilde{\chi}_{2\times1}\\ \tilde{\varphi}_{2\times1}\end{pmatrix}+mc^2\begin{pmatrix}\tilde{\varphi}_{2\times1}\\ -\tilde{\chi}_{2\times1}\end{pmatrix}+e\phi\begin{pmatrix}\tilde{\varphi}_{2\times1}\\ \tilde{\chi}_{2\times1}\end{pmatrix}. \tag{7.41}$$

We can make an attempt to examine Eq. 7.40 and Eq. 7.41 as a *pair* of two 2×1 matrix equations instead of a single 4×1 matrix equation. These equations would be

$$i\hbar\frac{\partial}{\partial t}\tilde{\varphi}=c\vec{\sigma}\bullet\vec{\pi}\tilde{\chi}+mc^2\tilde{\varphi}+e\phi\tilde{\varphi} \tag{7.42}$$

$$\text{and } i\hbar\frac{\partial}{\partial t}\tilde{\chi}=c\vec{\sigma}\bullet\vec{\pi}\tilde{\varphi}-mc^2\tilde{\chi}+e\phi\tilde{\chi}. \tag{7.43}$$

Now, an electron at rest has an energy $E_0 = mc^2$, which is ~0.511MeV. It is huge compared to energies involved in spectral transitions of an atom. It is therefore instructive to extract the factor $\exp\left(-i\frac{E_0}{\hbar}t\right)$ from $\left(\tilde{\varphi}(t),\tilde{\chi}(t)\right)$:

$$\begin{bmatrix}\tilde{\varphi}(t)\\ \tilde{\chi}(t)\end{bmatrix}=\exp\left(-i\frac{E_0}{\hbar}t\right)\begin{bmatrix}\varphi(t)\\ \chi(t)\end{bmatrix}. \tag{7.44}$$

The functions $\left(\varphi(t),\chi(t)\right)$ that appear on the right-hand side of Eq. 7.44 would also be time dependent, but only mildly so, since *most* of the time dependence is placed in the factor $\exp\left(-i\frac{E_0}{\hbar}t\right)$. Inserting Eq. 7.44 in Eqs. 7.41 gives the pair of equations

$$i\hbar\frac{\partial\varphi}{\partial t}=c\vec{\sigma}\bullet\vec{\pi}\chi+e\phi\varphi \tag{7.45a}$$

$$0\overset{\leftarrow}{\simeq}i\hbar\frac{\partial\chi}{\partial t}\overset{\rightarrow}{=}c\vec{\sigma}\bullet\vec{\pi}\varphi+\left(e\phi-2mc^2\right)\chi\simeq c\vec{\sigma}\bullet\vec{\pi}\varphi-2mc^2\chi. \tag{7.45b}$$

In obtaining Eq. 7.45b, we have made use of the fact that $\frac{\partial\chi}{\partial t}\approx 0$ since $\chi(t)$ is quite nearly independent of time, and also that the term with the coefficient $2mc^2$ is the dominant one. It then follows that

$$\frac{\vec{\sigma}\bullet\vec{\pi}}{2mc}\varphi\approx\chi, \tag{7.45c}$$

allowing an interpretation of χ as the *small* part and φ as the *large* part of the four-component wavefunction in Eq. 7.44. Using Eq. 7.45c in Eq. 7.45a gives a decoupled equation for the large component (using the result from Solved Problem P4.5), often referred to as the *Pauli equation*:

$$i\hbar\frac{\partial\varphi}{\partial t}\simeq\left[\frac{(\vec{\sigma}\bullet\vec{\pi})(\vec{\sigma}\bullet\vec{\pi})}{2m}+e\phi\right]\varphi=\left[\frac{\vec{\pi}\bullet\vec{\pi}+i\vec{\sigma}\bullet\vec{\pi}\times\vec{\pi}}{2m}+e\phi\right]\varphi. \tag{7.46}$$

The approximations involved in such decoupling of the large and the small components forewarn us about the weakness in the attempted separation of the large and the small components. There is a stronger way of achieving this, which is not only more rigorous but also insightful, by carrying out a series of transformations known as the Foldy–Wouthuysen (FW) transformations [5]. Prior to engaging ourselves with the FW transformations, we shall persist with the Pauli equation even if only tentatively.

$$\text{Now, }\frac{\vec{\pi}\bullet\vec{\pi}}{2m}=\frac{\pi^2}{2m}=\frac{\left(\vec{p}_c-\frac{e}{c}\vec{A}\right)\bullet\left(\vec{p}_c-\frac{e}{c}\vec{A}\right)}{2m}=\frac{\vec{p}_c^{\,2}}{2m}+\frac{i\hbar e}{2mc}\left(\vec{\nabla}\bullet\vec{A}+\vec{A}\bullet\vec{\nabla}\right)+\frac{e^2A^2}{2mc^2}. \tag{7.47}$$

Ignoring the quadratic term in the vector potential (please revisit Eq. 2.73 through Eq. 2.75 in Chapter 2, and refer to the discussion in Chapter 6, immediately after Eq. 6. 97),

$$\frac{\vec{\pi}\bullet\vec{\pi}}{2m}=\frac{\vec{p}_c^{\,2}}{2m}+\frac{i\hbar}{2m}2\vec{A}\bullet\vec{\nabla}=\frac{\vec{p}_c^{\,2}}{2m}+\frac{i\hbar e}{mc}\left(\frac{1}{2}\vec{B}\times\vec{r}\right)\bullet\vec{\nabla}=\frac{\vec{p}_c^{\,2}}{2m}-\frac{e}{mc}\left(\frac{1}{2}\vec{B}\bullet\vec{r}\times\left(-i\hbar\vec{\nabla}\right)\right), \tag{7.48a}$$

$$\text{or, }\frac{\vec{\pi}\bullet\vec{\pi}}{2m}=\frac{\vec{p}_c^{\,2}}{2m}-\frac{e}{2mc}\left(\vec{B}\bullet\vec{r}\times\vec{p}\right)=\frac{\vec{p}_c^{\,2}}{2m}-\frac{e}{2mc}\left(\vec{\ell}\bullet\vec{B}\right). \tag{7.48b}$$

Furthermore, for an arbitrary function $f(\vec{r})$

$$(\vec{\pi}\times\vec{\pi})f(\vec{r})=\left(\vec{p}_c-\frac{e}{c}\vec{A}\right)\times\left(\vec{p}_c-\frac{e}{c}\vec{A}\right)f(\vec{r})=\left(i\hbar\frac{e}{c}\right)\begin{bmatrix}\left(\vec{\nabla}\times\vec{A}\right)f(\vec{r})\\+\left(\vec{A}\times\vec{\nabla}\right)f(\vec{r})\end{bmatrix}, \tag{7.49a}$$

$$\text{i.e., }(\vec{\pi}\times\vec{\pi})f(\vec{r})=\left(i\hbar\frac{e}{c}\right)\left[\vec{\nabla}\times\left\{\vec{A}f(\vec{r})\right\}+\left(\vec{A}\times\vec{\nabla}f(\vec{r})\right)\right], \tag{7.49b}$$

therefore,

$$\frac{i}{2m}\vec{\sigma}\bullet(\vec{\pi}\times\vec{\pi})f(\vec{r})=\left(i\hbar\frac{e}{c}\right)\frac{i}{2m}\vec{\sigma}\bullet\begin{bmatrix}\left(\vec{\nabla}\times\vec{A}\right)f\\+\vec{\nabla}f\times\cancel{\vec{A}+\vec{A}\times\vec{\nabla}f}\end{bmatrix}=-\frac{e\hbar}{2mc}\vec{\sigma}\bullet\vec{B}(\vec{r})f(\vec{r}). \tag{7.50}$$

Using Eq. 7.48b and Eq. 7.50 in Eq. 7.44, we get

$$i\hbar\frac{\partial\varphi}{\partial t}=\begin{bmatrix}\frac{\vec{p}_c^{\,2}}{2m}-\frac{e}{2mc}\vec{\ell}\bullet\vec{B}\\-\frac{e\hbar}{2mc}\frac{2\vec{s}}{\hbar}\bullet\vec{B}(\vec{r})+e\phi\end{bmatrix}\varphi=\begin{bmatrix}\frac{\vec{p}_c^{\,2}}{2m}-\frac{e}{2mc}\vec{\ell}\bullet\vec{B}\\+\frac{e}{mc}\vec{s}\bullet\vec{B}+e\phi\end{bmatrix}\varphi. \tag{7.51}$$

Now, we know from the Solved Problem P4.3 that

$$\vec{s} = \frac{\hbar}{2}\vec{\sigma}, \tag{7.52}$$

and we can therefore conclude that $i\hbar\frac{\partial}{\partial t}\varphi = \left[\frac{p_c^{\,2}}{2m} - \frac{e}{2mc}\left(\vec{\ell} + 2\vec{s}\right)\cdot\vec{B} + e\phi\right]\varphi.$ (7.53)

We now define the orbital magnetic moment as $\vec{\mu}_\ell = -\mu_B\frac{g_\ell\vec{\ell}}{\hbar}$ (see Fig. 7.2), (7.54a)

with $\mu_B = \frac{|e|\hbar}{2mc}$ (called as the *Bohr Magneton*) (7.54b)

where $|e|$ denotes the magnitude of the electron charge

and $g_\ell = 1.$ (7.54c)

Also, we define the spin magnetic moment as $\vec{\mu}_s = -\mu_B\frac{g_s\vec{s}}{\hbar}$, (7.54d)

with $g_s = 2.$ (7.54e)

The ratio $\frac{g_s}{g_\ell} = 2$ (7.55)

is called the gyromagnetic ratio. The total magnetic moment is

$$\vec{\mu} = -\mu_B\frac{g_\ell\vec{\ell} + g_s\vec{s}}{\hbar} = -\mu_B\frac{\vec{\ell} + 2\vec{s}}{\hbar} = \frac{e}{2mc}\left(\vec{\ell} + 2\vec{s}\right). \tag{7.56}$$

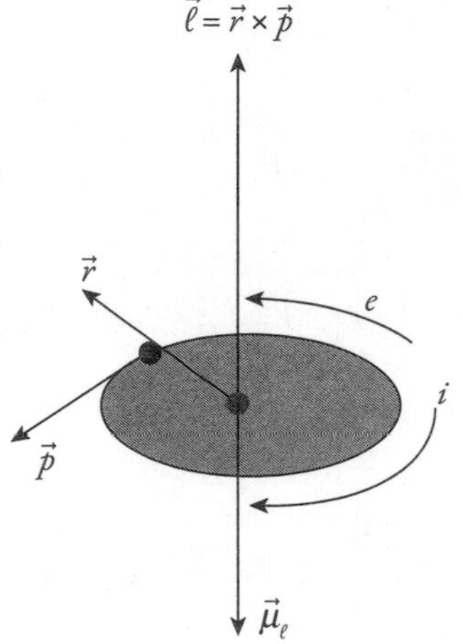

Fig. 7.2 This is a semi-classical model of an electron circling a proton in the hydrogen atom. The electron's instantaneous position with respect to the nucleus is $\vec{r}$, instantaneous kinetic momentum is $\vec{p}$, and the orbital angular momentum is $\vec{\ell}$. This motion can be likened to an electric current $i = \frac{e}{T} = e\frac{v}{2\pi r}$, where e is the electron's charge, v its speed, and T the time period for one full rotation. Associated with this current is then a magnetic dipole moment $\vec{\mu}_\ell = \frac{1}{c}iA\hat{A} = \frac{1}{c}\left(e\frac{v}{2\pi r}\right)\left(\pi r^2\right)\hat{A} = \frac{1}{c}\frac{evr}{2}\hat{A}$, i.e., $\vec{\mu}_\ell = \frac{e}{2c}\vec{r}\times\vec{v} = \frac{e\hbar}{2mc}\frac{\vec{\ell}}{\hbar}$.

and the Eq. 7.53 now becomes

$$i\hbar\frac{\partial\varphi}{\partial t} = \left[\frac{p_c^{\ 2}}{2m} - \vec{\mu}\bullet\vec{B} + e\phi\right]\varphi. \tag{7.57}$$

Funquest: Argue if modeling an electron to be revolving around the nucleus of an atom provides (or not!) a satisfactory idea of the quantum orbital angular momentum of the electron, and likewise of the idea of a particle rotating about its own axis accounts for its quantum spin angular momentum.

The electron spin magnetic moment was introduced in the early days of quantum theory prior to the relativistic formulation in an *ad hoc* manner in semi-empirical models aimed at the interpretation of the interaction of an atom with an applied magnetic field. The semi-empirical analysis was successful to a fair extent, but deficiencies in it had to be addressed. Dirac's relativistic theory does just that. The two-component Pauli equation (Eq. 7.57) provides for the two states of an electron observed in a Stern–Gerlach experiment (see Chapter 1). However, to determine expectation values to order $\frac{v^2}{c^2}$, all the four components in $\psi_{4\times 1}$ (Eqs. 7.38a,b) are required. Dirac equation describes both electrons (positive energy states) and the positrons ("negative energy" states). The solution has four components. The matrix vector operator $\vec{\alpha}$ has nonzero elements in the off-diagonal blocks. When the matrix operation on the four-component wavefunction is carried out, $\tilde{\varphi}_{2\times 1}$ and $\tilde{\chi}_{2\times 1}$ get superposed and the terminology "small" and "large" becomes obstructive. Operators that have nonzero elements in the off-diagonal positions mix the components of the wavefunction in the first two rows with those in the next two rows of the four-component wavefunction. Such operators are called as "odd" operators. It is the odd operators that make the Pauli two-component equation (Eq. 7.57) unsatisfactory. A systematic method to diminish the role of the odd operators is discussed in the next section.

7.2 Foldy–Wouthuysen Transformations

We now depart from the procedure laid out in the previous section which led us to the Pauli equation (Eq. 7.46). Instead, we now employ a systematic method, developed by several contributors including Dirac, Pryce, Newton, and Wigner, and by Foldy and Wouthuysen [7,8,9]. The FW transformation (FWT) method employs a series of transformations (brought about using operators S_i, $i = 1, 2, 3, \ldots$) of Dirac's relativistic quantum equation, Eq. 7.38:

$$\left[H\psi = i\hbar\frac{\partial\psi}{\partial t}\right] \underset{S_1}{\overset{FWT\ 1}{\rightarrow}} \left[H'\psi' = i\hbar\frac{\partial\psi'}{\partial t}\right] \underset{S_2}{\overset{FWT\ 2}{\rightarrow}} \\ \underset{S_2}{\overset{FWT\ 2}{\rightarrow}} \left[H''\psi'' = i\hbar\frac{\partial\psi''}{\partial t}\right] \underset{S_3}{\overset{FWT\ 3}{\rightarrow}} \left[H'''\psi''' = i\hbar\frac{\partial\psi'''}{\partial t}\right], \tag{7.58}$$

which systematically seek to diminish the role of the odd operators. The essential motivation for these transformations is to write the transformed Dirac equation in a form that is *relatively* free from odd operators.

The first FWT is effected by employing an operator S such that the wave function ψ is transformed to ψ'

$$\psi \to \psi' = e^{iS}\psi, \tag{7.59}$$

$$\text{correspondingly, } \psi = e^{-iS}\psi'. \tag{7.60}$$

In general, S may be time-dependent and does not commute with the time-derivative operator $\frac{\partial}{\partial t}$. The FWT is strategized to *choose* $S_i; i = 1,2,3$ such that the role of the "odd" operators is scaled down, by order $O\left(\frac{1}{mc^2}\right)$. Since the rest–mass–energy mc^2 is large (~0.511 MeV for an electron), this transformation is very gainful. When FWT is effected thrice, the role of the odd operators is diminished by $O\left(\frac{1}{m^3c^6}\right)$ resulting in a far more satisfactory theory than the approximation that led us to the two-component Pauli equation.

From Eq. 7.59, we see that

$$H\left(e^{-iS}\psi'\right) = H\psi = i\hbar\frac{\partial}{\partial t}\psi = \left(i\hbar\frac{\partial}{\partial t}\right)\left(e^{-iS}\psi'\right) = \left(i\hbar\frac{\partial}{\partial t}e^{-iS}\right)\psi' + \left(e^{-iS}\right)\left(i\hbar\frac{\partial\psi'}{\partial t}\right).$$

$$\text{Hence, } e^{+iS}He^{-iS}\psi' = i\hbar e^{+iS}\frac{\partial}{\partial t}e^{-iS}\psi' + i\hbar\frac{\partial}{\partial t}\psi',$$

$$\text{i.e., } \left[e^{+iS}He^{-iS} - i\hbar e^{+iS}\frac{\partial}{\partial t}e^{-iS}\right]\psi' = i\hbar\frac{\partial\psi'}{\partial t} = H'\psi'. \tag{7.61}$$

In other words, the Hamiltonian H transforms as

$$H \to H' = \left(e^{+iS}He^{-iS}\right) - \left(i\hbar e^{+iS}\frac{\partial}{\partial t}e^{-iS}\right). \tag{7.62}$$

To obtain the *Transformed Dirac–Hamiltonian* (TDH) H_1, where the subscript '1' denotes the first FWT brought about using the operator S (rather S_1), we first examine the transformation of an operator

$$\Omega \to \Omega' = e^{+iS}\Omega e^{-iS} = \lim_{\xi\to 1} e^{+i\xi S}\Omega e^{-i\xi S} = \lim_{\xi\to 1} F(\xi), \tag{7.63a}$$

in order to study the *first* of the two terms on the right hand side of Eq. 7.62.

$$\text{with } F(\xi) = e^{+i\xi S}\Omega e^{-i\xi S}. \tag{7.63b}$$

Using Eqs. 7.63a,b in Eq. 7.62, we get, on using the result of the Solved Problem P7.2, for the *first* TDH the following expression:

$$H' = \begin{bmatrix} H + i[S,H] + \frac{i^2}{2!}[S,[S,H]] + \\ \frac{i^3}{3!}[S,[S,[S,H]]] + \frac{i^4}{4!}[S,[S,[S,[S,H]]]] + \ldots \end{bmatrix} + e^{+iS}\left(-i\hbar\frac{\partial}{\partial t}\right)e^{-iS}. \tag{7.64}$$

The term $e^{+iS}\left(-i\hbar\frac{\partial}{\partial t}\right)e^{-iS}$ in the above expression also needs to be determined carefully, since $\frac{\partial S}{\partial t}$ does not commute with S. Expanding both the operators involving the exponential of $+iS$ and $-iS$ carefully, it results in an infinite series consisting of nested commutators:

$$\text{i.e., } H' = \begin{bmatrix} H + i[S,H] + \frac{i^2}{2!}[S,[S,H]] + \\ \frac{i^3}{3!}[S,[S,[S,H]]] + \\ \frac{i^4}{4!}[S,[S,[S,[S,H]]]] + \ldots \end{bmatrix} - \hbar\dot{S} - \frac{i\hbar}{2}[S,\dot{S}] + \frac{\hbar}{6}[S,[S,\dot{S}]] + \ldots \quad (7.65)$$

We rewrite the original Dirac Hamiltonian (Eq. 7.38) as

$$H = \beta mc^2 + c\vec{\alpha}\cdot\left(\vec{p} - \frac{e}{c}\vec{A}\right) + e\phi = \beta e_0 + \theta + \varepsilon, \quad (7.66a)$$

with the rest–mass–energy denoted by $e_0 = mc^2$, (7.66b)

the odd operator represented by $\theta = c\vec{\alpha}\cdot\left(\vec{p} - \frac{e}{c}\vec{A}\right)$, (7.66c)

and the interaction with the scalar potential represented by

$$\varepsilon = e\phi. \quad (7.66d)$$

We now choose the FWT operator S to be of the order $\bigcirc\left(\frac{1}{e_0}\right)$. Considering the fact that the Hamiltonian has a term in βe_0, it would only be the fifth term $\left[S,\left[S,\left[S,[S,H]\right]\right]\right]$ in Eq. 7.65 that would be of $\bigcirc\left(\frac{1}{e_0^{\,3}}\right)$. To retain terms of $\bigcirc\left(\frac{1}{e_0^{\,3}}\right)$, we may approximate $H \simeq \beta e_0$ in this *fifth* term, but in all the previous terms we must employ the full Hamiltonian (Eq. 7.66a). Truncating both the infinite series in Eq. 7.65 to $\bigcirc\left(\frac{1}{e_0^{\,3}}\right)$, we get the *first* TDH as

$$H' \simeq H + i[S,H] + \frac{i^2}{2!}[S,[S,H]] + \frac{i^3}{3!}[S,[S,[S,H]]] + \frac{i^4}{4!}\left[S,\left[S,\left[S,\left[S,\beta mc^2\right]\right]\right]\right] - \hbar\dot{S} - \frac{i\hbar}{2}[S,\dot{S}] + \frac{\hbar}{6}[S,[S,\dot{S}]]. \quad (7.67)$$

For the first FWT, the choice for the transformation operator is

$$S_1 = \frac{-i\beta\theta}{2e_0} = \frac{-i\beta\theta}{2mc^2} = \frac{-i\begin{bmatrix} 1_{2\times2} & 0_{2\times2} \\ 0_{2\times2} & -1_{2\times2} \end{bmatrix} c \begin{bmatrix} 0_{2\times2} & \vec{\sigma}_{2\times2} \\ \vec{\sigma}_{2\times2} & 0_{2\times2} \end{bmatrix} \bullet \left(\vec{p} - \frac{e}{c}\vec{A}\right)}{2mc^2}. \quad (7.68)$$

To determine H', we need each of the terms on the right-hand side of Eq. 7.67. The first term in H' is the full Dirac Hamiltonian, also employed in the second and the third terms, which are dealt with in the Solved Problem P7.3. The fourth term $\frac{i^3}{3!}\left[S,\left[S,\left[S,H\right]\right]\right]$ and the fifth term $\frac{i^4}{4!}\left[S,\left[S,\left[S,\left[S,\beta mc^2\right]\right]\right]\right]$ are obtained similarly.

We also need $\hbar\dot{S}$, $-\frac{i\hbar}{2}\left[S,\dot{S}\right]$ and $+\frac{\hbar}{6}\left[S,\left[S,\dot{S}\right]\right]$. For example,

$$\frac{-i\hbar}{2}\left[S,\dot{S}\right]_- = \frac{-i\hbar}{2}\left[S,\left(\frac{-i\beta\dot{\theta}}{2mc^2}\right)\right]_- = \frac{-i\hbar}{8m^2c^4}\left[\theta,\dot{\theta}\right]_-.$$

Compiling all the terms, we get $H' = \beta mc^2 + \varepsilon' + \theta'$, (7.69)

$$\text{where } \varepsilon' = \varepsilon + \frac{\beta\theta^2}{2mc^2} - \frac{\beta\theta^4}{8m^3c^6} - \frac{1}{8m^2c^4}\left[\theta,\left[\theta,\varepsilon\right]\right] - \frac{i\hbar}{8m^2c^4}\left[\theta,\dot{\theta}\right] \tag{7.70a}$$

$$\text{and } \theta' = \frac{\beta}{2mc^2}\left[\theta,\varepsilon\right] - \frac{\theta^3}{3m^2c^4} + \frac{i\hbar\beta\dot{\theta}}{2mc^2}. \tag{7.70b}$$

The leading "odd" term is now of the order $\bigcirc\left(\frac{1}{e_0}\right)$, where $e_0 = mc^2$ is the rest–mass–energy.

The second FWT can now be implemented. This would give the *second* TDH:

$$H'' = e^{+iS_2}H'e^{-iS_2} + e^{+iS_2}\left(-i\hbar\frac{\partial}{\partial t}\right)e^{-iS_2}. \tag{7.71}$$

The second transformation is effected using $S_2 = \frac{-i\beta\theta'}{2mc^2}$, (7.72)

which yields the second TDH

$$H'' = \beta mc^2 + \varepsilon' + \theta''. \tag{7.73}$$

A third FWT is now carried out to obtain the third TDH:

$$H''' = e^{+iS_3}H''e^{-iS_3} + e^{+iS_3}\left(-i\hbar\frac{\partial}{\partial t}\right)e^{-iS_3}, \tag{7.74}$$

using $S_3 = \frac{-i\beta\theta''}{2mc^2}$. (7.75)

The transformation is a bit tedious, but is implemented using essentially the same methods used above. The third TDH turns out to be

$$H''' = \beta mc^2 + \varepsilon'' \tag{7.76a}$$

$$\text{i.e., } H''' = \beta\left(mc^2 + \frac{\theta^2}{2mc^2} - \frac{\theta^4}{8m^3c^6}\right) + \varepsilon - \frac{1}{8m^2c^4}\left[\theta,\left[\theta,\varepsilon\right]\right] - \frac{i\hbar}{8m^2c^4}\left[\theta,\dot{\theta}\right]. \tag{7.76b}$$

To obtain the third TDH, we must obtain

$$\frac{\theta^2}{2mc^2} = \frac{\left(c\vec{\alpha}\bullet\left(\vec{p}-\frac{e}{c}\vec{A}\right)\right)^2}{2mc^2} = \frac{1}{2m}\left\{\vec{\alpha}\bullet\left(\vec{p}-\frac{e}{c}\vec{A}\right)\right\}\left\{\vec{\alpha}\bullet\left(\vec{p}-\frac{e}{c}\vec{A}\right)\right\} = \frac{1}{2m}\left\{\begin{pmatrix}\vec{0} & \vec{\sigma}\\ \vec{\sigma} & \vec{0}\end{pmatrix}\bullet\vec{\pi}\right\}\left\{\begin{pmatrix}\vec{0} & \vec{\sigma}\\ \vec{\sigma} & \vec{0}\end{pmatrix}\bullet\vec{\pi}\right\},$$

i.e.,

$$\frac{\theta^2}{2mc^2} = \frac{1}{2m}\begin{pmatrix} \vec{0} & \vec{\sigma}\cdot\vec{\pi} \\ \vec{\sigma}\cdot\vec{\pi} & \vec{0} \end{pmatrix}\begin{pmatrix} \vec{0} & \vec{\sigma}\cdot\vec{\pi} \\ \vec{\sigma}\cdot\vec{\pi} & \vec{0} \end{pmatrix} = \frac{1}{2m}\begin{pmatrix} \vec{\sigma}\cdot\vec{\pi}\,\vec{\sigma}\cdot\vec{\pi} & 0 \\ 0 & \vec{\sigma}\cdot\vec{\pi}\,\vec{\sigma}\cdot\vec{\pi} \end{pmatrix} = \frac{1}{2m}\vec{\sigma}\cdot\vec{\pi}\,\vec{\sigma}\cdot\vec{\pi}\,1_{4\times4}$$

$$\text{or, } \frac{\theta^2}{2mc^2} = \frac{1}{2m}\left\{\vec{\pi}^2 + i\vec{\sigma}\cdot\vec{\pi}\times\vec{\pi}\right\}1_{4\times4} = \frac{1}{2m}\left\{\vec{\pi}^2 - e\vec{\sigma}\cdot\vec{B}\right\}1_{4\times4}.$$

The above result is on account of the fact that

$$\vec{\pi}\times\vec{\pi} = \left(\vec{p} - \frac{e}{c}\vec{A}\right)\times\left(\vec{p} - \frac{e}{c}\vec{A}\right) = \vec{p}\times\vec{p} - \frac{e}{c}\left(\vec{p}\times\vec{A} + \vec{A}\times\vec{p}\right) + \frac{e^2}{c^2}\left(\vec{A}\times\vec{A}\right),$$

$$\text{i.e., } \vec{\pi}\times\vec{\pi} = \frac{ie\hbar}{c}\left(\vec{\nabla}\times\vec{A} + \vec{A}\times\vec{\nabla}\right) = \frac{ie\hbar}{c}\vec{B}$$

since, for an arbitrary function $f(\vec{r})$,

$$\left(\vec{\nabla}\times\vec{A} + \vec{A}\times\vec{\nabla}\right)f = \vec{\nabla}\times\left(\vec{A}f\right) + \left(\vec{A}\times\vec{\nabla}f\right) = \left(\vec{\nabla}\times\vec{A}\right)f \cancel{-\left(\vec{A}\times\vec{\nabla}f\right) + \left(\vec{A}\times\vec{\nabla}f\right)} = \vec{B}f.$$

$$\text{Thus, } \frac{\theta^2}{2mc^2} = \frac{\left(\vec{p} - \frac{e}{c}\vec{A}\right)^2}{2m} - \frac{e\hbar}{2mc}\vec{\sigma}\cdot\vec{B}. \tag{7.77a}$$

We also require

$$[\theta, \varepsilon] + i\hbar\dot{\theta} = c\left[\vec{\alpha}\cdot\left(\vec{p} - e\vec{A}\right), e\phi\right] + i\hbar c\frac{\partial}{\partial t}\left(\vec{\alpha}\cdot\vec{\pi}\right) = -ie\hbar c\left[\vec{\alpha}\cdot\vec{\nabla}, \phi\right] + i\hbar c\frac{\partial}{\partial t}\left(\vec{\alpha}\cdot\vec{\pi}\right).$$

We can simplify the first term on the right-hand side by recognizing that

$$\underline{\left[\vec{\alpha}\cdot\vec{\nabla}, \phi\right]f} = \left(\vec{\alpha}\cdot\vec{\nabla}\phi - \phi\vec{\alpha}\cdot\vec{\nabla}\right)f = \vec{\alpha}\cdot\vec{\nabla}(\phi f) - \phi\vec{\alpha}\cdot\vec{\nabla}f = \left(\vec{\alpha}\cdot\vec{\nabla}\phi\right)f + \cancel{\left(\vec{\alpha}\cdot\vec{\nabla}f\right)\phi} - \cancel{\phi\vec{\alpha}\cdot\vec{\nabla}f}$$
$$= \left(\vec{\alpha}\cdot\vec{\nabla}\phi\right)f.$$

$$\text{Therefore } [\theta, \varepsilon] + i\hbar\dot{\theta} = -ie\hbar c\vec{\alpha}\cdot\vec{\nabla}\phi + i\hbar c\frac{\partial}{\partial t}\left(\vec{\alpha}\cdot\vec{\pi}\right) = -ie\hbar c\vec{\alpha}\cdot\vec{\nabla}\phi + i\hbar c\frac{\partial}{\partial t}\left(\vec{\alpha}\cdot\left(\vec{p} - \frac{e}{c}\vec{A}\right)\right),$$

$$\text{i.e., } [\theta, \varepsilon] + i\hbar\dot{\theta} = -ie\hbar c\vec{\alpha}\cdot\vec{\nabla}\phi - ie\hbar\left(\vec{\alpha}\cdot\frac{\partial\vec{A}}{\partial t}\right) = i\hbar ce\vec{\alpha}\cdot\vec{E}, \tag{7.77b}$$

since $\vec{E} + \frac{1}{c}\frac{\partial\vec{A}}{\partial t} = -\vec{\nabla}\phi$.

Using Eq. 7.77b, we get

$$\frac{1}{8m^2c^4}\left[\theta, [\theta, \varepsilon] + i\hbar\dot{\theta}\right] = \frac{i\hbar ce}{8m^2c^4}\left[\theta, \vec{\alpha}\cdot\vec{E}\right],$$

i.e., $$\frac{1}{8m^2c^4}\left[\theta,[\theta,\varepsilon]+i\hbar\dot{\theta}\right]=\frac{ie\hbar}{8m^2c^2}\left[\vec{\alpha}\bullet\vec{\pi},\vec{\alpha}\bullet\vec{E}\right]=\frac{ie\hbar}{8m^2c^2}\left(\vec{\alpha}\bullet\vec{\pi}\ \vec{\alpha}\bullet\vec{E}-\vec{\alpha}\bullet\vec{E}\ \vec{\alpha}\bullet\vec{\pi}\right).$$

Hence,

$$\frac{1}{8m^2c^4}\left[\theta,[\theta,\varepsilon]+i\hbar\dot{\theta}\right]=\frac{ie\hbar}{8m^2c^2}\left\{\begin{pmatrix}\vec{0} & \vec{\sigma}\bullet\vec{\pi}\\ \vec{\sigma}\bullet\vec{\pi} & \vec{0}\end{pmatrix}\begin{pmatrix}\vec{0} & \vec{\sigma}\bullet\vec{E}\\ \vec{\sigma}\bullet\vec{E} & \vec{0}\end{pmatrix}-\begin{pmatrix}\vec{0} & \vec{\sigma}\bullet\vec{E}\\ \vec{\sigma}\bullet\vec{E} & \vec{0}\end{pmatrix}\begin{pmatrix}\vec{0} & \vec{\sigma}\bullet\vec{\pi}\\ \vec{\sigma}\bullet\vec{\pi} & \vec{0}\end{pmatrix}\right\},$$

i.e., $$\frac{1}{8m^2c^4}\left[\theta,[\theta,\varepsilon]+i\hbar\dot{\theta}\right]=\frac{ie\hbar}{8m^2c^2}\left\{\begin{pmatrix}\vec{\sigma}\bullet\vec{\pi}\ \vec{\sigma}\bullet\vec{E} & 0\\ 0 & \vec{\sigma}\bullet\vec{\pi}\ \vec{\sigma}\bullet\vec{E}\end{pmatrix}-\begin{pmatrix}\vec{\sigma}\bullet\vec{E}\ \vec{\sigma}\bullet\vec{\pi} & 0\\ 0 & \vec{\sigma}\bullet\vec{E}\ \vec{\sigma}\bullet\vec{\pi}\end{pmatrix}\right\}.$$

Therefore, $$\frac{1}{8m^2c^4}\left[\theta,[\theta,\varepsilon]+i\hbar\dot{\theta}\right]=\frac{ie\hbar}{8m^2c^2}\left(\vec{\sigma}\bullet\vec{\pi}\ \vec{\sigma}\bullet\vec{E}-\vec{\sigma}\bullet\vec{E}\ \vec{\sigma}\bullet\vec{\pi}\right)1_{4\times4},$$

or, $$\frac{1}{8m^2c^4}\left[\theta,[\theta,\varepsilon]+i\hbar\dot{\theta}\right]=\frac{ie\hbar}{8m^2c^2}\left(\vec{\pi}\bullet\vec{E}+i\vec{\sigma}\bullet\vec{\pi}\times\vec{E}-\vec{E}\bullet\vec{\pi}-i\vec{\sigma}\bullet\vec{E}\times\vec{\pi}\ \right)1_{4\times4}.$$

Furthermore,

$$\vec{\pi}\cdot\vec{E}-\vec{E}\cdot\vec{\pi}=\left(\vec{p}-\frac{e}{c}\vec{A}\right)\cdot\vec{E}-\vec{E}\cdot\left(\vec{p}-\frac{e}{c}\vec{A}\right)=\vec{p}\cdot\vec{E}-\vec{E}\cdot\vec{p},$$

and $$\vec{p}\cdot\vec{E}f=-i\hbar\vec{\nabla}\cdot\left(\vec{E}f\right)=-i\hbar\left(f\vec{\nabla}\cdot\vec{E}+\vec{E}\cdot\vec{\nabla}f\right)=f\vec{p}\cdot\vec{E}+\vec{E}\cdot\vec{p}f,$$

we therefore have

$$\vec{\pi}\cdot\vec{E}-\vec{E}\cdot\vec{\pi}=\vec{p}\cdot\vec{E}.$$

Also,

$$\vec{\pi}\times\vec{E}-\vec{E}\times\vec{\pi}=\left(\vec{p}-\frac{e}{c}\vec{A}\right)\times\vec{E}-\vec{E}\times\left(\vec{p}-\frac{e}{c}\vec{A}\right)=\vec{p}\times\vec{E}-\vec{E}\times\vec{p}-\frac{e}{c}\left(\vec{A}\times\vec{E}-\vec{E}\times\vec{A}\right),$$

and

$$\vec{p}\times\vec{E}f=-i\hbar\vec{\nabla}\times\left(\vec{E}f\right)=-i\hbar\left(f\vec{\nabla}\times\vec{E}-\vec{E}\times\vec{\nabla}f\right)=f\vec{p}\times\vec{E}-\vec{E}\times\vec{p}f,$$

from which it follows that

$$\vec{\pi}\times\vec{E}-\vec{E}\times\vec{\pi}=\vec{p}\times\vec{E}-2\vec{E}\times\vec{p}+2\frac{e}{c}\vec{E}\times\vec{A}=\vec{p}\times\vec{E}-2\vec{E}\times\vec{\pi}.$$

Therefore,

$$\frac{1}{8m^2c^4}\left[\theta,[\theta,\varepsilon]+i\hbar\dot{\theta}\right]=\frac{e\hbar^2}{8m^2c^2}\left(div\ \vec{E}\right)+\frac{ie\hbar^2}{8m^2c^2}\vec{\sigma}\bullet\text{curl}\ \vec{E}+\frac{e\hbar}{4m^2c^2}\vec{\sigma}\bullet\vec{E}\times\vec{\pi}. \tag{7.77c}$$

Compiling all the terms, we get the *third* TDH:

$$H''' = \left[\begin{array}{l} \beta\left(mc^2 + \dfrac{\left(\vec{p} - \frac{e}{c}\vec{A}\right)^2}{2m} - \dfrac{\vec{p}^4}{8m^3c^2}\right) + e\phi - \beta\dfrac{e\hbar}{2mc}\vec{\sigma}\bullet\vec{B} - \\ \dfrac{e\hbar^2}{8m^2c^2}\left(div\vec{E}\right) - \dfrac{ie\hbar^2}{8m^2c^2}\vec{\sigma}\bullet\text{curl }\vec{E} - \dfrac{e\hbar}{4m^2c^2}\vec{\sigma}\bullet\vec{E}\times\vec{\pi} \end{array}\right]. \tag{7.78}$$

We have obtained the above TDH using *three* FW transformations of the Dirac equation (Eq. 7.66a). The foundation of this method is the fully relativistic quantum equation that is Lorentz covariant. We know that considering the very high speed of light, nonrelativistic mechanics provides a fairly good approximation to the relativistic equation of motion. It is therefore not surprising that one expects the *dominant* part of Eq. 7.78 to be the nonrelativistic Hamiltonian

$$H_{NR} = \frac{\left(\vec{p} - \frac{e}{c}\vec{A}\right)^2}{2m} + e\phi. \tag{7.79}$$

The extra term $\dfrac{\vec{p}^4}{8m^3c^2}$ in Eq. 7.78 can be recognized as the correction due to the relativistic kinetic energy (RKE), given by

$$T_{K.E.}^{Rel} = E - mc^2 = \left(p^2c^2 + m^2c^4\right)^{\frac{1}{2}} - mc^2 = \left(m^2c^4\left(\frac{p^2c^2}{m^2c^4} + 1\right)\right)^{\frac{1}{2}} - mc^2,$$

$$\text{i.e., } T_{K.E.}^{Rel} = mc^2\left[\left(\frac{p^2}{m^2c^2} + 1\right)^{\frac{1}{2}} - 1\right] - mc^2\left[1 + \frac{1}{2}\left(\frac{p}{mc}\right)^2 - \frac{1}{8}\left(\frac{p}{mc}\right)^4 + .. - 1\right],$$

$$\text{or, } T_{K.E.}^{Rel} = \frac{p^2}{2m} - \frac{p^4}{8m^3c^2} + \ldots.. \tag{7.80}$$

The term mc^2 in Eq. 7.78 is only an additive constant term coming from the rest–mass–energy, and curl $\vec{E}$ in the last-but-one term is zero for the central field in the hydrogen atom. It is then natural to ask if the remaining terms in the TDH we obtained after three FW transformations can be treated using perturbation theory (Chapter 6) with H_{NR} treated as the unperturbed Hamiltonian, and other terms as perturbations. We shall learn in Chapter 8 that spectroscopic studies of atoms in the presence a magnetic field by George Eugene Uhlenbeck and Samuel Goudsmit required for their interpretation an *additional* magnetic moment possessed by an electron. This led to a *semi-empirical two-component* theory, *aka* Pauli equation, in which the single-component wavefunction is replaced by a two-component wavefunction (Eq. 7.45). The 4×4 matrix β has no role in the two-component theory that was employed historically as a compromise before the development of the relativistic theory by Dirac.

Ignoring therefore β in Eq. 7.78, amid various parallel developments in the theory of quantum mechanics, perturbative corrections to the nonrelativistic Schrödinger equation were attempted. The two-component Pauli theory involved the introduction of electron spin angular momentum $\vec{s} = \frac{\hbar}{2}\vec{\sigma}$ (Eq. 4.43 to Eq. 4.46). Its interaction with the magnetic field was treated perturbatively as magnetic-dipole (md) interaction, $\frac{e\hbar}{2mc}\vec{\sigma}\bullet\vec{B}$, which appears in Eq. 7.78. Different physicists introduced one or more *piecewise* perturbations. We skip the historical record and merely indicate the individual corrections as were attempted using perturbation theory. We *rewrite* the Hamiltonian in Eq. 7.78 as

$$H = \left[\begin{array}{l} \beta\left(\underbrace{mc^2}_{\substack{rest\\mass\\energy}} + \frac{\left(\vec{p} - \frac{e}{c}\vec{A}\right)^2}{2m} - \underbrace{\left(\frac{\vec{p}^4}{8m^3c^2}\right)}_{RKE}\right) + e\phi - \\ \underbrace{\beta\frac{e\hbar}{2mc}\vec{\sigma}\bullet\vec{B}}_{md} - \underbrace{\left(\frac{e\hbar^2}{8m^2c^2}\left(div\vec{E}\right)\right)}_{D} - \left(\underbrace{\frac{ie\hbar^2}{8m^2c^2}\vec{\sigma}\bullet\text{curl }\vec{E}}_{zero} + \underbrace{\frac{e\hbar}{4m^2c^2}\vec{\sigma}\bullet\vec{E}\times\vec{\pi}}_{so}\right) \end{array}\right]. \tag{7.81}$$

Funquest: Is it not awesome that the RKE term, the Darwin term, the spin-orbit coupling etc. have emerged neatly as a result of the FWT? Is there any need left to model the electron 'spin' wrongly as a particle rotating about an axis going through it?

The relativistic kinetic energy (RKE) correction can be treated as the $\frac{\vec{p}^4}{8m^3c^2}$ perturbation. A perturbative correction involving the divergence of the electric field was introduced by Darwin. It corresponds to the term $\frac{e\hbar^2}{8m^2c^2}\left(div\vec{E}\right)$, labeled by D in Eq. 7.81. It is also called as the *zitterbewegung* term. As a result of zitterbewegung, the electron cannot be localized; this raises the energy of the particle states. The zitterbewegung is caused by the interference between the positive and the negative energy components of the wavefunction. It *would vanish if the wave packet is a superposition of only positive or only negative energy solutions.* "Zitterbewegung" is eliminated when you take expectation values for wave packets made up completely of positive energy states (or completely negative energy states), as achieved by the FW transformation. The negative energy states correspond to positron states, and zitterbewegung is interpreted as a result of the spontaneous creation and destruction of electron–positron pairs. The FW transformation allows us to seek a decoupling between positive and negative energy states and find in a limiting process a description in terms of particles and antiparticles [9]. In the last term $\left(\frac{e\hbar}{4m^2c^2}\vec{\sigma}\bullet\vec{E}\times\vec{\pi}\right)$ in Eq. 7.81, $\vec{\pi}$ can be replaced by $\vec{p}$ since the term involving the vector potential would be weaker in magnitude by a *further* factor of c. In Chapter 8, we shall discuss in detail some of the major spectroscopic consequences of the magnetic dipole (md) term and the spin–orbit (so) interaction. In the next section, we determine corrections to the nonrelativistic energy eigenvalue of the Schrödinger equation for the hydrogen atom using perturbative treatment of the extra terms in Eq. 7.78.

7.3 Unraveling Relativistic Effects as Perturbations over Nonrelativistic Schrödinger Equation, and More

Treating Eq. 7.78 (or Eq. 7.81) as a Hamiltonian in which are added various perturbations helps us interpret the physical content of the Dirac equation (Eq. 7.38) as perturbations to Schrödinger nonrelativistic theory. The physical content in the terms in the TDH (Eq. 7.67, Eq. 7.71, and Eq. 7.74, i.e., Eq. 7.78) is essentially the same as in the original Dirac equation (Eq. 7.38). In the form in various terms appear in the third TDH, they are amenable easily using perturbation theory, which is the language commonly used, especially by experimentalists. We have already obtained in Chapter 5 the energy level of the nth discrete bound state of the nonrelativistic (Bohr–Schrödinger) hydrogen atom. From Eq. 5.47, it is given by

$$E_n = -\frac{\mu Z^2 e^4}{2\hbar^2}\frac{1}{n^2}. \tag{7.82}$$

The first-order perturbation correction to the energy of the nth eigenvalue due to the RKE, using Eq. 6.88, is

$$E_n^{(1)} = \left\langle n^{(0)} \middle| V \middle| n^{(0)} \right\rangle = \langle V \rangle. \tag{7.83}$$

First-Order Perturbative Correction due to Spin–Orbit Interaction

In order to determine the correction to the nonrelativistic energy value E_n due to the spin–orbit interaction term in Eq. 7.81, using $\vec{p}$ instead of $\vec{\pi}$ as explained in the paragraph after Eq. 7.81, we must determine the expectation value of the perturbation Hamiltonian

$$H'_{so} = \frac{e}{2m^2c^2}\vec{s}\bullet\vec{E}\times\vec{p} = \frac{e}{2m^2c^2}\vec{s}\bullet\left(-\frac{\partial V}{\partial r}\frac{\vec{r}}{r}\right)\times\vec{p} = -\frac{e}{2m^2c^2}\left(\frac{1}{r}\frac{\partial V}{\partial r}\right)\vec{s}\bullet\vec{\ell} \tag{7.84}$$

with

$$V = -\frac{Ze^2}{r}. \tag{7.85}$$

Hence, $$H'_{so} = \frac{e}{2\mu^2c^2}\left(\frac{Ze}{r^3}\right)\vec{s}\bullet\vec{\ell} = \frac{Ze^2}{2\mu^2c^2}\left(\frac{1}{r^3}\right)\vec{s}\bullet\vec{\ell}, \tag{7.86}$$

and the first-order perturbation correction due to it becomes

$$E_{n,so}^{(1)} = \left\langle n^{(0)} \middle| H'_{so} \middle| n^{(0)} \right\rangle = \frac{Ze^2}{2\mu^2c^2}\left\langle\frac{1}{r^3}\right\rangle\left\langle\vec{s}\bullet\vec{\ell}\right\rangle. \tag{7.87}$$

With the radial functions listed in Table 5.1 (Chapter 5), we find that

$$\left\langle\frac{1}{r^3}\right\rangle = \frac{Z^3}{n^3 a_0^3 \ell\left(\ell+\frac{1}{2}\right)(\ell+1)}, \tag{7.88}$$

where $a_0 = \dfrac{\hbar^2}{\mu e^2}$ is the Bohr radius.

The total angular momentum of the electron is

$$\vec{j} = \vec{\ell} + \vec{s}. \tag{7.89a}$$

Thus, the *so* term is interpreted as one that results from the coupling of spin angular momentum with the orbital angular momentum. This can be easily worked out using the methods in Chapter 4. We see that

$$j^2 = \vec{j}\cdot\vec{j} = \ell^2 + 2\vec{\ell}\cdot\vec{s} + s^2, \tag{7.89b}$$

and hence $\vec{\ell}\cdot\vec{s} = \frac{1}{2}\left[j^2 - \ell^2 - s^2\right].$ (7.89b)

The expectation values of the above operator is

$$\left\langle\vec{\ell}\cdot\vec{s}\right\rangle = \left\langle\frac{1}{2}\left[j^2 - \ell^2 - s^2\right]\right\rangle = \frac{\hbar^2}{2}\left[j(j+1) - \ell(\ell+1) - s(s+1)\right],$$

i.e., $\left\langle\vec{\ell}\cdot\vec{s}\right\rangle = \frac{\hbar^2}{2}\left[j(j+1) - \ell(\ell+1) - \frac{3}{4}\right].$ (7.90)

Using the results of Eq. 7.88 and Eq. 7.90 in Eq. 7.87, we get

$$\left\langle H'_{so}\right\rangle = \frac{Ze^2}{2\mu^2c^2}\left\{\frac{Z^3}{n^3{a_0}^3\ell\left(\ell+\frac{1}{2}\right)(\ell+1)}\right\}\left[\frac{\hbar^2}{2}\left\{j(j+1) - \ell(\ell+1) - s(s+1)\right\}\right], \tag{7.91a}$$

i.e., $\left\langle H'_{so}\right\rangle = \frac{Z^4e^2}{2\mu^2c^2}\frac{1}{n^2}\frac{\mu^3e^6\hbar^2}{\hbar^6}\frac{j(j+1) - \ell(\ell+1) - s(s+1)}{2n\ell\left(\ell+\frac{1}{2}\right)(\ell+1)}.$ (7.91b)

Hence, $E^{(1)}_{n,so} = \left\langle H'_{so}\right\rangle = \frac{\mu c^2}{2}(Z\alpha)^4\frac{1}{n^2}\frac{j(j+1) - \ell(\ell+1) - s(s+1)}{2n\ell\left(\ell+\frac{1}{2}\right)(\ell+1)},$

i.e., $E^{(1)}_{n,so} = \left\langle H'_{so}\right\rangle = -E_n(Z\alpha)^2\frac{\left\{j(j+1) - \ell(\ell+1) - \frac{3}{4}\right\}}{2n\ell\left(\ell+\frac{1}{2}\right)(\ell+1)},$ (7.91c)

where we have now written the energy of the nth bound state as

$$E_n = -\frac{\mu Z^2e^4}{2\hbar^2}\frac{1}{n^2} = -\frac{\mu c^2Z^2}{2}\left(\frac{e^2}{\hbar c}\right)^2\frac{1}{n^2} = -\frac{\mu c^2}{2}\left\{Z\left(\frac{e^2}{\hbar c}\right)\right\}^2\frac{1}{n^2} = -\frac{\mu c^2}{2}(Z\alpha)^2\frac{1}{n^2}, \tag{7.92}$$

where $\alpha = \frac{e^2}{\hbar c}.$ (7.93)

The universal constant α is of great fundamental importance; it is a dimensionless number whose value is $7.2973525693\times10^{-3}$. It is the inverse of 137.035999084. Its precision measurement and accurate calculation using rigorous theory is of great interest. It provides a measure of the strength of coupling between elementary charged particles and the electromagnetic radiation. Because of its importance in accounting for the splitting of the energy level resulting in what is called as the *fine structure* schematically shown in Fig. 7.3, α is called as the *fine structure constant*. Its importance can perhaps be gauged from Feynman's remark "*all good theoretical physicists put this number up on their wall and worry about it*".

As a result of the spin–orbit interaction, the energy level $E_{n\ell}$ splits into two, one for $j = \ell + \frac{1}{2}$ and the other for $j = \ell - \frac{1}{2}$. The first-order energy correction in Eq. 7.92 is then obtained by replacing $\ell = j \mp \frac{1}{2}$ for $j = \ell \pm \frac{1}{2}$, resulting in the splitting of the energy level as shown schematically in Fig. 7.3, since

for $j = \ell + \frac{1}{2}$ we have

$$j(j+1) - \ell(\ell+1) - \frac{3}{4} = \left(\ell + \frac{1}{2}\right)\left(\ell + \frac{3}{2}\right) - \ell(\ell+1) - \frac{3}{4} = \ell, \tag{7.94a}$$

and for $j = \ell - \frac{1}{2}$, we have

$$j(j+1) - \ell(\ell+1) - \frac{3}{4} = \left(\ell - \frac{1}{2}\right)\left(\ell + \frac{1}{2}\right) - \ell(\ell+1) - \frac{3}{4} = -\ell - 1. \tag{7.94b}$$

Funquest: Why do you think the value of the fine structure constant given above includes a number of places of decimal? Get more information about this from other sources.

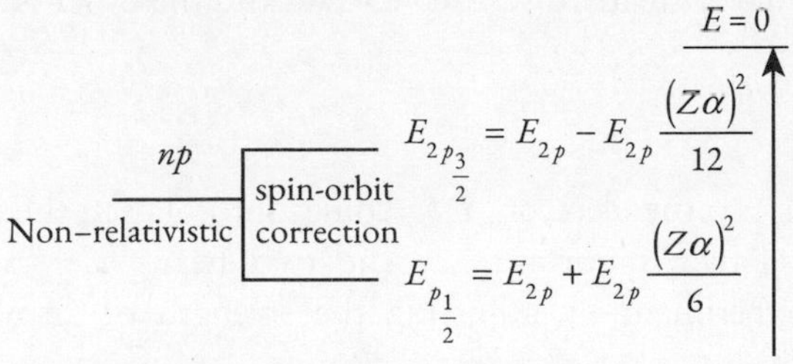

Fig. 7.3 Spin–orbit splitting of the 2*p* level in the hydrogen atom. Note that the $2p_{\frac{3}{2}}$ level is less negative, and the $2p_{\frac{1}{2}}$ level is more negative than the energy of the nonrelativistic 2*p* level.

Relativistic Kinetic Energy Correction

We shall now estimate the first-order perturbation correction due to the term RKE in Eq. 7.81. Using H'_{RKE} as the perturbation Hamiltonian, we see that

$$\Delta E^{(1)}_{RKE} = \left\langle H'_{RKE} \right\rangle = \left\langle -\frac{\vec{p}^4}{8\mu^3 c^2} \right\rangle = \left\langle -\left(\frac{\vec{p}^2}{2m}\right)^2 \frac{1}{2\mu c^2} \right\rangle = \frac{-1}{2\mu c^2}\left\langle \left(\frac{\vec{p}^2}{2\mu}\right)^2 \right\rangle, \tag{7.95a}$$

i.e., $$\Delta E^{(1)}_{RKE} = \frac{-1}{2\mu c^2}\left\langle (E_n - V)^2 \right\rangle = \frac{-1}{2\mu c^2}\left[E_n^{\,2} - 2E_n\langle V\rangle + \langle V^2\rangle\right]. \tag{7.95b}$$

Since the hydrogenic potential is given by Eq. 7.85, we must determine $\left\langle \frac{1}{r}\right\rangle$ and $\left\langle \frac{1}{r^2}\right\rangle$. Again, using the hydrogenic wavefunctions listed in Table 5.1 and using results from the Solved Problems P5.4 and P5.5 (Chapter 5), we find that

$$\left\langle \frac{1}{r}\right\rangle = \frac{1}{n^2 a_0} \tag{7.96a}$$

and $$\left\langle \frac{1}{r^2}\right\rangle = \frac{1}{\left(\ell + \frac{1}{2}\right) n^3 a_0^{\,2}}. \tag{7.96b}$$

Hence, $$\Delta E^{(1)}_{RKE} = -E_n\left(\frac{Z\alpha}{n}\right)^2\left[\frac{3}{4} - \frac{n}{\left(\ell + \frac{1}{2}\right)}\right]. \tag{7.97}$$

The RKE correction is different for different values of the orbital angular momentum quantum number. Even as the $2s$ and $2p$ energy levels of the hydrogen atom are degenerate due to the SO(4) symmetry discussed in Chapter 5, their energies are now separated on account of the relativistic kinetic energy correction $\Delta E^{(1)}_{RKE}$, as shown in Fig. 7.4. The intrinsically negative energy of the nth bound state becomes *more* negative, resulting in a downward energy shift, schematically shown in Fig. 7.4. The spin–orbit interaction splits the energy of the $2p$ state in Fig. 7.4 *further*, as already shown in Fig. 7.3.

In the Hamiltonian of Eq. 7.81, there are two terms that involve the electron spin. One of these is H'_{so}, which we have already discussed earlier, and the other is

$$H'_{md} = \frac{e\hbar}{2\mu c}\vec{\sigma}\bullet\vec{B}. \tag{7.98}$$

While H'_{so} is intrinsic to the electron, H'_{md} comes into play in the presence of an applied magnetic field. The relative importance of the two terms H'_{so} and H'_{md} can therefore be controlled in an experiment by adjusting the strength of an applied magnetic field.

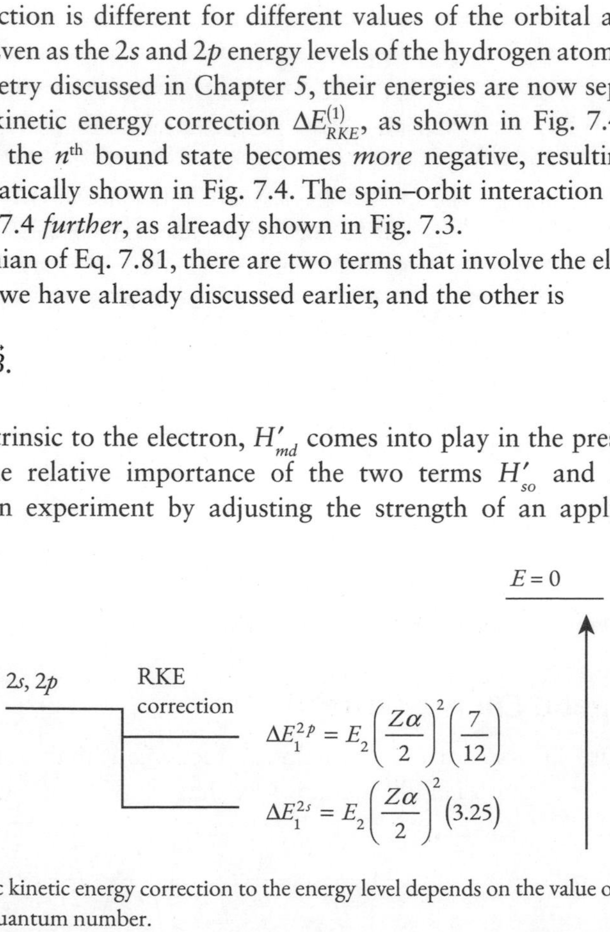

Fig. 7.4 The relativistic kinetic energy correction to the energy level depends on the value of the orbital angular momentum quantum number.

Depending on the interplay between the relative strengths of these two terms that can be treated using first-order perturbation theory, the experimental spectroscopic findings are extremely complex. We shall study these complexities in Chapter 8 by investigating the class of spectroscopies that come under the Zeeman effect. The only term in Eq. 7.81 that now remains to be treated perturbatively is the Darwin term

$$H'_D = -\frac{e\hbar^2}{8\mu^2c^2}\left(div\vec{E}\right) = -\frac{e\hbar^2}{8\mu^2c^2}\vec{\nabla}\bullet\left(\frac{-Ze^2}{r^2}\right)\hat{e}_r = \frac{\pi\hbar^2 Ze^2}{2\mu^2c^2}\delta^3(\vec{r}). \tag{7.99}$$

Since it involves the Dirac-δ function, the region near the origin is of interest in the determination of its expectation value. Since all wavefunctions with $\ell \geq 1$ go rapidly to zero close to the origin (see Fig. 5.1, Chapter 5), only $\ell = 0$ (s orbital) is of interest.

The Darwin correction in the first order therefore is

$$\Delta E_D^{(1)} = \frac{\pi\hbar^2 Ze^2}{2\mu^2c^2}\left\langle\psi_{n,\ell=0,m=0}\left|\delta^3(\vec{r})\right|\psi_{n,\ell=0,m=0}\right\rangle = \frac{\pi\hbar^2 Ze^2}{2\mu^2c^2}\left|\psi_{n,\ell=0,m=0}(r=0)\right|^2,$$

$$\text{i.e., } \Delta E_D^{(1)} = \frac{\hbar^2 Ze^2}{2\mu^2c^2}\frac{1}{n^3a_0^{\,3}}, \tag{7.100a}$$

$$\text{i.e., } \Delta E_D^{(1)} = \left(\frac{\mu e^2}{\hbar^2}\right)^3\frac{\hbar^2 Ze^2}{\mu^2c^2}\frac{1}{2n^3} = \left(\frac{e^8}{\hbar^4c^4}\right)\mu c^2\frac{1}{2n^3} = \alpha^4\mu c^2\frac{1}{2n^3} = -E_n\frac{(Z\alpha)^2}{n}, \tag{7.100b}$$

$$\text{since } \left|\psi_{n,\ell=0,m=0}(r=0)\right|^2 = \frac{1}{\pi n^3 a_0^{\,3}}. \tag{7.101}$$

Strictly speaking, we know that nature demands a four-component wavefunction. In the two-component approximation that was attempted earlier, it is only natural that some aspects of the physical information are lost. These details come from the coupling of the two pairs of the two-component functions that together constitute the four-component wavefunction. This coupling is due to the odd operators in the Dirac equation. The three FW transformations reduce the coupling between the two pairs of two-component functions, but do not eliminate the same. In the Darwin approximation, the neglect of the residual coupling can be approximately accounted for by pretending that the Coulomb interaction between the nucleus and the electron has a nonlocal character, as if the electron itself is delocalized over a spherical region of space whose radius is the Compton wavelength. The delocalization of the electron is referred to as *zitterbewegung*. Effectively, the neglect of the residual coupling is modeled by a zigzag delocalized motion of the electron. A perturbative treatment of this effect was proposed by Charles Galton Darwin (1887–1962), *grandson of Charles Robert Darwin (1809–1882), author of the theory of evolution of life.*

Adding the results of Eq. 7.91c, Eq. 7.97, and Eq. 7.100b, we find the first-order perturbation estimate of the correction to the Bohr–Schrödinger energy of the nth bound state (Eq. 7.92). The correction is

$$E_n^{(1)} = E_n^{(1)s\text{-}o} + E_n^{(1)RKE} + E_n^{(1)D} = E_n\frac{(Z\alpha)^2}{n^2}\left(\frac{n}{j+\frac{1}{2}} - \frac{3}{4}\right). \tag{7.102}$$

E_n being intrinsically negative, the energy levels get slightly more tightly bound. The energy eigenvalues of the relativistic hydrogen atom therefore is

$$E_{nj}^{Dirac} = E_n\left[1+\frac{(Z\alpha)^2}{n^2}\left(\frac{n}{j+\frac{1}{2}}-\frac{3}{4}\right)\right]. \tag{7.103}$$

Note that even if the individual corrections $E_n^{(1)s\text{-}o}$ and $E_n^{(1)RKE}$ depend on the orbital angular momentum quantum number, the sum (Eq. 7.102) depends on (n, j); it is independent of ℓ. The energy of the relativistic hydrogen atom's discrete bound states therefore depend on (n, j) and not on ℓ. The bound state spectral energy levels are shown schematically (not to scale) in Fig. 7.5. The second column in this figure illustrates how the Bohr–Schrödinger energy levels get corrected in Dirac theory.

We must add that nature requires a theory that is even larger than Dirac's. Willis Lamb and Robert Retherford detected in an experiment they carried out in 1947 a transition between the $^2S_{1/2}$ and $^2P_{1/2}$ levels of hydrogen, which are *degenerate* in Dirac's theory, as per Eq. 7.103. This energy shift is called as the Lamb–Retherford shift; or only as Lamb shift for short. It is shown schematically in the third column in Fig. 7.5. The Lamb shift results from the interaction of the hydrogen atom with vacuum fluctuations. It stimulated the development of the theory of quantum electrodynamics (QED), by R. P. Feynman, S. Tomonaga, J. Schwinger, and F. Dyson.

Furthermore, even as the fine structure that results in spin–orbit splitting results from the angular momentum coupling $\vec{j}_e = \vec{\ell}_e + \vec{s}_e$ between the electron's orbital and spin angular momentum, one must include consequences of a *further* angular momentum coupling between and the total angular momentum $\vec{j}_e$ of the electron, and the intrinsic spin angular momentum of the nucleus (proton), $\vec{j}_p$, which is also $\frac{1}{2}$. The electron–proton angular momentum coupling results in a net angular momentum,

$$\vec{F} = \vec{j}_e + \vec{j}_p;\ \left|j_e - j_p\right| \le F \le \left(j_e + j_p\right). \tag{7.104}$$

The additional structure that results from the electron–nucleus angular momentum coupling is called as the hyperfine structure. It is indicated (again, not to scale) in the fourth column in Fig. 7.5. We will discuss it further in Section 8.4 (Chapter 8).

Hydrogen being the most abundant atom in the entire universe, observation of the Doppler shift of the transition between the hyperfine split levels of the hydrogen atom in the ground state provides a temperature map of the universe. The coupling of the angular momentum of the proton and the electron results in the hyperfine doublet, $F = 1$ and $F = 0$ (shown schematically in Fig. 7.5). The $(F = 1) \rightarrow (F = 0)$ transition corresponds to an energy of $\sim 5.87 \times 10^{-6}$ eV. The frequency of this radiation is 1420.405751768 MHz, and it corresponds to the wavelength of 21.106114054160 cm, which can be measured very precisely using sophisticated radio-astronomy techniques. The 21 cm line was first detected in 1951 by Ewen and Purcell. Studies on the 21 cm line, often referred to as the "H I" line, have led to groundbreaking discoveries in astronomy. For example, map of the neutral hydrogen in our galaxy established its spiral structure. From the red-shift of the 21-cm line, one can estimate distribution of matter in the universe. Observations of the distribution of the intensity of H I line provides perceptive mapping of the cosmos. Spectroscopic consequences of the fine structure and that of the hyperfine structure are discussed in Chapter 8.

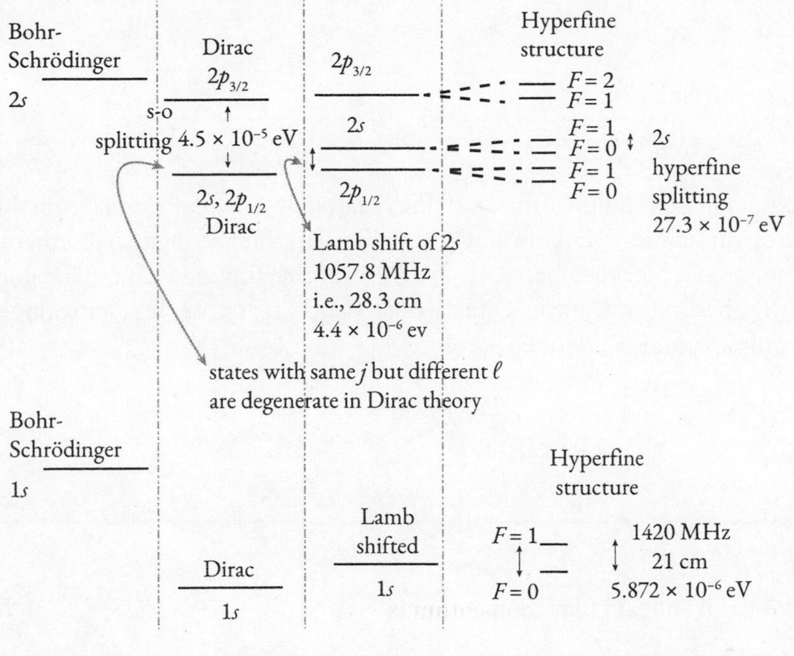

Fig. 7.5 Bohr–Schrödinger energy levels in the nonrelativistic energy spectrum of the hydrogen atom shift due to relativistic effects. RKE lowers the energy levels; spin–orbit interaction splits energy levels of $j = \ell \pm \frac{1}{2}$. The Darwin correction affects only the $\ell = 0$ states. The three corrections, s-o, RKE, and Darwin, are intrinsically present in Dirac theory; the FW transformations bring them to the surface and make them manifest. The energy levels are further shifted, due to factors not accounted for in Dirac theory. These additional shifts are the Lamb–Retherford shift and shift on account of the hyperfine structure.

While concluding this section, we reiterate the fact that the odd operators in the Dirac equation couple the two pairs of the two-component wavefunctions. The FW transformations only reduce this coupling, not eliminate it. Dirac equation has been obtained by factoring the relativistic invariant expressed in Eq. 7.12, and its solutions therefore accommodated both positive and negative energy states. The positive energy states describe the electron, and the existence of negative energy states predicted the electron's antiparticle, namely the positron. The positron was observed in a cloud chamber experiment conducted by Carl Anderson in 1932.

Funquest: Assuming there is extra terrestrial intelligence elsewhere in the cosmos, can you guess which spectral line would be most likely be used by intelligent aliens to establish communication with others in the cosmos?

7.4 Separation of the *Relativistic* Hydrogen Atom in Radial and Angular Parts

Since odd operators are present in the Dirac Hamiltonian (Eq. 7.38), it is not obvious that the angular part and the radial part of the relativistic four-component wavefunction are separable even for the Coulomb potential. The Dirac Hamiltonian for the Coulomb potential

$$e\phi = -\frac{Ze^2}{r} \tag{7.105}$$

in the hydrogen atom is

$$H_{DC} = \beta\mu c^2 + c\vec{\alpha}\bullet\vec{p} + e\phi(|\vec{r}|). \tag{7.106}$$

The orbital angular momentum and also the spin angular momentum both do not commute with the Dirac–Coulomb (DC) Hamiltonian. Hence, neither of them provides a good quantum number for the relativistic hydrogen atom. Toward recognizing good quantum numbers of the hydrogen atom in Dirac's relativistic theory, we first introduce two 4 × 4 matrix operators, ρ_1 and $\vec{\Xi}$, defined as

$$(\rho_1)_{4\times4} = \begin{pmatrix} 0_{2\times2} & 1_{2\times2} \\ 1_{2\times2} & 0_{2\times2} \end{pmatrix} \tag{7.107a}$$

and

$$\vec{\Xi}_{4\times4} = \begin{pmatrix} \vec{\sigma}_{2\times2} & 0_{2\times2} \\ 0_{2\times2} & \vec{\sigma}_{2\times2} \end{pmatrix}. \tag{7.107b}$$

Just as the Pauli spin angular momentum is

$$\vec{s}_{2\times2} = \frac{\hbar}{2}\vec{\sigma}_{2\times2}, \tag{7.108a}$$

the *Dirac spin* angular momentum is

$$\vec{s}_{4\times4} = \frac{\hbar}{2}\vec{\Xi}_{4\times4}. \tag{7.108b}$$

In terms of the Dirac spin, we see that

$$\rho_1\vec{\Xi} = \begin{pmatrix} 0 & 1 \\ 1 & 0 \end{pmatrix}\begin{pmatrix} \vec{\sigma} & 0 \\ 0 & \vec{\sigma} \end{pmatrix} = \begin{pmatrix} 0 & \vec{\sigma} \\ \vec{\sigma} & 0 \end{pmatrix} = \vec{\alpha}, \tag{7.109a}$$

$$\text{and } \rho_1\vec{\alpha} = \begin{pmatrix} 0 & 1 \\ 1 & 0 \end{pmatrix}\begin{pmatrix} 0 & \vec{\sigma} \\ \vec{\sigma} & 0 \end{pmatrix} = \begin{pmatrix} \vec{\sigma} & 0 \\ 0 & \vec{\sigma} \end{pmatrix} = \vec{\Xi}. \tag{7.109b}$$

Furthermore,

$$\vec{\Xi}\bullet\vec{p} = \begin{bmatrix} (\vec{\sigma}\bullet\hat{e}_r)(\vec{\sigma}\bullet\hat{e}_r) & 0 \\ 0 & (\vec{\sigma}\bullet\hat{e}_r)(\vec{\sigma}\bullet\hat{e}_r) \end{bmatrix}(\vec{\Xi}\bullet\vec{p}) = (\vec{\Xi}\bullet\hat{e}_r)(\vec{\Xi}\bullet\hat{e}_r)(\vec{\Xi}\cdot\vec{p}) = (\vec{\Xi}\bullet\hat{e}_r)\frac{(\vec{\Xi}\bullet\vec{r})(\vec{\Xi}\cdot\vec{p})}{r},$$

$$\text{hence } \vec{\Xi}\bullet\vec{p} = \vec{\Xi}\bullet\hat{e}_r\left[\hat{e}_r\bullet\vec{p} + \frac{i\vec{\Xi}\bullet\vec{r}\times\vec{p}}{r}\right] = \vec{\Xi}\bullet\hat{e}_r\left[\left(-i\hbar\frac{\partial}{\partial r}\right) + \frac{i}{r}(\vec{\Xi}\bullet\vec{\ell})\right]. \tag{7.110}$$

We have made use of $\vec{r}\bullet\vec{p} = -i\hbar r\hat{e}_r\bullet\vec{\nabla} = -i\hbar r\hat{e}_r\bullet\left(\hat{e}_r\frac{\partial}{\partial r} + \hat{e}_\theta\frac{1}{r}\frac{\partial}{\partial\theta} + \hat{e}_\varphi\frac{1}{r\sin\theta}\frac{\partial}{\partial\varphi}\right) = -i\hbar r\frac{\partial}{\partial r}$.

Likewise, using Eq. 7.110, we see that

$$\vec{\alpha}\bullet\vec{p} = (\rho_1\vec{\Xi})\bullet\vec{p} = \begin{pmatrix} 0 & 1 \\ 1 & 0 \end{pmatrix}\vec{\Xi}\bullet\vec{p} = \begin{pmatrix} 0 & 1 \\ 1 & 0 \end{pmatrix}(\vec{\Xi}\bullet\hat{e}_r)\left[\left(-i\hbar\frac{\partial}{\partial r}\right) + \frac{i}{r}(\vec{\Xi}\bullet\vec{\ell})\right], \tag{7.111a}$$

i.e., $\vec{\alpha}\cdot\vec{p} = \alpha_r\left[\left(-i\hbar\frac{\partial}{\partial r}\right) + \frac{i}{r}\left(\vec{\Xi}\cdot\vec{\ell}\right)\right].$ (7.111b)

We define an operator

$$K_{4\times4} = \beta\left(\hbar 1_{4\times4} + \vec{\Xi}_{4\times4}\cdot\vec{\ell}\right) = \hbar\beta + \beta\vec{\Xi}_{4\times4}\cdot\vec{\ell} = \begin{bmatrix} \vec{\sigma}\cdot\vec{\ell}+\hbar & 0 \\ 0 & -\vec{\sigma}\cdot\vec{\ell}-\hbar \end{bmatrix}, \tag{7.112}$$

but it can be written in alternative equivalent ways. We see that

$$\beta K_{4\times4} = \hbar + \vec{\Xi}_{4\times4}\cdot\vec{\ell}, \tag{7.113}$$

which gives $\vec{\Xi}_{4\times4}\cdot\vec{\ell} = \beta K_{4\times4} - \hbar.$ (7.114)

Therefore,

$$\vec{\alpha}\cdot\vec{p} = \left(\rho_1\vec{\Xi}\right)\cdot\vec{p} = \alpha_r\left[\left(-i\hbar\frac{\partial}{\partial r}\right) + \frac{i}{r}\left(\beta K_{4\times4} - \hbar\right)\right] = -i\alpha_r\left[\left(\hbar\frac{\partial}{\partial r}\right) + \frac{\hbar}{r} - \frac{\beta K_{4\times4}}{r}\right] \tag{7.115}$$

We sometimes use the subscript 2×2 and/or 4×4 sometimes, as in Eqs. 7.108a,b, Eq. 7.112, and Eq. 7.113, only to highlight the matrix character of the operators referred to. However, most often, these subscripts will be dropped; their presence must be inferred from the context.

Using Eq. 7.115, the Dirac–Coulomb Hamiltonian (Eq. 7.106) therefore is

$$H_{DC} = \beta\mu c^2 - ic\alpha_r\left(\hbar\frac{\partial}{\partial r} + \frac{\hbar}{r} - \frac{1}{r}\beta K\right) + e\phi(r). \tag{7.116}$$

Now,

$$\left(\vec{\sigma}\cdot\vec{\ell}\right)\left(\vec{\sigma}\cdot\vec{\ell}\right) = \ell^2 + i\vec{\sigma}\cdot\vec{\ell}\times\vec{\ell} = \ell^2 + i\vec{\sigma}\cdot\left(i\hbar\vec{\ell}\right) = \ell^2 - \hbar\vec{\sigma}\cdot\vec{\ell},$$

i.e., $\hbar\vec{\sigma}\cdot\vec{\ell} + \left(\vec{\sigma}\cdot\vec{\ell}\right)\left(\vec{\sigma}\cdot\vec{\ell}\right) = \ell^2.$ (7.117a)

Hence,

$$j^2 = \vec{j}\cdot\vec{j} = \left(\vec{s}+\vec{\ell}\right)\cdot\left(\vec{s}+\vec{\ell}\right) = s^2 + 2\left(\vec{s}\cdot\vec{\ell}\right) + \ell^2 = s^2 + \hbar\vec{\sigma}\cdot\vec{\ell} + \left[\hbar\vec{\sigma}\cdot\vec{\ell} + \left(\vec{\sigma}\cdot\vec{\ell}\right)\left(\vec{\sigma}\cdot\vec{\ell}\right)\right],$$

i.e.,

$$j^2 = \left(\vec{\sigma}\cdot\vec{\ell}\right)^2 + 2\hbar\vec{\sigma}\cdot\vec{\ell} + \frac{3\hbar^2}{4} = \left(\vec{\sigma}\cdot\vec{\ell}+\hbar\right)^2 - \frac{\hbar^2}{4}. \tag{7.117b}$$

Using Eq. 7.113,

$$j^2 = (\beta K)^2 - \frac{\hbar^2}{4}, \tag{7.117c}$$

$$\text{or}\ \left(j^2 + \frac{\hbar^2}{4}\right)1_{4\times4} = (\beta K)^2 = \begin{bmatrix} 1&0&0&0\\0&-1&0&0\\0&0&1&0\\0&0&0&-1 \end{bmatrix}\begin{bmatrix} K&0&0&0\\0&K&0&0\\0&0&K&0\\0&0&0&K \end{bmatrix} \times \begin{bmatrix} 1&0&0&0\\0&-1&0&0\\0&0&1&0\\0&0&0&-1 \end{bmatrix}\begin{bmatrix} K&0&0&0\\0&K&0&0\\0&0&K&0\\0&0&0&K \end{bmatrix},$$

i.e.,

$$\left(j^2 + \frac{\hbar^2}{4}\right)1_{4\times4} = K^2 1_{4\times4}. \tag{7.118}$$

The operator K^2 commutes with the Dirac–Coulomb Hamiltonian, and from the above equation we see that its eigenvalue in the basis of the eigenvectors of H_{DC} would be

$$\kappa^2\hbar^2 = j(j+1)\hbar^2 + \left(\frac{\hbar}{2}\right)^2 = \hbar^2\left[j^2 + j + \frac{1}{4}\right] = \hbar^2\left(j + \frac{1}{2}\right)^2. \tag{7.119}$$

Equation 7.119 thus provides another good quantum number, along with j, for the relativistic hydrogen atom in the Dirac model. This quantum number is the square root of κ^2, and the two possible values are

$$\kappa = \pm\left(j + \frac{1}{2}\right) \text{ for } j = \ell \mp \frac{1}{2}. \tag{7.120}$$

Essentially, the two possible values correspond to the two possible parity states of the atom. *Dirac parity* is represented by the operator P_D, which commutes with the Dirac–Coulomb operator. It is

$$P_D = \beta P = \begin{bmatrix} 1_{2\times2} & 0_{2\times2} \\ 0_{2\times2} & -1_{2\times2} \end{bmatrix} P, \tag{7.121a}$$

and
it has eigenvalues

$$\omega = +1 \tag{7.121b}$$

$$\text{and } \omega = -1. \tag{7.121c}$$

Using the eigenvalues of the Dirac-parity operator, we can write the quantum number κ as

$$\kappa = \left(j + \frac{1}{2}\right)\omega. \tag{7.122}$$

The operators $\{H, j^2, j_z, K\}$ provide a complete set of commuting operators and provide good quantum numbers. The Dirac–Coulomb eigenvalue (Eq. 7.103) is degenerate on account of the sign of $\omega = \pm1$; it is determined by (n, j), i.e., by $(n, |\kappa|)$ and *not* by (n, κ). This degeneracy prompts the search for an extra symmetry, similar to the SO(4) symmetry that accounts for the n^2-fold degeneracy (Eq. 5.85 in Chapter 5) of the *nonrelativistic* hydrogen atom. The residual degeneracy (and associated symmetry) in the energy state (n, j) need not be dismissed as *accidental*; its physical origin has been studied [11–13]. The SO(4) symmetry of the Dirac–Coulomb quantum problem is explained by *six generators* on introducing a pseudo-spin operator [14, 15]. We have already seen that the symmetry of the Dirac–Coulomb Hamiltonian is *broken* by vacuum polarization, which results in the Lamb shift (Fig. 7.5).

In Table 7.2, we present the quantum numbers κ and ω for various atomic orbitals.

Table 7.2 Quantum numbers for atomic states in the hydrogen atom. Note that the quantum number κ includes information about both j and parity.

Orbital	ℓ	*Parity*	κ	j	ω	$j+\frac{\omega}{2}$	$(-1)^{j+\frac{\omega}{2}}$
$s_{\frac{1}{2}}$	0	+1	−1	$\frac{1}{2}$	−1	0	+1
$p_{\frac{1}{2}}$	1	−1	+1	$\frac{1}{2}$	+1	1	−1
$p_{\frac{3}{2}}$	1	−1	−2	$\frac{3}{2}$	−1	1	−1
$d_{\frac{3}{2}}$	2	+1	+2	$\frac{3}{2}$	+1	2	+1
$d_{\frac{5}{2}}$	2	+1	−3	$\frac{5}{2}$	−1	2	+1
$f_{\frac{5}{2}}$	3	−1	+3	$\frac{5}{2}$	+1	3	−1
..	..	..	..	..	..	..	...

Using now the Dirac–Coulomb Hamiltonian (Eq. 7.116), we can determine both the energy eigenvalues E_{nj} and the four-component wavefunctions $u_{n\kappa m}$ of the stationary states of the hydrogen atom by solving the equation

$$\left[\beta\mu c^2 - ic\alpha_r\left(\hbar\frac{\partial}{\partial r}+\frac{\hbar}{r}-\frac{1}{r}\beta K\right)+e\phi(r)\right]u_{n\kappa m} = E_{n\kappa}u_{n\kappa m}. \tag{7.123a}$$

Replacing the operator K by its eigenvalue $\hbar\kappa$, the equation to be solved is

$$\text{i.e., } \left[\beta\mu c^2 - ic\alpha_r\left(\hbar\frac{\partial}{\partial r}+\frac{\hbar}{r}-\frac{1}{r}\beta\hbar\kappa\right)+e\phi(r)\right]u_{n\kappa m} = E_{n\kappa}u_{n\kappa m}. \tag{7.123b}$$

Operating on Eq. 7.123b by α_r,

$$\left[\alpha_r\beta\mu c^2 - ic\alpha_r\alpha_r\left(\hbar\frac{\partial}{\partial r}+\frac{\hbar}{r}-\frac{1}{r}\beta\hbar\kappa\right)+\alpha_r e\phi(r)\right]u_{n\kappa m} = E_{n\kappa}\alpha_r u_{n\kappa m}, \tag{7.124}$$

$$\text{i.e., } \begin{bmatrix} \mu c^2\alpha_r\beta u_{n\kappa m} - ic\hbar\frac{\partial}{\partial r}u_{n\kappa m} - ic\frac{\hbar}{r}u_{n\kappa m} \\ +i\frac{c\hbar\kappa}{r}\beta u_{n\kappa m} + e\phi(r)\alpha_r u_{n\kappa m} \end{bmatrix} = E_{n\kappa}\alpha_r u_{n\kappa m}. \tag{7.125}$$

Now, operating on Eq. 7.125 by β, we get

$$\begin{bmatrix} \mu c^2 \beta\alpha_r \beta u_{n\kappa m} - ic\hbar\beta\dfrac{\partial}{\partial r}u_{n\kappa m} - ic\dfrac{\hbar}{r}\beta u_{n\kappa m} \\ +i\dfrac{c\hbar\kappa}{r}\beta\beta u_{n\kappa m} + e\phi(r)\beta\alpha_r u_{n\kappa m} \end{bmatrix} = E_{n\kappa}\beta\alpha_r u_{n\kappa m}. \tag{7.126}$$

Since, $\beta\alpha_r = -\alpha_r\beta$, we have

$$\begin{bmatrix} -\mu c^2 \alpha_r \beta\beta u_{n\kappa m} - ic\hbar\beta\dfrac{\partial}{\partial r}u_{n\kappa m} - ic\dfrac{\hbar}{r}\beta u_{n\kappa m} \\ +i\dfrac{c\hbar\kappa}{r}u_{n\kappa m} - e\phi(r)\alpha_r\beta u_{n\kappa m} \end{bmatrix} = -E_{n\kappa}\alpha_r\beta u_{n\kappa m}. \tag{7.127}$$

Now, adding Eq. 7.125 and Eq. 7.127, we get

$$\begin{bmatrix} -\mu c^2 \alpha_r(1-\beta)u_{n\kappa m} - ic\hbar\dfrac{\partial}{\partial r}(1+\beta)u_{n\kappa m} - ic\dfrac{\hbar}{r}(1+\beta)u_{n\kappa m} \\ +i\dfrac{c\hbar\kappa}{r}(1+\beta)u_{n\kappa m} + e\phi(r)\alpha_r(1-\beta)u_{n\kappa m} \end{bmatrix} = E_{n\kappa}\alpha_r(1-\beta)u_{n\kappa m}, \tag{7.128}$$

and subtracting Eq. 7.127 from Eq. 7.125, we get

$$\begin{bmatrix} \mu c^2 \alpha_r(1+\beta)u_{n\kappa m} - ic\hbar\dfrac{\partial}{\partial r}(1-\beta)u_{n\kappa m} - ic\dfrac{\hbar}{r}(1-\beta)u_{n\kappa m} \\ +i\dfrac{c\hbar\kappa}{r}(1-\beta)u_{n\kappa m} + e\phi(r)\alpha_r(1+\beta)u_{n\kappa m} \end{bmatrix} = E_{n\kappa}\alpha_r(1+\beta)u_{n\kappa m}. \tag{7.129}$$

We shall now explore the possibility of expressing the 4 × 4 Dirac–Coulomb wavefunction as a pair of two 2 × 2 functions in which the radial part is separated from the angular part. The angular part of the wavefunction would be placed in the vector spherical harmonics,

$$\left(u_{n\kappa m}\right)_{4\times1} = \begin{bmatrix} \left(u_+\right)_{2\times1} \\ \left(u_-\right)_{2\times1} \end{bmatrix} = \begin{bmatrix} i\dfrac{\mathrm{P}(\mathrm{r})}{r}\Omega_{\kappa m}(\hat{r}) \\ \dfrac{\mathrm{Q}(\mathrm{r})}{r}\Omega_{-\kappa m}(\hat{r}) \end{bmatrix}, \tag{7.130}$$

where the vector spherical harmonics spinors have been defined in Chapter 4 (Eq. 4.120). The separation of u_+ and u_- in radial and angular parts is in anticipation of such a separation being possible; this is, however, subject to scrutiny. It turns out that such a separation indeed works out, as is shown below. It also enables writing the four-component differential equation merely in terms of two-component form. This simplification is achieved because of the following properties:

$$(1-\beta)u_{n\kappa m} = \left(\begin{bmatrix} 1_{2\times2} & 0_{2\times2} \\ 0_{2\times2} & 1_{2\times2} \end{bmatrix} - \begin{bmatrix} 1_{2\times2} & 0_{2\times2} \\ 0_{2\times2} & -1_{2\times2} \end{bmatrix}\right)\begin{bmatrix} \left(u_+\right)_{2\times1} \\ \left(u_-\right)_{2\times1} \end{bmatrix} = \begin{bmatrix} 0 & 0 \\ 0 & 2 \end{bmatrix}\begin{bmatrix} u_+ \\ u_- \end{bmatrix} = 2\begin{bmatrix} 0 \\ u_- \end{bmatrix}, \tag{7.131a}$$

and

$$(1+\beta)u_{n\kappa m} = \left(\begin{bmatrix} 1_{2\times2} & 0_{2\times2} \\ 0_{2\times2} & 1_{2\times2} \end{bmatrix} + \begin{bmatrix} 1_{2\times2} & 0_{2\times2} \\ 0_{2\times2} & -1_{2\times2} \end{bmatrix}\right)\begin{bmatrix} \left(u_+\right)_{2\times1} \\ \left(u_-\right)_{2\times1} \end{bmatrix} = \begin{bmatrix} 2 & 0 \\ 0 & 0 \end{bmatrix}\begin{bmatrix} u_+ \\ u_- \end{bmatrix} = 2\begin{bmatrix} u_+ \\ 0 \end{bmatrix}. \tag{7.131b}$$

Using Eqs. 7.131a,b, Eq. 7.128 takes the following form:

$$-\mu c^2\alpha_r 2\begin{bmatrix}0\\u_-\end{bmatrix}-ic\hbar\frac{\partial}{\partial r}2\begin{bmatrix}u_+\\0\end{bmatrix}-ic\frac{\hbar}{r}2\begin{bmatrix}u_+\\0\end{bmatrix}++i\frac{c\hbar\kappa}{r}2\begin{bmatrix}u_+\\0\end{bmatrix}+e\phi(r)\alpha_r 2\begin{bmatrix}0\\u_-\end{bmatrix}=E_{n\kappa}\alpha_r 2\begin{bmatrix}0\\u_-\end{bmatrix},$$

$$\text{i.e., } -\mu c^2\begin{bmatrix}\sigma_r u_-\\0\end{bmatrix}-ic\hbar\frac{\partial}{\partial r}\begin{bmatrix}u_+\\0\end{bmatrix}-ic\frac{\hbar}{r}\begin{bmatrix}u_+\\0\end{bmatrix}+i\frac{c\hbar\kappa}{r}\begin{bmatrix}u_+\\0\end{bmatrix}+e\phi(r)\begin{bmatrix}\sigma_r u_-\\0\end{bmatrix}=E_{n\kappa}\begin{bmatrix}\sigma_r u_-\\0\end{bmatrix},$$

and hence

$$-\mu c^2\sigma_r u_- - ic\hbar\frac{\partial}{\partial r}u_+ - ic\frac{\hbar}{r}u_+ + i\frac{c\hbar\kappa}{r}u_+ + e\phi(r)\sigma_r u_- = E_{n\kappa}\sigma_r u_-. \tag{7.132}$$

Likewise, Eq. 7.129 takes the following form:

$$\mu c^2\alpha_r 2\begin{bmatrix}u_+\\0\end{bmatrix}-ic\hbar\frac{\partial}{\partial r}2\begin{bmatrix}0\\u_-\end{bmatrix}-ic\frac{\hbar}{r}2\begin{bmatrix}0\\u_-\end{bmatrix}+i\frac{c\hbar\kappa}{r}2\begin{bmatrix}0\\u_-\end{bmatrix}+e\phi(r)\alpha_r 2\begin{bmatrix}u_+\\0\end{bmatrix}=E_{n\kappa}\alpha_r 2\begin{bmatrix}u_+\\0\end{bmatrix},$$

i.e.,

$$\mu c^2\begin{bmatrix}0\\\sigma_r u_+\end{bmatrix}-ic\hbar\frac{\partial}{\partial r}\begin{bmatrix}0\\u_-\end{bmatrix}-ic\frac{\hbar}{r}\begin{bmatrix}0\\u_-\end{bmatrix}+i\frac{c\hbar\kappa}{r}\begin{bmatrix}0\\u_-\end{bmatrix}+e\phi(r)\begin{bmatrix}0\\\sigma_r u_+\end{bmatrix}=E_{n\kappa}\begin{bmatrix}0\\\sigma_r u_+\end{bmatrix},$$

and hence

$$\mu c^2\sigma_r u_+ - ic\hbar\frac{\partial}{\partial r}u_- - ic\frac{\hbar}{r}u_- + i\frac{c\hbar\kappa}{r}u_- + e\phi(r)\sigma_r u_+ = E_{n\kappa}\sigma_r u_+. \tag{7.133}$$

We have already studied in Chapter 4 the response of the vector spherical harmonics to the operator $(\vec{\sigma}\cdot\hat{e}_r)$. From Eq. 4.134,

$$\sigma_r u_+ = \sigma_r\frac{P_{n\kappa}(r)\Omega_{\kappa m}(\hat{r})}{r}=\frac{P_{n\kappa}(r)}{r}\{\sigma_r\Omega_{\kappa m}(\hat{r})\}=\frac{P_{n\kappa}(r)}{r}\{-\Omega_{-\kappa m}(\hat{r})\}, \tag{7.134a}$$

$$\text{and } \sigma_r u_- = \sigma_r\frac{iQ_{n\kappa}(r)\Omega_{-\kappa m}(\hat{r})}{r}=\frac{iQ_{n\kappa}(r)}{r}\{\sigma_r\Omega_{-\kappa m}(\hat{r})\},$$

$$\text{i.e., } \sigma_r u_- = \frac{iQ_{n\kappa}(r)}{r}\{-\Omega_{\kappa m}(\hat{r})\}=-\frac{iQ_{n\kappa}(r)\Omega_{\kappa m}(\hat{r})}{r}. \tag{7.134b}$$

Therefore, using Eqs. 7.134a,b in Eq. 7.132, we get

$$\begin{bmatrix}-\mu c^2\left(-\dfrac{iQ_{n\kappa}(r)\cancel{\Omega_{\kappa m}(\hat{r})}}{r}\right)-ic\hbar\dfrac{\partial}{\partial r}\left(\dfrac{P_{n\kappa}(r)\cancel{\Omega_{\kappa m}(\hat{r})}}{r}\right)\\[2ex] -ic\dfrac{\hbar}{r}\left(\dfrac{P_{n\kappa}(r)\cancel{\Omega_{\kappa m}(\hat{r})}}{r}\right)\\[2ex] +i\dfrac{c\hbar\kappa}{r}\left(\dfrac{P_{n\kappa}(r)\cancel{\Omega_{\kappa m}(\hat{r})}}{r}\right)+e\phi(r)\left(-\dfrac{iQ_{n\kappa}(r)\cancel{\Omega_{\kappa m}(\hat{r})}}{r}\right)\end{bmatrix}=-E_{n\kappa}\left(\frac{iQ_{n\kappa}(r)\cancel{\Omega_{\kappa m}(\hat{r})}}{r}\right).$$

The cancelation of the vector spherical harmonics in the above equation is a consequence of the application of the properties presented in Eqs. 7.134a,b. On count of the elimination of the angular parts, we are left only with the radial functions and their derivatives. We can therefore replace the partial derivative operator $\frac{\partial}{\partial r}$ in Eq. 7.133 by $\frac{d}{dr}$ and write

$$\left[\begin{array}{l} -\mu c^2\left(\frac{Q_{n\kappa}(r)}{r}\right) - ic\hbar\frac{d}{dr}\left(\frac{P_{n\kappa}(r)}{r}\right) - ic\frac{\hbar}{r}\left(\frac{P_{n\kappa}(r)}{r}\right) \\ +i\frac{c\hbar\kappa}{r}\left(\frac{P_{n\kappa}(r)}{r}\right) - ie\phi(r)\left(\frac{Q_{n\kappa}(r)}{r}\right) \end{array}\right] = -iE_{n\kappa}\left(\frac{Q_{n\kappa}(r)}{r}\right),$$

or
$$\left[\begin{array}{l} -\mu c^2\left(\frac{Q_{n\kappa}(r)}{r}\right) - ic\hbar\frac{1}{r}\frac{dP}{dr} \cancel{+ic\hbar\frac{P}{r^2} - ic\frac{\hbar}{r}\left(\frac{P_{n\kappa}(r)}{r}\right)} \\ +i\frac{c\hbar\kappa}{r}\left(\frac{P_{n\kappa}(r)}{r}\right) - ie\phi(r)\left(\frac{Q_{n\kappa}(r)}{r}\right) \end{array}\right] = -iE_{n\kappa}\left(\frac{Q_{n\kappa}(r)}{r}\right),$$

$$\text{or, } -c\hbar\left[\frac{d}{dr}+\frac{\kappa}{r}\right]P_{n\kappa}(r) + \left(e\phi(r) - \mu c^2\right)Q_{n\kappa}(r) = E_{n\kappa}Q_{n\kappa}(r). \tag{7.135}$$

Similarly, using Eqs. 7.134a,b in Eq. 7.133, we get

$$\left[\begin{array}{l} -\mu c^2\left(\frac{P_{n\kappa}(r)}{r}\cancel{\{\Omega_{-\kappa m}(\hat{r})\}}\right) + ic\hbar\frac{\partial}{\partial r}\left(\frac{iQ_{n\kappa}(r)}{r}\cancel{\{\Omega_{-\kappa m}(\hat{r})\}}\right) + \\ ic\frac{\hbar}{r}\left(\frac{iQ_{n\kappa}(r)}{r}\cancel{\{\Omega_{-\kappa m}(\hat{r})\}}\right) - i\frac{c\hbar\kappa}{r}\left(\frac{iQ_{n\kappa}(r)}{r}\cancel{\{\Omega_{-\kappa m}(\hat{r})\}}\right) \\ -e\phi(r)\left(\frac{P_{n\kappa}(r)}{r}\cancel{\{\Omega_{-\kappa m}(\hat{r})\}}\right) \end{array}\right] = -E_{n\kappa}\left(\frac{P_{n\kappa}(r)}{r}\cancel{\{\Omega_{-\kappa m}(\hat{r})\}}\right),$$

i.e.,

$$-\mu c^2\frac{P_{n\kappa}(r)}{r} - c\hbar\frac{d}{dr}\left(\frac{Q_{n\kappa}(r)}{r}\right) - c\frac{\hbar}{r}\frac{Q_{n\kappa}(r)}{r} + \frac{c\hbar\kappa}{r}\frac{Q_{n\kappa}(r)}{r} - e\phi(r)\frac{P_{n\kappa}(r)}{r} = -E_{n\kappa}\frac{P_{n\kappa}(r)}{r},$$

or

$$-\mu c^2\frac{P_{n\kappa}(r)}{r} - c\hbar\frac{1}{r}\frac{dQ}{dr} \cancel{+c\hbar\frac{Q}{r^2} - c\frac{\hbar}{r}\frac{Q_{n\kappa}(r)}{r}} + \frac{c\hbar\kappa}{r}\frac{Q_{n\kappa}(r)}{r} - e\phi(r)\frac{P_{n\kappa}(r)}{r} = -E_{n\kappa}\frac{P_{n\kappa}(r)}{r},$$

$$\text{and hence } c\hbar\left(\frac{d}{dr}-\frac{\kappa}{r}\right)Q_{n\kappa}(r) + \left(e\phi(r) + \mu c^2\right)P_{n\kappa}(r) = E_{n\kappa}P_{n\kappa}(r). \tag{7.136}$$

The elimination of the vector spherical harmonics that has worked out is a ratification of our ansatz (Eq. 7.130) that the Dirac four-component wavefunction for the Coulomb potential is separable in radial and angular part. Equations 7.135 and 7.136 are coupled differential equations for $P_{n\kappa}(r)$ and $Q_{n\kappa}(r)$, which are respectively called as the large and the small part of the radial functions. The so-called small component becomes increasingly important with the number of protons in the nucleus. The electron probability density $u^{\dagger}_{n\kappa m}u_{n\kappa m}$ is distributed among both the large and the small parts of the (4-component) wave function, and the normalization condition is

$$\int_0^{\infty} r^2 dr \left\{ \left| \frac{P_{n\kappa}(r)}{r} \right|^2 + \left| \frac{Q_{n\kappa}(r)}{r} \right|^2 \right\} = \int_0^{\infty} dr \left\{ P_{n\kappa}(r)^2 + Q_{n\kappa}(r)^2 \right\} = 1. \qquad (7.137)$$

The probability densities, however, are very unevenly distributed in the functions $P_{n\kappa}(r)$ and $Q_{n\kappa}(r)$. When Z is small, the function $P_{n\kappa}(r)$ approximates to the nonrelativistic wavefunction and is called as the large part of the wavefunction. Readers may refer to other sources [10] for methods to obtain solutions to the coupled differential equations Eq. 7.135 and Eq. 7.136.

Solved Problems

P7.1: From Eq. 7.37, show that $i\hbar \frac{\partial}{\partial t}\psi_{4\times 1} = \left(c\vec{\alpha}\bullet\vec{\pi} + \beta mc^2 + e\phi\right)_{4\times 4}\psi_{4\times 1} = H\psi_{4\times 1}$.

Solution:

Expanding Eq. 7.37, we have

$$\left(i\hbar\left(\gamma^0\partial_0 + \gamma^1\partial_1 + \gamma^2\partial_2 + \gamma^3\partial_3\right) - \frac{e}{c}\left(\gamma^0 A_0 + \gamma^1 A_1 + \gamma^2 A_2 + \gamma^3 A_3\right) - mc\right)_{4\times 4}\psi_{4\times 1} - 0_{4\times 1},$$

From Eq. 7.3, we see that $\begin{bmatrix} A^0 \\ A^1 \\ A^2 \\ A^3 \end{bmatrix} = \begin{bmatrix} A_0 \\ -A_1 \\ -A_2 \\ -A_3 \end{bmatrix}$, and hence

$$\begin{pmatrix} i\hbar\left(\beta\frac{\partial}{\partial ct} + \beta\alpha_x\frac{\partial}{\partial x^1} + \beta\alpha_y\frac{\partial}{\partial x^2} + \beta\alpha_z\frac{\partial}{\partial x^3}\right) \\ -\frac{e}{c}\left(\beta A^0 - \beta\alpha_x A^1 - \beta\alpha_y A^2 - \beta\alpha_z A^3\right) - mc \end{pmatrix}_{4\times 4}\psi_{4\times 1} = 0_{4\times 1},$$

i.e., $\left(i\hbar\beta\frac{\partial}{\partial ct} + i\hbar\beta\vec{\alpha}\bullet\vec{\nabla} - \frac{e}{c}\beta\phi + \frac{e}{c}\beta\vec{\alpha}\bullet\vec{A} - mc\right)_{4\times 4}\psi_{4\times 1} = 0_{4\times 1}$, which gives us the result.

P7.2: Determine Ω' (Eq. 7.63a) in the limit $\xi \to 1$.

Solution:

Expanding $F(\xi)$ near $\xi = 0$, we see that

$$\Omega' = \lim_{\xi\to 1}\left\{\sum_{n=0}^{\infty}\frac{\xi^n}{n!}\left[\frac{\partial^n F(\xi)}{\partial \xi^n}\right]_{\xi=0}\right\} = \lim_{\xi\to 1}\left\{\begin{matrix} \frac{\xi^0}{0!}F_{\xi=0} + \frac{\xi}{1!}\left(\frac{\partial F}{\partial \xi}\right)_{\xi=0} + \frac{\xi^2}{2!}\left(\frac{\partial^2 F}{\partial \xi^2}\right)_{\xi=0} + \\ \frac{\xi^3}{3!}\left(\frac{\partial^3 F}{\partial \xi^3}\right)_{\xi=0} + \frac{\xi^4}{4!}\left(\frac{\partial^4 F}{\partial \xi^4}\right)_{\xi=0} \cdots \end{matrix}\right\},$$

i.e., $\Omega' = F_{\xi=0} + \left(\frac{\partial F}{\partial \xi}\right)_{\xi=0} + \frac{1}{2}\left(\frac{\partial^2 F}{\partial \xi^2}\right)_{\xi=0} + \frac{1}{6}\left(\frac{\partial^3 F}{\partial \xi^3}\right)_{\xi=0} + \frac{1}{24}\left(\frac{\partial^4 F}{\partial \xi^4}\right)_{\xi=0} + ..$

We note that the limit $\xi \to 1$ is sought after the partial derivatives of $F(\xi)$ are determined at $\xi = 0$.

For every integer n, we need $\left[\frac{\partial^n F(\xi)}{\partial \xi^n}\right]_{\xi=0}$.

We obtain $\frac{\partial F(\xi)}{\partial \xi} = \frac{\partial}{\partial \xi} e^{+i\xi S}\Omega e^{-i\xi S} = \left(iSe^{+i\xi S}\right)\Omega e^{-i\xi S} + e^{+i\xi S}\Omega\left(-iSe^{-i\xi S}\right) = ie^{+i\xi S}[S,\Omega]_- e^{-i\xi S}$,

$\frac{\partial^2 F(\xi)}{\partial \xi^2} = i^2 e^{+i\xi S}\left[S,[S,\Omega]\right]e^{-i\xi S}$; $\frac{\partial^3 F(\xi)}{\partial \xi^3} = i^3 e^{+i\xi S}\left[S,\left[S,[S,\Omega]\right]\right]e^{-i\xi S}$.

Thus, $\frac{\partial^n F(\xi)}{\partial \xi^n} = i^n e^{+i\xi S}\left[S,\left[S,\left[S,...[S,\Omega]\right]\right]\right]e^{-i\xi S}$, and hence $\left[\frac{\partial^n F(\xi)}{\partial \xi^n}\right]_{\xi=0} = i^n\left[S,\left[S,\left[S,...[S,\Omega]\right]\right]\right]$.

Thus, for $n = 0$, $\left[\frac{\partial^n F(\xi)}{\partial \xi^n}\right]_{\xi=0} = F_{\xi=0} = \Omega$,

for $n = 1$, $\left[\frac{\partial F(\xi)}{\partial \xi}\right]_{\xi=0} = i[S,\Omega]$; for $n = 2$, $\left[\frac{\partial^2 F(\xi)}{\partial \xi^2}\right]_{\xi=0} = i^2\left[S,[S,\Omega]\right]$;

for $n = 3$, $\left[\frac{\partial^3 F(\xi)}{\partial \xi^3}\right]_{\xi=0} = i^3\left[S,\left[S,[S,\Omega]\right]\right]$, etc.

P7.3: Show that (a) the second term in H' (Eq. 7.67) turns out to be $-\theta + \frac{\beta\theta^2}{mc^2} + \frac{1}{2mc^2}\beta[\theta,\varepsilon]_-$, and (b) the third term turns out to be $-\frac{\beta\theta^2}{2mc^2} - \frac{\theta^3}{2m^2c^4} - \frac{1}{8m^2c^4}\left[\theta,[\theta,\varepsilon]\right]$.

Solution:

The second term is $i[S,H]_- = i\left[\frac{-i\beta\theta}{2mc^2},\beta mc^2 + \theta + \varepsilon\right]_- = \left[\frac{\beta\theta}{2mc^2},\beta mc^2\right]_- + \left[\frac{\beta\theta}{2mc^2},\theta\right]_- + \left[\frac{\beta\theta}{2mc^2},\varepsilon\right]_-$,

i.e., $i[S,H]_- = \frac{1}{2}(\beta\theta\beta - \beta^2\theta) + \frac{1}{2mc^2}(\beta\theta^2 - \theta\beta\theta) + \frac{1}{2mc^2}(\beta\theta\varepsilon - \varepsilon\beta\theta)$.

Since $\beta\theta = -\theta\beta$ and $\beta\varepsilon = \varepsilon\beta$, $i[S,H]_- = \frac{1}{2}(-\beta\beta\theta - \beta^2\theta) + \frac{1}{2mc^2}(\beta\theta^2 + \beta\theta\theta) + \frac{1}{2mc^2}(\beta\theta\varepsilon - \beta\varepsilon\theta)$.

Therefore, $i[S,H]_- = -\theta + \frac{\beta\theta^2}{mc^2} + \frac{1}{2mc^2}\beta[\theta,\varepsilon]_-$.

The third term in H' is $\frac{i^2}{2}\left[S,[S,H]_-\right]_- = \frac{i}{2}\left[\begin{matrix} S,-\theta + \frac{\beta\theta^2}{mc^2} + \\ \frac{1}{2mc^2}\beta[\theta,\varepsilon]_- \end{matrix}\right]_- = \frac{i}{2}\left[\begin{matrix} \frac{-i\beta\theta}{2mc^2},-\theta + \frac{\beta\theta^2}{mc^2} + \\ \frac{1}{2mc^2}\beta[\theta,\varepsilon]_- \end{matrix}\right]_-$,

$\frac{i^2}{2}\left[S,[S,H]_-\right]_- = -\frac{\beta\theta^2}{2mc^2} - \frac{\theta^3}{2m^2c^4} - \frac{1}{8m^2c^4}\left[\theta,[\theta,\varepsilon]\right]$.

P7.4: A Minkowski space–time diagram is a geometric representation of motion in space–time. Time is usually plotted along the vertical axis as per common convention. Considering the propagation of light, the spatial distance traversed by light in time t, given by ct, is plotted orthogonal to the time axis. Thus, light on a Minkowski diagram travels at 45° to the time axis, along the surfaces of a 45° cone, called as the light cone. A point in space–time is a *world-point event*; its locus on the diagram is a *world-line*.

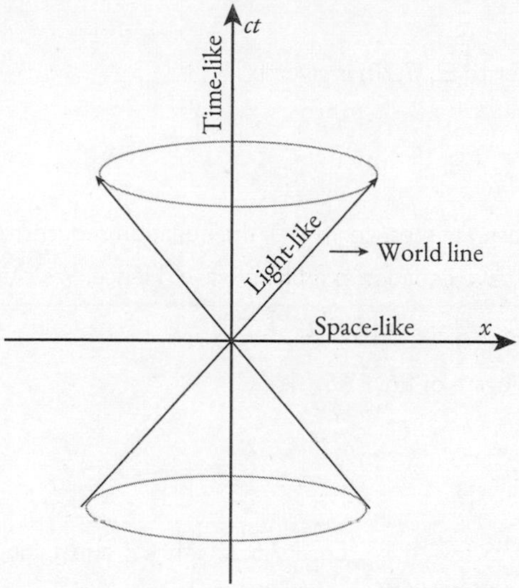

From Eq. 7.4a, the invariant interval between two events is $d^2 = g_{\mu\nu} x^\mu x^\nu$. Event coordinates for which $d^2 = c^2t^2 - |\vec{x}|^2 \gg 0$ belong to a region of space–time called *time-like*, and those for which $d^2 = c^2t^2 - |\vec{x}|^2 \ll 0$ belong to the space–time region called *space-like*. Events for which $d^2 = c^2t^2 - |\vec{x}|^2 = 0$ are on the light-cone whose surface defines a region known as *light-like*.

Physical processes take place in the *future light cone* or the forward light cone, for which $d^2 > 0$. The region $d^2 < 0$ defines the *past light cone*. Depending on where a person is located in the space–time continuum, one person's past can therefore be another's future. This has philosophical implications with regard to our perception of past, present, and future, and it even raises ontological issues. *QUESTION: An atomic clock is placed in a jet airplane. The clock measures a time interval of 3,600 s when the jet plane moves at a speed 400 m/s. How much larger a time interval does an identical clock held by an observer at rest on the ground measure?*

Solution:

Using Lorentz transformations (Eq. 7.1) for the time dilation (Eq. 7.2a), we have $\Delta t = \dfrac{\Delta\tau}{\sqrt{1-\beta^2}} \geq \Delta\tau$. Let us call the reference frame to be attached to the earth as frame S and the S frame to be the rest frame of the atomic clock. S' frame is the reference frame of the atomic clock that is present in the jet air plane. It is given that the proper time $\Delta\tau = 3600\text{s}$ as per the clock in the moving jet plane. The time interval in the rest frame (S frame) is $\Delta t \neq \Delta\tau$. We see that $\gamma = \frac{1}{\sqrt{1-\beta^2}} \simeq 1 + \frac{\beta^2}{2}$. From Eq. 7.2a, $\Delta t = \gamma\Delta\tau$. The difference in the intervals measured by the two clocks therefore is $\Delta t - \Delta\tau = (\gamma - 1)\Delta\tau \simeq \frac{\beta^2}{2}\Delta\tau = 3.2ns$.

P7.5: Show that the KG equation leads to a probability density that can possibly be negative and therefore it is unphysical.

Solution:

The KG equation (refer to Eq. 7.20) is given by

$$\left[\Box + \left(\frac{mc}{\hbar}\right)^2\right]\psi = 0. \qquad \text{(P7.5.1)}$$

When the rest mass vanishes, i.e. $m = 0$, the equation reduces to the wave equation.

The solution for wave equation is a plane wave. Hence, we can obtain a similar solution for the KG equation: $\psi = \exp\left(\pm ik_\mu x^\mu\right) = \exp\left(\pm i(k_0 t - \vec{k}.\vec{x})\right)$.

The complex conjugate of Eq. P7.5.1 is

$$\left[\Box + \left(\frac{mc}{\hbar}\right)^2\right]\psi^* = 0. \qquad \text{(P7.5.2)}$$

Multiplying Eq. P7.5.1 with ψ^*, Eq. P.7.5.2 with ψ^*, and taking the difference of the two, we get $\psi^*\Box\psi - \psi\Box\psi^* = 0$, i.e., $\partial_\mu\left(\psi^*\partial^\mu\psi - \psi\partial^\mu\psi^*\right) = 0$.

Therefore, $\frac{\partial}{\partial t}\left(\psi^*\frac{\partial\psi}{\partial t} - \psi\frac{\partial\psi^*}{\partial t}\right) - \vec{\nabla}\cdot\left(\psi^*\vec{\nabla}\psi - \psi\vec{\nabla}\psi^*\right) = 0$.

The probability current density *four* vector is $J^\mu = (J^0, \vec{J}) = (\rho, \vec{J})$, where

$\vec{J} = \frac{1}{2im}\left(\psi^*\vec{\nabla}\psi - \psi\vec{\nabla}\psi^*\right)$ and the density is $\rho = \frac{i}{2m}\left(\psi^*\frac{\partial\psi}{\partial t} - \psi\frac{\partial\psi^*}{\partial t}\right)$. We thus have the equation of continuity: $\frac{\partial\rho}{\partial t} + \vec{\nabla}\cdot\vec{J} = 0$.

For $\psi = \exp\left(-ik_\mu x^\mu\right) = \exp\left(-i\left(k_0 t - \vec{k}\cdot\vec{x}\right)\right)$,

$\rho = \frac{i}{2m}(-ik_0 - ik_0) = \frac{k_0}{m}$, since $k_\mu = (k_0, \vec{k}) = (E, \vec{k})$.

Using now $p_\mu p^\mu = \hbar^2 k_\mu k^\mu$ (and $\hbar = 1$), we get

$E^2 - p^2 = \left(k_0\right)^2 - \vec{k}\cdot\vec{k}$. Hence, $m^2 = \left(k_0\right)^2 - \vec{k}\cdot\vec{k}$; or $\left(k_0\right)^2 = \vec{k}\cdot\vec{k} + m^2$.

We are therefore led to the conclusion that $k_0 = \pm\sqrt{\vec{k}\cdot\vec{k} + m^2} = \pm E$. Therefore, the probability density turns out to be positive and also negative, which is unphysical.

P7.6: Prove the following properties of gamma matrices (Eq. 7.29a and Eq. 7.29b) using Clifford Algebra $[\gamma^\mu,\gamma^\nu] = \gamma^\mu\gamma^\nu + \gamma^\nu\gamma^\mu = 2g^{\mu\nu}$.

(i) $(\gamma^0)^2 = I$

(ii) $(\gamma^i)^2 = -I$

(iii) $[\gamma^0,\gamma^i] = 0$

(iv) $[\gamma^i,\gamma^j] = 0$

Solution:

(i) For $\mu = 0,\ \nu = 0$

$$[\gamma^0,\gamma^0] = \gamma^0\gamma^0 + \gamma^0\gamma^0 = 2g^{00}$$
$$\Rightarrow 2(\gamma^0)^2 = 2g^{00}$$
$$\Rightarrow (\gamma^0)^2 = I$$

(ii) For $\mu = i,\ \nu = i$

$$[\gamma^i,\gamma^i] = \gamma^i\gamma^i + \gamma^i\gamma^i = 2g^{ii}$$
$$\Rightarrow 2(\gamma^i)^2 = 2g^{ii}$$
$$\Rightarrow (\gamma^i)^2 = -I$$

(iii) For $\mu = 0,\ \nu = i$

$$[\gamma^0,\gamma^i] = 2g^{0i} = 0$$

(iv) For $\mu = i,\ \nu = j$ with $i \neq j$

$$[\gamma^i,\gamma^j] = 2g^{ij} = 0.$$

P7.7: Prove that $Tr(\gamma^0) = 0$ (see Eq. 7.28a).

Solution:

We see that

$$Tr(\gamma^i\gamma^0\gamma^i) = Tr\gamma^i(-\gamma^i\gamma^0) = -Tr(\gamma^i)^2\gamma^0 = Tr(\gamma^0).$$

Also,

$$Tr(\gamma^i\gamma^0\gamma^i) = Tr(\gamma^i\gamma^i\gamma^0) = -Tr(\gamma^0).$$

Taking the difference of the above two results, we get

$$2Tr(\gamma^0) = 0\text{, i.e., } Tr(\gamma^0) = 0.$$

Exercises

E7.1: Consider the KG equation in the presence of an EM field $A^\mu = (A^0,\vec{A})$. Treat the interaction term in the minimal coupling (Eq. 7.34b) scheme.

(a) Take the nonrelativistic limit and show that the KG equation reduces to the Schrödinger equation with minimal coupling.

(b) Comment on the charge and current densities in the presence of an EM field for a scalar wave function.

(c) Determine the reflection and transmission coefficient from an electrostatic potential having the form $\phi(z) = \begin{cases} 0, z < 0 \\ \phi_0, z > 0 \end{cases}$. What is the *Klein paradox?*

E7.2: What should be the minimum dimension of gamma matrices? Are those the only matrices that are appropriate to develop a relativistic formulation of quantum mechanics?

E7.3: Prove that (a) $[\alpha^i, \alpha^j]_+ = 2\delta^{ij}$ (b) $[\alpha^i, \beta]_+ = 0$

The form of β and α matrices are presented in Eq. 7.32b and Eq. 7.32c.

E7.4: The free particle positive energy Dirac solution is $\psi = N\begin{pmatrix} \phi \\ \frac{\vec{\sigma}\bullet\vec{p}}{E+m}\phi \end{pmatrix}\exp\left(ik_\mu \bullet x^\mu\right)$, $E = \sqrt{k^2 + m^2}$ and $\hbar = c = 1$. Assuming normalization $\psi^\dagger\psi = E/m$, obtain the normalization constant N. Show that for the negative energy solution it is also possible to choose normalization so that $\psi^\dagger\psi$ is positive.

E7.5: Define a conjugate operator $\overline{\psi} = \psi^\dagger(x)\gamma^0$.

(a) Obtain the Dirac equation that can be satisfied by $\overline{\psi}$ in the gamma matrix notation. (b) Show that the Dirac probability current $j^\mu = \overline{\psi}\gamma^\mu\psi$ satisfies the conservation law $\partial_\mu j^\mu = 0$.

E7.6: Use an alternative choice for the matrices $\vec{\alpha}$ and β in the Dirac equation as

$\vec{\alpha} = \begin{pmatrix} -\vec{\sigma} & 0 \\ 0 & \vec{\sigma} \end{pmatrix}$ and $\beta = \begin{pmatrix} 0 & 1 \\ 1 & 0 \end{pmatrix}$ and write the wavefunction as $\psi = \begin{pmatrix} \phi \\ \chi \end{pmatrix} e^{ip.x}$, $(\hbar = c = 1)$.

Obtain the equations satisfied by ϕ and χ. Verify if $E^2 = p^2 + m^2$.

E7.7: Show that the differential equations for ϕ and χ decouple if $m = 0$.

E7.8: Determine the explicit forms for ϕ and χ for the case $\vec{p} = p(\sin\theta, 0, \cos\theta)$ and $m = 0$, satisfying $\phi^\dagger\phi = \chi^\dagger\chi = 1$.

E7.9: What is *Kramer's recursion rule?*

E7.10: Determine the energy shift due to a perturbation $H'_{RKE} = \frac{-p^4}{8m^3c^2}$ (Eq. 7.78) for the $n = 3$ state of Hydrogen.

E7.11: Calculate the energies of all the Dirac levels (in eV) for the $n = 1,2,3$ states of the hydrogen atom. Do not include hyperfine splitting. Sketch the levels in a schematic diagram and show all possible electric dipole transitions.

E7.12: If the general form of the spin–orbit coupling for a particle of mass m and spin $\vec{S}$ moving in a potential $V(r)$ is $H'_{so} = \frac{1}{2m^2c^2}\vec{L}\bullet\vec{S}\frac{1}{r}\frac{dV}{dr}$, what is the effect of that coupling on the spectrum of an electron bound in isotropic three-dimensional isotropic harmonic oscillator?

E7.13: Sketch the energy level diagram for the three-dimensional isotropic harmonic oscillator inclusive of the spin–orbit coupling. Show the energy shifts due to the spin–orbit coupling for the lowest two states.

E7.14: Muons (mass: ~207 m_e) have a mean lifetime of ~2.2 μs in their rest frame and decay as per the following pathways:

$$\mu^- \to e^- + \bar{\nu}_e + \nu_\mu \quad ; \quad \mu^+ \to e^+ + \nu_e + \bar{\nu}_\mu$$

The mouns travel at ~99.4% the speed of light. If muon decay takes place in the upper atmosphere about 10 to 15 km above the earth's surface, how far can they travel? How come we can see the decay tracks on the earth's surface in a cloud chamber experiment? (See [16].)

References

[1] P. C. Deshmukh, *Foundations of Classical Mechanics* (Cambridge University Press, New Delhi, 2019).

[2] O. Klein, Quantentheorie und fünfdimensionale relativitätstheorie, *Zeitschrift für Physik*, 37:12, 895–906 (1926).

[3] W. Gordon, Der comptoneffekt nach der schrödingerschen theorie, *Zeitschrift f'ur Physik* 40:1–2, 117–133 (1926).

[4] P. A. M. Dirac, The Quantum Theory of the Electron. *Proc. Roc. Soc.* (Lond.) A117, 610–624 (1928).

[5] P. A. M. Dirac, On the Preponderance of Matter over Antimatter. *Proc. Roc. Soc.* (Lond.) A118, 351–361 (1928).

[6] J. D. Bjorken and S. D. Drell, *Relativistic Quantum Mechanics* (McGraw-Hill, New York, 1964).

[7] T. D. Newton and E. P. Wigner, Localized States for Elementary Systems. *Revs. Mod. Phys.* 21, 400 (1949).

[8] L. L. Foldy and S. A. Wouthuysen, On the Dirac Theory of Spin 1/2 Particles and Its Non-Relativistic Limit. *Phys Rev.*, 78:29 (1950).

[9] John P. Costella and Bruce H. J. McKellar, The Foldy-Wouthuysen Transformation, *Am. J. Phys.*, 63:12 (1995).

[10] W. R. Johnson, *Atomic Structure Theory – Lectures on Atomic Physics* (Springer, Berlin, 2007).

[11] J. P. Dahl and Th. Jorgensen, On the Dirac-Kepler Problem: The Johnson-Lippmann Operator, Supersymmetry and Normal-mode Representations, *Int. J. Quantum Chem.*, 53, 161–181 (1995).

[12] T. T. Khachidze and A. A. Khelashvili, The hidden symmetry of the Coulomb problem in relativistic quantum mechanics: From Pauli to Dirac, *Am. J. Phys.* 74, 628 (2006).

[13] H. Katsura and H. Aoki, Exact Supersymmetry in the Relativistic Hydrogen Atom in General Dimensions – Supercharge and the Generalized Johnson-Lippmann Operator, *J. Math. Phys.*, 47, 032301 (2006).

[14] Jing-Ling Chen, Dong-Ling Deng, and Ming-Guang Hu, SO(4) symmetry in the Relativistic Hydrogen Atom, *Phys. Rev.*, A 77, 034102 (2008).

[15] A. A. Stahlhofen, Comment on "SO(4) Symmetry in the Relativistic Hydrogen Atom," *Phys. Rev.*, A 78, 036101 (2008).

[16] Voma Uday Kumar, Gnaneswari Chitikela, Niharika Balasa, and P. C. Deshmukh, Revisiting Table-Top Demonstration of Relativistic Time-Dilation and Length-Contraction, *Bulletin of the Indian Association of Physics Teachers*, 11:6, 172–179 (2019).

8

Quantum Mechanics of Spectral Transitions

GMRT. http://www.ncra.tifr.res.in/ncra/gmrt/about-gmrt/goals-of-gmrt. Courtesy: NCRA , TIFR, India.

The GMRT (Giant Metrewave Radio Telescope) is an array of thirty fully steerable parabolic radio telescopes having a diameter of 45 meters for radio astronomical research at meter wavelengths. It is an extremely advanced facility set up by the National Centre for Radio Astrophysics at Khodad (near Pune, India).

Theories of the formation of structure in the Big-Bang Universe predict the presence of proto galaxies or proto clusters of galaxies made up of clouds of neutral hydrogen gas before their gravitational condensation into galaxies. It should in principle be possible to detect these through the well-known radio line emitted by neutral hydrogen at a frequency of 1420 MHz. (see Fig. 7.4 in Chapter 7). The line is, however, expected to be very weak and red-shifted to meter wavelengths because of the expansion of the universe between emission, billions of years ago, and detection at the present epoch. Visit: http://www.ncra.tifr.res.in/ncra/gmrt/about-gmrt/goals-of-gmrt (courtesy: NCRA , TIFR, India). (downloaded on 26 March 2023).

In the previous seven chapters, we have introduced the barebones of quantum mechanics. The methodology we reviewed equips us to study the physical universe we live in. Spectroscopy is a powerful experimental technique used to investigate matter. Basically, spectroscopy investigates a target system through its interaction with a probe, which can be electromagnetic radiation, or elementary or composite particles, or their combination. Our prototype of the target system is an atom. Atomic spectroscopy, or more generally atomic physics and quantum physics developed synchronously. These disciplines continue to interpenetrate each other to jointly push the frontiers of science, engineering, and technology.

8.1 Spectroscopic Oscillator Strengths

Spectroscopy literature is embellished by terminology from classical, semi-classical, and quantum-mechanical models. An all too pervasive term employed in the description of spectral transitions is the *oscillator strength* [1, 2]. It has its origins in the earliest explanatory models that used classical mechanics. The phenomenology pertaining to the spectroscopic oscillator strength is now comprehensively integrated into the fully quantum mechanical scheme. It is insightful to track the metamorphosis of the term *oscillator strength* from its earliest avatar in semi-classical physics to its current usage within the quantum models.

Let us consider a spectral transition in an atom which results from the absorption of electromagnetic energy. It is natural to inquire how much power is pumped by the electric field into the atomic system. In the classical model, each atomic electron is modeled as an oscillator bound elastically about an equilibrium position. This electron oscillator would be engaged in an interaction with an applied electromagnetic field. The time-dependent electric field is represented by $\hat{\varepsilon}E_0 e^{-i(\omega t+\theta)}$, E_0 being the amplitude of the electric field, $\hat{\varepsilon}$ the unit vector along which the field is polarized, and θ is a phase factor. The elastic restoring force on the electron is $-k\vec{r}(t)$, k being the spring constant of the model oscillator and $\vec{r}(t)$ the instantaneous position of the electron. Damping of this oscillator must be accounted for, as the electron would lose energy due to its coupling with unspecified degrees of freedom. The electron's motion may be hindered by other electrons, and it could also radiate away some of its energy when accelerated. The electron's dynamics is then described by the equation of motion (Chapter 5 of Reference [3]) for a damped and driven oscillator:

$$m\ddot{\vec{r}}(t) = -\left(m\omega_{0,s}^2\right)\vec{r} - \left(\Gamma_d\right)m\dot{\vec{r}} + \hat{\varepsilon}eE_0e^{-i(\omega t+\theta)}. \tag{8.1}$$

Funquest: Just what is damping? The electromagnetic interaction is conservative, so what is causing energy loss?

In Eq. 8.1, m and e are respectively the mass and charge of the electron, $\omega_{0,s}$ is its natural (circular) frequency of oscillation, and ω is the circular frequency of the electric field. Γ_d is an effective damping constant and, $\hat{\varepsilon}E_0e^{-i(\omega t+\theta)}$ is the driving force per unit charge. It is easy to see that the solution to this differential equation of motion for the damped and driven electron oscillator is

$$\vec{r}(t) = \hat{\varepsilon}\frac{e}{m}\frac{E_0}{\omega_{0,s}^2 - \omega^2 - i\Gamma_d\omega}e^{-i(\omega t+\theta)}. \tag{8.2}$$

We see that $\vec{r} \to 0$ as $E_0 \to 0$, i.e., the displacement of the electron in this model about its mean equilibrium position is essentially due to the applied electric field. This displacement results in an induced dipole moment. Notwithstanding the fact that this model is completely classical and allows for a simultaneous accurate description of the position and velocity of the electron, it provides a fruitful basis, as will soon be seen, to the notion of atomic photoionization cross section and its relationship to physical properties of the atom, such as its polarizability. The (induced) dipole moment of the electron is proportional to the applied electric field,

$$\vec{d} = \alpha\vec{E}. \tag{8.3}$$

The proportionality constant α is the dipole moment per unit field, known as the atomic polarizability. Clearly,

$$\alpha(\omega)=\frac{|e\vec{r}|}{|\vec{E}|}=\frac{e\left|\hat{\varepsilon}\frac{e}{m}\frac{E_0}{\omega_{0,s}^2-\omega^2-i\Gamma_d\omega}e^{-i(\omega t+\theta)}\right|}{\left|\hat{\varepsilon}E_0e^{-i(\omega t+\theta)}\right|}=\frac{e^2}{m}\frac{1}{\omega_{0,s}^2-\omega^2-i\Gamma_d\omega}, \tag{8.4a}$$

$$\text{i.e., } \alpha(\omega)\approx\frac{e^2}{m}\frac{1}{2\omega_{0,s}\left[\left(\omega_{0,s}-\omega\right)-i\frac{\Gamma_d\omega}{2\omega_{0,s}}\right]}, \tag{8.4b}$$

near the resonance frequency $\omega\approx\omega_{0,s}$. Also, one can see that

$$\frac{1}{\omega_{0,s}^2-\omega^2-i\Gamma_d\omega}=\frac{m}{e^2}\alpha(\omega). \tag{8.5}$$

The frequency dependence of the atomic polarizability is explicitly pointed out in the above relation. Using Eq. 8.2 and Eq. 8.5, we get for the instantaneous position of the electron

$$\vec{r}(t)=\hat{\varepsilon}\frac{E_0\alpha(\omega)}{e}e^{-i(\omega t+\theta)}. \tag{8.6}$$

The instantaneous dipole moment is then given by

$$\vec{d}(t)=e\vec{r}(t)=\hat{\varepsilon}E_0\alpha(\omega)e^{-i(\omega t+\theta)}. \tag{8.7}$$

The average power pumped into the atomic system by the electric field provides a measure of the photoionization probability. It is the time average of the rate at which work W is done by the electric field on the charge and is given by

$$\langle Q\rangle=\frac{1}{T}\int_0^T\frac{dW}{dt}dt=\frac{1}{T}\int_0^T\frac{\vec{F}\bullet\overrightarrow{dr}}{dt}dt, \tag{8.8a}$$

i.e.,

$$\langle Q\rangle=\frac{1}{T}\int_0^T\left[\left(\frac{\vec{F}}{e}\right)\bullet\left(e\dot{\vec{r}}\right)\right]dt. \tag{8.8b}$$

The physical quantity of interest is the real part

$$\langle Q\rangle_{\text{Real}}=\frac{1}{T}\int_0^T\left[\text{Re}\left(\frac{\vec{F}}{e}\right)\bullet\text{Re}\left(e\dot{\vec{r}}\right)\right]dt. \tag{8.9}$$

The scalar product in the integrand is

$$\text{Re}\left(\frac{\vec{F}}{e}\right)\bullet\text{Re}\left(e\dot{\vec{r}}\right)=\hat{\varepsilon}\frac{E_0}{2}\left[e^{-i(\omega t+\theta)}+e^{+i(\omega t+\theta)}\right]\bullet\hat{\varepsilon}\frac{E_0\omega}{2}\left[(-i)\alpha(\omega)e^{-i(\omega t+\theta)}+(+i)\alpha^*(\omega)e^{+i(\omega t+\theta)}\right],$$

i.e.,

$$\text{Re}\left(\frac{\vec{F}}{e}\right)\bullet\text{Re}\left(e\dot{\vec{r}}\right)=\frac{E_0^2\omega}{4}\left[-i\alpha(\omega)e^{-i2(\omega t+\theta)}+i\alpha^*(\omega)-i\alpha(\omega)+i\alpha^*(\omega)e^{+i2(\omega t+\theta)}\right]. \tag{8.10}$$

The first and the fourth terms on the right-hand side of Eq. 8.10 are time-dependent. It is of natural interest to examine what these two terms would contribute over long time intervals. The integration over time is from 0 to ∞, but the range of integration can be extended from −∞ to +∞ and recognizing that E_0 is zero for $t < 0$. We note that

$$\int_{t\to-\infty}^{t\to\infty} e^{\mp i2(\omega t+\theta)}dt = \pm\frac{\pi}{2}\delta(\omega). \tag{8.11}$$

This term, under spectral integration, can contribute only for $\omega = 0$, which corresponds to the static field (no oscillations). The second and the third terms on the right-hand side of Eq. 8.10 correspond to the sum of a complex number with its complex conjugate, and hence they are equal to twice its imaginary part. Thus, the contribution from the terms in the bracket on the right-hand side of Eq. 8.10 adds up to $2\,\mathrm{Im}\,\alpha(\omega)$. The average power pumped into the atomic system by the electric field is therefore proportional to the imaginary part of the frequency-dependent atomic dipole polarizability:

$$\langle Q\rangle_{\mathrm{Real}} = \frac{\omega E_0^2}{2}\mathrm{Im}\,\alpha(\omega). \tag{8.12}$$

Furthermore, since

$$\alpha(\omega) = \frac{e^2}{m}\frac{1}{\omega_{0,s}^2 - \omega^2 - i\Gamma_d\omega}\times\frac{\omega_{0,s}^2 - \omega^2 + i\Gamma_d\omega}{\omega_{0,s}^2 - \omega^2 + i\Gamma_d\omega},$$

we find that

$$\mathrm{Im}\left[\alpha(\omega)\right] = \frac{e^2}{m\left(\omega_{0,s}+\omega\right)^2}\frac{\Gamma_d\omega}{\left(\omega_{0,s}-\omega\right)^2 + \left(\frac{\Gamma_d\omega}{\left(\omega_{0,s}+\omega\right)^2}\right)^2}. \tag{8.13}$$

Close to the resonance energy, $\omega \simeq \omega_{0,s}$, we get

$$\mathrm{Im}\left[\alpha(\omega)\right] \simeq \frac{e^2}{m\left(2\omega_{0,s}\right)^2}\frac{\Gamma_d\omega}{\left(\omega_{0,s}-\omega\right)^2 + \left(\frac{\Gamma_d\omega}{\left(2\omega_{0,s}\right)^2}\right)^2} = \frac{e^2}{2m\omega_s}\frac{\frac{\Gamma_d}{2}\left(\frac{\omega}{\omega_s}\right)}{\left(\omega_{0,s}-\omega\right)^2 + \left(\frac{\Gamma_d}{2}\right)^2\left(\frac{\omega}{\omega_s}\right)^2},$$

i.e.,

$$\mathrm{Im}\left[\alpha(\omega\to\omega_s)\right] \simeq \frac{e^2}{2m\omega_s}\frac{\frac{\Gamma_d}{2}}{\left(\omega_{0,s}-\omega\right)^2 + \left(\frac{\Gamma_d}{2}\right)^2}. \tag{8.14}$$

Essentially, we have ignored the *difference* between unity and $\frac{\omega}{\omega_s}$ *compared* to unity, but not the said difference *itself* when it stands alone.

Hence, using Eq. 8.14 in Eq. 8.12, we get

$$\langle Q \rangle_{\text{Real}} = \frac{E_0^2 e^2}{4m} \frac{\frac{\Gamma_d}{2}}{\left(\omega_{0,s} - \omega\right)^2 + \left(\frac{\Gamma_d}{2}\right)^2}. \tag{8.15}$$

The average power pumped into the atomic system is now related to the frequency-dependent spectral distribution, called as the *oscillator strength* $\frac{df}{d\omega}$. It is defined such that its integral over all possible frequencies adds up to unity:

$$1 = \int_0^\infty \frac{df}{d\omega} d\omega = \int_{-\infty}^\infty \frac{df}{d\omega} d\omega, \tag{8.16a}$$

there being no transitions corresponding to negative frequencies. This choice of the definition of the oscillator strength involving the resolution of unity turns out to be very fruitful, since we have considered at this point exactly *1* electron in the photoabsorption process. Later, in Section 8.4, you will find that this can be readily extended to *N* electrons. Now, we also have the mathematical resolution of the unity in terms of the Dirac-δ function (more appropriately called as a functional or a generalized distribution), given by

$$1 = \int_{-\infty}^\infty \delta\left(\omega - \omega_{0,s}\right) d\omega. \tag{8.16b}$$

Comparing Eqs. 8.16a and 8.16b, we see that

$$\frac{df}{d\omega} = \delta\left(\omega - \omega_{0,s}\right). \tag{8.17}$$

Defining the oscillator strength this way enables one to borrow the physically appealing and intuitive formalism of the dynamics of the classical oscillator, and apply it fruitfully in the quantum description of atomic photoionization (Chapter 6). There are several representations of the Dirac-δ [4], of which the following form provides a forthright connect with the oscillator strength:

$$\delta\left(\omega - \omega_{0,s}\right) = \lim_{\Gamma_d \to 0} \frac{1}{\pi} \frac{\frac{\Gamma_d}{2}}{\left\{\left(\omega - \omega_{0,s}\right)^2 + \left(\frac{\Gamma_d}{2}\right)^2\right\}}. \tag{8.18}$$

The right-hand side of the above equation is the *Lorentzian representation* of the Dirac-δ function (see the Solved Problem P8.10). It has its highest value at $\omega = \omega_{0,s}$, and it quickly becomes small $(\to 0)$ as ω moves away from $\omega_{0,s}$. From Eq. 8.16 and Eq. 8.18, we get the following expression for the oscillator strength, expressed now in terms of the Dirac-δ function:

$$\frac{df}{d\omega} = \delta\left(\omega - \omega_{0,s}\right) = \lim_{\Gamma_d \to 0} \frac{1}{\pi} \frac{\frac{\Gamma_d}{2}}{\left\{\left(\omega - \omega_{0,s}\right)^2 + \left(\frac{\Gamma_d}{2}\right)^2\right\}}. \tag{8.19}$$

Using Eq. 8.15, we now get the average power pumped into the atomic system by the electric field in terms of the spectral distribution of the oscillator, namely the atomic oscillator strength:

$$\langle Q\rangle = \frac{\pi}{4}\frac{e^2E_0^2}{m}\frac{df}{d\omega}. \tag{8.20}$$

It is no surprise that the power pumped contains the Dirac-δ function, since one surely expects the power absorbed to peak at the resonance. It is also useful to bear in mind that this average power was related to the imaginary part of the atomic polarizability (Eq. 8.12). This property makes it possible to connect our formalism to macroscopic observables, since the atomic polarizability $\alpha(\omega)$ is proportional to the susceptibility $\chi(\omega)$ of the medium:

$$\chi(\omega) = \eta\alpha(\omega), \tag{8.21a}$$

η being the number of atoms per unit volume in a dilute atomic gas. In turn, the susceptibility provides the proportionality (Chapter 12 of Reference [3]) between the polarization (which is the dipole moment per unit volume) property of the medium and the applied electric field:

$$\vec{P}(\vec{r},t) = \chi(\omega)\vec{E}(\vec{r},t). \tag{8.21b}$$

The photoionization cross section σ (Eq. 6.120) that we obtained in Chapter 6 is just the average power pumped into the atomic oscillator system per unit intensity I of the electric field:

$$\sigma = \frac{\langle Q\rangle}{I}. \tag{8.22a}$$

In the semi-empirical treatment of atomic photoionization, the electromagnetic radiation absorbed by an atom is treated as a classical wave described by Maxwell's equations. In Chapter 6, we used Fermi's golden rule to obtain the photoionization cross section σ, in terms of $W_{i\to f}$, the transition rate (transition probability per unit time), given by Eq. 6.123, for an atomic electron to be ejected from a discrete bound state into a continuum state. The resulting photoionization cross section, per unit intensity of the electric field, with $\hbar\omega$ taking care of the energy absorbed in the process, is

$$\sigma = \frac{\langle Q\rangle}{I} = \frac{\hbar\omega}{I}W_{i\to f}. \tag{8.22b}$$

The primary physical quantity in the oscillator strength is the transition matrix element M, which we have used in Eqs. 6.132a,b of Chapter 6. We immediately recognize the presence of the momentum operator in it, since

$$M = \left\langle f \mid e^{i\vec{k}\bullet\vec{r}}\hat{\varepsilon}\bullet\vec{\nabla}\mid i\right\rangle = \frac{i}{\hbar}\left\langle f \mid e^{i\vec{k}\bullet\vec{r}}\hat{\varepsilon}\bullet\vec{p}\mid i\right\rangle. \tag{8.23a}$$

The exponential term in the matrix element can be expanded in powers of $\frac{r}{\lambda}$. For large wavelengths, i.e., low energies (up to ~6 keV) above the ionization threshold, one may approximate $e^{i\vec{k}\bullet\vec{r}} \approx 1$. For reasons that will soon become clear, this approximation is called as the *dipole approximation*, in which we set

$$M = \left\langle f \mid e^{i\vec{k}\bullet\vec{r}}\hat{\varepsilon}\bullet\vec{\nabla}\mid i\right\rangle = \frac{i}{\hbar}\left\langle f \mid p_\varepsilon \mid i\right\rangle. \tag{8.23b}$$

At shorter wavelengths the dipole approximation fails, and one must go beyond it [5, 6]. The reasons for the failure of the dipole approximation are sometimes rather complex, some of which would be alluded to later.

From the fundamental uncertainty relation between position and momentum, we now note that

$$\left[r_k, p_k^2\right] = r_k p_k p_k - p_k p_k r_k = \left[r_k, p_k\right] p_k + p_k \left[r_k, p_k\right] = 2i\hbar p_k. \tag{8.24}$$

Considering now the fact that the potential energy operator commutes with the position operator for each of the three Cartesian components $k = 1, 2, 3$, we have

$$[r_k, H_0] = [r_k, \frac{p^2}{2m}] = \frac{i\hbar}{m} p_k. \tag{8.25}$$

$$\text{Thus, } M = \frac{i}{\hbar}\langle f|p_\varepsilon|i\rangle = \frac{i}{\hbar}\left\langle f\left|\frac{m}{i\hbar}[r_\varepsilon, H]\right|i\right\rangle = \frac{m}{\hbar^2}(E_i - E_f)\langle f|r_\varepsilon|i\rangle, \tag{8.26a}$$

$$\text{and using } \hbar\omega_{if} = -\hbar\omega_{fi}, \text{ we get } (p_\varepsilon)_{fi} = \langle f|p_\varepsilon|i\rangle = im\omega_{fi}\langle f|r_\varepsilon|i\rangle = im\omega_{fi}(r_\varepsilon)_{fi}, \tag{8.26b}$$

$$\text{for the } \textit{components } p_\varepsilon = \hat{\varepsilon}\bullet\vec{p} \text{ and } r_\varepsilon = \hat{\varepsilon}\bullet\vec{r}. \tag{8.26c}$$

The matrix element of the momentum operator in Eq. 8.23b is therefore related to that of the position operator by the following relation:

$$\vec{p}_{fi} = im\omega_{fi}\vec{r}_{fi}. \tag{8.27}$$

The transition matrix element can be determined in either the *momentum form*, also called as the *velocity form* (since these two are scaled only by the mass), or equivalently the *position form*, also called as the length form [1, 2, 7], since position/distance has the dimension of length. There is also a third form, called as the acceleration form, for which we refer the readers to Reference [8]. It is the *length form* that gives the approximation $e^{i\vec{k}\bullet\vec{r}} \approx 1$ its name, viz., the *dipole approximation*, since the probability amplitude for transition from the initial bound state $|i\rangle$ to the continuum state $|f\rangle$ involves the matrix element of the *position* operator (Eq. 8.23b and Eq. 8.27); after all, it is directly proportional to the *dipole moment* operator, the dipole moment merely being the product of charge and *displacement*.

Funquest: In view of the spectroscopic transition selection rules that we obtain using the Wigner–Eckart theorem (Section 4.4, Chapter 4), what will be the selection rules in the dipole approximation? How would these change if we include just one more term beyond the dipole approximation in the power series expansion of $e^{i\vec{k}\bullet\vec{r}}$?

As mentioned earlier, however, terms ignored in the dipole approximation are often of importance, not merely at energies well above the ionization threshold, but sometimes at *much lower energies* [5,6]. At the dipole 'Cooper minimum' (Fig. 5.3a,b in Chapter 5), for example, higher multi-pole terms become important even at low energies relative to the dipole term, since the latter goes to zero. Nonetheless, the dipole approximation is in general very successful and accounts for a vast majority of spectral transitions. The equality (Eq. 8.27) of the length and velocity forms of the transition matrix elements is a property of the central field potential, since in obtaining it we have exploited the commutation of the potential energy operator with the position operator. In a many-electron atom, which we discuss in Chapter 9, the potential experienced by every electron is *not* local, and therefore it does not commute with the position operator,

$$\text{i.e.,} \left[\vec{r}, V\left(\vec{r},\vec{r}'\right)\right] \neq 0, \tag{8.28}$$

and Eq. 8.27 does not hold.

The transition matrix element can be factored into a *physical part* and a *geometrical part* using the Wigner–Eckart theorem (Section 4.4 of Chapter 4). In the dipole approximation, transitions can take place between initial and final states for which the orbital angular momentum differs, and the selection rule is $\delta\ell = \pm 1$. However, when this rule is satisfied, the photoionization cross section does not *always* diminish monotonically with energy, as one expects from Eq. 6.141. The cross section diminishes toward a Cooper minimum (Fig. 5.3a,b, Chapter 5), but then it may rise again [1, 2, 5, 6, 9]. The phenomenology that results is illustrated in the Fig. 8.1.

It should be clear from Fig. 5.3a,b (Chapter 5) and Fig. 8.1 that the Cooper minimum is a result of the discrete bound state in the photoionization transition matrix having one or more nodes. It would not occur in the case of those bound state orbitals that have no nodes in their radial wavefunctions. The number of nodes in the radial wavefunction is $n - \ell - 1$ (Eq. 5.49). Thus, photoionization from 1*s*, 2*p*, 3*d*, 4*f*, and 5*g* orbitals would not have a Cooper minimum. Fig. 5.3a,b (Chapter 5) and Fig. 8.1 show the change in the sign of the radial matrix element when it goes through the Cooper minimum. In the illustration seen in the Fig. 8.1, we see that the oscillator strength *first* rises immediately above the ionization threshold. Its maximum is not at the threshold, but it is at a slightly higher energy; it is therefore called as a *delayed* maximum. It is due to the centrifugal potential (see the discussion following Eq. 5.18 in Chapter 5) in the radial Schrödinger equation. The delayed maximum results from the *small-r* behavior of the radial functions (Fig. 5.1 in Chapter 5), which tells us that radial functions (including those in the continuum) go to zero more rapidly near the origin, for higher orbital angular momentum

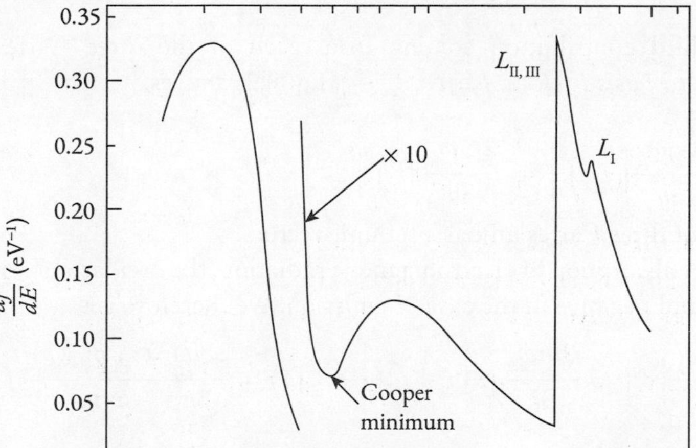

Fig. 8.1 The change in sign of the matrix elements (Eq. 8.26) makes the photoionization cross-section go through a zero at some energy, called as the *Cooper minimum*. It is illustrated here, with permission from Reference [2]. The *delayed maximum* in the photoionization cross section is also seen in this figure.

quantum number ℓ. Naturally, if the final continuum state in the photoionization matrix element has a high value of ℓ, it would not penetrate the atomic region where the initial state radial function of a discrete bound state from which photoionization originates, and thus have less overlap with the initial state wavefunction, resulting in a smaller value of the photoionization transition probability. As the energy of the continuum state would increase above the threshold, a continuum wavefunction would progressively *overcome* the centrifugal barrier, thereby increasing its overlap with the discrete bound initial state. The cross section therefore first increases above the threshold, as indicated by the oscillator strength shown in Fig. 8.1. Thus, phenomena such as (a) the occurrence of the Cooper minimum, (b) the centrifugal barrier effect, and (c) the opening up of an inner threshold, all cause a *departure* from the usual monotonic decrease of the photoionization cross section from the threshold with increasing photon energy. More complex correlation effects, such as an autoionization resonance [2], also result in dramatic fluctuations in the cross section.

Figure 8.2 shows an interesting illustration, from Reference [10], of the delayed maximum in photoionization from the $4f$ subshell of atomic mercury (Hg). These plots have been obtained using a theoretical method called as the relativistic random phase approximation (RRPA). It is a relativistic many-electron theory whose discussion is beyond the scope of this book. However, the figure is included here to raise an interesting feature of the photoionization cross section. The minimum seen in the photoionization cross section at ~160 eV in this figure is *not* a Cooper minimum, since the $4f$ radial function is nodeless. Rather, it is the result of competing oscillator strengths in $4f \rightarrow \varepsilon d$ and $4f \rightarrow \varepsilon g$ dipole channels (where ε stands for the final state of the photoelectron in the continuum), shown in the lower panel of Fig. 8.2.

The definition given in Eq. 8.19 of the oscillator strength in the *classical model* is modified in the *semi-classical* quantum model (SCQM) in which the oscillator strength for a one-electron atomic system ($N = 1$) is now *defined* by

$$f_{fi} \underset{definition}{\overset{QM,\, N=1}{=}} \frac{2m\omega_{fi}}{\hbar}\left|\vec{r}_{fi}\right|^2 = \frac{2mE_{fi}}{\hbar^2}\left|\vec{r}_{fi}\right|^2. \tag{8.29a}$$

With one-third contribution coming from each of the three Cartesian components $\left(\textit{corresponding to basis vectors } \hat{\varepsilon} \equiv \hat{\varepsilon}_x, \hat{\varepsilon}_y, \hat{\varepsilon}_z\right)$, Eq. 8.29a gives,

$$f_{fi}^{(r_\varepsilon)} \underset{definition}{\overset{QM}{=}} \frac{2m\omega_{fi}}{3\hbar}\left|\left(\vec{r}\cdot\hat{\varepsilon}\right)_{fi}\right|^2 = \frac{2m\omega_{fi}}{3\hbar}\left|\left(r_\varepsilon\right)_{fi}\right|^2, \tag{8.29b}$$

where $\hat{\varepsilon}$ is one of three Cartesian basis set unit vectors.

In the case of absorption of electromagnetic radiation, the oscillator strength is considered to be positive, and negative in the case of emission. We therefore get

$$f_{fi}^{(r_\varepsilon)} = \frac{2m\omega_{fi}}{3\hbar}\left|x_{fi}\right|^2 = \frac{2m\omega_{fi}}{3\hbar}\langle i \mid r_\varepsilon \mid f\rangle\langle f \mid r_\varepsilon \mid i\rangle = \frac{2m\omega_{fi}}{3\hbar}\frac{\langle i \mid p_\varepsilon \mid f\rangle}{im\omega_{if}}\langle f \mid r_\varepsilon \mid i\rangle,$$

$$\text{i.e., } f_{fi}^{(r_\varepsilon)} = +\frac{2i}{3\hbar}\langle i \mid p_\varepsilon \mid f\rangle\langle f \mid r_\varepsilon \mid i\rangle, \text{ since } \omega_{fi} = -\omega_{if}. \tag{8.30a}$$

$$\text{Likewise, } f_{fi}^{(r_\varepsilon)} = \frac{2m\omega_{fi}}{3\hbar}\left|\left(r_\varepsilon\right)_{fi}\right|^2 = \frac{2m\omega_{fi}}{3\hbar}\langle i \mid r_\varepsilon \mid f\rangle\langle f \mid r_\varepsilon \mid i\rangle = \frac{2m\omega_{fi}}{3\hbar}\langle i \mid r_\varepsilon \mid f\rangle\frac{\langle f \mid p_\varepsilon \mid i\rangle}{im\omega_{fi}},$$

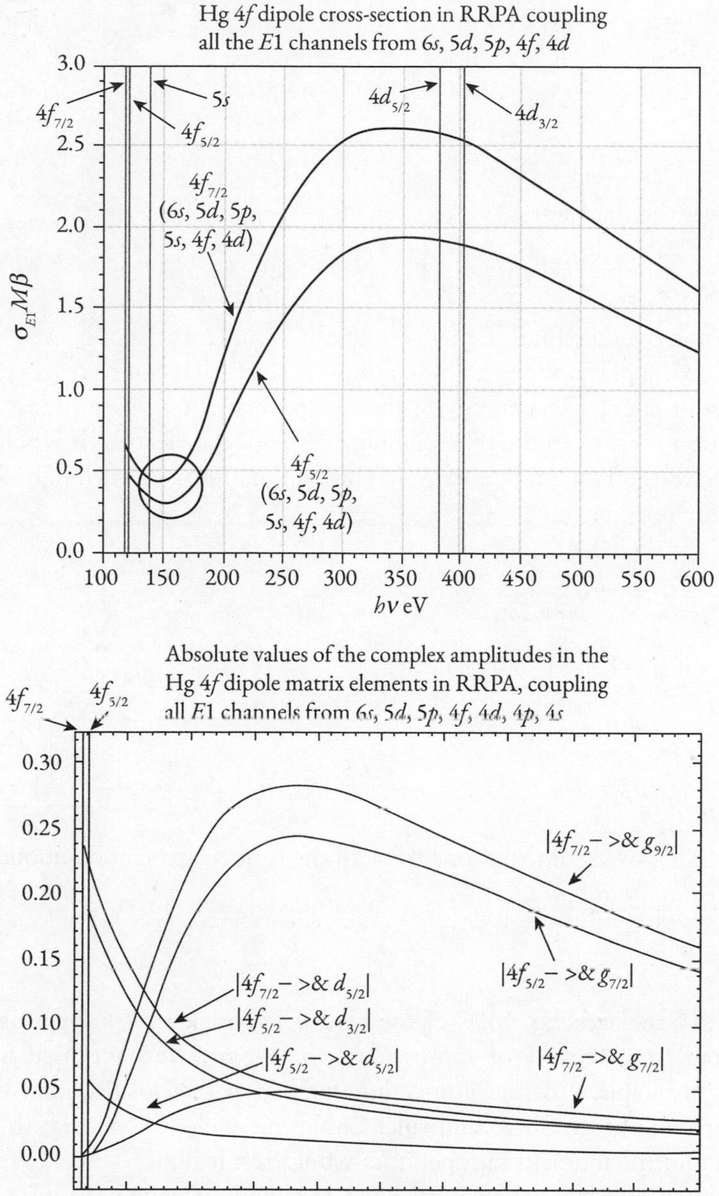

Fig. 8.2 Photoionization of atomic mercury from its $4f$ subshell. The upper panel shows that the cross sections go through a minimum a little above 150 eV, then goes through a delayed maximum, and eventually it decreases. The figure on the right-hand side shows the magnitudes of the complex matrix elements in the six relativistic dipole channels: $4f_{7/2} \to \varepsilon g_{9/2}$, $4f_{7/2} \to \varepsilon g_{7/2}$, $4f_{7/2} \to \varepsilon d_{5/2}$ and $4f_{5/2} \to \varepsilon g_{7/2}$, $4f_{5/2} \to \varepsilon d_{5/2}$, $4f_{5/2} \to \varepsilon d_{3/2}$. The continuum energy of the final state is represented by the symbol ε. Transition matrix amplitudes in channels ($4f \to g$) with $\delta\ell = +1$ show *delayed maximum* due to the centrifugal barrier effect, discussed in the text. These figures are results of theoretical calculations using the relativistic random phase approximation (RRPA) from Reference [10].

$$\text{i.e., } f_{fi}^{(r_\varepsilon)} = -\frac{2i}{3\hbar}\langle i \mid r_\varepsilon \mid f\rangle\langle f \mid p_\varepsilon \mid i\rangle. \tag{8.30b}$$

Adding half of Eq. 8.30a to that of Eq. 8.30b, we get

$$f_{fi}^{(r_\varepsilon)} = \frac{i}{3\hbar}\left[\langle i \mid p_\varepsilon \mid f\rangle\langle f \mid r_\varepsilon \mid i\rangle - \langle i \mid r_\varepsilon \mid f\rangle\langle f \mid p_\varepsilon \mid i\rangle\right]. \tag{8.31}$$

Summing over the final states,

$$\sum_f f_{fi}^{(r_\varepsilon)} = \frac{i}{3\hbar}\sum_f \left[\langle i \mid p_\varepsilon \mid f\rangle\langle f \mid r_\varepsilon \mid i\rangle - \langle i \mid r_\varepsilon \mid f\rangle\langle f \mid p_\varepsilon \mid i\rangle\right]. \tag{8.32}$$

The summation over the final state $|f\rangle$ essentially includes the sum over all discrete bound states, and also an integration over all the continuum states that can be reached as determined by conservation of energy. The energy of the electromagnetic radiation absorbed in single-photon absorption must overcome the binding energy of the electron it would eject; it must therefore be above the ionization energy I. This summation is accordingly described more fully by the following equivalence:

$$\sum_f |f\rangle\langle f| \equiv \sum_f^{discrete} |f\rangle\langle f| + \int_{E_f > I} dE_f, \tag{8.33a}$$

$$\text{with (left hand side being) } \sum_f |f\rangle\langle f| = 1. \tag{8.33b}$$

From Eq. 8.32, it follows that

$$\sum_f f_{fi}^{(r_\varepsilon)} = \frac{i}{3\hbar}\left[\langle i \mid p_\varepsilon r_\varepsilon - r_\varepsilon p_\varepsilon \mid i\rangle\right] = -\frac{i}{3\hbar}\left[\langle i \mid [r_\varepsilon, p_\varepsilon]_- \mid i\rangle\right] = -\frac{i}{3\hbar}(i\hbar) = \frac{1}{3}. \tag{8.34}$$

Finally, summing over contributions from all the three Cartesian components, we get for this *semi-classical quantum model*

$$\sum_f f_{fi}^{QM} = 1. \tag{8.35}$$

Equation 8.35 replaces Eq. 8.16 of the classical model; it is known as the *Thomas–Reiche–Kuhn (TRK) sum rule for the oscillator strengths*. We have used in the Eq. 8.34 the uncertainty principle, viz., the non-commutativity of the position and the momentum operators, in arriving at the TRK sum rule. Could one have asked what the commutation of the position and the momentum operators would result in, if one believed that Eq. 8.35 must correspond to the classical result in Eq. 8.16? Wouldn't one have discovered the non-commutativity of the position and momentum operators, and thereby the uncertainty principle? As has been pointed out in Reference [2], it was this reverse logic (in conjunction with various other spectroscopic analyses that Heisenberg was deeply involved with) that in fact led Heisenberg to develop the principle of uncertainty.

Replacement of Eq. 8.19 by the *definition* in Eq. 8.29 and the corresponding replacement of Eq. 8.16 by Eq. 8.35 have resulted from the modification of the classical model (a bound electron represented by a damped and driven oscillator) by the semi-classical *quantum model* (a bound electron a represented by a state vector $|i\rangle$). The intuitive classical idea and the vocabulary of using the oscillator is thus *extended*, albeit with appropriate modifications in quantum mechanics. The quantum model now developed is adequate for most purposes, but one must remember that strictly speaking it is not *fully* quantum mechanical; only the atomic system is quantized, not the electromagnetic radiation field it interacts with. It is for this reason that this model is called *semi-classical*. There are other situations, such as the description of a two-level atomic system interacting with a tuned laser, that need for their description a fully quantum model. In this, the radiation field is *also* quantized. The Jaynes–Cummings model of cavity electrodynamics (cQED) describes the fully quantum mechanical model [11].

From Eq. 8.29 and Eq. 8.35, we see that

$$\sum_f \omega_{fi} \left|\vec{r}_{fi}\right|^2 = \frac{\hbar}{2m}, \tag{8.36}$$

which is another sum rule. The TRK sum rule in fact belongs to a *family of sum rules* for $\sum_f \omega_{fi}^n \left|\vec{r}_{fi}\right|^2$. Equation 8.35 corresponds to $n = 0$, and Eq. 8.36 to $n = 1$. These terms are respectively referred to as "moment 0" and "moment 1." References [1, 2] should be read for further details. When the number of electrons is greater than 1, the quantum mechanical definition in Eq. 8.29 is summed over all the electrons in an N-electron atom, since each electron could participate in photoabsorption. Hence, we have the following result:

$$\sum_f f_{fi}^{(N)} \underset{\text{definition}}{\overset{N\text{-electron atom}}{=}} \sum_f \frac{2m\omega_{fi}}{\hbar} \left| \sum_{n=1}^{N} \vec{r}_{fi}^{(n)} \right|^2 = N. \tag{8.37}$$

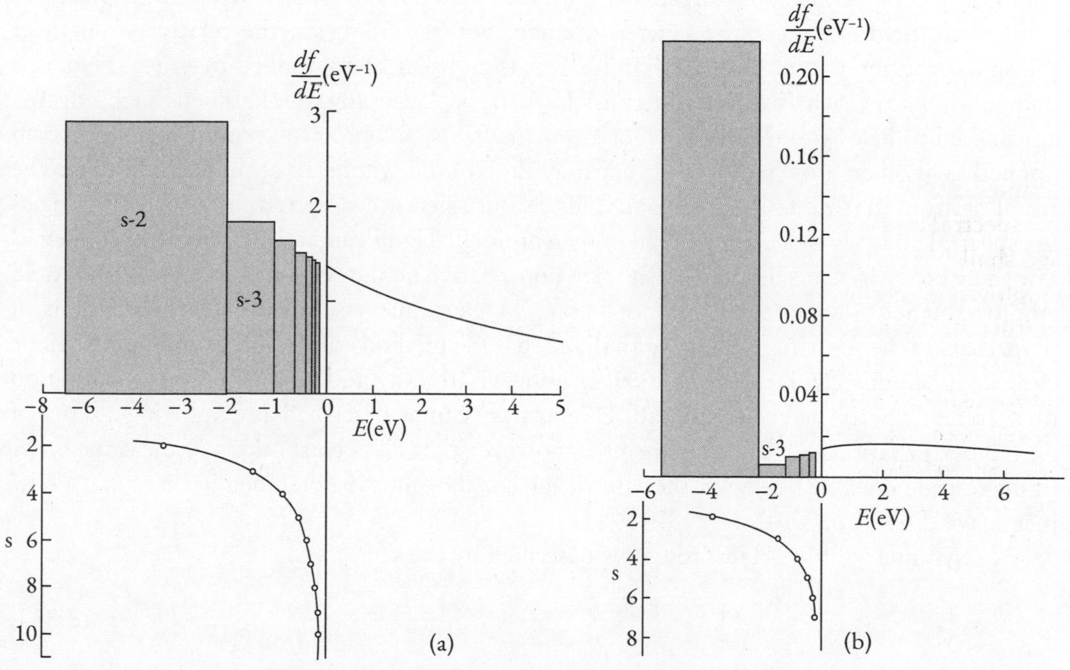

Fig. 8.3 Oscillator strength distributions (reproduced with permission from Ref. [2]) in the discrete and the continuous spectra of (a) H (theory) and (b) Li (experiment). Oscillator strengths for discrete transitions are plotted as histogram blocks of height $\frac{f_s}{\Delta E_s}$, where the base of each block is determined by the geometrical construction of tangents to the E_s versus s curve as shown in the lower part of (a).

The oscillator strengths are continuous across the ionization threshold as seen in Fig. 8.3a,b, reproduced from References [1, 2]. This property is of great utility in the development of the single-channel and multi-channel quantum defect theory [12, 13] (SC-QDT and MQDT). The Fig. 8.3b even has a Cooper minimum, which in this case is in the discrete part of the spectrum. However, most often it is above the ionization threshold. In a few cases, there can even be multiple Cooper minima [14]. Finally, we would like to point out that the ideas developed in this chapter are primarily based on quantum collision theory. After all, electron–ion collision is related to photoionization by the *time-reversal theory* [2, 15, and 16], discussed in Chapter 10. The notion of oscillator strength (and the *generalized oscillator strength*) is of great importance in theories of quantum collisions [1].

When an atom is placed in an applied electromagnetic field, the atomic structure is altered. One must then re-examine the atomic structure, and very often methods of perturbation theory (Chapter 6) are effective. We shall discuss this in the next section.

8.2 Stark–Lo Surdo Effect and Parity-Violating Spectroscopic Transitions

Spectroscopy of an atom placed in an electric field is known after Johannes Stark and Antonio Lo Surdo. The underlying modification of the atomic structure resulting from the interaction of an atom with an applied electric field is called as the Stark–Lo Surdo effect, or more commonly only as the Stark effect. On the other hand, spectroscopy of an atom placed in a magnetic field is named after Pieter Zeeman, but depending on the relative strength of the spin–orbit interaction (Eq. 7.84 and Fig. 7.4) and that of an applied magnetic field, it is named after Friedrich Paschen and Ernst E. A. Back. Generally, the Paschen–Back effect is considered to belong to the "family" of Zeeman effects. Taking into account the effect of an applied field on an atom requires the consideration of the symmetry of the Hamiltonian. The simplest atom that we take up for our study as our reference system to build methodologies for spectral analysis is of course the hydrogen atom. Its Hamiltonian is spherically symmetric. We shall begin by considering the simpler non-relativistic model; our model system would employ the Schrödinger equation, not Dirac. We can improvise on it by using methods of perturbation theory (Chapter 6). We shall see that the methods of perturbation theory can be cascaded, i.e., one can incorporate *further* improvisation by adding a (smaller) perturbation on an already perturbed Hamiltonian. This strategy can be described as employing a series of nested perturbations. One must remember, however, that all perturbations of the same order of magnitude must be treated at the same level, i.e., they must be included in the same "nest," just as we did in Eq. 7.102.

The symmetry of the Hamiltonian is described by the set

$$G = \left\{R_i;\ i = 1,2,3,....,h\right\} \tag{8.38a}$$

of all operators which commute with the Hamiltonian,

$$\text{i.e., for every } i,\ \left[R_i, H\right]_- = 0\,. \tag{8.38b}$$

The set G constitutes a mathematical group. It is called as the *Group of Schrödinger equation*. Since R_i commutes with H for every i, all of these operators can be diagonalized simultaneously with the Hamiltonian. Hence, the search for quantum eigenstates of the system can be made among the eigenstates of the symmetry operators (see the Solved Problem P1.8 in Chapter 1). This remarkable property provides a widely used strategy to solve quantum mechanical problems using group theory. This approach is fundamental to solving problems not only in atomic physics, but also in molecular and condensed matter physics, as well as in nuclear and particle physics. The matrix representations (denoted below by fat letters) of the symmetry operators commute with that of the Hamiltonian. Hence, considering the orthonormal eigenbasis of the symmetry operator,

$$[\mathbb{R},\mathbb{H}]_{i,j} = 0, \tag{8.39a}$$

$$\text{i.e.,} \sum_k \mathbb{R}_{ik}\delta_{ik}\mathbb{H}_{kj} - \sum_k \mathbb{H}_{ik}\mathbb{R}_{kj}\delta_{kj} = 0, \tag{8.39b}$$

$$\text{or } \mathbb{H}_{ij}\left(\mathbb{R}_{ii} - \mathbb{R}_{jj}\right) = 0. \tag{8.39c}$$

Hence, the condition $\mathbb{R}_{ii} \neq \mathbb{R}_{jj}$ ensures that $\mathbb{H}_{ij} = 0$. In other words, the Hamiltonian has no *non*zero matrix element between eigenstates of a symmetry operator R belonging to different eigenvalues of R. In other words, off-diagonal elements of H cannot connect functions of different symmetry. Furthermore, if the operator P_R that operates on the wavefunction space represents the symmetry operator R, then for the *stationary* eigenstates of the Hamiltonian one can see that

$$P_R\left(H\psi_n\right) = P_R\left(E\psi_n\right) \tag{8.40a}$$

$$\text{and } H\left(P_R\psi_n\right) = E\left(P_R\psi_n\right). \tag{8.40b}$$

We see from Eqs. 8.40a,b that the functions $P_R\psi_n$ and ψ_n are essentially degenerate. A perturbation, such as the application of an external electric and/or magnetic field, can remove the degeneracy of the quantum eigenstates – *if its symmetry is different from that of the Hamiltonian*. Depending on the complete symmetry of the Hamiltonian, the perturbation would break the symmetry partially or fully. One must remember that the Group of Schrödinger equation may consist of both *geometrical* symmetries (like rotations, reflections, and translations) and also *dynamical* symmetries; an example of the latter is the Pauli–Lenz dynamical symmetry (Eq. 5.67 of Chapter 5). A physical system may also have *discrete* symmetries (see Appendix A).

We shall first discuss the Stark effect. The potential experienced by an electron in the hydrogen atom placed in a constant electric field $\vec{\varepsilon} = \varepsilon\hat{e}_z$ is the sum of the Coulomb potential and $e\varepsilon z$, which is the potential energy of the interaction between the electron and the applied electric field. This effective potential V is schematically shown in Fig. 8.4. The z-axis is *chosen* to be the same as that of the applied electric field.

The full Hamiltonian H for the system is no longer represented by the nonrelativistic Hamiltonian H_0; it must now include the extra potential due to the interaction of the electron with the applied field. Under conditions that the extra potential can be treated using perturbation theory, it is given by

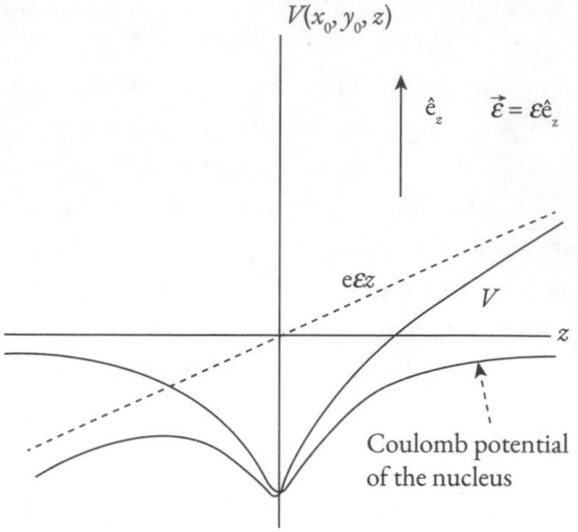

Fig. 8.4 Schematic diagram showing the effective potential experienced by an electron in the hydrogen atom placed in a constant electric field.

$$H = H_0 + H' = \left\{ \frac{\left(-i\hbar\vec{\nabla}\right)^2}{2m} - \frac{Ze^2}{r} \right\} + \lambda e\varepsilon z, \tag{8.41}$$

where λ is a perturbation order parameter, discussed in Chapter 6. Our final results will be, as expected, independent of this order parameter.

From Eq. 6.88 of Chapter 6, we know that the first order correction to the energy of the ground state of the hydrogen atom due to the perturbation term in Eq. 8.41 is

$$E^{(1)}_{n=1,\ell=0,m=0} = \left\langle \psi_{n=1,\ell=0,m=0} \middle| e\varepsilon z \middle| \psi_{n=1,\ell=0,m=0} \right\rangle = e\varepsilon \iiint d^3V \left\{ z \left| \psi_{n=1,\ell=0,m=0}\left(\vec{r}\right) \right|^2 \right\} = 0. \tag{8.42}$$

One really does not have to evaluate the above triple integral. Since it would involve integration over whole space, it must include integration over the z-coordinate from $-\infty$ to $+\infty$ and would go to zero, the integrand being an odd function of z. Since there is no first order perturbative correction to the energy of the ground state of the hydrogen atom, we conclude that there is no Stark effect on the ground state that is linear in ε. We proceed to determine the second order correction. From Eq. 6.92 of Chapter 6, it is

$$E^{(2)}_{n=1,\ell=0,m=0} = e^2\varepsilon^2 \sum_{\substack{n\neq 1\\ \ell,m}} \frac{\left| \left\langle \psi_{n,\ell,m} \middle| z \middle| \psi_{n=1,\ell=0,m=0} \right\rangle \right|^2}{E_1 - E_n} = -e^2\varepsilon^2 \sum_{\substack{n\neq 1\\ \ell,m}} \frac{\left| \left\langle \psi_{n,\ell,m} \middle| z \middle| \psi_{n=1,\ell=0,m=0} \right\rangle \right|^2}{E_n - E_1}. \tag{8.43}$$

Since for every n, $\left(E_n - E_1\right) > 0$, we conclude that the ground-state energy is *lowered*; it becomes *more* negative. Now, from Eq. 5.47 of Chapter 5, we know that the hydrogen atom's energy levels go as $\frac{1}{n^2}$, and for every $n > 2$, $\left(E_n - E_1\right) > \left(E_2 - E_1\right)$. Furthermore, the energy levels crowd in as the value of the principle quantum number increases, as shown in

Fig. 5.2. We may therefore extract $\frac{1}{E_2 - E_1}$ from the summation in Eq. 8.43 and obtain a *lower bound* for the 2nd order perturbation correction. This enables us to write the inequality

$$E^{(2)}_{n=1,\ell=0,m=0} > -e^2\varepsilon^2 \frac{1}{E_2 - E_1} \sum_{\substack{n\neq 1\\ \ell,m}} \left|\left\langle \psi_{n,\ell,m} \middle| z \middle| \psi_{n=1,\ell=0,m=0} \right\rangle\right|^2. \tag{8.44}$$

In order to obtain this lower bound, we need the sum

$$S = \sum_{n\neq 1,\ell,m} \left\langle \psi_{n=1,\ell=0,m=0} \middle| z \middle| \psi_{n,l,m} \right\rangle \left\langle \psi_{n,l,m} \middle| z \middle| \psi_{n=1,\ell=0,m=0} \right\rangle,$$

or, including the term for $n = 1$ (which adds *nothing* to the sum),

$$S = \sum_{n,\ell,m} \left\langle \psi_{n=1,\ell=0,m=0} \middle| z \middle| \psi_{n,\ell,m} \right\rangle \left\langle \psi_{n,\ell,m} \middle| z \middle| \psi_{n=1,\ell=0,m=0} \right\rangle. \tag{8.45}$$

Inclusion of $n = 1$ helps us recognize the unit operator sandwiched in Eq. 8.45, so we can immediately write

$$S = \left\langle \psi_{n=1,\ell=0,m=0} \middle| z^2 \middle| \psi_{n=1,\ell=0,m=0} \right\rangle. \tag{8.46a}$$

It is easy to evaluate the above sum. Due to symmetry, it is nothing but one-third the expectation value of the operator r^2 in the ground state of the hydrogen atom. Thus, using solution to the Solved Problem P5.4 (Chapter 5),

$$S = \frac{1}{3}\left\langle r^2 \right\rangle_{n=1,\ell=0,m=0} = \frac{a_0^2}{Z^2}, \tag{8.46b}$$

$$\text{and } E^{(2)}_{n=1,\ell=0,m=0} > -e^2\varepsilon^2 \frac{1}{E_2 - E_1}\frac{a_0^2}{Z^2}. \tag{8.47}$$

Since the applied field has symmetry about the z-axis, this problem can be solved using parabolic coordinates. The Schrödinger equation for the Coulomb potential could be solved in this coordinate system using the method of separation of variables (Chapter 5). From Eq. 8.43, we can obtain the *sensitivity* of the second order energy correction to the magnitude of the applied field:

$$\frac{\partial E^{(2)}_{n=1,\ell=0,m=0}}{\partial \varepsilon} = -\left[2e^2 \sum_{\substack{n\neq 1\\ \ell,m}} \frac{\left|\left\langle \psi_{n,\ell,m} \middle| z \middle| \psi_{n=1,\ell=0,m=0} \right\rangle\right|^2}{E_n - E_1} \right]\varepsilon. \tag{8.48}$$

Furthermore, since the dipole moment is

$$d = -\frac{\partial E^{(2)}_{n=1,\ell=0,m=0}}{\partial \varepsilon} = \bar{\alpha}\varepsilon, \tag{8.49a}$$

we immediately find the static atomic polarizability; it is given by

$$\bar{\alpha} = 2e^2 \sum_{\substack{n\neq 1\\ \ell,m}} \frac{\left|\left\langle \psi_{n,\ell,m} \middle| z \middle| \psi_{n=1,\ell=0,m=0} \right\rangle\right|^2}{E_n - E_1}. \tag{8.49b}$$

In terms of the static atomic polarizability, the 2nd order energy correction therefore is

$$E^{(2)}_{n=1,\ell=0,m=0} = -\frac{1}{2}\bar{\alpha}\varepsilon^2 = -\frac{1}{2}d\varepsilon. \tag{8.50}$$

The energy of a dipole in an electric field typically is the product of the dipole moment and the magnitude of the field, but Eq. 8.50 has a factor of *half*, since the dipole moment under consideration is *induced* by the applied field, which itself is proportional to the magnitude of the field. The expression in Eq. 8.50 bears this out. The above consideration can be extended for other systems, for example, a gas of molecules. The polarizability then turns out to be a tensor, rather than a scalar as for the hydrogen atom. As the density of the gas increases, different approximations can be developed, such as the Clausius–Mosotti relation at intermediate densities. This approach provides the first steps towards determining electric susceptibility of bulk matter.

We have determined the first order and second order Stark effect on hydrogen atom's ground state. We now inquire if and how the excited states of the hydrogen atom show the Stark effect. This can also be studied using perturbation theory, but the methods of Chapter 6 need to be improvised, since the energy states are now degenerate. We know from Chapter 5 that on account of the SO(4) symmetry of the Coulomb potential, all energy states with $n \geq 2$ have n^2 – *fold* degeneracy. This degeneracy is doubled due to the electron spin as we have learned in Chapter 7, but for now we persist with the nonrelativistic model. The four degenerate states $|n,\ell,m\rangle$ of the first excited state $(n = 2)$ are $\{|2,0,0\rangle,|2,1,-1\rangle,|2,1,0\rangle,|2,1,1\rangle\}$. There are therefore four linearly independent solutions which satisfy the time-independent Schrödinger equation with the same eigenvalue:

$$H_0\psi^{(0)}_{n=2,r}(\vec{r}) = E^{(0)}_{n=2}\psi^{(0)}_{n=2,r}(\vec{r}) \text{ for } r = 1,2,3,4. \tag{8.51a}$$

Without any loss of generality, we consider the four linearly independent functions to be orthonormalized,

$$\text{i.e., for every } r, s: \left\langle \psi^{(0)}_{n=2,s} \middle| \psi^{(0)}_{n=2,r} \right\rangle = \delta_{rs}, \tag{8.51b}$$

and similar orthogonality holds also with respect to normalized eigenfunctions corresponding to all *other* principal quantum numbers,

$$\text{i.e., for every } r, s: \left\langle \psi^{(0)}_{N\neq 2,s} \middle| \psi^{(0)}_{n=2,r} \right\rangle = 0 \text{ for } s = 1,2,3,\ldots,d_N. \tag{8.51c}$$

In writing the above result, we have recognized that the state with principal quantum number $N \neq 2$ is d_N – *fold* degenerate, with $d_N = N^2$.

The Schrödinger equation with the *full* Hamiltonian is

$$H\psi_{n=2,r}(\vec{r}) = E_{n=2,r}\psi_{n=2,r}(\vec{r}) \text{ for } r = 1,2,3,4. \tag{8.52}$$

The eigenvalues $E_{n=2,r}$ for different values of r may or may not be the same. That obviously includes the possibility that some of them are the same, others being different. Essentially, this means that the perturbation may remove the degeneracy either partially or fully (see Fig. 6.1

of Chapter 6). In general, the wavefunction in Eq. 8.52 for every value of r can be written as a superposition of (a) $\chi^{(0)}_{n=2,r}$, which itself is a mixture of the four degenerate unperturbed functions $\psi^{(0)}_{n=2,s}(\vec{r})$, with $r = 1,2,3,4$, but the mixing coefficients c_{rs} in general would be *different* for each pair (r, s), though in some special case they may turn out to be equal, and (b) various functions scaled by different powers of the perturbation order parameter λ resulting from the perturbation corrections to each $\psi_{n=2,r}(\vec{r})$, due to the perturbation interaction. Thus,

$$\text{for } r = 1,2,3,4,\ \psi_{n=2,r}(\vec{r}) = \lambda^0\chi^{(0)}_{n=2,r} + \lambda^1\psi^{(1)}_{n=2,r} + \lambda^2\psi^{(2)}_{n=2,r} + \ldots.. \tag{8.53}$$

$$\text{with } \chi^{(0)}_{n=2,r} = \sum_{s=1}^{4} c_{rs}\psi^{(0)}_{n=2,s}(\vec{r}). \tag{8.54}$$

With both r and s taking four values each, we need to determine the sixteen coefficients c_{rs}.

The energy eigenvalue in Eq. 8.52 of the full Hamiltonian can also be written as a power series in the perturbation order parameter, similar to Eq.8.53, i.e.,

$$E_{n=2,r}(\vec{r}) = \lambda^0 E^{(0)}_{n=2} + \lambda^1 E^{(1)}_{n=2,r} + \lambda^2 E^{(2)}_{n=2,r} + \ldots.. \tag{8.55}$$

The first order correction $\psi^{(1)}_{n=2,r}$ in Eq. 8.53 is expressible in terms of all the unperturbed functions which constitute a complete set of basis for an arbitrary function:

$$\text{For every } r = 1,2,3,4:\ \psi^{(1)}_{n=2,r} = \sum_{N}\sum_{s=1}^{d_N} a_{n=2,r;Ns}\psi^{(0)}_{Ns}. \tag{8.56}$$

Since one needs every basis function in the complete set that spans the space of the wavefunctions, the sum in Eq. 8.56 includes summation over *all* values of the principal quantum numbers. Up to the retention of the first order terms in the perturbation order parameter, Eq. 8.52 becomes:

For $r = 1,2,3,4$,

$$\left(H_0 + \lambda H_1\right)\left\{\lambda^0\chi^{(0)}_{n=2,r} + \lambda^1\psi^{(1)}_{n=2,r}\right\} = \left\{\lambda^0 E^{(0)}_{n=2} + \lambda^1 E^{(1)}_{n=2,r}\right\}\left\{\lambda^0\chi^{(0)}_{n=2,r} + \lambda^1\psi^{(1)}_{n=2,r}\right\}. \tag{8.57}$$

As we did in Chapter 6, we can now equate the coefficients corresponding to the same powers of λ on the two side of Eq. 8.57:

$$\text{Equating the coefficient of } \lambda^0:\ H_0\chi^{(0)}_{n=2,r} = E^{(0)}_{n=2}\chi^{(0)}_{n=2,r}, \tag{8.58}$$

wherein $\chi^{(0)}_{n=2,r}$ is given by Eq. 8.54.

$$\text{Likewise, for } \lambda^1:\ H_0\psi^{(1)}_{n=2,r} + H_1\chi^{(0)}_{n=2,r} = E^{(0)}_{n=2}\psi^{(1)}_{n=2,r} + E^{(1)}_{n=2,r}\chi^{(0)}_{n=2,r}, \tag{8.59}$$

wherein $\psi^{(1)}_{n=2,r}$ is given by Eq. 8.56.

Using Eq. 8.54 and Eq. 8.56 in Eq. 8.59, we get

$$\begin{bmatrix} H_0\left(\sum_N\sum_{s=1}^{d_N} a_{n=2,r;Ns}\psi_{Ns}^{(0)}\right)+ \\ H_1\left(\sum_{s=1}^{4} c_{rs}\psi_{n=2,s}^{(0)}(\vec{r})\right) \end{bmatrix} = \begin{bmatrix} E_{n=2}^{(0)}\left(\sum_N\sum_{s=1}^{d_N} a_{n=2,r;Ns}\psi_{Ns}^{(0)}\right)+ \\ E_{n=2,r}^{(1)}\left(\sum_{s=1}^{4} c_{rs}\psi_{n=2,s}^{(0)}(\vec{r})\right) \end{bmatrix}, \tag{8.60a}$$

$$\text{i.e.,}\quad \begin{bmatrix} \left(\sum_N\sum_{s=1}^{d_N} a_{n=2,r;Ns}E_N^{(0)}\psi_{Ns}^{(0)}\right)+ \\ H_1\left(\sum_{s=1}^{4} c_{rs}\psi_{n=2,s}^{(0)}(\vec{r})\right) \end{bmatrix} = \begin{bmatrix} E_{n=2}^{(0)}\left(\sum_N\sum_{s=1}^{d_N} a_{n=2,r;Ns}\psi_{Ns}^{(0)}\right)+ \\ E_{n=2,r}^{(1)}\left(\sum_{s=1}^{4} c_{rs}\psi_{n=2,s}^{(0)}(\vec{r})\right) \end{bmatrix}, \tag{8.60b}$$

$$\text{or}\quad \left(\sum_N\sum_{s=1}^{d_N} a_{n=2,r;Ns}\left(E_N^{(0)}-E_{n=2}^{(0)}\right)\psi_{Ns}^{(0)}\right)+\left(\sum_{s=1}^{4} c_{rs}\left(H_1-E_{n=2,r}^{(1)}\right)\psi_{n=2,s}^{(0)}(\vec{r})\right)=0. \tag{8.60c}$$

Projecting Eq. 8.60c on $\left\langle\psi_{n=2,u}^{(0)}\right|$ wherein u is some *particular* member of the set of $r = 1,2,3,4$, we get

$$\left(\sum_N\sum_{s=1}^{d_N} a_{n=2,r;Ns}\left\langle\psi_{n=2,u}^{(0)}\middle|\left(E_N^{(0)}-E_{n=2}^{(0)}\right)\middle|\psi_{Ns}^{(0)}\right\rangle\right)+\left(\sum_{s=1}^{4} c_{rs}\left\langle\psi_{n=2,u}^{(0)}\middle|\left(H_1-E_{n=2,r}^{(1)}\right)\middle|\psi_{n=2,s}^{(0)}\right\rangle\right)=0. \tag{8.61}$$

Since $\left\langle\psi_{n=2,u}^{(0)}\middle|\psi_{N\neq 2,s}^{(0)}\right\rangle = 0$ for every $N \neq 2$, while $\left(E_N^{(0)} - E_{n=2}^{(0)}\right) = 0$ for $N = n$, the first term above vanishes, and we are left with the following relation for every r, and for every u,

$$\sum_{s=1}^{4} c_{rs}\left[\left\langle\psi_{n=2,u}^{(0)}\middle|H_1\middle|\psi_{n=2,s}^{(0)}\right\rangle - E_{n=2,r}^{(1)}\delta_{us}\right] = 0. \tag{8.62}$$

We must now solve essentially linear homogeneous system of equations in four unknowns for every r. The condition for nontrivial solution is, for every r,

$$\left|\left\langle\psi_{n=2,u}^{(0)}\middle|H_1\middle|\psi_{n=2,s}^{(0)}\right\rangle - E_{n=2,r}^{(1)}\delta_{us}\right| = 0. \tag{8.63}$$

The solution is given by the roots $E_{n=2,r}^{(1)}$ for $r = 1,2,3,4$. The four roots may all be different from each other, or only some of them may be equal to each other, and others may be different. This would depend on whether the degeneracy is fully or partially removed. The residual degeneracy, if any, may possibly be removed by high order perturbation correction, or it may persist to all orders. Let us illustrate the solution for a particular value of r. For example, for $r = 2$, we use the root $E_{n=2,r=2}^{(1)}$, and get

$$\text{for every } u,\ \sum_{s=1}^{4} c_{2s}\left[\left\langle\psi_{n=2,u}^{(0)}\middle|H_1\middle|\psi_{n=2,s}^{(0)}\right\rangle - E_{n=2,r=2}^{(1)}\delta_{us}\right] = 0. \tag{8.64}$$

In particular, for $\underline{u = 1}$,

$$\begin{bmatrix} c_{21}\left\langle \psi^{(0)}_{n=2,1}\middle| H_1 \middle| \psi^{(0)}_{n=2,1}\right\rangle - E^{(1)}_{n=2,2} + \\ c_{22}\left\langle \psi^{(0)}_{n=2,1}\middle| H_1 \middle| \psi^{(0)}_{n=2,2}\right\rangle + c_{23}\left\langle \psi^{(0)}_{n=2,1}\middle| H_1 \middle| \psi^{(0)}_{n=2,3}\right\rangle + c_{24}\left\langle \psi^{(0)}_{n=2,1}\middle| H_1 \middle| \psi^{(0)}_{n=2,4}\right\rangle \end{bmatrix} = 0; \tag{8.65a}$$

for $\underline{u = 2}$,

$$\begin{bmatrix} c_{21}\left\langle \psi^{(0)}_{n=2,2}\middle| H_1 \middle| \psi^{(0)}_{n=2,1}\right\rangle - E^{(1)}_{n=2,2} + \\ c_{22}\left\langle \psi^{(0)}_{n=2,2}\middle| H_1 \middle| \psi^{(0)}_{n=2,2}\right\rangle + c_{23}\left\langle \psi^{(0)}_{n=2,2}\middle| H_1 \middle| \psi^{(0)}_{n=2,3}\right\rangle + c_{24}\left\langle \psi^{(0)}_{n=2,2}\middle| H_1 \middle| \psi^{(0)}_{n=2,4}\right\rangle \end{bmatrix} = 0; \tag{8.65b}$$

for $\underline{u = 3}$,

$$\begin{bmatrix} c_{21}\left\langle \psi^{(0)}_{n=2,3}\middle| H_1 \middle| \psi^{(0)}_{n=2,1}\right\rangle - E^{(1)}_{n=2,2} + \\ c_{22}\left\langle \psi^{(0)}_{n=2,3}\middle| H_1 \middle| \psi^{(0)}_{n=2,2}\right\rangle + c_{23}\left\langle \psi^{(0)}_{n=2,3}\middle| H_1 \middle| \psi^{(0)}_{n=2,3}\right\rangle + c_{24}\left\langle \psi^{(0)}_{n=2,3}\middle| H_1 \middle| \psi^{(0)}_{n=2,4}\right\rangle \end{bmatrix} = 0; \tag{8.65c}$$

and for $\underline{u = 4}$,

$$\begin{bmatrix} c_{21}\left\langle \psi^{(0)}_{n=2,4}\middle| H_1 \middle| \psi^{(0)}_{n=2,1}\right\rangle - E^{(1)}_{n=2,2} + \\ c_{22}\left\langle \psi^{(0)}_{n=2,4}\middle| H_1 \middle| \psi^{(0)}_{n=2,2}\right\rangle + c_{23}\left\langle \psi^{(0)}_{n=2,4}\middle| H_1 \middle| \psi^{(0)}_{n=2,3}\right\rangle + c_{24}\left\langle \psi^{(0)}_{n=2,4}\middle| H_1 \middle| \psi^{(0)}_{n=2,4}\right\rangle \end{bmatrix} = 0. \tag{8.65d}$$

We must now determine the matrix elements

$$\left\langle \psi^{(0)}_{n=2,u}\middle| H_1 \middle| \psi^{(0)}_{n=2,s}\right\rangle = \left\langle \psi_{nlm}\middle| e\varepsilon z \middle| \psi_{n'l'm'}\right\rangle, \tag{8.66}$$

with appropriate values of the quantum numbers. These can be determined from the eigenfunctions of the hydrogen atom, which we know from Chapter 5. From parity consideration, we see that all diagonal elements must be zero. Therefore, we need not determine these elements. We now examine

$$\left\langle \psi^{(0)}_{n=2,u}\middle| H_1 \middle| \psi^{(0)}_{n=2,s}\right\rangle = e\varepsilon \int_{\theta=0}^{\pi} \sin\theta d\theta \int_{\varphi=0}^{2\pi} d\varphi \int_0^{\infty} r^2 dr R_{n\ell}(r)(r\cos\theta) R_{n'\ell'}(r) Y^*_{\ell m}(\hat{r}) Y_{\ell' m'}(r). \tag{8.67a}$$

Now, from Appendix C, we know that $\cos\theta = P_1(\cos\theta) = \sqrt{\frac{4\pi}{3}} Y_{\ell=1,m=0}(\theta)$, (8.67b)

hence,

$$\left\langle \psi_{n\ell m}\middle| e\varepsilon z \middle| \psi_{n'\ell' m'}\right\rangle = \sqrt{\frac{4\pi}{3}} e\varepsilon \int_0^{\infty} r^3 dr R_{n\ell}(r) R_{n'\ell'}(r) \int_{\theta=0}^{\pi} \sin\theta d\theta \int_{\varphi=0}^{2\pi} d\varphi Y^*_{\ell m}(\hat{r}) Y_{\ell,m=0}(\theta) Y_{\ell' m'}(\hat{r}). \tag{8.68}$$

Integral of a product of three spherical harmonics (Gaunt integrals) are well documented, using which

$$\int_{\varphi=0}^{2\pi} d\varphi Y^*_{\ell m}(\hat{r}) Y_{\ell,m=0}(\theta) Y_{\ell' m'}(\hat{r}) = N \int_{\varphi=0}^{2\pi} d\varphi e^{i(m'-m)\varphi}. \tag{8.69}$$

The integral on the right-hand side vanishes unless $m' = m$. Of the four degenerate states $\{|2,0,0\rangle, |2,1,-1\rangle, |2,1,0\rangle, |2,1,1\rangle\}$ we began with, the only pair of functions for which $m' = m$ are $\{|2,0,0\rangle, |2,1,0\rangle\}$, i.e., $\{2s, 2p_0\}$. From the Wigner-Eckart theorem, we know that the matrix element $\left\langle \psi^{(0)}_{n=2,\ell,m} \middle| e\varepsilon z \middle| \psi^{(0)}_{n=2,\ell',m'} \right\rangle$ is nonzero only if $\delta\ell = \pm 1, \delta m = 0$. We therefore solve the following equation to get the roots:

$$\begin{bmatrix} -E^{(1)}_{n=2} & \left\langle \psi_{2p_0} \middle| H_1 \middle| \psi_{2s} \right\rangle \\ \left\langle \psi_{2s} \middle| H_1 \middle| \psi_{2p_0} \right\rangle & -E^{(1)}_{n=2} \end{bmatrix} \begin{bmatrix} c_1 \\ c_2 \end{bmatrix} = 0. \tag{8.70}$$

The roots are given by:

$$\begin{vmatrix} -E^{(1)}_{n=2} & \left\langle \psi_{2p_0} \middle| H_1 \middle| \psi_{2s} \right\rangle \\ \left\langle \psi_{2s} \middle| H_1 \middle| \psi_{2p_0} \right\rangle & -E^{(1)}_{n=2} \end{vmatrix} = 0, \tag{8.71a}$$

$$\text{i.e., } \left(E^{(1)}_{n=2}\right)^2 - \left\langle \psi_{2s} \middle| H_1 \middle| \psi_{2p_0} \right\rangle \left\langle \psi_{2p_0} \middle| H_1 \middle| \psi_{2s} \right\rangle = 0, \tag{8.71b}$$

$$\text{or, } E^{(1)}_{n=2} = \pm\sqrt{\left\langle \psi_{2s} \middle| H_1 \middle| \psi_{2p_0} \right\rangle \left\langle \psi_{2p_0} \middle| H_1 \middle| \psi_{2s} \right\rangle} = \pm\sqrt{\left|\left\langle \psi_{2s} \middle| H_1 \middle| \psi_{2p_0} \right\rangle\right|^2}. \tag{8.71c}$$

The matrix element in the above equation can be readily obtained using the wavefunctions from Chapter 5. It is:

$$\left\langle \psi_{2s} \middle| H_1 \middle| \psi_{2p_0} \right\rangle = e\varepsilon \int_{r=0}^{\infty} r^3 R_{2s}(r) R_{2p}(r) dr \oiint d\Omega Y^*_{\ell=0,m=0}(\hat{r}) \cos\theta Y_{\ell=1,m=0}(\hat{r}), \tag{8.72a}$$

$$\text{i.e., } \left\langle \psi_{2s} \middle| H_1 \middle| \psi_{2p_0} \right\rangle = \begin{bmatrix} 2\left(\frac{Z}{2a}\right)^{3/2} \frac{1}{\sqrt{3}} \left(\frac{Z}{2a}\right)^{3/2} e\varepsilon \times \int_{r=0}^{\infty} r^3 \left(1 - \frac{Zr}{2a}\right) e^{-Zr/2a} \left(\frac{Zr}{a}\right) e^{-Zr/2a} dr \\ \times \int_{\theta=0}^{\pi} \sin\theta d\theta \int_{\varphi=0}^{2\pi} d\varphi \left(\frac{1}{\sqrt{4\pi}}\right) (\cos\theta) \left(\sqrt{\frac{3}{4\pi}} \cos\theta\right) \end{bmatrix}, \tag{8.72b}$$

$$\text{or } \left\langle \psi_{2s} \middle| H_1 \middle| \psi_{2p_0} \right\rangle = \frac{Z^3}{12 a_0^{\,3}} e\varepsilon \times \int_{r=0}^{\infty} dr\, r^3 \left(\frac{Zr}{a_0}\right) \left(1 - \frac{Zr}{2a_0}\right) e^{-Zr/a_0} = -3\frac{e\varepsilon a_0}{Z}. \tag{8.72c}$$

Using Eq. 8.71c, we find the first order perturbation correction:

$$E^{(1)}_{n=2} = \pm\left|\left\langle \psi_{2s} \middle| H_1 \middle| \psi_{2p_0} \right\rangle\right| = \pm\frac{3e\varepsilon a_0}{Z}. \tag{8.73}$$

Out of the four degenerate states, one of the levels shifts by

$$\left[E^{(1)}_{n=2}\right]_1 = -\frac{3e\varepsilon a_0}{Z}, \tag{8.74a}$$

another by

$$\left[E^{(1)}_{n=2}\right]_2 = +\frac{3e\varepsilon a_0}{Z}, \tag{8.74b}$$

and the other two remain unchanged. We can now obtain the two new nondegenerate wavefunctions by determining the coefficients c_1 and c_2 in Eq. 8.70 with $E^{(1)}_{n=2}$ given by Eq. 8.74a for the first and by Eq. 8.74b for the second.

For the first root (Eq. 8.74a), we get

$$\left(+3\frac{e\varepsilon a_0}{Z}\right)c_1 + \left\langle \psi_{2p_0}\middle|H_1\middle|\psi_{2s}\right\rangle c_2 = 0, \tag{8.75a}$$

$$\text{i.e.,}\quad \left(+3\frac{e\varepsilon a_0}{Z}\right)c_1 + \left(-3\frac{e\varepsilon a_0}{Z}\right)c_2 = 0, \tag{8.75b}$$

$$\text{and hence } c_1 = c_2. \tag{8.76}$$

For the second root (Eq. 8.74b), we likewise get,

$$c_2 = -c_1. \tag{8.77}$$

Accordingly, the two nondegenerate wavefunctions that have first-order energy corrections $\pm\frac{3e\varepsilon a_0}{Z}$ are

$$\psi_{\mp} = \frac{1}{\sqrt{2}}\left(\psi_{2s} \mp \psi_{2p_0}\right). \tag{8.78}$$

The linear combinations do not have well-defined parity, and therefore the expectation value of the dipole moment in these states does not vanish. This dipole moment interacts with the applied field. The energy splitting is shown in Fig. 8.5. We see that the linear superposition ψ_- has diminished electron density above the *XY* plane (along positive *z*) and increased electron density below the *XY* plane (along negative *z*).

The degeneracy is thus partially lifted. The wavefunctions of the nondegenerate states are a superposition of 2*s* and 2*p*, and hence the orbital angular momentum quantum number (and thus also parity) is not a "good" quantum number for these states. However, the perturbed Hamiltonian has cylindrical symmetry,

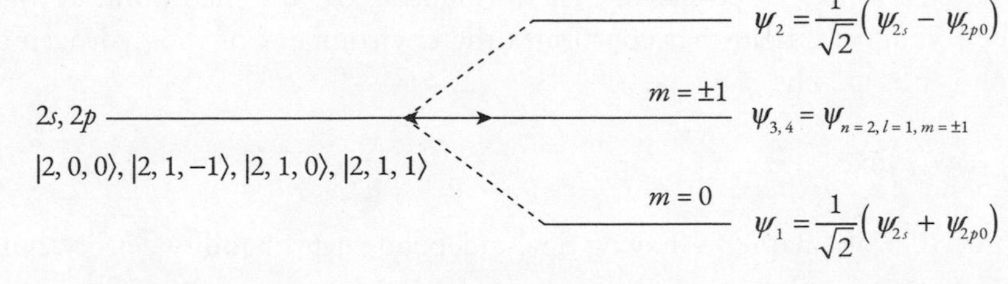

Fig. 8.5 First-order Stark effect for $n = 2$, the first excited state of the hydrogen atom. The energy splitting is $E^{(1)}_{n=2} = \mp\left(3ea_0 \times \varepsilon\right)$.

$$\text{i.e., } \left[H', \ell_z\right]_- = 0, \tag{8.79}$$

and hence m is a good quantum number. One must remember, however, from Fig. 7.5 of Chapter 7 that additional factors such as the fine structure and Lamb shift cause additional complexities due to which the 2s, 2p functions are only approximately degenerate.

We have treated the Stark effect using perturbation theory. However, there is a complexity associated with this since the perturbation potential $e\varepsilon z$ goes to infinity as $z \to \pm\infty$. This region cannot be wished away. After all, even the normalization integral of bound states requires integration over the whole space. As a result of this, the eigenspectrum of the full Hamiltonian inclusive of the Stark perturbation is not bounded. Its energy spectrum has a continuum that is unbound from both above and below. The energy eigenstates have discrete eigenspectrum of the main/unperturbed Hamiltonian H_0 and continuous eigenspectrum of the perturbation Hamiltonian into which the former is embedded.

The degeneracy of the discrete and continuum results in *configuration interaction*; the eigenstates are thus not strictly discrete or continuum; they are mixed. This makes the energy states metastable. They are in fact resonances whose lifetimes are inversely related to the widths of the states, as per Eq. 1.108. The lowest states have rather large lifetimes, but the higher excited states can decay through field ionization. The lowest states, however, may also decay if the field strengths are strong. Only for zero applied field the Coulomb potential has an infinite number of discrete bound states, but the slightest of applied field turns the bound states of the discrete spectrum into metastable resonances due to mixing with the continuum. The mixing of 2s and 2p states, often called *Stark mixing*, effectively diminishes the lifetime of the $n = 2$ state. That this state has a finite lifetime, let alone that it gets diminished by Stark mixing, is itself a curio if we continue to regard the bound states eigenspectrum to consist of sharp energy eigenvalues given by Eq. 5.47 of Chapter 5. Those energy states are sharp; they define *stationary* eigenstates having infinite lifetime resulting from their probability density being independent of time. From the standpoint of the energy–time uncertainty relation (Eq. 1.108 of Chapter 1), $\delta E = 0$ for sharp energy levels and only an infinite lifetime of the state would make the solution compatible with the energy–time uncertainty. We must remember, however, that the quantum problem we solved in Chapter 5 essentially considered a *model* Hamiltonian for an isolated hydrogen atom; it ignored coupling between the atom and its environment. Nature does not allow such separation of the atom and its environment, and an exchange of energy between the atom and the environment is possible. A mechanism to incorporate this exchange in a model Hamiltonian renders it non-Hermitian, with complex eigenvalues, as was determined first by Weisskopf and Wigner by integrating the quantized electromagnetic field in their formalism. On the incorporation of coupling with the electromagnetic field that constitutes the environment of the hydrogen atom, the wavefunction is not merely

$$\psi_n(\vec{r},t) = \psi(\vec{r})\exp\left(-\frac{i}{\hbar}E_n t\right), \tag{8.80a}$$

as given in Eq. 5.5b of Chapter 5, having time-independent probability density, but it is

$$\psi_n(\vec{r},t) = \psi(\vec{r})\exp\left(-\frac{i}{\hbar}E_n t\right)\exp\left(-\frac{1}{2}\gamma_n t\right). \tag{8.80b}$$

The wavefunction in Eq. 8.80b has a time-dependent probability density that diminishes as $\exp\left(-\gamma_n t\right)$; it represents the fact that the nth bound state has a finite, not infinite, lifetime, which is

$$\left(\Delta t\right)_n = \frac{1}{\gamma_n}. \tag{8.81}$$

After a time corresponding to this lifetime, the probability density of the nth eigenstate would diminish to $\frac{1}{e}$ times its value at $t = 0$. One can see that Eq. 8.80b effectively involves a complex eigenvalue corresponding to the non-Hermiticity of the Hamiltonian.

We have considered only one possible decay mechanism, from $n = 2$ to $n = 1$. For higher excited states, there are multiple decay pathways accessible to the system from transition to lower states through a spontaneous emission of photon. A spontaneous transition from an excited state to a lower state therefore has a width, and the emission line has a consequent spectral lineshape, which is Lorentzian. As we know, the wavefunctions for 2*s* and 2*p* are degenerate, and while the transition $2p \rightarrow 2s$ is favored by the selection rule (Eqs. 4.150a,b in Chapter 4) the transition from 2*s* isn't, being one from $j = 0$ to $j = 0$; see the discussion following Eqs. 4.151a,b in Chapter 4. The lifetime of the 2*p* state is of the order of 10^{-9} *s* while that of 2*s* is much longer, of the order of 10^{-1} *s*. The degenerate 2*s* and 2*p* states have a time-dependent probability density that involves a single energy value and a single decay rate. However, their Stark mixing (Eq. 8.78) splits the energy (Fig. 8.6) and the time-dependent phases (Eq. 8.80b) catalyze the dipole-favored transitions thereby effectively lowering the lifetime of the $n = 2$ state. In other words, in the presence of the applied electric field, the metastable 2*s* state acquires some character of the unstable 2*p* state due to Stark mixing resulting in a shortening of the lifetime of the $n = 2$ state via a radiative transition mediated by electrodynamic dipole interaction.

The transition $2s \rightarrow 1s$ violates the parity selection rule (Laporte's rule – see the discussion after Eqs. 4.152a,b in Chapter 4), but only because we considered the electrodynamical dipole interaction in the Wigner–Eckart selection rule. The origin of this selection rule is the fact that the QED Lagrangian that is involved in light–matter interactions commutes with parity. Nature, however, accommodates such a transition, albeit when effected by a different interaction, namely the weak interaction. Most of atomic and molecular spectroscopy is governed by the electrodynamics interaction, and the $2s \rightarrow 1s$ interaction effected by the weak interaction is *called* as parity violation, even if parity is *not* conserved in the weak interaction. The electrodynamical interaction is, however, seamlessly unified with the weak interaction, as explained in the Electro-weak Theory (EWT) developed by Sheldon Glashow, Abdus Salam, and Steven Weinberg. Salient features of the EWT are highlighted in Section A.2 of Appendix A.

Now, the parity of an atom consisting of N electrons with n_1 electrons having orbital angular momentum ℓ_i is

$$\Pi = -1^{\left(\sum_{i=1}^{N}\ell_i\right)}, \tag{8.82}$$

since the parity quantum number is multiplicative. Thus, it is the open shell electrons in an atom that determine the parity of an atom. For example, the ground state of bismuth atom,

$[Hg]6p^3\ {}^4S^{o}_{\frac{3}{2}}$, has odd parity, indicated by the superscript "o." The transition matrix element for the electric dipole "$E1$" operator $\vec{D}$ for a transition from an initial state $|i\rangle$ to a final state $|f\rangle$ may be written using the unitary parity operator $\hat{P}$ (Appendix A) as

$$T_{fi}^{(E1)} = \langle f|\vec{D}|i\rangle = \langle f|\hat{P}^{\dagger}\hat{P}\vec{D}|i\rangle = \langle \hat{P}f|\hat{P}\vec{D}|i\rangle, \tag{8.83a}$$

$$\text{i.e., } T_{fi}^{(E1)} = -\Pi_f\Pi_i\langle f|\vec{D}|i\rangle = -\Pi_f\Pi_i T_{fi}^{(E1)}, \tag{8.83b}$$

where Π_i and Π_f are the parities of respectively the initial and the final states. If the parity of the initial state and that of the final state is the same, we must conclude that $T_{fi}^{(E1)}$ is the negative of itself, i.e., it must essentially be zero. Electric dipole transition between states having the same parity is therefore forbidden, which is the Laporte's selection rule. Under the *EWT*, the interaction Hamiltonian consists of a part that is symmetric under parity, and another that is antisymmetric:

$$H = H_{(even)} + H_{(odd)}. \tag{8.84}$$

Under parity, this Hamiltonian is not invariant:

$$\hat{P}^{-1}H\hat{P} = \hat{P}^{-1}H_{(even)}\hat{P} + \hat{P}^{-1}H_{(odd)}\hat{P} = \left\{H_{(even)} - H_{(odd)}\right\} \neq H, \tag{8.85a}$$

$$\text{i.e., } \left[\hat{P},H\right]_{-} \neq 0. \tag{8.85b}$$

It is then obvious that the time evolution operator is also not invariant under parity transformation:

$$\hat{P}^{-1}U(t)\hat{P} \neq U(t), \tag{8.86a}$$

$$\text{i.e., } \left[\hat{P},U(t)\right]_{-} = \left[\hat{P},e^{-\frac{i}{\hbar}Ht}\right]_{-} \neq 0. \tag{8.86b}$$

The matrix element of the time evolution operator between an initial and a final state is then not equal to that in the corresponding parity-reversed states (denoted below by placing a tilde on top of the states' designation):

$$\langle \tilde{f}|U(t)|\tilde{i}\rangle = \langle f|\hat{P}^{\dagger}U(t)\hat{P}|i\rangle = \langle f|\hat{P}^{-1}U(t)\hat{P}|i\rangle \neq \langle f|U(t)|i\rangle, \tag{8.87a}$$

$$\text{where } |\tilde{i}\rangle = \hat{P}|i\rangle \text{ and } |\tilde{f}\rangle = \hat{P}|f\rangle. \tag{8.87b}$$

We see that the expectation value of the Hamiltonian in the parity-operated state is different from that in the original state:

$$\langle \tilde{a}|H|\tilde{a}\rangle = \langle a|\hat{P}^{\dagger}H\hat{P}|a\rangle = \langle a|\hat{P}^{-1}H\hat{P}|a\rangle \neq \langle a|H|a\rangle, \tag{8.88a}$$

$$\text{i.e., } E_{\tilde{a}} \neq E_{a}. \tag{8.88b}$$

The transition $6s\ ^2S_{1/2} \rightarrow 7s\ ^2S_{1/2}$ in ^{133}Cs [6–8] has been observed in experiments. It is commonly called as Atomic Parity Violation (APV) transition. This transition has been extensively investigated theoretically. Such transitions are also known as Parity Non-Conserving (PNC) transitions. Other similar transitions in atoms have also been studied. APV occurs due to the electroweak interaction effected by an exchange of Z^0 boson between an electron and the nucleus. It can occur also due to parity-violating inter-nuclear interactions. As a result of PNC, there occurs a small mixing of atomic states having opposite parity. Laporte's rule predicts $T_{fi}^{(E1)}$ (Eq. 8.83) to be zero, but PNC transition allows it. We illustrate this using a simple example, considering the 1*s* and 2*s* states of the hydrogen atom (Fig. 8.6), which have the same parity.

In the presence of the parity-violating interaction, H_{PNC}, the 2*s* state acquires a character from the opposite parity states. According to perturbation theory (following the argument that led us to Eq. 6.72 in Chapter 6), the parity-mixed 2*s* state is given by

$$|2\tilde{s}\rangle = |2s\rangle + \sum_o |o\rangle \frac{\langle o|H_{PNC}|2s\rangle}{\varepsilon_{2s} - \varepsilon_o}, \tag{8.89a}$$

where $|o\rangle$ represents odd parity intermediate states and ε_o are the energies of those states. The 1*s* state *also* acquires an odd parity admixture on account of parity violation. Hence,

$$|1\tilde{s}\rangle = |1s\rangle + \sum_o |o\rangle \frac{\langle o|H_{PNC}|1s\rangle}{\varepsilon_{1s} - \varepsilon_o}. \tag{8.89b}$$

The $E1_{\text{PNC}}$ transition amplitude between the parity mixed states $1\tilde{s}, 2\tilde{s}$ is then *nonzero*:

$$E1_{\text{PNC}} = \langle 2\tilde{s}|d|1\tilde{s}\rangle \neq 0, \tag{8.90a}$$

i.e.,

$$E1_{\text{PNC}} = \sum_I \left[\frac{\langle 2s|d|I\rangle\langle I|H_{\text{PNC}}|1s\rangle}{\varepsilon_{1s} - \varepsilon_I} + \frac{\langle 2s|H_{\text{PNC}}|I\rangle\langle I|d|1s\rangle}{\varepsilon_{2s} - \varepsilon_I} \right] \neq 0. \tag{8.90b}$$

$E1_{\text{PNC}}$ has been measured using extremely sensitive interferometry apparatus. Theoretical methods required to study this are also very sophisticated. Although we have considered hydrogen atom as a case study to show the role of parity violation in altering atomic transition selection rules, it is preferable to use heavier atoms, since the relevant observables scale as Z^4. An important advantage of EDM (see Appendix A), or $E1_{\text{PNC}}$, as probes of particle physics is that the atomic experiments are table-top experiments and far cheaper than accelerator-based experiments. For further details on this topic, we refer the reader to References [17–19].

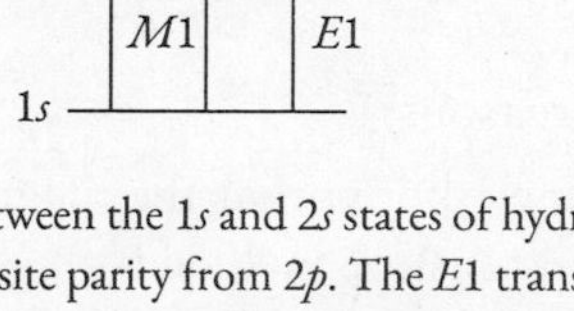

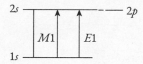

Fig. 8.6 Electric dipole ($E1$) transition between the 1*s* and 2*s* states of hydrogen is forbidden. However, due to parity violation, 2*s* acquires opposite parity from 2*p*. The $E1$ transition between 1*s* and parity mixed 2*s* is then allowed.

8.3 Zeeman-Paschen–Back Spectroscopies

The Zeeman effect was first observed in 1896. Atomic spectral lines were seen to be "split" when an atom was placed in an external magnetic field. From Eq. 6.97 of Chapter 6, we have

$$i\hbar\frac{\partial\psi(\vec{r},t)}{\partial t} = \left[-\frac{\hbar^2\vec{\nabla}^2}{2m} - \frac{ie\hbar}{mc}\vec{A}\cdot\vec{\nabla} + \frac{e^2}{2mc^2}A^2 - \frac{Ze^2}{r}\right]\psi(\vec{r},t). \tag{8.91}$$

A constant magnetic field $\vec{B}$ in which an atom is placed may be considered to be uniform over the atomic dimension. Choosing a Cartesian coordinate system to be so oriented that its z-axis is along the direction of the applied magnetic field, the vector potential of the electromagnetic field interacting with the atom is

$$\vec{A} = \frac{1}{2}\vec{B}\times\vec{r} = \frac{1}{2}B_z\hat{e}_z\times\left(x\hat{e}_x + y\hat{e}_y + z\hat{e}_z\right) = \frac{1}{2}B_z x\hat{e}_y - \frac{1}{2}B_z y\hat{e}_x. \tag{8.92}$$

We shall use the same approximation that was discussed following the Eq. 6.97 in the Chapter 6. It led us to ignore quadratic term in the vector potential, so we treat Eq. 8.91 using methods of perturbation theory with the perturbation Hamiltonian given by Eq. 6.99, viz.,

$$H'_\ell = -\frac{ie\hbar}{mc}\vec{A}\bullet\vec{\nabla} = -\frac{ie\hbar}{mc}\left(\frac{1}{2}\vec{B}\times\vec{r}\right)\bullet\vec{\nabla} = -\frac{ie\hbar}{2mc}\vec{B}\bullet\vec{r}\times\vec{\nabla} = \frac{e}{2mc}\vec{B}\bullet\vec{\ell} = -\vec{\mu}_\ell\cdot\vec{B}, \tag{8.93}$$

since, from Eqs. 7.54a,b,c we have

$$\vec{\mu}_\ell = -\mu_B\frac{g_\ell\vec{\ell}}{\hbar}, \tag{8.94a}$$

with the Bohr magneton given by $\mu_B = \dfrac{e\hbar}{2mc}$ (8.94b)

and $g_\ell = 1.$ (8.94c)

The Schrödinger equation to be solved perturbatively,

$$i\hbar\frac{\partial\psi(\vec{r},t)}{\partial t} = \left[-\frac{\hbar^2\vec{\nabla}^2}{2m} - \frac{Ze^2}{r} - \vec{\mu}_\ell\cdot\vec{B}\right]\psi(\vec{r},t), \tag{8.95}$$

must however be replaced by

$$i\hbar\frac{\partial\psi(\vec{r},t)}{\partial t} = \left[-\frac{\hbar^2\vec{\nabla}^2}{2m} - \frac{Ze^2}{r} - \vec{\mu}\cdot\vec{B} + \xi(r)\vec{s}\cdot\vec{\ell}\right]\psi(\vec{r},t), \tag{8.96}$$

with $\vec{\mu} = -\mu_B\dfrac{g_\ell\vec{\ell} + g_s\vec{s}}{\hbar} = -\mu_B\dfrac{\vec{\ell} + 2\vec{s}}{\hbar} = -\dfrac{e}{2mc}\left(\vec{\ell} + 2\vec{s}\right),$ (8.97)

on account of the following two reasons:

(i) from Eq. 7.56 of Chapter 7 we know that the magnetic moment due to the electron's intrinsic angular momentum must be included
and

(ii) having included the effect of the electron's spin angular momentum, we must include the spin–orbit interaction, which appears as the last term in the perturbed Hamiltonian in Eq. 8.97 with

$$\xi(r) = -\frac{e}{2m^2c^2}\left(\frac{1}{r}\frac{\partial V}{\partial r}\right), \tag{8.98}$$

as we know from Eq. 7.84 of Chapter 7.

The perturbation then consists of two terms (a) $-\vec{\mu}\cdot\vec{B}$, which can be controlled since $\vec{B}$ is the externally applied laboratory magnetic field, and (b) $\xi(r)\vec{s}\cdot\vec{\ell}$, which is *intrinsic* to the atomic system. As mentioned earlier in the discussion following Eq. 7.80 (Chapter 7), the perturbative term that includes the effect of the electron spin was introduced on an ad hoc basis by Uhlenbeck and Goudsmit using an *invalid* classical model of a spinning top. In a sense it was an appropriate correction, but only accidently so; it found its rationale only later in relativistic quantum mechanics, as discussed in Chapter 7.

The strategy to execute perturbative treatment of the terms (a) $-\vec{\mu}\cdot\vec{B}$ and (b) $\xi(r)\vec{s}\cdot\vec{\ell}$ then depends on their *relative* strengths:

(i) $\left|\vec{\mu}\cdot\vec{B}\right| \gg \left|\xi(r)\vec{s}\cdot\vec{\ell}\right|$ – this case is referred to as the *normal Zeeman effect* (NZE) or *strong field* $\left(\vec{B}_{external}\right)$ Zeeman effect.

(ii) $\left|\vec{\mu}\cdot\vec{B}\right| \gtrsim \left|\xi(r)\vec{s}\cdot\vec{\ell}\right|$ – this case is called as the *Paschen–Back* or *Paschen–Back–Zeeman* effect or *intermediate field* Zeeman effect.

(iii) $\left|\vec{\mu}\cdot\vec{B}\right| < \left|\xi(r)\vec{s}\cdot\vec{\ell}\right|$ – this case is referred to as the *anomalous Zeeman effect* or the *weak field* Zeeman effect.

All the three cases here belong to the *family* of Zeeman effects. Case (iii) is called anomalous only because the electron spin was not understood when it was experimentally studied, and it had then seemed like an anomaly. However, it must be remembered that the electron's intrinsic spin angular momentum is included in the total magnetic moment $\vec{\mu}$ (see Eq. 8.97), which is employed in the NZE (case (i)), and it is of course present in the spin–orbit interaction (case (ii)). Electron spin is therefore of importance in *all* the three cases. The term *strong* field only refers to the strength of the applied magnetic field relative to that in the other two cases; the field under consideration is, however, not strong enough to require consideration of the quadratic term $\left(\frac{e^2}{2mc^2}A^2\right)$ in the vector potential in Eq. 8.91.

Case (i): Normal Zeeman effect (or "strong" field Zeeman effect)

In this case, we ignore the spin–orbit interaction, and the perturbation Hamiltonian H'_{NZE} is given by $-\vec{\mu}\cdot\vec{B}$. We shall use Eq. 6.88 of Chapter 6 to obtain the first-order perturbation correction to the energy level:

$$\Delta E_{NZE} = \left\langle\psi_0\middle|H'_{NZE}\middle|\psi_0\right\rangle = \left\langle\psi_0\middle|\left(-\vec{\mu}\cdot\vec{B}\right)\middle|\psi_0\right\rangle. \tag{8.99}$$

At this point, we are confronted with a *choice*: which basis set should we use for the unperturbed vectors, $\left|\psi_0\right\rangle = \left|n,\ell,m_\ell,m_s\right\rangle$ or $\left|\psi_0\right\rangle = \left|n,\ell,j,m_j\right\rangle$? From a mathematical point of view, we can expand a wavefunction in the basis $\left\{\left|n,\ell,m_\ell,m_s\right\rangle\right\}$ as well as in the alternative basis $\left\{\left|n,\ell,j,m_j\right\rangle\right\}$. Our choice is, however, easily made, since the spin–orbit interaction is weak and ignorable in this case. We therefore employ the $\left\{\left|n,\ell,m_\ell,m_s\right\rangle\right\}$ basis for the unperturbed states. Therefore,

$$\Delta E_{\text{NZE}} = \left\langle n,\ell,m_\ell,m_s\right| \mu_B \frac{\left(\ell_z + 2s_z\right)B}{\hbar}\left|n,\ell,m_\ell,m_s\right\rangle = \mu_B B\left(m_\ell + 2m_s\right). \tag{8.100}$$

The perturbative correction to energy is determined by the *sum* $\left(m_\ell + 2m_s\right)$, regardless of the individual values of m_ℓ & m_s. Since the energy levels split, so do the spectral lines that are due to atomic transitions from one level to another. The dipole spectral line $n'p \rightarrow ns$ splits into three lines, and the spectral line $n'd \rightarrow np$ also splits into three lines. These lines are called as Lorentz triplet.

The splitting of the electric dipole spectral line $n'p \rightarrow ns$ is schematically shown in Fig. 8.7a. In an electric dipole transition, the m_s quantum number must be the same for the initial and the final states; transitions take place only for $\delta m_s = 0$. The six-fold degenerate $n'p$ state splits into five energy levels, since $\left(m_\ell = -1, m_s = \frac{1}{2}\right)$ and $\left(m_\ell = +1, m_s = -\frac{1}{2}\right)$ both have $m_\ell + 2m_s = 0$ and remain degenerate. The ns energy level splits into two levels, shown in Fig. 8.7a. The Lorentz triplets with $\delta M = -1,0,+1$ corresponding to $m_s = +\frac{1}{2}$ have exactly the same energies as the triplets in the transitions corresponding to $m_s = -\frac{1}{2}$. Only three lines are therefore seen in a spectrometer even if there actually are six transitions. Historically, these lines were studied before the theory of electron spin was formulated. The Lorentz triplets seen in the experiment were *explained away* with *no* reference to the electron spin. The spectral diagram in Fig. 8.7b seemed completely satisfactory, though it turned out to be fortuitous.

The electric dipole spectral line $n'd \rightarrow np$ also splits into Lorentz triplets. The ten $n'd$ degenerate levels split into equally spaced *seven* levels shown in the upper part of Fig. 8.8a. The degeneracy is only partially removed by the applied magnetic field since the values of $m_\ell + 2m_s$ are the same for the following $\left(m_\ell, m_s\right)$ pairs –

(a) $\left(0,\frac{1}{2}\right)$ and $\left(2,-\frac{1}{2}\right)$; which both give $m_\ell + 2m_s = 1$,

(b) $\left(-1,\frac{1}{2}\right)$ and $\left(1,-\frac{1}{2}\right)$; both give $m_\ell + 2m_s = 0$, and

(c) $\left(-2,\frac{1}{2}\right)$ and $\left(0,-\frac{1}{2}\right)$; for which both have $m_\ell + 2m_s = -1$.

Five of the seven levels corresponding to $m_s = +\frac{1}{2}$ are shown in the first column in Fig. 8.8a, and the other set of five levels corresponding to $m_s = -\frac{1}{2}$ in the next. Note that

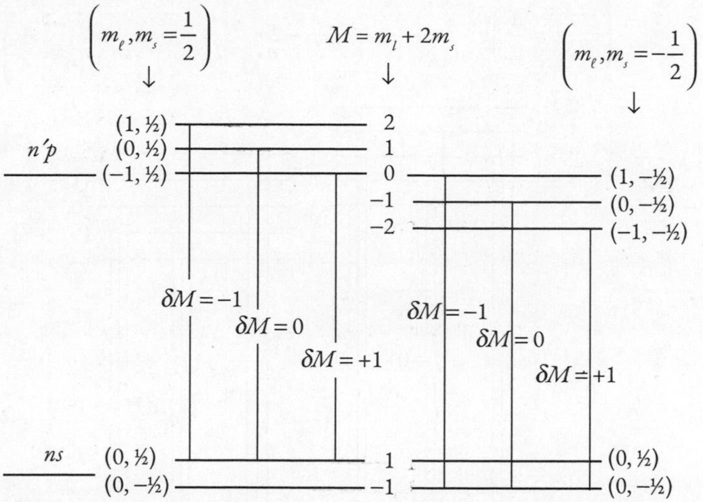

Fig. 8.7a The electric dipole transition $n'p \rightarrow ns$ splits into six transitions, three of which correspond to $m_s = +\frac{1}{2}$ and the other three to $m_s = -\frac{1}{2}$. Only three frequencies are, however, observed in the spectrometer since the splitting of the energy levels is equispaced.

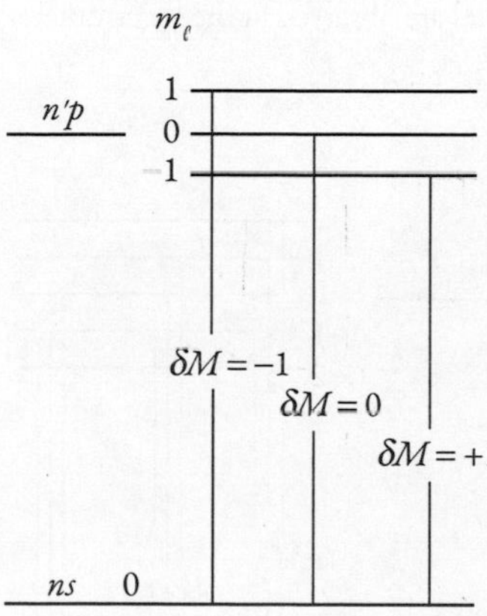

Fig. 8.7b The Lorentz triplet was historically *explained away* with no reference to the electron's spin.

the $n'd$ states with $M = -1,0,1$ are doubly degenerate, and the np state with $M = 0$ is also doubly degenerate. There are a total of nine transitions corresponding to $m_s = +\frac{1}{2}$, and another nine corresponding to $m_s = -\frac{1}{2}$. Both the sets of nine transitions are shown in Fig. 8.8a.

The splitting between the energy levels being equidistant; only three different spectral frequencies are observed in the spectrometer, resulting from both the sets of transitions. This allows, yet again, a fortuitous explanation of the Lorentz triplets without any reference to the electron spin, using the schematic diagram in Fig. 8.8b. In Figs. 8.7a,b and Figs. 8.8a,b, the transitions with $\delta M = 0$ are called as π-lines and those with $\delta M = \pm 1$ as σ-lines.

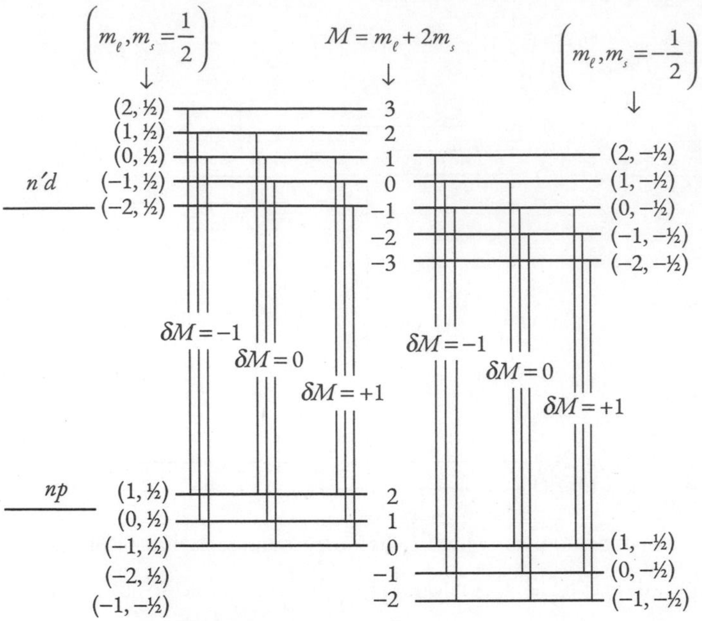

Fig. 8.8a The electric dipole transition $n'd \rightarrow np$ splits into nine transitions for $m_s = +\frac{1}{2}$ and also for $m_s = -\frac{1}{2}$. However, only three frequencies are observed in the spectrometer since the splitting of the energy levels is equispaced.

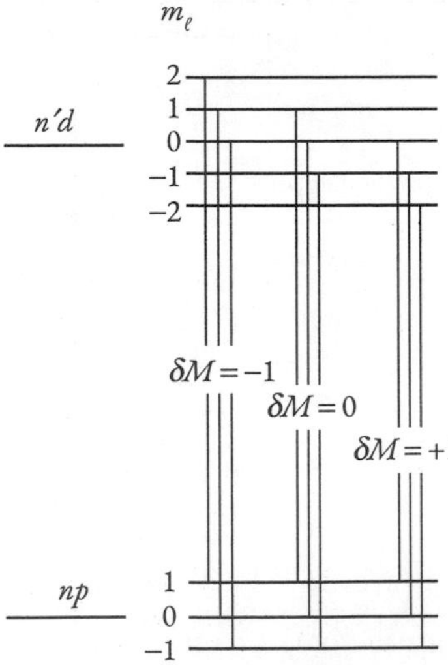

Fig. 8.8b The $n'd \rightarrow np$ electric dipole line also splits into three spectral lines that are observed, which could also be explained away with no reference to the electron's spin.

In the early days of quantum mechanics, the accidental explanation (as in Fig. 8.7b and Fig. 8.8b) of the Lorentz triplets seemed accidentally "normal," which is why this case is called as the NZE. The term is obviously misleading because it conceals the subtle role of the electron spin, but the terminology has remained in use due to its historical occurrence.

Case (ii): Paschen–Back–Zeeman effect (or "intermediate" field Zeeman effect)
Intrinsic properties of an atomic system are described correctly by quantum mechanics. These include *both* the orbital angular momentum and the spin angular momentum. It is the *external* applied field whose direction $\hat{B} = \hat{e}_z$ and magnitude $|\vec{B}|$ that we can control. We may diminish $|\vec{B}|$ till the dipolar coupling becomes comparable to the spin–orbit coupling, but remains slightly larger, i.e., $|\vec{\mu} \cdot \vec{B}| \gtrsim |\xi(r)\vec{s} \cdot \vec{\ell}|$. Unlike the case dealt with in the discussion on NZE, the spin–orbit perturbation Hamiltonian must now be explicitly considered. However, the dipolar coupling of the atom's magnetic moment with the external field is slightly larger than the spin–orbit coupling. Hence, the basis $\{|n,\ell,m_\ell,m_s\rangle\}$ is more appropriate than $\{|n,\ell,j,m_j\rangle\}$ to describe the unperturbed states $\{\psi^{(0)}\}$ to determine perturbative corrections to the energy spectrum using Eq. 6.88 of Chapter 6. This case is referred to as the Paschen–Back–Zeeman effect, or simply as Paschen–Back effect. We shall denote it as PB-ZE. It consists of two corrections:

(i) correction due to the atom's dipolar coupling with the applied magnetic field, i.e., due to the perturbation Hamiltonian $-\vec{\mu} \cdot \vec{B}$,
and
(ii) correction due to the spin–orbit coupling Hamiltonian, $\xi(r)\vec{s} \cdot \vec{\ell}$.

The *net* correction to the energy value from the first-order perturbation theory due to the *two* corrections mentioned here therefore is

$$\Delta E^{PB-ZE} = \Delta E_M^{PB-ZE} + \Delta E_{s-o}^{PB-ZE}, \tag{8.101}$$

where

$$\Delta E_M^{PB-ZE} = \langle n,\ell,m_\ell,s,m_s|\left(-\vec{\mu}\cdot\vec{B}\right)|n,\ell,m_\ell,s,m_s\rangle = \langle n,\ell,m_\ell,s,m_s|\left(\mu_B\frac{\vec{\ell}+2\vec{s}}{\hbar}\cdot\vec{B}\right)|n,\ell,m_\ell,s,m_s\rangle$$

$$\text{i.e., } \Delta E_M^{PB-ZE} = \mu_B B\left(m_\ell + 2m_s\right) \tag{8.102}$$

$$\text{and } \Delta E_{s-o;\ell\neq 0}^{PB-ZE} = \langle n,\ell,m_\ell,m_s|\xi(r)\vec{\ell}\cdot\vec{s}|n,\ell,m_\ell,m_s\rangle,$$

$$\text{i.e., } \Delta E_{s-o;\ell\neq 0}^{PB-ZE} = \langle n,\ell|\xi(r)|n,\ell\rangle\langle \ell,m_\ell,s,m_s|\vec{\ell}\cdot\vec{s}|\ell,m_\ell,s,m_s\rangle. \tag{8.103}$$

In Eq. 8.102, we have used the eigenvalue equations for ℓ_z and for s_z, and in Eq. 8.103, we have factored the radial and the angular integrals. Using now the solution of the Solved Problem P5.5 in Chapter 5, the radial integral is

$$\langle n,\ell|\xi(r)|n,\ell\rangle = \langle n,\ell|\left(\frac{e}{2m^2c^2}\frac{1}{r}\frac{\partial V}{\partial r}\right)|n,\ell\rangle = \frac{e}{2m^2c^2}\int_{r=0}^{\infty} r^2 dr\left[R_{n\ell}(r)\right]^2\left(\frac{Ze}{r^3}\right),$$

$$\text{i.e., } \langle n,\ell|\xi(r)|n,\ell\rangle = \frac{Ze^2\hbar^2}{2m^2c^2}\times\frac{Z^3}{a_0^3 n^3\ell\left(\ell+\frac{1}{2}\right)(\ell+1)}. \tag{8.104}$$

The angular matrix element is

$$\left\langle \ell,m_\ell,s,m_s \middle| \vec{\ell}\cdot\vec{s} \middle| \ell,m_\ell,s,m_s \right\rangle = \left\langle \ell,m_\ell,s,m_s \middle| \ell_x s_x + \ell_y s_y + \ell_z s_z \middle| \ell,m_\ell,s,m_s \right\rangle,$$

$$\text{i.e., } \left\langle \ell,m_\ell,s,m_s \middle| \vec{\ell}\cdot\vec{s} \middle| \ell,m_\ell,s,m_s \right\rangle = \left\langle \ell,m_\ell,s,m_s \middle| \ell_z s_z \middle| \ell,m_\ell,s,m_s \right\rangle = \left(m_\ell\hbar\right)\left(m_s\hbar\right), \quad (8.105)$$

which follows from the fact that the components of the orbital and spin angular momenta that are orthogonal to the direction of the magnetic field would involve the ladder operators that *raise* and *lower* the magnetic quantum numbers; and orthogonality with the vector in the dual conjugate space then gives zero.

Combining the result in Eqs. 8.104 and 8.105, we get

$$\Delta E^{PB-ZE}_{s-o;\ell\neq 0} = \frac{Ze^2\hbar^2}{2m^2c^2}\frac{Z^3\times\left(m_\ell m_s\right)}{a_0^3 n^3\ell\left(\ell+\frac{1}{2}\right)(\ell+1)} = \frac{Ze^2\hbar^2}{2m^2c^2}\frac{Z^3\times\left(m_\ell m_s\right)}{\left(\frac{\hbar^2}{me^2}\right)^3 n^3\ell\left(\ell+\frac{1}{2}\right)(\ell+1)},$$

$$\text{i.e., } \Delta E^{PB-ZE}_{s-o;\ell\neq 0} = \frac{Z^2\alpha^2}{2}\frac{me^4\times Z^2\times\left(m_\ell m_s\right)}{\hbar^2 n^3\ell\left(\ell+\frac{1}{2}\right)(\ell+1)} = \kappa_{n\ell} m_\ell m_s, \quad (8.106)$$

where we have employed the values of the Bohr radius a_0, and the fine-structure constant α. Essentially, the splitting in the energy levels is proportional to the product $m_\ell m_s$ of the magnetic quantum numbers, but the proportionality $\kappa_{n\ell}$ depends parametrically on the principal and the orbital angular momentum quantum numbers. The net *correction* (Eq. 8.101) in the first order due to the combined effect of the external magnetic field and the spin–orbit interaction is

$$\Delta E^{PB-ZE} = \mu_B B\left(m_\ell + 2m_s\right) + \kappa_{n\ell} m_\ell m_s, \quad (8.107)$$

so the energy $E_{n\ell m}$ shifts to

$$E'_{n\ell m} = E_{n\ell m} + \mu_B B\left(m_\ell + 2m_s\right) + \kappa_{n\ell} m_\ell m_s. \quad (8.108)$$

A spectroscopic transition $(n',\ell',m') \to (n,\ell,m)$ would therefore result in an energy *shift* due to the PB-ZE by

$$\delta E = \left[E_{n'\ell'm'} + \mu_B B\left(m'_\ell + 2m'_s\right) + \kappa_{n'\ell'} m'_\ell m'_s\right] - \left[E_{n\ell m} + \mu_B B\left(m_\ell + 2m_s\right) + \kappa_{n\ell} m_\ell m_s\right], \quad (8.109a)$$

$$\text{i.e., } \delta E = \left(E_{n'\ell'm'} - E_{n\ell m}\right) + \mu_B B\left(m'_\ell - m_\ell\right) + \left(\kappa_{n'\ell'} m'_\ell - \kappa_{n\ell} m_\ell\right) m_s, \quad (8.109b)$$

since the magnetic spin quantum numbers of the initial and final states in an electric dipole transition remain the same. The frequency shift would be $\frac{\delta E}{\hbar} = \delta\nu$.

Case (iii): Anomalous Zeeman effect (or "weak" field Zeeman effect)

The third case we shall consider corresponds to the external field having only a weak magnitude: $\left|\vec{\mu}\cdot\vec{B}\right| < \left|\xi(r)\vec{s}\cdot\vec{\ell}\right|$. The weakest term would be treated perturbatively, and the larger term, viz., the spin–orbit interaction, must now be included in the unperturbed Hamiltonian. In other words, we first obtain the eigenvalues and eigenfunctions of $\left[-\frac{\hbar^2\vec{\nabla}^2}{2m} - \frac{Ze^2}{r} - \vec{\mu}\cdot\vec{B} + \xi(r)\vec{s}\cdot\vec{\ell}\right]$, which would be regarded as the unperturbed Hamiltonian H_0, and use for our eigenbasis $\left\{\left|n,\ell,j,m_j\right\rangle\right\}$ and not $\left\{\left|n,\ell,m_\ell,m_s\right\rangle\right\}$. The coupling term $\left(-\vec{\mu}\cdot\vec{B}\right)$ in the Hamiltonian (Eq. 8.96) with the magnetic field would be treated using perturbation theory. In the case (i) also we had treated the term $\left(-\vec{\mu}\cdot\vec{B}\right)$ using first-order perturbation theory, but we had ignored the spin–orbit interaction term, since the external field was considered to be the most important term. In the present case, the external field being weak relative to the spin–orbit interaction, the latter is included in the unperturbed Hamiltonian generating the eigenbasis $\left\{\left|n,\ell,j,m_j\right\rangle\right\}$, and the first–order correction to energy eigenvalue therefore is

$$\Delta E_{\text{AZE}}^{\text{Weak-Field}} = \left\langle \ell s j m_j \left| \frac{\mu_B}{\hbar}\left(j_z + s_z\right)B \right| \ell s j m_j \right\rangle. \tag{8.110}$$

The algebra must proceed differently now compared to that in the case (i) since m_ℓ and m_s are no longer good quantum numbers. The subscript AZE stands for *anomalous Zeeman effect*, but only because in the early days of quantum mechanics the spin–orbit interaction that provides the unperturbed states was not known at that time. The terminology has stayed, even after a satisfactory understanding of the spin–orbit interaction as discussed in Chapter 7. The first-order perturbation energy correction in the spin–orbit coupled basis is

$$\Delta E_{\text{AZE}}^{\text{Weak-Field}} = \left\langle \ell s j m_j \left| \left[-\left(-\mu_B\frac{\vec{\ell}+2\vec{s}}{\hbar}\right)\cdot\vec{B}\right] \right| \ell s j m_j \right\rangle = \left\langle \ell s j m_j \left| \frac{\mu_B}{\hbar}\left(j_z + s_z\right)B \right| \ell s j m_j \right\rangle \tag{8.111}$$

and

$$\Delta E^{PB-ZE}_{s=0;\ell\neq 0} = \mu_B B m_j + \frac{\mu_B}{\hbar}B\left\langle \ell s j m_j \left| s_z \right| \ell s j m_j \right\rangle = \frac{Z^2\alpha^2}{2}\frac{me^4\times Z^2\times\left(m_\ell m_s\right)}{\hbar^2 n^3\ell\left(\ell+\frac{1}{2}\right)(\ell+1)} = \kappa_{n\ell}m_\ell m_s, \tag{8.112}$$

since $\vec{\ell} + 2\vec{s} = \vec{j} + \vec{s}$.

The first-order perturbation correction to energy when the magnetic field is weak requires the determination of the matrix element of j_z and also that of s_z. The former is straightforward, but the latter is problematic (if only slightly) since $\left|\ell s j m_j\right\rangle$ is not an eigenstate of s_z. Hence, $\vec{s}$ being a vector operator, we consider the matrix element of an operator having rank 1 and use the Wigner–Eckart theorem (Eq. 4.151 in Chapter 4):

$$\left\langle \alpha' j' m' \left| \vec{V}_q^{(k=1)} \right| \alpha j m \right\rangle = \frac{\left\langle \alpha' j' \left\| \vec{V}^{(k=1)} \right\| \alpha j \right\rangle}{\sqrt{2j'+1}} \times \left\langle (j1)j'm' \middle| j1mq \right\rangle. \tag{8.113}$$

We see that

$$\frac{\left\langle \alpha' j' m' \left| \vec{V}_q^{(k=1)} \right| \alpha j m \right\rangle}{\left\langle \alpha' j' m' \left| \vec{J}_q^{(k=1)} \right| \alpha j m \right\rangle} = \frac{\left\langle \alpha' j' \left\| \vec{V}^{(k=1)} \right\| \alpha j \right\rangle}{\left\langle \alpha' j' \left\| \vec{J}^{(k=1)} \right\| \alpha j \right\rangle}, \tag{8.114a}$$

$$\text{or } \langle \alpha' j' m' | \vec{V}_q^{(k=1)} | \alpha j m \rangle = \frac{\langle \alpha' j' \| \vec{V}^{(k=1)} \| \alpha j \rangle}{\langle \alpha' j' \| \vec{J}^{(k=1)} \| \alpha j \rangle} \langle \alpha' j' m' | \vec{J}_q^{(k=1)} | \alpha j m \rangle = C \langle \alpha' j' m' | \vec{J}_q^{(k=1)} | \alpha j m \rangle. \tag{8.114b}$$

Recognizing now that

$$\langle \alpha j m | \vec{V} \cdot \vec{J} | \alpha j m \rangle = \sum_{m'=-j}^{j} \langle \alpha j m | \vec{V} | \alpha j m' \rangle \cdot \langle \alpha j m' | \vec{J} | \alpha j m \rangle,$$

or using Eq. 8.114b, $\langle \alpha j m | \vec{V} \cdot \vec{J} | \alpha j m \rangle = \sum_{m'=-j}^{j} C \langle \alpha j m | \vec{J} | \alpha j m' \rangle \cdot \langle \alpha j m' | \vec{J} | \alpha j m \rangle,$

i.e., $\langle \alpha j m | \vec{V} \cdot \vec{J} | \alpha j m \rangle = C \langle \alpha j m | J^2 | \alpha j m \rangle = C\hbar^2 j(j+1)$

$$\text{which gives } C = \frac{\langle \alpha j m | \vec{V} \cdot \vec{J} | \alpha j m \rangle}{\hbar^2 j(j+1)}. \tag{8.115}$$

$$\text{Therefore, } \langle \alpha j m' | \vec{V}_q^{(k=1)} | \alpha j m \rangle = \frac{\langle \alpha j m | \vec{V} \cdot \vec{J} | \alpha j m \rangle}{\hbar^2 j(j+1)} \langle \alpha j m' | \vec{J}_q^{(k=1)} | \alpha j m \rangle, \tag{8.116}$$

$$\text{and for } m' = m \text{ we get } \langle \alpha j m | \vec{V}_q^{(k=1)} | \alpha j m \rangle = \frac{\langle \alpha j m | \vec{V} \cdot \vec{J} | \alpha j m \rangle}{\hbar^2 j(j+1)} \langle \alpha j m | \vec{J}_q^{(k=1)} | \alpha j m \rangle. \tag{8.117}$$

The above relation holds good for each of the three components $q = -1, 0, +1$, so the equality holds for the vector itself:

$$\langle \alpha j m | \vec{V} | \alpha j m \rangle = \frac{\langle \alpha j m | \vec{V} \cdot \vec{J} | \alpha j m \rangle}{\hbar^2 j(j+1)} \langle \alpha j m | \vec{J} | \alpha j m \rangle. \tag{8.118}$$

Hence, considering the component along $\hat{e}_z$,

$$\hbar^2 j(j+1) \langle \alpha j m | V_z | \alpha j m \rangle = \hbar^2 j(j+1) \langle \alpha j m | (\vec{V} \bullet \hat{e}_z) | \alpha j m \rangle = \langle \alpha j m | (\vec{V} \cdot \vec{J})(\vec{J} \bullet \hat{e}_z) | \alpha j m \rangle. \tag{8.119}$$

Our interest of course is in the matrix element of s_z, which is

$$\hbar^2 j(j+1) \langle \alpha j m | s_z | \alpha j m \rangle = \langle \alpha j m | (\vec{s} \cdot \vec{J}) J_z | \alpha j m \rangle = m\hbar \langle \alpha j m | (\vec{s} \cdot \vec{J}) | \alpha j m \rangle, \tag{8.120a}$$

$$\text{i.e., } \hbar^2 j(j+1) \langle \alpha j m | s_z | \alpha j m \rangle = m_j \hbar \left\langle \alpha j m_j \left| \left(\frac{J^2 - L^2 + S^2}{2} \right) \right| \alpha j m_j \right\rangle, \tag{8.120b}$$

$$\text{or } \hbar^2 j(j+1) \langle \alpha j m | s_z | \alpha j m \rangle = m_j \hbar \times \hbar^2 \frac{[j(j+1) - \ell(\ell+1) + s(s+1)]}{2}. \tag{8.120c}$$

We therefore have the important result

$$\langle \alpha j m_j | s_z | \alpha j m_j \rangle = m_j \hbar \times \frac{[j(j+1) - \ell(\ell+1) + s(s+1)]}{2j(j+1)}, \tag{8.121}$$

which enables us to obtain the first-order perturbation correction due to the AZE (Eq. 8.111):

$$\Delta E_{AZE}^{Weak-Field} = \mu_B B m_j + \frac{\mu_B}{\hbar} B m_j \hbar \times \frac{[j(j+1) - \ell(\ell+1) + s(s+1)]}{2j(j+1)}, \tag{8.122}$$

i.e., $$\Delta E_{AZE}^{Weak-Field} = \mu_B B m_j \left[1 + \frac{j(j+1) - \ell(\ell+1) + s(s+1)}{2j(j+1)}\right]. \tag{8.123}$$

We therefore write

$$\Delta E_{AZE}^{Weak-Field} = g\mu_B B m_j, \tag{8.124}$$

with $$g = \left[1 + \frac{j(j+1) - \ell(\ell+1) + s(s+1)}{2j(j+1)}\right], \tag{8.125}$$

called as the *Lande's g factor* (please visit *One hundred years ago Alfred Landé unriddled the Anomalous Zeeman Effect and presaged electron spin* by Horst Schmidt-Böcking et al. 2023 Phys. Scr. 98 014005). The energy splitting given by Eq. 8.123 depends on whether $j = \ell + s$ or $j = \ell - s$:

$$\Delta E_{AZE}^{Weak-Field} = \mu_B B m_j \left[1 + \frac{\left(\ell \pm \frac{1}{2}\right)\left(\ell \pm \frac{1}{2} + 1\right) - \ell(\ell+1) + s(s+1)}{2\left(\ell \pm \frac{1}{2}\right)\left(\ell \pm \frac{1}{2} + 1\right)}\right] = \mu_B B m_j \left[1 \pm \frac{1}{2\ell+1}\right]. \tag{8.126a}$$

Hence, when $j = \ell + \frac{1}{2}$, $$\Delta E_{AZE}^{Weak-Field} = \mu_B B m_j \left[\frac{2\ell+2}{2\ell+1}\right], \tag{8.126b}$$

and when $j = \ell - \frac{1}{2}$, $$\Delta E_{AZE}^{Weak-Field} = \mu_B B m_j \left[\frac{2\ell}{2\ell+1}\right]. \tag{8.126c}$$

The weak-field splitting of the spin–orbit split $2p_{3/2}$ and $2p_{1/2}$ levels of the sodium atom placed in a magnetic field are shown in Fig. 8.9.

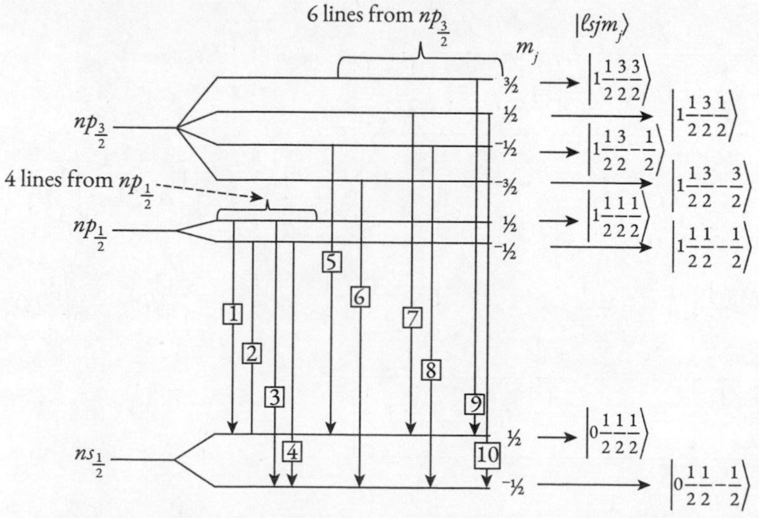

Fig. 8.9 The d_1 and d_2 lines of the sodium atom split into 10 lines.

As shown in Fig. 8.9, the energy level $2p_{3/2}$ splits into four levels corresponding to $m_j = \frac{3}{2}, \frac{1}{2}, -\frac{1}{2}, -\frac{3}{2}$, the level $2p_{1/2}$ splits into two levels corresponding to $m_j = \frac{1}{2}, -\frac{1}{2}$ and the level $2s_{1/2}$ also splits into two levels. The resulting eight states $\left|(\ell, s)\, j, m_j\right\rangle$ can be written in the $\left\{\left|(\ell, s)\, m_\ell, m_s\right\rangle\right\}$ basis using the Clebsch–Gordan coefficients:

$$\left|(\ell, s)\, j, m_j\right\rangle = \sum_{m_l=-\ell}^{\ell} \sum_{m_s=-1/2}^{1/2} \left|(\ell, s) m_\ell m_s\right\rangle \left\langle(\ell, s) m_\ell m_s \middle|(\ell, s)\, j,\, m_j\right\rangle. \tag{8.127}$$

Using the Clebsch–Gordan coefficients from Eqs. 4.96a,b and Eq. 4.97, the eight states $\left|(\ell, s)\, j, m_j\right\rangle$ in Fig. 8.9 are listed below, using α, β to denote $\left|m_s = \pm\frac{1}{2}\right\rangle$ and Y_ℓ^m to denote $\left|\ell, m_\ell\right\rangle$ states:

$\ell = 1, s = \frac{1}{2};\quad 2p_{\frac{3}{2}}\quad j = \frac{3}{2};\ m_j = -\frac{3}{2},\ -\frac{1}{2}, \frac{1}{2}, \frac{3}{2}$

$$2p_{\frac{3}{2}}\quad j = \frac{3}{2},\ m_j = \frac{3}{2} \qquad \left|\left(1, \frac{1}{2}\right) \frac{3}{2}, \frac{3}{2}\right\rangle = \alpha Y_1^1; \tag{8.128a}$$

$$2p_{\frac{3}{2}}\quad j = \frac{3}{2},\ m_j = \frac{1}{2} \qquad \left|\left(1, \frac{1}{2}\right) \frac{3}{2}, \frac{1}{2}\right\rangle = \sqrt{\frac{2}{3}}\alpha Y_1^0 + \frac{1}{\sqrt{3}}\beta Y_1^1; \tag{8.128b}$$

$$2p_{\frac{3}{2}}\quad j = \frac{3}{2},\ m_j = -\frac{1}{2} \qquad \left|\left(1, \frac{1}{2}\right) \frac{3}{2}, -\frac{1}{2}\right\rangle = \sqrt{\frac{2}{3}}\beta Y_1^0 + \sqrt{\frac{1}{3}}\alpha Y_1^{-1}; \tag{8.128c}$$

$$2p_{\frac{3}{2}}\quad j = \frac{3}{2},\ m_j = -\frac{3}{2} \qquad \left|\left(1, \frac{1}{2}\right) \frac{3}{2}, -\frac{3}{2}\right\rangle = \beta Y_1^{-1}. \tag{8.128d}$$

$\ell = 1, s = \frac{1}{2};\quad 2p_{\frac{1}{2}}\quad j = \frac{1}{2};\ m_j = -\frac{1}{2}, \frac{1}{2}$

$$2p_{\frac{1}{2}}\quad j = \frac{1}{2},\ m_j = \frac{1}{2} \qquad \left|\left(1, \frac{1}{2}\right) \frac{1}{2}, \frac{1}{2}\right\rangle = -\sqrt{\frac{1}{3}}\alpha Y_1^0 + \sqrt{\frac{2}{3}}\beta Y_1^1; \tag{8.129a}$$

$$2p_{\frac{1}{2}}\quad j = \frac{1}{2},\ m_j = -\frac{1}{2} \qquad \left|\left(1, \frac{1}{2}\right) \frac{1}{2}, -\frac{1}{2}\right\rangle = \sqrt{\frac{1}{3}}\beta Y_1^0 - \sqrt{\frac{2}{3}}\alpha Y_1^{-1}. \tag{8.129b}$$

$\ell = 0, s = \frac{1}{2};\quad 2s_{\frac{1}{2}}\quad j = \frac{1}{2};\ m_j = -\frac{1}{2}, \frac{1}{2}$

$$2s_{\frac{1}{2}}\quad j = \frac{1}{2};\ m_j = \frac{1}{2} \qquad \left|\left(0, \frac{1}{2}\right) \frac{1}{2}, \frac{1}{2}\right\rangle = \alpha Y_0^0; \tag{8.130a}$$

$$2s_{\frac{1}{2}} \quad j = \frac{1}{2};\ m_j = -\frac{1}{2} \qquad \left|\left(0,\ \frac{1}{2}\right)\frac{1}{2},\ -\frac{1}{2}\right\rangle = \beta Y_0^0. \tag{8.130b}$$

We can also use this example to illustrate one of the many applications of the Wigner–Eckart theorem (Chapter 4, Eqs. 4.151a,b). It is instructive to ask what would be the line intensities of various transitions shown in Fig. 8.9. For example, the probability amplitude for the dipole spectral transition number 5 in Fig. 8.9 using the Wigner–Eckart theorem is:

$$\left\langle j'=\frac{1}{2},m'=-\frac{1}{2}\right|T_q^{(k=1)}\left|j=\frac{3}{2},m=-\frac{1}{2}\right\rangle =$$

$$= \left[\frac{\left\langle \alpha, j'=\frac{1}{2}\left\|T_q^{(k)}\right\|\alpha, j=\frac{3}{2}\right\rangle}{\sqrt{\left(2\times\frac{3}{2}\right)+1}} \times \left\langle\left(j_1=\frac{3}{2}, j_2=k=1\right)m_1=-\frac{1}{2}, m_2=q\right|\left(j_1=\frac{3}{2}, j_2=k=1\right)j=\frac{1}{2}, m=-\frac{1}{2}\right\rangle\right]. \tag{8.131}$$

Likewise, the probability amplitude for the dipole spectral transition number 10 in Fig. 8.9 is

$$\left\langle\left(0\ \frac{1}{2}\right)\frac{1}{2}\ \frac{1}{2}\right|T_q^{(k)}\left|\left(1\ \frac{1}{2}\right)\frac{3}{2}\ \frac{1}{2}\right\rangle =$$

$$= \left[\frac{\left\langle \alpha, j'=\frac{1}{2}\left\|T_q^{(k)}\right\|\alpha, j=\frac{3}{2}\right\rangle}{\sqrt{\left(2\times\frac{3}{2}\right)+1}} \times \left\langle\left(j_1=\frac{3}{2}, j_2=k=1\right)m_1=\frac{1}{2}, m_2=q\right|\left(j_1=\frac{3}{2}, j_2=k=1\right)j=\frac{1}{2}, m=\frac{1}{2}\right\rangle\right]. \tag{8.132}$$

If we want to compare the intensities of the transitions of line number 5 and line number 10, then it is sufficient to examine the ratio of the modulus-square of the right-hand sides in Eq. 8.131 and Eq. 8.132. The actual determination of the reduced matrix elements in Eq. 8.131 and Eq. 8.132 is not required. The Clebsch–Gordan coefficients can be obtained readily from the tables given in Chapter 4, and we readily find that the transitions '5' and '10' in Fig. 8.9 would be of equal intensity.

Along with quantum collisions, which we shall study in Chapter 10, spectroscopy gives us the capability to explore and research the physical world around us. Internal structural details of the atom generate additional complexities in the spectra, over and above the *fine structure* we have discussed in this section. The intrinsic angular momentum of the nucleus (coming from that of the neutrons and protons) results in a corresponding magnetic moment that couples to the extra-nuclear magnetic moment of the electrons in an atom. This results in a *hyperfine structure* and an associated family of spectroscopies, which we study in the next section. Their applications include diverse areas such as the workings of atomic clocks, laser cooling of atoms, and astro-spectroscopy.

8.4 Hyperfine Structure Spectroscopy

We have come across the hyperfine structure of the hydrogen atom in Chapter 7 (Eq. 7.104). Fig. 7.4 merits reproduction in this chapter as well for immediate reference and is presented below in Fig. 8.10.

The hyperfine structure shown in the last column of Fig. 8.10 is due to the coupling of the proton's angular momentum, which we shall denote by $\vec{I}$, with that of the electron, $\vec{J}$. Both I and J are equal to ½. These two add up to a net angular momentum

$$\vec{F} = \vec{I} + \vec{J}, \tag{8.133}$$

with $F = 0,1$. The corresponding energy levels for the 1s state are separated by about 5.872×10^{-6} eV (Fig. 8.10). This constitutes the hyperfine structure of the 1s level of the hydrogen atom. The transition $F = 1 \rightarrow F = 0$ between the hyperfine split levels of the hydrogen atom provides us with an eye into the cosmos, where hydrogen is the most abundant atom. The temperatures out there are extremely low, and the hydrogen atoms are in their ground states $1s^2S_{1/2}$. Nonetheless, the density of atoms is so low in interstellar space that atoms that are slightly excited survive for long enough times without colliding with each other. The spectral transition $F = 1 \rightarrow F = 0$ corresponds to a wavelength of 21 cm (H1 line)

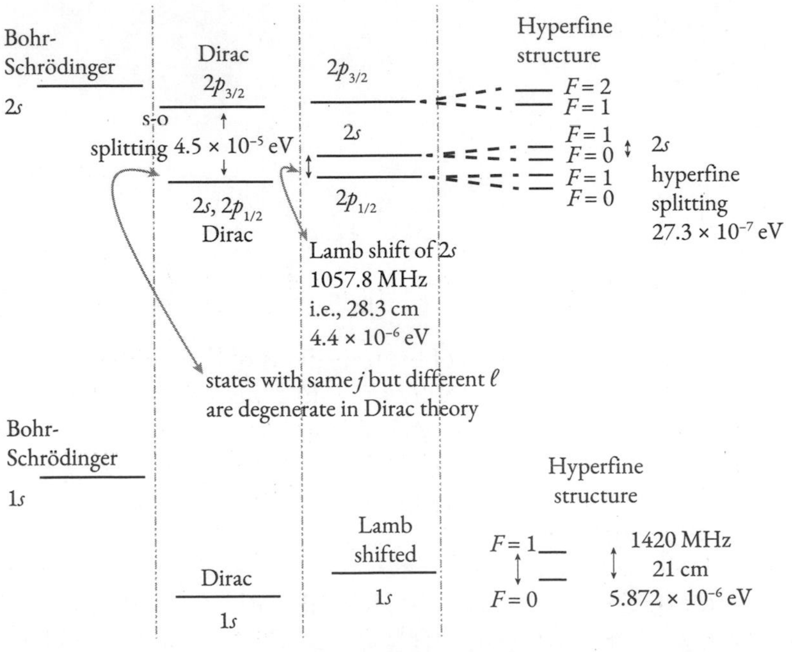

Fig. 8.10 Bohr–Schrödinger energy levels in the nonrelativistic energy spectrum of the hydrogen atom shift due to relativistic effects. RKE lowers the energy levels; spin–orbit interaction splits energy levels of $j = \ell \pm \frac{1}{2}$. The Darwin correction affects only the $\ell = 0$ states. The energy levels are further shifted due to factors not accounted for in Dirac theory. These additional shifts are the Lamb–Retherford shift, and shift on account of the hyperfine structure. This diagram is of course only schematic, not to scale.

and was predicted in 1944 by H. C. van de Hulst. It was first observed by Harold Ewen and Edward M. Purcell in 1951. Radio telescopes placed at various parts of the world are tuned in to the 21 cm waves; measurements of the Doppler shift in this line help determine velocities of atomic hydrogen in space and the location in space of where atomic hydrogen gas is found. The amount of hydrogen is estimated from the intensity of this line. Due to rotation of a galaxy, the 21 cm line appears with a double peak, one that is red-shifted and the other blue-shifted. Among its numerous applications, studies of Doppler shift of the H1 line provides information about *non*-Keplerian rotation curves (Fig. 8.7 and Fig. 8.8 in Reference [3]), indicating the presence of dark matter.

To examine the hyperfine structure of the hydrogen atom, let us denote the electron's angular momentum by $\vec{J}$ with that of the proton by $\vec{I}$. In the presence of an applied magnetic field $\vec{B}$ the *perturbation* term in Eq. 8.95 gets modified to

$$H' = g_e\mu_B\vec{J}\cdot\vec{B} - g_p\mu_p\vec{I}\cdot\vec{B} + h\vec{I}\bullet A\bullet\vec{J}, \tag{8.134}$$

where μ_B is the Bohr magneton (Eq. 8.94b) and μ_p is the nuclear magneton $\mu_p = \frac{e\hbar}{2M_pc}$; it is smaller than the Bohr magneton by the factor $\frac{\mu_e}{M_p}$, which is the ratio of the mass of the electron to that of the proton. g_e and g_p are the Lande g factors (Eq. 8.125) appropriate for the electron and the proton respectively. The term $H'_{hfs} = \vec{I}\bullet A\bullet\vec{J}$ is the perturbation Hamiltonian coming from the hyperfine interaction. In general, A has the character of a second rank tensor, called as the *hyperfine tensor*. Often it is a scalar as would be considered in the present discussion. As we have learned from Chapter 4, two alternative basis sets may be considered in the context of the angular momentum coupling $\vec{J} + \vec{I}$. The basis of the factor states $\left\{\left|JM_JIM_I\right\rangle\right\} \equiv \left\{\left|(JI)M_JM_I\right\rangle\right\}$ would be appropriate when the applied magnetic field is high, just as we have studied in the case of the strong-field Zeeman effect. When the applied magnetic field is weak, a linear superposition of these vectors would provide a more appropriate eigenbasis. We may express the hyperfine interaction in a form similar to that of the spin–orbit interaction (Eqs. 7.90 and 7.91a,b in Chapter 7):

$$\left\langle\vec{J}\bullet\vec{I}\right\rangle = \left\langle\frac{1}{2}\left[F^2 - J^2 - I^2\right]\right\rangle = \frac{\hbar^2}{2}\left[F(F+1) - J(J+1) - I(I+1)\right]. \tag{8.135}$$

We proceed to examine the matrix representation of the hyperfine interaction perturbation Hamiltonian in Eq. 8.134:

$$H'_{hfs} = hA\vec{I}\bullet\vec{J} = hA\left(I_zJ_z + I_xJ_x + I_yJ_y\right) = hA\left[I_zJ_z + \frac{1}{2}\left(I_+J_- + I_-J_+\right)\right], \tag{8.136}$$

where we have used the ladder operators (from Eq. 4.22a in Chapter 4). Using the techniques in the Solved Problem P4.3 in Chapter 4, it is now straightforward to obtain the following representation of the hyperfine interaction perturbation Hamiltonian in the factor basis $\left\{\left|M_JM_I\right\rangle\right\}$:

$$\mathbb{H}'_{hfs} = \begin{bmatrix} \mathbb{H}'_{hfs} & |\uparrow\uparrow\rangle & |\uparrow\downarrow\rangle & |\downarrow\uparrow\rangle & |\downarrow\downarrow\rangle \\ |\uparrow\uparrow\rangle & \begin{pmatrix} \frac{1}{2}g_e\mu_B B_0 \\ -\frac{1}{2}g_p\mu_p B_0 \\ +\frac{hA}{4} \end{pmatrix} & 0 & 0 & 0 \\ |\uparrow\downarrow\rangle & 0 & \begin{pmatrix} \frac{1}{2}g_e\mu_B B_0 \\ +\frac{1}{2}g_p\mu_p B_0 \\ -\frac{hA}{4} \end{pmatrix} & \left(\frac{hA}{2}\right) & 0 \\ |\downarrow\uparrow\rangle & 0 & \left(\frac{hA}{2}\right) & \begin{pmatrix} -\frac{1}{2}g_e\mu_B B_0 \\ -\frac{1}{2}g_p\mu_p B_0 \\ -\frac{hA}{4} \end{pmatrix} & 0 \\ |\downarrow\downarrow\rangle & 0 & 0 & 0 & \begin{pmatrix} -\frac{1}{2}g_e\mu_B B_0 \\ +\frac{1}{2}g_p\mu_p B_0 \\ +\frac{hA}{4} \end{pmatrix} \end{bmatrix} \tag{8.137}$$

The energy eigenvalues may be obtained by diagonalizing the above matrix. Hyperfine structure transitions that can occur are shown (following the discussion in Reference [20]) in the Fig. 8.11a and Fig. 8.11b for the strong-field and the weak-field case respectively.

Another important application of hyperfine structure spectroscopy provides us with precision time keeping through the working of the caesium $\left({}^{133}Cs\right)$ atomic clock. The spin–orbit split $[Xe]6p_{3/2} \rightarrow [Xe]6s^{1}\,{}^{2}S_{1/2}$ (D2 line: 852 nm) and $[Xe]6p_{1/2} \rightarrow [Xe]6s^{1}\,{}^{2}S_{1/2}$ (D1 line: 894 nm) lines of the caesium atom are further split due to the hyperfine structure. The intrinsic angular momentum of the nucleus of the ${}^{133}Cs$ atom is $I = \frac{7}{2}$, so the $6p_{3/2}$ level splits into four levels with $F = 5,4,3,2$ and both the $6p_{1/2}$ and the $6s_{1/2}$ levels split into two levels each with $F = 4,3$. The unit 1 second is *defined* by taking the fixed numerical value of the caesium frequency $\delta\nu$, the unperturbed ground-state hyperfine transition (Fig. 8.12) frequency of the ${}^{133}Cs$ atom, to be 9192631770 when expressed in the unit Hz. The hyperfine structure of the caesium atom thus provides a natural oscillator whose inverse frequency provides a unit for measuring time intervals. The time standard Coordinated Universal Time (UTC) is maintained using some 400 precision atomic clocks placed worldwide. Occasional variations in the earth's angular speed

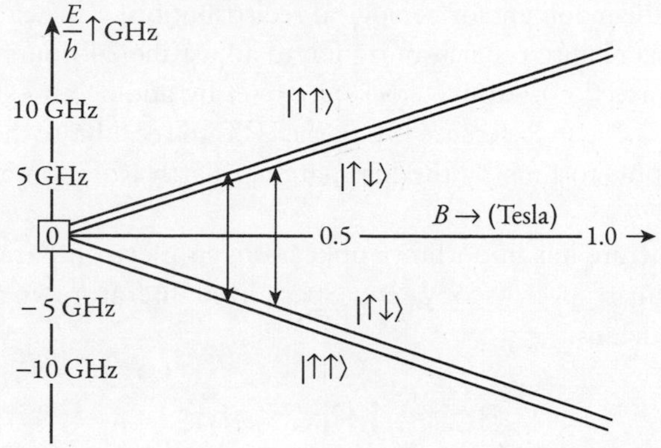

Fig. 8.11a At high fields, the base vectors $\left\{ JM_J IM_I \right\} \equiv \left\{ M_J M_I \right\}$ provide an eigenbasis of good quantum numbers. The system's quantum states can therefore be labeled by $\left\{ |\uparrow\uparrow\rangle, |\uparrow\downarrow\rangle, |\downarrow\uparrow\rangle, |\downarrow\downarrow\rangle \right\}$. Figure adapted from Reference [20].

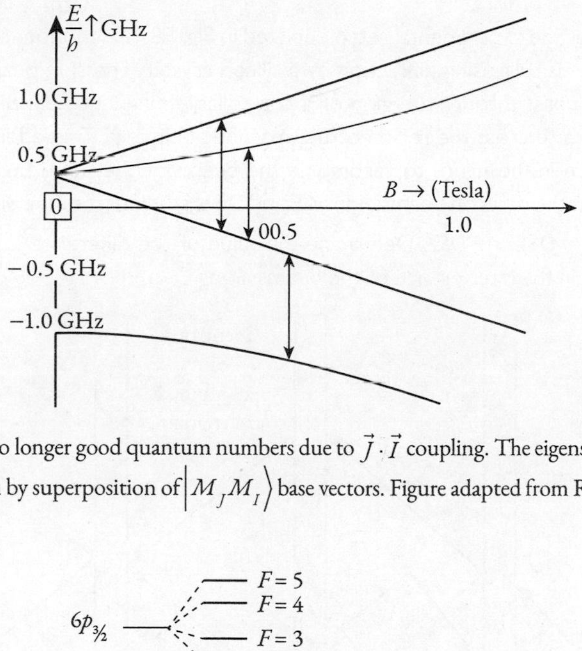

Fig. 8.11b $M_J M_I$ are no longer good quantum numbers due to $\vec{J} \cdot \vec{I}$ coupling. The eigenstates of the Hamiltonian are now given by superposition of $\left| M_J M_I \right\rangle$ base vectors. Figure adapted from Reference [20].

$6p_{3/2}$ ⸺ $\quad$ $F = 5$
$\qquad\qquad\qquad\qquad F = 4$
$\qquad\qquad\qquad\qquad F = 3$
$\qquad\qquad\qquad\qquad F = 2$

$6p_{1/2}$ ⸺ $\qquad F = 4$
$\qquad\qquad\qquad\qquad F = 3$

$$1\ \text{sec} = \frac{9192631770}{(\delta v \text{ in } Cs\ 6s_{1/2}: F = 4 \leftrightarrow F = 3)}$$

$6s_{1/2}$ ⸺ $\qquad F = 4$
$\qquad\qquad\qquad\qquad F = 3$ $\quad$ **9.192 631 770 GHz**

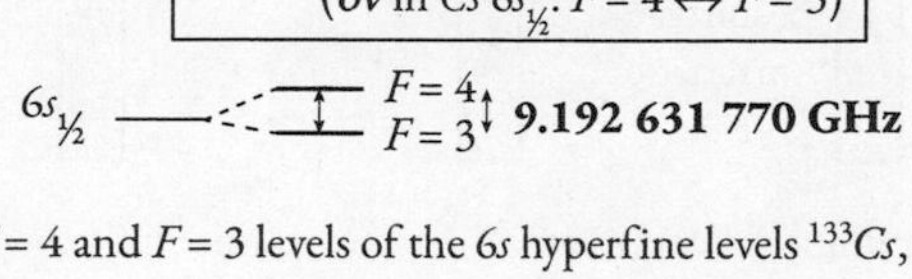

Fig. 8.12 Transition between $F = 4$ and $F = 3$ levels of the $6s$ hyperfine levels ^{133}Cs, which defines the unit of time.

due to tidal effects of the moon and/or geological redistribution of the earth's internal mass (as the earth's core evolves) require resynchronization to adjust the 24-hours day length (over and above the leap year correction). Atomic clocks are reset by adding an extra second, called the "leap second" (see Eq. 3.19c in Reference [3]) to the UTC to recalibrate them. These corrections are determined from time to time by the International Earth Rotation and Reference System Service (IERS).

The hyperfine structure has important applications in Electron Paramagnetic Resonance spectroscopy, Mössbauer spectroscopy, etc. Specialized literature in these areas may be consulted for further details.

Solved Problems

P8.1: Neutron is a charge-less baryon; it is not a fundamental Dirac particle. It has a net spin $s = \frac{1}{2}\hbar$, which comes from the quark model of the neutron. Though neutral, the magnetic moment of the neutron is $\mu_n \simeq -1.913\mu_N$, where $\mu_N = \frac{e\hbar}{2m_p}$ is the nuclear magneton. Consider the experimental setup depicted in Fig. P8.1. A monoenergetic beam of thermal neutrons is split using diffraction by a silicon crystal. The split beam along the path *AD* is made to pass through a region of a controllable magnetic field $\vec{B}$. The time duration for the beam to cross the region of the magnetic field is τ. Two additional crystals are used as shown in the figure to recombine the beams at the point *E* (Fig. P8.1). Interference between the neutron beams along *CE* and *DE* is studied from the *difference* in counts in the detectors *DS*1 and *DS*2. Determine the value of the difference $\delta|\vec{B}|$ to obtain *successive maxima* in the interference of the beams along *CE* and *DE*.

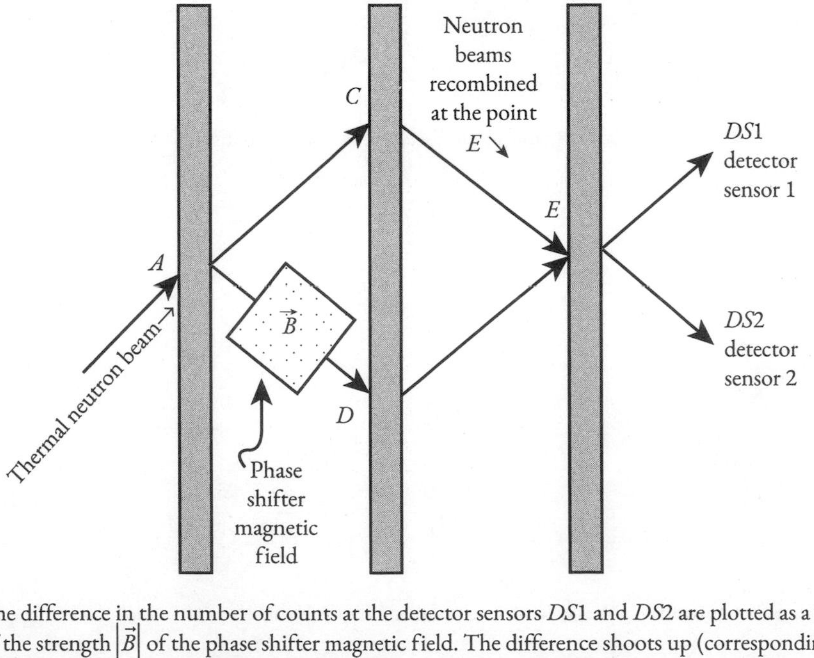

Fig. P8.1 The difference in the number of counts at the detector sensors *DS*1 and *DS*2 are plotted as a function of the strength $|\vec{B}|$ of the phase shifter magnetic field. The difference shoots up (corresponding to bright fringes) and goes to zero (dark fringes) as a function of the strength of the magnetic field.

Solution:

Time evolution operator due to the interaction of the neutron magnetic moment with the magnetic field in the path $A \to D$ over the period τ is

$$U(\tau) = \exp\left(-\frac{iH\tau}{\hbar}\right) = \exp\left(-\frac{i\left(-\vec{\mu}\bullet\vec{B}\right)\tau}{\hbar}\right) = \exp\left(i\frac{\left(\vec{\mu}\bullet\vec{B}\right)}{\hbar}\right)\tau = \exp\left(i\frac{\left(\gamma\vec{s}\bullet\vec{B}\right)}{\hbar}\right)\tau,$$

where γ is the gyromagnetic ratio. It would add a phase $\xi = \omega\tau$ to the beam component $A \to D$, where the circular frequency $\omega = \frac{\left(\vec{\mu}\bullet\vec{B}\right)}{\hbar} = \omega(B)$ is a linear function of the magnetic field strength. A change δB in the magnitude of the field strength results in a change $\delta\xi$ in the phase. To obtain successive maxima, we must change the magnetic field to attain $\delta\xi = 4\pi$, since the spin angular momentum of the neutron follows SU(2) algebra corresponding to spin–half (see Eq. 4.72b from Chapter 4). Results of an experiment in which $\delta\xi = 4\pi$ in successive maxima are reported by A.Werner, R.Colella, A.W.Overhauser, and C.F.Eagen (*Phys. Lett.*, 35:1053 (1972)). Isn't this a beautiful experiment that reveals the symmetry of the SU(2) group in which the unit operation requires a rotation through 4π?

P8.2: This problem involves continuation of the discussion on neutron interferometry that we began with in the above problem, P8.1. We shall now consider effects due to the gravitational field. We so we consider two arrangements of the experimental set up in Fig. P8.1. In the first setup, we shall arrange the apparatus such that the paths *ADE* and *ACE* that interfere at *E* are all in a horizontal plane (Fig. P8.2). Neutrons therefore experience essentially the same gravitational field as they go along the path *ADE* as along the path *ACE*. In the second setup, we lift the apparatus and turn it upward through an angle θ about the line *AD*. As a result of this, the segment *CE* is raised to *C′E′* as shown in Fig. P8.2. Comment on the conditions for constructive and destructive interference at the recombination point *E′* between the sub-beams *AC′E′* and *ADE′*.

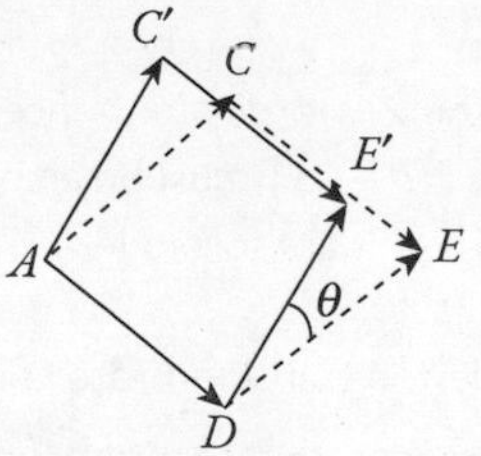

Fig. P8.2 In the first experiment with the neutron interferometer, we consider the case when the plane *ADECA* is horizontal. We then raise the arm *CE* to *C′E′* by turning the interferometer through an angle θ about the arm *AD*. The track *C′E′* is now at a height $DE\sin\theta$ higher than the arm *AD*. Note that $DE = DE'$.

Solution:

Neutrons passing through the arms *AC′* experience the same variation in the gravitational field as those which pass through *DE′*, but the neutrons which pass through the segment $C' \to E'$ are at a higher gravitational potential compared to those that pass through the segment $A \to D$.

The propagator (Chapter 2, Eq. 2.1a) for the path $\left.K\left(E',t_f;A,t_0\right)\right]_{via\ D} \neq \left.K\left(E',t_f;A,t_0\right)\right]_{via\ C'}$

since the phase $\frac{1}{\hbar}\int_{C'}^{E'} dtL(x,\dot{x}) \neq \frac{1}{\hbar}\int_{A}^{D} dtL(x,\dot{x})$;

the difference in the two phases is

$$\delta\phi = \frac{1}{\hbar}\int_{i}^{f} dt\, mg(DE\sin\theta) = \frac{1}{\hbar}mg(DE\sin\theta)\tau = \frac{1}{\hbar}mg(DE\sin\theta)\left[\frac{mC'E'}{p}\right]$$

since the time interval τ is nothing but the distance traversed divided by the speed, the latter being given by the momentum p divided by the mass m. Using the expression for the de Broglie wavelength, we get $\delta\phi == \frac{1}{\hbar}m^2g(DE\sin\theta)\left[\frac{\lambda}{h}C'E'\right]$. Clearly, the interference pattern is sensitive to the angle of inclination of the apparatus. Such an experiment is reported by S.A.Werner, R.Colella, and A.W.Overhauser (*Phys. Rev.* A21:1419 [1980]). The interference pattern is governed by the phase difference $\delta\phi$ and is very sensitive to the inclination θ whose sine appears in the expression for the phase difference. This experiment not only provides an experimental test of the SU(2) algebra of the spin-$\frac{1}{2}$ particles, but also a sensitive test of the equivalence between the inertial mass and the gravitational mass. An improvisation of this experiment that includes passage of the sub-beam along *AD* to pass through a region of the magnetic field (similar to the case in P8.1) would further highlight the sensitive dependence of the interference pattern on tiny variations in the geometrical setup and on the magnetic field. Such experiments are of great value in realizing the importance of geometrical (Pancharatnam–Berry) phase (discussed in Chapter 2) in the context of gravitational fields (gravitational geometric phase).

P8.3: Consider a particle in a one-dimensional square-well potential. Write down the transition matrix dipole moment element, $\mu_{ab} = ex_{ab} = \int \psi_b^*(ex)\psi_a dx$, where *a* and *b* represent the quantum number *n* (see Eqs. 3.60a,b from Chapter 3). Determine the transition selection rules.

Solution:

The operator *ex* is an odd function of the *x*-coordinate; hence, transition can take place only between wavefunctions of opposite parity. Hence, $|(a-b)|$ must be an odd number (examples: $2 \to 1$ and $1 \to 4$ are allowed transitions and $1 \to 3$ and $4 \to 2$ are forbidden).

P8.4: Using Fermi's golden rule for transition rate, obtain the differential decay rate for beta decay. Assume a short-range contact interaction of the form $V_\beta = g\delta(\vec{r}_i - \vec{r}_j)$ for the weak interaction potential and neglect any effect of coulomb interaction.

Solution:

From Fermi's golden rule (Eq. 6.123 and Eq. 6.131a–c, Chapter 6), the transition (decay) rate is $\lambda = \frac{2\pi}{\hbar}|M_{fi}|^2 \rho(E_f)$, where $\rho(E_f)$ is the density of the final states, $M_{fi} = \int \psi_f^* V_\beta \psi_i dV$ is the transition matrix element, V_β is the (weak) interaction potential, the initial state is $\psi_i = u_P$ and the final state is $\psi_f = u_D\varphi_\beta\eta_\nu$; u_P and u_D being the nuclear states; and the lepton states for the beta particle and the neutrino are φ_β and η_ν respectively.

The matrix element therefore is $M_{fi} = \int u_D^*\varphi_\beta^*\eta_\nu^* V_\beta u_P dV$.

Admitting plane wave solutions for β and ν,

we have $\varphi_\beta = \frac{1}{V^{1/2}} e^{i\vec{k}_e \bullet \vec{r}} = \frac{1}{V^{1/2}}\left(1 + i\vec{k}_e \bullet \vec{r} - \frac{(\vec{k}_e \bullet \vec{r})^2}{2} + \ldots\right) \approx \frac{1}{V^{1/2}}$ and similarly $\eta_\nu \approx \frac{1}{V^{1/2}}$.

Hence, $M_{fi} = \frac{g}{V}\int u_D^* u_P dV = \frac{g}{V} M'_{fi}$, where M'_{fi} is the nuclear matrix elements and the transition rate is $\lambda = \frac{2\pi}{\hbar V^2} g^2 \mid M'_{fi} \mid^2 \rho(E_f)$. The density of final states is $\rho(E_f) = \frac{dN}{dE_f}$, where $dN = \frac{4\pi}{(2\pi\hbar)^3} p^2 dp V$ is the number of states in the spatial volume V and the momentum volume $4\pi p^2 dp$.

The total $dN_{total} = dN_e dN_\nu = \left(\frac{4\pi}{(2\pi\hbar)^3}\right)^2 p_e^2 dp_e p_\nu^2 dp_\nu V^2$.

With $E_f = E_e + E_\nu = E_e + p_\nu c$, we have $p_\nu = \left(\frac{E_f - E_e}{c}\right)$ and $dp_\nu = \frac{dE_f}{c}$, at a given E_e.

Therefore, $dN_{total} = dN_e dN_\nu = \left(\frac{16\pi^2}{(2\pi\hbar)^6}\right) p_e^2 dp_e \left(\frac{E_f - E_e}{c}\right)^2 \frac{dE_f}{c} V^2$

and $\rho(E_f) = \frac{dN_{total}}{dE_f} = \left(\frac{16\pi^2}{(2\pi\hbar)^6 c^3}\right) p_e^2 dp_e \left(E_f - E_e\right)^2 V^2$.

Hence, the differential decay rate at a particular p_e is given by

$$d\lambda(p_e) = \frac{2\pi}{\hbar V^2} g^2 \mid M'_{fi} \mid^2 \left(\frac{16\pi^2}{(2\pi\hbar)^6 c^3}\right) p_e^2 dp_e \left(E_f - E_e\right)^2 V^2 = \frac{g^2 \mid M'_{fi} \mid^2}{2\pi^3 \hbar^7 c^3}\left(E_f - E_e\right)^2 p_e^2 dp_e.$$

P8.5: How many fine-structure levels will the 3*d* state of hydrogen split into? Determine the energies of these levels. What happens to these levels when the atom is placed in a weak magnetic field?

Solution:

From Chapter 7, Eq. 7.103, for the $3d$ state of the hydrogen atom, $n = 3, \ell = 2, s = 1/2$.

The total angular momentum can be $j = \ell + s, \ell + s - 1, \ldots |\ell - s|$. Therefore, $j = 3/2, 5/2$.

The required energies are $E_{nj}^{Dirac} = E_n\left[1 + \frac{(Z\alpha^2)}{n^2}\left(\frac{n}{j + 1/2} - \frac{3}{4}\right)\right]$.

Hence, $E\left(n = 3, j = \frac{3}{2}\right) = -\frac{13.6}{3^2}\left[1 + \frac{(1/137)^2}{3^2}\left(\frac{3}{2} - \frac{3}{4}\right)\right] = -1.5111178 \text{ eV}$

and $E\left(n = 3, j = \frac{5}{2}\right) = -\frac{13.6}{3^2}\left[1 + \frac{(1/137)^2}{3^2}\left(1 - \frac{3}{4}\right)\right] = -1.5111133 \text{ eV}$.

In a weak magnetic field, the energy level for each j splits further for $m_j = j, j-1, \ldots, -j+1, -j$.

P8.6: Deuterium atoms are kept in an external magnetic field. Considering the electron-nuclear hyperfine interactions, how many spectral lines in the ESR (electron spin resonance) spectrum does one expect?

Solution:

In the presence of a magnetic field, the electron state in deuterium has two energy levels corresponding to the two spin states denoted by $m_s = +1/2$ and $m_s = -1/2$. The nucleus of deuterium has a nuclear spin coming from that of a proton and a neutron; it is given by $I = 1$. Therefore, $m_I = I, (I-1), \ldots, -(I-1), -I = \{1, 0, -1\}$. Considering the interaction of the two m_s levels of the electron with the nucleus of deuterium, we get six energy levels as shown in the figure below.

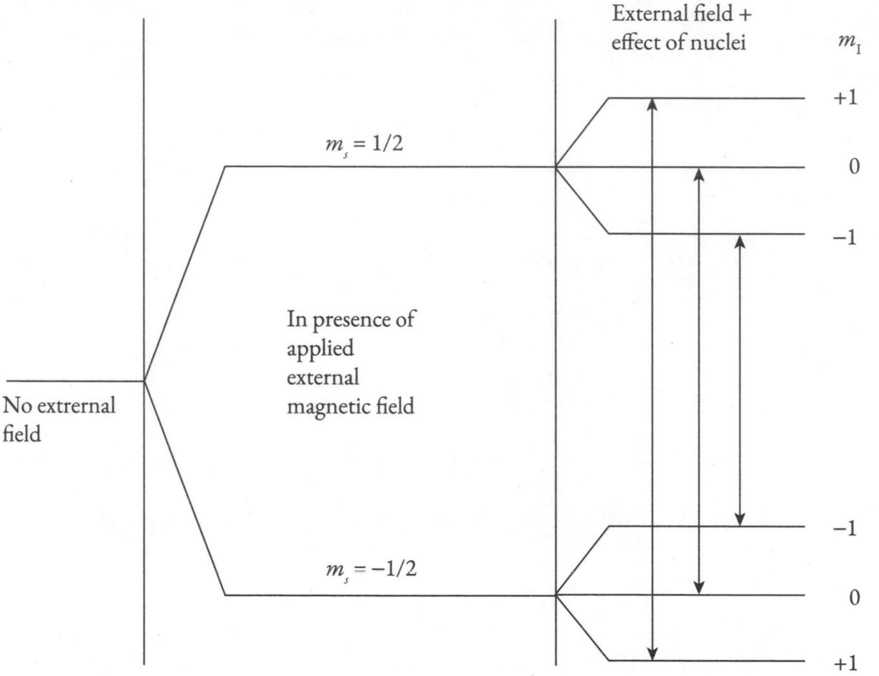

In the spectrum, we would expect to see three lines corresponding to the selection rules $\Delta m_s = \pm 1$ and $\Delta m_I = 0$, shown in the above figure.

P8.7: A proton is kept in a magnetic field of 14092 Gauss. Determine the splitting of the energy level of the proton due to the magnetic field and obtain the frequency of the incident radiation required to flip the spin state of the proton.

Solution:

Energy of the proton in the magnetic field is $E = \mu_z B$, where $\mu_z = g_I \mu_N m_I = \hbar \gamma_I m_I$, γ_I is the proton's gyromagnetic ratio; its value is 26750 rad/s/Gauss. Therefore, the energy of the nuclear levels is $-\hbar \gamma_I m_I B$ with $m_I = \pm 1/2$. The difference in energy levels is $\Delta E = \hbar \gamma B$, and for a spin–flip transition to occur, the energy of the incident radiation must be $h\nu = \hbar \gamma B$, where ν is the frequency of the incident radiation. Therefore, we have $\nu = \dfrac{\gamma B}{2\pi} = \dfrac{26750 \times 14092}{2\pi} = 60 MHz$. (A resonant transition between nuclear spin states is known as *Nuclear Magnetic Resonance (NMR)*, discovered by Isidor Isaac Rabi. He was awarded the Nobel Prize in Physics in 1944.)

P8.8: Show that the radiation from the transition between the two hyperfine levels of hydrogen Is ground state has a wavelength of 21 cm.

Solution:

From Eq. 8.135, the hyperfine splitting is

$\Delta E = \frac{2}{3}(Z\alpha^4)\left(\frac{m_e}{m_N}\right)(m_e c^2)g_N \frac{1}{n^3}(F(F+1)-3/2)$, where F is the *total* angular momentum (including the electron and the proton angular momentum). For the hydrogen atom in its ground state, the electron (spin + orbital) angular momentum is $J = 1/2$ and the proton has $I = 1/2$. Therefore, the *total* (electron + nucleus) angular momentum is $F = 0,1$. The energy difference between the two hyperfine levels determines the wavelength of the emitted radiation:

$$\Delta E|_{F=1} - \Delta E|_{F=0} = \frac{4}{3}(Z\alpha)^4\left(\frac{m_e}{m_N}\right)(m_e c^2)g_N\frac{1}{3} = \frac{4}{3}(1/137)^4\left(\frac{0.51}{938}\right)(0.51\times 10^6)(5.56) = 5.84\times 10^{-6}\,\text{eV}$$

Therefore, the wave length is: $\lambda = \frac{2\pi\hbar c}{\Delta E|_{F=1} - \Delta E|_{F=0}} = \frac{2\pi\times 1973}{5.84\times 10^{-6}}\mathring{A} \simeq 21.2\,\text{cm}.$

P8.9: (a) The spectral line of singly ionized calcium in a distant galaxy is measured in an experiment to be 410 nm. The wavelength of this spectral line in a *rest frame* is known to be 393.3 nm. Calculate the redshift and the recessional velocity. (b) According to the empirical Hubble's Law, the velocities of recession between our galaxy and other galaxies are directly proportional to the distance between them: $v = H_0 d$, where H_0 is the Hubble constant and d is the distance of the galaxy. If this galaxy is 163 Mpc away, calculate the Hubble constant. (c) From the above calculation, estimate the age of the universe.

Solution:

(a) The redshift is given by $z = \frac{\Delta\lambda}{\lambda} = \frac{410 - 393.3}{410} = 0.0407.$

The recessional velocity is $v = c \times z = 1200\,\text{km/s}.$

(b) Hubble's constant: $H_0 = \frac{v}{d} = \frac{1200\,\text{km/s}}{163\,\text{Mpc}} = 75\,\text{km/s/Mpc}.$

(c) Assuming that the universe has been expanding at a constant rate since the beginning, the age of the universe is given by $\frac{1}{H_0} = \frac{1}{75\,\text{km/s/Mpc}} = \frac{1}{75}\times 3.09\times 10^{19}\,\text{s} \simeq 13.03$ billion years.

***Note**: Edwin Hubble's observations showed the existence of galaxies beyond the Akash-Ganga (Milky-way) galaxy. In 1929, Hubble showed from his observation of the redshift of light coming from distant galaxies that the farther a galaxy is from us, the faster it recedes from us. This showed that the universe is expanding. The 2011 Nobel Prize in physics was awarded to Saul Perlmutter, Brian P. Schmidt, and Adam G. Riess "for the discovery of the accelerating expansion of the Universe through observations of distant supernovae." The source of this acceleration is not well understood, hence attributed to "dark energy". Within the framework of the general theory of relativity, it corresponds to a positive vacuum energy (positive value of the cosmological constant).*

P8.10: With reference to Eq. 8.18 through Eq. 8.20, show that the Lorentzian function is a representation of the Dirac-δ.

Solution:

First, it will be useful to revisit the Solved Problem P6.8 in Chapter 6. To demonstrate that the Lorentzian profile presents a Dirac-δ function, it is sufficient to show that the function $f(x) = \frac{1}{\pi}\frac{1}{1+x^2}$ is nonnegative, locally integrable, and $\int_{\mathbb{R}^n} f(x)dx = 1$.

It is readily seen that $f(x) \geq 0$ for all values of x,

$$\int_a^b f(x)dx = \frac{1}{\pi}\int_a^b \frac{1}{1+x^2}dx = \frac{1}{\pi}[\arctan(x)]_a^b = \frac{1}{\pi}[\arctan(b) - \arctan(a)] < \infty, \text{ since}$$

$\arctan(x) \in \left[-\frac{\pi}{2}, \frac{\pi}{2}\right]$. Therefore, the function is locally integrable.

We see that $\int_{\mathbb{R}^n} f(x)dx = \int_{-\infty}^{\infty} f(x)dx = \frac{1}{\pi}\int_{-\infty}^{\infty}\frac{1}{1+x^2}dx = \frac{1}{\pi}[\arctan(x)]_{-\infty}^{\infty} = \frac{1}{\pi}\left[\frac{\pi}{2} - \left(-\frac{\pi}{2}\right)\right] = 1$. Hence,

$$\lim_{\Gamma_d \to 0} f_{\Gamma_d/2}(\omega - \omega_{0,s}) = \delta(\omega - \omega_{0,s}) \Rightarrow \lim_{\Gamma_d \to 0}\frac{1}{\pi}\frac{\Gamma_d/2}{(\omega - \omega_{0,s})^2 + (\Gamma_d/2)^2} = \delta(\omega - \omega_{0,s}).$$

It is a good exercise to plot $L_\varsigma(\tilde{\omega})$ as a function of $\tilde{\omega}$.

P8.11: For our galaxy, the recessional velocity is given by 2,675 km/s, and its inclination is 82°. The wavelength of a H_α line is observed to be 6622.8Å at a radius of 4 kiloparsec. Calculate the rotational velocity, and using that determine the dynamical mass inside the radius of 4 kiloparsec.

Solution:

The rest frame wavelength of the H_α line is given by 6563Å. The line-of-sight velocity is given by $v_{LOS} = \frac{c(\lambda_{obs} - \lambda_{rest})}{\lambda_{rest}} = 3 \times 10^5 \text{ km/s} \times \frac{59.8}{6563} = 2733.5 \text{ km/s}$. The rotational velocity is given by $v_{rotational} = \frac{v_{LOS} - v_{recessional}}{\sin i}$, where i is the inclination angle. $v_{rotatonal} = \frac{2733.5 - 2675}{\sin(82)} = 59.07 \text{ km/s}$

The dynamical mass is given by

$$M_{dynamical} = \frac{(v_{rotation})^2 r}{G} = \frac{(59.07 \text{ km/s})^2 \times 4 \text{ kpc}}{1.33 \times 10^{21} \text{ km}^3 M_{sun}^{-1} \text{ s}^2} \times \frac{3.08 \times 10^{16} \text{ km}}{1 \text{kpc}} = 3.23 \times 10^9 M_{sun}.$$

Comments: In this problem, to calculate the dynamical mass, we used the Newton–Keplarian formula, where the $v_{rotation} = \sqrt{\frac{M_{dynamical} G}{r}}$ which predicts a decrease in the rotational speed as the radius from the center is increased. However, it is observed that instead of declining, the velocity curve flattens out with distance. To empirically account for this flattening, the hypothesis of non-luminous mass nicknamed "dark matter" has been proposed (see section 8.4, Chapter 8, in Reference [3]).

Exercises

E8.1: In the Solved Problem P8.6, we discussed the ESR spectrum of a deuterium atom in an external magnetic field. Similarly, considering the electron–nuclear interaction, how many lines in the ESR spectra does one expect for an atom of (a) hydrogen and (b) tritium?

E8.2: List all the degenerate states of the $n = 3$ level of the relativistic hydrogen atom. Consider an atom of hydrogen that is placed in an electric field, and determine the number of distinct energy sublevels that the $n = 3$ level splits into. Find the residual degeneracy, if any, of each sub-level. Numerically calculate the level shift for each case.

E8.3: Evaluate the spin–orbit interaction for a hydrogen-like *muonic atom* (in which there is a muon instead of an electron).

E8.4: The Lyman-α line is emitted when the atomic electron undergoes a transition from $n = 2$ level to the ground state. The spin–orbit interaction splits the Lyman-α line into a fine-structure doublet. Calculate the wavelengths of the two transitions in the Lyman-α doublet.

E8.5: In the previous problem, we studied the fine structure of the Lyman-α transition. What happens to the Lyman-α doublet if the hydrogen atom is placed in a magnetic field that is (a) weak and (b) strong?

E8.6: The Balmer-α line in the visible spectrum of the hydrogen atom results from the transition from $n = 3$ to $n = 2$ energy levels. How many such spectral lines can one expect to see considering the fine-structure splitting? What happens to these lines if the atom is kept in a weak magnetic field?

E8.7: Consider the transition from $3p_{3/2,1/2}$ to 3s that is responsible for the D1 and D2 lines of the sodium atom. Calculate the splitting of these spectral lines when the sodium atom is placed in a strong magnetic field.

E8.8: An absorption line of calcium has a wavelength of 3933 Å, but it is observed in a distant moving galaxy to have a wavelength of 4002 Å. How fast is this galaxy moving? Is it moving toward or away from us? Using the value of the Hubble's constant to be 73 km/s/Mpc, estimate how far away the galaxy is from us.

E8.9: The K line of calcium, which has a rest frame wavelength of 393.3 nm, is observed to be 394.6 nm when coming from a distant galaxy. Calculate the redshift and the recessional speed of the galaxy. How far away is the galaxy from us?

E8.10: Find the probability that an electron will be *removed* by an electric field from a spherical potential well with short-range forces in which the electron is in a bound s state. The electric field is assumed to be weak, such that $|e|\varepsilon \ll \hbar^2\kappa^3/m$, where m is the electron mass, $\kappa = \sqrt{(2m|E|)}/\hbar$, and E is the binding energy of the electron in the well.

References

[1] J. W. Cooper, Photoionization from Outer Atomic Subshells: A Model Study. *Physical Review* 128:2, 681–693 (1962); U. Fano and J. W. Cooper, Spectral Distribution of Atomic Oscillator Strengths. *Reviews of Modern Physics* 40, 441 (1968).

[2] U. Fano and A. R. P. Rau, *Atomic Collisions and Spectra* (Elsevier Academic Press, 1986).

[3] P. C. Deshmukh, *Foundations of Classical Mechanics* (Cambridge University Press, New Delhi, 2019).

[4] George B. Arfken and Hans J. Weber, *Mathematical Methods for Physicists*, 6th edition (Elsevier Academic Press, 2005).

[5] O. Hemmers, G. Fisher, P. Glans, D. L. Hansen, H. Wang, S. B. Whitfield, R. Wehlitz, J. C. Levin, I. A. Sellin, R. C. C. Perera, E. W. B. Dias, H. S. Chakraborty, P. C. Deshmukh, S. T. Manson, and D. W. Lindle, Beyond the Dipole Approximation: Angular-distribution Effects in Valence Photoemission. *J. Phys. B: At. Mol. Opt. Phys.* 30, L727 (1997).

[6] O. Hemmers, R. Guillemin, E. P. Kanter, B. Krassig, D. W. Lindle, S. H. Southworth, R. Wehlitz, J. Baker, A. Hudson, M. Lotrakul, D. Rolles, W. C. Stolte, I. C. Tran, A. Wolska, S.W. Yu, M. Ya. Amusia, K. T. Cheng, L.V. Chernysheva, W. R. Johnson, and S.T. Manson, Dramatic Nondipole Effects in Low-energy Photoionization: Experimental and Theoretical Study of Xe 5s. *Phys. Rev. Lett.* 91:5, 053002-1 (2003).

[7] A. F. Starace, Length and Velocity Formulas in Approximate Oscillator-strength Calculations. *Phys. Rev. A* 3:4, 1242–1245 (1971).

[8] S. Chandrasekhar, On the Continuous Absorption Coefficient of the Negative Hydrogen Ion. *Astrophysics Journal* 102, 223 (1945).

[9] S. T. Manson, Systematics of Zeros in Dipole Matrix Elements for Photoionizing Transitions: Nonrelativistic Calculations. *Phys. Rev. A* 31, 3698 (1985).

[10] Tanima Banerjee, Electron Correlation and Relativistic Effects in Atomic Multipole Transitions. Ph.D. thesis (Indian Institute of Technology Madras, India, 2008).

[11] Andrew D. Greentree, Jens Koch, and Jonas Larson, Fifty Years of Jaynes-Cummings. *Physics Journal of Physics B Atomic Molecular Physics* 46:22, 220201 (2013). Doi:10.1088/0953-4075/46/22/220201.

[12] M. J. Seaton, Quantum Defect Theory I. General Formulation. *Proc. Phys. Soc.* 88, 801 (1966).

[13] C. M Lee and W. R. Johnson, Scattering and Spectroscopy: Relativistic Multichannel Quantum-Defect Theory. *Phys. Rev. A* 22, 979 (1980).

[14] Aarthi Ganesan, Sourav Banerjee, Pranawa C. Deshmukh, and Steven T. Manson, Photoionization of Xe5s: Angular Distribution and Wigner Time Delay in the Vicinity of the Second Cooper Minimum. *Journal of Phys.B.: Atomic, Molec. and Opt. Phys.* (2020). DOI: https://doi.org/10.1088/1361-6455/abbe2e.

[15] P. C. Deshmukh and Sourav Banerjee, Time Delay in Atomic and Molecular Collisions and Photoionization/photodetachment. *International Reviews in Physical Chemistry*, 40:1, 127–153 (2020). https://doi.org/10.1080/0144235X.2021.1838805

[16] P. C. Deshmukh, S. Banerjee, A. Mandal, and S. T. Manson, Eisenbud–Wigner–Smith Time Delay in Atom–Laser Interactions. *Eur. Phys. J. Spec. Top.* (2021). https://doi.org/10.1140/epjs/s11734-021-00225-7.

[17] M. S. Safronova, D. Budker, D. DeMille, A. Derevianko, and Charles W. Clark, Search for New Physics with Atoms and Molecules. *Revs. Mod. Phys.*, 90:2, 025008 (2018).

[18] C. S. Wood, S. C. Bennett, D. Cho, B. P. Masterson, J. L. Roberts, C. E. Tanner, and C. E. Wieman, Measurement of Parity Nonconservation and an Anapole Moment in Cesium Science 275, 1759 (1997).

[19] George Toh, Amy Damitz, Carol E. Tanner, W. R. Johnson, and D. S. Elliott, Determination of the Scalar Polarizability of the Cesium 6s 2S1/2 – 7s 2S1/2 Transition and Implications for Atomic Parity Violation. *Phys. Rev. Lett.* 123, 073002 (2019).

[20] G. Denninger, Hyperfine Interaction. https://www3.pi2.uni-stuttgart.de/official/g.denninger/GKMR2005/PDF/hyperfine_interaction.pdf (downloaded on November 15, 2022).

9

The Many-Electron Atom

D. R. Hartree. B668, https://arkiv.dk/en/vis/5944851. Courtsey: Niels Bohr Archive.

It may well be that the high-speed digital computer will have as great an influence on civilization as the advent of nuclear power.

—Douglas R. Hartree

Statistics is intrinsic to quantum theory. Description of nature requires it. In classical physics, we take recourse to statistical analysis either when we are not interested in a detailed solution, or when we are unable to cope up with a large amount of data. Statistical methods enable us to extract important and useful information about the system to understand its physical properties. For example, rather than keeping track of kinetic energies of individual molecules, we determine their average, and benefit from the notion of *temperature*. The great debate between Albert Einstein and Niels Bohr in the late 1920s and in the 1930s eventually established (*especially after the work of John Bell in 1964, discussed in Chapter 11*) that the role of statistics in describing nature is far deeper than was suspected earlier. Statistics must be invoked to describe physical properties of nature even when we are dealing with a single particle, or for that matter even vacuum. In this chapter, we study another important manifestation of the role of statistics in quantum theory; this comes from the identity of particles in nature.

9.1 Symmetry Properties of Wavefunctions for Bosons and Fermions

If we have ten electrons, classical physics would regard the particles as distinguishable; the ten electrons could be *numbered* from one through ten, and one could, at least in principle, follow the dynamics of a particular, say the seventh, electron. Electrons are fundamental particles in nature. They are not amenable to their description by classical statistics. They must be described by quantum theory. In an N-electron system, it is impossible to tag any of these electrons by a number, such as the seventh. This impacts how the electrons occupy energy levels available to them.

The smallest *many*-particles system is a *pair* of identical particles. If we subject the pair of particles to an *interchange*, effected by the operator $\widehat{I}$, and then interchange them yet again, we must get the original state. We expect the double interchange to be the identity operator. We therefore write

$$\hat{I}\left\{\hat{I}\Psi\left(q_1,q_2\right)\right\} = \Psi\left(q_1,q_2\right), \tag{9.1}$$

where the two-particle state is represented by $\Psi\left(q_1,q_2\right)$, with $q_i, i = 1,2$ representing the coordinates of the two particles. These coordinates would include not only the three space coordinates but also the spin coordinate. We may call the two-particle wavefunction $\Psi\left(q_1,q_2\right)$ as a *geminal*, after the constellation "Gemini" whose prominent stars, Castor and Pollux (*Punarvasu nakshatra* in Hindu astronomy), appear as twins. The two particles (eg., electrons) that are at the center of our discussion are, however, not merely *identical*; they are *indistinguishable*, unlike Castor and Pollux. We have already seen in Chapter 7 that as a result of the constancy of the speed of light in every inertial frame of reference, we require a relativistic formulation of quantum mechanics. On requiring a quantum equation to be Lorentz covariant, we are led to a deep-seated property of fundamental particles: they have an intrinsic angular momentum, called as spin angular momentum (*see the discussion following Eq. 7.57, Chapter 7*). In Chapter 4, we have seen that a quantum angular momentum is a vector operator $\vec{j}$ such that $\vec{j}\bullet\vec{j} = j^2$ and $\vec{j}\bullet\hat{u} = j_u$ are *compatible* observables, represented by commuting operators. The unit vector $\hat{u}$ picks a direction in space; it is common to choose a Cartesian coordinate system with its z-axis along $\hat{u}$. The operators j^2 and $j_z = j_u$ therefore have common eigenkets, labeled by the eigenvalue $\left[\hbar^2 j\left(j+1\right)\right]$ of j^2, and $\left[\hbar m\right]$ of j_u. We know from Chapter 4 that the value of j is either half-integer or integer. Particles with half integer, $\left(\frac{1}{2},\frac{3}{2},\frac{5}{2},\ldots\right)$ intrinsic spin angular momentum are called *fermions* (after Enrico Fermi), and those with integer $(0,1,2,\ldots)$ intrinsic spin angular momentum are called *bosons* (after Satyendranath Bose). As mentioned earlier, the fact that identical particles in nature are indistinguishable (i.e., they cannot be tagged by numbers, or colors, or in any manner) makes it necessary for us to adopt two different schemes to account for how they are distributed in the energy states accessible to them. This can be stated in a *deceptively* simple manner: *Fermions are particles with half-integer spin. They are distributed in various energy states accessible to them according to a statistical law known as the Fermi–Dirac (FD) statistical distribution. Bosons are particles with integer spins; these occupy energy states accessible to them according to a different law, known as the Bose–Einstein (BE) statistical distribution.* Both FD and BE statistics are different from the classical distribution function, namely the Maxwell–Boltzmann (MB) statistics.

It is imperative now that two interchanges of identical particles being the identity, a single interchange would, at best, change the *phase* of the state function of the two identical particles:

$$\hat{I}\ \Psi(q_1,q_2) = \Psi(q_2,q_1) = e^{i\alpha}\Psi(q_1,q_2). \tag{9.2}$$

Furthermore, since $e^{i2\alpha} = +1$,

$$\text{we have } e^{i\alpha} = \pm 1, \text{ i.e., } \alpha = 0 \text{ or } \pi. \tag{9.3}$$

We have

$$\Psi(q_2,q_1) = \Psi(q_1,q_2) \tag{9.4a}$$

when $\alpha = 0$ (boson case, integer spin),

$$\text{and } \Psi(q_2,q_1) = -\Psi(q_1,q_2) \tag{9.4b}$$

when $\alpha = \pi$ (fermion case, half-integer spin).

The two-particles wavefunction is *symmetric* (Eq. 9.4a) in the case of bosons and *antisymmetric* (Eq. 9.4b) in the case of fermions. Figure 9.1 depicts the one-to-one relation between (i) the *spin* angular momentum being half-integer or integer, (ii) symmetry of the two-particle wavefunction and (iii) the statistical distribution law that accounts for how the particles occupy the energy states accessible to them. The simplicity depicted in Fig. 9.1 is dangerously *deceptive* even as it is incredibly rewarding and elegant. Feynman points out [1] that "*it appears to be one of the few places in physics where there is a rule which can be stated very simply, but for which no one has found a simple and easy explanation. The explanation is down deep in relativistic quantum mechanics.*" Likewise, Tomonaga [2] points out that "*relation between spin and statistics is apparent, but hard to understand.*" We shall merely state this relationship and move on to exploit it for our understanding of the many-electron atom. We refer the readers to References [2] and [3] for a study of the deep connection between intrinsic spin angular momentum and quantum statistics.

In the rest of this chapter, the two identical particles we shall consider is a pair of electrons. These belong, of course, to the family of fundamental particles in nature. From the solution of the Dirac equation (Chapter 7) we have seen that the electron spin is $s = \frac{1}{2}$; an electron is therefore a fermion. Let us write the two-electron (geminal) wavefunction in terms of a product of two *separate* one-electron wavefunctions, $u_i(q_k)$, which would represent the spin part *as well as* the spatial orbital part. The one-electron wavefunction $u_i(q_k)$ is called as *spin-orbital*. It has two components, as in Eq. 7.134a (Chapter 7). The argument q_k that appears

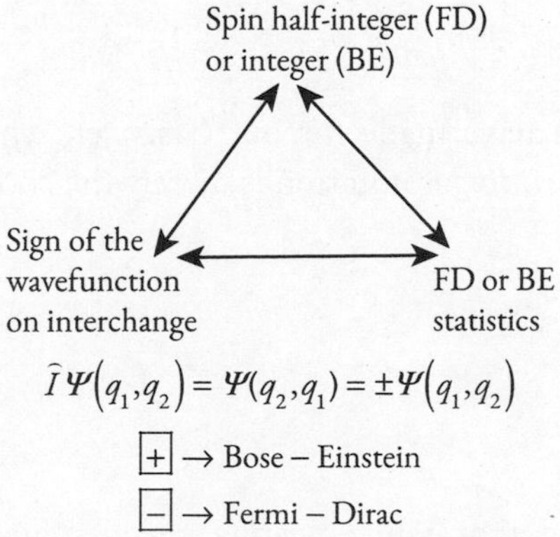

Fig. 9.1 The relationship between symmetry of the N-particle wavefunction, value of the intrinsic *spin* angular momentum quantum number, and the statistical distribution function.

in the wavefunction consists of four coordinates, viz., the three space coordinates and one spin coordinate. The argument of each spin-orbital, therefore is

$$q_k \equiv (\vec{r}_k, \zeta_k), \tag{9.5}$$

$\vec{r}_k$ being the position vector that represents the three space coordinates, and ζ_k the spin coordinate of the electron which is in the quantum state i. The *subscript i* of the spin-orbital represents a *set* of four *good quantum labels* (n, ℓ, m_ℓ, m_s). We have labeled the electron state by the (nonrelativistic) atomic hydrogenic quantum numbers n, ℓ, m_ℓ (used in Chapter 5), augmented by the inclusion of the spin eigenvalue $\left(m_s = \pm\frac{1}{2}\right)$ of the Z-component of the spin angular momentum operator $\vec{s}\bullet\hat{e}_z$. The electron spin angular momentum vector operator $\vec{s}_{2\times 2}$ is essentially the same as the one we have used in Chapter 7 (Eq. 7.108a). Thus,

$$i \equiv (n_i, \ell_i, m_{\ell i}, m_{si}). \tag{9.6a}$$

We could also use an *alternative* set of quantum numbers,

$$i \equiv (n_i, \ell_i, j_i, m_i), \tag{9.6b}$$

(with $\vec{j} = \vec{\ell} + \vec{s}$), if the spin-orbit coupling is strong. The set of vectors $\{|i\rangle\}$ therefore represents the basis $\{|n_i, \ell_i, m_{\ell i}, m_{si}\rangle\}$ or $\{|n_i, \ell_i, j_i, m_i\rangle\}$; both the basis sets being mathematically equivalent. However, one may be more suitable than the other in a given situation, as discussed in the Section 8.3 of Chapter 8.

In Eq. 9.6a, $-\ell_i < -\ell_i + 1 < .. < m_{\ell i} < .. < \ell_i - 1 < \ell_i$,

whereas in Eq. 9.6b, $-j_i < -j_i + 1 < .. < m_i < .. < j_i - 1 < j_i$.

We shall represent the spin-orbital by the *de Broglie–Schrödinger* notation or the equivalent *Dirac* notation,

$$u_i(q_j) \equiv u_i(\vec{r}_j, \zeta_j) \equiv \langle \vec{r}_j, \zeta_j \mid n_i, \ell_i, m_{\ell i}, m_{si}\rangle \equiv \langle j \mid i\rangle, \tag{9.7a}$$

$$\text{or, } u_i(q_j) \equiv u_i(\vec{r}_j, \zeta_j) \equiv \langle \vec{r}_j, \zeta_j \mid n_i, \ell_i, j_i, m_i\rangle \equiv \langle j \mid i\rangle, \tag{9.7b}$$

depending on the choice we have made for our basis set, whether that in Eq. 9.6a or in Eq. 9.6b. The *de Broglie–Schrödinger* notation is merely the coordinate representation of the Hilbert space vector

$$|i\rangle \equiv \left|n_i, \ell_i, m_{\ell_i}, \sigma_i\right\rangle \tag{9.8a}$$

$$\text{or, } |i\rangle \equiv |n_i, \ell_i, j_i, m_i\rangle. \tag{9.8b}$$

We must now take care that while writing the geminal wavefunction of the *pair* of electrons in terms of individual spin-orbitals of single electrons, we do not lose the *signature of indistinguishability* of the two electrons. Electrons being indistinguishable, the *two-electron* system can at best be described by just what it is – a *two-electron* wavefunction. This certainly does not mean that the "two-electron-system" becomes a fundamental particle! The fundamental particle is of course the individual electron. We therefore express the two-electron wavefunction as a *product* of two *single-electron* functions, but we must demand that the *form* of the product function must respect the indistinguishability of the two electrons. Electrons being fermions, the geminal wavefunction must be antisymmetric. The requirements of *separability* and *indistinguishability* are fulfilled satisfactorily by writing the geminal as an *antisymmetric combination* of the factor functions, which are the one-electron spin-orbitals. The following superposition of the product functions meets this requirement:

$$\Psi(q_1,q_2) \equiv \Psi_{i,j}(q_1,q_2) = \frac{1}{\sqrt{2}}\left[u_i(q_1)u_j(q_2) - u_i(q_2)u_j(q_1)\right]. \tag{9.9a}$$

This form can be written as a determinant:

$$\Psi(q_1,q_2) = \frac{1}{\sqrt{2}}\begin{vmatrix} u_i(q_i) & u_i(q_j) \\ u_j(q_i) & u_j(q_j) \end{vmatrix}. \tag{9.9b}$$

The factor $\frac{1}{\sqrt{2}}$ in Eqs. 9.9a,b is a normalization constant; we may assume, without any loss of generality, that the spin-orbitals are individually orthogonalized and normalized. The quantum labels i, j on the geminal wavefunction may be expanded, as in Eqs. 9.7a,b, or, for the sake of brevity they may be suppressed, as in Eqs. 9.9a,b. The pair of labels i and j must, however, be different from each other, at least in one of the four quantum numbers in the set, else the two-electron wavefunction vanishes (check Eq. 9.9a,b). They could of course be different in more than one quantum number, and possibly different in all the four. This property results from the antisymmetry of the two-electron wavefunction. This is an expression of a very general property of all fermions; it comes from their statistics. No two fermions occupy the same quantum state. Historically, in its first *avatar*, this property appeared as the *Pauli's exclusion principle*: all four quantum numbers of electrons in an atom are never the same. Even as Pauli is honored by naming the spinors (Chapters 4, 7, 8) after him, he had not anticipated the electron spin property when he formulated the *exclusion principle*. He had only stipulated at that point of time that there must be a fourth quantum number with a double value, and no two electrons would have the same set of four quantum numbers in an atom. The exclusion principle was inspired by the mechanism under discussion to fill up the atomic orbitals with electrons as per the periodic table's *aufbau* (German: "building up") principle. Later developments showed that this fourth quantum number found its rationale in Dirac's relativistic formulation of quantum mechanics, as discussed in Chapter 7 (see in particular the discussion following Eq. 7.57 and also that following Eq. 7.80).

The forms presented in Eq. 9.9a and Eq. 9.9b for the two-electron wavefunction can be readily generalized to N electrons ($N \geq 2$). The antisymmetry of the N-electron wavefunction requires it to change its sign every time there is an interchange of *any* pair of electrons in the system; after all, they are all identical and indistinguishable. This is completely in line with the relationship depicted in Figure 9.1. The N-electron wavefunction is therefore written as an $N \times N$ determinant, which is a straightforward generalization of Eq. 9.9b:

$$\psi^{(N)} = \frac{1}{\sqrt{N!}} \begin{vmatrix} u_1(1) & \cdots & \cdots & u_1(N) \\ u_2(1) & \cdots & \cdots & \cdots \\ \cdots & \cdots & \boxed{\langle j \mid i\rangle = u_i(q_j)} & \langle N \mid i\rangle \\ \cdots & \cdots & \cdots & \cdots \\ u_N(1) & \cdots & \cdots & u_N(N) \end{vmatrix}. \tag{9.10}$$

For clarity, we have placed a typical element $\langle j \mid i\rangle = u_i(q_j)$ in the ith row and j^{th} column of the above determinant in a box. The antisymmetry of the wavefunction and associated exclusion principle is beautifully manifest in the above $N \times N$ determinant. It vanishes when any two rows or columns are the same, and its sign *reverses* when any two rows or columns are interchanged. The form of the wavefunction in Eq. 9.10 is known as the *Slater determinant*, named after John C. Slater.

Each spin-orbital in the Slater determinant is a product of a two-component *spinor* from the spin vector space and an *orbital*, which is a solution of the nonrelativistic Schrödinger equation (Chapter 5). Using the hydrogenic quantum numbers we have

$$u_i(q_j) = \psi_{n_i,\ell_i,m_{\ell i}}(\vec{r}_j)\,\chi_{m_{si}}(\zeta_j). \tag{9.11}$$

It represents the *probability amplitude* that an electron at space-spin coordinate $q_j \equiv (\vec{r}_j,\zeta_j)$ is in the quantum state $|i\rangle \equiv |n_i,\ell_i,m_{\ell i},m_{si}\rangle$.

It is now instructive to discuss further the antisymmetric character of the two-electron (geminal) wavefunction by considering the effect of interchange of the indistinguishable electrons on the (a) spatial part of the wavefunction and (b) the spin part. Since the spin of each electron is $\frac{1}{2}$, we know from the angular momentum coupling algebra (Chapter 4) that the *total* spin angular momentum of the pair of electrons is

$$\vec{S} = \vec{s}_1 + \vec{s}_2 \text{ with } s_1 = \frac{1}{2} \text{ and } s_2 = \frac{1}{2}.$$

Hence, $S = 0$ (singlet, with $m_S = 0$), (9.12a)

or $S = 1$ (triplet, with $m_S = -1,0,+1$). (9.12b)

Now, in terms of a space part $\phi(\vec{r}_1,\vec{r}_2)$ and a spin part $\chi(\zeta_1,\zeta_2)$, the two-electron wavefunction $\psi(q_1,q_2)$ is

$$\psi(q_1,q_2) = \phi(\vec{r}_1,\vec{r}_2)\chi(\zeta_1,\zeta_2). \tag{9.13}$$

We know that $\psi(q_2,q_1) = -\psi(q_1,q_2)$, and the antisymmetry may come either from the spin part, or from the space-orbital part. In the former case, the two-electron wavefunction is a singlet (denoted below by the superscript S) and in the latter case it is a triplet (superscript T). Accordingly,

$$\psi^{(S)}(q_2,q_1) = \phi(\vec{r}_2,\vec{r}_1)\chi(\zeta_2,\zeta_1) = \phi(\vec{r}_1,\vec{r}_2)\left[-\chi(\zeta_1,\zeta_2)\right], \tag{9.14a}$$

i.e., the space-orbital part is symmetric,

$$\phi^{(S)}(\vec{r}_2,\vec{r}_1) = +\phi(\vec{r}_1,\vec{r}_2), \tag{9.14b}$$

and the spin part of the wavefunction is antisymmetric

$$\chi^{(S)}(\zeta_2,\zeta_1) = -\chi(\zeta_1,\zeta_2). \tag{9.14c}$$

The two electrons have antiparallel spins in this case. The symmetric space-orbital part of the singlet state is

$$\phi^{(S)}(\vec{r}_1,\vec{r}_2) = N\left[\varphi_1(\vec{r}_1)\varphi_2(\vec{r}_2) + \varphi_1(\vec{r}_2)\varphi_2(\vec{r}_1)\right]. \tag{9.14d}$$

The alternative way of getting the two-electron wavefunction $\psi(q_1,q_2)$ to be antisymmetric is

$$\psi^{(T)}(q_2,q_1) = \phi(\vec{r}_2,\vec{r}_1)\chi(\zeta_2,\zeta_1) = \left[-\phi(\vec{r}_1,\vec{r}_2)\right]\chi(\zeta_1,\zeta_2), \tag{9.15a}$$

i.e., the space-orbital part is antisymmetric

$$\phi^{(T)}(\vec{r}_2,\vec{r}_1) = -\phi(\vec{r}_1,\vec{r}_2), \tag{9.15b}$$

and the spin part is symmetric, i.e.,

$$\chi^{(T)}(\zeta_2,\zeta_1) = \chi(\zeta_1,\zeta_2). \tag{9.15c}$$

This represents the triplet state; the two electrons have parallel spins in this case. The antisymmetric space-orbital part is

$$\phi^{(T)}(\vec{r}_1,\vec{r}_2) = N\left[\varphi_1(\vec{r}_1)\varphi_2(\vec{r}_2) - \varphi_1(\vec{r}_2)\varphi_2(\vec{r}_1)\right]. \tag{9.15d}$$

Funquest: Between the two-electron singlet and the triplet state, which do you think has the lower energy (see the Solved Problem P9.6)?

In this section, we dealt primarily with two-electron states. In the next section, we discuss in detail the $N \geq 2$ case, considering the inter-electron interactions, and also the interaction of each electron with the Coulomb potential of the atom's nucleus. Solving the Schrödinger and/or the Dirac equation for such a system requires smart strategies.

9.2 Hartree–Fock Self-Consistent Field

An N-electron atomic system described by hydrogenic quantum numbers has an infinite-dimensional Hilbert space. However, only N of the *one-electron state vectors* in this Hilbert space are *occupied*. The Fermi–Dirac distribution function permits occupancy of one, and only one, electron per state. The occupation number of each state is therefore either one or zero. The *particular* set of N states that are occupied defines a *configuration*. A simple example would illustrate this. Consider an atom of Beryllium ($Z = 4$). The neutral Beryllium atom has four electrons, and we may consider two alternative configurations, $1s^2 2s^2$ and $1s^2 3s^2$. Both are valid; they represent different ways in which the four electrons occupy accessible one-electron states. The *occupation number* of the following four *one-electron* states $\left(|i\rangle \equiv \left|n_i, \ell_i, m_{\ell i}, m_{si}\right\rangle\right)$ is unity in the configuration $1s^2 2s^2$: $|1\rangle \equiv \left|1,0,0,\frac{1}{2}\right\rangle$, $|2\rangle \equiv \left|1,0,0,-\frac{1}{2}\right\rangle$, $|3\rangle \equiv \left|2,0,0,\frac{1}{2}\right\rangle$, and $|4\rangle \equiv \left|2,0,0,-\frac{1}{2}\right\rangle$.

For the configuration $1s^2 3s^2$, it is the following four states that are occupied: $|1\rangle \equiv \left|1,0,0,\frac{1}{2}\right\rangle$, $|2\rangle \equiv \left|1,0,0,-\frac{1}{2}\right\rangle$, $|3\rangle \equiv \left|3,0,0,\frac{1}{2}\right\rangle$, and $|4\rangle \equiv \left|3,0,0,-\frac{1}{2}\right\rangle$.

The Slater determinants for the two configurations are different; they therefore effectively spell out the occupation numbers 1 or 0 of the respective one-electron states. In a particular configuration, the occupation number of the *select* four one-electron states is *one*, and it is *zero* for all the other states. Considering the indistinguishability of the N electrons, there are $N!$ ways of permuting the electrons in the N quantum states. The N-electron Slater determinant is therefore a sum over the $N!$ number of permutations of the occupancy of the N-electrons in the *select* N number of *one-electron* states that belong to a given configuration. Each term in the summation must of course involve an appropriate *sign*, since odd number of interchanges of the electrons would reverse the sign of the product functions, while an even number of interchanges would leave it unchanged. Thus, using the permutation operator P and the parity $p = \pm 1$ of the permutation, the Slater determinant (Eq. 9.10) is completely equivalent to

$$\psi^{(N)}_{1,2,..N}\left(q_1,..,q_N\right) = \left[\frac{1}{\sqrt{N!}}\sum_{P=1}^{N!}(-1)^p P\right]\left\{u_1\left(q_1\right)u_2\left(q_2\right)..u_N\left(q_N\right)\right\}, \tag{9.16a}$$

$$\text{i.e., } \psi^{(N)}_{1,2,..N}\left(q_1,..,q_N\right) = \left[\frac{1}{\sqrt{N!}}\sum_{P=1}^{N!}(-1)^p P\right]\langle q_1|1\rangle\langle q_2|2\rangle..\langle q_N|N\rangle. \tag{9.16b}$$

The subscript i in the single-electron spin-orbitals $f_i\left(q_j\right)$ is the set of four "good" quantum numbers, and the arguments $\left(q_j\right)$ are the *set* of four coordinates (the spatial position vector $\vec{r}$ *and* the coordinate ζ in the spin-space). The identification of the $\left(q_j\right)^{\text{th}}$ coordinate is notional; indistinguishability of the identical particles is ensured by the *antisymmetrizer operator* which has $N!$ permutations built into it; it is defined as

$$A = \frac{1}{\sqrt{N!}}\sum_{P=1}^{N!}(-1)^p P. \tag{9.17}$$

The factor $\frac{1}{\sqrt{N!}}$ ensures normalization of the N-electrons wavefunction (assuming that the factor one-electron spin-orbitals are individually normalized). The antisymmetrizer operator "A" operates on the product $\left\{u_1\left(q_1\right)u_2\left(q_2\right)..u_N\left(q_N\right)\right\}$ of N *one*-electron wavefunctions and generates an antisymmetric, *normalized*, N-electron wavefunction that denotes a particular configuration. If one expands the sum in Eq. 9.16, there would be $N!$ terms to be written.

For a small atom like neon, this number would be 3628800, and for xenon we shall have $2.308437 \times 10^{+71}$ terms. Can you even imagine how much space it would take to write it? The antisymmetrizer operator and the determinant form of the N-electron wavefunction makes it possible to write it in a compact form. It is humbling, especially when you remind yourself that this is only for a *single* configuration of the N-electron state. A complete description would require a *superposition* of various configurations (as in the case of the Beryllium atom discussed above), i.e., a linear combination of a number of Slater determinants. This consideration is important when a number of different configurations are near-degenerate. A sum of different configurations, each represented by a Slater determinant, generates a *superposition* of Slater determinants. Such a superposition represents *configuration interaction*. The problem of obtaining what would be a correct *single* Slater determinant for a given configuration is itself an arduous task; it is augmented enormously on account of configuration interaction. In this chapter, we shall restrict ourselves to the consideration of a *single* Slater determinant, i.e., to determine the Slater determinant for a *particular* configuration. Complexities owing to *many-electron correlations* that require more than a single configuration constitute rich physics of atoms, molecules, and condensed matter. However, a single Slater determinant is often sufficient to provide many important properties of the N-electron system.

As one considers an atom or ion with more than one electron, one encounters intimidating difficulties. Unlike the case of the single-electron hydrogenic atom, exact analytical solutions cannot be obtained for a system of many electrons. The electron–electron term between every pair of electrons makes the solution *inseparable* in single-particle coordinates. Such a separation in one-electron coordinates is possible only in an *approximate* manner. The approximation that is employed in achieving this is therefore called as independent particle *approximation* (IPA) or independent particle *method* (IPM). The IPA seeks to express a many-electron wavefunction in terms of products of one-electron functions:

$$\psi^{(N)}_{1,2,..N}\left(q_1,..,q_N\right) \leftrightarrow \left\{u_1\left(q_1\right)u_2\left(q_2\right)..u_N\left(q_N\right)\right\}. \tag{9.18}$$

In its first implementation, the requirement of antisymmetry of the many-electron function was ignored. We shall, however, consider the many-electron wavefunction to be an *antisymmetrized* superposition of the product of one-electron wavefunctions, since the electrons are *indistinguishable* fermions. We shall therefore use Eq. 9.10 (equivalently Eqs. 9.16a,b) instead of the simple product in Eq. 9.18. The Slater determinant is in accordance with the Fermi–Dirac statistics and includes correlations between every pair of electrons due to the electron spin; the simple *un*-antisymmetrized form in Eq. 9.18 isn't. A product of N single-electron wavefunctions was first attempted around 1928 by D. R. Hartree employing (*un*-antisymmetrized) Eq. 9.18. Extension of this work by V. Fock [7, 8] employed Eq. 9.16, which is obviously better suited and will be discussed in the rest of this chapter.

In the context of the above discussion, we shall classify many-electron correlations into two kinds:

(i) *Electron spin correlations*, also known as *Fermi–Dirac*, or *statistical*, or as *exchange correlations* (for reasons discussed later in this chapter),
and
(ii) Coulomb correlations.

In his original work, Hartree developed a strategy known as the *self-consistent-field* (SCF) method to obtain the N-electron wavefunction. The augmentation of this method to include the electron spin statistics correlations (also known as exchange correlations) is what is known as the Hartree–Fock (HF) SCF method. It includes the spin correlations. We *define* Coulomb correlations as those correlations that are *left out of the HF SCF method*. Addressing *Coulomb correlations* would require a superposition of Slater determinants resulting from the *configuration interaction*. The HF SCF provides an excellent starting point for the study of N-electron systems within the framework of the IPA; the *spin correlations* are, however, contained in the antisymmetry of the wavefunction and hence included in the HF SCF approximation. The methodology not only is applicable to atoms/ions but is readily extended to study other many-body systems, including molecules, clusters, and solids.

The expression for the many-electron wavefunction (as in Eq. 9.16) is only formal. A struggle lies ahead to determine the one-electron functions. Exact analytical one-electron wavefunctions are available only for the single-electron hydrogenic atom/ions. The $N > 1$ case poses a challenge, since the two electrons interact not merely with the nucleus in the atom but also with each other, through their exchange (spin) correlation, *and* also through the Coulomb interaction. The following discussion would explain why this problem is rather vexing.

The N-electron Schrödinger equation to be solved is

$$H^{(N)}\psi^{(N)} = E^{(N)}\psi^{(N)}. \tag{9.19}$$

Setting up this equation presents a *Catch-22* situation: the solution can be obtained only after the eigenvalue equation (Eq. 9.19) is solved, but the very construction of the Hamiltonian whose eigenfunctions we need requires the solution for its construction! To appreciate this point, consider the two-electron Coulomb interaction term in the N-electron Hamiltonian,

$$H^{(N)}\left(q_1,q_2,..,q_N\right) = \sum_{i=1}^{N}\left(-\frac{1}{2}\nabla_i^2 - \frac{Z}{r_i}\right) + \sum_{i<j=1}^{N}\frac{1}{d_{ij}}, \tag{9.20}$$

$$\text{i.e., } H^{(N)}\left(q_1,q_2,..,q_N\right) = H_1 + H_2, \tag{9.21}$$

$$\text{with } H_1 = \sum_i f\left(\left|\vec{r}_i\right|\right) = \sum_i\left(-\frac{1}{2}\nabla_i^2 - \frac{Z}{r_i}\right), \tag{9.22a}$$

$$\text{and } H_2 = \sum_{i<j=1}^{N}\frac{1}{d_{ij}} = \sum_{i<j}\mathrm{v}\left(\left|\vec{r}_i - \vec{r}_j\right|\right). \tag{9.22b}$$

In the system of units we have employed, the electron charge and also its mass is of unit magnitude. The operator H_1 can be constructed without any difficulty; it consists of one-electron operators, the kinetic energy operator, and the operator corresponding to the potential energy of each electron in the Coulomb field of the nucleus. The gradient operator is subscripted by i since it must take partial derivatives of an operand specifically with respect to the components of $\vec{r}_i$. The hard part lies in the construction of the operator H_2, which consists of the two-electron Coulomb interaction operators. It cannot be defined in terms of an interaction between two classical point charges. An electron can only be described by its quantum mechanical probability density. The electron's charge times its probability density provides the charge density, in terms of which the electron-electron interactions must be described. The immediate riddle now lies in the fact that the probability density can be determined only in terms of the electron wavefunctions, which can be determined only after the N-electron equation is solved. Figure 9.2 exhibits the dilemma we have: the solution to the problem is necessary even to merely pose it!

Douglas R. Hartree (1897–1958) came up with the idea of getting SCF solutions to break the *Catch-22* conundrum.

> Douglas R. Hartree was helped by his father William Hartree in solving numerical problems as were involved in solving the SCF problem. It is no wonder that with his skills in solving numerical problems, D. R. Hartree designed a large differential analyzer, in 1935. The prototype for this machine was a small-scale calculator built from pieces of children's Meccano – which actually solved useful equations concerned with atomic theory in 1934! When John Eckert set up the Electronic Numerical Integrator and Computer (ENIAC) – *the first electronic general-purpose digital Turing-complete computer* – Hartree was asked to go to the USA to advice on its use. Hartree showed how to use ENIAC to calculate trajectories of projectiles. He predicted at Cambridge in 1946 that "*It may well be that the high-speed digital computer will have as great an influence on civilization as the advent of nuclear power.*" How truthful Hartree's vision has turned out to be!

Hartree's strategy to tackle the Catch-22 puzzle (Fig. 9.2) employs the *self-consistent-field* (SCF) method. It begins with "*guess*" wavefunctions to construct the Hamiltonian, no matter that the guess may be even grossly incorrect. In principle, the guess can be correct only by a lucky chance. Using the guess wavefunction to build the Hamiltonian, one can now *solve* the Schrödinger equation (Eq. 9.19). One must then inquire if the solution to the Schrödinger equation turns out to be the same as one that was guessed *initially* to build the Hamiltonian. The two sets of wavefunctions are compared to determine if they lie close to each other within a desired numerical convergence criterion. If the convergence fails, the trial functions must be varied, and the process iterated upon till self-consistency is attained. The meaning of *self-consistency* is now *self-evident*! Hartree's original method used IPA product functions (Eq. 9.18); the HF method employs the antisymmetrized sum (Eq. 9.16) of the products.

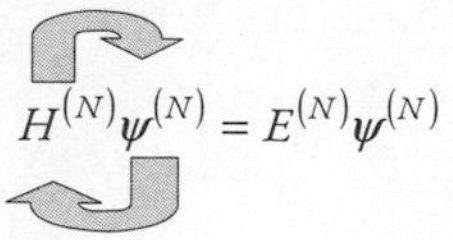

Fig. 9.2 The electron charge densities can be obtained only in terms of the electron wavefunctions, which can be obtained only after the differential equation is solved; but the Hamiltonian operator can be constructed only in terms of the electron wavefunctions!

Having been confronted with the difficulty in solving the two-electron problem, one may ask [9] "*how many bodies are required before we have a problem? G. E. Brown points out that this can be answered by a look at history. In eighteenth-century Newtonian mechanics, the three-body problem was insoluble. With the birth of relativity ... and ... quantum electrodynamics in 1930, the two- and one-body problems became insoluble. And within modern quantum field theory, the problem of zero bodies (vacuum) is insoluble. So, if we are out after exact solutions, no bodies at all are already too many!*"

Attainment of the desired self-consistency requires an iterative variation of the one-electron spin-orbitals. Variations in the orbitals, however, cannot be made randomly. The variation is subject to the following mandatory constraints:

(i) normalization of the spin-orbitals $\langle i|j\rangle = 1$ for $j = i$, (9.23a)

and

(ii) their orthogonality $\langle i|j\rangle = 0$ for $j \neq i$, (9.23b)

i and j being the collective complete *set* of good quantum numbers (Eqs. 9.8a,b) of the ith and j^{th} *occupied* single-particle states.

The HF strategy to seek self-consistent-field solutions to the N-electron Schrödinger equation is inspired by a very powerful principle that is well known in fundamental physics, namely the *principle of variation*. The SCF solutions are just exactly those as obtained from the variational principle; the correct SCF solutions are those as would make the expectation value of the N-electron Hamiltonian in the N-electron Slater determinant wavefunction a minimum. Accordingly,

$$\delta\langle \psi^{(N)} \mid H^{(N)} \mid \psi^{(N)}\rangle = 0. \tag{9.24}$$

We see from Eq. 9.21 that the expectation value $\langle \psi^{(N)} \mid \Omega \mid \psi^{(N)}\rangle$ of an operator Ω, with $\Omega = H_1$ and $\Omega = H_2$, is required. What is intimidating in obtaining the same is the fact that there are $N!$ terms in the N-electron determinantal wavefunction $\psi^{(N)}$. In fact, it appears *twice* in Eq. 9.24, sandwiching the operator. This is an arduous task even for moderate value of N. What makes it even more daunting is the fact that q being a set of four coordinates $(\vec{r},\zeta)$, *integration* over each q_i involves integration over the three space coordinates and summation over the spin coordinate. However, abundant simplification results on recognizing that the operator Ω (both H_1 and H_2) itself is *symmetric* under an arbitrary number of exchanges of the identical particles. To discover the anticipated simplification, we first write the expectation value as an integral, with one of the determinants written in an equivalent form using the antisymmetrizer operator (Eq. 9.17). Accordingly,

$$\langle \psi^{(N)} \mid \Omega \mid \psi^{(N)}\rangle = \int..\int dq_1..dq_N \left\{\psi^{(N)}\right\}^* \Omega \frac{1}{\sqrt{N!}}\sum_{P=1}^{N!}(-1)^p P\left\{u_1(q_1)..u_N(q_N)\right\}, \tag{9.25a}$$

$$\text{i.e., } \langle \psi^{(N)} \mid \Omega \mid \psi^{(N)}\rangle = \frac{1}{\sqrt{N!}}\sum_{P=1}^{N!}(-1)^p \int..\int dq_1..dq_N \left\{\psi^{(N)}\right\}^* \Omega\, P\left\{u_1(q_1)..u_N(q_N)\right\}. \tag{9.25b}$$

It is at this point that we take advantage of the symmetry of the operator Ω. Being symmetric under an interchange of every pair of electrons amid the N-electron system, Ω commutes with the permutation operator P. Hence,

$$\langle \psi^{(N)} \mid \Omega \mid \psi^{(N)} \rangle = \frac{1}{\sqrt{N!}} \sum_{P=1}^{N!} (-1)^p \int .. \int dq_1 .. dq_N \left\{\psi^{(N)}\right\}^* P\Omega \left\{u_1(q_1) .. u_N(q_N)\right\}. \tag{9.26}$$

Next, having moved the permutation operator from the right side of Ω to its left, we shall now take it further to the left, in fact all the way outside the integration symbol. However, in doing so we must *undo* the permutation it would bring about in $\left\{\psi^{(N)}\right\}^*$ that appears in the integrand in Eq. 9.26. Therefore,

$$\langle \psi^{(N)} \mid \Omega \mid \psi^{(N)} \rangle = \frac{1}{\sqrt{N!}} \sum_{P=1}^{N!} (-1)^p P \int .. \int dq_1 .. dq_N \left\{P^{-1} \left\{\psi^{(N)}\right\}^*\right\} \Omega \left\{u_1(q_1) .. u_N(q_N)\right\}. \tag{9.27a}$$

Effectively, we have $PP^{-1}\left\{\psi^{(N)}\right\}^*$, but that leaves $\psi^{(N)*}$ unchanged. The parity of the operator P^{-1} is of course the same as that of P, which is $(-1)^P$. It is odd or even depending on the number of interchanges of the indistinguishable particles under the permutation being respectively odd or even. Thus, we have

$$\langle \psi^{(N)} \mid \Omega \mid \psi^{(N)} \rangle = \frac{1}{\sqrt{N!}} \sum_{P=1}^{N!} (-1)^p P \int .. \int dq_1 .. dq_N \left\{(-1)^p \left\{\psi^{(N)}\right\}^*\right\} \Omega \left\{u_1(q_1) .. u_N(q_N)\right\}, \tag{9.27b}$$

i.e., $$\langle \psi^{(N)} \mid \Omega \mid \psi^{(N)} \rangle = \frac{1}{\sqrt{N!}} \sum_{P=1}^{N!} (-1)^{2p} P \int .. \int dq_1 .. dq_N \left\{\psi^{(N)}\right\}^* \Omega \left\{u_1(q_1) .. u_N(q_N)\right\},$$

or, $$\langle \psi^{(N)} \mid \Omega \mid \psi^{(N)} \rangle = \frac{1}{\sqrt{N!}} \sum_{P=1}^{N!} P \int .. \int dq_1 .. dq_N \left\{\psi^{(N)}\right\}^* \Omega \left\{u_1(q_1) .. u_N(q_N)\right\}. \tag{9.27c}$$

The value of the integral $\int .. \int dq_1 .. dq_N \left\{\psi^{(N)}\right\}^* \Omega \left\{u_1(q_1) .. u_N(q_N)\right\}$ on the right hand side of Eq. 9.27 does not change under any permutation; it returns exactly the same value under each of the $N!$ number of permutations. Essentially, it therefore gets added under the summation in Eq. 9.27c to itself $N!$ times. Hence,

$$\langle \psi^{(N)} \mid \Omega \mid \psi^{(N)} \rangle = \frac{1}{\sqrt{N!}} \left[N! \times \int .. \int dq_1 .. dq_N \left\{\psi^{(N)}\right\}^* \Omega \left\{u_1(q_1) .. u_N(q_N)\right\}\right],$$

i.e., $$\langle \psi^{(N)} \mid \Omega \mid \psi^{(N)} \rangle = \sqrt{N!} \int .. \int dq_1 .. dq_N \left\{\psi^{(N)}\right\}^* \Omega \left\{u_1(q_1) .. u_N(q_N)\right\}. \tag{9.28}$$

We now have $N!$ terms in $\left\{\psi^{(N)}\right\}^*$. This is a lot of terms, yet only a half of the number of terms in the integrand we originally had in Eq. 9.25a. We shall now show that *additional* simplification is achieved on examining the integrals on specific examination of the forms $\Omega = H_1$ and $\Omega = H_2$. First, we shall examine the one-electron operator term. Using Eq. 9.22a and Eq. 9.28, we have

$$\left\langle \psi^{(N)} \mid H_1 \mid \psi^{(N)} \right\rangle = \sqrt{N!} \int .. \int dq_1 .. dq_N \left\{\psi^{(N)}\right\}^* \left[\sum_{i=1}^{N} f(q_i)\right] \left\{u_1(q_1) ... u_N(q_N)\right\}, \tag{9.29a}$$

$$\text{i.e., } \left\langle \psi^{(N)} \mid H_1 \mid \psi^{(N)} \right\rangle = \int .. \int dq_1 .. dq_N \begin{Bmatrix} \left[\sum_{P=1}^{N!} (-1)^p P \left\{ u_1^*(q_1) .. u_N^*(q_N) \right\} \right] \\ \times \left[\sum_{i=1}^{N} f(q_i) \right] \left\{ u_1(q_1) .. u_N(q_N) \right\} \end{Bmatrix}. \tag{9.29b}$$

Note that the factor $\sqrt{N!}$ in Eq. 9.29a is canceled by $\frac{1}{\sqrt{N!}}$, which is present in the antisymmetrizer operator. Our next step is to (i) first identify the particular term (out of the $N!$ terms in $\sum_{P=1}^{N!} (-1)^p P \left\{ u_1^*(q_1) .. u_N^*(q_N) \right\}$) that corresponds to the identity permutation and write it explicitly, and (ii) pack the remaining $(N! - 1)$ terms in what will be referred to as the remainder R. Accordingly,

$$\left\langle \psi^{(N)} \mid H_1 \mid \psi^{(N)} \right\rangle = \int .. \int dq_1 .. dq_N \left\{ u_1^*(q_1) .. u_N^*(q_N) \right\} \left[\sum_{i=1}^{N} f(q_i) \right] \left\{ u_1(q_1) .. u_N(q_N) \right\} + R \tag{9.29c}$$

$$\text{with } R = \int .. \int dq_1 .. dq_N \left[\sum_{P=1}^{N!-1} (-1)^p P \left\{ u_1^*(q_1) .. u_N^*(q_N) \right\} \right] \left[\sum_{i=1}^{N} f(q_i) \right] \left\{ u_1(q_1) .. u_N(q_N) \right\}. \tag{9.30a}$$

There are N operators $f(q_i)$ with $(i = 1,2, \ldots ,N)$ in R, from which we shall now first separate out the term for $i = 1$:

$$\text{or, } R = \begin{bmatrix} \int .. \int dq_1 .. dq_N \left[\sum_{P=1}^{N!-1} (-1)^p P \left\{ u_1^*(q_1) .. u_1^*(q_N) \right\} \right] \left[f(q_1) \right] \left\{ u_1(q_1) .. u_N(q_N) \right\} + \\ \int .. \int dq_1 .. dq_N \left[\sum_{P=1}^{N!-1} (-1)^p P \left\{ u_1^*(q_1) .. u_N^*(q_N) \right\} \right] \left[\sum_{i=2}^{N} f(q_i) \right] \left\{ u_1(q_1) .. u_N(q_N) \right\} \end{bmatrix}. \tag{9.30b}$$

The identity permutation is already excluded from R; the first term on the right-hand side of Eq. 9.30b therefore has *at least* one interchange; say $q_1 \rightleftharpoons q_2$. We shall refer to this *term* as $T_{1\rightleftharpoons 2}$. Hence,

$$T_{1\rightleftharpoons 2} = -\int .. \int dq_1 .. dq_N \begin{pmatrix} \left\{ u_1^*(q_2) u_2^*(q_1) .. u_N^*(q_N) \right\} \times \\ \left[f(q_1) \right] \left\{ u_1(q_1) .. u_N(q_N) \right\} \end{pmatrix}. \tag{9.31a}$$

The minus sign on the right-hand side of the above equation takes care of the fact that with a *single interchange* the sign of the N-electron wavefunction is reversed. We shall, however, presently find that this sign is inconsequential in this particular case, as $T_{1\rightleftharpoons 2}$ turns out to be zero.

Since the N integrations are independent of each other,

$$T_{1\rightleftharpoons 2} = -\begin{Bmatrix} \left[\int dq_1 u_2^*(q_1) f(q_1) u_1(q_1) \right] \times \\ \left[\int dq_2 u_1^*(q_2) u_2(q_2) \right] \left[\prod_{j=3}^{N} \int dq_j u_j^*(q_j) u_j(q_j) \right] \end{Bmatrix}, \tag{9.31b}$$

where $\prod_{j=3}^{N}$ represents the product of integrals for $j = 3$ through N. Clearly,

$$\left[\prod_{j=3}^{N}\int dq_j u_j^*(q_j)u_j(q_j)\right] = \left[\prod_{j=3}^{N}\int dz u_j^*(z)u_j(z)\right] = 1, \tag{9.32a}$$

but owing to the orthogonality of the one-electron spin-orbitals,

$$\int dq_2 u_1^*(q_2)u_2(q_2) = \int dy u_1^*(y)u_2(y) = 0, \tag{9.32b}$$

$$\text{and hence } T_{1\rightleftharpoons 2} = 0. \tag{9.32c}$$

In Eqs. 9.32a,b we have deliberately used an arbitrary symbols (z) for the integration variable, since it is dummy. It is easy to apply the same technique to discover that all the remaining terms in R would also be zero. From Eq. 9.29c we therefore get

$$\langle\psi^{(N)} \mid H_1 \mid \psi^{(N)}\rangle = \left[\sum_{i=1}^{N}\right]\left[\int..\int\begin{pmatrix} dq_1 .. dq_N\{u_1^*(q_1)..u_N^*(q_N)\} \\ \times f(q_i)\{u_1(q_1)..u_N(q_N)\}\end{pmatrix}\right], \tag{9.33a}$$

$$\text{i.e., } \langle\psi^{(N)} \mid H_1 \mid \psi^{(N)}\rangle = \sum_{i=1}^{N}\left\{\begin{matrix}\left[\int dq_i u_1^*(q_i)f(q_i)u_1(q_i)\right] \\ \times\left[\prod_{j\neq i}\int dq_j u_j^*(q_j)u_j(q_j)\right]\end{matrix}\right\}, \tag{9.33b}$$

$$\text{and hence, } \langle\psi^{(N)} \mid H_1 \mid \psi^{(N)}\rangle = \sum_{i=1}^{N}\int dq_i u_i^*(q_i)f(q_i)u_i(q_i), \tag{9.34a}$$

$$\text{i.e., } \langle\psi^{(N)} \mid H_1 \mid \psi^{(N)}\rangle = \sum_{i=1}^{N}\int dz u_i^*(z)f(z)u_i(z) = \sum_{i=1}^{N}\langle i|f|i\rangle. \tag{9.34b}$$

The use of different integration variable symbols in Eqs. 9.34a,b reinforces the equivalence of alternative notations. The dummy labels can be replaced by any other label; they are completely arbitrary. The subscripts i which characterize the sets of four good quantum labels (Eq. 9.8) also get summed over, but the labels can *only* run over the N sets of one-electron states that are *occupied* in the specific configuration corresponding to the Slater determinant. Eq. 9.34b is a tremendous simplification over Eq. 9.29a. In Eq. 9.34a,b, we have a sum over one-electron integrals. It provides the expectation values of the one-electron operators in the single-particle spin-orbitals. In the Dirac notation, the result in Eq. 9.34b appears compact and elegant.

For the two-electron operator, we have

$$\langle\psi^{(N)} \mid H_2 \mid \psi^{(N)}\rangle = \sqrt{N!}\int..\int dq_1 .. dq_N\left[\begin{matrix}\{\psi^{(N)}\}^*\left\{\sum_{i<j}v\left(\left|\vec{r}_i - \vec{r}_j\right|\right)\right\} \\ \times\{u_1(q_1)...u_N(q_N)\}\end{matrix}\right]. \tag{9.35a}$$

Canceling the factor $\sqrt{N!}$ in Eq. 9.35a with that in the denominator of the antisymmetrizer, just as we did earlier in the case of the one-electron operator, we have

$$\left\langle \psi^{(N)} \mid H_2 \mid \psi^{(N)} \right\rangle = \int .. \int dq_1 .. dq_N \left(\begin{array}{l} \left[\sum_{P=1}^{N!} (-1)^p P \left\{ u_1^*(q_1) .. u_N^*(q_N) \right\} \right] \times \\ \left[\frac{1}{2} \sum_{j=1, j\neq i}^{N} \sum_{i=1}^{N} \frac{1}{d_{ij}} \right] \left\{ u_1(q_1) ... u_N(q_N) \right\} \end{array} \right). \tag{9.35b}$$

This time around, we shall separate out the $N!$ terms under the permutations on the right-hand side of the above equation in three parts:

(i) a term corresponding to the identity permutation, which we shall represent by ι (please pay attention to the notation used),

(ii) a term corresponding to just *one* interchange, say $i \rightleftharpoons j$, which we shall represent by $E_{i\rightleftharpoons j}$, and

(iii) the rest of the terms, which we shall collectively refer to as the remainder R.

Thus, $$\left\langle \psi^{(N)} \mid H_2 \mid \psi^{(N)} \right\rangle = \iota - E_{i\rightleftharpoons j} + R, \tag{9.36}$$

with $$\iota = \int .. \int dq_1 .. dq_N \left(\begin{array}{l} \left\{ u_1^*(q_1) .. u_N^*(q_N) \right\} \times \\ \left[\frac{1}{2} \sum_{\substack{j=1,\\ j\neq i}}^{N} \sum_{i=1}^{N} \frac{1}{d_{ij}} \right] \left\{ u_1(q_1) ... u_N(q_N) \right\} \end{array} \right), \tag{9.37a}$$

$$E_{i\rightleftharpoons j} = \int .. \int dq_1 .. dq_N \left(\begin{array}{l} \left\{ u_1^*(q_1) u_2^*(q_2) .. u_j^*(q_i) u_i^*(q_j) .. u_N^*(q_N) \right\} \times \\ \left[\frac{1}{2} \sum_{\substack{j=1,\\ j\neq i}}^{N} \sum_{i=1}^{N} \frac{1}{d_{ij}} \right] \left\{ u_1(q_1) ... u_N(q_N) \right\} \end{array} \right), \tag{9.37b}$$

and $$R = \int .. \int dq_1 .. dq_N \left(\begin{array}{l} \left[\sum_{P=3}^{N!} (-1)^p P \left\{ u_1^*(q_1) .. u_N^*(q_N) \right\} \right] \times \\ \left[\frac{1}{2} \sum_{\substack{j=1,\\ j\neq i}}^{N} \sum_{i=1}^{N} \frac{1}{d_{ij}} \right] \left\{ u_1(q_1) ... u_N(q_N) \right\} \end{array} \right). \tag{9.37c}$$

The minus sign behind the term $E_{i\rightleftharpoons j}$ in Eq. 9.36 is because of the fact that there is exactly one interchange in that term compared to the identity term in ι. We now consider the term $E_{i\rightleftharpoons j}$ in further detail. We shall denote the *one-over-distance* term by $\frac{1}{d_{rs}}$, since it impacts only the integration over q_r and q_s. Accordingly,

$$E_{i\rightleftharpoons j} = \frac{1}{2} \sum_{\substack{j=1,\\ j\neq i}}^{N} \sum_{i=1}^{N} J_{ij} = \frac{1}{2} \sum_{\substack{j=1,\\ j\neq i}}^{N} \sum_{i=1}^{N} \iint dq_r dq_s u_j^*(q_r) u_i^*(q_s) \frac{1}{d_{rs}} u_i(q_r) u_j(q_s), \tag{9.38a}$$

$$\text{i.e., } E_{i\rightleftharpoons j} = \frac{1}{2}\sum_{\substack{j=1,\\ j\neq i}}^{N}\sum_{i=1}^{N}\langle ji|v|ij\rangle. \tag{9.38b}$$

Integration over all *other* coordinates q_ℓ, with $\ell \neq i, \ell \neq j$, will not involve the Coulomb interaction. These integrations produce integrals of the type we had considered in Eq. 9.32a, giving unity. We shall denote the two-center integral on the right-hand side of Eq. 9.38a as

$$J_{ij} = \iint dq_r dq_s u_j^*(q_r)u_i^*(q_s)\frac{1}{d_{rs}}u_i(q_r)u_j(q_s). \tag{9.38c}$$

It is called as the *exchange* integral. The term R in Eq. 9.37c will have more than one interchange, so the product of integrals over independent degrees of freedom has orthogonalities, exactly as in Eq. 9.32b. These orthogonalities render R equal to zero. Finally, the term ι (Eq. 9.37a), by reasoning that is even *simpler* than employed in the analysis of the term $E_{1\rightleftharpoons 2}$, is

$$\iota = \frac{1}{2}\sum_{\substack{j=1,\\ j\neq i}}^{N}\sum_{i=1}^{N} I_{ij} = \frac{1}{2}\sum_{\substack{j=1,\\ j\neq i}}^{N}\sum_{i=1}^{N}\iint dq_r dq_s u_i^*(q_r)u_j^*(q_s)\frac{1}{d_{rs}}u_i(q_r)u_j(q_s). \tag{9.39a}$$

$$\text{i.e., } \iota = \frac{1}{2}\sum_{\substack{j=1,\\ j\neq i}}^{N}\sum_{i=1}^{N}\langle ij|v|ij\rangle. \tag{9.39b}$$

We shall denote the two-center integral on the right-hand side of Eq. 9.39a as

$$I_{ij} = \iint dq_r dq_s u_i^*(q_r)u_j^*(q_s)\frac{1}{d_{rs}}u_i(q_r)u_j(q_s). \tag{9.39c}$$

It is called as the *Coulomb* integral. We now have a very compact result:

$$\langle \psi^{(N)} \mid H_2 \mid \psi^{(N)}\rangle = \frac{1}{2}\sum_{\substack{j=1,\\ j\neq i}}^{N}\sum_{i=1}^{N}\left[I_{ij} - J_{ij}\right] = \frac{1}{2}\sum_{\substack{j=1,\\ j\neq i}}^{N}\sum_{i=1}^{N}\left[\langle ij|v|ij\rangle - \langle ji|v|ij\rangle\right], \tag{9.40a}$$

$$\text{i.e., } \langle \psi^{(N)} \mid H_2 \mid \psi^{(N)}\rangle = \frac{1}{2}\sum_{j=1}^{N}\sum_{i=1}^{N}\left[\langle ij|v|ij\rangle - \langle ji|v|ij\rangle\right]. \tag{9.40b}$$

In writing the form Eq. 9.40b, we have removed the condition $j \neq i$ in Eq. 9.40a, since the two terms in the rectangular bracket above *cancel* each other exactly for $j = i$; together they contribute only a zero to the sum. The arithmetic cancelation of the Coulomb term with the exchange term also addresses an essentially physical problem, viz., the effect of *self-interaction* of each electron; it is explicitly annulled in the HF method. The *exchange* integral, J_{ij}, has its origin in the antisymmetry of the many-electron wavefunction, and therefore the Fermi–Dirac statistical correlations are also called as *exchange* correlations. Like the two-center Coulomb integrals (Eq. 9.39c), the exchange integrals (Eq. 9.38c) also consist of matrix elements of the very same Coulomb interaction operator $\frac{1}{distance}$ in products of two spin-orbitals; however, the term J_{ij} comes from the *exchange* $i \rightleftharpoons j$ in the complex conjugate part of the two-center integrals. It is for this reason that the *Coulomb interaction* that is compelled by the antisymmetry of the wavefunction that is *specifically* called as the *exchange* interaction. It is a matrix element of essentially the *one-over-distance* Coulomb interaction, but the exchange term has no classical analogue. It has its origin in the indistinguishability of the fermions. Combining the results of Eq. 9.34a,b and Eq. 9.40a,b, we have

$$E^{(N)} = \left\langle \psi^{(N)} \mid H \mid \psi^{(N)} \right\rangle = \left\langle \psi^{(N)} \mid H_1 \mid \psi^{(N)} \right\rangle + \left\langle \psi^{(N)} \mid H_2 \mid \psi^{(N)} \right\rangle, \tag{9.41a}$$

i.e.,

$$E^{(N)} = \left\{ \begin{array}{l} \sum_{i=1}^{N} \int dq u_i^*(q) f(q) u_i(q) \\ +\frac{1}{2}\sum_{j=1}^{N}\sum_{i=1}^{N} \iint dq_r dq_s \left[\begin{array}{l} u_i^*(q_r) u_j^*(q_s) \frac{1}{d_{rs}} u_i(q_r) u_j(q_s) - \\ u_j^*(q_r) u_i^*(q_s) \frac{1}{d_{rs}} u_i(q_r) u_j(q_s) \end{array} \right] \end{array} \right\}. \tag{9.41b}$$

We express it compactly using the Dirac notation as

$$E^{(N)} = \left\langle \psi^{(N)} \mid H \mid \psi^{(N)} \right\rangle = \sum_{i=1}^{N} \langle i|f|i\rangle + \frac{1}{2}\sum_{j=1}^{N}\sum_{i=1}^{N}\left[\langle ij|v|ij\rangle - \langle ji|v|ij\rangle\right]. \tag{9.41c}$$

The variational principle from Eq. 9.24 now reads as

$$\delta\left\{ \sum_{i=1}^{N} \langle i|f|i\rangle + \frac{1}{2}\sum_{j=1}^{N}\sum_{i=1}^{N}\left[\langle ij|v|ij\rangle - \langle ji|v|ij\rangle\right] \right\} = 0, \tag{9.42a}$$

$$\text{i.e., } \delta \left(\begin{array}{l} \sum_{i=1}^{N} \int dq u_i^*(q) f(q) u_i(q) \\ +\frac{1}{2}\sum_{j=1}^{N}\sum_{i=1}^{N} \iint dq_1 dq_2 \left[\begin{array}{l} u_i^*(q_1) u_j^*(q_2) \frac{1}{d_{12}} u_i(q_1) u_j(q_2) - \\ u_j^*(q_1) u_i^*(q_2) \frac{1}{d_{12}} u_i(q_1) u_j(q_2) \end{array} \right] \end{array} \right) = 0. \tag{9.42b}$$

The forms in Eq. 9.42a and Eq. 9.42b are of course equivalent; the former has elegance from its compactness, while the latter is helpful because of its transparency; the spin-orbitals and the integration variables are explicitly manifest in it.

The integration variables are arbitrary (dummy); they get integrated out. The Coulomb integral in Eq. 9.39c can be rewritten as

$$I_{ij} = \iint dq_1 dq_2 u_i^*(q_1) u_j^*(q_2) \frac{1}{d_{12}} u_i(q_1) u_j(q_2). \tag{9.43a}$$

Separating the integration over the three continuous space coordinates and the discrete summation over the space coordinate, we get

$$I_{ij} = \left[\begin{array}{l} \left\{ \iint dV_1 dV_2 u_i^*(\vec{r}_1) u_j^*(\vec{r}_2) \frac{1}{d_{12}} u_i(\vec{r}_1) u_j(\vec{r}_2) \right\} \\ \times \left(\sum_{\zeta_1} \sum_{\zeta_2} \langle \zeta_1 \mid \sigma_i \rangle^* \langle \zeta_2 \mid \sigma_j \rangle^* \langle \zeta_1 \mid \sigma_i \rangle \langle \zeta_2 \mid \sigma_j \rangle \right) \end{array} \right], \tag{9.43b}$$

i.e.,

$$I_{ij} = \left[\begin{array}{l} \left(\sum_{\zeta_1} \langle \sigma_i | \zeta_1 \rangle \langle \zeta_1 \mid \sigma_i \rangle \right) \left(\sum_{\zeta_2} \langle \sigma_j | \zeta_2 \rangle \langle \zeta_2 \mid \sigma_j \rangle \right) \\ \times \iint dV_1 dV_2 \left\{ u_i^*(\vec{r}_1) u_j^*(\vec{r}_2) \frac{1}{d_{12}} u_i(\vec{r}_1) u_j(\vec{r}_2) \right\} \end{array} \right], \tag{9.43c}$$

and hence,

$$I_{ij} = \iint dV_1 dV_2 u_i^*(\vec{r}_1) u_j^*(\vec{r}_2) \frac{1}{d_{12}} u_i(\vec{r}_1) u_j(\vec{r}_2). \tag{9.43d}$$

Likewise, the exchange integral is

$$J_{ij} = \iint dq_1 dq_2 u_j^*(q_1) u_i^*(q_2) \frac{1}{d_{12}} u_i(q_1) u_j(q_2), \tag{9.44a}$$

$$\text{i.e., } J_{ij} = \left\{ \begin{array}{l} \sum_{\zeta_1} \sum_{\zeta_2} \langle j | \zeta_1 \rangle \langle i | \zeta_2 \rangle \langle \zeta_1 | i \rangle \langle \zeta_2 | j \rangle \\ \times \iint dV_1 dV_2 u_j^*(\vec{r}_1) u_i^*(\vec{r}_2) \frac{1}{d_{12}} u_i(\vec{r}_1) u_j(\vec{r}_2) \end{array} \right\}, \tag{9.44b}$$

$$J_{ij} = \left[\begin{array}{l} \left(\sum_{\zeta_1} \langle j | \zeta_1 \rangle \langle \zeta_1 | i \rangle \right) \left(\sum_{\zeta_2} \langle i | \zeta_2 \rangle \langle \zeta_2 | j \rangle \right) \\ \times \iint dV_1 dV_2 u_j^*(\vec{r}_1) u_i^*(\vec{r}_2) \frac{1}{d_{12}} u_i(\vec{r}_1) u_j(\vec{r}_2) \end{array} \right], \tag{9.44c}$$

and hence,

$$J_{ij} = \delta_{\sigma_i,\sigma_j} \iint dV_1 dV_2 u_j^*(\vec{r}_1) u_i^*(\vec{r}_2) \frac{1}{d_{12}} u_i(\vec{r}_1) u_j(\vec{r}_2). \tag{9.44d}$$

The integration over the set of four coordinates in Eq. 9.34 for the single-electron operators can also be similarly separated into integration over the continuous space coordinates and discrete summation over the spin coordinates. Thus,

$$\langle\psi^{(N)} \mid H \mid \psi^{(N)}\rangle = \left\{ \begin{matrix} \sum_{i=1}^{N} \int dV u_i^*(\vec{r}) f(\vec{r}) u_i(\vec{r}) \\ +\frac{1}{2} \sum_i \sum_j \left[\begin{matrix} \iint dV_1 dV_2 u_i^*(\vec{r}_1) u_j^*(\vec{r}_2) \frac{1}{d_{12}} u_i(\vec{r}_1) u_j(\vec{r}_2) \\ -\delta(\sigma_i, \sigma_j) \\ \times \iint dV_1 dV_2 u_j^*(\vec{r}_1) u_i^*(\vec{r}_2) \frac{1}{d_{12}} u_i(\vec{r}_1) u_j(\vec{r}_2) \end{matrix} \right] \end{matrix} \right\}. \tag{9.45}$$

On interchanging $(\vec{r}_1)$ with $(\vec{r}_2)$ in the exchange term (since $d_{12} = d_{21}$), we may rewrite the expectation value of the N-electron Hamiltonian in the N-electron antisymmetrized single-configuration wavefunction as

$$\langle\psi^{(N)} \mid H \mid \psi^{(N)}\rangle = \left\{ \begin{matrix} \sum_{i=1}^{N} \int dV u_i^*(\vec{r}) f(\vec{r}) u_i(\vec{r}) \\ +\frac{1}{2} \sum_i \sum_j \left[\begin{matrix} \iint dV_1 dV_2 u_i^*(\vec{r}_1) u_j^*(\vec{r}_2) \frac{1}{d_{12}} u_i(\vec{r}_1) u_j(\vec{r}_2) \\ -\delta(\sigma_i, \sigma_j) \\ \times \iint dV_1 dV_2 u_j^*(\vec{r}_2) u_i^*(\vec{r}_1) \frac{1}{d_{12}} u_i(\vec{r}_2) u_j(\vec{r}_1) \end{matrix} \right] \end{matrix} \right\}. \tag{9.46}$$

The advantage in interchanging the dummy labels under integration to rewrite Eq. 9.45 as Eq. 9.46 will become clear shortly.

The variational problem now requires us to determine the extremum of the left-hand side of Eq. 9.45, subject to the constraints imposed by the conditions stated in Eq. 9.23a and Eq. 9.23b.

Using now Lagrange's method (see the Solved Problem P9.7) of variational multipliers [10], λ_{ij}, the condition that $\langle\psi^{(N)} \mid H \mid \psi^{(N)}\rangle$ is an "extremum," subject to the constraints (Eqs. 9.23a,b), is then expressed as

$$0 = \delta \left[\begin{matrix} \langle\psi^{(N)} \mid H \mid \psi^{(N)}\rangle \\ +\left(\sum_{i=1}^{N} \lambda_{ii} \int dV u_i^*(\vec{r}) u_i(\vec{r}) \right) \\ +\sum_{i>j} \delta_{m_{s_i}, m_{s_j}} \left\{ \lambda_{ij} \int dV u_i^*(\vec{r}) u_j(\vec{r}) + \lambda_{ji}^* \int dV u_j^*(\vec{r}) u_i(\vec{r}) \right\} \end{matrix} \right], \tag{9.47}$$

$$\text{i.e., } 0 = \delta\left[\langle\psi^{(N)} \mid H \mid \psi^{(N)}\rangle\right] + \delta[M], \tag{9.48}$$

$$\text{with } M = \left[\begin{matrix} \sum_{i=1}^{N} \lambda_{ii} \int dV u_i^*(\vec{r}) u_i(\vec{r}) + \\ \sum_{i>j} \delta_{m_{s_i}, m_{s_j}} \left(\lambda_{ij} \int dV u_i^*(\vec{r}) u_j(\vec{r}) + \lambda_{ji}^* \int dV u_j^*(\vec{r}) u_i(\vec{r}) \right) \end{matrix} \right]. \tag{9.49}$$

While seeking the variations as required in Eq. 9.48, we shall now make an approximation. It is known as the *frozen orbital approximation* according to which, variations in the single particle orbitals are made *one* at a time, which is to say that all of the *other* $(N-1)$ orbitals are considered "frozen" during the consideration of the variation in a particular orbital. This turns out to be an excellent approximation; it is very satisfactory toward accounting for many gross properties of atomic structure and dynamics. Surely, one might raise questions about the utility of this approximation, since the variation in *any* orbital would, in fact, change the corresponding electron charge density distribution and thereby impact the charge distribution of all the other orbitals in the many-electron system. This cascading effect generates Coulomb *correlations*. These result in a time-dependent upsurge of density fluctuations in the entire many-electron atomic field. Only the *static average* of the charge density is considered in the HF SCF method. This amounts to pretending that all the other orbitals are frozen; this characteristic feature of the HF SCF method is therefore aptly called as the *frozen orbital approximation*. Under the frozen orbital approximation, we now consider the variations required in Eq. 9.48 due to changes in a single, say the kth orbital; keeping all *other* orbitals frozen. We consider the variation in the term M first:

$$\delta[M]=\begin{bmatrix}\lambda_{kk}\int dV\left\{\delta u_k^*(\vec{r})\right\}u_k(\vec{r})+\lambda_{kk}\int dV u_k^*(\vec{r})\left\{\delta u_k(\vec{r})\right\}\\+\sum_j\delta_{\sigma_k,\sigma_j}\left(\lambda_{kj}\int dV\left\{\delta u_k^*(\vec{r})\right\}u_j(\vec{r})+\lambda_{jk}^*\int dV u_j^*(\vec{r})\left\{\delta u_k(\vec{r})\right\}\right)\end{bmatrix}. \tag{9.50}$$

The first two terms provide a measure of the overlap between the *change* in a basis function, and that same function. Obviously, a *change* in a function can have no overlap with the original function itself; else it wouldn't be a *change* at all! This is similar to unit vectors in polar coordinate systems. Any *change* in a unit vector $\hat{u}$ is always orthogonal to it (see, for example, Chapter 2 of Reference [11]):

$$\hat{u}\bullet\delta\hat{u}=0. \tag{9.51}$$

$$\text{Hence, }\delta[M]=\sum_j\delta_{\sigma_k,\sigma_j}\left(\lambda_{kj}\int dV_1\left\{\delta u_k^*(\vec{r}_1)\right\}u_j(\vec{r}_1)+\lambda_{jk}^*\int dV_1 u_j^*(\vec{r}_1)\left\{\delta u_k(\vec{r}_1)\right\}\right). \tag{9.52}$$

For later convenience in writing the right-hand side of Eq. 9.52, we have now used an integration label V_1 instead of V (which is dummy, anyway).

Using Eq. 9.46 and 9.52,

$$0=\delta\int dV_1\left\{\delta u_k^*(\vec{r}_1)\right\}\begin{bmatrix}f(\vec{r}_1)u_i(\vec{r}_1)+\\\left\{\frac{1}{2}\sum_j\int dV_2\frac{u_j^*(\vec{r}_2)}{d_{12}}\begin{pmatrix}u_k(\vec{r}_1)u_j(\vec{r}_2)-\\\delta(\sigma_i,\sigma_j)u_k(\vec{r}_2)u_j(\vec{r}_1)\end{pmatrix}\right\}\\+\sum_j\delta(\sigma_k,\sigma_j)\lambda_{kj}u_j(\vec{r}_1)\end{bmatrix}+\{c.c.\}. \tag{9.53}$$

The term $\{c.c.\}$ in the above equation is the exact *complex conjugate* of the previous one. In obtaining the above form, we have taken advantage of the labels interchange which we carried out in rewriting Eq. 9.45 in the form Eq. 9.46, and have also used Eq. 9.52. It is because of the relabeling of the dummy integration labels that we could factor out common terms in Eq. 9.53.

Funquest: In obtaining the complex conjugate term explicitly in Eq. 9.53, which we have left for you as an exercise, have you recognized that you would need to make use of the Hermiticity of the one-electron operators so that

$$\int dV u_k^*(\vec{r}) f(\vec{r})\{\delta u_k(\vec{r})\} = \int dV \{\delta u_k(\vec{r})\}\{f(\vec{r}) u_k^*(\vec{r})\}\ ?$$

We now take cognizance of the fact that the variation in the factor $\{\delta u_k^*(\vec{r}_1)\}$ in the integrand of Eq. 9.53 (and correspondingly the variation in the factor $\{\delta u_k(\vec{r}_1)\}$ in the complex conjugate term) is totally *arbitrary*. The relationship must hold good for any *unsystematic* variation in the one-electron orbital, namely the chosen k^{th} orbital. Hence, the *necessary and sufficient condition* for Eq. 9.53 to be valid for *arbitrary* variation $\{\delta u_k^*(\vec{r}_1)\}$ is (within the frozen orbital approximation) that the *other* factor in the integrand (in Eq. 9.53) is zero, i.e.,

$$\begin{bmatrix} f(\vec{r}_1)u_k(\vec{r}_1) \\ +\left[\sum_j\left\{\int dV_2 \dfrac{u_j^*(\vec{r}_2)}{r_{12}}\left(u_k(\vec{r}_1)u_j(\vec{r}_2) - \delta(\sigma_k,\sigma_j)u_k(\vec{r}_2)u_j(\vec{r}_1)\right)\right\}\right] \\ +\sum_j \delta(\sigma_k,\sigma_j)\lambda_{kj}u_j(\vec{r}_1) \end{bmatrix} = 0, \tag{9.54a}$$

$$\text{i.e.,}\quad \left\{\begin{matrix} f(\vec{r}_1)u_k(\vec{r}_1) + \\ \left[\sum_j\left\{\int dV_2 \dfrac{u_j^*(\vec{r}_2)}{r_{12}}\left(u_k(\vec{r}_1)u_j(\vec{r}_2) - \delta(\sigma_k,\sigma_j)u_k(\vec{r}_2)u_j(\vec{r}_1)\right)\right\}\right] \end{matrix}\right\} = -\sum_j \delta(\sigma_k,\sigma_j)\lambda_{kj}u_j(\vec{r}_1). \tag{9.54b}$$

Equation 9.54a,b is known as the *single-particle HF equation*. It has emerged as the *necessary and sufficient condition* that the *expectation value of the N-electron Hamiltonian in the N-electron single-configuration Slater determinantal antisymmetric wavefunction is an extremum*, within the framework of the *frozen orbital approximation*, and subject to the *constraint of orthonormality of the single-particle orbitals*. Essentially, in the *HF SCF Independent Particle Approximation* (HF SCF IPA), each electron in an atom/ion consisting of the N electrons is considered to experience a potential determined by the central field nuclear attraction, and a *static* average potential determined by the remaining $(N - 1)$ electrons. This approximation is considered to be a part of the IPA family even if statistical (Fermi–Dirac, exchange) correlations are incorporated in it. The Coulomb *correlations* are ignored in the HF SCF scheme; this is in fact only a tautological remark, since *Coulomb correlations are defined to be just those that are left out of the HF SCF approximation.*

N^2 Lagrange's variational parameters λ_{ij} for $i, j = 1, .., N$ were introduced in Eq. 9.47, which we examine carefully now. From the composition of the factors that the variational parameters multiply, you would notice that

$$\lambda_{kj} = \lambda^*_{jk}, \tag{9.55a}$$

i.e., the matrix $[\lambda] = [\lambda]^\dagger$. (9.55b)

In other words, the Lagrange's variational multipliers constitute a *self-adjoint* matrix $\left[\lambda_{ij}\right]$. We can now diagonalize this Hermitian matrix using a unitary transformation. Let C be the unitary transformation that diagonalizes the matrix of the Lagrange multipliers:

$$[C]^\dagger [\lambda][C] = [\lambda']_{\text{diagonal}} = \left[\lambda'_{ij}\delta_{ij}\right]. \tag{9.56}$$

The same matrix would also effect a transformation of the basis of the orbitals, giving us new orbitals that we may represent by placing a *prime* on it:

$$u'_i(\vec{r}) = \sum_{k=1}^{N} c^\dagger_{ik}\, u_k(\vec{r}). \tag{9.57}$$

The single-particle HF SCF equation, Eq. 9.54, can then be *rewritten* in terms of the primed orbitals $\{u'_k(\vec{r})\}$. The corresponding primed Lagrange multipliers $\{\lambda'_{ij}\}$ are members of a diagonal matrix, and hence nonzero only for $i = j$. Now, the form of an equation is always preserved under a unitary transformation when every term in that equation is subjected to the same transformation. The HF SCF equation therefore takes a very elegant isomorphic form, when subjected to the transformation that diagonalizes the matrix of Lagrange multipliers. It is (see the Solved Problem P9.9):

$$\left\{\begin{matrix} f(\vec{r}_1)u_k(\vec{r}_1) + \\ \sum_j \left[\int dV_2 \dfrac{u^*_j(\vec{r}_2)}{r_{12}}\left(u_k(\vec{r}_1)u_j(\vec{r}_2) - \delta(\sigma_k,\sigma_j)u_k(\vec{r}_2)u_j(\vec{r}_1)\right)\right]\end{matrix}\right\} = -\lambda_{kk}u_k(\vec{r}_1), \tag{9.58a}$$

i.e.,

$$\left\{\begin{matrix} f(\vec{r}_1)u_k(\vec{r}_1) + \\ \sum_j \left[\int dV_2 \dfrac{u^*_j(\vec{r}_2)}{r_{12}}\left(u_k(\vec{r}_1)u_j(\vec{r}_2) - \delta(\sigma_k,\sigma_j)u_k(\vec{r}_2)u_j(\vec{r}_1)\right)\right]\end{matrix}\right\} = \varepsilon_k u_k(\vec{r}_1), \tag{9.58b}$$

where $\varepsilon_k = -\lambda_{kk}$. (9.59)

Equations 9.58a,b is referred to as the HF equation in the *diagonal* form. There are N such coupled integro-differential equations. These are amenable to iterative numerical solutions. The single-particle orbitals are varied iteratively to solve the N coupled integro-differential equations till self-consistency within a chosen convergence factor is attained. The meaning of the term *self-consistent-field* is now self-evident. As mentioned earlier, the first self-consistent-field method that was introduced by Hartree did *not* include the exchange terms.

The HF method that we have discussed here includes the exchange correlations. The use of a single Slater determinant addresses the Fermi–Dirac exchange correlations, but not the Coulomb correlations, since the *frozen orbital approximation* has been applied. In order to consider the effect of the *Coulomb correlations*, more than one Slater determinantal wavefunction corresponding to different configurations must be used. The beauty of quantum physics, and how it developed synchronously with atomic physics, is displayed magnificently in connections between the mathematical framework of quantum theory and properties of nature as revealed in careful experiments. The intrinsic structure of the mathematical principle of variation that governs the development of the HF SCF method, likewise, connects directly to experiments in atomic physics. We shall find in the next section that the Lagrange multipliers, introduced in Eq. 9.47 as variational multipliers to address the constraints summarized in Eqs. 9.23a,b, turn out to be associated with *measurable* atomic quantum properties.

9.3 Koopmans' Theorem

Multiplying Eq. 9.58 by $u_k^*(\vec{r}_1)$ and integrating over the whole space, we get

$$\left\{\begin{array}{l}\int dV_1 u_k^*(\vec{r}_1) f(\vec{r}_1) u_k(\vec{r}_1) \\ +\sum_j \left[\int dV_1 \int dV_2 \dfrac{u_k^*(\vec{r}_1) u_j^*(\vec{r}_2)}{r_{12}} \left(\begin{array}{l} u_k(\vec{r}_1) u_j(\vec{r}_2) \\ -\delta(m_k, m_j) u_k(\vec{r}_2) u_j(\vec{r}_1)\end{array}\right)\right]\end{array}\right\} = \varepsilon_k \int dV_1 u_k^*(\vec{r}_1) u_k(\vec{r}_1), \quad (9.60a)$$

i.e., using the notation from Eq. 9.41 and Eq. 9.42,

$$\langle k \mid f \mid k\rangle + \sum_{j=1}^{N} [\langle kj \mid v \mid kj\rangle - \langle kj \mid v \mid jk\rangle] = \varepsilon_k. \quad (9.60b)$$

Summing over k,

$$\sum_{k=1}^{N} n_k \langle k \mid f \mid k\rangle + \sum_{k=1}^{N} n_k \sum_{j=1}^{N} n_j [\langle kj \mid v \mid kj\rangle - \langle kj \mid v \mid jk\rangle] = \sum_{k=1}^{N} \varepsilon_k, \quad (9.60c)$$

where n_k and n_j are the occupation numbers respectively of the kth and the jth single-particle states. The occupation numbers are, of course, unity, when the state is occupied, and zero otherwise.

Note that

$$\left\langle \psi^{(N)} \middle| H \middle| \psi^{(N)} \right\rangle + \left\langle \psi^{(N)} \middle| H_2 \middle| \psi^{(N)} \right\rangle = \sum_{k=1}^{N} \varepsilon_k, \quad (9.60d)$$

$$\text{or,} \left\langle \psi^{(N)} \middle| H_1 \middle| \psi^{(N)} \right\rangle + 2\left\langle \psi^{(N)} \middle| H_2 \middle| \psi^{(N)} \right\rangle = \sum_{k=1}^{N} \varepsilon_k, \quad (9.60e)$$

which shows that the energy resulting from the sum of the diagonal Lagrange multipliers on the right-hand side of Eq. 9.60e exceeds the expectation value of the N-electron Hamiltonian in the N-particle Slater determinant. The difference comes from the two-electron term $\left\langle \psi^{(N)} \middle| H_2 \middle| \psi^{(N)} \right\rangle$.

Now, we know that the expectation value of the N-electron Hamiltonian in the antisymmetrized wavefunction is given by Eq. 9.41c. If we now *remove* one electron from this N-electron system, say the one in the ℓth spin-orbital, we may write an expression for the residual $(N-1)$ electron system similar to Eq. 9.46. While doing so, we shall use the *frozen orbital approximation*; i.e., we assume that the remaining $(N-1)$ spin orbitals are exactly the same as in the original N-electron system. For the $(N-1)$ electron system, the energy functional corresponding to the expectation value of the $(N-1)$ electron Hamiltonian in the $(N-1)$ electron antisymmetrized wavefunction can therefore be written as

$$E_{n_\ell=0}^{(N-1)} = \sum_{\substack{r=1 \\ r\neq\ell}}^{N} \langle r|f|r\rangle + \frac{1}{2}\sum_{\substack{r=1 \\ r\neq\ell}}^{N}\sum_{\substack{s=1 \\ s\neq\ell}}^{N}\left[\langle rs|v|rs\rangle - \langle sr|v|rs\rangle\right]. \tag{9.61}$$

The summation index on the right-hand side of the above equation runs from 1 through N, but excludes ℓ. It corresponds to the $(N-1)$ electron system with a *hole* (vacancy) in the ℓth state. Subtracting Eq. 9.61 from Eq. 9.41c, we get

$$E^{(N)} - E_{n_\ell=0}^{(N-1)} = \left[\begin{array}{l} \left\{\begin{array}{l} \sum_{i=1}^{N} \langle i|f|i\rangle \\ +\frac{1}{2}\sum_{j=1}^{N}\sum_{i=1}^{N}\left[\langle ij|v|ij\rangle - \langle ji|v|ij\rangle\right] \end{array}\right\} \\ -\left\{\begin{array}{l} \sum_{\substack{r=1 \\ r\neq\ell}}^{N} \langle r|f|r\rangle \\ +\frac{1}{2}\sum_{\substack{r=1 \\ r\neq\ell}}^{N}\sum_{\substack{s=1 \\ s\neq\ell}}^{N}\left[\langle rs|v|rs\rangle - \langle sr|v|rs\rangle\right] \end{array}\right\} \end{array}\right]. \tag{9.62}$$

The difference $\left(E^{(N)} - E_{n_\ell=0}^{(N-1)}\right)$ therefore has $\langle \ell|f|\ell\rangle$, coming from a difference in the one-electron integrals, and also a difference that comes from the two-electron integrals. On account of the frozen orbital approximation, the summation over r, and over s, in the energy functional $E_{n_\ell=0}^{(N-1)}$ is over essentially the very *same* set of quantum numbers as the corresponding summations in $E^{(N)}$, *except for* $i=\ell$ or $j=\ell$. Hence, every term in the energy functional $E_{n_\ell=0}^{(N-1)}$ gets canceled by the corresponding term in $E^{(N)}$, except for a few terms. In particular, the terms in $E^{(N)}$ that survive are (i) the one-electron integrals $\langle \ell|f|\ell\rangle$, (ii) the two-electron terms when i takes *all* the N values and $j=\ell$, and (iii) the two-electron terms when j takes *all* the N values and $i=\ell$. Hence,

$$E^{(N)} - E_{n_\ell=0}^{(N-1)} = \left\{\begin{array}{l} \langle \ell|f|\ell\rangle \\ +\frac{1}{2}\sum_{i=1}^{N}\left[\langle i\ell \mid v \mid i\ell\rangle - \langle \ell i \mid v \mid i\ell\rangle\right] \\ +\frac{1}{2}\sum_{j=1}^{N}\left[\langle \ell j \mid v \mid \ell j\rangle - \langle j\ell \mid v \mid \ell j\rangle\right] \end{array}\right\}. \tag{9.63}$$

Note that under the summation index i in the *Coulomb* term in Eq. 9.63, ℓ is the *second* index, whereas under the summation index j, it is the *first* index. In the exchange terms these positions are *reversed*. In the second summation in Eq. 9.63, we may interchange the indices $\ell \rightleftharpoons j$ *within* the square bracket (i.e., for *both* the Coulomb and the exchange terms) since the *difference* between the Coulomb and the exchange terms would not change on interchanging the quantum *labels*. The manipulation of these labels renders Eq. 9.63 in the following form:

$$E^{(N)} - E^{(N-1)}_{n_\ell=0} = \left\{ \begin{array}{l} \langle \ell | f | \ell \rangle + \\ \dfrac{1}{2}\sum_{i=1}^{N} [\langle i\ell \mid v \mid i\ell \rangle - \langle \ell i \mid v \mid i\ell \rangle] \\ +\dfrac{1}{2}\sum_{j=1}^{N} [\langle j\ell \mid v \mid j\ell \rangle - \langle \ell j \mid v \mid j\ell \rangle] \end{array} \right\}. \tag{9.64}$$

We immediately recognize that the *two* summations in Eq. 9.64 are exactly equal to each other. After all, i and j are only dummy labels! Hence, using Eq. 9.60b, we get

$$E^{(N)} - E^{(N-1)}_{n_\ell=0} = \langle \ell | f | \ell \rangle + \sum_{k=1}^{N} [\langle \ell k \mid v \mid \ell k \rangle - \langle k\ell \mid v \mid \ell k \rangle] = \varepsilon_\ell. \tag{9.65}$$

The left-hand side can be interpreted as the energy required for the ionization of the N-electron atom in its ℓth spin-orbital, i.e., the ionization energy to create a vacancy in that orbital, also called as the *ionization potential*. It is the binding energy of an electron in the ℓth spin-orbital. The frozen orbital approximation is fundamental to this interpretation. One may therefore expect some disagreement between the experimentally measured ionization potential and that we get from Eq. 9.65, known as Koopmans' (*not* Koopman's) theorem [12]. It is named after Tjalling Charles Koopmans (1910–1985), who was a student of Hendrik Kramers. Koopmans was a joint winner of the 1975 Nobel *Memorial* Prize in *Economic Sciences*, for his work on optimum allocation of resources.

Funquest: Explicitly write all the terms in Eq. 9.62 for the $1s^2 2s^2$ configuration of the Beryllium atom ($Z = 4$). Obtain an expression for $\left(E^{(4)} - E^{(3)}_{n_{1s\uparrow}=0}\right)$ explicitly writing step by step all the terms in Eq. 9.62 through Eq. 9.65.

The Koopmans' theorem imparts a physical and measurable interpretation to the parameters $\lambda_{\ell\ell} = -\varepsilon_\ell$. Hitherto treated merely as Lagrange's variational parameters, they have turned out to be the ionization potentials, albeit within the limitation of the frozen orbital approximation. The ionization potential of an atom is naturally sensitive to its physico-chemical environment, making it a very useful characteristic physical property of a quantum system to study. Highly sophisticated experimental techniques such as the Electron Spectroscopy for Chemical Analysis (ESCA), etc., have been developed to study the same.

Let us now write the HF SCF Eq. 9.58b in the following form:

$$\left\{\begin{array}{l} f(\vec{r}_1)u_k(\vec{r}_1)+ \\ \left[\sum_j \int dV_2 \frac{u_j^*(\vec{r}_2)u_j(\vec{r}_2)}{r_{12}}\right] u_k(\vec{r}_1) \\ -\sum_j \delta(\sigma_k,\sigma_j)\left[\int dV_2 \left\{\frac{u_j^*(\vec{r}_2)u_j(\vec{r}_1)}{r_{12}} u_k(\vec{r}_2)\right\}\right] \end{array}\right\} = \varepsilon_k u_k(\vec{r}_1). \tag{9.66}$$

We shall now rewrite the integrals in the above equation using a compact notation. We define

$$v_{el}(1) = \sum_j \left[\int dV_2 \frac{u_j^*(\vec{r}_2)u_j(\vec{r}_2)}{r_{12}}\right],$$

$$\text{i.e., } v_{el}(1) = \sum_j \left[\int dV_2 \frac{\rho_j(\vec{r}_2)}{r_{12}}\right] = \int dV_2 \frac{\left\{\sum_j \rho_j(\vec{r}_2)\right\}}{r_{12}} = \int dV_2 \frac{\rho(\vec{r}_2)}{r_{12}}, \tag{9.67}$$

$$\text{where } \rho(\vec{r}_2) = \sum_\sigma \rho_\sigma(\vec{r}_2) = \sum_\sigma \left[\sum_j \left\{u_{j\sigma}^*(\vec{r}_2)u_{j\sigma}(\vec{r}_2)\right\}\right] \tag{9.68}$$

$$\text{with } \rho_\sigma(\vec{r}_2) = \sum_j u_{j\sigma}^*(\vec{r}_2)u_{j\sigma}(\vec{r}_2) = \sum_j \left|u_{j\sigma}(\vec{r}_2)\right|^2. \tag{9.69}$$

Equation 9.66 takes the following form:

$$\left[\begin{array}{l} f(\vec{r}_1)u_k(\vec{r}_1) + v_{el}(1)\, u_k(\vec{r}_1) \\ -\sum_j \delta(\sigma_k,\sigma_j)\left\{\int dV_2 \frac{u_j^*(\vec{r}_2)u_k(\vec{r}_2)}{r_{12}}\right\} u_j(\vec{r}_1) \end{array}\right] = \varepsilon_k u_k(\vec{r}_1). \tag{9.70}$$

Multiplying Eq. 9.66 by the spin part $\left\langle \zeta_1 \middle| \frac{1}{2},\sigma_k \right\rangle = \chi_{\frac{1}{2},\sigma_k}(\zeta_1)$ of the wavefunction, we get

$$\left[\begin{array}{l} f(\vec{r}_1)\chi_{\frac{1}{2},\sigma_k}(\zeta_1)u_k(\vec{r}_1) \\ +\left\{\left[\sum_j \int dV_2 \frac{u_j^*(\vec{r}_2)u_j(\vec{r}_2)}{r_{12}}\right] \chi_{\frac{1}{2},\sigma_k}(\zeta_1)\right\} u_k(\vec{r}_1) \\ -\sum_j \delta(\sigma_k,\sigma_j)\left[\int dV_2 \left\{\frac{u_j^*(\vec{r}_2)u_j(\vec{r}_1)}{r_{12}} u_k(\vec{r}_2)\right\} \chi_{\frac{1}{2},\sigma_k}(\zeta_1)\right] \end{array}\right] = \varepsilon_k u_{\alpha_k}(\vec{r}_1)\chi_{\frac{1}{2},\sigma_k}(\zeta_1). \tag{9.71}$$

You would recognize that it is only the last term in the above equation (the set of exchange terms on the left hand side) that comes in the way of casting Eq. 9.71 as an eigenvalue equation. The essential reason for this is the fact that the exchange interactions in this last term involve non-locality.

We *define* a new operator, called the *Coulomb operator*, as

$$V^{dc} = \sum_j V_j^{dc}(q_1), \tag{9.72a}$$

$$\text{such that } V_j^{dc}(q_1) = \int dq_2 \frac{u^*_j(q_2)u_j(q_2)}{r_{12}} = \int dV_2 \frac{\rho_j^{dc}(\vec{r}_2)}{r_{12}}. \tag{9.72b}$$

The numerator in the above integrand

$$u_j^*(\vec{r}_2)u_j(\vec{r}_2) = \rho_j^{dc}(\vec{r}_2), \tag{9.73}$$

is called as the *Coulomb* (or direct) *charge density*. It is just the probability density multiplied by the electron's charge. In Eq. 9.72b, the variable "2" gets integrated out; hence, the operator V_j^{dc} depends only on (q_1). In writing Eq. 9.72b, we have used the fact that the sum $\sum_{\zeta_2} \langle \sigma_j \mid \zeta_2 \rangle \langle \zeta_2 \mid \sigma_j \rangle$ is unity.

Also, we *define* another operator, called as the *exchange operator*, as

$$V^{ex}(q_1) = \sum_j V_j^{ex}(q_1), \tag{9.74a}$$

by *stating* its effect on an *arbitrary function*, say $\phi(q_1)$, of (q_1):

$$V_j^{ex}(q_1)\phi(q_1) = \left[\int dq_2 \frac{u_j^*(q_2)\phi(q_2)}{r_{12}}\right] u_j(q_1). \tag{9.74b}$$

Eq. 9.74b is the *characteristic defining property* of the *exchange operator*. If the arbitrary function $\phi(q_1)$ is chosen to be the one-electron spin-orbital $u_k(q_1)$, we have

$$V_j^{ex}(q_1)u_k(q_1) = \left[\int dq_2 \frac{u_j^*(q_2)u_k(q_2)}{r_{12}}\right] u_j(q_1). \tag{9.74c}$$

Equation 9.74c is not an eigenvalue equation. By separating the summation over the spin coordinate and integration over the space coordinates in it, we get

$$V_j^{ex}(q_1)u_k(q_1) = \delta(\sigma_j, \sigma_k)\left[\int dV_2 \frac{u_j^*(\vec{r}_2)u_k(\vec{r}_2)}{r_{12}}\right] u_j(\vec{r}_1)\chi_{\frac{1}{2},\sigma_j}(\zeta_1). \tag{9.74d}$$

We now *define* 'exchange density' as

$$\rho_{j,k}^{ex}(\vec{r}_2) = u_j^*(\vec{r}_2)u_k(\vec{r}_2). \tag{9.75}$$

Equation 9.75 is similar to Eq. 9.73; both have the dimensions of probability density, and when multiplied by the electron's charge, they give charge densities. However, recognizing $\rho_{j,k}^{ex}$ as a charge density requires reconciliation with the antisymmetric nature of the many-electron wavefunction resulting from Fermi–Dirac statistics. There is *no* classical analogue for this term.

We can now write Eq. 9.74d as

$$V_j^{ex}(q_1)u_k(q_1) = \delta(\sigma_j,\sigma_k)\left[\int dV_2 \frac{\rho_{j,k}(\vec{r}_2)}{r_{12}}\right]u_j(\vec{r}_1)\chi_{\frac{1}{2},\sigma_j}(\zeta_1). \tag{9.76}$$

Equation 9.71 can now be rewritten as

$$f(\vec{r}_1)u_k(q_1) + V^{dc}(q_1)u_k(q_1) - V^{ex}(q_1)u_k(q_1) = \varepsilon_k u_k(q_1), \tag{9.77a}$$

or, more compactly as

$$\left[f(q_1) + V^{HF}(q_1)\right]u_k(q_1) = \varepsilon_k u_k(q_1), \tag{9.77b}$$

with the *Hartree-Fock potential operator* V^{HF} *defined* as

$$V^{HF}(q_1) = V^{dc}(q_1) - V^{ex}(q_1). \tag{9.78}$$

It seems a bit outlandish that introduction of the operator V^{ex} has enabled us to write the HF equation in a form (Eq. 9.77b) that appears to be an eigenvalue equation, but the reason is merely the peculiar definition (Eqs. 9.74a–d) of the exchange operator that we have chosen. Notwithstanding the pretentious *eigenvalue-like* appearance of Eqs. 9.77a,b, the *global* integration in the exchange density makes it essentially *nonlocal*. In the next section, we introduce an approximation to the nonlocal exchange density which provides some simplification, but of course at a cost.

> ***Funquest:*** If dipole transition matrix elements (Eq. 8.27, Chapter 8) are studied using the Hartree-Fock approximation, would the results in the length gauge be identical to those in the velocity gauge?

9.4 Local Density Approximations

The atomic HF SCF scheme has a certain amount of "electron–electron correlation" built into it since it takes into account the *antisymmetry* of the many-electron wavefunction. This correlation originates from the *identity* of electrons; an exchange of one electron with another results in a configuration that is indistinguishable from the former, but the sign of the wavefunction must be reversed. The electron wavefunction is described by the Fermi–Dirac statistics. Exchange correlations are incorporated in the HF SCF method and are also equivalently referred to as the Fermi–Dirac (or sometimes simply "Fermi") correlations, or as Pauli correlations, since the Pauli Exclusion Principle is also governed by essentially the same phenomenology. Often, these correlations are alternatively called as *spin correlations*, since they result from the intrinsic spin angular momentum of the electron. We have seen that the HF method includes the *N!* permutations of the *N* identical, indistinguishible, electrons. The HF single-electron, two-component, spin-orbitals constitute the *N*-electron Slater determinant.

Only one *set* of superpositions of the *product of one-electron wavefunctions* is included in the HF method. This set comes from permutations of the electrons *within a single configuration* consisting of a particular set of N spin-orbitals that are occupied, and the rest are not. Even if a single product of N number of one-electron spin-orbitals is considered in the HF SCF approximation, the spin correlations are fully included in it. The HF method is said to belong to the family of the *Independent Particle Approximation* (IPA), inspite of its inclusion of the "spin correlations," since the methodology is based on products of a particular set of N single-electron wavefunctions.

The two-electron exchange integral J_{ij} (Eq. 9.44a) corresponding to the spin/exchange correlations is very cumbersome to evaluate. It scales up rapidly with the number of electrons. An approximation to this term is therefore often employed. The exchange potential (Eqs. 9.74a–d) that results from the exchange integral is *nonlocal*. An approximation to this term is therefore often employed. The position operator does not commute with the operator for the exchange potential in the Hamiltonian (see Eq. 8.28), as a result of which the spectroscopic matrix element in the length gauge and in the velocity gauge (also called as momentum gauge) turn out to produce different results; see Eq. 8.27 (Chapter 8) and the discussion following it. Spectroscopic results in the HF model are therefore gauge-dependent [13, 14]. A number of approximations to the exchange terms have been developed, which trace their origins to the method introduced by John Slater. These include Slater's method rooted in the 'Local Density Approximation' (LDA) to the exchange energy, and also the 'Generalized Gradient Approximation' (GGA), which accounts for both the local density, and its gradient [15–20]. However, unlike the HF SCF, the LDA and GGA approximations suffer from self-interaction errors.

One can see that the HF single particle equation (Eq. 9.71) for spin up, if $m_{sk} = +\frac{1}{2}$ (or spin down, if $m_{sk} = -\frac{1}{2}$), corresponds to a Schrödinger-type equation for an electron in a potential field generated by the nucleus and other electrons whose charge density is ρ, with

$$\rho = \rho_{\uparrow} + \rho_{\downarrow} = \sum_{m_s} \rho_{m_s} = \sum_{m_s}\left(\sum_i u^*_{i,m_{si}} u_{i,m_{si}}\right) = \sum_{m_s}\left(\sum_i \left|u_{i,m_{si}}\right|^2\right), \tag{9.79a}$$

$$\text{i.e., } \rho_{\uparrow} = \sum_i \left|u_{i,\uparrow}\right|^2, \text{ and } \rho_{\downarrow} = \sum_i \left|u_{i,\downarrow}\right|^2, \tag{9.79b}$$

except that the charge of a single electron corresponding to ↑ (or ↓, as the case may be) is reduced from Eq. 9.75. Once this charge is *scooped* out, thereby creating a hole (called as the "Fermi" or "exchange correlation" hole). The remaining charge density integrates to the charge of (N –1) electrons. The Fermi hole may be considered to have a spherical shape with radius r_F, occupying a volume $\frac{4}{3}\pi r_F^3$, such that

$$\frac{4}{3}\pi r_F^3 \times \left|\rho_{\uparrow}\right| = |e|, \tag{9.80a}$$

which is the charge of a single electron. The radius of the Fermi hole therefore is

$$r_F = \left(\frac{3}{4\pi}\frac{|e|}{\left|\rho_{\uparrow}\right|}\right)^{1/3}. \tag{9.80b}$$

The exchange hole does not have a sharp boundary. The charge density of electrons with the same spin (↑ or ↓) as that of the electron under consideration would grow with distance away from the center of the Fermi hole. The exchange potential energy term in the HF equation (Eq. 9.71) represents potential energy at the position of the electron at $\vec{r}_1$ generated by an *exchange* charge density $-e\sum_{k=1}^{N}\frac{u_j^*(\vec{r}_1)u_k^*(\vec{r}_2)u_k(\vec{r}_1)u_j(\vec{r}_2)}{u_j^*(\vec{r}_1)u_j(\vec{r}_1)}$, located at $\vec{r}_2$. This depends on both $\vec{r}_1$ and $\vec{r}_2$. The probability that an electron at $\vec{r}_1$ is in the jth state being $\frac{u_j^*(\vec{r}_1)u_j(\vec{r}_1)}{\sum_{\ell=1}^{N}u_\ell^*(\vec{r}_1)u_\ell(\vec{r}_1)}$; it is used to provide a measure of the *weight* of its contribution to the net *exchange* charge density $-e\frac{\left\{\sum_{j=1}^{N}\sum_{k=1}^{N}u_j^*(\vec{r}_1)u_k^*(\vec{r}_2)u_k(\vec{r}_1)u_j(\vec{r}_2)\right\}}{\sum_{\ell=1}^{N}u_\ell^*(\vec{r}_1)u_\ell(\vec{r}_1)}$. The exchange energy is expressible as a function of a density. This *function of a function* is called as the *density functional*. The density can be obtained using alternative ways; not necessarily from the many-body wavefunctions. This approach led to the development of an elegant many-body electronic structure theory known as the 'Density Functional Theory' (DFT), developed by Kohn and Sham in 1965 [18]. The DFT has emerged as a powerful tool to develop computational methods to obtain many-body electronic structure [19,20] in atoms, molecules, and solids.

Funquest: If dipole transition matrix elements (Eq. 8.27) are studied using the LDA to the Hartree-Fock approximation, would the results in the length gauge be identical to those in the velocity gauge?

To conclude this chapter, we reiterate that as a result of Coulomb correlations, the net wavefunction of the *N*-electron system must be written as a linear superposition of two (or more) Slater determinants. A complete description may well require a superposition of an infinite alternative Slater determinants, each corresponding to a different "configuration" that spells out the occupancy number (1 or 0) of the possible one-electron spin-orbitals. An iterative self-consistent-field can then be generated as before. Such a scheme, which includes superposition of Slater determinants for different configurations, is called as the Multi-Configuration Hartree-Fock method (MCHF), and/or Configuration Interaction (CI), Many-Body Perturbation Theory, and the Coupled Cluster methods [21]. The MCHF/CI methods take (partial) account of the Coulomb correlations that are left out of the HF formulation. Other many-body methods include the Many-Body Perturbation Theory (MBPT) and the Coupled-Cluster Theory [22]. Unfortunately, there is no formalism that can be developed, even in principle, which can include the Coulomb correlations *completely*. This is because of the fact that a many-body problem is simply not amenable to *exact* solutions. The MCHF/CI is one of the several approximate methods that have been developed to include the Coulomb correlations in one's analysis. Other common many-body approximations are the random phase approximation (RPA), Feynman–Goldstone diagrammatic perturbation theory, etc.

The HF, MCHF, RPA, and other methods mentioned above are primarily based on the Schrödinger equation. However, the Schrödinger equation is nonrelativistic; it is not Lorentz covariant and does not therefore accommodate consequences of the fact that the speed of light is finite and constant in every inertial frame of reference. Relativistic effects are sometimes included, if only partially, in methods based on the Schrödinger equation by using perturbative corrections that model some of the relativistic effects (see Chapter 7). However, such an approach is often treacherous, as it is not easy to take care that all perturbative terms of equal importance are properly incorporated on an equal footing. A safer approach is to base the many-electron formalism on Dirac's relativistic equation (as in Chapter 7) rather than the Schrödinger equation. Yet again, a self-consistent-field many-electron formalism based on the Dirac equation can be built, analogous to the HF method, and the resulting relativistic scheme is then called as the Dirac–Hartree–Fock (DHF) method [23, 24]. Elements of the Slater determinent are then four-components relativistic one-electron wavefunctions (bispinors), such as in Eq. 7.124 of Chapter 7. Many-body relativistic methods can subsequently be developed to address the Coulomb correlations, such as the Relativistic Multi-Configuration Dirac (Hartree) Fock (MCDHF/MCDF) methods [23], the relativistic random phase approximation (RRPA) [25], etc. Spectacular advances in obtaining solutions to the HF SCF equations have been made; the literature is vast. Computational resources employed now include not merely very powerful classical computing, but also implementations using qubits (see Chapter 11, and Appendix E) on a quantum computer [26].

Funquest: Write the relativistic Slater determinant wavefunction for the ground state of an atom of Beryllium atom ($Z = 4$) by explicitly writing all the quantum numbers, all the four components, and all the space-spin coordinates.

Solved Problems

P9.1: Determine whether the following functions are symmetric or antisymmetric under the exchange of position.

(a) $\psi(x_1,x_2) = 4(x_1 - x_2)^2 + \dfrac{2}{x_1x_2}$

(b) $\psi(x_1,x_2) = \dfrac{\sin(x_1 - x_2)}{\cos(x_1 - x_2) + 5x_1x_2}$

(c) $\psi(x_1,x_2) = \dfrac{1}{x_2 + 3}e^{-|x_1|}$

Solution:

(a) The function $\psi(x_1,x_2)$ is symmetric, since $\psi(x_1,x_2) = \psi(x_2,x_1)$.

(b) Antisymmetric, since $\psi(x_1,x_2) = -\psi(x_2,x_1)$.

(c) Neither symmetric nor antisymmetric, since

$$\psi(x_2,x_1) = \frac{1}{x_1 + 3}e^{-|x_2|} \neq \pm\psi(x_1,x_2).$$

P9.2: The abundance of the lithium isotopes Li^7 and Li^6 are 92.41% and 7.59% respectively. Are Li^7 and Li^6 nuclei classified as bosons or fermions?

Solution:

For Li^7, $Z = 3$ (three protons).

Number of neutrons: $N_n = A–Z = 4$.

Both protons and neutrons are spin-½ particles. The total spin for Li^7 can be $(3+4)\times\frac{1}{2}=\frac{7}{2}$; hence, it is a fermion.

In the case of Li^6, $Z = 3$ and $N_n = 3$. Its total spin can be $(3+3)\times\frac{1}{2}=3$; hence, it is a boson.

P9.3: A system consisting of three identical bosons which are non-interacting in nature move in a common external one-dimensional harmonic oscillator potential. Obtain the energy levels and wave functions of the ground state, the first excited state, and the second excited state of the system.

Solution:

For the 1D harmonic oscillator, the energy is $\varepsilon_n = \left(n+\frac{1}{2}\right)\hbar\omega$ (Eq. 3.77, Chapter 3). The ground-state energy is $\varepsilon_0 = \frac{1}{2}\hbar\omega$.

For the ground state of the particular bosonic system, the particles can be occupied in the lowest energy level as shown.

The total energy of the non-interacting set of three bosons therefore is $3\varepsilon_0$, and its wavefunction is the product of the three wavefunctions of individual bosons: $\psi(x_1,x_2,x_3) = \phi_0(x_1)\phi_0(x_2)\phi_0(x_3)$, where $\phi_0(x_i)$; $i = 1,2,3$ stands for the individually normalized wavefunctions.

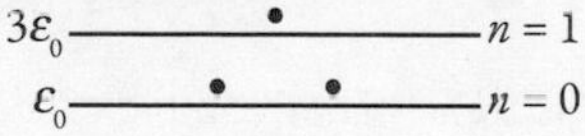

Configuration of the *first* excited state.

Total energy of the system is $2\varepsilon_0 + 3\varepsilon_0 = 5\varepsilon_0$.

The symmetric wavefunction is written as

$$\psi(x_1,x_2,x_3) = \frac{1}{\sqrt{3}}\left(\phi_0(x_1)\phi_0(x_2)\phi_1(x_3) + \phi_0(x_1)\phi_0(x_3)\phi_1(x_2) + \phi_0(x_2)\phi_0(x_3)\phi_1(x_1)\right)$$

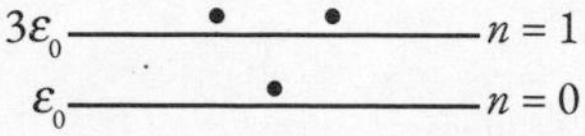

Configuration of the *second* excited state.

Total energy of the system is $\varepsilon_0 + \left(2\times 3\varepsilon_0\right) = 7\varepsilon_0$.

The composite wavefunction is

$$\psi(x_1,x_2,x_3) = \frac{1}{\sqrt{3}}\left(\phi_0(x_1)\phi_1(x_2)\phi_1(x_3) + \phi_0(x_2)\phi_1(x_3)\phi_1(x_1) + \phi_0(x_3)\phi_1(x_1)\phi_1(x_2)\right)$$

P9.4: Illustrate the Koopmans theorem for a three-electron system in the *frozen orbital approximation*.

Solution:

The energy of a three-electron system is (using Eq. 9.41c):

$$E(\psi^{(3)}) = \langle 1|f|1\rangle + \langle 2|f|2\rangle + \langle 3|f|3\rangle + (J_{12} - K_{21}) + (J_{13} - K_{31}) + (J_{23} - K_{32})$$

With the help of an external agency (light for example) if the electron in state "2" is removed so that $n_2 = 0$, $n_1 = 1$, and $n_3 = 1$, then within the *frozen orbital approximation* the remaining 2-electron system is

$$E(\psi^{(2)}) = \langle 1|f|1\rangle + \langle 3|f|3\rangle + (J_{13} - K_{31})$$

Hence,

$$E(\psi^{(3)}) - E(\psi^{(2)}) = \langle 2|f|2\rangle + (J_{12} - K_{21}) + (J_{23} - K_{32}) = \varepsilon_2 = -\lambda_{22}.$$

$\langle i|f|i\rangle + \sum_j \left[\langle ij|v|ij\rangle - \langle ji|v|ij\rangle\right] = \varepsilon_k$ (Eq. 9.60b), ε_2 being the ionization potential or binding energy for the state $n = 2$ given by the negative of the Lagrange's multiplier.

P9.5: Show that the Coulomb and the exchange integrals are both real.

Solution:

The Coulomb and exchange integral are presented in Eq. 9.39c and Eq. 9.38c.

The Coulomb integral is

$$I_{ij} = \iint dq_r dq_s u_i^*(q_r) u_j^*(q_s) \frac{1}{d_{rs}} u_i(q_r) u_j(q_s)$$

where the inverse distance $\frac{1}{d_{rs}}$ is real. The complex conjugate of the Coulomb integral is $I_{ij}^* = \iint dq_r dq_s u_i^*(q_r) u_j^*(q_s) \frac{1}{d_{rs}} u_i(q_r) u_j(q_s) = I_{ij}$, hence real.

The exchange integral is $J_{ij} = \iint dq_r dq_s u_j^*(q_r) u_i^*(q_s) \frac{1}{d_{rs}} u_i(q_r) u_j(q_s)$

and $J_{ij}^* = \iint dq_r dq_s u_i^*(q_r) u_j^*(q_s) \frac{1}{d_{rs}} u_j(q_r) u_i(q_s)$. On interchanging the dummy variables, one can see that $J_{ij} = J_{ij}^*$, and hence real.

P9.6(a): Obtain the matrix representation of the Coulomb interaction in the basis $\left\{\varphi_1(\vec{r}_1)\varphi_2(\vec{r}_2), \varphi_2(\vec{r}_1)\varphi_1(\vec{r}_2)\right\}$ of two-electron product states and show that the it is diagonalized in the basis of the singlet and the triplet states (Eq. 9.14 and Eq. 9.15), viz., $\left\{\phi^{(S)}(\vec{r}_1,\vec{r}_2), \phi^{(T)}(\vec{r}_1,\vec{r}_2)\right\}$.

P9.6(b): Determine the energy difference between the singlet and the triplet states, find which has lower energy, and discuss why this is so.

Solution:

(a) The basis of a pair of electrons is $\left\{\varphi_1(\vec{r}_1)\varphi_2(\vec{r}_2), \varphi_2(\vec{r}_1)\varphi_1(\vec{r}_2)\right\}$ and the Coulomb interaction is $\hat{U} = \frac{1}{|\vec{r}_i - \vec{r}_j|} = \frac{1}{r_{ij}}$. Accordingly, the matrix representation of the Coulomb interaction is:

$$U = \begin{pmatrix} U_{11} & U_{12} \\ U_{21} & U_{22} \end{pmatrix} = \begin{pmatrix} \langle 1|U|1\rangle & \langle 1|U|2\rangle \\ \langle 1|U|2\rangle & \langle 2|U|2\rangle \end{pmatrix}$$

The matrix element $U_{11} = \iint d^3\vec{r}_1 d^3\vec{r}_2 \varphi_1^*(\vec{r}_1)\varphi_2^*(\vec{r}_2)\frac{1}{r_{12}}\varphi_1(\vec{r}_1)\varphi_2(\vec{r}_2) = J$ (Coulomb integral) and

$U_{12} = \iint d^3\vec{r}_1 d^3\vec{r}_2 \varphi_1^*(\vec{r}_1)\varphi_2^*(\vec{r}_2)\frac{1}{r_{12}}\varphi_1(\vec{r}_2)\varphi_2(\vec{r}_1) = K$ (Exchange integral).

We see that $U_{21} = \iint d^3\vec{r}_1 d^3\vec{r}_2 \varphi_1^*(\vec{r}_2)\varphi_2^*(\vec{r}_1)\frac{1}{r_{12}}\varphi_1(\vec{r}_1)\varphi_2(\vec{r}_2) = K^*$

and $U_{22} = \iint d^3\vec{r}_1 d^3\vec{r}_2 \varphi_1^*(\vec{r}_2)\varphi_2^*(\vec{r}_1)\frac{1}{r_{12}}\varphi_1(\vec{r}_2)\varphi_2(\vec{r}_1) = J^*$. Therefore,

$$U = \begin{pmatrix} U_{11} & U_{12} \\ U_{21} & U_{22} \end{pmatrix} = \begin{pmatrix} J & K \\ K^* & J^* \end{pmatrix}.$$

From the problem P9.5, we have seen that both the Coulomb and the exchange integrals are real. Therefore:

$$U = \begin{pmatrix} J & K \\ K & J \end{pmatrix}.$$

The matrix U can be diagonalized: $T_{2\times2}U_{2\times2}T^{\dagger}_{2\times2} = U^{diagonal}_{2\times2}$, where $T_{2\times2}$ is an unitary matrix: $T_{2\times2} = \frac{1}{\sqrt{2}}\begin{pmatrix} 1 & -1 \\ 1 & 1 \end{pmatrix}$. The unitary operator transforms the basis as follows:

$$T_{2\times2}\begin{pmatrix} \varphi_1(\vec{r}_1)\varphi_2(\vec{r}_2) \\ \varphi_2(\vec{r}_1)\varphi_1(\vec{r}_2) \end{pmatrix} = \frac{1}{\sqrt{2}}\begin{pmatrix} \varphi_1(\vec{r}_1)\varphi_2(\vec{r}_2) - \varphi_1(\vec{r}_2)\varphi_2(\vec{r}_1) \\ \varphi_1(\vec{r}_1)\varphi_2(\vec{r}_2) + \varphi_1(\vec{r}_2)\varphi_2(\vec{r}_1) \end{pmatrix} = \begin{pmatrix} \phi^{Triplet} \\ \phi^{Singlet} \end{pmatrix}.$$

For a two-electron system total spin is $\vec{S} = \vec{s}_1 + \vec{s}_2$; $S = 0,1$. For the singlet, $S = 0, m_s = 0$. The spin part of the wavefunction is antisymmetric and the space part must be symmetric. For $S = 1, m_s = -1,0,1$ (triplet), the space part of the wavefunction must be symmetric.

(b) Diagonalization of the Coulomb interaction matrix:

$$T_{2\times2}\begin{pmatrix} J & K \\ K^* & J^* \end{pmatrix}T_{2\times2} = \frac{1}{2}\begin{pmatrix} J-K-K+J & J+K-K-J \\ J-K+K-J & J+K+K+J \end{pmatrix} = \begin{pmatrix} J-K & 0 \\ 0 & J+K \end{pmatrix}.$$

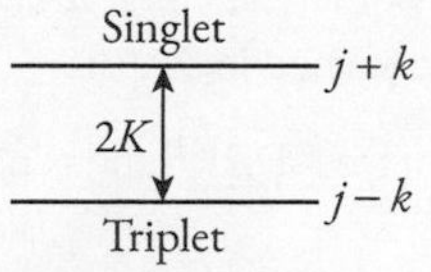

The diagonal elements represent energy eigenvalues of the triplet and singlet states. The difference in energy is $2K$.

Electrons in the triplet states keep themselves far apart to prevent the spatial wavefunction in Eq. 9.15d go to zero. In the singlet state, separation between the two electrons is relatively less. The triplet state of a pair of electrons therefore has lower energy, electron-electron Coulomb interaction being repulsive.

P9.7: Show that the condition to obtain extremum of a function of two variables, say $f(x,y)$, subject to a constraint that $g(x,y) = k$, where k is a constant, turns out to be the same as the condition to obtain the extremum of the function $\left[f(x,y) - \lambda g(x,y)\right]$, where the variational multiplier λ may be determined using *other* physical characteristics of the problem.

Solution:

For the function $f(x,y)$ to be an extremum: $df = 0$:

$$\frac{\partial f}{\partial x}dx + \frac{\partial f}{\partial y}dy = 0.$$

Changes in x and y being independent, $\frac{\partial f}{\partial x} = 0$ and $\frac{\partial f}{\partial y} = 0$.

The constraint $g(x,y) = k$ implies that the increment $dg(x,y) = 0$:

i.e., $\frac{\partial g}{\partial x}dx + \frac{\partial g}{\partial y}dy = 0$. From the above relations, it follows that

$$\frac{\left(\frac{\partial f}{\partial x}\right)}{\left(\frac{\partial g}{\partial x}\right)} = \lambda = \frac{\left(\frac{\partial f}{\partial y}\right)}{\left(\frac{\partial g}{\partial y}\right)}$$

and hence

$$\left(\frac{\partial f}{\partial x}\right) - \lambda\left(\frac{\partial g}{\partial x}\right) = 0 \text{ and } \left(\frac{\partial f}{\partial y}\right) - \lambda\left(\frac{\partial g}{\partial y}\right) = 0.$$

In other words, obtaining the extremum of the function $f(x,y)$ with the constraint $g(x,y) = k$ is equivalent to seeking the extremum of the function $f(x,y) - \lambda g(x,y)$.

P9.8: Compose a computer algorithm for the calculation of Helium atom using the local density approximation.

Solution:

Step 1: Solve the Hydrogen atom as discussed in P6.9 using Numerov's method. Obtain the auxiliary function $\chi_{E,\ell}(r) = rR(r)$, after solving the radial part of the Schrödinger equation as given in P6.9.2.

Step 2: Solve for the Hartree direct potential ($V^{dc}(r)$) with the help of Eq. 9.72b and Eq. 9.73. The integration in Eq. 9.72 can be accomplished using a numerical integration subroutine. The term $\rho(r')$ in Eq. 9.73 may be constructed using the wavefunction obtained in Step 1.

Step 3: The exchange potential in the Hartree–Fock approximation in the LDA is

$$V^{ex}(r) = -\left[\frac{3\left|\chi_{E,\ell}(r)\right|^2}{2\pi^2 r^2}\right]^{1/3}. \qquad \text{(P9.8.1)}$$

Step 4: Solve the Schrödinger equation for the Helium atom using the discussion in P6.9. The equation to be solved is

$$-\frac{\hbar^2}{2\mu}\frac{d^2\chi_{E,\ell}(r)}{dr^2}+\left\{V(r)+\frac{\hbar^2}{2\mu}\frac{\ell(\ell+1)}{r^2}\right\}\chi_{E,\ell}(r)=E\chi_{E,\ell}(r) \tag{P9.8.2}$$

where $V(r)=\frac{2}{r}+V^{dc}(r)-V^{ex}(r)$. Obtain the wavefunction and energy after solving P 9.8.2 using numerical procedures.

Step 5: Using the wavefunction obtained in Step 4, repeat the steps from step 2 onwards. The $\rho(r')$ in Step 2 in the new iteration may now be constructed using the wavefunction obtained in Step 4.

Step 6: Repeat the iteration until E (in Eq. P9.8.2) in consecutive iterations is the same within a numerical convergence factor.

P9.9: Rewrite the single-particle Hartree–Fock self-consistent-field equation (Eq. 9.54) using the orbitals obtained on employing the transformations given in Eq. 9.56 and Eq. 9.57.

Solution:

The Hartree–Fock equation in Eq. 9.54b is

$$\left\{\begin{array}{l} f(\vec{r}_1)u_k(\vec{r}_1)+ \\ \left[\left[\sum_j\left\{\int dV_2\frac{u_j^*(\vec{r}_2)}{r_{12}}\left(u_k(\vec{r}_1)u_j(\vec{r}_2)-\delta(\sigma_k,\sigma_j)u_k(\vec{r}_2)u_j(\vec{r}_1)\right)\right\}\right]\right]\end{array}\right\}=-\sum_j\delta(\sigma_k,\sigma_j)\lambda_{kj}u_j(\vec{r}_1)$$

The transformation in Eq. 9.57 is $u_i'(\vec{r})=\sum_{k=1}^{N}c_{ik}^{\dagger}\,u_k(\vec{r})=\sum_{k=1}^{N}c_{ki}^{*}\,u_k(\vec{r})$.

Use the inverse unitary transformation $u_j(\vec{r})=\sum_{l=1}^{N}c_{jl}\,u_l'(\vec{r})$ and substitute every unprimed wavefunction using an expansion in terms of the primed orbitals in the HF equation. Use the orthonormality of the single-particle wavefunctions and diagonalize the matrix of the variational multipliers, $[C^{\dagger}][\lambda][C]=[\lambda^{diagonal}]=\left[\lambda_{ij}\delta_{ij}\right]$ to get the HF equation in the *diagonal* form (Eqs. 9.58a,b).

Exercises

E9.1: Three non-interacting particles are confined to move in a one-dimensional infinite potential well of length a, described as

$$V(x)=\begin{cases}0, 0<x<a\\ \infty, \text{other values of } x\end{cases}.$$

Determine the energy and wave function of the ground, first and the second excited states when the three particles are (a) spin-less and distinguishable with their masses $m_1<m_2<m_3$; (b) identical bosons; (c) identical spin-1/2 particles; and (d) distinguishable spin-1/2 particles.

E9.2: Referring to P9.3 with an additional condition that all the three particles are in the same spin state, obtain the energy and the wave function of the ground, first, and the second excited state.

E9.3: Find the ground state energy and the wave function of a system of N non-interacting identical particles that are confined to a one-dimensional infinite well when the particles are (a) bosons and (b) spin-1/2 fermions.

E9.4: Neglecting the spin–orbit interaction find the energy levels and the wave functions of the three lowest states of a two-electron atom.

E9.5: Write the Slater determinant for the ground state of the carbon atom. If you expand this determinant, how many terms would be there?

E9.6: Four particles are to be distributed among five energy states a, b, c, d, and e. Enumerate all possible ways for such a distribution if the particles are (a) fermion, (b) bosons, and (c) classical particles. Also find the probability of finding four particles in a particular states.

E9.7: A *parafermion* is a particle for which the maximum occupancy of any given single-particle state is k, which is an integer greater than zero. (For $k = 1$, parafermions are just the fermions; and for $k = \infty$ the parafermions are what we recognize as the bosons.) Consider a system of n particles having single-particle energy is ε. The Hamiltonian of the system is $H = \varepsilon n$. Obtain the occupation function $n(\mu,T)$ where μ is the chemical potential and T is the absolute temperature. What is n when $\mu = -\infty$, $\mu = \varepsilon$, and when $\mu = +\infty$? Show that the distribution function $n(\mu, T)$ reduces to the Fermi and Bose distributions in the appropriate limits.

E9.8: Show that the shortest path between two points in Euclidean xy plane is a straight line.

E9.9: Obtain the optical path in an atmosphere where the velocity v increases linearly with height y: $v(y) = y/b$, $b > 0$. At the surface of a black hole (called its event horizon), the gravitational force is so huge that the velocity of light tends to zero (i.e., $v = 0$ at $y = 0$). Use Fermat's principle to show that light will not escape completely from the black hole unless the path is perpendicular to the event horizon.

E9.10: Consider two parallel coaxial wire circles to be connected by a surface of minimum area that is generated by a revolving curve $y(x)$ about the x-axis. The curve is required to pass through fixed endpoints (x_1, y_1) and (x_2, y_2) as shown in Fig. E9.10.1. Using variational calculus, choose the curve $y(x)$ so that the area of the resulting surface will be the least.

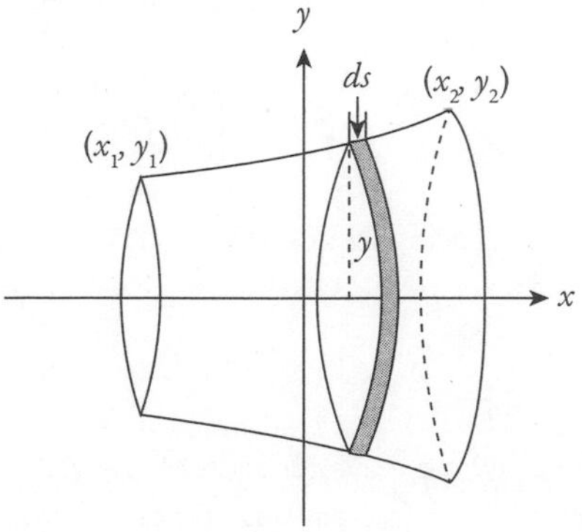

Fig. E9.10.1 A soap film suspended between the wire circles.

E9.11: Show that the Hatree and Hatree–Fock equations are identical for the ground state of helium (refer Eq. 9.19 and Eq. 9.20). Determine the Hatree–Fock potential for such a system.

E9.12: Consider a system having N interacting electron, the Hamiltonian of which is given by $H^N(q_1,q_2,\ldots,q_N)=\sum_{i=1}^{N}\left(-\frac{1}{2}\nabla_i^2-\frac{Z}{r_i}\right)+\sum_{i<j=1}^{N}\frac{1}{d_{ij}}$. Prove that $[H,L]=0$, where $\vec{L}=\sum_i\vec{L}_i$ is the total angular momentum of the system.

E9.13: The Hartree–Fock method is applicable not only for fermions but also for bosons. Consider a system consisting of N spin-0 particles in one dimension having spin-orbital coordinates x_i (for spin-0 particles, only the orbital coordinate matters). If the bosons interact via a δ-function potential, the Hamiltonian is written as $H=-\sum_{i=1}^{N}\frac{\partial^2}{\partial x_i^2}+g\sum_{i>j}^{N}\delta(x_i-x_j)$. This means that the bosons experience an infinitely large interaction in an infinitely short range. Although this problem can be solved exactly, consider the Hartree–Fock approximation here and show by the minimization of the energy functional

$E=\frac{\langle\psi|H|\psi\rangle}{\langle\psi|\psi\rangle}$ with respect to ϕ that the Hartree–Fock equation is

$$\left[-\frac{\partial^2}{\partial x^2}-g(N-1)\,|\,\varphi(x)\,|^2\right]\varphi(x)=\varepsilon\varphi(x).$$

E9.14: Consider solving the ground state of helium atom using the density referring to the P9.8. Rather than solving the Hartree direct potential ($V^{dc}(r)$) by calculating the integrals in Eq. 9.72b and Eq. 9.73, find it using the Poisson's equation. Construct an ordinary differential equation to solve the $V^{dc}(r)$ using Numerov's technique.

References

[1] Richard P. Feynman, *The Feynman Lectures in Physics*, Vol. 3, p. 4–3.

[2] Sin-Itiro Tomonaga, *The Theory of Spin.* Translated into English by Takeshi Oka (The University of Chicago Press, 1997).

[3] V. B. Beretetskii, E. M. Lifshitz, and L. P. Pitaevskii, *Relativistic Quantum Theory* (Pergamon Press, 1974).

[4] D. R. Hartree, The Wave Mechanics of an Atom with a Non-Coulomb Central Field. Part II. Some Results and Discussion Mathematical Proceedings of the Cambridge Philosophical Society. 24:111–132 (1928) DOI: 10.1017/S0305004100011920.

[5] D. R. Hartree, *Calculation of Atomic Structure* (Wiley, 1952).

[6] D. R Hartree and W. Hartree, Self-consistent Field with Exchange for Beryllium. *Proc. Royal Soc. Lond. A.* 150:869, 9 (1935). doi:9.1098/rspa.1935.0085.

[7] V. A. Fock, Self Consistent Field. *Z. Phys.* 62:11, 795 (1930). doi:9.1007/BF01330439.

[8] J. C. Slater, *Quantum Theory of Atomic Structure* (McGraw-Hill, 1960).

[9] R. D. Mattuck, *A Guide to Feynman Diagrams in the Many-Body Problem: Second Edition* (McGraw-Hill, 1976).

[10] G. B. Arfken and H. J. Weber, *Mathematical Methods for Physicists* (Elsevier Academic Press, 7th edition, 2013).

[11] P.C. Deshmukh, *Foundations of Classical Mechanics* (Cambridge University Press, 2019).

[12] T. C. Koopmans, Über die Zuordnung von Wellenfunktionen und Eigenwerten zu den Einzelnen Elektronen Eines Atoms. *Physica* 1:104 (1933).

[13] S. Chandrasekhar, On the Continuous Absorption Coefficient of the Negative Hydrogen Ion. *Astrophysics Journal* 102, 223 (1945).

[14] A. Starace, Length and Velocity Formulas in Approximate Oscillator Strength Calculations. *Phys. Rev. A*, 3:4, 1242 (1971).

[15] J. C. Slater, Theory of Complex Spectra. *Phys.Rev.* 34:1293 (1929).

[16] J. C. Slater, A Simplification of the Hartree-Fock Method. *Phys. Rev.* 81:3, 385 (1951).

[17] I. Lindgren and K. Schwartz, Analysis of the Electronic Exchange in Atoms. *Phys. Rev. A* 5, 542 (1972).

[18] W. Kohn and L. J. Sham, Self-consistent Equation including Exchange and Coulomb Affects. *Phys. Rev.* 140, A1133 (1965).

[19] Axel D. Becke, Perspective: Fifty Years of Density-functional Theory in Chemical Physics. *J. Chem. Phys.* 140, 18A301 (2014). https://doi.org/9.1063/1.4869598.

[20] R. O. Jones, Density Functional Theory, Its Origins, Rise to Prominence, and Future. *Revs. Mod. Phys.* 87:3, 897 (2015).

[21] C.F Fischer, T. Brage, P Joensson, *Computational Atomic Structure, An MCHF Approach* (Routledge, 2022).

[22] I. Shavitt and R. J. Bartlett, *Many-Body Methods in Chemistry and Physics: MBPT and Coupled-Cluster Theory* (Cambridge University Press, Cambridge, 2009).

[23] F. C Smith and W. R Johnson, Relativistic Self-consistent Fields with Exchange. *Phys. Rev* 160 (1967).

[24] I. P. Grant, *Relativistic Quantum Theory of Atoms and Molecules* (Springer, 2007).

[25] W. R. Johnson and C. D. Lin, Multi-Channel Relativistic Random Phase Approximation for the Photoionization of Atoms, *Phys. Rev.* A20, 964 (1979).

[26] Google AI quantum and collaborators, Frank Arute, Kunal Arya, et al., Hartree-Fock on a Superconducting Qubit Quantum Computer. *SCIENCE* 369:6507, 1084–1089 (August 28, 2020). DOI: 10.1126/science.abb9811

10

Quantum Collisions

Niels Bohr. B24, https://arkiv.dk/en/vis/5894685. Courtesy: Niels Bohr Archive.

How wonderful that we have met with a paradox. Now we have some hope of making progress

—Niels Bohr

Probing an elementary or composite particle, an atom, an ion, a molecule, a cluster, or, for that matter any quantum system, involves studying how the target responds to a query – an investigation that is made using a probe that could be electromagnetic radiation (a photon or a beam of light that is incident on the target), or impinging the target with some particles, such as electrons, positrons, neutrons, α-particles, or composites of the same such as atoms, ions, and molecules. Typically, the investigation is called "spectroscopy" when the probe is electromagnetic radiation that is absorbed, scattered or reemitted by the target, which is the object of investigation. When the probe is some other particle, it is usually called "collisions." The quantum mechanics of spectroscopy and collisions is, however, very closely and seamlessly related, as we shall uncover in this chapter. Niels Bohr's remark quoted above in the frame of a diagram depicting collisions signposts the fact that studies of quantum collisions have resulted in many landmark developments in physics.

10.1 Partial Wave Analysis, Faxen–Holstmark Equation for Potential Scattering

Very commonly, quantum collisions are analyzed using what is called as partial wave analysis, which is a mathematical framework developed by Hilding Faxen and Johan Holtsmark in 1927 [1–4]. The Faxen–Holtsmark theory aims at investigating a target T using a monoenergetic incident beam of projectile particles P. Figures 10.1a–d schematically illustrate the scattering process. A monoenergetic beam of incident particles represented by a wavepacket in Fig. 10.1a enters an apparatus containing the target atoms. The polar axis of this coordinate frame is the Z-axis of a Cartesian coordinate system that is chosen to be along the direction of incidence of the probe particles – also referred to as projectiles. Information about the target is obtained from the interaction between the projectiles and the target.

The *probability* of encounter between the incident and the target particles is of primary interest in the scattering formalism. A measure of the encounter probability is provided by the ratio

$$\Upsilon = \frac{\textit{number of scattering events per unit time per unit scatterer}}{\textit{flux of the incident particles with respect to the target}}. \tag{10.1}$$

The flux of the incident particles is

$$\Phi_P = \frac{N_P}{S}, \tag{10.2}$$

where N_P is the number of probing particles reaching the target *per unit time*, and S is the effective cross-sectional area of the incident beam that enters the apparatus through the collimator. Clearly, the dimensions of the flux Φ_P are $T^{-1}L^{-2}$. Let us denote the number of particles P that interact with the *target per unit time* by N_I. This is a fraction of N_A, which is the number of particles A reaching the target per unit time. If P is the probability that an incident particle interacts with the target and thereby gets *removed from the incident flux by scattering*, then

$$P \times N_P = N_I. \tag{10.3}$$

N_P and N_I both have dimension of T^{-1}; i.e., inverse time. N_I would be proportional to the incident flux Φ_P and also to n_T, which is the number of target particles that intercept the beam. Thus, N_I is proportional to the product $n_T\Phi_P$. Let us denote the constant of proportionality by σ. We therefore write

$$N_I = \sigma n_T \Phi_P \tag{10.4}$$

or

$$\sigma = \frac{N_I}{n_T\Phi_P} = \frac{PN_P}{n_T\Phi_P}, \tag{10.5}$$

which is just the ratio Υ defined in Eq. 10.1. The dimensions of σ are given by the square of length, L^2, i.e., area. σ is therefore called as the *scattering cross section* and is usually measured in units of Mega-barn, denoted as Mb. The barn is defined to be 10^{-24} cm^2, so 1 Mb is 10^{-18} cm^2. From Eq. 10.1 and Eq. 10.5, we see that the scattering cross section σ provides a measure of the intrinsic propensity of interaction between the scatterer and the probe, independent of the actual intensity of the incident particles and the number of scattering particles.

The distance "b" shown in Fig. 10.1a is called as the impact-parameter; it is the closest distance at which the projectile would pass the target's center if it were *not* to interact. Figure 10.1b shows a section of the incident beam that is transverse to the direction of its propagation. Due to the interaction between the projectiles and the target particles, all particles in the annular ring between radius b and $b+db$ get scattered through the shaded portion shown in Fig. 10.1c, which is a curved annular strip on the surface of a sphere of radius r, measured from the scatterer's center. It subtends an angle $d\theta$ at the center. A tiny area $\delta S = r^2\delta\Omega$ on this strip is shown in Fig. 10.1d. The basis set of unit vectors $\left\{\hat{e}_r, \hat{e}_\theta, \hat{e}_\varphi\right\}$ shown in Fig. 10.1d is the one that belongs to the spherical polar coordinate system (see, for example, Chapter 2 of Reference [5]).

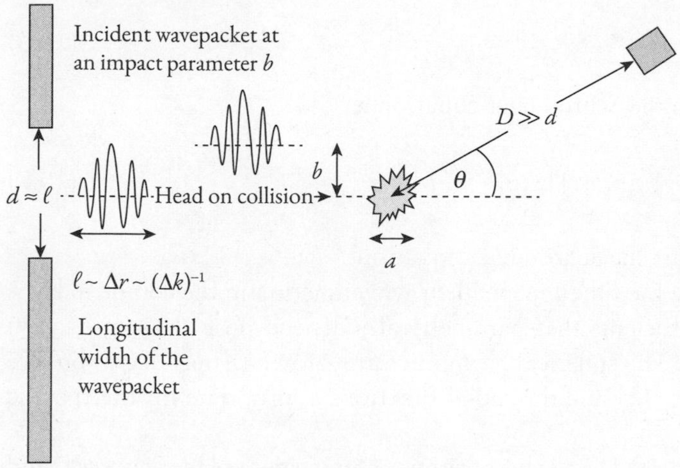

Fig. 10.1a Schematic diagram showing scattering of an incident beam of probe particles by a target scatterer. We assume that the transverse width d is about the same as the wavepacket's longitudinal width.

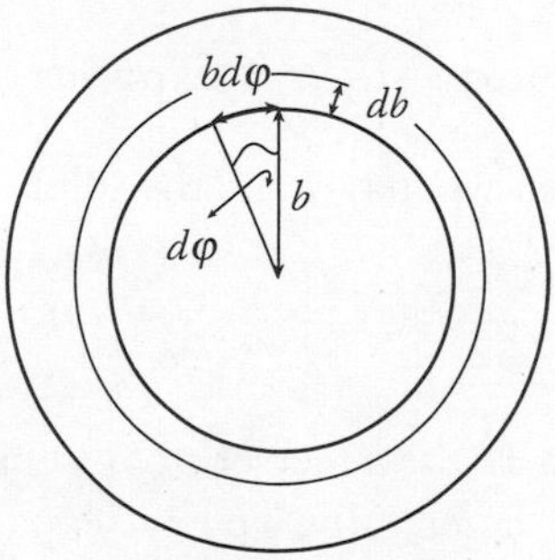

Fig. 10.1b Different parts of the incident particles are at different impact parameter distances from the scatterer's center. The scattering process is considered to have cylindrical symmetry about the Z-axis, independent of the azimuthal angle φ.

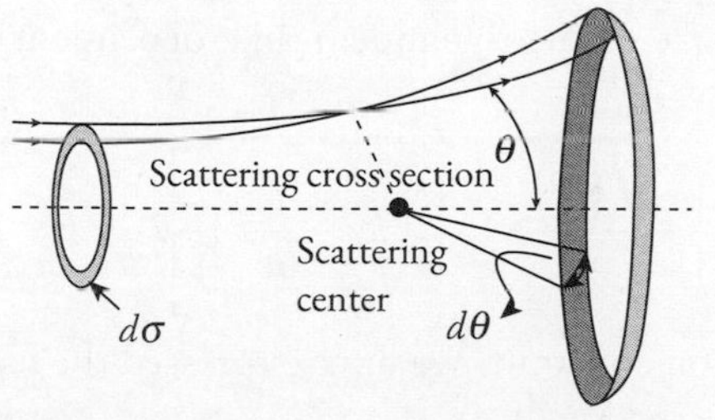

Fig. 10.1c The shaded ring on the left is in a flat plane, perpendicular to the direction of incidence. It is the same ring shown in Fig. 10.1b. The shaded strip on the right is on the surface of a sphere centered at the scatterer's center at a distance r.

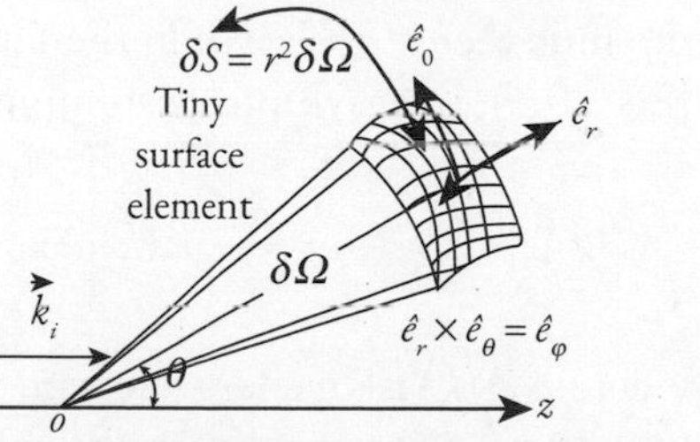

Fig. 10.1d A tiny surface element δS on the shaded curved strip (on the spherical surface) shown in Fig. 10.1c subtends a solid angle $\delta\Omega$ at the center of the scatterer.

The beam of incident particles is not expected to be strictly monoenergetic; even if mostly so. It is considered to be represented by a surface of constant phase, which is a traveling plane wave that enters the apparatus through a collimator. The propagation of this plane wave is along its wave vector $\vec{k}_i = k\hat{e}_z$. The energy of the monoenergetic particle in the incident beam is $E = \dfrac{\hbar^2 k^2}{2m}$, m being the mass of the projectile particle. The incident beam, however, has a slight energy spread, and the wave vector may be only centered around k, and have a width Δk.

We shall consider the spread Δk in the magnitude of the wave vector later. For the time being, we shall consider strict monoenergetic incident waves traveling along $\vec{k}_i = k\hat{e}_z$ (Fig. 10.1d). We can now set up the Schrödinger equation for our system of interacting probe particles and the target particles. If we consider the temporal evolution of the wavefunction to be completely determined by the *dynamical phase* (Eqs. 1.114a,b) described by

$$\exp\left(-i\frac{E}{\hbar}t\right) = \exp(-i\omega t), \tag{10.6}$$

then we may write the solution to the Schrödinger equation as

$$\psi_{\vec{k}_i}^{SE}(\vec{r},t) = \left[\sum_{\ell=0}^{\infty}\sum_{m=-\ell}^{\ell} c_{\ell m}(k) Y_{\ell m}(\hat{e}_r) R_{\ell m}(k,r)\right]\exp\left(-i\frac{E}{\hbar}t\right), \tag{10.7}$$

where the superscript *SE* represents the Schrödinger equation.

Essentially, we have expanded the time-independent wavefunction in the complete basis of spherical harmonics with coefficients that parametrically depend on k (i.e., on energy) and the radial functions $R_{\ell m}(k,r)$. The spherical harmonics are known (Appendix C) but we must determine the coefficients $c_{\ell m}(k)$ and the radial functions appropriate for the specific target potential.

The solution can also be interpreted as a superposition of an incident plane wave ψ_{inc} and a scattered spherical wave ψ_{scat}. The amplitude of the scattered wave in general is anisotropic, hence it must be scaled by a function $f(\hat{\Omega})$, wherein $\hat{\Omega}$ denotes a unit vector in the direction of scattering. It could be in the forward direction, along $\hat{e}_z$, or opposite to it, or, for that matter, in any other direction in the three-dimensional space. This direction would be indicated by the unit vector $\hat{e}_r$ shown in the Fig. 10.1d. Inclusive of the dynamical time-dependent phase (Eq. 10.6), the asymptotic solution is

$$\psi_{\vec{k}_i}^{+}(\vec{r},t;r\to\infty) \to A(k)\left[\exp\left\{i\left(\vec{k}_i\bullet\vec{r} - \frac{E}{\hbar}t\right)\right\} + \frac{f(\hat{\Omega})}{r}\exp\left\{i\left(kr - \frac{E}{\hbar}t\right)\right\}\right], \tag{10.8}$$

wherein $A(k)$ is an energy dependent normalization constant. We have expressed the incident part and the scattered part of the wavefunction respectively as

$$\psi_{inc} = \exp\left(i\vec{k}_i\bullet\vec{r}\right) \tag{10.9a}$$

$$\text{and } \psi_{scat} = \frac{f(\hat{\Omega})}{r}\exp(ikr). \tag{10.9b}$$

Funquest: Suggest various alternative pairs of basis functions whose linear superpositions would express the general solution to the Schrödinger equation for potential scattering.

The amplitude of the scattered wave must diminish with distance away from the center of the target, since the surface area of a sphere which the scattered wave (in Eq. 10.8) crosses as time progresses would increase as the square of the radius of that sphere. The factor $\frac{1}{r}$ in Eq. 10.9b takes care of this, since the intensity of a wave goes as the square of the amplitude. The angle dependent function $f(\hat{\Omega})$ is called as the *scattering amplitude*. We shall call Eq. 10.8 as *phenomenological* solution since

- the surface of constant phase of the wavefront represented by the first term $\exp\left\{i\left(\vec{k}_i\bullet\vec{r} - \frac{E}{\hbar}t\right)\right\}$ in Eq. 10.8 is a plane incident wave traveling along the positive Z-axis. The phase of this surface is $\eta_{inc} = \vec{k}_i \bullet \vec{r} - \omega t$, i.e., $\eta_{inc} = kz - \omega t$, which means that on the surface of *constant* phase, we have $0 = d\eta_{inc} = k_i dz - \omega dt$, which guarantees that $\frac{dz}{dt} = \frac{\omega}{k} \rangle 0$, i.e., the z-coordinate of the traveling surface increases with time, and

- the surface of constant phase of the wavefront represented by the second term $\exp\left\{i\left(kr - \frac{E}{\hbar}t\right)\right\}$ in Eq. 10.8 is a spherical wave traveling radially *outward* from the scatterer's center. The phase of this surface being $\eta_{scat} = kr - \omega t$, on the surface of *constant* phase, we have $0 = d\eta_{scat} = kdr - \omega dt$, which guarantees that $\frac{dr}{dt} = \frac{\omega}{k} > 0$, i.e., the r-coordinate of the traveling surface *increases* with time.

We shall refer to the form of the wavefunction in Eq. 10.8 as *phenomenological*; it has not been obtained by solving the Schrödinger equation, but it has been built to *describe the phenomenon of scattering*. The time-independent part of Eq. 10.8 is

$$\psi^{+}_{\vec{k}_i}(\vec{r}; r \to \infty) = A(k)\left[\psi_{inc} + \psi_{scat}\right] = A(k)\left[\exp\left(i\vec{k}_i \bullet \vec{r}\right) + \frac{f\left(\hat{\Omega}\right)}{r}\exp(ikr)\right]. \tag{10.10}$$

In Eq. 10.8 and Eq. 10.10, we have placed a superscript plus sign (+) on the wavefunction. This is to denote the *scattering* boundary condition, called as the *outgoing wave boundary condition*; it has a spherical *outgoing* wave in the solution. In Section 10.3, we shall contrast this boundary condition with its time-reversed process. Our focus in this chapter is the quantum mechanics of electron-ion/atom scattering whose time-reversed process is atomic/molecular *photoionization* (or *photodetachment* of a negative ion). In the case of *photoionization/photodetachment*, we shall employ the *ingoing wave boundary condition* and denote the wavefunction as $\psi^{-}_{\vec{k}_i}(\vec{r}; r \to \infty)$, i.e., with a minus (−) sign on the wavefunction as a superscript. The time-reversal symmetry is discussed in Appendix A. Our discussion follows the treatment in References [3, 4], salient features of which can also be found in References [5, 9, 10].

Figures 10.2a,b show multiple pathways to form a quantum system and for its decay. These processes are reversible. The techniques we describe in the present chapter are of general utility in atomic and molecular physics, nuclear and particle physics, etc. However, the backdrop for our discussion would be primarily the electron-atom scattering. It serves as a fertile context to illustrate methods of quantum collision theory. Figure 10.2c depicts the scenario represented by Eq. 10.10 (better by Eq. 10.8, which includes the time-dependence). Our prototype target is an atom whose potential is spherically symmetric. Its wavefunction is therefore separable in radial and angular coordinates. The angular part of the wavefunction would be the spherical harmonics; we have already studied them (Chapter 4, and Appendix C). We shall therefore focus our attention on the determination of the radial part of the Schrödinger wavefunction in the scattering phenomena.

Considering the symmetry of the scattering process (Fig. 10.2c) about the axis along which the incident beam impinges upon the target, we expect the radial functions in Eq. 10.7 to be independent of the magnetic quantum number m. We know from Eq. 5.17a–d of Chapter 5

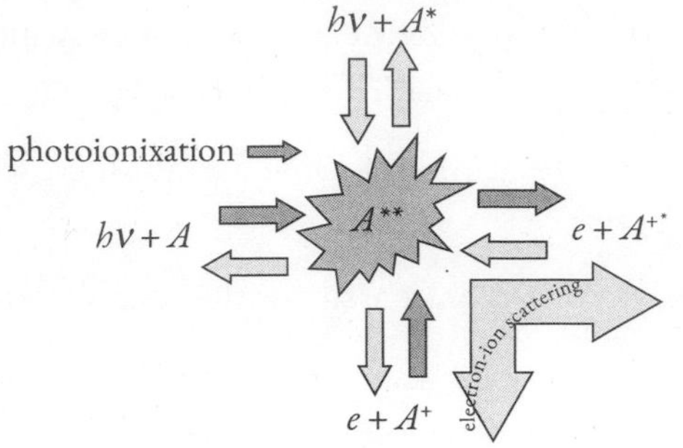

Fig. 10.2a This schematic diagram showing multiple pathways is based on Fig. 1.1 from Reference [4]. $h\nu$ denotes a photon, e denotes an electron, A denotes a neutral atom and $A+$ denotes a positive ion. An asterisk denotes an excited state.

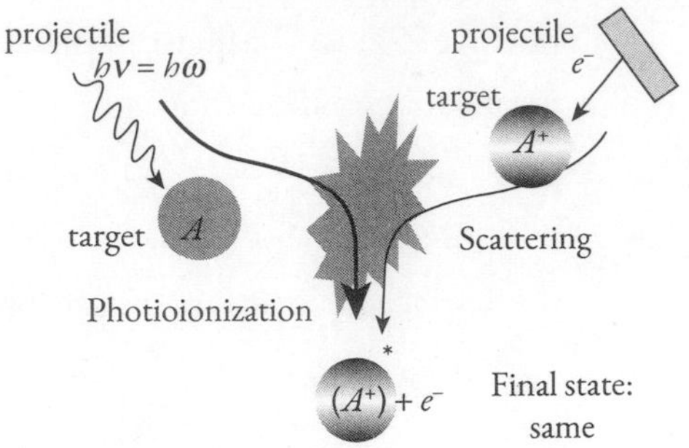

Fig. 10.2b The final state of photoionization $\left(h\nu + A \rightarrow A^{+^*} + e\right)$ and electron-ion scattering $\left(A^+ + e \rightarrow A^{+^*} + e\right)$, are the same, even though the initial ingredients of the reaction are different. These two processes are related by time-reversal symmetry.

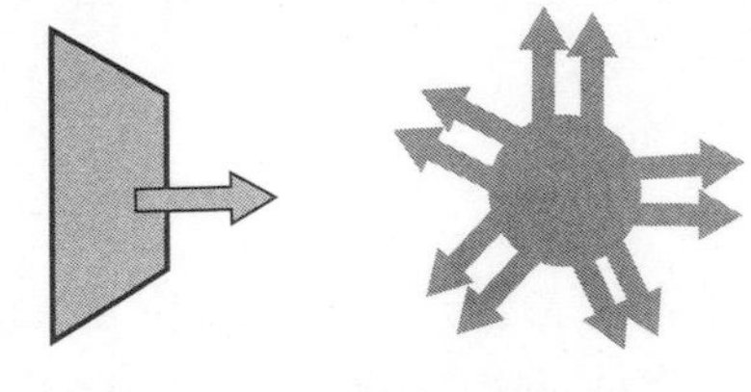

Fig. 10.2c Schematic diagram showing a plane wavefront incident on a target scattering center. The scattered wave would travel radially outward after scattering, but the scattering yield can be anisotropic.

that the radial function $R_{k,\ell}(r)$ for a particular value of ℓ would be a solution of the radial differential equation

$$R''_{k\ell}(r) + \frac{2}{r}R'_{k\ell}(r) - \frac{\ell(\ell+1)}{r^2}R_{k\ell}(r) + \frac{2\mu}{\hbar^2}\left[E - V(r)\right]R_{k\ell}(r) = 0. \tag{10.11}$$

The wavefunction inclusive of the angular part is

$$\psi(\vec{r}) = R_{\varepsilon\ell}(r)Y_\ell^m(\hat{r}), \tag{10.12}$$

where ε denotes the energy $(E > 0)$ of the scattering states. If we now introduce a new function $y_{\varepsilon\ell}(r)$ such that

$$R_{\varepsilon\ell}(r) = \frac{y_{\varepsilon\ell}(r)}{r}, \tag{10.13}$$

then the differential equation that the function $y_{\varepsilon\ell}(r)$ must satisfy is

$$\left[-\frac{\hbar^2}{2m}\frac{d^2}{dr^2} + \left\{V(r) + \frac{1}{2m}\frac{\ell(\ell+1)}{r^2}\right\} - E\right]y_{\varepsilon\ell}(r) = 0. \tag{10.14}$$

The object of our investigation is to study the target potential, so the strategy we adopt is to compare the solution to Eq. 10.14 *with* and *without* the potential $V(r)$; we would then expect that the difference in the solutions will subsume information about the potential. The *zero potential* case is a special case of Eq. 10.14; it is also spherically symmetric since it is zero in *all* the directions. For this case, Eq. 10.14 reduces to

$$\left[-\frac{\hbar^2}{2m}\frac{d^2}{dr^2}+\frac{1}{2m}\frac{\ell(\ell+1)}{r^2}-E\right]y_{\varepsilon\ell}(r)=0. \tag{10.15}$$

The above equation has a very simple solution when $\ell = 0$ (s orbitals). We shall therefore first obtain the solution for the *s* orbital. We shall then explore if we can discover a *method of mathematical induction* to obtain the solution for $\ell + 1$ from that for ℓ. The motivation to look for such a method is obvious. Beginning with the solution for $\ell = 0$, one could then build the solution for $\ell = 1$, and continue the induction to obtain solution for *any* value of ℓ. The differential equation to be solved for $\ell = 0$ is

$$\left[\frac{d^2}{dr^2}+k^2\right]y_{\varepsilon,\ell=0}(r)=0, \tag{10.16}$$

where the energy being essentially kinetic is written as

$$E=\frac{p^2}{2m}=\frac{(\hbar k)^2}{2m} \tag{10.17a}$$

$$\text{with } k=\frac{p}{\hbar}=\frac{\sqrt{2mE}}{\hbar}. \tag{10.17b}$$

We can immediately write the solution to the differential equation Eq. 10.16. It is

$$y_{\varepsilon,\ell=0}(r)=rR_{\varepsilon,\ell=0}(r)=Ne^{\perp ikr}; \tag{10.18}$$

it contains sine and cosine functions. The radial function $R_{\varepsilon\ell}(r)$ requires a subsequent division by r (as per Eq. 10.13), so we must discard the cosine terms in Eq. 10.18, for it blows up at the origin. Physically acceptable solution for the *s* orbital therefore is

$$R_{\varepsilon,\ell=0}(r)=N\frac{\sin(kr)}{r}. \tag{10.19}$$

Normalizing this function on the $\frac{k}{2\pi}$ scale using the Dirac-δ function, using a procedure similar to that we used in the Solved Problem P1.10 in Chapter 1, we get

$$\int_0^\infty R_{k'\ell}R_{k\ell}r^2dr=\delta\left(\frac{k'-k}{2\pi}\right)=2\pi\delta(k'-k), \tag{10.20a}$$

$$\text{i.e., } \int_0^\infty \sin(k'r)\sin(kr)dr=\frac{2\pi}{N^2}\delta(k'-k). \tag{10.20b}$$

Now,

$$\int_0^\infty \sin(k'r)\sin(kr)dr=\frac{\pi}{2}\delta(k'-k),$$

from which we conclude that $N = 2$. The normalized radial wavefunction for the $\ell = 0$ component of the free-particle $(V = 0)$ therefore is

$$R_{\varepsilon,\ell=0}(r) = \frac{2\sin(kr)}{r}. \tag{10.21}$$

Our interest is of course in obtaining the solution for an arbitrary value of ℓ, not just for the special case $\ell = 0$. We shall therefore obtain the solution for an arbitrary value of the orbital angular momentum quantum number ℓ, but only for the zero potential. Subsequently, we shall proceed to obtain the solution for the target potential $V(r) \neq 0$. In order to get the solution $R_{k,\ell}(r)$ for arbitrary ℓ, we take advantage of the fact that regardless of the nature of the potential, and also regardless of the value of the energy, the small-r solution is given by Eq. 5.26 of Chapter 5, viz.,

$$R_{k\ell}(r \to 0) \sim r^{\ell}. \tag{10.22}$$

Hence, for $0 \leq r < \infty$, we write the solution as

$$R_{k\ell}(r) = N' r^{\ell} \xi_{k\ell}(r) \tag{10.23}$$

and enquire what would be the condition on the function $\xi_{k\ell}(r)$ which has now been introduced. To determine this condition, we substitute Eq. 10.23 in Eq. 10.11, but with $V = 0$, which is

$$R''_{k\ell}(r) + \frac{2}{r} R'_{k\ell}(r) + \left\{k^2 - \frac{\ell(\ell+1)}{r^2}\right\} R_{k\ell}(r) = 0. \tag{10.24}$$

We need the first and the second derivatives of Eq. 10.23, which are given respectively by

$$R'_{k,\ell}(r) = N'\left[\ell r^{\ell-1}\xi_{k,l}(r) + r^{\ell}\xi'_{k\ell}(r)\right], \tag{10.25a}$$

and

$$R''_{k,\ell}(r) = N'\left[\ell(\ell-1)r^{\ell-2}\xi_{k,\ell}(r) + \ell r^{\ell-1}\xi'_{k,\ell}(r) + \ell r^{\ell-1}\xi'_{k,\ell}(r) + r^{\ell}\xi''_{k,\ell}(r)\right],$$

i.e., $$R''_{k,\ell}(r) = N'\left[\ell(\ell-1)r^{\ell-2}\xi_{k,\ell}(r) + 2\ell r^{\ell-1}\xi'_{k,\ell}(r) + r^{\ell}\xi''_{k,\ell}(r)\right]. \tag{10.25b}$$

Using Eq. 10.23, Eq. 10.25a, and Eq. 10.25b in Eq. 10.24, we get

$$\begin{bmatrix} r^{\ell}\xi''_{k,\ell}(r) + 2\ell r^{\ell-1}\xi'_{k,\ell}(r) + \ell(\ell-1)r^{\ell-2}\xi_{k,\ell}(r) + \\ \frac{2}{r}\left\{\ell r^{\ell-1}\xi_{k,\ell}(r) + r^{\ell}\xi'_{k,\ell}(r)\right\} + \left\{k^2 - \frac{\ell(\ell+1)}{r^2}\right\} r^{\ell}\xi_{k,\ell}(r) \end{bmatrix} = 0. \tag{10.26}$$

Rearranging Eq. 10.26 and dividing by r^{ℓ}, we get

$$\xi''_{k,\ell}(r) + (2\ell+1)r^{-1}\xi'_{k,\ell}(r) + k^2\xi_{k,\ell}(r) = 0. \tag{10.27}$$

The condition on the function $\xi_{k\ell}(r)$ that was introduced in Eq. 10.23 therefore is that it must satisfy the above second-order differential equation. If we can find the solution to Eq. 10.27, we would have found the solution to the radial differential equation for the special case $V = 0$, and for arbitrary ℓ. We are primarily looking for a solution to the radial Schrödinger equation using the *method of mathematical induction*. Therefore we differentiate Eq. 10.27 with respect to r one *more* time to get a *second*-order differential equation for $\xi'_{k,\ell}(r)$. The highest order of the derivative of $\xi_{k,\ell}(r)$ that would appear in the resulting equation would be 3. The resulting equation is

$$\xi'''_{k,\ell}(r) + \frac{2\ell+1}{r}\xi''_{k,\ell}(r) + \frac{k^2 - 2(\ell+1)}{r^2}\xi'_{k,\ell}(r) = 0. \tag{10.28}$$

We now *propose* a recursion relation,

$$\xi'_{k,\ell}(r) = r\xi_{k,\ell+1}(r). \tag{10.29}$$

The recursion relation suggested above provides the function $\xi_{k,\ell+1}$ from (the derivative of) $\xi_{k,\ell}$. The multiplying factor r balances the dimensions on the two sides of Eq. 10.29. We find that

$$\xi''_{k,\ell}(r) = r\xi'_{k,\ell+1}(r) + \xi_{k,\ell+1}(r) \tag{10.30a}$$

$$\text{and } \xi'''_{k,\ell}(r) = r\xi''_{k,\ell+1}(r) + \xi'_{k,\ell+1}(r) + \xi'_{k,\ell+1}(r) = r\xi''_{k,\ell+1}(r) + 2\xi'_{k,\ell+1}(r). \tag{10.30b}$$

It can be easily verified now that Eq. 10.28 is satisfied on substituting Eq. 10.29 and Eqs. 10.30a,b in it, thereby ratifying our proposal for the recursion relation. This is a huge step, because we can now obtain the radial function for an arbitrary ℓ starting with the solution in Eq. 10.21 for $\ell = 0$ using Eq. 10.23, and the recursion relation (Eq. 10.29). We can therefore write

$$\xi_{k,\ell=0}(r) = \frac{R_{k,\ell=0}(r)}{N'}, \tag{10.31a}$$

and for an arbitrary ℓ we therefore have

$$\xi_{k,\ell}(r) = \left(\frac{1}{r}\frac{d}{dr}\right)^{\ell}\xi_{k,\ell=0}(r) = \left(\frac{1}{r}\frac{d}{dr}\right)^{\ell}\frac{R_{k,\ell=0}(r)}{N'} = \frac{1}{N'}\left(\frac{1}{r}\frac{d}{dr}\right)^{\ell}\frac{2\sin(kr)}{r}. \tag{10.31b}$$

Exploiting the property that we can always multiply the solution of a differential equation with respect to r by an energy-dependent (i.e., k-dependent) parameter, we write the solution to the radial Schrödinger equation (albeit for $V = 0$) along with a multiplicative factor $\frac{(-1)^{\ell}}{k^{\ell}}$, included for later convenience, as

$$R_{k,\ell}(r) = N'r^{\ell}\left\{\frac{1}{N'}\left(\frac{1}{r}\frac{d}{dr}\right)^{\ell}\frac{2\sin(kr)}{r}\right\} = \frac{(-1)^{\ell}2}{k^{\ell}}r^{\ell}\left\{\left(\frac{1}{r}\frac{d}{dr}\right)^{\ell}\frac{\sin(kr)}{r}\right\}, \text{ for } \ell = 0,1,2,3,.. \tag{10.32}$$

Essentially, $\xi_{k,\ell}(r)$ is obtained by operating on $\frac{\sin(kr)}{r}$ by the operator $\left(\frac{1}{r}\frac{d}{dr}\right)$ a total of ℓ number of times. Operating just once gives us

$$\left(\frac{1}{r}\frac{d}{dr}\right)\frac{\sin(kr)}{r} = \frac{1}{r}\left\{\frac{k\cos(kr)}{r} - \frac{\sin(kr)}{r^2}\right\} \underset{r\to\infty}{=} \frac{1}{r}\frac{k\cos(kr)}{r} = \frac{(-1)k}{r}\frac{\sin\left(kr - \frac{\pi}{2}\right)}{r}.$$

We see that the effect of the operator $\left(\frac{1}{r}\frac{d}{dr}\right)$ on the operand is to regenerate the operand with a slight difference: the phase of the sinusoidal function drops by $\frac{\pi}{2}$, and the result is scaled by $\frac{(-1)k}{r}$. When we do this ℓ times, the result is

$$\left(\frac{1}{r}\frac{d}{dr}\right)^{\ell}\frac{\sin(kr)}{r}=\frac{(-1)^{\ell}k^{\ell}}{r^{\ell}}\;\frac{\sin\left(kr-\frac{\ell\pi}{2}\right)}{r}. \tag{10.33}$$

The multiplicative factor $\frac{(-1)^{\ell}}{k^{\ell}}$ included in Eq. 10.32 gets canceled, and the result is

$$R_{k,\ell}(r)=\frac{y_{\varepsilon\ell}(r)}{r}=\frac{2\sin\left(kr-\frac{\ell\pi}{2}\right)}{r}\text{ for }\ell=0,1,2,3,.. \tag{10.34}$$

The simplicity of this result is appealing; the solution for an arbitrary value of the orbital angular momentum quantum number has exactly the same form as that for the *s*-orbital, except for the accumulation of a phase-shift of $\frac{\pi}{2}$ for each successive value of ℓ.

We have now obtained the solution $y_{\varepsilon\ell}(r)$ to Eq. 10.15, which is the special case corresponding to $V = 0$ of the general case that is of our primary interest, viz. Eq. 10.14. We now consider $V(r) \neq 0$, but one whose influence diminishes rapidly as one goes away from the scattering center. Beyond a certain distance, we consider the potential to be therefore weak, as shown in Fig. 10.3. For such a potential, the solution $y_{\varepsilon,\ell=0}(r)$ would be only *slightly* different from that in Eq. 10.18; we may therefore write its asymptotic solution with a slight modification as

$$y_0(k,r;r\to\infty)=F_0(k,r)\exp(\pm ikr), \tag{10.35a}$$

where $F_0(k,r)$ is a *slowly* varying function of r. Note that the $\ell = 0$ quantum number has been written as a subscript.

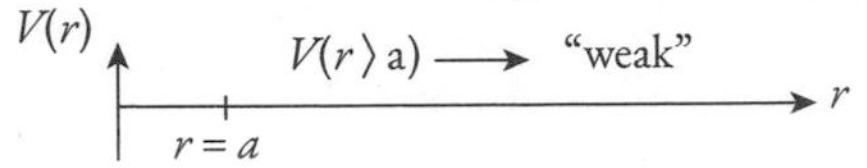

Fig. 10.3 A physical potential becomes weak as one moves away from its source. In particular, we consider potentials that asymptotically fall *faster* than the Coulomb potential.

Accordingly, $y_0'(k,r)=\pm ikF_0(k,r)\exp(\pm ikr)+F_0'(k,r)\exp(\pm ikr)$, (10.35b)

and

$$y_0''(k,r)=\left[\begin{array}{r}(\pm ik)^2F_0(k,r)\exp(\pm ikr)\pm ikF_0'(k,r)\exp(\pm ikr)+\\+F_0''(k,r)\exp(\pm ikr)\pm ikF_0'(k,r)e^{\pm ikr}\end{array}\right],$$

i.e., $y_0''(k,r)=(\pm ik)^2F_0(k,r)\exp(\pm ikr)\pm 2ikF_0'(k,r)\exp(\pm ikr)+F_0''(k,r)\exp(\pm ikr)$. (10.35c)

Putting Eqs. 10.35a,b,c in Eq. 10.14, we get

$$\begin{bmatrix} \cancel{(\pm ik)^2 F_0(k,r)\exp(\pm ikr)} \pm 2ikF_0'(k,r)\exp(\pm ikr) + F_0''(k,r)\exp(\pm ikr) + \\ \cancel{k^2 F_0(k,r)\exp(\pm ikr)} - V(r)F_0(k,r)\exp(\pm ikr) \end{bmatrix} = 0, \quad (10.36)$$

i.e., $\pm 2ikF_0'(k,r)\exp(\pm ikr) + F_0''(k,r)\exp(\pm ikr) - V(r)F_0(k,r)\exp(\pm ikr) = 0.$ (10.37)

We may now ignore $\frac{F_0''(k,r)}{F_0(k,r)}$ relative to $\frac{F_0'(k,r)}{F_0(k,r)}$, since $F_0(k,r)$ is only a *slowly* varying function of r. From Eq. 10.37, we get

$$\cancel{\frac{F_0''(k,r)}{F_0(k,r)}} \pm 2ik\frac{F_0'(k,r)}{F_0(k,r)} = V(r), \quad (10.38a)$$

i.e., $\frac{F_0'(k,r)}{F_0(k,r)} \simeq \frac{1}{\pm 2ik}V(r).$ (10.38b)

Integration of Eq. 10.38b gives

$$F_0(k,r) = \exp\left(\frac{1}{\pm 2ik}\int V(r)dr\right). \quad (10.39)$$

We note that if $V(r)$ is the Coulomb potential, the integral in the exponent on the right-hand side of Eq. 10.39 would be

$$\int V(r)dr = \int \frac{1}{r}dr = \ell n(r), \quad (10.40)$$

which does *not* allow $F_0(k,r)$ to become independent of r. Our analysis, however, is satisfactory for any potential $V(r)$ that asymptotically falls *faster* than the Coulomb potential. The case of the Coulomb potential, however, has an exact solution; we have already dealt with it fully in Section 5.4 of Chapter 5. We therefore conclude that for any potential

$$V(r; r \to \infty) \sim r^{-(1+\varepsilon)} \text{ with } \varepsilon > 0, \quad (10.41)$$

the asymptotic approximation

$$y_0(k,r;r \to \infty) = F_0(k,r)\exp(\pm ikr) = B_0^{(1)}(k)\exp(+ikr) + B_0^{(2)}(k)\exp(-ikr) \quad (10.42)$$

is satisfactory. Combining the results in Eq. 10.33 and Eq. 10.42, we see that for (a) an arbitrary value of the orbital angular momentum and (b) for potentials that fall faster than the Coulomb potential, the asymptotic solution to Eq. 10.14 is

$$y_\ell(k,r) \underset{r\to\infty}{\to} c_1\cos\left(kr - \frac{\ell\pi}{2}\right) + c_2\sin\left(kr - \frac{\ell\pi}{2}\right), \quad (10.43a)$$

i.e., $y_\ell(k,r) \underset{r\to\infty}{\to} A_\ell(k)\sin\left(kr - \frac{\ell\pi}{2} + \delta_\ell(k)\right),$ (10.43b) wherein $\delta_\ell(k)$ is a k-dependent (i.e., energy-dependent) phase shift that may be different for different orbital angular momentum quantum numbers. The solution to the radial Schrödinger equation (Eq. 10.11) therefore is

$$R_{\varepsilon\ell}(r; r\to\infty) \sim \frac{y_{\varepsilon\ell}(r)}{r} \sim A_\ell(k)\frac{1}{r}\sin\left(kr - \frac{\ell\pi}{2} + \delta_\ell(k)\right), \tag{10.44a}$$

$$\text{i.e., } R_{\varepsilon\ell}(r; r\to\infty) \simeq A_\ell(k)\frac{1}{2ir}\left\{\exp\left(i\left(kr - \frac{\ell\pi}{2} + \delta_\ell(k)\right)\right) - \exp\left(-i\left(kr - \frac{\ell\pi}{2} + \delta_\ell(k)\right)\right)\right\}. \tag{10.44b}$$

Comparing Eq. 10.34 with Eqs. 10.44a,b we see that the complete effect of the potential $V(r)$ is to produce the phase shift $\delta_\ell(k)$ in the radial solution, relative to that for $V = 0$. Therefore, $\delta_\ell(k)$ is called as the scattering phase shift. *It has all the information about the scattering potential.* Using the asymptotic radial wavefunction, the *time-independent* part of the solution (Eq. 10.7) to the scattering problem therefore is

$$\psi^{SE}_{\vec{k}_i}(\vec{r}; r\to\infty) = \sum_{\ell=0}^{\infty}\sum_{m=-\ell}^{\ell}\left[\begin{array}{l} c_{\ell m}(k)\, Y_{\ell m}(\hat{e}_r)\times \\ A_\ell(k)\frac{1}{2ir}\left\{\exp\left(i\left(kr - \frac{\ell\pi}{2} + \delta_\ell(k)\right)\right) - \exp\left(-i\left(kr - \frac{\ell\pi}{2} + \delta_\ell(k)\right)\right)\right\}\end{array}\right]. \tag{10.45}$$

Funquest: On what does the scattering phase shift depend parametrically? What is it independent of?

The angular basis set employed consists of the spherical harmonics, and the linearly independent base pair that has been employed for the radial function is $\{\exp(+ikr), \exp(-ikr)\}$. The phenomenological solution (Eq. 10.8, Eq. 10.10) to the scattering problem can also be expanded in essentially the same linearly independent basis set. We shall therefore be able to compare the coefficients of corresponding base functions and thereby determine $f(\hat{\Omega})$.

The scattered part ψ_{scat} (Eq. 10.10b) is already explicitly written in the basis $\{\exp(+ikr), \exp(-ikr)\}$. We now proceed to write the incident part ψ_{inc} (Eq. 10.10a) in the same basis. Toward that goal, we first take advantage of the symmetry of the incident wave about the Z-axis and introduce variables $\rho = kr$ and $\mu = \cos\theta$. We expand the incident wavefunction in the basis of product of the spherical harmonics (Appendix C) and the spherical Bessel functions, and take advantage of the symmetry about the Z-axis. Identifying the Legendre polynomial $P_\ell(\mu)$ as $\sqrt{\frac{4\pi}{2\ell+1}}Y_\ell^{m=0}(\hat{\Omega})$, we can write the incident wavefunction as

$$\exp(i\rho\mu) = \sum_{\ell=0}^{\infty}\sum_{m=-\ell}^{+\ell} c_{\ell,m}Y_\ell^m(\hat{r})\, j_\ell(\rho) = \sum_{\ell=0}^{\infty} a_\ell P_\ell(\mu) j_\ell(\rho) = \sum_{\ell=0}^{\infty} i^\ell(2\ell+1)P_\ell(\mu) j_\ell(\rho), \tag{10.46}$$

where we have used the result of the Solved Problem P10.1.

Using the result of Eq. 10.46 in the phenomenological solution (Eq. 10.8 and Eq. 10.10) to the scattering problem, we get

$$\psi^{+}_{\vec{k}_i}(\vec{r};r\to\infty)=A(k)\left[\begin{array}{l}\left\{\sum_{\ell=0}^{\infty}\sum_{m=-\ell}^{\ell}i^{\ell}\sqrt{4\pi(2\ell+1)}\dfrac{e^{i\left(kr-\frac{\ell\pi}{2}\right)}-e^{-i\left(kr-\frac{\ell\pi}{2}\right)}}{2ikr}Y_{\ell,m}(\theta,\phi)\delta_{m0}\right\}+\\ \dfrac{f(\hat{\Omega})}{r}e^{ikr}\end{array}\right], \tag{10.47a}$$

$$\text{i.e., }\psi^{+}_{\vec{k}_i}(\vec{r};r\to\infty)=A(k)\left[\left\{\sum_{\ell=0}^{\infty}\frac{e^{i\left(kr-\frac{\ell\pi}{2}\right)}-e^{-i\left(kr-\frac{\ell\pi}{2}\right)}}{2ikr}(2\ell+1)P_{\ell}(\mu)i^{\ell}\right\}+\frac{f(\hat{\Omega})}{r}e^{ikr}\right]. \tag{10.47b}$$

The above relationship expresses the solution to the scattering problem in the basis $\{\exp(+ikr),\exp(-ikr)\}$, just as in Eq. 10.45. The coefficients of the basis functions in Eq. 10.45 and Eq. 10.47 must be respectively equal to each other. Equating the coefficients of $\exp(-ikr)$ we get

$$A(k)\frac{(-1)\times\exp\left(-i\left(-\frac{\ell\pi}{2}\right)\right)i^{\ell}\sqrt{4\pi(2\ell+1)}Y_{\ell,m}(\theta,\varphi)\delta_{m0}}{2ikr} \tag{10.48a}$$

$$=-c_{lm}(k)A_{\ell}(k)\frac{\exp\left(-i\left(-\frac{\ell\pi}{2}+\delta_{\ell}(k)\right)\right)}{2ir}Y_{\ell m}(\hat{e}_r),$$

$$\text{and hence } c_{\ell m}(k)=\frac{A(k)}{kA_{\ell}(k)}i^{\ell}\sqrt{4\pi(2\ell+1)}\exp(i\delta_{\ell}(k))\delta_{m0}. \tag{10.48b}$$

We can now use the above value of $c_{\ell m}$ while equating the coefficients of $\exp(+ikr)$. While doing so, we also note that $Y_{\ell}^{m=0}(\hat{\Omega})$ is nothing but $\sqrt{\frac{2\ell+1}{4\pi}}P_{\ell}(\mu)$ and that $i^{\ell}e^{-i\frac{\ell\pi}{2}}$ is unity. We therefore get

$$\sum_{\ell=0}^{\infty}\left[\begin{array}{l}\dfrac{\cancel{A(k)}}{k\cancel{A_{\ell}(k)}}i^{\ell}(2\ell+1)\exp(i\delta_{\ell}(k))P_{\ell}(\mu)\times\\ \cancel{A_{\ell}(k)}\dfrac{1}{2ir}\exp\left(i\left(-\dfrac{\ell\pi}{2}+\delta_{\ell}(k)\right)\right)\end{array}\right]-\cancel{A(k)}\left\{\sum_{\ell=0}^{\infty}\frac{e^{i\left(-\frac{\ell\pi}{2}\right)}}{2ikr}(2\ell+1)P_{\ell}(\mu)i^{\ell}\right\}=\cancel{A(k)}\frac{f(\hat{\Omega})}{r},$$

$$\text{i.e., }\sum_{\ell=0}^{\infty}\left[\frac{(2\ell+1)}{2ikr}\exp(i2\delta_{\ell}(k))P_{\ell}(\mu)i^{\ell}\exp\left(-i\frac{\ell\pi}{2}\right)\right]-\left\{\sum_{\ell=0}^{\infty}\frac{i^{\ell}\exp\left(-i\frac{\ell\pi}{2}\right)}{2ikr}(2\ell+1)P_{\ell}(\mu)\right\}=\frac{f(\hat{\Omega})}{r}$$

$$\text{or, } f(k,\theta) = \frac{1}{2ik}\sum_{\ell=0}^{\infty}(2\ell+1)\left[\exp\left(2i\delta_\ell(k)\right)-1\right]P_\ell(\cos\theta), \tag{10.49a}$$

$$\text{or, } f(k,\theta) = \frac{1}{k}\sum_{\ell=0}^{\infty}(2\ell+1)\exp\left(i\delta_\ell(k)\right)\sin\delta_\ell(k)P_\ell(\cos\theta), \tag{10.49b}$$

$$\text{or, } f(k,\theta) = \sum_{\ell=0}^{\infty}(2\ell+1)\left[\frac{\exp\left(2i\delta_\ell(k)\right)-1}{2ik}\right]P_\ell(\cos\theta), \tag{10.49c}$$

$$\text{and } a_\ell(k) = \frac{\exp\left(2i\delta_\ell(k)\right)-1}{2ik}, \tag{10.50}$$

which is called as the partial wave amplitude corresponding to the ℓth partial wave,

$$\text{and } S_\ell(k) = \exp\left(2i\delta_\ell(k)\right), \tag{10.51}$$

which is called as the S-matrix element (see below the Solved Problems P10.7 and P10.8). It is the diagonal element in the angular momentum representation of what is called as the scattering operator. This formalism is based on methods developed earlier by Rayleigh. The above result is known as the Faxen–Holtzmark equation for the energy and polar-angle-dependent scattering amplitude $f\left(\hat{\Omega}\right)$, which was introduced in Eq. 10.10. It is independent of the azimuthal angle φ due to the axial symmetry of the incident wavefunction. The phenomenological solution in Eq. 10.8 and Eq. 10.10 that we used to obtain the Faxen–Holtsmark relation for the scattering amplitude is referred to as the scattering solution as per *outgoing wave boundary condition*. This terminology will become clear in the next section.

10.2 Time-Reversal Symmetry – *Ingoing* Wave and *Outgoing* Wave Boundary Conditions

The role of symmetry is fundamental in interpreting the laws of nature. We have discussed salient features of this in Appendix A. That the same laws that govern nuclear and particle physics apply to atomic physics is only a striking expression of the unifying beauty of physics. We have already seen that the electroweak theory that accounts for nuclear decay also accounts for parity violating low-energy transitions in atomic physics. We have discussed this in Section 8.2 of Chapter 8. In this section, we shall see how the time reversal symmetry connects the solution to quantum collision problem with that of atomic photoionization. This is really amazing, since the input to a quantum collision experiment is very different from that to an atomic photoionization experiment. In particular, we have an electron and a positively charged ion in the initial state of an elastic or non-elastic electron-ion scattering. The final state would also have an electron and an ion, in the ground or an excited state. In the photoionization of a *neutral atom* (or photodetachment of a negative ion, such as Li^-), a *photon* is absorbed, and an electron that initially belonged to the atom (or ion) in a *bound state* is knocked out into the continuum. Note that the ingredients of the two processes - *ion-electron collision and photoionization* - are different, but in their final states *both* the processes have an ion and an electron. The electron knocked out of an atom/ion in photoionization/photodetachment is referred to as the photoelectron. It escapes to the asymptotic region as a free particle. Figures 10.2a,b and 10.4 show that the final state in scattering is the same as that of atomic photoionization (or photodetachment of a negative ion), though the initial states are different; even the *constituents* in their initial states are different. In the final state of photoionization (photodetachment) one has a photoelectron that propagates *away* from the reaction zone as a free particle. In the initial state of this process, it exists only as an integral part of the atom in which it is in a bound state; it has no existence as an independent electron in the initial state. Figure 10.4 shows that the photoelectron in the photoionization (photodetachment) process can be regarded as the time-reversed electron in the scattering process. In order to map the photoionization (photodetachment) process to the scattering phenomenology, one can, however, depict the electron's *initial* state in the former process by *ingoing* spherical waves (Fig. 10.4, rightmost panel). It shows that the electron flux through a spherical surface in any given ingoing direction exactly cancels that from the opposite side, representing thereby that there is no free electron in the initial state of photoionization (photodetachment). Photoionization is therefore often referred to as *half-scattering*. We shall discuss the boundary conditions as applied to photoionization. However, with regard to its quantum mechanical formalism, the discussion would also describe photodetachment of a negative ion.

Funquest: In atomic photoionization, do you think that the photoelectron is knocked out of an atom instantly when a photon (whose energy is more than the binding energy of the electron) is absorbed by the atom?

From Eq. 10.46, the incident wavefunction $\exp(ikz)$ is

$$\psi_{inc}(\vec{r};r\to\infty) = \sum_{\ell} i^{\ell}(2\ell+1)P_{\ell}(\cos\theta)\ \frac{e^{i\left(kr-\frac{\ell\pi}{2}\right)} - e^{-i\left(kr-\frac{\ell\pi}{2}\right)}}{2ikr}, \tag{10.52a}$$

$$\text{i.e., } \psi_{inc}(\vec{r};r\to\infty) = \sum_{\ell}(2\ell+1)\ P_{\ell}(\cos\theta)\frac{e^{ikr} - e^{-ikr}e^{+i\frac{\ell\pi}{2}}e^{+i\frac{\ell\pi}{2}}}{2ikr}. \tag{10.52b}$$

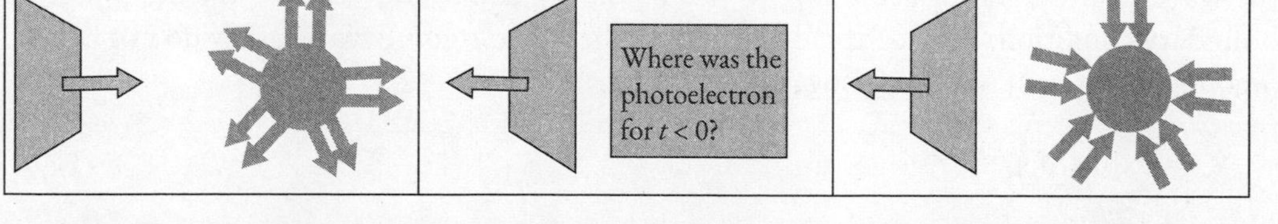

Fig. 10.4 The left-most panel depicts a scattering experiment in which a monoenergetic beam of incident particles impinging on a target is scattered away in various directions by the target. The right-most panel shows a photoelectron that escapes away to the asymptotic region as a free particle, but since it did not exist as a free particle in the initial state, it must be depicted by *incoming* spherical waves, in which the incoming flux from any direction is canceled from that coming in from the opposite side.

Using now (a) $e^{i\ell\pi} = \left(e^{i\pi}\right)^{\ell} = (-1)^{\ell}$, (10.53a)

and (b) $P_{\ell}(\cos\theta)(-1)^{\ell} = P_{\ell}(-\cos\theta)$, (10.53b) we may write

$$\psi_{inc}\left(\vec{r};r\to\infty\right)=\frac{1}{2ikr}\sum_{\ell}(2\ell+1)\left[P_{\ell}(\cos\theta)e^{ikr}-P_{\ell}(-\cos\theta)e^{-ikr}\right],\tag{10.54a}$$

i.e.,

$$\psi_{inc}\left(\vec{r};r\to\infty\right)=\frac{1}{2ikr}\sum_{\ell}(2\ell+1)\,P_{\ell}\left(\cos\theta\right)\left[e^{ikr}-(-1)^{\ell}e^{-ikr}\right].\tag{10.54b}$$

Likewise, we rewrite the asymptotic form of the total wavefunction (Eq.10.45) as

$$\psi_{\vec{k}_i}^{SE}\left(\vec{r};r\to\infty\right)=\sum_{\ell}c_{\ell}i^{\ell}(2\ell+1)\,P_{\ell}(\cos\theta)\frac{\sin\left(kr-\dfrac{\ell\pi}{2}+\delta_{\ell}\right)}{kr},\tag{10.55a}$$

or as $\psi_{\vec{k}_i}^{SE}\left(\vec{r};r\to\infty\right)=\dfrac{1}{2ikr}\sum_{\ell}c_{\ell}(2\ell+1)\left[P_{\ell}\left(\cos\theta\right)e^{+i(\delta_{\ell}+kr)}-P_{\ell}\left(-\cos\theta\right)e^{-i(\delta_{\ell}+kr)}\right].$ (10.55b)

in which δ_{ℓ} is the phase-shift caused by the scattering potential.

The scattered wave ψ_{scat} is essentially the difference

$$\psi_{scat}\left(\vec{r};r\to\infty\right)=\psi_{\vec{k}_i}^{SE}\left(\vec{r};r\to\infty\right)-\psi_{inc}\left(\vec{r};r\to\infty\right),\tag{10.56}$$

and this requires making a specific choice of the normalization c_{ℓ} in Eqs. 10.55a,b since we must have

$$\frac{f\left(\hat{\Omega}\right)}{r}\exp\left(ikr\right)=\left\{\begin{array}{l}\dfrac{1}{2ikr}\sum_{\ell}c_{\ell}(2\ell+1)\left[P_{\ell}\left(\cos\theta\right)e^{+i(\delta_{\ell}+kr)}-P_{\ell}\left(-\cos\theta\right)e^{-i(\delta_{\ell}+kr)}\right]\\ -\dfrac{1}{2ikr}\sum_{\ell}(2\ell+1)\,P_{\ell}\left(\cos\theta\right)\left[e^{ikr}-(-1)^{\ell}e^{-ikr}\right]\end{array}\right\}.\tag{10.57}$$

The left-hand side has only a component along $\exp(ikr)$. Hence the constant c_{ℓ} must have such a value that would make the coefficient of $\exp(-ikr)$ vanish. Normalization of the wavefunction is thus according to this boundary condition, which is the outgoing wave boundary condition. The scattered waves are spherical *outgoing* waves; they do not have any *ingoing* component. On choosing

$$c_{\ell}=\exp\left(+i\delta_{\ell}\right)\tag{10.58}$$

for this boundary condition, the scattered wavefunction is given by

$$\psi_{scat}\left(\vec{r};r\to\infty\right)\to\left(\psi_{\vec{k}_i}^{SE}-\psi_{inc}\right)=\frac{e^{ikr}}{r}\sum_{\ell}(2\ell+1)\,P_{\ell}\left(\cos\theta\right)\left(\frac{\exp\left(2i\delta_{\ell}\right)-1}{2ik}\right)\tag{10.59}$$

with the scattering amplitude given by Eq. 10.49. Accordingly, the time-dependent solution to the scattering problem is given by

$$\psi_T^+(\vec{r},t;r\to\infty)=\exp\left(i(kz-\omega t)\right)+\frac{\exp\left(i(kr-\omega t)\right)}{r}\sum_\ell (2\ell+1)\,P_\ell(\cos\theta)\left(\frac{\exp(2i\delta_\ell)-1}{2ik}\right). \tag{10.60}$$

We have placed a superscript "+" on the wavefunction to indicate the choice made in Eq. 10.58 corresponding to the outgoing wave boundary condition. As mentioned earlier, photoionization is *time-reversed* scattering, and is commonly referred to as "half-scattering." The term *motion reversal* rather than *time reversal* was preferred by Wigner, since the time-reversal relationship involves complex conjugation of the wavefunction in addition to $t\to -t$ under the *anti-unitary operator* Θ for time reversal symmetry, as discussed in Appendix A. The role of the time-reversal symmetry that enables interpretation of photoionization and photodetachment in terms of half scattering would become further emphasized in the next section in the framework of the *reciprocity theorem* (Eqs. 10.150a,b and Fig. 10.8). To apply the boundary condition according to the ingoing wave boundary condition, we therefore choose the complex conjugate of the choice that was made in Eq. 10.58, as per Eq. A.27a,b (from Appendix A):

$$c_\ell=\exp(-i\delta_\ell). \tag{10.61}$$

Using Eq. 10.61 in the total wavefunction from Eq. 10.55b, we get

$$\psi^-(\vec{r};r\to\infty)=$$

$$\frac{1}{2ikr}\sum_\ell \exp(-i\delta_\ell)(2\ell+1)\left[P_\ell(\cos\theta)\exp\left(i(kr+\delta_\ell)\right)-P_\ell(-\cos\theta)\exp\left(-i(kr+\delta_\ell)\right)\right], \tag{10.62a}$$

i.e.,

$$\psi^-(\vec{r};r\to\infty)=\frac{1}{2ikr}\sum_\ell (2\ell+1)\left[P_\ell(\cos\theta)\exp(ikr)-P_\ell(-\cos\theta)\exp\left(-i(kr+2\delta_\ell)\right)\right]. \tag{10.62b}$$

A superscript "-" is now placed on the wavefunction to indicate the *ingoing wave* boundary condition. On subtracting now the incident plane wave (Eq. 10.54a) from the total wavefunction, we get the *scattered part*

$$\psi_{scat}(\vec{r};r\to\infty)=-\frac{\exp(-ikr)}{r}\left\{\frac{1}{2ik}\sum_\ell (2\ell+1)\left[\exp(-2i\delta_\ell)-1\right]P_\ell(-\cos\theta)\right\}. \tag{10.63}$$

The total wavefunction written as a sum of the incident plane wave and the scattered part is

$$\psi_{Tot}(\vec{r};r\to\infty)=\exp\left(+i\vec{k}\bullet\vec{r}\right)-\frac{\exp(-ikr)}{r}\left\{\frac{1}{2ik}\sum_\ell (2\ell+1)\left[\exp(-2i\delta_\ell)-1\right]P_\ell(-\cos\theta)\right\} \tag{10.64}$$

We now make use of the time-reversal operator (Appendix A, Section 3), for which

$$\Theta\vec{r}\Theta^{-1}=\vec{r}\quad\text{(commutation with the position operator)} \tag{10.65}$$

and

$$\theta \vec{p} \theta^{-1} = -\vec{p} \quad \text{(anti-commutation with the momentum operator).} \tag{10.66}$$

As discussed in Appendix A, time reversal involves complex conjugation and

$$t \to -t. \tag{10.67}$$

Photoionization/photodetachment corresponds to *half* of time-reversed *scattering* (Fig. 10.2a,b). Note the similarity with nuclear β-decay in which the electron escaping a decaying nucleus does not exist as a *free* particle in the initial state. The photoelectron's direction of escape defines the orientation of the Z-axis, now reversed, on account of Eq. 10.66. Fig. 10.5a,b,c shows the *defining direction* in photoionization. Accordingly, Eq. 10.64 should be modified by noting that

$$P_\ell(-\cos\theta) = P_\ell(\cos(\theta + \pi)), \tag{10.68}$$

in correspondence with a reversal of the Z-axis.

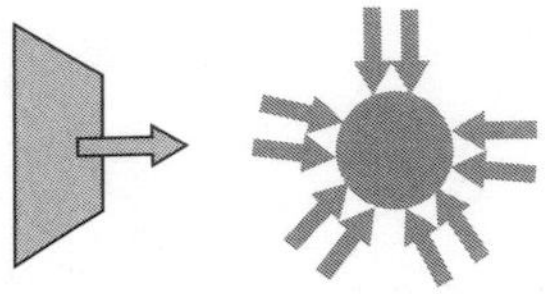

Fig. 10.5a The Z-axis is assumed pointed from left to right. This figure shows the plane wave moving along the positive Z-axis.

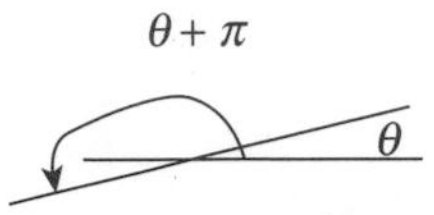

Fig. 10.5b The polar angle must be measured with respect to positive Z-axis, which in Fig. 10.5c is shown to be along the exit channel defined by the direction in which the photoelectron escapes.

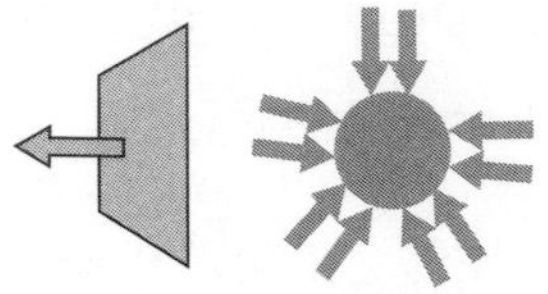

Fig. 10.5c This figure shows the plane wave moving from right to left.
$\cos(\theta + \pi) = -\cos\theta$
$P_\ell(-\cos\theta) = P_\ell(\cos(\theta + \pi))$

This reversal of the Z-axis comes from the fact that in a collision experiment, it is the *entrance channel* that is *unique*, and the quantum description considers the direction of the incident plane wave to be moving along the positive Z-axis. In photoionization, it is the *exit channel* which is unique hence all orientations must be made with reference to this direction, which is the direction in which the photoelectron escapes. This direction provides the unique reference axis in the photoionization process with respect to which the polar angle is measured. Using now the dynamical phase $\exp(-i\omega t)$ corresponding to a stationary state wavefunction, along with Eq. 10.68, the ingoing wavefunction can be obtained from Eq. 10.64 to get:

$$\psi_{Tot}^{\ *} = \exp\left(+i\left(\vec{k}\bullet\vec{r} - \omega t\right)\right) + \frac{\exp(-i(kr + \omega t))}{r}\left\{\frac{1}{2ik}\sum_{\ell=0}^{\infty}(2\ell+1)\left[\exp(2i\delta_\ell) - 1\right]P_\ell(\cos(\theta))\right\} \tag{10.69}$$

with the understanding that the polar angle θ is now measured with reference to the Z-axis along which the photoelectron escapes. The terms $\exp\left(i\vec{k}\bullet\vec{r}\right)$ and the dynamical phase $\exp\left(-i\frac{E}{\hbar}t\right)$ of a stationary state are invariant under complex-conjugation *and* $t \to -t$. The first term in Eq. 10.69 corresponds to the photoelectron escaping along the exit direction which defines the $+Z$ axis, with reference to which the polar angle is measured. The second term has a spherical *ingoing* wave, travelling towards the centre. It corresponds to flux travelling in all opposite directions, cancelling each other.

The choice (see Eq. 10.61) $c_\ell = \exp\left(-i\delta_\ell(k)\right)$ cancels the outgoing spherical waves in the scattered solution, as must happen in the photoionization event, the exit channel being unique. Not surprisingly, the coefficient chosen to represent the boundary condition for photoionization is the complex conjugate of that employed to describe the collision process. Figure 10.5a–c depicts the time-reversed (rather, motion reversed) relationship between the solution for quantum collision and that for photoionization effected by the time reversal operator Θ (see Appendix A, Section 3). This result has important consequences in the study of photoionization and also photodetachment of a negative ion. Ingoing wave boundary conditions must be employed to determine photoionization/photodetachment transition amplitudes. These are important for the investigations of photoelectron angular distribution and also in the study of time delay in photoionization. Since photoionization is only *half* scattering, a corresponding adjustment is made in the expression of Eisenbud–Wigner–Smith (EWS) time delay (or 'Wigner time delay') in photoionization [6–8], discussed in the next Section.

10.3 Optical Theorem, Reciprocity Theorem, and Wigner Time Delay

A fundamental theorem in the theory of quantum collisions bears a strong similarity with optical phenomena; it relates the scattering cross section corresponding to the removal of incident flux of the projectiles by an obstacle that would have *otherwise* propagated in the *forward* direction. This theorem, known as the optical theorem, enunciates a conservation principle about the intensity redistribution brought about on account of scattering by a target potential of an incident beam of probing projectiles. The intensity reorganization due to scattering is quantified in terms of the flux redistribution obtained from the probability current density vector (Eqs. 3.3a,b in Chapter 3), given by

$$\vec{j}(\vec{r}) = \frac{\hbar}{2mi}\left[\psi^*(\vec{r})\vec{\nabla}\psi(\vec{r}) - \psi(\vec{r})\vec{\nabla}\psi^*(\vec{r})\right] = \mathrm{Re}\left\{\frac{\hbar}{mi}\,\psi^*(\vec{r})\vec{\nabla}\psi(\vec{r})\right\}. \tag{10.70}$$

Component of the current density vector along a direction $\left(\hat{e}_r\right)$ is given by

$$\vec{j}(\vec{r})\cdot\hat{e}_r = \mathrm{Re}\left\{\frac{\hbar}{mi}\,|A(k)|^2\left(e^{-i\vec{k}_i\bullet\vec{r}} + \frac{f^*\left(\hat{\Omega}\right)e^{-ikr}}{r}\right)\frac{\partial}{\partial r}\left(e^{i\vec{k}_i\bullet\vec{r}} + \frac{f\left(\hat{\Omega}\right)e^{ikr}}{r}\right)\right\}, \tag{10.71a}$$

where we have used Eq. 10.8 for the wavefunction corresponding to the asymptotic scattering solution. It can be broken down into three parts, corresponding to the incident component, the scattered component, and the interference term as

$$\vec{j}(\vec{r})\cdot\hat{e}_r = \left\{\vec{j}_{\text{inc}}(\vec{r}) + \vec{j}_{\text{scat}}(\vec{r}) + \vec{j}_{\text{int}}(\vec{r})\right\}\cdot\hat{e}_r. \tag{10.71b}$$

Let us first examine the term corresponding to the incident beam. It is given by

$$\vec{j}_{inc}(\vec{r}) = \text{Re}\left\{\frac{\hbar}{mi}A(k)^* e^{-ik_i z}\left(\hat{e}_x\frac{\partial}{\partial x} + \hat{e}_y\frac{\partial}{\partial y} + \hat{e}_z\frac{\partial}{\partial z}\right)A(k)e^{ik_i z}\right\} = |A(k)|^2\frac{\hbar\vec{k}_i}{m} = |A(k)|^2\vec{v}_i, \tag{10.72}$$

and its flux through an elemental area δS orthogonal to the direction of incidence is

$$\delta\phi_{\delta S}^{inc} = \vec{j}(\vec{r})\cdot\delta S\hat{e}_z = |A(k)|^2 v_i\delta S = |A(k)|^2\frac{\delta z}{\delta t}\delta S = |A(k)|^2 v_i\delta S = |A(k)|^2\frac{\delta V}{\delta t}. \tag{10.73}$$

If we choose normalization corresponding to unit probability density, then $A(k) = 1$ so that the incident wavefunction is $\psi_{inc} = e^{i\vec{k}_i\cdot\vec{r}}$, for which $\psi^*\psi = 1$. It corresponds to a probability current density vector of $\vec{j}_{inc}(\vec{r}) = \frac{\hbar\vec{k}_i}{m} = \vec{v}_i$. From Fig. 10.6, we see that it would provide for a density of one particle per unit volume; i.e., one particle crossing unit area in unit time at velocity $\vec{v}_i$. For arbitrary normalization, the incident flux per unit area is

the $\delta\phi_{inc} = |A(k)|^2 v_i$. (10.74)

Our discussion in this section follows that in Reference [3] closely even if not wholly. The current density vector term corresponding to the scattered wavefunction is

$$\vec{j}_{scat}(\vec{r}) = \text{Re}\left[|A(k)|^2\frac{\hbar}{mi}\left\{\frac{f^*(\hat{\Omega})}{r}e^{-ikr}\right\}\left(\hat{e}_r\frac{\partial}{\partial r} + \hat{e}_\theta\frac{1}{r}\frac{\partial}{\partial\theta} + \hat{e}_\varphi\frac{1}{r\sin\theta}\frac{\partial}{\partial\varphi}\right)\left\{\frac{f(\hat{\Omega})}{r}e^{ikr}\right\}\right]. \tag{10.75}$$

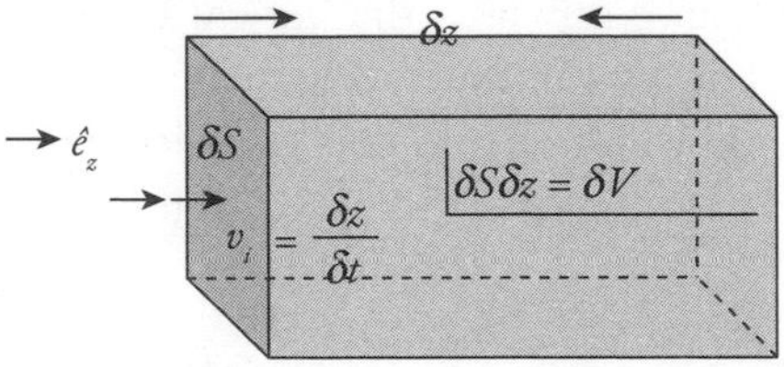

Fig. 10.6 Estimation of the correspondence between the probability current density and the density of projectile particles.

Now, the terms $\frac{1}{r}\frac{\partial}{\partial\theta}\left\{\frac{f(\hat{\Omega})}{r}e^{ikr}\right\}$ and $\frac{1}{r\sin\theta}\frac{\partial}{\partial\varphi}\left\{\frac{f(\hat{\Omega})}{r}e^{ikr}\right\}$ both have a term in $1/r$ and another in $1/r^2$, of which we need to only retain the former since the region of interest for our analysis is the asymptotic region, $r\to\infty$. Retaining only the leading terms, we have

$$\vec{j}_{scat}(\vec{r};r\to\infty)\simeq \text{Re}\left[|A(k)|^2\frac{\hbar}{mi}\left\{\frac{f^*(\hat{\Omega})}{r}e^{-ikr}\right\}\hat{e}_r(ik)\left\{\frac{f(\hat{\Omega})}{r}e^{ikr}\right\}\right]=|A(k)|^2\frac{\hbar k}{m}\frac{|f(\hat{\Omega})|^2}{r^2}\hat{e}_r. \tag{10.76}$$

The scattered flux through the elemental area $\delta S\hat{e}_r$ (Fig. 10.1d) is given by

$$\delta\phi_{scat,through\ \delta S}=\vec{j}_{scat}(\vec{r})\cdot\delta S\hat{e}_r\approx|A(k)|^2\frac{\hbar k}{m}\frac{|f(\hat{\Omega})|^2}{r^2}\hat{e}_r\cdot\delta S\hat{e}_r=|A(k)|^2\frac{\hbar k}{m}\frac{|f(\hat{\Omega})|^2}{r^2}r^2\delta\Omega. \tag{10.77}$$

A measure of the scattering probability is given by the ratio of the term for the scattered flux, given in Eq. 10.77 to the incident flux, given in Eq. 10.74. We see that:

$$\delta\sigma\ (through\ \delta S)=\frac{\delta\phi^{scattered}_{\delta S=r^2\delta\Omega}}{\delta\phi^{incident}_{unit\ area}}=\frac{\left\{|A(k)|^2\frac{\hbar k}{m}\frac{|f(\hat{\Omega})|^2}{r^2}r^2\delta\Omega\right\}}{\left\{|A(k)|^2 v_i\right\}}=|f(\hat{\Omega})|^2\delta\Omega. \tag{10.78}$$

From Eq. 10.5, we know that this quantity must have the dimension of area; it is the scattering cross section. From the above relation, we see that the differential scattering cross section per unit angle is given by

$$\frac{d\sigma}{d\Omega}=\lim_{\delta\Omega\to 0}\frac{\delta\sigma}{\delta\Omega}=|f(\hat{\Omega})|^2. \tag{10.79a}$$

This definition of *differential cross section* is independent of the normalization $A(k)$. We have used this expression of the differential cross section in Chapter 5 in the context (Eqs. 5.167a,b and Eqs. 5.169a,b) of the continuum eigenfunctions of the Coulomb potential, reconnecting with which would be instructive. The total scattering cross section is obtained from integration over the solid angle:

$$\sigma_{tot}=\iint\frac{d\sigma}{d\Omega}d\Omega=\int_{\theta=0}^{\pi}\int_{\varphi=0}^{2\pi}|f(\hat{\Omega})|^2\sin\theta d\theta d\varphi. \tag{10.79b}$$

Funquest: What is the significance of the term 'differential' in the definition of the 'differential cross section'?

It is often useful to analyze the differential scattering cross section and the total cross section in terms of the partial wave scattering phase shifts for the different orbital angular momentum quantum numbers. This scheme is illustrated below in the Solved Problem P10.2. The utility of this procedure is appreciated in the interpretation of the Ramsauer–Townsend effect, which we visit below in the Solved Problem P10.5. We now proceed to examine the connection of the scattering cross section with the flux along the direction $\hat{\Omega}$ resulting from the interference term. To determine this contribution, we examine the component of the probability current density vector coming from the part $\vec{j}_{\text{int}}$ (from Eq. 10.71b) along the unit vector $\hat{e}_r$:

$$\vec{j}_{\text{int}}(\vec{r})\cdot\hat{e}_r=\text{Re}\left[\frac{\hbar}{mi}|A(k)|^2\left\{e^{-i\vec{k}_i\cdot\vec{r}}\frac{\partial}{\partial r}\left(\frac{f(\hat{\Omega})e^{ikr}}{r}\right)+\frac{f^*(\hat{\Omega})e^{-ikr}}{r}\frac{\partial}{\partial r}\left(e^{i\vec{k}_i\cdot\vec{r}}\right)\right\}\right]. \tag{10.80a}$$

Considering the asymptotic region $r \to \infty$ and ignoring terms of $O\left(\frac{1}{r^2}\right)$ with respect to those of $O\left(\frac{1}{r}\right)$, we get

$$\vec{j}_{\text{int}}(\vec{r})\cdot\hat{e}_r = \text{Re}\left[\frac{\hbar}{mi}\,|A(k)|^2\left\{e^{-i\vec{k}_i\cdot\vec{r}}(ik)\frac{f(\hat{\Omega})e^{ikr}}{r} + \frac{f^*(\hat{\Omega})e^{-ikr}}{r}(ik\cos\theta)e^{i\vec{k}_i\cdot\vec{r}}\right\}\right], \quad (10.80\text{b})$$

$$\text{i.e., } \vec{j}_{\text{int}}(\vec{r})\cdot\hat{e}_r = \text{Re}\left[\frac{\hbar k}{m}\,|A(k)|^2\left\{\frac{f(\hat{\Omega})e^{ikr(1-\cos\theta)}}{r} + \cos\theta\frac{f^*(\hat{\Omega})e^{-ikr(1-\cos\theta)}}{r}\right\}\right]. \quad (10.80\text{c})$$

Now, a realistic physical incident beam cannot be strictly monoenergetic; it would contain waves whose wavelengths spread over a certain range corresponding to which an integration over the wave number from k to $k + \Delta k$ would be necessitated. The amplitudes of the wave $|A(k)|$ may not, however, be much different for these different wavenumbers, so the primary integration over k in Eq. 10.80c is

$$\int_k^{k+\Delta k} e^{\pm ik'r(1-\cos\theta)}dk' = \left.\frac{e^{\pm ik'r(1-\cos\theta)}}{\pm ir(1-\cos\theta)}\right]_k^{k+\Delta k} = \left(\frac{e^{\pm i(k+\Delta k)r(1-\cos\theta)} - e^{\pm ikr(1-\cos\theta)}}{\pm ir(1-\cos\theta)}\right). \quad (10.81)$$

In the asymptotic region $r \to \infty$, the above terms would go to zero, unless $\theta \simeq 0$. Essentially, we see that in the asymptotic region, which is specifically the one of our interest, the interference terms (Eqs. 10.80a–c) are of importance only in the neighborhood of *forward scattering*, i.e., for $\theta \simeq 0$. This is similar to an optical effect: when light falls on an obstacle, the intensity of light that would have continued to propagate in the forward direction is diminished; it is scattered away by the obstacle in various directions. Now, for stationary states, the probability density of the wavefunction is independent of time, hence from the equation of continuity (Eq. 3.2 in Chapter 3) we get

$$0 = \frac{\partial\rho}{\partial t} = \iiint\left(\vec{\nabla}\cdot\vec{j}(\vec{r})\right)dV = \oiint\vec{j}(\vec{r})\cdot\hat{e}_r r^2 d\Omega = \oiint\left\{\begin{matrix}\vec{j}_{inc}(\vec{r}) + \\ \vec{j}_{scat}(\vec{r}) + \vec{j}_{\text{int}}(\vec{r})\end{matrix}\right\}\cdot\hat{e}_r r^2 d\Omega, \quad (10.82\text{a})$$

$$\text{i.e., } 0 = \oiint\left\{\vec{j}_{scat}(\vec{r}) + \vec{j}_{\text{int}}(\vec{r})\right\}\cdot\hat{e}_r r^2 d\Omega, \quad (10.82\text{b})$$

since the surface integral over a closed surface of the incident flux is evidently zero.

$$\text{Now, } \oiint\vec{j}_{scat}(\vec{r})\cdot\vec{dS} = \frac{\hbar k}{m}|A(k)|^2\oiint\left|f(\hat{\Omega})\right|^2 d\Omega = \frac{\hbar k}{m}|A(k)|^2\,\sigma_{total}, \quad (10.83\text{a})$$

and on recognizing that the integration over the polar angle would be of consequence only in the neighborhood of a small angle $\Delta\theta$ about $\theta = 0$, the direction of *forward scattering* (Fig. 10.7), we see that $\oiint\vec{j}_{\text{int}}(\vec{r})\cdot\hat{e}_r r^2 d\Omega = \int_{\theta=0}^{\theta=0+\Delta\theta}\sin\theta d\theta\int_{\varphi=0}^{2\pi}d\varphi\vec{j}_{\text{int}}(\vec{r})\cdot\hat{e}_r r^2,$

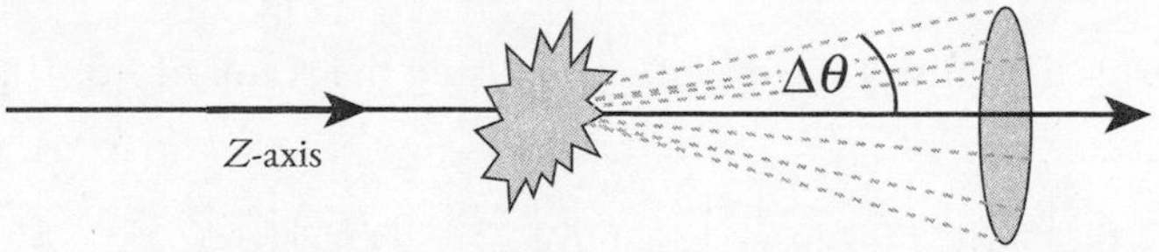

Fig. 10. 7 The scatterer scoops out flux from the incident beam that would have continued its travel along the forward direction and spreads it in all the directions.

$$\text{i.e.,} \oiint \vec{j}_{\text{int}}(\vec{r})\cdot\hat{e}_r r^2 d\Omega = 2\pi \int_{\theta=0}^{\theta=0+\Delta\theta} \sin\theta d\theta \text{Re}\left[\frac{\hbar k}{m}\left|A(k)\right|^2 \left\{\begin{matrix}\dfrac{f(\hat{\Omega})e^{ikr(1-\cos\theta)}}{r} + \\ \cos\theta\dfrac{f^*(\hat{\Omega})e^{-ikr(1-\cos\theta)}}{r}\end{matrix}\right\}\right] r^2. \tag{10.83b}$$

Using Eq. 10.83a and Eq. 10.83b in Eq. 10.82b, we see that the factor $\frac{\hbar k}{m}\left|A(k)\right|^2$ drops out, and we get

$$0 = \sigma_{tot} + 2\pi \int_{\theta=0}^{\theta=0+\Delta\theta} \sin\theta d\theta \text{Re}\left\{\frac{f(\hat{\Omega})e^{ikr(1-\cos\theta)}}{r} + \cos\theta\frac{f^*(\hat{\Omega})e^{-ikr(1-\cos\theta)}}{r}\right\} r^2. \tag{10.84a}$$

However, the scattering amplitude $f(\hat{\Omega})$ may not vary much from the *forward scattering amplitude* $f(\theta = 0)$ within the narrow range $\Delta\theta$ of the polar angle, we may therefore write

$$0 = \sigma_{tot} + 2\pi \text{Re}\left\{f(0)re^{ikr}\int_{\theta=0}^{\theta=0+\Delta\theta} \sin\theta d\theta e^{-ikr\cos\theta} + f^*(0)re^{-ikr}\int_{\theta=0}^{\theta=0+\Delta\theta} \sin\theta d\theta e^{+ikr\cos\theta}\right\}. \tag{10.84b}$$

The integration is over the polar angle θ in the range $0 < \theta \le \Delta\theta$ with $\Delta\theta \ne 0$, though small. It is easiest to carry out this integration by changing the variable $\theta \to \mu$ such that $\cos\theta = \mu$; $-\sin\theta d\theta = d\mu$. The result is

$$0 = \sigma_{total} + 2\pi \text{Re}\left\{f(0)re^{ikr}\left(\frac{e^{-ikr} - e^{-ikr\cos\Delta\theta}}{-ikr}\right) + f^*(0)re^{-ikr}\left(\frac{e^{ikr} - e^{ikr\cos\Delta\theta}}{ikr}\right)\right\}, \tag{10.85a}$$

$$\text{i.e.,}\ 0 = \sigma_{total} + 2\pi \text{Re}\left\{f(0)\left(\frac{1}{-ik} - \frac{e^{ikr(1-\cos\Delta\theta)}}{-ik}\right) + f^*(0)\left(\frac{1}{ik} - \frac{e^{-ikr(1-\cos\Delta\theta)}}{ik}\right)\right\}. \tag{10.85b}$$

We would need to integrate *further* over a narrow range of the wavenumbers corresponding to deviation from absolute monochromaticity of the incident beam of projectiles. The integrals required for the analysis of Eqs. 10.85a,b are

$$\int_k^{k+\Delta k} e^{\pm ik'r(1-\cos\Delta\theta)}dk' = \frac{e^{\pm i(k+\Delta k)r(1-\cos\Delta\theta)} - e^{\pm ikr(1-\cos\Delta\theta)}}{\pm ir(1-\cos\Delta\theta)};\ \Delta\theta \ne 0. \tag{10.86}$$

The numerator consists of the trigonometric functions sine and cosine, which both are of modulo unity. The denominator becomes large asymptotically (since $\Delta\theta \ne 0$).

The oscillatory terms are thus of the order $O\left(\frac{1}{r}\right)$. Using the result of Eq. 10.86 in Eq. 10.85b, we get

$$0 = \sigma_{total} + 2\pi \mathrm{Re}\left\{ f(0)\left[\frac{i}{k} + O\left(\frac{1}{r}\right)\right] + f^{*}(0)\left[\frac{-i}{k} + O\left(\frac{1}{r}\right)\right]\right\}. \tag{10.87}$$

Ignoring now the oscillatory terms, we get a result that is simple, but very powerful: The total scattering cross section (Eq. 10.79b) is given by

$$\sigma_{total} = \frac{4\pi}{k}\left[\mathrm{Im} f(0)\right], \tag{10.88}$$

called as the *Optical Theorem* (Bohr–Peierls–Placzek relation). It states that *the total scattering cross section is equal to* $4\pi/k$ *times the imaginary part of the forward scattering (complex) amplitude*. It is independent of the normalization $A(k)$. We have obtained this result in the framework of quantum collision physics, but the name of the theorem is borrowed from optics since it corresponds to the fact that the loss of intensity of light in the "shadow" of an obstacle in the forward direction results from the scattering of the incident beam by the target. Likewise, in a collision, we have the conservation of intensity of the incident beam of projectiles, measured in terms of the probability flux. The conservation principle abridged in the optical theorem is manifest in the interconnections between Eqs. 10.79a,b, Eq. 10.82a,b and Eq. 10.88: the flux in the incident beam that would have continued to move on in the *forward* (Fig. 10.7) direction is scattered by the target, and it is redistributed in various directions. The scattered intensity is not necessarily isotropic, though it may be so in special cases.

We have considered a slight departure from monochromaticity of the incident beam and carried out integration (Eq. 10.86) over wavenumbers in the range Δk. This was for the restricted purpose of unearthing the importance (Eqs. 10.84a,b, 10.85a,b) of forward scattering in our analysis. Nonetheless, we have *primarily* employed (in Eq. 10.72) scattering of a *monoenergetic* beam of incident projectiles. This was an idealization, and realistically we must employ a wavepacket that is a superposition of a number of different wavelengths. The *phase* velocities of different frequencies in the wavepacket may be different, resulting in a distortion of the shape of the wavepacket as it propagates at a *group* velocity. We shall denote the incident wavepacket by an integral in the three-dimensional k-space as:

$$\Phi_{inc}(\vec{r},t) = \frac{1}{(2\pi)^{3/2}} \iiint d^3\vec{k}\left\{A(\vec{k})\, e^{+i\left(\vec{k}\cdot\vec{r} - \omega(k)t\right)}\right\}, \tag{10.89a}$$

$$\text{i.e., } \Phi_{inc}(\vec{r},t) = \frac{1}{(2\pi)^{3/2}} \iiint d^3\vec{k}\left\{\left|A(\vec{k})\right| e^{i\alpha(\vec{k})} e^{+i\vec{k}\cdot\vec{r}} e^{-i\omega(\vec{k})t}\right\}, \tag{10.89b}$$

$$\text{or, } \Phi_{inc}(\vec{r},t) = \frac{1}{(2\pi)^{3/2}} \iiint d^3\vec{k}\left|A(\vec{k})\right| e^{i\beta(\vec{k})}, \tag{10.89c}$$

$$\text{with } \beta(\vec{k}) = \vec{k}\cdot\vec{r} - \omega(\vec{k})t + \alpha(\vec{k}), \tag{10.90}$$

and the k-dependent and time-independent *complex* coefficients $A(\vec{k})$ have been written in terms of its real amplitude, and phase, as

$$A(\vec{k}) = \left|A(\vec{k})\right| e^{i\alpha(\vec{k})}. \tag{10.91a}$$

It can be determined from the form of the wavepacket at $t = 0$:

$$A(\vec{k}) = \frac{1}{(2\pi)^{3/2}} \iiint d^3\vec{r} \left\{ \Phi_{incident}(\vec{r},0) e^{-i\vec{k}\cdot\vec{r}} \right\}, \tag{10.91b}$$

which is the wavefunction in the momentum space.

Consideration of scattering of an incident wavepacket (Eqs. 10.89a–c), rather than that of a monochromatic incident wave, *revalidates* the optical theorem (10.79a,b) as will be discussed below (Eq. 10.121a,b). In general, the k-dependence of the circular frequency $\omega(k)$ is not linear. Each individual wave $\frac{1}{(2\pi)^{3/2}} A(\vec{k}) e^{+i(\vec{k}\cdot\vec{r} - \omega(k)t)}$ in the wavepacket travels at the phase velocity

$$v_\phi = \frac{\omega(k)}{k} = \frac{E(k)/\hbar}{k} = \frac{(\hbar k)^2/2m}{\hbar k} = \frac{\hbar k}{2m}, \tag{10.92}$$

and the wavepacket travels at the group velocity

$$\left.\frac{d\omega(k)}{dk}\right]_{k_i} = \frac{\hbar k_i}{m} = v_i. \tag{10.93}$$

The normalization of the wavepacket is represented by the fact that

$$\iiint d^3\vec{r} \left|\Phi_{incident}(\vec{r},0)\right|^2 = 1 = \iiint d^3\vec{k} \left|A(\vec{k})\right|^2. \tag{10.94}$$

The oscillatory terms due to different phases would mutually destroy various different components in the incident wavepacket. Therefore, the condition for the incident wavepacket $\left|\Phi_{inc}(\vec{r},t)\right|$ to be large enough is that the phase β in Eq. 10.90 must *not change* much with respect to k (in the k-space), in order to avoid the possible cancellations. Simply stated, this condition is

$$\left[\vec{\nabla}_k \beta(k)\right]_{\vec{k}=\vec{k}_i} = \vec{0}, \tag{10.95}$$

where the gradient is in the k-space. In one dimension, this condition is:

$$0 = \left.\frac{d\beta(k)}{dk}\right]_{k_i} = z - \left[\frac{d\omega(k)}{dk}\right]_{k_i} t + \left[\frac{d\alpha(k)}{dk}\right]_{k_i},$$

i.e., $z = -\left[\frac{d\alpha(k)}{dk}\right]_{k_i} + \left[\frac{d\omega(k)}{dk}\right]_{k_i} t.$

Hence the condition for Eq. 10.95 to hold in three dimensions is

$$\vec{r}(t) = -\left[\vec{\nabla}_k \alpha\left(\vec{k}\right)\right]_{\vec{k}_i} + \left[\vec{\nabla}_k \omega\left(\vec{k}\right)\right]_{\vec{k}_i} t. \tag{10.96}$$

Equation 10.96 has exactly the same form as that of a classical kinematic equation $\left\{\vec{r}(t) = \vec{r}_0 + \vec{v}_i\left(t - t_0\right)\right\}$ for a particle that starts out its motion at time t_0 from a point $\vec{r}_0$ at a velocity $\vec{v}_i$ with

$$\vec{r}_0 = -\left[\vec{\nabla}_k \alpha\left(\vec{k}\right)\right]_{\vec{k}_i} \tag{10.97}$$

and from Eq. 10.93, we have $\vec{v}_i = \left[\vec{\nabla}_k \omega\left(\vec{k}\right)\right]_{\vec{k}_i} = \frac{\hbar \vec{k}_i}{m}$. (10.98)

In our analysis, $\vec{r}_0$ has its origin in the $\vec{k}$-dependence (Eq. 10.97) of the phase $\alpha\left(\vec{k}\right)$, and $\vec{v}_i$ originates from the $\vec{k}$-dependence of $\omega\left(\vec{k}\right)$. The functions $\alpha\left(\vec{k}\right)$ and $\omega\left(\vec{k}\right)$ in the vicinity of the mean incident wave vector $\vec{k}_i$ are given by

$$\alpha\left(\vec{k}\right) = \alpha\left(\vec{k}_i\right) + \left[\vec{\nabla}_k \alpha(\vec{k})\right]_{\vec{k}_i} \bullet \left(\vec{k} - \vec{k}_i\right) + \ldots = \alpha\left(\vec{k}_i\right) + \left[-\vec{r}_0\right] \bullet \left(\vec{k} - \vec{k}_i\right) + \ldots. \tag{10.99}$$

Now, $\omega(k) = \frac{E(k)}{\hbar} = \frac{\hbar^2 k^2}{\hbar 2m} = \frac{\hbar k^2}{2m}$,

and the group velocity of the wavepacket is $\vec{v}_i = \vec{\nabla}_k \omega\left(\vec{k}\right)\Big]_{\vec{k}_i} = \vec{\nabla}_k\left(\frac{\hbar k^2}{2m}\right)\Big]_{\vec{k}_i} = \frac{\hbar \vec{k}_i}{m}$. This makes $\vec{v}_i \bullet \vec{k}_i = \frac{\hbar k_i^2}{m} = 2\omega\left(k_i\right)$, and hence

$$\omega\left(\vec{k}\right) = \omega\left(\vec{k}_i\right) + \left[\nabla_k \omega\left(\vec{k}\right)\right]_{\vec{k}_i} \bullet \left(\vec{k} - \vec{k}_i\right) + . = \omega\left(\vec{k}_i\right) + \vec{v}_i \bullet \vec{k} - \vec{v}_i \bullet \vec{k}_i + . = -\omega\left(\vec{k}_i\right) + \vec{v}_i \bullet \vec{k} + .. \tag{10.100}$$

Using Eq. 10.99 and Eq. 10.100 in Eq. 10.89b, we get

$$\Phi_{inc}\left(\vec{r},t\right) = (2\pi)^{-3/2}\, e^{i\left(\vec{r}_0 \cdot \vec{k}_i + \alpha(k_i) + \omega(k_i)t\right)} \left[\iiint d^3\vec{k} \left|A\left(\vec{k}\right)\right| e^{i\vec{k}\cdot\left(\vec{r} - \vec{v}_i t - \vec{r}_0\right)}\right], \tag{10.101}$$

The phase $\exp\left\{i\left(\alpha\left(\vec{k}_i\right) + \vec{r}_0 \bullet \vec{k}_i\right)\right\}$ is inconsequential, dropping which we get

$$\Phi_{inc}\left(\vec{r},t\right) = (2\pi)^{-3/2}\, e^{i\omega(k_i)t} \left[\iiint d^3\vec{k} \left|A\left(\vec{k}\right)\right| e^{i\vec{k}\cdot\left(\vec{r} - \vec{v}_i t - \vec{r}_0\right)}\right], \tag{10.102}$$

The free wavepacket centered at the point $\vec{r}_0$ at time t_0 would therefore retain its shape at time t, centered at $\vec{r} = \vec{r}_0 + \vec{v}_i\left(t - t_0\right)$, but the condition for this is that the higher-order terms in Eq. 10.99 and Eq. 10.100 can be ignored. We therefore check out this condition by examining the *next* term in the expansion (Eq. 10.100a), which is $\left[\frac{d^2\omega(\vec{k})}{dk^2}\right]_{k_i}(k-k_i)^2$.

We enquire if this term is indeed small. The second derivative of $\omega(k)=\frac{\hbar k^2}{2m}$ with respect to k being $\frac{\hbar}{m}$, we must examine if $\frac{\hbar}{m}(k-k_i)^2 t \ll 1$, i.e., if $\frac{\hbar}{m}(\Delta k)^2 t \ll 1$. This amounts to questioning if $\frac{D}{k_i} \ll (\Delta r)^2$, since $t = D/v_i$ and $\Delta k \approx (\Delta r)^{-1}$. Typically in a laboratory experiment, $k_i^{-1} \sim 10^{-8}$ cm, the longitudinal width of the wavepacket along the direction of incidence is $\Delta r \sim 10^{-1}$ cm and $D \sim 10^2$ cm (Fig. 10.1a). Hence the required condition is in fact satisfied, and we may write, using Eq. 10.102:

$$\Phi_{inc}(\vec{r},t) = \exp\left\{i\frac{E(\vec{k}_i)}{\hbar}(t-t_0)\right\}\left\{\Phi_{inc}\left(\vec{r}-\vec{v}_i(t-t_0),t_0\right)\right\}. \tag{10.103}$$

Let us now consider a free particle wave packet incident on the scatterer target impinging at an impact parameter $\vec{b}$, rather than head-on (Fig. 10.1a). The representation of the wavepacket in Eq. 10.89a gets only slightly modified to

$$\Phi_{\vec{b}}(\vec{r},t) = (2\pi)^{-3/2}\iiint d^3\vec{k}\left[\left|A(\vec{k})\right|\exp\left\{i\left\{\vec{k}\cdot(\vec{r}-\vec{b})-\omega(\vec{k})t\right\}\right\}\right], \tag{10.104a}$$

$$\text{i.e., } \Phi_{\vec{b}}(\vec{r},t) = (2\pi)^{-3/2}\iiint d^3\vec{k}\left\{\left|A(\vec{k})\right|e^{-i\vec{k}\cdot\vec{b}}e^{+i\vec{k}\cdot\vec{r}}e^{+i\omega(\vec{k}_i)t}e^{-i\vec{k}\cdot\vec{v}_i t}\right\}, \tag{10.104b}$$

which can be written as

$$\Phi_b(\vec{r},t) = e^{i\xi}\chi\left(\vec{r}-\vec{b}-\vec{v}_i t\right), \tag{10.105}$$

where the phase angle ξ is independent of k,

$$\text{and } \chi\left(\vec{r}-\vec{b}-\vec{v}_i t\right) = (2\pi)^{-3/2}\iiint d^3\vec{k}\left[\left|A(\vec{k})\right|e^{+i(\vec{k}-\vec{k}_i)\cdot(\vec{r}-\vec{b}-\vec{v}_i t)}\right] \tag{10.106}$$

determines the *shape* of the wavepacket. Since the shape function is multiplied only by a phase factor to give the wavepacket $\Phi_{\vec{b}}$, we conclude that when the wavepacket is normalized, the *shape* function is square integrable to unity:

$$\iiint d^3\vec{s}\;|\chi(\vec{s})|^2 = 1. \tag{10.107}$$

While Eqs. 10.104a–c represents the incident wavepacket, the solution to the complete scattering problem may be written by using the asymptotic form

$$\psi^+_{\vec{k}_i}(\vec{r};r\to\infty)\underset{r\to\infty}{\to} A(k)\left\{e^{i\vec{k}_i\cdot\vec{r}} + \frac{f(\hat{\Omega})}{r}e^{ikr}\right\}.$$

Now, we may write the wavenumber $k = \left|\vec{k}\right|$ about the mean k_i as

$$k = k_i + \hat{k}_i \cdot \left(\vec{k} - \vec{k}_i\right) + \ldots \tag{10.108}$$

and use it to write the complete solution as

$$\Psi^+_{\vec{b}}\left(\vec{r},t\right) \underset{r\to\infty}{\to} \left[\Phi_{\vec{b}}\left(\vec{r},t\right) + \Psi^{scat}_{\vec{b}}\left(\vec{r},t\right)\right], \tag{10.109}$$

i.e.,

$$\Psi^+_{\vec{b}}\left(\vec{r},t\right) \underset{r\to\infty}{\to} \left[\Phi_{\vec{b}}\left(\vec{r},t\right) + (2\pi)^{-3/2} \iiint d^3\vec{k} \left\{ A\left(\vec{k}\right) e^{-i\vec{k}\cdot\vec{b}} \frac{f\left(\vec{k},\hat{\Omega}\right)}{r} e^{ikr} e^{-i\omega(k)t} \right\}\right]. \tag{10.110}$$

We see that the scattering term in the above equation vanishes due to cancellation of the oscillatory terms for the distant past (much before the projectile-target interaction) when the k-dependent amplitude peaks around a given wavenumber in the incident wavepacket. Such cancellation would, however, not occur in the distant future (well after the interaction) when the scattered flux is observed in a detector (where the phase of the integrand in Eq. 10.110 can be stationary). We now write, for the latter case, the complex scattering amplitude in terms of its magnitude and phase as

$$f\left(\vec{k},\hat{\Omega}\right) = \left|f\left(\vec{k},\hat{\Omega}\right)\right| e^{i\Lambda(\vec{k},\hat{\Omega})} \simeq \left|f\left(\vec{k}_i,\hat{\Omega}\right)\right| e^{i\Lambda(\vec{k},\hat{\Omega})}. \tag{10.111}$$

Expanding now the phase angle $\Lambda\left(\vec{k},\hat{\Omega}\right)$ about the point $\vec{k}_i$ in the k-space,

$$\Lambda\left(\vec{k},\hat{\Omega}\right) \simeq \Lambda\left(\vec{k}_i,\hat{\Omega}\right) + \left[\vec{\nabla}_k \Lambda\left(\vec{k}_i,\hat{\Omega}\right)\right]_{\vec{k}=\vec{k}_i} \cdot \left(\vec{k} - \vec{k}_i\right), \tag{10.112}$$

$$\text{i.e., } \Lambda\left(\vec{k},\hat{\Omega}\right) \simeq \Lambda\left(\vec{k}_i,\hat{\Omega}\right) + \vec{\rho}\left(\hat{\Omega}\right) \cdot \left(\vec{k} - \vec{k}_i\right), \tag{10.113}$$

$$\text{with } \left[\vec{\nabla}_k \Lambda\left(\vec{k}_i,\hat{\Omega}\right)\right]_{\vec{k}=\vec{k}_i} = \vec{\rho}\left(\hat{\Omega}\right); \ \left|\vec{\rho}\left(\hat{\Omega}\right)\right| \ll \Delta r. \tag{10.114}$$

As a result of the k-dependence of the phase in the complex scattering amplitude, the scattered wave undergoes a time delay, given by

$$\tau = \frac{\left(\hat{k}_i \cdot \vec{\rho}\right)}{v_i} = \frac{\hat{k}_i \cdot \left[\vec{\nabla}_k \Lambda\left(\vec{k}_i,\hat{\Omega}\right)\right]_{\vec{k}=\vec{k}_i}}{v_i}, \tag{10.115}$$

since the numerator gives the displacement in the effective center of the scattered wavepacket due to the interaction in the scattering region.

The component of $\left[\vec{\nabla}_k \Lambda\left(\vec{k}_i,\hat{\Omega}\right)\right]_{\vec{k}=\vec{k}_i}$ along the unit vector $\hat{k}_i$ is the *directional derivative* of the phase with respect to k, and we get

$$\tau = \frac{1}{v}\frac{d\Lambda}{dk} = \frac{\hbar}{v}\frac{d\Lambda}{d(\hbar k)} = \frac{\hbar}{v}\frac{d\Lambda}{dp} = \frac{\hbar}{v}\frac{d\Lambda}{dE}\frac{p}{m} = \hbar\frac{d\Lambda}{dE}. \tag{10.116}$$

The scattering phase shift (from Eq. 10.59, Eq. 10.60) in the ℓth partial wave is $2\delta_\ell(k)$. The corresponding time delay in it is therefore is $2\hbar\frac{d\delta_\ell(k)}{dE}$. On the other hand, consideration of photoionization (also photodetachment) as *half-scattering* (Fig. 10.2a,b) renders the corresponding time delay to be *half* of what it is in collisions, and hence it is given by $\hbar\frac{d\delta_\ell(k)}{dE}$. The position of the wavepacket at an instant t is given by

$$r \simeq v_i t - v_i\tau = v_i t - \left(\vec{\rho}\bullet\vec{k}_i\right) = z - \left(\vec{\rho}\bullet\vec{k}_i\right). \tag{10.117}$$

It must be noted that the time *delay* may also be negative, whence it is called as time *advancement*. We note that the time delay τ has its origin in the term $\vec{\nabla}_k\Lambda\left(\vec{k}_i,\hat{\Omega}\right)$ where Λ is the phase factor in the scattering amplitude (Eq. 10.110); it corresponds to the energy dependence of the phase of the scattering amplitude. It was studied extensively by Eisenbud [6], Wigner [7], and Smith [8] and is known as the EWS time-delay, or simply as the *Wigner time delay*. Through the time-reversal symmetry, it has important applications in the dynamics of photoionization and photodetachment [4, 5, 9, 10].

The solution to the scattering problem with outgoing wave boundary condition can now be written as

$$\Psi^+_{\vec{b}}(\vec{r},t) \underset{r\to\infty}{\rightarrow} \left[\Phi_{\vec{b}}(\vec{r},t) + f\left(\vec{k}_i,\hat{\Omega}\right)\frac{e^{i\{k_i r-\omega(\vec{k}_i)t\}}}{r}e^{-i\vec{k}_i\cdot\vec{b}}\chi\left(r\hat{k}_i - \vec{v}_i t + \vec{\rho}(\hat{\Omega}) - \vec{b}\right)\right]. \tag{10.118}$$

The probability density in the scattered part of the solution is

$$\left|\Psi^{+\,scat}_{\vec{b}}(\vec{r},t)\right|^2 = \left|f\left(\vec{k}_i,\hat{\Omega}\right)\right|^2\frac{1}{r^2}\left|\chi\left(\vec{\rho}(\hat{\Omega}) + \hat{k}_i r - \vec{v}_i t - \vec{b}\right)\right|^2, \tag{10.119}$$

from which we obtain the probability of scattering along the direction $\hat{\Omega}$ as

$$P_b\left(\hat{\Omega}\right) = \int_0^\infty r^2 dr\left|\Psi^{+\,scat}_{\vec{b}}(\vec{r},t)\right|^2 = \left|f\left(\vec{k}_i,\hat{\Omega}\right)\right|^2\int_0^\infty r^2 dr\frac{1}{r^2}\left|\chi\left(\vec{\rho}(\hat{\Omega}) + \hat{k}_i r - \vec{v}_i t - \vec{b}\right)\right|^2, \tag{10.120a}$$

$$\text{i.e., } P_b\left(\hat{\Omega}\right) = \left|f\left(\vec{k}_i,\hat{\Omega}\right)\right|^2\int_{-\infty}^{\infty} dz\left|\chi\left(\vec{\rho}(\hat{\Omega}) + \hat{k}_i\left(r - v_i t\right) - \vec{b}\right)\right|^2 \tag{10.120b}$$

since $\vec{v}_i = \hat{k}_i v_i$ and we have changed the integration variable from 'r' to 'z', incident projectiles being along the Cartesian positive Z-axis.

The differential cross section is then obtained by integrating over all the different impact parameters:

$$\frac{d\sigma}{d\Omega} = \iint d^2\vec{b}P_b\left(\hat{\Omega}\right) = \left|f\left(\vec{k}_i,\hat{\Omega}\right)\right|^2\int_{-\infty}^{\infty} dz\iint d^2\vec{b}\left|\chi\left(\vec{\rho}(\hat{\Omega}) + \hat{k}_i z - \vec{b}\right)\right|^2. \tag{10.121a}$$

The integral in the above equation is essentially over the whole space, and from Eq. 10.107 we know that it is unity. Hence,

$$\frac{d\sigma}{d\Omega} = \left|f\left(\vec{k}_i, \hat{\Omega}\right)\right|^2 \tag{10.121b}$$

just as in Eq. 10.79a. The optical theorem thus holds good for a description of the incident beam by a realistic wavepacket consisting of particles having a spread over various energies, just as well as for an ideally monoenergetic incident beam of particles.

Funquest: What is the value of the scattering differential cross section and also the total cross section for the Coulomb potential? Explain your answer.

We shall now proceed to *define* the *scattering operator*, $\widehat{S}$. We shall find that the conservation principle expressed in the optical theorem requires that the operator $\widehat{S}$ is unitary [11] (see also the Solved Problem P10.8). For an arbitrary coordinate-system, we denote the direction of incidence of the projectile particles by the unit vector $\hat{n}$. The asymptotic solution to the complete collision problem for scattering along a direction represented by the unit vector $\hat{n}'$ is

$$\psi(\vec{r}; r \to \infty) \to \exp\left(ikr\hat{n}\bullet\hat{n}'\right) + \frac{f\left(\hat{n},\hat{n}'\right)\exp(ikr)}{r}. \tag{10.122}$$

In the above equation, $f\left(\hat{n},\hat{n}'\right)$ is the scattering amplitude and we have omitted the mention of the arbitrary overall normalization constant in Eq. 10.122. The directions of incidence and of scattering are shown in Fig. 10.8. Typically, we can choose a coordinate system whose polar axis is either along $\hat{n}$ or along $\hat{n}'$. In the present discussion, we shall choose the latter. Now, any linear combination of solutions of the kind expressed in Eq. 10.122 *for different directions of incidence* would also be a solution to the Schrödinger differential equation, hence we write the solution as

$$\Psi(\vec{r}; r \to \infty) \to \iint F(\hat{n})\exp\left(ikr\ \hat{n}\bullet\hat{n}'\right)d\Omega + \frac{\exp(ikr)}{r}\iint F(\hat{n})f\left(\hat{n},\hat{n}'\right)d\Omega. \tag{10.123a}$$

Now, $\left(kr\hat{n}\bullet\hat{n}'\right)$ changes very rapidly in the asymptotic $r \to \infty$ region with slight changes in $\hat{n}\bullet\hat{n}'$, hence $\iint F(\hat{n})\exp\left(ikr\hat{n}\bullet\hat{n}'\right)d\Omega$ is largely determined by the condition $\hat{n}\bullet\hat{n}' = \pm 1$, i.e., $\hat{n} = \pm\hat{n}'$ whence $F(\hat{n}) \simeq F\left(\pm\hat{n}'\right)$. We emphasize that the superposition in Eq. 10.123a maintains the direction of the scattered wave unique; it fixes the polar axis. The integration is essentially over different directions of incidence. With the polar angle θ measured with respect to the direction $\hat{n}'$, the superposition is given by

$$\Psi\left(\vec{r}; r \to \infty\right) \to 2\pi\int_{\theta=0}^{\pi} \sin\theta d\theta F(\hat{n})\exp\left(ikr\ \hat{n}\bullet\hat{n}'\right) + \frac{\exp(ikr)}{r}\iint f\left(\hat{n},\hat{n}'\right)F(\hat{n})d\Omega \tag{10.123b}$$

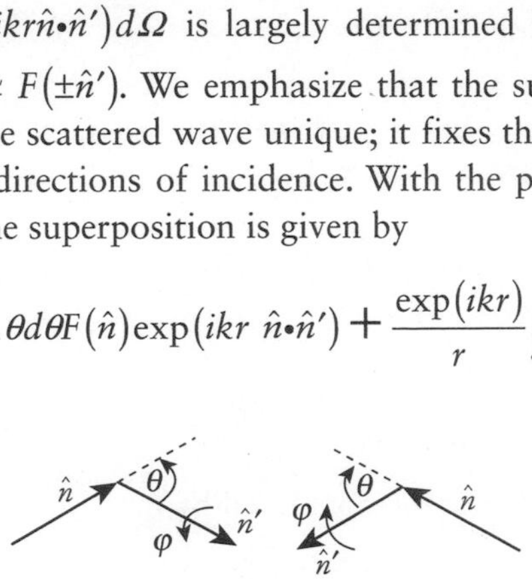

Fig. 10.8 Illustration of the reciprocity theorem. It enables the application of the outgoing wave boundary conditions in scattering, and the ingoing wave boundary conditions in photoionization and photodetachment.

$$\text{i.e., } \Psi(\vec{r};r\to\infty)\to 2\pi\left.\frac{F(\hat{n})\exp(ikr\cos\theta)}{ikr}\right|_{\cos\theta=1}^{\cos\theta=-1}+\frac{\exp(ikr)}{r}\iint f(\hat{n},\hat{n}')F(\hat{n})d\Omega.$$

$$\text{or, } \Psi(\vec{r};r\to\infty)\to\frac{2\pi i}{k}\left[\left\{\begin{matrix}\dfrac{F(-\hat{n}')\exp(-ikr)}{r}-\\ \dfrac{F(\hat{n}')\exp(ikr)}{r}\end{matrix}\right\}+\frac{k}{2\pi i}\frac{\exp(ikr)}{r}\iint f(\hat{n},\hat{n}')F(\hat{n})d\Omega\right].$$

We may drop the multiplier $\frac{2\pi i}{k}$ since it can be absorbed in the overall normalization of the wavefunction, and write the solution as

$$\Psi(\vec{r};r\to\infty)\to\left\{\begin{matrix}\dfrac{F(-\hat{n}')\exp(-ikr)}{r}-\\ \dfrac{F(\hat{n}')\exp(ikr)}{r}\end{matrix}\right\}+\frac{k}{2\pi i}\frac{\exp(ikr)}{r}\iint f(\hat{n},\hat{n}')F(\hat{n})d\Omega,$$

$$\text{i.e., } \Psi(\vec{r};r\to\infty)\to\frac{F(-\hat{n}')\exp(-ikr)}{r}-\frac{\exp(ikr)}{r}\left\{F(\hat{n}')-\frac{k}{2\pi i}\iint f(\hat{n},\hat{n}')F(\hat{n})d\Omega\right\}. \tag{10.124}$$

Along with the time-dependent dynamical phase (Eq. 1.115 in Chapter 1), the wavefunction is

$$\Psi(\vec{r},t;r\to\infty)\to\overbrace{\frac{F(-\hat{n}')\exp(-i(kr+\omega t))}{r}}^{\text{ingoing spherical wave}}-\overbrace{\frac{\exp(i(kr-\omega t))}{r}}^{\text{outgoing spherical wave}}\left\{\begin{matrix}F(\hat{n}')-\\ \dfrac{k}{2\pi i}\iint f(\hat{n},\hat{n}')F(\hat{n})d\Omega\end{matrix}\right\}; \tag{10.125}$$

it is a superposition of ingoing and outgoing spherical waves. We introduce an operator $\mathfrak{F}$ (note the *script* font) defined by its *effect* on an arbitrary function F which depends only on a direction $\hat{n}'$:

$$\mathfrak{F}F(\hat{n}')=\frac{1}{4\pi}\iint f(\hat{n},\hat{n}')F(\hat{n})d\Omega. \tag{10.126}$$

The result of this operation requires integration over all directions $\hat{n}$ in space, weighted by a factor $f(\hat{n},\hat{n}')$ which depends on the direction $\hat{n}'$ (the argument of the operand F) and also the direction $\hat{n}$ over which the integration is carried out. Using this operator, Eq. 10.124 simplifies to

$$\Psi(\vec{r};r\to\infty)\to\frac{\exp(-ikr)}{r}F(-\hat{n}')-\frac{\exp(ikr)}{r}(1+2ik\mathfrak{F})F(\hat{n}'). \tag{10.127}$$

A new operator

$$\mathfrak{s}=1+2ik\mathfrak{F} \tag{10.128}$$

called the *scattering operator* was introduced by Heisenberg in 1943. He rebuilt a formulation of quantum mechanics in which this operator played a vital role. Using the scattering operator, Eq. 10.127 becomes

$$\Psi(\vec{r};r \to \infty) \to \frac{\exp(-ikr)}{r}F(-\hat{n}') - \frac{\exp(ikr)}{r}\mathfrak{s}F(\hat{n}'). \tag{10.129}$$

On the inclusion of the dynamical phase, Eq. 10.125 becomes

$$\Psi(\vec{r},t;r \to \infty) \to \frac{\exp(-i(kr+\omega t))}{r}F(-\hat{n}') - \frac{\exp(i(kr-\omega t))}{r}\mathfrak{s}F(\hat{n}'). \tag{10.130}$$

From Eq. 10.129 and Eq. 10.130 we see that $\left\langle \frac{F(-\hat{n}')}{r} \middle| \frac{F(-\hat{n}')}{r} \right\rangle$ provides a measure of the intensity of the *ingoing* waves, and $\left\langle \frac{F(\hat{n}')}{r} \middle| \mathfrak{s}^\dagger \mathfrak{s} \middle| \frac{F(\hat{n}')}{r} \right\rangle$ provides a measure of the intensity of the *outgoing* waves. Conservation of the ingoing and outgoing flux is then expressed by the *unitarity of the scattering operator*:

$$\mathfrak{s}\mathfrak{s}^\dagger = 1. \tag{10.131}$$

We shall see now how this property of the scattering operator provides, in fact, the optical theorem (Eq. 10.88). It follows from the definition in Eq. 10.128 of the scattering operator that

$$\mathfrak{s}\mathfrak{s}^\dagger = (1+2ki\mathfrak{F})(1-2ki\mathfrak{F}^\dagger) = 1 + 2ki(\mathfrak{F}-\mathfrak{F}^\dagger) + 4k^2(\mathfrak{F}\mathfrak{F}^\dagger), \tag{10.132}$$

and furthermore the unitarity of the scattering operator implies that

$$(\mathfrak{F}-\mathfrak{F}^\dagger) = 2ki\mathfrak{F}\mathfrak{F}^\dagger. \tag{10.133}$$

We observe at this point that in Eq. 10.126, the integration is over the unprimed direction $\hat{n}$, which is the *first* index in the factor $f(\hat{n},\hat{n}')$ in the integrand. The Hermitian adjoint of Eq. 10.126 is

$$\mathfrak{F}^\dagger F(\hat{n}') = \frac{1}{4\pi}\iint f^*(\hat{n}',\hat{n}'')F(\hat{n}'')d\Omega'' \tag{10.134}$$

in which the integration is over the double-primed variable, which is now the *second* index in the factor $f^*(\hat{n}',\hat{n}'')$. From Eq. 10.128 and Eq. 10.134, we get

$$(\mathfrak{F}-\mathfrak{F}^\dagger)F(\hat{n}') = \frac{1}{4\pi}\iint f(\hat{n},\hat{n}')F(\hat{n})d\Omega - \frac{1}{4\pi}\iint f^*(\hat{n}',\hat{n}'')F(\hat{n}'')d\Omega'', \tag{10.135}$$

i.e., $2ki\mathfrak{F}\mathfrak{F}^\dagger F(\hat{n}') = 2ki\mathfrak{F}\{\mathfrak{F}^\dagger F(\hat{n}')\} = 2ki\mathfrak{F}\left\{\frac{1}{4\pi}\iint f^*(\hat{n}',\hat{n}'')F(\hat{n}'')d\Omega''\right\}$. (10.136a)

Hence, $\iint f(\hat{n},\hat{n}')F(\hat{n})d\Omega - \iint f^*(\hat{n}',\hat{n}'')F(\hat{n}'')d\Omega'' = 2ki\mathfrak{F}\iint f^*(\hat{n}',\hat{n}'')F(\hat{n}'')d\Omega''$.
(10.136b)

Since the operator $\mathfrak{F}$ on the right-hand side would operate only on the operand function $F(\hat{n}'')$, $f^*(\hat{n}',\hat{n}'')$ being treated only as a scalar multiplier,

$$\iint f(\hat{n},\hat{n}')F(\hat{n})d\Omega - \iint f^*(\hat{n}',\hat{n}'')F(\hat{n}'')d\Omega'' = 2ki\left[\iint f^*(\hat{n}',\hat{n}'')\{\mathfrak{F}F(\hat{n}'')\}d\Omega''\right]. \quad (10.136c)$$

i.e., using Eq. 10.126,

$$\iint f(\hat{n},\hat{n}')F(\hat{n})d\Omega - \iint f^*(\hat{n}',\hat{n}'')F(\hat{n}'')d\Omega'' = 2ki\left[\iint f^*(\hat{n}',\hat{n}'')\frac{1}{4\pi}\iint f(\hat{n},\hat{n}'')F(\hat{n})d\Omega d\Omega''\right]. \quad (10.136d)$$

Dropping now the double-prime on the *dummy* variable in the second term on the left hand, we get

$$\iint\{f(\hat{n},\hat{n}') - f^*(\hat{n}',\hat{n})\}F(\hat{n})d\Omega = \iint\left\{\frac{ki}{2\pi}\iint f^*(\hat{n}',\hat{n}'')f(\hat{n},\hat{n}'')d\Omega''\right\}F(\hat{n})d\Omega. \quad (10.136e)$$

The integrals on both sides of the above integral are definite integrals in the same angular space. Their integrands are therefore equal, and hence

$$f(\hat{n},\hat{n}') - f^*(\hat{n}',\hat{n}) = \frac{ki}{2\pi}\iint f^*(\hat{n}',\hat{n}'')f(\hat{n},\hat{n}'')d\Omega''. \quad (10.137)$$

For the case of scattering in the forward direction, we have $\hat{n}' = \hat{n}$ and hence

$$f(\hat{n},\hat{n}) - f^*(\hat{n},\hat{n}) = \frac{ki}{2\pi}\iint f^*(\hat{n},\hat{n}'')f(\hat{n},\hat{n}'')d\Omega'' = \frac{ki}{2\pi}\iint\left|f(\hat{n},\hat{n}'')\right|^2 d\Omega'', \quad (10.138)$$

$$\text{or } 2i\,\mathrm{Im}\left[f(\hat{n},\hat{n})\right] = \frac{ki}{2\pi}\iint\left|f(\hat{n},\hat{n}'')\right|^2 d\Omega'' = \frac{ki}{2\pi}\sigma_{total}, \quad (10.139)$$

which we see, on using Eq. 10.88 and Eqs. 10.121a,b, to be essentially the *optical theorem*. In the immediate context, it has resulted from the *unitarity of the scattering operator* (Eq. 10.131). We find that this property corresponds to the conservation of flux in the scattered wave with respect to that in the incident wave.

Let us now construct the time-reversed solution corresponding to the wavefunction in Eq. 10.130. As discussed in Appendix A (Eqs. A.27a,b), this involves complex conjugation, along with the discrete reversal of sign of time, $t \rightarrow -t$. Therefore, first taking the complex conjugate of Eq. 10.130, we get

$$\Psi^*(\vec{r},t;\, r\rightarrow\infty) \rightarrow \frac{e^{+i(kr+\omega t)}}{r}F^*(-\hat{n}') - \frac{e^{-i(kr-\omega t)}}{r}\mathfrak{s}^*F^*(\hat{n}') \quad (10.140)$$

and now letting $t \rightarrow -t$

$$\Psi^*(\vec{r},-t;\, r\rightarrow\infty) \rightarrow \frac{e^{+i(kr-\omega t)}}{r}F^*(-\hat{n}') - \frac{e^{-i(kr+\omega t)}}{r}\mathfrak{s}^*F^*(\hat{n}'). \quad (10.141)$$

We note that the space part of the time-reversed function is

$$\xi = \left\{ \frac{\exp(ikr)}{r} F^*(-\hat{n}') - \frac{\exp(-ikr)}{r} \mathfrak{s}^* F^*(\hat{n}') \right\}. \tag{10.142}$$

The scattering process is therefore appropriately represented by the function ξ as well. Let us now denote the effect of the operator $\mathfrak{s}^*$ on $F^*(\hat{n}')$ by $-\Phi(-\hat{n}')$, i.e.,

$$\mathfrak{s}^* F^*(\hat{n}') = -\Phi(-\hat{n}'). \tag{10.143}$$

Essentially, the above relation provides a *definition* of $-\Phi(-\hat{n}')$.

Hence, $F^*(\hat{n}') = (\mathfrak{s}^*)^{-1} \mathfrak{s}^* F^*(\hat{n}') = (\mathfrak{s}^*)^{-1}\{-\Phi(-\hat{n}')\} = -(\mathfrak{s}^*)^{-1} \Phi(-\hat{n}')$. (10.144a)

The scattering operator being unitary, we can write the above result also as

$$F^*(\hat{n}') = -(\mathfrak{s}^*)^{\dagger} \Phi(-\hat{n}') = -\tilde{\mathfrak{s}}\Phi(-\hat{n}'), \tag{10.144b}$$

wherein the symbol $\tilde{\mathfrak{s}}$ denotes the transpose of $\mathfrak{s}$. Now, since the parity operator P *reverses* the sign of the argument of its operand, we note that

$$F^*(-\hat{n}') = PF^*(\hat{n}') = P\{-\tilde{\mathfrak{s}}\Phi(-\hat{n}')\} = -P\tilde{\mathfrak{s}}\Phi(-\hat{n}') = -P\tilde{\mathfrak{s}}P\Phi(\hat{n}'). \tag{10.145}$$

Using Eq. 10.143 and Eq. 10.145 in Eq. 10.142 we get

$$\xi = \left\{ \frac{e^{+ikr}}{r} \{-P\tilde{\mathfrak{s}}P\Phi(\hat{n}')\} - \frac{e^{-ikr}}{r} \{-\Phi(-\hat{n}')\} \right\}, \tag{10.146}$$

$$\text{i.e., } \xi = \left\{ \frac{e^{-ikr}}{r} \Phi(-\hat{n}') - \frac{e^{+ikr}}{r} (P\tilde{\mathfrak{s}}P)\Phi(\hat{n}') \right\}. \tag{10.147}$$

Comparison of Eq. 10.147 and Eq. 10.129 gives the following correspondence:

$$\left[\underbrace{\frac{\exp(-ikr)}{r} F(-\hat{n}') - \frac{\exp(ikr)}{r} \mathfrak{s}F(\hat{n}')}_{\text{Eq. 10.129}} \right] \leftrightarrow \left[\underbrace{\frac{e^{-ikr}}{r} \Phi(-\hat{n}') - \frac{e^{+ikr}}{r} (P\tilde{\mathfrak{s}}P)\Phi(\hat{n}')}_{\text{Eq. 10.147}} \right]. \tag{10.148}$$

The wave functions in Eq. 10.129 and in Eq. 10.147 are essentially expressed in the base pair $\{\exp(ikr), \exp(-ikr)\}$, consisting of spherical ingoing and outgoing waves. From the coefficient of $\exp(-ikr)$, we find the correspondence: $F \leftrightarrow \Phi$; i.e., the two functions differ *only* in how they are denoted.

From the coefficient of $\exp(-ikr)$ we identify

$$P\tilde{\mathfrak{s}}P = \mathfrak{s}. \tag{10.149}$$

Simultaneous operations of (i) transposition and (ii) parity (inversion) leaves the scattering operator invariant, i.e.,

$$\mathfrak{s}(\hat{n},\hat{n}') = \mathfrak{s}(-\hat{n}',-\hat{n}). \tag{10.150a}$$

Note that the arguments of the scattering operator $\mathfrak{s}$ on the two sides of Eq. 10.150a are (i) interchanged (transposed) and (ii) reversed. Equivalently, we have the following property for the scattering amplitude:

$$f(\hat{n},\hat{n}') = f(-\hat{n}',-\hat{n}). \tag{10.150b}$$

The result in Eq. 10.150a,b is known as the *Reciprocity Theorem* [11]. The equality of time-reversed *scattering* processes comes from Eq. 10.150a,b. Along with Fig. 10.2a,b, Fig. 10.4, and Fig. 10.5a,b,c, we interpret photoionization (and photodetachment) as *half-scattering* (see also the discussion on Eq. 10.116a,b). It arises essentially from time-reversal symmetry in the scattering process.

10.4 Lippmann–Schwinger Integral Equation, Green Functions, Born Approximations

A powerful method of solving differential equations with boundary conditions employs an inverse approach in which one unscrambles integral equations. It is particularly well adapted to solve problems in quantum collisions described by the Schrödinger equation $\left\{\frac{\left(-i\hbar\vec{\nabla}\right)^2}{2m} + V(\vec{r})\right\}\psi(\vec{r}) = E\psi(\vec{r})$. To minimize writing symbols that only represent fundamental constants, we shall express the interaction potential in terms of a *reduced potential*

$$U(\vec{r}) = \frac{2m}{\hbar^2}V(\vec{r}) \tag{10.151a}$$

and the energy eigenvalue of the Schrödinger equation in terms of

$$k^2 = \frac{2mE}{\hbar^2}. \tag{10.151b}$$

Using the reduced potential, the differential equation

$$\left(\vec{\nabla}^2 + k^2\right)\psi_{\vec{k}}(\vec{r}) = U(\vec{r})\psi_{\vec{k}}(\vec{r}) \tag{10.152}$$

essentially represents the Schrödinger equation. It is a differential equation, subject to boundary conditions determined by physical conditions. It has the form

$$\Omega\sigma(\vec{r}) = \xi(\vec{r}) \tag{10.153}$$

in which the target of our inquiry is to determine $\sigma(\vec{r})$, Ω being a linear operator, and the unknown function $\sigma(\vec{r})$ appears, however, in the function $\xi(\vec{r})$ that appears on the right-hand side. The *source* of inhomogeneity is the function $\xi(\vec{r})$; if it were zero then one has the homogeneous equation

$$\Omega\sigma(\vec{r}) = 0. \tag{10.154}$$

In the absence of the potential in the Schrödinger equation, one has the homogeneous equation for a free particle,

$$\left(\vec{\nabla}^2 + k^2\right)\phi_{\vec{k}}(\vec{r}) = 0. \tag{10.155}$$

The operator $\left(\vec{\nabla}^2 + k^2\right)$ is known as the Helmholtz operator. Our interest is in determining the solution $\psi_{\vec{k}}(\vec{r})$ corresponding to a specific *source*, subject to boundary conditions appropriate to the physical conditions. The technique we adopt is the Green function method, known after the British mathematician-physicist George Green (1793 – 1841) who developed it in the context of problems in electrodynamics. It is useful in solving partial differential equations when one can solve the inhomogeneous equation

$$\Omega G_0(\vec{r},\vec{r}_0) = \delta^3(\vec{r} - \vec{r}_0) \tag{10.156a}$$

in which the *source* of inhomogeneity is the Dirac-δ (generalized) function ('distribution'), Ω being the operator in Eq. 10.153, which is the inhomogeneous equation to be solved. Employing the Helmholtz operator for the operator Ω, we have

$$\left(\vec{\nabla}^2 + k^2\right)G_0(k,\vec{r},\vec{r}_0) = \delta^3(\vec{r} - \vec{r}_0). \tag{10.156b}$$

We can now verify in a straightforward manner that

$$\psi_{\vec{k}_i}(\vec{r}) = \phi_{\vec{k}_i}(\vec{r}) + \iiint d^3\vec{r}'\, G_0(k,\vec{r},\vec{r}')U(\vec{r}')\psi_{\vec{k}_i}(\vec{r}') \tag{10.157}$$

is a solution to Eq. 10.152, the inhomogeneous equation that is of our primary interest. In order to first check the veracity of this claim, we operate on the right-hand side of Eq. 10.157 by the Helmholtz operator:

$$\left(\vec{\nabla}^2 + k^2\right)\begin{bmatrix}\phi_{\vec{k}_i}(\vec{r}) + \\ \iiint d^3\vec{r}'\, G_0(k,\vec{r},\vec{r}')U(\vec{r}')\psi_{\vec{k}_i}(\vec{r}')\end{bmatrix} = \left(\vec{\nabla}^2 + k^2\right)\iiint d^3\vec{r}'\, G_0(k,\vec{r},\vec{r}')U(\vec{r}')\psi_{\vec{k}_i}(\vec{r}'), \tag{10.158a}$$

i.e.,

$$\left(\vec{\nabla}^2 + k^2\right)\begin{bmatrix}\phi_{\vec{k}_i}(\vec{r}) + \\ \iiint d^3\vec{r}'\, G_0(k,\vec{r},\vec{r}')U(\vec{r}')\psi_{\vec{k}_i}(\vec{r}')\end{bmatrix} = \iiint d^3\vec{r}'\left\{\left(\vec{\nabla}^2 + k^2\right)G_0(k,\vec{r},\vec{r}')\right\}U(\vec{r}')\psi_{\vec{k}_i}(\vec{r}'), \tag{10.158b}$$

or,

$$\left(\vec{\nabla}^2 + k^2\right)\begin{bmatrix}\phi_{\vec{k}_i}(\vec{r}) + \\ \iiint d^3\vec{r}'\, G_0(k,\vec{r},\vec{r}')U(\vec{r}')\psi_{\vec{k}_i}(\vec{r}')\end{bmatrix} = \iiint d^3\vec{r}'\,\delta^3(\vec{r} - \vec{r}_0)U(\vec{r}')\psi_{\vec{k}_i}(\vec{r}') = U(\vec{r})\psi_{\vec{k}_i}(\vec{r}), \tag{10.158c}$$

which is just Eq. 10.152, fulfilling our objective. The equation 10.157 is tautological, since the wavefunction appears in the integrand on the right-hand side. It recasts the Schrödinger equation as an inverse problem using integration, but the boundary conditions that are required to solve the differential equation must be incorporated by *choosing* $G_0(k,\vec{r},\vec{r}')$ appropriately. We shall discover that this inverse approach enables us to develop effective approximation methods to solve complex problems of potential scattering. The function $G_0(k,\vec{r},\vec{r}_0)$ is known as the Green function; it is a solution to the Dirac-δ inhomogeneity and in the present case it is referenced to the momentum k (in units of $\hbar$). We note that the Dirac-δ being a generalized function (often called as *distribution*), the Green function also is not, as such, a *function*; it is specific to each set of boundary conditions. An examination of Eq. 10.157 tells us that apart from the additive function $\phi_{\vec{k}_i}(\vec{r})$, the solution to the differential equation is obtained by integration over whole space of an integrand, which consists of a product of the Green function $G_0(k,\vec{r},\vec{r}')$ and the inhomogeneity term $U(\vec{r}')\psi_{\vec{k}}(\vec{r}')$, the integration (dummy) variable being $\vec{r}'$. The unprimed parameter $\vec{r}$ is the specific argument of the wavefunction on the left-hand side of Eq. 10.157. The Green function is thus the consequential term (hence called as the *kernel*) for a given inhomogeneity term which determines the wavefunction amplitude $\psi_k(\vec{r})$ at the point $\vec{r}$, which we shall refer to as the *field point*. Essentially, the Green function *propagates* the *influence* of a source at $\vec{r}'$ to the specific field point $\vec{r}$, but the net function at the field point requires integration over source points $\vec{r}'$, over the entire space, in the spirit of the principle of superposition. The kernel is therefore also called as the *propagator* or as the *influence function*. Equation 10.157 thus represents a cause-effect relationship, schematically sketched in Fig. 10.9.

For simplicity, omitting the momentum label k, we now write the Fourier representation of the Green function

$$G_0(\vec{R}) = G_0(k,\vec{R}) = G_0(k,\vec{r}-\vec{r}') \tag{10.159a}$$

$$\text{where } \vec{R} = \vec{r} - \vec{r}' \tag{10.159b}$$

as

$$G_0(\vec{R}) = \frac{1}{(2\pi)^3}\iiint d^3\vec{k}' g(\vec{k}')e^{i\vec{k}'\cdot\vec{R}}. \tag{10.160}$$

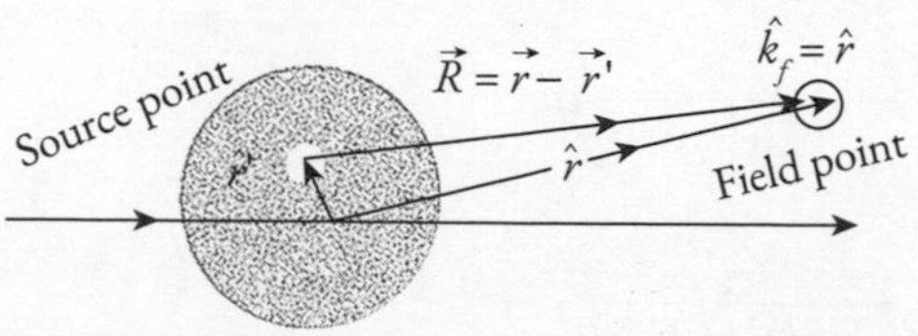

Fig. 10.9 Schematic representation of the cause-effect relationship depicting the influence at the field point of the extended source of inhomogeneity in the differential equation. Whole space integration over the extended region of the source must be carried out to get the cumulative effect at the field point.

Using Eq. 10.160 and the Fourier transform (see the Solved Problem P1.10 in Chapter 1) of the Dirac-δ function in Eq. 10.156b, we get

$$\left(\vec{\nabla}^2 + k^2\right)\left\{\frac{1}{(2\pi)^3}\iiint d^3\vec{k}' g\left(\vec{k}'\right) e^{i\vec{k}'\cdot\vec{R}}\right\} = \left\{\frac{1}{(2\pi)^3}\iiint d^3\vec{k}' e^{i\vec{k}'\cdot\vec{R}}\right\}, \tag{10.161a}$$

$$\text{i.e., } \left(\vec{\nabla}^2 + k^2\right)\iiint d^3\vec{k}' g\left(\vec{k}'\right) e^{i\vec{k}'\cdot\vec{R}} = \iiint d^3\vec{k}' e^{i\vec{k}'\cdot\vec{R}}, \tag{10.161b}$$

$$\text{i.e., } \iiint d^3\vec{k}' g\left(\vec{k}'\right)\left(\vec{\nabla}^2 + k^2\right) e^{i\vec{k}'\cdot\vec{R}} = \iiint d^3\vec{k}' e^{i\vec{k}'\cdot\vec{R}}, \tag{10.161c}$$

$$\text{or, } \iiint d^3\vec{k}' g\left(\vec{k}'\right)\left(k^2 - k'^2\right) e^{i\vec{k}'\cdot\vec{R}} = \iiint d^3\vec{k}' e^{i\vec{k}'\cdot\vec{R}}. \tag{10.161d}$$

The integration on both sides of Eq. 10.161d being definite integrals in essentially the same k-space, the corresponding integrands must be equal. Hence,

$$g\left(\vec{k}'\right) = \frac{1}{k^2 - k'^2} = \left(\frac{1}{k - k'}\right)\left(\frac{1}{k + k'}\right) = \frac{-1}{(k + k')(k' - k)}. \tag{10.162}$$

Using the above result now in Eq. 10.160, we have the following expression for the Green function:

$$G_0(\vec{R}) = \frac{-1}{(2\pi)^3}\iiint d^3\vec{k}' \frac{1}{(k + k')(k' - k)} e^{i\vec{k}'\cdot\vec{R}}. \tag{10.163a}$$

The integration volume element in the k-space is $d^3\vec{k} \equiv k^2 dk \sin\theta d\theta d\phi$, wherein the polar angles in the momentum space are taken with reference to a polar axis which may be taken to be along the unit vector $\hat{R} = \vec{R}/|\vec{R}|$. Azimuthal symmetry about this axis yields a factor of 2π, using which the Green function becomes

$$G_0\left(\vec{R}\right) = \frac{-1}{(2\pi)^2}\iint k'^2 dk' \sin\theta' d\theta' \frac{1}{(k + k')(k' - k)} e^{i\vec{k}'\cdot\vec{R}}, \tag{10.163b}$$

$$\text{i.e., } G_0\left(\vec{R}\right) = \frac{-1}{(2\pi)^2}\int_{k'=0}^{\infty} \frac{k'^2}{(k + k')(k' - k)} dk' \int_{\theta'=0}^{\pi} \sin\theta' d\theta' e^{ik'R\cos\theta'},$$

$$\text{i.e., } G_0\left(\vec{R}\right) = \frac{-1}{(2\pi)^2}\int_{k'=0}^{\infty} \frac{k'^2}{(k + k')(k' - k)} dk' \int_{\cos\theta'=-1}^{\cos\theta'=1} d(\cos\theta') e^{ik'R(\cos\theta')},$$

$$\text{i.e., } G_0\left(\vec{R}\right) = \frac{-1}{(2\pi)^2}\int_{k'=0}^{\infty} \frac{k'}{(k + k')(k' - k)} \frac{e^{ik'R} - e^{-ik'R}}{iR} dk',$$

$$\text{or, } G_0\left(\vec{R}\right) = \frac{-1}{(2\pi^2 R)}\int_{k'=0}^{\infty} \frac{k' \sin(k'R)}{(k + k')(k' - k)} dk'. \tag{10.163c}$$

Now, in the k-space, the integration variable has the range $0 \le k' < \infty$, but for the determination of the integral it is expeditious to exploit the fact that the integrand is an even function of k', since $k'\sin(k'R)$ is a product of two odd functions. This allows the integral to be *mathematically* equal to *half* of the integration range *doubled,* to $(-\infty,+\infty)$:

$$G_0\left(\vec{R}\right) = \frac{-1}{4\pi^2 R}\int_{k'=-\infty}^{\infty} \frac{k'\sin(k'R)}{(k+k')(k'-k)}dk', \tag{10.164a}$$

$$\text{i.e., } G_0(\vec{R}) = \frac{-1}{2\pi^2 R}\frac{1}{2}\int_{k'=-\infty}^{\infty}\left[\frac{1}{2}\left\{\frac{1}{(k'+k)}+\frac{1}{(k'-k)}\right\}\right]\frac{\exp(ik'R)-\exp(-ik'R)}{2i}dk', \tag{10.164b}$$

$$\text{or, } G_0(\vec{R}) = I_1 + I_2, \tag{10.165a}$$

where

$$I_1 = \frac{-1}{16\pi^2 Ri}\int_{k'=-\infty}^{\infty}\left\{\frac{\exp(ik'R)}{(k'+k)}+\frac{\exp(ik'R)}{(k'-k)}\right\}dk', \tag{10.165b}$$

and

$$I_2 = \frac{1}{16\pi^2 Ri}\int_{k'=-\infty}^{\infty}\left\{\frac{\exp(-ik'R)}{(k'+k)}+\frac{\exp(-ik'R)}{(k'-k)}\right\}dk'. \tag{10.165c}$$

The integral representation of the Green function over the range $(-\infty,+\infty)$ is a fruitful ploy, since it enables us to use contour integration in the complex-k' plane to evaluate it, notwithstanding the presence of the poles in the integrands (Eqs. 10.164a,b and 10.165a,b). The complex-k' plane includes the single point at infinity, so the integration in Eq. 10.164 can be carried over a closed semi-circle of infinite radius that goes through the single point at infinity, circumventing the singularities at $k' = \pm k$. Cauchy's theorem enables assigning a value to the integral: given the Laurent expansion of a function of $f(z)$ about a point z_0,

$$f(z) = \sum_{n=-\infty}^{+\infty} a_n (z - z_0)^n, \tag{10.166}$$

the contour integral $\oint_C f(z)dz$ along a closed path C that encircles an isolated singularity at z_0 is given by

$$\frac{1}{2\pi i}\oint_C f(z)dz = a_{-1}, \tag{10.167}$$

which is the coefficient of $\frac{1}{z - z_0}$. For the evaluation of the integral I_1 (Eq. 10.165b), the semi-circle is selected using Jordan's lemma to be in the *upper* half (Fig. 10.10) of the complex k'-plane, since the integrand in I_1 has $\exp(ik'R)$ in the numerator, and $R > 0$. A suitable contour for the evaluation of the integral in Eq. 10.165 is thus a semi-circle in the upper half of the complex k'-plane. Circumvention of the poles is possible, but it can be done in various alternative ways. Specific choice of the contour is dictated by the boundary condition. For the description of potential scattering, our choice shall be governed by the *outgoing* wave boundary condition, discussed in Section 10.1.

Four different contours can be chosen in accordance with the boundary conditions, by flexing very slightly the part of the path of integration along the real axis to circumvent (include, or exclude) the poles at $k' = \pm k$. These are shown in Fig. 10.11.

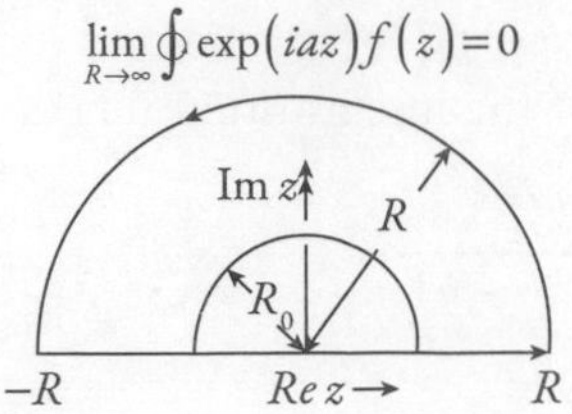

Fig. 10.10 For a function $f(z)$ that is analytic in the upper half of the complex plane in the region $|z| > R_0$, and $\lim_{R\to\infty} |f(z)| \to 0$ then $\lim_{R\to\infty} \oint \exp(iaz) f(z) = 0$ (for $(a > 0)$; the path integral being over the large semicircle. When $a < 0$, the integration must be over a similar semi-circle in the lower half of the complex plane. This is a widely used result in complex analysis known as the Jordan's lemma.

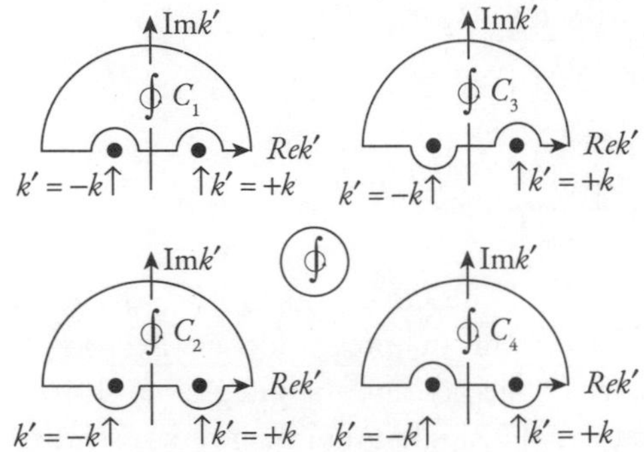

Fig. 10.11 The closed paths C_1, C_2, C_3 and C_4, show four different ways to choose the path of integration for the evaluation of the integral I_1 (Eqs. 10.165a,b). The choice C_1 gives zero, and the remaining three choices correspond to different boundary conditions listed in Table 10.1.

As shown in Fig. 10.11, there are four alternative paths which are semi-circular in the upper half of the complex k'-plane which circumvent the poles. The path C_1 excludes both the poles, while the path C_2 includes both of them. The path C_3 includes the pole at $-k$ but excludes that at $+k$, while the path C_4 excludes the pole at $-k$ but includes that at $+k$. The circumvention of the poles on the real axis is effected by encircling them over semi-circular paths having infinitesimal radius. Results of the contour integration I_1 using Eq. 10.167 are compiled in Table 10.1.

For the evaluation of I_2, the semi-circle is in the lower half (Fig.10.12) of the complex k'-plane, since the integrand in I_2 has $\exp(-ik'R)$ in the numerator, instead of $\exp(+ik'R)$. Again, the path of integration may circumvent the poles on the real k'-axis in four alternative ways, depicted in Fig. 10.12. Results of the contour integration I_2 using Eq. 10.167 are compiled in Table 10.2.

Our interest is in solving the integral equation of potential scattering for the *outgoing* boundary condition. We shall denote the Green function for the outgoing wave boundary condition with a superscript "+." As seen from Tables 10.1 and 10.2, it is given by

$$G_0^+(\vec{R}) = \frac{-1}{16\pi^2 Ri} I_1(C_4) + \frac{1}{16\pi^2 Ri} I_2(C_3), \tag{10.168a}$$

Table 10.1 Values of the contour integrals I_1 over the paths shown in Fig. 10.11 (infinite semi-circle in the upper half of the complex k'-plane). The time-dependent dynamical phase $\exp(-i\omega t)$ is central to the interpretation of the solution as "ingoing" or "outgoing" wave.

$I_1(C_1)=0$	$I_1(C_3)=2\pi i\exp(-ikR)$ **Ingoing wave boundary condition**
$I_1(C_2)=2\pi i\{\exp(ikR)+\exp(-ikR)\}$ **Ingoing wave plus outgoing wave boundary condition**	$I_1(C_4)=2\pi i\exp(ikR)$ **Outgoing wave boundary condition**

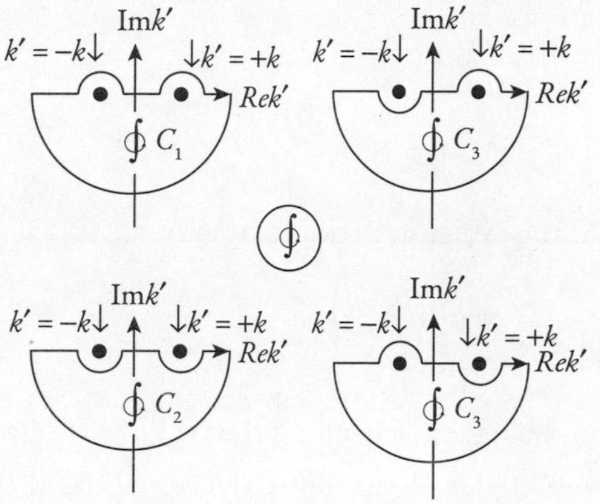

Fig. 10.12 Four different paths of integration for the evaluation of the integral I_2 (Eqs. 10.165a,c). The choice C_2 gives zero; the remaining three choices correspond to different boundary conditions listed in Table 10.2.

where $I_1(C_4)$ denotes the value of the integral I_1 along the path C_4 and $I_2(C_3)$ denotes the integral I_2 along the path C_3. Hence, the Green function with outgoing wave boundary condition is

$$G_0^+(\vec{R}) = -1\left\{\frac{2\pi i\exp(ikR)}{16\pi^2 Ri}+\frac{2\pi i\exp(ikR)}{16\pi^2 Ri}\right\} = -\frac{\exp(ikR)}{4\pi R}. \tag{10.168b}$$

Table 10.2 Values of the contour integrals I_2 over the paths shown in Fig. 10.12 (infinite semi-circle in the lower half of the complex k'-plane). The time-dependent dynamical phase $\exp(-i\omega t)$ is central to the interpretation of the solution as "ingoing" or "outgoing" wave.

$I_2(C_1)=-2\pi i\{\exp(ikR)+\exp(-ikR)\}$ **Ingoing wave plus outgoing wave boundary condition**	$I_2(C_3)=-2\pi i\exp(ikR)$ **Outgoing wave boundary condition**
$I_2(C_2)=0$	$I_2(C_4)=-2\pi i\exp(-ikR)$ **Ingoing wave boundary condition**

There is an alternative to circumventing the poles, by *departing* from the real axis at the poles. The complementary method is to carry out the contour integration *without* leaving the real axis, but *displace* the poles by an infinitesimal amount ε' (Fig. 10.13). One can then take the limit $\varepsilon' \to 0$. Displacing the poles to $k' = \pm(k + i\varepsilon')$, we have

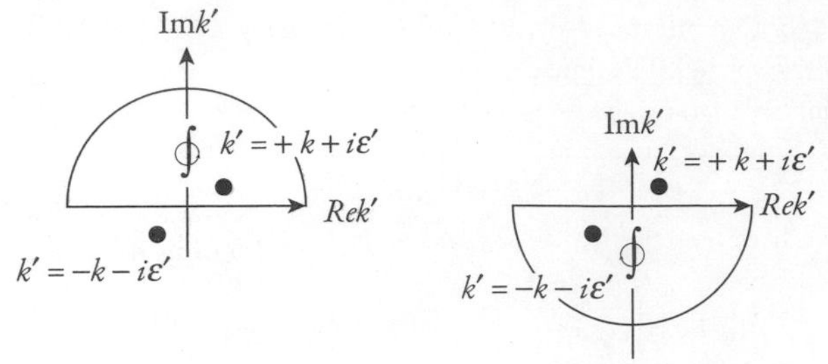

Fig. 10.13 This contour integration path includes the entire real axis. The poles are avoided by displacing them infinitesimally and then taking the limit as the magnitude of the displacement tends to zero.

$$k'^2 = (k^2 + i2k\varepsilon' + \varepsilon'^2) \simeq (k^2 + i\varepsilon) \tag{10.169a}$$

$$\text{with } \varepsilon = 2k\varepsilon'. \tag{10.169b}$$

Instead of Eq. 10.163a, the Green function may now be written as

$$G_0^+(\vec{R}) = \frac{-1}{(2\pi)^3} \lim_{\varepsilon \to 0^+} \iiint d^3\vec{k}' \frac{\exp\left(i\vec{k}' \cdot \vec{R}\right)}{k'^2 - k^2 - i\varepsilon}, \tag{10.170}$$

with the path of integration being entirely along the real k' axis, apart from the semi-circle at infinity (Fig. 10.13). The boundary condition that needs to be specified is indicated again by the superscript "+" placed in Eq.10.170.

Equation 10.157 inclusive of the Green function chosen specifically with reference to the boundary conditions is known as the *Lippmann–Schwinger equation for potential scattering*. It is equivalent to the Schrödinger's differential equation together with the boundary condition, the latter being included in the specific choice of the Green function. For the *outgoing* wave boundary condition, the Lippmann–Schwinger equation is

$$\psi^+_{\vec{k}_i}(\vec{r}) = \phi_{\vec{k}_i}(\vec{r}) + \iiint d^3\vec{r}'\, G_0^+(k, \vec{r}, \vec{r}')\, U(\vec{r}')\, \psi^+_{\vec{k}_i}(\vec{r}'), \tag{10.171a}$$

$$\text{i.e., } \psi^+_{\vec{k}_i}(\vec{r}) = \phi_{\vec{k}_i}(\vec{r}) - \frac{1}{4\pi} \iiint d^3\vec{r}' \left\{ \frac{\exp(ikR)}{R} \right\} U(\vec{r}')\, \psi^+_{\vec{k}_i}(\vec{r}'). \tag{10.171b}$$

Since in a collision experiment the detector is placed well away from the scattering target, we examine the distance R between the source point and the field point (Fig. 10.9) in the asymptotic limit. This distance is given by

$$R = \left\{(\vec{r} - \vec{r}') \cdot (\vec{r} - \vec{r}')\right\}^{\frac{1}{2}} = r\left\{1 + \frac{1}{2}\frac{r'^2 - 2rr'\cos\theta}{r^2} + \frac{\frac{1}{2}\left(\frac{1}{2} - 1\right)}{1 \times 2}\frac{(r'^2 - 2rr'\cos\theta)^2}{r^4} - \ldots\right\}, \tag{10.172a}$$

$$\text{i.e., } R = r - \hat{r} \cdot \vec{r}' + \frac{r'^2}{2r} - \frac{r'^4}{8r^3} + \frac{r'^3\cos\theta}{2r^2} - \frac{r'^2\cos^2\theta}{2r} .. = r - \hat{r} \cdot \vec{r}' + \frac{r'^2}{2r}(1 - \cos^2\theta) - \frac{r'^4}{8r^3} + \frac{r'^3\cos\theta}{2r^2} .., \tag{10.172b}$$

$$\text{or, } R = r - \hat{r}\bullet\vec{r}' + \frac{r'^2\sin^2\theta}{2r} - \frac{r'^4}{8r^3} + \frac{r'^3\cos\theta}{2r^2} .. = r - \hat{r}\bullet\vec{r}' + \frac{(\hat{r}\times\vec{r}')^2}{2r} + ... \quad (10.172c)$$

Hence, we have

$$\frac{\exp(ikR)}{R} \simeq \frac{\exp\left(ik\left(r-\hat{r}\bullet\vec{r}'+\frac{r'^2\sin^2\theta}{2r}\right)\right)}{r-\hat{r}\bullet\vec{r}'+\frac{r'^2\sin^2\theta}{2r}} = \frac{\exp(ikr)\times\exp(-ik(\hat{r}\bullet\vec{r}'))\times\exp\left(ik\left(\frac{r'^2\sin^2\theta}{2r}\right)\right)}{r\left(1-\frac{\hat{r}\bullet\vec{r}'}{r}+\frac{r'^2\sin^2\theta}{2r^2}\right)}, \quad (10.173a)$$

$$\text{i.e., } \frac{\exp(ikR)}{R} \simeq \frac{\exp(ikr)}{r}\exp(-ikr'\cos\theta)\exp\left(ik\left(\frac{r'^2\sin^2\theta}{2r}\right)\right)\left(1-\frac{\hat{r}\bullet\vec{r}'}{r}+\frac{r'^2\sin^2\theta}{2r^2}\right)^{-1}, \quad (10.173b)$$

$$\text{or, } \frac{\exp(ikR)}{R} \simeq \frac{\exp(ikr)}{r}\exp(-ikr'\cos\theta)\left\{\begin{matrix} 1+\frac{ikr'^2\sin^2\theta}{2r}+ \\ +\frac{1}{2}\left(\frac{ikr'^2\sin^2\theta}{2r}\right)^2+.. \end{matrix}\right\}\left(\begin{matrix} 1+\frac{r'\cos\theta}{r} \\ -\frac{r'^2\sin^2\theta}{2r^2}+.. \end{matrix}\right). \quad (10.173c)$$

In the asymptotic limit, we therefore get

$$\left\{\frac{\exp(ikR)}{R};\ r\to\infty\right\} \simeq \left\{\frac{\exp(ikr)}{r}\exp\left(-i\vec{k}_f\bullet\vec{r}'\right)\right\}, \quad (10.174)$$

where $k\hat{r} = k\hat{k}_f = \vec{k}_f$ (Fig. 10.9). (10.175)

In the asymptotic limit, the Lippmann–Schwinger equation can therefore be written as

$$\psi^+_{\vec{k}_i}(\vec{r}) = (2\pi)^{-\frac{3}{2}}\exp\left(i\vec{k}_i\bullet\vec{r}\right) - \frac{1}{4\pi}\frac{\exp(ikr)}{r}\iiint d^3\vec{r}'\left\{\exp\left(-i\vec{k}_f\bullet\vec{r}'\right)\right\}U(\vec{r}')\psi^+_{\vec{k}_i}(\vec{r}'), \quad (10.176a)$$

$$\text{or, } \psi^+_{\vec{k}_i}(\vec{r}) = (2\pi)^{-\frac{3}{2}}\left\{e^{i\vec{k}_i\bullet\vec{r}} + \frac{e^{ikr}}{r}f\left(\hat{k}_i,\hat{k}_f\right)\right\}, \quad (10.176b)$$

where we recognize the scattering amplitude (Eq. 10.9b) as

$$f\left(\hat{k}_i,\hat{k}_f\right) = -2\pi^2\iiint d^3\vec{r}'\phi_{\vec{k}_f}(\vec{r}')U(\vec{r}')\psi^+_{\vec{k}_i}(\vec{r}') = -2\pi^2\left\langle\phi_{\vec{k}_f}\middle|U\middle|\psi^+_{\vec{k}_i}\right\rangle = -\left(\frac{2\pi}{\hbar}\right)^2 m\left\langle\phi_{\vec{k}_f}\middle|V\middle|\psi^+_{\vec{k}_i}\right\rangle. \quad (10.177)$$

Above, we have the *integral representation of the scattering amplitude* which directly lends itself to an interpretation in terms of a transition matrix element:

$$f\left(\hat{k}_i,\hat{k}_f\right) = -\left(\frac{2\pi}{\hbar}\right)^2 mT_{fi}, \quad (10.178)$$

wherein

$$T_{fi} = \left\langle \phi_{\vec{k}_f} \middle| V \middle| \psi^{+}_{\vec{k}_i} \right\rangle \tag{10.179}$$

is called as the *transition matrix element*.

In terms of the transition matrix element, the differential cross section for scattering is then given by

$$\frac{d\sigma}{d\Omega} = m^2 \left(\frac{2\pi}{\hbar} \right)^4 \left| T_{fi} \right|^2. \tag{10.180}$$

As mentioned earlier, the present formalism of the problem of potential scattering in the framework of the Lippmann–Schwinger equation provides an opportunity to develop a series of approximations in an iterative manner. The approximations are known as the *Born series of approximations*. An approximation is necessitated by the fact that the Lippmann–Schwinger equation expresses the wavefunction on the left-hand side (Eq. 10.157) in terms of an integral, whose integrand on the right-hand side albeit has the *unknown* function; it is *just* what we want to know:

$$\overset{?}{\overbrace{\psi^{+}_{\vec{k}_i}(\vec{r})}} = \phi_{\vec{k}_i}(\vec{r}) + \iiint d^3\vec{r}'\, G^{+}_{0}(k,\vec{r},\vec{r}')\, U(\vec{r}')\, \overset{?}{\overbrace{\psi^{+}_{\vec{k}_i}(\vec{r}')}}. \tag{10.181}$$

Since $\psi^{+}_{\vec{k}_i}$ in the integrand is unknown, we *presume*, to 0th order approximation, that it is the same as the incident wave. We shall denote the 0th order solution by placing (0) as a superscript on the wavefunction:

$$\psi^{+(0)}_{\vec{k}_i}(\vec{r}) = \phi_{\vec{k}_i}(\vec{r}) + 0 = (2\pi)^{-3/2} \exp\left(i\vec{k}_i \cdot \vec{r}\right). \tag{10.182}$$

This solution is not helpful; it ignores the scattering potential. One can, however, make incremental headway by improvising on the 0th approximation. Step by step superior solutions to the potential scattering problem can be constructed by employing the nth order approximation in the *integrand* for the $(n+1)^{\text{th}}$ order solution ($n = 0,1,2,3,..$), as organized in Table 10.3. The cascading iterative approximations organized in Table 10.3 constitute the *Born approximation series*.

Table 10.3 The series of Born approximations

0th order solution	$\psi^{+(0)}_{\vec{k}_i}(\vec{r}) = \phi_{\vec{k}_i}(\vec{r}) + 0 = (2\pi)^{-3/2} \exp\left(i\vec{k}_i \cdot \vec{r}\right)$
1st order solution	$\psi^{+(1)}_{\vec{k}_i}(\vec{r}) = \phi_{\vec{k}_i}(\vec{r}) + \iiint d^3\vec{r}'\, G^{+}_{0}(k,\vec{r},\vec{r}')\, U(\vec{r}')\, \psi^{+(0)}_{\vec{k}_i}(\vec{r}')$
2nd order solution	$\psi^{+(2)}_{\vec{k}_i}(\vec{r}) = \phi_{\vec{k}_i}(\vec{r}) + \iiint d^3\vec{r}'\, G^{+}_{0}(k,\vec{r},\vec{r}')\, U(\vec{r}')\, \psi^{+(1)}_{\vec{k}_i}(\vec{r}')$
...	...
n^{th} order solution	$\psi^{+(n)}_{\vec{k}_i}(\vec{r}) = \phi_{\vec{k}_i}(\vec{r}) + \iiint d^3\vec{r}'\, G^{+}_{0}(k,\vec{r},\vec{r}')\, U(\vec{r}')\, \psi^{+(n-1)}_{\vec{k}_i}(\vec{r}')$

It is now instructive to write the 2nd-order solution elaborately:

$$\psi^{+(2)}_{\vec{k}_i}(\vec{r}) = \begin{bmatrix} \phi_{\vec{k}_i}(\vec{r}) + \\ \iiint d^3\vec{r}'\, G_0^+(k,\vec{r},\vec{r}')U(\vec{r}') \begin{Bmatrix} \phi_{\vec{k}_i}(\vec{r}) + \\ \iiint d^3\vec{r}''\, G_0^+(k,\vec{r},\vec{r}'')U(\vec{r}'')\psi^{+(0)}_{\vec{k}_i}(\vec{r}'') \end{Bmatrix} \end{bmatrix}, \quad (10.183a)$$

$$\text{i.e., } \psi^{+(2)}_{\vec{k}_i}(\vec{r}) = \begin{bmatrix} \phi_{\vec{k}_i}(\vec{r}) + \\ +\iiint d^3\vec{r}'\, G_0^+(k,\vec{r},\vec{r}')U(\vec{r}')\phi_{\vec{k}_i}(\vec{r}') \\ +\iiint d^3\vec{r}' \iiint d^3\vec{r}''\, G_0^+(k,\vec{r},\vec{r}')U(\vec{r}')G_0^+(k,\vec{r},\vec{r}'')U(\vec{r}'')\phi_{\vec{k}_i}(\vec{r}'') \end{bmatrix}. \quad (10.183b)$$

Note that in the 2nd-order term, the potential U appears maximum twice under the integrand. By extension, we see that in the nth order term, the potential appears a maximum of n times. We therefore develop a compact notation, and rewrite the 2nd-order solution as

$$\psi^{+(2)}_{\vec{k}_i}(\vec{r}) = \phi_{\vec{k}_i}(\vec{r}) + \iiint d^3\vec{r}'\{K_1(k,\vec{r},\vec{r}')\}\phi_{\vec{k}_i}(\vec{r}') + \iiint d^3\vec{r}''\{K_2(k,\vec{r},\vec{r}'')\}\phi_{\vec{k}_i}(\vec{r}''), \quad (10.184a)$$

where the kernels of the integrals are

$$K_1(k,\vec{r},\vec{r}') = G_0^+(k,\vec{r},\vec{r}')U(\vec{r}'), \quad (10.184b)$$

$$K_2(k,\vec{r},\vec{r}'') = \iiint d^3\vec{r}'\, G_0^+(k,\vec{r},\vec{r}')U(\vec{r}')G_0^+(k,\vec{r},\vec{r}'')U(\vec{r}''). \quad (10.184c)$$

We use the kernels to *define* $\phi_i(\vec{r}); i = 0,1,2$ as

$$\phi_0(\vec{r}) = \phi_{\vec{k}_i}(\vec{r}), \quad (10.185a)$$

$$\phi_1(\vec{r}) = \iiint d^3\vec{r}'\{K_1(k,\vec{r},\vec{r}')\}\phi_{\vec{k}_i}(\vec{r}'), \quad (10.185b)$$

$$\phi_2(\vec{r}) = \iiint d^3\vec{r}''\{K_2(k,\vec{r},\vec{r}'')\}\phi_{\vec{k}_i}(\vec{r}''), \text{ etc.} \quad (10.185c)$$

Using now the newly defined functions $\phi_0(\vec{r})$, $\phi_1(\vec{r})$ and $\phi_2(\vec{r})$, we can write the 2nd-order solution as

$$\psi^{+(2)}_{\vec{k}_i}(\vec{r}) = \phi_0(\vec{r}) + \phi_1(\vec{r}) + \phi_2(\vec{r}). \quad (10.186a)$$

By extension, we may write the nth order solution as

$$\psi^{+(n)}_{\vec{k}_i}(\vec{r}) = \sum_{m=0}^{n} \phi_m(\vec{r}), \quad (10.186b)$$

and the complete solution as an infinite series,

$$\psi^+_{\vec{k}_i}(\vec{r}) = \sum_{m=0}^{\infty} \phi_m(\vec{r}). \tag{10.187}$$

The nth order Born approximation to the scattering amplitude therefore is given by (for $n = 1, 2$):

$$f_{B_1}(\hat{k}_i, \hat{k}_f) = -2\pi^2 \left\langle \phi_{\vec{k}_f} \middle| U \middle| \psi^{+(0)}_{\vec{k}_i} \right\rangle = -2\pi^2 \left\langle \phi_{\vec{k}_f} \middle| U \middle| \phi_{\vec{k}_i} \right\rangle, \tag{10.188a}$$

$$f_{B_2}(\hat{k}_i, \hat{k}_f) = -2\pi^2 \left\langle \phi_{\vec{k}_f} \middle| U \middle| \psi^{+(1)}_{\vec{k}_i} \right\rangle = -2\pi^2 \left\langle \phi_{\vec{k}_f} \middle| U \middle| \phi_{\vec{k}_i} \right\rangle - 2\pi^2 \left\langle \phi_{\vec{k}_f} \middle| U G_0^+ U \middle| \phi_{\vec{k}_i} \right\rangle. \tag{10.188b}$$

The nth order scattering amplitude is given by

$$f_{B_n}(\hat{k}_i, \hat{k}_f) = \sum_{j=1}^{n} \overline{f}_{B_j}, \tag{10.189a}$$

$$\text{where } \overline{f}_{B_j} = -2\pi^2 \left\langle \phi_{\vec{k}_f} \middle| U G_0^+ U \ldots. G_0^+ U \middle| \phi_{\vec{k}_i} \right\rangle \tag{10.189b}$$

is a multiple scattering amplitude (Fig. 10.14), in which the interaction potential U appears j times, and G_0^+ appears $(j-1)$ times.

$$\text{Note that } \overline{f}_{B_1} = f_{B_1}(\hat{k}_i, \hat{k}_f). \tag{10.190}$$

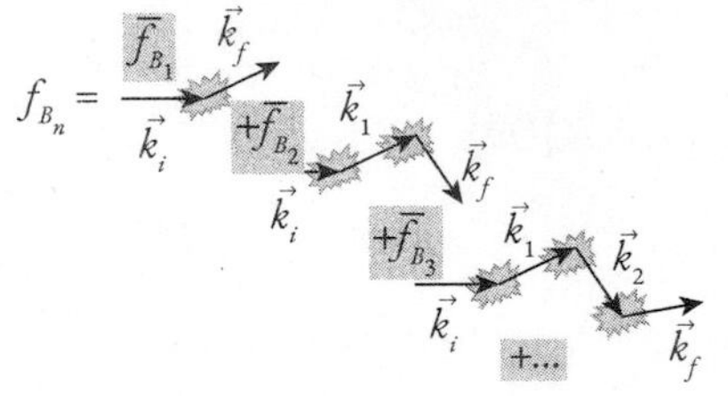

Fig. 10.14 Schematic representation of the Born series of multiple scattering (Eqs. 10.189a,b). Adapted from Chapter 8 of Reference [3].

The Born series (Eq. 10.189a,b) represents multiple scattering by the target potential. We depict it schematically in Fig. 10.14.

From Eq. 10.88a, the scattering amplitude in the 1st-order Born approximation is

$$f_{B_1}(k, \vec{\Delta}) = -2\pi^2 \iiint d^3\vec{r}\, \phi^*_{\vec{k}_f}(\vec{r}) U(\vec{r}) \phi_{\vec{k}_i}(\vec{r})$$

$$= -\frac{2\pi^2}{(2\pi)^{3/2}(2\pi)^{3/2}} \iiint d^3\vec{r} \exp\left(-i\vec{k}_f \bullet \vec{r}\right) U(\vec{r}) \exp\left(+i\vec{k}_i \bullet \vec{r}\right),$$

$$\text{i.e. } f_{B_1}(k, \vec{\Delta}) = -\frac{1}{4\pi} \iiint d^3\vec{r}\; U(\vec{r}) \exp\left(+i\vec{\Delta} \bullet \vec{r}\right) = -\frac{m}{2\pi\hbar^2} \iiint d^3\vec{r}\; V(\vec{r}) e^{+i\vec{\Delta}\bullet\vec{r}} = -\frac{m}{2\pi\hbar^2} \tilde{V}(\vec{\Delta}), \tag{10.191}$$

where $\vec{\Delta} = \vec{k}_i - \vec{k}_f$ (10.192)

is the momentum transfer,

and $\tilde{V}(\vec{\Delta}) = \iiint d^3\vec{r}\, V(\vec{r}) e^{+i\vec{\Delta}\cdot\vec{r}}$ (10.193)

is the Fourier transform of the scattering potential. In other words, *scattering amplitude in the first Born approximation is proportional to the Fourier transform of the potential.* The differential scattering cross section is

$$\left.\frac{d\sigma(\hat{\Omega})}{d\Omega}\right]_{B1} = \left|f_{B1}\left(\hat{\Omega} = \hat{k}_f\right)\right|^2. \tag{10.194}$$

From Eqs. 10.193 and 10.194 we see that the differential cross section remains the same if the *sign* of the interaction is reversed; it remains the same for attractive and repulsive interactions. The total cross section in the first Born approximation is

$$\sigma_{B1}^{Total} = \iint d\Omega \left.\frac{d\sigma(\hat{\Omega})}{d\Omega}\right]_{B1} = \int_{\varphi=0}^{2\pi} d\varphi \int_{\theta=0}^{\pi} \sin\theta d\theta \left|f_{B1}\left(\hat{\Omega} = \hat{k}_f\right)\right|^2. \tag{10.195a}$$

For a spherically symmetric scattering potential, it is convenient to choose a coordinate system that has $\hat{e}_z = \hat{\Delta}$, the direction in which the momentum transfer takes place. The total scattering cross section in the first Born approximation in this case is then given by

$$\sigma_{B1}^{Total} = 2\pi \int_{\theta=0}^{\pi} \sin\theta d\theta \left|f_{B1}\left(\hat{\Omega} = \hat{k}_f\right)\right|^2. \tag{10.195b}$$

In elastic scattering, $\left|\vec{k}_i\right| = \left|\vec{k}_f\right|$ and hence the magnitude Δ of their difference (Fig. 10.15a) is in the range $0 \le \Delta \le 2k$, corresponding to the range $0 \le \theta \le \pi$ of the angle of scattering. From Fig. 10.15a, we see that

$$\Delta = \left|\vec{k}_i - \vec{k}_f\right| = 2k\sin\frac{\theta}{2} \tag{10.196a}$$

$$\text{and } \Delta^2 = 4k^2\sin^2\frac{\theta}{2} = 2k^2(1-\cos\theta). \tag{10.196b}$$

For a spherically symmetric potential, the integral in Eq. 10.191 giving the scattering amplitude in the first Born approximation is easily determined. It is given by

$$f_{B_1}(k,\Delta) = -\frac{1}{4\pi}(2\pi)\int_{r=0}^{\infty} r^2U(r)dr \int_{\theta=0}^{\pi} \sin\theta d\theta e^{+ir\Delta\cos\theta} = -\frac{1}{2}\int_{r=0}^{\infty} r^2U(r)dr \int_{\mu=-1}^{+1} d(\cos\theta) e^{+ir\Delta(\cos\theta)}, \tag{10.197a}$$

$$\text{i.e., } f_{B_1}(k,\Delta) = \left\{-\frac{1}{2}\int_{r=0}^{\infty} r^2U(r)\left(\frac{e^{+ir\Delta} - e^{-ir\Delta}}{ir\Delta}\right)dr\right\} = -\frac{1}{\Delta}\int_{r=0}^{\infty} rU(r)\sin(r\Delta)dr. \tag{10.197b}$$

Let us now study the scattering amplitude and the differential cross section for the Yukawa potential

$$U(\vec{r}) = -\frac{U_0}{\alpha}\frac{\exp(-\alpha r)}{r} = -\zeta\frac{\exp(-\alpha r)}{r}, \tag{10.198}$$

where $a = \frac{1}{\alpha}$ is the "range" of the potential; it is the distance at which its strength drops to $1/e$ of that at the origin, and the parameter α is adjustable. We may vary it keeping the ratio $\left(U_0/\alpha\right) = \zeta$ unchanged. In the limit $\alpha \to 0$, the scattering amplitude (Eq. 10.197b) in the first Born approximation for the Yukawa potential turns out to be

$$f_{B_1}(k,\Delta) = \frac{\zeta}{\Delta}\int_{r=0}^{\infty}\exp(-\alpha r)\sin(r\Delta)dr = \frac{\zeta}{\Delta^2+\alpha^2} = \frac{\zeta}{4k^2\sin^2\frac{\theta}{2}+\alpha^2}, \tag{10.199}$$

and the differential scattering cross section is

$$\left.\frac{d\sigma}{d\Omega}\right]_{B_1} = \left|f_{B_1}(k,\Delta)\right|^2 = \left|\frac{\zeta}{4k^2\sin^2\frac{\theta}{2}+\alpha^2}\right|^2. \tag{10.200}$$

$$\sigma_{B1}^{Total} = \iint d\Omega \left.\frac{d\sigma(\hat{\Omega})}{d\Omega}\right]_{B1} = \int_{\varphi=0}^{2\pi}d\varphi\int_{\theta=0}^{\pi}\sin\theta d\theta\left|f_{B1}\left(\hat{\Omega}=\hat{k}_f\right)\right|^2. \tag{10.200a}$$

From Eqs. 10.196a,b, we have $\sin\theta d\theta = \frac{\Delta d\Delta}{k^2}$ and this allows us to express the total cross section in Eq. 10.195b as

$$\text{i.e., } \sigma_{B1}^{Total} = \frac{2\pi}{k^2}\int_{\Delta=0}^{2k}\left|f_{B1}(\Delta)\right|^2\Delta d\Delta = \frac{2\pi}{k^2}\zeta^2\int_{\Delta=0}^{2k}\frac{\Delta d\Delta}{\left(\Delta^2+\alpha^2\right)^2}, \tag{10.200b}$$

$$\text{i.e., } \sigma_{B1}^{Total} = \frac{4\pi\zeta^2}{\left(4k^2+\alpha^2\right)}. \tag{10.200c}$$

In the high energy limit $(k \to \infty)$,

$$\sigma_{B1}^{Total}(k\to\infty) = \frac{\pi\zeta^2}{k^2} = \frac{\pi\hbar^2\zeta^2}{2mE}. \tag{10.201}$$

We thus find that the total scattering cross section in the first Born approximation is inversely proportional to the energy of the incident particles in the high energy limit. The differential cross section is independent of the sign of the potential; hence it is the same for "attractive" and "repulsive" potentials. In the limit $\alpha \to 0$ (but ζ remaining the same), the Yukawa potential becomes the same as the Coulomb. The differential cross section in the first Born approximation in this limit becomes

$$\left.\frac{d\sigma}{d\Omega}\right]_{B_1}^{Coulomb} = \left|f_{B_1}(k,\Delta)\right|^2 = \left|\frac{\pi\zeta^2}{4k^2\sin^2\frac{\theta}{2}}\right|^2, \tag{10.202}$$

in complete agreement with Eq. 5.169b of Chapter 5, which we had obtained by *solving* the Schrödinger equation for the Coulomb potential.

It is remarkable that the differential scattering cross section in the first Born approximation for the Coulomb potential is essentially the same as we get from the Schrödinger equation (Eq. 5.169); we had already noted in Chapter 5 that the latter agrees *also* with the classical expression for Rutherford scattering (Appendix 8A of Chapter 8 in Ref. [13]). We note that the Fourier transform (Eq. 10.193) of the Yukawa potential is negligible beyond $r \simeq \frac{1}{\alpha} = a$. From the Fourier transform of the potential we see that it is appreciable for $\Delta \lesssim \alpha$, i.e., to small angles, for which $\Delta \simeq k\theta$. Most of the scattering therefore occurs in the forward direction (Fig. 10.15b), within a cone having an angle $\theta \lesssim \frac{\alpha}{k}$. As the angle of scattering Δ increases, the differential scattering cross section diminishes (Fig. 10.15c). One must remember that approximations are extremely useful, but there is a price to be paid. In the present case, we find that the scattering amplitude (Eqs. 10.197a,b) in the first Born approximation being essentially a real number, its imaginary part is zero:

$$\operatorname{Im} f_{B1}(\theta = 0) = 0. \tag{10.203}$$

The optical theorem therefore does not apply. Thus the unitarity of the S operator and associated conservation of flux is not satisfied in the Born approximation. The scattering amplitude includes the potential term to the first power, but the scattering cross section has a quadratic term in the potential, which is the reason behind Eq. 10.203 and the failure of the optical theorem. A perturbative expansion of the scattering amplitude and the cross section

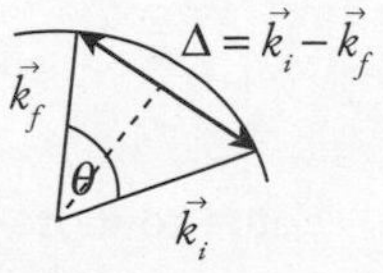

Fig. 10.15a Change in momentum in collision.

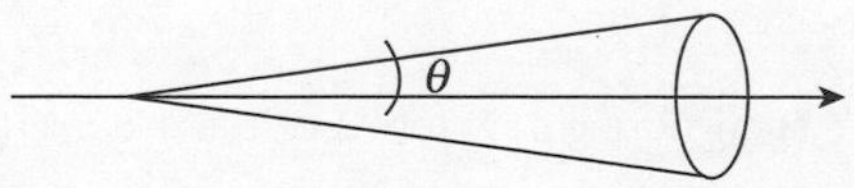

Fig. 10.15b Most of the scattering takes place in a small angular cone along the forward direction.

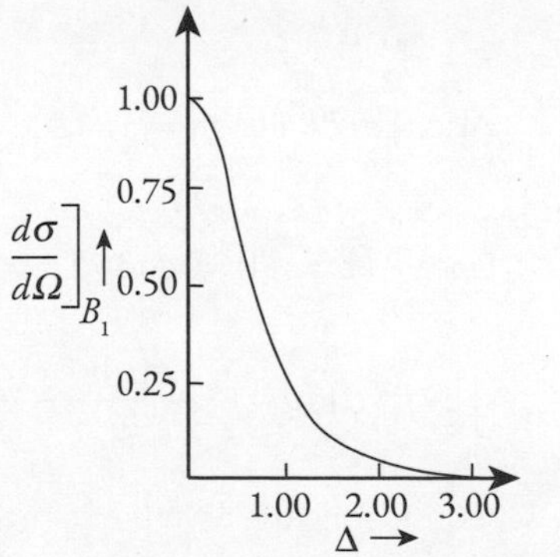

Fig. 10.15c Angular anisotropy of the differential cross section in the first Born approximation. Adapted from Chapter 8 of Reference [3].

allows comparison of terms in corresponding order of the perturbation parameter, and this enables us to identify the following *non-linear* nature of the optical theorem:

$$\sigma_{B1}^{Total} = \iint d\Omega \left| f_{B1}(\hat{\Omega}) \right|^2 = \frac{4\pi}{k} \mathrm{Im} f_{B2}(\theta = 0). \tag{10.204}$$

We expect the 1st-order Born approximation to be effective when the scattered wave is not significantly different from the incident wave, as one would expect when the scattering potential is very weak and/or the incident energy is high enough; this would justify Eq. 10.188a. In this context, *high energy* would mean an energy of the incident projectiles that is significantly higher than the interaction energy between the incident particle and the scattering potential. Presence of many electrons in the target, however, results in many-electron correlations, as a result of which the Born approximation breaks down even at high energies [12–14], especially in the angular distribution of the scattering yield. Barring such effects, the Born approximation is very useful at high energies to predict gross features of quantum collisions and is hence widely used.

Solved Problems

P10.1: Determine $\int_{-1}^{+1} e^{i\rho\mu} P_{\ell'}(\mu) d\mu$ and find the coefficient a_ℓ. Hint: Since a_ℓ must be independent of the distance, consider the asymptotic region and ignore terms of order $O(\rho^{-2})$ relative to those of order $O(\rho^{-1})$.

Solution:

$$\int_{-1}^{+1} e^{i\rho\mu} P_\ell(\mu) d\mu = \sum_{\ell'=0}^{\infty} a_{\ell'} \left[\int_{-1}^{+1} P_\ell(\mu) P_{\ell'}(\mu) d\mu \right] j_{\ell'}(\rho) = \sum_{\ell'=0}^{\infty} a_{\ell'} \left[\frac{2}{2\ell'+1} \delta_{\ell'\ell} \right] j_{\ell'}(\rho)$$

Hence, $\int_{-1}^{+1} e^{i\rho\mu} P_\ell(\mu) d\mu = a_\ell \left[\frac{2}{2\ell+1} \right] j_\ell(\rho)$.

Also, $\int_{-1}^{+1} e^{i\rho\mu} P_\ell(\mu) d\mu = \int_{-1}^{+1} P_\ell(\mu) e^{i\rho\mu} d\mu = \left[P_\ell(\mu) \frac{e^{i\rho\mu}}{i\rho} \right]_{-1}^{+1} - \int_{-1}^{+1} P'_\ell(\mu) \frac{e^{i\rho\mu}}{i\rho} d\mu$, where $P'_\ell(\mu) = \frac{d}{d\mu} P_\ell(\mu)$.

Since $P_\ell(\mu = +1) = 1$ and $P_\ell(\mu = -1) = (-1)^\ell P_\ell(\mu = 1) = (-1)^\ell$,

$$\int_{-1}^{+1} e^{i\rho\mu} P_\ell(\mu) d\mu = \frac{e^{i\rho} - (-1)^\ell e^{-i\rho}}{i\rho} - \frac{1}{i\rho} \int_{-1}^{+1} P'_\ell(\mu) e^{i\rho\mu} d\mu.$$

Considering the asymptotic region, and ignoring now terms of the order $O(\rho^{-2})$ relative to those of order $O(\rho^{-1})$, we get

$$a_\ell \left[\frac{2}{2\ell+1} \right] j_\ell(\rho) = \frac{e^{i\rho} - (-1)^\ell e^{-i\rho}}{i\rho} = e^{i\frac{\ell\pi}{2}} \left(\frac{e^{i\rho} e^{-i\frac{\ell\pi}{2}} - e^{i\frac{\ell\pi}{2}} e^{-i\rho}}{i\rho} \right) = e^{i\frac{\ell\pi}{2}} \left(\frac{e^{i\left(\rho - \frac{\ell\pi}{2}\right)} - e^{-i\left(\rho - \frac{\ell\pi}{2}\right)}}{i\rho} \right),$$

since $(-1)^{\ell} = \left(e^{i\pi}\right)^{\ell} = e^{i\ell\pi}$. Finally, noting that $j_{\ell}(\rho;\rho \to \infty) = \dfrac{\sin\left(\rho - \dfrac{\ell\pi}{2}\right)}{\rho}$, we get $a_{\ell} = i^{\ell}(2\ell + 1)$. We have used the asymptotic limit $\rho \to \infty$, but that does not impact the result for a_{ℓ} since it is independent of distance.

P10.2: Obtain an expression for the total scattering cross section in terms of the partial waves phase shifts.

Solution:

We shall use Eq. 10.49(b) and Eq. 10.79(a).

$$\sigma_{tot} = \iint \frac{d\sigma}{d\Omega} d\Omega = \int_{\theta=0}^{\pi} \int_{\varphi=0}^{2\pi} |f(\Omega)|^2 \sin\theta d\theta d\varphi = 2\pi \int_{\theta=0}^{\pi} |f(\Omega)|^2 \sin\theta d\theta.$$

$$\text{Hence, } \sigma_{tot} = 2\pi \int_{\theta=0}^{\pi} \sum_{\ell=0}^{\infty} \sum_{\ell'=0}^{\infty} (2\ell+1)(2\ell'+1) P_{\ell}(\cos\theta) P_{\ell'}(\cos\theta) \exp\left(i\left(\delta_{\ell} - \delta_{\ell'}\right)\right) \sin\delta_{\ell} \sin\delta_{\ell'} \sin\theta d\theta.$$

$$\text{Since } \int_0^{\pi} P_{\ell}(\cos\theta) P_{\ell'}(\cos\theta) \sin\theta d\theta = \frac{2}{2\ell+1}\delta_{\ell\ell'}, \text{ we get } \sigma_{tot} = \frac{4\pi}{k^2} \sum_{\ell=0}^{\infty} (2\ell+1)\sin^2\delta_{\ell} = \sum_{\ell=0}^{\infty} \sigma_{\ell},$$

where the cross section for the ℓth partial wave is $\sigma_{\ell}(k) = \dfrac{4\pi}{k^2}(2\ell+1)\sin^2\delta_{\ell}(k)$.

P10.3: How important is it to sum over the orbital angular momentum ℓ up to ∞ in obtaining the scattering cross section in terms of the partial waves?

Solution:

Consider scattering of particles having incident momentum $\vec{p}$ arriving at the scattering center at an impact parameter (shortest distance from the scattering center) "b" and assume that the scattering potential has practically no effect beyond a range "a." The classical angular momentum of the incident particle is $\vec{L} = \vec{b} \times \vec{p}$. Making the semi-classical approximation $\hbar\ell \simeq \sqrt{\hbar^2\ell(\ell+1)}$ (which improves with the value of the orbital angular momentum), we have $\left|\vec{L}\right| = \left|\vec{b} \times \vec{p}\right| = bp \simeq \hbar\ell$. Since we expect practically no scattering when $b > a$, we see that $\ell_{max} \le \dfrac{ap}{\hbar}$; i.e., $\ell_{max} \le ka$. Often a few partial waves are sufficient, depending on the accuracy one seeks, although in some sophisticated computations hundreds of partial waves can be used. The diminishing importance of higher orbital angular momenta in partial wave analysis is due the centrifugal barrier effect, discussed in Chapter 5.

P10.4: Consider an experiment involving elastic scattering of neutrons. The scattering phase shifts measured in this experiment are $\delta_0 = 95°, \delta_1 = 72°, \delta_3 = 60°, \delta_4 = 18°, \delta_5 = 5°$, and all higher phase shifts being negligible. Find the total scattering cross section.

Solution:

$$\sigma_{tot} = \frac{4\pi}{k^2}\left(\sin^2\delta_0 + 3\sin^2\delta_1 + 5\sin^2\delta_2 + 7\sin^2\delta_3 + 9\sin^2\delta_4 + 11\sin^2\delta_5\right) = \frac{4\pi}{k^2}10.702$$

P10.5: In the early 1920s, Ramsauer and Towsend independently discovered that the cross section for scattering of electrons at some energies from gaseous atoms dropped significantly at some energies. Electrons simply seem to go through the target, as if it were transparent. Explain the Ramsauer–Townsend effect.

Solution:

From P10.3 we know that only a few partial waves with the smallest values of the orbital angular momentum are expected to contribute to scattering. At very low energies, only the s-wave $(\ell = 0)$ is important. The Ramsauer–Townsend effect is the effective disappearance of scattering when $\delta_{\ell=0}(k) = n\pi$, since $\sigma_\ell = \sin^2 \delta_\ell(k)$. [Please refer to S.G. Kukolich, Demonstration of Ramsauer–Townsend Effect in Xenon. *Am. J. Phys.* 36:8, 701–703 (1968).]

P10.6: A particle of mass m is scattered from a spherical repulsive potential of radius R, with $V(r) = V_0$ for $r \le R$. The potential is zero for $r > R$. Determine the total cross section in the limit of low energies.

Solution:

From Eq. 10.191: $f_{B_1}(k,\vec{\Delta}) = -\dfrac{m}{2\pi\hbar^2}\iiint d^3\vec{r}V(\vec{r})e^{i\vec{\Delta}.\vec{r}}$, where $\vec{\Delta} = \vec{k}_i - \vec{k}_f$.

Using Eq. 10.197b,

$$f_{B_1}(k,\vec{\Delta}) = -\frac{2m}{\Delta\hbar^2}\int_0^\infty rV(r)\sin(\Delta r)dr = -\frac{2m}{\Delta\hbar^2}\int_0^R rV_0\sin(\Delta r)dr = -\frac{2mV_0}{\hbar^2\Delta}\left[\frac{\sin(\Delta R)}{\Delta^2} - \frac{R\cos(\Delta R)}{\Delta}\right].$$

In the limit $\Delta R \to 0$, we get isotropic cross section $\dfrac{d\sigma}{d\Omega} = |f(\Omega)|^2 = \dfrac{4m^2V_0^2R^6}{9\hbar^4}$ and the total cross section is

$$\sigma_{tot} = \iint \frac{d\sigma}{d\Omega}d\Omega = \frac{16\pi m^2V_0^2R^6}{9\hbar^4}.$$

P10.7: With reference to the Solved Problem P3.5 (Chapter 3), express the coefficients B and F in terms of A and G, i.e., express the amplitudes of the outgoing waves in terms of those of the incoming waves (note that "ingoing" and "incoming" waves mean the same thing in the present context).

Solution:

We must find the 2 × 2 matrix $\mathbb{S}$ such that $\begin{pmatrix} B \\ F \end{pmatrix} = \begin{pmatrix} S_{11} & S_{12} \\ S_{21} & S_{22} \end{pmatrix}\begin{pmatrix} A \\ G \end{pmatrix}$.

From the solution of P3.5 (Chapter 3), we have

$$\begin{bmatrix} A \\ B \end{bmatrix} = \begin{bmatrix} \dfrac{e^{2ika}\left(2\cosh 2\kappa a + i\gamma_- \sinh 2\kappa a\right)}{2} & \dfrac{i\gamma_+}{2}\sinh 2\kappa a \\ -\dfrac{i\gamma_+}{2}\sinh 2\kappa a & \dfrac{e^{-2ika}\left(2\cosh 2\kappa a - i\gamma_- \sinh 2\kappa a\right)}{2} \end{bmatrix}\begin{bmatrix} F \\ G \end{bmatrix}.$$

Re-arranging: $$\begin{bmatrix} B \\ F \end{bmatrix} = \begin{bmatrix} \dfrac{i\gamma_+ \sinh 2\kappa a}{e^{-2ika}\left(2\cosh 2\kappa a - i\gamma_- \sinh 2\kappa a\right)} & \dfrac{2}{e^{-2ika}\left(2\cosh 2\kappa a - i\gamma_- \sinh 2\kappa a\right)} \\ \dfrac{2}{e^{-2ika}\left(2\cosh 2\kappa a - i\gamma_- \sinh 2\kappa a\right)} & \dfrac{i\gamma_+ \sinh 2\kappa a}{e^{2ika}\left(2\cosh 2\kappa a - i\gamma_- \sinh 2\kappa a\right)} \end{bmatrix} \begin{bmatrix} A \\ G \end{bmatrix},$$

Therefore, $$\mathbb{S} = \begin{bmatrix} \dfrac{i\gamma_+ \sinh 2\kappa a}{e^{-2ika}\left(2\cosh 2\kappa a - i\gamma_- \sinh 2\kappa a\right)} & \dfrac{2}{e^{-2ika}\left(2\cosh 2\kappa a - i\gamma_- \sinh 2\kappa a\right)} \\ \dfrac{2}{e^{-2ika}\left(2\cosh 2\kappa a - i\gamma_- \sinh 2\kappa a\right)} & \dfrac{i\gamma_+ \sinh 2\kappa a}{e^{2ika}\left(2\cosh 2\kappa a - i\gamma_- \sinh 2\kappa a\right)} \end{bmatrix}.$$

P10.8: With reference to the above problem, P10.7, show that the S-matrix is unitary.

Solution:

For the S-matrix in P10.7, we have $$S^\dagger = \begin{bmatrix} \dfrac{i\gamma_+ \sinh 2\kappa a}{e^{-2ika}\left(2\cosh 2\kappa a - i\gamma_- \sinh 2\kappa a\right)} & \dfrac{2}{e^{-2ika}\left(2\cosh 2\kappa a - i\gamma_- \sinh 2\kappa a\right)} \\ \dfrac{2}{e^{-2ika}\left(2\cosh 2\kappa a - i\gamma_- \sinh 2\kappa a\right)} & \dfrac{i\gamma_+ \sinh 2\kappa a}{e^{-2ika}\left(2\cosh 2\kappa a - i\gamma_- \sinh 2\kappa a\right)} \end{bmatrix}.$$

Hence $S^\dagger S = I$; the S-matrix is unitary.

P10.9: Prove that Maxwell's equations are invariant under time reversal.

Solution:

$$\vec{\nabla}\bullet\vec{E}(\vec{r},t) = \frac{\rho(\vec{r},t)}{\varepsilon_0} \Rightarrow \vec{\nabla}\bullet\vec{E}(\vec{r},-t) = \frac{4\pi\rho(\vec{r},-t)}{\varepsilon_0}, \text{ since } \vec{E}(\vec{r},-t) = \vec{E}(\vec{r},t) \text{ and } \rho(\vec{r},-t) = \rho(\vec{r},t).$$

$$\vec{\nabla}\bullet\vec{B}(\vec{r},t) = 0 \Rightarrow \vec{\nabla}\bullet\vec{B}(\vec{r},-t) = -\vec{\nabla}\bullet\vec{B}(\vec{r},t) = -0 = 0 \text{ since } \vec{B}(\vec{r},-t) = -\vec{B}(\vec{r},t).$$

Likewise, $$\vec{\nabla}\times\vec{B}(\vec{r},-t) = \mu_0\vec{J}_f(\vec{r},-t) + \mu_0\varepsilon_0\frac{\partial\vec{E}(\vec{r},-t)}{\partial t},$$

since $\vec{B}(\vec{r},-t) = -\vec{B}(\vec{r},t)$; $\dfrac{\partial}{\partial(-t)} = -\dfrac{\partial}{\partial t}$; $\vec{E}(\vec{r},-t) = \vec{E}(\vec{r},t)$; $\vec{J}_f(\vec{r},-t) = -\vec{J}_f(\vec{r},t)$.

Finally, time reversal symmetry of the equation for the curl of the electric field results from the fact that

$$\vec{\nabla}\times\vec{E}(\vec{r},t) = \vec{\nabla}\times\vec{E}(\vec{r},-t) \text{ and } -\frac{\partial\vec{B}(\vec{r},-t)}{\partial(-t)} = -\frac{\partial\vec{B}(\vec{r},t)}{\partial t}.$$

P10.10: Prove that the Lorentz's force is invariant under time reversal.

Solution:

$$\vec{F}(\vec{r},t) = q\left(\vec{E}(\vec{r},t) + \vec{v}\times\vec{B}(\vec{r},t)\right) \Rightarrow \vec{F}(\vec{r},-t) = q\left(\vec{E}(\vec{r},-t) + (-\vec{v})\times\vec{B}(\vec{r},-t)\right) = q\left(\vec{E}(\vec{r},t) + (-\vec{v})\times\left(-\vec{B}(\vec{r},t)\right)\right).$$

P10.11: Prove that under time reversal an electromagnetic wave reverses its direction of propagation, but the polarization remains unchanged.

Solution:

Poynting vector is given by $\vec{E}(\vec{r}, -t) \times \vec{B}(\vec{r}, -t) = \vec{E}(\vec{r}, t) \times \left(-\vec{B}(\vec{r}, t)\right) = -\vec{E}(\vec{r}, -t) \times \vec{B}(\vec{r}, t)$

Therefore, under time reversal an electromagnetic wave reverses its direction of propagation, but the polarization remains unchanged.

P10.12: Prove that (a) $\Theta P \Theta^{-1} = -P$, (b) $\Theta X \Theta^{-1} = X$, and (c) $\Theta J \Theta^{-1} = -J$, where P, X, J are respectively the momentum, position, and angular momentum operators.

Solution:

(a) Θ anti-commutes with the momentum operator P (refer Appendix A)

i.e., $\Theta P = -P\Theta$. Hence $\Theta P \Theta^{-1} = -P\Theta\Theta^{-1} = -P$.

(b) Θ commutes with position operator X (Appendix A), i.e., $\Theta X = X\Theta$.

Hence, $\Theta X \Theta^{-1} = X\Theta\Theta^{-1} = X$.

(c) Θ anti-commutes with angular momentum operator J (Appendix A).

Hence, $\Theta J \Theta^{-1} = -J\Theta\Theta^{-1} = -J$.

P10.13: Show that $\Theta^{-1} \dfrac{P^2}{2m} \Theta = -\dfrac{P^2}{2m}$.

Solution:

Θ anti-commutes with the Hamiltonian H (refer Appendix A)

i.e., $H\Theta = -\Theta H$. Hence $\Theta^{-1} H \Theta = -\Theta^{-1}\Theta H = -H$. Consequently, $\Theta^{-1} \dfrac{P^2}{2m} \Theta = -\dfrac{P^2}{2m}$.

Exercises

E10.1: With reference to Solved Problem P3.5, obtain the scattering matrix for its *time reversed* state.

$$\psi^*(x) = \begin{Bmatrix} A^* e^{-ikx} + B^* e^{ikx} (x < -a) \\ C^* e^{-\kappa x} + D^* e^{\kappa x} (-a < x < a) \\ F^* e^{-ikx} + G^* e^{ikx} (a < x) \end{Bmatrix}$$

E10.2: Obtain the quantum mechanical expression for s-wave cross section for scattering from a hard sphere of radius R.

E10.3: Determine the Born approximation to the differential and total cross section for scattering of a particle of mass m by the δ-function potential $V(r) = g\delta^3(r)$.

E10.4: Particles are scattered from the potential $V(r) = \dfrac{g}{r^2}$, where g is a positive constant. Obtain the radial wave equation and determine its regular solution.

E10.5: Use the Born approximation to discuss qualitatively the scattering by a crystal lattice of identical atoms.

E10.6: Determine the scattering cross section for a low-energy particle from a potential given by $V = -2V_0$ for $r < a$, $V = 0$ for $r > a$.

E10.7: Using the optical theorem and the Born approximation amplitude, obtain an expression for the total scattering cross section for a complex potential.

E10.8: Determine the s-wave phase shift, as a function of wave number k, for a spherically symmetric potential that is infinitely repulsive inside a sphere of radius r_0 and vanishes outside.

E10.9: Using the Born approximation, determine the differential cross section $\frac{d\sigma}{d\Omega}$ for a central field Gaussian potential of the form $V(r) = \frac{V_0}{\sqrt{4\pi}} e^{\frac{-r^2}{4a^2}}$.

E10.10: A (point) particle is scattered by a second particle with a rigid core; that is, the scattering potential is $V(r) = 0$ for $r > a$ and $V(r) = \infty$ for $r < a$. The energy of the scattered particles satisfies $ka = 1$. Find the expression for δ_ℓ.

E10.11: A particle of mass μ and momentum $p = \hbar k$ is scattered by the potential $V(r) = \frac{V_0 a}{r} e^{\frac{-r}{a}}$, where V_0 and $a > 0$ are real constants (Yukawa potential). Using the Born approximation, find the differential cross section and total cross section.

E10.12: A beam of particles of mass m and energy E propagates along the z-axis of a coordinate system and scatters from a potential $V = v$ if $|x| \le L, |y| \le L$ *and* $|z| \le L$, and zero otherwise; v is a small constant energy. Determine the differential scattering cross section in the Born approximation.

E10.13: Consider scattering of two identical spin-less particles of mass m. The interaction potential depends on the distance r between the particles and is given by $V(r) = -\frac{h}{16m}\left(\frac{1}{R} + \frac{1}{a}\right)^2$; $r < R$ and the potential is zero for $r > R$. R and a are constants, with $R << a$.

Determine the phase shift $\delta_0(k)$ in the low-energy limit. Given $kR << \frac{R}{a} << 1$ and $k = \sqrt{\frac{2\mu E}{\hbar^2}}$, where E is the energy in the center-of-mass frame and μ is the reduced mass.

E10.14: A particle of mass m is scattered by a central potential $V(r) = -\frac{\hbar^2}{ma^2}\frac{1}{\cosh^2(r/a)}$, where "a" is a constant. Given that the equation $\frac{d^2y}{dx^2} + k^2 y + \frac{2}{\cosh^2 x} y = 0$ has solutions $y = e^{\pm ikz}(\tanh x \mp ik)$, determine the s-wave contribution to the total scattering cross section at energy E.

References

[1] H. Faxén und J. P. Holtsmark, Beitrag zur Theorie des Durchganges langsamer Elektronen durch Gase. *Zeitschrift für Physik*. 45, 307–324 (1927).

[2] Ta-You Wu and Takashi Ohmura, *Quantum Theory of Scattering* (Dover Publications, 2014).

[3] C. J. Joachain, *Quantum Collision Theory* (North-Holland Publishing Co., Amsterdam, 1975).

[4] U. Fano and A. R. P. Rau, *Theory of Atomic Collisions and Spectra* (Academic Press. 1986).

[5] P. C. Deshmukh, D. Angom, and A. Banik, *Symmetry in Electron-atom Collision and Photoionization Process* (Invited article in DST-SERC-School publication, Narosa, 2011).

[6] L. Eisenbud, Ph. D. thesis (unpublished), Princeton University, 1948.

[7] E. P. Wigner, Lower Limit for the Energy Derivative of the Scattering Phase Shift. *Phys. Rev.* 98:145 (1955). doi:10.1103/PhysRev.98.145.

[8] F. T. Smith, Lifetime Matrix in Collision Theory. *Phys. Rev.* 118: 349 (1960). doi:10.1103/PhysRev.118.349.

[9] P. C. Deshmukh and S. Banerjee, Time Delay in Atomic and Molecular Collisions and Photoionisation/Photodetachment. *Int. Revs in Phys. Chem.* 40, 27–153 (2020) doi: 10.1080/0144235X.2021.1838805.

[10] P. C. Deshmukh, S. Banerjee, A. Mandal, and S. T. Manson, Eisenbud–Wigner–Smith Time Delay in Atom–Laser Interactions. *Eur. Phys. J. Spec. Top* (2021) https://doi.org/10.1140/epjs/s11734-021-00225-7.

[11] L. D. Landau and E. M. Lifshitz, *Quantum Mechanics – Non-Relativistic Theory* (3rd Edition, Pergamon Press Ltd, 1977).

[12] P. C. Deshmukh *Foundations of Classical Mechanics* (Cambridge University Press, 2019).

[13] E. W. B. Dias, H. S. Chakraborty, P. C. Deshmukh, S. T. Manson, O. Hemmers, G. Fisher, P. Glans, D. L. Hansen, H. Wang, S. B. Whitfield, D. W. Lindle, R. Wehlitz, J. C. Levin, I. A. Sellin, R. C. C. Perera, Breakdown of the Independent Particle Approximation in High-Energy Photoionization. *Phys. Rev. Lett.* 78:24, 4553–4556 (1997).

[14] D. L. Hansen, O. Hemmers, H. Wang, D. W. Lindle, P. Focke, I. A. Sellin, C. Heske, H. S. Chakraborty, P. C. Deshmukh and S. T. Manson, Validity of the Independent Particle Approximation: The Exception, Not the Rule. *Phys.Rev. A 60*, R2641–2644 (1999).

11

Introduction to Quantum Information and Quantum Computing

Albert Einstein and Neils Bohr. B189, https://arkiv.dk/en/vis/5905821. Courtesy: Niels Bohr Archive.

When Einstein died, his greatest rival, Bohr, found for him words of moving admiration. When a few years later Bohr in turn died, someone took a photograph of the blackboard in his study. There's a drawing on it. A drawing of the 'light-filled box' in Einstein's thought experiment. To the very last, the desire to challenge oneself and understand more. And to the very last: doubt.

—Carlo Rovelli

In this chapter we shall study how quantum entanglement is tested in a laboratory experiment, and how it empowers us to boost computing powers to unprecedented levels. We develop our notion of *reality* from our day-to-day experiences. Quantum theory is largely counterintuitive because it conflicts with our naïve and untutored perceptions of position and momentum. It accounts for physical events in the universe with enduring cogency; but it demands reconciliation with the principle of uncertainty and a consequent statistical description of nature. Quantum theory has impacted science, technology, and also human lifestyle, notwithstanding the fact that relations such as Eq. 1.105 (Chapter 1) and Eq. 3.3 (Chapter 3) characterize quantum theory as essentially *probabilistic*. Einstein's famous quote "*God does not play dice*" grossly undervalues his unease, and also his insight, in quantum physics. Bohr's proverbial response "*it is not your job to tell God what to do*," on the other hand, underscores not just his confidence in quantum theory but also his extraordinary insight in an exhaustive discernment of the laws of nature. The previous ten chapters are inspired by the triumph of quantum mechanics. In the present chapter, we revisit a few elements of the Bohr–Einstein deliberations and also the works of John Bell three decades later, which provided a methodology, based on which experiments could be performed to obtain clarity on the *probability conundrum* in quantum theory. In the meantime, intellectual churning over half a century since the Bohr–Einstein debates led to a deeper understanding of the *principle of superposition* and *entanglement*. The Bohr–Einstein debates immortalized the fifth Solvay conference held in 1927 [1], but continued through subsequent years, with two major publications in 1935 – one by Einstein, Podolsky, and Rosen [2], and the other by Bohr [3] – providing major landmarks. In-depth analysis of the works triggered by Einstein and Bohr, and later by John Bell [4], has now emerged as a robust cornerstone of quantum information science and quantum computing, heralding the second quantum revolution.

11.1 EPR Paradox; Bell's Inequalities

Probabilistic considerations are involved even in classical phenomena; a coin that is tossed has half a chance of landing obverse, or reverse. Einstein's concerns were not quite about the role of probability and uncertainty *per se* in quantum theory, but with the analysis and interpretation of probability (see the Solved Problems P11.8, P11.9, *and* P11.10 for mind-boggling consequences of nature's probabilistic character). His disquiet with quantum mechanics amounted to a deeper question: Did quantum theory provide a complete theory of nature? Is anything missing in quantum theory? If anything is missing in the theory, could it suggest that the nature of probability in quantum mechanics is intrinsically the same as in classical physics? After all, the reason in classical physics for a coin that is flipped has only half a chance of landing obverse or reverse, is that all the forces on the coin – including the exact torque on it when it is flipped, and the forces due to the atmosphere on the coin – are not known. We shall refer to the integrated *missing* information as the *hidden variables*. If this information is plugged in, it would be plausible in principle – *even if enormously cumbersome* – in a classical theory to predict exactly how the coin would land.

A genius that he was, Einstein inquired if there was any missing information that made quantum theory incomplete; if only that missing information could be added, quantum theory would be deterministic. One may appreciate this concern by considering a paradoxical situation represented in Fig. 11.1. It schematically depicts a diatomic molecule having a net spin angular momentum $S = 0$ fragment into two individual atoms a_L and a_R, each having spin $s = \frac{1}{2}$, with the former flying away to the left, and the latter toward the right. One can in principle carry out measurements using a Stern-Gerlach type apparatus on the two atoms a_L and a_R when they are meters, or kilometres, or even light years, apart. If the projection of the spin angular momentum of a_L, along the magnetic field in the +Z direction is measured and found to be $\uparrow$, then that of a_R must be $\downarrow$. If one now measures the projection of the spin angular momentum of a_R, along the magnetic field in the +X direction, and finds it to be $\uparrow$, then that of a_L must be $\downarrow$. Such independent measurements on well-separated atoms, if possible, would provide information about *orthogonal* components of the spin angular momentum. This is clearly incompatible with Eq. 4.13a,b. It is then astonishing that nature prevents measurements of orthogonal components of the angular momentum of such fragmented particles *even* if they are *well separated*. Einstein described this as counterintuitive *spooky action at a distance*; it requires non-locality of a physical theory to

Fig. 11.1 After the explosion of a diatomic molecule, atom a_L speeds away to the left, and a_R to the right.

account for the phenomenology. Einstein–Podolsky–Rosen (EPR) argued that such spooky action could only be enabled by hidden variables that are not an integral part of the quantum theory. EPR therefore concluded that quantum theory is incomplete. In the context of the example we have considered, the EPR conundrum refers to the incompleteness of quantum theory that takes no cognizance of possible hidden variables. These hidden variables could account for the spooky action at a distance that limits simultaneous and accurate information about canonically conjugate variables. Bohr's counter to the proposal of hidden variables was his conviction that quantum entanglement was intrinsic to the description of natural phenomena; it would dispense with the notion of hidden variables that was invoked by EPR. Einstein proposed various gedanken experiments to support the possibility of how hidden variables would make quantum theory complete; but Bohr successfully found loopholes in Einstein's arguments. The Solvay conference of 1927 is famous for the Bohr-Einstein debates, which continued for many years.

In the EPR paradox described above, we have admitted in our analysis information about a physical property of a_R by *inferring* it from a measurement on a_L, and not directly from a measurement on a_R itself. A theory that allows this is said to endorse *counterfactual definiteness*. It amounts to the assumption that not only the measurement we carry out has a definite answer, but also a measurement that has *not* been executed. A dramatic way to flaunt the idea of counterfactual definiteness is to ask [2] *if the moon is there when nobody is looking at it*.

Trusting *counterfactual definiteness* to obtain simultaneous and complete information about canonically conjugate physical properties (such as position and momentum, or orthogonal components of angular momentum) of both particles such as a_L and a_R makes the Heisenberg uncertainty principle paradoxical. Concomitantly, the corresponding element of intrinsic probabilistic character of nature would be rendered redundant. However, nature limits accessible information to be no more than that can be *actually* measured; it does not allow us to *conjecture* additional information by invoking parameters based on supplementary principles, no matter how strong those are within their restricted domains. This aspect of nature is correctly described by quantum theory on the basis of entanglement, not hidden variables.

It is necessary to *test* the assumption that physical reality consists not only of the measurement that we carry out but also of the measurement that we do not. The inability to simultaneously determine orthogonal components of spin angular momentum of each of the two atoms can then at best be accounted for by invoking the principle of non-locality of reality (since the atoms a_L and a_R are separated by an arbitrary distance, even galaxies apart) or by hidden variables. It then becomes appropriate to ask if (a) *counterfactual definiteness* is satisfied by the laws of nature *and* (b) reality is *non-local*. Can both (a) and (b) be correct? It is intriguing to inquire if quantum theory satisfies *counterfactual definiteness and also* admits a perception of reality governed by the tenets of locality. If (a) is to be correct and one *still cannot* get simultaneous accurate information about conjugate physical properties of the fragmented pieces a_L and a_R, one would have to admit that there are hidden variables that prevent access to such information; this would make quantum theory incomplete. Alternatively, one would have to conclude that (b) is correct, i.e., reality is non-local; but the possibility of *non-local hidden variables* may persist. This is mindboggling, since it requires *entanglement* between the separated particles that is maintained at all times, regardless of their separation. As a consequence of this, a measurement on one fragment impacts accessible information about the other particle instantly. This is *not* to be construed as an influence that possibly *travels* faster than light between the separated particles. Rather, it is a property of nature that remains applicable at all instants of time between the entangled particles and is accounted *completely* by quantum theory; information *does not* have to *travel* – *it is just there* at all times, regardless of the separation between the entangled particles, no matter how large, thereby attributing a counter-intuitive 'non-locality' to 'reality'. One might call this as "*Reality-2.0*".

After the EPR and Bohr papers [3, 4] in 1935, there was no significant advance toward resolving the Einstein–Bohr debate for nearly three decades, even as quantum physics spearheaded progress in quantum devices (e.g., semiconductors, lasers). Fundamentally different perceptions of reality that propelled the Bohr–Einstein debate did not come in the way of the success of quantum mechanics. Development of relativistic quantum mechanics and of quantum field theory eclipsed the diverse perspectives of Bohr and Einstein, till John Bell [5] in 1964 formulated a *mathematical inequality* that would help determine the nature of probability and correlation in quantum mechanics, to examine if quantum theory was incomplete, as EPR suspected. Essentially, Bell's inequality provides a mathematical test, based on experiments, that determines if a theory is both counterfactual and local. John Bell's paper *On the Einstein–Podolsky–Rosen Paradox* proposed a mathematical inequality between physical quantities based on predictions of probabilities of correlations between disjoint events that would be satisfied if the probabilities are determined on the basis of *counterfactual definiteness and local hidden variables*.

It turns out, as discussed subsequently, that results of actual experiments do not satisfy Bell's inequality, thereby refuting the argument that quantum theory is incomplete. The experiments which established this are the 2022-Nobel-Prize winning works of Alain Aspect, John Clauser, and Anton Zeilinger. That reality has a non-local character (possibly inclusive of non-local hidden variables) is admittedly counterintuitive, but it has been tested repeatedly in several sophisticated experiments. Quantum theory argues that the sum total of unknowns in physical processes in nature – *based on the idea of counterfactual definiteness and spatially bound reality* – cannot be described in terms of local causality (aka *factorizability*). In other words, quantum theory *cannot be both local and counterfactual;* ***it is not incomplete, as Einstein contrived.*** One must, however, reconcile with the fact that no set of locally available unknowns can make the outcome of an experiment deterministic.

The arguments of Einstein and Bohr are subtle and deep, so is the work of John Bell. Our goal in this chapter is a modest one, to provide only a first introduction to the profound subject of entanglement, since it provides the basis for computation using qubits instead of the classical bits. In the remaining chapter, our focus is on introduction to quantum computation. We therefore illustrate Bell inequality using a toy example. Infallible phraseology of Bell's theorem, however, requires a more rigorous framework than what we adopt in this chapter; we recommend that the reader refers to primary literature for further details. Accordingly, we consider an unpretentious physical situation based on the principles of locality and counterfactual definiteness.

Figures 11.2 and 11.3 are based on Reference [6]. Using these figures, we demonstrate a contrafactual definite situation. The sets D, H, and V (Fig. 11.2a) together constitute the *universal* set in our example. In Fig. 11.2b, the sets are displaced so that each of the three sets has a part that intersects (overlaps) with the other two, and also a part that does not. Denoting now the complement of the sets H, V, and D respectively by H', V', and D' we find the following inequality between the areas of overlaps:

$$(VD') \cup (H'D) \geq (VH'). \tag{11.1}$$

In Eq. 11.1, we have used a compact notation for the intersection of two sets V and H', $V \cap H' = VH'$, etc., and the union of sets is represented by $\cup$. Eq. 11.1 tells us that if we were to play a blind-folded game 'pin the nose' (which is to place a pin or a marker *blindfolded* on a face-picture) on Fig. 11.2b, the probability of pinning the areas indicated by the left hand side is greater than or equal to the probability of pinning the area indicated on the right side. Now, VH' has a part that intersects with D, and also a part that does not. Equation 11.1 does not speak about whether VH' overlaps with D, or with D'. Likewise, whether or not $H'D$ overlaps with V or V' is not referred to, nor does it refer to whether or not VD' overlaps with H or H'. What is *not measured* is not spoken about here, but counterfactual definiteness allows us to include that information as an intrinsic part of our local reality.

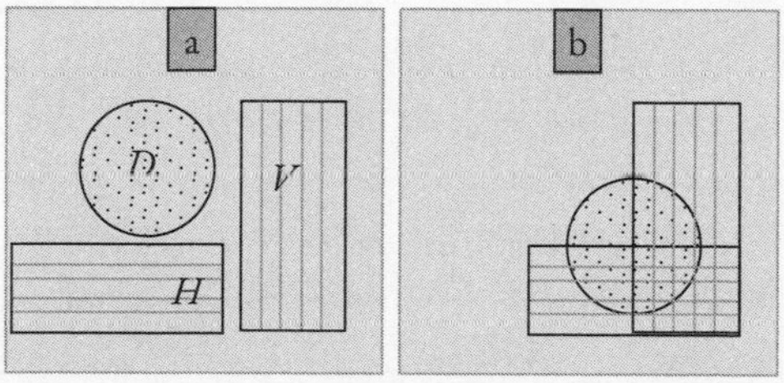

Fig. 11.2 The panel [a] on the left shows Venn diagrams for a set H (rectangle with horizontal stripes), a set V (rectangle with vertical stripes), and a third set D (a dotted circle). The panel [b] on the right shows these three sets displaced and arranged such that parts of D, H and V overlap with each other, constituting partial intersections that we shall denote by $\cap$.

In other words, under the assumption of *locality* and *counterfactual definiteness*, Eq. 11.1 is completely equivalent to

$$(VHD' \cup VH'D') \cup (VH'D \cup V'H'D) \geq VH'D \cup VH'D', \tag{11.2}$$

$$\text{since } VD' \equiv (VHD') \cup (VH'D'), \tag{11.3a}$$

$$H'D \equiv (VH'D) \cup (V'H'D), \tag{11.3b}$$

$$\text{and } VH' \equiv (VH'D) \cup (VH'D'). \tag{11.3c}$$

A theory that does *not* satisfy the inequality (Eq. 11.2) cannot be both *local* and *counterfactual*. The inequalities expressed in Eq. 11.2 and Eq. 11.3 present, in a simplified way, Bell's inequality that must be satisfied by all theories that are *both* local and counterfactual. The underlying correlation presented in the inequality (Eqs. 11.2 and 11.3a,b,c) that would be of importance to us is that it represents a quantitative comparison of possible correlation between *three events* (H, V, and D), each of which when *paired* with either of the remaining *two* has *two* different outcomes (whether the pair intersects/overlaps, or does not). The underlying logic is both *local* and *counterfactual*.

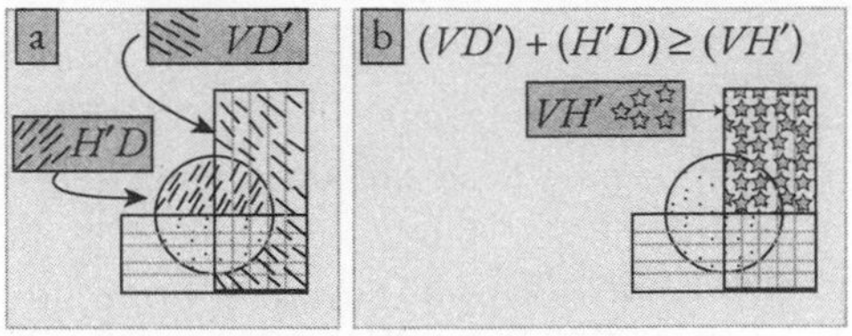

$$(VHD' \cup VH'D') \cup (VH'D \cup V'H'D) \geq VH'D \cup VH'D'$$

Fig. 11.3 The sets H' and D' denote the complements of H and D respectively. The intersection $H'D = H' \cap D$ of H' and D is shaded with dashes sloping left to right upward, while the intersection $VD' = V \cap D'$ of V and D' is shaded with dashes left to right sloping downward. The intersection $VH' = V \cap H'$ of V and H' is shown with stars (asterisks).

Figure 11.4 depicts another situation in which we have three circles labeled H, V, and D, each partially intersecting (overlapping) with each of the other two. Seven segments in this figure are labeled "1" through "7" by the letters of the sets that intersect (overlap); the complement of each of the three sets being denoted by a prime, just as in the previous example. Each of these seven segments is *double* valued in *two* ways. A segment that is a part of V may overlap with H, or it may not overlap with H; but the same segment may also overlap with D or it may not overlap with D. We shall interpret the overlap as a measure of the probability of correlation between events that label the overlapping segment. From Fig. 11.4, once again invoking the probability of blind-folded 'pin the nose' (explained below Eq. 11.1) on its various parts, we see that

$$('4' + '7') + ('6' + '3') \geq ('6' + '7'), \tag{11.4a}$$

$$\text{i.e., } (VHD' \cup VH'D') \cup (VH'D \cup V'H'D) \geq (VH'D \cup VH'D'), \tag{11.4b}$$

$$\text{or } \mathrm{Prob}(VD') + \mathrm{Prob}(H'D) \geq \mathrm{Prob}(VH'). \tag{11.4c}$$

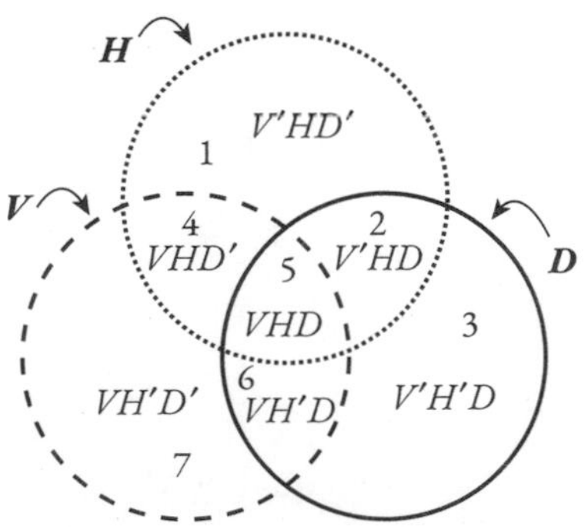

Fig. 11.4 Each of the three circles H (dotted line for circumference), V (dashed line for circumference), and D (continuous line for circumference) partly overlaps with the other two circles. The circle H has a part that overlaps with circle V and a part that does not. It also has a part that overlaps with the circle D and a part that does not. The circles V and D also have similar relationships with the other two circles.

We shall see below that quantum correlations violate this inequality. This would lead us to the conclusion that quantum theory cannot be both local and counterfactual. In order to test the inequality in Eqs. 11.4a,b, we shall now set up an experiment somewhat similar to that described in Fig. 11.1, but different in details. We describe this experiment with reference to Figs. 11.5a,b. The experiment we envisage has a singlet state of a pair of spin-half fermions break up, with one fragment observed by Alice at a distant location and the other observed by Bob in the opposite direction (Fig. 11.5a). Each fragment has spin half $\left(s = \frac{1}{2}\right)$ and may be in one of the two quantum eigen states with $m_s = \pm\frac{1}{2}$, i.e., m_s is ↑ (up), or ↓ (down) with reference to the direction of a magnetic field in a Stern–Gerlach (SG) apparatus, as per our discussion in Chapters 1 and 4. We consider three alternative basis sets in which an arbitrary state vector of the spin-half system can be expressed. We shall denote the three basis sets as V (whose axis of quantization is at 0° with respect to the laboratory Z-axis), H (whose axis of quantization is at 90° with respect to the Z-axis), and D (whose axis of quantization is at 45° with respect to the Z-axis). Incidentally, the two characters, Alice and Bob, were invented in 1977 by three computer scientists, (Ronald) Rivest, (Adi) Shamir, and (Leonard) Adleman, at MIT, to describe certain functions in the implementation of public key cryptography. These three scientists are commonly referenced together as RSA, the first letters of their last names.

Alice and Bob may individually pick the SG basis sets randomly, with no reference to the choice made by the other, and tabulate their findings in various experiments they perform. We inquire if the correlations in the findings of Alice and Bob are explicable by hidden variables (as Einstein would wish) or by quantum entanglement (as Bohr claimed). Toward that end, we develop the following book-keeping procedure that will enable us map our findings to test the inequality (Eq. 11.4) in correlations in the outcome of the experiments depicted in the Fig. 11.5a,b:

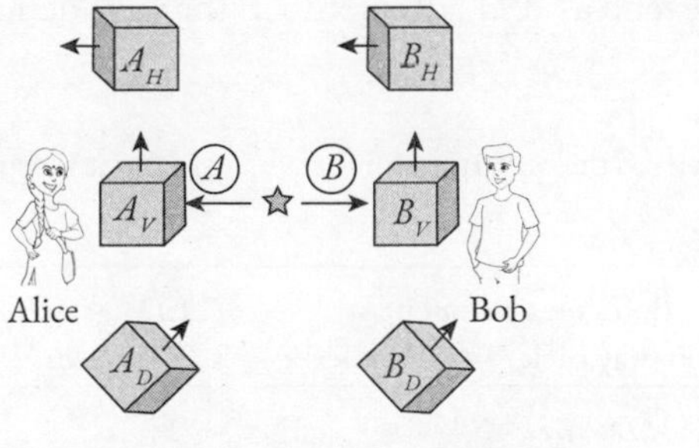

Fig. 11.5a The top row, middle, and the bottom row depict three pairs of Stern–Gerlach (SG) basis sets, with Alice and Bob. The direction of the SG magnetic field (axis of quantization) in the three pairs of apparatus is different. Alice and Bob may choose to perform experiment in any of the three basis sets of SG apparatus.

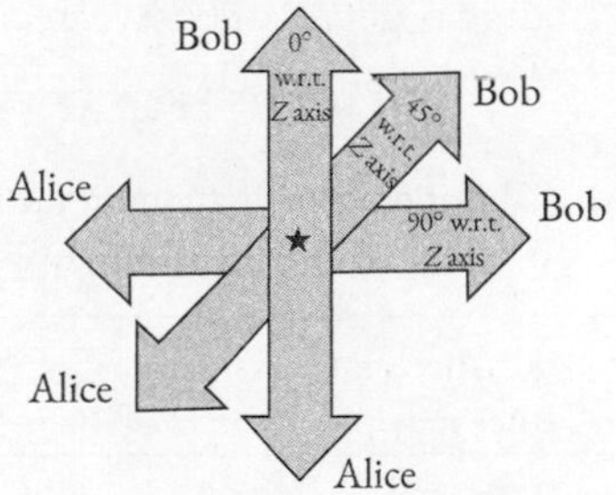

Fig. 11.5b When Alice chooses V, Bob may choose H or V or D. They pick their apparatus without consulting each other. Likewise, when Alice chooses H, Bob may independently choose H or V or D, etc. Their findings are examined after all experiments are performed [2].

- If Alice gets spin up ↑ and Bob gets spin down ↓ we *record* the event as "1."
- If Alice gets spin up ↑ and Bob *also* gets spin up ↑, we *record* the event as "0."

We shall refer to this *Book-Keeping Convention* as *BKC*. What makes the correlations interesting and useful for our analysis is the fact that the orientations (*V*, *D*, or *H*) of the magnetic field chosen by Alice and Bob are totally independent of the other's choice. For example, if Alice picks the orientation *V* and records ↑, Bob may pick the orientation *V* or *H* or *D*. We know that whichever orientation of the magnetic field Bob picks, he may record either ↑ or ↓, which are the only two possibilities. When *both* Alice and Bob choose the same orientation, the result would always be recorded as "0" according to the *BKC*. We focus our attention on those cases in which they choose *different* orientations. Note that just as we had three circles in Fig. 11.4, we have three alternative orientations of the SG basis sets for the two observers (Alice and Bob) with each getting one of the two possible outcomes (↓ or ↑) in their respective experiments. There are therefore $2^3 = 8$ configurations under consideration. These are catalogued in the last three columns of Table 11.1 as "1" or "0" according to the *BKC* we have adopted for three cases of interest, represented by (i) *A*:*D*, *B*:*H*; i.e., Alice (A) picked the *D* orientation and Bob (B) picked the *H* orientation, (ii) *A*:*H*, *B*:*V* and (iii) *A*:*V*, *B*:*D*. Merely interchanging the roles of Alice and Bob would not give us any new correlation.

The symbols "1" and "0" represent a binary choice; we could equally well use other indicators, such as Yes/No, On/Off, etc. It is clear that for any of the three choices of orientations of the magnetic field that Alice can make, the outcome of her measurement is either ↑ or ↓ with reference to magnetic field in Alice's SG apparatus. Correspondingly, Bob could choose any of the three orientations, and the outcome of his measurement would also be either ↑ or ↓, with reference to the axis of quantization in his apparatus. We shall now examine the *correlation* between Alice's and Bob's findings. The correlations would be statistical, our aim is to determine if they can be accounted for by hidden variables. We device the following book-keeping method to record the correlations:

- When Alice chooses *V* (i.e., at 0°) and gets ↑ *and* Bob chooses *D* (i.e., at 45°) and he also gets ↑ then we record the event as *VD*′. Remember that *D*′ denotes the *complement* of D.

Table 11.1 Tabulation of correlations in the measurement of the quantum number m_s carried out independently by Alice (A) and Bob (B).

	Alice's measurement **Alice may pick *D* or *H* or *V***	**Bob's measurement** **Bob may pick *D* or *H* or *V***	***A*:*D*** ***B*:*H***	***A*:*H*** ***B*:*V***	***A*:*V*** ***B*:*D***
*P*1	*D*↑ *H*↑ *V*↑	*D*↓ *H*↓ *V*↓	1	1	1
*P*2	*D*↑ *H*↑ *V*↓	*D*↓ *H*↓ *V*↑	1	0	0
*P*3	*D*↑ *H*↓ *V*↑	*D*↓ *H*↑ *V*↓	0	0	1
*P*4	*D*↑ *H*↓ *V*↓	*D*↓ *H*↑ *V*↑	0	1	0
*P*5	*D*↓ *H*↑ *V*↑	*D*↑ *H*↓ *V*↓	0	1	0
*P*6	*D*↓ *H*↑ *V*↓	*D*↑ *H*↓ *V*↑	0	0	1
*P*7	*D*↓ *H*↓ *V*↑	*D*↑ *H*↑ *V*↓	1	0	0
*P*8	*D*↓ *H*↓ *V*↓	*D*↑ *H*↑ *V*↑	1	1	1

- When Alice chooses H (i.e., at 90°) and gets ↑ *and* Bob chooses D (i.e., at 45°) and he also gets ↑ then we record the event as HD'.
- When Alice chooses D (i.e., at 45°) and gets ↑ *and* Bob chooses H (i.e., at 90°) and he also gets ↑ then we record the event as DH'.
- In the notation (such as VH') that we have adopted here, the *first* letter (V in this case) corresponds to Alice's measurement and the second (H') to Bob's measurement.

We are interested in examining if the correlations indicated in the last three columns of Table 11.1 are explicable in terms of Bell's inequality (Eqs. 11.5a,b), as Einstein wished, or it requires quantum entanglement, as Bohr claimed. We have seen here that if the correlations between Alice's findings and those of Bob are governed by the principles of locality and counterfactual definiteness, then as per Eq. 11.4c, the probability of the records VD', $H'D$ and VH' would satisfy the following relation:

$$\text{Prob}(VD') + \text{Prob}(H'D) \geq \text{Prob}(VH'), \tag{11.5a}$$

or, equivalently

$$\text{Prob}(VD') + \text{Prob}(DH') \geq \text{Prob}(VH'), \tag{11.5b}$$

since the order in which the letters are written is not of any relevance in Fig. 11.4. Reminding ourselves that the prime represents the *complement* of a set, Bell's inequality (Eqs. 11.5a,b) would have us expect that

$$\text{Prob}(\text{Alice}, V\uparrow; \text{Bob}, D\uparrow) + \text{Prob}(\text{Alice}, D\uparrow; \text{Bob}, H\uparrow) \geq \text{Prob}(\text{Alice}, V\uparrow; \text{Bob}, H\uparrow), \tag{11.6a}$$

or, in other words

$$\text{Prob}(A\uparrow @0°, B\uparrow @45°) + \text{Prob}(A\uparrow @45°, B\uparrow @90°) \geq \text{Prob}(A\uparrow @0°, B\uparrow @90°). \tag{11.6b}$$

We will now check out if prediction of quantum theory bears out the Bell's inequality in Eqs. 11.6a,b. From quantum theory we know that prior to a measurement, a spin-half system is in a state of superposition of the base pair $\left|\frac{1}{2},\frac{1}{2}\right\rangle$ and $\left|\frac{1}{2},-\frac{1}{2}\right\rangle$. We consider SG apparatus (discussed in Chapter 1) in which the axis of quantization is along the Z-axis of a laboratory Cartesian frame of reference. We shall indicate the base pair corresponding to this orientation of the magnetic field by placing a subscript L (for 'laboratory') on the base vectors: $\left\{\left|\frac{1}{2},\frac{1}{2}\right\rangle_{(L)}, \left|\frac{1}{2},-\frac{1}{2}\right\rangle_{(L)}\right\}$. We shall simplify the notation and write this base pair as $\{\phi_{L+}, \phi_{L-}\}$. In the experimental situation described in Fig. 11.5, the axis of quantization in Alice's experiment could be in general different from the orientation of the field in Bob's experiment. We shall denote the angles that Alice's and Bob's magnetic field axis of quantization make with the laboratory Z-axis as θ_A and θ_B respectively. We shall label the eigenbasis appropriate for Alice's experiment by the letter A, and Bob's by the letter B. We thus have two more basis sets – $\{\phi_{A+}, \phi_{A-}\}$ and $\{\phi_{B+}, \phi_{B-}\}$. We thus have three alternative base pairs, subscripted respectively by the letters L, A, and B. We now write the antisymmetrized *direct product* of the quantum state vectors of the fragments A (measured by Alice) and B (measured by Bob) in the base pair L.

$$|\psi_{AB}\rangle = \sqrt{\frac{1}{2}}\left(|\phi^A_{L+}\rangle|\phi^B_{L-}\rangle - |\phi^A_{L-}\rangle|\phi^B_{L+}\rangle\right). \tag{11.7}$$

The state vectors denoting the fragment with Alice are *super*-scripted by the letter A, and those denoting the fragment with Bob by the letter B.

It is instructive to ask if we can write the above product state of the two fragments as

$$|\psi_{AB}\rangle \stackrel{?}{=} \sqrt{\frac{1}{2}}\left(|\phi_{A+}\rangle|\phi_{B-}\rangle - |\phi_{A-}\rangle|\phi_{B+}\rangle\right). \tag{11.8}$$

The short answer to this question must be "no" since the orientations of the magnetic fields in Alice's and Bob's SG apparatus are, in general, different. However, in some special case the equality in Eq. 11.8 may hold. We shall seek further insight into this by writing the factor states $\{|\phi^A_{L+}\rangle, |\phi^A_{L-}\rangle\}$ in Eq. 11.7 in terms of base pairs $\{\phi_{A+}, \phi_{A-}\}$:

$$|\phi^A_{L+}\rangle = \cos\frac{\theta_A}{2}|\phi_{A+}\rangle - \sin\frac{\theta_A}{2}|\phi_{A-}\rangle, \tag{11.9a}$$

$$\text{and } |\phi^A_{L-}\rangle = \cos\frac{\theta_A}{2}|\phi_{A-}\rangle + \sin\frac{\theta_A}{2}|\phi_{A+}\rangle. \tag{11.9b}$$

The choice of the coefficients in the above superpositions as cosine and sine functions ensures that the sum of their modulus squared is properly normalized. Their relative phases are chosen such that they are orthogonal as well:

$$\left\langle \begin{pmatrix} \cos\frac{\theta_A}{2}\langle\phi_{A+}| - \\ \sin\frac{\theta_A}{2}\langle\phi_{A-}| \end{pmatrix} \middle| \begin{pmatrix} \cos\frac{\theta_A}{2}|\phi_{A-}\rangle + \\ \sin\frac{\theta_A}{2}|\phi_{A+}\rangle \end{pmatrix} \right\rangle = \cos\frac{\theta_A}{2}\sin\frac{\theta_A}{2} - \sin\frac{\theta_A}{2}\cos\frac{\theta_A}{2} = 0. \tag{11.10}$$

The base pair $\{|\phi^A_{L+}\rangle, |\phi^A_{L-}\rangle\}$ is, of course, orthonormal. The choice of using *half-angles* as arguments of the sine and cosine is in accordance with the fact that spin-half particles follow SU(2) algebra: as explained in the discussion following Eq. 4.48d in Chapter 4, rotation angles corresponding to the spin-half system are *twice* as large compared to those generated by ordinary integer orbital angular momentum.

Likewise, we express the vectors in Bob's basis $\{|\phi^B_{L+}\rangle, |\phi^B_{L-}\rangle\}$ in terms of the base pair $\{\phi_{B+}, \phi_{B-}\}$:

$$|\phi^B_{L+}\rangle = \cos\frac{\theta_B}{2}|\phi_{B+}\rangle - \sin\frac{\theta_B}{2}|\phi_{B-}\rangle, \tag{11.11a}$$

$$\text{and } |\phi^B_{L-}\rangle = \cos\frac{\theta_B}{2}|\phi_{B-}\rangle + \sin\frac{\theta_B}{2}|\phi_{B+}\rangle. \tag{11.11b}$$

Since

$$\left\langle \left(\begin{matrix} \cos\frac{\theta_B}{2}\langle\phi_{B+}| - \\ \sin\frac{\theta_B}{2}\langle\phi_{B-}| \end{matrix} \right) \middle| \left(\begin{matrix} \cos\frac{\theta_B}{2}|\phi_{B-}\rangle + \\ \sin\frac{\theta_B}{2}|\phi_{B+}\rangle \end{matrix} \right) \right\rangle = \cos\frac{\theta_B}{2}\sin\frac{\theta_B}{2} - \sin\frac{\theta_B}{2}\cos\frac{\theta_B}{2} = 0, \tag{11.12}$$

we see that the base pair $\left\{|\phi^B_{L+}\rangle, |\phi^B_{L-}\rangle\right\}$ is also orthogonal. Using now Eqs. 11.9a,b and Eqs. 11.11a,b in Eq. 11.7, we get

$$|\psi_{AB}\rangle = \sqrt{\frac{1}{2}} \left\{ \begin{matrix} \cos\frac{\theta_A}{2}\cos\frac{\theta_B}{2}|\phi_{A+}\rangle|\phi_{B-}\rangle + \cos\frac{\theta_A}{2}\sin\frac{\theta_B}{2}|\phi_{A+}\rangle|\phi_{B+}\rangle \\ -\sin\frac{\theta_A}{2}\cos\frac{\theta_B}{2}|\phi_{A-}\rangle|\phi_{B-}\rangle - \sin\frac{\theta_A}{2}\sin\frac{\theta_B}{2}|\phi_{A-}\rangle|\phi_{B+}\rangle \\ -\cos\frac{\theta_A}{2}\cos\frac{\theta_B}{2}|\phi_{A-}\rangle|\phi_{B+}\rangle + \cos\frac{\theta_A}{2}\sin\frac{\theta_B}{2}|\phi_{A-}\rangle|\phi_{B-}\rangle \\ -\sin\frac{\theta_A}{2}\cos\frac{\theta_B}{2}|\phi_{A+}\rangle|\phi_{B+}\rangle + \sin\frac{\theta_A}{2}\sin\frac{\theta_B}{2}|\phi_{A+}\rangle|\phi_{B-}\rangle \end{matrix} \right\}, \tag{11.13a}$$

$$\text{i.e., } |\psi_{AB}\rangle = \sqrt{\frac{1}{2}} \left\{ \begin{matrix} \left(\cos\frac{\theta_A}{2}\cos\frac{\theta_B}{2} + \sin\frac{\theta_A}{2}\sin\frac{\theta_B}{2}\right)|\phi_{A+}\rangle|\phi_{B-}\rangle \\ +\left(\cos\frac{\theta_A}{2}\sin\frac{\theta_B}{2} - \sin\frac{\theta_A}{2}\cos\frac{\theta_B}{2}\right)|\phi_{A+}\rangle|\phi_{B+}\rangle \\ +\left(\cos\frac{\theta_A}{2}\sin\frac{\theta_B}{2} - \sin\frac{\theta_A}{2}\cos\frac{\theta_B}{2}\right)|\phi_{A-}\rangle|\phi_{B-}\rangle \\ -\left(\sin\frac{\theta_A}{2}\sin\frac{\theta_B}{2} + \cos\frac{\theta_A}{2}\cos\frac{\theta_B}{2}\right)|\phi_{A-}\rangle|\phi_{B+}\rangle \end{matrix} \right\}. \tag{11.13b}$$

Hence,

$$|\psi_{AB}\rangle = \sqrt{\frac{1}{2}} \left\{ \begin{matrix} \left(\cos\frac{\theta_A - \theta_B}{2}\right)|\phi_{A+}\rangle|\phi_{B-}\rangle - \left(\sin\frac{\theta_A - \theta_B}{2}\right)|\phi_{A+}\rangle|\phi_{B+}\rangle \\ -\left(\sin\frac{\theta_A - \theta_B}{2}\right)|\phi_{A-}\rangle|\phi_{B-}\rangle - \left(\cos\frac{\theta_A - \theta_B}{2}\right)|\phi_{A-}\rangle|\phi_{B+}\rangle \end{matrix} \right\}, \tag{11.14}$$

and we note that *when* $\theta_A = \theta_B$, we get the right-hand side of Eq. 11.8. Using a slightly more revealing notation to depict whether Alice and Bob measure spin ↑ or ↓, we may write Eq. 11.14 equivalently as

$$|\psi_{AB}\rangle = \sqrt{\frac{1}{2}} \left\{ \begin{matrix} \left(\cos\frac{\theta_A - \theta_B}{2}\right)|A\uparrow\rangle|B\downarrow\rangle - \left(\sin\frac{\theta_A - \theta_B}{2}\right)|A\uparrow\rangle|B\uparrow\rangle \\ -\left(\sin\frac{\theta_A - \theta_B}{2}\right)|A\downarrow\rangle|B\downarrow\rangle - \left(\cos\frac{\theta_A - \theta_B}{2}\right)|A\downarrow\rangle|B\uparrow\rangle \end{matrix} \right\}. \tag{11.15}$$

We see that the probability that both Alice and Bob measure spin up or spin down is the same, given by

$$P_{same} = \frac{1}{2}\sin^2\left(\frac{\theta_A - \theta_B}{2}\right), \tag{11.16a}$$

while either of them measuring spin up and the other measuring spin down is also equal to each other, which is

$$P_{different} = \frac{1}{2}\cos^2\left(\frac{\theta_A - \theta_B}{2}\right). \tag{11.16b}$$

We denote the orientations of the magnetic field in the SG apparatus of Alice and Bob if $\theta = 0°$ by V, if $\theta = 45°$ by D, and if $\theta = 90°$ by H. An event in which Alice employs V and records spin ↑ while Bob employs H and records ↓ will be recorded as $\left(\text{Alice}, V\uparrow; \text{Bob}, H\downarrow\right)$. In the notation of Eq. 11.6, it is (VH). On the other hand, an event in which Alice employs V and records spin ↑ while Bob employs H and records ↑ will be recorded as $\left(\text{Alice}, V\uparrow; \text{Bob}, H\uparrow\right)$; it corresponds to (VH'), since H' denotes the *complement* of the set H.

Using Eqs. 11.16a,b in Eq. 11.6b, we ask if

$$\frac{1}{2}\sin^2\left(\frac{(45-0)°}{2}\right) + \frac{1}{2}\sin^2\left(\frac{(90-45)°}{2}\right) \geq \frac{1}{2}\sin^2\left(\frac{(90-0)°}{2}\right), \tag{11.17a}$$

$$\text{i.e., if } \frac{1}{2}\sin^2(22.5°) + \frac{1}{2}\sin^2(22.5°) \geq \frac{1}{2}\sin^2(45°), \tag{11.17b}$$

$$\text{or, if } 0.1464466093 \geq 0.25, \tag{11.17c}$$

which is obviously not the case! We therefore conclude that the prediction of Bell's inequality (which is based on the principles of locality and contrafactual definiteness) is violated by the statistical interpretation (Eqs. 11.16a,b) of probability in quantum theory. In other words, quantum theory cannot be both local and counterfactual. One must therefore conclude that the spooky action at a distance predicted by quantum theory cannot be accounted for by local hidden variables; rather, it represents *entanglement* between the distant particles which are with Alice and Bob, irrespective of the distance between them. A number of experiments have been carried out to test Bell's inequality, but every result violates it. Experiments therefore reaffirm that quantum entanglement is completely independent of the distance between the entangled species; that natural phenomena do not occur in a manner that would be consistent with local hidden variables that have well-defined values regardless of whether they are measured or not.

All experimental tests conducted so far have concluded that quantum mechanics is a *fundamental* theory of nature that *cannot* be built from components of classical physics. The notion of local reality that dominates our thinking because of acquaintance with classical physics is not consistent with natural phenomena. Nature demands the quantum theory at a *fundamental* level to account for its character; entanglement occurs in nature, and its explanation cannot be broken down into classical fragments. Bell-type inequalities apply to correlations in probable outcome of experiments that are accounted for by factorable local conditions. However, experiments conducted in disjoint spaces require non-factorability for their explanation. It brings us to a re-interpretation of 'reality' which we may call as 'Reality-2.0'.

It would allude to non-locality of the physical reality for entangled quantum systems. One might be tempted to say that quantum physics is strange, but it is more appropriate to say is that it is nature that is strange; quantum theory provides her correct description. Physical phenomena occur in nature, not in theories. A theory only describes natural phenomena, it is not obliged to reason it out in the sense of causality. Quantum physics provides the correct description of natural phenomena, whereas classical physics is intrinsically inconsistent, since it requires position and momentum of an object whose measurements are incompatible.

Having presented salient aspects of the EPR paradox and Bell's inequalities, we hasten to add that a rigorous analysis requires addressing loopholes in the arguments; and there are two kinds of loopholes that specialists scrutinize – the *locality loophole* and the *detection loophole*. Experiments [7–10] conducted after carefully blocking the loopholes have also concluded that Bell-type inequalities are violated in natural processes, providing robust credence to entanglement as a fundamental property in nature. Since Bell-type inequalities establish that the joint assumptions of *hidden-variables* and the principle of *local causation* contradict the outcome of experiments, Henry Stapp has proclaimed [11] that Bell's theorem is the "most profound discovery of science." The Nobel prize (2022) awarded to Alan Aspect, John Clauser and Anton Zeilinger celebrates the experimental demonstration of violation of Bell inequalities and pioneering quantum information science.

Bell's formulation of the mathematical inequality represents a *family of conditions* on *correlations* between probabilistic results of experiments performed in disjoint physical spaces. The probability of obtaining a result of an experiment is estimated in Bell's theorem based on the principle of local causality. Correlations in the outcome of the experiment can then be compared with predictions of quantum theory. This provides a mechanism to understand the role of probability in quantum theory. The role of probability in quantum theory is therefore *fundamental*; it recognizes that description of physical processes in nature requires an *intrinsically* statistical tackle. Thereby, quantum theory accounts for the mindboggling phenomenon of entanglement between objects irrespective of the distance between them; it explains counterintuitive outcomes of experiments as has been unfailingly tested time and time again in sophisticated laboratories. We shall not, however, attempt an exhaustive review of this field. Instead, we proceed to discuss how entanglement enables processing quantum information and quantum computation. The Solved Problem P11.1 illustrates the enormous capacity of a quantum computer over that of a classical machine.

11.2 Universal Quantum Gates, Deutsch–Jozsa Algorithm

That nature is correctly described by quantum mechanics no matter how weird it seems was the reason for Richard Feynman to say in his keynote address "Simulating Physics with Computers", at the MIT Physics of Computation Conference [12]: "*Nature isn't classical, dammit, and if you want to make a simulation of nature, you'd better make it quantum mechanical, and by golly it's a wonderful problem, because it doesn't look so easy.*" Along with other works [13–15] in the early 1980s, Feynman's aforementioned inspirational remark foreshows an outstanding innovation in computational strategy. Even as quantum devices (e.g., semiconductors) were being used in the fabrication of computers, computational algorithms were based on the *classical* binary Boolean logic; hence referred to as 'classical' computing. The enterprising ideas that led to the initiatives in the 1980s mentioned above triggered the developments in 'quantum' computing in which the basic unit of information is the quantum superposition of the *binary digits* (bits) "0" and "1," rather than the alternatives "on" (represented by the binary digit '1') or "off" (represented by '0') of an electrical switch. After all, the "on" or "off" only represent *classically exclusive alternatives*, like "alive or dead," "white or black," "up↑ or down↓," and "high or low." The basic unit of quantum computation therefore is the *inclusive* superposition – called *qubit* – of the alternatives "1" and "0," rather than the *exclusive* alternatives "on" or "off" states of a classical electrical switch. Quantum computational logic, founded on the principles of entanglement discussed in Section 11.1, rather than Boolean algebra, would simulate nature correctly since it is in conformity with the spirit of Young's double slit experiment that demands for its explanation an *inclusive superposition* of the classically exclusive alternatives – *namely Slit σ, Slit ξ from the Fig. 2.5a in Chapter 2*. It is the principle of superposition that inspires the description of the physical state of a system by a vector that can be expressed as a linear superposition of eigenvectors (Chapter 1), with coefficients whose modulus square give the probability of being in a particular eigenstate, as discussed in Chapter 1 (Born interpretation).

Major scintillating milestones in the development of computers include (a) the works of Charles Babbage (1791–1871), who came up with the idea of digital programmable machines, (b) Claude E. Shannon (1916–2001), who figured out that the symbolic logic developed by George Boole (1815–1864) could be applied to electrical circuits, (c) the works of Maurice Karnaugh (1924–2022) who developed powerful graphical techniques to simplify Boolean expressions into a *minimal sum of products*, (d) the outstanding work of Alan Turing (1912–1954), etc. Turing's work was in some sense a response to David Hilbert's (1862–1943) quest: *Is there a general algorithm mathematicians can follow to determine if a mathematical statement is provable?* The *Church–Turing thesis* that *a computing problem can be solved on any computer we can hope to build, if and only if, it can be solved on the Turing machine* revolutionized the entire field of information science. These developments come under the domain of classical computation in which a computer breaks down complex information (such as numbers, text, and instructions to process the same) into bits and processes, the same using a multitude of rapid "on" or "off" switching processes.

A quantum computer processes and stores information digitally, but information is encoded in a *superposition* of two digits, "0" and "1." We shall call the classical bits as *cbits*, and denote them briefly as 'cb'. Since each cbit can be in either "on" (i.e., "1") or "off" (i.e., "0") state, if one has 8 cbits, these could be set in any *one* of the 2^8 states. If one had 80 cbits, they can be set in any *one* of the 2^{80} states. If we, however, employ 80 qubits instead of classical bits, then one can employ them to represent a *superposition* of 2^{80} states. Using the *qubit* instead of the classical bit as the basic unit of *information* has revolutionized computing power. Gigantic metamorphosis in this field is underway even as these lines are being written and read. Since physical states are described by vectors in a Hilbert space, henceforward we shall identify a single qubit (1 qubit, or 1qb) as a superposition of orthonormal states $|1\rangle$ and $|0\rangle$:

$$\left|\psi_{1\text{qb}}\right\rangle = \alpha_0|0\rangle + \alpha_1|1\rangle, \tag{11.18a}$$

where α_0 and α_1 are in general complex numbers. Equivalently,

$$\left|\psi_{1\text{qb}}\right\rangle = \rho_0 e^{i\varphi_0}|0\rangle + \rho_1 e^{i\varphi_1}|1\rangle, \tag{11.18b}$$

$$\text{or, } \left|\psi_{1\text{qb}}\right\rangle = e^{i\varphi_0}\left[\rho_0|0\rangle + \rho_1 e^{i(\varphi_1-\varphi_0)}|1\rangle\right], \tag{11.18c}$$

where ρ_i, φ_i; $i = 0,1$ are four real numbers, representing two amplitudes and two phases.

A qubit *by itself* also contains just one bit of information, just like a switch being on or off. A single qubit can therefore communicate no more than a single bit of classical information. This result is known as Holevo's theorem, after the Russian scientist Alexander Holevo (born: 1943). However, one can employ *entanglement* between two or more (i.e., multiple) qubits to attain computational capabilities *above* (in fact, well above) that of classical computers. This point would be elaborated in Section 11.3.

There are many physical systems that can be used as candidates for qubits. Designing a physically operational quantum computer is, however, extremely difficult due to challenges inherent in implementing and processing entangled qubits. Any physical system that has *two well-defined states* amenable to superposition (Eq. 11.18) can be a candidate as a qubit. To be useful, it must be small enough to fabricate a device having a practical size. The superposition must be robust and inert to *any inadvertent* interaction with its environment that would destroy the superposition; after all a measurement (intended, or not) induces a system in a state of superposition to collapse into an eigenstate. Criteria proposed by DiVincenzo [16] are usually invoked to build a scalable quantum computer. These consist of five requirements for the construction of the quantum computer itself, and another two in order to employ it for communication. We encourage the readers to refer to more specialized sources on quantum information and computing science for details. Essentially, processing quantum information requires

(i) a physical well-defined qubit system that is scalable;
(ii) conveniently implementable qubit initialization method;
(iii) the qubit states to have significant life times to enable manipulation for computation;
(iv) implementable universal gate operations between the qubits;
(v) measurement capability that is typically qubit specific.

Table 11.2 Some candidates for a physical qubit, and their quantum states that are superposed in the respective qubit.

	Physical system	$\|1\rangle$	$\|0\rangle$
1	Spin system (e.g., electron, or quantum dot, or atomic spin in an optical lattice)	$\|\downarrow\rangle \equiv \|\text{down}\rangle$	$\|\uparrow\rangle \equiv \|\text{up}\rangle$
2	Photon polarization	Vertical polarization	Horizontal polarization
3	Photon number (as in cavity quantum electrodynamics experiment)	Single photon state	Vacuum
4	Josephson junction (superconducting flux)	Anticlockwise current	Clockwise current
5	Josephson junction (energy of superconducting phase)	Ground state	1st excited state
6	Ion trap	Ground and excited state of an ion	

We omit a detailed discussion of the DiVincenzo criteria. Some physical candidates that can be employed as qubits are reported in Table 11.2. Both in terms of information storage capacity and the time required for information processing, quantum superposition offers incredible advantage over classical computers. The Solved Problem P11.1 demonstrates this. Some applications are mentioned toward the end of this chapter, but even the media covers updates practically on a daily basis.

Just like the classical bits are manipulated using gate operations such as the *OR*-gate, *AND*-gate, *NAND*-gate, *EXOR*-gate, and *EXNOR*-gate, qubits are manipulated using *quantum gates*. We shall discuss *universal quantum gate operations*; these are universal in the sense that they represent operations no matter which physical system (such as those listed in Table 11.2) is used as a qubit.

A qubit (Eq. 11.18) is a superposition of $|0\rangle$ and $|1\rangle$ with complex coefficients α_0 and α_1. We thus work in a vector space over the *complex* numbers having dimension 2; i.e., vector space over real numbers having the dimension 4. However, we shall be dealing with normalized states such that

$$\left|\alpha_0\right|^2 + \left|\alpha_1\right|^2 = 1, \tag{11.19}$$

and this constraint makes one of the four real numbers redundant. Furthermore, we can choose the zero of the angles that represent the phases in Eqs. 11.18b,c such that $\varphi_0 = 0$, which allows us to represent a qubit by just two real numbers. We hasten to add that quantum theory requires a complex vector space; it cannot be factored into a tensor product of two real vector spaces [17]. It will soon be found advantageous to choose the required two real numbers to be the polar angle θ and the azimuthal angle φ of the spherical polar coordinate system [Chapter 2 of Reference 18]. We therefore represent a qubit as

$$\left|\psi_{1qb}\right\rangle = \cos\frac{\theta}{2}|0\rangle + e^{i\varphi}\sin\frac{\theta}{2}|1\rangle, \tag{11.20}$$

with $0 \leq \theta \leq \pi$ and $0 \leq \varphi < 2\pi$. This representation of a qubit is often referred to as the *standard representation*, and the orthonormal basis $\{|0\rangle, |1\rangle\}$ in which we have expressed the arbitrary qubit as the *computational basis*. We construct a unit sphere (i.e., a sphere of unit radius) and represent a qubit by a point on its surface (Fig. 11.6). The choice of trigonometric functions cosine and sine as coefficients ensures normalization of the state. As a result of choosing the half-polar angle $\left(\frac{\theta}{2}\right)$, the vectors $|0\rangle$ and $|1\rangle$ that are orthogonal in the Hilbert space appear as antipodal points on the polar (Z-) axis of the sphere, named as the *Bloch sphere* after Felix Bloch, who contributed one of the first detailed understanding of a superposition of spin-half particles in the context of magnetic resonance spectroscopy. It is a usual practice to place the vector $|0\rangle$ at the "North pole" (i.e., on the positive Z-axis) of the Bloch sphere and the vector $|1\rangle$ at the "South pole". The opposite convention can also be chosen, and adhered to. The antipodal appearance on the Bloch sphere of the orthogonal states $|0\rangle$ and $|1\rangle$ is no mystery: after all, the orthogonality is in accordance with the rules of algebra of vectors in the Hilbert space; it is not in the three-dimensional space in which we have employed the spherical polar coordinate system. The Bloch sphere is in the space $\mathbb{R}^3$, while the qubit is in the complex vector space $\mathbb{C}^2$. Such a qubit is an element of the SU(2) group, whereas rotations in $\mathbb{R}^3$ are elements of SO(3). As discussed in Chapter 4, these groups are homomorphic, not isomorphic. In SU(2), we require a rotation through 4π to generate the identity (see the discussion below Eq. 4.48d in Chapter 4).

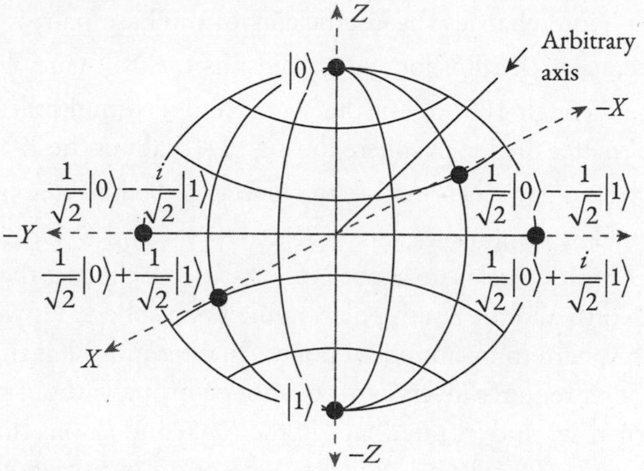

Fig. 11.6 Representation of a qubit according to Eq. 11.20 on a Bloch sphere. Qubit states on the intersection of the Cartesian coordinates and the Bloch sphere are explicitly labeled.

Qubit manipulation from one state to another is achieved using *universal quantum gates*; it can be represented as rotation of the Bloch sphere about an appropriate axis representing a unitary transformation of the qubit state vector.

It is most convenient to represent the qubit in Eq. 11.18a using a matrix representation of the base vectors

$$|0\rangle \to \begin{bmatrix} 1 \\ 0 \end{bmatrix};\ |1\rangle \to \begin{bmatrix} 0 \\ 1 \end{bmatrix}, \tag{11.21a}$$

corresponding to which, an arbitrary qubit such as that in Eq. 11.18a is represented by a matrix having one column and two rows:

$$|\psi_{1qb}\rangle \to \alpha_0 \begin{bmatrix} 1 \\ 0 \end{bmatrix} + \alpha_1 \begin{bmatrix} 0 \\ 1 \end{bmatrix} = \begin{bmatrix} \alpha_0 \\ 0 \end{bmatrix} + \begin{bmatrix} 0 \\ \alpha_1 \end{bmatrix} = \begin{bmatrix} \alpha_0 \\ \alpha_1 \end{bmatrix}. \tag{11.21b}$$

Qubit manipulation by quantum gates can then be represented using matrix algebra. For example, the Pauli matrices σ_x, σ_y and σ_z (defined in Solved Problem P4.3 of Chapter 4) perform the following quantum-gate operations:

$$\sigma_x|\psi_{1qb}\rangle = \sigma_x\left[\alpha_0|0\rangle + \alpha_1|1\rangle\right] \to \boxed{\begin{bmatrix} 0 & 1 \\ 1 & 0 \end{bmatrix}\begin{bmatrix} \alpha_0 \\ \alpha_1 \end{bmatrix} = \begin{bmatrix} \alpha_1 \\ \alpha_0 \end{bmatrix}} \to |\psi_{1qb}\rangle' = \alpha_1|0\rangle + \alpha_0|1\rangle, \tag{11.22a}$$

$$\sigma_y|\psi_{1qb}\rangle = \sigma_y\left[\alpha_0|0\rangle + \alpha_1|1\rangle\right] \to \boxed{\begin{bmatrix} 0 & -i \\ i & 0 \end{bmatrix}\begin{bmatrix} \alpha_0 \\ \alpha_1 \end{bmatrix} = i\begin{bmatrix} -\alpha_1 \\ \alpha_0 \end{bmatrix}} \to |\psi_{1qb}\rangle' = -i\alpha_1|0\rangle + i\alpha_0|1\rangle, \tag{11.22b}$$

and

$$\sigma_z|\psi_{1qb}\rangle = \sigma_z\left[\alpha_0|0\rangle + \alpha_1|1\rangle\right] \to \boxed{\begin{bmatrix} 1 & 0 \\ 0 & -1 \end{bmatrix}\begin{bmatrix} \alpha_0 \\ \alpha_1 \end{bmatrix} = \begin{bmatrix} \alpha_0 \\ -\alpha_1 \end{bmatrix}} \to |\psi_{1qb}\rangle' = \alpha_0|0\rangle - \alpha_1|1\rangle. \tag{11.22c}$$

The Pauli-X gate (σ_x) interchanges the coefficients of the base pair; it is therefore referred to as the Pauli-Not gate, or as the flip-flop gate, or just as the Not-gate. The Pauli-Y gate (σ_y) rotates the Bloch sphere through 180° about the *Y*-axis and also multiplies the base vectors by $\pm i$, and the Pauli-*Z* gate rotates the Bloch sphere through 180° about the *Z*-axis; it flips the phase of $|1\rangle$ since the azimuthal angle is measured with respect to the half-*ZX* plane (Fig. 11.7). The quantum gates are denoted by symbols such as $-\boxed{X}-$, $-\boxed{Y}-$ and $-\boxed{Z}-$, shown in this figure. The Pauli-X gate derives its nomenclature as the *NOT* gate from the fact that it reverses ("nots") the base vectors at the North and the South poles on the Boch sphere. However, it does not map each point on the Bloch sphere to its antipodal point. An operation that maps every point on a sphere to its antipodal point requires inversion; it cannot be brought about by a unitary rotation.

Qubit states get turned around (rotated) about the *X*, *Y*, and *Z* axis through π respectively under the Pauli-X, -*Y*, and -*Z* gates. The identity operation, represented by the 2 × 2 unit matrix, is referred to as the Pauli-I gate. One can see that the *Y* gate can be effectively written as $-iZX$ and that the *Z* gate only overturns the phase of the South pole on the Bloch sphere. Rotations of the Bloch sphere through arbitrary angles are represented by the so-called Rotation gates. An anticlockwise rotation through an arbitrary angle θ about an axis along the direction $\hat{u}$, looking *toward* the origin along that axis, is denoted by $-\boxed{R_{\hat{u}}(\theta)}-$ and represented by the 2 × 2 matrix

$$R_{\hat{u}}(\theta) = \exp\left(-i\frac{\theta}{2}\hat{u}\bullet\vec{\sigma}\right). \tag{11.23}$$

Note the *half*-angle in Eq. 11.23 for reasons already discussed earlier; it stems from the homomorphism between the SU(2) and SO(3) groups. For $\theta = \pi$, the rotation gate is a Pauli-gate, but up to a phase; for example,

$$R_X(\pi) = \exp\left(-i\frac{\pi}{2}\hat{e}_x\bullet\vec{\sigma}\right) = -i\sigma_X \equiv -iX. \tag{11.24}$$

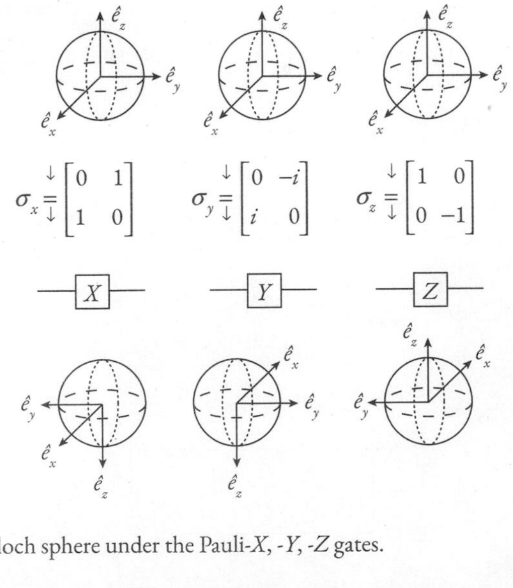

Fig. 11.7 Rotation of the Bloch sphere under the Pauli-*X*, -*Y*, -*Z* gates.

For applications in developing computational algorithms using qubits, it is expeditious to employ Pauli Power gates $-\boxed{A^t}-$ *where A is X or Y or Z* defined as

$$A^t = \exp\left(-i\frac{\pi}{2}t(A - I)\right). \tag{11.25}$$

Depending on A being X, or Y, or Z, the Pauli Power gates are called respectively as the *Pauli X (or Y or Z) power gate.*

We have discussed operations on a single qubit; the quantum gates discussed above are single-qubit quantum gates. For ready reference, some important single-qubit quantum gates are compiled in Table 11.3, along with their common notations and matrix representations.

Table 11.3 provides only a partial, illustrative, list of single-qubit quantum gates. There are other gates, such as the fractional phase shift gate, V-gate, and pseudo-Hadamard gate, for which we refer the reader to specialized literature. We hurriedly turn now to one of the first applications of information science, viz. the Deutsch–Jozsa algorithm [13, 20].

We shall consider only its *simplest* demonstration to exhibit the advantage of performing quantum computing with qubits over classical computing with cbits. Toward this goal, Deutsch raised an interesting question: what is the minimal number of steps required to determine if a function-box that admits one of two possible inputs (such as "0" or "1") and outputs also "0" or "1" is *sensitive* or *insensitive* to the input?

Deutsch's inquiry was motivated by his interest in the "Many Worlds Interpretation" (MWI) of quantum mechanics that was advanced by Hugh Everett in the mid-1950s. Everett was unaware of a similar viewpoint advanced earlier by none other than Schrödinger himself, who had raised difficulties with the Born–Bohr interpretation of quantum mechanics, according to which a measurement induces collapse of a system in a state of superposition into an eigenstate. Everett's MWI was upheld by Bryce DeWitt, who wrote: "*every quantum transition taking place in every star, in every galaxy, in every remote corner of the universe is splitting our local world on Earth into myriad copies of itself.*" David Deutsch worked with DeWitt in the 1970s and carried the conviction that the MWI provided the correct understanding of the quantum universe. Deutsch's quantum algorithm, which we describe below, was motivated by his belief that quantum computing would *ratify* the MWI.

We shall *not* delve any further on the many-worlds interpretation of quantum mechanics. Instead, we focus on Deutsch's question: what is the minimal number of steps required to determine if a function box 'f' that admits one of two possible inputs (such as "0" or "1") and outputs also "0" or "1" is *sensitive* or *insensitive* to the input? This question is easily clarified with reference to Table 11.4.

When the f function-box delivers an output that *depends* on the input, it is called "balanced"; when it does not, it is called "constant". The first and the fourth columns of Table 11.4 correspond to the f function-box being "constant". The second and the third columns correspond to the f function-box being "balanced". One may therefore inquire if the f function-box is balanced or constant by studying the output in a *single* measurement on the f function-box. Deutsch–Jozsa (DJ) algorithm employs information manipulation using

Table 11.3 A few commonly employed single-qubit quantum gates are listed here.

Sr. No.	Name of the single qubit quantum gate	Symbol	Matrix representation	Remarks
1	Pauli $-X$ gate	$-\boxed{X}-$	$\begin{bmatrix} 0 & 1 \\ 1 & 0 \end{bmatrix}$	Rotation of the Bloch sphere through π about the X-axis.
2	Pauli $-Y$ gate	$-\boxed{Y}-$	$\begin{bmatrix} 0 & -i \\ i & 0 \end{bmatrix}$	Rotation of the Bloch sphere through π about the Y-axis.
3	Pauli $-Z$ gate	$-\boxed{Z}-$	$\begin{bmatrix} 1 & 0 \\ 0 & -1 \end{bmatrix}$	Rotation of the Bloch sphere through π about the Z-axis.
4	X, Y, Z Rotation gate $\hat{u} = \hat{e}_X\left(X\text{-rotation}\right)$ $\hat{u} = \hat{e}_Y\left(Y\text{-rotation}\right)$ $\hat{u} = \hat{e}_Z\left(Z\text{-rotation}\right)$	$-\boxed{R_{\hat{u}}\left(\theta\right)}-$	$\exp\left(-i\frac{\theta}{2}\hat{u}\bullet\vec{\sigma}\right)$ $\vec{\sigma}$ is the Pauli vector matrix.	Rotation of the Bloch sphere through an arbitrary angle Θ about an axis along $\hat{u}$ as per the convention defined in the text.
5	Pauli X Power gate	$-\boxed{X^t}-$	$\exp\left(+i\frac{\pi}{2}t\right)R_X\left(t\pi\right)$	Rotation of the Bloch sphere through an arbitrary angle $t\pi$ about the X-axis.
6	Pauli Y Power gate	$-\boxed{Y^t}-$	$\exp\left(+i\frac{\pi}{2}t\right)R_Y\left(t\pi\right)$	Repeated operations by the Y gate
7	Pauli Z Power gate	$-\boxed{Z^t}-$	$\exp\left(+i\frac{\pi}{2}t\right)R_Z\left(t\pi\right)$	Repeated operations by the Z gate
8	$\sqrt{X} = \sqrt{NOT}$ Square root of Not gate	$-\boxed{\sqrt{X}}-$ or as $-\boxed{\sqrt{\neg}}-$, using the notation for classical cbit gate.	$\frac{1}{2}\begin{bmatrix} 1+i & 1-i \\ 1-i & 1+i \end{bmatrix}$	Check out its square
9	Quantum coin flip gate	$-\boxed{QCF}-$ (compare two successive operations of the $-\boxed{QCF}-$ gate with $-\boxed{\sqrt{X}}-$)	$\frac{1}{\sqrt{2}}\begin{bmatrix} 1 & -1 \\ 1 & 1 \end{bmatrix}$	Passing a signal through one QCF gate randomizes the state, but putting two QCF gates in a row yields a deterministic result. It has been called as a machine that first scrambles eggs and then unscrambles them [19].
10	Phase shift gate	$-\boxed{P\left(\theta\right)}-$ or $-\boxed{R_\theta}-$	$\begin{bmatrix} 1 & 0 \\ 0 & \exp\left(i\theta\right) \end{bmatrix}$	This gate changes the phase of the pure state $\left\|1\right\rangle$ relative to that of the state $\left\|0\right\rangle$.
11	Hadamard gate	$-\boxed{H}-$	$\frac{1}{\sqrt{2}}\begin{bmatrix} 1 & 1 \\ 1 & -1 \end{bmatrix}$	Named after the French mathematician Jacques Salomon Hadamard (1865 to 1963).

Table 11.4 Listed here are all the alternatives that have to be considered when an *f* function-box accepts only two inputs "0" or "1", and has only two possible outputs, "0" or "1".

Output → **Input ↓**	**Constant**	**Balanced**	**Balanced**	**Constant**
0	0	0	1	1
1	0	1	0	1
	Output is "0" regardless of the input, i.e., it is independent of the input.	Output is the same as the input; thus depends on the input.	Output is opposite to what the input is; thus output depends on the input.	Output is "1" regardless of the input, i.e., it is independent of the input.

quantum gates to achieve this, whereas if classical algorithm is to be used, one must carry out measurements with the *f* function-box minimally *twice*. Sure enough, the DJ method requires *supplementary resources*, but the fact that the *f* function-box is used only *once* flaunts the reward of using quantum rather than classical logic.

The supplementary resource required in the DJ algorithm involves the use of the single-qubit Hadamard gate (11th row in Table 11.3) and a *two-qubit gate* called as the Control-*NOT* (often written as *CNOT*, or *C-NOT*, or *CX*; with "*X*" referencing the Pauli-Not) gate. A two-qubits (2-qubit) state vector is expressed in terms of the quantum state of each of the two qubits which constitute it, referring either of them as the *first* qubit and the other as the *second* qubit. Each of these is in a state of superposition of $|0\rangle$ and $|1\rangle$, or equivalently in a state of superposition of $|\text{off}\rangle$ or $|\text{on}\rangle$; corresponding to a cat being in a superposed state of being $|\text{dead}\rangle$ or $|\text{alive}\rangle$. The complete basis for a two-qubit system is therefore $\{|00\rangle, |01\rangle, |10\rangle, |11\rangle\}$, with the first digit identifying the state of the first qubit, and the next describing the state of the second. A general 2-qubit state is a linear superposition of these four base vectors. Extending the practice of denoting a single qubit by a matrix having two rows and one column (Eqs. 11.21a,b), we shall denote a 2-qubit quantum state vector by a matrix having four rows and one column:

$$\left|\psi_{2\text{qb}}\right\rangle = \begin{bmatrix} a|00\rangle + b|01\rangle \\ + c|10\rangle + d|11\rangle \end{bmatrix} \leftrightarrow \begin{bmatrix} a \\ b \\ c \\ d \end{bmatrix}, \tag{11.26a}$$

or equivalently as

$$\left|\psi_{2\text{qb}}\right\rangle = \begin{bmatrix} \alpha_{00}|00\rangle + \alpha_{01}|01\rangle \\ + \alpha_{10}|10\rangle + \alpha_{11}|11\rangle \end{bmatrix} \leftrightarrow \begin{bmatrix} \alpha_{00} \\ \alpha_{01} \\ \alpha_{10} \\ \alpha_{11} \end{bmatrix}. \tag{11.26b}$$

It is *of course* possible for one or more of the four coefficients in the above superposition to be zero. A fundamental question that we must ask is if the 2-qubit state is expressible as a tensor product (i.e., a direct product, *such as* that in Eq. 4.75 of Chapter 4) of two individual 1qb vectors. To illustrate the importance of this question, let us consider the following normalized 2-qubit state in which $\alpha_{01} = 0$ and also $\alpha_{10} = 0$:

$$\left|\Phi^+\right\rangle = \sqrt{\frac{1}{2}}\{|00\rangle + |11\rangle\}. \tag{11.27a}$$

We now ask if $\left|\Phi^+\right\rangle$ can be written as a tensor product of two separate one-qubit vectors (Fig. 11.8):

$$\left|\Phi^+\right\rangle \overset{?}{=} \left(a|0\rangle + b|1\rangle\right)\left(c|0\rangle + d|1\rangle\right). \tag{11.27b}$$

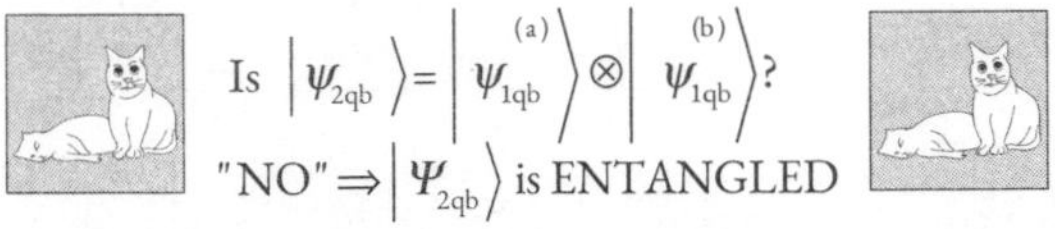

Fig. 11.8 The criterion to determine if a 2-qubit state vector represents an entangled 2-qubit state (i.e., a Bell state) or not requires examining whether or not the 2-qubit vector is expressible as a tensor product of separate 1qb vectors.

For the right-hand sides of Eq. 11.27b to correspond to the right-hand side of Eq. 11.27a, one must have $ad = 0$ and also $bc = 0$. However, if $a = 0$, then the coefficient of $|00\rangle$ in Eq. 11.27b would also vanish and if $d = 0$ then the coefficient of $|11\rangle$ must also vanish. It is therefore impossible to express $\left|\Phi^+\right\rangle$ as a tensor product of two single qubit state vectors. Consideration of $bc = 0$ would lead us to the same conclusion. A 2-qubit state that cannot be factored into a product of single-qubit vectors is called as an *entangled state*, also known as *Bell state*. Such a 2-qubit state vector is maximally correlated (i.e., maximally entangled). The Solved Problem P11.11 illustrates how a Bell state can be produced.

We consider two single qubits, which we shall refer to as $|A1\rangle$ and $|B2\rangle$, given by

$$|A1\rangle = c_0^{A1}|0\rangle + c_1^{A1}|1\rangle \tag{11.28a}$$

and $$|B2\rangle = c_0^{B2}|0\rangle + c_1^{B2}|1\rangle. \tag{11.28b}$$

Between the above two qubits, the following *four* (no more, and no fewer) Bell states can be formed:

$$\left|\beta_{s+}\right\rangle = \sqrt{\frac{1}{2}}\left(\left|0^{A1}0^{B2}\right\rangle + \left|1^{A1}1^{B2}\right\rangle\right) = \sqrt{\frac{1}{2}}\left(|00\rangle + |11\rangle\right), \tag{11.29a}$$

$$\left|\beta_{s-}\right\rangle = \sqrt{\frac{1}{2}}\left(\left|0^{A1}0^{B2}\right\rangle - \left|1^{A1}1^{B2}\right\rangle\right) = \sqrt{\frac{1}{2}}\left(|00\rangle - |11\rangle\right), \tag{11.29b}$$

$$\left|\beta_{d+}\right\rangle = \sqrt{\frac{1}{2}}\left(\left|0^{A1}1^{B2}\right\rangle + \left|1^{A1}0^{B2}\right\rangle\right) = \sqrt{\frac{1}{2}}\left(|01\rangle + |10\rangle\right), \tag{11.30a}$$

and

$$\left|\beta_{d-}\right\rangle = \sqrt{\frac{1}{2}}\left(\left|0^{A1}1^{B2}\right\rangle - \left|1^{A1}0^{B2}\right\rangle\right) = \sqrt{\frac{1}{2}}\left(|01\rangle - |10\rangle\right). \tag{11.30b}$$

We have denoted Bell states by $|\beta\rangle$ subscripted by $s\pm$ or $d\pm$, with s and d denoting whether the two single bits in each term are the same (both $|0\rangle$ or both $|1\rangle$) or *d*ifferent (one $|0\rangle$ and the other $|1\rangle$) 1qb states. The subscript $\pm$ denotes the relative phases of the superposed 2-qubit states. The qubits are distinguishable (and hence with two observers, Alice and Bob), notwithstanding the fact that radial function of the bound states of an atom go to zero only at infinite distance (e.g., Fig. 5.4 and Fig. 5.6). The superscripts $A1$ and $B2$ on the right-hand sides in Eqs. 11.29a,b and 11.30a,b denote that Alice has qubit number 1 and Bob has qubit number 2. Such a detailed designation is, however, not required; one may just remember that in the tensor product states, the first qubit is with Alice (*ladies first*) and the second with Bob. The superscripts $A1$ and $B2$ are therefore usually suppressed, as also indicated in the expressions at extreme right in the Eqs. 11.29a,b and11.30a,b. The Bell states $\left|\beta_{s\pm}\right\rangle$ and $\left|\beta_{d\pm}\right\rangle$ are commonly denoted in literature also by

$$\left|\beta_{s\pm}\right\rangle \leftrightarrow \left|\Phi^{\pm}\right\rangle \tag{11.31a}$$

and

$$\left|\beta_{d\pm}\right\rangle \leftrightarrow \left|\Psi^{\pm}\right\rangle. \tag{11.31b}$$

Regardless of the notation, the four alternative superpositions presented on the right-hand sides of Eqs. 11.29a,b and 11.30a,b are the only possible 2-qubit Bell states. The idea of maximally entangled "Bell" states is readily extended to a multi-qubit (i.e., n-qubit, including $n > 2$) system. In general, an *entangled* multi-qubit is one that cannot be factored as a tensor (direct) product of lower qubits. It must be remembered that the location of the physical qubits that participate in a Bell state are not commented upon in their definitions. In other words, they could be located in the same laboratory, or separated by an arbitrary distance, even galaxies apart. This point would come up in the discussion on teleportation in Section 11.3. Having now introduced the 2-qubit quantum states (entangled, or factorable), we proceed to discuss the *CNOT* (Controlled Not) 2-qubit gate.

The CX gate operates on a 2-qubit state $\left|\psi_{2qb}\right\rangle$ to generate a transformed state $\left|\psi_{2qb}\right\rangle'$:

$$(\mathrm{CX})\left|\psi_{2qb}\right\rangle = \left|\psi_{2qb}\right\rangle'. \tag{11.32}$$

The CX gate is *defined* by its characteristic operation on a 2-qubit state, of which one is regarded as the CONTROL qb (considered as the *first* of the distinguishable two qubits) and the other as the TARGET qb (considered as the *second* qb); it performs the Pauli *NOT* operation on the target qb if the CONTROL qb is in the state $|1\rangle$, i.e., at the South pole of the Bloch sphere. In as much as the single qb gates are represented by 2×2 matrices, the 2-qubit gates are represented by 4×4 matrices. On writing a general 2-qubit vector in the form in Eq. 11.26b, the defining character of the *CX* gate is represented by the following matrix equation:

$$\left|\psi_{2qb}\right\rangle' \rightarrow \begin{bmatrix} 1 & 0 & 0 & 0 \\ 0 & 1 & 0 & 0 \\ 0 & 0 & 0 & 1 \\ 0 & 0 & 1 & 0 \end{bmatrix} \begin{bmatrix} \alpha_{00} \\ \alpha_{01} \\ \alpha_{10} \\ \alpha_{11} \end{bmatrix} = \begin{bmatrix} \alpha_{00} \\ \alpha_{01} \\ \alpha_{11} \\ \alpha_{10} \end{bmatrix}. \tag{11.33}$$

The first two rows in the 1-column matrix which represents the 2-qubit state vector represent the control qb, and the next two rows represent the target qb. Matrix multiplication in the above equation bares out the fact that the target qb is reversed when the control qb is in state $|1\rangle$. The condition of reversing the target qb only when the control qb is in the state "1" would remind you of the Boolean truth table for the classical 2-input *XOR* gate:

$$XOR \leftrightarrow \begin{pmatrix} X & Y & Z \\ 0 & 0 & 0 \\ 0 & 1 & 1 \\ 1 & 0 & 1 \\ 1 & 1 & 0 \end{pmatrix}. \tag{11.34}$$

Unlike the classical *OR* gate which provides an output if either or both of the inputs are "1," the Boolean *XOR* gate provides an output when one of the two inputs is "1" but only if the other is "0," which is why it is called as the "Exclusive *OR*" (*XOR*) gate. *Y* is left unchanged in column *Z* when *X* has "0"; only when *X* has "1" do the values of *Y* get negated in the column *Z*. The 2-qubit *CX* gate performs similar logic. The *CX* quantum gate is therefore denoted as shown in the panel on the right in Fig. 11.9, using the same symbol $(-\oplus-)$ that is used for the classical *XOR* gate.

The DJ algorithm uses a 2-qubit *CX* gate and employs the *f-CX* function box instead of the *f* function-box. In addition, it uses two 1qb Hadamard gates, H_1 and H_2, as shown in Fig. 11.10, but the important thing is that the *f* function-box is used *only once*. Two single qubits, both in pure states $|1\rangle$ (i.e., the South pole on the Bloch sphere of each qb), are fed (at stage "*a*" in Fig. 11.10) respectively to two Hadamard gates H_1 and H_2. The output of each of the two Hadamard gates is then input to the 2-qubit *f-CX* function box. The first qubit's states are subscripted by "1," and the second qubit states by "2". We examine the physical states of both the qubits at each stage of the operations, marked as "*a*," "*b*" and "*c*" in this figure.

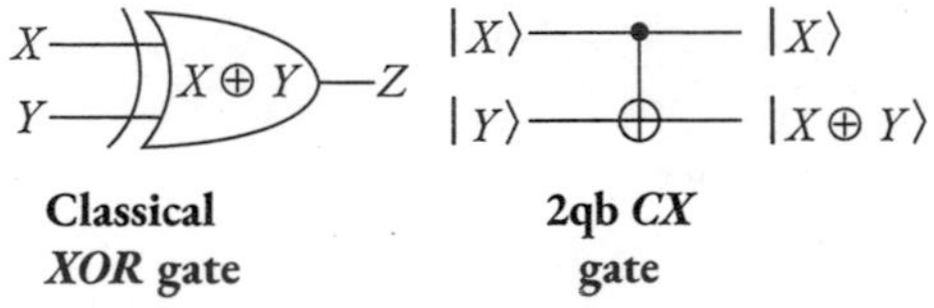

Fig. 11.9 The 2-qubit Control-Not gate is denoted using the symbol $-\oplus-$ that denotes the classical *XOR* (exclusive *OR*) gate.

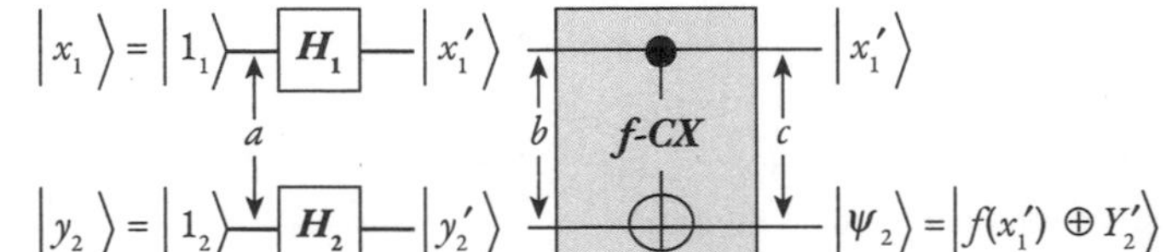

Fig. 11.10 In the DJ algorithm, two single qubits each in pure state $|1\rangle$ are input to two 1qb Hadamard gates H_1 and H_2. The outputs from the two *H* gates are fed to the 2-qubit *f-CX* function box.

At stage "*a*," the two input 1qb pure states are obviously not entangled. Their tensor product is operated on by the two Hadamard gates *independent* of each other. Using the matrix representation for the Hadamard gate given in Table 11.3 (11th row), the operation we have at stage "*a*" therefore is:

$$H_1H_2|\psi_{in}\rangle = H_1H_2|1_11_2\rangle = H_1|1_1\rangle H_2|1_2\rangle, \tag{11.35a}$$

$$\text{i.e., } H_1H_2|\psi_{in}\rangle = \frac{1}{\sqrt{2}}\left(|0_1\rangle - |1_1\rangle\right)\frac{1}{\sqrt{2}}\left(|0_2\rangle - |1_2\rangle\right), \tag{11.35b}$$

$$\text{or } H_1H_2|\psi_{in}\rangle = \frac{1}{2}\left(|0_10_2\rangle - |1_10_2\rangle - |0_11_2\rangle + |1_11_2\rangle\right). \tag{11.35c}$$

The 2-qubit state vector written as a sum of the four terms in Eq. 11.35c is obviously not an entangled 2-qubit; it is in fact produced from the tensor product of state vectors for individual qubits "1" and "2". It is not any of the four possible Bell states (Eqs. 11.29a,b and 11.30a,b). We started out with pure states at stage "*a*" but the input to the *f-CX* function box consists of *all the four* possible combinations of the two qubits. At stage "*b*" in the Fig. 11.10, we have

$$|x_1'y_2'\rangle = \frac{1}{2}\left(|0_10_2\rangle - |1_10_2\rangle - |0_11_2\rangle + |1_11_2\rangle\right) \tag{11.36}$$

On the left-hand side of Eq. 11.36, the subscripts "1" and "2" represent the state vectors of the first and the second qubit. They include combinations of the possibility of the first qubit to be in the state "0" or "1" with the second qubit in "0" or "1". The modulus of the probability amplitudes of all combinations are all equal, but the + and − symbols in the superposition specify which combinations are in phase and which are not. The two un-entangled qubits $|x_1'\rangle$ and $|y_2'\rangle$ are now input to the *f-CX* function box. Both the qubits have every possible base vector from the 2-qubit basis $\{|0_10_2\rangle, |1_10_2\rangle, |0_11_2\rangle, |1_11_2\rangle\}$. Since the *f-CX* function box is used instead of the *f* function box, the output of the *f-CX* function box at stage "*c*" in Fig. 11.10 consists of the state $|x_1'\rangle$ and $|\psi_2\rangle = |f(x_1') \oplus y_2'\rangle$, instead of $|x'\rangle$ and $|x' \oplus y'\rangle$ as we had in Fig. 11.9. Since $|x_1'\rangle$ is given by Eq. 11.36, in order to find $|\psi_2\rangle$, we must determine

$$f\text{-CX}|0_10_2\rangle = |0_1\ f(0)_X \oplus 0_Y\rangle, \tag{11.37a}$$

$$f\text{-CX}|0_11_2\rangle = |0_1\ f(0)_X \oplus 1_Y\rangle, \tag{11.37b}$$

$$f\text{-CX}|1_10_2\rangle = |1_1\ f(1)_X \oplus 0_Y\rangle, \tag{11.37c}$$

$$\text{and } f\text{-CX}|1_11_2\rangle = |1_1\ f(1)_X \oplus 1_Y\rangle. \tag{11.37d}$$

The subscripts *X* and *Y* on the right-hand sides of Eqs. 11.37a–d correspond to the first and the second columns of the *XOR* truth table in Eq. 11.34. The values of $f(0)_X$ and $f(1)_X$, however, depend on whether the *f* function box is *constant* or *balanced*, which is what we wish to determine. The values of *Z* in the last column would be the *same* or *different* from those of $f(0)_X$ and $f(1)_X$ which would appear in the *first column* in the *XOR* truth table, depending respectively on whether the second column (i.e., value of *Y*) has "0" or "1". The *XOR* results are organized in Figs. 11.11a,b and compiled in Eqs. 11.38a–d. Under the "*NOT*" operation, we know that "0" and "1" get interchanged, but since we do not know the function *f*, we denote the result of the *NOT* operation on $f(0)$ by $\overline{f(0)}$ and that on $f(1)$ by $\overline{f(1)}$; these values of course can only be either "0" or "1". The notation of denoting the opposite of $f(0)$ by $\overline{f(0)}$, and similarly denoting by $\overline{f(1)}$ the opposite of $f(1)$, has been used in Figs. 11.11a,b, and in all of the analysis that follows.

Using the 2-qubit CX instruction (Fig. 11.9) we have

$$f\text{-CX}\left|0_1 0_2\right\rangle = \left|0_1 f(0)_2\right\rangle, \tag{11.38a}$$

$$f\text{-CX}\left|0_1 1_2\right\rangle = \left|0_1 \overline{f(0)}_2\right\rangle, \tag{11.38b}$$

$$f\text{-CX}\left|1_1 0_2\right\rangle = \left|1_1 f(1)_2\right\rangle, \tag{11.38c}$$

$$\text{and } f\text{-CX}\left|1_1 1_2\right\rangle = \left|1_1 \overline{f(1)}_2\right\rangle. \tag{11.38d}$$

Collecting all the terms from Eqs. 11.38a–d, we get

$$f\text{-CX}\left[\frac{1}{2}\begin{pmatrix}\left|0_1 0_2\right\rangle - \left|1_1 0_2\right\rangle - \left|0_1 1_2\right\rangle \\ +\left|1_1 1_2\right\rangle\end{pmatrix}\right] = \frac{1}{2}\begin{bmatrix}\left|0_1 f(0)_2\right\rangle - \left|0_1 \overline{f(0)}_2\right\rangle - \left|1_1 f(1)_2\right\rangle \\ +\left|1_1 \overline{f(1)}_2\right\rangle\end{bmatrix}. \tag{11.39}$$

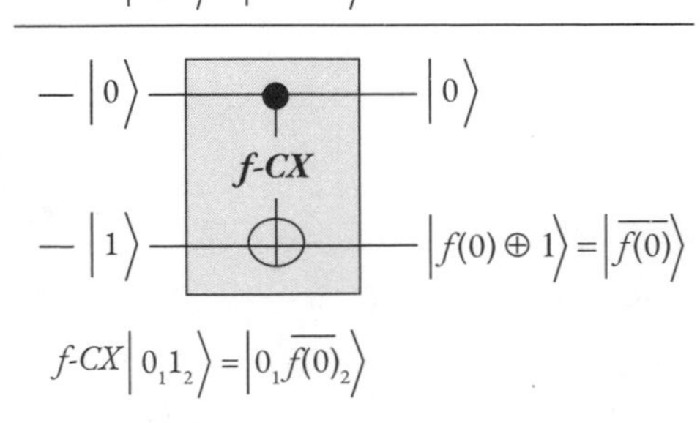

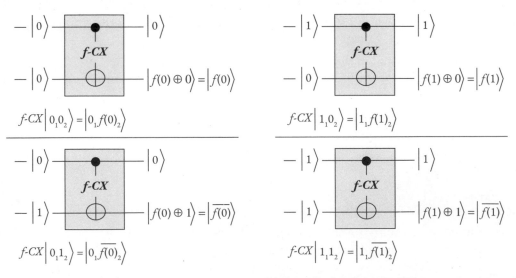

Fig. 11.11a Depicting the operations in Eqs. 11. 38a,b.

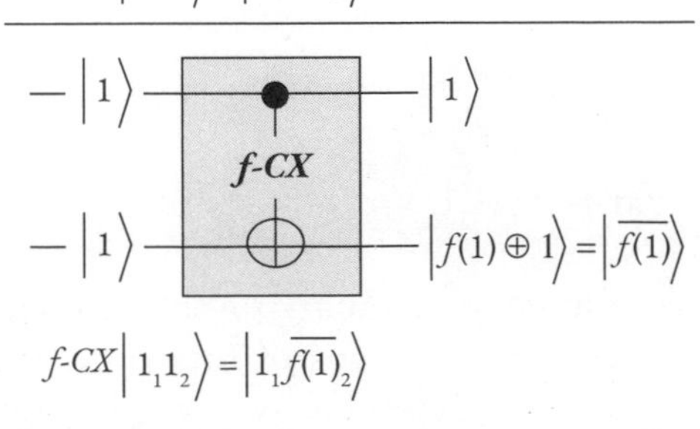

Fig. 11.11b Depicting the operations in Eqs. 11. 38c,d.

The output from the function *f-CX* function box at stage "*c*" (Fig. 11.10) therefore has the following two qubits:

$$\left|x_1'\psi_2\right\rangle = \left|x_1' f\left(x_1'\right) \oplus y_2'\right\rangle = \frac{1}{2}\left(\left|0_1 f(0)_2\right\rangle - \left|1_1 f(1)_2\right\rangle - \left|0_1 \overline{f(0)}_2\right\rangle + \left|1_1 \overline{f(1)}_2\right\rangle\right). \quad (11.40)$$

Now, if the *f* function box is *constant*, $f(0) \overset{\rightarrow}{\underset{\leftarrow}{=}} f(1)$ which also guarantees that $\overline{f(0)} \overset{\rightarrow}{\underset{\leftarrow}{=}} \overline{f(1)}$.

On the other hand, if the *f* function box is *balanced*, we have $f(1) = \overline{f(0)}$ and $f(0) = \overline{f(1)}$. For the case that when the function box is *constant*, we shall therefore have:

$$\left|x_1'\psi_2\right\rangle = \left|x_1'\psi_{2c}\right\rangle = \frac{1}{2}\left(\left|0_1 f(0)_2\right\rangle - \left|1_1 f(0)_2\right\rangle - \left|0_1 \overline{f(0)}_2\right\rangle + \left|1_1 \overline{f(0)}_2\right\rangle\right). \quad (11.41a)$$

We have added a subscript "*c*" and have written the left-hand side as $\left|\psi_{2c}\right\rangle$ to remind us that this corresponds to the assumption that the function *f* box is *constant*. We can see that the state $\left|\psi_{2c}\right\rangle$ factorizes as

$$\left|x_1'\psi_{2c}\right\rangle = \left|x_1' f\left(x_1'\right) \oplus y_2'\right\rangle = \frac{1}{2}\left[\left|0_1\right\rangle\left(\left|f(0)_2\right\rangle - \left|\overline{f(0)}_2\right\rangle\right) + \left|1_1\right\rangle\left(\left|\overline{f(0)}_2\right\rangle - \left|f(0)_2\right\rangle\right)\right] \quad (11.41b)$$

$$\text{or, } \left|x_1'\psi_{2c}\right\rangle = \left|x_1' f\left(x_1'\right) \oplus y_2'\right\rangle = \frac{1}{2}\left[\left(\left|0_1\right\rangle - \left|1_1\right\rangle\right)\left(\left|f(0)_2\right\rangle - \left|f(0)_2\right\rangle\right)\right]. \quad (11.41c)$$

We denote the 1st qubit vector, $\sqrt{\frac{1}{2}}\left(\left|0_1\right\rangle - \left|1_1\right\rangle\right)$, as $\left|X_{1-}\right\rangle$ and find that

$$\left|x_1'\psi_{2c}\right\rangle = \sqrt{\frac{1}{2}}\left[\left|X_{1-}\right\rangle\left(\left|f(0)_2\right\rangle - \left|\overline{f(0)}_2\right\rangle\right)\right]. \quad (11.42)$$

Likewise, denoting the vector $\sqrt{\frac{1}{2}}\left(\left|0_1\right\rangle + \left|1_1\right\rangle\right)$ as $\left|X_{1+}\right\rangle$, for the case that the function *f* box is *balanced*, we get

$$\left|x_1'\psi_{2b}\right\rangle = \frac{1}{2}\left[\left(\left|0_1\right\rangle + \left|1_1\right\rangle\right)\left(\left|f(0)_2\right\rangle - \left|\overline{f(0)}_2\right\rangle\right)\right] = \sqrt{\frac{1}{2}}\left[\left(\left|X_{1+}\right\rangle\right)\left(\left|f(0)_2\right\rangle - \left|\overline{f(0)}_2\right\rangle\right)\right]. \quad (11.43)$$

Note that for this case we have added a subscript "*b*" (for *balanced*) on the 2-qubit state vector. Comparing Eq. 11.42 with Eq. 11.43, we see that the second qubit is in the state $\frac{1}{\sqrt{2}}\left(\left|f(0)_2\right\rangle - \left|\overline{f(0)}_2\right\rangle\right)$ in *both* the cases – i.e., regardless of the function box being 'balanced' or 'constant'. We do not know this state at this stage, since we are yet to determine whether the function *f* box is constant or balanced. The state of the second qubit is therefore insensitive to the nature of the function box. However, when the function *f* box is *constant*, the first qubit is in the state $\left|X_{1-}\right\rangle$, and it is in the state $\left|X_{1+}\right\rangle$ if the function *f* box is *balanced*. If we can therefore determine if the first qubit is in the state $\left|X_{1+}\right\rangle$ or $\left|X_{1-}\right\rangle$, we can determine if the function f box is *constant* or *balanced*. The two states $|X_{1-}\rangle$ and $|X_{1+}\rangle$ are easily distinguished by operating on them by a single qubit Hadamard gate. The Hadamard gate has the property that when it operates on a pure state, it gives a mixed state, but when it operates again on that mixed state, it recovers the original pure state. This is readily exhibited by the following matrix equations:

$$H|0\rangle = \frac{1}{\sqrt{2}}\begin{bmatrix}1 & 1\\ 1 & -1\end{bmatrix}\begin{bmatrix}1\\ 0\end{bmatrix} = \frac{1}{\sqrt{2}}\begin{bmatrix}1\\ 1\end{bmatrix} = \left(\frac{1}{\sqrt{2}}|0\rangle + \frac{1}{\sqrt{2}}|1\rangle\right) = |X_+\rangle, \tag{11.44a}$$

$$\text{and } H|1\rangle = \frac{1}{\sqrt{2}}\begin{bmatrix}1 & 1\\ 1 & -1\end{bmatrix}\begin{bmatrix}0\\ 1\end{bmatrix} = \frac{1}{\sqrt{2}}\begin{bmatrix}1\\ -1\end{bmatrix} = \frac{1}{\sqrt{2}}|0\rangle - \frac{1}{\sqrt{2}}|1\rangle = |X_-\rangle. \tag{11.44b}$$

Operating on $|X_\pm\rangle$ by the Hadamard operator, we have

$$H|X_+\rangle = \frac{1}{\sqrt{2}}\begin{bmatrix}1 & 1\\ 1 & -1\end{bmatrix}\left\{\frac{1}{\sqrt{2}}\begin{bmatrix}1\\ 1\end{bmatrix}\right\} = \frac{1}{2}\begin{bmatrix}2\\ 0\end{bmatrix} = \begin{bmatrix}1\\ 0\end{bmatrix} = |0\rangle, \tag{11.45a}$$

$$\text{and } H|X_-\rangle = \frac{1}{\sqrt{2}}\begin{bmatrix}1 & 1\\ 1 & -1\end{bmatrix}\left\{\frac{1}{\sqrt{2}}\begin{bmatrix}1\\ -1\end{bmatrix}\right\} = \frac{1}{2}\begin{bmatrix}0\\ 2\end{bmatrix} = \begin{bmatrix}0\\ 1\end{bmatrix} = |1\rangle. \tag{11.45b}$$

An SG (Stern–Gerlach) type experiment on the output of the Hadamard gate would determine the state of the vector and distinguish between the states $|0\rangle$ and $|1\rangle$, and thus identify whether the first qubit in Eq. 11.42 is $|X_+\rangle$ or $|X_-\rangle$; thereby identifying the f function box as *constant* or *balanced*. That we have employed extra resources does not undermine the fact that the function f box has been used *only once*. Using qubits instead of the cbits, we could input a superposition of all the four base vectors $\{|00\rangle, |01\rangle, |10\rangle, |11\rangle\}$ of a 2-qubit state and thereby figure out if the function box is *constant* or *balanced* in a *single* query on the box. A classical algorithm would require a minimum of *two* queries on the function box. The DJ algorithm was the first one that demonstrated that quantum algorithm can achieve what a classical one cannot. In the Solved Problem P11.17, you will see how the DJ algorithm can be extended to n-qubits. Such capability is now called as *quantum supremacy*. Tangible benefits of quantum supremacy are astounding; they are poised to bring in the second quantum revolution in many fields: cyber-security, cryptography, computational quantum chemistry, drug-design, artificial intelligence, machine learning, optimization problems, climatology, financial markets, communication, etc. Awe-inspiring advances in each of these areas draw their potency from *entanglement* between qubits, a few prominent aspects of which are illustrated and discussed in the next section.

11.3 No Cloning, No Go Theorems, and Quantum Teleportation

Prior to discussing what an algorithm that employs qubits can do, we first rule out what it cannot, in the spirit of a *no go* theorem. A no go theorem typically describes a situation that is untenable. In fact, the Bell's theorem which we discussed in Section 11.1 is a *no go* theorem;

it rules out compatibility of predictions of quantum theory with local hidden variables. We first ascertain that *cloning* is not possible using qubits. Cloning involves duplication, making a copy of an object. This requires, of course, the original object to be copied, whatever it is, and a machine that would produce the copy. Also, ancillary material on which the copy would be made is required; that is, if cloning is possible at all – which is just what we are inquiring. In order to appreciate this theorem, we quickly remind ourselves of what is expected of a *copier*. A photocopying machine, for example, accepts an arbitrary picture as input, whether it is that of the Khardungla Pass, or that of the river Saraswathi, and produces a copy of the input on an ancillary plain white sheet of paper. It could also make a copy of a *combination* of two pictures (Fig. 11.12), and produce the original and a copy of the original. Our *expectation* from a quantum copier is intuitively driven by our experience with a photocopier which essentially clones an object into a duplicate, using an ancillary sheet on which the copy is produced.

Going by our intuitive thinking about a copier, we expect a quantum cloning machine to admit an arbitrary quantum state, along with an ancillary state on which it would be copied, and deliver the original unhampered, along with its exact copy on the ancillary state (Fig. 11.13). In as much as the photocopier duplicates an arbitrary image, or a combination of images, we expect the quantum cloning machine to be capable of producing an exact copy of an arbitrary quantum state. We represent this expectation by the following equation, in which $\hat{\Omega}_C$ represents the *quantum cloning operator* (for the cloning machine), $|\psi\rangle$ is an *arbitrary* state vector in the Hilbert space that represents a physical system, and $|\psi^A\rangle$ is the ancillary state on which the copy would be produced:

$$\hat{\Omega}_C\left(|\psi\rangle \otimes |\psi^A\rangle\right) = |\psi\rangle \otimes |\psi\rangle. \tag{11.46}$$

The symbol $\otimes$ between two state vectors denotes their tensor product. We explore the possibility of cloning a state that is an arbitrary superposition $\left(\alpha|\xi\rangle + \beta|\phi\rangle\right)$ of two states $|\xi\rangle$ and $|\phi\rangle$. For Eq. 11.46 to hold for

$$|\psi\rangle = \left(\alpha|\xi\rangle + \beta|\phi\rangle\right), \tag{11.47}$$

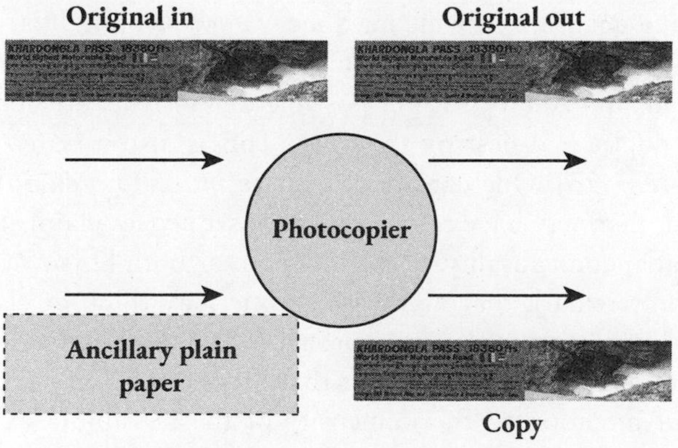

Fig. 11.12 A photocopier accepts as input any document or image along with an ancillary blank sheet of plain paper and delivers the original unaltered and a copy of the same.

we require

$$\Omega_c\left[\left(\alpha|\xi\rangle+\beta|\phi\rangle\right)\otimes|\psi^A\rangle\right]=\left(\alpha|\xi\rangle+\beta|\phi\rangle\right)\otimes\left(\alpha|\xi\rangle+\beta|\phi\rangle\right), \tag{11.48a}$$

$$\text{i.e., } \alpha\Omega_c\left(|\xi\rangle\otimes|\psi^A\rangle\right)+\beta\Omega_c\left(|\phi\rangle\otimes|\psi^A\rangle\right)=\begin{bmatrix}\alpha^2|\xi\rangle\otimes|\xi\rangle+\alpha\beta|\xi\rangle\otimes|\phi\rangle+\\ \beta\alpha|\phi\rangle\otimes|\xi\rangle+\beta^2|\phi\rangle\otimes|\phi\rangle\end{bmatrix}, \tag{11.48b}$$

$$\text{or } \alpha|\xi\rangle\otimes|\xi\rangle+\beta|\phi\rangle\otimes|\phi\rangle=\begin{bmatrix}\alpha^2|\xi\rangle\otimes|\xi\rangle+\alpha\beta|\xi\rangle\otimes|\phi\rangle+\\ \beta\alpha|\phi\rangle\otimes|\xi\rangle+\beta^2|\phi\rangle\otimes|\phi\rangle\end{bmatrix}. \tag{11.48c}$$

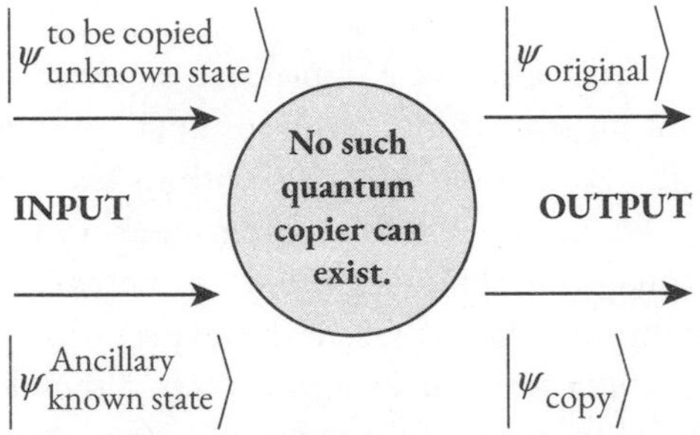

Fig. 11.13 We expect a quantum cloning machine to accept an arbitrary quantum state to be copied as input and copy it on an ancillary state and produce the original along with a copy of the original.

We have only used the linearity of the cloning operator Ω_c, and of course its desired cloning capability (Eq. 11.46).

Comparing the two sides of Eq. 11.48c, we see that the requirement reduces to $\alpha^2 = \alpha$ and $\beta^2 = \beta$, and that the cross terms must vanish, i.e., $\alpha\beta = 0 = \beta\alpha$. The superposition in Eq. 11.47 is, however, destroyed if either $\alpha = 0$ or $\beta = 0$, and of course if both α and β are zero. In other words, it is not possible to have a quantum copier that clones an arbitrary quantum state. The conclusion we have been led to is called as the *no cloning* theorem [21, 22]. It states that a quantum copying machine *cannot* be built. It is, as mentioned earlier, a *no go* theorem. The inverse operation is also impossible: if we have two copies of an arbitrary unknown quantum state, it is not possible to construct a machine that would save just one of the two copies and destroy the other. This is also a *no go* theorem, known as the *no deleting theorem* [23]. One can think of it as *inverse no-cloning* theorem. Another recent *no go* theorem that may have far reaching consequences claims that a post-quantum theory, if any, must abandon causality, or purification, or both [23]. We shall not attempt to elaborate this, but only remark that even though one may think of classical physics as an approximation to which quantum theory approaches in a limiting manner (e.g., $\hbar \to 0$), the Reference [24] is a *no-go* theorem for theories that decohere to quantum mechanics.

We now turn our attention to the immensity of the possibilities that open up due to quantum entanglement. We begin with a discussion on teleportation. The Cambridge dictionary [25] provides the following meaning of teleportation: "*the act of traveling by an imaginary very fast form of transport that uses special technology or special mental powers, or of causing someone or something to travel in this way; teleportation is probably among the more viable of science fiction's wild dream.*" Such usage of the term "*teleportation*," or some variant thereof, is misleading. It prejudices us, and results in an incorrect understanding of *teleportation*. The discussion that follows would clarify that teleportation might appear to be spooky, but it does not have anything to do with travel at all. It is *always there* between entangled qubits, and it is independent of time. It is not imaginary; it is very *real* in the sense of *reality* ('Reality-2.0') as we understand from the Bell's theorem, discussed in Section 11.1. It does not involve transport of any physical object, rather, it *teleports quantum information*. It does not belong to the sphere of science fiction; it very much is an integral part of *mainstream science, engineering, and technology*.

To appreciate teleportation, consider an arbitrary qubit

$$\left|\psi_{a,1}\right\rangle = a_0\left|0_{a,1}\right\rangle + a_1\left|1_{a,1}\right\rangle \tag{11.49}$$

which Alice is in possession of. The subscript "a" refers to the fact that the qubit is with Alice and the subscript "1" refers to this qubit as the qubit number "1". We shall refer to this qubit as qb1. We shall employ two additional qubits as a supplementary resources available to Alice and Bob in the initial state; these shall be referred to as qb2 and qb3. The coefficients a_0 and a_1 in the superposition (Eq. 11.49) are the complex *probability amplitudes* that the physical state of the system is respectively in the state $|0\rangle$ and $|1\rangle$. Alice would not know the values of (a_0, a_1); – any measurement on the state would simply induce the collapse of the system into one of the two possible eigenstates, akin to performing a 'which-way' (Young's) double-slit experiment (Chapters 2 and 3). Direct measurement of these coefficients is therefore not possible. Teleportation involves communication of information about the qubit (Eq. 11.49) by Alice to Bob, who is considered to be at an arbitrary remote location. Coming to think of it, Alice would not be able to tell this to Bob even if he were in the *same* laboratory with Alice, *since Alice herself would not know of the coefficients* in Eq. 11.49. The central challenge lies in just how communication of information about the *unknown* mixing coefficients (a_0, a_1) in the qubit can be achieved. In some sense, it is like asking Alice to read out a poem to Bob without reading any of its constituent stanzas; also asking Bob to listen to the whole poem without listening to any part of the composition! It must be noted that this situation is grossly different from photocopying a poem from a piece of paper and handing it over to Bob without reading it; the information that is to be communicated is not classical, it is a physical state that is a *superposition* of possible eigenstates – it is a qubit which cannot be cloned!

Our goal therefore is to find a mechanism for Bob to obtain information about the superposition state $\left|\psi_{a,1}\right\rangle$ regardless of his whereabouts. In the following, we demonstrate just how quantum entanglement enables this, albeit using additional resources. We need to put into service, as a resource available to Alice and Bob in the initial state, two *other* qubits that are paired in *any* one of the four Bell states (Eqs. 11.29a,b and 11.30a,b). Furthermore, we shall see that success of the method we present does not depend on just where Bob is; he could be sharing Alice's laboratory, or be elsewhere – even in a different building in the city, or in another country, or for that matter in a different galaxy.

If one wonders how information must travel from Alice to Bob, it would be natural to expect that this cannot happen faster than c, the speed of light. However, one must recognize that entanglement between two qubits in a Bell state *does not require* information to *travel* from one qubit to the other. Limiting value of the speed of light therefore is irrelevant; entanglement is a fundamental property of nature that describes the non-local *reality* correctly as discussed in Section 11.1. Information in quantum entanglement does not *travel per se* from one qubit to the other; it is *just there*, between the entangled qubits. It is sustained even if the participating qubits are separated, and move away from each other, no matter at *what speed* and *how far*, until a measurement decoheres the state (i.e., the entanglement is destroyed). An interaction of the qubit with any probe (intended or inadvertent – such as an impurity in the apparatus with which the qubit interacts) amounts to a measurement; it results in the collapse of the system that is initially in a state of superposition into one of its eigenstates. The method we discuss below would work no matter in which particular Bell state the qubits qb2 and qb3 are initially entangled in, though the specific steps one must take differ in details. Furthermore, in *addition* to the qubits "2" and "3" in a Bell state, we shall need an extra resource that is also *crucial*, but we shall come to that a little bit later. *At this juncture, we underscore the fact that the independence of the correlation between entangled qubits with respect to their physical whereabouts is central to the understanding of Bell's theorem discussed in Section 11.1.* We now proceed with the consideration that the additional resource Alice and Bob have with them consists of qb2 (which is with Alice) and qb3 (with Bob) in the Bell state $\left|\beta_{s+}\right\rangle$ (Eq. 11.29b). We denote this Bell state by slightly expanding the notation in Eq. 11.29b to specify the qubit numbers "2" and "3," and also include the subscripts "*a*" (for Alice) and "*b*" (for Bob):

$$\left|\beta_{2,3,s+}\right\rangle = \sqrt{\frac{1}{2}}\left(\left|0_{a2}0_{b3}\right\rangle + \left|1_{a2}1_{b3}\right\rangle\right). \tag{11.50}$$

We shall now describe a mechanism to teleport the quantum information contained in the qubit described in Eq. 11.49 from Alice to Bob. For the present illustration, we use $\left|\beta_{2,3,s+}\right\rangle$ but a consistent methodology exists also for all other Bell pairs. As long as the entanglement between qb2 and qb3 is not broken, we admit the possibility that Bob moves spatially away at this juncture, *along* with qb3 that he remains in possession of. Only an interaction (intended or not) of the Bell state $\left|\beta_{2,3,s+}\right\rangle$ with a probe can (and does) destroy the correlation. It would induce the Bell state to collapse into one of the possible eigenstates $\{|00\rangle,|01\rangle,|10\rangle,|11\rangle\}$ of the two qubits. Alice's strategy at this juncture is to entangle the qubits qb1 and qb2, both of which are in her physical possession. Alice's experiment on qb1 and qb2 impacts qb3 in a unique manner, since qb2 and qb3 are initially in a Bell state (Eq. 11.50). After all, two qubits are distinguishable, even if they may be entangled.

Consider the possibility that Alice entangles qb1 and qb2 in a Bell state by employing the 2-qubit *CX* gate, using qb1 as the control qubit, and qb2 as the target qubit. Prior to the application of this *CX* operation, the 3qb state of the qubits qb1, qb2, and qb3 is

$$\left|\Psi_{'1','2','3'}\right\rangle = \left(a_0\left|0_{a,1}\right\rangle + a_1\left|1_{a,1}\right\rangle\right) \otimes \left(\sqrt{\frac{1}{2}}\left(\left|0_{a,2}0_{b,3}\right\rangle + \left|1_{a,2}1_{b,3}\right\rangle\right)\right), \tag{11.51a}$$

$$\text{i.e., } \left|\Psi_{'1','2','3'}\right\rangle = \sqrt{\frac{1}{2}}\left(a_0\left|0_1 0_2 0_3\right\rangle + a_0\left|0_1 1_2 1_3\right\rangle + a_1\left|1_1 0_2 0_3\right\rangle + a_1\left|1_1 1_2 1_3\right\rangle\right). \tag{11.51b}$$

On operating by the CX gate (Eq. 11.33), the 3qb state $\left|\Psi_{'1','2','3'}\right\rangle$ is transformed to

$$\left|\Psi_{'1','2','3'}\right\rangle' = \sqrt{\frac{1}{2}}\begin{pmatrix} a_0\left|0_{a,1}\right\rangle\left|0_{a,2}0_{b,3}\right\rangle + a_0\left|0_{a,1}\right\rangle\left|1_{a,2}1_{b,3}\right\rangle \\ +a_1\left|1_{a,1}\right\rangle\left|1_{a,2}0_{b,3}\right\rangle + a_1\left|1_{a,1}\right\rangle\left|0_{a,2}1_{b,3}\right\rangle \end{pmatrix}, \tag{11.52a}$$

$$\text{i.e., } \left|\Psi_{'1','2','3'}\right\rangle' = \begin{Bmatrix} \sqrt{\frac{1}{2}}a_0\left|0_{a,1}\right\rangle\left(\left|0_{a,2}0_{b,3}\right\rangle + \left|1_{a,2}1_{b,3}\right\rangle\right) \\ +\sqrt{\frac{1}{2}}a_1\left|1_{a,1}\right\rangle\left(\left|1_{a,2}0_{b,3}\right\rangle + \left|0_{a,2}1_{b,3}\right\rangle\right) \end{Bmatrix}. \tag{11.52b}$$

At this stage Alice would apply the Hadamard (H) operation on qb1. The qb1 (Eq. 11.49) would remain in its original state when the CX gate was applied to get $\left|\Psi_{'1','2','3'}\right\rangle'$; after all qb1 was used as the *control* qb, but subsequent application of the H-gate on qb1 would now change its state, according to the universal Hadamard gate representation:

$$\frac{1}{\sqrt{2}}\begin{bmatrix} 1 & 1 \\ 1 & -1 \end{bmatrix}\begin{bmatrix} a_0 \\ a_1 \end{bmatrix} = \frac{1}{\sqrt{2}}\begin{bmatrix} a_0 + a_1 \\ a_0 - a_1 \end{bmatrix}. \tag{11.53}$$

Now that qb1 has changed from the superposition in Eq. 11.49 to that in Eq. 11.53, the 3qb state $\left|\Psi_{'1','2','3'}\right\rangle'$ in Eq. 11.52b changes *correspondingly* to

$$\left|\Psi_{'1','2','3'}\right\rangle'' = \sqrt{\frac{1}{2}}\begin{Bmatrix} a_0\dfrac{\left|0_{a,1} + 1_{a,1}\right\rangle}{\sqrt{2}}\left(\left|0_{a,2}0_{b,3}\right\rangle + \left|1_{a,2}1_{b,3}\right\rangle\right) \\ +a_1\dfrac{\left|0_{a,1} - 1_{a,1}\right\rangle}{\sqrt{2}}\left(\left|1_{a,2}0_{b,3}\right\rangle + \left|0_{a,2}1_{b,3}\right\rangle\right) \end{Bmatrix}, \tag{11.54a}$$

$$\text{i.e., } \left|\Psi_{'1','2','3'}\right\rangle'' = \frac{1}{2}\begin{Bmatrix} a_0\left|0_{a,1}0_{a,2}\right\rangle\left|0_{b,3}\right\rangle + a_0\left|1_{a,1}0_{a,2}\right\rangle\left|0_{b,3}\right\rangle \\ +a_0\left|0_{a,1}1_{a,2}\right\rangle\left|1_{b,3}\right\rangle + a_0\left|1_{a,1}1_{a,2}\right\rangle\left|1_{b,3}\right\rangle \\ +a_1\left|0_{a,1}1_{a,2}\right\rangle\left|0_{b,3}\right\rangle - a_1\left|1_{a,1}1_{a,2}\right\rangle\left|0_{b,3}\right\rangle \\ +a_1\left|0_{a,1}0_{a,2}\right\rangle\left|1_{b,3}\right\rangle - a_1\left|1_{a,1}0_{a,2}\right\rangle\left|1_{b,3}\right\rangle \end{Bmatrix}, \tag{11.54b}$$

$$\text{or, } \left|\Psi_{'1','2','3'}\right\rangle'' = \frac{1}{2}\left\{\begin{array}{l} \left|0_{a,1}0_{a,2}\right\rangle\left(a_0\left|0_{b,3}\right\rangle + a_1\left|1_{b,3}\right\rangle\right) \\ +\left|1_{a,1}0_{a,2}\right\rangle\left(a_0\left|0_{b,3}\right\rangle - a_1\left|1_{b,3}\right\rangle\right) \\ +\left|0_{a,1}1_{a,2}\right\rangle\left(a_0\left|1_{b,3}\right\rangle + a_1\left|0_{b,3}\right\rangle\right) \\ +\left|1_{a,1}1_{a,2}\right\rangle\left(a_0\left|1_{b,3}\right\rangle - a_1\left|0_{b,3}\right\rangle\right) \end{array}\right\}. \tag{11.54c}$$

The 3qb state state $\left|\Psi_{'1','2','3'}\right\rangle''$ is a superposition of four terms, given in the Eq. 11.54c. Each of the four terms consists of a tensor product of qb1 and qb2, which both are with Alice, while qb3 is with Bob. Being in physical possession of the distinguishable qubits qb1 and qb2, Alice can make a measurement on each, and determine if she has $\left|0_{a,1}0_{a,2}\right\rangle$, or $\left|1_{a,1}0_{a,2}\right\rangle$, or $\left|0_{a,1}1_{a,2}\right\rangle$, or $\left|1_{a,1}1_{a,2}\right\rangle$. This would pin down exactly which of the four possible superposition of $\left\{\left|0_{b,3}\right\rangle, \left|1_{b,3}\right\rangle\right\}$ is with Bob, and specify whether it is from the first, or second, or third, or the fourth row on the right-hand side of Eq. 11.54c.

Table 11.5 lists the four possibilities. Depending on which of the four ($\left|0_{a,1}0_{a,2}\right\rangle$, or $\left|1_{a,1}0_{a,2}\right\rangle$, or $\left|0_{a,1}1_{a,2}\right\rangle$, or $\left|1_{a,1}1_{a,2}\right\rangle$) 2-qubit state of qb1 and qb2 is with Alice, Bob can apply an appropriate gate, indicated in the last column of Table 11.5 on qb3, to determine the original state (Eq. 11.49) of qb1. Alice would need to use a conventional (i.e., classical) communicational channel to inform Bob the states of qb1 and qb2 (in the first column, Table 11.5). She could send a messenger, or make a phone call, or flash laser beams for Bob to observe and interpret according to a pre-arranged code. Bob can then implement the gate operation on qb3, as per the last column in Table 11.5, and thereby determine the original superposition state in qb1 as existed in Eq. 11.49. We thus have a complete solution to the teleportation challenge. A quantum superposition is communicated by Alice to a remote Bob, albeit using a classical communication channel after entangling qb1 with qb2.

Table 11.5 The third column shows the state of the third qubit which is with Bob corresponding to each of the four possible states of the first two qubits "1" and "2" with Alice. The last column apprises us about just what experiment Bob must perform on qb3 to get the qubit state in Eq. 11.49. It is left as an exercise for the reader to verify that the gate operations described in the last column on the qb3 state described in the second column yields the qubit state described in Eq. 11.49.

Sr. No.	Alice has ↓	Bob's qb3 ↓	Bob can apply this ↓ gate on qb'3'
1	$\left\|0_{a,1}0_{a,2}\right\rangle$	$\left(a_0\left\|0_{b,3}\right\rangle + a_1\left\|1_{b,3}\right\rangle\right)$	Identity
2	$\left\|1_{a,1}0_{a,2}\right\rangle$	$\left(a_0\left\|0_{b,3}\right\rangle - a_1\left\|1_{b,3}\right\rangle\right)$	Pauli-*Z* gate
3	$\left\|0_{a,1}1_{a,2}\right\rangle$	$\left(a_0\left\|1_{b,3}\right\rangle + a_1\left\|0_{b,3}\right\rangle\right)$	Pauli-*X* gate
4	$\left\|1_{a,1}1_{a,2}\right\rangle$	$\left(a_0\left\|1_{b,3}\right\rangle - a_1\left\|0_{b,3}\right\rangle\right)$	Pauli-*ZX* gate

It is superfluous to add that the communication using the classical channel mentioned in the last step above cannot be executed at a speed faster than that of light; hence at no stage are we in conflict with the limits on transport of information imposed by the special theory of relativity. That said, we hasten to add that the methodology we have employed to communicate quantum information about the state of the qb1 (with Alice in Eq. 11.49) to Bob has essentially used *entanglement* in two ways:

- The procedure harnessed the Bell state $\left|\beta_{2,3,s+}\right\rangle$, in Eq. 11.50.
- The methodology involved operations by quantum gates, in getting Eqs. 11.54a,b,c from Eqs. 11.52, and for the gate operations Bob must perform as indicated in the last column of Table 11.5.

The correlation between entangled qubits in Bell states does not involve information that has to be *transferred* from one qubit to the other. Rather, the quantum information *exists* between the entangled qubits, irrespective of their physical locations and/or relative movement, till either qubit interacts with any agency. It is not anything that *travels* "instantly" – the quantum information between entangled qubits is just *there*, drawing its charisma from a *non-local reality* ("REALITY 2.0") that takes away the mystery about *spooky* action at a distance. This fundamental hallmark of all implications of the quantum entanglement has for its basis the Bell's theorem (Section 11.1). Note also that qb1 does not remain in the original state of superposition given in Eq. 11.49. It has to be operated upon to obtain the first column in Table 11.5, and cannot therefore remain in the original mixed state. Teleportation communicates quantum information from Alice to Bob, but the process involves collapse of the superposition state into one of the four possible eigenstates of the tensor product $|\mathrm{qb1}\rangle|\mathrm{qb2}\rangle$. The quantum information in Eq. 11.49 is teleported, but the qb1 does not stay in that superposition state any longer. This is in contrast with *classical* cloning in which one gets a copy from a copying machine along with the original, which remains unchanged (Fig. 11.12).

As mentioned earlier, variants of the above method can surely be implemented. The supplementary resource of having qb2 and qb3 in a Bell state can in fact be any one of the four different maximally entangled qubits given in Eqs. 11.29a,b and 11.30a,b, though we used $\left|\beta_{2,3,s+}\right\rangle$ in the example deliberated upon.

In the example that follows, we consider a different Bell state of qb2 and qb3 to be available to Alice and Bob as an initial condition:

$$\left|\beta_{2,3,d-}\right\rangle = \sqrt{\frac{1}{2}}\left(\left|0_2 1_3\right\rangle - \left|1_2 0_3\right\rangle\right). \tag{11.55}$$

This Bell state is from Eq. 11.30b, with the notation again slightly expanded, to include the qubit numbers qb2 and qb3. The 3qb state of the qubits qb1, qb2, and qb3 now is

$$\left|\Psi_{'1','2','3'}\right\rangle = \left(a_0\left|0_{a,1}\right\rangle + a_1\left|1_{a,1}\right\rangle\right)\left(\sqrt{\frac{1}{2}}\left(\left|0_{a,2}1_{b,3}\right\rangle - \left|1_{a,2}0_{b,3}\right\rangle\right)\right), \tag{11.56a}$$

$$\text{i.e., } \left|\Psi_{'1','2','3'}\right\rangle = \sqrt{\frac{1}{2}}\left(a_0\left|0_1 0_2 1_3\right\rangle - a_0\left|0_1 1_2 0_3\right\rangle + a_1\left|1_1 0_2 1_3\right\rangle - a_1\left|1_1 1_2 0_3\right\rangle\right). \tag{11.56b}$$

As before, we may admit the possibility that Bob spatially moves away from the original location to a different location, carrying the qb3 along with him. We have already seen that the speed at which he moves away, nor the spatial distance to which he does, are both irrelevant. The only thing that matters is that the entanglement between qb2 and qb3 is preserved. From the previous illustration, we know that Alice would have to entangle qb1 with qb2 and inform Bob just how she has entangled qb1 and qb2, using a classical communication channel. Bob would then be able to determine the quantum information in the superposition given in Eq. 11.49. Alice could of course pair qb1 and qb2 in any one of the four Bell states, and Bob would not know in which of these four states she has entangled qb1 with qb2, till she uses a classical communication channel to share that information with him. For the sake of illustration, we consider the case when Alice entangles qb1 and qb2 in Bell state $\left|\beta_{1,2,d-}\right\rangle$, given in Eq. 11.30b. Non-zero contributions to the probability amplitude $\left\langle\beta_{1,2,d-}\middle|\Psi_{'1','2','3'}\right\rangle$ can come only from those terms on the right-hand side of Eq. 11.56b in which the states of qb1 and qb2 correspond to those in Bell state $\left|\beta_{1,2,d-}\right\rangle$. From Eq. 11.55 and Eq. 11.56b, we see that

$$\left\langle\beta_{1,2,d-}\middle|\Psi_{'1','2','3'}\right\rangle = n\left\{a_0\left|0_3\right\rangle + a_1\left|1_3\right\rangle\right\}, \tag{11.57}$$

where n is an appropriate normalization constant. Since qb3 is with Bob, he would already be in possession of just the quantum information in Eq. 11.49 that Alice intended to teleport; except that he would not know what he has *unless* Alice informs him using a classical channel that she entangled qb1 and qb2 in the Bell state $\beta_{1,2,d-}$. Alice could certainly entangle qb1 and qb2 in a different Bell pair, for example in the state $\left|\beta_{1,2,d+}\right\rangle$, given in Eq. 11.30a, whence the state of qb3 with Bob would be

$$\left\langle\beta_{1,2,d+}\middle|\Psi_{'1','2','3'}\right\rangle = n\left\{a_0\left|0_3\right\rangle - a_1\left|1_3\right\rangle\right\}, \tag{11.58}$$

which is of course different from that in Eq. 11.49. Nonetheless, on knowing over a classical communication channel which particular Bell state Alice has entangled qb1 with qb2, Bob can perform an appropriate single-qubit gate operation on qb3 to determine the original state in Eq. 11.49, thereby completing teleportation of the quantum information.

Teleportation is thus communication of quantum information using a classical communication channel. One may ask if there are challenges in communicating *classical* information using quantum entanglement. Indeed this involves overcoming certain constraints, since a single qubit (qb) cannot communicate information beyond that in a single classical bit (cbit). This constraint is established as a theorem in information science, known after Alexander Holevo, as mentioned in Section 11.1. The constraint can be circumvented using *dense coding*, discussed in the next section.

11.4 Dense Coding and Applications of Quantum Computing

Holevo's theorem [26] recognizes an upper limit, often called as *Holevo-bound* or *Holevo-limit*, on the amount of information that can be carried by qubits. We shall not prove this theorem. However, considering how deep-seated its consequences are in the field of information science, we shall ruminate on its simplest upshot: a single qubit (qbit, or qb) cannot communicate information about any more than a single classical bit (cbit, or cb). In order to savor the quintessence of this deep-seated theorem, and to appreciate how *dense coding* enables us *circumvent* the Holevo-bound, we shall discuss an example of dense coding in its simplest form. Depending on the context and in its more general form, it is also referred to as *superdense coding*.

The Holevo-bound dares us to find a mechanism for Alice to communicate information to Bob about *two* classical bits (not their quantum superposition) using a *single* qubit. The two *classical* bits with Alice can only constitute one of the four alternatives: $|0_1\rangle|0_2\rangle$, or $|0_1\rangle|1_2\rangle$, or $|1_1\rangle|0_2\rangle$, or $|1_1\rangle|1_2\rangle$. If the two classical bits were two switches, the four possibilities just mentioned correspond *respectively* to the switches being off,off; off-on; on-off; and on-on. The subscript on each bit denotes the bit number. Dense coding does not break the Holevo-bound; it dodges it using quantum entanglement as an ancillary crutch. Alice and Bob would use two qubits, each of which is in a state of superposition of two (other) classical bits. Alice would have one of these qubits with her, and Bob would have the other. These two qubits are in *one* of the four possible Bell states (Eqs. 11.29a,b and 11.30a,b). We shall denote this particular Bell state as $|\phi_{ab}\rangle$. The subscript ab denotes the fact that one of the qubits in the Bell pair is with Alice, the other with Bob. We shall presume that the particular Bell state under our attention is

$$|\phi_{ab}\rangle = |\beta_{1,2,s+}\rangle. \tag{11.59}$$

which (Eq. 11.29a) is $\sqrt{\frac{1}{2}}\left(|0_a0_b\rangle + |1_a1_b\rangle\right)$. Alice can perform a single-gate operation on her qb, qb1, which is in this entanglement. Alice of course has the choice to perform the identity operation (i.e., do nothing, just leave it alone). In general, Alice's single-gate operation Ω on qb1 would *transform* Bell state $|\phi_{ab}\rangle$:

$$\Omega|\phi_{ab}\rangle = |\phi_{ab}\rangle'. \tag{11.60}$$

Depending on which particular combination out of the four possibilities (1) $|0_1\rangle|0_2\rangle$, or (2) $|0_1\rangle|1_2\rangle$, or (3) $|1_1\rangle|0_2\rangle$, or (4) $|1_1\rangle|1_2\rangle$ is with Alice, she would perform one of the operations (1) $\mathbb{I}$ (identity), or (2) $\mathbb{X}$, or (3) $\mathbb{Z}$, or (4) $i\mathbb{Y}$ *respectively*, where $\mathbb{X}$, $\mathbb{Y}$ and $\mathbb{Z}$ are respectively the matrix representations of the Pauli *X*, *Y* and *Z* gates (Table 11.3). Using the matrix representations of the single-qubit Pauli gates, it is easily verified that the state $|\phi_{ab}\rangle'$ is given by the last column in Table 11.6 when $|\phi_{ab}\rangle$ is $|\beta_{a,b,s+}\rangle$.

Alice would now *send* her qubit to Bob. This is a crucial step in the context of the Holevo-limit. We note that only a *single qubit* is communicated by Alice to Bob. It is under this constraint that the mechanism of dense coding is to be appreciated. We note that the information about the two classical bits with Alice is coded in the Bell state $|\phi_{ab}\rangle'$, and Bob is

Table 11.6 Transformation of the state $|\phi_{ab}\rangle \to |\phi_{ab}\rangle'$ under the single-qb gate operations performed by Alice on the qb in her possession.

Sr. No.	Alice has ↓	Alice applies this ↓ gate	$\vert\phi_{ab}\rangle'$
1	$\vert 0_{a,1}\rangle\vert 0_{a,2}\rangle$	Identity, $\mathbb{I}$	$\vert\beta_{a,b,s+}\rangle$
2	$\vert 0_{a,1}\rangle\vert 1_{a,2}\rangle$	$\mathbb{X}$	$\vert\beta_{a,b,d+}\rangle$
3	$\vert 1_{a,1}\rangle\vert 0_{a,2}\rangle$	$\mathbb{Z}$	$\vert\beta_{a,b,s-}\rangle$
4	$\vert 1_{a,1}\rangle\vert 1_{a,2}\rangle$	$i\mathbb{Y}$	$\vert\beta_{a,b,d-}\rangle$

now in possession of both the qubits entangled in this Bell state. He can perform Control-Not operation on the Bell state $|\phi_{ab}\rangle'$. He would use the qubit which he received from Alice as the *control qb*, and his qb as the *target qb*. As a result of this operation, Bob gets a new 2-qubit state, which we denote by $|\phi_{ab}\rangle''$:

$$CX|\phi_{ab}\rangle' = |\phi_{ab}\rangle''. \tag{11.61}$$

The results would be different depending on the four possible pairs of the classical bits originally with Alice. The result of the operation described in Eq. 11.61 is summarized in the Table 11.7. We note that the state $|\phi_{ab}\rangle''$ is factorable, and the factors are also given in Table 11.7.

Since the state $|\phi_{ab}\rangle''$ is factorable, Bob can conduct separate measurements on each of the participating distinguishable qubit, *without* affecting the other. First, he can measure the state of the qubit that was originally with him. From Table 11.7, we see that if Bob finds his qb to be in the state $|0_b\rangle$, then the two cbits whose information Alice was to communicate

Table 11.7 Transformation of the 2-qubit state described by $CX|\phi_{ab}\rangle' \to |\phi_{ab}\rangle''$ under the 2-qubit Control-Not gate operation performed by Bob on the 2-qubit state $|\phi_{ab}\rangle'$.

Sr. No.	Alice has ↓	$CX\vert\phi_{ab}\rangle'$	$\vert\phi_{ab}\rangle''$
1	$\vert 0_{a,1}\rangle\vert 0_{a,2}\rangle$	$CX\vert\beta_{a,b,s+}\rangle$	$\sqrt{\frac{1}{2}}\left(\vert 0_a 0_b\rangle + \vert 1_a 0_b\rangle\right) = \sqrt{\frac{1}{2}}\left(\vert 0_a\rangle + \vert 1_a\rangle\right)\vert 0_b\rangle$
2	$\vert 0_{a,1}\rangle\vert 1_{a,2}\rangle$	$CX\vert\beta_{a,b,d+}\rangle$	$\sqrt{\frac{1}{2}}\left(\vert 1_a 1_b\rangle + \vert 0_a 1_b\rangle\right) = \sqrt{\frac{1}{2}}\left(\vert 1_a\rangle + \vert 0_a\rangle\right)\vert 1_b\rangle$
3	$\vert 1_{a,1}\rangle\vert 0_{a,2}\rangle$	$CX\vert\beta_{a,b,s-}\rangle$	$\sqrt{\frac{1}{2}}\left(\vert 0_a 0_b\rangle - \vert 1_a 0_b\rangle\right) = \sqrt{\frac{1}{2}}\left(\vert 0_a\rangle - \vert 1_a\rangle\right)\vert 0_b\rangle$
4	$\vert 1_{a,1}\rangle\vert 1_{a,2}\rangle$	$CX\vert\beta_{a,b,d-}\rangle$	$\sqrt{\frac{1}{2}}\left(-\vert 1_a 1_b\rangle + \vert 0_a 1_b\rangle\right) = \sqrt{\frac{1}{2}}\left(-\vert 1_a\rangle + \vert 0_a\rangle\right)\vert 1_b\rangle$

to him are either $\left|0_{a,1}\right\rangle\left|0_{a,2}\right\rangle$ (first row of Table 11.7) or $\left|1_{a,1}\right\rangle\left|0_{a,2}\right\rangle$ (third row); and if he finds his qb to be in the state $\left|1_b\right\rangle$, the two cbits are either $\left|0_{a,1}\right\rangle\left|1_{a,2}\right\rangle$ (second row) or $\left|1_{a,1}\right\rangle\left|1_{a,2}\right\rangle$ (fourth row). Bob is now in *partial* possession of the information about the two cbits with Alice. He now has to figure out which one of *two* possible pairs of cbits is with Alice. This is a healthy advance from the *four* possibilities he had to worry about at the beginning. The residual two alternatives differ only in the *relative* phase ($\pm$) of the states $\left|0_a\right\rangle$ and $\left|1_a\right\rangle$ in the *first* factor qubit in the last column of Table 11.7. The final step therefore is to use the Hadamard gate on the first factor qubit in the last column of the Table 11.7, which is the qubit Bob received from Alice. The Hadamard gate transforms a pure state into a mixed state, and vice versa. When it operates on a mixed state $\frac{1}{\sqrt{2}}(|0\rangle + |1\rangle)$ with *in-phase* superposition, it outputs the state $|0\rangle$, and when it operates on the out-of-phase superposition $\frac{1}{\sqrt{2}}(|0\rangle - |1\rangle)$, it outputs $|1\rangle$. By determining the final output to be $|1\rangle$ or $|0\rangle$, Bob can then figure out whether the superposition involved was *in-phase* or *out-of-phase*, and thereby make the final distinction between $\left|0_{a,1}\right\rangle\left|0_{a,2}\right\rangle$ and $\left|1_{a,1}\right\rangle\left|0_{a,2}\right\rangle$, or between $\left|0_{a,1}\right\rangle\left|1_{a,2}\right\rangle$ and $\left|1_{a,1}\right\rangle\left|1_{a,2}\right\rangle$, that remained to be done. The highlight of the strategy described above is that a single qubit is sent by Alice to Bob, which enables him to possess the state $\left|\phi_{ab}\right\rangle'$ amenable for subsequent quantum gate operations. The algorithm described above uses quantum entanglement to side-step the Holevo-limit, and is therefore called as *dense coding*.

There are alternative ways of achieving dense coding. For example, one can use maximally entangled three qubits (3qb), which are an extension from the 2-qubit Bell state to 3qb systems. A fully correlated 3qb system is not factorable in terms of tensor products of lower qubits; neither a direct product of the three factor qubits, nor that of one qubit with a Bell state of the other two. Such a 3qb state is called as a W state, after Wolfgang Dür [27]. It is written as

$$W_n = \frac{1}{\sqrt{2+2n}}\left\{|100\rangle + \sqrt{n}e^{i\gamma}|010\rangle + \sqrt{n+1}e^{i\delta}|001\rangle\right\}. \tag{11.62}$$

In the above superposition the vector $|(\text{qb}1)(\text{qb}2)(\text{qb}3)\rangle$ denotes the states of the qb1 (with Alice), and qb2 and qb3 (both with Bob). *For example*, for $n = 1, \gamma = 0, \delta = 0$, the W state is

$$W_1 = \frac{1}{\sqrt{4}}\left\{|100\rangle + |010\rangle + \sqrt{2}|001\rangle\right\}. \tag{11.63}$$

We shall outline the procedure to achieve dense coding using the 3qb state W_1. We know that the three qubits can remain entangled regardless of their physical locations, so we consider W_1 consisting of qb1 with Alice, and qb2 and qb3 with Bob. Depending on which pair of cbits she has, Alice would perform a different single-qubit gate operation Ω on qb1 which transforms the 3qb state:

$$\Omega|W_1\rangle = |W_1\rangle'. \tag{11.64}$$

Possible results of this operation are presented in Table 11.8.

The four 3qb states in the last column in Table 11.8 are orthogonal to each other. Alice can now send the transformed qubit (qb1) in her possession to Bob. Subsequently Bob can carry out measurements corresponding to 3qb projection operators, and thereby determine which of the four possible pairs of cbits Alice has. Yet again, a single qubit is sent by Alice to Bob, but it has in it coded information about which of the four pairs of cbits she has, thus dodging the Holevo limit.

In this chapter, we have restricted ourselves only to provide a flavor of the vast field of quantum information science. Much of our discussion has involved a maximum of three qubits, but the Solved Problems P11.17, P11.18, and P11.19 would provide a flavor of generalizations to higher number of qubits can be made. Our coverage is far from being exhaustive; for the latter, readers are referred to specialized literature [28, 29, for example] on entanglement and quantum computing that heralds the second quantum revolution. Applications of quantum computing concern not only the search algorithms and for prime factorization, but also cybersecurity, quantum chemistry and drug development, fabrication of more lasting miniature batteries, modelling of finance management systems, management of traffic, search for exoplanets, synthesizing images of black holes using a multitude of ground-based and space telescopes, and in the areas of artificial intelligence, machine learning, deep learning, etc. Quantum chemistry on a quantum computer has now become a viable option. It opens up exciting opportunities to tackle challenging problems in important areas such as drug designing, etc. Methods of 'second quantization' are used to map a quantum chemistry Hamiltonian into qubit operators. The method of second quantization is also important to solve many other problems in physics, such as the study of electron correlations in atomic, molecular, and condensed matter physics, etc. We provide a brief introduction to second quantization in Appendix D, and to quantum chemistry on a quantum computer using the Variational Quantum Eigensolver (VQE) in Appendix E, with a limited objective only to induct readers into these expanding vistas.

Table 11.8 Transformation of the 3qb state $\Omega|W_1\rangle = |W_1\rangle'$ under the single qubit operation Ω performed by Alice on qb1.

Sr. No.	Alice has ↓	$\boldsymbol{\Omega}|\boldsymbol{W}_1\rangle$	$|\boldsymbol{W}_1\rangle'$
1	$|0_{a,1}\rangle|0_{a,2}\rangle$	$I|W_1\rangle$	$\frac{1}{\sqrt{4}}\{|100\rangle+|010\rangle+\sqrt{2}|001\rangle\}$
2	$|0_{a,1}\rangle|1_{a,2}\rangle$	$X|W_1\rangle$	$\frac{1}{\sqrt{4}}\{|000\rangle+|110\rangle+\sqrt{2}|101\rangle\}$
3	$|1_{a,1}\rangle|0_{a,2}\rangle$	$Z|W_1\rangle$	$\frac{1}{\sqrt{4}}\{-|100\rangle+|010\rangle+\sqrt{2}|001\rangle\}$
4	$|1_{a,1}\rangle|1_{a,2}\rangle$	$iY|W_1\rangle$	$\frac{1}{\sqrt{4}}\{|000\rangle-|110\rangle-\sqrt{2}|101\rangle\}$

Supplementary Notes

Introduction to Programming in Python

Python is a high-level easy to learn programming language; its syntax is easy to read and remember.

(1) For a beginner's guide please visit website: https://www.python.org/doc/.
(2) *Python Programming for School Students: Suitable for Class X to XII Students* – by Mahendra Verma.
(3) *Practical Numerical Computing Using Python: Scientific & Engineering Applications* – by Mahendra Verma.

Introduction to Qiskit

Qiskit is an open-source SDK (Software Development Kit) for working with quantum computers at the level of pulses, circuits, and application modules. Please visit https://qiskit.org/.

Running a code on IBM Quantum Processor and IBM-Q Composer:

To run our quantum circuit on an actual quantum processor from IBM you can create an account at https://quantum-computing.ibm.com. Once you create your account, you will receive a 128-character long API token. Make a note of it, as you would need it to perform calculations on IBM's quantum processors.

Solved Problems

P11.1: Consider the world's fastest supercomputer, the Japanese built Fugaku supercomputer. It has a peak speed of 442.010 petaflops. How does this compare to a 64-qubit quantum computer? What happens if you double the number of qubits in the quantum computer? Discuss your answer in terms of (a) time taken for equivalent calculations, (b) space required to store the data.

Solution:

(a) The Fugaku Supercomputer is a 64-bit, 7.6 million-core supercomputer, which can perform around 4.4×10^{17} floating-point operations per second. An n-qubit quantum computer, on the other hand, can store 2^n states and perform operations on them simultaneously. Now consider our 64-qubit quantum computer: The 64-qubit quantum computer can operate on 2^{64} states simultaneously. Assuming it takes one second to complete one operation, let us calculate the speed of our system.

Number of states available $= 2^{64} = 1.8 \times 10^{19}$ states.

Effective number of operations per second = time taken for one operation × number of states on which this operation is performed simultaneously $= 1 \times 2^{64} = 1.8 \times 10^{19}$ operations per second.

The ratio of the speed of the 64-qubit quantum computer to that of the Fugaku supercomputer is $1.8 \times 10^{19} \div 4.4 \times 10^{17} \approx 41$.

This means that the supercomputer will take ~41 seconds to complete the same amount of work that a 64-qubit quantum computer performs in one second. If we double our qubit count to 128 qubits,

Number of states available $= 2^{128} = 3.4 \times 10^{38}$ states.

Effective number of operations per second $= 1 \times 2^{128} = 3.4 \times 10^{38}$ operations per second.

Similarly, the speed of the 128-qubit quantum computer as compared with the Fugaku supercomputer is $3.4 \times 10^{38} \div 4.4 \times 10^{17} = 7.4 \times 10^{20}$. Accordingly, a calculation that takes one second on a 128-bit quantum computer, will take 7.4×10^{20} seconds on the Fugaku supercomputer. In comparison, the current age of the universe is approximately 4.3×10^{17} seconds.

(b) We can easily draw up similar arguments about the space requirements.

A 64-qubit register can store 2×2^{64} bits of data $= 3.6 \times 10^{19}$ bits $= 4.5 \times 10^{18}$ bytes $= 4.5$ million TB.

Size of 1 TB microSD card $= 15\,\text{mm} \times 11\,\text{mm} \times 1\,\text{mm} = 1.65 \times 10^{-7}\,\text{m}^3$.

Thus, the space taken by 4.5 million microSD cards would be $0.76\,\text{m}^3$.

This is the size of a box which is $1\,\text{m} \times 1\,\text{m} \times 76\,\text{cm}$. Now, if we double the number of qubits to 128, then we can store 2×2^{128} bits of data (2^{126} bytes), we would require a space equal to $7 \times 10^{18}\,\text{m}^3$, or one-third the volume of moon! In comparison, a 127-qubit quantum chip can be made to have a size of approximately $25\,\text{mm} \times 30\,\text{mm} \times 3.5\,\text{mm}$!!

P11.2: (a) In a quantum circuit the number of input and output qubits is the same. Justify this statement.

(b) Quantum gates are unitary and hence reversible. Let us consider a state $|0\rangle$ to which we apply a Hadamard gate. Is the output deterministic? Show that the Hadamard gate is logically reversible.

Solution:

(a) A qubit or a system of qubits changes its state by going through a series of transformations. A unitary transformation is described by a matrix U with complex entries. Unitary $(UU^{\dagger} = 1)$ can be satisfied only if the number of input and output qubits is equal. Unlike classical gates, quantum gates are reversible; the gate's inputs can be reconstructed from its outputs.

(b) Applying a Hadamard gate to a pure state $|0\rangle$ gives the superposed state $\frac{1}{\sqrt{2}}\left(|0\rangle + |1\rangle\right)$. The output is *deterministic* even if it is a superposition. The Hadamard gate is logically reversible as we can apply another Hadamard gate to the superposed state which returns pure input state.

P11.3: (a) Find the antipodal state $|\psi^{\perp}\rangle$ of a general quantum gate $|\psi\rangle = \cos\left(\frac{\theta}{2}\right)|0\rangle + e^{i\varphi}\sin\left(\frac{\theta}{2}\right)|1\rangle$. Furthermore, show that $\langle\psi|\psi^{\perp}\rangle = 0$.

(b) Does applying a Pauli-X gate negate the above state $|\psi\rangle$?

Solution:

(a) Antipodal polar coordinates of the point (θ, φ) are $(\pi - \theta, \varphi + \pi)$.

$$|\psi^{\perp}\rangle = \cos\left(\frac{\pi-\theta}{2}\right)|0\rangle + e^{i(\varphi+\pi)}\sin\left(\frac{\pi-\theta}{2}\right)|1\rangle = \sin\left(\frac{\theta}{2}\right)|0\rangle - e^{i\varphi}\cos\left(\frac{\theta}{2}\right)|1\rangle.$$

We see that $\langle\psi|\psi^{\perp}\rangle = \sin\left(\frac{\theta}{2}\right)\cos\left(\frac{\theta}{2}\right) - \sin\left(\frac{\theta}{2}\right)\cos\left(\frac{\theta}{2}\right) = 0.$

(b) Pauli-X gate, also known as the NOT gate, negates pure states; it does not reverse the sign of a mixed state.

P11.4: Prove that $R_y\left(\frac{\pi}{2}\right)R_z(\pi)e^{i\frac{\pi}{2}} = H$.

Solution:

$$R_x(\theta) = \cos\left(\frac{\theta}{2}\right)I - i\sin\left(\frac{\theta}{2}\right)X = \cos\left(\frac{\theta}{2}\right)\begin{pmatrix}1 & 0\\ 0 & 1\end{pmatrix} - i\sin\left(\frac{\theta}{2}\right)\begin{pmatrix}0 & 1\\ 1 & 0\end{pmatrix},$$

i.e., $R_x(\theta) = \begin{pmatrix}\cos\left(\frac{\theta}{2}\right) & -i\sin\left(\frac{\theta}{2}\right)\\ -i\sin\left(\frac{\theta}{2}\right) & \cos\left(\frac{\theta}{2}\right)\end{pmatrix}$.

Likewise, $R_y(\theta) = \cos\left(\frac{\theta}{2}\right)I - i\sin\left(\frac{\theta}{2}\right)Y = \begin{pmatrix}\cos\left(\frac{\theta}{2}\right) & -\sin\left(\frac{\theta}{2}\right)\\ \sin\left(\frac{\theta}{2}\right) & \cos\left(\frac{\theta}{2}\right)\end{pmatrix}$.

The proof follows readily from matrix multiplication.

$$R_z(\theta) = \cos\left(\frac{\theta}{2}\right)I - i\sin\left(\frac{\theta}{2}\right)Z = \cos\left(\frac{\theta}{2}\right)\begin{pmatrix}1 & 0\\ 0 & 1\end{pmatrix} - i\sin\left(\frac{\theta}{2}\right)\begin{pmatrix}1 & 0\\ 0 & -1\end{pmatrix} = \begin{pmatrix}\exp\left(-i\frac{\theta}{2}\right) & 0\\ 0 & \exp\left(i\frac{\theta}{2}\right)\end{pmatrix}$$

$$R_y\left(\frac{\pi}{2}\right)R_z(\pi)e^{i\frac{\pi}{2}} = \frac{1}{\sqrt{2}}\begin{pmatrix}1 & -1\\ 1 & 1\end{pmatrix}\begin{pmatrix}-i & 0\\ 0 & i\end{pmatrix}\left(\cos\frac{\pi}{2} + i\sin\frac{\pi}{2}\right) = \frac{1}{\sqrt{2}}\begin{pmatrix}1 & 1\\ 1 & -1\end{pmatrix}.$$

P11.5: Show that $e^{i\pi/2}R_x\left(\frac{\pi}{2}\right)R_z\left(\frac{\pi}{2}\right)R_x\left(\frac{\pi}{2}\right) = H$.

Solution:

$$e^{i\pi/2}\left(\frac{1}{\sqrt{2}}\right)^3\begin{pmatrix}\cos\left(\frac{\pi}{4}\right) & -i\sin\left(\frac{\pi}{4}\right)\\ -i\sin\left(\frac{\pi}{4}\right) & \cos\left(\frac{\pi}{4}\right)\end{pmatrix}\begin{pmatrix}e^{-i\pi 3/4} & 0\\ 0 & e^{i\pi/4}\end{pmatrix}\begin{pmatrix}\cos\left(\frac{\pi}{4}\right) & -i\sin\left(\frac{\pi}{4}\right)\\ -i\sin\left(\frac{\pi}{4}\right) & \cos\left(\frac{\pi}{4}\right)\end{pmatrix} =$$

$$= i\left(\frac{1}{2}\right)^3\begin{pmatrix}1 & -i\\ -i & 1\end{pmatrix}\begin{pmatrix}1-i & 0\\ 0 & 1+i\end{pmatrix}\begin{pmatrix}1 & -i\\ -i & 1\end{pmatrix} = H.$$

P11.6: Show that there exists unitary gates A, B and C such that $ABC = I$ and $U = e^{i\alpha}AXBXC$, given the decomposition $U = e^{i\alpha}R_z(\beta)R_y(\gamma)R_z(\delta)$. *Hint*: You can choose $A = R_z\left(\frac{\delta-\beta}{2}\right)$, $B = R_y\left(\frac{-\gamma}{2}\right)R_z\left(\frac{-\delta-\beta}{2}\right)$, and $C = R_z(\beta)R_y\left(\frac{\gamma}{2}\right)$.

Solution:

Performing the operations in the sequence A first, B next, and C last is mathematically equivalent to operating on a quantum state by the operator CBA. Using $XR_z(\theta)X = R_z(-\theta)$ and $XR_y(\theta)X = R_y(-\theta)$, we have

$$CXBXA = R_z(\beta)R_y\left(\frac{\gamma}{2}\right)XR_y\left(\frac{-\gamma}{2}\right)X\bullet XR_z\left(\frac{-\delta-\beta}{2}\right)XR_z\left(\frac{\delta-\beta}{2}\right),$$

$$CXBXA = R_z(\beta)R_y\left(\frac{\gamma}{2}\right)R_y\left(\frac{\gamma}{2}\right)R_z\left(\frac{\delta+\beta}{2}\right)R_z\left(\frac{\delta-\beta}{2}\right).$$

Therefore, $CXBXA = R_z(\beta)R_y(\gamma)R_z(\delta)$. Apply controlled-$U$ gate (which operates on a target qubit of and only if the control qubit is $|1\rangle$) on arbitrary state $a|0\rangle + b|1\rangle$.

$$\text{Controlled-}U\ |0\rangle\left(a|0\rangle + b|1\rangle\right) = e^{i\alpha}|0\rangle \otimes C\bullet B\bullet A\left(a|0\rangle + b|1\rangle\right) = |0\rangle \otimes C\bullet B\bullet A\left(a|0\rangle + b|1\rangle\right),$$

$$\text{Controlled-}U\ |1\rangle\left(a|0\rangle + b|1\rangle\right) = |1\rangle \otimes e^{i\alpha}C\bullet B\bullet A\left(a|0\rangle + b|1\rangle\right) = |1\rangle \otimes U\left(a|0\rangle + b|1\rangle\right).$$

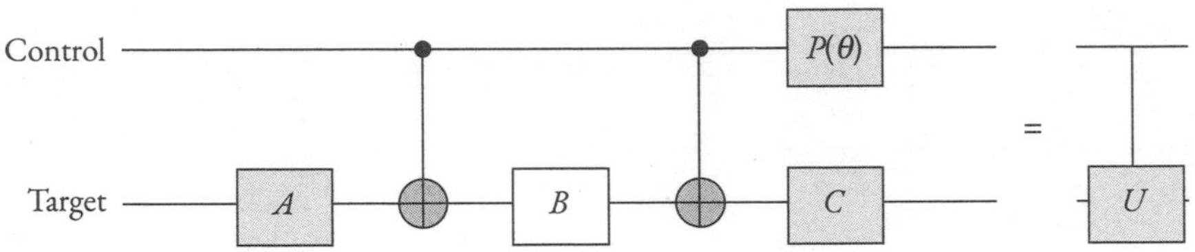

P11.7: Consider the gates (i) H (ii) $G_A = Z\sqrt{Y}$ and (iii) $G_B = \sqrt{Z}\sqrt{X}\sqrt{Z}$. Sketch their operations on an arbitrary state, $|\psi\rangle$, using the Bloch sphere representation. From the final position of the qubit state on the Bloch sphere, can we conclude that $H = G_A = G_B$?

Solution:

(i) The Hadamard operation on an arbitrary state, $|\psi\rangle = \alpha_0|0\rangle + \alpha_1|1\rangle$.

$$H|\psi\rangle = \frac{1}{\sqrt{2}}\begin{pmatrix}1 & 1\\ 1 & -1\end{pmatrix}\begin{pmatrix}\alpha_0\\ \alpha_1\end{pmatrix}$$

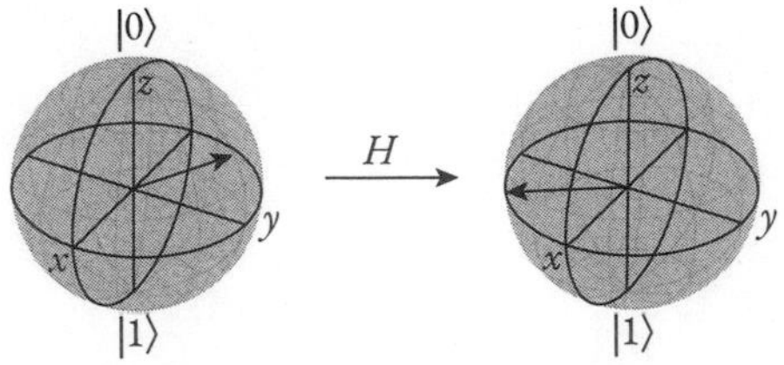

(ii) Pauli-Z gate - π rotation about Z axis in the anticlockwise direction.
$\sqrt{Y}$ gate - $\pi/2$ rotation about the y axis in the anticlockwise direction

$$\sqrt{Y}Z|\psi\rangle = \left(\frac{1+i}{2}\right)\begin{pmatrix}1 & -1\\ 1 & 1\end{pmatrix}\begin{pmatrix}1 & 0\\ 0 & -1\end{pmatrix}\begin{pmatrix}\alpha_0\\ \alpha_1\end{pmatrix} = \left(\frac{e^{i\pi/4}}{\sqrt{2}}\right)\begin{pmatrix}1 & 1\\ 1 & -1\end{pmatrix}\begin{pmatrix}\alpha_0\\ \alpha_1\end{pmatrix} = e^{i\pi/4}H|\psi\rangle$$

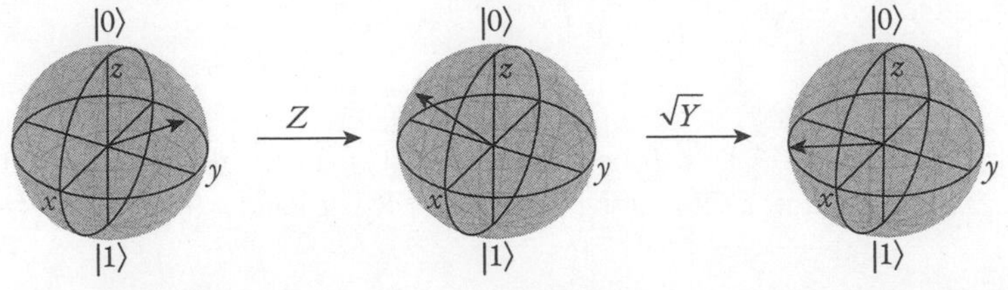

(iii) $\sqrt{Z}$ gate - $\pi/2$ rotation about the z axis in the anticlockwise direction

$\sqrt{X}$ gate - $\pi/2$ rotation about the x axis in the anticlockwise direction

$$\sqrt{Z}\sqrt{X}\sqrt{Z}|\psi\rangle = \begin{pmatrix} 1 & 0 \\ 0 & i \end{pmatrix}\begin{pmatrix} 1+i & 1-i \\ 1-i & 1+i \end{pmatrix}\begin{pmatrix} 1 & 0 \\ 0 & i \end{pmatrix}\begin{pmatrix} \alpha_0 \\ \alpha_1 \end{pmatrix} = \left(\frac{e^{i\pi/4}}{\sqrt{2}}\right)\begin{pmatrix} 1 & 1 \\ 1 & -1 \end{pmatrix}\begin{pmatrix} \alpha_0 \\ \alpha_1 \end{pmatrix} = e^{i\pi/4}H|\psi\rangle$$

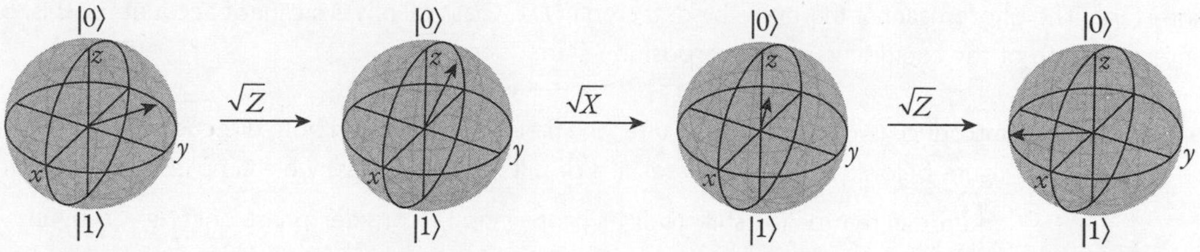

The orientation of the vector on the Bloch sphere looks the same state for H, G_A and G_B gates, but the gates $G_A = G_B = e^{i\pi/4}H$ introduce an additional phase which is not manifest on the two-parameter (θ, φ) Bloch sphere (Eq. 11.20 and Fig. 11.6). The additional phase $\xi = \pi/4$ cannot be represented on the Bloch sphere. While $H^2 = I$, $G_A^8 = G_B^8 = I$ (Idempotent), the Gates G_A and G_B are *Hadamard-like* gates, and are reversible; they do not recover the original state when operated on the input twice in succession.

P11.8: Mach Zehnder Interferometer (MZI) consists of a photon source (S), two 50:50 beam splitters ($BS1$ and $BS2$), two mirrors ($M1$ and $M2$) and two detectors ($D1$ and $D2$) as shown in the figure given below. If an incident photon is in the x direction, which of the two detectors would register it? Given: The vector of the photon moving horizontally is $|x\rangle = \begin{pmatrix} 1 \\ 0 \end{pmatrix}$ and that of the photon moving vertically is $|y\rangle = \begin{pmatrix} 0 \\ 1 \end{pmatrix}$.

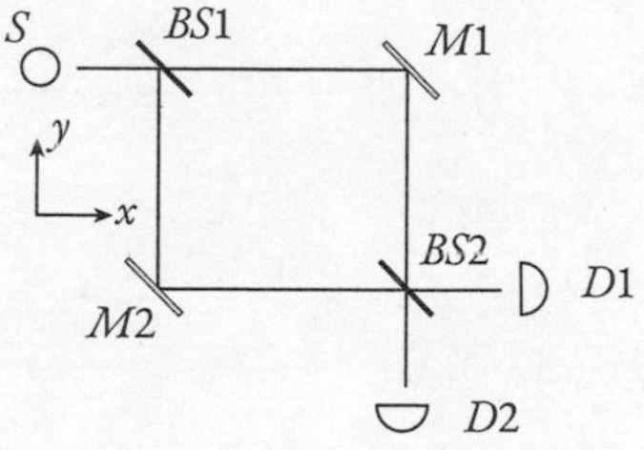

Operators representing a beam splitter: $BS = \frac{1}{\sqrt{2}}\begin{pmatrix} 1 & i \\ i & 1 \end{pmatrix}$; for interaction with glass: $A(\varphi_x, \varphi_y) = \begin{pmatrix} e^{i\varphi_x} & 0 \\ 0 & e^{i\varphi_y} \end{pmatrix}$; and for a mirror, $M = \begin{pmatrix} 0 & 1 \\ 1 & 0 \end{pmatrix}$.

Solution:

When a photon traveling in the x-direction passes through the beam splitter, the result is

$$BS|x\rangle = \frac{1}{\sqrt{2}}\begin{pmatrix} 1 & i \\ i & 1 \end{pmatrix}\begin{pmatrix} 1 \\ 0 \end{pmatrix} = \frac{1}{\sqrt{2}}\begin{pmatrix} 1 \\ i \end{pmatrix} = \frac{1}{\sqrt{2}}(|x\rangle + i|y\rangle).$$

The resultant is a superposed state. When this state passes through a mirror, the Pauli-X gate rotates the state through π radians around the x axis. $MBS|x\rangle = \frac{1}{\sqrt{2}}\begin{pmatrix} 0 & 1 \\ 1 & 0 \end{pmatrix}\begin{pmatrix} 1 \\ i \end{pmatrix} = \frac{1}{\sqrt{2}}(i|x\rangle + |y\rangle).$

Once it passes through the second beam splitter, we get $BSMBS|x\rangle = \frac{1}{2}\begin{pmatrix} 1 & i \\ i & 1 \end{pmatrix}\begin{pmatrix} i \\ 1 \end{pmatrix} = i|x\rangle$.

Therefore, the probability that the photon will be detected in the detector placed in the x direction is one. To reach $D1$, both paths experience one reflection and so arrive in phase with each other with their phases shifted by 90°. The paths to $D2$, however, are 180° out of phase and destructively interfere. The photon cannot therefore be detected in $D2$. Classical physics cannot account for this; Born interpretation of the coefficients in a superposition state does.

P11.9: Introduce two glass slabs (G) after the first beam splitter in both directions as shown in the figure below. If the two glasses are of different thickness, which implies that each would add a different phase shift to light propagation, (consider phase shifts given by $\varphi_x = \frac{\pi}{2}$ and $\varphi_y = \frac{\pi}{4}$), what is the probability of detecting the photon in the x and y directions respectively?

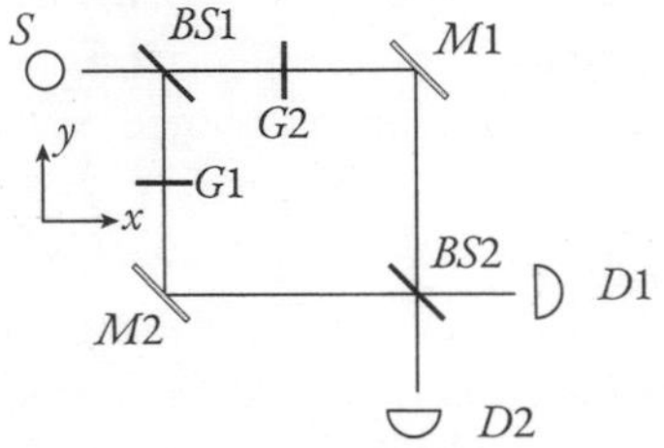

Solution:

When two glasses are introduced along x and y direction respectively,

$A(\varphi_x, \varphi_y)BS|x\rangle = \frac{1}{\sqrt{2}}\begin{pmatrix} e^{i\varphi_x} & 0 \\ 0 & e^{i\varphi_y} \end{pmatrix}\begin{pmatrix} 1 \\ i \end{pmatrix} = \frac{1}{\sqrt{2}}\left(e^{i\varphi_x}|x\rangle + ie^{i\varphi_y}|y\rangle\right)$. This is now followed by interaction with the mirror, $MA(\varphi_x, \varphi_y)BS|x\rangle = \frac{1}{\sqrt{2}}\begin{pmatrix} 0 & 1 \\ 1 & 0 \end{pmatrix}\begin{pmatrix} e^{i\varphi_x} \\ ie^{i\varphi_y} \end{pmatrix} = \frac{1}{\sqrt{2}}\begin{pmatrix} ie^{i\varphi_y} \\ e^{i\varphi_x} \end{pmatrix}$.

Once light passes through the second beam splitter, we shall have

$$BSMA(\varphi_x, \varphi_y)BS|x\rangle = \frac{1}{2}\begin{pmatrix} 1 & i \\ i & 1 \end{pmatrix}\begin{pmatrix} i\exp(i\varphi_y) \\ \exp(i\varphi_x) \end{pmatrix} = \frac{1}{2}\begin{pmatrix} i\left(\exp(i\varphi_x) + \exp(i\varphi_y)\right) \\ \left(\exp(i\varphi_x) - \exp(i\varphi_y)\right) \end{pmatrix}.$$

For glasses with different thickness, $\varphi_x = \frac{\pi}{2}$ and $\varphi_y = \frac{\pi}{4}$, the probability that the photon will be detected in $D1$ is 0.85355. Probability that the photon will be detected in $D2$ is 0.14645.

P11.10: The figure below shows a quantum circuit that simulates MZI. Find the probability of detecting $|0\rangle$ at the output.

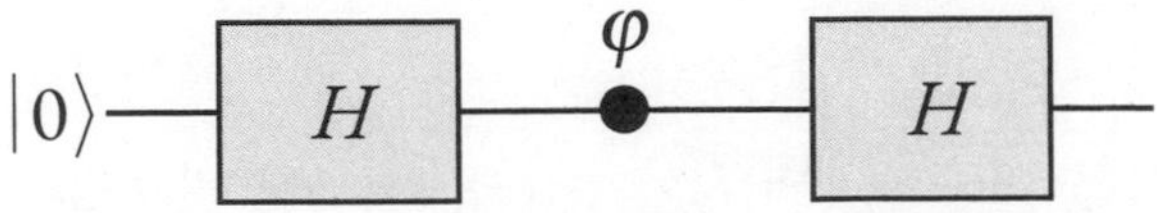

Solution:

Apply the Hadamard gate to the $|0\rangle$ state: $H|0\rangle = \frac{1}{\sqrt{2}}\begin{pmatrix} 1 & 1 \\ 1 & -1 \end{pmatrix}\begin{pmatrix} 1 \\ 0 \end{pmatrix} = \frac{1}{\sqrt{2}}\begin{pmatrix} 1 \\ 1 \end{pmatrix} = \frac{1}{\sqrt{2}}(|0\rangle + |1\rangle)$.

Apply now the phase shift gate: $P(\theta)H|0\rangle = \frac{1}{\sqrt{2}}\begin{pmatrix} 1 & 0 \\ 0 & e^{i\varphi} \end{pmatrix}\begin{pmatrix} 1 \\ 1 \end{pmatrix} = \frac{1}{\sqrt{2}}\begin{pmatrix} 1 \\ e^{i\varphi} \end{pmatrix} = \frac{1}{\sqrt{2}}(|0\rangle + e^{i\varphi}|1\rangle)$.

Finally, operating by the Hadamard gate:

$$HP(\theta)H|0\rangle = \frac{1}{2}\begin{pmatrix} 1 & 1 \\ 1 & -1 \end{pmatrix}\begin{pmatrix} 1 \\ e^{i\varphi} \end{pmatrix} = \frac{1}{2}\begin{pmatrix} 1+e^{i\varphi} \\ 1-e^{i\varphi} \end{pmatrix} = \left(\frac{1+e^{i\varphi}}{2}\right)|0\rangle + \left(\frac{1-e^{i\varphi}}{2}\right)|1\rangle.$$

Probability of detecting the state $|0\rangle$ is $P_0(\varphi) = |\langle 0|HP(\theta)H|0\rangle|^2 = \frac{1+\cos\varphi}{2}$.

Probability of detecting the state $|1\rangle$ is $P_1(\varphi) = |\langle 1|HP(\theta)H|1\rangle|^2 = \frac{1-\cos\varphi}{2}$.

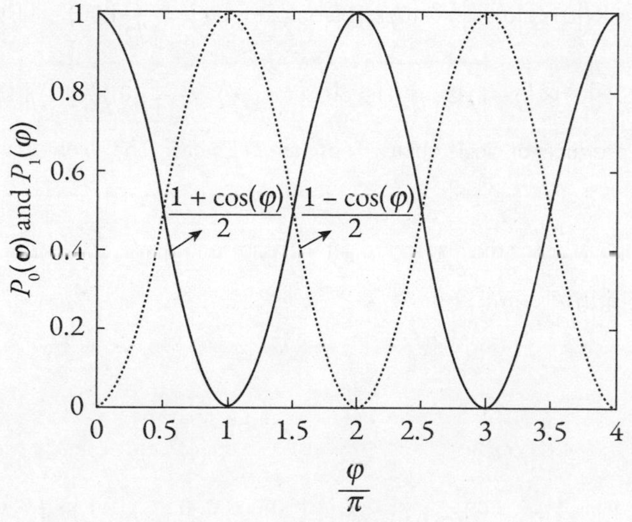

P11.11: What are the outputs of the quantum circuit shown in the figure below, when the inputs are: $|00\rangle$, $|01\rangle$, $|10\rangle$ and $|11\rangle$?

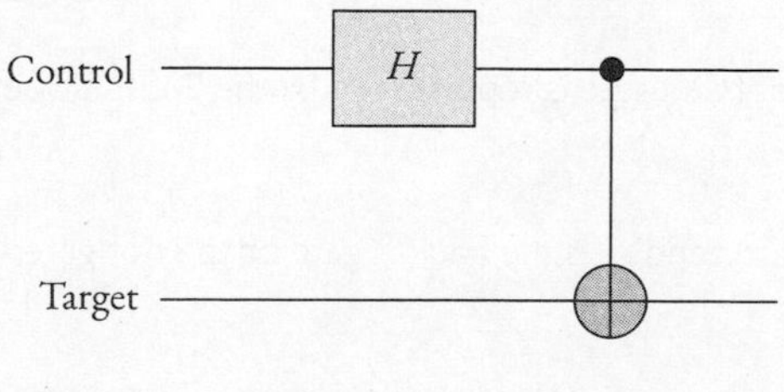

Solution:

Consider the input $|\overset{C}{0}\overset{T}{0}\rangle$, C represents the control qubit input and T represents the target qubit input.

Apply Hadamard gate: $H|0\rangle = \frac{1}{\sqrt{2}}\begin{pmatrix} 1 & 1 \\ 1 & -1 \end{pmatrix}\begin{pmatrix} 1 \\ 0 \end{pmatrix} = \frac{1}{\sqrt{2}}\begin{pmatrix} 1 \\ 1 \end{pmatrix} = \frac{1}{\sqrt{2}}(|0\rangle + |1\rangle)$.

For 2 qubits as shown in the circuit, the input after applying the Hadamard gate will be

$H|00\rangle = \frac{1}{\sqrt{2}}(|00\rangle + |10\rangle)$. Now, perform the *C-NOT* operation,

$$CXH|00\rangle = \frac{1}{\sqrt{2}} CX(|0\rangle \otimes |0\rangle + |1\rangle \otimes |0\rangle)\, CXH|00\rangle = \frac{1}{\sqrt{2}}\begin{pmatrix} 1 & 0 & 0 & 0 \\ 0 & 1 & 0 & 0 \\ 0 & 0 & 0 & 1 \\ 0 & 0 & 1 & 0 \end{pmatrix}\begin{pmatrix} 1 \\ 0 \\ 1 \\ 0 \end{pmatrix} = \frac{1}{\sqrt{2}}\begin{pmatrix} 1 \\ 0 \\ 0 \\ 1 \end{pmatrix}.$$

Therefore, the output is $|\psi'\rangle = \frac{1}{\sqrt{2}}(|00\rangle + |11\rangle)$

Similarly, for the input, $\left|\overset{CT}{01}\right\rangle$, $CXH|01\rangle = \frac{1}{\sqrt{2}} CX(|0\rangle \otimes |1\rangle + |1\rangle \otimes |1\rangle)$,

The output is, $|\psi'\rangle = \frac{1}{\sqrt{2}}(|01\rangle + |10\rangle)$.

For the input, $\left|\overset{CT}{10}\right\rangle$, $CXH|10\rangle = \frac{1}{\sqrt{2}} CX(|0\rangle \otimes |0\rangle - |1\rangle \otimes |0\rangle)$. The output is $|\psi'\rangle = \frac{1}{\sqrt{2}}(|00\rangle - |11\rangle)$.

For the input $\left|\overset{CT}{11}\right\rangle$, $CXH|11\rangle = \frac{1}{\sqrt{2}} CX(|0\rangle \otimes |1\rangle - |1\rangle \otimes |1\rangle)$. The output is $|\psi'\rangle = \frac{1}{\sqrt{2}}(|01\rangle - |10\rangle)$.

The outputs cannot be factored as products of single qubits. These are *Bell states*. The given circuit is a way of obtaining entangled states.

P11.12: Show that the unitary operator for the Hadamard gate is represented as

$$\Omega = \frac{1}{\sqrt{2}}\left[|0\rangle\langle 0| + |1\rangle\langle 0| + |0\rangle\langle 1| - |1\rangle\langle 1|\right].$$

Solution:

Let $|0\rangle = \begin{pmatrix} 1 \\ 0 \end{pmatrix}$ and $|1\rangle = \begin{pmatrix} 0 \\ 1 \end{pmatrix}$

$$\Omega = \frac{1}{\sqrt{2}}\left[\begin{pmatrix} 1 \\ 0 \end{pmatrix}(1\ \ 0) + \begin{pmatrix} 0 \\ 1 \end{pmatrix}(1\ \ 0) + \begin{pmatrix} 1 \\ 0 \end{pmatrix}(0\ \ 1) - \begin{pmatrix} 0 \\ 1 \end{pmatrix}(0\ \ 1)\right] = \frac{1}{\sqrt{2}}\left[\begin{pmatrix} 1 & 0 \\ 0 & 0 \end{pmatrix} + \begin{pmatrix} 0 & 0 \\ 1 & 0 \end{pmatrix} + \begin{pmatrix} 0 & 1 \\ 0 & 0 \end{pmatrix} - \begin{pmatrix} 0 & 0 \\ 0 & 1 \end{pmatrix}\right].$$

$$\therefore \Omega = \frac{1}{\sqrt{2}}\begin{pmatrix} 1 & 1 \\ 1 & -1 \end{pmatrix} = H, \text{ which is the Hadamard gate.}$$

P11.13: Show that the controlled-not (CX) gate is represented by the 2-qubit operator

$$\Omega = |0\rangle\langle 0| \otimes I + |1\rangle\langle 1| \otimes X,$$

where I is the identity operator and X is the Pauli-X gate or the not gate.

Solution:

$$\Omega = \begin{pmatrix} 1 \\ 0 \end{pmatrix}(1\ \ 0) \otimes \begin{pmatrix} 1 & 0 \\ 0 & 1 \end{pmatrix} + \begin{pmatrix} 0 \\ 1 \end{pmatrix}(0\ \ 1) \otimes \begin{pmatrix} 0 & 1 \\ 1 & 0 \end{pmatrix} = \begin{pmatrix} 1 & 0 \\ 0 & 0 \end{pmatrix} \otimes \begin{pmatrix} 1 & 0 \\ 0 & 1 \end{pmatrix} + \begin{pmatrix} 0 & 0 \\ 0 & 1 \end{pmatrix} \otimes \begin{pmatrix} 0 & 1 \\ 1 & 0 \end{pmatrix}$$

$$\text{i.e., } \Omega = \begin{pmatrix} 1 & 0 & 0 & 0 \\ 0 & 1 & 0 & 0 \\ 0 & 0 & 0 & 0 \\ 0 & 0 & 0 & 0 \end{pmatrix} + \begin{pmatrix} 0 & 0 & 0 & 0 \\ 0 & 0 & 0 & 0 \\ 0 & 0 & 0 & 1 \\ 0 & 0 & 1 & 0 \end{pmatrix} = \begin{pmatrix} 1 & 0 & 0 & 0 \\ 0 & 1 & 0 & 0 \\ 0 & 0 & 0 & 1 \\ 0 & 0 & 1 & 0 \end{pmatrix} = CX, \text{ which is the CNOT gate.}$$

P11.14: (a) Consider a two-qubit state where the second (target) qubit is initially in the pure state $|0\rangle$, and the first (control) cubit is in the pure state $|1\rangle$. Determine if the *CNOT* gate operating on the two-qubit states results in cloning.

(b) Consider a two-qubit pure state where the second (target) qubit is initially in the pure state $|0\rangle$, and the first (control) cubit is also in the pure state $|0\rangle$ state. Determine if the *CNOT* gate operating on the two-qubit states results in cloning.

(c) Does the above result make the *CNOT* operator a quantum cloning operator?

Solution:

(a) A copy of the first qubit is produced.

(b) Two copies of $|0\rangle$ are produced.

(c) The *CNOT* gate is not a quantum cloning operator since copies of only pure states are produced in the above examples; not that of a state of superposition.

P11.15: Check whether $I \otimes \sigma_i \overset{?}{=} \sigma_i \otimes I$, where I is the identity and σ_i are the Pauli matrices.

Solution:

$$I \otimes \sigma_1 = \begin{pmatrix} 1 & 0 \\ 0 & 1 \end{pmatrix} \otimes \begin{pmatrix} 0 & 1 \\ 1 & 0 \end{pmatrix} = \begin{pmatrix} 0 & 1 & 0 & 0 \\ 1 & 0 & 0 & 0 \\ 0 & 0 & 0 & 1 \\ 0 & 0 & 1 & 0 \end{pmatrix}, \text{ whereas } \sigma_1 \otimes I = \frac{1}{\sqrt{2}} \begin{pmatrix} 0 & 0 & 1 & 0 \\ 0 & 0 & 0 & 1 \\ 1 & 0 & 0 & 0 \\ 0 & 1 & 0 & 0 \end{pmatrix},$$

$$I \otimes \sigma_2 = \begin{pmatrix} 1 & 0 \\ 0 & 1 \end{pmatrix} \otimes \begin{pmatrix} 0 & -i \\ i & 0 \end{pmatrix} = \begin{pmatrix} 0 & -i & 0 & 0 \\ i & 0 & 0 & 0 \\ 0 & 0 & 0 & -i \\ 0 & 0 & i & 0 \end{pmatrix}, \ \sigma_2 \otimes I = \begin{pmatrix} 0 & -i \\ i & 0 \end{pmatrix} \otimes \begin{pmatrix} 1 & 0 \\ 0 & 1 \end{pmatrix} \frac{1}{\sqrt{2}} \begin{pmatrix} 0 & 0 & -i & 0 \\ 0 & 0 & 0 & -i \\ i & 0 & 0 & 0 \\ 0 & i & 0 & 0 \end{pmatrix},$$

$$\text{and } I \otimes \sigma_3 = \begin{pmatrix} 1 & 0 \\ 0 & 1 \end{pmatrix} \otimes \begin{pmatrix} 1 & 0 \\ 0 & -1 \end{pmatrix} = \begin{pmatrix} 1 & 0 & 0 & 0 \\ 0 & -1 & 0 & 0 \\ 0 & 0 & 1 & 0 \\ 0 & 0 & 0 & -1 \end{pmatrix}, \ \sigma_3 \otimes I = \begin{pmatrix} 1 & 0 \\ 0 & -1 \end{pmatrix} \otimes \begin{pmatrix} 1 & 0 \\ 0 & 1 \end{pmatrix} = \frac{1}{\sqrt{2}} \begin{pmatrix} 1 & 0 & 0 & 0 \\ 0 & 1 & 0 & 0 \\ 0 & 0 & -1 & 0 \\ 0 & 0 & 0 & -1 \end{pmatrix}.$$

Therefore, in general, $I \otimes \sigma_i \neq \sigma_i \otimes I$.

P11.16: Bell states can be transformed into each other under local unitary transformations. Prove the following relations:

(a) $|\beta_{s-}\rangle = (I \otimes \sigma_3)|\beta_{s+}\rangle = (\sigma_3 \otimes I)|\beta_{s+}\rangle$

(b) $|\beta_{d+}\rangle = (I \otimes \sigma_1)|\beta_{s+}\rangle = (\sigma_1 \otimes I)|\beta_{s+}\rangle$

(c) $|\beta_{d-}\rangle = (I \otimes -i\sigma_2)|\beta_{s+}\rangle = (i\sigma_2 \otimes I)|\beta_{s+}\rangle$

where, σ_i are the Pauli matrices.

Solution:

(a) $$|\beta_{s+}\rangle = \frac{1}{\sqrt{2}}(|0\rangle\otimes|0\rangle + |1\rangle\otimes|1\rangle) = \frac{1}{\sqrt{2}}\left[\begin{pmatrix}1\\0\end{pmatrix}\otimes\begin{pmatrix}1\\0\end{pmatrix} + \begin{pmatrix}0\\1\end{pmatrix}\otimes\begin{pmatrix}0\\1\end{pmatrix}\right] = \frac{1}{\sqrt{2}}\left[\begin{pmatrix}1\\0\\0\\0\end{pmatrix} + \begin{pmatrix}0\\0\\0\\1\end{pmatrix}\right] = \frac{1}{\sqrt{2}}\begin{pmatrix}1\\0\\0\\1\end{pmatrix}$$

$$(I\otimes\sigma_3)|\beta_{s+}\rangle = \frac{1}{\sqrt{2}}\begin{pmatrix}1&0&0&0\\0&-1&0&0\\0&0&1&0\\0&0&0&-1\end{pmatrix}\begin{pmatrix}1\\0\\0\\1\end{pmatrix} = \frac{1}{\sqrt{2}}\begin{pmatrix}1\\0\\0\\-1\end{pmatrix} = \frac{1}{\sqrt{2}}(|0\rangle\otimes|0\rangle - |1\rangle\otimes|1\rangle)$$

Similarly, $$(\sigma_3\otimes I)|\beta_{s+}\rangle = \frac{1}{\sqrt{2}}\begin{pmatrix}1&0&0&0\\0&1&0&0\\0&0&-1&0\\0&0&0&-1\end{pmatrix}\begin{pmatrix}1\\0\\0\\1\end{pmatrix} = \frac{1}{\sqrt{2}}\begin{pmatrix}1\\0\\0\\-1\end{pmatrix} = \frac{1}{\sqrt{2}}(|0\rangle\otimes|0\rangle - |1\rangle\otimes|1\rangle)$$

$$\therefore |\beta_{s-}\rangle = (I\otimes\sigma_3)|\beta_{s+}\rangle = (\sigma_3\otimes I)|\beta_{s+}\rangle$$

(b) Similarly, $(I\otimes\sigma_1)|\beta_{s+}\rangle = \frac{1}{\sqrt{2}}(|0\rangle\otimes|1\rangle + |1\rangle\otimes|0\rangle)$, $(\sigma_1\otimes I)|\beta_{s+}\rangle = \frac{1}{\sqrt{2}}(|0\rangle\otimes|1\rangle + |1\rangle\otimes|0\rangle)$

$$\therefore |\beta_{d+}\rangle = (I\otimes\sigma_1)|\beta_{s+}\rangle = (\sigma_1\otimes I)|\beta_{s+}\rangle$$

(c) $(I\otimes -i\sigma_2)|\beta_{s+}\rangle = \frac{1}{\sqrt{2}}(|0\rangle\otimes|1\rangle - |1\rangle\otimes|0\rangle)$, $(i\sigma_2\otimes I)|\beta_{s+}\rangle = \frac{1}{\sqrt{2}}(|0\rangle\otimes|1\rangle - |1\rangle\otimes|0\rangle)$

$$\therefore |\beta_{d-}\rangle = (I\otimes -i\sigma_2)|\beta_{s+}\rangle = (i\sigma_2\otimes I)|\beta_{s+}\rangle$$

P11.17: In Section 11.2 we dealt with Deutsch-Jozsa algorithm which took a single qubit, $|1_1\rangle$, as the input for the first register and $|1_2\rangle$ as the input for the second register. Generalize the algorithm so for n qubits input.

Solution:

1. The first register is an n-qubit state, each of the qubit initialized to $|1\rangle$. The second register is a single qubit state, initialized to $|1\rangle$: $|\psi_0\rangle = |1\rangle^{\otimes n}|1\rangle$.
2. The notation $|1\rangle^{\otimes n}$ is a compact way to represent $|1\rangle\otimes|1\rangle\otimes|1\rangle\otimes|1\rangle\otimes|1\rangle\cdots$ *n times*. Another equivalent notation is $|1^{\otimes n}\rangle$.
3. Apply a Hadamard gate to each of the qubits: $|\psi_1\rangle = \frac{1}{\sqrt{2^{n+1}}}\sum_{x=0}^{2^n-1}|x\rangle(|0\rangle - |1\rangle)$, where $|x\rangle$ is the xth term in the expansion of the product $\prod_n(|0\rangle - |1\rangle)$.
4. Apply the quantum function box on $|\psi_1\rangle$ to obtain $|\psi_2\rangle$:

$$|\psi_2\rangle = \frac{1}{\sqrt{2^{n+1}}}\sum_{x=0}^{2^n-1}|x\rangle(|f(x)\rangle - |1\oplus f(x)\rangle) = \frac{1}{\sqrt{2^{n+1}}}\sum_{x=0}^{2^n-1}(-1)^{f(x)}|x\rangle(|0\rangle - |1\rangle).$$

5. We now apply the Hadamard gate to each qubit in the first register, to obtain

$$|\psi_3\rangle = \frac{1}{2^n}\sum_{x=0}^{2^n-1}(-1)^{f(x)}\left[\sum_{z=0}^{2^n-1}(-1)^{x\cdot z}|z\rangle\right](|0\rangle - |1\rangle) = \frac{1}{2^n}\sum_{z=0}^{2^n-1}\left[\sum_{x=0}^{2^n-1}(-1)^{f(x)}(-1)^{x\cdot z}\right]|z\rangle(|0\rangle - |1\rangle),$$

where $x \bullet z$ is the sum of bitwise product, and evaluates to 1 in our case.

6. Measure the state of the n-qubits of the first register. The probability of measuring $|1\rangle^{\otimes n}$ is given by $\left|\frac{1}{2^n}\sum_{x=0}^{2^n-1}(-1)^{f(x)+1}\right|^2$. This evaluates to 1, if $f(x)$ is constant, and to 0 if $f(x)$ is balanced.

P11.18: Verify by mathematical induction that $H^{\otimes n}[i,j] = \frac{1}{\sqrt{2^n}}(-1)^{i\bullet j}$, where $i, j \in [0, 2^n)$.

Solution:

For the case $n = 1$, we have $H^{\otimes 1}[i,j] = \frac{1}{\sqrt{2^1}}(-1)^{i\bullet j}$, where $i, j \in [0, 2^1)$. For $n = k-1$, let's assume that $H^{\otimes k-1}[i,j] = \frac{1}{\sqrt{2^{k-1}}}(-1)^{i\bullet j}$, where $i, j \in [0, 2^{k-1})$. For $n = k$, we have $H^{\otimes k} = H \otimes H^{\otimes k-1} = \frac{1}{\sqrt{2}}\begin{pmatrix}1 & 1\\ 1 & -1\end{pmatrix} \otimes H^{\otimes k-1} = \frac{1}{\sqrt{2}}\begin{pmatrix}H^{\otimes k-1} & H^{\otimes k-1}\\ H^{\otimes k-1} & -H^{\otimes k-1}\end{pmatrix}$. Note that $H^{\otimes k-1}$ is a $2^{k-1} \times 2^{k-1}$ matrix. Now, for $i, j \in [0, 2^{k-1})$, $H^{\otimes k}[i,j] = \frac{1}{\sqrt{2}}H^{\otimes k-1}[i', j'] - \frac{1}{\sqrt{2^k}}(-1)^{i'\bullet j'}$.

Here, $i' = i_{1:}$, $j' = j_{1:}$ and $i_0 = 0$, $j_0 = 0$,

$$\begin{aligned} i \bullet j &= (i_0 \wedge j_0) \oplus (i_1 \wedge j_1) \oplus (i_2 \wedge j_2) \oplus \cdots \oplus (i_{n-1} \wedge j_{n-1}) = (0 \wedge 0) \oplus (i_1 \wedge j_1) \oplus (i_2 \wedge j_2) \oplus \cdots \oplus (i_{n-1} \wedge j_{n-1}) \\ &= 0 \oplus (i_1 \wedge j_1) \oplus (i_2 \wedge j_2) \oplus \cdots \oplus (i_{n-1} \wedge j_{n-1}) = i' \bullet j' \end{aligned}$$

thus, giving us, $H^{\otimes k}[i,j] = \frac{1}{\sqrt{2^k}}(-1)^{i\bullet j}$.

Similarly, for $i \in [2^{k-1}, 2^k), j \in [0, 2^{k-1})$ and $i \in [0, 2^{k-1}), j \in [2^{k-1}, 2^k)$, $H^{\otimes k}[i,j] = \frac{1}{\sqrt{2^k}}(-1)^{i'\bullet j'}$, where $i' = i_{1:}$, $j' = j_{1:}$ and $i_0 = 1, j_0 = 0$, and $i_0 = 0, j_0 = 1$ respectively, we can show that $i \bullet j = i' \bullet j'$. Finally, for $i, j \in [2^{k-1}, 2^{k-1})$, $H^{\otimes k}[i,j] = \frac{-1}{\sqrt{2^k}}(-1)^{i'\bullet j'}$, where $(i_0, j_0) = (1,1)$. In this case, we get $i \bullet j = 1 \oplus i' \bullet j'$, thus giving us $(-1)^{i\bullet j} = (-1)^{1\oplus i'\bullet j'} = (-1)^{1+i'\bullet j'} = -(-1)^{i'\bullet j'}$. Therefore, $H^{\otimes k}[i,j] = \frac{1}{\sqrt{2^k}}(-1)^{i\bullet j}$. Combining all the cases, for $n = k$, we get $H^{\otimes k}[i,j] = \frac{1}{\sqrt{2^k}}(-1)^{i\bullet j}$.

P11.19: The Greenberger–Horne–Zeilinger (GHZ) State is an n-qubit entangled state defined as $|\mathrm{GHZ}\rangle = \frac{|0\rangle^{\otimes n} + |1\rangle^{\otimes n}}{\sqrt{2}}$ for $n > 2$. Build a circuit for a 3-qubit GHZ state using Qiskit and measure it to verify that the GHZ state is indeed an entangled state.

Solution:

```
# Import Qiskit SDK.

from qiskit import QuantumRegister
from qiskit import ClassicalRegister
from qiskit import QuantumCircuit
from matplotlib import pyplot as plotter

# Simulation backend
from qiskit import Aer
# To perform simulation
from qiskit import execute
# For plotting out results
from qiskit.tools.visualization import plot_histogram
# Declaring a quantum register with 3 qubits
qr = QuantumRegister(3)

# Declaring a classicalregister with 3 bits
cr = ClassicalRegister(3)

# Declaring a quantumcircuit using the registers declared above
qc = QuantumCircuit(qr, cr)

# Applying a Hadamard gate on qubit 0
qc.h(qr[0])

# Applying CX gates to obtain our GHZ state
qc.cx(qr[0],qr[1])
qc.cx(qr[0],qr[2])

# Add a barrier to separate the input from the measurement
qc.barrier(qr)

# Apply the measurement gates
qc.measure(qr,cr)

# Prepare the simulation backend.
sim = Aer.get_backend("qasm_simulator")

# Let's simulate the circuit using execute(…)
res = execute(qc, backend = sim).result()

# Finally to plot the output we use plot_histogram(...)
plot_histogram(res.get_counts(qc))
plotter.show()
```

P11.20: Write a Qiskit program that simulates the quantum circuit given in problem P11.11.

Solution:

```
# Importing Qiskit SDK
from qiskit import QuantumRegister
from qiskit import ClassicalRegister
from qiskit import QuantumCircuit

# Get the value of pi
from numpy import pi

# For plotting our results
from qiskit.tools.visualization import plot_histogram
from matplotlib import pyplot as plotter

# Simulation backend
from qiskit import Aer
from qiskit import execute

# Prepare our quantum and classical registers
qreg_q = QuantumRegister(2, 'q')
creg_c = ClassicalRegister(2, 'c')

# Prepare the circuit.
circuit = QuantumCircuit(qreg_q, creg_c)

# By default, the qubits are in |0& state.
# Use a not gate to flip the qubit(s) to |1& state.
# Ex: circuit.x(qreg_1[1]) flips qubit 1 to |1& state.

# Apply Hadamard gate on Qubit 1
circuit.h(qreg_q[0])

# Apply C-NOT on Qubit 2 with Qubit 1 as control.
circuit.cx(qreg_q[0], qreg_q[1])

# Add a barrier in our circuit
circuit.barrier(qreg_q[0], qreg_q[1])

# Add measurement gates
circuit.measure(qreg_q[0], creg_c[0])
circuit.measure(qreg_q[1], creg_c[1])
```

```
# Prepare our simulation backend
sim = Aer.get_backend( "qasm_simulator" )

# Get the result
res = execute( circuit, backend = sim ).result()

# Generate the histogram of the output
plot_histogram( res.get_counts( circuit ) )

# Plot the histogram
plotter.show()
```

P11.21: Write a program to encode the circuit in problem P11.10, and verify the output for the following values of φ. (a) $\varphi = \pi/2$ (b) $\varphi = \pi/4$ (c) $\varphi = \pi$.

Solution:

```
# Importing Qiskit SDK
from qiskit import QuantumRegister
from qiskit import ClassicalRegister
from qiskit import QuantumCircuit

# Get the value of pi
from numpy import pi

# For plotting our results
from qiskit.tools.visualization import plot_histogram
from matplotlib import pyplot as plotter

# Simulation backend
from qiskit import Aer
from qiskit import execute

# Prepare our quantum and classical registers
qreg_q = QuantumRegister(1, 'q')
creg_c = ClassicalRegister(1, 'c')

# Prepare the circuit.
circuit = QuantumCircuit(qreg_q, creg_c)

# Apply Hadamard gate on the qubit.
circuit.h(qreg_q[0])

# We now apply a phase gate.
# A phase of pi/2 is implemented by s gate.
# Similarly, for pi/4 it is t gate
```

```
# and z gate perform the pi flip.
circuit.s(qreg_q[0])

# Finally, apply the second hardamard gate.
circuit.h(qreg_q[0])

# Add a barrier in our circuit
circuit.barrier(qreg_q[0])

# Add measurement gate
circuit.measure(qreg_q[0], creg_c[0])

# Prepare our simulation backend
sim = Aer.get_backend( "qasm_simulator" )

# Get the result
res = execute( circuit, backend = sim ).result()

# Generate the histogram of the output
plot_histogram( res.get_counts( circuit ) )

# Plot the histogram
plotter.show()
```

Exercises

E11.1: Prove that the matrix ρ describes the density operator if and only if $Tr[\rho] = 1$ and ρ is positive and semi-definite.

E11.2: Density operator of a pure state and mixed state are $\rho = \sum_i |\psi_i\rangle\langle\psi_i|$ and $\rho = \sum_i p_i |\psi_i\rangle\langle\psi_i|$ respectively, where p_i forms a valid probability distribution and $|\psi_i\rangle$ is a valid pure state. Show that:

(a) $\rho^2 = \rho$ for pure state, $\rho^2 \neq \rho$ for mixed state.

(b) $Tr[\rho^2] = 1$ for a pure state, $Tr[\rho^2] < 1$ for mixed state.

E11.3: Find the density operators of the following states, and classify the following states as pure or mixed states.

(a) All four Bell states

(b) $|\psi\rangle = \dfrac{|0\rangle \otimes |1\rangle + |1\rangle \otimes |0\rangle + |1\rangle \otimes |1\rangle}{\sqrt{3}}$

E11.4: Find the reduced density matrix of the states given in problem P11.16. Show that the partial traces $Tr_A(|\psi\rangle\langle\psi|)$ and $Tr_B(|\psi\rangle\langle\psi|)$ have the same eigen values.

E11.5: Prove the Schmidt decomposition theorem: For a pure state $|\psi\rangle$ of a bipartite composite state system, *AB*, there exists orthonormal states $|\phi\rangle_A$ for system *A*, and $|\phi\rangle_B$ for a system *B*, such that $|\psi\rangle = \sum_i \lambda_i |\phi\rangle_A |\phi\rangle_B$, where, λ_i are non-negative real numbers satisfying $\sum_i \lambda_i^2 = 1$, also known as the Schmidt co-efficient.

E11.6: With reference to E11.5, find the Schmidt decomposition of the following states;

(a) $|\psi_1\rangle = \frac{1}{2}(|00\rangle - |01\rangle + |10\rangle - |11\rangle)$

(b) $|\psi_2\rangle = \frac{1}{2}(|00\rangle + |01\rangle + |10\rangle - |11\rangle)$

References

[1] https://en.wikipedia.org/wiki/Solvay_Conference#Fifth_conference (downloaded on 15 January 2022).

[2] N. David Mermin, Is the Moon There When Nobody Looks? Reality and the Quantum Theory. *Physics Today* 38: 4, 38 (1985); https://doi.org/10.1063/1.880968.

[3] A. Einstein, B. Podolsky and N. Rosen, Can Quantum-Mechanical Description of Physical Reality Be Considered Complete? *Phys. Rev.* 47, 777–780 (1935).

[4] N. Bohr. Can Quantum-Mechanical Description of Physical Reality Be Considered Complete? *Phys. Rev.* 48, 696–702 (1935).

[5] J. Bell, On the Einstein-Podolsky-Rosen Paradox. *Physics* 1:3, 195–290 (1964).

[6] David Harrison, Bell's Inequalities and Quantum Correlations. *Am. J. Phys.* 50:9, 811–816 (1982).

[7] B. Hansen et al., Loophole-free Bell Inequality Violation Using Electron Spins Separated by 1.3 Kilometers. *Nature* 536, 682–686 (2015).

[8] M. Giustina et al., Significant Loophole-free Test of Bell's Theorem with Entangled Photons. *Phys. Rev. Lett.* 115, 250401 (2015).

[9] L. K. Shalm et al., Strong Loophole-free Test of Local Realism. *Phys. Rev. Lett.* 115, 250402 (2015).

[10] John F. Clauser, Michael A. Horne, Abner Shimony, and Richard A. Holt, Proposed Experiment to Test Local Hidden-Variable Theories. *Phys. Rev. Lett.* 23, 880 (1969); *Erratum Phys. Rev. Lett.* 24, 549 (1970).

[11] Andrew Whitaker, *John Bell and the Most Profound Discovery of Science*. https://physicsworld.com/a/john-bell-profound-discovery-science/.

[12] R. P. Feynman, Simulating Physics with Computers. *Int. J. Theor. Phys.* 21, 467 (1982) https://doi.org/10.1007/BF02650179.

[13] David Deutsch, Quantum Theory, the Church–Turing Principle and the Universal Quantum Computer, *Proceedings of the Royal Society of London. A. Mathematical and Physical Sciences A400*, 97–117 (1985). https://royalsocietypublishing.org/doi/pdf/10.1098/rspa.1985.0070.

[14] P. Benioff, The Computer as a Physical System: A Microscopic Quantum Mechanical Hamiltonian Model of Computers as Represented by Turing Machines. *J. Stat. Phys.* 22, 563 (1980).

[15] P. Benioff, Quantum Mechanical Models of Turing Machines That Dissipate No Energy. *Phys. Rev. Lett.* 48, 1581 (1982).

[16] David P. DiVincenzo. The Physical Implementation of Quantum Computation. *Fortschritte der Physik* 48:9–11, 771–783 (2000). arXiv:quant-ph/0002077. Bibcode:2000ForPh..48..771D. doi:10.1002/1521-3978(200009)48:9/11@771::AID-PROP771&3.0.CO;2-E.

[17] M-O Renou, D. Trillo, M. Weilenmann, et al., Quantum Theory Based on Real Numbers Can Be Experimentally Falsified. *Nature* 600, 625 (2021).

[18] P. C. Deshmukh, *Foundations of Classical Mechanics* (Cambridge University Press, 2019).

[19] Brian Hayes, Computing Science: Square Root of Not, *American Scientist* 83:4, 304–08 (1995). http://www.jstor.org/stable/29775474.

[20] David Deutsch and Richard Jozsa, Rapid Solutions of Problems by Quantum Computation, *Proceedings of the Royal Society of London A*. 439, 553–558 (1992); doi:10.1098/rspa.1992.0167.

[21] W. Wootters, and W. Zurek, A Single Quantum Cannot Be Cloned, *Nature* 299, 802–803 (1982). https://doi.org/10.1038/299802a0.

[22] D. Dieks, Communication by EPR Devices, *Physics Letters A* 92, 271–272 (1982).

[23] A. K. Pati and S. L. Braunstein, Impossibility of Deleting an Unknown Quantum State. *Nature* 404, 164 (2000).

[24] C. M. Lee and J. H. Selby, A No-Go Theorem for Theories That Decohere to Quantum Mechanics. *Proc. R. Soc. A* 474, 20170732 (2018). http://dx.doi.org/10.1098/rspa.2017.0732.

[25] https://dictionary.cambridge.org/dictionary/english/teleportation (downloaded on February 21, 2022).

[26] Alexander S. Holevo, Bounds for the Quantity of Information Transmitted by a Quantum Communication Channel. *Problems of Information Transmission* 9, 177–183 (1973).

[27] Wolfgang Dür, Guifré Vidal, and Ignacio Cirac, Three Qubits Can Be Entangled in Two Inequivalent Ways. *Phys Rev. A* 62, 062314 (2000).

[28] M. A. Nielson and I. L. Chuang, *Quantum Computation and Quantum Information* (Cambridge University Press, 2013).

[29] John Preskill, *Quantum Computation*. California Institute of Technology Lecture Notes for Ph 219 (1998). http://theory.caltech.edu/~preskill/ph219/ph219_2022.html.

Appendix A

Symmetry of the Hamiltonian

P. Dirac and E. Wigner. AIP Emilio Segrè Visual Archives, Physics Today Collection.

A symmetry transformation can be represented on the Hilbert space of physical states by an operator that is either linear and unitary, or anti-linear and anti-unitary.
—Eugene P. Wigner

A.1 Continuous, Dynamical, and Discrete Symmetries

The role of symmetry in natural laws has been underscored by many, notably by Eugene P. Wigner [1]. Symmetry plays a huge role in our understanding of the laws of nature. Plato (427–327 BCE) recognized that the only regular solids, whose faces are polygons having identical faces meeting at identical solid angles, are the (i) tetrahedron, (ii) cube, (iii) octahedron, (iv) dodecahedron, and (v) icosahedron. Their symmetry properties are celebrated in Euler's theorem

$$V - E + F = 2, \tag{A.1}$$

where V is the number of vertices, E the number of edges, and F the number of faces. In 1905, Einstein recognized the symmetry in Maxwell's laws of electrodynamics and formulated the special theory of relativity. This symmetry resulted from the mind-boggling invariance of the speed of light in all inertial frames of reference. It required for its account a non-Euclidean four-dimensional (flat) space–time continuum [Chapter 13 of Reference 2]. Maxwell's theory of electrodynamics engendered an unimaginable constancy of the speed of light in all inertial frames of reference, regardless of their mutual motion. Einstein harbored length-contraction and time-dilation to account for the *invariance* of the speed of light, which is a fundamental *symmetry* principle. About ten years later, Einstein interpreted gravity in terms of an *equivalence principle*, inspired yet again by symmetry. A few years later, in 1918, Emmy Noether established that each *conservation law* is in essence an equivalent expression of an underlying *symmetry*. Rudimentary introduction to some of these ideas is now found even in undergraduate texts, such as Reference [2]. In this chapter, we discuss the role of symmetry in atomic structure and dynamics.

The hallmark of quantum theory is the recognition of employing operators to represent dynamical variables of classical mechanics. The simplest symmetry operation is translation through an infinitesimal displacement in homogeneous space. We represent it by the operator $\tau\left(\overrightarrow{dx'}\right)$; from a position eigenket, it furnishes a displaced one:

$$\tau\left(\overrightarrow{dx'}\right) \mid \vec{x'}\rangle = \mid \vec{x'} + \overrightarrow{dx'}\rangle. \tag{A.2}$$

Its effect on an arbitrary state $\mid \alpha\rangle$ in the Hilbert space is

$$\mid \alpha\rangle \rightarrow \mid \alpha\rangle' = \tau\left(\overrightarrow{dx'}\right) \mid \alpha\rangle. \tag{A.3a}$$

As discussed in Chapter 1 (Eq. 1.65d), the operator for infinitesimal displacement in homogeneous space is

$$\tau\left(\overrightarrow{dx}\right) = 1 - i\frac{\vec{p}}{\hbar}\bullet\overrightarrow{dx}. \tag{A.3b}$$

Preservation of the metric norm of the state vector requires the displacement vector to be unitary, with the desired properties listed in Eqs. 1.57a–d (from Chapter 1). The generator of translations in homogeneous space is the operator for linear momentum (Eq. 1.65d), which is conserved under translations. The displacement operators constitute a mathematical group; it is *abelian* since its generators satisfy the property

$$\left[p_i, p_j\right]_- = 0; \quad i, j = 1,2,3. \tag{A.4}$$

In contrast, the symmetry operators that effect rotational displacements in isotropic space (Chapter 4) constitute a *non-abelian* group (Eq. 4.18a).

The 3 × 3 transformation matrices $\mathbb{R}$ corresponding to rotations (Eqs. 4.10a,b,c; Eq. 4.12) and similar matrices $\mathbb{P}$ that represent inversions/reflections are unimodular orthogonal transformations whose determinant is ±1. For both the matrices $\mathbb{R}$ and $\mathbb{P}$, the matrix transpose is equal to its inverse,
but

$$|\mathbb{R}| = +1, \tag{A.5a}$$

while

$$|\mathbb{P}| = -1. \tag{A.5b}$$

Together, rotations and parity constitute the group $O(3)$. The rotation group has the *special* property that the determinant of its matrix is $|\mathbb{R}| = +1$ and hence denoted by $SO(3)$, the letter S standing for this special property and O for the orthogonal character of the transformation matrix. The argument n in the notation $SO(n)$ denotes the number of generators of the symmetry group, given by $\frac{n(n-1)}{2}$. For $n = 3$, we have $\frac{n(n-1)}{2} = 3$, which are the three components $\{J_x, J_y, J_z\}$ of the angular momentum operator.

The algebra of the generators of rotations is embodied in the very definition of the quantum angular momentum vector operator $\vec{J}$. We have discussed it in Chapter 4. Here we revisit only a few specific properties briefly to underscore symmetry characteristics associated with the angular momentum. Its characteristic defining property is easily stated in terms of the commutation law for any two of its Cartesian components (Eq. 4.18a):

$$\left[J_k, J_\ell\right]_- = i\hbar\varepsilon_{k\ell m}J_m, \tag{A.6a}$$

$$\text{where } \varepsilon_{k\ell m} = \begin{cases} +1 \text{ for even permutations of } k,\ell,m \\ -1 \text{ for odd permutations of } k,\ell,m \\ 0 \text{ if any two of the three indices are equal} \end{cases}. \tag{A.6b}$$

As discussed in Chapter 4 (Eq. 4.48b), the operator for infinitesimal rotation $\left(\overrightarrow{\delta\xi}\right)$ in isotropic space is

$$U_R\left(\overrightarrow{\delta\xi}\right) = 1 - \frac{i}{\hbar}\vec{j}\bullet\overrightarrow{\delta\xi}. \tag{A.6c}$$

The continuity of transformations in the rotation group makes it a classic example of what are known as *Lie groups*, named after the Norwegian mathematician Sophus Lie. The characteristic feature of a Lie group is that the symmetry transformations that it consists of vary *smoothly*, continuously. Lie groups are therefore of monumental importance in differential calculus. The abstract algebraic, geometrical, and topological structure of the Lie groups unifies the mathematics of different topics like Bessel functions and spherical harmonics. The $SO(3)$ group of rotations is homomorphic with $SU(2)$, which consists of all unitary, unimodular, 2×2 matrices. States of particles with half-odd-integer spin require a rotation through 4π to return to the original. The Lie algebra of angular momentum operators includes rules for addition of angular momenta, using the Wigner $3j$ functions (Clebsch–Gordan coefficients), and the Wigner–Eckart theorem, which separates a transition matrix element into a physical part and a geometric part. Several important aspects of this have been dealt with in Chapter 4.

Apart from the symmetries mentioned above, a Hamiltonian may have what is known as *dynamical* symmetry; it is structured in the details of the potential that describes specific interactions that governs the Lagrangian of the system, from which the Hamiltonian is developed. A classic example of this symmetry is the Coulomb interaction between the proton and electron in the hydrogen atom. The specific form of the interaction, namely the fact that it diminishes as the inverse of the distance between the two particles, is important for an *additional* symmetry, over and above the spherical symmetry that is responsible for the conservation of angular momentum. As a result of this extra symmetry, the Laplace–Runge–Lenz vector is conserved. The Pauli–Lenz symmetry operator (Section 5.2 in Chapter 5) expands the set of hydrogen atom's symmetry operators, albeit within the subspace of bound states, from the $SO(3)$ group to $SO(4)$. The additional symmetry makes it possible to solve the Schrödinger equation for the hydrogen atom not only in the spherical polar coordinates but also, for example, in the parabolic coordinates (Section 5.3 in Chapter 5). The $SO(4)$ symmetry of the hydrogen atom, rather than $SO(3)$, is of fundamental importance to the understanding of the degeneracies of its eigenstates. The $SO(4)$ group consists of six operators, since in this case $\frac{n(n-1)}{2} = 6$, namely $\{J_x, J_y, J_z\}$ and $\{A_x, A_y, A_z\}$, which respectively are the three components of the vector angular momentum operator, and the three components of the Pauli–Lenz vector operator. The rank of the $SO(3)$ group is 1, J^2 being its Casimir operator. The rank of the $SO(4)$ group, on the other hand, is 2, and its Casimir operators are C_1 and C_2 (Eqs. 5.72a,b in Chapter 5; also References [3, 4]).

Translational displacements in homogeneous space, rotations in isotropic space, and temporal evolution of the physical state of a system are *continuous* spatio-temporal symmetry transformations. The space transformations are translations or rotations, and the temporal transformation we have considered hitherto involve only a gradual and continuous variations. In the spirit of the Noether's theorem, symmetries in (i) continuous translation in homogeneous space, (ii) continuous rotation in isotropic space, and (iii) continuous temporal transformation are associated *respectively* with the conservation of linear momentum, angular momentum, and energy.

Continuous time variation $t \to t' = (t \pm \delta t)$ with $\delta t \to 0$ must be contrasted with the *discrete* time-reversal, $t \to -t$, discussed below in Section A.3.

In relativistic mechanics, the symmetry principle of importance is the Poincaré transformation in the (Minkowski) space–time continuum. A special case of this that leaves the origin fixed is the Lorentz covariance, brought about by the Lorentz transformation. It leaves the speed of light invariant. Curved space–time, as required in general relativity, involves projective symmetries that leave its geodesic structure invariant. The metric structure of the manifold is preserved under these spatio-temporal transformations. The connection between symmetries and associated invariance principles is indeed far reaching.

We discuss parity (Eq. A.5b), the reflection symmetry, further in the next section.

A.2 Parity Nonconservation in Physical Phenomena

Unlike rotation, parity (Eq. A.5b) is a *discrete* symmetry. There are three discrete symmetries in the *standard model of physics*; *together* they constitute an invariance principle for all natural phenomena in the physical universe. Each symmetry operation may not be individually associated with a conserved physical quantity. However, under the combined symmetry operations viz., (i) C (charge conjugation), (ii) P (parity), and (iii) T (*discrete* time-reversal) a Lorentz invariant quantum field theory is invariant. In the present section, we restrict ourselves to a brief discussion on the parity operation, since it is of particular interest in atomic spectroscopy discussed in Section 8.2 of Chapter 8. Specialized literature on the standard model of physics may be referred to for details.

Parity consists of mirror *reflection*, in which only one of the three spatial Cartesian axes is reversed, or the *inversion*, in which all the three axes are reversed. Thus, under the parity operator, which we shall denote by $\hat{P}$, a Cartesian coordinate frame $\{X, Y, Z\}$ is transformed into $\{-X, Y, Z\}$ (reflection), or into $\{-X, -Y, -Z\}$ (inversion). In fact, the latter is *also* a mirror reflection. It is nicely interpreted in the spherical polar coordinate system in which a point whose coordinates are $\{r,\theta,\varphi\}$ goes to $\{r,\pi-\theta,\varphi+\pi\}$. If this operation is carried out twice in succession, one would of course recover the original state. The parity operator $\hat{P}$ has the following properties:

$$[\vec{r},\hat{P}]_+ = 0 \text{ (anti-commutation with the position operator)} \tag{A.7a}$$

and

$$[\vec{p},\hat{P}]_+ = 0 \text{ (anti-commutation with the momentum operator).} \tag{A.7b}$$

Accordingly, for an arbitrary state vector $|\alpha\rangle$, we shall have

$$\langle\alpha|\ \hat{P}^\dagger\ \vec{r}\ \hat{P}\ |\alpha\rangle = -\langle\alpha|\vec{r}|\alpha\rangle, \tag{A.8a}$$

$$\text{i.e., } \hat{P}^\dagger\ \vec{r}\ \hat{P} = -\vec{r}. \tag{A.8b}$$

Likewise, for the momentum operator, we have $\hat{P}^\dagger\vec{p}\hat{P} = -\vec{p}$. (A.8c)

Consequently, the parity operator commutes with the operator $\vec{J}$ for angular momentum:

$$[\vec{J},\hat{P}]_- = 0, \tag{A.9a}$$

$$\text{i.e., } \hat{P}^\dagger\vec{J}\hat{P} = \vec{J}. \tag{A.9b}$$

Also, the parity operator would conserve probability, i.e., it would leave the norm of an arbitrary state vector invariant. Hence,

$$\hat{P}^\dagger\hat{P} = 1. \tag{A.10}$$

The parity operator therefore is unitary, Hermitian, and it is its own inverse:

$$\hat{P}^2 = 1. \tag{A.11}$$

Accordingly, eigenvalues of the parity operator are ± 1. The plus sign corresponds to *even* parity and the minus sign to *odd* parity. Thus, for a particle represented by a wavefunction $\psi(\vec{r})$, we have

$$\hat{P}\psi(\vec{r}) = \pm\psi(-\vec{r}). \tag{A.12a}$$

i.e.,

$$\hat{P}\psi(r,\theta,\varphi) = \pm\psi(r,\pi-\theta,\varphi+\pi). \tag{A.12b}$$

The spherical harmonics (Appendix C) have the following property

$$Y_{\ell m}(\pi-\theta,\varphi+\pi) = (-1)^\ell\, Y_{\ell m}(\theta,\varphi), \tag{A.13}$$

and hence parity of an electron in a central field atomic Hamiltonian (Chapter 5) is $+1$ or -1, depending on the orbital angular momentum quantum number ℓ being even or odd. For an atom in the presence of an applied field (such as the electric field studied in Section 8.2

of Chapter 8), the orbital angular momentum ceases to be a good quantum number. Parity is then not well-defined in such cases.

One would commonly expect that a physical process that occurs in nature would take place also under the parity operation. This expectation was, however, challenged around 1950 by what is referred to as the $\tau - \theta$ puzzle. Two particles τ and θ discovered in the cosmic showers were found to be identical in terms of charge, mass, spin, and lifetime, but were found to decay differently, both via the weak interaction:

$$\tau^{+} \rightarrow \pi^{+} + \pi^{+} + \pi^{-}, \tag{A.14a}$$

but

$$\theta^{+} \rightarrow \pi^{+} + \pi^{0}. \tag{A.14b}$$

The parity of pions, which are products of the above decay processes, was determined to be odd. Thus, if the spin of the decaying particle is ℓ, the parity of τ^{+} would be $(-1)^{\ell+1}$, whereas that of θ^{+} would be $(-1)^{\ell}$. One would then conclude that the particles τ and θ, otherwise identical, have opposite parity. This led to the conclusion that (a) either τ and θ are different particles, or, (b) conservation of parity is violated in *weak* interaction. It needed the courage of Tsung-Dao Lee and Chen Ning Yang [5] to propose that it must be the latter, i.e., parity is *not* conserved in weak interaction. This was confirmed in the famous experiment on β-decay done by Madam Wu and collaborators [6].

We shall first briefly summarize a few salient features of *fundamental interactions in nature* in order to place our discussion in its proper context. Details pertaining to these aspects are often found in books on particle physics, but they have a place, even if only a brief one, in the context of atomic physics. It is this unifying beauty of physics that we allude to. It would lead us to understand how parity violation impacts *atomic* processes. It is common knowledge that there are four fundamental interactions in nature: gravity, electromagnetic, strong interaction, and weak interaction. The much familiar *gravitational* interaction occurs between particles having mass. It has infinite range. The *electromagnetic* interaction occurs between particles having electric charge. Like gravity, this interaction also has an infinite range. The *weak* interaction affects all particles; it has a short range. The *strong* interaction binds *quarks* together and binds nucleons within the nucleus of an atom. The *coupling constant* provides a measure of the strength of the interaction. Relative strengths are referenced to the strength of the *strong* interaction, which is considered to be of the order 10^{0} at a distance of 1 fm. It is 10^{-2} (more correctly, the *inverse* of the *fine structure constant*) for the electromagnetic interaction. It is $\sim 10^{-6}$ for weak interaction and 10^{-39} for the gravitational interaction. Unity of the *weak* and the *electromagnetic* interactions into the synthesized term "*electroweak*" interaction is recognized in the *Standard Model* of physics.

The strong interaction is mediated by *gluons*, which are massless. Gluons interact with quarks and other gluons through a property that is analogous to the electric charge but is essentially different from it; it is called *color charge*. There are three types of color charge (±red, ±green, ±blue). The theory that accounts for quark–gluon interactions is called quantum chromo-dynamics (QCD). The range of the strong interaction is very short; it is of the order of 10^{-15} m, i.e., 1 fm (femtometer, named earlier as one fermi, after Enrico Fermi). The range of the weak interaction is about two orders of magnitude shorter, but in the electro-weak theory (EWT), it is *unified* with the electromagnetic interaction which has infinite range. The electroweak interaction is a unified description of the weak interaction and the electromagnetic interaction. Sheldon Glashow, Abdus Salam, and Steven Weinberg were jointly awarded the 1979 Nobel Prize in Physics for developing the EWT. The EWT is one of the primary cornerstones of the standard model of physics. It is extensively discussed in literature on the standard model and on particle physics.

The electroweak unification (EWT) enables spectroscopic transitions in atoms which violate parity conservation. This is uncommon, since most of atomic spectroscopy is governed by the electromagnetic interaction which conserves parity. Considering the importance of parity-violating transitions in atomic physics, we discuss a few salient features of the weak interaction in this section. Parity-violating atomic transitions are discussed in Section 8.2 of Chapter 8.

At very short distances, ~10^{-18} m, the *weak* interaction and the *electromagnetic* interaction have comparable strengths. The weak interaction, however, gets radically feebler as the distance increases. As the distance increases by an order of magnitude, it becomes ~10,000 times weaker than the electromagnetic interaction. This difference is attributed to what is perceived as strength of the interaction; it is in fact a *combined* effect of the mass of the boson that mediates the interaction *and* the distance of the interaction. The bosons that mediate the weak interaction are the W^+, W^-, and the Z^0. The mass of $W^\pm$ bosons is ~80.38 GeV c^{-2} and that of Z^0 is ~91.19 GeV c^{-2}, whereas photons, which mediate the electromagnetic interaction, are massless. The $W^\pm$ bosons have plus and minus charge but are otherwise identical; they are antiparticles of each other. The Z^0 particle is neutral; it is its own antiparticle. Fundamental interactions are described by gauge symmetries described by a group. $SU(3)$ is the gauge group for strong nuclear interactions, $SU(2)$ for weak nuclear interactions, and $U(1)$ for electromagnetic interactions. The bosons that mediate the gauge interactions correspond to the generators of the gauge group. The gauge group $SU(2)$ for weak interactions has three generators, so the weak nuclear force has three gauge bosons, $W^\pm$ and Z^0.

Now, as mentioned here, parity is not conserved in the weak interaction. On account of the unification of the weak interaction with the electromagnetic interaction, one would expect that parity would be violated in some of the atomic processes as well, even if they are predominantly governed by the electromagnetic interaction. The EWT is sometimes (though rarely) also called as quantum flavor-dynamics (QFD), since the primary particles that are involved in the electroweak interaction are the quarks (which constitute particles such as neutron and proton), which come in six types, called *flavors*: strange, charm, top, bottom, up, and down. An example of the weak interaction is a down quark in a neutron changing into an up quark, resulting in the neutron transformation into a proton, along with the emission of an electron and an antineutrino (more specifically, an *electron-antineutrino*). This process is depicted using its *Feynman diagram* in Fig. A.1. In this diagram, the arrow of time is from bottom to the top. Destruction (annihilation) of particles is indicated by an arrow *into* the vertex, and creation of particles by an arrow *out* of the vertex. For an antiparticle, the convention is just the opposite.

Radioactive β-decay (the experiment of Madam Wu and collaborators, mentioned earlier) of the cobalt nucleus is described by

$$ {}_{Z=27}^{60}\mathrm{Co} \rightarrow {}_{Z=28}^{60}\mathrm{Ni} + e^- + \overline{\nu}_e + 2\gamma. \tag{A.15}$$

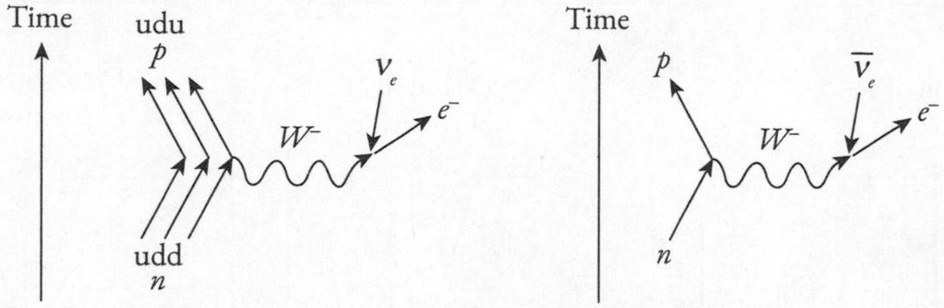

Fig. A.1 A neutron (n) is a spin-half baryon. It consists of three quarks, up, down, down. The weak interaction mediated by W^- boson transforms it into a proton (p) which consists of the three quarks up, down, up. An electron and an antineutrino (electron-antineutrino) are emitted.

The Ni isotope is in fact produced in an excited state, which quickly emits two photons, denoted by 2γ in Eq. A.15. Using an experimental technique (called as the Rose–Gorter method) the directions at which photons exit the Ni nucleus provide information about the orientation of the magnetic moment of the nucleus. In Wu's experiment, anisotropy was found in the rate at which β particles were detected - when the Co nuclear magnetic moment was polarized in opposite directions. Most of the β-emission yield (~70%) takes place in the direction opposite to that of the spin moment. By polarizing the magnetic moment of the Co nuclei in the opposite direction, the direction of the magnetic moment could be swapped. This amounts to reversing the direction of the spin. Anisotropy in the β-decay yield was then measured again, in what is now equivalent to the parity reversed (mirror image) space. Seventy percent of the β emission yield in the parity-reversed experiment, however, occurred in the same direction as the spin magnetic moment, establishing the breakdown of parity conservation. This experiment was performed in two different arrangements of the apparatus that effectively *mirrored* each other. It could be concluded from this experiment that the direction in which β particles were preferentially emitted in radioactive decay, relative to that of the spin moment, is not the same in the mirror image space. In other words, parity was not conserved (Fig. A.2a) in the weak interaction which results in β-decay. In a spherical polar coordinate system in which the polar angle θ is measured with respect to the Cartesian Z-axis (defined to be along the magnetic field in this experiment) and the azimuthal angle φ with respect to the ZX half-plane, the differential rates for β-decay in the directions for (θ, φ) and $(\pi - \theta, \varphi + \pi)$ do not have identical dependence on the polar angle. Figure A.2b, showing neutrino helicity, illustrates parity nonconservation. Helicity of a particle is defined as $\hat{\sigma} \bullet \hat{p}$ where $\hat{\sigma}$ and $\hat{p}$ are unit vectors respectively in the directions of the intrinsic spin angular momentum vector and its linear momentum. Thus, the helicity is +1 for a particle whose intrinsic angular momentum and linear momentum are parallel to each other and −1 when they are antiparallel. Helicity is a well-defined Lorentz-invariant quantity for a massless particle traveling at the speed of light. On the other hand, a Dirac particle with mass would be in a state of superposition of the two helicity states and would travel at a speed less than that of light. Now, the helicity of a particle would be reversed by providing a Lorentz boost to the frame of reference from which the observation is carried out and the particle under observation is thereby overtaken by the observer. On overtaking, the direction of the particle's intrinsic spin angular momentum vector would seem to be invariant, but its linear momentum would appear to be reversed. This would therefore be equivalent to

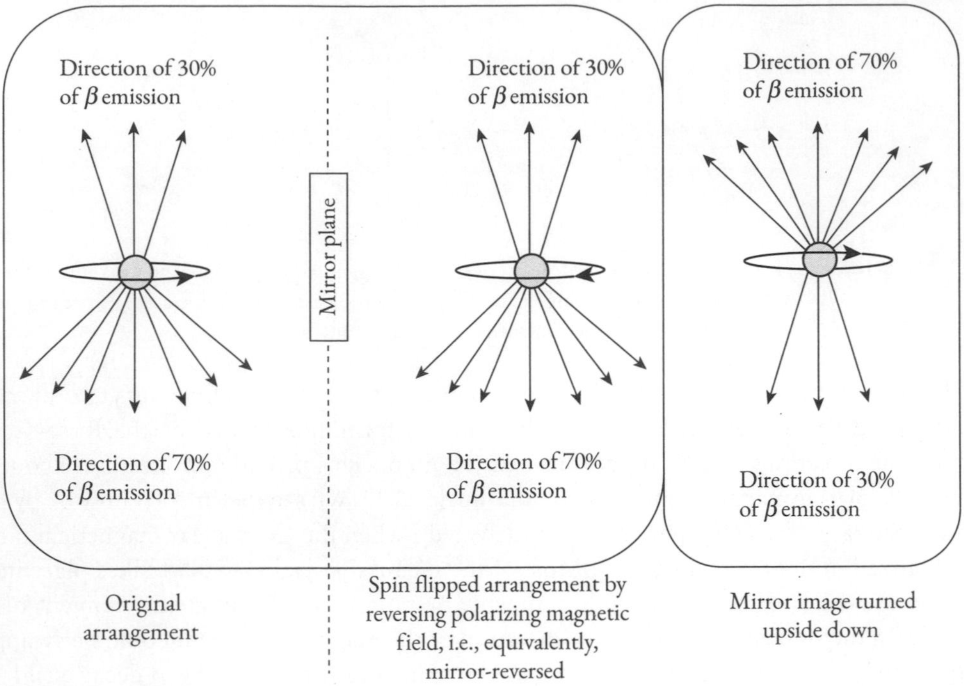

Fig. A.2a Parity violation in β-decay experiment.

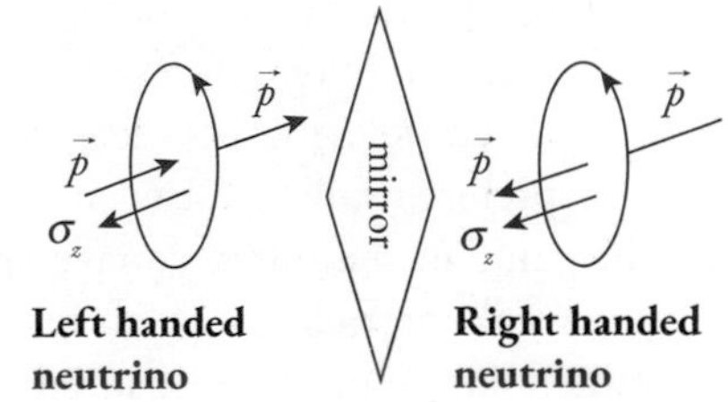

Fig. A.2b A mirror image of the left-handed neutrino would produce the right-handed neutrino; but the latter has not been ever detected.

observing the particle from a left-handed coordinate frame of reference after the boost, if it was observed from a right-handed frame prior to the boost. Observations before and after the boost are therefore akin to those carried out before and after a parity operation. Upon reflection (Fig. A.2b), the direction of the intrinsic angular momentum of a neutrino remains unchanged, it being an axial vector. The direction of the linear momentum vector, which is a polar vector, however, reverses. Under parity, a right-handed neutrino would result from a left-handed neutrino, but a right-handed neutrino has never been found. Parity violation was found to be maximal in β-decay; i.e., a mirror image of this physical process is never detected. Being a fermion, a neutrino and an antineutrino have opposite intrinsic parities. Neutrinos are *always* found to be left-handed, and antineutrinos *always* right-handed. Incidentally, from this, one can also argue that a massless neutrino cannot possess a magnetic dipole moment, since no mechanism can exist that could flip its spin direction.

The weak interaction treats right-handed particles differently from those that are left-handed. Neither a right-handed neutrino nor a left-handed antineutrino has ever been detected. Parity is not conserved in the weak interactions because of the helicity property mentioned earlier. Expectation value of the helicity operator would be zero if parity was conserved, but it is not. A *different* property closely related to helicity is chirality; it distinguishes a neutrino *with-mass* from an antineutrino. For massless particles, helicity and chirality are the same.

Parity violation, theorized by Lee and Yang and detected by Madam Wu in 1956 led to a speculation, proposed by Lev Landau in 1957, that the *combined* operation of charge conjugation *and* parity (CP) would correspond to a conservation principle. Charge conjugation refers to a *particle* ↔ *antiparticle* interchange. The CP symmetry was also, however, found to be violated in 1964, albeit *indirectly*, in the experiments conducted by James Cronin and Val Fitch. *Direct* violation of CP symmetry was found toward the end of the last century in experiments conducted at Fermilab and at CERN. There also are supplementary cases of great interest in which CP violation finds a distinctive manifestation, such as in the BaBar experiment at SLAC National Accelerator Laboratory, Stanford University, and in the Belle collaboration experiment at the KEK accelerator in Japan.

Violation of CP symmetry prompted the recovery of a conservation principle by including, along with charge conjugation (C) and parity (P), the operation for time-reversal (T). An intricate issue related to CP-, and hence T-, violation is at the bottom of baryogenesis theory, which is aimed at identifying an early process in the evolution of the universe that would account for the matter–antimatter asymmetry. The CPT symmetry is an important consideration in Andrei Sakharov's analysis of baryogenesis. One hoped that the combined CPT symmetry would correspond to a fundamental principle. Extensive specialized literature is available on the CPT symmetry, but some semi-technical synopses (for example Reference [7]) are also available. That C, P, and T symmetries are *together* satisfied in all physical processes has emerged as an important cornerstone of the Standard Model of physics. We see that the symmetries associated with *continuous* and *discrete* transformations are associated with conservation laws that are respectively additive and multiplicative. Important contributions to the development of the CPT theorem were made by Schwinger (1951), Lüders and Pauli (1954), and also by our hero of quantum entanglement (Chapter 11) - John Bell (1951).

Most of atomic spectroscopy involves the electrodynamic interaction between an electron's charge and the electromagnetic field. Parity is conserved in this interaction, since the electrodynamics Lagrangian is invariant under parity. On account of the EW unification, parity is, however, *not* conserved in some atomic transitions, discussed in Section 8.2 of Chapter 8.

The T-symmetry connects solutions of quantum collisions with those of atomic photoionization/photodetachment. This connection is discussed in Chapter 11, where it plays a fundamental role toward the understanding of the time-delay in photoionization/photodetachment processes. The time-reversal symmetry is discussed in the next section.

A.3 Time-Reversal Symmetry: Search for Electron Dipole Moment

It is conjectured from voluminous studies that matter was created at the Big Bang. Cosmic Microwave Background Anisotropy (CMBA) studies using space probes provide powerful tools to investigate distribution of matter in the universe. One would expect that matter and antimatter would be created in equal measure. Matter would also be annihilated along with an equal amount of antimatter, and energy would be created in this process. That we, and also the physical matter in the universe around us, exist at all, not annihilated by antimatter, is therefore mysterious. If laws of nature were to be the same for matter and for antimatter, CP symmetry would be secure; if not, it is violated. In the experiments performed by Cronin and Fitch it was found that CP symmetry is broken. The absence of adequate explanation of the observed matter–antimatter imbalance makes one wonder if there is physics *B*eyond the *Standard Model*, referred as BSM, and what it may be. The weak interaction accounts for CP violation but is not enough to account for the estimated *excess* of matter over antimatter. Sophisticated experiments are in progress to search for additional sources of CP violation. In order to recover a conservation principle, T-violation (T stands for time-reversal) is then predicted in the SM to accompany CP violation, toward conserving *combined* CPT symmetry. One may also ask if the matter–antimatter symmetry involves conversion of antimatter into matter. Such a process would require violation of time-reversal symmetry.

The T-symmetry compensates for CP violation to restore a conserved quantity, namely the CPT symmetry. Likewise, TC symmetry is violable and compensated for by the P symmetry, and PT symmetry is violable and compensated for by C symmetry. The conservation of the *combined* CPT symmetry, i.e., the *product* of the three discrete symmetries, however, provides a robust conservation principle. Thus, violation of any pair of the three symmetries is compensated by that of the third. This has been tested in a large number of very sophisticated and accurate measurements. Even the Lamb shift in hydrogen and anti-hydrogen has been found recently to be identical, as is expected from CPT symmetry. T-violation (equivalent to CP violation) has been detected, for example, in the decay of B-mesons in experiments. The CPT symmetry is therefore an important tenet of the SM, but it does not fully account for the matter–antimatter asymmetry. Also, the origin of T-violation is not comprehensively explained in the SM. Even if the matter–antimatter asymmetry and the CPT theorem are irresistibly engaging and constitute fundamental topics of pronounced importance to the SM and BSM, our focal interest in the present Section lies in understanding the role of T symmetry in atomic dynamics. The CPT symmetry is of great consequence in particle physics, where it is discussed extensively. It is, however, of substantial importance also in atomic physics [8–10]. We shall now discuss how the time-reversal symmetry connects the solutions of the Schrödinger equation with *outgoing* wave boundary conditions to those with *ingoing* wave boundary conditions [11–13]. The former are employed in atomic quantum collisions, and the latter in atomic photoionization/photodetachment, discussed in Chapter 10.

In general, a symmetry operator Γ can be expressed as

$$\Gamma = 1 - \frac{i}{\hbar}\varepsilon G, \tag{A.16}$$

where G is a Hermitian generator of Γ. We have already seen that the operators for translational symmetry in homogeneous space (Eq. A.3b) and for rotations in isotropic space (Eq. A.6c) have the above form.

We know that the classical equations of motion are symmetric as $t \to -t$. We quantify this statement by considering two identical particles, represented by I and II. For particle I, let

$\vec{r}_I(t=0)$ and $\vec{p}_I(t=0)$ represent the initial position and momentum, whereas for the particle II, $\vec{r}_{II}(t=0)=\vec{r}_I(t=0)$, but $\vec{p}_{II}(t=0)=-\vec{p}_I(t=0)$. At a later time $t>0$, *if* $\vec{r}_{II}(t)=\vec{r}_I(-t)$ and $\vec{p}_{II}(t)=-\vec{p}_I(-t)$, then motion has time-reversal symmetry [14]. This is illustrated in Fig. A.3.

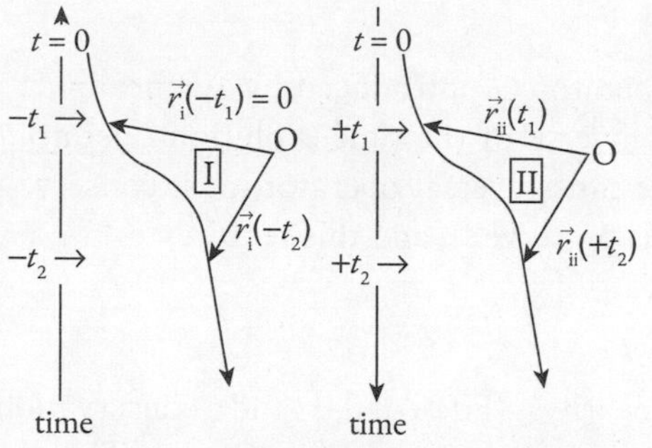

Fig. A.3 Depiction of the time-reversal symmetry in classical laws of motion.

Time-reversal symmetry in classical mechanics is related to the fact that the classical equation of motion, such as Newton's second law, involves the *second* derivative of the position vector with respect to time. Consequently, $t \rightarrow -t$ happens twice and leaves the law invariant. Hamilton's equations involve only the first-order derivative with respect to time, but there are two equations, and the equation for $\dot{p}$ has a minus sign while that for $\dot{q}$ does not. Hamilton's equations are therefore invariant under time-reversal, not surprisingly, since they contain the same physics as in Newton's second law.

An interesting case is motion of a charged particle in a magnetic field. If one writes its equation of motion as

$$m\frac{d^2\vec{r}}{dt^2} = q\frac{d\vec{r}}{dt} \times \vec{B}, \tag{A.17}$$

one might suspect that time-reversal symmetry is broken, since only the first derivative with respect to time appears on the right. However, this is not a problem since $\vec{B} \rightarrow -\vec{B}$ as $t \rightarrow -t$, as can be seen from Biot–Savart law of electrodynamics. More generally, Maxwell's equations are invariant under time-reversal, which originates from the fact that as $t \rightarrow -t$ (i) charge density remains invariant, while the current density reverses its sign, and (ii) electric field remains invariant, but the magnetic field reverses its sign.

Laws of physics are quantum, not classical. As one would expect, time-reversal symmetry in the context of quantum considerations requires special, expectedly counter-intuitive, and totally new considerations. Let us denote the quantum operator for time-reversal by Θ and apply it to the fundamental principle of quantum mechanics, represented by the uncertainty relation,

$$\left[\vec{r}_i, \vec{p}_j\right]_- = i\hbar\delta_{ij}. \tag{A.18}$$

This gives

$$\Theta\left[\vec{r}_i, \vec{p}_i\right]\Theta^\dagger = -\left[\vec{r}_i, \vec{p}_i\right] = \Theta\left(i\hbar\delta_{ij}\right)\Theta^\dagger, \tag{A.19a}$$

$$\text{i.e., } -\left(i\hbar\delta_{ij}\right) = \Theta\left(i\hbar\delta_{ij}\right)\Theta^{\dagger}. \tag{A.19b}$$

It immediately follows that the sign on the two sides of the above equation can be reconciled if

$$\Theta i = -i\Theta, \tag{A.20a}$$

i.e., if the time-reversal operator Θ, introduced by Wigner in 1932, is *antilinear*. It is now obvious that it is not the inverse of the time-evolution operator, $U\left(t,t_0\right)$, which is a linear operator. Furthermore, the time-reversal operator must conserve probability (i.e., the norm of the time reversed state is preserved), and this requires

$$\Theta^{\dagger}\Theta = 1. \tag{A.20b}$$

The properties 'antilinearity' (Eq. A.20a) and 'conservation of probability' (A.20b) together identify time-reversal as an *anti-unitary* operation. The operation by the time-reversal operator on two arbitrary state vectors $|\alpha\rangle$ and $|\beta\rangle$ is therefore represented by the relations

$$\Theta|\alpha\rangle = |\tilde{\alpha}\rangle \text{ and } \Theta|\beta\rangle = |\tilde{\beta}\rangle, \tag{A.21a}$$

$$\text{with } \Theta\{a|\alpha\rangle + b|\beta\rangle\} = \{a^{*}\Theta|\alpha\rangle + b^{*}\Theta|\beta\rangle\}, \tag{A.21b}$$

$$\text{and } \langle\tilde{\beta}|\tilde{\alpha}\rangle = \langle\beta|\alpha\rangle^{*}. \tag{A.21c}$$

We now contrast the properties of the operator for time-evolution and that for time-reversal. The time-evolution of an arbitrary state vector $|\alpha\rangle$ in the Hilbert space over an infinitesimal time interval δt gives us the quantum state $|\alpha,t;t_0\rangle$ at time t, from the state $|\alpha,t_0\rangle$ at time t_0, i.e.,

$$|\alpha,t_0+\delta t;t_0\rangle = U\left(t_0+\delta t,t_0\right)|\alpha,t_0\rangle. \tag{A.22a}$$

Setting $t_0 = 0$, we may write this simply as

$$|\alpha,\delta t;0\rangle = U(\delta t,0)|\alpha,0\rangle = \left(1 - i\frac{H}{\hbar}\delta t\right)|\alpha,0\rangle. \tag{A.22b}$$

Let us now denote the time-reversed state of $|\alpha\rangle$ by $|\alpha_R\rangle$:

$$|\alpha_R\rangle = \Theta|\alpha\rangle. \tag{A.23}$$

We expect the state reached on time-*evolution* through a time-interval δt of the time-*reversed* state, to be the same as that reached on time-*reversal* of the time-*evolved* (through $-\delta t$) state, i.e., we expect

$$\left[1 - i\frac{H}{\hbar}\delta t\right]\Theta|\alpha\rangle = \Theta\left[1 - i\frac{H}{\hbar}(-\delta t)\right]|\alpha\rangle. \tag{A.24}$$

Since the state $|\alpha\rangle$ is arbitrary, we are led to the operator equivalence

$$\left[1 - i\frac{H}{\hbar}\delta t\right]\Theta = \Theta\left[1 - i\frac{H}{\hbar}(-\delta t)\right], \tag{A.25a}$$

$$\text{i.e., } \Theta - i\frac{H\Theta}{\hbar}\delta t = \Theta - \Theta i\frac{H}{\hbar}(-\delta t), \tag{A.25b}$$

$$\text{i.e., } iH\Theta\delta t = \Theta iH(-\delta t), \tag{A.25c}$$

$$\text{or } iH\Theta = -\Theta iH. \tag{A.25d}$$

Now, contrary to the property described by Eq. A.20a, *if* the time-reversal operator were linear, then we would be led to the conclusion that $[\Theta,H]_+ = 0$, rather than $[\Theta,H]_- = 0$. For a moment, if we *assume* that the time-reversal operator *anti*-commutes with the Hamiltonian, then we would have

$$H\Theta|E_n\rangle = -\Theta H|E_n\rangle = -\Theta E_n|E_n\rangle = -E_n\Theta|E_n\rangle, \tag{A.26}$$

which essentially would make $|\Theta|E_n\rangle\rangle$ an eigenstate of the Hamiltonian belonging to the eigenvalue $-E_n$. If we applied this result for the consideration of a free-particle, which can have only positive energy states, this would be obviously paradoxical. Our tentative assumption is therefore fallacious, and we conclude *yet again* that the time-reversal operator is *antilinear*, as already seen in Eq. A.20. Evidently, it commutes with the Hamiltonian. The above discussion reinforces the conclusion that the time-reversal operator is anti-unitary. As a result of this, the quantum time reversed state involves (i) complex conjugation along with (ii) $t \to -t$, i.e., the time reversed state would be described by

$$\psi_R(\vec{r},t) = \Theta[\psi(\vec{r},t)] = \psi^*(\vec{r},-t), \tag{A.27a}$$

$$|\psi_R(t)\rangle = \Theta|\psi(t)\rangle. \tag{A.27b}$$

That complex conjugation and $t \to -t$ must go together under time-reversal operation is seen by recognizing that if $\psi(\vec{r},t)$ is a solution of the Schrödinger equation, then $\psi(\vec{r},-t)$ is not a solution, but $\psi^*(\vec{r},-t)$ is.

In particular, at $t = 0$,

$$|\psi_R(0)\rangle = \Theta|\psi(0)\rangle. \tag{A.28}$$

The initial state (at time zero) is mapped by the time-reversal operator into the initial state of the time-reversed motion. The time-reversal operator is more aptly called as '*motion reversal*' (preferred by Wigner) operator, but the term time-reversal is rather widely used. The operator has the following properties:

$$\Theta\vec{r}\Theta^{-1} = \vec{r} \text{ (commutation with the position operator)}, \tag{A.29a}$$

$$\Theta\vec{p}\Theta^{-1} = -\vec{p} \text{ (anti-commutation with the momentum operator),} \tag{A.29b}$$

$$\text{and } \Theta^{\dagger}\vec{J}\Theta^{-1} = -\vec{J} \text{ (anti-commutation with the angular momentum).} \tag{A.29c}$$

These properties must be compared with corresponding properties of the parity operation (Eq. A.7–A.9). The role of time-reversal symmetry needs very careful consideration; it is nontrivial. Time itself is not an observable – it cannot be represented by a Hermitian operator. In the quantum theory, time is therefore treated only as a parameter, as discussed in Chapter 1. Even if time itself is not an observable, *time-delay* (see Eq. 10.115) is a measurable quantity [15, 16], and is represented by a Hermitian operator.

We must now ask how one may detect violation of the time-reversal symmetry. We must identify an *observable* which may arise necessarily from time-reversal violation. A quantum description of the electron accommodates a cloud of virtual particles around it resulting from vacuum polarization, and an asymmetry in the charge distribution of the virtual particles would give rise to electron's electric dipole moment, written as eEDM. The nEDM, neutron's electric dipole moment, is also a subject of abundant investigations. Such a dipole moment is conjectured, but hitherto not detected in an experiment. If an experiment succeeds in ascertaining the eEDM, it would be a manifestation of both parity-violation and violation of the time-reversal symmetry. We therefore explore if an electron has, or hasn't, a permanent electric dipole moment. We investigate what would happen if it does have, even if it has not been seen in an experiment as yet. Even a classical electric dipole does not need to have a spatial expanse; the multipole expansion of a charge distribution already allows for a 'point' dipole, just as the net charge of the dipole can of course be zero. Now, the intrinsic permanent electric dipole of a charged particle, assuming it exists, must be parallel to its intrinsic spin angular momentum vector. In fact, it absolutely *has* to be so, since

(i) a rotation about $\vec{d}$ would leave the charge distribution invariant, and
(ii) it is only a rotation about $\vec{J}$ that can achieve this.

We therefore recognize the *parallel* orientation, $\vec{J} \nearrow\nearrow \vec{d}$, to be a natural law. Now, we inquire if this law is conserved under parity and under time-reversal. We have

$$\hat{P}^{\dagger}\vec{J}\hat{P} = \vec{J} \qquad \& \qquad P\vec{d}P^{-1} = -\vec{d} \text{ (Fig. A.4a),} \tag{A.30a}$$

and also

$$\Theta^{\dagger}\vec{J}\Theta^{-1} = -\vec{J} \qquad \& \qquad \Theta\vec{d}\Theta^{-1} = \vec{d} \text{ (Fig. A.4b).} \tag{A.30b}$$

Only one of the two properties, $\vec{J}$ or $\vec{d}$, reverses sign under parity, and so also under time-reversal. It would therefore seem that the law $\vec{J} \nearrow\nearrow \vec{d}$ would remain valid only if $\vec{d} = \vec{0}$, or if parity transformation and time-reversal transformation, both produce a new particle that is otherwise identical to the previous one, except that for the new particle we would have $\vec{J} \nearrow\swarrow \vec{d}$ instead of $\vec{J} \nearrow\nearrow \vec{d}$. Such a particle, along with the original particle, would constitute a degenerate pair of particles, but no such degenerate pairs of particles exist. Now, if an experiment reveals that $\vec{d} \neq \vec{0}$, the law $\vec{J} \nearrow\nearrow \vec{d}$ could remain invariant if P-symmetry is

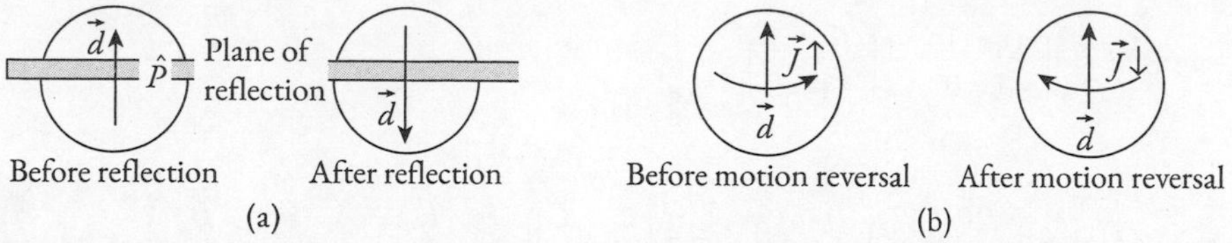

Fig. A.4 Under parity transformation, the angular momentum remains invariant, but the electric dipole moment would reverse sign. Under the time-reversal operation, just the opposite would happen.

violated (to address the dilemma in Fig. A.4a), and also, if T-symmetry is violated, so as to address the dilemma in Fig. A.4b.

In other words, discovery of eEDM would provide evidence for T-violation as much as for P-violation. The latter has been well established in EWT, but T-violation is a topic of intense current research. The SM of particle physics is very successful, and it predicts a non-zero eEDM. Its value has an upper bound, which is $|d_e| < 2.9 \times 10^{-38}\, e$ cm.

Discovering such a tiny dipole moment is therefore exceptionally challenging, but it would be very rewarding since it would demonstrate T-violation (as much as P-violation). Moreover, if the experimental value turns out to be different from what is predicted by the SM, it would open up the doors for physics *b*eyond the *s*tandard *m*odel, BSM. Reliable atomic many-body theories are therefore needed to calculate the EDM, and hypersensitive experiments need to be performed to detect it in the laboratory.

Let us build the time-reversal operator for spin-½ particles. Since it is anti-unitary, let us write it as a product of a unitary operator U with the complex-conjugation operator K:

$$\Theta = UK. \tag{A.31}$$

Since the time reversal operator reverses the sign of the angular momentum (Eq. A.29c), for $\vec{\sigma}$ (Pauli spin angular momentum vector operator), we must have

$$\Theta\vec{\sigma}\Theta^{-1} = UK\vec{\sigma}K^{-1}U^{-1} = U(K\vec{\sigma}K)U^{-1} = -\vec{\sigma}. \tag{A.32}$$

$$\text{Now, } K\sigma_x K = K\begin{pmatrix} 0 & 1 \\ 1 & 0 \end{pmatrix}K = K\begin{pmatrix} 0 & K \\ K & 0 \end{pmatrix} = \begin{pmatrix} 0 & K^2 \\ K^2 & 0 \end{pmatrix} = \sigma_x, \tag{A.33a}$$

$$\text{and likewise, } K\sigma_y K = -\sigma_y, \tag{A.33b}$$

$$\text{and } K\sigma_z K = \sigma_z. \tag{A.33c}$$

Our task now is to determine an operator U such that

$$U\sigma_x U^{-1} = -\sigma_x \;;\; U\sigma_y U^{-1} = +\sigma_y \;;\; U\sigma_z U^{-1} = -\sigma_z. \tag{A.34}$$

From the above, we see that U can only change the phase of σ_y, it must therefore have the form $U = \exp(i\xi)\sigma_y$. Since ξ is only a phase, it can be chosen according to our convenience. A convenient choice is $\xi = \frac{\pi}{2}$ which gives

$$U = \exp\left(i\frac{\pi}{2}\right)\begin{pmatrix} 0 & 1 \\ -1 & 0 \end{pmatrix} = \begin{pmatrix} 0 & 1 \\ -1 & 0 \end{pmatrix}, \tag{A.35}$$

and

$$\Theta^2 = UKUK = \begin{pmatrix} 0 & 1 \\ -1 & 0 \end{pmatrix} K \begin{pmatrix} 0 & 1 \\ -1 & 0 \end{pmatrix} K = \begin{pmatrix} 0 & K \\ -K & 0 \end{pmatrix}\begin{pmatrix} 0 & K \\ -K & 0 \end{pmatrix} = \begin{pmatrix} -K^2 & 0 \\ 0 & -K^2 \end{pmatrix} = -1_{2\times 2}. \tag{A.36}$$

Obtaining the time-reversal *twice* of a state reverses its sign! We can verify easily that for any other angle ξ also (i.e., even when $\xi \neq \frac{\pi}{2}$), we always have $\Theta^2 = -1_{2\times 2}$. It is indeed remarkable that the time-reversed state of a time-reversed state does not give us the same state, but gives one whose sign is changed. This is similar to the SU(2) algebra in which we found that a rotation through 2π reversed the sign of a spin-½ state. This peculiar property is of great importance in understanding degeneracy associated with the time-reversal symmetry.

Let us *tentatively* assume that the time-reversed state $\Theta|\psi\rangle$ gives the same quantum state as $|\psi\rangle$, i.e., $\Theta|\psi\rangle$ is *not* linearly independent of $|\psi\rangle$. In other words, we have assumed that Θ only scales a Hilbert scale vector $|\psi\rangle$ by an insignificant phase, say ζ.

This assumption allows us to write $\Theta|\psi\rangle = \exp(i\zeta)|\psi\rangle$ (A.37a)

which would give

$$\Theta^2|\psi\rangle = \Theta\exp(i\zeta)|\psi\rangle = \exp(-i\zeta)\Theta|\psi\rangle = \exp(-i\zeta)\exp(+i\zeta)|\psi\rangle = +|\psi\rangle, \tag{A.37b}$$

contradicting $\Theta^2 = -1$.

$\Theta|\psi\rangle$ is therefore linearly independent of $|\psi\rangle$ and we therefore have *degenerate* eigenstates of the Hamiltonian. This degeneracy is called as *Kramers degeneracy* [17, 18], understood in terms of *Kramers Theorem*. When a spin-half system has time-reversal symmetry, then its *every energy eigenstate* is *doubly degenerate*. Along with the dynamical symmetry (Chapter 5), this is another example of having degeneracy which is *not* assigned to the *geometrical* symmetry of the Hamiltonian. This property has great importance in the spectroscopy of condensed matter, since a crystal field cannot remove the Kramers degeneracy, the electric field being invariant under time-reversal (see the Solved Problem P10.9). The limited objective of this Appendix is to provide a framework for a few applications discussed especially in Chapters 8 and 10. For a detailed discussion on chirality, etc., the reader must refer to specialized literature.

References

[1] E. P. Wigner, *Group Theory and Its Application to the Quantum Mechanics of Atomic Spectra* (Academic Press 1959).

[2] P. C. Deshmukh, *Foundations of Classical Mechanics* (Cambridge University Press, New Delhi, 2019).

[3] P. C. Deshmukh, Aarthi Ganesan, N. Shanthi, Blake Jones, James Nicholson, and A. Soddu, The "Accidental" Degeneracy of the Hydrogen Atom Is No Accident! *Canadian Journal of Physics* 10.1139/cjp-2014-0300.

[4] P. C. Deshmukh, Aarthi Ganesan, Sourav Banerjee, and Ankur Mandal, Accidental Degeneracy of the Hydrogen Atom and Its Non-accidental Solution in Parabolic Coordinates. *Canadian Journal of Physics* 99:10 853–860 (2021). https://doi.org/10.1139/cjp-2020-0258.

[5] T. D. Lee and C. N. Yang, Question of Parity Conservation in Weak Interactions. *Phys. Rev.* 104, 254 (1956).

[6] C. S. Wu, E. Ambler, R. W. Hayward, D. D. Hoppes, and R. P. Hudson, Experimental Test of Parity Conservation in β-decay. *Phys. Rev.* 105:4 1413–1415 (1957).

[7] P. C. Deshmukh and J. Libby, Symmetry Principles and Conservation Laws in Atomic and Subatomic Physics. *Resonance* 15, 832 (September 2010), and *Resonance* 15, 926 (October 2010).

[8] M. S. Safronova, D. Budker, D. DeMille, A. Derevianko, and Charles W. Clark, Search for New Physics with Atoms and Molecules. *Revs. Mod. Phys.*, 90:2 025008 (2018).

[9] C. S. Wood, S. C. Bennett, D. Cho, B. P. Masterson, J. L. Roberts, C. E. Tanner, C. E. Wieman, Measurement of Parity Nonconservation and an Anapole Moment in Cesium. *Science* 275, 1759 (1997).

[10] Gworge Toh, Amy Damitz, Carol E. Tanner, W. R. Johnson, and D. S. Elliott, Determination of the Scalar Polarizability of the Cesium 6s 2S1/2 – 7s 2S1/2 Transition and Implications for Atomic Parity Violation. *Phys. Rev. Lett.* 123, 073002 (2019).

[11] G. Breit and H. A. Bethe, Ingoing Waves in Final State of Scattering Problems. *Phys. Rev.* 93, 888 (1954); http://prola.aps.org/pdf/PR/v93/i4/p888_1.

[12] U. Fano and A. R. P. Rau, *Theory of Atomic Collisions and Spectra* (Academic Press, 1986).

[13] Pranawa C. Deshmukh, Dilip Angom, and Alak Banik, *Symmetry in Electron-Atom Collisions and Photoionization Process*, DST-SERC School at the Birla Institute of Technology, Pilani, January 9–28, 2011.

[14] J. M. Domingos, Time Reversal in Classical and Quantum Mechanics. *Int. J. Theor. Phys.* 18:3, 213 (1979).

[15] P. C. Deshmukh and S. Banerjee, Time Delay in Atomic and Molecular Collisions and Photoionization/photodetachment. *International Reviews in Physical Chemistry* 40:1, 127–153 (2020) https://doi.org/10.1080/0144235X.2021.1838805.

[16] P. C. Deshmukh, S. Banerjee, A. Mandal, and S. T. Manson, Eisenbud–Wigner–Smith Time Delay in Atom–Laser Interactions. *Eur. Phys. J. Spec. Top* (2021); https://doi.org/10.1140/epjs/s11734-021-00225-7.

[17] H. A. Kramers, Théorie générale de la rotation paramagnétique dans les cristaux, *Proceedings Koninklijke Akademie van Wetenschappen* 33: 959–972 (1930).

[18] E. Wigner, Über die Operation der Zeitumkehr in der Quantenmechanik, *Nachr. Akad. Ges. Wiss. Göttingen* 31, 546–559 (1932). http://www.digizeitschriften.de/dms/img/?PPN=GDZPPN002509032.

Appendix B

Schrödinger, Heisenberg, and Dirac "Pictures" of Quantum Dynamics

Dirac, Heisenberg, and Schrödinger. https://repository.aip.org/islandora/object/nbla%3A292226, Bohr Library & Archives (aip.org) D7.

A consistent description of the physical state of a system is provided by its wavefunction whose time-evolution is described by the Schrödinger equation

$$H\psi_{\alpha}(\vec{r},t) = i\hbar\frac{\partial}{\partial t}\psi_{\alpha}(\vec{r},t). \tag{B.1}$$

The wavefunction $\psi_{\alpha}(\vec{r},t)$ is the coordinate representation of the state vector $|\alpha,t\rangle$ in the Hilbert space that describes the system. Unitary transformations of an orthonormal basis set that spans the Hilbert space amount to rotation of the basis set. An example of this is the unitary transformation from $\{|j_1,m_1,j_2,m_2\rangle\}$ basis to $\{|j_1,j_2,j,m\rangle\}$ using the Clebsch–Gordan coefficients that we discussed in Chapter 4. Such unitary transformations preserve the norm of the state vectors, and alternative basis sets connected by them provide mathematically equivalent descriptions of quantum mechanics. Preference of using one basis over another is dictated by algebraic elegance.

Alternative descriptions of the temporal evolution of a physical system are possible using *generalized* rotations that leave the physics invariant, but alter the description of the temporal evolution of the system. Whereas the *Schrödinger picture* describes temporal evolution of a physical system using methodologies in which the operators are independent of time, and all the time-dependence appears in the wavefunctions, in the Heisenberg *picture* all the time-dependence is contained in the operators, while the wavefunctions are considered independent of time. In the *Dirac picture* (also called as the *Interaction picture*), both the wavefunctions and the operators are considered to be time-dependent. The three pictures are essentially equivalent, and transformations from any one of them to any other are effected using *generalized* rotations brought about by unitary transformations described below.

The time-evolution operator in the Schrödinger picture is given by Eq. 1.86b (Chapter 1). If the reference time t_0 is considered to be zero, the time evolution of a physical state is described by

$$\psi(\vec{r},t) = \exp\left(-i\frac{H}{\hbar}t\right)\psi(\vec{r},0), \tag{B.2}$$

where the exponential operator is

$$\exp\left(-i\frac{H}{\hbar}t\right) = 1 - i\frac{H}{\hbar}t + \frac{1}{2!}\left(i\frac{H}{\hbar}t\right)^2 - \frac{1}{3!}\left(i\frac{H}{\hbar}t\right)^3 + .. \tag{B.3}$$

which gives the dynamical phase (Eq. 1.114b in Chapter 1) of a stationary state:

$$\psi_S(\vec{r},t) = e^{-i\frac{E}{\hbar}t}\ \psi_S(\vec{r},0). \tag{B.4}$$

Operators like q, p, H are independent of time in the Schrödinger picture. To be specific, we shall rewrite the wavefunction in the Schrödinger equation, in the Schrödinger picture, with a subscript S:

$$H\psi_S(\vec{r},t) = i\hbar\frac{\partial}{\partial t}\psi_S(\vec{r},t). \tag{B.5}$$

The operator that effects the generalized rotation from the Schrödinger to the Heisenberg picture is $O_H = e^{i\frac{H}{\hbar}t}$. Its Hermitian adjoint is $O_H^\dagger = e^{-i\frac{H}{\hbar}t}$. It transforms a Schrödinger picture operator Ω_S to the corresponding Heisenberg picture operator Ω_H according to the following relation:

$$\Omega_H = e^{i\frac{H}{\hbar}t}\Omega_S e^{-i\frac{H}{\hbar}t}. \tag{B.6}$$

While operators transform as per Eq. B.6, the transformation of a wavefunction $\psi_S(\vec{r},t)$ in the Schrödinger picture to corresponding wavefunction $\psi_H(\vec{r},t)$ in the Heisenberg picture is described by

$$\psi_H(\vec{r},t) = O_H\psi_S(\vec{r},t) = \exp\left(i\frac{H}{\hbar}t\right)\exp\left(-i\frac{E}{\hbar}t\right)\psi_S(\vec{r},0) = \psi_S(\vec{r},0), \tag{B.7}$$

i.e., the Heisenberg picture wavefunction is independent of time; *at all times it is nothing but the wavefunction in the Schrödinger picture at the initial time*. An operator (Eq. B.6) in the Heisenberg picture is manifestly time-dependent; its time dependence is

$$\frac{\partial\Omega_H}{\partial t} = \left(\frac{\partial}{\partial t}\exp\left(i\frac{H}{\hbar}t\right)\right)\Omega_S\exp\left(-i\frac{H}{\hbar}t\right) + \exp\left(i\frac{H}{\hbar}t\right)\Omega_S\left(\frac{\partial}{\partial t}\exp\left(-i\frac{H}{\hbar}t\right)\right),$$

i.e., $$\frac{\partial\Omega_H}{\partial t} = \left(\frac{iH}{\hbar}\exp\left(i\frac{H}{\hbar}t\right)\right)\Omega_S\exp\left(-i\frac{H}{\hbar}t\right) + \exp\left(i\frac{H}{\hbar}t\right)\Omega_S\left(\frac{-iH}{\hbar}\right)\exp\left(-i\frac{H}{\hbar}t\right),$$

i.e., $$\frac{\partial\Omega_H}{\partial t} = \left\{\frac{iH}{\hbar}\right\}\left\{\exp\left(i\frac{H}{\hbar}t\right)\Omega_S\exp\left(-i\frac{H}{\hbar}t\right)\right\} - \left\{\exp\left(i\frac{H}{\hbar}t\right)\Omega_S\exp\left(-i\frac{H}{\hbar}t\right)\right\}\left\{\frac{iH}{\hbar}\right\},$$

i.e., $$\frac{\partial\Omega_H}{\partial t} = \left\{\frac{iH}{\hbar}\right\}\{\Omega_H\} - \{\Omega_H\}\left\{\frac{iH}{\hbar}\right\} = \frac{i}{\hbar}[H,\Omega_H]_-,$$

or, $i\hbar\frac{\partial \Omega_H}{\partial t} = \left[\Omega_H, H\right]_-$.

When the Heisenberg picture operator has an explicit time-dependence as well, its equation of motion is

$$i\hbar\frac{d\Omega_H}{dt} = \left[\Omega_H, H\right]_- + i\hbar\frac{\partial \Omega_H}{\partial t}. \tag{B.8}$$

The Dirac (interaction) picture is especially useful when the Hamiltonian of a system consists of a solvable part and another that isn't. For example,

$$H = H_0 + H_1, \tag{B.9}$$

of which H_0 represents the solvable part (which, for book-keeping purpose, may be referred to as the *s*-part) and H_1 is the part that produces the difficulty in solving the Schrödinger equation with the *full* Hamiltonian H. Since H_1 produces the said difficulty, we may refer to it, again merely for the purpose of book-keeping, as the *d*-part.

The *d*-part, H_1, may even include an explicit time-dependence (such as that considered in Section 8.3 of Chapter 8). Even in situations where methods based on the perturbation theory breakdown, one can use the Dirac picture to develop excellent approximation methods using the form of the Hamiltonian given by Eq. B.9.

The operator $O = e^{i\frac{H_0}{\hbar}t}$ brings about transformation of a Schrödinger picture operator Ω_S to that in the Dirac picture, written as

$$\Omega_I(t) = e^{i\frac{H_0}{\hbar}t}\Omega_S e^{-i\frac{H_0}{\hbar}t}. \tag{B.10}$$

The corresponding transformation of the wavefunction is

$$\psi_I(\vec{r},t) = e^{i\frac{H_0}{\hbar}t}\psi_S(\vec{r},t). \tag{B.11}$$

Note that in the -

(i) - Schrödinger picture, the wavefunctions ψ_S are time-dependent and the operators Ω_S are independent of time,
(ii) - Heisenberg picture, it is the wavefunctions ψ_H that are independent of time while the operators Ω_H are time-dependent,
(iii) - Dirac (Interaction) picture, the operators Ω_I, and the wavefunctions ψ_I are *both* time-dependent.

The inverse transformation from the Dirac picture wavefunction to the Schrödinger picture wavefunction is

$$e^{-i\frac{H_0}{\hbar}t}\psi_I(\vec{r},t) = \psi_S(\vec{r},t), \tag{B.12}$$

and hence the Schrödinger equation is

$$\left[H_0 + H_1\right]\psi_S(\vec{r},t) = i\hbar\frac{\partial}{\partial t}\left\{e^{-i\frac{H_0}{\hbar}t}\psi_I(\vec{r},t)\right\},$$

i.e., $H_0\psi_S(\vec{r},t)+H_1\psi_S(\vec{r},t)=i\hbar\left(-i\frac{H_0}{\hbar}\right)e^{-i\frac{H_0}{\hbar}t}\psi_I(\vec{r},t)+i\hbar e^{-i\frac{H_0}{\hbar}t}\frac{\partial}{\partial t}\psi_I(\vec{r},t),$

i.e., $\cancel{H_0\psi_S(\vec{r},t)}+H_1\psi_S(\vec{r},t)=\cancel{H_0\psi_S(\vec{r},t)}+i\hbar e^{-i\frac{H_0}{\hbar}t}\frac{\partial}{\partial t}\psi_I(\vec{r},t),$ (B.13)

or, $H_1 e^{-i\frac{H_0}{\hbar}t}\psi_I(\vec{r},t)=i\hbar e^{-i\frac{H_0}{\hbar}t}\frac{\partial}{\partial t}\psi_I(\vec{r},t).$ (B.14)

Operating on both sides of the above equation by $e^{+i\frac{H_0}{\hbar}t}$, we get

$$\left(e^{+i\frac{H_0}{\hbar}t}H_1 e^{-i\frac{H_0}{\hbar}t}\right)\psi_I(\vec{r},t)=i\hbar\frac{\partial}{\partial t}\psi_I(\vec{r},t),$$

i.e., $(H_I(t))\psi_I(\vec{r},t)=i\hbar\frac{\partial}{\partial t}\psi_I(\vec{r},t),$ (B.15)

where $H_I(t)$ is the transform of the operator H_1 to the interaction (Dirac) picture. Equation B.15 has a "form" just like the Schrödinger equation. Even as the operator on the left-hand side in Eq. B.15 is the transformation of only the d-part (H_1) of the Hamiltonian, the transformation itself is brought about using the s-part operator (H_0) as per Eq. B.10 and Eq. B.11. When the *d*-part is zero, the interaction picture wavefunction also becomes independent of time and the Heisenberg and the Dirac pictures become identical. One can also see that the expectation values of an operator in the three representations are all equal:

$\langle\Omega\rangle_S=\langle\Omega\rangle_H=\langle\Omega\rangle_I.$ (B.16)

The utility of the Dirac picture emerges from the description of the *time-evolution* operator $U(t,t_0)$, which describes the wavefunction at a time $t\neq t_0$ from its value at a reference time t_0, which has the following properties:

- $\psi_I(t)=U(t,t_0)\psi_I(t_0)$ (B.17a)
- $\psi_I(t)=U(t,t_1)U(t_1,t_0)\psi_I(t_0)$ (B.17b)

It therefore follows that

- $U(t_1,t_3)=U(t_1,t_2)U(t_2,t_3)$ (*closure* property) (B.18a)
- $1=U(t,t)=U(t,t_0)U(t_0,t)$ (existence of *identity*) (B.18b)
- $U^{-1}(t_0,t)=U(t,t_0)$ (existence of *inverse*) (B.18c)
- $\langle\psi_I(t)|\psi_I(t)\rangle=\langle\psi_I(t_0)|\psi_I(t_0)\rangle$ (preservation of the *norm*)

i.e., $\langle \psi_I(t) | \psi_I(t) \rangle = \langle \psi_I(t_0) | U^\dagger(t,t_0) U(t,t_0) | \psi_I(t_0) \rangle$

or, $U^\dagger(t,t_0) = U^{-1}(t,t_0)$ (*unitarity* of the time-evolution operator). (B.18d)

Since the differential equation

$$H_I(t)\, U(t,t_0)\psi_I(t_0) = i\hbar \frac{\partial}{\partial t} U(t,t_0)\psi_I(t_0)$$

holds good for an arbitrary state $\psi_I(t_0)$, we are led to the following operator identity

$$i\hbar \frac{\partial}{\partial t} U(t,t_0) = H_I(t) U(t,t_0). \tag{B.19}$$

It represents the equation of motion for the time-development operator. Now,

$$\psi_I(\vec{r},t) = e^{+i\frac{H_0}{\hbar}t} \psi_S(\vec{r},t) = e^{+i\frac{H_0}{\hbar}t} e^{-i\frac{H}{\hbar}(t-t_0)} \psi_S(\vec{r},t_0) = e^{+i\frac{H_0}{\hbar}t} e^{-i\frac{H}{\hbar}(t-t_0)} e^{-i\frac{H_0}{\hbar}t_0} \psi_I(\vec{r},t_0), \tag{B.20}$$

which essentially describes the time evolution in the Dirac picture from time t_0 to t. Hence

$$U(t,t_0) = e^{+i\frac{H_0}{\hbar}t} e^{-i\frac{H}{\hbar}(t-t_0)} e^{-i\frac{H_0}{\hbar}t_0}, \tag{B.21}$$

which is the formal solution to Eq. B.18. In general,

$$[H, H_0]_- \neq 0, \tag{B.22}$$

hence the order in which the operators are treated in Eq. B.21 is very important. Using Dyson's chronological operator, powerful approximations are developed using an integral equation for the time development operator. We ask the reader to refer to other sources for a detailed discussion on fruitful methods that are available to study many-electron correlations using the Dyson chronological operator.

Appendix C

Spherical Harmonics

Watercolor caricature of the French mathematician Adrien-Marie Legendre by French artist Julien-Léopold Boilly, 1820 (Public Domain). https://en.wikipedia.org/wiki/Adrien-Marie_Legendre.

We know from Chapter 4 that the *orbital* angular momentum operators ℓ^2 and ℓ_z commute with each other and can therefore be simultaneously diagonalized in their common eigenbasis. From Eqs. 4.57a,b, these operators are given by

$$\ell^2 = -\hbar^2 \left\{ \frac{1}{\sin\theta}\frac{\partial}{\partial\theta}\left(\sin\theta\frac{\partial}{\partial\theta}\right) + \frac{1}{\sin^2\theta}\frac{\partial^2}{\partial\varphi^2} \right\}, \tag{C.1}$$

$$\text{and } \ell_z = -i\hbar\frac{\partial}{\partial\varphi}. \tag{C.2}$$

The representation of the simultaneous eigenvectors $|\ell m\rangle$ of ℓ^2 and ℓ_z in the coordinate space in various equivalent notations is

$$\langle \hat{e}_r | \ell m\rangle = \langle \theta,\varphi | \ell m\rangle = Y_{\ell m}(\theta,\varphi) = Y_{\ell m}(\hat{e}_r), \tag{C.3}$$

where $\hat{r}$ denotes the unit vector along the position vector $\vec{r} = \overrightarrow{OP}$ of an arbitrary point P in space whose polar and azimuthal angles respectively are θ and φ, shown in Fig. C.1.

The functions $Y_{\ell m}(\theta,\varphi)$ are called as spherical harmonics. They satisfy the following equations:

$$\ell_z Y_{\ell m}(\theta,\varphi) = i\hbar\frac{\partial}{\partial\varphi}Y_{\ell m}(\theta,\varphi), \tag{C.4}$$

$$\text{and } \ell^2 Y_{\ell m}(\theta,\varphi) = -\hbar^2 \left\{ \frac{1}{\sin\theta}\frac{\partial}{\partial\theta}\left(\sin\theta\frac{\partial}{\partial\theta}\right) + \frac{1}{\sin^2\theta}\frac{\partial^2}{\partial\varphi^2} \right\} Y_{\ell m}(\theta,\varphi). \tag{C.5}$$

Writing the eigenvalue equation for ℓ^2 as

$$\ell^2 Y_{\ell m}(\theta,\varphi) = \lambda\hbar^2 Y_{\ell m}(\theta,\varphi), \tag{C.6}$$

we find that the differential equation to be solved for the spherical harmonics is

$$\left[\left\{ \frac{1}{\sin\theta}\frac{\partial}{\partial\theta}\left(\sin\theta\frac{\partial}{\partial\theta}\right) + \frac{1}{\sin^2\theta}\frac{\partial^2}{\partial\varphi^2} \right\} + \lambda \right] Y_{\ell m}(\theta,\varphi) = 0. \tag{C.7}$$

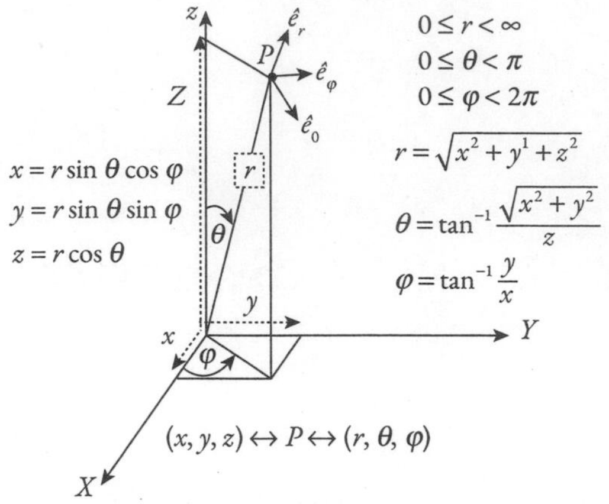

Fig. C.1 This figure shows the relationship between the Cartesian coordinate system and the spherical polar coordinates. The three Cartesian unit vectors $\left\{\hat{e}_x, \hat{e}_y, \hat{e}_z\right\}$ are constant vectors along respective Cartesian axes. The unit vectors $\left\{\hat{e}_r, \hat{e}_\theta, \hat{e}_\varphi\right\}$ of the spherical polar coordinate system at a typical point P are as shown in the figure. They point respectively in the direction of the *increasing* radial distance r, polar angle θ, and the azimuthal angle φ. The notation is the same as in Reference [1].

From Chapter 4, we already know that the eigenvalue of the operator ℓ^2 is $\ell(\ell+1)\hbar^2$. Hence, we shall set $\lambda = \ell(\ell+1)$ hereafter. The polar angle θ and the azimuthal angle φ are independent degrees of freedom. Hence, we seek a solution to the above differential equation using the method of separation of variables and factorize the spherical harmonics into a function of only the polar angle, and another of the azimuthal angle alone:

$$Y_{\ell m}(\theta,\varphi) = \Theta(\theta)\Phi(\varphi). \tag{C.8}$$

Insertion of Eq. C.8 in C.7 provides a neat separation of the partial differential equation C.7,

$$-\frac{1}{\Phi}\frac{d^2\Phi}{d\varphi^2} = \frac{\sin^2\theta}{\Theta}\left[\frac{1}{\sin\theta}\frac{d}{d\theta}\left(\sin\theta\frac{d\Theta}{d\theta}\right) + \ell(\ell+1)\Theta\right], \tag{C.9}$$

where the left-hand side depends only on the azimuthal angle and the right-hand side only on the polar angle. We may therefore set each side to be a constant, and choose this constant to be m^2 wherein m is to be determined. It would soon be seen that this choice turns out to be a particularly convenient one. The two independent ordinary differential equations to be now solved are

$$\frac{d^2\Phi}{d\varphi^2} + m^2\Phi(\varphi) = 0, \tag{C.10}$$

and $$\frac{1}{\sin\theta}\frac{d}{d\theta}\left(\sin\theta\frac{d\Theta}{d\theta}\right) + \left(\ell(\ell+1) - \frac{m^2}{\sin^2\theta}\right)\Theta = 0. \tag{C.11}$$

The boundary condition $\Phi(\varphi) = \Phi(\varphi + 2\pi)$ determines the solution of Eq. C.10 to be

$$\Phi(\varphi) = N_\varphi \exp(im\varphi), \tag{C.12}$$

where N_φ is an arbitrary constant and $m = 0, \pm1, \pm2, \pm3, \ldots$, i.e., zero, or a positive or negative integer, thus ratifying the choice of the constant of separation we had made to be a *convenient* one.

It is now expedient to solve the polar equation by introducing an auxiliary variable

$$\omega = \cos\theta; \tag{C.13a}$$

correspondingly

$$d\omega = -\sin\theta d\theta, \tag{C.13b}$$

$$\text{and } P(\omega) = \Theta(\theta). \tag{C.13c}$$

The differential equation corresponding to Eq. C.11 satisfied by $P(\omega)$ is

$$(1-\omega^2)\frac{d^2P_{\ell m}(\omega)}{d\omega^2} - 2\omega\frac{dP_{\ell m}(\omega)}{d\omega} + \left(\ell(\ell+1) - \frac{m^2}{1-\omega^2}\right)P_{\ell m}(\omega) = 0, \tag{C.14a}$$

$$\text{or, } \frac{d}{d\omega}\left[(1-\omega^2)\frac{dP_{\ell m}(\omega)}{d\omega}\right] - \frac{m^2}{1-\omega^2}P_{\ell m}(\omega) + \ell(\ell+1)P_{\ell m}(\omega) = 0. \tag{C.14b}$$

For $m = 0$, the equation reduces to

$$(1-\omega^2)\frac{d^2P_\ell(\omega)}{d\omega^2} - 2\omega\frac{dP_\ell(\omega)}{d\omega} + \ell(\ell+1)P_\ell(\omega) = 0, \tag{C.15a}$$

$$\text{i.e., } \frac{d}{d\omega}\left[(1-\omega^2)\frac{dP_\ell(\omega)}{d\omega}\right] + \ell(\ell+1)P_\ell(\omega) = 0. \tag{C.15b}$$

$$\text{i.e., } \frac{d^2P_\ell(\omega)}{d\omega^2} - \frac{2\omega}{(1-\omega^2)}\frac{dP_\ell(\omega)}{d\omega} + \frac{\ell(\ell+1)}{(1-\omega^2)}P_\ell(\omega) = 0. \tag{C.15c}$$

Equations C.15a,b are known as the Legendre's differential equation, and Eqs. C.14a,b as the *associated* Legendre differential equation, after the French mathematician Adrien-Marie Legendre (1752–1833). Equation C.15 remains invariant as $\theta \to (\pi - \theta)$, and correspondingly $\omega \to -\omega$, so the functions $P(\omega)$ are symmetric or anti-symmetric with respect to the XY-plane. We see that the Legendre differential equation has regular singular points at $\omega = \pm1$. It has a power series solution

$$P_\ell(\omega) = \sum_{k=0}^{\infty} c_k \omega^k, \tag{C.16}$$

substituting which in the Legendre differential equation one obtains the recursion relation

$$c_{k+2} = \frac{k(k+1) - \ell(\ell+1)}{(k+1)(k+2)}c_k = \frac{(k+\ell+1)(k-\ell)}{(k+1)(k+2)}c_k. \tag{C.17}$$

$$\text{Since } \frac{c_{k+2}}{c_k} \underset{k\to\text{large}}{\rightarrow} 1, \tag{C.18}$$

we must require that, in order to have physically acceptable solutions, the power series must terminate for some maximum value $k_{max} = \ell$ of the integer k. Thence, higher power terms would not blow up in the entire range $-1 \le \omega \le +1$. The physically acceptable solutions are the so-called Legendre *polynomials* in (i) either the even powers of ω *and* $c_1 = 0$, when $\ell = 0,2,4,6,..$ or, (ii) odd powers of ω *and* $c_2 = 0$, when $\ell = 1,3,5,7,...$

The solution is given by

$$P_\ell(\omega) = \left\{ \begin{array}{l} c_0\left[1 - \ell(\ell+1)\dfrac{\omega^2}{2!} + \ell(\ell+1)(\ell-2)(\ell+3)\dfrac{\omega^4}{4!} - ...\right] \\ + c_1\left[\omega - (\ell-1)(\ell+2)\dfrac{\omega^3}{3!} + (\ell-1)(\ell+2)(\ell-3)(\ell+4)\dfrac{\omega^5}{5!} - ...\right] \end{array} \right\}. \tag{C.19}$$

For both even or odd integer value of ℓ, the solution can be compactly written as

$$P_\ell(\omega) = K_\ell \sum_{r=0}^{k}\left[(-1)^r \frac{(2\ell - 2r)!}{(\ell - r)!r!}\frac{\omega^{\ell-2r}}{(\ell-2r)!}\right], \tag{C.20}$$

where K_ℓ is a normalization constant. The polynomials are conveniently expressed popularly as the Rodrigues' formula, after the French amateur mathematician Olinde Rodrigues (1795–1851):

$$P_\ell(\omega) = \frac{K_\ell}{\ell!}\left(\frac{d}{d\omega}\right)^\ell (\omega^2 - 1)^\ell. \tag{C.21}$$

The choice

$$K_\ell = \frac{1}{2^\ell} \tag{C.22a}$$

normalizes the polynomials giving

$$P_\ell(1) = 1. \tag{C.22b}$$

The first few polynomials for $\ell = 0,1,2,3,4,..$ are commonly denoted by their spectroscopic symbols, which *respectively* are $s, p, d, f, g, ...$ These are given by

$$P_{\ell=0}(\omega) = 1, \tag{C.23a}$$

$$P_{\ell=1}(\omega) = \omega, \tag{C.23b}$$

$$P_{\ell=2}(\omega) = \frac{1}{2}(3\omega^2 - 1), \tag{C.23c}$$

$$P_{\ell=3}(\omega) = \frac{1}{2}(5\omega^2 - 3\omega), \tag{C.23d}$$

$$P_{\ell=4}(\omega) = \frac{1}{8}\left(35\omega^4 - 30\omega^2 + 3\right), \text{ etc.} \tag{C.23e}$$

Differentiation Eq. C.15a m-times, and denoting the mth-order derivative operator as

$$\frac{d^m}{d\omega^m} = \frac{d^{m-1}}{d\omega^{m-1}}\frac{d}{d\omega} \tag{C.24}$$

as $\Omega_{m-1,1}$ we get

$$\Omega_{m-1,1}\left[\left(1-\omega^2\right)\frac{d^2P_\ell(\omega)}{d\omega^2}\right] - \Omega_{m-1,1}\left[2\omega\frac{dP_\ell(\omega)}{d\omega}\right] + \ell(\ell+1)\Omega_{m-1,1}\left[P_\ell(\omega)\right] = 0. \tag{C.25}$$

Using the Leibnitz formula,

$$\frac{d^m}{d\omega^m}\left[f(\omega)g(\omega)\right] = \sum_{r=0}^{m}\binom{m}{r}\frac{d^rf}{d\omega^r}\frac{d^{m-r}g}{d\omega^{m-r}} = \sum_{r=0}^{m}\frac{m!}{r!(m-r)!}\frac{d^rf}{d\omega^r}\frac{d^{m-r}g}{d\omega^{m-r}}, \tag{C.26}$$

we get

$$\left(1-\omega^2\right)\frac{d^2y_{\ell m}(\omega)}{d\omega^2} - 2\omega(m+1)\frac{dy_{\ell m}(\omega)}{d\omega} + (\ell-m)(\ell+m+1)y_{\ell m}(\omega) = 0, \tag{C.27}$$

$$\text{where } y_{\ell m}(\omega) = \frac{d^mP_\ell(\omega)}{d\omega^m}. \tag{C.28}$$

One can continue the analysis in the same manner and deduce that

$$P_{\ell m}(\omega) = \left(1-\omega^2\right)^{\frac{m}{2}}y(\omega) = \left(1-\omega^2\right)^{\frac{m}{2}}\frac{d^m}{d\omega^m}P_\ell(\omega) \tag{C.29}$$

is a solution of the Associated Legendre differential equation.

Using Eq. C.12 and Eq. C.29, the normalized spherical harmonics (Eq. 6.8) are thus given by

$$Y_{\ell m}(\theta,\varphi) = \left\{(-1)^m\sqrt{\frac{2\ell+1}{4\pi}\frac{(\ell-m)!}{(\ell+m)!}}\right\}P_\ell(\cos\theta)\exp(im\varphi), \text{ when } m \geq 0, \tag{C.30a}$$

$$\text{and } Y_{\ell m}(\theta,\varphi) = (-1)^{|m|}Y_{\ell,|m|}^{*}(\theta,\varphi), \text{ when } m < 0. \tag{C.30b}$$

Inclusive of the Legendre polynomials (Eqs. C.23a–e), the first few normalized spherical harmonics are given by

$$Y_{00}(\theta,\varphi) = \frac{1}{\sqrt{4\pi}}, \tag{C.31a}$$

$$Y_{10}(\theta,\varphi) = \sqrt{\frac{3}{4\pi}}\cos\theta, \tag{C.31b}$$

$$Y_{1,\pm1}(\theta,\varphi) = \mp\sqrt{\frac{3}{8\pi}}\sin\theta\exp(\pm im\varphi), \tag{C.31c}$$

$$Y_{2,0}(\theta,\varphi) = \sqrt{\frac{5}{16\pi}}(3\cos^2\theta - 1), \tag{C.31d}$$

$$Y_{2,\pm1}(\theta,\varphi) = \mp\sqrt{\frac{15}{8\pi}}\sin\theta\cos\theta\exp(\pm i\varphi), \tag{C.31e}$$

$$Y_{2,\pm2}(\theta,\varphi) = \mp\sqrt{\frac{15}{32\pi}}\sin^2\theta\exp(\pm 2i\varphi), \text{ etc.} \tag{C.31f}$$

The spherical harmonic function (Eq. C.31b) for $\ell = 1$, $m = 0$ is often called as the p_z orbital. It is shown in Fig. C.2. It represents the angular part of the atomic orbital, *at an arbitrary distance a_0 from the centre*. The radial dependence of the atomic orbital is shown in Fig. 5.3. The upper lobe is assigned a plus sign and the lower lobe the minus sign, since $\cos\theta$ is positive above the XY-plane, and negative below it. The spherical harmonics $Y_{\ell,m}(\theta,\varphi)$ appear in the description of the probability amplitude $R_{E,\ell}(r)Y_{\ell,m}(\theta,\varphi)$. The probability density would involve $\left|Y_{\ell,m}(\theta,\varphi)\right|^2$, and for $\ell = 1$, $m = 0$ it would scale as $\cos^2\theta$. A polar plot of this function would therefore *also* have a figure of 8, but the lobes would not be perfect spheres. They would be stretched out along the Z-axis, and both the lobes would have the plus sign. One must, however, remember that unless any external field is applied to break the spherical symmetry of the potential, the Z-axis provides *only* a reference direction, called as the *axis of quantization*.

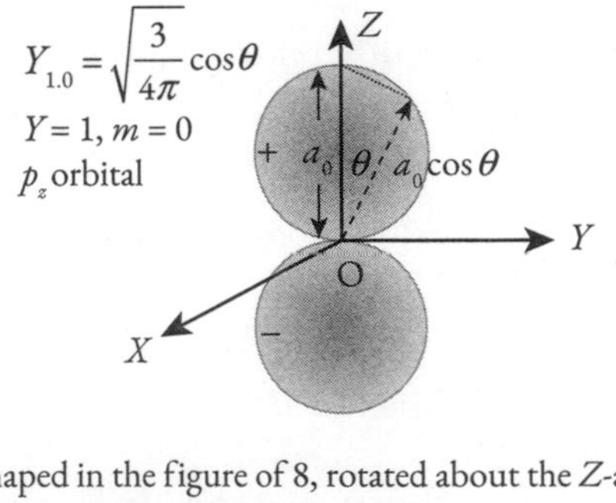

Fig. C.2 The atomic p_z orbital is shaped in the figure of 8, rotated about the Z-axis through one full circle. The lobes above and below the XY-plane are perfectly spherical. Since the diameter of a circle subtends an angle of 90° at any point on the circumference, the side of the triangle shown by dashed lines scales as $a_0\cos\theta$, as shown, corresponding to the Legendre polynomial for $\ell = 1$, $m = 0$.

Spherical harmonics for $m \neq 0$ obviously have both a real part and an imaginary part, but their superpositions can be constructed to provide a linearly independent alternative basis. The resulting real superpositions can then be sketched just as in Fig. C.2.

For example,

$$\frac{1}{\sqrt{2}}\left(Y_{\ell=1,m=1} \pm Y_{\ell=1,m=-1}\right) = p_{\pm} \tag{C.32}$$

gives real functions that can be plotted as functions of θ, φ at any given distance a_0. Similar superpositions that yield real functions can be constructed for higher values of the orbital angular momentum; many of these are commonly tabulated in other easily accessible sources.

Reference

[1] P. C. Deshmukh, *Foundations of Classical Mechanics* (Cambridge University Press, New Delhi, 2019).

APPENDIX D

Occupation Number Formalism
Second Quantization

Pascual Jordan. B525, https://arkiv.dk/en/vis/5941470. Courtesy: Niels Bohr Archive.

We might say that the three operators $a^{\dagger}$, a and $n = a^{\dagger}a$ correspond respectively to the Creator (Brahma), the Destroyer (Shiva), and the Preserver (Vishnu) in Hindu mythology.

—J. J. Sakurai in *Advance Quantum Mechanics*

The aforementioned remark by Sakurai may be taken in a lighter vein. When we consider matter $\rightleftarrows$ energy conversion, it does become essential to introduce operators for particle creation and annihilation. We will, however, introduce them in this appendix with a *limited* objective, to indicate their efficacy in going *beyond* the Hartree–Fock (HF) self-consistent field (SCF) method (Chapter 9). It would prepare the readers to tackle problems involving a many-electron system going *beyond* the single-particle approximation, also referred to as the Independent Particle Approximation (IPA). The mathematical machinery we employ to achieve this is the occupation number formalism. *Also*, it will familiarize the reader with basic tools introduced in Appendix E to solve the quantum mechanical many-electron problem on a *quantum computer* (Chapter 11). Much of the occupation number formalism was developed by Jordan and Wigner [1, 2].

D.1 CREATION AND ANNIHILATION OPERATORS

In Chapter 1, we introduced "quantization" as a mathematical framework to describe the laws of nature in a consistent and successful manner. Essentially, it encompassed dispensing with the classical description of a system in terms of the dynamical variables q and p, and replacing them by *operators* q_{op} and p_{op}. These *operate* on wavefunctions, which are coordinate representations of state vectors in a Hilbert space. The resulting mathematical contraption initially seemed abstract, but unlike the classical description, it provides a suitable and beneficial description of nature. Quantization leads to discrete energy eigen-spectra when boundary conditions on the differential (Schrödinger) equation are appropriate for *bound-states* of the physical system under study, as well as to an energy continuum in the case of *unbound* states. When a bound state and a continuum state are degenerate, and both are accessible to the system, one has meta-stable states (resonances) which may decay (autoionization) into separate fragments of the system. Such a process can also be described as annihilation of a particle in a discrete level and creation of the same in a continuum eigenstate. In particular, the description of atomic photoionization (Section 6.3, Chapter 6) as well as that of an autoionization resonance benefits from a reformulation in terms of annihilation and creation operators.

In Chapter 9, we have discussed the HF theory to address the *N*-electron problem. It provides solutions within the framework of the IPA since its basic element is the Slater determinant (SD) written as a product of a *particular set* of *N* number of $u_i\left(q_j\right)$, the *single-electron* spin orbitals (Eqs. 9.7a,b). The HF SCF solutions ignore the Coulomb correlations since the frozen orbital approximation (see the discussion above Eq. 9.50) is employed. Correlated dynamics of the *N*-electron system is addressed in the HF SCF method only to the extent of the application of the Fermi–Dirac statistics, required by the half-integer spin of the electrons (Chapters 4 and 7) – i.e., spin correlations are included in the HF SCF method; however, the Coulomb correlations are not.

Theories that address correlations are aimed at unfolding the *collecting description* of the many-electron systems, even if only in terms of single-electron wavefunctions (spin-orbitals); after all, the fundamental units of matter are the elementary particles – the electrons in the present case. The collective description of a many-particle system must go *beyond* the IPA. A single SD, having resulted from the frozen orbital approximation, cannot describe the *N*-electron system correctly. A *particular set* (which would be our *reference* state) of *N* number of spin-orbitals cannot be adequate to describe the true natural state of the *N*-electron system. One must therefore employ a *superposition* of SDs in which different *sets* of *N* spin-orbitals are used to generate the (antisymmetrized) product wavefunctions. The different SDs would have the same number of electrons, but they would be made up with a *different set* of *N* spin-orbitals. The different SDs can be obtained by annihilating one or more electrons from the "previous" (reference) set of spin-orbitals and creating an equal number in another set, giving a different SD. This brings us to the annihilation and creation operators that operate on occupation number vectors. For reasons that would become clear later, the occupation number formalism is also called as the method of *second quantization*.

In the context of addressing many-electron correlations beyond the HF method, second quantization does not involve any fundamentally novel *concept*, such as the uncertainty principle. Rather, it provides an alternative mathematical structure that involves a *compact notation* and a *convenient framework* to describe the many-electron structure, as well as its interaction with various probes (i.e., dynamics). It provides a platform to develop powerful mathematical methods to address *correlations* that go *beyond* the single-particle approximations.

Elements of an SD (Eq. 9.10) are the *one-electron* spin-orbitals which are solutions of the *N*-integro-differential HF equation (Eqs. 9.58a,b). In an SD, there are *N* different *subscripts* of the spin-orbitals. The set of *N* subscripts (common to elements in all columns in a given row of a SD) stands for a particular way in which the *N* electrons occupy accessible one-electron spin-orbitals, one in each, zero in the rest, a total of *N* spin-orbitals being occupied. The vacant spin-orbitals are also possible solutions of the same Hamiltonian; only their occupation numbers are zero in the reference SD. The select spin-orbitals whose occupation number is "1" characterize a particular *configuration*; it is unique for each SD, as discussed in the beginning of Section 9.2 of Chapter 9.

A given configuration is therefore made up of *one* of the infinite ways of choosing N number of *one-electron spin-orbitals* that are chosen for the SCF scheme to generate the SD. We have illustrated alternative SDs for the beryllium ($Z=4$) atom in Chapter 9 characterizing the one-electron spin-orbitals by the $\left\{\left|n_\alpha,\ell_\alpha,m_{\ell,\alpha},m_{s,\alpha}\right\rangle\right\}$ set of quantum numbers. We may represent the set of four quantum numbers collectively by a single symbol, thereby introducing a compact notation $\left|n_\alpha,\ell_\alpha,m_{\ell,\alpha},m_{s,\alpha}\right\rangle \leftrightarrow |\alpha\rangle$. Another example would clarify this further. We consider an atom of magnesium (Z = 12) in its ground state: $1s^2 2s^2 2p^6 3s^2$, but this time we employ a different basis, viz. $\left\{\left|n_\alpha,\ell_\alpha,j_\alpha,m_{j\alpha}\right\rangle\right\}$. In this case, our compact notation is $\left|n_\alpha,\ell_\alpha,j_\alpha,m_{j\alpha}\right\rangle \leftrightarrow |\alpha\rangle$. A specific reference configuration of the magnesium atom under consideration is $1s^2_{\frac{1}{2}}2s^2_{\frac{1}{2}}2p^2_{\frac{1}{2}}2p^4_{\frac{3}{2}}3s^2_{\frac{1}{2}}$. We shall refer to the state of the Mg atom corresponding to this configuration as $\left\langle q_1q_2\cdots q_{12}\middle|\psi_A\right\rangle$. Essentially, this configuration declares just *which* one-electron spin-orbitals are *occupied*. The 12 one-electron states described by the vectors $\left|n_\alpha,\ell_\alpha,j_\alpha,m_{j\alpha}\right\rangle$ that are occupied in $\left\langle q_1q_2\cdots q_{12}\middle|\psi_A\right\rangle$ are

$$\left|1,0,\frac{1}{2},\pm\frac{1}{2}\right\rangle,\left|2,0,\frac{1}{2},\pm\frac{1}{2}\right\rangle,\left|2,1,\frac{1}{2},\pm\frac{1}{2}\right\rangle,\left|2,1,\frac{3}{2},\left(\pm\frac{1}{2}\ \&\ \pm\frac{3}{2}\right)\right\rangle,\left|3,0,\frac{1}{2},\pm\frac{1}{2}\right\rangle.$$

One can also consider an alternate configuration of the magnesium atom, such as $1s^2_{\frac{1}{2}}2s^2_{\frac{1}{2}}2p^2_{\frac{1}{2}}2p^4_{\frac{3}{2}}4s^2_{\frac{1}{2}}$. We shall refer to this state by $\left\langle q_1q_2\cdots q_{12}\middle|\psi_B\right\rangle$. The 12 one-electron states that are occupied in $\left\langle q_1q_2\cdots q_{12}\middle|\psi_B\right\rangle$ are

$$\left|1,0,\frac{1}{2},\pm\frac{1}{2}\right\rangle,\left|2,0,\frac{1}{2},\pm\frac{1}{2}\right\rangle,\left|2,1,\frac{1}{2},\pm\frac{1}{2}\right\rangle,\left|2,1,\frac{3}{2},\left(\pm\frac{1}{2}\ \&\ \pm\frac{3}{2}\right)\right\rangle,\left|4,0,\frac{1}{2},\pm\frac{1}{2}\right\rangle.$$

Both the configurations, $\left|\psi_A\right\rangle$ and $\left|\psi_B\right\rangle$, have 12 electrons, but *configured* differently. The occupation number of the states $\left|3,0,\frac{1}{2},+\frac{1}{2}\right\rangle$ and $\left|3,0,\frac{1}{2},-\frac{1}{2}\right\rangle$ is "1" in $\left|\psi_A\right\rangle$ and "0" in $\left|\psi_B\right\rangle$, while the occupation number of $\left|4,0,\frac{1}{2},+\frac{1}{2}\right\rangle$ and $\left|4,0,\frac{1}{2},-\frac{1}{2}\right\rangle$ is "0" in $\left|\psi_A\right\rangle$ and "1" in $\left|\psi_B\right\rangle$. The SDs for the configurations $\left|\psi_A\right\rangle$ and $\left|\psi_B\right\rangle$ are different, but in addition to these two there can in fact be an infinite number of alternative configurations that differ from each other in the occupation numbers of the infinite possible one-electron spin-orbitals.

The wavefunction of the Mg atom must therefore be written as a linear combination of *all* such SDs:

$$\left|\psi^{CI}\right\rangle=\sum_{i=1}^{n}a_i\left|\psi_i^{SD}\right\rangle, \tag{D.1}$$

where $n\to\infty$. It can be restricted to a finite number to include only those configurations that are nearly degenerate and correspond to the same total angular momentum. The superscript "CI"

on the left stands for "configuration interaction." It is impossible to do a calculation in which $n \to \infty$. Different choices of truncating the summation to a finite number represent *different ways* of doing a multi-configuration Hartree–Fock (MCHF) calculation [3] for an N-electron system. These correspond to *different approximations*. The infinity is an overwhelming idea; its subsets may have finite number of elements, but also may have *infinite* terms. Such subsets are classified using sophisticated methods of many-body theory. An example of a *subset* of all possible configurations having an infinity of terms is the random phase approximation (RPA) [4, 5]; it contains an infinite number of configurations, but only those which belong to a particular *type*, classified by *many-electron correlations* theory.

For a given configuration represented by an SD, one therefore has a unique set consisting of N number of *occupied* spin-orbitals. Out of an infinite *ordered* set $|\alpha_1, \alpha_2, \alpha_3, \alpha_4, \ldots\rangle$, a particular set of N one-electron states are occupied, and the rest *vacant*. The physical content of an SD is completely equivalent to a set of occupation numbers $|n_1, n_2, n_3, n_4, \ldots\rangle$ such that n_i gives the occupation number of the one-electron spin-orbital $|\alpha_i\rangle$, n_i being either "0" or "1" and $\sum_{i=1}^{\infty} n_i = N$, the total number of electrons. In the case of bosons too, one can develop a similar formalism with reference to an *ordered* set of single-boson states, but in this case (i) the coordinate representation of an ordered set $|\alpha_1, \alpha_2, \alpha_3, \alpha_4, \ldots\rangle$ describing the N-boson $\left(\sum_{i=1}^{\infty} n_i = N\right)$ system is *symmetric*, not antisymmetric, under an interchange of two bosons, (ii) the number n_i of bosons in any state $|\alpha_i\rangle$ is arbitrary: $0 \le n_i \le N$. The state vectors $|n_1, n_2, n_3, n_4, \ldots\rangle$ constitute an occupation number vector space which represents the state of a multi-particle system of fermions or bosons. We shall develop the algebra of the occupation vector space in Section D.2. In the present section, we restrict ourselves to only recognize that we must expect the creation and annihilation operators to have *different* commutation properties for fermions and for bosons.

We begin with a set $\{\alpha_1, \alpha_2, \alpha_3, \alpha_4, \ldots\}$ whose members are sequenced as per a pre-determined select *order* of α_i which represents the eigenvalues of a CSCO (Section 1.3 of Chapter 1). The *good* quantum numbers α_i would be determined by the symmetry group of the Hamiltonian; it is different for different systems, such as an atom, a molecule, or a crystal. An occupation number vector would be written as $|n_1, n_2, n_3, n_4, \ldots\rangle$, where n_k is the number of particles in the state α_k. Fundamental particles being either fermions or bosons, we must build our algebra for these two types of particles; the occupation numbers n_k for a system of fermions can only be "0" or "1," for bosons we have $0 \le n_i \le N$. We may construct an arbitrary occupation number state by operating on a *vacuum* state, defined as

$$|\psi_v^{(0)}\rangle = |0, 0, \ldots 0, 0, \ldots\rangle, \tag{D.2}$$

by an appropriate set of *particle creation operators* that populate the system. The subscript v denotes the vacuum. A single particle in the γth state, for example, has

$$\psi_\gamma^{(1)} = |0, 0, \ldots, n_\gamma = 1, 0, \ldots\rangle. \tag{D.3}$$

Consider a 3-fermion system in its lowest state thus constructed from the fermion *vacuum* with *creation* operators $(a_i^\dagger)$ for the ith single-fermion state:

$$|n_1 = 1, n_2 = 1, n_3 = 1\rangle = |1,1,1\rangle = a_1^\dagger a_2^\dagger a_3^\dagger \left|\psi_v^{(0)}\right\rangle. \tag{D.4}$$

If we now want to describe the *reduction* of the occupation number of the $i = 2$ state from "1" to "0," we must ask what would be the effect of an *annihilation* operator (a_2) on the occupation number vector $a_1^\dagger a_2^\dagger a_3^\dagger \left|\psi_v^{(0)}\right\rangle$. Since one cannot annihilate a particle that is not even there, it is intuitive to define

$$a_2 \left|\psi_v^{(0)}\right\rangle = 0. \tag{D.5}$$

Now, we need the vector $a_2 a_1^\dagger a_2^\dagger a_3^\dagger \left|\psi_v^{(0)}\right\rangle$, and to use Eq. D.5 to obtain it, we must first inquire how to interchange the position of the operator a_2 with *other* creation and annihilation operators on its right. The commutation properties must be consistent with the *interchange* of particles defined in Eq. 9.2 through Eqs. 9.4a,b of Chapter 9; these rules must therefore be *different* for fermions and for bosons. The statistical properties of the identical particles (whether fermions or bosons) must be integrated in the commutation properties of the creation and annihilation operators. We shall introduce the mathematical framework of occupation number vector space in the next section.

D.2 Occupation Number States Formalism

The Schrödinger equation for N identical particles, *whether fermions or bosons*, is

$$i\hbar \frac{\partial}{\partial t} \Psi(x_1, x_2, \ldots, x_N, t) = H\Psi(x_1, x_2, \ldots, x_N, t), \tag{D.6}$$

where H is the N-particle Hamiltonian, and we express the N-particle wavefunction to be written as a product of one-particle functions:

$$\Psi(x_1, \ldots, x_N, t) = \sum_{\alpha_1'} \sum_{\alpha_2'} \cdots \sum_{\alpha_N'} C(\alpha_1', \alpha_2', \ldots, \alpha_N', t)\, \psi_{\alpha_1'}(x_1)\, \psi_{\alpha_2'}(x_2) \ldots \psi_{\alpha_N'}(x_N), \tag{D.7}$$

$\psi_{\alpha_i'}(x_i)$ being the single-particle wavefunction. In Eq. D.7, the subscript represents the *set* of eigenvalues of CSCO and (x_i) represents all the coordinates (including spin) of that particle. The multiple summations over all the *labels* (quantum states) accommodate the configurations corresponding to particles at various coordinates being in *different* quantum states: the coordinates are only notional. We shall closely follow Ref. [6]. The particles being identical, it is impossible to distinguish between them. The possibility that a particle at coordinate q_i is in the state $|\alpha_j\rangle$ and that at q_j is in the state $|\alpha_i\rangle$ cannot be distinguished from the possibility that the particle at coordinate q_i is in the state $|\alpha_i\rangle$ and that at q_j in the state $|\alpha_j\rangle$, respectively. One must therefore employ a linear superposition of all possible interchanges between the N identical particles, with due respect to the interchange resulting in a symmetric or antisymmetric combination, for bosons and fermions respectively:

$$\Psi(x_1, x_2, \ldots, x_i, \ldots x_j, \ldots x_N, t) = \pm\Psi(x_1, x_2, \ldots, x_j, \ldots x_i, \ldots x_N, t). \tag{D.8}$$

The plus sign in Eq. D.8 is appropriate for bosons, and the minus for fermions. The factor $\left[\psi_{\alpha_1'}(x_1)\psi_{\alpha_2'}(x_2)\ldots\psi_{\alpha_N'}(x_N)\right]$ in Eq. D.7 is insensitive to the actual position of any single-particle spin-orbital; the necessary and sufficient condition corresponding to Eq. D.8 therefore is that the *expansion coefficients* in Eq. D.7 are *correspondingly* symmetric or antisymmetric with respect to interchange of corresponding quantum numbers. Hence, under an interchange $i \rightleftarrows j$ of the quantum labels,

$$C(\alpha_1,\ldots,\alpha_i,\ldots,\alpha_j,\ldots,\alpha_N,t) = \pm C(\alpha_1,\ldots,\alpha_j,\ldots,\alpha_i,\ldots,\alpha_N,t), \tag{D.9}$$

the plus sign being appropriate for bosons, and the minus sign for fermions. Using Eq. D.7 in the Schrödinger equation (Eq. D.6), we get

$$i\hbar\frac{\partial}{\partial t}C(\alpha_1,\ldots,\alpha_N,t) = \sum_{\alpha_1'}\sum_{\alpha_2'}\cdots\sum_{\alpha_N'}\left[\begin{array}{l} C(\alpha_1',\alpha_2',\ldots,\alpha_N',t)\times \\ \int dx_1\int dx_2\ldots\int dx_N \left\{\begin{array}{l}\psi_{\alpha_1}(x_1)^\dagger\psi_{\alpha_2}(x_2)^\dagger \\ \ldots\psi_{\alpha_N}(x_N)^\dagger\end{array}\right\} H \left\{\begin{array}{l}\psi_{\alpha_1'}(x_1)\psi_{\alpha_2'}(x_2) \\ \ldots\psi_{\alpha_N'}(x_N)\end{array}\right\}\end{array}\right], \tag{D.10}$$

The N-particle Hamiltonian (for both fermions and the bosons) has the form

$$H = \sum_{k=1}^{N} T(x_k) + \frac{1}{2}\sum_{k=1}^{N}\sum_{\ell=1}^{N}{}' V(x_k,x_\ell), \tag{D.11}$$

where the prime (') on the *second* summation notation denotes the exclusion $j \neq i$. It consists of one-particle operators T (such as the kinetic energy operator for each particle) and two-particle operators V, representing interaction (whatever) between the identical particles. Writing the one-particle and the two-particle terms separately, we have

$$\begin{aligned} i\hbar\frac{\partial}{\partial t}C(\alpha_1,\ldots,\alpha_N,t) = & \sum_{k=1}^{N}\sum_{\alpha_1'}\sum_{\alpha_2'}\cdots\sum_{\alpha_N'}\left[\begin{array}{l} C(\alpha_1',\alpha_2',\ldots,\alpha_N',t)\times\int dx_1\int dx_2\ldots\int dx_N \\ \left\{\begin{array}{l}\psi_{\alpha_1}(x_1)^\dagger\psi_{\alpha_2}(x_2)^\dagger \\ \ldots\psi_{\alpha_N}(x_N)^\dagger\end{array}\right\} T(x_k) \left\{\begin{array}{l}\psi_{\alpha_1'}(x_1)\psi_{\alpha_2'}(x_2) \\ \ldots\psi_{\alpha_N'}(x_N)\end{array}\right\}\end{array}\right] \\ & + \\ & \frac{1}{2}\sum_{k=1}^{N}\sum_{\ell=1}^{N}{}'\sum_{\alpha_1'}\sum_{\alpha_2'}\cdots\sum_{\alpha_N'}\left[\begin{array}{l} C(\alpha_1',\alpha_2',\ldots,\alpha_N',t)\times\int dx_1\int dx_2\ldots\int dx_N \\ \left\{\begin{array}{l}\psi_{\alpha_1}(x_1)^\dagger\psi_{\alpha_2}(x_2)^\dagger \\ \ldots\psi_{\alpha_N}(x_N)^\dagger\end{array}\right\} V(x_k,x_\ell) \left\{\begin{array}{l}\psi_{\alpha_1'}(x_1)\psi_{\alpha_2'}(x_2) \\ \ldots\psi_{\alpha_N'}(x_N)\end{array}\right\}\end{array}\right]. \end{aligned} \tag{D.12}$$

Carrying out integration over the *one-particle terms*, *and* using the orthogonality and normalization of the single-particle spin-orbitals, we get

$$i\hbar\frac{\partial}{\partial t}C(\alpha_1,\alpha_2,\ldots,\alpha_N,t) = \\ \sum_{\alpha_k'}\sum_{k=1}^{N}C(\alpha_1,\ldots,\alpha_k',\ldots,\alpha_N,t)\int dx_k\psi_{\alpha_k}(x_k)^\dagger T(x_k)\psi_{\alpha_k'}(x_k) + \\ +\sum_{\alpha_1'}\cdots\sum_{\alpha_N'}C(\alpha_1',\ldots,\alpha_N',t)\int dx_1\ldots\int dx_N\begin{bmatrix}\psi_{\alpha_1}(x_1)^\dagger\ldots\psi_{\alpha_N}(x_N)^\dagger\times \\ \left\{\frac{1}{2}\sum_{k=1}^{N}\sum_{\ell=1}^{N}{}'V(x_k,x_\ell)\right\}\begin{Bmatrix}\psi_{\alpha_1'}(x_1)\ldots \\ \psi_{\alpha_N'}(x_N)\end{Bmatrix}\end{bmatrix}. \tag{D.13}$$

Now, working with the two-particle terms, we get

$$i\hbar\frac{\partial}{\partial t}C(\alpha_1,\alpha_2,\ldots,\alpha_N,t) = \sum_{\alpha_k'}\sum_{k=1}^{N}\begin{Bmatrix}C(\alpha_1,\ldots,\alpha_{k-1},\alpha_{k'},\alpha_{k+1},\ldots,\alpha_N,t)\times \\ \times\int dx_k\psi_{\alpha_k}(x_k)^\dagger T(x_k)\psi_{\alpha_k'}(x_k)\end{Bmatrix}, \\ +\frac{1}{2}\sum_{k=1}^{N}\sum_{\ell=1}^{N}{}'\sum_{\alpha_k'}\sum_{\alpha_\ell'}\begin{Bmatrix}C(\alpha_1,\ldots,\alpha_{k-1},\alpha_k',\alpha_{k+1},\ldots,\alpha_{\ell-1},\alpha_\ell',\alpha_{\ell+1},\ldots,\alpha_N,t)\times \\ \times\int dx_k\int dx_l\begin{Bmatrix}\psi_{\alpha_k}(x_k)^\dagger \\ \psi_{\alpha_\ell}(x_\ell)^\dagger\end{Bmatrix}V(x_k,x_\ell)\begin{Bmatrix}\psi_{\alpha_k'}(x_k) \\ \psi_{\alpha_\ell'}(x_\ell)\end{Bmatrix}\end{Bmatrix} \tag{D.14}$$

or,

$$i\hbar\frac{\partial}{\partial t}C(\alpha_1,\alpha_2,\ldots,\alpha_N,t) = \sum_{\gamma}\sum_{k=1}^{N}\begin{Bmatrix}C(\alpha_1,\ldots,\alpha_{k-1},\gamma,\alpha_{k+1},\ldots,\alpha_N,t)\times \\ \times\int dx_k\psi_{\alpha_k}(x_k)^\dagger T(x_k)\psi_\gamma(x_k)\end{Bmatrix} \\ +\frac{1}{2}\sum_{k=1}^{N}\sum_{\ell=1}^{N}{}'\sum_{\gamma}\sum_{\gamma'}\begin{Bmatrix}C(\alpha_1,\ldots,\alpha_{k-1},\gamma,\alpha_{k+1},\ldots,\alpha_{\ell-1},\gamma',\alpha_{\ell+1},\ldots,\alpha_N,t)\times \\ \times\int dx_k\int dx_l\begin{Bmatrix}\psi_{\alpha_k}(x_k)^\dagger \\ \psi_{\alpha_\ell}(x_\ell)^\dagger\end{Bmatrix}V(x_k,x_\ell)\begin{Bmatrix}\psi_\gamma(x_k) \\ \psi_{\gamma'}(x_\ell)\end{Bmatrix}\end{Bmatrix}. \tag{D.15}$$

In Eq. D.15, we have relabeled the summation label α_k' in the one-particle operators as γ, and the labels α_k' and α_ℓ' in the two-particle operators respectively as γ and γ'.

There is a specific differential equation for each set of quantum numbers $\{\alpha_1,\alpha_2,\ldots,\alpha_N\}$, and there are infinite sets of such *coupled* differential equations with time-dependent coefficients.

Let us discuss the case of an N-electron system. Electrons being fermions, an interchange $i \rightleftarrows j$ would reverse the sign of the coefficient C (Eq. D.9):

$$C(\alpha_1,\alpha_2,\ldots,\alpha_i,\ldots,\alpha_j\ldots,\alpha_N,t) = -C(\alpha_1,\alpha_2,\ldots,\alpha_j,\ldots,\alpha_i\ldots,\alpha_N,t). \tag{D.16}$$

The coefficient $C(\alpha_1, \alpha_2, \ldots, \alpha_i, \ldots, \alpha_j \ldots, \alpha_N, t)$ is unique to a particular SD and is declared by the N quantum labels $(\alpha_1, \alpha_2, \ldots, \alpha_i, \ldots, \alpha_j \ldots, \alpha_N)$ that appear in it. It can be represented by a coefficient $f(n_1, \ldots, n_i, \ldots n_\infty)$, where n_i is the number of electrons in the state $|\alpha_i\rangle$. Accordingly, we have the correspondence

$$C(\alpha_1, \alpha_2, \ldots, \alpha_i, \ldots, \alpha_j \ldots, \alpha_N, t) \equiv f(n_1, \ldots, n_i, \ldots n_\infty, t), \tag{D.17}$$

where the fermion occupation numbers

$$n_i = 0 \text{ or } 1, \tag{D.18a}$$

$$\text{and } \sum_i n_i = N. \tag{D.18b}$$

The most general many-electron wavefunction (Eq. D.7) can therefore be written as the following summation over SDs:

$$\Psi(x_1, \ldots, x_N, t) = \sum_{n_1=0}^{1} \cdots \sum_{n_i=0}^{1} \cdots \sum_{n_\infty=0}^{1} f(n_1, \ldots, n_i, \ldots n_\infty, t)\, \Phi_{n_1, \ldots, n_i, \ldots n_\infty}(x_1, \ldots, x_N), \tag{D.19}$$

where the SD is

$$\Phi_{n_1, \ldots, n_i, \ldots n_\infty}(x_1, \ldots, x_N) = \frac{1}{\sqrt{N!}} \begin{vmatrix} u_{\alpha_1}(x_1) & \ldots & u_{\alpha_1}(x_N) \\ \ldots & u_{\alpha_i}(x_j) & u_{\alpha_i}(x_N) \\ u_{\alpha_N}(1) & \ldots & u_{\alpha_N}(x_N) \end{vmatrix}, \tag{D.20}$$

and

$$f(n_1, \ldots n_i, \ldots, n_j, \ldots, \ldots n_\infty, t) \equiv C(\alpha_1 < \ldots < \alpha_i < \ldots < \alpha_j < \ldots < \alpha_N, t), \tag{D.21a}$$

and, effectively,

$$f(n_1, n_2, \ldots n_i, \ldots, n_j, \ldots, \ldots n_\infty, t) \equiv \tilde{C}(n_1, n_2, \ldots n_i, \ldots, n_j, \ldots, \ldots n_\infty, t). \tag{D.21b}$$

In terms of the occupation number vectors, the state vector (Eqs. D.7 and D.19) is

$$|\Psi(t)\rangle = \sum_{n_1=0}^{1} \cdots \sum_{n_i=0}^{1} \cdots \sum_{n_\infty=0}^{1} f(n_1, \ldots n_i, \ldots n_\infty, t) |n_1, \ldots n_i, \ldots n_\infty\rangle. \tag{D.22}$$

In order to control and manipulate the occupation numbers, we define fermion creation and annihilation operators, using the idea we introduced in Eq. D.4. These are defined by writing an occupation number state vector as

$$|n_1 n_2 \ldots n_i \ldots n_j \ldots n_\infty\rangle = (a_1^\dagger)^{n_1} (a_2^\dagger)^{n_2} \ldots (a_i^\dagger)^{n_i} \ldots (a_j^\dagger)^{n_j} \ldots (a_\infty^\dagger)^{n_\infty} |0\rangle, \tag{D.23}$$

where $|0\rangle$ is the vacuum state. The effect of a fermion *annihilation operator* a_k on the occupation number state is

$$a_k|n_1 \ldots n_i \ldots n_j \ldots n_\infty\rangle = a_k\left[\left(a_1^\dagger\right)^{n_1} \ldots \left(a_i^\dagger\right)^{n_i} \ldots \left(a_j^\dagger\right)^{n_j} \ldots \left(a_\infty^\dagger\right)^{n_\infty}\right]|0\rangle. \tag{D.24}$$

To make use of the property illustrated in Eq. D.5, namely that one cannot annihilate a particle that is not there, we must define how to *move* the annihilation operator a_k across the string of creation operators on its right in Eq. D.24. For fermions there is a phase change at every interchange between a pair of them. Hence,

$$a_k|n_1 \ldots n_i \ldots n_j \ldots n_\infty\rangle = (-1)^{S_k}\left(a_1^\dagger\right)^{n_1} \ldots \left(a_i^\dagger\right)^{n_i} \ldots a_k\left(a_k^\dagger\right)^{n_k} \ldots \left(a_\infty^\dagger\right)^{n_\infty}|0\rangle, \tag{D.25}$$

$$\text{where } S_k = \sum_{i=1}^{k-1} n_i. \tag{D.26}$$

Accordingly,

$$a_k|\ldots n_k \ldots\rangle = (-1)^{S_k}\left(n_k\right)^{\frac{1}{2}}|\ldots(n_k - 1)\ldots\rangle \text{ when } n_k = 1, \tag{D.27a}$$

$$\text{and } a_k|\ldots n_k \ldots\rangle = 0 \text{ when } n_k = 0. \tag{D.27b}$$

Likewise, for the fermion *creation operator* $a_k^\dagger$, we have

$$a_k^\dagger|\ldots n_k \ldots\rangle = (-1)^{S_k}\left(n_k + 1\right)^{\frac{1}{2}}|\ldots(n_k + 1)\ldots\rangle \text{ when } n_k = 0, \tag{D.28}$$

$$\text{and } a_k^\dagger|\ldots n_k \ldots\rangle = 0 \text{ when } n_k = 1. \tag{D.29}$$

In both cases, i.e., when a one-fermion state is vacant or filled,

$$a_k^\dagger a_k|\ldots n_k \ldots\rangle = n_k|\ldots n_k \ldots\rangle,\ n_k = 0,1. \tag{D.30}$$

The fermion creation and the annihilation operators are therefore so defined as to satisfy the following anti-commutation rules:

$$\left[a_k, a_m^\dagger\right]_+ = a_k a_m^\dagger + a_m^\dagger a_k = \delta_{mk}, \tag{D.31a}$$

$$\left[a_k^\dagger, a_m^\dagger\right]_+ = a_k^\dagger a_m^\dagger + a_m^\dagger a_k^\dagger = 0, \tag{D.31b}$$

$$\text{and } \left[a_k, a_m\right]_+ = a_k a_m + a_m a_k = 0. \tag{D.31c}$$

The above anti-commutation rules capture the Fermi–Dirac statistics obeyed by particles having half-integer intrinsic spin angular momentum. We now ask if Eq. D.15 would remain valid under *simultaneous* interchange of quantum labels (such as $i \rightleftarrows j$) on its *both* sides.

It may be tempting to assume that the answer to this question is yes, but that would not be correct, since the interchange under consideration is *not* merely that of *dummy* summation labels. We observe that on the right-hand side of Eq. D.14 (or Eq. D.15), in the terms corresponding to the single-particle operators, α_k' appears once extra and α_k appears once less. Furthermore, in the two-particle terms, α_k' and α_ℓ' appear once extra, while α_k and α_ℓ appear once less. Remember that we have the label γ on the right-hand side in the one-electron terms in place of α_k, which appears on the left. Likewise, we have the labels γ and γ' on the right-hand side in the two-electron terms respectively in place of α_k' and α_ℓ', which appear on the left-hand side. The relabeling is only for the purpose of highlighting the above-mentioned difference in the quantum labels on the two sides of the equation. The difference in the occupancy is easily represented by scaling the one-particle terms on the right-hand side by the factor $\left(\sqrt{n_{\alpha_k}+1}\,\delta_{n_{\alpha_k},0}\right)\left(\sqrt{n_{\alpha_{k'}}}\,\delta_{n_{\alpha_{k'}},1}\right)$ and the two-particle terms on the right-hand side by $\left(\sqrt{n_{\alpha_k}+1}\,\delta_{n_{\alpha_k},0}\sqrt{n_{\alpha_{k'}}}\,\delta_{n_{\alpha_{k'}},1}\right)\left(\sqrt{n_{\alpha_\ell}+1}\,\delta_{n_{\alpha_\ell},0}\sqrt{n_{\alpha_{\ell'}}}\,\delta_{n_{\alpha_{\ell'}},1}\right)$, where n_α is the number of fermions in the state $|\alpha\rangle$, subject to Eqs. D.18a,b. We must also get the correct phase corresponding to the shift in the occupation numbers, which is attained on using the annihilation and creation operators. The phase (Eq. D.26) is obviously affected by the occupation numbers in the *previous* one-particle states. The critical term of our focus is $\alpha_k' \rightarrow \alpha_k$ in the one-electron terms and $\alpha_k', \alpha_\ell' \rightarrow \alpha_k, \alpha_\ell$ in the two-electron terms. The phase factor would therefore be, for $\alpha_k' < \alpha_k$, $(-1)^{n_{\alpha_k'+1}+n_{\alpha_k'+2}+\ldots+n_{\alpha_k-1}}$, and for $\alpha_k' > \alpha_k$, it is $(-1)^{n_{\alpha_k+1}+n_{\alpha_k+2}+\ldots+n_{\alpha_k'-1}}$.

In general, let us consider the one-particle operator term in which the *occupancy* shifts from a single-particle state j to the state i. The one-electron term therefore has the form

$$\begin{aligned}\delta_{n_i'0}\delta_{n_j'1}\left(\sqrt{n_i'+1}\right)\left(\sqrt{n_j'}\right)(-1)^{S_j-S_i}\left|\ldots n_i'+1\ldots n_j'-1\ldots n_\infty'\right\rangle &= \\ &= a_i^\dagger a_j\left|\ldots n_i'\ldots n_j'\ldots n_\infty'\right\rangle.\end{aligned} \tag{D.32}$$

In the case of the two-particle operator terms, the occupancy shifts from k,ℓ to i,j, but while using Eq. D.26 (and D.28) to track the phase, we must remember that calculation of S_ℓ using Eq. D.26 needs care. With $i < j < k < \ell$, $n_k = 1$ *prior* to the reduction in the occupation of the kth state, but $n_k = 0$ *after* the reduction in the occupation of the kth state. The calculation of S_ℓ would be different in these two cases by a factor of (−1). The Schrödinger equation (Eq. D.6) therefore takes the following expression in the occupation number formalism, with the state vector given by Eq. D.23:

$$i\hbar\frac{\partial}{\partial t}\left|\psi(t)\right\rangle = H\left|\psi(t)\right\rangle, \tag{D.33}$$

with

$$H = \sum_{i,j} a_i^\dagger \langle i|T|j\rangle a_j + \frac{1}{2}\sum_{i,j,k,\ell} a_i^\dagger a_j^\dagger \langle ij|V|k\ell\rangle a_\ell a_k, \tag{D.34a}$$

or, equivalently,

$$H = \sum_{i,j} \langle i|T|j\rangle a_i^\dagger a_j + \frac{1}{2}\sum_{i,j,k,\ell} \langle ij|V|k\ell\rangle a_i^\dagger a_j^\dagger a_\ell a_k. \quad \text{(D.34b)}$$

Note that the *ordering* of the indices ℓ, k in the annihilation operators is of importance. It is the result of the factor (-1) phase difference discussed above, and the Eq. D.31c.

The formalism for bosons is similar, but the rules for the creation and annihilation operators for bosons, *corresponding* to Eqs. D.31a,b,c, are different, since a many-boson wavefunction remains symmetric under an exchange of a pair of bosons. The rules appropriate for boson creation and annihilation operators are

$$\left[b_k, b_m^\dagger\right]_- = b_k b_m^\dagger - b_m^\dagger b_k = \delta_{mk}, \quad \text{(D.35a)}$$

$$\left[b_k^\dagger, b_m^\dagger\right]_- = b_k^\dagger b_m^\dagger - b_m^\dagger b_k^\dagger = 0, \quad \text{(D.35b)}$$

$$\text{and } \left[b_k, b_m\right]_+ = b_k b_m - b_m b_k = 0. \quad \text{(D.35c)}$$

We can now follow a procedure similar to that adopted for the case of fermions, but with due respect to the fact that the rules D.35a,b,c must now be used instead of D.31a,b,c, and the occupation number of the ith bosonic state can be $0 \le n_i \le N$. We would then arrive at the Schrodinger equation in the occupation number formalism for bosons, and find that it has the same form as that for fermions, but with the Hamiltonian given by

$$H = \sum_{i,j} \langle i|T|j\rangle b_i^\dagger b_j + \frac{1}{2}\sum_{i,j,k,\ell} \langle ij|V|k\ell\rangle b_i^\dagger b_j^\dagger b_k b_\ell. \quad \text{(D.36)}$$

The ordering of the indices in the last term is not important now on account of Eq. D.35c.

We now introduce "field" operators

$$\hat{\psi}(x) = \sum_i \psi_i(x) c_i \quad \text{(D.37a)}$$

and

$$\hat{\psi}^\dagger(x) = \sum_j \psi_j^\dagger(x) c_j^\dagger, \quad \text{(D.37b)}$$

where $\psi_i(x)$ is the wavefunction and $c_i, c_j^\dagger$ are respectively the annihilation and the creation operators (Eqs. D.31a,b,c for fermions and Eqs. D.35a,b,c for bosons). The reason they are called 'field' operators is that they can be defined for every point x, thereby generating a field. Equations D.37a,b constitute *field quantization* since the scalar fields are expressed in terms of *operators*. Of course, for a particle having spin η, we have $(2\eta + 1)$ components, and we may therefore write

$$\psi_i(x) = \begin{bmatrix} \psi_i(x)_1 \\ \cdots \\ \psi_i(x)_{2\eta+1} \end{bmatrix} \equiv \psi_i(x)_\alpha, \alpha = 1, \ldots, (2\eta + 1). \quad \text{(D.38)}$$

In particular, electrons have spin-½, and hence for the electrons we have

$$\psi_i(x) = \begin{bmatrix} \psi_i(x)_1 \\ \psi_i(x)_2 \end{bmatrix} \equiv \psi_i(x)_\alpha, \alpha = 1,2. \tag{D.39}$$

From the boson and fermion commutation/anti-commutation properties (Eqs. D.35a,b,c and Eqs. D.31a,b,c), it follows that

$$\left[\hat{\psi}_\alpha(x), \hat{\psi}^\dagger_\beta(x')\right]_\mp = \sum_i \psi_i(x)_\alpha \psi_i(x')_\alpha = \delta_{\alpha\beta}\delta(x - x'), \tag{D.40a}$$

$$\left[\hat{\psi}_\alpha(x), \hat{\psi}_\beta(x')\right]_\mp = 0, \tag{D.40b}$$

$$\text{and } \left[\hat{\psi}^\dagger_\alpha(x), \hat{\psi}^\dagger_\beta(x)\right]_\mp = 0, \tag{D.40c}$$

with the *upper* sign being applicable for bosons, and the *lower* sign for fermions. An N-particle operator is

$$J^{(N)} = \sum_{i=1}^{N} J\left(x_{i,op}\right). \tag{D.41}$$

The corresponding operator, using field quantized operator terminology (Eqs. D.37a,b) is

$$\hat{J}^{(N)} = \sum_{i,j=1}^{N} \langle j|J|i\rangle c_j^\dagger c_i = \sum_{i,j} \iiint d^3x \psi_j(x)^\dagger J(x) \psi_i(x) c_j^\dagger c_i, \tag{D.42a}$$

$$\text{i.e., } \hat{J}^{(N)} = \iiint d^3(x) \hat{\psi}^\dagger(x) J(x) \hat{\psi}(x). \tag{D.42b}$$

In particular, the field-quantized Hamiltonian therefore takes the form

$$H = \begin{bmatrix} \iiint d^3(x)\hat{\psi}^\dagger(x) T(x) \hat{\psi}(x) + \\ \frac{1}{2}\iiint d^3x \iiint d^3x' \hat{\psi}^\dagger(x) \hat{\psi}^\dagger(x') V(x,x') \hat{\psi}(x') \hat{\psi}(x) \end{bmatrix}. \tag{D.43}$$

We recall (Chapter 1) that "quantization" required us to employ *operators* in place of classical dynamical variables. For example, the prescription to quantize a classical dynamical variable is $J(x) \to J_{op}\left(x_{op}\right)$. In field quantization (Eqs. D.37a,b), we have used operators a *second* time now; hence, the method of occupation number formalism is also called as *second quantization*.

References

[1] P. Jordan, Zur Quantenmechanik der Gasentartung. *Zeits. f. Phys.* 44:473–480 (1927).

[2] P. Jordan and E. Wigner, Über das Paulische Äquivalenzverbot. Zeitschriftfür Physik. *Zeits. f. Phys.* 47, 631 (1928).

[3] C. Fischer, T. Brage, and P. Jönsson, *Computational Atomic Structure: An MCHF Approach* (Taylor & Francis, Inc., 1997).

[4] D. Bohm and D. Pines, A Collective Description of Electron Interactions: III. Coulomb Interactions in a Degenerate Electron Gas. *Phys. Rev.* 92:609 (1953).

[5] P. C. Deshmukh and S. T. Manson, Photoionization of Atomic Systems Using the Random-Phase Approximation Including Relativistic Interactions. *Atoms* 10, 71 (2022) https://doi.org/10.3390/atoms10030071.

[6] A. L. Fetter and J. D. Walecka, *Quantum Theory of Many Particle Systems* (Dover Publications, 2003).

Appendix E

Electron Structure Studies with Qubits

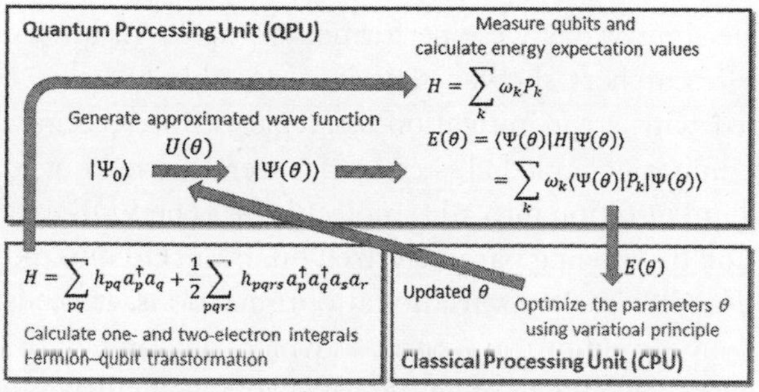

A schematic view of the VQE-based quantum chemical calculation.

Included with permission from: Kenji Sugisaki, Takumi Kato, Yuichiro Minato, Koji Okuwaki and Yuji Mochizuki, Variational quantum eigensolver simulations with the multireference unitary coupled cluster ansatz: a case study of the C2v quasi-reaction pathway of beryllium insertion into a H_2 molecule, *Phys. Chem. Chem. Phys.*, 24, 8439 (2022).

Accurate prediction of chemical and material properties from first principles quantum chemistry is a challenging task on traditional computers. Recent developments in quantum computation offer a route toward highly accurate solutions with polynomial cost.

—Jarrod R. McClean, Ryan Babbush, Peter J. Love, and Alán Aspuru-Guzik, *J. Phys. Chem. Lett.* 5, 24:4368–4380 (2014), https://pubs.acs.org/doi/10.1021/jz501649m.

Information processing employing mathematical modeling using qubits has enormous potential in drug development for clinical trials against dreadful diseases. There are many different ways in which a molecule can be folded to optimize a chemical reaction. Designing smart materials for emerging technologies also requires mathematical simulations, most efficiently implemented on a quantum computer. In this appendix, we provide a cursory introduction to the young and expanding field of electron structure studies with qubits.

The original meaning of the term *quantum supremacy* proposed by John Preskill in 2012 was intended to describe *the point where quantum computers can do things that classical computers cannot*. It is often interpreted as the *demonstrated* and *quantified* ability to process *any problem faster* on a quantum computer than on a classic computer. The term *quantum advantage* is also much in vogue; it is used to describe attainment of a quantum computational algorithm in solving *a real-world problem faster* than on a classical computer. Platforms for the development of quantum computing architecture include (a) *quantum gate–based* (Chapter 11) and (b) *quantum annealing–based* approach (which employs optimization techniques akin to those in operations research). Industry giants such as Google, Honeywell, IBM, and Intel use the quantum gate–based platform, while D-wave systems employs quantum annealing.

The *quantum phase estimation* (QPE) was the first algorithm that was proposed to solve the Schrödinger equation on a quantum computer. It is a fully quantum algorithm to obtain eigenvalues of a Hamiltonian, but it requires rather sophisticated hardware and employs the inverse quantum Fourier transform (IQFT) method. It is a multipurpose program that is a part of other quantum algorithms, including Shor's algorithm. A full configuration-interaction computation that provides variationally the best wavefunction has been carried out using the QPE algorithm [1]. However, QPE requires a very large number of qubits.

An alternative approach employs the *variational quantum eigensolver* (VQE), which utilizes quantum *and* classical resources to solve quantum eigenvalue problems [2]. Of specific interest is the study of many-electron correlations. The Hartree–Fock method addresses exchange (Fermi–Dirac, statistical) correlations, but not the Coulomb correlations, as discussed in Chapter 9. In order to tackle the Coulomb correlations, it is natural to expect computing using qubits to be valuable, but one must map the many-electron Hamiltonian in terms of qubits. In recent years, a Hartree–Fock problem has actually been solved on a superconducting qubit quantum computer [3]. *Google AI Quantum collaborators* performed a VQE simulation of two intermediate-scale chemistry problems. The simulations were performed on up to 12 qubits, involving up to 72 two-qubit gates. The researchers showed that it is possible to achieve chemical accuracy when VQE is combined with error mitigation strategies. More recently, VQE simulations with the multi-reference unitary-coupled cluster ansatz were carried out to study the quasi-reaction pathway of beryllium insertion into a H_2 molecule [4]. The VQE uses a QPU to prepare approximate wavefunction by utilizing parameterized quantum circuits and evaluating energy expectation values, and a CPU for the variational optimizations. It needs a parametrized quantum circuit, defined by an empirical "ansatz" which determines the accuracy of the wavefunction and energy. The development of new ansatz provides challenging opportunities. There are two major parts of the VQE: (a) computations on QPU – quantum processing unit, and (b) computations on CPU – classical processing unit.

The primary strategy involves mapping the second quantized many-electron Hamiltonian and the wavefunction into the corresponding qubit entities, applying *fermion–qubit transformations*. Information of the energy expectation value obtained from the QPU is transferred to the CPU, and the CPU carries out variational optimization of the parameters and convergence check. If the variational calculation did not converge, a set of information of the revised parameters is returned to the QPU. The QPU executes evaluation of the energy expectation value using the new parameters. These procedures are iterated until convergence.

Ansatz need to be balanced out; they are *chemistry inspired* and alternatively inspired by *hardware efficiency*. In chemistry-inspired ansatz, ground-state trial wavefunction of a molecule is built from operators generating single- and double-excitation configurations which are obtained from a HF SCF wavefunction determined in advance, using a classical computer.

The subject is vast; in this appendix we have a modest goal to only provide an introduction to mapping the Hamiltonian for many *indistinguishable* electrons (fermions) to a Hamiltonian in terms of *distinguishable* qubits. Techniques for such mappings are (1) Jordan–Wigner, often abbreviated as "JW" [5, 6], (2) Parity [7], and (3) Bravyi–Kitaev [8, 9]. We shall illustrate only the essence of the JW scheme in which one transforms the electronic structure Hamiltonian for the indistinguishable electrons, to that for distinguishable qubits, using second quantization methods, and constructing an Ising-type Hamiltonian. The method is well adapted to execute coupled cluster calculations with single and double excitations. The method relies on using universal quantum gates; any operation on a quantum computer can be reduced to operations by a set of quantum gates such as those we discussed in Chapter 11. An arbitrary unitary operation can be expressed as a finite sequence of universal quantum gates. Decomposition of an arbitrary quantum gate into a product of elementary single qubit quantum gates and CNOT gate is possible. Our strategy therefore is to (1) use second quantized Hamiltonian – creation and annihilation operators – and (2) map the Hamiltonian using Pauli matrices, which are naturally adapted to manipulation by quantum gates, as discussed in Chapter 11.

The second quantized many-electron operator is, from Eqs. D.34a,b (Appendix D),

$$H = \sum_{r,s} h_{rs} f_r^\dagger f_s + \frac{1}{2}\sum_{r,s,t,u} h_{rstu} f_r^\dagger f_s^\dagger f_u f_t. \tag{E.1}$$

We have denoted the electron creation and annihilation operators by $f^\dagger$ and f respectively only to emphasize that they operate on *f*ermion occupation number states. Written out more fully by writing the summation over the spin variables α, β explicitly, the Hamiltonian is

$$\begin{aligned} H = &\sum_{i,\alpha}\sum_{j,\beta} f_{i\alpha}^\dagger \langle i\alpha|h|j\beta\rangle f_{j\beta} + \\ &\frac{1}{2}\sum_{i,\alpha}\sum_{j,\beta}\sum_{k,\gamma}\sum_{\ell,\delta} f_{i\alpha}^\dagger f_{j\beta}^\dagger \langle i\alpha, j\beta|v|\ell\delta, k\gamma\rangle f_{k\gamma} f_{\ell\delta}, \end{aligned} \tag{E.2}$$

or, equivalently, using the integrals we used in Chapter 9 (Eqs. 9.38a,b,c and Eqs. 9.39a,b,c) as

$$\begin{aligned} H = &\sum_{i,\alpha}\sum_{j,\beta} f_{i\alpha}^\dagger \left(\int \psi_{i\alpha}^*(q) h(q) \psi_{j\beta}(q)\, dq\right) f_{j\beta} + \\ &+\frac{1}{2}\sum_{i,\alpha}\sum_{j,\beta}\sum_{k,\gamma}\sum_{\ell,\delta} f_{i\alpha}^\dagger f_{j\beta}^\dagger \left(\iint dq\,dq' \begin{Bmatrix}\psi_{i\alpha}^*(q)\times \\ \psi_{j\beta}^*(q)\end{Bmatrix} v(q,q') \begin{Bmatrix}\psi_{l\delta}(q')\times \\ \psi_{k\gamma}(q')\end{Bmatrix}\right) f_{k\gamma} f_{l\delta}. \end{aligned} \tag{E.3}$$

We now discuss how to map the creation and annihilation operators into Pauli matrices using JW transformations. This is achieved by defining qubit creation and annihilation operators, $Q^\dagger$ and Q. Qubits are distinguishable, like numerable spins in a one-dimensional Ising–Lenz model of ferromagnetism (Fig. E.1).

The JW transformation seeks to establish a relationship between the creation and annihilation operator $f^\dagger$ and f for the indistinguishable fermions and the creation and annihilation operators $Q^\dagger$ and Q for the distinguishable qubits. Using these relations, we can recast the Hamiltonian in Eqs. E.1–E.3 in terms of the Pauli matrices $\sigma_i, i = 1,2,3$ (Chapter 4) in a form

$$H = \sum_{j=1}^{N} \alpha_j P_j = \sum_{j=1}^{N} \alpha_j \left(\Pi_i \sigma_i^j\right), \tag{E.4}$$

wherein the summation j is over various terms in the Hamiltonian, $\alpha_j \leftrightarrow \alpha_j\left(h_{rs}, h_{rstu}\right)$ are scalars corresponding to the one- and two-center integrals in the Eq. E.1, and i denotes a qubit index on which the Pauli operator acts. For the fermion operators, we have (Eqs. D.31a,b,c in Appendix D):

$$\left[f_j^\dagger, f_k^\dagger\right]_+ = 0; \quad \left[f_j, f_k\right]_+ = 0; \quad \left[f_j^\dagger, f_k\right]_+ = \delta_{jk}. \tag{E.5}$$

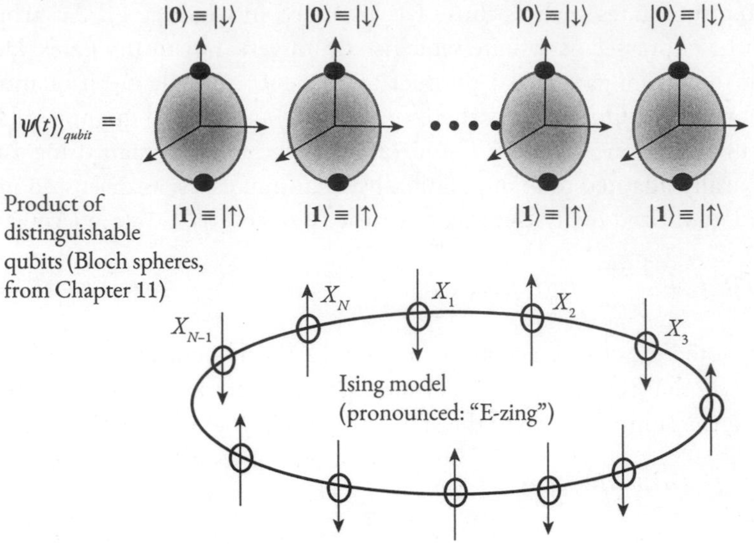

Fig. E.1 Distinguishable qubits, numbered like spins in an Ising model.

The effect of the electron creation and annihilation operators on occupied, and vacant, number states,

$$f_j^{\dagger}|0\rangle = |1\rangle; \quad f_j^{\dagger}|1\rangle = 0; \quad f_j|0\rangle = 0; \quad f_j|1\rangle = |0\rangle, \tag{E.6}$$

is analogous to that of the electron spin raising and lowering operators (Eqs. 4.22a,b,c and Solved Problem P4.3 in Chapter 4):

$$\sigma_j^{+}|\downarrow\rangle = |\uparrow\rangle; \quad \sigma_j^{+}|\uparrow\rangle = 0; \quad \sigma_j^{-}|\downarrow\rangle = 0; \quad \sigma_j^{-}|\uparrow\rangle = |\downarrow\rangle, \tag{E.7}$$

since from Eqs. 4.18a,b,c (Chapter 4) we have, setting the Dirac constant (i.e., *reduced* Planck's constant) to unity,

$$\vec{s} = \frac{1}{2}\vec{\sigma} \text{ and } \left[s_i, s_j\right]_- = \varepsilon_{ijk} i s_k, \tag{E.8}$$

$$\text{and } s_j^{\pm} = s_j^x \pm i s_j^y. \tag{E.9}$$

Also, we have

$$s_j^+ s_j^- = \left\{\left(s_j^x\right)^2 - i\left[s_j^x, s_j^y\right]_- + \left(s_j^y\right)^2\right\} = \left(\frac{1}{2} + s_j^z\right). \tag{E.10}$$

$$\text{Note that } \left(f_j^{\dagger}\right)^2 = 0; \quad \left(f_j\right)^2 = 0; \quad \left(\sigma_j^{\pm}\right)^2 = 0. \tag{E.11}$$

$$\text{Now, we have } \left[s_j^+, s_j^-\right]_+ = 1 \tag{E.12a}$$

just like $\left[f_j^\dagger, f_j\right]_+ = 1.$ (E.12b)

The Eqs. E.10, E.11, and E.12a,b suggest that it would be wonderful if we can define *qubit creation and annihilation operators*, similar to electron creation and annihilation operators, by associating them with the Pauli spin *raising and lowering operators* respectively. However, there is a difficulty, since for different qubits, corresponding to different sites in Fig. E.1, the rules are different, as can be seen from the fact that

Whereas $\left[f_j^\dagger, f_k\right]_+^{j\neq k} = 0,$ (E.13a)

we have $\left[s_j^+, s_j^-\right]_-^{j\neq k} = 0;$ (E.13b)

i.e., we have the *anti*-commutator in Eq. E.13a but the commutator in Eq. E.13b. The JW transformation compensates for this difference by defining the qubit creation and annihilation operators as follows:

$$Q_j^+ = f_j^\dagger \exp\left(-i\pi\sum_{k=1}^{j-1} f_k^\dagger f_k\right) = f_j^\dagger\left[\mathop{\Pi}_{k=1}^{j-1}\exp\left(-i\pi f_k^\dagger f_k\right)\right] = f_j^\dagger P_j, \tag{E.14a}$$

and

$$Q_j^- = f_j \exp\left(+i\pi\sum_{k=1}^{j\ 1} f_k^\dagger f_k\right) = f_j\left[\mathop{\Pi}_{k=1}^{j-1}\exp\left(+i\pi f_k^\dagger f_k\right)\right] = f_j P_j, \tag{E.14b}$$

which give

$$Q_j^+ Q_j^- = f_j^\dagger f_j. \tag{E.15}$$

The factor P_j in Eqs. E.14a,b is a phase factor that addresses the discrepancy in Eqs. E.13a,b. With reference to the qubits laid out in accordance with the Ising model (Fig. E.1), we see that the qubit creation and destruction operators acquire a *non-local* character. The phase is +1 if the number of $f_j \equiv Q_j^- \otimes Z_{j-1}^{\rightarrow}$ occupied modes *till* j is even, and −1 if it is odd. The inverse relationship between the electron creation and annihilation operators and the qubit creation and annihilation operators is given by

$f_j^\dagger \equiv Q_j^+ \otimes Z_{j-1}^{\rightarrow},$ (E.16a)

and $f_j \equiv Q_j^- \otimes Z_{j-1}^{\rightarrow}$, with (E.16b)

$$Z_i^{\rightarrow} \equiv \sigma_i^z \otimes \sigma_{i-1}^z \otimes \ldots \sigma_1^z \otimes \sigma_0^z\,. \tag{E.17}$$

The operator $Z_i^{\rightarrow}$ is the parity operator; its eigenvalues are ±1.

We see that the phase factor is

$$P_j = \prod_{k=1}^{j-1} \sum_{n=0}^{\infty} \frac{(i\pi)^n}{n!} \left(f_k^\dagger f_k\right)^n = \prod_{k=1}^{j-1} \left[1 + \sum_{n=1}^{\infty} \frac{(i\pi)^n}{n!} \left(f_k^\dagger f_k\right)^n\right], \tag{E.18}$$

and using the idempotence property

$$\left(f_k^\dagger f_k\right)^n = \left(f_k^\dagger f_k\right), \tag{E.19}$$

we have

$$P_j = \prod_{k=1}^{j-1} \left[1 + \left\{\exp(i\pi) - 1\right\}\left(f_k^\dagger f_k\right)\right] = \prod_{k=1}^{j-1} \left(1 - 2f_k^\dagger f_k\right). \tag{E.20}$$

We shall also use the correspondence between

$$s_j^+ s_j^- - \frac{1}{2} = s_j^z \tag{E.21a}$$

$$\text{and } f_j^\dagger f_j - \frac{1}{2} = Q_j^+ Q_j^- - \frac{1}{2}. \tag{E.21b}$$

$$\text{Since } \left(1 - 2f_k^\dagger f_k\right)^2 = 1, \tag{E.22a}$$

$$\text{we get } \left(P_j\right)^2 = \left[\prod_{k=1}^{j-1} \left(1 - 2f_k^\dagger f_k\right)\right]^2 = 1, \tag{E.22b}$$

and hence

$$P_j P_{j+1} = \left\{\exp\left(+i\pi \sum_{k=1}^{j-1} f_k^\dagger f_k\right)\right\}\left\{\exp\left(+i\pi \sum_{i=1}^{j} f_i^\dagger f_i\right)\right\}, \tag{E.23a}$$

$$\text{i.e., } P_j P_{j+1} = \left\{\exp\left(+i\pi \sum_{k=1}^{j-1} f_k^\dagger f_k\right)\right\}\left\{\exp\left(+i\pi \sum_{i=1}^{j-1} f_i^\dagger f_i\right)\right\} \exp\left(+i\pi f_j^\dagger f_j\right),$$

$$\text{or, } P_j P_{j+1} = \exp\left(+i\pi f_j^\dagger f_j\right) = \left(1 - 2f_j^\dagger f_j\right). \tag{E.23b}$$

Likewise, we see that

$$\left(Q_j^+\right)\left(Q_{j+1}^-\right) = f_j^\dagger P_j P_{j+1} f_{j+1} = f_j^\dagger \left(1 - 2f_j^\dagger f_j\right) f_{j+1} = f_j^\dagger f_{j+1}. \tag{E.24}$$

It can be seen, using Eq. E.21a, that we have the correspondence

$$f_j^\dagger \leftrightarrow \sigma_j^+ \otimes \sigma_{j-1}^z \otimes \sigma_{j-2}^z \otimes \ldots \otimes \sigma_1^z \tag{E.25a}$$

and

$$f_j \leftrightarrow \sigma_j^- \otimes \sigma_{j-1}^z \otimes \sigma_{j-2}^z \otimes \ldots \otimes \sigma_1^z, \tag{E.25b}$$

and we employ the convention

$$\sigma^z\left|\uparrow\right\rangle = -1\left|\uparrow\right\rangle. \tag{E.26}$$

Using the JW transformation, the electronic Hamiltonian (Eq. E.1–3) is transformed to the qubit Hamiltonian, and the occupation number vectors

$$\underbrace{\left|O_{n-1}O_{n-2}\cdots O_3O_2O_1O_0\right\rangle}_{\substack{\boxed{j=0,1,2,3,\ldots,n-2,n-1} \\ n\text{ fermions occupation number state}}};\quad \underbrace{O_j \in \{0,1\}}_{\text{fermion occupation numbers}} \tag{E.27}$$

are mapped into qubit states

$$\underbrace{\left|q_{n-1}\right\rangle \otimes \left|q_{n-2}\right\rangle \otimes \ldots \otimes \left|q_0\right\rangle}_{\substack{\boxed{j=0,1,2,3,\ldots,n-2,n-1} \\ n\text{ qubits state}}};\quad \underbrace{\left|q_j\right\rangle \in \{qubit\}}_{\substack{\text{Bloch sphere superposition} \\ \text{of } |0\rangle\ \&\ |1\rangle;\ |\uparrow\rangle\ \&\ |\downarrow\rangle}}, \tag{E.28}$$

to solve the electron structure problem on a quantum computer using the VQE. Correlations beyond the Hartree–Fock method can be tackled using, for example, the unitary coupled-cluster singles and doubles (UCCSD) method, by including operators of the following form [10]:

$$\hat{T}_{CCSD} = \hat{T}_1 + \hat{T}_2 = \sum_{\substack{i\in virtual \\ \alpha\in occupied}} t_i^{\alpha} a_i^{\dagger} a_{\alpha} + \sum_{\substack{i,j\in virtual \\ \alpha,\beta\in occupied}} t_{ij}^{\alpha\beta} a_i^{\dagger} a_j^{\dagger} a_{\beta} a_{\alpha} \tag{E.29a}$$

and

$$\hat{T}_{UCCSD} = \left(\hat{T}_1 - \hat{T}_1^{\dagger}\right) + \left(\hat{T}_2 - \hat{T}_2^{\dagger}\right). \tag{E.29b}$$

The VQE method relies on *first* preparing the qubit Hamiltonian on a classical computer, making an ansatz to represent a trial wavefunction $\left|\psi(\vec{\theta})\right\rangle$, and then obtaining the extremum of

$$E \leq \frac{\left\langle \psi(\vec{\theta})\middle|H_{el}\middle|\psi(\vec{\theta})\right\rangle}{\left\langle \psi(\vec{\theta})\middle|\psi(\vec{\theta})\right\rangle}, \tag{E.30}$$

just as in the Hartree–Fock theory, but now using a quantum computer [11–13].

References

[1] Kenji Sugisaki, Chikako Sakai, Kazuo Toyota, Kazunobu Sato, Daisuke Shiomi, and Takeji Takui, Quantum Algorithm for Full Configuration Interaction Calculations without Controlled Time Evolutions, *J. Phys. Chem. Lett.* 12, 45:11085–11089 (2021). https://doi.org/10.1021/acs.jpclett.1c03214.

[2] Kenji Sugisaki, Takumi Kato, Yuichiro Minato, Koji Okuwaki, and Yuji Mochizuki, Variational Quantum Eigensolver Simulations with the Multireference Unitary Coupled Cluster Ansatz: A Case Study of the C2v Quasi-reaction Pathway of Beryllium Insertion into a H_2 Molecule. *Phys. Chem. Chem. Phys.* 24, 8439 (2022).

[3] Frank Arute, Kunal Arya, et al. (81 authors), Twelve-qubit Quantum Computing for Chemistry. *Science* 369 (6507):1084–1089 (August 2020). DOI: 10.1126/science.abb9811.

[4] Kenji Sugisaki, Takumi Kato, Yuichiro Minato, et al., Towards Accurate Description of Chemical Reaction Energetics by Using Variational Quantum Eigensolver: A Case Study of the C2v Quasi-Reaction Pathway of Beryllium Insertion to H_2 Molecule. DOI:10.33774/chemrxiv-2021-w7n78.

[5] P. Jordan, Zur Quantenmechanik der Gasentartung. *Zeits. f. Phys.* 44, 473–480 (1927).

[6] P. Jordan and E. Wigner, Über das Paulische Aquivalenzverbot. *Z. Phys.* 47(9–10): 631–651 (1928). http://dx.doi.org/10.1007/BF01331938.

[7] S. Parity Bravyi, J. M. Gambetta, A. Mezzacapo, and K. Temme, Tapering of Qubits to Simulate Fermionic Hamiltonians. *arXiv* (2017), 1701.08213.

[8] S. B. Bravyi and A. Y. Kitaev, Fermionic Quantum Computation. *Ann. Phys.* 298:1, 210–226 (2002). doi:10.1006/aphy.2002.6254.

[9] Jacob T. Seeley, Martin J. Richard, and Peter J. Love, The Bravyi-Kitaev Transformation for Quantum Computation of Electronic Structure. *J. Chem. Phys.* 137, 224109 (2012); https://doi.org/10.1063/1.4768229.

[10] Dmitry A. Fedorov, Bo Peng, Niranjan Govind, and Yuri Alexeev, VQE Method: A Short Survey and Recent Developments. arXiv:2103.08505v2 [quant-ph] (30 Aug 2021).

[11] S. McArdle and S. Endo, Quantum Computational Chemistry. *Rev. Mod. Phys.* 92:1, 015003 (2020). doi:10.1103/revmodphys.92.015003.

[12] Y. Cao, J. Romero, J. P. Olson, M. Degroote, P. D. Johnson, M. Kieferova, I. D. Kivlichan, T. Menke, B. Peropadre, N. P. D. Sawaya, S. Sim, L. Veis, and A. Aspuru-Guzik, Quantum Chemistry in the Age of Quantum Computing. *Chem. Rev.* 119(19), 10856–10915 (2019). doi:10.1021/acs.chemrev.8b00803.

[13] M. Cerezo, A. Arrasmith, R. Babbush, S. C. Benjamin, S. Endo, K. Fujii, J. R. McClean, K. Mitarai, X. Yuan, L. Cincio, and P. J. Coles, Variational Quantum Algorithms. arXiv:2012.09265 (2020).

Index